Plant Genetics and Molecular Breeding

Plant Genetics and Molecular Breeding

Special Issue Editor
Pedro Martínez-Gómez

MDPI • Basel • Beijing • Wuhan • Barcelona • Belgrade

Special Issue Editor
Pedro Martínez-Gómez
CEBAS-CSIC
Centro de Edafología y Biología Aplicada del Segura
Department of Plant Breeding
Spain

Editorial Office
MDPI
St. Alban-Anlage 66
4052 Basel, Switzerland

This is a reprint of articles from the Special Issue published online in the open access journal *International Journal of Molecular Sciences* (ISSN 1422-0067) from 2018 to 2019 (available at: https://www.mdpi.com/journal/ijms/special_issues/plant_genetic_breeding)

For citation purposes, cite each article independently as indicated on the article page online and as indicated below:

LastName, A.A.; LastName, B.B.; LastName, C.C. Article Title. *Journal Name* **Year**, *Article Number*, Page Range.

ISBN 978-3-03921-175-3 (Pbk)
ISBN 978-3-03921-176-0 (PDF)

Cover image courtesy of Pedro Martínez-Gómez. Original Painting by Juana Serrano.

Contents

About the Special Issue Editor

Pedro Martínez-Gómez is an international reference for Prunus Breeding and Biotechnology. The topic of his research is the use of integrated genetic, genomic, proteomic, transcriptomic, and epigenetic approaches to Prunus breeding. He has a degree in Agricultural Sciences (University of Lleida, Spain), a Master of Science in Plant Breeding (International Centre for Advanced Mediterranean Agronomic Studies, Zaragoza, Spain) and a PhD in Fruit Genetics and Breeding (University of Murcia, Spain). He spent three years as a Post-Graduate Researcher in the Department of Pomology at the University of California, Davis, working on peach and almond breeding and biotechnology with professors Thomas Gradziel. After this period, Dr. Martínez-Gómez started a tenure-track process in the Department of Plant Breeding at CEBAS-CSIC of Murcia, leading a new laboratory for molecular markers application to fruit breeding. Currently, he is working as a Scientific Researcher at CEBAS-CSIC in Murcia, Spain. He is an international reference in Prunus Breeding and Biotechnology, with 120 papers published in international journals included in the SCI and more than 2400 citations in the Web of Science. In August 2006, he received in Seoul (Korea) the "Miklos Faust International Award", an International Award of the American Society for Horticultural Science (ASHS) and the International Society for Horticultural Science (ISHS), for its contribution to the international diffusion of the research in fruit science. In January 2012, he received in Teheran (Iran) the "Khwarizmi International Award" organized by the Iranian Research Organization for Science and Technology (IROST) and the "FAO Medal Award" from the Food and Agriculture Organization for his contribution to the research in fruit trees in developing countries.

Preface to "Plant Genetics and Molecular Breeding"

The development of new plant varieties is a long and tedious process involving the generation of large seedling populations for the selection of the best individuals. In addition, the requirements for new varieties must be anticipated several years (between 5, in the case of horticultural crops, and 15, in the case of fruit crops) in advance, as this is the average time from the original cross to the release of a new pre-crop variety. A critical decision is the choice of the parents to be used. Subsequent crosses can be of the complementary type (when we cross two varieties with complementary characteristics to obtain a new variety that integrates the good features of both varieties) or of the transgressive type, (when two varieties with good features are crossed in order to obtain a progeny performance better than that of the parents). While the ability of breeders to generate large populations is almost unlimited, the management, phenotyping (genetic studies), and selection of these seedlings are the main factors limiting the generation of new cultivars. In this context, molecular (DNA) studies for the development of marker-assisted selection (MAS) strategies are particularly useful when the evaluation of a character is expensive, time-consuming, or with long juvenile periods. In addition, proteomic (proteins and enzymes), transcriptomic (RNA), and epigenetic (DNA methylation and histone modifications) studies are being applied to breeding programs.

These integrated approaches have been classically performed for the genetic characterization of plant material. However, at this moment, suitable markers to be applied in the selection of agronomical traits are essential for the clarification of the mentioned genomic studies, and more efficient markers should be developed. This Special Issue, "Plant Genetics and Molecular Breeding", presents a total of 34 articles with 33 original research articles and 1 review article broadly covering the field of Genetics and Molecular Plant Breeding from a multidisciplinary perspective. Manuscripts are focused on the integration of different phenotypical and molecular tools for the analysis of different traits with a high interest in breeding. This multidisciplinary studies were performed mainly in cereal crops including rice and wheat, industrial crops including oilseed rape, sugarcane, soybean, sesame, and minor oil crops, vegetable crops including cabbage, tomato, broccoli, chickpea, and cucumber, and ornamental crops including Chrysanthemum, Paeonia, rose, and Aechmea. Some studies were carried out in fruit trees, such as kiwi and almond, and in crops with environmental and medical uses.

The papers published in this Special Issue report highly novel results and plausible and testable new models for the integrative analysis of the different approaches applied to plant breeding, including genetic (phenotyping and transmission of agronomic characters), physiology (flowering, ripening, organ development), genomic (DNA regions responsible for the different agronomic characters), transcriptomic (gene expression analysis of the characters), proteomic (proteins and enzymes involved in the expression of the characters), metabolomic (secondary metabolites), and epigenetic (DNA methylation and histone modifications) approaches. The final objective of these studies is the development of new MAS strategies linked to the most important agronomic traits. In this context, these integrated approaches have been applied to the analysis of different agronomic traits with great interest in breeding related to the increase of the production (nutrient use efficiency, yield, pollen development, plant development, cytoplasmic male sterility, elongated internode, abortive buds), the increase of the quality (grain quality, starch composition, phenolic acids, leaf colour, flower colour, polyunsaturated fatty acids, plant architecture, flower development, floral scent), and the reduction of the production costs of plants with biotic (nematode resistance) and abiotic (flowering time, drought resistance, waterlogging resistance, salt stress, heat, heavy-metal) stress tolerance.

The assayed molecular approaches at genomic, transcriptomic, proteomic, metabolomic, and epigenetic levels together with an increasingly accurate phenotyping will facilitate the breeding of new climate-resilient varieties resistant to abiotic and biotic stress, with suitable productivity and quality, to extend the adaptation and viability of the current varieties.

Pedro Martínez-Gómez
Special Issue Editor

International Journal of
Molecular Sciences

Editorial

Editorial for Special Issue "Plant Genetics and Molecular Breeding"

Pedro Martínez-Gómez

Department of Plant Breeding, CEBAS-CSIC, P.O. Box 164, 30100 Espinardo, Murcia, Spain;
pmartinez@cebas.csic.es; Tel.: +34-968-396-200

Received: 21 May 2019; Accepted: 29 May 2019; Published: 30 May 2019

The development of new plant varieties is a long and tedious process involving the generation of large seedling populations to select the best individuals. In addition, the requirements for new varieties must be anticipated several years in advance (between five in the case of horticultural crops and 15 in the case of fruit crops), depending on the average time from the original cross to the release of a new pre-crop variety. One important decision is the choice of which parents to use. Subsequent crosses can include the complementary type (the cross of two varieties with complementary characteristics to obtain a new variety that integrates the positive aptitudes of both varieties) or the transgressive type, in which two varieties are crossed with positive aptitudes to obtain progeny performance that supersedes either parent [1]. Although the ability of breeders to generate large populations is almost unlimited, the management, phenotyping (genetic studies), and selection of seedlings are the main factors limiting the generation of new cultivas [2]. In this context, molecular (DNA) studies for the development of marker-assisted selection (MAS) strategies are particularly useful when the evaluation of character is expensive, time-consuming, or comprises long juvenile periods. In addition, proteomic (proteins and enzymes), transcriptomic (RNA), and epigenetic (DNA methylation and histone modifications) studies are being applied to breeding programs [3].

Integrated approaches have been classically performed for the genetic characterization of plant material [4]. However, the current development of suitable markers applicable in the selection for agronomical traits must be used for the clarification of the above-mentioned genomic studies and the development of more efficient markers. In this context, this Special Issue, "Plant Genetics and Molecular Breeding", presents a total of 34 articles with 33 original research articles and one review article broadly covering the field of genetics and molecular plant breeding from a multidisciplinary perspective (Table 1). Manuscripts focus on the integration of different phenotypical and molecular tools for the analysis of different traits with a high interest in breeding. These multidisciplinary studies have been performed predominantly in cereal crops including rice [5–13] and wheat [14]; industrial crops including oilseed rape [15–17], sugarcane [18,19], soybean [20], sesame [21], and minor oil crops [22]; vegetable crops including cabbage [23,24], tomato [25], broccoli [26], chickpea [27], and cucumber [28]; ornamental crops including *Chrysanthemum* [29,30], *Paeomia* [31], rose [32] and *Aechmea* [33]; fruit and nut trees such as kiwi [34] and almond [35]; and crops with environmental [36,37] and medical uses [38].

Papers published in this Special Issue has reported high novelty results as well as plausible and testable new models for the integrative analysis of the different approaches applicable to plant breeding, including genetic (phenotyping and transmission of agronomic characters), physiology (flowering, ripening, organ development), genomic (DNA regions responsible for different agronomic characters), transcriptomic (gene expression analysis of the characters), proteomic (proteins and enzymes involved in the expression of the characters), metabolomic (secondary metabolites) and epigenetic (DNA methylation and histone modifications) approaches (Table 1). The objective of these studies is the development of new MAS strategies linked to the most important agronomic traits. In this context, these integrated approaches have been applied to the analysis of different agronomic traits with a focus on breeding related to an increase in production (nutrient use efficiency, yield, pollen development, plant development, cytoplasmic male sterility, elongated internode, abortive buds), increase in quality (grain quality, starch composition, phenolic acids, leaf colour, flower colour, polyunsaturated fatty acids, plant architecture, flower development, floral scent), and the reduction of costs with biotic (nematode resistance) and abiotic (flowering time, drought resistance, waterlogging resistance, salt stress, heat tolerance, heavy metal tolerance) stress tolerance (Table 1).

Table 1. Contributors to the Special Issue "Plant Genetics and Molecular Breeding".

Crop	Species	Trait	Integrated Approaches	Reference
Cereal	Rice	Nutrient use efficiency	Physiology, Genomics	Jewel et al. [5]
			Physiology, Genomics, Transcriptomics	Ali et al. [6]
		Grain yield	Genomics, Transcriptomics, Transformation	Fu et al. [7]
			Genetics, Genomics, Transcriptomics,	Zhang et al. [8]
		Starch accumulation	Physiology, Genomics, Transcriptomics	Zha et al. [9]
		Starch composition	Genomics, Transcriptomics, Transformation	Jiang et al. [10]
		Pollen development	Physiology, Genomics, Transcriptomics	Sun et al. [11]
		Endorpesm development	Genomics, Transcriptomics, Transformation	Wang et al. [12]
		Plant development	Genomics, Transcriptomics, Transformation	Xue et al. [13]
	Wheat	Drought stress	Physiology, Genomics	Bhatta et al. [14]
Industrial	Oilseed rape	Cytoplasmic male esterility	Physiology, Genomics, Transcriptomics	Ding et al. [15]
		Nematode resistance	Genomics, Transcriptomics, Transformation	Zhong et al. [16]
		Heavy metal tolerance	Genomics, Transcriptomics, Metabolomics	Pan et al. [17]
	Sugarcane	Stem borer resistance	Physiology, Genomics, Transformation	Gao et al. [18]
		Plant growth	Genomics, Transcriptomics, Genetic Transformation	Wang et al. [19]
	Soybean	Branching	Physiology, Genomics, Transcriptomics	Shim et al. [20]
	Sesame	Yield	Genetics, Genomics, Transcriptomics	Zhou et al. [21]
	Oil crops	Polyunsaturated fatty acids	Physiology, Genomics, Transcriptomics	Wu et al. [22]
Vegetable	Cabbage	Cytoplasmic male sterility	Genomics, Transcriptomics, Proteomics	Han et al. [23]
		Leaf colour	Genetics, Genomics, Transcriptomics	Liu et al. [24]
	Tomato	Elongated internode	Genomics, Transcriptomics, Transformation	Sun et al. [25]
	Broccoli	Abortive buds	Genetics, Genomics, Transcriptomics,	Shu et al. [26]
	Chickpea	Heat tolerance	Genetics, Genomics, Transcriptomics	Paul et al. [27]
	Cucumber	Drought stress	Physiology, Genomics, Transcriptomics	Wang et al. [28]
Ornamental	*Chysanthemun*	Flower development	Physiology, Genomics, Transcriptomics	Yang et al. [29]
		Salt stress	Genomics, Transcriptomics, Transformation	He et al. [30]
	Paeonia	Flower colour	Genomics, Transcriptomics, Transformation	Zhang et al. [31]
	Rose	Flower coluor	Genomics, Transcriptomics, Transformation	Sui et al. [32]
	Aechmea	Plant architecture	Genomics, Transcriptomics, Transformation	Lei et al. [33]
Fruit tree	Kiwi fruit	Waterlogging resistance	Physiology, Genomics, Transcriptomics	Pan et al. [34]
	Almond	Flowering time	Physiology, Genomics, Epigenetics	Prudencio et al. [35]
Environmental	Desert moss	Drought tolerance	Genomics, Transcriptomics, Transformation	Li et al. [36]
	Wintersweet	Floral scent	Physiology, Genomics, Transcriptomics,	Li et al. [37]
Medical	*Salvia*	Phenolic acids	Transcriptomics, Transformation, Metabolomics	Wang et al. [38]

Overall, the 34 contributions published in this Special Issue (Table 1) illustrate the advances in the field of plant genetics and molecular breeding as well as the different integrated approaches necessary for plant breeding programs of the 21st century. The application of massive sequencing methodologies ("deep-sequencing") of the genome (DNA-Seq) [14,17,21], transcriptome (RNA-Seq) [15,19,20,28,37], and proteome [23], focused on lowering the costs of sequencing technologies, has been also widely reported in this Special Issue. These methodologies allow for broader knowledge of the complete genome and transcriptome, respectively. Currently, this application is of great interest to breeding programs, considering the high number of plants species with reference genomes [39].

To conclude, we assert that human activities are producing a significant increase in global temperatures, a phenomenon referred to as climate change. According to the "Intergovernmental Panel on Climate Change (IPCC) Fourth Assessment Report", the average global temperature has increased by 0.74 °C over the last century and is expected to rise between 1.1 °C and 6.0 °C before 2100 [40]. Climate change is affecting all life processes on earth, including food crop production. Increases in temperature are modifying the growth stages of plants, especially those in temperate zones that are adapted to seasonal changes in solar radiation, temperature, and water availability. These molecular approaches at genomic, transcriptomic, proteomic, metabolomic and epigenetic levels, together with an increasingly accurate phenotyping, will facilitate the breeding of new climate-resilient varieties resistant to abiotic stress with a suitable productivity and quality to extend the adaptation and viability of current varieties.

Acknowledgments: This study has been supported by Grants Nº 19879/GERM/15 of the Seneca Foundation of the Region of Murcia and Nº RTI2018-095556-B-I00 of the Spanish Ministry of Science.

Conflicts of Interest: The author declares no conflict of interest.

Abbreviations

DNA-Seq	DNA sequencing
MAS	Marker-assisted selection
RNA-Seq	RNA sequencing

References

1. Hayward, M.D.; Bosemark, N.O.; Romagosa, I. *Plant Breeding. Principles and Prospects*; Springer: Berlin, Germany, 1993; 450p.
2. Martínez-Gómez, P.; Prudencio, A.S.; Gradziel, T.M.; Dicenta, F. The delay of flowering time in almond: A review of the combined effect of adaptation, mutation and breeding. *Euphytica* **2017**, *213*, 197. [CrossRef]
3. Martínez-Gómez, P.; Sánchez-Pérez, R.; Rubio, M. Clarifying omics concepts, challenges and opportunities for Prunus breeding in the post-genomic era. *OMICS J. Int. Biol.* **2012**, *16*, 268–283. [CrossRef] [PubMed]
4. Zeinalabedini, M.; Majourhat, K.; Kayam-Nekoui, M.; Grigorian, V.; Torchi, M.; Dicenta, F.; Martínez-Gómez, P. Comparison of the use of morphological, protein and DNA markers in the genetic characterization of Iranian wild Prunus species. *Sci. Hort.* **2008**, *16*, 268–283. [CrossRef]
5. Jewel, Z.A.; Ali, J.; Mahender, A.; Hernandez, J.; Pang, Y.; Li, Z. Identification of Quantitative Trait Loci Associated with Nutrient Use Efficiency Traits, Using SNP Markers in an Early Backcross Population of Rice (*Oryza sativa* L.). *Int. J. Mol. Sci.* **2019**, *20*, 900. [CrossRef] [PubMed]
6. Ali, J.; Jewel, Z.A.; Mahender, A.; Anandan, A.; Hernadez, J.; Li, Z. Molecular Genetics and Breeding for Nutrient Use Efficiency in Rice. *Int. J. Mol. Sci.* **2018**, *19*, 1762. [CrossRef]
7. Fu, X.; Zhou, M.; Chen, M.; Shen, L.; Zhu, Y.; Wang, J.; Zhu, L.; Gao, Z.; Dong, G.; Guo, L.; et al. Enhanced Expression of QTL *qLL9/DEP1* Facilitates the Improvement of Leaf Morphology and Grain Yield in Rice. *Int. J. Mol. Sci.* **2019**, *20*, 866. [CrossRef]
8. Zhang, Z.H.; Zhu, Y.Z.; Wang, S.L.; Fan, Y.Y.; Zhuang, J.Z. Importance of the Interaction between Heading Date Genes *Hd1* and *Ghd7* for Controlling Yield Traits in Rice. *Int. J. Mol. Sci.* **2019**, *20*, 516. [CrossRef]
9. Zha, K.; Xie, H.; Ge, M.; Wang, Z.; Si, W.; Gu, L. Expression of Maize MADS Transcription Factor *ZmES22* Negatively Modulates Starch Accumulation in Rice Endosperm. *Int. J. Mol. Sci.* **2019**, *20*, 483. [CrossRef]
10. Jiang, J.Z.; Kuo, C.H.; Chen, B.H.; Chen, M.K.; Li, C.S.; Ho, S.L. Effects of *OsCDPK1* on the Structure and Physicochemical Properties of Starch in Developing Rice Seeds. *Int. J. Mol. Sci.* **2018**, *19*, 3247. [CrossRef]
11. Sun, L.; Xiang, X.; Tang, Z.; Yu, P.; Wen, X.; Wang, H.; Abbas, A. *OsGPAT3* Plays a Critical Role in Anther Wall Programmed Cell Death and Pollen Development in Rice. *Int. J. Mol. Sci.* **2018**, *19*, 4017. [CrossRef]
12. Wang, H.; Zhang, Y.; Sun, L.; Xu, P.; Tu, R.; Meng, S.; Wu, W.; Anis, G.B.; Hussain, K.; Riaz, A.; et al. *WB1*, a Regulator of Endosperm Development in Rice, Is Identified by a Modified MutMap Method. *Int. J. Mol. Sci.* **2018**, *19*, 2159. [CrossRef]
13. Xue, M.; Long, Y.; Zhao, Z.; Huang, G.; Huang, K.; Zhang, T.; Jiang, Y.; Yuan, Q.; Pei, X. Isolation and Characterization of a Green-Tissue Promoter from Common Wild Rice (*Oryza rufipogon* Griff.). *Int. J. Mol. Sci.* **2018**, *19*, 2009. [CrossRef] [PubMed]
14. Bhatta, M.; Morgounov, A.; Belamkar, V.; Baenziger, S. Genome-Wide Association Study Reveals Novel Genomic Regions for Grain Yield and Yield-Related Traits in Drought-Stressed Synthetic Hexaploid Wheat. *Int. J. Mol. Sci.* **2018**, *19*, 3011. [CrossRef]
15. Ding, B.; Hao, M.; Mai, D.; Zaman, Q.U.; Sang, S.; Wang, H.; Wang, W.; Fu, L.; Cheng, H.; Hu, Q. Transcriptome and Hormone Comparison of Three Cytoplasmic Male Sterile Systems in *Brassica napus*. *Int. J. Mol. Sci.* **2018**, *19*, 4022. [CrossRef]

16. Zhong, X.; Zhou, Q.; Cui, N.; Cai, D.; Tang, G. BvcZR3 and BvHs1^{pro-1} Genes Pyramiding Enhanced Beet Cyst Nematode (*Heterodera schachtii* Schm.) Resistance in Oilseed Rape (*Brassica napus* L.). *Int. J. Mol. Sci.* **2019**, *20*, 1740. [CrossRef] [PubMed]

17. Pan, Y.; Zhu, M.; Wang, S.; Ma, G.; Huang, X.; Qiao, C.; Wang, R.; Xu, X.; Liang, Y.; Lu, K.; et al. Genome-Wide Characterization and Analysis of Metallothionein Family Genes That Function in Metal Stress Tolerance in *Brassica napus* L. *Int. J. Mol. Sci.* **2018**, *19*, 2181. [CrossRef]

18. Gao, S.; Yang, Y.; Guo, J.; Su, Y.; Wu, Q.; Wang, C.; Que, Y. Particle Bombardment of the cry2A Gene Cassette Induces Stem Borer Resistance in Sugarcane. *Int. J. Mol. Sci.* **2018**, *19*, 1692. [CrossRef]

19. Wang, L.; Liu, F.; Zhang, X.; Wang, W.; Sun, T.; Chen, Y.; Dai, M.; Yu, S.; Xu, L.; Su, Y.; et al. Expression Characteristics and Functional Analysis of the *ScWRKY3* Gene from Sugarcane. *Int. J. Mol. Sci.* **2018**, *19*, 4059. [CrossRef]

20. Shim, S.; Ha, J.; Kim, M.Y.; Choi, M.S.; Kang, S.T.; Jeong, S.C. *GmBRC1* is a Candidate Gene for Branching in Soybean (*Glycine max* (L.) Merrill). *Int. J. Mol. Sci.* **2019**, *20*, 135. [CrossRef] [PubMed]

21. Zhou, R.; Dossa, K.; Li, D.; Yu, J.; You, J.; Wei, X.; Zhang, X. Genome-Wide Association Studies of 39 Seed Yield-Related Traits in Sesame (*Sesamum indicum* L.). *Int. J. Mol. Sci.* **2018**, *19*, 2794. [CrossRef]

22. Wu, P.; Zhang, L.; Feng, T.; Lu, W.; Zhao, H.; Li, J.; Lü, S. A Conserved Glycine Is Identified to be Essential for Desaturase Activity of IpFAD2s by Analyzing Natural Variants from *Idesia polycarpa*. *Int. J. Mol. Sci.* **2018**, *19*, 3932. [CrossRef]

23. Han, F.; Zhang, X.; Yang, L.; Zhuang, M.; Zhang, Y.; Li, Z.; Fang, Z.; Lv, H. iTRAQ-Based Proteomic Analysis of Ogura-CMS Cabbage and Its Maintainer Line. *Int. J. Mol. Sci.* **2018**, *19*, 3180. [CrossRef]

24. Liu, Y.; Yu, H.; Han, F.; Li, Z.; Fang, Z.; Yang, L.; Zhuang, M.; Lv, H.; Liu, Y.; Li, Z.; et al. Differentially Expressed Genes Associated with the Cabbage Yellow-Green-Leaf Mutant in the *ygl-1* Mapping Interval with Recombination Suppression. *Int. J. Mol. Sci.* **2018**, *19*, 2936. [CrossRef]

25. Sun, X.; Shu, J.; Ali Mohamed, A.M.; Deng, X.; Zhi, X.; Bai, J.; Cui, Y.; Lu, X.; Du, Y.; Wang, X.; et al. Identification and Characterization of EI (*Elongated Internode*) Gene in Tomato (*Solanum lycopersicum*). *Int. J. Mol. Sci.* **2019**, *20*, 2204. [CrossRef]

26. Shu, J.; Zhang, L.; Liu, Y.; Li, Z.; Fang, Z.; Yang, L.; Zhuang, M.; Zhang, Y.; Lv, H. Normal and Abortive Buds Transcriptomic Profiling of Broccoli ogu Cytoplasmic Male Sterile Line and Its Maintainer. *Int. J. Mol. Sci.* **2018**, *19*, 2501. [CrossRef]

27. Paul, P.J.; Samineni, S.; Thudi, M.; Sajja, S.B.; Rathore, A.; Das, R.R.; Khan, A.W.; Chaturvedi, S.K.; Lavanya, G.R.; Varshney, R.K.; et al. Molecular Mapping of QTLs for Heat Tolerance in Chickpea. *Int. J. Mol. Sci.* **2018**, *19*, 2166. [CrossRef]

28. Wang, M.; Jiang, B.; Peng, Q.; Liu, W.; He, X.; Liang, Z.; Lin, Y. Transcriptome Analyses in Different Cucumber Cultivars Provide Novel Insights into Drought Stress Responses. *Int. J. Mol. Sci.* **2018**, *19*, 2067. [CrossRef]

29. Yang, Y.; Sun, M.; Yuan, C.; Han, Y.; Zhang, T.; Cheng, T.; Wang, J.; Zhang, Q. Interactions between WUSCHEL- and CYC2-like Transcription Factors in Regulating the Development of Reproductive Organs in *Chrysanthemum morifolium*. *Int. J. Mol. Sci.* **2019**, *20*, 1276. [CrossRef]

30. He, L.; Wu, Y.H.; Zhao, Q.; Wang, B.; Liu, Q.L.; Zhang, L. Chrysanthemum *DgWRKY2* Gene Enhances Tolerance to Salt Stress in Transgenic Chrysanthemum. *Int. J. Mol. Sci.* **2018**, *19*, 2062. [CrossRef]

31. Zhang, X.; Xu, Z.; Yu, X.; Zhao, L.; Zhao, M.; Han, X.; Qi, S. Identification of Two Novel R2R3-MYB Transcription factors, *PsMYB114L* and *PsMYB12L*, Related to Anthocyanin Biosynthesis in *Paeonia suffruticosa*. *Int. J. Mol. Sci.* **2019**, *20*, 1055. [CrossRef]

32. Sui, X.; Zhao, M.; Zhao, L.; Han, X. RrGT2, A Key Gene Associated with Anthocyanin Biosynthesis in *Rosa rugosa*, Was Identified Via Virus-Induced Gene Silencing and Overexpression. *Int. J. Mol. Sci.* **2018**, *19*, 4057. [CrossRef]

33. Lei, M.; Li, Z.Y.; Wang, L.B.; Fu, Y.L.; Ao, M.F.; Xu, L. Constitutive Expression of *Aechmea fasciata SPL14* (*AfSPL14*) Accelerates Flowering and Changes the Plant Architecture in *Arabidopsis*. *Int. J. Mol. Sci.* **2018**, *19*, 2085. [CrossRef] [PubMed]

34. Pan, D.L.; Wang, G.; Wang, T.; Jia, Z.H.; Guo, Z.R.; Zhang, J.Y. *AdRAP2.3*, a Novel Ethylene Response Factor VII from *Actinidia deliciosa*, Enhances Waterlogging Resistance in Transgenic Tobacco through Improving Expression Levels of *PDC* and *ADH* Genes. *Int. J. Mol. Sci.* **2019**, *20*, 1189. [CrossRef]

35. Prudencio, A.S.; Werner, O.; Martínez-García, P.J.; Dicenta, F.; Ros, R.M.; Martínez-Gómez, P. DNA Methylation Analysis of Dormancy Release in Almond (*Prunus dulcis*) Flower Buds Using Epi-Genotyping by Sequencing. *Int. J. Mol. Sci.* **2018**, *19*, 3542. [CrossRef]

36. Li, X.; Gao, B.; Zhang, D.; Liang, Y.; Liu, X.; Zhao, J.; Zhang, J.; Wood, A.J. Identification, Classification, and Functional Analysis of *AP2/ERF* Family Genes in the Desert Moss *Bryum argenteum. Int. J. Mol. Sci.* **2018**, *19*, 3673. [CrossRef]

37. Li, Z.; Jiang, Y.; Liu, D.; Ma, J.; Li, J.; Li, M.; Sui, S. Floral Scent Emission from Nectaries in the Adaxial Side of the Innermost and Middle Petals in *Chimonanthus praecox. Int. J. Mol. Sci.* **2018**, *19*, 3278. [CrossRef]

38. Wang, B.; Niu, J.; Huang, Y.; Liu, Y.; Zhou, W.; Hu, S.; Li, L.; Wang, D.; Wang, S.; Cao, X.; et al. Molecular Characterization and Overexpression of *SmJMT* Increases the Production of Phenolic Acids in *Salvia miltiorrhiza. Int. J. Mol. Sci.* **2018**, *19*, 3788. [CrossRef]

39. Aranzana, M.J.; Decroocq, V.; Dirlewanger, E.; Eduardo, I.; Gao, Z.S.; Gasic, K.; Iezzoni, A.; Peacce, C.; Prieto, H.; Tao, R.; et al. *Prunus* genetics and applications after de novo genome sequencing: achievements and prospects. *Hort. Res.* **2019**, *6*, 58. [CrossRef]

40. IPCC. *Climate Change 2007: The Physical Science Basis. Contribution of Working Group I to the Fourth Assessment Report of the Intergovernmental Panel on Climate Change;* Cambridge University Press: Cambridge, UK; New York, NY, USA, 2007.

International Journal of
Molecular Sciences

Article

Identification of Quantitative Trait Loci Associated with Nutrient Use Efficiency Traits, Using SNP Markers in an Early Backcross Population of Rice (*Oryza sativa* L.)

Zilhas Ahmed Jewel [1,†], Jauhar Ali [1,*,†], Anumalla Mahender [1,†], Jose Hernandez [2], Yunlong Pang [1,3] and Zhikang Li [4]

1   Rice Breeding Platform, International Rice Research Institute (IRRI), Los Baños, Laguna 4031, Philippines; jeweluplb@gmail.com (Z.A.J.); m.anumalla@irri.org (A.M.); y.pang@sdau.edu.cn (Y.P.)
2   College of Agriculture, University of the Philippines Los Baños, Laguna 4031, Philippines; joehernandez56@gmail.com
3   College of Agronomy, Shandong Agricultural University, Taian 271018, China
4   Institute of Crop Sciences, Chinese Academy of Agricultural Sciences, Beijing 100081, China; zhkli1953@126.com
*   Correspondence: J.Ali@irri.org; Tel.: +63-2580-5600 (ext. 2541)
†   These authors contributed equally to this work.

Received: 24 October 2018; Accepted: 23 January 2019; Published: 19 February 2019

Abstract: The development of rice cultivars with nutrient use efficiency (NuE) is highly crucial for sustaining global rice production in Asia and Africa. However, this requires a better understanding of the genetics of NuE-related traits and their relationship to grain yield. In this study, simultaneous efforts were made to develop nutrient use efficient rice cultivars and to map quantitative trait loci (QTLs) governing NuE-related traits in rice. A total of 230 BC_1F_5 introgression lines (ILs) were developed from a single early backcross population involving Weed Tolerant Rice 1, as the recipient parent, and Hao-an-nong, as the donor parent. The ILs were cultivated in field conditions with a different combination of fertilizer schedule under six nutrient conditions: minus nitrogen (−N), minus phosphorus (−P), (−NP), minus nitrogen phosphorus and potassium (−NPK), 75% of recommended nitrogen (75N), and NPK. Analysis of variance revealed that significant differences ($p < 0.01$) were noted among ILs and treatments for all traits. A high-density linkage map was constructed by using 704 high-quality single nucleotide polymorphism (SNP) markers. A total of 49 main-effect QTLs were identified on all chromosomes, except on chromosome 7, 11 and 12, which are showing 20.25% to 34.68% of phenotypic variation. With further analysis of these QTLs, we refined them to four top hotspot QTLs (QTL harbor-I to IV) located on chromosomes 3, 5, 9, and 11. However, we identified four novel putative QTLs for agronomic efficiency (AE) and 22 QTLs for partial factor productivity (PFP) under −P and 75N conditions. These interval regions of QTLs, several transporters and genes are located that were involved in nutrient uptake from soil to plant organs and tolerance to biotic and abiotic stresses. Further, the validation of these potential QTLs, genes may provide remarkable value for marker-aided selection and pyramiding of multiple QTLs, which would provide supporting evidence for the enhancement of grain yield and cloning of NuE tolerance-responsive genes in rice.

Keywords: nutrient use efficiency; quantitative trait loci (QTLs), molecular markers; agronomic efficiency; partial factor productivity

Int. J. Mol. Sci. **2019**, *20*, 900

1. Introduction

Rice (*Oryza sativa* L.) is one of the most prominent staple food crops in the world and it has significantly contributed to global food security [1]. By 2050, rice production has to be increased by more than 60% to meet the rapid increase in food demand [2,3]. Approximately 90% of rice cultivation is carried out in irrigated and lowland rainfed ecosystems in Asia, where we still face problems such as decreasing arable land and freshwater availability, and increased labor costs and biotic and abiotic stress factors that impede production and productivity. Further, the drastic rise in fertilizer costs drives the search for suitable rice cultivars that are efficient in grain yield production [4–8]. Nutrient fertilizers are one of the most predominant factors that influence the genetic enhancement of grain yield productivity in most agricultural regions [6].

In the last four decades, the amount of nitrogen (N) applied to crops has risen 12 to 104 teragrams per year (Tg year^{-1}), and 30%–50% of the N has been harvested in the grain and the remaining 70%–50% of the fertilizer has been lost through a combination of leaching, denitrification, volatilization, surface runoff, and microbial consumption [9]. Therefore, the excessive amount and long-term use of fertilizers cause severe environmental pollution, and this is one of the significant costly inputs for poor farmers [10–14]. Hence, the identification of crops with higher NuUE that are less dependent on NPK fertilizers (nitrogen, phosphorous, and potassium) is crucial for the sustainability of agriculture. Kant et al. [15] estimated that an increase of 1% in NuUE could save about US$ 1.1 billion annually. The Green Revolution period witnessed a modest increase of 13% in harvested rice area in Asia, whereas grain yield production more than doubled, from 240 to 513 million tons, with consumption of fertilizers high in NPK increasing from 6.7 to 69.0 million tons. According to the FAO electronic database, particularly in Asian countries, China, Sri Lanka, Indonesia, and India had the highest average fertilizer consumption of 565.2 kg ha^{-1}, 251.7 kg ha^{-1}, 185.141 kg ha^{-1}, and 157.48 kg ha^{-1}, respectively, from 2005 to 2014 (http://ricestat.irri.org:8080/wrsv3/entrypoint.htm#). Therefore, the identification of superior cultivars with improved nutrient use efficiency (NuUE) is an essential target research area for plant breeders, for attaining higher grain yield and also reducing production costs [16].

In the improvement of the tolerance of rice cultivars of single biotic and abiotic stresses, tremendous progress has been achieved [8,17–21]. However, these tolerant varieties have not been able to succeed to attain higher grain yields under both the irrigated and rainfed ecosystems. To overcome these conditions, breeding aspects need to focus on developing multiple-stress-tolerant cultivars (MSTC) and improving nutrient use efficiency, which are vital for the sustainability of grain yield productivity, which could have a more significant impact on higher yield, especially under low-fertilizer-input conditions, besides being more beneficial to poor farmers. Several critical components of morphological, physiological, and agronomic traits are reducing grain yield and decreasing biomass content under nutrient-deficient conditions. These traits become altered due to changes in their molecular mechanism and physiological pathways, leading to their susceptibility [22]. The exploitation of rice genetic resources, using advanced genomic technologies, is essential for the identification of rice cultivars with NuUE for increasing crop grain yield under low-input conditions.

Developing rice varieties with stress tolerance and NuUE through conventional breeding approaches is extremely slow because of several factors that influence the molecular genetics and physiological mechanisms underlying low-input tolerance, the complexity of NuUE traits, and the absence of efficient breeding selection criteria [23–25]. The combination of advanced molecular marker-assisted breeding and conventional breeding platforms could speed up the breeding procedure for varietal development and identifying trait-associated QTLs. Over the past two decades, several types of traditional molecular markers, such as restriction fragment length polymorphism (RFLP), randomly amplified polymorphic DNA (RAPD), amplified fragment length polymorphism (AFLP) and simple sequence repeats (SSRs), have been used for QTL identification for NuUE-related traits under low-input conditions of nitrogen [26–30], phosphorus [7,31–38], and potassium [39,40]. However, these molecular markers have some disadvantages vis-à-vis SNP markers, such as partial chromosome coverage, labor-intensiveness, low resolution, and higher cost. Among these various molecular markers, SNP markers have a comprehensive range of applications in the construction of genetic maps, uniform distribution, and cloning of QTLs. Their clear advantages, such as high density and accurate and reliable approaches, have been used to perform high-throughput genotyping based on SNPs across the genome [41–44]. In a genomics era, high-throughput genotyping with next-generation sequencing (NGS) and various array-based SNP detection platforms have become excellent tools for the dissecting of complex traits and identification of trait-associated genes and alleles in rice [45–48]. As compared to NGS, a SNP array can be used for many samples within a short period; it has a low cost and analysis of data interpretation is relatively easy [47,49]. The recently developed SNP array has been successfully used in diversity studies and genome-wide association studies (GWAS), and in the identification of numerous QTLs and genes in rice [43,49–51].

Over the years, the identification of QTLs for varied agro-morphological traits related to NuUE, by using different populations, hardly resulted in any significant impact regarding rice crop improvement and the development of varieties. The QTLs were identified related to NPK use efficiency using recombinant inbred lines (RILs) [34,48,52–54], backcross inbred lines (BILs) [55], doubled haploids (DHs) [39,56–59], BC$_2$F$_3$ [33], introgression lines (ILs) [60,61], and chromosome segment substitution lines (CSSLs) [35] populations. These QTLs have a smaller effect, and many of them exhibit significant epistatic and QTL x Environment interactions, making them less amenable to breeding programs.

Several QTLs may correspond to known genes in the N or P metabolic pathway, for example, Qyd-2b for N deficiency tolerance was located in the vicinity of the gene encoding cytosolic glutamine synthetase (GS1) [62], and Qyd-3b and Qpn-3 were nearby the genes for glutamate dehydrogenase (GDH2) [26]. Interestingly, Qyd-12 was detected only in low-P conditions and was co-localized with a major QTL, Pup1, on chromosome 12, which was reported to be involved in P absorption [63]. These results indicate that the QTLs specially detected under single N or P deficiency conditions may be involved in different pathways of N and P metabolism. The tightly linked markers have breeding potential in pyramiding elite genes and QTLs for N and P use efficiency.

Although N and P use efficient QTLs and linked traits were reported earlier in different genetic backgrounds of mapping populations, to date, there are no reports of different combinations of NPK fertilizers and their response to the foremost NuUE traits, such as grain yield response, agronomic efficiency, and partial factor productivity in rice. In the present study, taking advantage of the rapid development of SNP technologies, we used a 6K SNP array to genotype a BC$_1$F$_5$ introgression line (IL) derived from a cross between Weed Tolerant Rice 1 (WTR-1), as the recipient parent, and Hao-an-nong (HAN), as the donor parent. In this study, we identified the NuUE-related QTLs by subjecting the population to varying nutrient rates and understanding the performance of yield and its components.

2. Results

2.1. Phenotypic Variation of NuUE Traits and Their Correlation among Traits

A total of 230 BC$_1$F$_5$ ILs were evaluated in six different NPK combinations (–N, –P, –NP, –NPK, 75N and NPK) during the dry season of 2014 in the experimental fields of International Rice Research Institute (IRRI). The key traits of NuUE were analyzed in the total of ILs and were compared with those of the parents and checks (Figure 1). The descriptive statistics for five traits as grain yield (GY), 1000-grain weight (1000-Gwt), percentage of spikelet fertility (PSPF), biomass yield (BY), filled grains per plant (FGN) mean values and testing of significance are presented in Tables 1 and 2. The highest GY range of 19.29–50.26 g/plant was observed in NPK, followed by 20.82–48.83 g/plant in 75N, and 12.81–42.95 g/plant in –NPK, and the lowest GY range of 13.89–34.96 g/plant was detected in –N conditions. Among these six nutrient conditions, 18 ILs in NPK, 23 ILs in 75N, 4 ILs in –P, and 1 IL in both –NPK and –NP had more than 40 g/plant. In –N conditions, the lowest GY was observed (34.96 g/plant). The best performing ILs under each of the six nutrient conditions with the highest grain yields were *GSR IR2-1-RF6-NU7-NU2-NU76-NU96* -WTR 1-RF6 in NPK that gave 50.26 g/plant;

and 42.95 g/plant (*GSR IR2-1-RF6-NU7-NU3-NU82-NU97*-WTR 1-RF6) in −NPK; 48.83 g/plant (*GSR IR2-1-Y17-NU2-NU5-NU6-NU9*-WTR 1-Y17) in 75N; 34.96 g/plant (*GSR IR2-1-L3-NU1-NU1-NU1-NU1*-WTR 1-LI3) in −N; 40.94 g/plant (*GSR IR2-1-RF6-NU4-NU9-NU14-NU66*-WTR 1-RF6) in −P; and 40.43 g/plant (*GSR IR2-1-RF6-NU7-NU2-NU77-NU94*-WTR 1-RF6) in −NP conditions, respectively. Across all six nutrient conditions, the lowest GYs were observed in WTR 1-RF13 (*GSR IR2-1-RF13-NU2-NU8-NU4-NU12*), WTR 1-RF14 (*GSR IR2-1-RF14-NU2-NU4-NU6-NU3*) with 11.85 g/plant; 12.43 g/plant detected in −NP; and followed by WTR 1-LI3 (*GSR IR2-1-L3-NU1-NU1-NU3-NU5*) and WTR 1-LI12 (*GSR IR2-1-L12-NU1-NU4-NU7-NU5*) with 12.86 g/plant and 12.81 g/plant in −NPK conditions, respectively. Under −NP and −NPK conditions, the recipient parent (Weed Tolerant Rice 1) gave 19.87 g/plant and 20.94 g/plant, respectively, while the donor parent (Hao-an-nong) gave 15.73 g/plant and 16.02 g/plant, respectively.

After further analysis of ILs with higher grain yield in each NuUE condition, we selected the first ten highest GY lines, and GY ranged from 28.86 to 50.26 g/plant. Interestingly, we identified promising lines, WTR 1-RF6 (*GSR IR2-1-RF6-NU7-NU3-NU82-NU97*), that showed higher GY in four different nutrient conditions: −N (29.11 g/plant), −P (38.40 g/plant), −NP (31.90 g/plant), and −NPK (42.95 g/plant). Similarly, introgression line WTR 1-RF6 (*GSR IR2-1-RF6-NU7-NU2-NU77-NU94*) had higher GY in three different NuUE conditions, −NP (40.43 g/plant), −NPK (31.43 g/plant), and NPK (42.76 g/plant); and WTR 1-RF6 (*GSR IR2-1-RF6-NU7-NU2-NU76-NU96*) in −P (40.09 g/plant), −NP (33.09 g/plant), and NPK (50.26 g/plant) conditions. Likewise, WTR 1-Y17 (*GSR IR2-1-Y17-NU2-NU5-NU6-NU8* and *GSR IR2-1-Y17-NU2-NU5-NU6-NU9*) had higher GY in 75N (42.00 g/plant and 48.83 g/plant, respectively), −NP (35.73 g/plant and 33.41 g/plant, respectively), and −NPK (33.29 g/plant and 34.75 g/plant, respectively) conditions, The results for NuUE key traits, such as GY (−NP, −NPK), FGN (75N, −NP, −NPK), and BY (NPK, 75N, −N, −P, −NP, −NPK), had a high level of phenotypic variation with a coefficient of variation (CV) of above 20%. The correlations of five traits in six different nutrient conditions were identified to show significant relationships between the respective traits. GY was found to be significantly and positively correlated with BY; FGN in NPK; FGN in 75N; GWT, BY, and FGN in −P; PSPF, BY, and FGN in −N; and BY and FGN in −NP ($p < 0.01$) conditions. However, GY showed itself to be negatively correlated with 1000-Gwt in NPK and with BY in 75N conditions. Among the six nutrient conditions, GY was negatively correlated with BY under 75N. However, the remaining GWT, PSPF and FGN traits followed with significant positive correlation between the traits.

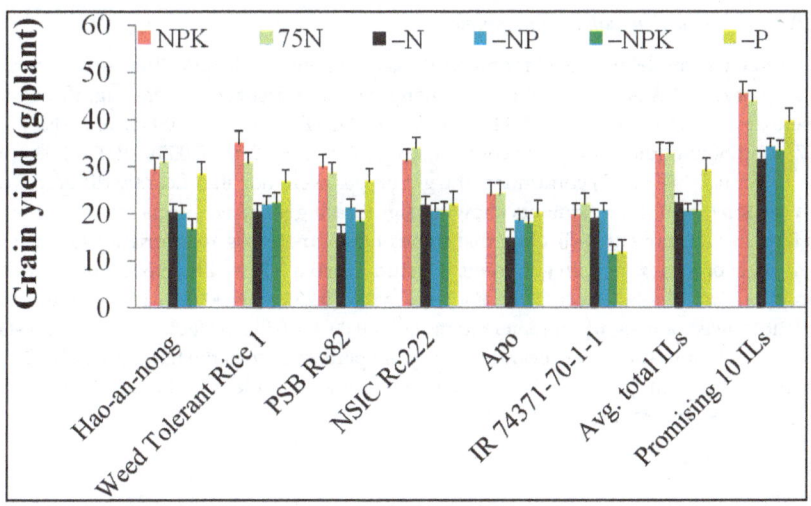

Figure 1. Grain yield performances of introgression lines, parents and checks in all six nutrient conditions.

Table 1. Statistical description of nutrient use efficiency-introgression lines for critical traits under six NPK combinations.

Traits	NuUE Condition	Mean ± Std. Error of Mean	Range (Min–Max)	SD	Variance (S^2)	CV%
GY	NPK	32.48 ± 0.34	19.29–50.26	5.21	27.19	16.04
	75N	32.58 ± 0.35	20.82–48.83	5.44	29.69	16.70
	–N	21.61 ± 0.24	13.89–34.96	3.79	14.37	17.54
	–P	26.74 ± 0.31	17.57–42.19	4.78	22.86	17.88
	–NP	20.41 ± 0.29	11.86–40.43	4.49	20.20	22.00
	–NPK	20.49 ± 0.29	12.81–42.95	4.54	20.66	22.16
1000 Gwt	NPK	27.35 ± 0.13	19.20–31.75	2.05	4.21	7.50
	75N	27.35 ± 0.14	15.18–39.50	2.24	5.03	8.19
	–N	26.41 ± 0.11	20.65–30.60	1.82	3.31	6.89
	–P	27.13 ± 0.16	14.80–33.75	2.52	6.36	9.29
	–NP	26.88 ± 0.14	15.60–31.10	2.14	4.58	7.96
	–NPK	27.16 ± 0.12	21.50–33.35	1.83	3.37	6.76
PSPF	NPK	87.02 ± 0.28	68.05–95.24	4.34	18.84	4.99
	75N	87.14 ± 0.24	72.39–94.35	3.76	14.15	4.32
	–N	89.24 ± 0.29	68.50–96.59	4.44	19.78	4.98
	–P	87.60 ± 0.27	72.95–95.95	4.21	17.75	4.81
	–NP	89.93 ± 0.28	70.52–97.37	4.30	18.54	4.78
	–NPK	89.88 ± 0.26	77.27–97.75	4.09	16.76	4.55
FGN	NPK	1518.90 ± 17.95	956.83–2621.83	273.41	74,756.36	18.00
	75N	1402.42 ± 19.23	783.00–3177.50	293.04	85,876.01	20.90
	–N	1012.30 ± 11.88	686.50–1660.83	180.97	32,750.56	17.88
	–P	1318.12 ± 16.54	786.50–2251.17	252.01	63,512.83	19.12
	–NP	947.57 ± 13.26	445.33–1704.50	202.11	40,848.47	21.33
	–NPK	1040.31 ± 14.50	565.83–1675.83	220.94	48,814.58	21.24
BY	NPK	79.72 ± 1.14	35.92–137.23	17.43	303.83	21.86
	75N	210.24 ± 1.29	41.52–30,066.06	1968.81	3,876,231.42	936.46
	–N	50.47 ± 1.48	23.46–336.63	22.54	508.41	44.66
	–P	69.26 ± 1.18	38.31–144.27	18.07	326.83	26.09
	–NP	45.01 ± 0.84	18.16–89.43	12.90	166.49	28.66
	–NPK	47.29 ± 0.91	18.09–116.50	13.95	194.736	29.50

GY—grain yield; 1000-Gwt—1000-grain weight; PSPF—percentage of spikelet fertility; FGN—filled grains per plant; BY—biomass yield (BY); CV—coefficient of variance; SD—standard deviation.

2.2. ANOVA and Interaction with the Environment

The fixed effect model was used for the analysis of variance with Satterthwaite denominator, and the summary of ANOVA is given in Table 2. Among the six nutrient conditions, significant genotypic effects were observed under –N ($F = 1.44$, $p = 0.0005$), –NP ($F = 1.76$, $p < 0.0001$), and –NPK ($F = 2.47$, $p = 0.0007$) conditions at the 1% level. In contrast, under 75N ($F = 1.2$, $p = 0.053$), –P ($F = 1.38$, $p = 0.0658$), and NPK ($F = 1.07$, $p = 0.4293$) conditions, the genotypes were not significantly different from each other. It was noted that the treatments showing significant genotypic effects are those that lack the nitrogen fertilizer component, indicating that some of these materials are nutrient use efficient. In the significant effect of fertilizers with genotype, the study used a −2 log-likelihood ratio test and found a significant environmental effect at the 1% level ($\chi^2 = 15.23$, $p = 0.0001$). However, genotype and environment showed non-significant interactions ($\chi^2 = 0.00$, $p = 0.9992$) (Table 3). This suggests that the different fertilizer applications significantly affect yield performance of the ILs as a whole. On the other hand, the non-significant genotypes by environment interactions indicate the consistent performance of genotypes across environments.

Table 2. ANOVA for the testing of significance of genotype effect per fertilizer condition.

S. No.	Environment	Degrees of Freedom	Sum of Squares	Mean Squares	F-Value	Satterthwaite Denominator	Pr (>F)
1	NPK	230	12,493.16	54.56	1.07	34.38	0.4293
2	75N	230	12,736.71	55.62	1.20	451.78	0.0530
3	−N	230	6540.20	28.56	1.44	451.61	0.0005 ***
4	−P	230	9273.85	40.50	1.38	62.36	0.0658
5	−NP	230	9288.00	40.56	1.76	229.00	0.0000 ***
6	−NPK	230	9503.51	41.50	2.47	37.76	0.0007 ***

Significant codes: 0 '***', 0.001 '**', 0.01 '*' 0.0.

Table 3. Testing for significance of fertilizer and its combined effect with genotype using −2 log-likelihood ratio test.

Effect	Model	AIC	BIC	Log-Likelihood	Chi Square	Degrees of Freedom	Pr (>Chisq)
Environment	1	17,071.23	18,457.21	−8301.62			
	2	17,058.00	18,449.91	−8294.00	15.2284	1	0.0001 ***
Genotype X Environment	3	17,056.00	18,441.98	−8294.00			
	4	17,058.00	18,449.91	−8294.00	0	1	0.9992

AIC—Akaike information criterion; BIC—Bayes information criterion; YLD—Yield; Deg—Degrees of freedom; Env—Environment; Rep—Replication; Blck—Block; Model 1: YLD~1 + Deg + (1|Env) + (1|Rep:Env) + (1|Rep:Blck:Env) + (1|Deg:Env). Model 2: YLD~1 + Deg + (1|Rep:Env) + (1|Rep:Blck:Env) + (1|Deg:Env). Model 3: YLD~1 + Deg + (1|Env) + (1|Rep:Env) + (1|Rep:Blck:Env) + (1|Deg:Env). Model 4: YLD~1 + Deg + (1|Env) + (1|Rep:Env) + (1|Rep:Blck:Env).

2.3. Analysis of Agronomic Efficiency (AE) and Partial Factor Productivity (PFP)

In the NuUE study, agronomic use efficiency and partial factor productivity were calculated using the fertilizer application rate of NPK (160–50–50 kg ha^{-1}). The computation using the equations (i to iii) revealed 15.06–34.11 kg of grain for each kg of N derived from an NPK fertilizer given in 3.2:1:1 ratio, respectively. However, using this equation, we found 117 ILs with >15 kg of grain per kg^{-1} N (NPK), along with 33 ILs (−P) and 161 ILs (75N). These were also compared with parents and we identified 42 ILs, 23 ILs, and 79 ILs that had a higher AE with respect to each equation (Table 4). Similarly, analysis of partial factor productivity helped to identify 16 ILs (NPK), 4 ILs (−P), and 151 ILs (75N) that had a >50 kg of grain per kg^{-1} N (NPK) condition. In comparison with parents, 25 ILs (NPK), 61 ILs (−P), and 117 ILs (75N) showed more than 50 kg of grain per kg^{-1} N in each condition.

Table 4. Determination of NPK fertilizer efficiency in ILs under experiment on NuUE.

Agronomic Efficiency (AE)			
AE applied nitrogen	AE formula	>15 kg grain kg^{-1} nitrogen applied	>Parents
AE(N) = grain yield (N fertilized–0NPK unfertilized) in kg ha^{-1}/Fertilizer N in kg ha^{-1}	$AE_{(N)} = (Y_{NPK} - Y_{0NPK}) \div F_N$	117	74
	$AE(N) = (Y_{NK} - Y_{0NPK}) \div F_N$	33	28
	$AE(N) = (Y_{75N} - Y_{0NPK}) \div F_{75N}$	161	86
Partial Factor Productivity (PFP)			
PFP Applied Nitrogen	PFP formula	>50 kg grain kg^{-1} nitrogen applied	>Parents
PFP(N) = grain yield N fertilized in kg ha^{-1}/Fertilizer N in kg ha^{-1}	$PFP_{(N)} = Y_{(+NPK)} \div F_N$	16	25
	$PFP_{(N)} = Y_{(-P)} \div F_N$	4	61
	$PFP_{(N)} = Y_{(75N)} \div F_N$	151	117

2.4. Construction of Linkage Map and Segregation of SNP Markers

A total of 704 high-quality polymorphic SNP markers, with an average for each chromosome of 58 SNP markers, were used to genotype the 230 ILs, and a high-density genetic linkage map was developed covering 1526.8 cM, with an average of 127.2 cM per SNP marker (Table 5). The logarithm of the odds (LOD) thresholds and explained phenotypic variance for six nutrient conditions ranged from 2.52% to 17.76% and 5.87% to 34.68%, respectively. Using single marker analysis (SMA), a total of 261 QTLs were mapped on all 12 chromosomes, with an average of 21 QTLs for yield attributed to seven key component traits of NuUE in rice. Among the 12 chromosomes, the highest number of QTLs was located on chromosome 6 (26 QTLs), whereas the lowest number of QTLs was observed on chromosome 12 (6 QTLs), and all QTLs are listed in Table S1. The hotspot QTL regions are shown in Figures 2 and 3. The majority of the QTLs were associated with 1000-Gwt, and PSPF traits had a negative additive effect of 61.9% and 61.5%, indicating that alleles from the recipient parent WTR-1 contributed to increasing phenotype.

Table 5. Distribution of 704 polymorphic single nucleotide polymorphism (SNP) markers distributed in across the 12 chromosomes, with their average distance, genome size, coverage percentage, genetic distance, and physical distance per cM.

S. No.	Chr	Marker No.	Average Distance (Kb)	Genome Size (Kb)	Genome Size (Gramene)	Coverage Percentage	Genetic Distance (cM)	Physical Distance per (Kb)
1	Chr01	76	564.0	42,492.4	43,270.92	98.20	181.8	238.01
2	Chr02	45	797.4	35,401.9	35,937.25	98.51	157.9	227.59
3	Chr03	72	497.7	35,824.4	36,413.81	98.38	166.4	218.83
4	Chr04	84	405.0	33,864.4	35,502.69	95.39	129.6	273.94
5	Chr05	50	479.8	29,100.3	29,958.43	97.14	122.3	244.96
6	Chr06	73	422.7	30,809.5	31,248.78	98.59	124.4	251.20
7	Chr07	74	393.9	28,942.5	29,697.62	97.46	118.6	250.40
8	Chr08	43	654.5	27,809.9	28,443.02	97.77	121.1	234.87
9	Chr09	41	521.7	21,348.9	23,012.72	92.77	93.5	246.13
10	Chr10	43	464.0	19,635.6	23,207.28	84.61	83.8	276.94
11	Chr11	58	492.2	28,312.7	29,021.10	97.56	117.9	246.15
12	Chr12	45	603.7	27,023.4	27,531.85	98.15	109.5	251.43

Chr—chromosome; Kb—kilo base pairs; cM—CentiMorgan.

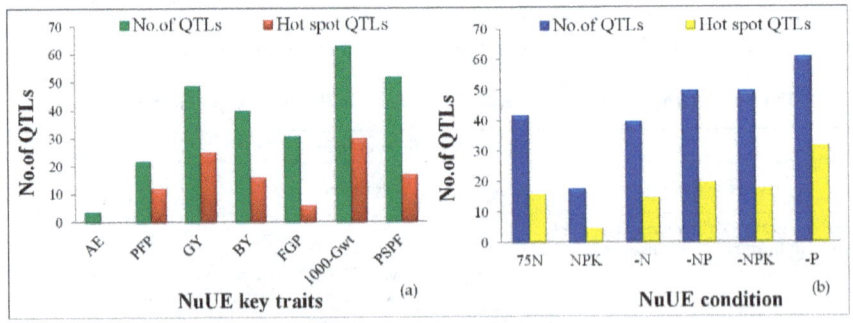

Figure 2. Distribution of trait-wise (**a**); and NPK combinations (**b**), associated with a total number of quantitative trait loci (QTLs) and hotspot QTLs in different nutrient conditions in rice.

Out of the 261 QTLs identified, 62.5% (163 QTLs) had a positive additive effect contributed by WTR-1, and the remaining 37.5% followed with the negative allele from donor parent Hao-an-nong. QTLs with >20% phenotypic variation explained (PVE) were determined as major QTLs with an LOD score of >9.50, while others were considered as minor QTLs. In summary, a total of 261 QTLs for seven traits under six different combinations of NPK nutrients were identified, out of which 49 were major QTLs and 212 were minor QTLs (Table S1). Out of these NuUE conditions, a large number of QTLs were observed in –P (61 QTLs), –NP, and –NPK (50 QTLs), and followed with 75N (42 QTLs) and –N (40 QTLs), and the fewest were detected in NPK conditions with 18 QTLs. The PVE of significant major QTLs ranged from 20.25% to 34.68% at 9.55 to 17.76 LOD value, whereas, for minor QTLs, the explained PV ranged from 5.87% to 19.98% at 2.52 to 9.49 LOD value. The distribution of all major and minor QTLs, represented on all 12 chromosomes, is shown in Figure 3, and the hotspot QTLs (>5 QTLs), identified over nine chromosomes (chr01, chr02, chr03, chr04, chr05, chr07, chr08, chr09, and chr11), were linked to six essential nutrient traits (PFP, GY, BY, FGP, 1000-Gwt, and PSPF) in rice.

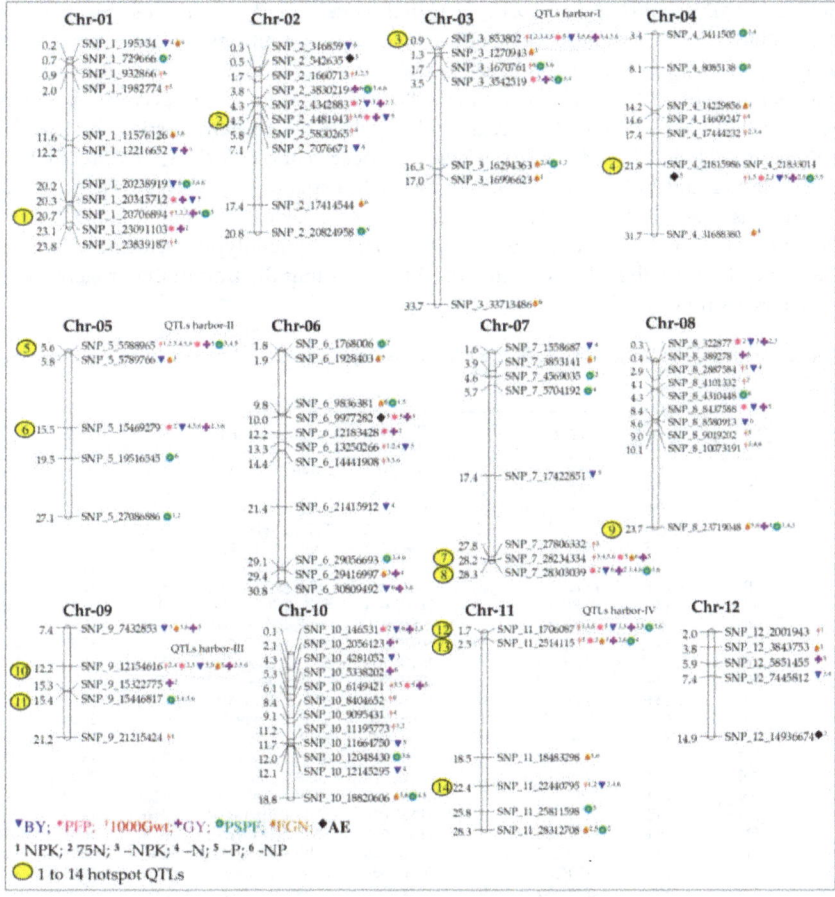

Figure 3. Linkage map of 261 QTLs distributed on 12 chromosomes, with respective polymorphic markers and colors depicting the QTLs governing crucial different nutrient traits.

Out of the 704 polymorphic markers, SNP_2_4481943 had the highest LOD value recorded (17.76), which explained PV of 34.68%, followed by two other markers, SNP_10_6149421 (LOD 16.28) and SNP_8_10073191 (LOD 15.31), which explained PV of 32.32% and 30.73%, respectively. Interestingly, these QTLs were significantly associated with 1000-Gwt under —NPK conditions, and their negative value of the allele was exhibited from donor parent Hao-an-nong. In contrast, the lowest LOD (2.52) was recorded in SNP_6_9836381, which explained PV of 5.87%, followed by three other SNP markers, SNP_12_14936674 (2.55), SNP_12_7445812 (2.55), and SNP_12_5851455 (2.53), which explained PV of 5.92%, 5.92%, and 5.89%, respectively. These four QTLs were linked with different key traits such as FGN, AE, BY, and GY under –NP, 75N, –N, and –NPK conditions. The lowest LOD value of two markers (SNP_6_9836381 and SNP_12_14936674) was contributed by the negative allele from donor parent Hao-an-nong, and two other markers (SNP_12_7445812 and SNP_12_5851455) had a positive allele from recipient parent WTR-1.

2.5. QTLs for NuUE Traits

2.5.1. Agronomic Efficiency (AE)

Agronomic efficiency (AE) was determined in the experiment on NuUE, and we found four QTLs (*qAE_2.1*, *qAE_4.1*, *qAE_6.1* and *qAE_12.1*) that were located on chromosomes 2, 4, 6, and 12, respectively (Table 6). For the agronomic efficiency of applied nitrogen in terms of −P conditions (applied nitrogen and potassium fertilizer), three QTLs were detected with an LOD score and phenotypic variation of 2.77, 4.01, and 4.52 and 6.43%, 9.17%, and 10.27%, respectively, with a positive additive effect of 3.16, 2.13, and 2.28, which indicates that the progeny carried the trait from the recipient parent WTR-1. QTL *qAE_6.1* consisted of a peak marker (SNP_6_9977282) that was also associated with GY and PFP under −P conditions, with explained phenotypic variation of 17.6%. In 75N conditions, one QTL (*qAE_12.1*) was detected with a peak marker (SNP_12_14936674) located on 14936674 bp of chromosome 12, whereas LOD score and phenotypic variation were 2.55% and 5.92%, respectively, since the additive value of −2.8 shows that the trait in the progeny came from donor parent Hao-an-nong.

Table 6. Putative QTLs identified for AE and PFP for six different nutrient conditions using 230 nutrient use efficient ILs.

S. No.	NuUE Condition [a]	Trait [b]	QTLs [c]	Chr	Position (bp) [d]	Peak Marker [e]	LOD Value	PVE% [f]	Additive Effect
1	−P	AE	qAE_2.1	2	542,635	SNP_2_542635	2.77	6.43	3.16
2	−P	AE	qAE_4.1	4	21,815,986	SNP_4_21815986	4.01	9.17	2.13
3	−P	AE	qAE_6.1	6	9,977,282	SNP_6_9977282	4.52	10.27	2.28
4	75N	AE	qAE_12.1	12	14,936,674	SNP_12_14936674	2.55	5.92	−2.80
5	−P	PFP	qPFP_1.1	1	20,345,712	SNP_1_20345712	8.64	18.71	2.87
6	−P	PFP	qPFP_2.2	2	4,481,943	SNP_2_4481943	11.68	24.44	−3.11
7	−P	PFP	qPFP_3.1	3	853,802	SNP_3_853802	8.93	19.28	3.05
8	−P	PFP	qPFP_4.1	4	21,833,014	SNP_4_21833014	10.59	22.44	3.03
9	−P	PFP	qPFP_5.1	5	5,588,965	SNP_5_5588965	3.39	7.81	2.10
10	−P	PFP	qPFP_6.1	6	9,977,282	SNP_6_9977282	8.07	17.60	2.71
11	−P	PFP	qPFP_7.1	7	28,234,334	SNP_7_28234334	7.35	16.16	2.82
12	−P	PFP	qPFP_8.2	8	8,437,588	SNP_8_8437588	9.90	21.14	−2.89
13	−P	PFP	qPFP_9.2	9	12,154,616	SNP_9_12154616	8.73	18.89	3.15
14	−P	PFP	qPFP_10.2	10	6,149,421	SNP_10_6149421	12.15	25.28	−3.16
15	−P	PFP	qPFP_11.1	11	1,706,087	SNP_11_1706087	5.93	13.25	2.29
16	75N	PFP	qPFP_1.2	1	23,091,103	SNP_1_23091103	5.89	13.17	3.45
17	75N	PFP	qPFP_2.1	2	4,342,883	SNP_2_4342883	9.44	20.25	−3.99
18	75N	PFP	qPFP_3.2	3	3,542,519	SNP_3_3542519	7.32	16.09	4.16
19	75N	PFP	qPFP_4.1	4	21,833,014	SNP_4_21833014	7.60	16.66	3.68
20	75N	PFP	qPFP_5.2	5	15,469,279	SNP_5_15469279	9.78	20.91	−4.05
21	75N	PFP	qPFP_6.2	6	12,183,428	SNP_6_12183428	4.46	10.14	2.92
22	75N	PFP	qPFP_7.2	7	28,303,039	SNP_7_28303039	7.21	15.89	4.04
23	75N	PFP	qPFP_8.1	8	322,877	SNP_8_322877	7.09	15.64	−3.50
24	75N	PFP	qPFP_9.1	9	12,154,616	SNP_9_12154616	7.87	17.19	4.23
25	75N	PFP	qPFP_10.1	10	146,531	SNP_10_146531	9.13	19.68	−3.92
26	75N	PFP	qPFP_11.2	11	2,514,115	SNP_11_2514115	3.66	8.41	2.57

[a] NPK condition trait: −P (negative phosphorus) and 75N (75% of nitrogen). [b] Trait name: AE (agronomic efficiency), PFP (partial factor productivity). [c] Name of identified QTL. [d] Nucleotide position (bp) of the SNP detected on each chromosome. [e] Peak marker of identified QTL. [f] Explanation of phenotypic variation.

2.5.2. Partial Factor Productivity (PFP)

A total of 22 QTLs were identified across all chromosomes, except chromosome 12, under two different nutrient conditions (−P and 75N). Eleven QTLs were associated with PFP with LOD scores ranging from 3.39 (SNP_5_5588965) to 12.15 (SNP_10_6149421), which explained phenotypic variation from 7.81% to 25.28% in −P conditions. Among these 11 QTLs, the highest LOD value of three peak markers, SNP_2_4481943 (LOD 11.68), SNP_8_8437588 (LOD 9.9), and SNP_10_6149421 (LOD 12.15), was contributed by the Hao-an-nong allele on chromosomes 2, 8, and 10, and eight QTLs remained on chromosomes 1, 3, 4, 5, 6, 7, 9, and 11 that had a positive additive value, indicating the contribution of recipient parent WTR-1. Under –P conditions, one of the peak markers (SNP_1_20345712) was located at the 20345712 bp position on chromosome 1, which is associated with BY and GY, with an LOD score of 4.01 and 8.63, explaining a PV of 9.17% and 18.71%, respectively (Table 6). Similarly, we discovered 11 significant QTLs for PFP under 75N conditions. Of the 11 QTLs detected, four (*qPFP_2.1, qPFP_5.2, qPFP_8.1,* and *qPFP_10.1*) were contributed by the donor parent allele of Hao-an-nong, and the remaining seven QTLs by the WTR-1 allele. An average LOD value of 7.22 and PV of 15.82% were explained by 11 QTLs. The two highest peak markers were SNP_5_15469279 and SNP_2_4342883, located on chromosomes 2 and 5, respectively, which explained PV of 20.25% and 20.91%.

2.5.3. Grain Yield (GY)

Forty-nine QTLs were identified for GY. Each QTL explained 5.89%~25.28% of the phenotypic variation, with respective LOD scores of 2.53~12.15. Among the total QTLs of GY, nine QTLs showed more than 20% of the phenotypic variance and were located on chromosome 2, 4, 5, 8, and 10. Four GY QTLs (*qGY_2.4, qGY_4.2, qGY_8.4,* and *qGY_10.5*) were detected in –P, two QTLs (*qGY_5.4, qGY_10.4*) in –NP and one QTL (*qGY_5.3*) significantly expressed under –NPK condition. Similarly, two GY QTLs (*qGY_2.2* and *qGY_5.2*) were identified in 75N nutrient condition. The peak marker of SNP_5_15469279 was located on chromosome 5, and it was associated with GY under three different nutrient conditions, such as 75N, –NPK, and –NP. Whereas LOD and phenotypic variations of this marker was recorded 9.78, 9.97, and 9.92 and 20.91%, 21.26%, and 21.18%, respectively, since additive value shows negative allele contributing from a donor parent Hao-an-nong.

2.5.4. Biomass Yield (BY)

A total of 40 QTLs were discovered in six nutrient conditions, except for NPK. Under deficiency of –N, –P, –NP, and –NPK conditions, 39 QTLs were linked with BY on all 12 chromosomes. Chromosomes 1, 2, 5, 8, 10, and 11 had more than three QTLs controlling BY. The contributions of these QTLs showed phenotypic variation, ranging from 5.99% to 17.63%, and their LOD values ranged from 2.57 to 8.08, respectively. These QTLs had the largest effects, with negative additive values showing them to be contributed by donor parent Hao-an-nong to the progeny. Under 75N conditions, one QTL (*qBY_11.3*), located on chromosome 11, at the position of 22440795 bp, had a PVE of 7.37% and it was contributed by the Hao-an-nong allele.

2.5.5. Percentage of Spikelet Fertility (PSPF)

In total, 52 QTLs were identified for six nutrient conditions and they were distributed on all chromosomes except for chromosome 12. These QTLs explained phenotypic variation ranging from 5.98% to 29.76%. Of these 52 QTLs, 6 QTLs for 75N, 3 QTLs for NPK, 10 QTLs for –P, and 11 QTLs for each nutrient condition (–N, –NP, and –NPK) were significantly expressed, with an LOD score ranging from 2.57 to 14.73. Under –N deficiency conditions, four QTLs (*qPSPF_3.3, qPSPF_6.4, qPSPF_8.2,* and *qPSPF_9.1*) and another three QTLs (*qPSPF_3.4, qPSPF_6.5,* and *qPSPF_9.2*) for —NPK conditions were recorded with PVE of >20%. The peak markers were located on chromosomes 3, 6, 8, and 9 with the favorable allele contribution from WTR-1.

2.5.6. 1000-Gwt

Out of the 261 QTLs, 63 were identified in 1000-Gwt for six nutrient (–N, –P, –NP, –NPK, 75N, and NPK) conditions. These QTLs together explained an average phenotypic variation ranging from 5.95% to 34.68% and an LOD score of 2.56 to 17.76. Among these QTLs, 24 (38%) were from donor parent Hao-an-nong, which was lower than WTR-1's contribution with 39 QTLs (62%). The QTLs *q1000Gwt_2.4, q1000Gwt_8.5,* and *q1000Gwt_10.1* had an LOD score greater than the threshold of 15 and explained PV of 17.76%, 15.31%, and 16.26%, respectively, under –NPK conditions. The three highest LOD values (17.76, 15.31, and 16.28) were observed on chromosomes 2, 8, and 10 and were contributed by the WTR-1 allele

2.5.7. Filled Grain Number (FGN)

A total of 31 QTLs had an LOD score of 2.52 to 7.59 (higher than the threshold level) and were located across all 12 chromosomes identified under six different nutrient conditions. The QTLs had phenotypic variation ranging from 5.87% to 15.84%. The QTLs *qFGN_4.2*, *qFGN_5.1*, and *qFGN_7.1* on chromosomes 4, 5, and 7 had a large additive effect (−118.1, −127.97, and −139.93), accounting for 7.16%, 10.17%, and 7.25% of the phenotypic variation, respectively. The alleles for increasing FGN came from both Hao-an-nong (10 QTLs) and WTR-1 (21 QTLs). Only two QTLs (*qFGN_1.1* and *qFGN_3.4*) controlling FGN on chromosomes 1 and 3 were detected, and explained 10.23% and 8.82% of the phenotypic variation under −N conditions, respectively. Out of 31 QTLs, 10 were contributed with an allele from Hao-an-nong, and the remaining 21 QTLs carried an allele from recipient parent WTR-1.

2.5.8. Hotspot QTLs for Multiple Traits

Among the total of 261 QTLs, 106 were identified as promising QTLs located over 14 hotspot regions of nine chromosomes (1, 2, 3, 4, 5, 7, 8, 9, and 11), and each chromosome had more than five QTLs contributing to single SNP peak markers (Figure 2a,b). The QTLs identified under each of the six nutrient conditions showed 32 QTLs identified in −P conditions, and these were significantly linked with six traits (PFP, GY, BY, FGP, 1000-Gwt, and PSPF) that explained PV ranging from 5.99% to 27.9% with LOD values of 2.57 to 13.64. Subsequently, several QTLs were identified under five nutrient conditions, −N (15 QTLs), −NP (20 QTLs), −NPK (18 QTLs), 75N (16 QTLs), and NPK (5 QTLs), and these were discovered across nine chromosomes. On each chromosome, the hotspot regions were identified that contributed to different traits under six different nutrient conditions. The peak marker SNP_1_20706894 on chromosome 1 was found to be linked to five QTLs associated with three traits, 1000-Gwt (75N, NPK, and −NPK), GY (−N), and PSPF (−P) conditions. Similarly, five QTLs located on chromosomes 2, 9, and 11, with peak markers SNP_2_4481943, SNP_9_15446817, and SNP_11_2514115, respectively, were associated with five traits: 1000-Gwt, PFP, BY, GY, and PSPF. Further analysis of the remaining hotspot regions, chr04 (eight QTLs), chr05 (seven QTLs), chr07 (eight QTLs-SNP_7_28303039 and six QTLs-SNP_7_28234334), chr08 (six QTLs) and chr11 (six QTLs), showed association with six prominent critical traits: 1000-Gwt, PFP, BY, GY, PSPF, and FGN. One to 14 hotspot QTLs per chromosome were classified (Table S1 and Figure 3) for the discovery of the top hotspot QTLs (a QTL that contributes to ≥10 QTLs). Out of these 14 hotspot QTLs, four QTLs were located on chromosome 3 (harbor-I QTLs), 11 QTLs on chromosome 5 (harbor-II QTLs), and 10 QTLs on chromosome 9 (harbor-III QTLs) and chromosome 11 (harbor-IV QTLs) at the nucleotide positions of 853,802; 5,588,965; 12,154,616; and 1,706,087 bp, respectively. The top hotspot regions of each chromosome were designated as QTLs harbor-I to -IV. While analyzing these top QTLs, harbor-I to -IV, we identified a positive allele contributed from recipient parent WTR-1 (Figure 3). Of these harbor-I to -IV QTLs, 16 QTLs were expressed on chr03, chr05, chr09, and chr11 for −P; nine QTLs on chr03, chr05, and chr11 for −NPK; seven QTLs on chr03, chr05, chr09, and chr11 for −NP; six QTLs on chr03, chr09, and chr11 for −N; five QTLs on chr03, chr05, and chr09 for 75N, and two QTLs on chr03 and chr05 for NPK, respectively. As a result, the top four hotspot QTLs explained phenotypic variation ranging from 6.09% to 26.97% and LOD scores ranging from 2.62 to 13.11.

2.5.9. Fine-Tuning of QTLs Harbor-I to -IV

The top four hotspot QTLs of chromosomes 3, 5, 9, and 11 were refined through QTARO databases. The first QTL (harbor-I) was identified on chromosome 3, between peak markers SNP_3_853802 and SNP_3_16294363, covering a span of 15.44 Mbp regions. In this region, nine genes (*pez1*, *OsIRO3*, *Mit*, *OsApx1*, *RPN10*, *OsFRDL1*, *OsMTP8.1*, *OsGS1; 2*, and *OsPT2*) and four QTLs (*qRFWw3*, *n-p3*, *qDLR3*, and *qZNT-3*), documented based on earlier reports, were associated with soil stress tolerance mechanism-related traits, for uptake and translocation from roots to shoots, for various nutrients, such as iron, cadmium, phosphate, and manganese (Table S2). Two transcription factors (*IDEF2* and *OsHsfA4a*) and one gene (*OsZIP5*), identified in the region spanning 21.5 Mbp on chromosome 5 (QTL harbor-II); two QTLs (*qLBI-9* and *qALSRL-9*) and one gene (*OsSTR1*), located on chromosome 9 covering a 13.78 Mbp region (QTL harbor-III); and two QTLs (*qDLR11* and *qRRE-11*), identified on chromosome 11 covering a region of 26.62 Mbp (QTL harbor-IV), were identified. Of these eight QTLs, QTL harbors distributed on three chromosomes (3, 9, and 11) accounted for PV ranging from 3.47% (*qDLR3*) to 18.3% (*qALSRL-9*) and LOD value ranging from 2.64 (*qRRE-11*) to 15.23 (*qn-p3*).

3. Discussion

NuUE is a complex trait influenced by several factors, and it is considered a vital trait to improve rice grain yield productivity in marginal and rainfed lowland areas. Rice breeding programs need to incorporate NuUE traits as this helps resource-poor farmers to save on fertilizer, and allows maximization of valuable resources to increase profitability. To date, several morphological and agronomic traits have been identified, and they can be used as indicators of nitrogen (N) [52,64–66], phosphorus (P) [7,38,67–70], and potassium (K) [16,39,40,71–73] deficiency tolerance under both field and hydroponic conditions. Using different genetic backgrounds of mapping populations, such as RILs, NILs, BILs, ILs, CSSLs, DHs, and BC$_2$F$_4$, there are several reports on QTL identification with independent studies of low-N, -P, and -K conditions. The previously identified QTLs were mapped on different chromosomes using low-density linkage maps that were constructed by PCR-based molecular markers, such as RFLP, AFLP, SSR, and STS [26,29,31,34,39,48,52,55,58,67,74–76]. However, in this study, we used SNP markers to identify the QTLs.

For the identification of QTLs, single-marker analysis, simple interval mapping, and composite interval mapping are widely used methods [77,78]. The development of nutrient use efficient selective ILs, through an early generation backcross breeding approach, allows both the detection of QTLs for NuUE traits and the simultaneous development of promising breeding materials into varieties. This is the strength of selective introgression breeding first proposed by Tanksley [78], and later demonstrated for rice by Li et al. [79]. However, for QTL detection in selective introgression backcross breeding populations, we could use single marker analysis (SMA) and association mapping with single nucleotide polymorphism (SNP) markers, and use a higher threshold of LOD value of 2.5 to declare a strong QTL.

3.1. Analysis of Critical NuUE Traits

The phenotypic evaluation of 230 BC$_1$F$_5$ ILs varied significantly in six different NPK nutrient combinations: N, –P, –NP, –NPK, 75N, and NPK. Grain yield is a quantitative trait and a highly complex character for all crops [80], and is the essence of any breeding program. Various morphological and physiological plant traits contribute to grain yield. Yield components are inter-related with each other, indicating a complex chain of relationships, which is highly influenced by the environment [81]. The breeding strategy in rice mainly depends on the degree of associated characters as well as the magnitude and nature of variation [82,83]. Based on a higher grain yield level, 18 ILs for NPK, 23 ILs for 75N, 4 ILs for –P, and 1 IL for both –NPK and –NP had more than 40 g/plant, and in –N conditions the lowest GY was observed (34.96 g/plant).

An understanding of the correlations between traits is of great importance in breeding programs, especially if the selection of one of them is impaired by low heritability or difficulties of measurement and identification [84]. Information on trait correlations has been helpful as a basis for selection in breeding programs. Ashura [85] showed a positive correlation between the number of filled grains per panicle, number of panicles per plant, and 1000-grain weight for grain yield. A correlation study enables breeders to understand the major traits for which selection can be based on population improvement. GY had a significant positive correlation with BY and FGN in NPK; FGN in 75N; GWT, BY, and FGN in –P; PSPF, BY, and FGN in –N; and BY and FGN in –NP (*p* < 0.01) conditions. A negative correlation was recorded for GY with 1000-Gwt in NPK and for BY in 75N conditions. Under all six nutrient conditions, GY showed a negative relationship with BY and the remaining traits, GWT, PSPF, and FGN, had a significant positive correlation between the traits. Highly correlated traits were indicated to have a common genetic basis, suggesting that these eight relative phenotypic traits could be used for the evaluation of P-deficiency tolerance in BILs at the seedling stage [86].

3.2. Promising Traits of AE-Associated QTLs

The application of AE and PFP traits was essential to plant nutrients for an optimum quantity and right proportion, through the correct method and time of application, and this is the key to increased and sustained crop production [87]. These two traits are important for the measurement of nutrient use efficiency, and this also provides an integrative index that quantifies total economic output relative to the use of all the nutrient resources in the system [88]. In 1996, Cassman et al. [10] defined PFP and AE and showed that they could be increased by raising the amount, uptake, and use of available nutrients, and by increasing the efficiency with which applied nutrients are taken up by the crop and used to produce grain. In this study, we identified 42 ILs with higher AE in NPK, 23 ILs in –P, and 79 ILs in 75N, as compared with their parents, and they were explicit in >15 kg grain kg^{-1} N conditions. Yoshida [89] and Cassman et al. [10], respectively, estimated AE to be 15–25 kg kg^{-1} and 15–20 kg kg^{-1} in the dry season in farmers' fields in the Philippines. In a similar way, Wen-Xia et al. [90] mentioned the AE in two kinds of rice: Jinzao had an AE range of 8.02–20.14 kg grain kg^{-1} N and Shanyou63 had an AE range of 3.40–18.37 kg grain kg^{-1}, differing with N management. In hybrid rice, AE for applied P was 5.2 kg grain kg^{-1} P and for applied K was 11.8 kg grain kg^{-1} K, where the fertilizer application rate for NPK was 200–75–200 and 200–150–200 kg ha^{-1}, respectively; AE for applied P in non-hybrid rice was 2.3 kg grain kg^{-1} P and 4.7 kg grain kg^{-1} P, where the fertilizer rate was the same [81].

Four novel QTLs were identified for AE (*qAE_2.1*, *qAE_4.1*, *qAE_6.1*, and *qAE_12.1*) on chromosomes 2, 4, 6, and 7, which explained phenotypic variation ranging from 5.93% to 10.27%. For these four AE QTLs, each SNP marker position was analyzed at both sides of the 500 kp (up- and down-stream of 1 Mb) regions on the chromosome (Table S3). With the respective positions of the 1 Mb region on chromosome 2, seven genes were identified influencing diverse functional roles as a defense to pathogen resistance and physiological mechanisms. Of the seven genes revealed, *BiP3* for BLB resistance [91]; *OsHPL3* for multiple biotic disease resistance, such as brown planthopper, striped stem borer, and BLB [92]; *BiP1* (*Os06g0622700*) for increased seed storage protein, starch content in the endosperm, and also stress responses [93,94]; *OsHPR1* (*Os02g0104700*) involved in photo-respiratory metabolism [95]; and *OsNOA1* regulating chlorophyll biosynthesis, plastid development, and Rubisco formation in a temperature-dependent manner [95], were identified on chromosome 2. Similarly, on chromosome 6, genes *Pi9* and *Pi2* were noticed [96], which are related to blast disease resistance, along with another three QTLs (*qPHw6-2*, *qSFWd6*, *amy6-1*) related to drought tolerance and amylose content [97,98]. Interestingly, a major QTL for P deficiency tolerance that was previously mapped on chromosome 12 was found to be closely associated with peak marker SNP_12_14936674 [31], and in the same region of markers was linked with drought tolerance QTLs [99].

In PFP analysis, 16 ILs in NPK, 4 ILs in –P, and 151 ILs in 75N conditions yielded >50 kg grain kg^{-1} N and, in comparison with parents, 25 ILs (NPK), 61 ILs (–P), and 117 ILs (75N) yielded more than 50 kg grain kg^{-1} N in each condition. Amanullah et al. [100] showed, with a maize crop, a PFP of applied N of 36.62 kg grain kg^{-1} N and an AE of applied N of 22.49 kg grain kg^{-1} N. After applying DAP and SSP in fields, the AE of two fertilizer applications resulted in 13.01 and 13.71 kg grain kg^{-1} P, and PFP resulted in 63.58 and 61.92 kg grain kg^{-1} P. In this study, 22 QTLs for PFP traits were identified under 75N (11 QTLs) and –P (11 QTLs) conditions, with a PV ranging from 8.41% to 20.91% and 7.81% to 25.28%, respectively. Out of these 22 QTLs, six QTLs at SNP marker positions were analyzed on both sides of 500 kb (up- and down-stream of 1Mb) regions, which were explained by PV of more than 20% on chromosomes 2, 4, 5, 8, and 10. Five genes and six QTLs were identified in the 1 Mb region associated with PFP traits. On chromosome 2, three genes were identified as

OsWRKY71 [101], *Os4CL3* [102], and *OsWRKY45* [103,104], which were functionally related to defense signaling molecules, such as SA, Me, JA and lignin biosynthesis. Two QTLs (*qDSR_8* and *qALRR_8*) on chromosome 8 [105,106] and three QTLs (*qRRE_10*, *qRFW_10*, and *qRCCL_10*) [107,108] and one gene, *OsAT1/Spl18* [109], located on chromosome 10 revealed having tolerance of aluminum and alkaline stress, and also showed blast disease and ultraviolet-B resistance in rice. On chromosome 5, gene *OsWRKY45* was found to be closely associated with SNP_5_15469279, and it is involved in multiple biotic and abiotic stress tolerance/resistance, such as BLB, blast, sheath blight, drought, salinity, and cold [103,104,110].

3.3. Categorizing NuUE QTLs and Related Traits

Averages of 21.75 QTLs were distributed across all 12 chromosomes. Out of 261 QTLs, the highest numbers were located on chromosome 3 (28 QTLs), chromosome 11 (27 QTLs), chromosome 6 (26 QTLs), chromosome 1 (24 QTLs), and chromosomes 5, 8, and 10 (23 QTLs), and the remaining chromosomes 2, 4, 7, and 9 had a range of 18 to 22 QTLs. All the identified QTLs had PVE ranging from 5.87% to 34.68% and an LOD threshold range of 2.52–17.76. The lowest number of QTLs was observed on chromosome 12, with PVE from 5.89% to 9.49% and LOD value of 2.6 to 11.82. Based on the different nutrient conditions, the largest number of QTLs was detected under P deficiency (–P: 61 QTLs), and the lowest number of QTLs was identified for NPK (18 QTLs) conditions. For other nutrient conditions, the numbers were as follows: –NP (50 QTLs), –NPK (50 QTLs), –N (40 QTLs), and 75N (42 QTLs). A comprehensive literature survey revealed that a majority of the QTLs associated with low phosphorus were reported on chromosomes 1, 2, and 12 (recently reviewed by Mahender et al. [7], Ali et al. [20], and van de Wiel et al. [70], whereas, for low nitrogen conditions, the majority of morpho-physiological trait-linked QTLs in rice were located on chromosomes 3, 5, and 8 [34,52–54,76]. Under six different nutrient conditions, 24.1% of the QTLs were associated with 1000-Gwt, followed by 19.9% for PSPF, 18.7% for GY, 15.3% for BY, 11.8% for FGP, 8.4% for PFP, and 1.5% for AE (Figure 3).

3.4. Consistency and Comparisons of Major QTLs across Different Genetic Backgrounds

In this study, a total of 261 QTLs were identified using a 6K SNP array-based genetic linkage map analysis, in 230 BC_1F_5 introgression lines, that were tested under six different nutrient conditions. Out of 261 QTLs, 49 major QTLs were found with more than 20% PVE and LOD threshold value ranging from 9.44 to 17.76 distributed across all chromosomes, except for chromosomes 7, 11, and 12. Of these 49 QTLs, the highest number (27 QTLs) was associated with 1000-Gwt and the lowest number (6 QTLs) with PFP. For other traits, nine QTLs were associated with GY and seven QTLs with PSPF, which were detected across five nutrient conditions: 75N, –P, –N, –NP, and –NPK. Interestingly, out of these 261 QTLs, 14 QTLs were consistently detected under more than four nutrient treatments, indicating that these were critical traits that were controlled by similar genes under different nutrient conditions. The consistencies of QTLs were identified on nine chromosomes: 1, 2, 3, 5, 7, 8, 9, 10, and 11.

On chromosome 2, SNP_2_4342883 was linked with four QTLs (*qPFP*, *qBY*, *qGY*, and *q1000Gwt*), and SNP_5_15469279 was linked with three QTLs (*qPFP*, *qBY*, and *qGY*) on chromosome 5, which were expressed under four nutrient conditions: –P, –NP, –NPK, and 75N. On chromosome 7, SNP_7_28303039 was linked with four QTLs (*qPFP*, *qBY*, *qGY*, and *qPSPF*) and, at another location on the same chromosome, SNP_7_28234334 was associated with four QTLs (*q1000gwt*, *qPFP*, *qFGN*, and *qGY*) that were identified in –N, –NP, –NPK, and 75N conditions. In a similar way, on chromosome 8, SNP_8_23719048 was linked with three QTLs (*qFGN*, *qGY*, and *qPSPF*); on chromosome 10, SNP_10_18820606 was linked with two QTLs (*qFGN* and *qPSPF*); on chromosome 11 with six QTLs (*q1000gwt*, *qPFP* *qBY*, *qGY*, *qPSPF*, and *qFGN*); and, on chromosome 9, SNP_9_12154616 was linked with four QTLs (*q1000gwt*, *qPFP*, *qBY*, and *qFGN*), which were consistently recorded in all nutrient deficiency conditions, such as –N, –P, –NP, and –NPK, respectively. Under –N, –P, –NPK, and 75N conditions, on chromosome 1 (SNP_1_20706894), three QTLs (*q1000gwt*, *qGY*, and *qPSPF*) were identified. Further, two QTLs (*q1000gwt*, and *qBY*) on chromosome 11 (SNP_11_22440795) and five QTLs (*q1000gwt*, *qPFP*, *qGY*, *qPSPF*, and *qFGN*) on the same chromosome were consistently expressed in −N, −P, −NP, −NPK, and 75N conditions. However, five QTLs (*q1000Gwt*, *qPFP*, *qBY*, *qFGN*, and *qGY*) located on chromosome 3 (SNP_3_853802) and four QTLs (*q1000Gwt*, *qPFP*, *qGY*, and *qPSPF*) located on chromosome 5 (SNP_5_5588965) were detected consistently under six NuUE conditions: −N, −P, −NP, −NPK, 75N, and NPK. For all 14 of these QTLs, the WTR-1 alleles were significantly associated with key component traits in NuUE, but, in the case of chromosome 10, associated with two QTLs; these were contributed by the Hao-an-nong (HAN) allele in −N, −P, −NP and, −NPK conditions. These results indicated that the promising yield-related traits might share a similar genetic basis under different combinations of NPK treatments.

In previous studies, numerous mapping populations have been used to identify the QTLs associated with NPK deficiency tolerance traits in rice [26,28,29,31,34,48,52,55,58,61,67,74–76,111]. For each chromosome level, we identified seven or more QTLs located on chromosomes 2, 5, 8, and 10 and PVE ranged from 20.25% to 34.68%, and they were significantly associated with 1000-Gwt, PFP, FY, and PSPF under nutrient-deficient conditions of −P, −N, −NP, and −NPK. Several researchers in separate studies mentioned that chromosomes 2, 5, 8, and 10 were markedly associated with several agronomic traits, such as GY, nitrogen use efficiency (NUE), agronomic nitrogen use efficiency (agNUE), PSPF, nitrogen absorption ability (NAA), nitrogen content in shoots (NCS), harvest index (HI), root dry weight (RDW), shoot dry weight (SDW), and number of tillers (NT) QTLs under −N conditions [34,52,54,76]. Similarly, under low-P conditions, 36 QTLs were reported on the same chromosomes 2, 5, 8, and 10, respectively. These chromosomes were linked with phosphorus deficiency tolerance traits, such as phosphorus uptake (PUP), phosphorus use efficiency (PUE), phosphorus use efficiency for grain yield (PUEg), RDW, relative root length (RRL), panicle number per plant (PNPP), phosphors translocation (PT), phosphors translocation efficiency (PTE), spikelet fertility (SPF), 1000-Gwt, BY, relative root volume (RRV), and relative plant height (RPH), mapped in the different genetic backgrounds of mapping populations [48,55,56,59–61,67]. However, the major QTL, *Pup1*, was located on chromosome 12 at 54.5 cM, which explained PV of 78.8%, contributing to enhancing the uptake capacity of P from soils [31,55], and, further, the specified region of the 278-kbp sequence was significantly directly linked with P deficiency tolerance [112]. In this study, six QTLs were located on chromosome 12, with PVE ranging from 5.89% to 9.49%. Among these, one QTL for AE (*qAE_12.1*) was very close to this region (59.7cM), where *Pup1* is located on chromosome 12. Several researchers identified the same region as contributing toward tolerance of several biotic and abiotic stresses, such as drought, cold [97,113–116], and aluminum toxicity [116] in rice.

Interestingly, the identified associated QTLs governing seven traits (1000-Gwt, FGN, PSPF, BY, GY, AE, and PFP) on chromosome 6, under phosphorus-deficient conditions, significantly contributed to PVE ranging from 7.51% to 27.49%. Earlier independent studies of Ni et al. [31] and Wissuwa et al. [55] mapped major QTLs for RTA, RSDW, and RRDW on the same chromosome 6, and a group of P-responsive genes and transcription factors were also located in this region [113], which may confer tolerance of P deficiency [117]. However, in comparison to *Pup1* on chromosome 12, major QTLs and P-responsive genes were mostly located on chromosome 6, which indicates that both have independent genes and regulatory pathways [118].

In the deficiency of −N, −P, −NP, and −NPK conditions, four QTLs for GY were identified in −P (*qGY_2.4*, *qGY_4.2*, *qGY_8.4*, and *qGY_10.5*), two QTLs in −NP (*qGY_5.4* and *qGY_10.4*), and one QTL in −NPK (*qGY_5.3*), which were detected on different chromosomes (2, 4, 5, 8, and 10) from the analysis of 49 major QTLs. These QTLs were explained by their PV ranges from 20.25% to 25.28%. This was in contrast to Yue et al. [52], who reported three QTLs for 1000-Gwt on chromosome 3 and 7 and another three QTLs for GY located on chromosomes 1, 4, and 9, and the phenotypic variation explained by these QTLs ranged from 4.93% to 26.73% and 5.73% to 6.80%, respectively. However, in the present study, on chromosome 4, peak marker SNP_4_21833014 was found to be close to *qGYP-4* and was also flanked by RM273–RM241 [52]. In the present study of the genomic region of SNP markers from SNP_1_195334 to SNP_1_23839187 on chromosome 1, five QTLs were identified (*qBY*, *qGY*, *qFGN*, *qPSPF*, and *q1000Gwt*), which explained PV ranging from 7.18% to 26.83% under N-deficient conditions. However, in the same genomic region of chromosome 1, several QTLs governing relative grain yield, grain weight, 1000-Gwt, and nitrate transporter, such as *NRT 2.1* and *OsNRT2.3,b* are significantly involved in nitrogen deficiency tolerance and improving N uptake and enhancement of grain yield under nitrogen deficiency [119–123]

3.5. QTL Hotspots

Association with several traits within a single genomic region has immense potential value for the development of desired target trait enhancement through breeding and marker-assisted selection (MAS) applications. A single gene may affect more than two traits through pleiotropy or with closely linked genomic loci [124,125]. In the present study, 14 hotspot QTLs were identified to be located on chromosomes 1, 2, 3, 4, 5, 7, 8, 9, and 11, which had PVE ranges from 5.99% to 34.68% (Figure 3). Of these 14, we revealed the top four QTL harbor regions on chromosomes 3, 5, 9, and 11, and were designated as QTLs harbor-I to -IV. Each of these QTL harbor regions contained more than 10 QTLs on the same genomic regions of SNP markers, and significantly carried the positive allele from recipient parent WTR-1. The top four hotspot QTLs (a total of 45 QTLs) located on chromosomes 3, 5, 9, and 11 had PVE ranging from 6.09% to 26.97% and LOD values of 2.62 to 13.11. The four chromosomal (3, 5, 9, and 11) regions were shared by nitrogen (N) [34,52], phosphorus (P) [39,48,55,59–61], and potassium (K) [39] deficiency QTLs reported in different mapping populations in rice.

Significant reports exist on chromosomes 3, 6, and 11 involved in phosphorus deficiency tolerance in rice. Using BILs derived from inter-specific crosses of *Oryza sativa* L. X *O. rufipogon* Griff. [38], F$_3$ lines from the crosses between P-deficiency-tolerant variety NERICA10 and sensitive variety Hitomebore [126], and 271 introgression lines [60], identified QTL clusters for SDW, BY, and P uptake on chromosomes 3 and 11, which showed high significance for P-deficiency tolerance. Comparison with previous studies showed a majority of the candidate genes for P-deficiency tolerance (*Pup1*) and *PSTOL1* located on chromosome 12 [63,112,118,127] and P-responsive genes (such as *OsPTF1*) on chromosome 6 [117]. Our present study revealed that a major region of chromosomes 3, 5, 9, and 11 might provide novel loci to discover candidate genes related to P-deficiency tolerance, and these genes/alleles might play a co-regulated role for P-deficiency tolerance. These significant results would be more helpful to breeders and biotechnologists to understand the genetic and molecular basis of the physiological mechanisms of P-deficiency tolerance in rice.

In addition, we compared the presently identified QTL harbor regions of chromosomes 3, 5, 9, and 11 with previously reported genes and loci by using Q-TARO (http://qtaro.abr.affrc.go.jp/), and those are significantly associated with soil stress tolerance mechanisms, such as uptake and transportation of various nutrient elements from the soil to roots and shoots [128–135]. In QTL harbor-I on chromosome 3 (32.86 Mb) and QTL harbor-II (21.5 Mb) on chromosome 5 regions, nine genes and four QTLs were identified, along with IDEF2, *OsHsfA4a*, and *OsZIP* [131,136,137], which are involved in Fe homeostasis, phosphate uptake and translocation, distribution of Zn content from roots to shoots, and tolerance of cadmium stress (Table S2). Four QTLs (*qRFWw3*, *n-p3*, *qDLR3*, and *qZNT-3*) associated with multiple traits of low nitrogen and zinc toxicity tolerance showed resistance to alkaline stress [26,98,106,138]. QTL harbor-III (13.78 Mb) and QTL harbor-IV (26.62 Mb) were located on chromosomes 9 and 11. Two QTLs (*qLBI-9* and *qALSRL-9*) and one gene (*OsSTR1*) located on chromosome 9 were associated with mycorrhizal formation and resistance to Fe and Al toxicity [105,133,139], whereas chromosome 11 contained two QTLs (*qDLR11* and *qRRE-11*) responsible for tolerance of Al toxicity and alkaline stress [106,140].

According to the physical positions of one of the QTL hotspot regions, SNP marker SNP_3_853802 was associated with the highest number of 14 QTLs that were located on chromosome 3, and were distributed as five QTLs in –P (*q1000gwt_3*, *qPFP_3*, *qBY_3*, *qFGN_3*, and *qGY_3*), two QTLs in –N (*q1000gwt_3* and *qGY_3*), two QTLs in –NP (*qBY_3* and *qGY_3*), three QTLs in -NPK (*q1000gwt_3*, *qBY_3*, and *qGY_3*), and one QTL in 75N and NPK (*q1000gwt_3*). Hạnh et al. [141] identified three hotspot regions with response to low-nitrogen QTLs, which were flanked by RM265-RM165 on chromosome 1, RM3199-RM514 on chromosome 3, and RM080-RM281 on chromosome 8, which were significantly associated with the traits as total fresh weight of leaf blades (FW), RDW, SDW, nitrogen concentration in sheaths plus stem (NS), nitrogen concentration in leaf blades (NL), plant height (PH), chlorophyll content index (CCI), NUE, physiological nitrogen use efficiency (pNUE), and agNUE in RIL populations of IR64 and Azucena. In another study, Senthilvel et al. [58], using doubled haploid (DH) lines of IR64/Azucena in a pot experiment with three N doses (native, 0 kg/ha^{-1}; normal, 100 kg/ha^{-1}; and high, 200 kg/ha^{-1}), identified seven main-effect QTLs that were associated with NUE traits and plant grain yield on chromosome 3 [58]. In similar studies, with two different N fertilizer rates (0 N and $130–135 \text{ kg N ha}^{-1}$), in field conditions with RILs derived from two *Oryza sativa*–ssp. *indica* rice varieties (Zhenshan97/Minghui63), Wei et al. [76] reported a major QTL for grain yield (*qRGY3*) on chromosome 3, and another two QTLs (*qRGY7* and *qRGY11*) on chromosomes 7 and 11, detected with PV of 10.8% in 2006 and 16.0% in 2007. Moreover, three QTLs (*qRBM9-1*, *qRBM9-2*, and *qRBM10*) for biomass yield on chromosomes 9 and 10 together explained 33.6% of the total phenotypic variation. However, in corroboration with Senthilvel et al. [58] and Wei et al. [76], QTL mapping studies on DH and RIL populations revealed that chromosome 3 holds a promising trait and was associated with nitrogen use efficiency (NUE) in rice. In P deficiency, several QTLs have been reported on chromosome 3 [55,60,61], on chromosome 5 [48,59,61], on chromosome 9 [59–61], and on chromosome 11 [48,60,61], which explained PV ranging from 4.4% to 16.6% in different mapping populations of BILs, ILs, RILs, and BC_2F_3 in rice. Under low K, Wu et al. [39] identified a total 21 QTLs, related to PH, TN, SDW, RDW, relative potassium concentration in plant (RKC), relative potassium use efficiency (RKUE), and relative potassium uptake (RKUP) on chromosomes 2, 3, 7, and 8, that collectively explained PV from about 8% to about 15% using DH populations. On chromosome 3, *qFGN_3.6* significantly associated with RTN and other traits, such as PH, SDW, RDW, and RKUP, were flanked by RG179-RG403 and RZ284-RZ394, contributing PV of 9.7% to 14.4% [39]. Thus, it is suggested that the identified QTL harbor regions support the reported NPK QTLs, and are refined with chromosomal regions of genes and QTLs that were involved in multiple stress tolerance and transportation of nutrient elements from soil to shoots and grain, which could be valuable information for the understanding of their genetic and physiological mechanisms in future molecular breeding programs in rice.

4. Materials and Methods

4.1. Plant Materials

A BC_1F_5 mapping population contained 230 introgression lines derived from a cross between Weed Tolerant Rice 1 (WTR-1), as the recipient parent, and Hao-an-nong (HAN), as the donor parent. The field experiments were carried out at the experimental farm of the International Rice Research Institute (IRRI), Los Baños, Laguna, Philippines (14.11° N, 121.15° E), during the 2014 dry season (DS). Seeds of 230 ILs, parents, and four checks (PSB Rc82, NSIC Rc222, Apo, and IR74371-70-1-1) were sown in a seedling nursery bed, and 21-day-old seedlings were transplanted with a single seedling per hill. The early generation of the backcross population (BC_1F_2) was grown in one generation under low-input, rainfed, and irrigation conditions during the 2011 wet season (WS), and was followed by four consecutive generations over the DS and WS during 2012 and 2013. Phenotypic screening for efficient selection was practiced based on higher grain yield under six nutrient conditions. Subsequently, this led to the development of 230 ILs. The detailed of the breeding strategies are described in Jewel et al [142]. The NuUE phenotyping experiment was laid out in an alpha lattice design with two replications, using a plot size of two rows × 12 plants/row, and a spacing distance of $0.2 × 0.2$ m. NPK nutrients in the form of urea, superphosphate, and muriate of potash were applied at 160, 50, and 50 kg ha^{-1} in the DS, and at 90, 30, and 30 kg ha^{-1} in the WS, respectively. The checks and parents were replicated in all six nutrient conditions. The NPK fertilizers were applied five times in splits, as in basal level, and at 20, 36, 54, and 72 days after transplanting (DAT), respectively, against –N, –P, –NP, –NPK, 75N, and NPK conditions (Figure 4). Field management, including pest control, weeding, and irrigation, followed IRRI's standard experimental farm practices to avoid adverse effects on grain quality.

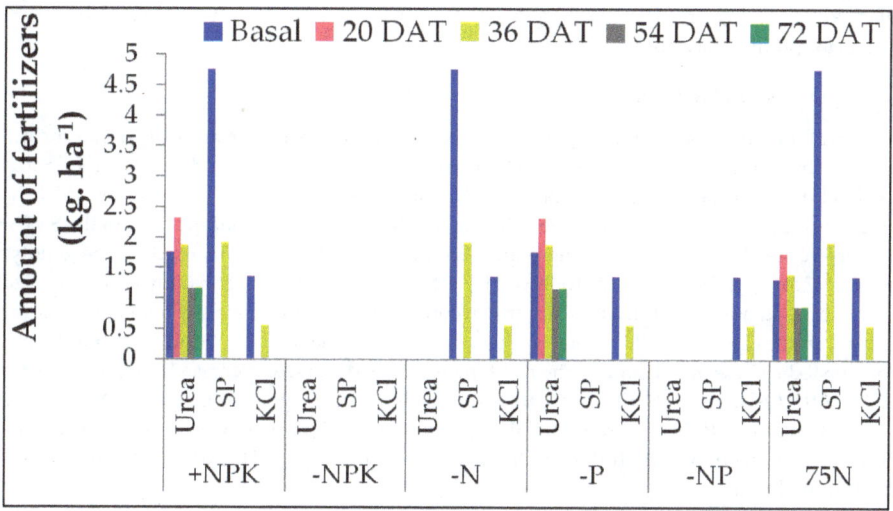

Figure 4. Application of fertilizers in six NuUE conditions with five splits.

4.2. Measurements of Agronomic and Yield-Attributed Traits

Each IL was selected in the middle row of plants, with three replications, and the plants were randomly sampled from each plot for phenotypic evaluation of six agronomic and yield-related traits. The six respective nutrient conditions, –N (PK fertilizer), –P (NK fertilizer), –NP (K fertilizer), –NPK (zero fertilizer), 75N (75% NPK fertilizer), and NPK (all nutrients), were recorded for seven prominent NuUE traits: agronomic efficiency (AE), partial factor productivity (PFP), grain yield (GY), biomass yield (BY), filled grains per plant (FGN), 1000-grain weight (1000-Gwt), and percentage of spikelet fertility (PSPF). Data analysis of the key phenotypic traits was performed with R software [143], and SPSS 17.0 software (IBM, Armonk, NY, USA) was used for the analysis of descriptive means, analysis of variance (ANOVA), Pearson correlation coefficient, and regression analysis, which were calculated among the ILs.

4.3. Calculation of Agronomic Efficiency (AE) and Partial Factor Productivity (PFP)

The increase in grain yield for each kg of fertilizer applied is known as AE. Using the formulas below, we calculated AE and PFP along with nitrogen fertilizer use condition. AE (kg kg^{-1}) = grain yield (N fertilized–N unfertilized) in kg ha^{-1}/Fertilizer N in kg ha^{-1} [144]. AE was computed for all the nutrient use efficient ILs, under six different nutrient conditions. The recommended fertilizer application rate was NPK = 160–50–50 kg ha^{-1}. Among them, three possible AE formulas were depicted:

(i) $AE_{(N)} = (Y_{NPK} - Y_{0NPK})/F_N$

(ii) $AE_{(N)} = (Y_{NK} - Y_{0NPK})/F_N$

(iii) $AE_{(75N)} = (Y_{(75N)} - Y_{0NPK})/F_{75N}$

Similarly, PFP of applied nitrogen is also called nitrogen use efficiency. PFP (N) = YN (crop yield with applied N (kg ha^{-1})/amount of fertilizer N applied (kg ha^{-1}) [144]. Based on applied NPK fertilizer, the performance of the selected and fixed nutrient use efficient ILs, across six different levels of nutrient conditions, was determined by the PFP of applied nitrogen using the following formula:

(iv) $PFP_{(N)} = Y_{(+NPK)}/F_N$

(v) $PFP_{(N)} = Y_{(-P)}/F_N$

(vi) $PFP_{(N)} = Y_{(75N)}/F_N$

Y_{+NPK} = crop yield with applied NPK fertilizer (kg ha^{-1}); F_{NPK} = amount of fertilizer NPK applied (kg ha^{-1}); F_N = fertilizer N in kg ha^{-1}; AE = agronomic efficiency applied nitrogen; PFP = partial factor productivity applied nitrogen.

4.4. Genotyping via 6K SNP Array

A total of 230 ILs of plant genomic DNA were extracted from leaf tissue using the cetyl trimethylammonium bromide (CTAB) method [145] and quantified by a NanoDrop 8000 spectrophotometer (ThermoFisher Scientific, Waltham, MA, USA). The concentration of DNA was adjusted to 50 ng μL^{-1}, and approximately 200 ng of DNA from each genotype used in the SNP array. The complete set of the BC_1F_5 mapping population, along with the parents used for the genotyping with 6K SNP array technology, was followed by DNA quantification, incubation, and hybridization of bead chips, staining, and image scanning according to the manufacturer's instructions for the Illumina Infinium assay, and this work was conducted in the Genotyping Services Laboratory of the International Rice Research Institute. The resulting intensity data were processed using the genotyping module V2011.1 of Genome Studio software (Illumina Inc., San Diego, CA, USA) for SNP calling. The generated genotypic data indicated that 704 high-quality SNP markers were identified and further used in the construction of high-density linkage maps for NuUE QTLs, related to agronomic and yield-attributed traits.

4.5. Mapping of QTLs and Hotspot Regions for NuUE

The mapping of QTLs was carried out with IciMapping software (QTL IciMapping version 4.0) [146] using a single marker analysis (SMA) method. The permutation method was used to obtain an empirical threshold for claiming QTLs based on 1000 runs of randomly shuffling the trait values [147], and the logarithm of odds (LOD) value threshold for claiming QTLs was 2.5. The genetic distance (cM) between SNP marker positions was changed to physical distance (kb), with 1 cM equal to 260 kbp [148,149]. For seven promising critical agronomic and yield-related traits under six NuUE conditions, significant markers on the same chromosome, linkage disequilibrium (LD), and phenotypic trait data were calculated by the genetics package in R software [150]. A graphical representation of the linkage map was constructed using MapChart software [151]. Further, for fine-tuning of hotspot QTLs, we used the Q-TARO database (http://qtaro.abr.affrc.go.jp/) for the analysis of previously published co-localization of QTLs and genes related to soil stress tolerance in the rice genome.

5. Conclusions

In many plant breeding programs, the development of rice varieties with NuUE is being considered as a means to reduce the usage of NPK fertilizers by improving their use efficiency, thereby improving grain yield productivity. In the present study, selective introgression lines derived from Weed Tolerant Rice 1, as the recipient parent, and Hao-an-nong, as the donor parent, enabled us to identify a large number of QTLs (261 putative QTLs) linked with NuUE traits, of which 49 QTLs showed high PVE ranging from 20.25% to 34.68%. Among them, 22 QTLs reported as novel QTLs were responsible for PFP and AE traits. They were identified among the promising top four hotspot QTLs (QTLs harbor I-IV), which comprised more than ten QTLs associated with critical NuUE traits, such as 1000-Gwt, PFP, BY, FGN, GY, and PSPF. The hotspot regions of QTLs, expressed across all six NuUE conditions, suggested an underlying uniform basis of genetic mechanisms, contributing to the tolerance of these traits and associated with tightly linked genes or QTLs, or pleiotropic regulations. Corroboration with several research efforts, of the earlier discovered QTLs for NUE traits, revealed that most of the genomic region was confirmed and their positions precisely confirmed by significant SNP peak markers with high LOD and PVE values, and this could have potential value for introgression of the target using MAS. The list of QTLs and refined hotspot regions will facilitate further validation in systematic breeding for specific adaptability under low-input conditions, and suggest that hotspot genomic regions could be used as targets for a superior understanding of the NuUE mechanism and for improving NuUE traits in rice. In the future, identified promising harbor QTLs would be useful for developing elite introgression breeding lines comprising positive QTLs via marker-assisted selection, and also using them for carrying them forward by a pyramiding approach over the ILs, with desirable QTLs within the same population. Further, this will lead to fine-mapping and molecular cloning of the critical loci that will be useful for enhancing grain yield and quality under low-input fertilizer management conditions.

Supplementary Materials: Supplementary materials can be found at http://www.mdpi.com/1422-0067/20/4/900/s1.

Author Contributions: J.A. and Z.A.J. conducted a phenotypic experiment; Z.A.J., J.A., J.H., Y.P., and A.M. analyzed genotypic data and mapped the QTLs; Z.A.J., J.A., and A.M. arranged the outline of the content in the manuscript and prepared a draft article; J.A. and Z.L. helped conceive the basic idea, gave suggestions, corrected the entire article, and improved on the prospects in breeding programs. All authors read and approved the research article.

Funding: The authors would like to thank and acknowledge the Bill & Melinda Gates Foundation (BMGF) for providing a research grant to ZL for the Green Super Rice Project under ID OPP1130530. We would also like to thank the Department of Science and Technology (DOST)-Philippine Council for Agriculture, Aquatic and Natural Resources Research and Development (PCCARRD), Philippines, for providing funds to JA through the University of the Philippines, Los Baños (UPLB), under the improved resource use efficient (iRUE) rice project.

Acknowledgments: The authors would like to thank the anonymous three reviewers from the International Journal of Molecular Science and two internal reviewers from at the International Rice Research Institute for their valuable suggestions and constructive comments that helped improve this manuscript.

Conflicts of Interest: The authors declare that the research review was conducted in the absence of any commercial or economic associations that could be construed as potential conflicts of interest.

References

1. Godfray, H.C.; Beddington, J.R.; Crute, I.R.; Haddad, L.; Lawrence, D.; Muir, J.F.; Pretty, J.; Robinson, S.; Thomas, S.M.; Toulmin, C. Food security: The challenge of feeding 9 billion people. *Science* **2010**, *327*, 812–818. [CrossRef] [PubMed]

2. FAO. The State of Food Insecurity in the World 2009. Available online: http://www.fao.org/tempref/docrep/fao/012/i0876e/i0876e_flyer.pdf (accessed on 30 October 2009).

3. Tilman, D.; Balzer, C.; Hill, J.; Befort, B.L. Global food demand and the sustainable intensification of agriculture. *Proc. Natl. Acad. Sci. USA* **2011**, *108*, 20260–20264. [CrossRef] [PubMed]

4. Edgerton, M.D. Increasing crop productivity to meet global needs for feed, food, and fuel. *Plant Physiol.* **2009**, *149*, 7–13. [CrossRef] [PubMed]

5. Naresh, R.K.; Gupta, R.K.; Ashok, K.; Singh, B.; Prakash, S.; Kumar, S.; Rathi, R.C. Direct-seeding and reduced-tillage options in the rice-wheat system of the Western Indo-Gangetic Plains. *Int. J. Agric. Sci.* **2011**, *7*, 197–208.

6. Bindraban, P.S.; Dimkpa, C.; Nagarajan, L.; Roy, A.; Rabbinge, R. Revisiting fertilisers and fertilisation strategies for improved nutrient uptake by plants. *Biol. Fertil. Soils* **2015**, *51*, 897–911. [CrossRef]

7. Mahender, A.; Anandan, A.; Pradhan, S.K.; Singh, O.N. Traits-related QTLs and genes and their potential applications in rice improvement under low phosphorus condition. *Arch. Agron. Soil Sci.* **2017**, *64*, 449–464. [CrossRef]

8. Ali, J.; Xu, J.L.; Gao, Y.M.; Ma, X.F.; Meng, L.J.; Wang, Y.; Pang, Y.L.; Guan, Y.S.; Xu, M.R.; Revilleza, J.E.; et al. Harnessing the hidden genetic diversity for improving multiple abiotic stress tolerance in rice (*Oryza sativa* L.). *PLoS ONE* **2017**, *12*, e0172515. [CrossRef]

9. Raun, W.R.; Johnson, G.V. Improving nitrogen use efficiency for cereal production. *Agron. J.* **1999**, *91*, 357–363. [CrossRef]

10. Cassman, K.G.; Gines, G.C.; Dizon, M.A.; Samson, M.I.; Alcanter, J.M. Nitrogen use efficiency in tropical lowland rice systems: Contribution from indigenous and applied nitrogen. *Field Crop. Res.* **1996**, *47*, 1–12. [CrossRef]

11. Giles, J. Nitrogen study fertilizes fears of pollution. *Nature* **2005**, *433*, 791. [CrossRef]

12. Vitousek, P.M.; Naylor, R.; Crews, T.; David, M.B.; Drinkwater, L.E.; Holland, E.; Johnes, P.J.; Katzenberger, J.; Martinelli, L.A.; Matson, P.A.; et al. Agriculture. Nutrient imbalances in agricultural development. *Science* **2009**, *19*, 324, 1519–1520. [CrossRef] [PubMed]

13. Hirel, B.; Tétu, T.; Lea, P.J.; Dubois, F. Improving nitrogen use efficiency in crops for sustainable agriculture. *Sustainability* **2011**, *3*, 1452–1485. [CrossRef]

14. Song, K.; Xue, Y.; Zheng, X.; Weiguang, L.; Qiao, H.; Qin, Q.; Yang, J. Effects of the continuous use of organic manure and chemical fertilizer on soil inorganic phosphorus fractions in calcareous soil. *Sci. Rep.* **2017**, *7*, 1164. [CrossRef] [PubMed]

15. Kant, S.; Bi, Y.; Rothstein, S.J. Understanding plant response to nitrogen limitation for the improvement of crop nitrogen use efficiency. *J. Exp. Bot.* **2011**, *62*, 1499–1509. [CrossRef] [PubMed]

16. Fageria, N.K.; Baligar, V.C.; Li, Y.C. Nutrient uptake and use efficiency by tropical legume cover crops at varying pH of an Oxisol. *J. Plant Nutr.* **2014**, *37*, 294–311. [CrossRef]

17. Das, G.; Rao, G.J.N. Molecular marker assisted gene stacking for biotic and abiotic stress resistance genes in an elite rice cultivar. *Front. Plant Sci.* **2015**, *6*, 1–18. [CrossRef] [PubMed]

18. Anumalla, M.; Roychowdhury, R.; Geda, C.K.; Mazid, M.; Rathoure, A.K. Utilization of plant genetic resources and diversity analysis tools for sustainable crop improvement with special emphasis on rice. *J. Adv. Res.* **2015**, *3*, 1155–1175.

19. Pang, Y.; Chen, K.; Wang, X.; Wang, W.; Xu, J.; Ali, J.; Li, Z. Simultaneous improvement and genetic dissection of salt tolerance of rice (*Oryza sativa* L.) by designed QTL pyramiding. *Front. Plant Sci.* **2017**, *8*, 1275. [CrossRef]

20. Ali, J.; Jewel, Z.A.; Mahender, A.; Anandan, A.; Hernandez, J.; Li, Z. Molecular genetics and breeding for nutrient use efficiency in rice. *Int. J. Mol. Sci.* **2018**, *19*, 1762. [CrossRef]

21. Jewel, Z.; Ali, J.; Pang, Y.; Anumalla, M.; Acero, B.; Hernandez, J.; Xu, J.; Li, Z. Developing Green Super Rice Varieties with High Nutrient Use Efficiency by Phenotypic Selection Under Varied Nutrient Conditions. *Preprints* **2018**. [CrossRef]

22. Novoa, R.; Loomis, R.S. Nitrogen and plant production. *Plant Soil* **1981**, *58*, 177–204. [CrossRef]

23. Baligar, V.C.; Fageria, N.K.; He, Z.L. Nutrient use efficiency in plants. *Commun. Soil Sci. Plant Anal.* **2001**, *32*, 921–950. [CrossRef]

24. McDonald, G.; William, B.; Chunyuan, H.; David, L. Nutrient use efficiency. In *Genomics and Breeding for Climate-Resilient Crops*; Springer: Berlin/Heidelberg, Germany, 2013; Volume 2, pp. 333–393.

25. Stahl, A.; Pfeifer, M.; Frisch, M.; Wittkop, B.; Snowdon, R.J. Recent genetic gains in nitrogen use efficiency in oilseed rape. *Front. Plant Sci.* **2017**, *8*, 963. [CrossRef] [PubMed]

26. Lian, X.; Xing, Y.; Yan, H.; Xu, C.; Li, X.; Zhang, Q. QTLs for low nitrogen tolerance at seedling stage identified using a recombinant inbred line population derived from an elite rice hybrid. *Appl. Genet.* **2005**, *112*, 85–96. [CrossRef] [PubMed]

27. Senthilvel, S.; Govindaraj, P.; Arumugachamy, S.; Latha, R.; Malarvizhi, P.; Gopalan, A.; Maheswaran, M. Mapping genetic loci associated with nitrogen use efficiency in rice (*Oryza sativa*) L. In Proceedings of the 4th International Crop Science Congress, Brisbane, Australia, 26 September–1 October 2004.

28. Srividya, A.; Vemireddy, L.R.; Hariprasad, A.S.; Jayaprada, M.; Sakile, S.; Puram, V.R.R.; Anuradha, G.; Siddiq, E.A. Identification and mapping of landrace derived QTL associated with yield and its components in rice under different nitrogen levels and environments. *Int. J. Plant Breed. Genet.* **2010**, *4*, 210–227. [CrossRef]

29. Tong, H.H.; Chen, L.; Li, W.P.; Mei, H.; Xing, Y.; Yu, X.; Xu, X.; Zhang, S.; Luo, L. Identification and characterization of quantitative trait loci for grain yield and its components under different nitrogen fertilization levels in rice (*Oryza sativa* L.). *Mol. Breed.* **2011**, *28*, 495–509. [CrossRef]

30. Vinod, K.K.; Heuer, S. Approaches towards nitrogen- and phosphorus-efficient rice. *AoB Plants* **2012**, *28*, 1–18. [CrossRef]

31. Ni, J.J.; Wu, P.; Senadhira, D.; Huang, N. Mapping QTLs for phosphorus deficiency tolerance in rice (*Oryza sativa* L.). *Theor. Appl. Genet.* **1998**, *97*, 1361–1369. [CrossRef]

32. Wissuwa, M.; Ae, N. Further characterization of two QTLs that increase phosphorus uptake of rice (*Oryza sativa* L.) under phosphorus deficiency. *Plant Soil* **2001**, *237*, 275–286. [CrossRef]

33. Lang, N.T.; Buu, B.C. Mapping QTLs for phosphorus deficiency tolerance in rice (*Oryza sativa* L.). *Omonrice* **2006**, *14*, 1–9.

34. Cho, Y.I.; Jiang, W.Z.; Chin, J.H.; Piao, Z.; Cho, Y.G.; McCouch, S.; Koh, H.J. Identification of QTLs associated with physiological nitrogen use efficiency in rice. *Mol. Cell* **2007**, *23*, 72–79.

35. Shimizu, A.; Kato, K.; Komatsu, A.; Motomura, K.; Ikehashi, H. Genetic analysis of root elongation induced by phosphorus deficiency in rice (*Oryza sativa* L.): Fine QTL mapping and multivariate analysis of related traits. *Appl. Genet.* **2008**, *117*, 987–996. [CrossRef] [PubMed]

36. Wang, Y.; Sun, Y.J.; Chen, D.Y.; Yu, S.B. Analysis of quantitative trait loci in response to nitrogen and phosphorus deficiency in rice using chromosomal segment substitution lines. *Acta Agron. Sin.* **2009**, *35*, 580–587. [CrossRef]

37. Chin, J.H.; Gamuyao, R.; Dalid, C.; Bustamam, M.; Prasetiyono, J.; Moeljopawiro, S.; Wissuwa, M.; Heuer, S. Developing rice with high yield under phosphorus deficiency: Pup1 sequence to application. *Plant Physiol.* **2011**, *156*, 1202–1216. [CrossRef] [PubMed]

38. Luo, X.D.; Liu, J.; Dai, L.F.; Zhang, F.T.; Wan, Y.; Xie, J.K. Linkage map construction and QTL identification of P-deficiency tolerance in *Oryza rufipogon* Griff. at early seedling stage. *Euphytica* **2017**, *213*, 96. [CrossRef]

39. Wu, P.; Ni, J.J.; Luo, A.C. QTLs underlying rice tolerance to low-potassium stress in rice seedlings. *Crop Sci.* **1998**, *38*, 1458–1462. [CrossRef]

40. Fageria, N.K.; Dos Santos, A.B.; De Moraes, M.F. Yield, potassium uptake, and use efficiency in upland rice genotypes. *Commun. Soil Sci. Plant Anal.* **2010**, *41*, 2676–2684. [CrossRef]

41. Feltus, F.A.; Wan, J.; Schulze, S.R.; Estill, J.C.; Jiang, N.; Paterson, A.H. An SNP resource for rice genetics and breeding based on subspecies *indica* and *japonica* genome alignments. *Genome Res.* **2004**, *14*, 1812–1819. [CrossRef]

42. Huang, Y.F.; Poland, J.A.; Wight, C.P.; Jackson, E.W.; Tinker, N.A. Using genotyping-by-sequencing (GBS) for genomic discovery in cultivated oat. *PLoS ONE* **2014**, *9*, e102448. [CrossRef]

43. Yu, H.; Xie, W.; Li, J.; Zhou, F.; Zhang, Q. A whole-genome SNP array (RICE6K) for genomic breeding in rice. *Plant Biotechnol. J.* **2014**, *12*, 28–37. [CrossRef]

44. Nadeem, M.A.; Nawaz, M.A.; Shahid, M.Q.; Doğan, Y.; Comertpay, G.; Yıldız, M.; Hatipoğlu, R.; Ahmad, F.; Alsaleh, A.; Labhane, N.; et al. DNA molecular markers in plant breeding: Current status and recent advancements in genomic selection and genome editing. *Biotechnol. Biotechnol. Equip.* **2018**, *32*, 261–285. [CrossRef]

45. Kumar, S.; Banks, T.W.; Cloutier, S. SNP Discovery through next-generation sequencing and its applications. *Int. J. Plant Genom.* **2012**, *2012*, 1–15. [CrossRef] [PubMed]

46. Xu, F.; Wang, W.; Wang, P.; Jun, L.M.; Chung Sham, P.; Wang, J. A fast and accurate SNP detection algorithm for next-generation sequencing data. *Nat. Commun.* **2012**, *3*, 1258. [CrossRef] [PubMed]

47. Thomson, M.J. High-throughput SNP genotyping to accelerate crop improvement. *Plant Breed. Biotechnol.* **2014**, *2*, 195–212. [CrossRef]

48. Wang, K.; Cui, K.; Liu, G.; Xie, W.; Yu, H.; Pan, J.; Huang, J.; Nie, L.; Shah, F. Identification of quantitative trait loci for phosphorus use efficiency traits in rice using a high density SNP map. *BMC Genet.* **2014**, *15*, 155. [CrossRef] [PubMed]

49. Chen, H.; Xie, W.; He, H.; Yu, H.; Chen, W.; Li, J.; Yu, R.; Yao, Y.; Zhang, W.; He, Y.; et al. A high-density SNP genotyping array for rice biology and molecular breeding. *Mol. Plant* **2014**, *7*, 541–553. [CrossRef] [PubMed]

50. Zhao, K.; Tung, C.W.; Eizenga, G.C.; Wright, M.H.; Ali, M.L.; Price, A.H.; Norton, G.J.; Islam, M.R.; Reynolds, A.; Mezey, J.; et al. Genome-wide association mapping reveals a rich genetic architecture of complex traits in *Oryza sativa*. *Nat. Commun.* **2011**, *2*, 467. [CrossRef]

51. Kurokawa, Y.; Noda, T.; Yamagata, Y.; Angeles-Shim, R.; Sunohara, H.; Uehara, K.; Furuta, T.; Nagai, K.; Jena, K.K.; Yasui, H.; et al. Construction of a versatile SNP array for pyramiding useful genes of rice. *Plant Sci.* **2016**, *242*, 131–139. [CrossRef]

52. Feng, Y.; Zhai, R.-R.; Lin, Z.-C.; Cao, L.-Y.; Wei, X.-H.; Cheng, S.-H. Quantitative trait locus analysis for rice yield traits under two nitrogen levels. *Rice Sci.* **2015**, *22*, 108–115.

53. Zhou, Y.; Tao, Y.; Tang, D.; Wang, J.; Zhong, J.; Wang, Y.; Yuan, Q.; Yu, X.; Zhang, Y.; Wang, Y.; et al. Identification of QTL associated with nitrogen uptake and nitrogen use efficiency using high throughput genotyped CSSLs in rice (*Oryza sativa* L.). *Front. Plant Sci.* **2017**, *8*, 1166. [CrossRef]

54. Dai, G.J.; Cheng, S.H.; Hua, Z.T.; Zhang, M.L.; Jiang, H.B.; Feng, Y.; Shen, X.H.; Su, Y.A.; He, N.; Ma, Z.B.; et al. Mapping quantitative trait loci for nitrogen uptake and utilization efficiency in rice (*Oryza sativa* L.) at different nitrogen fertilizer levels. *Genet. Mol. Res.* **2015**, *8*, 10404–10414. [CrossRef] [PubMed]

55. Wissuwa, M.; Yano, M.; Ae, N. Mapping of QTLs for phosphorus-deficiency tolerance in rice (*Oryza sativa* L.). *Appl. Genet.* **1998**, *97*, 777–783. [CrossRef]

56. Feng, M.; Xianwu, Z.; Guohu, M.I.; Ping, H.E.; Lihuang, Z.; Fusuo, Z. Identification of quantitative trait loci affecting tolerance to low phosphorus in rice (*Oryza sativa* L.). *Chin. Sci. Bull.* **2000**, *45*, 520–525.

57. Fang, P.; Wu, P. QTL × N-level interaction for plant height in rice (*Oryza sativa* L.). *Plant Soil* **2001**, *236*, 237–242. [CrossRef]

58. Senthilvel, S.; Vinod, K.K.; Malarvizhi, P.; Maheswaran, M. QTL and QTL environment effects on agronomic and nitrogen acquisition traits in rice. *J. Integr. Plant Biol.* **2008**, *50*, 1108–1117. [CrossRef] [PubMed]

59. Ping, M.U.; Huang, C.; Li, J.X.; Liu, L.F.; Li, Z.C. Yield trait variation and QTL mapping in a DH population of rice under phosphorus deficiency. *Acta Agron. Sin.* **2008**, *34*, 1137–1142.

60. Li, J.; Xie, Y.; Dai, A.; Liu, L.; Li, Z. Root and shoot traits responses to phosphorus deficiency and QTL analysis at seedling stage using introgression lines of rice. *J. Genet. Genom.* **2009**, *36*, 173–183. [CrossRef]

61. Xiang, C.; Ren, J.; Zhao, X.-Q.; Ding, Z.-S.; Zhang, J.; Wang, C.; Zhang, J.-W.; Joseph, C.A.; Zhang, Q.; Pang, Y.-L.; et al. Genetic dissection of low phosphorus tolerance related traits using selected introgression lines in rice. *Rice Sci.* **2015**, *22*, 264–274.

62. Yamaya, T.; Obara, M.; Nakajima, H.; Sasaki, S.; Hayakawa, T.; Sato, T. Genetic manipulation and quantitative trait loci mapping for nitrogen recycling in rice. *J. Exp. Bot.* **2002**, *53*, 917–925. [CrossRef]

63. Wissuwa, M.; Wegner, J.; Ae, N.; Yano, M. Substitution mapping of Pup1: A major QTL increasing phosphorus uptake of rice from a phosphorus-deficient soil. *Appl. Genet.* **2002**, *105*, 890–897. [CrossRef] [PubMed]

64. Duan, Y.H.; Zhang, Y.L.; Ye, L.T.; Fan, X.R.; Xu, G.H.; Shen, Q.R. Responses of rice cultivars with different nitrogen use efficiency to partial nitrate nutrition. *Ann. Bot.* **2007**, *99*, 1153–1160. [CrossRef] [PubMed]

65. Wu, L.; Yuan, S.; Huang, L.; Sun, F.; Zhu, G.; Li, G.; Fahad, S.; Peng, S.; Wang, F. Physiological mechanisms underlying the high-grain yield and high-nitrogen use efficiency of elite rice varieties under a low rate of nitrogen application in China. *Front. Plant Sci.* **2016**, *7*, 1024. [CrossRef] [PubMed]

66. Nguyen, H.T.T.; Dang, D.T.; Pham, C.V.; Bertin, P. QTL mapping for nitrogen use efficiency and related physiological and agronomical traits during the vegetative phase in rice under hydroponics. *Euphytica* **2016**, *212*, 473–500. [CrossRef]

67. Shimizu, A.; Yanagihara, S.; Kawasaki, S.; Ikehashi, H. Phosphorus deficiency-induced root elongation and its QTL in rice (*Oryza sativa* L.). *Appl. Genet.* **2004**, *109*, 1361–1368. [CrossRef]

68. Rose, T.J.; Wissuwa, M. Rethinking internal phosphorus utilization efficiency: A new approach is needed to improve PUE in grain crops. *Adv. Agron.* **2012**, *116*, 185–217.

69. Wissuwa, M.; Kretzschmar, T.; Rose, T.J. From promise to application: Root traits for enhanced nutrient capture in rice breeding. *J. Exp. Bot.* **2016**, *67*, 3605–3615. [CrossRef]

70. Wiel van de, C.C.M.; van der Linden, C.G.; Scholten, O.E. Improving phosphorus use efficiency in agriculture: Opportunities for breeding. *Euphytica* **2016**, *207*, 1–22. [CrossRef]

71. Wang, Y.; Wu, W.H. Genetic approaches for improvement of the crop potassium acquisition and utilization efficiency. *Curr. Opin. Plant Biol.* **2015**, *25*, 46–52. [CrossRef]

72. Xue, X.; Lu, J.; Ren, T.; Li, L.; Yousaf, M.; Cong, R.; Li, X. Positional difference in potassium concentration as diagnostic index relating to plant K status and yield level in rice (*Oryza sativa* L.). *Soil Sci. Plant Nutr.* **2016**, *62*, 31–38. [CrossRef]

73. Carmeis Filho, A.C.A.; Crusciol, C.A.C.; Guimarães, T.M.; Calonego, J.C.; Mooney, S.J. Impact of amendments on the physical properties of soil under tropical long-term no till conditions. *PLoS ONE* **2016**, *11*, e0167564. [CrossRef] [PubMed]

74. Tong, H.H.; Mei, H.W.; Yu, X.Q.; Xu, X.Y.; Li, M.S.; Zhang, S.Q.; Luo, L.J. Identification of related QTLs at late developmental stage in rice (*Oryza sativa* L.) under two nitrogen levels. *Acta Genet. Sin.* **2006**, *33*, 458–467. [CrossRef]

75. Piao, Z.; Li, M.; Li, P.; Zhang, C.; Wang, H.; Luo, Z.; Lee, J.; Yang, R. Bayesian dissection for genetic architecture of traits associated with nitrogen utilization efficiency in rice. *Afr. J. Biotechnol.* **2009**, *8*, 6834–6839.

76. Wei, D.; Cui, K.; Ye, G.; Pan, J.; Xiang, J.; Huang, J.; Nie, L. QTL mapping for nitrogen-use efficiency and nitrogen deficiency tolerance traits in rice. *Plant Soil* **2012**, *359*, 281–295. [CrossRef]

77. Tanksley, S.D. Mapping polygenes. *Annu. Rev. Genet.* **1993**, *27*, 205–233. [CrossRef] [PubMed]

78. Tanksley, S.D.; Young, N.D.; Paterson, A.H.; Bonierbale, M.W. RFLP mapping in plant breeding: New tools for an old science. *Nat. Biotechnol.* **1989**, *7*, 257–264. [CrossRef]

79. Li, Z.K.; Fu, B.Y.; Gao, Y.M.; Xu, J.L.; Ali, J.; Lafitte, H.; Jiang, Y.Z.; Rey, J.; Vijayakumar, C.; Maghirang, R.; et al. Genome-wide ILs and their use in genetic and molecular dissection of complex phenotypes in rice (*Oryza sativa* L.). *Plant Mol. Biol.* **2005**, *59*, 33–52. [CrossRef]

80. Ali, J.; Xu, J.L.; Gao, Y.; Fontanilla, M.; Li, Z.K. Breeding for yield potential and enhanced productivity across different rice ecologies through green super rice (GSR) breeding strategy. In *International Dialogue on Perceptions and Prospects of Designer Rice*; Muralidharan, K., Siddiq, E.A., Eds.; Society for the Advancement of Rice Research, Directorate of Rice Research: Hyderabad, India, 2013; pp. 60–68.

81. Rao, T.N. Improving Nutrient Use Efficiency: The Role of Beneficial Management Practices. In *Better Crops-India*; IPNI–India Program 133: Gurgaon, India, 2007; Volume 1, pp. 6–7.

82. Prasad, B.; Patwary, A.K.; Biswas, P.S. Genetic variability and selection criteria in fine rice (*Oryza sativa* L.). *Pak. J. Biol. Sci.* **2001**, *4*, 1188–1190.

83. Zahid, M.A.; Akhatar, M.; Sabar, N.; Zaheen, M.; Tahir, A. Correlation and path analysis studies of yield and grain traits in Basmati rice (*Oryza sativa* L.). *Asian J. Plant Sci.* **2006**, *5*, 643–645.

84. Cruz, C.D.; Regazzi, A.J. *Biometric Models Applied to Genetic Improvement*, 2nd ed.; Universidade Federal de Viçosa: Viçosa, Brazil, 1997; 390p.

85. Ashura, L. Inter-relationship between yield and some selected agronomic characters in rice. *Afr. Crop Sci. J.* **1998**, *6*, 83–88.

86. Lebreton, C.; LazicJancic, V.; Steed, A.; Pekic, S.Q.S. Identification of QTL for drought responses in maize and their use in testing causal relationships between traits. *J. Exp. Bot.* **1995**, *46*, 853–865. [CrossRef]

87. Cisse, L.; Amar, B. The importance of phosphatic fertilizer for increased crop production in developing countries. In Proceedings of the AFA 6th International Annual Conference, Cairo, Egypt, 31 January–2 February 2000.

88. Yadav, R.L. Assessing on-farm efficiency and economics of fertilizer N, P and K in rice-wheat systems of India. *Field Crop Res.* **2003**, *18*, 39–51. [CrossRef]

89. Yoshida, S. *Fundamentals of Rice Crop Science*; IRRI: Los Baños, Philippines, 1981; 269p.

90. Xie, W.X.; Wang, G.H.; Zhang, Q.C.; Guo, H.C. Effects of nitrogen fertilization strategies on nitrogen use efficiency in physiology, recovery, and agronomy and redistribution of dry matter accumulation and nitrogen accumulation in two typical rice cultivars in Zhejiang. *China J. Zhejiang Univ. Sci. B* **2007**, *8*, 208–216. [CrossRef] [PubMed]

91. Park, C.J.; Bart, R.; Chern, M.; Canlas, P.E.; Bai, W.; Ronald, P.C. Over-expression of the endoplasmic reticulum chaperone BiP3 regulates XA21-mediated innate immunity in rice. *PLoS ONE* **2010**, *5*, e9262. [CrossRef] [PubMed]

92. Ye, N.; Yang, G.; Chen, Y.; Zhang, C.; Zhang, J.; Peng, X. Two hydroxyl pyruvate reductases encoded by OsHPR1 and OsHPR2 are involved in photo respiratory metabolism in rice. *J. Integr. Plant Biol.* **2014**, *56*, 170–180. [CrossRef]

93. Yasuda, H.; Hirose, S.; Kawakatsu, T.; Wakasa, Y.; Takaiwa, F. Over-expression of BiPhas inhibitory effects on the accumulation of seed storage proteins in endosperm cells of rice. *Plant Cell Physiol.* **2009**, *50*, 1532–1543. [CrossRef]

94. Wakasa, Y.; Yasuda, H.; Oono, Y.; Kawakatsu, T.; Hirose, S.; Takahashi, H.; Hayashi, S.; Yang, L.; Takaiwa, F. Expression of ER quality control-related genes in response to changes in BiP1 levels in developing rice endosperm. *Plant J.* **2011**, *65*, 675–689. [CrossRef]

95. Yang, Q.; He, H.; Li, H.; Tian, H.; Zhang, J.; Zhai, L.; Chen, J.; Wu, H.; Yi, G.; He, Z.H.; et al. NOA1 functions in a temperature-dependent manner to regulate chlorophyll biosynthesis and Rubisco formation in rice. *PLoS ONE* **2011**, *6*, e20015. [CrossRef]

96. Qu, S.; Liu, G.; Zhou, B.; Bellizzi, M.; Zeng, L.; Dai, L.; Han, B.; Wang, G.L. The broad-spectrum blast resistance gene *Pi9* encodes a nucleotide-binding site–leucine-rich repeat protein and is a member of a multigene family in rice. *Genetics* **2006**, *172*, 1901–1914. [CrossRef]

97. Amarawathi, Y.; Singh, R.; Singh, A.K.; Singh, V.P.; Mohapatra, T.; Sharma, T.R.; Singh, N.K. Mapping of quantitative trait loci for basmati quality traits in rice (*Oryza sativa* L.). *Mol. Breed.* **2008**, *21*, 49–65. [CrossRef]

98. Cui, K.; Huang, J.; Xing, Y.; Yu, S.; Xu, C.; Peng, S. Mapping QTLs for seedling characteristics under different water supply conditions in rice (*Oryza sativa*). *Physiol. Plant.* **2008**, *132*, 53–68. [CrossRef] [PubMed]

99. Bernier, J.; Kumar, A.; Ramaiah, V.; Spaner, D.; Atlin, G. A large-effect QTL for grain yield under reproductive-stage drought stress in upland rice. *Crop Sci.* **2007**, *47*, 507–517. [CrossRef]

100. Amanullah, M.A.; Almas, L.K.; Amanullaj, J.; Zahir, S.; Rahman, H.; Khalil, S.K. Agronomic efficiency and profitability of P-fertilizers applied at different planting densities of maize in Northwest. *Pak. J. Plant Nutr.* **2012**, *35*, 331–341. [CrossRef]

101. Liu, X.; Bai, X.; Wang, X.; Chu, C. OsWRKY71, a rice transcription factor, is involved in rice defense response. *J. Plant Physiol.* **2007**, *164*, 969–979. [CrossRef] [PubMed]

102. Gui, J.; Shen, J.; Li, L. Functional characterization of evolutionarily divergent 4-coumarate: Coenzyme a ligases in rice. *Plant Physiol.* **2011**, *157*, 574–586. [CrossRef]

103. Tao, Z.; Liu, H.; Qiu, D.; Zhou, Y.; Li, X.; Xu, C.; Wang, S. A pair of allelic WRKY genes play opposite roles in rice-bacteria interactions. *Plant Physiol.* **2009**, *151*, 936–948. [CrossRef] [PubMed]

104. Tao, Z.; Kou, Y.; Liu, H.; Li, X.; Xiao, J.; Wang, S. OsWRKY45 alleles play different roles in abscisic acid signalling and salt stress tolerance but similar roles in drought and cold tolerance in rice. *J. Exp. Bot.* **2011**, *62*, 4863–4874. [CrossRef] [PubMed]

105. Nguyen, V.T.; Nguyen, B.D.; Sarkarung, S.; Martinez, C.; Paterson, A.H.; Nguyen, H.T. Mapping of genes controlling aluminum tolerance in rice: Comparison of different genetic backgrounds. *Mol. Genet. Genom.* **2002**, *267*, 772–780.

106. Qi, D.; Guo, G.; Lee, M.C.; Zhang, J.; Cao, G.; Zhang, S.; Suh, S.C.; Zhou, Q.; Han, L. Identification of quantitative trait loci for the dead leaf rate and the seedling dead rate under alkaline stress in rice. *J. Genet. Genom.* **2008**, *35*, 299–305. [CrossRef]

107. Sato, T.; Ueda, T.; Fukuta, Y.; Kumagai, T.; Yano, M. Mapping of quantitative trait loci associated with ultraviolet-B resistance in rice (*Oryza sativa* L.). *Appl. Genet.* **2003**, *107*, 1003–1008. [CrossRef]

108. Xue, Y.; Wan, J.; Jiang, L.; Wang, C.; Liu, L.; Zhang, Y.M.; Zhai, H. Identification of quantitative trait loci associated with aluminum tolerance in rice (*Oryza sativa* L.). *Euphytica* **2006**, *150*, 37–45. [CrossRef]

109. Mori, M.; Tomita, C.; Sugimoto, K.; Hasegawa, M.; Hayashi, N.; Dubouzet, J.G.; Ochiai, H.; Sekimoto, H.; Hirochika, H.; Kikuchi, S. Isolation and molecular characterization of a Spotted leaf 18 mutant by modified activation-tagging in rice. *Plant Mol. Biol.* **2007**, *63*, 847–860. [CrossRef] [PubMed]

110. Shimono, M.; Koga, H.; Akagi, A.; Hayashi, N.; Goto, S.; Sawada, M.; Kurihara, T.; Matsushita, A.; Sugano, S.; Jiang, C.J.; et al. Rice WRKY45 plays important roles in fungal and bacterial disease resistance. *Mol. Plant Pathol.* **2012**, *13*, 83–94. [CrossRef] [PubMed]

111. Feng, Y.; Cao, L.Y.; Wu, W.M.; Shen, X.H.; Zhan, X.D.; Zhai, R.R.; Wang, R.C.; Chen, D.B.; Cheng, S.H. Mapping QTLs for nitrogen-deficiency tolerance at seedling stage in rice (*Oryza sativa* L.). *Plant Breed.* **2010**, *129*, 652–656. [CrossRef]

112. Heuer, S.; Lu, X.; Chin, J.H.; Tanaka, J.P.; Kanamori, H.; Matsumoto, T.; De Leon, T.; Ulat, V.J.; Ismail, A.M.; Yano, M.; et al. Comparative sequence analyses of the major quantitative trait locus phosphorus uptake 1 (Pup1) reveal a complex genetic structure. *Plant Biotechnol. J.* **2009**, *7*, 456–457. [CrossRef] [PubMed]

113. Babu, R.C.; Nguyen, B.D.; Chamarerk, V.; Shanmugasundaram, P.; Chezhian, P.; Jeyaprakash, P.; Ganesh, S.K.; Palchamy, A.; Sadasivam, S.; Sarkarung, S.; et al. Genetic analysis of drought resistance in rice by molecular markers: Association between secondary traits and field performance. *Crop Sci.* **2003**, *43*, 1457–1469. [CrossRef]

114. Andaya, V.C.; Mackill, D.J. Mapping of QTLs associated with cold tolerance during the vegetative stage in rice. *J. Exp. Bot.* **2003**, *54*, 2579–2585. [CrossRef]

115. Peng, J.W.; Liu, Q.; Rong, X.M.; Zhu, H.M.; Xie, G.X.; Tang, G.R. Effects of different rational ratio of N, P, K fertilizer and amount of N fertilizer on photosynthesis character and yield of rice. *J. Hunan Agric. Univ. Nat. Sci.* **2004**, *30*, 123–127.

116. Wu, P.; Liao, C.Y.; Hu, B.; Yi, K.K.; Jin, W.Z.; Ni, J.J.; He, C. QTLs and epistasis for aluminum tolerance in rice (*Oryza sativa* L.) at different seedling stages. *Appl. Genet.* **2000**, *100*, 1295–1303. [CrossRef]

117. Yi, K.; Wu, Z.; Zhou, J.; Du, L.; Guo, L.; Wu, Y.; Wu, P. OsPTF1, a novel transcription factor involved in tolerance to phosphate starvation in rice. *Plant Physiol.* **2005**, *138*, 2087–2096. [CrossRef]

118. Gamuyao, R.; Chin, J.H.; Pariasca-Tanaka, J.; Pesaresi, P.; Catausan, S.; Dalid, C.; Slamet-Loedin, I.; Tecson-Mendoza, E.M.; Wissuwa, M.; Heuer, S. The protein kinase Pstol1 from traditional rice confers tolerance of phosphorus deficiency. *Nature* **2012**, *488*, 535–539. [CrossRef] [PubMed]

119. Moncada, P.; Martinez, C.P.; Borrero, J.; Chatel, M.; Gauch, H.; Guimaraes, E.; Tohme, J.; McCouch, S.R. Quantitative trait loci for yield and yield components in an *Oryza sativa* × *Oryza rufipogon* BC$_2$F$_2$ population evaluated in an upland environment. *Appl. Genet.* **2001**, *102*, 41–52. [CrossRef]

120. Araki, R.; Hasegawa, H. Expression of rice (*Oryza sativa* L.) genes involved in high-affinity nitrate transport during the period of nitrate induction. *Breed. Sci.* **2006**, *56*, 295–302. [CrossRef]

121. Katayama, H.; Mori, M.; Kawamura, Y.; Tanaka, T.; Mori, M.; Hasegawa, H. Production and characterization of transgenic plants carrying a high-affinity nitrate transporter gene (OsNRT2.1). *Breed. Sci.* **2009**, *59*, 237–243. [CrossRef]

122. Fu, Q.; Zhang, P.; Tan, L.; Zhu, Z.; Ma, D.; Fu, Y.; Zhan, X.; Cai, H.; Sun, C. Analysis of QTLs for yield-related traits in Yuanjiang common wild rice (*Oryza rufipogon* Griff.). *J. Genet. Genom.* **2010**, *37*, 147–157. [CrossRef]

123. Ogawa, S.; Valencia, M.O.; Lorieux, M.; Arbelaez, J.D.; McCouch, M.; Ishitani, M.; Selvaraj, M.G. Identification of QTLs associated with agronomic performance under nitrogen-deficient conditions using chromosome segment substitution lines of a wild rice relative, *Oryza rufipogon*. *Acta Physiol. Plant.* **2016**, *38*, 103. [CrossRef]

124. Hittalmani, S.; Shashidhar, H.E.; Bagali, P.G.; Ning, H.; Sidhu, J.S.; Singh, V.P.; Khush, G.S. Molecular mapping of quantitative trait loci for plant growth, yield and yield related traits across three diverse locations in a doubled haploid rice population. *Euphytica* **2002**, *125*, 207–214. [CrossRef]

125. Pelgas, B.; Bousquet, J.; Meirmans, P.G.; Ritland, K.; Isabel, N. QTL mapping in white spruce: Gene maps and genomic regions underlying adaptive traits across pedigrees, years and environments. *BMC Genom.* **2011**, *12*, 145. [CrossRef]

126. Koide, Y.; Pariasca-Tanaka, J.; Rose, T.; Fukuo, A.; Konisho, K.; Yanagihara, S.; Fukuta, Y.; Wissuwa, M. QTLs for phosphorus-deficiency tolerance detected in upland NERICA varieties. *Plant Breed.* **2013**, *132*, 259–265. [CrossRef]

127. Mukherjee, A.; Sarkar, S.; Chakraborty, A.S.; Yelne, R.; Kavishetty, V.; Biswas, T.; Mandal, N.; Bhattacharyya, S. Phosphate acquisition efficiency and phosphate starvation tolerance locus (PSTOL1) in rice. *J. Genet.* **2014**, *93*, 683–688. [CrossRef]

128. Yokoshom, K.; Yamaji, N.; Ueno, D.; Mitani, N.; Ma, J.F. OsFRDL1 is a citrate transporter required for efficient translocation of iron in rice. *Plant Physiol.* **2009**, *149*, 297–305. [CrossRef] [PubMed]

129. Ai, P.; Sun, S.; Zhao, J.; Fan, X.; Xin, W.; Guo, Q.; Yu, L.; Shen, Q.; Wu, P.; Miller, A.J.; et al. Two rice phosphate transporters, OsPht1;2 and OsPht1;6, have different functions and kinetic properties in uptake and translocation. *Plant J.* **2009**, *57*, 798–809. [CrossRef] [PubMed]

130. Zheng, L.; Ying, Y.; Wang, L.; Wang, F.; Whelan, J.; Shou, H. Identification of a novel iron regulated basic helix-loop-helix protein involved in Fe homeostasis in *Oryza sativa*. *BMC Plant Biol.* **2010**, *11*, 166. [CrossRef] [PubMed]

131. Lee, S.; Jeong, H.J.; Kim, S.A.; Lee, J.; Guerinot, M.L.; An, G. OsZIP5 is a plasma membrane zinc transporter in rice. *Plant Mol. Biol.* **2010**, *73*, 507–517. [CrossRef] [PubMed]

132. Ishimaru, Y.; Kakei, Y.; Shimo, H.; Bashir, K.; Sato, Y.; Nishizawa, N.K. A rice phenolic efflux transporter is essential for solubilizing precipitated apoplasmic iron in the plant stele. *J. Biol. Chem.* **2011**, *286*, 24649–24655. [CrossRef] [PubMed]

133. Gutjahr, C.; Radovanovic, D.; Geoffroy, J.; Zhang, Q.; Siegler, H.; Chiapello, M.; Casieri, L.; An, K.; An, G.; Guiderdoni, E.; et al. The half-size ABC transporters STR1 and STR2 are indispensable for mycorrhizal arbuscule formation in rice. *Plant J.* **2012**, *69*, 906–920. [CrossRef] [PubMed]

134. Chen, Z.; Fujii, Y.; Yamaji, N.; Masuda, S.; Takemoto, Y.; Kamiya, T.; Yusuyin, Y.; Iwasaki, K.; Kato, S.; Maeshima, M.; et al. Mn tolerance in rice is mediated by MTP8.1, a member of the cation diffusion facilitator family. *J. Exp. Bot.* **2013**, *64*, 4375–4387. [CrossRef]

135. Funayama, K.; Kojima, S.; Tabuchi-Kobayashi, M.; Sawa, Y.; Nakayama, Y.; Hayakawa, T.; Yamaya, T. Cytosolic glutamine synthetase1;2 is responsible for the primary assimilation of ammonium in rice roots. *Plant Cell Physiol.* **2013**, *54*, 934–943. [CrossRef]

136. Ogo, Y.; Kobayashi, T.; Nakanishi Itai, R.; Nakanishi, H.; Kakei, Y.; Takahashi, M.; Toki, S.; Mori, S.; Nishizawa, N.K. A novel NAC transcription factor, IDEF2, that recognizes the iron deficiency-responsive element 2 regulates the genes involved in iron homeostasis in plants. *J. Biol. Chem.* **2008**, *19*, 13407–13417. [CrossRef]

137. Shim, D.; Hwang, J.U.; Lee, J.; Lee, S.; Choi, Y.; An, G.; Martinoia, E.; Lee, Y. Orthologs of the class A4 heat shock transcription factor HsfA4a confer cadmium tolerance in wheat and rice. *Plant Cell* **2009**, *21*, 4031–4043. [CrossRef]

138. Dong, Y.J.; Ogawa, T.; Kamiunten, H.; Lin, D.Z.; Cheng, S.H.; Terao, H.; Matsuo, M. Detection of QTLs for zinc toxicity tolerance in rice (*Oryza sativa* L.). *Rice Genet. Newsl.* **2004**, *21*, 33–36.

139. Wan, J.L.; Zhai, H.Q.; Wan, J.M.; Yasui, H.; Yoshimura, A. Detection and analysis of QTLs for some traits associated with tolerance to ferrous iron toxicity in rice (*Oryza sativa* L.), using recombinant inbred lines. *Rice Genet. Newsl.* **2003**, *20*, 55–57.

140. Xue, Y.; Jiang, L.; Su, N.; Wang, J.K.; Deng, P.; Ma, J.F.; Zhai, H.Q.; Wan, J.M. The genetic basis and fine-mapping of a stable quantitative-trait loci for aluminium tolerance in rice. *Planta* **2007**, *227*, 255–262. [CrossRef] [PubMed]

141. Hanh, N.T.T.; Cuòng, P.V.; Pierre, B. Rice nitrogen use efficiency: Genetic dissection. *J. Sci. Dev.* **2013**, *11*, 814–825.

142. Jewel, Z.A.; Ali, J.; Pang, Y.; Mahender, A.; Acero, B.; Hernandez, J.; Xu, J.; Li, Z.K. Developing green super rice varieties with high nutrient use efficiency by phenotypic selection under varied nutrient conditions. *Crop J.* **2019**, in press.

143. R Core Team. *R: A Language and Environment for Statistical Computing*; R Foundation for Statistical Computing: Vienna, Austria, 2013.

144. Dobermann, A.R. *Nitrogen Use Efficiency-State of the Art*; Agronomy-Faculty Publications: Lincoln, NE, USA, 2005; 316p.

145. Murray, M.G.; Thampson, W.F. Rapid isolation of high molecular weight plant DNA. *Nucl. Acids Res.* **1980**, *8*, 4321–4325. [CrossRef]

146. Wang, J.; Li, H.; Zhang, L.; Meng, L. *Users' Manual of QTL IciMapping*; The Quantitative Genetics Group, Institute of Crop Science, Chinese Academy of Agricultural Sciences (CAAS): Beijing, China; Genetic Resources Program, International Maize and Wheat Improvement Center (CIMMYT): Mexico City, Mexico, 2014.

147. Churchill, G.A.; Doerge, R.W. Empirical threshold values for quantitative trait mapping. *Genetics* **1994**, *138*, 963–971.

148. Chen, M.; Presting, G.; Barbazuk, W.B.; Goicoechea, J.L.; Blackmon, B.; Fang, G.; Kim, H.; Frisch, D.; Yu, Y.; Sun, S.; et al. An integrated physical and genetic map of the rice genome. *Plant Cell* **2002**, *14*, 537–545. [CrossRef]

149. Tiwari, S.; Krishnamurthy, S.L.; Kumar, V.; Singh, B.; Rao, A.; Mithra, S.V.A.; Rai, V.; Singh, A.K.; Singh, N.K. Mapping QTLs for salt tolerance in rice (*Oryza sativa* L.) by bulked segregant analysis of recombinant inbred lines using 50K SNP chip. *PLoS ONE* **2016**, *11*, e0153610. [CrossRef]

150. Warnes, G.; Leisch, F. Package genetics: Population Genetics. 2015. Available online: https://cran.r-project.org/web/packages/genetics/genetics.pdf (accessed on 17 May 2016).

151. Voorrips, R.E. MapChart: Software for the Graphical Presentation of Linkage Maps and QTLs. *J Hered.* **2002**, *93*, 77–78. [CrossRef]

International Journal of
Molecular Sciences

Review

Molecular Genetics and Breeding for Nutrient Use Efficiency in Rice

Jauhar Ali [1,*,†], Zilhas Ahmed Jewel [1,†], Anumalla Mahender [1,†], Annamalai Anandan [2], Jose Hernandez [3] and Zhikang Li [4]

1   Rice Breeding Platform, International Rice Research Institute (IRRI), Los Baños, Laguna 4031, Philippines; jeweluplb@gmail.com (Z.A.J.); m.anumalla@irri.org (A.M.)
2   ICAR-National Rice Research Institute, Cuttack, Odisha 753006, India; anandanau@yahoo.com
3   Institute of Crop Science, College of Agriculture and Food Science, University of the Philippines Los Baños, Laguna 4031, Philippines; joehernandez56@gmail.com
4   Institute of Crop Sciences, Chinese Academy of Agricultural Science, Beijing 100081, China; zhkli1953@126.com
*   Correspondence: J.Ali@irri.org; Tel.: +63-2580-5600 (ext. 2541)
†   These authors contributed equally to this work.

Received: 25 April 2018; Accepted: 1 June 2018; Published: 14 June 2018

Abstract: In the coming decades, rice production needs to be carried out sustainably to keep the balance between profitability margins and essential resource input costs. Many fertilizers, such as N, depend primarily on fossil fuels, whereas P comes from rock phosphates. How long these reserves will last and sustain agriculture remains to be seen. Therefore, current agricultural food production under such conditions remains an enormous and colossal challenge. Researchers have been trying to identify nutrient use-efficient varieties over the past few decades with limited success. The concept of nutrient use efficiency is being revisited to understand the molecular genetic basis, while much of it is not entirely understood yet. However, significant achievements have recently been observed at the molecular level in nitrogen and phosphorus use efficiency. Breeding teams are trying to incorporate these valuable QTLs and genes into their rice breeding programs. In this review, we seek to identify the achievements and the progress made so far in the fields of genetics, molecular breeding and biotechnology, especially for nutrient use efficiency in rice.

Keywords: NPK fertilizers; agronomic traits; molecular markers; quantitative trait loci

1. Introduction

Global rice production increased by three-fold over the past three decades despite rice production constraints and rising input costs. Rice is a nutritionally important cereal crop and staple food of Asia. There is an urgent need for developing high-yielding, nutritious, resource use-efficient and multi-stress-tolerant rice varieties to keep up with the tremendous human population growth, especially in Asia, where rice remains the primary source of caloric intake. The yields of rice grain had seen remarkable improvement during the green revolution and post-green revolution. This increase in yield was primarily achieved through high-input-responsive varieties requiring more chemical fertilizers and pesticides and under an ample supply of irrigation water. This kind of approach that predominated over the past three to four decades now stands exhausted amidst our hope to raise productivity per se sustainably. We are now finding that yields are fast approaching a theoretical limit set by the crop's efficiency in harnessing applied inputs. In exploratory managed experimental plots, N fertilizer retrieval in a single year averaged 65% for maize, 57% for wheat and 46% for rice [1,2]. Alterations in the scale of farming operations and management practices such as tillage, seeding, weed and pest control, irrigation and harvesting usually resulted in on-farm variation (lower nutrient use efficiency) and did not accurately reflect the efficiencies obtained in the experimental plot. N recovery efficiency on average ranges from 20–30% for farmer-managed fields under rainfed conditions, from 30–55% under irrigated conditions [3,4] and rarely exceeds 50%.

Over the years, the rice varieties bred did not improve in nutrient absorption and were not developed to maximize nutrient absorption, but they have the capacity to use less than 50% of the applied nutrients. Breeding rice cultivars with improved nutrient use efficiency (NuUE) is becoming a prerequisite for lowering production costs. Such cultivars with NuUE protects the environment by reducing fertilizer application, decreasing the rate of nutrient application losses to ecosystems, decreasing input costs and improving rice yield with a guarantee for sustainability in agriculture while maintaining soil and ground water quality. On the other hand, improvement of NuUE is an essential prerequisite for expanding crop production into marginal lands with low nutrient availability. In light of high energy costs and progressively unpredictable resources, future agricultural systems with concern for improving yield productivity need to be more fruitful and efficient, especially considering fertilizer and irrigation water. In this context, the identification and development of rice varieties with superior grain yield under low input conditions have therefore become a high breeding priority [5]. Even though significant genotypic differences in nitrogen use efficiency exist in rice, genetic selection for this trait has not been carried out systematically [6–8]. This may be primarily because of the complexity involved in the overall phenotype and its evaluation and the non-availability of genetic tools to use. However, with the recent use of high-throughput single nucleotide polymorphism (SNP) markers with ease and high precision, this area of research needs improvement for better understanding [9–11].

Genetic and physiological traits often change with the interaction with environmental variables. Plants are efficient in the absorption and use of nutrients in controlled environments. Therefore, there is a need for a systematic breeding program to develop cultivars with high NuUE and water use efficiency (WUE) [12,13]. The traits involved, particularly nutrient absorption, transport, use and mobilization, should be identified to enhance NuUE and coupled with best management practices for sustainable agriculture.

Use of the wild species of *Oryza* and native landraces becomes imperative for exploiting the untapped reservoir of useful QTLs and genes, especially to broaden the genetic basis of rice and to enrich existing varieties [14,15]. Genetic selection and plant breeding techniques helped to develop rice varieties that are resistant to pests, diseases and adverse environmental conditions such as drought, submergence and salinity. However, for improving NuUE in rice crop, a proper genetic selection approach is necessary. Superior N-efficient genotypes are required as evidenced from the low recovery of N fertilizer, associated economic and environmental concerns and the lack of adoption of more efficient N management strategies [16,17]. Nitrogen use efficiency (NUE) mostly depends on interactions and the use of the nutrient in a proper way, water availability, light intensity, disease pressure and genotype, which could also be improved through appropriate genetic manipulation [6]. Plant ability to absorb and use nutrients under various environmental and ecological conditions is largely influenced by the genetic makeup and physiological components [12]. There are two major approaches to understand NuUE. First, the nutrient deficiency stress triggers a response of plants to it, which may lead to the identification of the processes affecting it. It would help us to understand how to sustain plants under low nutrient inputs. The second approach would be to exploit genetic variability (both natural and induced) through innovative molecular breeding schemes.

Molecular linkage genetic maps and quantitative trait locus (QTL) mapping technologies are helpful for estimating the number and position of the loci governing genetic variation using different types of segregating and fixed populations. Characterizing these loci to their map positions in the genome, as well as their phenotypic effects and epistatic interactions with other QTLs and loci [18–21] has enabled us to explore the genetic loci associated with complex traits such as drought, salinity, disease, NuUE and insect resistance in crop plants [18,22–29]. The rapid advancement in genome sequencing technologies and marker-aided breeding approaches has resulted in a change in breeding methods, providing new opportunities [5]. Association mapping is a method used to identify genes and QTLs underlying quantitatively inherited variation based on a diverse set of fixed lines. It allows the discovery of QTLs/genes using historical phenotypic data and eventually leads to identifying gene functions, under used alleles and allele combinations that can be useful for crop improvement [30,31]. Genome-wide association mapping depends on the strength of linkage disequilibrium (LD) across a diverse population besides identifying the relationships between markers and traits of agronomic and evolutionary interest [32,33].

Understanding the genetic basis of agronomic, physiological and morphological traits in rice is critical for developing new and improved rice varieties. Rice breeders can use this information to select parental lines for hybridization and screen segregating populations (Figure 1). Recently, researchers have been gaining access to the enormous online wealth of genomic and plant breeding resources, including high-quality genome sequences [34–36], dense SNP maps [37–39], extensive germplasm collections and public databases of genomic information [35,36,39–41]. In this review, we have attempted to gather all the necessary information on QTLs related to N, P and K for the benefit of breeders involved in developing rice varieties with NuUE for sustainable agriculture.

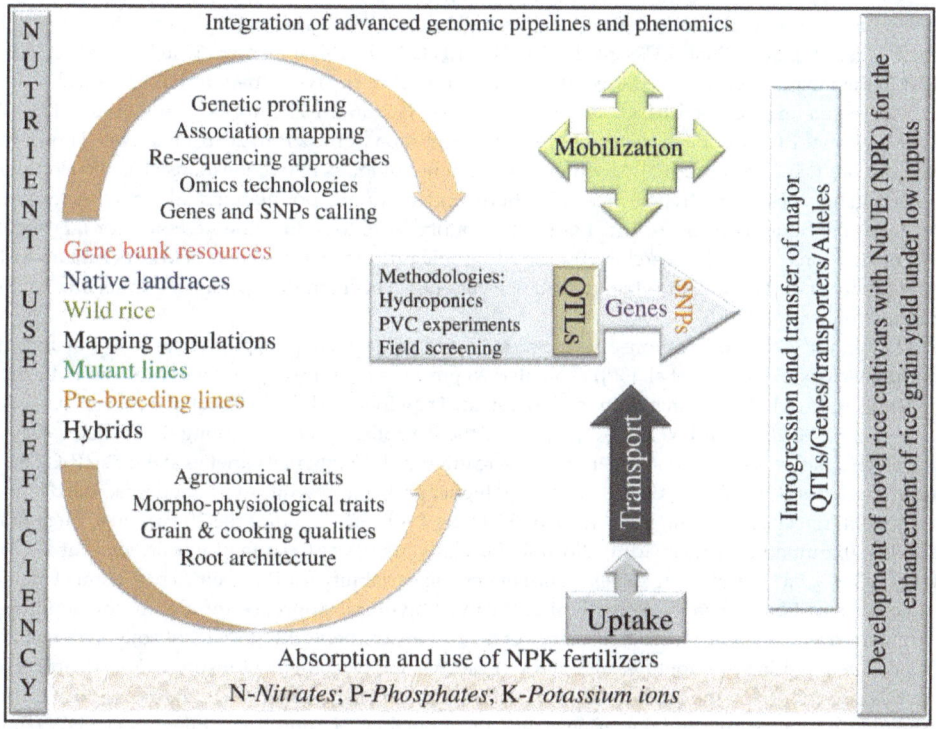

Figure 1. Integrated breeding and genomic approaches for improvement of rice cultivars superior in nutrient use efficiency (NuUE).

2. Screening Protocols and Breeding Efforts for Traits Related to Nutrient Use Efficiency

The literature is replete with NuUE screening protocols, especially for varieties, and very few are available for the systematic breeding of varieties with NuUE. Most of these NuUE studies use minus plots for different nutrients under study [42]. Research plots in institutions practice using omission or minus plots for any given target nutrient under study. Furthermore, researchers have always used natural sites with nutrient deficiencies for screening for any given nutrient such as the Pangil and Tiaong locations in the Philippines for P and Zn_ deficiency conditions, respectively.

2.1. Phosphorus

Deficiency of phosphorus is widespread in tropical and temperate acid soils. Screening and breeding for low phosphorus-tolerant (LPT) genotypes are some of the primary criteria for improving the use efficiency of P fertilizers. Worldwide, one-third of cultivable lands lack P in the soil to meet the requirement for ideal plant growth and development [43]. To avoid these stressful conditions, P is applied widely as an artificial fertilizer for improving grain yield for the burgeoning global population. The inconsistent use of fertilizers severely reduces income, and extreme conditions may cause environmental pollution [44]. Therefore, to overcome this crisis, the identification and improvement of P-efficient rice genotypes adapted to low-P soils would be a favorable solution for the enhancement of grain yield [45]. Developing P-efficient genotypes started with breeders involved in developing upland rice genotypes in an inadvertent manner. On the other hand, the mega-variety of India Swarna is a widely adaptable and popular variety among farmers perhaps because among its necessary traits is P responsiveness, as it possesses the *Pup1* QTL. Therefore, breeders should give more emphasis to developing lines tolerant of P_ deficiency with high grain yield. Krishnamurthy et al. [46] identified six LPT genotypes as Rasi, IET5854, IET14554, PRH122, IET15328 and IET17467, based on grain yield in field experiments at the Directorate of Rice Research in Hyderabad, India. Fageria et al. [47] reported seven lines (CAN 5164, CAN 4097, CAN 5170, IR3646-8-1-2, CAN 4137, A8-391 and IAC-47) at the National Rice and Bean Research Center of Embrapa in Brazil. In 2015, Saito et al. [48] found two varieties (Mudgo and DJ123) based on aboveground biomass at two locations. The development of such genotypes from diverse rice collections and mapping populations, along with cautious screening methodologies, is essential at the laboratory level to reduce the necessity for large-scale field evaluations. Several researchers used hydroponic nutrient solution and field experiments with different doses of P fertilizer to characterize rice varieties. This identified promising traits involved in tolerance of low P [48–58].

For grain yield and response to a graded level of applied phosphorus in low soil fertility conditions, Krishnamurthy et al. [50]) evaluated 28 pre-release promising rice varieties and hybrids at the Directorate of Rice Research farm in Hyderabad. They followed the protocol of 0–60 kg P_2O_5 ha^{-1} (i.e., 0, 10, 20, 30 40, 50 and 60 kg P_2O_5 ha^{-1}) for the P application rate. Among the 28 rice varieties, four distinct patterns were identified in response to grain yield. Eight rice varieties at 0–10 kg P_2O_5 ha^{-1} and six varieties at 20–30 kg P_2O_5 ha^{-1} exhibited higher grain yield, while five varieties recorded higher grain yield in responses at higher P rates of 50–60 kg P_2O_5 ha^{-1}. Out of the 28 varieties, three lines (IET 17190, Sumati and Rajavadlu) did not show any significant change in grain yield at 0–10 or 50–60 kg P_2O_5 ha^{-1}, indicating the existence of genetic variability for P-use efficiency. Chin et al. [58] suggested a soil-based screening method as the most favorable approach for identifying genotypes with tolerance of P_ deficiency. Aluwihare et al. [53] experimented with Ultisol soils, without any application of fertilizer for four decades at Rice Research and Development Institute (RRDI), Sri Lanka, and this also confirmed the absence of P [58]. At P0 and P30 (30 mg/kg P_2O_5) conditions, during early vegetative, late vegetative and flowering stages, plant height (PH), number of tillers (NT), SDW (shoot dry weight), SPC (shoot P concentration), SPU (shoot P uptake) and PUE (P use efficiency) were found to be the major indicators for P_ deficiency tolerance (PDT). Among the total genotypes, 13 were considered as highly tolerant, 13 as moderate and 4 as sensitive to P_ deficiency based on SDW and P use efficiency under P0 conditions. Cancellier et al. [59] and Fageria et al. [60] elucidated that plant height is a vital morphological trait for PDT screening as it significantly correlates with dry weight and yield. Panigrahy et al. [61] identified four low P-tolerant and four susceptible mutants by screening 300-ethane methane sulfonate (EMS)-induced (Nagina 22 [N22]) mutants under low-P field conditions.

However, experimentations at the gene expression level were carried out in controlled test tube, Petri plate or potted conditions with different rates of nutrients, which often included the zero condition (control) for less than a month's duration [13,62,63]. Li et al. [64] carried out expression profile studies using a DNA chip by subjecting rice at 6, 24 and 72 h under low-P stress and compared to a control treatment under normal P conditions. The study showed that genes directly involved in phosphorus absorption and use did not change significantly in transcription in rice shoots, relating to the inadequate low-P treatment. At 72 h under low phosphorus limitation, rice shoots did not develop severe phosphorus stress [65].

Specific genotypes known for their susceptibility to nutrient deficiency stress are useful for selection purposes, especially for different target nutrients. P_ deficiency tolerance was identified in a rice population derived from a cross between P-inefficient *japonica* cultivar "Nipponbare" and P-efficient *indica* landrace "Kasalath" [65].

On the other hand, several traits were studied to understand the phenotyping behavior of plants for precision screening and to progress in breeding activities. Root dry weight (RDW) is an important feature for evaluating the selection index for low-P tolerance in rice. Li et al. [49] reported that, at the seedling stage, dry weight had a significant genotypic variation (19.60%) in both standard and low-P conditions. TDW correlated with RRDW (relative root dry weight), RPH (relative plant height), RPUP (relative total P uptake), RSPA (relative shoot P accumulation), RPUE (relative P use efficiency) and RPC (relative P concentration) at $p < 0.01$. Several key morphological and physiological traits such as plant height, number of tillers, shoot root length, relative shoot and root dry weight and leaf age and root-attributed traits such as root diameter, root hair number and number of roots were used for screening and identifying tolerant genotypes under P_ deficiency conditions [61,66–70]. Increasing the productivity of grain yield under P_ deficiency conditions, increasing P taken up from the soil and improving the dry matter of internal use of P help to enhance the number of panicles and grain productivity [53,71]. Relative tiller dry weight (RTW), shoot dry weight and plant dry weight used as better screening criteria for identifying genotypes tolerant of low-P stress, especially RTW being sensitive, proved to be a reliable screening test. In recent days, image analysis has been becoming popular in high-throughput screening. Chen et al. [72] established an accurate, fast and operable method for diagnosing the crop nutrition status of NPK deficiencies in the color and shape of leaf parameters using a static scanning technology (SST) and hierarchical method in a pot experiment.

2.2. Nitrogen

Nitrogen fertilizer is an essential element for many aspects to improve grain yield, grain quality, flowering time and root development for extracting water and other nutrient elements from the soil [73,74]. On the other hand, the application of N is not uniform in all geographic regions of nations worldwide [75]. Several morphological and agronomic factors were found to influence the deficiency or high rates of N. Higher rates of N fertilizer consumption repeatedly led to environmental pollution and decreased nitrogen use efficiency (NUE) [76]. Therefore, the immediate focus should be to exploit the available variability in the use efficiency of rice cultivars through classical plant breeding methods and advanced biotechnological approaches to increase NUE in rice. Numerous research efforts have been conducted with different rates of N fertilizer in field experiments and hydroponic nutrient solution, and this was correlated with N use-efficient genotypes and higher grain yield (GY) parameters [77–79]. Chaturvedi [80] conducted a field experiment with different treatments of N fertilizer at the Agricultural Research Station in Chhattisgarh, India. Using an application of sulfur-containing nitrogenous fertilizer (Super Net) has significantly increased the grain yield and grain nitrogen content in hybrid rice variety Proagro 6207. Manzoor et al. [81] directed an experiment with nine different N rates (i.e., 0, 50, 75, 100, 125, 150, 175, 200 and 225 kg ha^{-1}) at the Rice Research Institute in Lahore, Pakistan, with Super basmati. Interestingly, at 200 kg N ha^{-1} and above, yield-attributed traits declined, and higher grain yield, number of grains per panicle, 1000-grain weight, number of tillers and panicle length significantly improved at 175 kg N ha^{-1}.

Likewise, Swamy et al. [82] evaluated ten rice genotypes under recommended rates of nitrogen (100 kg N ha^{-1}) and deficient N as no external nitrogen (i.e., N0) in a treatment grown in field conditions at Indian Institute of Rice Research (IIRR), Hyderabad. They found that 14% of root length (RL) decreased significantly under N_ deficiency. Haque and Haque [83] detected higher grain yield (5.36 t ha^{-1}) in 60 kg N ha^{-1}, and the highest NUE (344.50 kg grain kg^{-1} N) was recorded for BU dhan 1 at six different N rates (0, 20, 40, 60, 80 and 100 kg N ha^{-1}); they found an intermediate rate of N as economical and environment-friendly.

Employing a hydroponic experiment, Nguyen et al. [74] determined the effect of N supply in low and excess NH_4NO_3 concentration in Yoshida nutrient solution using three rice cultivars: IR64 (*Oryza sativa* ssp. *indica*), Azucena (*O. sativa* ssp. *japonica*) and TOG7105 (*O. glaberrima*). The rate of absorption of NUE (aNUE) and agronomic NUE (agNUE) decreased significantly, although at a gradual pace as the N supply increased, and physiological NUE (pNUE) declined progressively upon lowering the N supply.

To minimize N application and to use available N more efficiently, agronomic practices still need to be standardized. Nitrogen use efficiency is a complex trait and is associated with different components such as pNUE, aNUE, agNUE [77,78,84] and alteration in morpho-agronomic and physiological traits such as plant height, tiller number, grain yield, dry weight of shoots and roots, spikelet number, number of filled grains per panicle, 1000-grain weight, the leaf color chart (LCC) and chloroplasts [25,26,74,81,83,85–91] in rice. Alteration of the main traits was influenced by the response of N fertilizers, which may enhance the availability of N, which can lead to higher photo-assimilates and dry matter accumulation [80,92]. Therefore, considering the absorption, physiological and agronomic NUEs associated with morpho-agronomic traits will help to attain the balance between high grain yield and the eco-friendly nature of farm systems, which would be useful in developing crops with superior NUE.

2.3. Potassium

The availability of K in the soil is insufficient in developing countries, and it plays a significant role in crop grain yield and quality [93]. From 2012–2016, K fertilizer consumption globally increased from 28.6 Mt (K_2O) to 33.2 Mt (K_2O) [94]. Notably, East and South Asia are promising agricultural areas consuming 44.9% of the world K fertilizer, which is not adequate for improving grain yield under deficiency of K. The price of K fertilizers increased rapidly from 2003 (USD 165 per ton) to 2013 (USD 595 per ton) [94]. Therefore, the identification of K use efficiency (KUE) in rice is essential and needs to be used in developing genotypes with higher grain yield for K-deficient conditions. Dobermann et al. [95] mentioned that, as compared with other cereal crops, rice acquires 56–112 kg of K from soils in each harvest of yield of 4–8 t ha^{-1}, and yearly K demand for irrigated rice would be 9–15 × 106 tons by 2025. In physiological aspects, K is involved in many functions related to regulating osmotic potential, transporting assimilates, root development for uptaking water and nutrients, reducing the frequency of diseases, drought tolerance and photosynthetic activity [96–99]. Under different rates of K fertilizer (0, 25, 50, 75 and 100 kg ha^{-1}), Mehdi et al. [100] evaluated the response of rice cultivars in saline-sodic soil during 2005 and achieved the highest paddy yield (3.24 t ha^{-1}) and straw yield (3.92 t ha^{-1}) at 100 kg K_2O ha^{-1}. Similarly, Fageria et al. [101] elucidated lowland rice grain yield varying from 5.88–6.24 t ha^{-1} with an application of 125 kg ha^{-1} in different years. Analysis of six upland rice genotypes evaluated in a greenhouse under natural soil of 200 mg K kg^{-1} revealed that K uptake in shoot and grain and the K use efficiency ratio (KUER) were significantly and positively associated with grain yield [101], whereas, compared with grain, K concentration and uptake were higher in shoots. Arif et al. [102] conducted a pot experiment with three genotypes in a rain-protected wire house at the University of Agriculture in Faisalabad using hydroponic nutrient solution with different K rates of 0, 30, 60, 90 and 120 kg ha^{-1}, respectively. Among the three genotypes, IR6 (low KUE), Super basmati (medium KUE), genotype 99509 (high KUE), the highest thousand grain weight (TGW) (IR6), grain yield (g pot^{-1}) (Super basmati, 99509), number of panicles and tillers per pot (Super basmati) were recorded at optimum rates of 60 kg ha^{-1}. Earlier reports revealed that a higher rate of K influences increases in yield-attributed traits [103–106]. The increase in yield with an optimum rate of K plays a crucial role in increased N use and increasing chlorophyll synthesis and translocation of assimilates to reproductive parts [107]. Recently, Islam et al. [108] compared the application of K fertilizer between 40 and 80 kg ha^{-1} in a randomized complete block design. The significant ($p < 0.05$) increases in grain and straw yield in the treatment with K application rates of 40 and 80 kg ha^{-1} were 54% and 68% in the dry season and 39% and 45% in the wet season from 2003–2010 in field experiments at the Bangladesh Rice Research Institute farm. Hence, improving uptake, transport and translocation of K efficiency in shoots and rice grain is possible for identifying superior genotypes to further enhance grain yield by proper management practices.

3. Identification and Use of QTLs Related to Nutrient Use Efficiency

Developing rice varieties with multiple tolerance is possible provided large-effect QTLs/genes are available and exploited with innovative molecular breeding approaches. The number of reported QTLs is unwaveringly increasing day by day, but still, very few are applied in breeding programs. Obtaining more data that validate QTLs/genes in different genetic backgrounds and environments is a prerequisite for their large-scale application. In rice, there is an attempt to bring a few large-effect QTLs that confer tolerance of submergence, drought, salinity and P deficiency together through molecular marker-assisted breeding. *Pup1* is the best model for exploiting the NuUE QTLs currently being used, for which molecular markers are now available and evaluated in different genetic backgrounds under field conditions [5].

3.1. QTLs Related to Nitrogen Use Efficiency

Among the essential nutrient elements, nitrogen is the most important one for rice growth in natural ecosystems. The green revolution, which was a breakthrough in agricultural production to secure human nutrition in the past century, depended mainly on fertilizer application and high-yielding modern varieties [109–113]. In this context, nitrogen use-efficient crop varieties are of great concern. Further, genes and QTLs related to agronomy for NUE are presented in Tables 1 and 2. Deeper understanding of the molecular basis of NUE would enable us to provide valuable information for crop improvement through biotechnological approaches. Recent advances in genomics and proteomics approaches such as subtractive hybridization, differential display and microarray techniques are transforming our approach to identify the candidate genes that play a crucial role in the regulation of NUE [4,7,114–118]. In addition, marker-trait association for NUE through quantitative real-time polymerase chain reaction (RT-PCR) technology is being used [119–121]. The identification of potential candidate genes/proteins will serve as biomarkers in the regulation of NUE for screening genotypes for their nitrogen responsiveness. This will help to optimize nitrogen inputs in agriculture.

The modern rice varieties were all selected earlier for higher N uptake to obtain maximum grain yields. Conversely, the biggest problem with the increased N supply often leads to a decrease in N use efficiency. This is mainly due to high N uptake before flowering, but is also due to low N uptake during the reproductive growth phase and incomplete N translocation from vegetative plant parts to the grains [15,178]. Sustainable agriculture requires developing crop varieties with high yield potential and less dependency on heavy applications of N and P fertilizer. Similar to P, N has no systematic breeding program and screening protocol. The genotypes were screened either in nutrient minus fields or under solution culture.

In recent years, heavy nitrogen fertilization during panicle development has been popular in China to improve population dynamics and increase grain yield [179]. Panicle fertilization was adopted to increase grain yield and N recovery efficiency at IRRI [180]. Nitrogen use efficiency positively correlates with photosynthetic characteristics. The measures for promoting photosynthetic function and delaying senescence of leaves may indirectly enhance N absorption and use of rice and ultimately increase NUE. Some research efforts had been devoted to developing genotypes that use N more efficiently. This highly complicated objective requires an in-depth understanding of the genetic basis of N assimilation and N use at different developmental stages. The QTLs underlying related traits toward the late developmental stage in rice at two different nitrogen rates were investigated using a population of chromosome segment substitution lines (CSSLs) derived from a cross between Teqing and Lemont. A total of 31 QTLs referencing five traits, especially plant height, panicle number per plant, chlorophyll content, shoot dry weight and grain yield per plant, were detected. Under the normal nitrogen (150 kg/h^{-1} N fertilizer) rate, three QTLs were identified for each trait, and the under low nitrogen (0N) rate, five, four, five and two QTLs were detected for plant height, panicle number per plant, chlorophyll content and shoot dry weight, respectively. Most of the QTLs were located on chromosomes 2, 3, 7, 11 and 12 [166].

Table 1. Rice genes/QTLs governing key agronomic traits, the protein encoded, level of allele expression and their possible use in breeding programs.

S. No.	Traits	Name of QTL	Encoded Protein	Nature of Allele Suitable for Use in Breeding Programs	References
1	Grain number	Gn1a	Cytokinin oxidase	Low expression	[122]
2	Grain number and strong culm	dep1	PEBP-like domain protein	Loss of function	[123]
3	Grain number	WFP	OsSPL14	High expression	[124]
4	Grain number, low tiller number, and strong culm	Ipa	OsSPL14	High and ectopic expression	[125]
5	Grain size	gs3	Transmembrane protein	Loss of function	[126]
6	Grain size and filling	gw2	RING-type ubiquitin E3 ligase	Loss of function	[127]
7	Grain size	qSW5/GW5	Unknown	Loss of function	[128]
8	Grain filling	GIF1	Cell wall invertase	Restricted expression in the ovular vascular trace	[129]
9	Heading date	Hd1	CONSTANS-like protein	Loss-of-function allele leads to late heading	[130]
10	Heading date	Hd6	Subunit of protein kinase	Loss-of-function allele leads to early heading	[131]
11	Heading date	Hd3a	FT-like	Low expression leads to late heading	[132–134]
12	Heading date	Ehd1	B-type response regulator	Loss-of-function allele leads to late heading	[135]
13	Grain number, plant height and heading date	Ghd7	CCT domain protein	Functional allele	[136]
14	Days to heading	DTH8	CCT domain protein	Functional allele	[137]
15	Plant height	sd1	Gibberellin 20 oxidase	Loss of function	[138]
16	Lodging resistance	SCM2	F-box protein	High expression	[139]
17	Disease resistance	pi21	Proline-rich protein	Loss of function	[140]
18	Disease resistance	Pb1	CC-NBS-LRR protein	Functional allele	[141]
19	Salt tolerance	SKC1	HKT-type transporter	Gain of function	[142]
20	Cold tolerance	qLTG3-1	GRP and LTP domain	Functional allele	[143]
21	Submerge tolerance	Sub1A	ERF-related factor	Gain of function	[144]
22	Internode elongation under submergence conditions	SK2	ERF-related factor	Gain of function	[145]
23	Cadmium accumulation	OsHMA3	Putative heavy metal transporter	Functional allele	[146]
24	Seed shattering	sh4	Myb3 transcription factor	Loss of function	[147]
25	Seed shattering	qSH1	BEL1-like homeobox protein	Low expression in abscission layer between panicle and spikelet	[148]
26	Prostrate growth	PROG1	Zinc finger transcription factor	Loss of function	[149,150]
27	Disease resistance	RHBV	NS3 protein	Favorable gene or QTL alleles	[151]
28	Phosphorus uptake	Pup1	OsPupK46-2	High expression	[57]
29	Deep rooting	DRO1	Auxin signaling pathway	Functional allele	[152]

Table 2. Quantitative trait loci identified for traits related to nitrogen, phosphorus and potassium use efficiency in rice.

Entry		Phosphorus				
S. No.	Traits	Population	Cross	No. of QTLs		Reference
				M	E	
1	Phosphorus uptake, plant dry weight, tiller number; phosphorus use efficiency	NILs	*Nipponbare/Kasalath*	8	-	[65]
2	Relative tillering ability, relative shoot dry weight, relative root dry weight	RILs	*IR20/IR55178*	4	-	[153]
3	Phosphorus uptake, tiller number	NIL	*Nipponbare/Kasalath*	1 (Pup)	-	[154]
4	Root elongation, shoot dry weight, relative phosphorus content, relative Fe content	F_8	*Gimbozu/Kasalath*	6	-	[155]
5	Relative root length, relative shoot length, relative shoot dry weight, relative root dry weight	BILs	*OM2395/AS996*	1	-	[156]
6	Root elongation under phosphorus deficiency	CSSLs	*Nipponbare/Kasalath CSSL29*	1	-	[157]
7	Plant height, maximum root length, root number, root volume, root fresh weight, root dry weight, shoot dry weight, total dry weight, root/shoot dry weight ratio	ILs	*Yuefa/IRAT109*	24	29	[63]
8	Relative root length, relative root dry weight, relative shoot dry weight, relative total dry weight, relative root-shoot ratio of dry weight	BC_2F_4	*Shuhui 527/Minghui 86*	48	-	[158]
9	Total aboveground biomass, harvest index, P use efficiency for grain yield based on P accumulation in grains, P harvest index, P translocation, P translocation efficiency, P total aboveground P uptake, P use efficiency for biomass accumulation, P use efficiency for grain yield, P use efficiency for straw dry weight based on P accumulation in straw	RILs	*Zhenshan 97/Minghui 63*	36	-	[159]
10	Root dry weight, relative shoot dry weight, relative total dry weight	DHs	*ZYQ8/JX17*	6	-	[160]
		Nitrogen				
1	Plant height	DHs	*IR64/Azucena*	10	-	[161]
2	Rubisco, total leaf nitrogen, soluble protein content	BILs	*Nipponbare/Kasalath*	15	-	[162]
3	N uptake (NUP), grain yield, biomass yield, N use efficiency (NUE)	CSSLs	*9311/Nipponbare*	13		[118]
4	Toot system architecture, NDT, and morphological and physiological traits	CSSLs	*Curinga/IRGC105491*	13		[163]
5	Twelve physiological and agronomic traits	RILs	*IR64/Azucena*	63		[27]
6	Glutamine synthetase, glutamate synthase	BILs	*Nipponbare/Kasalath*	13	-	[164]
7	Glutamine synthetase, panicle number per plant, panicle weight	NILs	*Koshihikari/Kasalath*	1	-	[164]
8	Total grain nitrogen, total shoot nitrogen, nitrogen uptake, nitrogen use efficiency, nitrogen translocation efficiency	F_3	*Basmati370/ASD16*	43	-	[165]
9	Root dry weight, shoot dry weight, biomass	RILs	*Zhenshan97/Minghui 63*	52	103	[166]
10	Plant height, panicle number per plant, chlorophyll content, shoot dry weight	CSSLs	*Teqing/Lemont*	31	-	[167]
11	Total grain number, total leaf nitrogen, total shoot nitrogen, nitrogen uptake, specific leaf nitrogen	RILs	*IR69093-4-3-2/IR72*	32	-	[168]
12	Root length, root thickness, root biomass, biomass, etc.	RILs	*Bala/Azucena*	17	-	[169]
13	Relative root dry weight, spikelet number per panicle, spikelet fertility, 1000-grain weight	ILs	*Shuhui 527 × Minghui 86*	48	-	[170]
14	Total grain number, total leaf nitrogen, total shoot nitrogen, physiological nitrogen-use efficiency, biomass	RILs	*Dasanbyeo/TR22183*	20	58	[170]
15	Total plant nitrogen, nitrogen-use efficiency	DHs	*IR64/Azucena*	16	-	[171]

Table 2. *Cont.*

S. No.	Traits	Population	Cross	No. of QTLs M	No. of QTLs E	Reference
	Entry	Phosphorus				
16	Total plant nitrogen, nitrogen dry matter production efficiency, nitrogen grain production efficiency, total grain number	RIL	Dasanbyeo/TR22183	28	23	[172]
17	Grain yield per plant, biomass, harvest index, etc.	RILs	IR64/INRC10192	46	-	[173]
18	Plant height, root dry weight, shoot dry weight, chlorophyll content, root length, biomass	RILs	R9308/Xieqingzao B	7	-	[161]
19	Grain yield per plant, grain number per panicle	RILs	Zhenshan 97/HR5	19	11	[174]
20	Number of panicles per plant, number of spikelets per panicle, number of filled grains per panicle, grain density per panicle	RILs	Xieqingzao B/Zhonghui 9308	52	-	[175]
21	Nitrogen deficiency tolerance and nitrogen-use efficiency	RILs	Zhenshan 97 and Minghui 63	12		[176]
	Potassium					
1	Plant height, tiller number, shoot and root oven-dry weight	DHs	IR64/Azucena.	4	-	[177]

M = main-effect QTLs; E = epistatic QTLs.

Based on the use of two N supply levels, 5 mg N L^{-1} for low N and 40 mg N L^{-1} [167] for high N, QTLs for plant height in rice were mapped onto the Restriction Fragment Length Polymorphism (RFLP) linkage map of a doubled-haploid population derived from a cross between IR64 and Azucena. Two QTLs, one on chromosome 1 and the other on chromosome 8, were detected at high N levels (40 mg N L^{-1}) in soil-based nutrient solution culture experiments. Furthermore, a total of eight QTLs were identified at low N level and located on chromosomes 1, 2, 3, 4, 5 and 6, whereas the QTL flanked by molecular markers RZ730 and RZ801 on chromosome 1 was identified in all experimental conditions. The hypothesis suggests that the genotype showing higher N efficiency under low N level may carry the gene(s) for higher N efficiency. This study demonstrated that the effects of low N stress on plant height lessened. In the present study, the female parent IR64 was found to have a relatively higher N efficiency than the male parent Azucena under low N levels due to its lesser decline in plant height than Azucena. Furthermore, some of the QTLs associated with plant height were detected only at low N levels and might have some relationship with N efficiency [162]. QTL analysis was related to N and P tolerance traits such as root length at the seedling stage, productive panicles, seed setting ratio and yield. A few QTLs out of these were found to be located on similar chromosomal sections that showed the genes associated with the N or P metabolism pathway [181,182]. QTLs for rice panicle number and grain yield were detected under low nitrogen (N0) and low phosphorus (P0) conditions and helped to analyze the genetic basis of tolerance of soil nutrient deficiency. A total of 125 CSSLs with relatively few introgression segments were derived from *japonica* cultivar Nipponbare within the genetic background of *indica* cultivar 93–11. These were screened using an augmented design in field experiments with regular fertilization (NF), low nitrogen (N0) and low phosphorus (P0) treatments. Grain yield and panicle number per plant were measured for each CSSL, and their relative values based on regular fertilization treatment considered as the measurement for tolerance of the nutrient deficiency. Both regular fertilization and low phosphorus treatments showed adverse effects on grain yield and panicle number. The different responses observed among the CSSLs refer to the deficiency of nitrogen or phosphorus. The relative traits had a significantly negative correlation with the traits in the regular fertilizer treatment. Cultivar 93–11 showed higher tolerance of low-nutrient stresses than Nipponbare. The negative allelic effects of 38 QTLs were contributed by Nipponbare under nitrogen and phosphorus deficiency stresses. Out of these, 26 QTLs were responsible for yield and panicle number, and the remaining 12 QTLs specified the relative traits. Five QTLs were identified in common under both stresses. Moreover, 81% of the QTLs were specifically detected only in low nitrogen (N0) or phosphorus (P0) conditions. These different QTLs suggest that the response to limiting nitrogen and phosphorus conditions was regulated by various sets of genes in rice [168].

The application of N fertilizer is of particular importance for cultivating high-yielding rice. However, heavy nitrogen fertilizer uses with high loss of nitrogen in rice-growing areas have led to low N recovery rates and environmental pollution. Grain yields are used as an indicator of NUE since it is difficult to evaluate the amount of plant-available N from the soil or any source of N inputs, including fertilizer application and N fixation [183]. Genotypes with high NUE are those cultivars that produce high grain yields with the application of N, while those that do not yield well are genotypes with low NUE. Cultivars with high NUE have the ability to take up N and efficiently use it to produce grains [184]. The relative weight of root, shoot and plant under two different N treatments could reveal the cultivars showing tolerance of low N stress. The QTLs identified for relative performance were distinctive from those for root, shoot and plant weight detected under the two N treatment conditions [182].

The study of Tong et al. [174] revealed a correlation with path analysis indicating that spikelet fertility percentage had the most significant contribution to grain yield per plant at the 300-and 150-kg urea ha^{-1} rates, but filled grains per panicle contributed a strong positive relationship with grain yield per plant at the N0 level. Six of 15 QTLs identified with main effects were detected for each trait except SFP. Clusters of main-effect QTLs associated with several key traits were observed on chromosomes 1, 2, 3, 5, 7 and 10, respectively. The main-effect QTLs (*qGYPP-4b* and *qGNPP-12*) were identified at the N0 rate only, which explained 10.9% and 10.2% of the total phenotypic variation explained (PVE). The identification of genomic regions associated with yield and its components at different nitrogen rates will be useful in marker-assisted selection for improving the NUE of rice. The NUE-related trait in rice is so complex that different results were obtained in previous publications because of various experimental conditions, methods and materials. The main-effect QTL (M-QTL), epistatic QTL (E-QTL) and QTL × environment (Q × E) interactions of six traits were investigated using a fully-saturated simple sequence repeat (SSR) linkage map. Obara et al. [185] found a QTL region associated with panicle number and panicle weight on chromosome 2 that contains a regulator gene (*GS1*) for glutamine synthetase activity. The selected rice plants based on this QTL region showed superiority in tillering ability, panicle number and total panicle weight under low N rates.

Several researchers identified main-effect QTLs on chromosome 3 [171], chromosome 6 [186] and chromosomes 2 and 9 [170] by using doubled haploids and Recombinant Inbred Lines (DHs and RILs) populations.

Among these QTLs, one QTL was identified as being associated with the number of grains per panicle under low N rate, and it was located in a similar region to the *Pup1* locus on chromosome 12, thus encouraging the use of *Pup1* materials for testing low-N tolerance [5]. Recently, in a hydroponic experiment with CSSLs, Zhou et al. [118] identified a total 23 QTLs, with seven QTLs for N uptake (NUP) located on different chromosomes (2, 3, 6, 8, 10 and 11), with phenotypic variation (PV) ranging from 3.16–13.99%. Six QTLs for N use efficiency were located on chromosomes 2, 4, 6 and 10 and had explained PV ranging from 3.76–12.34%, respectively. The remaining 10 QTLs were responding to grain yield (GY) and biomass yield (BY). With the results of correlation analysis, Zhou et al. [118] suggested that both NUP and NUE had large effects on grain yield. Previous reports of Dong et al. [187,188] showed the NUP trait more closely associated with grain yield than NUE. NUE and NUP trait-linked QTLs are highly useful for improving grain yield under low-input conditions.

3.2. Phosphorus Use Efficiency and Related QTLs

Phosphorus is one of the essential macro-nutrients required for plant growth and development. Low availability of phosphorus in a variety of soils, especially in the tropics, often limits rice grain yields [189], along with the lack of available P sources locally in many countries. The higher importation and transportation costs of P fertilizers frequently prevent resource-poor farmers, especially in developing countries, from applying P to their deficient farmlands. Thus, developing rice cultivars with improved tolerance of P deficiency may therefore be a cost-effective solution to this problem. Rose and Wissuwa [45], optimistic that breeding for poor soil with high P uptake and high PUE needs to be developed and to maximize crop grain yield in such low-input systems, noticed that continuous cropping of poor soil is often related to poverty. It is also important to breed efficient crops. A combination of both P uptake and P internal nutrient efficiency is equally desirable for high-input systems, whereas it would facilitate a reduction in fertilizer rates without yield compensation. Dobermann and Fairhurst [190] reported in rice that P fertilizer use efficiency is only ~25%, which suggests considerable scope for improvement.

Several researchers have identified genes and QTLs governing agronomic traits related to nutrient use efficiency, and these are shown in Tables 1 and 2 and are represented in Figure 2 with the respective NPK QTLs located on 12 chromosomes associated with morpho-physiological traits under low-input conditions. The *Pup1* gene responsible for phosphorus uptake was identified and characterized by Chin et al. [57] (Table 1). Quantitative trait loci for P deficiency tolerance were identified in a rice population derived from a cross between P-inefficient *japonica* cultivar Nipponbare and P-efficient *indica* landrace Kasalath [65]. Tolerance of P deficiency was primarily caused by genotypic differences in P uptake; internal PUE had a negligible effect, and even phosphorus content changed slightly within 72 h in the shoots under low phosphorus stress, but phosphorus content decreased rapidly at 24 h in the roots [62].

Several studies were carried out to understand the genetics of tolerance of phosphorus deficiency in crops, and they identified several QTLs associated with it [54–66,154,156,191]. Su et al. [192] reported that 39 QTLs were associated with panicle number and weight of dry matter, chosen as the indices of P deficiency tolerance in wheat (*Triticum aestivum* L.).

The QTLs related to root traits, panicle number and seed set percentage were reported in rice [66,153,156]. Yield component traits such as panicle number and seed-setting percentage could be used as selection indices for P deficiency tolerance in rice [192]. However, only a few reports are available for the QTL mapping of grain yield and its components for P_ deficiency tolerance.

A significant QTL for P uptake was mapped to a 13.2-cM interval on the long arm of chromosome 12 flanked by markers C443-G2140. The position was estimated to be at 54.5-cM, a 3-cM distance from marker C443. Additional minor QTLs were found on chromosomes 2, 6 and 10 [155]. However, the first evidence supporting the presence of a significant QTL for P_ deficiency tolerance came from a study by Ni et al. [154].

A doubled-haploid population was derived from a cross between P_ deficiency-tolerant *japonica* rice IRAT109 and P deficiency-sensitive *japonica* rice Yuefu [193]. A total of 116 lines were evaluated for yield per plant and its component traits under P deficiency and normal conditions. There were significant differences in seed-setting percentage, panicle number per plant and yield per plant for the doubled haploid DH population between the two conditions, whereas there was no significant difference in 1000-grain weight and grain number per panicle. The results indicated that seed-setting percentage, panicle number per plant and yield per plant were easily influenced by P_ deficiency. Restricted fragment length polymorphism (RFLP) and simple sequence repeat (SSR) markers were used to cover 1535-cM of the rice genome to discover a total of 17 QTLs for plant yield and its components (1000-grain weight, seed-setting %, panicle number per plant, grain number per panicle) under P deficiency conditions. These QTLs explained from 2.65–20.78% of the phenotypic variance, with 12 QTLs showing higher than 10%. For 1000-grain weight, one QTL was detected, which had an logarithm of the odds LOD score of 5.13 and high contribution of PV (14.38%). Five QTLs were linked with seed-setting percentage, and three QTLs were linked with panicle number per plant [193]. Out of these five, three SP QTLs (*qSP2*, *qSP5* and *qSP11)* contributed more than 10%, and the three QTLs for panicle number per plant had high general contributions of more than 17%. Two QTLs (*qPN10* and *qPN12*) had an opposite additive effect. For grain number per panicle, four QTLs were detected, two of which (*qGN6* and *qGN7*) had high general contributions and positive effects. Four additive QTLs were found on chromosomes 2, 3, 6 and 7, which explained 4.77–13.55% of the phenotypic variance, for yield per plant. Three of them, *qYP3*, *qYP6* and *qYP7*, had high general contributions of more than 10% [194].

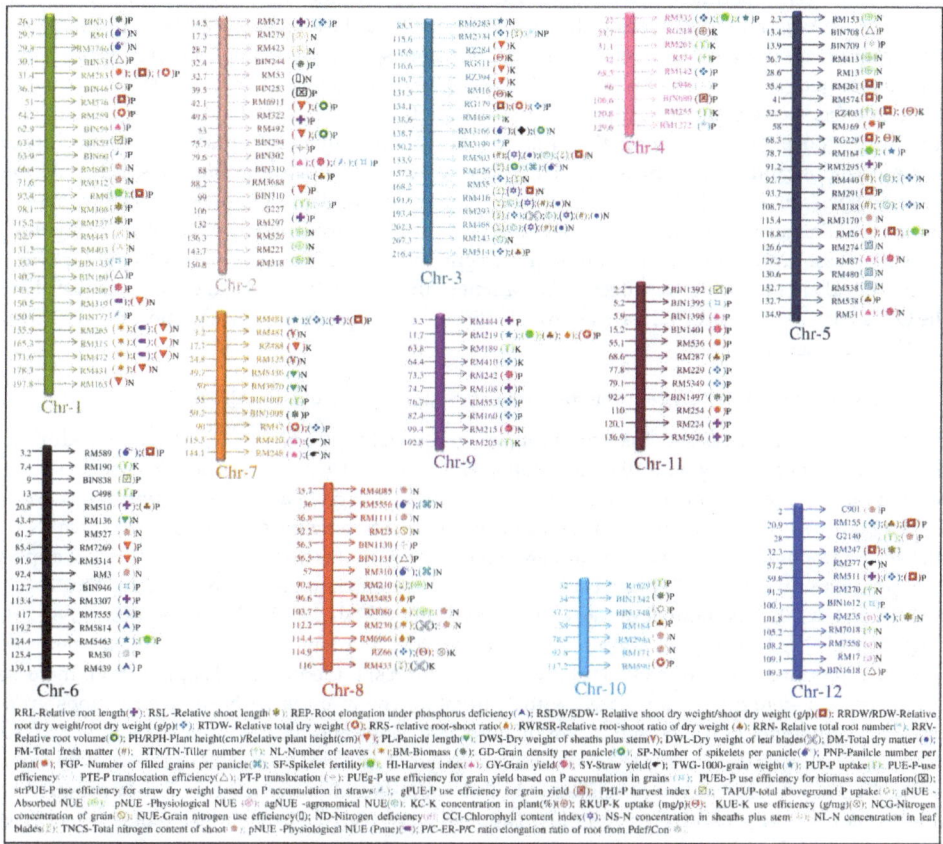

Figure 2. Diagram of 12 chromosomes with reported nutrient use efficiency (NuUE)-NPK QTLs linked to markers associated with the respective traits were identified through marker assisted selection (MAS) breeding approaches in a low-NPK environment using diverse mapping populations of rice.

3.3. Potassium Use Efficiency and Related QTLs

Among the essential elements, potassium is necessary for plant growth. It is the activator of many enzymes in plants and the osmotic regulator of cell solute potential, and it plays a significant role in plant growth and metabolism. In rice, increased application of K fertilizer significantly improves grain and milling quality, such as increasing the percentages of brown rice, milled rice and head milled rice; reducing chalkiness; and enhancing grain protein content [194]. Fageria et al. [101] reported on K uptake and the use efficiency of upland rice under Brazilian conditions. They conducted a greenhouse experiment with the K rate as zero (natural soil level) and 200 mg K kg^{-1} of soil with the objective of evaluating the influence of K on grain yield, K uptake and their use efficiency, especially for six upland rice genotypes grown on a Brazilian Oxisol. Shoot dry weight and grain yield were significantly influenced by K rate and genotype treatments. The potassium concentration in the shoot was about six-fold greater than that of the grain, across two K rates and six genotypes. However, the K use efficiency ratio (KUER) was about 6.5-times higher in the grain than in the shoot, over two K rates and six genotypes. Potassium uptake in shoot and grain and KUER were significantly and positively associated with grain yield. Besides these, soil Ca, K, base saturation, acidity saturation, Ca saturation, K saturation, Ca/K ratio and Mg/K ratio showed a significant influence on the K application rate.

A greenhouse experiment was conducted at four levels of saline water irrigation (tap water and 2, 4 and 6 dS m^{-1} of salinity) and four different methods of K application (spraying with distilled water as the control, application of potassium on soil, potassium spraying and application of potassium on soil plus spraying). The purpose was to study the efficiency of potassium spraying and use in the soil and their effect on yield and its components under salinity stress. The results showed that grain yield, number of shoots, 100-seed weight, tiller number, dry root weight and K uptake in seeds and shoots decreased significantly with increasing salinity. The best method of K application was soil intake plus spraying [195]. In an investigation of a DH population consisting of 123 lines derived from *indica* variety IR64 and *japonica* variety Azucena under a hydroponic experiment, Wu et al. [177] identified three QTLs associated with shoot and root dry weight under K-deficient conditions. These same three QTLs were also influencing the effect on K content in the plant (KC), K uptake and K use efficiency. The QTLs individually had PVE ranging from 8–15% and were positioned on chromosomes 2, 3, 5 and 8 in K_ deficiency conditions.

4. Effect of Nutrient Use Efficiency across Medium- and Long-Duration Rice

Singh et al. [6] assessed the variability in grain yield and N use of 10 medium-duration (119 ± 4 days after seeding) and 10 long-duration (130 ± 4 DAS) genotypes. These genotypes showed varying rates of acquisition and use of soil and fertilizer N. Significant diversity within genotypes was found in grain yield and N uptake, efficiency and partitioning parameters (physiological N use efficiency, agronomic N use efficiency, apparent recovery, partial factor productivity (PFP) of applied N, N productivity index and N harvest index). The N use-efficient genotypes were IR54790-B-B-38, BG380-2 and BG90-2 (medium duration) and IR3932-182-2-3-3-2, IR54853-B-B-318 and IR29723-88-2-3-3 (long duration), producing high grain yields at both low and high rates of N, whereas inefficient genotypes produced low grain yields at low N rates, but responded well to N application. Increases in grain yields were highly correlated with N uptake. The grain yield-N uptake relationship for individual genotypes indicated significant differences in slope and the grain yield obtained with soil N (GY0). Significant differences in GY0 were due to genotypic variation in N uptake and efficiency of use. The N harvest index was related to both N uptake and use efficiency. The N productivity index, which integrated both GY0 and PFP of applied N, provided a better ranking of rice genotypes. The performance levels of efficient and inefficient genotypes over a range of soil and fertilizer N supply were consistent across three seasons of trials.

5. QTLs for Both Low Nitrogen and Phosphorus Stress

Eight QTLs explained panicle number per plant under the three treatments. Five of the QTLs were identified under the low-nitrogen treatment, and three were identified under the low-phosphorus treatment. The alleles from Nipponbare at all the QTLs_ had adverse effects on panicle number (decreasing it by 42.6–62.9%). No common QTLs were identified for panicle number under both low-N and low-P stresses. A total of 18 QTLs for yield per plant were detected in three treatments [175]. Located on chromosome 4, a QTL (*Qyd-4c*) was identified in all treatments with relatively higher phenotypic variance explained (58.2%, 55.2% and 88.1%) under normal, low-N and low-P conditions, respectively. The authors detected another four QTLs (*Qyd-3a*, *Qyd-4a*, *Qyd-7a* and *Qyd-10*) in two treatments. The rest of the 13 QTLs were identified in only low-nitrogen or low-phosphorus treatments. Regarding relative yield, two and three QTLs were identified in different N and P treatments, respectively, of which *Qryd-7a* was a common QTL, suggesting that the CSSL containing the *Qryd-7a* locus was sensitive to both N and P_ deficiency stresses [127,172]. QTL *Qyd-4a* was located in the same chromosomal region as the QTL for dry weight of seedling root [167]. The authors conjectured this substitution region to be associated with root response to nutrient stresses, probably containing genes for regulating nutrient absorption and consequently affecting yield per plant in rice. Root elongation gets hit by either N or P_ deficiency [126,167,172], resulting in various nutrition assimilation in plants. Several QTLs from this study correspond to known genes in the N or P metabolic pathway. For example, *Qyd-2b* for N_ deficiency tolerance was located near the gene encoding cytosolic glutamine synthetase (*GS1*), and *Qyd-3b* and *Qpn-3* were nearby the genes for glutamate dehydrogenase (*GDH2*) [182]. Furthermore, *Qyd-12* was detected only under low-P conditions, and it co-localized with a significant QTL (*Pup1*) on chromosome 12, which was involved in P absorption [154]. These results indicate that the QTLs specifically detected under single N or P_ deficiency conditions may be involved in different pathways of N and P metabolism. Their tightly linked markers have breeding potential in pyramiding elite QTLs for N and P use efficiency.

Tolerance of low nitrogen stress conditions is a highly desired characteristic for sustainable crop production. The genetic components associated with low N tolerance in rice at the seedling stage, including main QTL effects, epistatic QTL effects and QTL by environment interactions, using a population of 239 RILs derived from a cross between popular Zhenshan 97 and Minghui 63, were studied [182] in solution culture. Root, shoot and plant weight over two N treatments were measured and the relative weight of the two treatments for each trait considered as measurements for low-N tolerance. Four to eight QTLs with main effects were detected for each of the nine traits. Very few QTLs were detected in both low and normal nitrogen conditions, and interestingly, most of the QTLs for the relative measurements were distinct from those for traits under the two nitrogen treatments, indicating very little commonality in the genetic basis of the traits and their relative performance under low and normal nitrogen conditions. In rice, some agronomic traits involving effective tiller number, spikelet fertility percentage and grain yield were studied under low nitrogen stress [166,170,185,196]. Two main-effect QTLs with large contribution rates were detected at the N0 rate. One of them affecting grain number per plant was detected at the interval RM117-RM101 on chromosome 12, accounting for 10.2% of the total phenotypic variance. There was no significant interaction between this M-QTL and environmental factors. This QTL is from the same region as a QTL (*Pup1*) related to phosphorus uptake [156]. Zhao et al. [33] reported that single segment substitution lines (SSSLs) each having a single chromosome segment derived from a donor under the same genetic background as the recipient parent were developed in rice by advanced backcrossing and genome-assisted selection. The QTLs for 22 essential traits were detected in rice with 32 SSSLs by a randomized block design in two to four cropping seasons. However, the QTLs controlling grain weight, grain length, the ratio of grain length to width and heading date were relatively stable. Fifty-nine QTLs were detected and distributed on chromosomes 1, 2, 3, 4, 6, 7, 8, 10 and 11, of which 18 were detected more than twice. Only 30.5% of the QTLs were repeatedly identified across different cropping seasons. Mostly the QTLs governing important agronomic traits showed small additive effects and instability. The stable QTLs usually had larger additive effects and were less affected by environment. With recent successful achievements in the Green Super Rice (GSR) project, efforts were made for highly adaptive rice cultivars with higher grain yield under low-input conditions [13,196–201]. About a 10% yield increase was obtained in elite GSR rice cultivars as compared with the local check variety NSIC Rc222 under multiple abiotic stress tolerance and low-input conditions, without compromising grain yield and quality [200]. Further progress in the genetic regulation of NuUE of GSR cultivars may provide valuable materials to understand the molecular and physiological pathways for the improvement of yield and grain quality under low-input conditions.

6. Agronomic Efficiency and Partial Factor Productivity QTLs

There is a significant increase in grain yield for each kg of fertilizer applied, termed agronomic efficiency (AE). Efficient fertilizer use is defined as maximum returns per unit of fertilizer applied [202]. According to Yadav [202], PFP and AE are useful measures of NUE, as they provide a basis for an integrative index that quantifies total economic output relative to the use of all nutrient resources in the system. Cassman et al. [203] defined PFP and AE to increase by increasing the amount, uptake and use of available nutrients and further by increasing the efficiency of applied nutrients that are taken up by the crop and used to produce grain.

Several researchers have studied AE and PFP in rice and other cereal crops. Dobermann [204] reported cereal crops in terms of AE of 10–30 kg grain kg^{-1} N, where >30 kg grain kg^{-1} is found in a well-managed system or at a low rate of N use or low soil N supply and for PFP 40–70 kg grain kg^{-1} N, with >70 kg^{-1} at low rates of N or in a well-managed efficient system. Wen-xia et al. [205] reported AE in two kinds of rice, one being Jinzao, with AE ranging from 8.02–20.14 kg grain kg^{-1} N, and the second one being Shanyou63, with an AE range of 3.4–18.37 kg grain kg^{-1} N absorbed. Yoshida [206]) estimated AE to be 15–25 kg grain kg^{-1} N, and Cassman et al. [203] reported AE at 15–20 kg grain kg^{-1} N in the dry season in farmers' fields in the Philippines.

Amanullah et al. [207] declared that in maize, PFP for applied N was 36.62 kg grain kg^{-1} N and AE for applied N was 22.49 kg grain kg^{-1} N, using DAP and SSP in the field for the AE of two fertilizer applications, resulting in 13.01 and 13.71 kg grain kg^{-1} P, and PFP resulting in 63.58 and 61.92 kg grain kg^{-1} P. Rao [208] reported AE for applied K in hybrid cotton to be 8.8 kg grain kg^{-1} K, where the application rate of the fertilizer is NPK at 200-150-100 kg ha^{-1}, and for non-hybrid cotton, 5.9 kg grain kg^{-1} K at the same rate of fertilizer application. In hybrid rice, AE for applied P was 5.2 kg grain kg^{-1} P and 11.8 kg grain kg^{-1} K with a fertilizer application rate of NPK of 200-75-200 and 200-150-200 kg ha^{-1}, respectively. The AE for applied P in non-hybrid rice was 2.3 kg grain kg^{-1} P and 4.7 kg grain kg^{-1} P, where the fertilizer rate was the same. Rao [208] in another study showed that only the application of P (N and K as blanket doses) gave AE for non-hybrid rice of 4.2–15.6 kg grain kg^{-1} P and 5.9–11.4 kg grain kg^{-1} P, where the P application rate was 75 and 150 kg ha^{-1} and plant spacing was 12.5 × 10 cm and 10 × 10 cm, respectively.

The application of a unit of fertilizer is economical if the increase in crop yield due to the quantity of fertilizer added is higher than the cost of the fertilizer used. However, if a unit of fertilizer does not increase the grain yield enough to pay for its cost, then its application will not be considered economical and will not be profitable even after a constant increase in grain yield [209]. The application of essential plant nutrients in optimum split dosages and proportion, dispensed to plants in an appropriate method and timing, is the key to increased and sustained crop production.

7. Conclusions

Improving global rice yield productivity under low-input conditions is the main challenge for plant breeders and molecular biologists to develop/improve appropriate rice cultivars. Improving NuUE (nutrient use efficiency) is a key component from an agronomic, economic and environmental viewpoint. Despite the highly complex nature of NuUE in rice, recent trends in molecular marker-assisted selection and advanced biotechnological tools can accelerate the dissecting of the polygenic nature of complex traits. Apart from several breeding and agronomic strategies, balanced N, P and K nutrient elements are required to maintain soil fertility, uptake and transportation from soil to grain to produce higher grain yield with nutrient quality traits. The combined genomic and phenomic studies are valuable to distinguish the QTL and gene responses to NPK acquisition and transportation identified, and very few of them are strongly used with the target trait of interest in plant breeding programs. So far, plenty of QTLs have been identified in diverse genetic backgrounds with significant PVE under different treatment doses of NPK. By using this QTL information, better NuUE genotypes can be developed suitable for resource-poor farmers. Further, by employing these rapid developments, an integrative SNP array with innovative techniques such as Next-generation sequencing (NGS) and Genotyping by sequencing (GBS)technologies, high-density and SNP linkage maps and molecular breeding approaches are feasible solutions for identifying cultivars with superior NuUE by incorporating them into breeding cycles and understanding the molecular genetics and physiological mechanisms of N, P and K status in plants under different fertilizers or deficiency conditions. However, a combined holistic approach requires different aspects of work in the pipeline and omic technologies for its implementation in modern NuUE breeding programs.

Author Contributions: J.A., Z.A.J. and A.M. worked on outlining the contents in the manuscript and prepared the draft article. A.A., A.M., J.A. and Z.L. contributed to the screening methodologies associated with aspects of molecular and genomic regions. J.A. and Z.L. conceived of the basic idea, gave suggestions, corrected the entire article and improved the prospects for breeding programs. All the authors read and approved the review article.

Acknowledgments: The authors would like to thank and acknowledge the Bill & Melinda Gates Foundation (BMGF) for providing a research grant to Z.L. for the Green Super Rice project under ID OPP1130530. We would also like to thank the Department of Agriculture (DA) of the Philippines for providing funds to J.A. under the Next-Gen project.

Conflicts of Interest: The authors declare that the research review was conducted in the absence of any commercial or economic associations that could be construed as a potential conflict of interest.

References

1. Ladha, J.K.; Pathak, H.; Krupnik, T.J.; Six, J.; Kessel, C.V. Efficiency of fertilizer nitrogen in cereal production: Retrospects and prospects. *Adv. Agron.* **2005**, *87*, 85–156.
2. Xu, X.; Liu, X.; He, P. Yield Gap, Indigenous Nutrient Supply and Nutrient Use Efficiency for Maize in China. *PLoS ONE* **2015**, *10*, e0140767. [CrossRef] [PubMed]
3. Roberts, T.L. Improving Nutrient Use Efficiency. *Turk. J. Agric. For.* **2008**, *32*, 177–182.
4. Liu, Z.; Zhu, C.; Jiang, Y.; Tian, Y.; Yu, J.; An, H.; Tang, W.; Sun, J.; Tang, J.; Chen, G.; et al. Association mapping and genetic dissection of nitrogen use efficiency-related traits in rice (*Oryza sativa* L.). *Funct. Integr. Genom.* **2016**, *16*, 323–333. [CrossRef] [PubMed]
5. Vinod, K.K.; Heuer, S. Approaches towards nitrogen- and phosphorus-efficient rice. *AoB Plants* **2012**, *28*, 1–18. [CrossRef] [PubMed]
6. Singh, U.; Ladha, J.K.; Castillo, E.G.; Punzalam, G.; Tirol-Padre, A.; Duqueza, M. Genotypic variation in nitrogen use efficiency in medium and long-duration rice. *Field Crops Res.* **1998**, *58*, 35–53. [CrossRef]
7. Han, M.; Okamoto, M.; Beatty, P.H.; Rothstein, S.J.; Good, A.G. The Genetics of Nitrogen Use Efficiency in Crop Plants. *Annu Rev. Genet.* **2015**, *49*, 269–289. [CrossRef] [PubMed]
8. Van Bueren, E.T.L.; Struik, P.C. Diverse concepts of breeding for nitrogen use efficiency, a review. *Agron. Sustain. Dev.* **2017**, *37*, 50. [CrossRef]
9. Chen, H.; Xie, W.; He, H.; Yu, H.; Chen, W.; Li, J.; Yu, R.; Yao, Y.; Zhang, W.; He, Y.; et al. A high-density SNP genotyping array for rice biology and molecular breeding. *Mol. Plant* **2014**, *7*, 541–553. [CrossRef] [PubMed]
10. Thomson, M.J.; Singh, N.; Dwiyanti, M.S.; Wang, D.R.; Wright, M.H.; Perez, F.A.; DeClerck, G.; Chin, J.H.; Malitic-Layaoen, G.A.; Juanillas, V.M.; et al. Large-scale deployment of a rice 6 K SNP array for genetics and breeding applications. *Rice* **2017**, *10*, 40. [CrossRef] [PubMed]
11. Feng, B.; Chen, K.; Cui, Y.; Wu, Z.; Zheng, T.; Zhu, Y.; Ali, J.; Wang, B.; Xu, J.; Zhang, W.; et al. Genetic Dissection and Simultaneous Improvement of Drought and Low Nitrogen Tolerances by Designed QTL Pyramiding in Rice. *Front. Plant Sci.* **2018**, *9*, 306. [CrossRef] [PubMed]
12. Baligar, V.C.; Fageria, N.K.; Hea, Z.L. Nutrient Use Efficiency in Plants. *Commun. Soil Sci. Plant Anal.* **2001**, *32*, 7–8. [CrossRef]
13. Ali, J.; Xu, J.L.; Gao, Y.M.; Fontanilla, M.A.; Li, Z.K. Green super rice (GSR) technology: An innovative breeding strategy-achievements & advances. In Proceedings of the 12th SABRAO Congress-Plant Breeding towards 2025: Challenges in a Rapidly Changing World, Chiang Mai, Thailand, 13–16 January 2012; pp. 16–17.
14. Kole, C.; Muthamilarasan, M.; Henry, R.; Edwards, D.; Sharma, R.; Abberton, M.; Batley, J.; Bentley, A.; Blakeney, M.; Bryant, J.; et al. Application of genomics-assisted breeding for generation of climate resilient crops: Progress and prospects. *Front. Plant Sci.* **2015**, *6*, 563. [CrossRef] [PubMed]
15. Stein, J.C.; Yu, Y.; Copetti, D.; Zhang, L.; Zhang, C.; Chougule, K.; Gao, D.; Iwata, A.; Goicoechea, J.L.; Wei, S.; et al. Genomes of 13 domesticated and wild rice relatives highlight genetic conservation, turnover and innovation across the genus *Oryza*. *Nat. Genet.* **2018**, *50*, 285–296. [CrossRef] [PubMed]
16. Broadbent, F.E.; De Datta, S.K.; Laureles, E.V. Measurement of nitrogen utilization efficiency in rice genotypes. *Agron. J.* **1987**, *79*, 786–791. [CrossRef]
17. Singh, U.; Cassman, K.G.; Ladha, J.K. *Innovative Nitrogen Management Strategies for Lowland Rice Systems*; Fragile Lives in Fragile Ecosystems, International Rice Research Institute, P.O. Box 933: Manila, Philippines, 1995; pp. 229–254.

18. Xiao, J.H.; Li, J.M.; Yuan, L.P.; Yuan, S.R. Identification of QTLs affecting traits of agronomic importance in a recombinant inbred population derived from a sub-specific rice cross. *Theor. Appl. Genet.* **1996**, *92*, 230–244. [CrossRef] [PubMed]
19. Wisser, R.J.; Sun, Q.; Hulbert, S.H.; Kresovich, S.; Nelson, R.J. Identification and Characterization of Regions of the Rice Genome Associated with Broad-Spectrum, Quantitative Disease Resistance. *Genetics* **2015**, *169*, 2277–2293. [CrossRef] [PubMed]
20. Bocianowski, J. Epistasis interaction of QTL effects as a genetic parameter influencing estimation of the genetic additive effect. *Genet. Mol. Biol.* **2013**, *36*, 93–100. [CrossRef] [PubMed]
21. Zhu, H.; Liu, Z.; Fu, X.; Dai, Z.; Wang, S.; Zhang, G.; Zeng, R.; Liu, G. Detection and characterization of epistasis between QTLs on plant height in rice using single segment substitution lines. *Breed. Sci.* **2015**, *65*, 192–200. [CrossRef] [PubMed]
22. Li, Z.S.; Pinson, R.M.; Stansel, J.W.; Park, D. Identification of two major genes and quantitative trait loci (QTLs) for heading date and plant height in cultivated rice (*Oryza sativa* L.). *Theor. Appl. Genet.* **1995**, *91*, 371–381. [CrossRef] [PubMed]
23. Kang, H.J.; Cho, Y.G.; Tlee, Y.; Eun, M.Y.; Shim, J.U. QTL mapping of genes conferring days to heading, culm length and panicle length based on molecular map of rice (*Oryza sativa* L.). *RDA J. Crop Sci.* **1998**, *40*, 55–61.
24. Yamamoto, T.; Yonemaru, J.; Yano, M. Towards the Understanding of Complex Traits in Rice: Substantially or Superficially? *DNA Res.* **2009**, *16*, 141–154. [CrossRef] [PubMed]
25. Zhao, K.; Tung, C.W.; Eizenga, G.C.; Wright, M.H.; Ali, M.L.; Price, A.H.; Norton, G.J.; Islam, M.R.; Reynolds, A.; Mezey, J.; et al. Genome-wide association mapping reveals a rich genetic architecture of complex traits in *Oryza sativa*. *Nat. Commun.* **2011**, *2*, 467. [CrossRef] [PubMed]
26. Nongpiur, R.C.; Singla-Pareek, S.L.; Pareek, A. Genomics Approaches for Improving Salinity Stress Tolerance in Crop Plants. *Curr. Genom.* **2016**, *17*, 343–357. [CrossRef] [PubMed]
27. Nguyen, H.T.T.; Dang, D.T.; Pham, C.V.; Bertin, P. QTL mapping for nitrogen use efficiency and related physiological and agronomical traits during the vegetative phase in rice under hydroponics. *Euphytica* **2016**, *212*, 473–500. [CrossRef]
28. Wang, H.; Qin, F. Genome-Wide Association Study Reveals Natural Variations Contributing to Drought Resistance in Crops. *Front. Plant Sci.* **2017**, *30*, 1110. [CrossRef] [PubMed]
29. Yadav, M.K.; Aravindan, S.; Ngangkham, U.; Subudhi, H.N.; Bag, M.K.; Adak, T.; Munda, S.; Samantary, S.; Jena, M. Correction: Use of molecular markers in identification and characterization of resistance to rice blast in India. *PLoS ONE* **2017**, *12*, e0179467. [CrossRef] [PubMed]
30. Flint-Garcia, S.A.; Thuillet, A.C.; Yu, J.; Pressoir, G.; Romero, S.M.; Mitchell, S.E.; Doebley, J.; Kresovich, S.; Goodman, M.M.; Buckler, E.S. Maize association population: A high-resolution platform for quantitative trait locus dissection. *Plant J.* **2005**, *44*, 1054–1064. [CrossRef] [PubMed]
31. Ersoz, E.S.; Yu, J.; Buckler, E.S. Applications of linkage disequilibrium and association mapping in maize. *Mol. Genet. Approach Maize Improv.* **2009**, *63*, 173–195.
32. Clark, R.M.; Schweikert, G.; Toomajian, C.; Ossowski, S.; Zeller, G.; Shinn, P.; Warthmann, N.; Hu, T.T.; Fu, G.; Hinds, D.A.; et al. Common sequence polymorphisms shaping genetic diversity in *Arabidopsis thaliana*. *Science* **2007**, *317*, 338–342. [CrossRef] [PubMed]
33. Zhao, F.M.; Zhu, H.T.; Ding, X.H.; Zeng, R.Z.; Zhang, Z.L.; Zhang, G. Detection of QTLs for Important Agronomic Traits and Analysis of Their Stabilities Using SSSLs in Rice. *Agric. Sci. China* **2007**, *6*, 769–778. [CrossRef]
34. Goff, S.A.; Ricke, D.; Lan, T.H.; Presting, G.; Wang, R.; Dunn, M.; Glazebrook, J.; Sessions, A.; Oeller, P.; Varma, H.; et al. A draft sequence of the rice genome (*Oryza sativa* L. ssp *japonica*). *Science* **2002**, *296*, 92–100. [CrossRef] [PubMed]
35. Yu, J.; Hu, S.; Wang, J.; Wong, G.K.; Li, S.; Liu, B.; Deng, Y.; Dai, L.; Zhou, Y.; Zhang, X.; et al. A draft sequence of the rice genome (*Oryza sativa* L. ssp *indica*). *Science* **2002**, *296*, 79–92. [CrossRef] [PubMed]
36. Du, H.; Yu, Y.; Ma, Y.; Gao, Q.; Cao, Y.; Chen, Z.; Ma, B.; Qi, M.; Li, Y.; Zhao, X.; et al. Sequencing and de novo assembly of a near complete *indica* rice genome. *Nat. Commun.* **2017**, *8*, 15324. [CrossRef] [PubMed]
37. McNally, K.L.; Childs, K.L.; Bohnert, R.; Davidson, R.M.; Zhao, K.; Ulat, V.J.; Zeller, G.; Clark, R.M.; Hoen, D.R.; Bureau, T.E.; et al. Genome wide SNP variation reveals relationships among landraces and modern varieties of rice. *Proc. Natl. Acad. Sci. USA* **2009**, *106*, 12273–12278. [CrossRef] [PubMed]

38. Huang, X.; Wei, X.; Sang, T.; Zhao, Q.; Feng, Q.; Zhao, Y.; Li, C.; Zhu, C.; Lu, T.; Zhang, Z.; et al. Genome wide association studies of 14 agronomic traits in rice landraces. *Nat. Genet.* **2010**, *42*, 961–967. [CrossRef] [PubMed]

39. Ebana, K.; Yonemaru, J.; Fukuoka, S.; Iwata, H.; Kanamori, K.; Namiki, N.; Nagasaki, H.; Yano, M. Genetic structure revealed by a whole-genome single nucleotide polymorphism survey of 5 diverse accessions of cultivated Asian rice (*Oryza sativa* L.). *Breed. Sci.* **2010**, *60*, 390–397. [CrossRef]

40. Zhao, K.; Aranzana, M.J.; Kim, S.; Lister, C.; Shindo, C.; Tang, C.; Toomajian, C.; Zheng, H.; Dean, C.; Marjoram, P.; et al. An *Arabidopsis* example of association mapping in structured samples. *PLoS Genet.* **2007**, *3*, e4. [CrossRef] [PubMed]

41. Agrama, H.A.; Yan, W.; Jia, M.; Fjellstrom, R.; McClung, A.M. Genetic structure associated with diversity and geographic distribution in the USDA rice world collection. *Nat. Sci.* **2010**, *2*, 247–291. [CrossRef]

42. Ali, J.; Xu, J.; Ismail, A.M.; Fu, B.Y.; Vijaykumar, C.H.M.; Gao, Y.M.; Domingo, J.; Maghirang, R.; Yu, S.B.; Gregorio, G.; et al. Hidden diversity for abiotic and biotic stress tolerances in the primary gene pool of rice revealed by a large backcross breeding program. *Field Crops Res.* **2006**, *97*, 66–76. [CrossRef]

43. MacDonald, G.K.; Bennett, E.M.; Potter, P.A.; Ramankutty, N. Agronomic phosphorus imbalances across the world's croplands. *Proc. Natl. Acad. Sci. USA* **2007**, *108*, 3086–3091. [CrossRef] [PubMed]

44. Cordell, D.; Drangert, J.O.; White, S. The story of phosphorus global food security and food for thought. *Glob. Environ. Chang.* **2009**, *19*, 92–305. [CrossRef]

45. Rose, T.J.; Wissuwa, M. Rethinking internal phosphorus utilization efficiency: A new approach is needed to improve PUE in grain crops. *Adv. Agron.* **2012**, *116*, 185–217.

46. Krishnamurthy, P.; Sreedevi, B.; Ram, T.; Padmavathi, G.; Kumar, R.M.; Rao, P.R; Rani, N.S.; Latha, P.C.; Singh, S.P. Evaluation of rice genotypes for phosphorus use efficiency under soil mineral stress conditions. *Oryza* **2010**, *47*, 29–33.

47. Fageria, N.K.; Morais, O.P.; Baligar, V.C.; Wrigh, R.J. Response of rice cultivars to phosphorus supply on an Oxisol. *Fertilizer Res.* **1988**, *16*, 195–206. [CrossRef]

48. Saito, K.; Vandamme, E.; Segda, Z.; Fofana, M.; Ahouanton, K. A screening protocol for vegetative-stage tolerance to phosphorus deficiency in upland rice. *Crop Sci.* **2015**, *55*, 1223–1229. [CrossRef]

49. Li, Z.K.; Fu, B.Y.; Gao, Y.M.; Xu, J.L.; Ali, J.; Lafitte, H.R.; Jiang, Y.Z.; Rey, J.D.; Vijayakumar, C.H.; Maghirang, R.; et al. Genome-wide ILs and Their Use in Genetic and Molecular Dissection of Complex Phenotypes in Rice (*Oryza sativa* L.). *Plant Mol. Biol.* **2005**, *59*, 33–52. [CrossRef] [PubMed]

50. Panigrahy, M.; Rao, D.N.; Sarla, N. Molecular mechanisms in response to phosphate starvation in rice. *Biotechnol. Adv.* **2009**, *27*, 389–397. [CrossRef] [PubMed]

51. Wissuwa, M.; Kretzschmar, T.; Rose, T.J. From promise to application: Root traits for enhanced nutrient capture in rice breeding. *J. Exp. Bot.* **2016**, *67*, 3605–3615. [CrossRef] [PubMed]

52. Vejchasarn, P.; Lynch, J.P.; Brown, K.M. Genetic Variability in Phosphorus Responses of Rice Root Phenotypes. *Rice* **2016**, *9*, 29. [CrossRef] [PubMed]

53. Aluwihare, Y.C.; Ishan, M.; Chamikara, M.D.M.; Weebadde, C.K.; Sirisena, D.N.; Samarasinghe, W.L.G.; Sooriyapathirana, S.D.S.S. Characterization and Selection of Phosphorus Deficiency Tolerant Rice Genotypes in Sri Lanka. *Rice Sci.* **2016**, *23*, 184–195. [CrossRef]

54. Mahender, A.; Anandan, A.; Pradhan, S.K.; Singh, O.N. Traits-related QTLs and genes and their potential applications in rice improvement under low phosphorus condition. *Arch. Agron. Soil Sci.* **2017**, *64*, 449–464. [CrossRef]

55. Yugandhar, P.; Veronica, N.; Panigrahy, M.; Nageswara Rao, D.; Subrahmanyam, D.; Voleti, S.R.; Mangrauthia, S.K.; Sharma, R.P.; Sarla, N. Comparing Hydroponics, Sand, and Soil Medium to Evaluate Contrasting Rice Nagina 22 Mutants for Tolerance to Phosphorus Deficiency. *Crop Sci.* **2017**, *57*, 1–9. [CrossRef]

56. Chithrameenal, K.; Vellaikumar, S.; Ramalingam, J. Identification of rice (*Oryza sativa* L.) genotypes with high phosphorus use efficiency (PUE) under field and hydroponic conditions. *Indian Res. J. Genet. Biotechnol.* **2017**, *9*, 23–37.

57. Chin, J.H.; Gamuyao, R.; Dalid, C.; Bustamam, M.; Prasetiyono, J.; Moeljopawiro, S.; Wissuwa, M.; Heuer, S. Developing rice with high yield under phosphorus deficiency: *Pup1* sequence to application. *Plant Physiol.* **2011**, *156*, 1202–1216. [CrossRef] [PubMed]

58. Sirisena, D.N.; Wanninayake, W.M.N. Identification of promising rice varieties for low fertile soils in the low country intermediate zone in Sri Lanka. *Ann. Sri Lanka Dep. Agric.* **2014**, *14*, 95–105.

59. Cancellier, E.L.; Brandao, D.R.; Silva, J.; Santos, M.M.; Fidelis, R.R. Phosphorus use efficiency of upland rice cultivars on Cerrado soil. *Ambience* **2012**, *8*, 307–318.

60. Fageria, N.K.; Knupp, A.M.; Moraes, M.F. Phosphorus Nutrition of Lowland Rice in Tropical Lowland Soil. *Commun. Soil Sci. Plant Anal.* **2013**, *44*, 2932–2940. [CrossRef]

61. Panigrahy, M.; Nageswara Rao, D.; Yugandhar, P.; Sravan Raju, N.; Krishnamurthy, P.; Voleti, S.R.; Ashok Reddy, G.; Mohapatra, T.; Robin, A.; Singh, A.K.; et al. Hydroponic experiment for identification of tolerance traits developed by rice Nagina 22 mutants to low-phosphorus in field condition. *Arch. Agron. Soil Sci.* **2014**, *60*, 565–576. [CrossRef]

62. Wu, P.; Ma, L.; Hou, X. Phosphate starvation triggers distinct alterations of genome expression in *Arabidopsis* roots and leaves. *Plant Physiol.* **2003**, *132*, 1260–1271. [CrossRef] [PubMed]

63. Li, J.; Xie, Y.; Dai, A.; Liu, L.; Li, Z. Root and shoot traits responses to phosphorus deficiency and QTL analysis at seedling stage using ILs of rice. *J. Genet. Genom.* **2009**, *36*, 173–183. [CrossRef]

64. Li, L.; Qiu, X.; Li, X.; Wang, S.; Zhang, Q.; Lian, X.M. Transcriptomic analysis of rice responses to low phosphorus stress. *Chin. Sci. Bull.* **2010**, *55*, 251–258. [CrossRef]

65. Wissuwa, M.; Yano, M.; Ae, N. Mapping of QTLs for phosphorus-deficiency tolerance in rice (*Oryza sativa* L.). *Theor. Appl. Genet.* **1998**, *97*, 777–783. [CrossRef]

66. Wissuwa, M.; Ae, N. Further characterization of two QTLs that increase phosphorus uptake of rice (*Oryza sativa* L.) under phosphorus deficiency. *Plant Soil.* **2001**, *237*, 275–286. [CrossRef]

67. Guo, Y.; Lin, W.; Shi, Q.; Liang, Y.; Chen, F.; He, H.; Liang, K. Screening methodology for rice (*Oryza sativa*) genotypes with high phosphorus use efficiency at their seedling stage. *J. Appl. Ecol.* **2002**, *13*, 1587–1591.

68. Yuan, H.; Liu, D. Signaling components involved in plant responses to phosphate starvation. *J. Integr. Plant Biol.* **2008**, *50*, 849–859. [CrossRef] [PubMed]

69. Chin, J.H.; Lu, X.; Haefele, S.M.; Gamuyao, R.; Ismail, A.; Wissuwa, M.; Heuer, S. Development and application of gene-based markers for the major rice QTL *Phosphate uptake 1*. *Theor. Appl. Genet.* **2010**, *120*, 1073–1086. [CrossRef] [PubMed]

70. Ramaekers, L.; Remans, R.; Rao, I.M.; Blair, M.W.; Vanderleyden, J. Strategies for improving phosphorus acquisition efficiency of crop plants. *Field Crops Res.* **2010**, *117*, 169–176. [CrossRef]

71. Wissuwa, M.; Mazzola, M.; Picard, C. Novel approaches in plant breeding for rhizosphere-related traits. *Plant Soil* **2009**, *321*, 409–430. [CrossRef]

72. Chen, L.; Lin, L.; Cai, G.; Sun, Y.; Huang, T.; Wang, K.; Deng, J. Identification of Nitrogen, Phosphorus, and Potassium Deficiencies in Rice Based on Static Scanning Technology and Hierarchical Identification Method. *PLoS ONE* **2014**, *9*, e113200. [CrossRef] [PubMed]

73. Place, G.A.; Sims, J.L.; Hall, U.L. Effects of nitrogen and phosphorous on the growth yield and cooking characteristics of rice. *Agron. J.* **1970**, *62*, 239–241. [CrossRef]

74. Nguyen, H.T.T.; Pham, C.V.; Bertin, P. The effect of nitrogen concentration on nitrogen use efficiency and related parameters in cultivated rices (*Oryza sativa* L. subsp. *indica* and *japonica* and *O. glaberrima* Steud.) in hydroponics. *Euphytica* **2014**, *198*, 137–151. [CrossRef]

75. Vitousek, P.M.; Naylor, R.; Crews, T.; David, M.B.; Drinkwater, L.E.; Holland, E.; Johnes, P.J.; Katzenberger, J.; Martinelli, L.A.; Matson, P.A.; et al. Agriculture. Nutrient imbalances in agricultural development. *Science* **2009**, *324*, 1519–1520. [CrossRef] [PubMed]

76. Bouwman, A.F.; Boumans, L.J.M.; Batjes, N.H. Emissions of N_2O and NO from fertilised fields: Summary of available measurement data. *Glob. Biogeochem. Cycles* **2002**, *16*, 6-1–6-13. [CrossRef]

77. Samborski, S.; Kozak, M.; Azevedo, R.A. Does nitrogen uptake affect nitrogen uptake efficiency or vice versa? *Acta Physiol. Plant.* **2008**, *30*, 419–420. [CrossRef]

78. Li, Y.; Yang, X.; Ren, B.; Shen, Q.; Guo, S. Why nitrogen use efficiency decreases under high nitrogen supply in rice (*Oryza sativa* L.) seedlings. *J. Plant Growth Regul.* **2012**, *31*, 47–52. [CrossRef]

79. Singh, H.; Verma, A.; Ansari, M.A.; Shukla, A. Physiological response of rice (*Oryza sativa* L.) genotypes to elevated nitrogen applied under field conditions. *Plant Signal. Behav.* **2015**, *9*, e29015. [CrossRef] [PubMed]

80. Chaturvedi, I. Effect of Nitrogen Fertilizers on Growth, Yield and Quality of Hybrid Rice (*Oryza sativa* L.). *J. Cent. Eur. Agric.* **2005**, *6*, 611–618.

81. Manzoor, Z.; Awan, T.H.; Zahid, M.A.; Faiz, F.A. Response of rice crop (Super Basmati) to different nitrogen levels. *J. Anim. Plant Sci.* **2006**, *16*, 1–2.

82. Swamy, K.N.; Kondamudi, R.; Kiran, T.V.; Vijayalakshmi, P.; Rao, Y.V.; Rao, P.R.; Subrahmanyam, D.; Voleti, S.R. Screening for nitrogen use efficiency with their root characteristics in rice (*Oryza* spp.) genotypes. *Ann. Biol. Sci.* **2015**, *3*, 8–11.

83. Haque, M.A.; Haque, M.M. Growth, Yield and Nitrogen Use Efficiency of New Rice Variety under Variable Nitrogen Rates. *Am. J. Plant Sci.* **2016**, *7*, 612–622. [CrossRef]

84. Kumagai, E.; Araki, T.; Kubota, F. Effects of nitrogen supply restriction on gas exchange and photosystem 2 function in flag leaves of a traditional low-yield cultivar and a recently improved high-yield cultivar of rice (*Oryza sativa* L.). *Photosynthetica* **2007**, *45*, 489–495. [CrossRef]

85. Maske, N.S.; Borkar, S.L.; Rajgire, H.J. Effects of Nitrogen Levels on Growth, Yield and Grain Quality of Rice. *J. Soil Crop* **1997**, *7*, 83–86.

86. Peng, S.; Cassman, K.G.; Virmani, S.S.; Sheehy, J.; Khush, G.S. Yield potential trends of tropical rice since the release of IR8 and the challenge of increasing rice yield potential. *Crop Sci.* **1999**, *39*, 1552–1559. [CrossRef]

87. Lawlor, D.W. Carbon and nitrogen assimilation in relation to yield: Mechanisms are the key to understanding production systems. *J. Exp. Bot.* **2002**, *53*, 773–787. [CrossRef] [PubMed]

88. Yang, J.; Peng, S.; Zhang, Z.; Wang, Z.; Visperas, R.M.; Zhu, Q. Grain and dry matter yields and portioning of assimilates in Japonica/Indica hybrid rice. *Crop Sci.* **2002**, *42*, 766–772. [CrossRef]

89. Ahmed, M.; Islam, M.M.; Paul, S.K.; Khulna, B. Effect of Nitrogen on Yield and Other Plant Characters of Local, T. Aman Rice Var. Jatai. *Res. J. Agric. Biol. Sci.* **2005**, *1*, 158–161.

90. Hamaoka, N.; Uchida, Y.; Tomita, M.; Kumagai, E.; Araki, T.; Ueno, O. Genetic variations in dry matter production, nitrogen uptake, and nitrogen use efficiency in the AA genome *Oryza* species grown under different nitrogen conditions. *Plant Prod. Sci.* **2013**, *16*, 107–116. [CrossRef]

91. Yogendra, N.D.; Kumara, B.H.; Chandrashekar, N.; Prakash, N.B.; Anantha, M.S.; Shashidhar, H.E. Real-time nitrogen management in aerobic rice by adopting leaf color chart (LCC) as influenced by silicon. *J. Plant Nutr.* **2017**, *40*, 1277–1286. [CrossRef]

92. Mandal, N.N.; Chaudhry, P.P.; Sinha, D. Nitrogen, phosphorus and potash uptake of wheat (var. Sonalika). *Environ. Ecol.* **1992**, *10*, 297.

93. Wang, Y.; Wu, W.H. Genetic approaches for improvement of the crop potassium acquisition and utilization efficiency. *Curr. Opin. Plant Biol.* **2015**, *25*, 46–52. [CrossRef] [PubMed]

94. FAO. *Current World Fertilizer Trends and Outlook to 2016*; Food and Agriculture Organization of the United Nations: Rome, Italy, 2012.

95. Dobermann, A.; Cassman, K.G.; Mamaril, C.P.; Sheehy, J.E. Management of phosphorus, potassium and sulfur in intensive irrigated lowland rice. *Field Crops Res.* **1998**, *56*, 113–138. [CrossRef]

96. Xiaoe, Y.; Romheld, V.; Marschner, H.; Baligar, V.C.; Martens, D.C. Shoot photosynthesis and root growth of hybrid and conventional rice cultivars as affected by N and K levels in the root zone. *Pedosphere* **1997**, *7*, 35–42.

97. Epstein, E.; Bloom, A.J. *Mineral Nutrition of Plants: Principles and Perspectives*, 2nd ed.; Sinauer Associates: Sunderland, MA, USA, 2005.

98. Fageria, N.K.; Dos Santos, A.B.; Moreira, A.; Moraes, M.F. Potassium soil test calibration for lowland rice on an inceptisol. *Commun. Soil Sci. Plant Anal.* **2010**, *41*, 2595–2601. [CrossRef]

99. Grzebisz, W.; Gransee, A.; Szczepaniak, W. The effects of potassium fertilization on water-use efficiency in crop plants. *J. Pant Nutr. Soil Sci.* **2013**, *176*, 355–374. [CrossRef]

100. Mehdi, S.M.; Sarfraz, M.; Hafeez, M. Response of rice advanced line PB-95 to potassium in saline sodic soil. *Pak. J. Biol. Sci.* **2007**, *10*, 2938–2939.

101. Fageria, N.K.; Dos Santos, A.B.; De Moraes, M.F. Yield, Potassium Uptake, and Use Efficiency in Upland Rice Genotypes. *Commun. Soil Sci. Plant Anal.* **2010**, *41*, 2676–2684. [CrossRef]

102. Arif, M.; Arshad, M.; Asghar, H.N.; Basara, S.M.A. Response of rice (*Oryza sativa*) genotypes varying in K use efficiency to various levels of potassium. *Int. J. Agric. Biol.* **2010**, *12*, 926–930.

103. Kalita, U.; Ojha, N.J.; Talukdar, M.C. Effect of levels and time of potassium application on yield and yield attributes of upland rice. *J. Potassium Res.* **1993**, *11*, 203–206.

104. Dunn, D.; Stevens, G. Rice potassium nutrition research progress (Missouri). *Better Crops* **2005**, *89*, 15–17.

105. Awan, T.H.; Manzoor, Z.; Safdar, M.E.; Ahmad, M. Yield response of rice to dynamic use of potassium in traditional rice growing area of Punjab. *Pak. J. Agric. Sci.* **2007**, *44*, 130–135.

106. Bahmaniar, M.A.; Ranjbar, G.A. Effects of nitrogen and potassium fertilizers on rice (*Oryza sativa* L.) genotypes processing characteristics. *Pak. J. Biol. Sci.* **2007**, *10*, 1829–1834. [PubMed]

107. Sarkar, R.K.; Malik, G.C. Effect of foliar spray of KNO_3 and Ca $(NO_3)^2$ on grass pea (*Lathyrus sativus* L.) grown in rice fallows. *Lathyrus Lathyrism Newslett.* **2001**, *2*, 47–48.

108. Islam, A.; Saha, P.K.; Biswas, J.C.; Saleque, M.A. Potassium Fertilization in Intensive Wetland Rice System: Yield, Potassium Use Efficiency and Soil Potassium Status. *Int. J. Agric. Pap.* **2016**, *1*, 7–21.

109. De, D.S.K.; Broadbent, F.E. Development changes related to nitrogen-use efficiency in rice. *Field Crops Res.* **1993**, *34*, 47–56.

110. William, R.R.; Johnson, G.V. Improving nitrogen use efficiency for cereal production. *Agron. J.* **1999**, *91*, 357–363.

111. Gregard, A.; Gelanger, G.; Michaud, R. Nitrogen use efficiency and morphological characteristics of timothy populations selected for low and high forage nitrogen concentrations. *Crop Sci.* **2000**, *40*, 422–429. [CrossRef]

112. Anil, K.; Nidhi, G.; Atul, K.G.; Vikram, S.G. Identification of Biomarker for Determining Genotypic Potential of Nitrogen-Use-Efficiency and Optimization of the Nitrogen Inputs in Crop Plants. *J. Crop Sci. Biotechnol.* **2009**, *12*, 183–194.

113. Pingali, P.L. Green revolution: Impacts, limits, and the path ahead. *Proc. Natl. Acad. Sci. USA* **2012**, *109*, 12302–12308. [CrossRef] [PubMed]

114. Beatty, P.H.; Shrawat, A.K.; Carroll, R.T.; Zhu, T.; Good, A.G. Transcriptome analysis of nitrogen-efficient rice over-expressing alanine aminotransferase. *Plant Biotechnol. J.* **2009**, *7*, 562–576. [CrossRef] [PubMed]

115. Kant, S.; Bi, Y.; Rothstein, S.J. Understanding plant response to nitrogen limitation for the improvement of crop nitrogen use efficiency. *J. Exp. Bot.* **2011**, *62*, 1499–1509. [CrossRef] [PubMed]

116. Kabir, G. Genetic approaches of increasing nutrient use efficiency especially nitrogen in cereal crops—A review. *J. Bio-Sci.* **2014**, *22*, 111–125. [CrossRef]

117. Rose, T.J.; Kretzschmar, T.; Waters, D.L.E.; Balindong, J.L.; Wissuwa, M. Prospects for Genetic Improvement in Internal Nitrogen Use Efficiency in Rice. *Agronomy* **2017**, *7*, 70. [CrossRef]

118. Zhou, Y.; Tao, Y.; Tang, D.; Wang, J.; Zhong, J.; Wang, Y.; Yuan, Q.; Yu, X.; Zhang, Y. Identification of QTL Associated with Nitrogen Uptake and Nitrogen Use Efficiency Using High Throughput Genotyped CSSLs in Rice (*Oryza sativa* L.). *Front. Plant Sci.* **2017**, *8*, 1166. [CrossRef] [PubMed]

119. Sinha, S.K.; Sevanthi, V.A.M.; Chaudhary, S.; Tyagi, P.; Venkadesan, S.; Rani, M.; Mandal, P.K. Transcriptome Analysis of Two Rice Varieties Contrasting for Nitrogen Use Efficiency under Chronic N Starvation Reveals Differences in Chloroplast and Starch Metabolism-Related Genes. *Genes (Basel)* **2018**, *11*, E206. [CrossRef] [PubMed]

120. Duan, Y.H.; Zhang, Y.L.; Ye, L.T.; Fan, X.R.; Xu, G.H.; Shen, Q.R. Responses of Rice Cultivars with Different Nitrogen Use Efficiency to Partial Nitrate Nutrition. *Ann. Bot.* **2007**, *99*, 1153–1160. [CrossRef] [PubMed]

121. Fan, X.; Xie, D.; Chen, J.; Lu, H.; Xu, Y.; Ma, C.; Xu, G. Over-expression of OsPTR6 in rice increased plant growth at different nitrogen supplies but decreased nitrogen use efficiency at high ammonium supply. *Plant Sci.* **2014**, *227*, 1–11. [CrossRef] [PubMed]

122. Ashikari, M.; Sakakibara, H.; Lin, S.; Yamamoto, T.; Takashi, T.; Nishimura, A.; Angeles, E.R.; Qian, Q.; Kitano, H.; Matsuoka, M. Cytokinin oxidase regulates rice grain production. *Science* **2005**, *309*, 741–745. [CrossRef] [PubMed]

123. Huang, X.; Qian, Q.; Liu, Z.; Sun, H.; He, S.; Luo, D.; Xia, G.; Chu, C.; Li, J.; Fu, X. Natural variation at the DEP1 locus enhances grain yield in rice. *Nat. Genet.* **2009**, *41*, 494–497. [CrossRef] [PubMed]

124. Miura, K.; Ikeda, M.; Matsubara, A.; Song, X.J.; Ito, M.; Asano, K.; Matsuoka, M.; Kitano, H.; Ashikari, M. OsSPL14 promotes panicle branching and higher grain productivity in rice. *Nat. Genet.* **2010**, *42*, 545–549. [CrossRef] [PubMed]

125. Jiao, Y.; Wang, Y.; Xue, D.; Wang, J.; Yan, M.; Liu, G.; Dong, G.; Zeng, D.; Lu, Z.; Zhu, X.; Qian, Q.; Li, J. Regulation of OsSPL14 by OsmiR156 defines ideal plant architecture in rice. *Nat. Genet.* **2010**, *42*, 541–544. [CrossRef] [PubMed]

126. Song, X.J.; Huang, W.; Shi, M.; Zhu, M.Z.; Lin, H.X. A QTL for rice grain width and weight encodes a previously unknown RING-type E3 ubiquitin ligase. *Nat. Genet.* **2007**, *39*, 623–630. [CrossRef] [PubMed]

127. Shomura, A.; Izawa, T.; Ebana, K.; Ebitani, T.; Kanegae, H.; Konishi, S.; Yano, M. Deletion in a gene associated with grain size increased yields during rice domestication. *Nat. Genet.* **2008**, *40*, 1023–1028. [CrossRef] [PubMed]

128. Weng, J.; Gu, S.; Wan, X.; Gao, H.; Guo, T.; Su, N.; Lei, C.; Zhang, X.; Cheng, Z.; Guo, X.; et al. Isolation and initial characterization of GW5,a major QTL associated with rice grain width and weight. *Cell Res.* **2008**, *18*, 1199–1209. [CrossRef] [PubMed]

129. Wang, E.; Wang, J.; Zhu, X.; Hao, W.; Wang, L.; Li, Q.; Zhang, L.; He, W.; Lu, B.; Lin, H.; et al. Control of rice grain-filling and yield by a gene with a potential signature of domestication. *Nat. Genet.* **2008**, *40*, 1370–1374. [CrossRef] [PubMed]

130. Yano, M.; Katayose, Y.; Ashikari, M.; Yamanouchi, U.; Monna, L.; Fuse, T.; Baba, T.; Yamamoto, K.; Umehara, Y.; Nagamura, Y.; et al. *Hd1*, a major photoperiod sensitivity quantitative trait locus in rice, is closely related to the Arabidopsis flowering time gene *CONSTANS*. *Plant Cell* **2000**, *12*, 2473–2483. [CrossRef] [PubMed]

131. Takahashi, Y.; Shomura, A.; Sasaki, T.; Yano, M. Hd6, a rice quantitative trait locus involved in photoperiod sensitivity, encodes the subunit of protein kinase CK2. *Proc. Natl. Acad. Sci. USA* **2001**, *98*, 7922–7927. [CrossRef] [PubMed]

132. Kojima, S.; Takahashi, Y.; Kobayashi, Y.; Monna, L.; Sasaki, T.; Araki, T.; Yano, M. Hd3a, a rice ortholog of the Arabidopsis FT gene, promotes transition to flowering downstream of Hd1 under shortday conditions. *Plant Cell Physiol.* **2002**, *43*, 1096–1105. [CrossRef] [PubMed]

133. Izawa, T.; Oikawa, T.; Sugiyama, N.; Tanisaka, T.; Yano, M.; Shimamoto, K. Phytochrome mediates the external light signal to repress FT orthologs in photoperiodic flowering of rice. *Genes Dev.* **2002**, *16*, 2006–2020. [CrossRef] [PubMed]

134. Tamaki, S.; Matsuo, S.; Wong, H.L.; Yokoi, S.; Shimamoto, K. Hd3a protein is a mobile flowering signal in rice. *Science* **2007**, *316*, 1033–1036. [CrossRef] [PubMed]

135. Doi, K.; Izawa, T.; Fuse, T.; Yamanouchi, U.; Kubo, T.; Shimatani, Z.; Yano, M.; Yoshimura, A. Ehd1, a B-type response regulator in rice, confers short-day promotion of flowering and controls FT-like gene expression independently of Hd1. *Genes Dev.* **2004**, *18*, 926–936. [CrossRef] [PubMed]

136. Xue, W.; Xing, Y.; Weng, X.; Zhao, Y.; Tang, W.; Wang, L.; Zhou, H.; Yu, S.; Xu, C.; Li, X.; Zhang, Q. Natural variation in Ghd7 is an important regulator of heading date and yield potential in rice. *Nat. Genet.* **2008**, *40*, 761–767. [CrossRef] [PubMed]

137. Wei, X.; Xu, J.; Guo, H.; Jiang, L.; Chen, S.; Yu, C.; Zhou, Z.; Hu, P.; Zhai, H.; Wan, J. DTH8 suppresses flowering in rice, influencing plant height and yield potential simultaneously. *Plant Physiol.* **2010**, *153*, 1747–1758. [CrossRef] [PubMed]

138. Sasaki, A.; Ashikari, M.; Ueguchi-Tanaka, M.; Itoh, H.; Nishimura, A.; Swapan, D.; Ishiyama, K.; Saito, T.; Kobayashi, M.; Khush, G.S.; et al. Green revolution: A mutant gibberellin-synthesis gene in rice. *Nature* **2002**, *416*, 701–702. [CrossRef] [PubMed]

139. Ookawa, T.; Hobo, T.; Yano, M.; Murata, K.; Ando, T.; Miura, H.; Asano, K.; Ochiai, Y.; Ikeda, M.; Nishitani, R.; et al. New approach for rice improvement using a pleiotropic *QTL* gene for lodging resistance and yield. *Nat. Commun.* **2010**, *1*, 132. [CrossRef] [PubMed]

140. Fukuoka, S.; Saka, N.; Koga, H.; Ono, K.; Shimizu, T.; Ebana, K.; Hayashi, N.; Takahashi, A.; Hirochika, H.; Okuno, K.; Yano, M. Loss of function of a proline-containing protein confers durable disease resistance in rice. *Science* **2009**, *325*, 998–1001. [CrossRef] [PubMed]

141. Hayashi, N.; Inoue, H.; Kato, T.; Funao, T.; Shirota, M.; Shimizu, T.; Kanamori, H.; Yamane, H.; Hayano-Saito, Y.; Matsumoto, T. Durable panicle blast-resistance gene Pb1 encodes an atypical CC-NBS-LRR protein and was generated by acquiring a promoter through local genome duplication. *Plant J.* **2010**, *64*, 498–510. [CrossRef] [PubMed]

142. Ren, Z.H.; Gao, J.P.; Li, L.G.; Cai, X.L.; Huang, W.; Chao, D.Y.; Zhu, M.Z.; Wang, Z.Y.; Luan, S.; Lin, H.X. A rice quantitative trait locus for salt tolerance encodes a sodium transporter. *Nat. Genet.* **2005**, *37*, 1141–1146. [CrossRef] [PubMed]

143. Fujino, K.; Sekiguchi, H.; Matsuda, Y.; Sugimoto, K.; Ono, K.; Yano, M. Molecular identification of a major quantitative trait locus, qLTG3-1, controlling low-temperature germinability in rice. *Proc. Natl. Acad. Sci. USA* **2008**, *105*, 12623–12628. [CrossRef] [PubMed]

144. Xu, K.; Xu, X.; Fukao, T.; Canlas, P.; Maghirang-Rodriguez, R.; Heuer, S.; Ismail, A.M.; Bailey-Serres, J.; Ronald, P.C.; Mackill, D.J. *Sub1A* is an ethylene-response-factor-like gene that confers submergence tolerance to rice. *Nature* **2006**, *442*, 705–708. [CrossRef] [PubMed]

145. Hattori, U.Y.; Nagai, K.; Furukawa, S.; Song, X.J.; Kawano, R.; Sakakibara, H.; Wu, J.; Matsumoto, T.; Yoshimura, A.; Kitano, H.; Matsuoka, M. The ethylene response factors SNOKEL1 and SNOKEL2 allow rice to adapt to deep water. *Nature* **2009**, *460*, 1026–1030. [CrossRef] [PubMed]

146. Ueno, D.; Koyama, E.; Kono, I.; Ando, T.; Yano, M.; Ma, J.F. Identification of a novel major quantitative trait locus controlling distribution of Cd between roots and shoots in rice. *Plant Cell Physiol.* **2009**, *50*, 2223–2233. [CrossRef] [PubMed]

147. Li, C.; Zhou, A.; Sang, T. Rice domestication by reducing shattering. *Science* **2006**, *311*, 1936–1939. [CrossRef] [PubMed]

148. Konishi, S.; Izawa, T.; Lin, S.Y.; Ebana, K.; Fukuta, Y.; Sasaki, T.; Yano, M. An SNP caused loss of seed shattering during rice domestication. *Science* **2006**, *312*, 1392–1396. [CrossRef] [PubMed]

149. Tan, L.; Li, X.; Liu, F.; Sun, X.; Li, C.; Zhu, Z.; Fu, Y.; Cai, H.; Wang, X.; Xie, D.; Sun, C. Control of a key transition from prostrate to erect growth in rice domestication. *Nat. Genet.* **2008**, *40*, 1360–1364. [CrossRef] [PubMed]

150. Jin, J.; Huang, W.; Gao, J.P.; Yang, J.; Shi, M.; Zhu, M.Z.; Luo, D.; Lin, H.X. Genetic control of rice plant architecture under domestication. *Nat. Genet.* **2008**, *40*, 1365–1369. [CrossRef] [PubMed]

151. Romero, L.E.; Lozano, I.; Garavito, A.; Carabali, S.J.; Triana, M.; Villareal, N.; Reyes, L.; Duque, M.C.; Martinez, C.P. Major QTLs Control Resistance to Rice Hoja Blanca Virus and Its Vector *Tagosodes orizicolus*. *G3 (Bethesda)* **2014**, *4*, 133–142. [CrossRef] [PubMed]

152. Uga, Y.; Yamamoto, E.; Kanno, N.; Kawai, S.; Mizubayashi, T.; Fukuoka, S. A major QTL controlling deep rooting on rice chromosome 4. *Sci. Rep.* **2013**, *3*, 3040. [CrossRef] [PubMed]

153. Ni, J.J.; Wu, P.; Senadhira, D.; Huang, N. Mapping QTLs for phosphorus deficiency tolerance in rice (*Oryza sativa* L.). *Theor. Appl. Genet.* **1998**, *97*, 1361–1369. [CrossRef]

154. Wissuwa, M.; Wegner, J.; Ae, N.; Yano, M. Substitution mapping of *Pup1*: A major QTL increasing phosphorus uptake of rice from a phosphorus-deficient soil. *Theor. Appl. Genet.* **2002**, *105*, 890–897. [PubMed]

155. Shimizu, A.; Yanagihara, S.; Kawasaki, S.; Ikehashi, H. Phosphorus deficiency-induced root elongation and its QTL in rice (*Oryza sativa* L.). *Theor. Appl. Genet.* **2004**, *109*, 1361–1368. [CrossRef] [PubMed]

156. Lang, N.T.; Buu, B.C. Mapping QTLs for phosphorus deficiency tolerance in rice (*Oryza sativa* L.). *Omonrice* **2006**, *14*, 1–9.

157. Shimizu, A.; Kato, K.; Komatsu, A.; Motomura, K.; Ikehashi, H. Genetic analysis of root elongation induced by phosphorus deficiency in rice (*Oryza sativa* L.): Fine QTL mapping and multivariate analysis of related traits. *Theor. Appl. Genet.* **2008**, *117*, 987–996. [CrossRef] [PubMed]

158. Chao, X.; Jie, R.; Xiu-qin, Z.; Zai-song, D.; Jing, Z.; Chao, W.; Jun-wei, Z.; Joseph, C.A.; Qiang, Z.; et al. Genetic Dissection of Low Phosphorus Tolerance Related Traits Using Selected Introgression Lines in Rice. *Rice Sci.* **2015**, *22*, 264–274. [CrossRef]

159. Wang, K.; Cui, K.; Liu, G.; Xie, W.; Yu, H.; Pan, J.; Huang, J.; Nie, L.; Shah, F. Identification of quantitative trait loci for phosphorus use efficiency traits in rice using a high density SNP map. *BMC Genet.* **2014**, *15*, 155. [CrossRef] [PubMed]

160. Feng, M.; Xianwu, Z.; Guohua, M.; He, P.; Zhu, L.; Zhang, F. Identification of quantitative trait loci affecting tolerance to low phosphorus in rice (*Oryza sativa* L.). *Chin. Sci. Bull.* **2000**, *45*, 519–525.

161. Fang, P.; Wu, P. QTL × N-level interaction for plant height in rice (*Oriza sativa* L.). *Plant Soil* **2001**, *236*, 237–242. [CrossRef]

162. Ishimaru, K.; Kobayashi, N.; Ono, K.; Yano, M.; Ohsugi, R. Are contents of Rubisco, soluble protein and nitrogen in flag leaves of rice controlled by the same genetics? *J. Exp. Bot.* **2001**, *52*, 1827–1833. [CrossRef] [PubMed]

163. Ogawa, S.; Valencia, M.O.; Lorieux, M.; Arbelaez, J.D.; McCouch, M.; Ishitani, M.; Selvaraj, M.G. Identification of QTLs associated with agronomic performance under nitrogen-deficient conditions using chromosome segment substitution lines of a wild rice relative, *Oryza rufipogon*. *Acta Physiol. Plant.* **2016**, *38*, 103. [CrossRef]

164. Obara, M.; Kajiura, M.; Fukuta, Y.; Yano, M.; Hayashi, M.; Yamaya, T.; Sato, T. Mapping of QTLs associated with cytosolic glutamine synthetase and NADH- glutamate synthase in rice (*Oryza sativa* L.). *J. Exp. Bot.* **2001**, *52*, 1209–1217. [PubMed]

165. Senthilvel, S.; Govindaraj, P.; Arumugachamy, S.; Latha, R.; Malarvizhi, P.; Gopalan, A.; Maheswaran, M. Mapping genetic loci associated with nitrogen use efficiency in rice (*Oryza sativa* L.). In Proceedings of the 4th International Crop Science Congress, Brisbane, Australia, 26 September–1 October 2004.

166. Tong, H.H.; Mei, H.W.; Yu, X.Q.; Xu, X.Y.; Li, M.S.; Zhang, S.Q.; Luo, L.J. Identification of Related QTLs at Late Developmental Stage in Rice (*Oryza sativa* L.) Under Two Nitrogen Levels. *Acta Genet. Sin.* **2006**, *33*, 458–467. [CrossRef]

167. Wang, Y.; Sun, Y.J.; Chen, D.Y.; Yu, S.B. Analysis of Quantitative Trait Loci in Response to Nitrogen and Phosphorus Deficiency in Rice Using Chromosomal Segment Substitution Lines. *Acta Agron. Sin.* **2009**, *35*, 580–587. [CrossRef]

168. Laza, M.R.; Kondo, M.; Ideta, O.; Barleen, E.; Imbe, T. Identification of quantitative trait loci for d13C and productivity in irrigated lowland rice. *Crop Sci.* **2006**, *46*, 763–773. [CrossRef]

169. MacMillan, K.; Emrich, K.; Piepho, H.P.; Mullins, C.E.; Price, A.H. Assessing the importance of genotype × environment interaction for root traits in rice using a mapping population II: Conventional QTL analysis. *Theor. Appl. Genet.* **2006**, *113*, 953–964. [CrossRef] [PubMed]

170. Cho, Y.I.; Jiang, W.Z.; Chin, J.H.; Piao, Z.; Cho, Y.G.; McCouch, S.; Koh, H.J. Identification of QTLs associated with physiological nitrogen use efficiency in rice. *Mol. Cell* **2007**, *23*, 72–79.

171. Senthilvel, S.; Vinod, K.K.; Malarvizhi, P.; Maheswaran, M. QTL and QTL × environment effects on agronomic and nitrogen acquisition traits in rice. *J. Integr. Plant Biol.* **2008**, *50*, 1108–1117. [CrossRef] [PubMed]

172. Piao, Z.; Li, M.; Li, P.; Zhang, C.; Wang, H.; Luo, Z.; Lee, J.; Yang, R. Bayesian dissection for genetic architecture of traits associated with nitrogen utilization efficiency in rice. *Afr. J. Biotechnol.* **2009**, *8*, 6834–6839.

173. Srividya, A.; Vemireddy, L.R.; Hariprasad, A.S.; Jayaprada, M.; Sakile, S.; Puram, V.R.R.; Anuradha, G.; Siddiq, E.A. Identification and mapping of landrace derived QTL associated with yield and its components in rice under different nitrogen levels and environments. *Int. J. Plant Breed. Genet.* **2010**, *4*, 210–227. [CrossRef]

174. Tong, H.H.; Chen, L.; Li, W.P.; Mei, H.; Xing, Y.; Yu, X.; Xu, X.; Zhang, S.; Luo, L. Identification and characterization of quantitative trait loci for grain yield and its components under different nitrogen fertilization levels in rice (*Oryza sativa* L.). *Mol. Breed.* **2011**, *28*, 495–509. [CrossRef]

175. Yue, F.; Rong-rong, Z.; Ze-chuan, L.; Li-yong, C.; Xing-hua, W.; Shi-hua, C. Quantitative trait locus analysis for rice yield traits under two nitrogen levels. *Rice Sci.* **2015**, *22*, 108–115. [CrossRef]

176. Wei, D.; Cui, K.; Ye, G.; Pan, J.; Xiang, J.; Huang, J.; Nie, L. QTL mapping for nitrogen-use efficiency and nitrogen-deficiency tolerance traits in rice. *Plant Soil* **2012**, *359*, 281–295. [CrossRef]

177. Wu, P.; Ni, J.J.; Luo, A.C. QTLs underlying Rice Tolerance to Low-Potassium Stress in Rice Seedlings. *Crop Sci.* **1998**, *38*, 1458–1462. [CrossRef]

178. Senaratne, R.; Ratnasinghe, D.S. Nitrogen fixation and beneficial effects of some grain legumes and green-manure crops on rice. *Biol. Fer. Soils* **1995**, *19*, 49–54. [CrossRef]

179. Lin, X.Q.; Zhou, W.J.; Zhu, D.F.; Zhang, Y. Effect of water management on photosynthetic rate and water use efficiency of leaves in paddy rice. *Chin. J. Rice Sci.* **2004**, *18*, 333–338, (in Chinese with English abstract).

180. Peng, S.; Cassman, K.G. Upper thresholds of nitrogen uptake rates and associated nitrogen fertilizer efficiencies in irrigated rice. *Agron. J.* **1998**, *90*, 178–185. [CrossRef]

181. Yamaya, T.; Obara, M.; Nakajima, H.; Sasaki, S.; Hayakawa, T.; Sato, T. Genetic manipulation and quantitative trait loci mapping for nitrogen recycling in rice. *J. Exp. Bot.* **2002**, *53*, 917–925. [CrossRef] [PubMed]

182. Lian, X.; Xing, Y.; Yan, H.; Xu, C.; Li, X.; Zhang, Q. QTLs for low nitrogen tolerance at seedling stage identified using a recombinant inbred line population derived from an elite rice hybrid. *Theor. Appl. Genet.* **2005**, *112*, 85–96. [CrossRef] [PubMed]

183. De, M.; Velk, P.L.G. The role of Azolla cover in improving the nitrogen use efficiency of lowland rice. *Plant Soil* **2004**, *263*, 311–321.

184. Ladha, J.K.; Kirk, G.J.D.; Bennett, J.; Peng, S.; Reddy, C.K.; Reddy, P.M.; Singh, U. Opportunities for increased nitrogen use efficiency from improved lowland rice germplasm. *Field Crops Res.* **1998**, *56*, 41–71. [CrossRef]

185. Obara, M.; Sato, T.; Sasaki, S.; Kashiba, K.; Nagano, A.; Nakamura, I.; Ebitani, T.; Yano, M.; Yamaya, T. Identification and characterization of a QTL on chromosome 2 for cytosolic glutamine synthetase content and panicle number in rice. *Theor. Appl. Genet.* **2004**, *110*, 1–11. [CrossRef] [PubMed]

186. Shan, Y.H.; Wang, Y.; Pan, X.B. Mapping of QTLs for nitrogen use efficiency and related traits in rice (*Oryza sativa* L.). *Agric. Sci. China* **2005**, *4*, 721–727.

187. Dong, G.C.; Wang, Y.L.; Zhang, Y.F.; Chen, P.; Yang, L.; Huang, J.; Zuo, B. Characteristics of yield and yield components in conventional *indica* rice cultivars with different nitrogen use efficiencies for grain output. *Acta Agron. Sin.* **2006**, *32*, 1511–1518.

188. Dong, G.C.; Wang, Y.; Yu, X.F. Differences of nitrogen uptake and utilization of conventional rice varieties with different growth duration. *Sci. Agric. Sin.* **2011**, *44*, 4570–4582.

189. Sanchez, P.A.; Salinas, J.G. Low-input technology for managing oxisols and ultisols in tropical America. *Adv. Agron.* **1981**, *34*, 279–406.

190. Dobermann, A.; Fairhurst, T. *Rice: Nutrient Disorders & Nutrient Management*; Potash & Phosphate Institute, Potash & Phosphate Institute of Canada, and International Rice Research Institute: Singapore; Los Baños, Philippines, 2000.

191. Su, J.Y.; Xiao, Y.M.; Li, M.; Liu, Q.; Li, B.; Tong, Y.; Jia, J.; Li, Z. Mapping QTLs for phosphorus-deficiency tolerance at wheat seedling stage. *Plant Soil* **2006**, *281*, 25–36. [CrossRef]

192. Liu, Y.; Li, Z.C.; Mi, G.H.; Zhang, H.L.; Mu, P.; Wang, X. Screen and identification for tolerance to low-phosphorus stress of rice germplasm (*Oryza sativa* L.). *Acta Agron. Sin.* **2005**, *31*, 238–242. (in Chinese with English abstract)

193. Ping, M.U.; Huang, C.; Li, J.X.; Liu, L.F.; Li, Z.C. Yield Trait Variation and QTL Mapping in a DH Population of Rice Under Phosphorus Deficiency. *Acta Agron. Sin.* **2008**, *34*, 1137–1142.

194. Liu, L.J.; Chang, E.H.; Fan, M.M.; Wang, Z.Q.; Yang, J.C. Effects of Potassium and Calcium on Root Exudates and Grain Quality During Grain Filling. *Acta Agron. Sin.* **2011**, *37*, 661–669.

195. Torkashv, M.; Vahed, S. The efficiency of potassium fertilization methods on the growth of rice (*Oryza sativa* L.) under salinity stress. *Afr. J. Biotechnol.* **2011**, *10*, 15946–15952.

196. Ali, J.; Franje, N.J.; Revilleza, J.E.; Acero, B. *Breeding for Low-Input Responsive Green Super Rice (GSR) Varieties for Rainfed Lowlands of Asia and Africa. University Library*; University of the Philippines at Los Baños: Los Baños, Philippines, 2016.

197. Yorobe, J.M.; Ali, J.; Pede, V.; Rejesus, R.M.; Velarde, O.P.; Wang, W. Yield and income effects of rice varieties with tolerance of multiple abiotic stresses: The case of green super rice (GSR) and flooding in the Philippines. *Agric. Econ.* **2016**, *47*, 1–11. [CrossRef]

198. Wu, L.; Yuan, S.; Huang, L.; Sun, F.; Zhu, G.; Li, G.; Fahad, S.; Peng, S.; Wang, F. Physiological Mechanisms Underlying the High-Grain Yield and High-Nitrogen Use Efficiency of Elite Rice Varieties under a Low Rate of Nitrogen Application in China. *Front. Plant Sci.* **2016**, *7*, 1024. [CrossRef] [PubMed]

199. Wang, F.; Peng, S. Yield potential and nitrogen use efficiency of China's super rice. *J. Integr. Agric.* **2017**, *16*, 1000–1008. [CrossRef]

200. Ali, J.; Xu, J.L.; Gao, Y.; Fontanilla, M.; Li, Z.K. Breeding for yield potential and enhanced productivity across different rice ecologies through green super rice (GSR) breeding strategy. In *International Dialogue on Perception and Prospects of Designer Rice*; Muralidharan, K., Siddiq, E.A., Eds.; Society for the Advancement of Rice Research, Directorate of Rice Research: Hyderabad, India, 2013; pp. 60–68.

201. Mortvedt, J.J.; Murphy, L.S.; Follett, R.H. *Fertilizer Technology and Application*; Meister Publishing Co.: Willoughby, OH, USA, 2001.

202. Yadav, R.L. Assessing on-farm efficiency and economics of fertilizer N., P and K in rice-wheat systems of India. *Field Crops Res.* **2003**, *18*, 39–51. [CrossRef]

203. Cassman, K.G.; Gines, G.C.; Dizon, M.A.; Samson, M.I.; Alceantara, J.M. Nitrogen use efficiency in tropical lowland rice systems: Contributions from indigenous and applied nitrogen. *Fields Crops Res.* **1996**, *47*, 1–12. [CrossRef]

204. Dobermann, A.R. *Nitrogen Use Efficiency-State of the Art*; Agronomy-Faculty Publications: Lincoln, NE, USA, 2005; p. 316.

205. Wen-xia, X.; Guang-huo, W.; Qi-chun, Z.; Guo, H.C. Effects of nitrogen fertilization strategies on nitrogen use efficiency in physiology, recovery, and agronomy and redistribution of dry matter accumulation and nitrogen accumulation in two typical rice cultivars in Zhejiang. *China J. Zhejiang Univ. Sci. B* **2007**, *8*, 208–216.

206. Yoshida, S. *Fundamentals of Rice Crop Science*; IRRI: Los Baños, Laguna, Philippines, 1981; 269p.

207. Amanullah; Muhammad, A.; Almas, L.K.; Amanullaj, J.; Zahir, S.; Rahman, H.; Khalil, S.K. Agronomic Efficiency and Profitability of P-Fertilizers Applied at Different Planting Densities of Maize in Northwest. *Pak. J. Plant Nutr.* **2012**, *35*, 331–341. [CrossRef]

Int. J. Mol. Sci. **2018**, *19*, 1762

208. Rao, T.N. Improving nutrient use efficiency: The role of beneficial management practices. In *Better Crops-India*; IPNI–India Program 133: Gurgaon, India, 2007; Volume 1, pp. 6–7.

209. Singh, D.P. Vermiculture biotechnology and biocomposting. In *Environmental Microbiology and Biotechnology*; Singh, D.P., Dwivedi, S.K., Eds.; New Age International (P) Limited Publishers: Darya Ganj, New Delhi, 2004; pp. 97–112.

International Journal of
Molecular Sciences

Article

Enhanced Expression of QTL *qLL9/DEP1* Facilitates the Improvement of Leaf Morphology and Grain Yield in Rice

Xue Fu [†], Jing Xu [†], Mengyu Zhou, Minmin Chen, Lan Shen, Ting Li, Yuchen Zhu, Jiajia Wang, Jiang Hu, Li Zhu, Zhenyu Gao, Guojun Dong, Longbiao Guo, Deyong Ren, Guang Chen, Jianrong Lin, Qian Qian * and Guangheng Zhang *

State Key Laboratory of Rice Biology, China National Rice Research Institute, Hangzhou 310006, China;
fuxuezsy@163.com (X.F.); xujing87@126.com (J.X.); zhoumy@mail.sustc.edu.cn (M.Z.); jiashf@126.com (M.C.);
baishushenlan@126.com (L.S.); 13258376009@163.com (T.L.); zyc1205926704@sina.com (Y.Z.);
wgwangjiajia@163.com (J.W.); hujiang588@163.com (J.H.); zhuli05@caas.cn (L.Z.);
zygao2000@hotmail.com (Z.G.); dongguojun@caas.cn (G.D.); guolongbiao@caas.cn (L.G.);
rendeyongsd@163.com (D.R.); chenguang0066@126.com (G.C.); ljr1970@hotmail.com (J.L.)
* Correspondence: qianqian188@hotmail.com (Q.Q.); zhangguangheng@126.com (G.Z.);
  Tel.: +86-571-63371418 (Q.Q.); +86-571-63370211 (G.Z.)
† These authors contributed equally to this work.

Received: 15 January 2019; Accepted: 13 February 2019; Published: 17 February 2019

Abstract: In molecular breeding of super rice, it is essential to isolate the best quantitative trait loci (QTLs) and genes of leaf shape and explore yield potential using large germplasm collections and genetic populations. In this study, a recombinant inbred line (RIL) population was used, which was derived from a cross between the following parental lines: hybrid rice Chunyou84, that is, *japonica* maintainer line Chunjiang16B (CJ16); and *indica* restorer line Chunhui 84 (C84) with remarkable leaf morphological differences. QTLs mapping of leaf shape traits was analyzed at the heading stage under different environmental conditions in Hainan (HN) and Hangzhou (HZ). A major QTL *qLL9* for leaf length was detected and its function was studied using a population derived from a single residual heterozygote (RH), which was identified in the original population. *qLL9* was delimited to a 16.17 kb region flanked by molecular markers C-1640 and C-1642, which contained three open reading frames (ORFs). We found that the candidate gene for *qLL9* is allelic to *DEP1* using quantitative real-time polymerase chain reaction (qRT-PCR), sequence comparison, and the clustered regularly interspaced short palindromic repeat-associated Cas9 nuclease (CRISPR/Cas9) genome editing techniques. To identify the effect of *qLL9* on yield, leaf shape and grain traits were measured in near isogenic lines (NILs) NIL-*qLL9*[CJ16] and NIL-*qLL9*[C84], as well as a chromosome segment substitution line (CSSL) CSSL-*qLL9*[KASA] with a Kasalath introgressed segment covering *qLL9* in the Wuyunjing (WYJ) 7 backgrounds. Our results showed that the flag leaf lengths of NIL-*qLL9*[C84] and CSSL-*qLL9*[KASA] were significantly different from those of NIL-*qLL9*[CJ16] and WYJ 7, respectively. Compared with NIL-*qLL9*[CJ16], the spike length, grain size, and thousand-grain weight of NIL-*qLL9*[C84] were significantly higher, resulting in a significant increase in yield of 15.08%. Exploring and pyramiding beneficial genes resembling *qLL9*[C84] for super rice breeding could increase both the source (e.g., leaf length and leaf area) and the sink (e.g., yield traits). This study provides a foundation for future investigation of the molecular mechanisms underlying the source–sink balance and high-yield potential of rice, benefiting high-yield molecular design breeding for global food security.

Keywords: *Oryza sativa* L.; leaf shape; yield trait; molecular breeding; hybrid rice

1. Introduction

Rice leaf morphogenesis and its spatial extension posture are important components of ideal plant architecture, which play a significant role in the photosynthetic efficiency and grain yield [1,2]. During the grain-filling stage, the top three leaves, particularly the flag leaf, are the main carbohydrate sources that were transported to panicles and grains for yield formation. The establishment of better source-to-sink biomass allocation would greatly contribute to the improvement of rice yield potential [3,4]. Therefore, using molecular genetic techniques to modulate the top three leaf morphology and improve the photosynthesis rate so as to balance the relationship with the grain sink will effectively achieve a high yield in rice.

The polarity development of leaves along the adaxial–abaxial, the medial–lateral, and the apical–basal axis determines the construction of the three-dimensional spatial morphology of the leaf. About 40 genes related to leaf morphogenesis have been cloned, and studies have mainly focused on leaf width, length, and rolling. Rice leaf width is mainly related to the number of veins and the distance between veins, which are regulated by the following aspects: microRNA shear-related genes, such as *OsDCL1* [5] and *GIF1* [6]; cell division-related genes, such as *SRL2* [7], *SLL1/RL9* [8,9], *OsCCC1* [10], and *OsWOX4* [11]; *NAL2 /NAL3* [12], *SLL1* [8], *OsCD1* [13], and genes related to coding transcription factors and cellulases; *NAL1/LSCHL4* [14,15], *NAL7* [16], *TDD1* [17], *OsCOW1* [18], *OsARF19* [19], and genes associated with auxin synthesis and metabolism; and genes including *NAL9* encoding an ATP-dependent Clp protease proteolytic subunit [20]. In a series of cloned rolling genes in rice, the type I genes, such as *ADL1* [21], *OsAGO7* [22], *OsAGO1a* [23], *SLL1* [8], and *RL9* [9], regulate the unbalanced development of different tissues in the adaxial/abaxial side, which affects the curl degree of blades. The type II genes are associated with the development of bulliform cells in the adaxial side, and changes in the size or amount affect the curl degree of the blade. Adaxially and abaxially rolled leaves appear as favored by various genes, such as *ACL1* [24], *LC2/OsVIL3* [25,26], *OsCOW1/NAL7* [16,18], *OsCD1/NRL1/ND1/sle1/DNL1* [13,14,27–29], *OsHox32* [30], *OsMYB103L* [31], *OsZHD1* [32], *REL1* [33], *REL2* [34], *RL14* [35], *ROC5* [36], *CLD1* [37], *SRL1* [38] *SFL1* [39], *SLL2* [40], *YABBY1* [41], and *LRRK1* [42]. The type III genes, such as *SLL1* [8], *RL9* [9], *SRL2* [7], *AVB* [43], and *OsSND2* [44], control the development of sclerenchyma in the abaxial side and affect the curl degree of blades. The genes of type IV, such as *CFL1* [45], include those with an abnormal cuticle development, leading to leaf curl. The genes of type V, such as *CVD1*, regulate commissural veins (CVs), and the lack of CV in *cvd1* mutant is the main cause of leaf curl [46].

The length, width, and area are the three traits determining the shape and size of a leaf, which are quantitatively inherited. Using a DH (doubled haploid) population, Li et al. detected two major QTLs for the flag leaf length, which were located on chromosome 4 and chromosome 8, respectively [47]. Yan et al. studied the genotype–environment interaction of eight plant morphological traits using a DH population and mapped seven QTLs on chromosomes 1, 2, 3, 4, 6, 9, and 10 related to the length of the flag leaf [48]. Farooq et al. identified three leaf length QTLs on chromosomes 1, 2, and 4 using IR64 derived introgression lines [49]. Although the important role of leaf traits in plant ideotype in rice has attracted great attention, the cloning of QTLs for leaf length is rarely documented.

The coordinated balance of source and sink is an essential component to ensure a high yield in rice. Notably, genetic populations used for leaf traits were generally among those used for yield traits, where QTLs for leaf traits were frequently located in regions in which QTLs for yield traits were detected [50–58]. The major leaf width QTL *qFLW4/LSCHL4/SPIKE* allelic to *NAL1* is related to both leaf morphology and yield traits [15,58,59]. Similarly, the pleiotropic effect on leaf morphology regulation was also found in the cloning of rice grain-shaped QTL. Large-grain alleles in the *GS2* locus simultaneously increase leaf length [60,61]. However, in rice high-yield breeding, studies have yet to be conducted on the influence of the leaf shape alleles/QTLs from different donors in the interaction of molecular level between the regulation of leaf morphogenesis and yield formation. As reported, up to 50% of the variation in panicle weight depended on the variation in leaf size [50]. The co-location of QTLs/genes for source–sink traits in rice could increase the source while expanding the sink, provided that the QTLs have the same effect direction for both traits, which will be invaluable genetic resources for breeding high-yield varieties [62].

In this study, QTL mapping for leaf length and width of the top three leaves in rice was performed at two different environments using a recombinant inbred lines (RILs) set derived from the cross between maintainer line and restorer line of the *indica–japonica* super hybrid rice Chunyou 84, followed by validation and fine mapping. The target major QTL *qLL9* controlling leaf length was delimitated into a 16.17 kb interval between markers C-1640 and C-1642 on chromosome 9, using a residual heterozygote identified from the RIL population, which were segregated at *qLL9* with high homogenous backgrounds. Then, gene cloning, functional analysis, and breeding potential assessments of *qLL9* were conducted.

2. Results

2.1. Analysis of Leaf Morphology in the RIL Population and Their Parental Cultivars

The morphology of the top three leaves of RIL parents Chunjiang 16B (CJ16) and Chunhui 84 (C84) was investigated and significant differences were found for leaf length, width, and area (Figure 1a–c). Compared with CJ16, the top three leaves length of C84 were 42.2%, 40.2%, and 46.6% longer, respectively (Figure 1d). Similarly, the leaf width of C84 was significantly wider than that of CJ16 (Figure 1e). Thus, a much larger leaf area was found in C84, that is, 2.7 times the flag leaf and 2.2 times both the second and third leaves (Figure 1f). In RIL population, the leaf traits of the top three leaves were all continuously distributed with large variations and transgressive segregation, showing a typical pattern of quantitative variation at both Hangzhou and Hainan experimental sites (Figure 2), which were suitable for QTL mapping. Furthermore, we observed that the leaf traits of RILs in Hangzhou generally tended to higher values, while those in Hainan tended to lower ones.

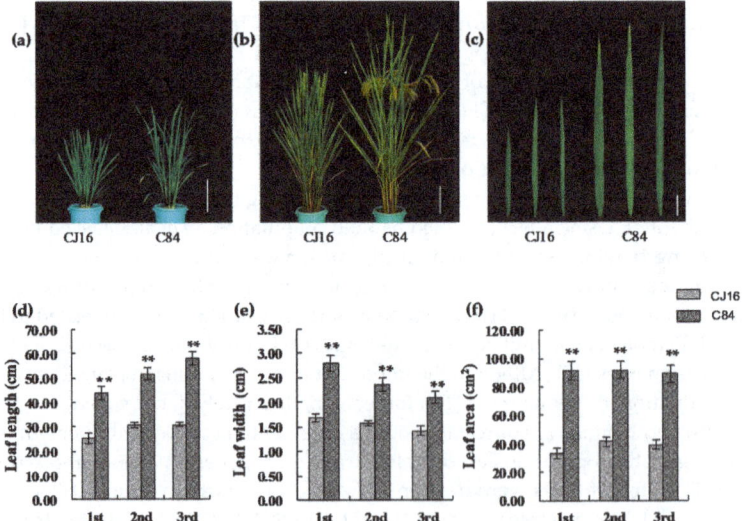

Figure 1. The leaf shape of parents of recombinant inbred lines (RILs). (**a**) Plant morphology at tillering stage; bar = 18 cm. (**b**) Plant morphology at heading stage; bar = 18 cm. (**c**) The top three leaves' shape of CJ16 and C84. From left to right are the first leaf, the second leaf, and the third leaf, respectively; bar = 5 cm. (**d**) Comparison of the top three leaves' length between CJ16 and C84. (**e**) Comparison of the top three leaves' width between CJ16 and C84. (**f**) Comparison of the top three leaves' area between CJ16 and C84. Data are represented as mean ± SD (*n* = 11). Asterisks represent significant difference determined by Student's *t*-test at *p*-value <0.01 (**), *p*-value <0.05 (*).

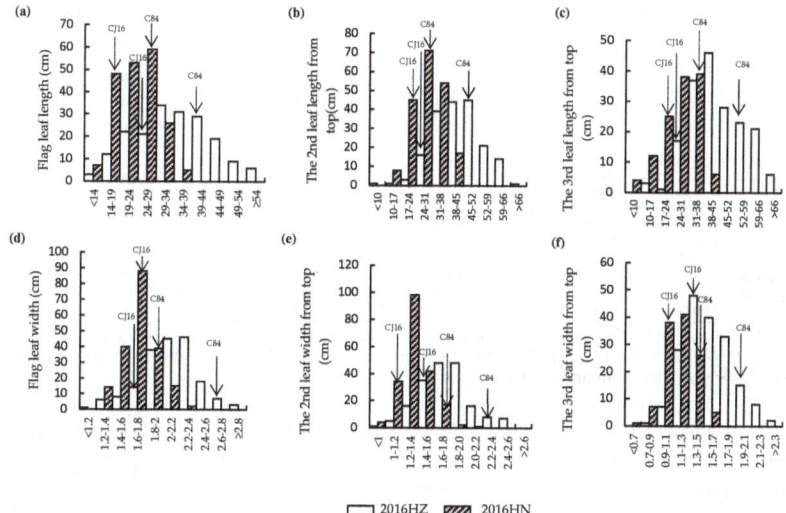

Figure 2. Frequency distributions of leaf traits in CJ16/C84 RILs. (**a**) Flag leaf length; (**b**) the second leaf length from top; (**c**) the third leaf length from top; (**d**) flag leaf width; (**e**) the second leaf width from top; (**f**) the third leaf width from top. HZ: Hangzhou; HN: Hainan.

2.2. Correlation Analysis and QTL Mapping

The correlation analysis for leaf traits in the RIL population showed a low correlation between Hangzhou and Hainan for the length of the second and the third leaf. However, the flag leaf length and width, and the second and third leaf width, were significantly positively correlated between the two environments. This could be the result of the *indica–japonica* subspecies differentiation of the parents and the distinct temperature and light conditions of Hangzhou and Hainan (Table 1).

QTL mapping was performed for the leaf length and width of the top three leaves (Table 2 and Figure 3). The results showed that a total of 27 QTLs were detected in the two environments, which were distributed on chromosomes 1, 2, 3, 5, 6, 9, 10, and 11. In Hangzhou, nine QTLs were detected, including one QTL for the flag leaf length, three QTLs for the second leaf length, and five QTLs for the third leaf length, which explained phenotypic variance in the range of 7.76%–32.41 %. In Hainan, 18 QTLs were detected, including two QTLs for the flag length, two QTLs for the flag leaf width, five QTLs for the second leaf length, four QTLs for the second leaf width, two QTLs for the third leaf length, and three QTLs for the third leaf width, which explained phenotypic variance in the range of 1.85%–30.63%. Among them, three leaf length QTLs, namely, the flag leaf length QTL *qFLL9*, the second leaf length QTL *qSLL9*, and the third leaf length QTL *qTLL9*, were simultaneously mapped in RM3700-B9-11 interval on chromosome 9 across the two environments, with the enhancing alleles all from C84, explaining phenotypic variance ranging from 19.19% to 32.41%, which agreed with the significantly positive correlation of flag leaf length between Hangzhou and Hainan (Table 1). Meanwhile, even though no significant correlation was found for the leaf length of the second and third leaf, consistent QTLs across both locations were also detected, that is, *qSLL6* and *qTLL2*, which could be because of their larger genetic effects and/or less sensitivity to environmental variation. In addition, we noted QTLs for the width of the second and third leaf were also detected in the interval between RM3700 and B9-11, but with the increasing alleles coming from CJ16. These results prompt us to mainly focus on the major region flanked by RM3700 and B9-11 on chromosomes 9, which showed stable effects on leaf morphological development, and named *qLL9*.

Table 1. Correlation analysis on leaf traits in recombinant inbred lines (RILs) derived from the cross of CJ16 and C84.

	HZ						HN				
	FLL	FLW	SLL	SLW	TLL	TLW	FLL	FLW	SLL	SLW	TLL
HZ-FLL											
HZ-FLW	−0.078										
HZ-SLL	0.833 **	−0.013									
HZ-SLW	−0.178 *	0.627 **	−0.072								
HZ-TLL	0.525 **	0.000	0.746 **	−0.023							
HZ-TLW	−0.203 **	0.597 **	−0.073	0.699 **	0.086						
HN-FLL	0.341 **	−0.317 **	0.271 **	−0.202 *	0.336 **	−0.252 **					
HN-FLW	−0.250 **	0.389 **	−0.298 **	0.382 **	−0.287 **	0.246 **	0.076				
HN-SLL	0.225 **	−0.268 **	0.135	−0.137	0.195 *	−0.226 **	0.873 **	0.229 **			
HN-SLW	−0.418 **	0.390 **	−0.510 **	0.441 **	−0.395 **	0.361 **	−0.021	0.754 **	0.160		
HN-TLL	0.000	−0.182 *	−0.098	−0.087	0.007	−0.132	0.642 **	0.254 **	0.813 **	0.313 **	
HN-TLW	−0.405 **	0.246 **	−0.477 **	0.337 **	−0.355 **	0.300 **	0.088	0.624 **	0.206 *	0.760 **	0.358 **

FLL: flag leaf length; FLW: flag leaf width; SLL: the second leaf length; SLW: the second leaf width; TLL: the third leaf length; TLW: the third leaf width. Data are represented as mean $\pm$ SD ($n = 3$). Asterisks represent significant difference determined by Student's t-test at p-value < 0.01 (**), p-value < 0.05 (*).

Table 2. Locations of quantitative trait loci (QTLs) for leaf traits in the RIL population.

Trait	QTL	Interval		Peak Position		Additive Effect		Explained Phenotypic Variance (%)	
		Marker 1	Marker 2	HZ	HN	HZ	HN	HZ	HN
FLL	qFLL9	RM3700	B9-11	61.1	56.87	−7.1707	−6.8682	32.41	30.63
	qFLL10	RM1375	H10-3		20.09		3.3926		5.09
FLW	qFLW3	H3-11	H3-12		152.3		0.0648		6.6
	qFLW6	RM3496	H6-3		27.14		−0.0839		16.07
SLL	qSLL2-1	2-30.2-a	2-42.1-a	36.2		−2.8643		8.98	
	qSLL2-2	H2-1	2-30.2-a		23.9		−1.3914		5.29
	qSLL6	wgw1	RM3183	59.4	60.1	3.5234	−4.2842	14.97	17.29
	qSLL9	RM3700	B9-11	60.1	57.6	−6.4659	−6.5431	29.62	20.85
	qSLL9-2	H9-3	H9-4		12.18		−0.7557		8.8
	qSLL10	RM1375	H10-3		27.11		3.9058		5.91
SLW	qSLW1	RM1247	H1-1		24.2		−0.0759		10.69
	qSLW6	wgw1	RM3183		59.35		−0.0727		10.31
	qSLW9	RM3700	B9-11		59.2		0.0839		18.39
	qSLW11	H11-9	H11-10		71.4		−0.0329		1.85
TLL	qTLL2	H2-1	2-30.2-a	13.0	19.44	−3.6555	−0.9935	13.10	8.99
	qTLL5	RM430	RM18751	70.3		−2.5759		8.43	
	qTLL6	wgw1	RM3183	35.2		−4.8389		15.65	
	qTLL9	RM3700	B9-11	59.1	61.2	−6.0826	−4.8326	21.87	19.19
	qTLL10	RM5689	RM1375	28.6		2.3523		7.76	
	qTLW1	H1-1	RM8111		22.41		−0.0669		10.79
TLW	qTLW6	wgw1	RM3183		59.35		−0.0684		13.84
	qTLW9	RM3700	B9-11		61.31		0.0834		16.28

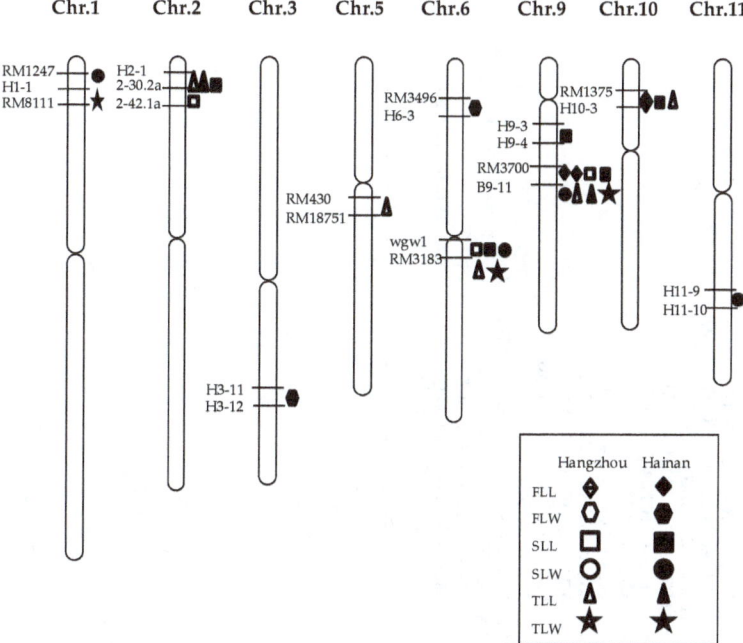

Figure 3. Locations of quantitative trait loci (QTLs) for leaf traits in the genetic map. FLL: flag leaf length; FLW: flag leaf width; SLL: the second leaf length; SLW: the second leaf width; TLL: the third leaf length; TLW: the third leaf width.

2.3. Fine Mapping and Leaf Shape Characterization of qLL9

According to the flanking markers RM3700 and B9-11 on chromosome 9, one residual heterozygote with a heterozygous segment covering the interval was identified from the RILs, from which a large population was derived. Nine representative near isogenic lines (NILs) with introgressions covering different portions of the target region were identified using DNA markers in the target interval for QTL fine mapping (Table 3; Figure 4a). Then, combined with their flag leaf length, *qLL9* was delimitated into a 16.17 kb region between markers C9-1640 and C9-1642 (Figure 4a).

To further clarify the effect of *qLL9* on leaf morphology, we compared the leaf traits in a near-isogenic line set of NIL-*qLL9*[CJ16] and NIL-*qLL9*[C84], and a chromosome segment substitution line (CSSL)-*qLL9*[KASA] and its recurrent parent Wuyunjing (WYJ) 7. In both groups, significant differences for the flag leaf were only found for the length, but not the width (Figure 5), while both the leaf length and width variation were significant for the second and third leaf from the top (Figure S1), which were highly consistent with the findings in our QTL primary mapping (Table 2). Obviously, compared with NIL-*qLL9*[CJ16], the longer flag leaf length of NIL-*qLL9*[C84] resulted in a larger flag leaf area (Figure 5d–f). However, the opposite allelic effects on leaf length and width of the second and third leaf from the top even made the leaf area variation not significant (Supplementary Figure 1). Similarly, after introgression of the Kasalath allele at *qLL9* in WYJ 7, CSSL-*qLL9*[KASA] resulted in a 1.5 times longer flag leaf from 29.4–43.5 cm^2 (Figure 5g–i).

On the other hand, it is noted that the flag leaf epidermal cells showed no significant difference in either the size or number in unit area between NIL-*qLL9*[CJ16] and NIL-*qLL9*[C84], which suggested the variation of flag leaf length could be mainly attributed to the increase of the total cell number (Figure 6). In summary, *qLL9* influenced the leaf length and area variation by regulating the development of leaf cells.

Table 3. Primers for QTL fine mapping, quantitative real-time polymerase chain reaction (qRT-PCR), vector construction, and gene editing.

Primer	Forward (5′-3′)	Reverse (5′-3′)	Experiment
RM3700	AAATGCCCATGCACAAC	TTGTCAGATTGTCACCAGGG	Fine mapping
C9-1594	CCTGTACACTGTAGGCCTGT	GGTGTCAAAGTACATAGGCCC	
C9-1635	GGTCGAAAGGAAGGAGAGCT	CTAGCCCTGCCTCGTTGTAA	
C9-1638	GTGTGTGTGTGTGTGTGT	TCATAGTACATGCCCTCCGT	
C9-1640	ATAAGTCCATATTGCCCACCTC	AAGCTTCTGGATCGTTAACAGG	
C9-1642	GTACCCTCCTCCGATGACAC	TTGTGGAGGACGAGAAGGTG	
C9-1715	GGTGGCGAGAAGAATTTGCA	TTTCGCCTCTCACTGACCTT	
B9-11	TCTTACGAATAGGCCCTTGG	AGAGCCCACAACACTTGTGC	
Actin	ATCCATCTTGGCATCTCTCAGC	CACAATGGATGGGCCAGACT	qRT-PCR
LOC_Os09g26960	CTGAGCCTCGCCAATCTG	CGAAGATCTCTCCATGCTC	
LOC_Os09g26970	CAAACATCTGGGCTTGGTCT	TCTAAGCAACCTGCCCAATC	
LOC_Os09g26980	ATTGATGTGAAAAGGCCAAGACT	CACCTTAAGCCCAAGGTTGTAG	
LOC_Os09g26999	GTAGCTGCAAAGCCAAGCTG	TTGAAGCAGCTGGAGCAAC	
POs26999	GGCCAGTGCCAAGCTTAAGGGAAGTTGGCCGCCTGCC	AGGGTCTTGCAGATCTCTCCACACGCAGCACGCCCAAC	Vector construction
Os09g26999-g++/g−	GGCAGGTGGTGATGGAGGCGCCG	AAACCGGCGCCTCCATCACCACC	Detection of target mutations
Os09g26999-JC	CGCGGATTTATACCCACCAC	CGCTCACCTTGAGGAAGGT	
Hyg-F1	GCTGTTATGCGGCCATTGTC	GACGTCTGTCGAGAAGTTTC	
Cas9-F2/pC1300-R2	ACCAGACACGAGACGACTAA	ATCGGTGCGGGCCTCTTC	
T3	ATCGGTGCGGGCCTCTTC		

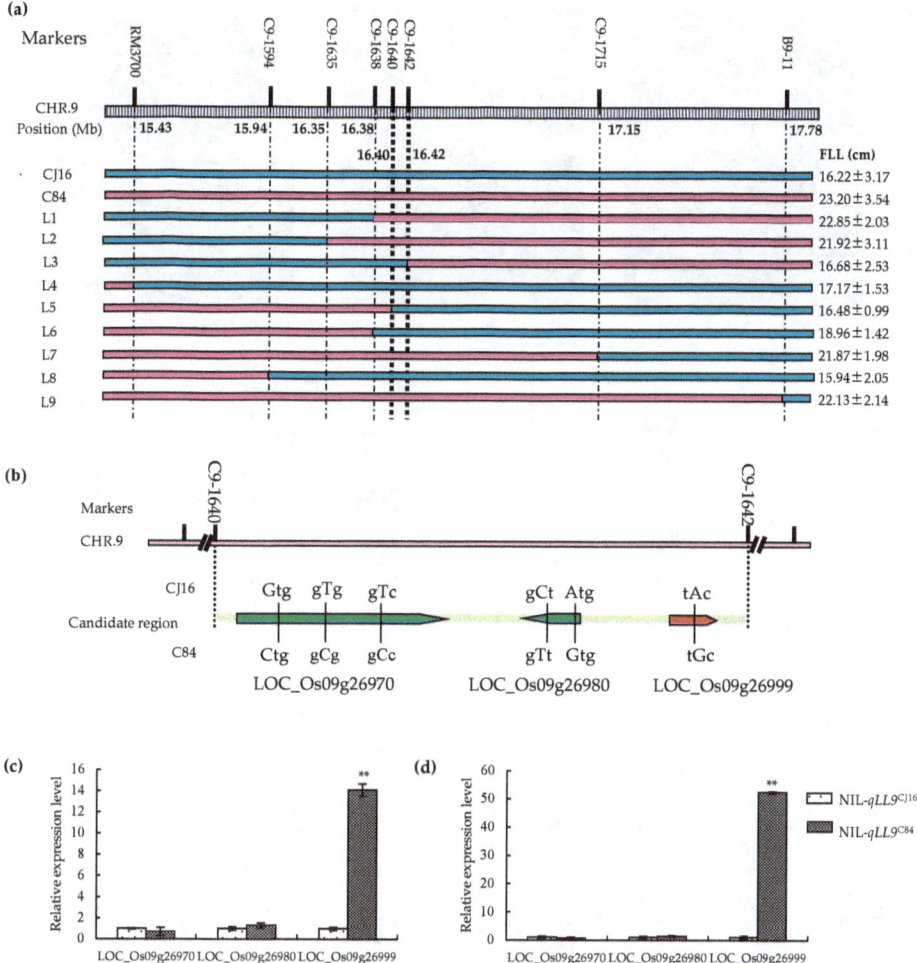

Figure 4. Map-based cloning of *qLL9* and expression analysis of candidate genes. (**a**) High-resolution mapping of *qLL9*. Numbers on the map indicate the physical distance on chromosome 9. Nine recombinant plants were used to refine the candidate region to 16.17 kb region by substitution mapping, in which green and pink rectangles indicate the homozygous CJ16 genotype and homozygous C84 genotype, respectively. Flag leaf length (FLL) values were obtained from the corresponding selfed progenies and represented as mean ± SD (*n* = 12). (**b**) Predicted open reading frames and sequence difference of the *qLL9* between CJ16 and C84 are shown. Relative expression of three candidate genes *LOC_09g26970*, *LOC_09g26980*, and *LOC_09g26999* of near isogenic line (NIL)-*qLL9*[CJ16] and NIL-*qLL9*[C84] at tillering stage (**c**) and heading stage (**d**) by quantitative real-time polymerase chain reaction (qRT-PCR). Data are mean ± SD (*n* = 3). Asterisks represent significant difference determined by Student's *t*-test at *p*-value < 0.01 (**).

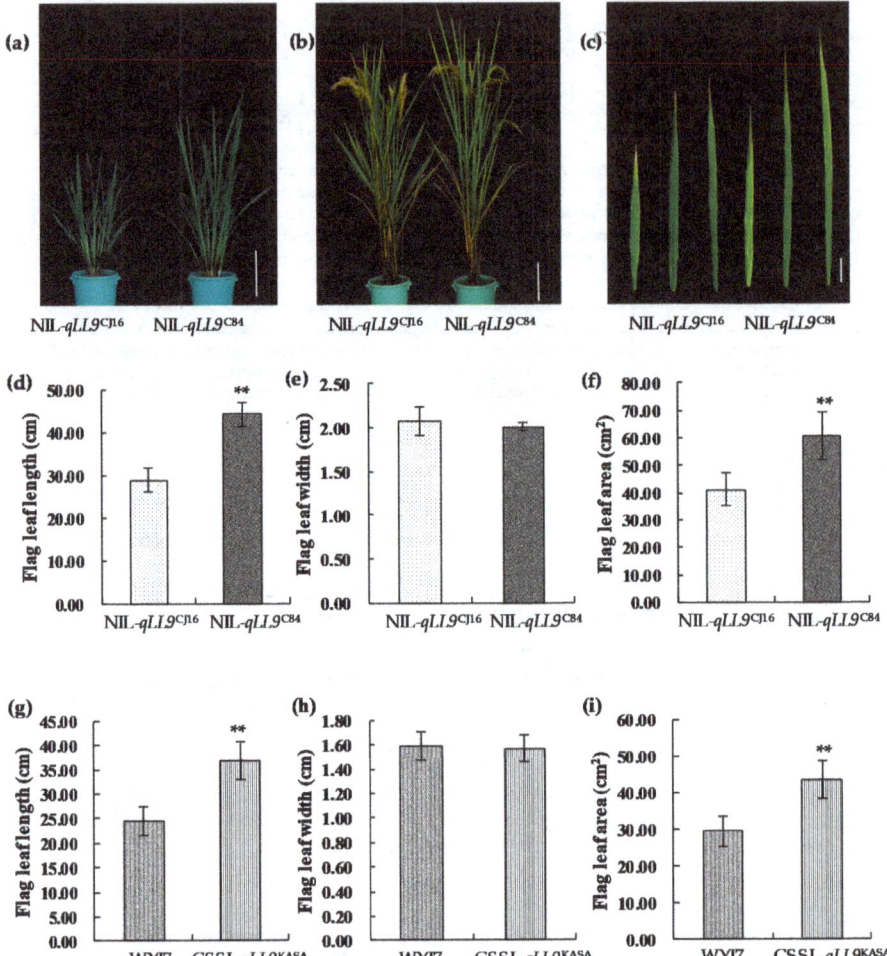

Figure 5. Phenotypes of near-isogenic lines and chromosome segment substitution lines. (**a**) Plant morphology at tillering stage; bar = 18 cm. (**b**) Plant morphology at heading stage; bar = 18 cm. (**c**) The top three leaf shape of NIL-*qLL9*^{CJ16} and NIL-*qLL9*^{C84}. From left to right are the first leaf, the second leaf, and the third leaf, respectively; bar = 5 cm. Comparison between the NIL-*qLL9*^{CJ16} and NIL-*qLL9*^{C84} for the flag leaf length (**d**), the flag leaf width (**e**), and the flag leaf area (**f**). Comparison between Wuyunjing (WYJ) 7 and chromosome segment substitution line (CSSL)-*qLL9*^{KASA} for the flag leaf length (**g**), the flag leaf width (**h**), and the flag leaf area (**i**). Data are mean ± SD (*n* = 15). Asterisks represent significant difference determined by Student's *t*-test at *p*-value < 0.01 (**).

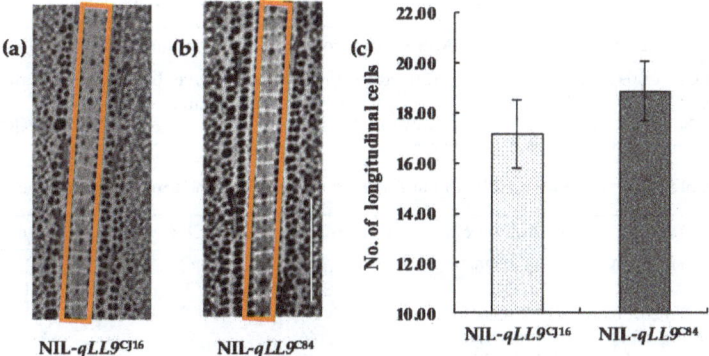

Figure 6. Histological cell morphology of NIL-*qLL9*^CJ16 and NIL-*qLL9*^C84. (**a,b**) Comparison of cytological morphological characteristics in the orange boxes between the two genotypes of the NIL set; bar = 200 μm. (**c**) Number of longitudinal cells in NIL-*qLL9*^CJ16 and NIL-*qLL9*^C84. Data are mean ± SD (*n* = 15) and no significant difference was found.

2.4. Determination of the Candidate Gene

According to the database of Rice Genome Annotation Project (http://rice.plantbiology.msu.edu), three open reading frames (ORFs) were predicted in the target region defined by C9-1640 and C9-1642 (Figure 4b and Table 4), namely, LOC_Os09g26970 encoding cytochrome P450 protein, LOC_Os09g26980 encoding retrotransposon protein, and LOC_Os09g26999 encoding G protein γ subunit (a cloned gene *DEP1*) [63]. By sequencing and qRT-PCR for the three candidates, six non-synonymous single nucleotide polymorphisms (SNPs) were found across the three candidate genes (Figure 4b and Table 5). No significant difference was detected in the LOC_Os09g26970 and LOC_Os09g26980 expression between NIL-*qLL9*^CJ16 and NIL-*qLL9*^C84 at both the tillering and heading stages, but the expression of LOC_Os09g26999 in NIL-*qLL9*^C84 was significantly increased by 14.1 times and 52.3 times, respectively, compared with NIL-*qLL9*^CJ16 at the two stages (Figure 4c,d). Therefore, LOC_Os09g26999 could be the best potential candidate for *qLL9*. In addition, we found that *qLL9* was allelic to the known gene *DEP1* [63]. We inferred that the difference of amino acid of *qLL9* would have an influence on the gene expression and would thus affect the leaf morphological development.

It has been reported that *DEP1* is mainly related to panicle development [63]. To further verify its correlation with leaf development, the CRISPR/Cas9 gene knockout technique was adopted in the genetic background of Nipponbare (Figure 7). Two knockout plants of LOC_Os09g26999, namely mutant 1 (MT-1) and mutant 2 (MT-2), were screened in T0 generation (Figure 8; Figure S2). After continuous selfing and marker assay, homozygous positive transgenic lines in T2 generation were obtained and used for investigating the flag leaf morphology. The results showed no significant difference in the flag leaf width between the knockout and wild-type plants. However, the flag leaf length and area were significantly decreased in both knockout lines. (Figure 8b–e). The results showed that *qLL9* was mainly responsible for the leaf length development, and the loss-function of *qLL9* could lead to a reduction in leaf length, and thereby leaf area.

To further explore the causal factor for differential expression of *LOC_Os09g26999* observed in the NIL set, we compared the promoter sequence of 2.0 kb upstream of the ATG between them, and found nine variations (Table 6). Then, the activity difference of the two types of promoters was compared by dual luciferase reporter assay. The result showed that the LUC-C84 promoter had 3.9 times higher expression of reporter genes than the LUC-CJ16 promoter (Figure 9). That is to say, the promoter activity of *qLL9* also played a significant role in the gene expression and leaf-trait variation.

Table 4. Annotated genes included in the 16.17kb region for *qLL9*.

Gene ID	Annotation from the Rice Genome Annotation Project
LOC_Os09g26970	Retrotransposon protein, Putative, Unclassified, Expressed
LOC_Os09g26980	Cytochrome P450, Putative, Expressed
LOC_Os09g26999	Gγ subunit; Dense and Erect Panicle1; DENSE PANICLE 1

Table 5. The position of SNP and amino acid variation for the three candidate genes.

Locus Name	Position on chr.9	Position on gene	CJ16	C84	CJ16	C84
LOC_Os09g26970	16393266	397	G	C	Val	Leu
LOC_Os09g26970	16393717	848	T	C	Val	Ala
LOC_Os09g26970	16395923	3054	T	C	Val	Ala
LOC_Os09g26980	16403850	4146	G	A	Ala	Val
LOC_Os09g26980	16404094	3902	T	C	Met	Cys
LOC_Os09g26999	16414735	3182	A	G	Tyr	Cys

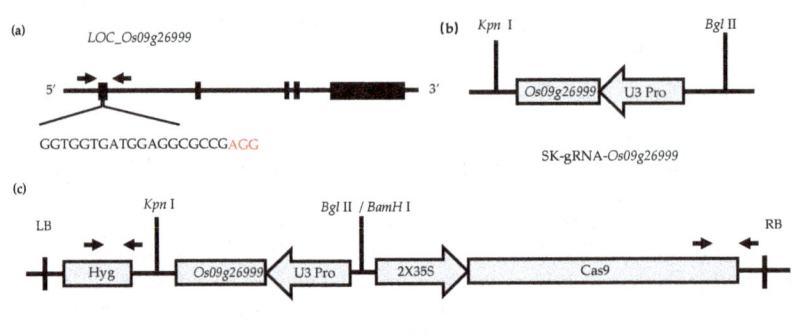

Figure 7. CRISPR/Cas9 vector construction. (**a**) Primer sequence on *LOC_Os09g26999*. (**b**) The intermediate vector SK-gRNA contains the U3 promotor and sgRNA scaffold. (**c**) Binary vector pC1300-Cas9 contains the 2 × 35S promotor and a Cas9 protein. SK-gRNA-*Os09g26999* are digested with *Kpn* I and *Bgl* II, respectively, and cloned into pC1300-Cas9 (digested with *Kpn* I and *BamH* I) by a one-step ligation.

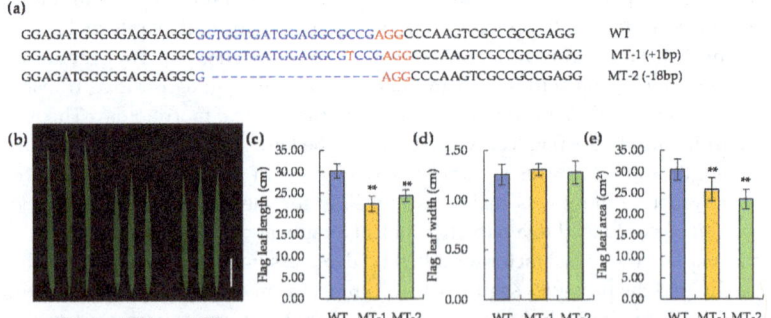

Figure 8. Sequence of target loci in CRISPR/Cas9 transgenic plants and phenotypic comparison between the knockout and wild-type (WT) plants. (**a**) Sequence variation of target loci in two transgenic plants. (**b**) Flag leaf variation; bar = 5 cm. Difference in (**c**) flag leaf length (cm), (**d**) flag leaf width (cm), and (**e**) flag leaf area. WT, Nipponbare; MT-1, Mutant 1; MT-2, Mutant 2. Data are represented as mean ± SD (*n* = 15). Asterisks represent significant difference determined by Student's *t*-test at *p*-value <0.01 (**).

Table 6. The position of nine SNPs in the promoter of LOC_Os09g26999.

Position	CJ16	C84
−1341	A	C
−1255	G	C
−951	C	G
−906	T	G
−629	G	T
−503	A	C
−493	G	C
−486	T	G
−31	TG	T

(a)

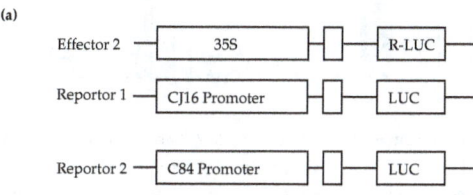

(b)

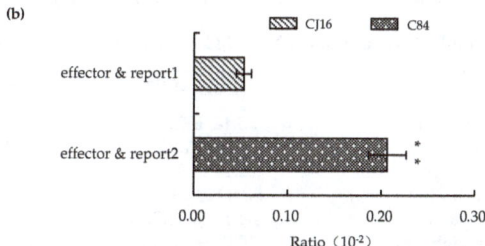

Figure 9. Comparison of the promoter activity of LOC_Os09g26999 between CJ16 and C84. (**a**) Diagram of carrier construction. (**b**) Ratio of promoter activity. Data are represented as mean ± SD ($n = 3$). Asterisks represent significant difference determined by Student's *t*-test at *p*-value <0.01 (**).

2.5. qLL9 Affecting the Yield Traits

To verify whether *qLL9* affected the yield formation, we firstly investigated its effects on grain shape using NIL-*qLL9*[CJ16] and NIL-*qLL9*[C84] plants (Figure 10). We found that the grain length, width, and thickness of NIL-*qLL9*[C84] are all slightly larger than those of NIL-*qLL9*[CJ16] by 6.66%, 2.55%, and 3.37%, respectively.

Then, we compared the other yield component traits between the NIL-*qLL9*[CJ16] and NIL-*qLL9*[C84] plants. There was no significant difference between them in the number of productive panicles per plant, the number of primary branches per panicle, the number of second branches per panicle, and the number of grains per panicle (Table 7). However, the thousand-grain weight of NIL-*qLL9*[C84] was significantly higher than that of NIL-*qLL9*[CJ16] (22.99 g and 20.79 g, respectively), while the seed setting rate of NIL-*qLL9*[C84] was lower (69.71% and 75.80%, respectively). Even so, the yield per plant of NIL-*qLL9*[C84] was 16.59% higher than that of NIL-*qLL9*[CJ16]. That is, the increase in the yield per plant was mainly attributed to the increase in thousand-grain weight. Further, yield measurement in plots also showed that the NIL-*qLL9*[C84] could yield more grains than NIL-*qLL9*[CJ16], increasing by 15.08%. As for the actual yield per hectare, NIL-*qLL9*[C84] could increase by 991.67 kg compared with NIL-*qLL9*[CJ16]. These results showed that the C84 allele at *qLL9* significantly increases the grain size, thousand-grain weight, and grain yield in the field production.

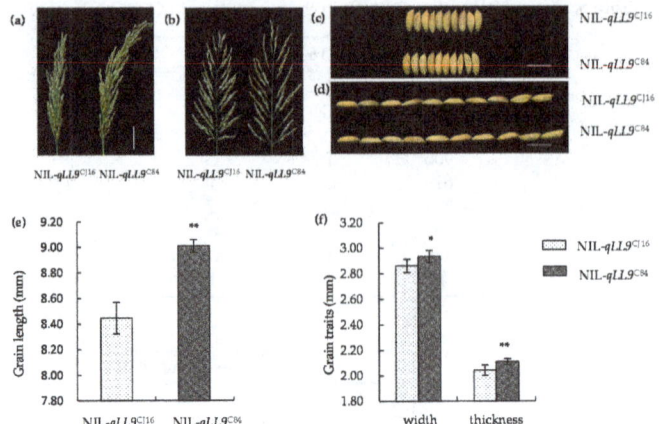

Figure 10. Panicle and grain morphology of NIL-*qLL9*^{CJ16} and NIL-*qLL9*^{C84}. (**a,b**) Panicle morphology at heading stage; bars = 4 cm. (**c,d**) Grain morphology at repining stage; bar = 1 cm. (**e**) Grain length. (**f**) Grain width and thickness. Data are represented as mean $\pm$ SD (n = 10). Asterisks represent significant difference determined by Student's t-test at p-value <0.01 (**), p-value <0.05 (*).

Table 7. Yield traits of NIL-*qLL9*^{CJ16} and NIL-*qLL9*^{C84}.

Trait	NIL-*qLL9*^{CJ16}	NIL-*qLL9*^{C84}
Panicle length (cm)	20.44 $\pm$ 0.84	26.7 $\pm$ 0.83 **
Panicles per plant	9.25 $\pm$ 1.42	9.42 $\pm$ 1.26
Number of primary branches	22.11 $\pm$ 2.18	21.4 $\pm$ 2.42
Number of secondry branches	80.00 $\pm$ 15.51	77.60 $\pm$ 7.84
Grains per panicle	404.11 $\pm$ 71.70	355.17 $\pm$ 34.09
1000-grain weight (g)	20.79 $\pm$ 0.74	22.99 $\pm$ 0.45 **
Seed setting rate (%)	75.80 $\pm$ 0.02	69.71 $\pm$ 0.02 **
Yield per plant (g)	27.93 $\pm$ 4.41	32.56 $\pm$ 8.19
Actual yield per plot (kg/48 m^2)	31.52 $\pm$ 3.20	36.28 $\pm$ 3.27 *
Actual yield change (%)	–	15.08

Data are represented as mean $\pm$ SD. Asterisks represent significant difference determined by Student's t-test at p-value <0.01(**), p-value <0.05 (*).

3. Discussion

In this study, the main effect QTL *qLL9* related to the leaf length was positioned by RIL population. *LOC_Os09g26999* was identified as the target gene through sequence comparison, expression analysis, and CRISPR-Cas9 gene editing technology. It is allelic to the known spike-shaped gene *DEP1/EP/qPE9-1* [63–65]. The plants of NILs and CSSLs carrying different alleles of *qLL9* showed great differences in leaf, spike, and yield. All results indicated that QTL *qLL9* has a pleiotropic function in rice. In addition to regulating spike development, *qLL9* is also plays a key role in the development of leaf morphology, grain shape, yield, and other traits.

Previous studies have demonstrated that QTL *qLL9* candidate gene *LOC_Os09g26999* encodes a cysteine-rich region [66]. Huang et al found that the replacement of a 637 bp stretch from the fifth exon of *LOC_Os09g26999* with a 12 bp sequence results in erect panicle architecture because of the early termination of translation [63]. Due to the differences of *LOC_Os09g26999* relative expression and promoter activity in CJ16 and C84, the promoter sequences and regulatory elements were compared by website (https://sogo.dna.affrc.go.jp). We found that nine nucleotide differences were observed among the parental lines, eight of which were located in cis-acting elements possibly by mediating enhancer activity depending on upstream region G/C mutation at 1255 bp upstream of ATG, which is consistent with site II transcriptional core sequence regulatory elements (TGGGCC[CJ16] to TGGCCC[C84]). It plays an important role in the specific expression of *proliferating cell nuclear antigen (PCNA)* gene in rice meristem [67,68]. We speculate that G/C mutation may be through mediation of enhancer activity dependent on far upstream regions. Comparison of CDS showed that a single base substitution event is observed from A to G at 3,182 bp downstream of ATG in *LOC_Os09g26999* between CJ16 and C84, leading to the substitution of amino acids from tyrosine to cysteine. This event might lead to changes in the structure of the γ subunit, affecting the signal transduction of G protein. Further work is needed to determine whether the difference expression of *LOC_Os09g26999* is due to changes in protein structure or differences in promoter activity.

The leaf morphological development in rice is regulated by the size and number of leaf epidermal cells. In *rot4-1D* mutant, the reduction in the number of leaf longitudinal cells induces a short leaf [69]. Tsuge et al revealed that the *Arabidopsis thaliana rotundifolia3* leaf mutant has the same number of cells as the wild type, but with reduced cell elongation in the leaf-length direction [70]. Rice flag leaf width QTL *qFLW7*, homologous to Arabidopsis *LNG1*, regulates the longitudinal growth of cells and its overexpression results in elongated leaves [71]. In this study, no significant difference was observed in the cell size per unit area of the leaf epidermis between NIL-*qLL9*[C84-] and NIL-*qLL9*[CJ16], whereas the total number of cells increased leading to leaf longer. The expression levels of eight cell cycle-related genes in NILs were analyzed at the heading stage (Figure 11 and Supplementary Table S1), which showed that the expression of *MCM5* plays an important role in the initiation and extension of DNA replication in the G1 phase, which was significantly up-regulated in NIL-*qLL9*[C84] compared to NIL-*qLL9*[CJ16] [72]. Plant Class A cyclin (Cyclin) *CYCA2;3* and *CYCA2;2* [73,74] could identify and interact with different cyclin-dependent kinases, which were also remarkably up-regulated in NIL-*qLL9*[C84]. Therefore, *qLL9* from *indica* rice C84 could improve the DNA replication efficiency of leaf tissue cells, accelerate cell division, and promote leaf elongation, which also corresponded to the fact that the morphology of the leaf epidermal cells of NIL-*qLL9*[C84] was unchanged while the total number of the cells increased. These findings indicated that *qLL9* may affect plant morphological development by participating in the regulation of cell division cycle. Notably, the regulatory mechanisms of different genes on leaf morphological development are different, thus their interactions in leaf morphogenesis should be studied in details in the future.

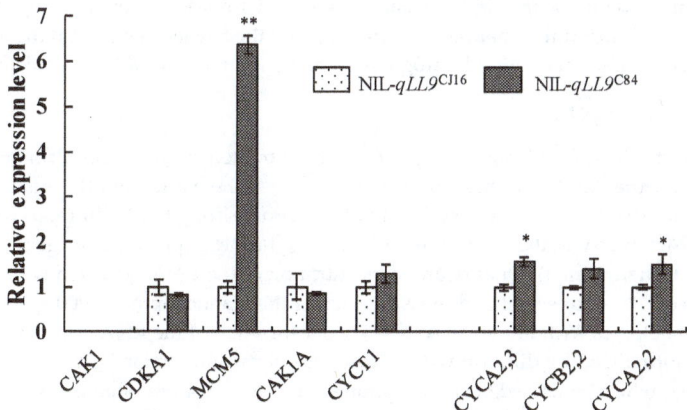

Figure 11. Comparison of eight cell cycle related gene expression between NIL-*qLL9*[CJ16] and NIL-*qLL9*[C84] at the heading stage by qRT-PCR, and the data are represented as mean ± SD (*n* = 3); asterisks represent significant difference determined by Student's t-test at p-value <0.01(**), p-value <0.05 (*).

4. Materials and Methods

4.1. RIL Population and Field Trial

The RIL population consisting of 188 lines was derived from the cross of the *japonica* rice Chunjiang 16B (CJ16) as the female parent and the *indica* rice Chunhui84 (C84) as the male parent, which are the maintainer and restorer lines of the commercial intersubspecific hybrid rice Chunyou84. The rice population was tested at experimental fields of the China Rice Research Institute located in Hangzhou, Zhejiang, and Lingshui, Hainan, during May–October 2016 and November 2016–April 2017, respectively. Twenty-five-day-old seedlings were transplanted at a hill spacing of 20 cm × 20 cm with three replications. In each replication, one line was grown in three-row plots with six plants per row. The block was managed in accordance with conventional field management, and diseases, insects, and weeds were controlled [75]. The leaf morphology and yield traits were investigated at the heading stage and mature stage. In each trial, data of the three replications were averaged for each line and used for data analysis.

4.2. Statistical and Genetic Analysis

The linkage map of the RILs consisted of sixty-nine simple sequence repeats (SSR) and eighty-nine sequence tagged site (STS) DNA markers. QTL analysis was conducted using QTL Network 2.1. Critical *F* values for genome-wise type I error were calculated with 1000 permutation tests and used for claiming a significant event. A significant level of $p < 0.005$ was set for candidate interval selection, putative QTL detection, and QTL effect estimation. The proportion of phenotypic variance (R^2) explained by a single main QTL for a given trait in a given population was calculated by Markov Chain Monte Carlo algorithm. In the genome scan, a testing window of 10 centimorgan (cM), filtration window of 10 cM, and walk speed of 1 cM were chosen. The naming of QTLs was based on the method of McCouch et al. [75].

4.3. Map-Based Cloning and Candidate-Gene Promoter Activity for qLL9

Using flanking markers RM3700 and B9-11 of *qLL9* obtained from QTL mapping, residual heterozygotes (RHs) were screened from the RIL population, which were segregated at *qLL9* with high homogenous genetic background. Six pairs of molecular markers were developed in this target interval (Table 3). By substitution mapping, we investigated the flag leaf length and the new developed marker genotypes of 2307 RH-derived introgression lines, and then the *qLL9* was fine mapped.

4.4. RNA Extraction and qRT-PCR

Total RNA of NIL-*qLL9*[CJ16] and NIL-*qLL9*[C84] was extracted from the penultimate leaves at the tillering stage and the flag leaf at heading stage to analyze the expression difference of candidate genes (Tables 3 and 4). We used the AxyPrepTM Multisource Total RNA Miniprep Kit (Axygen) to extract total RNA, which was then retro-transcribed using PrimeScriptTM RT Reagent Kit with gDNA Eraser (Takara, Dalian, China). Quality and concentration of the RNA extracted were checked with electrophoresis on 1% agarose gel and measured using the Nanodrop ND-2000 spectrophotometer (NanoDrop Technologies, Wilmington, CA, USA). Concentration of the RNA samples used for cDNA synthesis was normalized by dilution with RNase-free ultra-pure water. qRT-PCR assays of 20 μL reaction volumes, which contained 0.5 μL of synthesized cDNA, 0.4 μM of gene-specific primers, and 10 μL of SYBR® Premix Ex Taq™ (Takara), were conducted using ABI 7500 Real-time PCR System (Applied Biosystems, Foster, CA, USA). Following the manufacturer's instruction, the qRT-PCR conditions were set up as follows: denaturing at 95 °C for 30 s, then 40 cycles of 95 °C for 5 s, 55 °C for 30 s, and 72 °C for 30 s. To standardize the quantification of gene expression, we used the rice *Ubiquitin* (*UBQ*) gene (Os03g0234200, http://rapdb.dna.affrc.go.jp/) as an internal control.

According to the sequence shown in http://rice.plantbiology.msu.edu/index.shtml, the primer POs26999-F/R (Table 3) was used to amplify LOC_Os09g26999 promoter of CJ16 and C84, which were constructed into *pGreenII0800-LUC* using homologous recombination [76]. Positive clones were screened by colony PCR and sequencing and were named as *proCJ16-LUC* and *proC84-LUC*. The plasmids and the internal reference (R-LUC) were transformed into the protoplasts of rice variety 93-11 by 40% PEG-3350 (Sigma, St. Louis, MO, USA) solution-mediated transformation [77]. The dual-luciferase reporter gene detection kit (Promega Company, Madison, WI, USA) was used for detection and analysis of the promoter activity.

4.5. CRISPR/Cas9 Transgene Analysis

The target sequence of the potential candidate LOC_Os09g26999 refers to a previous study [78]. Synthetic primers Os26999-g++ and Os26999-g−−were used to make the target site adapter (Table 3). This adapter was connected to an SK-gRNA carrier, and positive intermediate carrier SK-gRNA-Os26999 was screened out. pC1300-2×35S::Cas9-gOs26999 final expression carrier was constructed by means of enzyme digestion-joining method. *Agrobacterium tumefaciens* EHA105 was transformed through positive cloning. *Agrobacterium tumefaciens*-mediated gene transfer experiments were carried out with the background of Nipponbare [79]. The result was analyzed via the sequential decoding method (http://dsdecode.scgene.com/) to identify transgenic positive plants. At heading stage, 15 wild-type and 15 mutant plants were selected to measure the length, width, and area of the flag leaf.

4.6. Construction of Near Isogenic Lines and Chromosome Segment Substitution Lines and Trait Measurement

Using flanking markers C-1640 and C-1642 (Table 3), a set of near isogenic lines (NILs) for *qLL9* was identified from RH-derived segregating populations with a higher homogenous background, named NIL-*qLL9*[CJ16] and NIL-*qLL9*[C84]. At the same time, one chromosome segment substitution line (CSSL) for *qLL9* was obtained with Kasalath (KASA) as the donor parent and Wuyunjing 7 (WYJ 7) as the recurrent parent through one cross followed by six continuous backcrosses and one self-crossing, named CSSL-*qLL9*[KASA].

The leaf length, width, and area of the NIL and CSSL set were investigated at the heading stage, while the panicle length, the number of primary and secondary branches, the number of grains per panicle, grain length, grain width, grain thickness, thousand-grain weight, and setting percentage were scored from 10 randomly selected main panicles at maturity. Then, Student's *t*-test was adopted to analyze the phenotypic difference between the two genotypic groups in each set.

The grain yield was measured using the NIL set NIL-*qLL9*[CJ16] and NIL-*qLL9*[C84], of which each was planted in three 48 m^2 plots, at a hill spacing of 20 cm × 20 cm. At maturity, three points in each plot were randomly harvested, with 30 hills per point, to determine the grain yield, which is then converted into the grain yield per hectare.

4.7. Morphological Observation on Leaf Epidermal Cytology

The commercially available transparent nail polish without color is selected, which is conducive to the transparency of microscopic materials. The flag leaves of tested rice plants at the heading stage were sampled and painted with the nail polish evenly at the same part for 10 min air-dry. When an open mouth exists at the end of the coating layer of the nail polish, the dried coating is torn with a transparent tape and placed on the fragment. Then, the cover slip was covered. The filter paper was covered by the blunt end of the dissecting needle, and the fragment was gently pressed to make a temporary filling piece. Under the electron microscopy, the cell size and number of 15 leaves of a single plant were analyzed [80].

5. Conclusions

Our work identified the genetic contribution of *qLL9* to both flag leaf morphologic development and yield formation. Improving the photosynthetic efficiency and coordinating the source–sink interaction through leaf morphogenesis are the premise to increase the rice yield and to establish the ideal plant type. Identification and utilization of the QTLs with beneficially pleiotropism for both leaf shape and grain yield would greatly contribute to molecular breeding of superior rice. For example, the QTL *qLSCHL4/NAL1* is correlated with leaf shape and involved in regulating the development of chlorophyll content, grain number and grain weight [15]. QTL *qFLW7*[9311] related to flag leaf width, could improve leaf shape and grain traits, remarkably increase rice yield in the field [71]. Similarly, *GS2* could increase both grain weight/size and leaf length and thus grain yield [60,61]. According to the main objectives of rice molecular design breeding, future studies should systematically analyze the source-sink relationship and the genetic network of plant type establishment, directional polymerize the *qLL9, NAL1, qFLW7, GS2*, and other beneficial genes from *indica* and *japonica* subspecies of rice to improve the yield potential and establish the ideal plant architecture in molecular breeding of superior rice.

Supplementary Materials: Supplementary materials can be found at http://www.mdpi.com/1422-0067/20/4/866/s1.

Author Contributions: Data curation, X.F., J.X., M.Z., T.L., Y.Z., and J.W.; Formal analysis, X.F., T.L., and D.R.; Funding acquisition, G.Z.; Investigation, X.F., J.X., M.Z., M.C., L.S., Y.Z., J.W., and J.H.; Methodology, J.H., L.Z., Z.G., L.G., D.R., G.C., and G.Z.; Project administration, Q.Q. and G.Z.; Resources, G.D. and J.L.; Software, M.Z. and Z.G.; Supervision, Q.Q. and G.Z.; Writing—original draft, X.F. and J.X.; Writing—review & editing, G.Z.

Funding: This work was supported by grants from the National Natural Science Foundation of China (31770195, 3186114300 and 31570184); National Key Research and Development Program (2016YFD0101801); and CAAS Science and Technology Innovation Program (CAAS-XTCX2016009).

Conflicts of Interest: The authors declare no conflict of interest.

References

1. Heath, D.V.; Gregory, F.G. The constancy of the mean net assimilation rate and its ecological importance. *Ann. Bot. Lond.* **1938**, *2*, 811–818. [CrossRef]
2. Donald, C.M. The breeding of crop ideotypes. *Euphytica* **1968**, *17*, 385–403. [CrossRef]
3. Mason, T.G.; Maskell, E.J. Studies on the transport of carbohydrates in the cotton plant. I. A study of diurnal variation in the carbohydrates of leaf, bark, and wood, and of the effects of ringing. *Ann. Bot.-Lond.* **1928**, *42*, 189–253. [CrossRef]
4. Hasson, A.; Blein, T.; Laufs, P. Leaving the meristem behind: The genetic and molecular control of leaf patterning and morphogenesis. *C. R. Biol.* **2010**, *333*, 350–360. [CrossRef] [PubMed]
5. Liu, B.; Li, P.C.; Li, X.; Liu, C.Y.; Cao, S.Y.; Chu, C.C.; Cao, X.F. Loss of function of *OsDCL1* affects micro RNA accumulation and causes developmental defects in rice. *Plant Physiol.* **2005**, *139*, 296–305. [CrossRef] [PubMed]
6. Wang, E.; Xu, X.; Zhang, L.; Zhang, H.; Lin, L.; Wang, Q.; Li, Q.; Ge, S.; Lu, B.R.; Wang, W.; et al. Duplication and independent selection of cell-wall invertase genes GIF1 and OsCIN1 during rice evolution and domestication. *BMC Evol. Biol.* **2010**, *10*, 108–120. [CrossRef] [PubMed]
7. Liu, X.F.; Li, M.; Liu, K.; Tang, D.; Sun, M.F.; Li, Y.F.; Shen, Y.; Du, G.J.; Cheng, Z.K. *Semi-Rolled Leaf2* modulates rice leaf rolling by regulating abaxial side cell differentiation. *J. Exp. Bot.* **2016**, *67*, 2139–2150. [CrossRef] [PubMed]
8. Zhang, G.H.; Xu, Q.; Zhu, X.D.; Qian, Q.; Xue, H.W. *SHALLOT-LIKE1* is a KANADI transcription factor that modulates rice leaf rolling by regulating leaf abaxial cell development. *Plant Cell* **2009**, *21*, 719–735. [CrossRef]
9. Yan, C.J.; Yan, S.; Zhang, Z.q.; Liang, G.H.; Lu, J.F.; Gu, M.H. Genetic analysis and gene fine mapping for a rice novel mutant (*rl9(t)*), with rolling leaf character. *Chin. Sci. Bull.* **2006**, *51*, 63–69. [CrossRef]
10. Kong, X.Q.; Gao, X.H.; Sun, W.; An, J.; Zhao, Y.X.; Zhang, H. Cloning and functional characterization of a cation–chloride cotransporter gene *OsCCC1*. *Plant Mol. Biol.* **2011**, *75*, 567–578. [CrossRef]
11. Tian, X.; Zhang, D.B. cDNA cloning and expression analysis of *OsWOX4* in rice. *J. Shanghai Jiaotong Univ. Agric. Sci.* **2011**, *29*, 8–14.

12. Cho, S.H.; Yoo, S.C.; Zhang, H.T.; Pandeya, D.; Koh, H.J.; Hwang, J.Y.; Kim, G.T.; Paek, N.C. The rice narrow leaf2 and narrow leaf3 loci encode WUSCHEL-related homeobox 3A *OsWOX3A*, and function in leaf, spikelet, tiller and lateral root development. *New Phytol.* **2013**, *198*, 1071–1084. [CrossRef] [PubMed]

13. Luan, W.J.; Liu, Y.Q.; Zhang, F.X.; Song, Y.L.; Wang, Z.Y.; Peng, Y.K.; Sun, Z.X. *OsCD1* encodes a putative member of the cellulose synthase-like D sub-family and is essential for rice plant architecture and growth. *Plant Biotechnol. J.* **2010**, *9*, 513–524. [CrossRef] [PubMed]

14. Hu, J.; Zhu, Li.; Zeng, D.L.; Gao, Z.Y.; Guo, L.B.; Fang, Y.X.; Zhang, G.H.; Dong, G.J.; Yan, M.X.; Liu, J.; Qian, Q. Identification and characterization of *NARROW AND ROLLED LEAF 1*, a novel gene regulating leaf morphology and plant architecture in rice. *Plant Mol. Biol.* **2010**, *73*, 283–292. [CrossRef] [PubMed]

15. Zhang, G.H.; Li, S.Y.; Wang, L.; Ye, W.J.; Zeng, D.L.; Rao, Y.C.; Peng, Y.L.; Hu, J.; Yang, Y.L.; Xu, J.; et al. *LSCHL4* from japonica cultivar, which is allelic to *NAL1*, increases yield of indica super rice 93-11. *Mol. Plant* **2014**, *7*, 1350–1364. [CrossRef] [PubMed]

16. Fujino, K.; Matsuda, Y.; Ozawa, K.; Nishimura, T.; Koshiba, T.; Fraaije, M.W.; Sekiguchi, H. *NARROW LEAF 7* controls leaf shape mediated by auxin in rice. *Mol. Genet. Genom.* **2008**, *279*, 499–507. [CrossRef] [PubMed]

17. Sazuka, T.; Kamiya, N.; Nishimura, T.; Ohmae, K.; Sato, Y.; Imamura, K.; Nagato, Y.; Koshiba, T.; Nagamura, Y.; Ashikari, M.; et al. A rice *tryptophan deficient dwarf* mutant, *tdd1*, contains a reduced level of indole acetic acid and develops abnormal flowers and organless embryos. *Plant J.* **2009**, *60*, 227–241. [CrossRef]

18. Woo, Y.M.; Park, H.J.; Su'udi, M.; Yang, J.I.; Park, J.J.; Back, K.; Park, Y.M.; An, G. *Constitutively wilted 1*, a member of the rice YUCCA gene family, is required for maintaining water homeostasis and an appropriate root to shoot ratio. *Plant Mol. Biol.* **2007**, *65*, 125–136. [CrossRef]

19. Zhang, S.Z.; Wu, T.; Liu, S.J.; Liu, X.; Jiang, L.; Wan, J.M. Disruption of *OsARF19* is Critical for Floral Organ Development and Plant Architecture in Rice (*Oryza sativa* L.). *Plant Mol. Biol. Rep.* **2016**, *34*, 748–760. [CrossRef]

20. Li, W.; Wu, C.; Hu, G.C.; Xing, Li.; Qian, W.J.; Si, H.M.; Sun, Z.X.; Wang, X.C.; Fu, Y.P.; Liu, W.Z. Characterization and fine mapping of a novel rice narrow leaf mutant nal9. *J. Integr. Plant Biol.* **2013**, *55*, 1016–1025. [CrossRef]

21. Hibara, K.; Obara, M.; Hayashida, E.; Abe, M.; Ishimaru, T.; Satoh, H.; Itoh, J.; Nagato, Y. The *ADAXIALIZED LEAF1* gene functions in leaf and embryonic pattern formation in rice. *Dev. Biol.* **2009**, *334*, 345–354. [CrossRef] [PubMed]

22. Shi, Z.Y.; Wang, J.; Wan, X.S.; Shen, G.Z.; Wang, X.; Zhang, J.L. Over-expression of rice *OsAGO7* gene induces upward curling of the leaf blade that enhanced erect-leaf habit. *Planta* **2007**, *226*, 99–108. [CrossRef] [PubMed]

23. Li, L.; Xue, X.; Zou, S.M.; Chen, Z.X.; Zhang, Y.F.; Li, Q.Q.; Zhu, J.K.; Ma, Y.Y.; Pan, X.B.; Pan, C.H. Suppressed expressed of AGO1a leads to adaxial leaf rolling in rice. *Chin. J. Rice Sci.* **2013**, *273*, 223–230.

24. Li, L.; Shi, Z.Y.; Li, L.; Shen, G.Z.; Wang, X.Q.; An, L.S.; Zhang, J.L. Overexpression of *ACL1 abaxially curled leaf 1*, increased bulliform cells and induced abaxial curling of leaf blades in rice. *Mol. Plant* **2010**, *35*, 807–817. [CrossRef] [PubMed]

25. Wang, J.; Hu, J.; Qian, Q.; Xue, H.W. LC2 and OsVIL2 promote rice flowering by photoperoid-induced epigenetic silencing of *OsLF. Mol. Plant* **2013**, *62*, 514–527. [CrossRef] [PubMed]

26. Zhao, S.Q.; Hu, J.; Guo, L.B.; Qian, Q.; Xue, H.W. *Rice leaf inclination 2*, a VIN3-like protein, regulates leaf angle through modulating cell division of the collar. *Cell Res.* **2010**, *20*, 935–947. [CrossRef] [PubMed]

27. Li, M.; Xiong, G.Y.; Li, R.; Cui, J.J.; Tang, D.; Zhang, B.C.; Pauly, M.; Cheng, Z.K.; Zhou, Y.H. Rice cellulose synthase-like D4 is essential for normal cell-wall biosynthesis and plant growth. *Plant J.* **2009**, *60*, 1055–1069. [CrossRef]

28. Yoshikaw, T.; Eiguchi, M.; Hibara, K.I.; Ito, J.I.; Nagato, Y. Rice *SLENDER LEAF 1* gene encodes cellulose synthase-like D4 and is specifically expressed in M-phase cells to regulate cell proliferation. *J. Exp. Bot.* **2013**, *647*, 2049–2061. [CrossRef]

29. Ding, Z.Q.; Lin, Z.F.; Li, Q.; Wu, H.; Xiang, C.Y.; Wang, J.F. *DNL1*, encodes cellulose synthase-like D4, is a major QTL for plant height and leaf width in rice (*Oryza sativa* L.). *Biochem. Biophys. Res. Commun.* **2014**, *457*, 133–140. [CrossRef]

30. Li, Y.Y.; Shen, A.; Xiong, W.; Sun, Q.L.; Luo, Q.; Song, T.; Li, Z.L.; Luan, W.J. Overexpression of *OsHox32* results in pleiotropic effects on plant type architecture and leaf development in rice. *Rice* **2016**, *9*, 46. [CrossRef]

31. Yang, C.H.; Li, D.Y.; Liu, X.; Ji, C.J.; Hao, L.L.; Zhao, X.F.; Li, X.B.; Chen, C.Y.; Cheng, Z.K.; Zhu, L.H. OsMYB103L, an R2R3-MYB transcription factor, influences leaf rolling and mechanical strength in rice (*Oryza sativa* L.). *BMC Plant Biol.* **2014**, *14*, 158. [CrossRef] [PubMed]

32. Xu, Y.; Wang, Y.H.; Long, Q.Z.; Huang, J.X.; Wang, Y.L.; Zhou, K.N.; Zheng, M.; Sun, J.; Chen, H.; Chen, S.H.; et al. Overexpression of *OsZHD1*, a zinc finger homeodomain class homeobox transcription factor, induces abaxially curled and drooping leaf in rice. *Planta* **2009**, *239*, 803–816. [CrossRef] [PubMed]

33. Chen, Q.L.; Xie, Q.J.; Gao, J.; Wang, W.Y.; Sun, B.; Liu, B.H.; Zhu, H.T.; Peng, H.F.; Zhao, H.B.; Liu, C.H.; et al. Characterization of *Rolled and Erect Leaf 1* in regulating leave morphology in rice. *J. Exp. Bot.* **2015**, *66*, 6047–6058. [CrossRef] [PubMed]

34. Yang, S.Q.; Li, W.Q.; Miao, H.; Gan, P.F.; Qiao, L.; Chang, Y.L.; Shi, C.H.; Chen, K.M. REL2, A Gene Encoding An Unknown Function Protein which Contains DUF630 and DUF632 Domains Controls Leaf Rolling in Rice. *Rice* **2016**, *9*, 37. [CrossRef] [PubMed]

35. Fang, L.K.; Zhao, F.M.; Cong, Y.F.; Sang, X.C.; Du, Q.; Wang, D.Z.; Li, Y.F.; Ling, Y.H.; Yang, Z.L.; He, G.H. Rolling-leaf14 is a 2OG-Fe II, oxygenase family protein that modulates rice leaf rolling by affecting secondary cellwall formation in leaves. *Plant Biotechnol. J.* **2012**, *10*, 524–532. [CrossRef] [PubMed]

36. Zou, L.P.; Sun, X.H.; Zhang, Z.G.; Liu, P.; Wu, J.X.; Tian, C.J.; Qiu, J.L.; Lu, T.G. Leaf Rolling Controlled by the Homeodomain Leucine Zipper Class IV Gene Roc5 in Rice. *Plant Physiol.* **2011**, *156*, 1589–1602. [CrossRef] [PubMed]

37. Li, W.Q.; Zhang, M.J.; Gan, P.F.; Qiao, L.; Yang, S.Q.; Miao, H.; Wang, G.F.; Zhang, M.M.; Liu, W.T.; Li, H.F.; et al. *CLD1/SRL1* modulates leaf rolling by affecting cell wall formation, epidermis integrity and water homeostasis in rice. *Plant J.* **2017**, *92*, 904–923. [CrossRef]

38. Xiang, J.J.; Zhang, G.H.; Qian, Q.; Xue, H.W. *Semi-rolled leaf1* encodes a putative glycosylphosphatidylinositol-anchored protein and modulates rice leaf rolling by regulating the formation of bulliform cells. *Plant Physiol.* **2012**, *159*, 1488–1500. [CrossRef]

39. Alamin, M.; Zeng, D.D.; Qin, R.; Sultana, H.; Jin, X.L.; Shi, C.H. Characterization and fine mapping of *SFL1*, a gene controlling *Screw Flag Leaf* in Rice. *Plant Mol. Biol. Rep.* **2017**, *35*, 491–503. [CrossRef]

40. Zhang, J.J.; Wu, S.Y.; Jiang, L.; Wang, J.L.; Zhang, X.; Guo, X.P.; Wu, C.Y.; Wan, J.M. A detailed analysis of the leaf rolling mutant sll2 reveals complex nature in regulation of bulliform cell development in rice (*Oryza sativa* L.). *Plant Biol.* **2015**, *17*, 437–448. [CrossRef]

41. Dai, M.Q.; Zhao, Y.; Ma, Q.; Hu, Y.F.; Hedden, P.; Zhang, Q.F.; Zhou, D.X. The rice *YABBY1* gene is involved in the feedback regulation of gibberellin metabolism. *Plant Physiol.* **2007**, *144*, 121–133. [CrossRef] [PubMed]

42. Zhou, Y.; Wang, D.; Wu, T.; Yang, Y.; Liu, C.; Yan, L.; Tang, D.; Zhao, X.; Zhu, Y.; Lin, J.; et al. LRRK1, a receptor-like cytoplasmic kinase, regulates leaf rolling through modulating bulliform cell development in rice. *Mol. Breed.* **2018**, *38*, 48. [CrossRef]

43. Ma, L.; Sang, X.C.; Zhang, T.; Yu, Z.Y.; Li, Y.F.; Zhao, F.M.; Wang, Z.W.; Wang, Y.T.; Yu, P.; Wang, N.; et al. *ABNORMAL VASCULAR BUNDLES* regulates cell proliferation and procambium cell establishment during aerial organ development in rice. *New Phytol.* **2017**, *213*, 275–286. [CrossRef] [PubMed]

44. Ye, Y.F.; Wu, K.; Chen, J.F.; Liu, Q.; Wu, Y.J.; Liu, B.M.; Fu, X.D. OsSND2, a NAC family transcription factor, is involved in secondary cell wall biosynthesis through regulating MYBs expression in rice. *Rice* **2018**, *11*, 36. [CrossRef] [PubMed]

45. Wu, R.H.; Li, S.B.; He, S.; Waßmann, F.; Yu, C.; Qin, G.; Schreiber, L.; Qu, L.J.; Gu, H. CFL1, a WW domain protein, regulates cuticle development by modulating the function of HDG1, a class IV homeodomain transcription factor, in rice and *Arabidopsis*. *Plant Cell* **2011**, *23*, 3392–3411. [CrossRef] [PubMed]

46. Jing, W.; Cao, C.J.; Shen, L.K.; Zhang, H.S.; Jing, G.Q.; Zhang, W.H. Characterization and fine mapping of a rice leaf-rolling mutant deficient in commissural veins. *Crop Sci.* **2017**, *57*, 2595–2604. [CrossRef]

47. Li, S.G.; He, P.; Wang, Y.P.; Li, H.Y.; Chen, Y.; Zhou, K.D.; Zhu, L.H. Genetic analysis and gene mapping of the leaf traits in rice *Oryza sativa* L. *Acta Agron. Sin.* **2000**, *26*, 261–265.

48. Yan, J.Q.; Zhu, J.; He, C.X.; Benmoussa, M.; Wu, P. Molecular marker-assisted dissection of genotype × environment interaction for plant type traits in rice (*Oryza sativa* L.). *Crop Sci.* **1999**, *39*, 538–544. [CrossRef]

49. Farooq, M.; Tagle, A.G.; Santos, R.E.; Ebron, L.A.; Fujita, D.; Kobayashi, N. Quantitative trait loci mapping for leaf length and leaf width in rice cv. IR64 derived lines. *J. Integr. Plant Biol.* **2010**, *52*, 578–584. [CrossRef]

50. Li, Z.K.; Pinson, S.R.M.; Stansel, J.W.; Paterson, A. Genetic dissection of the source-sink relationship affecting fecundity and yield in rice (*Oryza sativa* L.). *Mol. Breed.* **1998**, *4*, 419–426. [CrossRef]

51. Mei, H.W.; Luo, L.J.; Ying, C.S.; Wang, Y.P.; Yu, X.Q.; Guo, L.B.; Paterson, A.H.; Li, Z.K. Gene actions of QTLs affecting several agronomic traits resolved in a recombinant inbred rice population and two testcross populations. *Theor. Appl. Genet.* **2005**, *110*, 649–659. [CrossRef] [PubMed]

52. Cui, K.H.; Peng, S.; Xing, Y.; Yu, S.; Xu, C.; Zhang, Q. Molecular dissection of the genetic relationships of source, sink and transport tissue with yield traits in rice. *Theor. Appl. Genet.* **2003**, *106*, 649–658. [CrossRef] [PubMed]

53. Yue, B.; Xue, W.Y.; Luo, L.J.; Xing, Y.Z. QTL Analysis for flag leaf characteristics and their relationships with yield and yield traits in rice. *Acta Agron. Sin.* **2006**, *33*, 824–832. [CrossRef]

54. Jiang, S.k.; Zhang, X.J.; Wang, J.Y.; Chen, W.F.; Xu, Z.J. Fine mapping of the quantitative trait locus *qFLL9* controlling flag leaf length in rice. *Euphytica* **2010**, *176*, 341–347. [CrossRef]

55. Wang, P.; Zhou, G.L.; Cui, K.H.; Li, Z.K.; Yu, S.B. Clustered QTL for source leaf size and yield traits in rice (*Oryza sativa* L.). *Mol. Breed.* **2012**, *29*, 99–113. [CrossRef]

56. Wang, P.; Zhou, G.L.; Yu, H.H.; Yu, S.B. Fine mapping a major QTL for flag leaf size and yield-related traits in rice. *Theor. Appl. Genet.* **2011**, *123*, 1319–1330. [CrossRef]

57. Ding, X.P.; Li, X.K.; Xiong, L.Z. Evaluation of near-isogenic lines for drought resistance QTL and fine mapping of a locus affecting flag leaf width, spikelet number, and root volume in rice. *Theor. Appl. Genet.* **2011**, *123*, 815–826. [CrossRef]

58. Chen, M.L.; Luo, J.; Shao, G.N.; Wei, X.J.; Tang, S.Q.; Sheng, Z.H.; Song, J.; Hu, P.S. Fine mapping of a major QTL for flag leaf width in rice, *qFLW4*, which might be caused by alternative splicing of *NAL1*. *Plant Cell Rep.* **2012**, *31*, 863–872. [CrossRef]

59. Fujita, D.; Trijatmiko, K.R.; Tagle, A.G.; Sapasap, M.V.; Koide, Y.; Sasaki, K.; Tsakirpaloglou, N.; Gannaban, R.B.; Nishimura, T.; Yanagihara, S.; et al. *NAL1* allele from a rice landrace greatly increases yield in modern indica cultivars. *Proc. Natl. Acad. Sci. USA* **2013**, *110*, 20431–20436. [CrossRef]

60. Hu, J.; Wang, Y.X.; Fang, Y.X.; Zeng, L.J.; Xu, J.; Yu, H.P.; Shi, Z.Y.; Pan, J.J.; Zhang, D.; Kang, S.J.; et al. A Rare allele of *GS2* enhances grain size and grain yield in rice. *Mol. Plant* **2015**, *8*, 1455–1465. [CrossRef]

61. Duan, P.; Ni, S.; Wang, J.M.; Zhang, B.L.; Xu, R.; Wang, Y.X.; Chen, H.Q.; Zhu, X.D.; Li, Y.H. Regulation of *OsGRF4* by OsmiR396 controls grainsize and yield in rice. *Nat. Plants* **2015**, *2*, 15203. [CrossRef] [PubMed]

62. Quarrie, S.A.; Pekic, S.; Radosevic, Q.; Rancic, R.D.; Kaminska, A.; Barnes, J.D.; Leverington, M.; Ceoloni, C.; Dodig, D. Dissecting a wheat QTL for yield present in a range of environments: From the QTL to candidate genes. *J. Exp. Bot.* **2006**, *57*, 2627–2637. [CrossRef] [PubMed]

63. Huang, X.Z.; Qian, Q.; Liu, Z.B.; Sun, H.Y.; He, S.Y.; Luo, D.; Xia, G.M.; Chu, C.C.; Li, J.Y.; Fu, X.D. Natural variation at the *DEP1* locus enhances grain yield in rice. *Nat. Genet.* **2009**, *41*, 494–497. [CrossRef] [PubMed]

64. Wang, J.Y.; Nakazaki, T.; Chen, S.Q.; Chen, W.F.; Saito, H.; Tsukiyama, T.; Okumoto, Y.; Xu, Z.J.; Tanisaka, T. Identification and characterization of the erect-pose panicle gene *EP* conferring high grain yield in rice (*Oryza sativa* L.). *Theor. Appl. Genet.* **2009**, *119*, 85–91. [CrossRef] [PubMed]

65. Zhou, Y.; Zhu, J.Y.; Li, Z.Y.; Yi, C.D.; Liu, J.; Zhang, H.G.; Tang, S.Z.; Gu, M.H.; Liang, G.H. Deletion in a quantitative trait gene *qpe9-1* associated with panicle erectness improves plant architecture during rice domestication. *Genetics* **2009**, *183*, 315–324. [CrossRef] [PubMed]

66. Mao, S.; Lu, G.H.; Yu, K.H.; Bo, Z.; Chen, J.H. Specific protein detection using thermally reduced graphene oxide sheet decorated with gold nanoparticle-antibody conjugates. *Adv. Mater* **2010**, *22*, 3521–3526. [CrossRef] [PubMed]

67. Kosugi, S.; Suzuka, I.; Ohashi, Y. Two of three promoter elements identified in a rice gene for proliferating cell nuclear antigen are essential for meristematic tissue-specific expression. *Plant J.* **1995**, *7*, 877–886. [CrossRef]

68. Zhao, M.Z.; Sun, J.; Xiao, Z.Q.; Cheng, F.; Xu, H.; Tang, L.; Chen, W.F.; Xu, Z.J.; Xu, Q. Variations in *DENSE AND ERECT PANICLE 1* (*DEP1*) contribute to the diversity of the panicle trait in high-yielding japonica rice varieties in northern China. *Breed. Sci.* **2016**, *66*, 599–605. [CrossRef]

69. Narita, N.N.; Moore, S.; Horiguchi, G.; Kubo, M.; Demura, T.; Fukuda, H.; Goodrich, J.; Tsukaya, H. Overexpression of a novel small peptide ROTUNDIFOLIA4 decreases cell proliferation and alters leaf shape in Arabidopsis thaliana. *Plant J.* **2004**, *38*, 399–713. [CrossRef]

70. Tsuge, T.; Tsukaya, H.; Uchimiya, H. Two independent and polarized processes of cell elongation regulate leaf blade expansion in *Arabidopsis thaliana* (L.) Heynh. *Development* **1996**, *122*, 1589–1600.

71. Xu, J.; Wang, L.; Wang, Y.X.; Zeng, D.L.; Zhou, M.Y.; Fu, X.; Ye, W.J.; Hu, J.; Zhu, L.; Ren, D.Y.; et al. Reduction of *OsFLW7* expression enhanced leaf area and grain production in rice. *Sci. Bull.* **2017**, *62*, 1631–1633. [CrossRef]

72. Snyder, M.; He, W.; Zhang, J.J. The DNA replication factor MCM5 is essential for stat1-mediated transcriptional activation. *Proc. Natl. Acad. Sci. USA* **2005**, *102*, 14539–14544. [CrossRef] [PubMed]

73. Imai, K.K.; Ohashi, Y.H.; Tsuge, T.; Yoshizumi, T.; Matsui, M.; Oka, A.; Aoyama, T. The a-type cyclin CYCA2;3 is a key regulator of ploidy levels in *Arabidopsis* endoreduplication. *Plant Cell* **2006**, *18*, 382–396. [CrossRef] [PubMed]

74. Boudolf, V.; Lammens, T.; Boruc, J.; Leene, J.V.; Daele, H.V.D.; Maes, S.; Isterdael, G.V.; Russinova, E.; Kondorosi, E.; Witters, E.; et al. CDKB1;1 Forms a functional complex with CYCA2;3 to suppress endocycle onset. *Plant Physiol.* **2009**, *150*, 1482–1493. [CrossRef] [PubMed]

75. Zhou, M.Y.; Song, X.W.; Xu, J.; Fu, X.; Li, T.; Zhu, Y.C.; Xiao, X.Y.; Mao, Y.J.; Zeng, D.L.; Hu, J.; et al. Construction of genetic map and mapping and verification of grain traits QTLs using recombinant inbred lines derived from a cross between indica C84 and japonica CJ16B. *Chin. J. Rice Sci.* **2018**, *32*, 207–218.

76. Shen, L.; Hua, Y.F.; Fu, Y.P.; Li, J.; Liu, Q.; Jiao, X.Z.; Xin, G.W.; Wang, J.J.; Wang, X.C.; Yan, C.J.; et al. Rapid generation of genetic diversity by multiplex CRISPR/Cas9 genome editing in rice. *Sci. China Life Sci.* **2017**, *60*, 506–515. [CrossRef] [PubMed]

77. Ma, X.L.; Zhang, Q.Y.; Zhu, Q.L.; Liu, W.; Chen, Y.; Qiu, R.; Wang, B.; Yang, Z.F.; Li, H.Y.; Lin, Y.R.; et al. A robust CRISPR/Cas9 system for convenient, high-efficiency multiplex genome editing in monocot and dicot plants. *Mol. Plant* **2015**, *8*, 1274–1284. [CrossRef]

78. Hellens, R.P.; Allan, A.C.; Friel, E.N.; Bolitho, K.; Grafton, K.; Templeton, M.D.; Karunairetnam, S.; Gleave, A.P.; Laing, W.A. Transient expression vectors for functional genomics, quantification of promoter activity and RNA silencing in plants. *Plant Methods* **2005**, *1*, 13. [CrossRef]

79. You, M.K.; Lim, S.H.; Kim, M.J.; Jeong, Y.S.; Lee, M.G.; Ha, S.H. Improvement of the fluorescence intensity during a flow cytometric analysis for rice protoplasts by localization of a green fluorescent protein into chloroplasts. *Int. J. Mol. Sci.* **2015**, *16*, 788–804. [CrossRef]

80. Ishimaru, K.; Shirota, K.; Higa, M.; Kawamitsu, Y. Identification of quantitative trait loci for adaxial and abaxial stomatal frequencies in *Oryza sativa*. *Plant Physiol. Biochem.* **2001**, *39*, 173–177. [CrossRef]

Article

Importance of the Interaction between Heading Date Genes *Hd1* and *Ghd7* for Controlling Yield Traits in Rice

Zhen-Hua Zhang, Yu-Jun Zhu, Shi-Lin Wang, Ye-Yang Fan and Jie-Yun Zhuang *

State Key Laboratory of Rice Biology and Chinese National Center for Rice Improvement, China National Rice Research Institute, Hangzhou 310006, China; zhangzhenhua@caas.cn (Z.-H.Z.); yjzhu2013@163.com (Y.-J.Z.); 15621566500@163.com (S.-L.W.); fanyeyang@caas.cn (Y.-Y.F.)
* Correspondence: zhuangjieyun@caas.cn; Tel.: +86-571-6337-0369

Received: 3 January 2019; Accepted: 23 January 2019; Published: 26 January 2019

Abstract: Appropriate flowering time is crucial for successful grain production, which relies on not only the action of individual heading date genes, but also the gene-by-gene interactions. In this study, influences of interaction between *Hd1* and *Ghd7* on flowering time and yield traits were analyzed using near isogenic lines derived from a cross between *indica* rice cultivars ZS97 and MY46. In the non-functional *ghd7*^{ZS97} background, the functional *Hd1*^{ZS97} allele promoted flowering under both the natural short-day (NSD) conditions and natural long-day (NLD) conditions. In the functional *Ghd7*^{MY46} background, *Hd1*^{ZS97} remained to promote flowering under NSD conditions, but repressed flowering under NLD conditions. For *Ghd7*, the functional *Ghd7*^{MY46} allele repressed flowering under both conditions, which was enhanced in the functional *Hd1*^{ZS97} background under NLD conditions. With delayed flowering, spikelet number and grain weight increased under both conditions, but spikelet fertility and panicle number fluctuated. Rice lines carrying non-functional *hd1*^{MY46} and functional *Ghd7*^{MY46} alleles had the highest grain yield under both conditions. These results indicate that longer growth duration for a larger use of available temperature and light does not always result in higher grain production. An optimum heading date gene combination needs to be carefully selected for maximizing grain yield in rice.

Keywords: flowering time; gene-by-gene interaction; *Hd1*; *Ghd7*; rice; yield trait

1. Introduction

Flowering time is a pivotal factor in the adaption of cereals to various ecogeographic environments and agricultural practices, which is controlled by an intricate genetic network. Florigens are at the core of the network, which are encoded by *Hd3a* and *RFT1* in rice [1,2]. The expression of *Hd3a* and *RFT1* are regulated by two important pathways mediating by *Hd1* and *Ehd1*, respectively [3]. *Hd1* has dual functions, which enhances florigen genes expressions under short-day (SD) conditions but inhibits florigen genes expressions under long-day (LD) conditions. The function conversion of *Hd1* is related to *PhyB*, *Se5*, *Ghd7* and *Ghd8* [4–8]. Function loss of any of these genes attenuates the conversion and maintains *Hd1* as an activator under any day-length conditions. *Ehd1* activates florigen genes expressions to promote flowering under both the SD and LD conditions [9]. *Ehd1* likely acts as a signal integrator, and its expression is regulated by many genes [3]. Recent studies revealed that *Hd1* represses expression of *Ehd1* through interaction with *Ghd7* or *DTH8* [6–8].

Flowering time is closely related to the grain yield for crop, owing to its key role in maintaining an appropriate balance between full use of resources and avoidance of environmental stresses. Many heading date (HD) genes were reported to affect yield traits, and their natural variations have been used in rice breeding, such as *Ghd7* [10], *DTH8*/*Ghd8* [11,12], *Hd1* [13,14],

OsPRR37/Ghd7.1/DTH7/Hd2 [15–17], *RFT1* [18] and *OsMADS51* [19,20]. Abiotic stresses during flowering, such as high temperature, low temperature, and drought, can pose a serious threat to spikelet fertility and consequently induce yield loss. The relationship between HD gene and abiotic stress has been given attention in recent years. The *Ehd1-Hd3a/RFT1* pathway responses stress signals mediated by *Ghd7* [21], *OsABF* [22] or *OsMADS51* [20]. They integrate low temperature, high temperature, and drought signals, respectively, into HD pathway, which induce or repress floral transition to avoid flowering in the stress environments. Moreover, *Ghd7* and other four HD genes, including *Ghd2* [23], *OsHAL3* [24], *OsWOX13* [25] and *OsJMJ703* [26], were found to be involved in drought or salt tolerance during vegetative phase.

When the pleiotropic effects of individual HD genes on yield traits have become recognized, the role of gene-by-gene interaction remains to be explored. In the present study, influences of *Hd1* and *Ghd7* on HD and yield traits were analyzed using near isogenic lines (NILs) and NIL-F$_2$ populations derived from a cross between *indica* rice cultivars Zhenshan 97 (ZS97) and Milyang 46 (MY46). Our results showed that *Hd1* and *Ghd7* could independently promote and repress flowering, respectively, whereas the flowering-repressor function of *Hd1* under natural long-day (NLD) conditions required functional *Ghd7*. With delayed flowering, spikelet number and grain weight increased under both natural short-day (NSD) and NLD conditions, but the spikelet fertility and panicle number fluctuated. Rice lines with genotype of *hd1Ghd7* produced the highest grain yield under both conditions.

2. Results

2.1. Effects of Hd1 and Ghd7 on Heading Date

In this study, effects of *Hd1* and *Ghd7* on HD were investigated using three populations derived from the rice cross ZS97/MY46//MY46///MY46. ZS97 carries functional *Hd1* and non-functional *ghd7*, whereas MY46 carries non-function *hd1* and functional *Ghd7* [14,17]. The three populations included two NIL populations, namely R1-NIL and R2-NIL, and one NIL-F$_2$ population namely R2-F$_2$ (Figure 1). Each NIL population comprised all the four homozygous genotypic combinations of *Hd1* and *Ghd7*, i.e., *hd1*[MY46]*ghd7*[ZS97], *Hd1*[ZS97]*ghd7*[ZS97]; *hd1*[MY46]*Ghd7*[MY46] and *Hd1*[ZS97]*Ghd7*[MY46]. The NIL-F$_2$ population consisted of all the nine genotypic combinations, i.e., *hd1*[MY46]*ghd7*[ZS97], *Hd1*[heterozygous]*ghd7*[ZS97], *Hd1*[ZS97]*ghd7*[ZS97], *hd1*[MY46]*Ghd7*[heterozygous], *Hd1*[heterozygous]*Ghd7*[heterozygous], *Hd1*[ZS97]*Ghd7*[heterozygous], *hd1*[MY46]*Ghd7*[MY46], *Hd1*[heterozygous]*Ghd7*[MY46], and *Hd1*[ZS97]*Ghd7*[MY46]. The R1-NIL population was tested under both the NSD and NLD conditions, and the R2-F$_2$ and R2-NIL populations were tested in NLD conditions only. All the rice materials matured in seasons that are appropriate for rice growth.

The R1-NIL population consisted of 10, 7, 12, and 20 lines of *hd1*[MY46]*ghd7*[ZS97], *Hd1*[ZS97]*ghd7*[ZS97], *hd1*[MY46]*Ghd7*[MY46], and *Hd1*[ZS97]*Ghd7*[MY46], respectively. In the genetic background tested by whole-genome resequencing and marker analysis, this population was segregated at *Hd16* but homozygous at all the remaining 11 cloned quantitative trait loci (QTL) for HD, including *OsMADS51*, *DTH2*, *OsMADS50/DTH3*, *Hd6*, *Hd17*, *RFT1*, *Hd3a*, *OsPRR37/Ghd7.1/DTH7/Hd2*, *Hd18*, *DTH8/Ghd8* and *Ehd1*. The effects of *Hd1* and *Ghd7* on HD were tested under NSD conditions in Lingshui from Dec. 2016 to Apr. 2017 (16LS) and from Dec. 2017 to Apr. 2018 (17LS), and under NLD conditions in Hangzhou from May to Sep. in 2017 (17HZ).

Highly significant effects ($p < 0.0001$) of *Hd1* and *Ghd7* on HD were detected in all the three trials (Table 1). In the two trials under NSD conditions (16LS and 17LS), the functional *Hd1*[ZS97] and *Ghd7*[MY46] alleles promoted and delayed flowering, respectively, no matter whether its counterpart was functional or non-functional (Figure 2a,b). In 16LS and 17LS, the proportion of phenotypic variance explained (R^2) were estimated to be 80.74% and 75.69% for *Hd1*, and 5.79% and 6.50% for *Ghd7*, respectively. The interaction between *Hd1* and *Ghd7* was non-significant in the 17LS trial and significant in the 16LS trial with a small R^2 of 1.30%. Overall, *Hd1* and *Ghd7* largely act additively in regulating HD under NSD conditions.

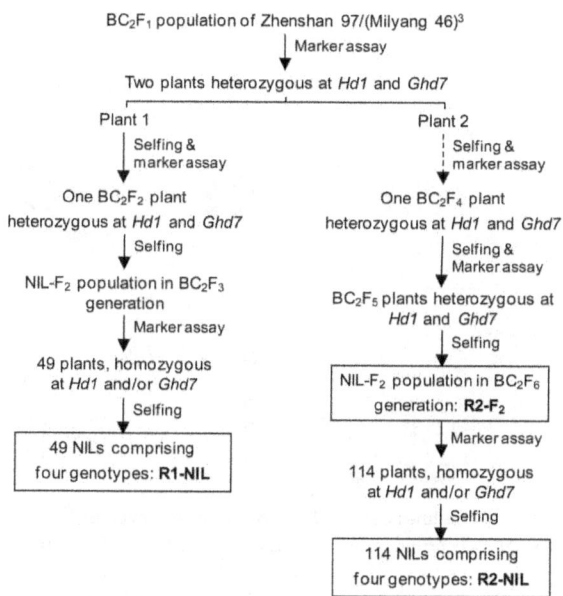

Figure 1. Development of the rice populations used in this study.

Table 1. The effects of *Hd1* and *Ghd7* on heading date and six yield traits.

Population	Trial	Trait	Hd1			Ghd7			Hd1 × Ghd7		
			P	*A*	R^2%	*P*	*A*	R^2%	*P*	I-effect	R^2%
R1-NIL	16LS	HD	<0.0001	10.09	80.74	<0.0001	0.51	5.79	<0.0001	−1.30	1.30
	17LS	HD	<0.0001	7.95	75.69	<0.0001	0.77	6.50	0.2586		
		NP	0.5940			0.0014	−0.44	7.51	0.0103		
		NSP	<0.0001	7.57	35.99	<0.0001	4.35	20.36	0.0438		
		NGP	<0.0001	7.60	41.05	<0.0001	3.62	17.38	0.1490		
		SF	0.0015	0.94	10.18	0.7543			0.1061		
		TGW	<0.0001	0.99	51.36	<0.0001	0.32	11.31	0.0479		
		GY	<0.0001	3.26	45.02	0.0122			0.0575		
	17HZ	HD	<0.0001	−3.30	3.03	<0.0001	6.08	56.54	<0.0001	3.06	16.43
		NP	<0.0001	0.72	4.93	<0.0001	−1.08	20.84	0.3167		
		NSP	0.0442			<0.0001	5.45	13.87	0.2970		
		NGP	0.1371			0.0677			0.3892		
		SF	0.7773			0.0198			0.8471		
		TGW	0.9515			<0.0001	1.03	38.44	0.0002	0.46	3.49
		GY	0.4677			0.8271			0.5233		
R2-NIL	18HZ	HD	<0.0001	−2.34	6.60	<0.0001	6.15	62.57	<0.0001	4.20	28.27
		NP	0.8453			0.0304			<0.0001	−0.45	9.97
		NSP	0.0253			<0.0001	9.50	43.27	<0.0001	4.07	7.81
		NGP	0.8697			<0.0001	4.05	16.34	0.0610		
		SF	<0.0001	0.99	3.10	<0.0001	−3.25	46.28	<0.0001	−1.80	14.12
		TGW	0.5604			<0.0001	0.33	24.38	<0.0001	0.29	18.28
		GY	0.0097	0.66	2.68	0.6200			<0.0001	−1.14	7.33

16LS, the trial conducted under natural short-day (NSD) conditions in Lingshui from Dec. 2016 to Apr. 2017; 17LS, the trial conducted under NSD conditions in Lingshui from Dec. 2017 to Apr. 2018; 17HZ, the trial conducted under the natural long-day (NLD) conditions in Hangzhou from May to Sep. in 2017; 18HZ, the trial conducted under the NLD conditions in Hangzhou from Apr. to Aug. in 2018. HD, heading date; NP, number of panicles per plant; NSP, number of spikelets per panicle; NGP, number of grains per panicle; SF, spikelet fertility (%); TGW, 1000-grain weight (g); GY, grain weight per plant (g). *A*, additive effect of replacing a Zhenshan 97 allele with a Milyang 46 allele. R^2%, proportion of phenotypic variance explained by the QTL effect. I-effect, positive value: parental type < recombinant type; negative value: parental type > recombinant type.

Int. J. Mol. Sci. **2019**, 20, 516

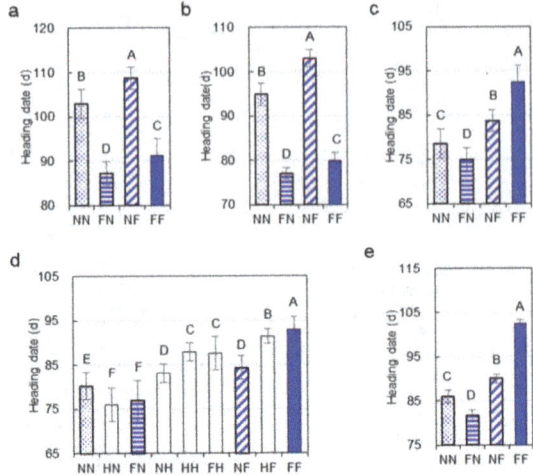

Figure 2. Heading date of rice lines classified based on the genotype of *Hd1* and *Ghd7*. (**a**) R1-NIL population under the NSD conditions in the 16LS trial. (**b**) R1-NIL population under the NSD conditions in the 17LS trial. (**c**) R1-NIL population under the NLD conditions the 17HZ trial. (**d**) R2-F$_2$ population under the NLD conditions in the 17HZ trial. (**e**) R2-NIL population under the NLD conditions in the 18HZ trial. NN, *hd1*^{MY46}*ghd7*^{ZS97}; HN, *Hd1*^{heterozygous}*ghd7*^{ZS97}; FN, *Hd1*^{ZS97}*ghd7*^{ZS97}; NH, *hd1*^{MY46}*Ghd7*^{heterozygous}; HH, *Hd1*^{heterozygous}*Ghd7*^{heterozygous}; FH, *Hd1*^{ZS97}*Ghd7*^{heterozygous}; NF, *hd1*^{MY46}*Ghd7*^{MY46}; HF, *Hd1*^{heterozygous}*Ghd7*^{MY46}; FF, *Hd1*^{ZS97}*Ghd7*^{MY46}. Data are presented in mean ± sd. Bars with different letters are significantly different at $p < 0.01$ based on Duncan's multiple range tests.

In the 17HZ trial under NLD conditions, the effects of *Hd1*, *Ghd7* and their interaction were all highly significant ($p < 0.0001$). The R^2 were estimated to be 3.03% for *Hd1*, 56.54% for *Ghd7*, and 16.43% for the interaction between the two genes (Table 1). Compared with NILs having the *hd1*^{MY46}*ghd7*^{ZS97} genotype, those having the *Hd1*^{ZS97}*ghd7*^{ZS97} genotypes flowered earlier by 3.51 d; compared with NILs having the *hd1*^{MY46}*Ghd7*^{ZS97} genotype, those having the *Hd1*^{ZS97}*Ghd7*^{MY46} genotype flowered later by 8.75 d (Figure 2c; Table 2). These indicated that *Hd1* regulates flowering dependent on *Ghd7* under NLD conditions, and its flowering-repressor activity requires the functional allele of *Ghd7*. For *Ghd7*, it delays flowering regardless of genotype of *Hd1* but its effect is enhanced by *Hd1*. HD was longer by 5.24 d in lines of *hd1*^{MY46}*Ghd7*^{MY46} than *hd1*^{MY46}*ghd7*^{ZS97}, whereas it was longer by 17.49 d in lines of *Hd1*^{ZS97}*Ghd7*^{MY46} than of *Hd1*^{ZS97}*ghd7*^{ZS97} (Table 2).

Table 2. Heading date and six yield traits of the four homozygous genotypes of *Hd1* and *Ghd7*.

Population	Trial	Group	HD	NP	NSP	NGP	SF	TGW	GY
R1-NIL	17LS	FN	87.2 ± 2.6 ^{Dd}	11.8 ± 1.1 ^{ABb}	79.3 ± 4.4 ^{Cc}	69.3 ± 4.7 ^{Cd}	87.4 ± 3.1 ^{Bb}	25.4 ± 1.1 ^{Cd}	20.7 ± 2.8 ^{Cc}
		FF	91.2 ± 3.8 ^{Cc}	11.6 ± 1.1 ^{ABb}	87.9 ± 6.9 ^{Bb}	77.6 ± 5.9 ^{Bc}	88.3 ± 2.6 ^{ABb}	27.0 ± 1.0 ^{Bc}	24.0 ± 2.9 ^{Bb}
		NN	103.0 ± 3.3 ^{Bb}	12.6 ± 0.9 ^{Aa}	92.0 ± 5.8 ^{Bb}	83.3 ± 5.5 ^{Bb}	90.6 ± 2.4 ^{Aa}	28.2 ± 0.9 ^{Ab}	29.4 ± 2.6 ^{Aa}
		NF	108.6 ± 2.5 ^{Aa}	11.0 ± 0.8 ^{Bb}	108.2 ± 7.2 ^{Aa}	96.7 ± 7.0 ^{Aa}	89.3 ± 2.4 ^{ABab}	28.9 ± 1.0 ^{Aa}	29.9 ± 3.0 ^{Aa}
	17HZ	FN	75.0 ± 1.6 ^{Dd}	15.5 ± 1.2 ^{ABa}	107.6 ± 8.6 ^{ABb}	89.0 ± 5.7 ^{Aab}	82.8 ± 3.9 ^{Aa}	22.8 ± 0.8 ^{Cb}	30.4 ± 1.5 ^{Aa}
		FF	78.5 ± 2.1 ^{Cc}	16.1 ± 1.2 ^{Aa}	100.1 ± 6.8 ^{Bc}	83.5 ± 4.0 ^{Ab}	83.6 ± 3.9 ^{Aa}	23.8 ± 0.8 ^{BCb}	30.5 ± 1.8 ^{Aa}
		NN	83.7 ± 1.5 ^{Bb}	14.5 ± 1.2 ^{BCb}	112.6 ± 9.1 ^{Aab}	89.8 ± 9.1 ^{Aab}	79.7 ± 5.7 ^{Aa}	24.9 ± 1.1 ^{ABa}	30.9 ± 3.1 ^{Aa}
		NF	92.5 ± 1.6 ^{Aa}	13.1 ± 1.4 ^{Cc}	115.0 ± 10.7 ^{Aa}	91.3 ± 11.8 ^{Aa}	79.6 ± 8.5 ^{Aa}	25.8 ± 1.2 ^{Aa}	29.9 ± 4.5 ^{Aa}
R2-NIL	18HZ	FN	81.6 ± 1.2 ^{Dd}	12.6 ± 1.0 ^{Aa}	115.4 ± 5.1 ^{Dd}	105.3 ± 4.9 ^{Bb}	91.3 ± 1.6 ^{Aa}	25.0 ± 0.3 ^{Cc}	31.8 ± 2.9 ^{ABb}
		FF	86.0 ± 1.4 ^{Cc}	11.7 ± 0.9 ^{BCbc}	120.4 ± 6.7 ^{Cc}	107.6 ± 5.9 ^{Bb}	89.3 ± 2.1 ^{Bb}	25.6 ± 0.4 ^{Bb}	30.8 ± 2.9 ^{Bbc}
		NN	90.0 ± 1.0 ^{Bb}	12.2 ± 1.2 ^{ABab}	131.5 ± 8.8 ^{Bb}	113.5 ± 7.9 ^{Aa}	86.4 ± 2.1 ^{Cc}	25.7 ± 0.6 ^{Bb}	33.3 ± 2.7 ^{Aa}
		NF	102.5 ± 0.7 ^{Aa}	11.3 ± 1.2 ^{Cc}	142.7 ± 7.5 ^{Aa}	115.7 ± 6.7 ^{Aa}	81.1 ± 2.7 ^{Dd}	26.3 ± 0.5 ^{Aa}	29.8 ± 3.2 ^{Bc}

FN, *Hd1*^{ZS97}*ghd7*^{ZS97}; FF, *Hd1*^{ZS97}*Ghd7*^{MY46}; NN, *hd1*^{MY46}*ghd7*^{ZS97}; NF, *hd1*^{MY46}*Ghd7*^{MY46}. Values are mean ± sd. Uppercase and lowercase letters following the values represent significant differences at $p < 0.01$ and $p < 0.05$, respectively, based on Duncan's multiple range tests.

2.2. Expressions of Genes Involved in the Photoperiod Pathway

The transcript levels of *Hd1*, *Ghd7*, *Ehd1*, *Hd3a* and *RFT1* at 2 h after sunrise were examined in seven-week-old rice lines in the R1-NIL population grown in the 17LS and 17HZ trials (Figure 3). In the 17LS trial under NSD conditions (Figure 3a), expression of *Hd1* and *Ghd7* was not affected by each other. The *Ehd1* expression was also not affected by either *Hd1* or *Ghd7*. For florigen genes, the expression of *Hd3a* was 7.87 times larger in lines of *Hd1^{ZS97}ghd7^{ZS97}* than *hd1^{MY46}ghd7^{ZS97}*, and 12.46 times larger in lines of *Hd1^{ZS97}Ghd7^{MY46}* than *hd1^{MY46}Ghd7^{MY46}*. These results indicate that *Hd1* promotes *Hd3a* expression regardless of *Ghd7* function, which was in accordance with that *Hd1* promotes flowering regardless of *Ghd7* function under NSD conditions. In addition, *Hd1* was also found to promote *RFT1* in the *Ghd7* background. At the same time, slightly repression of *Hd3a* by *Ghd7* was detected in the *hd1* background. These were consistent with the small effect of *Ghd7* under NSD conditions.

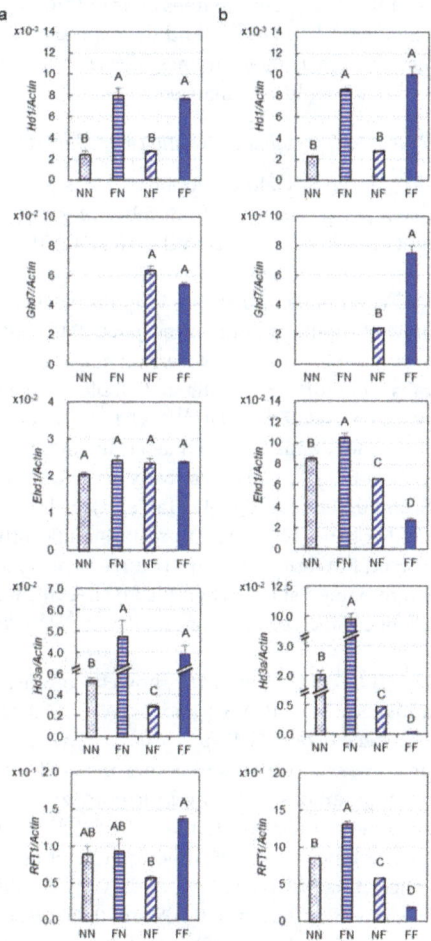

Figure 3. Transcript levels of five heading date genes in the R1-NIL population. (**a**) Under the NSD conditions in Lingshui. (**b**) Under the NLD conditions in Hangzhou. Data are presented in mean ± s. e. m (*n* = 3). Bars with different letters are significantly different at *p* < 0.01 based on Duncan's multiple range tests.

In the 17HZ trial conducted under NLD conditions (Figure 3b), expression of *Hd1* was not affected by *Ghd7*, but *Hd1* up-regulated *Ghd7* expression. The *Ghd7* expression was 2.12 times larger in lines of $Hd1^{ZS97}Ghd7^{MY46}$ than $hd1^{ZS97}Ghd7^{MY46}$. The expression of *Ehd1* in lines of $Hd1^{ZS97}ghd7^{ZS97}$ was 1.24 times as large as that in lines of $hd1^{MY46}ghd7^{ZS97}$, but the expression in lines of $Hd1^{ZS97}Ghd7^{MY46}$ was only 0.42 times as large as that in lines of $hd1^{MY46}Ghd7^{MY46}$. These suggest that *Hd1* significantly represses *Ehd1* expression in the *Ghd7* background. For florigen genes, the expressions of *Hd3a* and *RFT1* in lines of $Hd1^{ZS97}ghd7^{ZS97}$ were 4.86 and 1.55 times as large as that in lines of $hd1^{MY46}ghd7^{ZS97}$, indicating *Hd1* promotes expressions of florigen genes in the *ghd7* background. However, *Hd1* was converted to severely repress the florigen gene expressions in the *Ghd7* background. The expressions of *Hd3a* and *RFT1* in lines of $Hd1^{ZS97}Ghd7^{MY46}$ were only 0.07 and 0.32 times as large as those in lines of $hd1^{MY46}Ghd7^{MY46}$. In the meantime, significant repression of the *Ehd1*, *Hd3a* and *RFT1* expressions by *Ghd7* were detected in both the *Hd1* and *hd1* background, and the effect were larger in the *Hd1* background. The expressions of the three genes in lines of $hd1^{MY46}Ghd7^{MY46}$ were 0.77, 0.24 and 0.68 times as large as those in lines of $hd1^{MY46}ghd7^{ZS97}$; and the expressions in lines of $Hd1^{ZS97}Ghd7^{MY46}$ were 0.26. 0.004 and 0.14 times as large as those in lines of $Hd1^{ZS97}ghd7^{ZS97}$. These agreed with that flowering-repressor function of *Ghd7* could be enhanced by *Hd1*.

2.3. Influence of Hd1 and Ghd7 on Yield Traits and Its Relationship with HD

Grain yield per plant (GY), and five yield components traits including number of panicles per plant (NP), number of spikelets per panicle (NSP), number of grains per panicle (NGP), spikelet fertility (SF), 1000-grain weight (TGW), were measured in the R1-NIL population grown in the 17LS and 17HZ trials.

In the 17LS trial under NSD conditions, *Hd1* showed significant effects ($p < 0.01$) on all the six yield traits except NP; and *Ghd7* showed significant influences ($p < 0.01$) on all the six yield traits except SF and GY (Table 1). Interaction between the two genes were all non-significant at $p < 0.01$. Relationships between HD and the yield traits were further investigated (Table 2). The lines of $Hd1^{ZS97}ghd7^{ZS97}$ had the shortest HD, followed by $Hd1^{ZS97}Ghd7^{MY46}$, $hd1^{MY46}ghd7^{ZS97}$ and $hd1^{MY46}Ghd7^{MY46}$. Significant differences ($p < 0.05$) were detected for all the five yield determinants among the four genotypic groups. Three of the traits, NSP, NGP, and TGW, were positively correlated with HD, having correlation coefficients (r) of 0.823, 0.828, and 0.614, respectively (Table S1). Values of these three traits increased with delayed heading. On the other hand, NP and SF were not significantly correlated with HD. For GY, the values increased with delayed flowering among the three genotypic groups having the shortest to third shortest HD, and then remained stable when the HD became longer. Consequently, the two genotypic groups having the longest and second longest HD, $hd1^{MY46}Ghd7^{MY46}$ and $hd1^{MY46}ghd7^{ZS97}$, had little difference on GY.

In the 17HZ trial under NLD conditions, *Hd1* showed significant effects only on NP; and *Ghd7* showed significant influences on NP, NSP, and TGW ($p < 0.0001$). Significant interaction between the two genes was detected on TGW ($p < 0.001$). The interaction acted for increasing the values of the recombinant types, which was in accordance with the epistasis on HD. The HD and six yield traits were also compared among the four homozygous genotype groups (Table 2). The lines of $Hd1^{ZS97}ghd7^{ZS97}$ had the shortest HD, followed by $hd1^{MY46}ghd7^{ZS97}$, $hd1^{MY46}Ghd7^{MY46}$ and $Hd1^{ZS97}Ghd7^{MY46}$. Significant differences ($p < 0.05$) among the four genotypic groups were detected on four yield determinants, including NP, NGP, NSP, and TGW. Variations of TGW and NSP were positively correlated with HD, having r values of 0.708 and 0.355, respectively (Table S1). The two traits tended to increase with delayed heading. Similar tendency was observed for NGP though it was not significantly correlated with HD. Conversely, NP was negatively correlated with HD ($p < 0.05$), having r value of -0.670. SF also appeared to decrease with delayed heading though no significant difference was observed. Consequently, the largest value of GY in the four genotypic groups was observed for $hd1^{MY46}Ghd7^{MY46}$ which had the second longest HD.

2.4. Validation of the Influences of Hd1 and Ghd7 on HD and Yield Traits under NLD Conditions

The relationship between *Hd1* and *Ghd7* was further analyzed using the R2-F_2 population, which was segregated at *Hd1* and *Ghd7* loci but homozygous at all the remaining 12 cloned flowering QTL mentioned above. The 775 plants of this population were grown in Hangzhou in 2017 under NLD conditions. Significant effects were identified for both genes. The additive effect, dominance effect and R^2 were estimated to be 1.89 d, -0.89 d and 6.4% for *Hd1*, and 6.04 d, 1.91 d and 59.3% for *Ghd7*, respectively. The plants were classified into nine genotypic groups based on the *Hd1* and *Ghd7* alleles, and the HD values were compared (Figure 2d). *Hd1* promoted flowering in the *ghd7* background, but delayed heading when the genotype of *Ghd7* was functional or heterozygous. *Ghd7* delayed flowering regardless of the genotype of *Hd1* but its effect was enhanced by the functional *Hd1* allele.

Plants that were homozygous at *Hd1* and/or *Ghd7* were selected from the R2-F_2 population and selfed. The resultant R2-NIL population, consisting of 29, 26, 29, and 30 lines of $hd1^{MY46}ghd7^{ZS97}$, $Hd1^{ZS97}ghd7^{ZS97}$, $hd1^{MY46}Ghd7^{MY46}$, and $Hd1^{ZS97}Ghd7^{MY46}$, respectively, was tested in Hangzhou in 2018 under NLD conditions. Both the *Hd1* and *Ghd7*, as well as their interaction, had highly significant effects ($p < 0.0001$) on HD (Table 1), which were similar to those observed previously under NLD condition. $Hd1^{ZS97}$ promoted and repressed flowering in the $Ghd7^{MY46}$ and $ghd7^{ZS97}$ backgrounds, respectively, while $Ghd7^{MY46}$ delayed flowering regardless of the *Hd1* function (Figure 2e).

GY and five yield components traits were also measured in the R2-NIL population. *Hd1* showed significant effects on SF ($p < 0.0001$) and GY ($p < 0.01$), and *Ghd7* exhibited highly significant effects on NSP, NGP, SF and TGW ($p < 0.0001$) (Table 1). Highly significant epistatic effects of the two genes were detected on all the traits except NGP ($p < 0.0001$). For NSP and TGW, the interactions acted for increasing the values of the recombinant types, which were consistent with the epistasis on HD. For NP, SF, and GY, the opposite direction was found. The relationships between HD and the yield traits were further analyzed (Table 2). Lines of $Hd1^{ZS97}ghd7^{ZS97}$ had the shortest HD, followed by $hd1^{MY46}ghd7^{ZS97}$, $hd1^{MY46}Ghd7^{MY46}$, and $Hd1^{ZS97}Ghd7^{MY46}$. Significant differences were detected for all the yield traits among the four genotypic groups. NSP, NGP and TGW were positively correlated with HD ($p < 0.05$), having *r* values of 0.806, 0.507 and 0.672, respectively (Table S1). Values of these traits increased with delayed heading. On the other hand, SF and NP were negatively correlated with HD ($p < 0.05$), having *r* values of −0.855 and −0.349, respectively. SF decreased with delayed heading; compared with lines having the shortest HD, SF in lines having the third longest, the second longest, and the longest HD decreased by 1.9%, 4.9% and 10.2%, respectively. Similar tendency was observed for NP. Consequently, lines in the $hd1^{MY46}Ghd7^{MY46}$ genotypic group having the second longest HD produced the highest GY.

3. Discussion

The bi-functional action of *Hd1* has been well recognized, promoting flowering under SD conditions and inhibiting flowering under LD conditions [27]. Recent studies revealed that flowering repressing function of *Hd1* is dependent on *Ghd7* [6,7]. In the present study, this relationship between *Hd1* and *Ghd7* was confirmed. Under NSD conditions, *Hd1* always up-regulated expressions of the two florigen genes (Figure 3) and promoted flowering regardless of *Ghd7* genotype (Figure 2). Under NLD conditions, *Hd1* still promoted flowering (Figure 2) by up-regulating florigen genes in the *ghd7* background (Figure 3). In the *Ghd7* background, however, *Hd1* was found to up-regulate *Ghd7*, and down-regulate *Ehd1* and florigen genes, consequently leading to late flowering. For *Ghd7*, its flowering-repressor action was observed under both NSD and NLD conditions regardless of *Hd1* function. Taken together, our results suggest that *Hd1* and *Ghd7* could promote and repress flowering independently, whereas flowering-repressor function of *Hd1* under LD conditions requires the functional *Ghd7*.

Among the four homozygous genotypic combinations of *Hd1* and *Ghd7*, the *Hd1ghd7* group exhibited the shortest HD under NLD conditions. Compared to *Hd1ghd7*, heading was delayed by 3.4–4.3 d and 7.5–8.7 d in the *hd1ghd7* and *hd1Ghd7* groups, respectively. Strikingly, HD in the *Hd1Ghd7* group was delayed by 16.1–20.9 d, owing to the genetic interaction between *Hd1* and *Ghd7* under NLD condition. This is likely the reason the *Hd1Ghd7* genotype was hardly carried by early season *indica* cultivars grown in middle-lower regions of the Yangtze River and South China regions [28] and *japonica* cultivars in northeast China [29], where early flowering is essential to ensure sufficient grown period for late season *indica* cultivars or secure a harvest before cold weather approaches.

It is generally accepted that long growth duration is associated with high-yielding production in rice [29,30], if varieties are harvested before cold weather approaches. A larger number of HD genes were found to have pleiotropic effects on yield traits, and their late-flowering alleles were frequently used to enhance grain yield mainly by increasing spikelet number and partially by increasing grain weight [10–20]. As expected, NSP and TGW gradually increased with delayed flowering under both the NSD and NLD conditions in this study. However, SF and NP tended to decrease under NLD conditions when the HD has become relatively long. As a consequence of trade-off among different yield components, rice lines having the *hd1Ghd7* genotype which had the second longest HD produced the highest grain yield, rather than the lines having the *Hd1Ghd7* genotype which had the longest HD. These results indicate that longer growth duration for a more use of available temperature and light does not always result in higher grain production.

Spikelet sterility is a key determinant of grain yield and frequently used as an indicator for stress tolerance. Two alternative explanations could be given to the decrease of spikelet sterility with delayed flowering. Firstly, alteration of time of flowering causes some loss of seasonal adaptability of rice. Secondly, *Ghd7* and *Hd1* participate in the stress tolerance of rice. *Ghd7* has been found to respond to multiple abiotic stress, such as high temperature, low temperature, and drought. Moreover, overexpression of *Ghd7* increases drought sensitivity, whereas knock-down of *Ghd7* enhances drought tolerance [21]. Our study showed that *Ghd7* expression was dramatically up-regulated in the *Hd1* background. This may be a reason that caused low SF in lines of *Hd1Ghd7*. Moreover, alteration of SF by *Hd1* was also observed in the *ghd7* background (Table 2), suggesting *Hd1* could be involved in stress response independently.

Panicle number is generally recognized as an unstable trait among yield traits. Few genes were reported to have pleiotropic effects on flowering time and panicle number [21,31,32]. *Ghd7* is found to regulate panicle number in a density-dependent manner. It decreases and increases panicle number at normal field condition and low-density conditions, respectively, though it always suppresses flowering time [21]. In the NIL populations used in our study, negative correlation between NP and HD was detected in both trials conducted under NLD conditions at normal planting density (Table 2, Table S1). The lines of *Hd1Ghd7* with the longest HD always produced the least NP (Table 2), indicating that combination of *Hd1* and *Ghd7* could cause decrease of panicle number under NLD conditions.

Although late-flowering alleles of flowering genes generally increase spikelet number, their influences on panicle number and spikelet sterility are not necessarily positive. Thus, an optimum HD genes combination needs to be carefully selected for maximizing grain yield in rice. In the present study, lines carried *hd1* and *Ghd7* alleles from MY46 produced the highest grain yield in both trials conducted in Hangzhou (Table 2) where is in the middle-lower region of the Yangtze River. Among the 14 middle-season *indica* rice cultivars tested by Wei et al [28], MY46 is one the 10 cultivars having the combination of non-functional *hd1* and functional *Ghd7*. These indicate that this combination could have undergone intensive artificial selection and play a significant role in the adaption of middle-season rice.

4. Materials and Methods

4.1. Plant Material

Three rice populations segregating at both the *Hd1* and *Ghd7* loci were used in this study. The developing process was illustrated in Figure 1 and described below. One F$_9$ plant of ZS97/MY46 was crossed with MY46 for two generations. Two BC$_2$F$_1$ plants which were heterozygous at both the *Hd1* and *Ghd7* loci were identified and selfed. In one of the two BC$_2$F$_2$ populations produced, a plant which was heterozygous for both the genes was identified and selfed. The resultant BC$_2$F$_3$ population was assayed with functional or closely linked DNA markers for the two genes. A total of 49 plants which were homozygous at *Hd1* and/or *Ghd7* loci were identified and selfed. One NIL population namely R1-NIL, comprising all the four homozygous genotypic combinations of *Hd1* and *Ghd7*, was constructed.

Another BC$_2$F$_2$ population was advanced to the BC$_2$F$_4$ generation. A BC$_2$F$_4$ plant which was heterozygous for both the genes was identified. In the resultant BC$_2$F$_5$ population, plants which were heterozygous for both the genes were selected and selfed. A NIL-F$_2$ population in the BC$_2$F$_6$ generation, namely R2-F$_2$ population, was constructed. A total of 114 plants which were homozygous at *Hd1* and/or *Ghd7* loci were selected and selfed. One NIL population namely R2-NIL, which consisted of all the four homozygous genotypic groups, was constructed.

4.2. Field Experiments and Phenotyping

The rice populations were tested in the experimental stations of the China National Rice Research Institute located at either Hangzhou or Lingshui. During the period of floral transition in the rice materials tested, day length in Hangzhou and Lingshui were corresponding to NLD and NSD conditions, respectively [14]. In all the trials, the planting density was 16.7 cm × 26.7 cm. Field management followed the normal agricultural practice. For NIL sets, the experiments followed a randomized complete block design with two replications. In each replication, one line was grown in a single row of ten plants. HD was recorded for each plant. At maturity, five middle plants in each row were harvested in bulk and measured for six yield traits, including NP, NSP, NGP, SF (%), TGW (g) and GY (g). Of which TGW was evaluated using fully filled grain followed the procedure reported by Zhang et al. [33].

4.3. DNA Marker Genotyping and Quantitative Real-time PCR Analysis

For population development and QTL mapping, total DNA was extracted using 2 cm-long leaf sample following the method of Zheng et al. [34]. PCR amplification was performed according to Chen et al. [35]. The products were visualized on 6% non-denaturing polyacrylamide gels using silver staining or on 2% agarose gels using Gelred staining. Three DNA markers were used, including functional marker Si9337 for *Hd1*, functional marker Se9153 and closely linked marker RM5436 for *Ghd7* [10,17].

For expression analysis, penultimate leaves of rice lines in the R1-NIL population were harvested at 7:00 am in 17HZ and 9:00 am in 17LS, 2 h after sunrise. Total RNA was extracted using RNeasy Plus Mini Kit (QIAGEN, Hilden, German). First-strand cDNA was synthesized using ReverTra AceR Kit (Toyobo, Osaka, Japan). Quantitative real-time PCR was performed on Applied Biosystems 7500 using SYBR qPCR Mix Kit (Toyobo, Osaka, Japan) according to the manufacturer's instructions. *Actin1* was used as the endogenous control. The data were analyzed according to the $2^{-\Delta Ct}$ method. Three biological replicates and three technical replicates were used. The primers were selected from previous studies [10,20,36].

4.4. Data Analysis

For the NIL-F$_2$ population, QTL analysis was performed with single marker analysis in Windows QTL Cartgrapher 2.5 [37]. For the NIL populations, two-way ANOVA was conducted to test the main and epistatic effects. Duncan's multiple range test was used to examine the phenotypic differences among genotypic groups. The analysis was performed using the SAS procedure GLM [38].

Supplementary Materials: Supplementary materials can be found at http://www.mdpi.com/1422-0067/20/3/516/s1.

Author Contributions: J.-Y.Z. conceived and designed the experiments. Z.-H.Z. and Y.Y.F. performed laboratory experiments. Z.-H.Z., Y.-J.Z. and S.-L.W. performed the field experiments. Z.-H.Z. and J.-Y.Z. analyzed the data and drafted the manuscript. Z.-H.Z. and J.-Y.Z. revised the manuscript. All authors read and approved the final manuscript.

Funding: This work was supported by the National Natural Science Foundation of China (31571637).

Conflicts of Interest: The authors declare no conflict of interest.

Abbreviations

NSDs	Natural short-day conditions
NLDs	Natural long-day conditions
SD	Short-day conditions
LD	Long-day conditions
NIL	Near isogenic lines
ZS97	Zhenshan 97
MY46	Milyang 46
HD	Heading date
QTL	Quantitative trait locus
R^2	The proportion of phenotypic variance explained
GY	Grain yield per plant
NP	Number of panicles per plant
NSP	Number of spikelets per panicle
NGP	Number of grains per panicle
SF	Spikelet fertility
TGW	1000-grain weight
r	Correlation coefficient

References

1. Tamaki, S.; Matsuo, S.; Wong, H.L.; Yokoi, S.; Shimamoto, K. Hd3a protein is a mobile flowering signal in rice. *Science* **2007**, *316*, 1033–1036. [CrossRef] [PubMed]
2. Komiya, R.; Yokoi, S.; Shimamoto, K. A gene network for long-day flowering activates *RFT1* encoding a mobile flowering signal in rice. *Development* **2009**, *136*, 3443–3450. [CrossRef] [PubMed]
3. Hori, K.; Matsubara, K.; Yano, M. Genetic control of flowering time in rice: Integration of Mendelian genetics and genomics. *Theor. Appl. Genet.* **2016**, *129*, 2241–2252. [CrossRef] [PubMed]
4. Izawa, T.; Oikawa, T.; Sugiyama, N.; Tanisaka, T.; Yano, M.; Shimamoto, K. Phytochrome mediates the external light signal to repress *FT* orthologs in photoperiodic flowering of rice. *Genes Dev.* **2002**, *16*, 2006–2020. [CrossRef] [PubMed]
5. Ishikawa, R.; Aoki, M.; Kurotani, K.; Yokoi, S.; Shinomura, T.; Takano, M.; Shimamoto, K. Phytochrome B regulates *Heading date 1* (*Hd1*)-mediated expression of rice florigen *Hd3a* and critical day length in rice. *Mol. Genet. Genomics* **2011**, *285*, 461–470. [CrossRef]
6. Nemoto, Y.; Nonoue, Y.; Yano, M.; Izawa, T. *Hd1*, a CONSTANS ortholog in rice, functions as an *Ehd1* repressor through interaction with monocot-specific CCT-domain protein Ghd7. *Plant J.* **2016**, *86*, 221–233. [CrossRef]
7. Du, A.; Tian, W.; Wei, M.; Yan, W.; He, H.; Zhou, D.; Huang, X.; Li, S.; Ouyang, X. The DTH8-Hd1 module mediates day-length-dependent regulation of rice flowering. *Mol. Plant* **2017**, *10*, 948–961. [CrossRef]
8. Zhang, Z.; Hu, W.; Shen, G.; Liu, H.; Hu, Y.; Zhou, X.; Liu, T.; Xing, Y. Alternative functions of Hd1 in repressing or promoting heading are determined by Ghd7 status under long-day conditions. *Sci. Rep.* **2017**, *7*, 5388. [CrossRef]
9. Doi, K.; Izawa, T.; Fuse, T.; Yamanouchi, U.; Kubo, T.; Shimatani, Z.; Yano, M.; Yoshimura, A. *Ehd1*, a B-type response regulator in rice, confers short-day promotion of flowering and controls *FT-like* gene expression independently of *Hd1*. *Genes Dev.* **2004**, *18*, 926–936. [CrossRef]

10. Xue, W.; Xing, Y.; Weng, X.; Zhao, Y.; Tang, W.; Wang, L.; Zhou, H.; Yu, S.; Xu, C.; Li, X.; et al. Natural variation in *Ghd7* is an important regulator of heading date and yield potential in rice. *Nat. Genet.* **2008**, *40*, 761–767. [CrossRef]

11. Wei, X.; Xu, J.; Guo, H.; Jiang, L.; Chen, S.; Yu, C.; Zhou, Z.; Hu, P.; Zhai, H.; Wan, J. *DTH8* suppresses flowering in rice, influencing plant height and yield potential simultaneously. *Plant Physiol.* **2010**, *153*, 1747–1758. [CrossRef] [PubMed]

12. Yan, W.H.; Wang, P.; Chen, H.X.; Zhou, H.J.; Li, Q.P.; Wang, C.R.; Ding, Z.H.; Zhang, Y.S.; Yu, S.B.; Xing, Y.Z.; et al. A major QTL, *Ghd8*, plays pleiotropic roles in regulating grain productivity, plant height, and heading date in rice. *Mol. Plant* **2011**, *4*, 319–330. [CrossRef] [PubMed]

13. Endo-Higashi, N.; Izawa, T. Flowering time genes *Heading date 1* and *Early heading date 1* together control panicle development in rice. *Plant Cell Physiol.* **2011**, *52*, 1083–1094. [CrossRef] [PubMed]

14. Zhang, Z.-H.; Wang, K.; Guo, L.; Zhu, Y.-J.; Fan, Y.-Y.; Cheng, S.-H.; Zhuang, J.-Y. Pleiotropism of the photoperiod-insensitive allele of *Hd1* on heading date, plant height and yield traits in rice. *PLoS ONE* **2012**, *7*, e52538. [CrossRef] [PubMed]

15. Yan, W.; Liu, H.; Zhou, X.; Li, Q.; Zhang, J.; Lu, L.; Liu, T.; Liu, H.; Zhang, C.; Zhang, Z.; et al. Natural variation in *Ghd7.1* plays an important role in grain yield and adaptation in rice. *Cell Res.* **2013**, *23*, 969–971. [CrossRef] [PubMed]

16. Gao, H.; Jin, M.; Zheng, X.-M.; Chen, J.; Yuan, D.; Xin, Y.; Wang, M.; Huang, D.; Zhang, Z.; Zhou, K.; et al. *Days to heading 7*, a major quantitative locus determining photoperiod sensitivity and regional adaptation in rice. *Proc. Natl. Acad. Sci. USA* **2014**, *111*, 16337–16342. [CrossRef] [PubMed]

17. Zhang, Z.-H.; Cao, L.-Y.; Chen, J.-Y.; Zhang, Y.-X.; Zhuang, J.-Y.; Cheng, S.-H. Effects of *Hd2* in the presence of the photoperiod-insensitive functional allele of *Hd1* in rice. *Biol. Open* **2016**, *5*, 1719–1726. [CrossRef] [PubMed]

18. Zhu, Y.-J.; Fan, Y.-Y.; Wang, K.; Huang, D.-R.; Liu, W.-Z.; Ying, J.-Z.; Zhuang, J.-Y. *Rice Flowering Locus T 1* plays an important role in heading date influencing yield traits in rice. *Sci. Rep.* **2017**, *7*, 4918. [CrossRef]

19. Chen, J.-Y.; Guo, L.; Ma, H.; Chen, Y.-Y.; Zhang, H.-W.; Ying, J.-Z.; Zhuang, J.-Y. Fine mapping of *qHd1*, a minor heading date QTL with pleiotropism for yield traits in rice (*Oryza sativa* L.). *Theor. Appl. Genet.* **2014**, *127*, 2515–2524. [CrossRef]

20. Chen, J.-Y.; Zhang, H.-W.; Zhang, H.-L.; Ying, J.-Z.; Ma, L.-Y.; Zhuang, J.-Y. Natural variation at *qHd1* affects heading date acceleration at high temperatures with pleiotropism for yield traits in rice. *BMC Plant Biol.* **2018**, *18*, 112. [CrossRef]

21. Weng, X.; Wang, L.; Wang, J.; Hu, Y.; Du, H.; Xu, C.; Xing, Y.; Li, X.; Xiao, J.; Zhang, Q. *Grain number, plant height, and heading date7* is a central regulator of growth, development, and stress response. *Plant Physiol.* **2014**, *164*, 735–747. [CrossRef] [PubMed]

22. Zhang, C.; Liu, J.; Zhao, T.; Gomez, A.; Li, C.; Yu, C.; Li, H.; Lin, J.; Yang, Y.; Liu, B.; et al. A drought-inducible transcription factor delays reproductive timing in rice. *Plant Physiol.* **2016**, *171*, 334–343. [CrossRef] [PubMed]

23. Liu, J.; Shen, J.; Xu, Y.; Li, X.; Xiao, J.; Xiong, L. *Ghd2*, a *CONSTANS*-like gene, confers drought sensitivity through regulation of senescence in rice. *J. Exp. Bot.* **2016**, *67*, 5785–5798. [CrossRef] [PubMed]

24. Su, L.; Shan, J.X.; Gao, J.P.; Lin, H.X. OsHAL3, a blue light-responsive protein, interacts with the floral regulator Hd1 to activate flowering in rice. *Mol. Plant* **2016**, *9*, 233–244. [CrossRef] [PubMed]

25. Minh-Thu, P.T.; Kim, J.S.; Chae, S.; Jun, K.M.; Lee, G.S.; Kim, D.E.; Cheong, J.J.; Song, S.I.; Nahm, B.H.; Kim, Y.K. A WUSCHEL homeobox transcription factor, OsWOX13, enhances drought tolerance and triggers early flowering in rice. *Mol. Cells* **2018**, *41*, 781–798. [PubMed]

26. Song, T.; Zhang, Q.; Wang, H.; Han, J.; Xu, Z.; Yan, S.; Zhu, Z. *OsJMJ703*, a rice histone demethylase gene, plays key roles in plant development and responds to drought stress. *Plant Physiol. Biochem.* **2018**, *132*, 183–188. [CrossRef]

27. Yano, M.; Katayose, Y.; Ashikari, M.; Yamanouchi, U.; Monna, L.; Fuse, T.; Baba, T.; Yamamoto, K.; Umehara, Y.; Nagamura, Y.; et al. *Hd1*, a major photoperiod sensitivity quantitative trait locus in rice, is closely related to the Arabidopsis flowering time gene *CONSTANS*. *Plant Cell* **2000**, *12*, 2473–2484. [CrossRef]

28. Wei, X.-J.; Xu, J.-F.; Jiang, L.; Wang, J.-J.; Zhou, Z.-L.; Zhai, H.-Q.; Wan, J.-M. Genetic analysis for the diversity of heading date of cultivated rice in China. *Acta Agron. Sin.* **2012**, *38*, 10–12. [CrossRef]

29. Ye, J.; Niu, X.; Yang, Y.; Wang, S.; Xu, Q.; Yuan, X.; Yu, H.; Wang, Y.; Wang, S.; Feng, Y.; et al. Divergent *Hd1*, *Ghd7*, and *DTH7* alleles control heading date and yield potential of *japonica* rice in northeast China. *Front. Plant Sci.* **2018**, *9*, 35. [CrossRef]

30. Hu, Y.; Li, S.; Xing, Y. Lessons from natural variations: Artificially induced heading date variations for improvement of regional adaptation in rice. *Theor. Appl. Genet.* **2018**. [CrossRef]

31. Wang, Q.; Zhang, W.; Yin, Z.; Wen, C.K. Rice CONSTITUTIVE TRIPLE-RESPONSE2 is involved in the ethylene-receptor signalling and regulation of various aspects of rice growth and development. *J. Exp. Bot.* **2013**, *64*, 4863–4875. [CrossRef] [PubMed]

32. Xu, Q.; Saito, H.; Hirose, I.; Katsura, K.; Yoshitake, Y.; Yokoo, T.; Tsukiyama, T.; Teraishi, M.; Tanisaka, T.; Okumoto, Y. The effects of the photoperiod-insensitive alleles, *se13*, *hd1* and *ghd7*, on yield components in rice. *Mol. Breed.* **2014**, *33*, 813–819. [CrossRef] [PubMed]

33. Zhang, H.-W.; Fan, Y.-Y.; Zhu, Y.-J.; Chen, J.-Y.; Yu, S.-B.; Zhuang, J.-Y. Dissection of the *qTGW1.1* region into two tightly-linked minor QTLs having stable effects for grain weight in rice. *BMC Genet.* **2016**, *17*, 98–107. [CrossRef] [PubMed]

34. Zheng, K.L.; Huang, N.; Bennett, J.; Khush, G.S. *PCR-Based Marker-Assisted Selection in Rice Breeding*; IRRI Discussion Paper Series No. 12; International Rice Research Institute: Los Banos, CA, USA, 1995.

35. Chen, X.; Temnykh, S.; Xu, Y.; Cho, Y.G.; McCouch, S.R. Development of a microsatellite framework map providing genome-wide coverage in rice. *Theor. Appl. Genet.* **1997**, *95*, 553–567. [CrossRef]

36. Dong, Q.; Zhang, Z.-H.; Wang, L.-L.; Zhu, Y.-J.; Fan, Y.-Y.; Mou, T.-M.; Ma, L.-Y.; Zhuang, J.-Y. Dissection and fine-mapping of two QTL for grain size linked in a 460-kb region on chromosome 1 of rice. *Rice* **2018**, *11*, 44. [CrossRef] [PubMed]

37. Wang, S.; Basten, C.J.; Zeng, Z.-B. *Windows QTL Cartographer 2.5*; Department of Statistics, North Carolina State University: Raleigh, NC, USA, 2012.

38. SAS Institute Inc. *SAS/STAT User's Guide*; SAS Institute: Cary, NC, USA, 1999.

Article

Expression of Maize MADS Transcription Factor *ZmES22* Negatively Modulates Starch Accumulation in Rice Endosperm

Kangyong Zha †, Haoxun Xie †, Min Ge, Zimeng Wang, Yu Wang, Weina Si and Longjiang Gu *

National Engineering Laboratory of Crop Stress Resistance breeding, Anhui Agricultural University, Hefei 230036, China; kangyongzha929@163.com (K.Z.); xhx521xz@163.com (H.X.); 13215616815@163.com (M.G.); wzmhanchang@163.com (Z.W.); wangyu20180712@163.com (Y.W.); weinasi@ahau.edu.cn (W.S.)
* Correspondence: longjianggu@163.com; Tel.: +86-0551-65786021
† These authors contributed equally to this work.

Received: 31 December 2018; Accepted: 17 January 2019; Published: 23 January 2019

Abstract: As major component in cereals grains, starch has been one of the most important carbohydrate consumed by a majority of world's population. However, the molecular mechanism for regulation of biosynthesis of starch remains elusive. In the present study, *ZmES22*, encoding a MADS-type transcription factor, was modestly characterized from maize inbred line B73. *ZmES22* exhibited high expression level in endosperm at 10 days after pollination (DAP) and peaked in endosperm at 20 DAP, indicating that *ZmES22* was preferentially expressed in maize endosperm during active starch synthesis. Transient expression of *ZmES22* in tobacco leaf revealed that ZmES22 protein located in nucleus. No transactivation activity could be detected for ZmES22 protein via yeast one-hybrid assay. Transformation of overexpressing plasmid 35S::*ZmES22* into rice remarkedly reduced 1000-grain weight as well as the total starch content, while the soluble sugar was significantly higher in transgenic rice lines. Moreover, overexpressing *ZmES22* reduced fractions of long branched starch. Scanning electron microscopy images of transverse sections of rice grains revealed that altered expression of *ZmES22* also changed the morphology of starch granule from densely packed, polyhedral starch granules into loosely packed, spherical granules with larger spaces. Furthermore, RNA-seq results indicated that overexpressing *ZmES22* could significantly influence mRNA expression levels of numerous key regulatory genes in starch synthesis pathway. Y1H assay illustrated that ZmES22 protein could bind to the promoter region of *OsGIF1* and downregulate its mRNA expression during rice grain filling stages. These findings suggest that *ZmES22* was a novel regulator during starch synthesis process in rice endosperm.

Keywords: *Zea mays* L.; MADS transcription factor; *ZmES22*; starch

---

1. Introduction

Maize (*Zea mays* L.) is one of the most widely grown crop world-wide, as well as a critical model for various biological researches, especially for endosperm development [1]. Starch is the major component of maize grains, which accounted up to 71% on a dry weight basis. Therefore, comprehensive understanding of the molecular mechanism for regulation of starch synthesis will facilitate increase in yield to feed growing population.

Starch is composed of two major components, known as amylose and amylopectin. The process of starch biosynthesis has been reported to be under finely regulated by numerous genes, which mainly encoded multiple subunits or isoforms of four enzymes: ADP-glucose pyrophosphorylase (AGPase), starch synthase (SS), starch branching enzyme (SBE), and starch debranching enzyme (DBE) [2,3]. At the initial stage of starch synthesis, glucose-1-phosphate, together with ATP, are converted to

ADP-glucose (ADPG) via AGPase. In the developing endosperm, ADPG is mainly produced in the cytosol and transferred into amyloplast through an adenylate translocator, BT1 [4]. Afterwards, the synthesis of starch is furthered by chain elongation by transferring ADPG to the nonreducing end of a glucan primer. The amylose chain elongation is completed by granule-bound starch synthase I (GBSSI), whereas, amylopectin chains are elongated by a soluble form of starch synthase (SSI, SSII, SSIII, and SSIV). α-1,6-Glucosidic linkages is then introduced by starch branching enzyme (BEI and BEII) and finally, fine structure of amylopectin is achieved through removal of unnecessary branches by starch debranching enzymes (ISA and Pullulanase). Mutants defective in any key genes exhibited apparent abnormal characters of starch in reserve organs. Mutations in *OsAGPL2*, one of the large subunits of AGPase, caused severe defects in grain filling and starch synthesis [5]. Loss-of-function mutations occurred in *OsBT1* gene, which encoded an ADPG translocator, resulted in a remarkable reduction in grain weight than wild type [4]. Deficiency of *OsSSIIa* lead to a chalky interior appearance and the endosperm of the mutant lines are mainly consisted of loosely packed, spherical starch granules with larger air spaces [5]. *Grain Incomplete Filling 1* (*OsGIF1*), encoding a cell-wall invertase, was of great importance in regulation of sucrose unloading from phloem into cells of reserve organs. Mutant lines of *OsGIF1* showed severe defects in grain filling and in turn reduced the grain weight to 70% of wild type rice at 30 days after pollination (DAP) [6].

Since starch biosynthesis and accumulation are critical determinants for both grain quality and production, key transcriptional regulators, including several transcription factors (TFs), have also been demonstrated to play an important regulatory role in starch synthesis. Null mutants of *OsBZIP58* seeds exhibited altered starch composition as well as morphological defects with apparent white belly region [7]. SUSIBA2, a WRKY family transcription factor, could directly bind to the promoter of *pISA1* gene to regulate its expression, thus affecting the synthesis of starch in barley [8]. Additionally, one of AP2 family of transcription factors, SERF1, negatively regulates rice grain filling, and genetic mutations could enhance the starch synthesis process of rice [9]. *ZmbZIP91* was proved to be a key regulator of the starch synthesis by directly binding to ACTCAT elements in the promoters of starch synthesis genes [10]. The inhibition of *ZmDof3* led to defects of the kernel phenotype with decreased starch content and a partially patchy aleurone layer [1]. Altered expression of transcription factors, causing abnormal features in reserve organs, could provide profound implications in understanding the molecular mechanisms that control starch biosynthesis. Despite these research highlights, a comprehensive understanding of factors that regulate the expression of genes in network of starch synthesis remains largely unknown, especially in maize. Hence, screening and identification of key transcription factors involved in starch synthesis will be of great importance in breeding of high-yielding crops.

In previous studies, a total of 2298 transcription factors were identified and further examined using RNA-seq dataset from 18 representative tissues from maize [11], which provided profound clues regarding to the relationship between development and dynamic expression profiles of key transcription factors. With an emphasis on endosperm-specificity, we identified 36 transcription factors that were preferentially highly expressed in maize endosperm [12]. The mRNA expression profiles of one gene, encoding a typical MADS transcription factor (GRMZM2G159397, designated as *ZmES22*), were further confirmed via qRT-PCR assays. To test if this gene was related to starch synthesis, *ZmES22* was cloned from maize inbred line B73. Afterwards, molecular properties and biological functions were modestly comprehensively characterized in transgenic rice lines. Overexpressing *ZmES22* in rice significantly reduced 1000-grain weight as well as hindered starch accumulation. Besides, altered expression of *ZmES22* in transgenic rice also changed the starch structure and morphology of starch granules. Furthermore, RNA-seq analysis demonstrated that numerous key regulatory genes in starch synthesis were differentially expressed compared to that in WT plants. Yeast one hybrid assay revealed that ZmES22 could bind to the promoter of *OsGIF1* and downregulated its expression during grain filling process. This study illustrated that *ZmES22* could be a newfound transcription factor, which negatively regulated starch synthesis in rice endosperm.

2. Results

2.1. Sequence Analysis and Construction of Phylogenetic Tree for ZmES22 Homologues

As one of the largest transcription factor family in eukaryote, MADS-box proteins has been characterized by its important roles in a variety of aspects during plant growth and development [13,14]. To test if *ZmES22* were related to starch synthesis, this gene was firstly cloned from maize inbred line B73. *ZmES22* contained an open reading frame (ORF) of 723 bp and encoded a protein of 240 amino acids with a predicted molecular weight (Mw) of 27,903 Da and an isoelectric point (pI) of 8.92. Pfam analysis of ZmES22 revealed that the deduced protein sequence consisted of four conserved domains, namely the MADS-box domain (MADS-box), intervening (I), K-box domain, and the C terminal domain (Figure S1). In order to find homologs of *ZmES2*, blastp program was explored for protein sequence of ZmES22 to search against protein database for *Zea mays*, *Oryza sativa*, and *Arabidopsis thaliana*, respectively. Afterwards, pairwise amino acid distances were calculated using MEGA7 with Jones–Taylor–Thornton (JTT) model, and genes with diversity less than 0.8 were retained according to empirical experience. A total of 16 genes, including 6 from *Zea mays*, 5 from *Oryza sativa*, and 5 from *Arabidopsis thaliana* were identified, respectively (Tables S1 and S2). Phylogenetic tree was constructed using the conserved MADS domain, and clear orthologous relationship could be observed between *ZmES22* and *OsMADS7* (Figure 1 and Figure S2). Furthermore, the Multiple EM for Motif Elicitation (MEME) motif website search program was explored to identify the conserved motifs for all 16 homologues. Great majority homologues contained Motif 1, Motif 2, Motif 3, and Motif 4, indicating these motifs were probably evolutionary conserved (Figure 1). While, presence or absence for remained motifs was more variable.

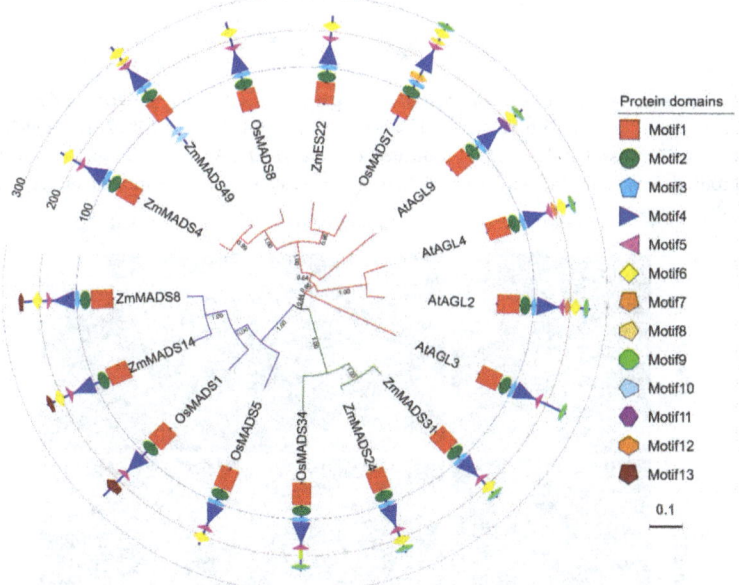

Figure 1. Phylogenetic tree of *ZmES22* homologous genes from maize, rice and Arabidopsis. Phylogenic tree of homologous genes of *ZmES22* from maize, rice and Arabidopsis MADS proteins, which was constructed using conserved MADS domain with MEGA7 software via Neighbor-joining method. Bootstrap value was indicated at each branch point. Gene IDs and predicted functions are listed in Supplementary Table S1.

2.2. Expression Profiles and Subcellular Localization of ZmES22

qRT-PCR assays were performed to investigate expression profiles of *ZmES22*. In line with previous transcriptome analysis, compared with nutritive organs, such as root, stem and leaf, *ZmES22* exhibited higher relative expression levels in reproductive organs (Figure 2). Intriguingly, significantly higher expression level of *ZmES22* was observed in endosperm than embryo at 10 DAP, and mRNA expression of *ZmES22* peaked in endosperm at 20 DAP. To ascertain the location of *ZmES22* protein, coding sequence of *ZmES22* was inserted into empty vector 35S::GFP. Afterwards, 35S::*ZmES22*-GFP construct and 35S::GFP were efficiently transfected tobacco leaf cells separately via *Agrobacterium* infiltration (Figure 3). Green fluorescence of 35S::GFP could be observed throughout the cell, whereas, green fluorescence of ZmES22-GFP fusion protein appeared only in nucleus (Figure 3), illustrating that ZmES22 protein functioned in nucleus. However, yeast one-hybrid assay demonstrated that ZmES22 protein did not have transcriptional activity in yeast cells (Figure S3). These results indicated that *ZmES22* may be involved in the regulation of development of endosperm with the help of other proteins.

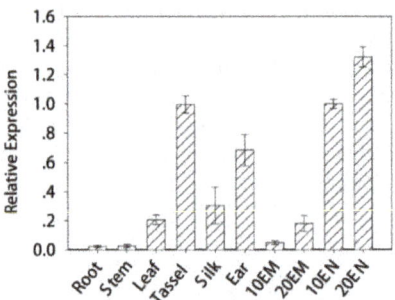

Figure 2. Expression pattern of ZmES22 across diverse tissues. Expression patterns of *ZmES22* in root, stem, leave, tassel, silk, ear, embryo and endosperm was quantified via qRT-PCR. The developmental stage of the embryo and endosperm is indicated by 10 and 20 DAP. Maize *Actin1* was used as the internal control. Error bars are standard deviations of three technical repeats and two biological repeats.

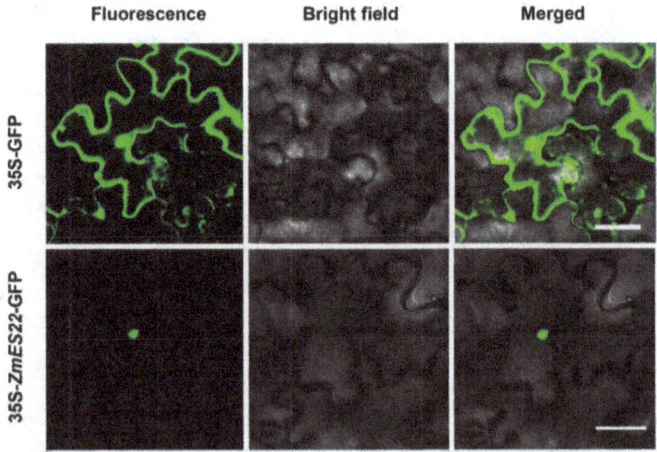

Figure 3. Subcellular localization of ZmES22 in tobacco. The 35S::*ZmES22*-GFP fusion construct and 35S::GFP vector were transiently expressed in tobacco epidermal cells and examined by a confocal laser scanning biological microscope, respectively. Bars = 50 μm.

2.3. Analysis of Agronomic Characters of ZmES22 Overexpression Transgenic Rice

To illustrate the function of *ZmES22*, twelve independent rice lines, which overexpressed *ZmES22* under the drive of CaMV 35S promoter, were obtained via *Agrobacterium* mediated transformation. qRT-PCR assays revealed that *ZmES22* expressed at distinct levels in transgenic rice lines, among which L8, L9, and L10 exhibited significantly higher expression (Figure S4). Therefore, these three transgenic rice lines was selected for further research. Compared to wild type (WT) plants, overexpression rice lines exhibited no visible difference during both the vegetative and reproductive stages, with similar plant height as well as panicle architecture (Figure S5). After maturation, agronomic traits, including grain length, grain width, grain thickness, and 1000-grain weight were minutely characterized for both transgenic rice lines and WT plants. There was no significant change in either grain length or grain width between transgenic plants and WT plants (Figure 4A,B). Nevertheless, grain thickness was dramatically decreased in overexpressed rice lines (Figure 4C, Student's *t*-test, *p*-value = 4.8×10^{-5}). Accordingly, 1000-grain weight of transgenic plants were significantly depleted by 3.88 g than that of WT plants (Figure 4D, Student's *t*-test, *p*-value = 1.8×10^{-8}). Additionally, total starch content, apparent amylose content (AAC) and soluble sugar content of both transgenic rice lines and WT plants were measured according to previously reported methods. Surprisingly, compared to WT plants, both total starch content and AAC were significantly reduced (Figure 5A,B, Student's *t*-test, *p*-value = 0.02), whereas, the content of soluble sugar in transgenic rice lines were significantly increased by 38% than that of WT plants (Figure 5C, Student's *t*-test, *p*-value = 8.4×10^{-4}). In particular, content of soluble sugar was two times larger in transgenic line L9 than that in WT plants. These results revealed that overexpression of *ZmES22* gene could significantly block starch biosynthesis process in endosperm of rice.

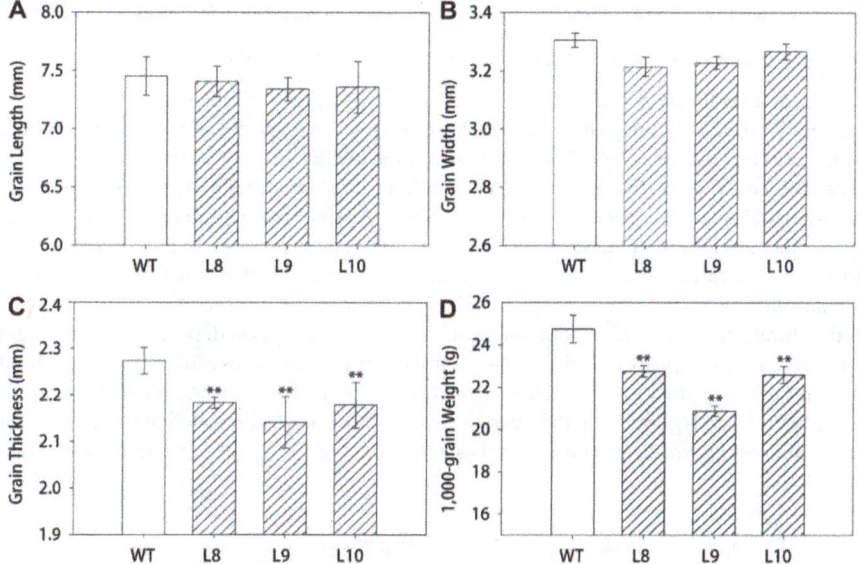

Figure 4. Agronomic characters of seeds from transgenic rice lines that overexpressed *ZmES22*. Grain agronomic characters including grain length (**A**), grain width (**B**), grain thickness (**C**), and 1000-grain weight (**D**) were minutely measured. Data are presented as mean ± SD of three replicates. L: transgenic lines of *ZmES22* seeds; WT: wild-type plants (Zhonghua 11), Student's *t*-test, ** *p*-value < 0.01.

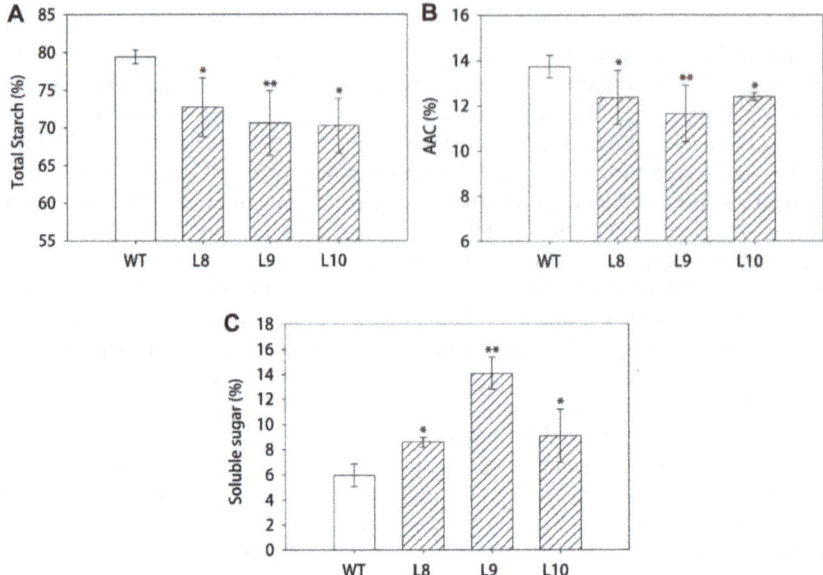

Figure 5. Overexpression of *ZmES22* in rice altered the starch composition. (**A**) Total starch content in rice endosperm. (**B**) Apparent amylose content (AAC) in rice endosperm. (**C**) Soluble sugar content in rice endosperm. Data are presented as mean ± SD of three replicates. L: transgenic lines of *ZmES22* seeds; WT: wild-type plants (Zhonghua 11), Student's *t*-test, * *p*-value < 0.05, ** *p*-value < 0.01.

2.4. Overexpression of ZmES22 Influences Starch Structure in Transgenic Rice

Both amylopectin blue value and the maximum absorption wavelength reflect the ability of amylopectin binding to iodine. Therefore, different BV and kmax can provide indicators for the basic distinction of starch structure [15]. To detect whether the relative content of amylose and amylopectin were altered by overexpression of *ZmES22*, BV and kmax of starch from both *ZmES22* overexpression rice seeds and WT plants were determined accordingly. As shown in Figure 6A,B, both BV and kmax of amylose and amylopectin in three transgenic lines were significantly smaller than that of WT plants (Student's *t*-test, *p*-value = 2.2×10^{-16}). Furthermore, morphology of starch granules was examined via scanning electron microscopy (SEM) [16]. SEM images of transverse sections of rice grains revealed that both central and dorsal endosperms were filled with densely packed, polyhedral starch granules in both transgenic rice and WT seeds, while ventral endosperm of transgenic rice seeds exhibited an apparent abnormity with a visible chalky region (Figure 6C), which was mainly consisted of loosely packed, spherical starch granules with larger air spaces. These results indicated that the overexpressing *ZmES22* could change starch structure as well as influence morphology of starch granules in transgenic rice lines.

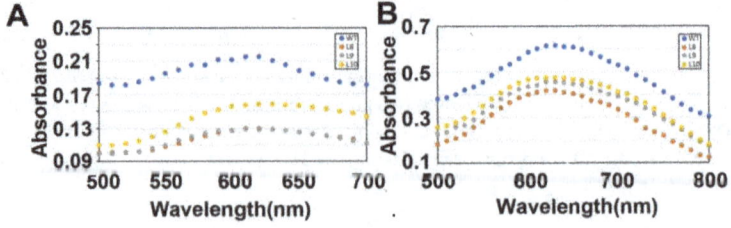

Figure 6. *Cont.*

C

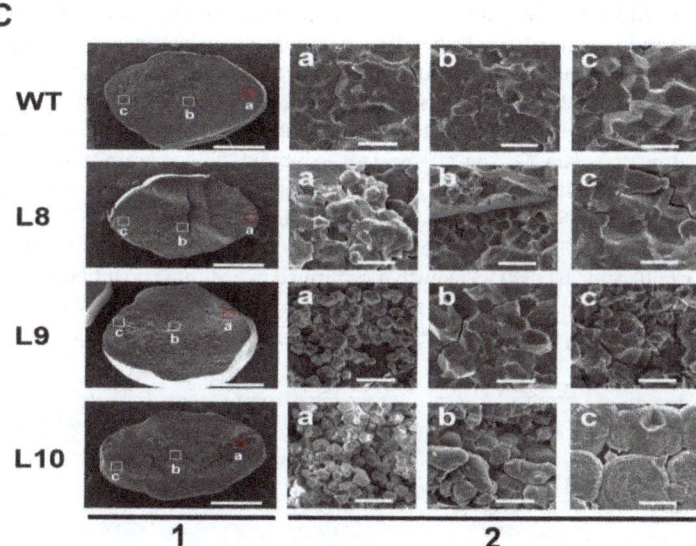

Figure 6. Blue value (BV), maximum absorbance (kmax) and scanning electron microscopy (SEM) images of the transverse sections of transgenic rice seeds. (**A**) BV at 600 nm and kmax represent the ability to combine with iodine. (**B**) BV at 680 nm and kmax represent the ability to combine with iodine. (**C**) Cross-sections of mature seeds are shown in (1). SEM of the ventral area of mature endosperm is shown in a of (2) and indicated by a red square in (1). Bars: 1 mm in (**1**); 10 μm in (2) a: dorsal; b: center; c: belly.

2.5. Overexpression of ZmES22 Influence Expression Profiles of Numerous Starch Synthesis Related Genes at 20 DAP Endosperm

To further explore the molecular basis of *ZmES22* in regulation of starch synthesis, expression profiles of 17 genes, which were preferentially expressed in developing endosperms and were demonstrated to be involved in starch synthesis, were compared between transgenic rice lines with WT plants at different developmental stages (3, 6, 10, and 20 DAP, Figure 7). The results illustrated that, except for *OsISA2*, *OsSSI*, and *OsSSIIa*, great majority of characterized starch synthesis related genes were downregulated depending on the individual genes when compared to WT plants (Figure 7). Interestingly, expression levels of *OsBEI* and *OsPUL* exhibited similar tendency that they were remarkably upregulated as grains got maturity (Figure 7). As is described previously, *ZmES22* was highly expressed in 20 DAP endosperm, therefore, we proposed that genes differentially expressed in transgenic rice plants at 20 DAP endosperm might be potentially key regulators. Nevertheless, no significant changes could be observed among all of tested genes at 20 DAP endosperms (Figure 7). In order to further investigate possible regulation by *ZmES22*, 20-DAP seeds for both overexpression rice lines and WT plants were collected for RNA-seq analysis, each was repeated with two biological replicates (Table 1, Figure 8A,B). Collectively, 1902 differentially expressed genes (DEGs), consisting of 986 upregulated and 916 downregulated genes in overexpression rice lines (Figure 8C), were determined with the following criteria:(1) the minimum fold-change of gene expression was 2.0; (2) the maximum adjusted *p* value was 0.05. In order to validate the RNA-seq data, 10 DEGs, including 5 upregulated and 5 downregulated genes, were randomly selected for quantitative real-time PCR analysis, and the results illustrated that RNA-seq data are of satisfactory quality (Figure 8D). To analyze the functional enrichments of the DEGs, both Gene Ontology (GO) and Kyoto Encyclopedia of Genes and Genomes (KEGG) analysis were performed using R package ClusterProfiler. DNA metabolic process, response to stress and carbohydrate metabolic process are the three mostly enriched GO items (Figure S6). Moreover, six pathways were significantly enriched in KEGG analysis (Figure 9), including starch and sucrose metabolism pathway (Figure S7), galactose metabolism, phenylalanine metabolism, plant hormone signal transduction pathway (Figure S8), etc. Interestingly, one gene, named *GIF1* (*Os04g0413500*), which was reported to be a key regulator to rice grain-filling and yield, was significantly enriched in carbohydrate metabolic process as well as starch and sucrose metabolism pathway in KEGG. The *gif1* mutant exhibited slower grain-filling rate and showed markedly more grain chalkiness than wild-type plants [6]. In the present study, *GIF1* gene was downregulated as much as 8-fold in overexpression rice lines compared with WT plants (Fisher's exact test, $p = 5.0 \times 10^{-5}$). The relative mRNA expression level of *OsGIF1* in four different developmental endosperms (3, 6, 10, and 20 DAP) were further confirmed by real-time quantitative PCR (Figure S9A). Because overexpression *ZmES22* lead to similar phenotype as *gif1* mutant, we therefore wonder if ZmES22 could bind to the promoter of *GIF1* and negatively regulate its expression?

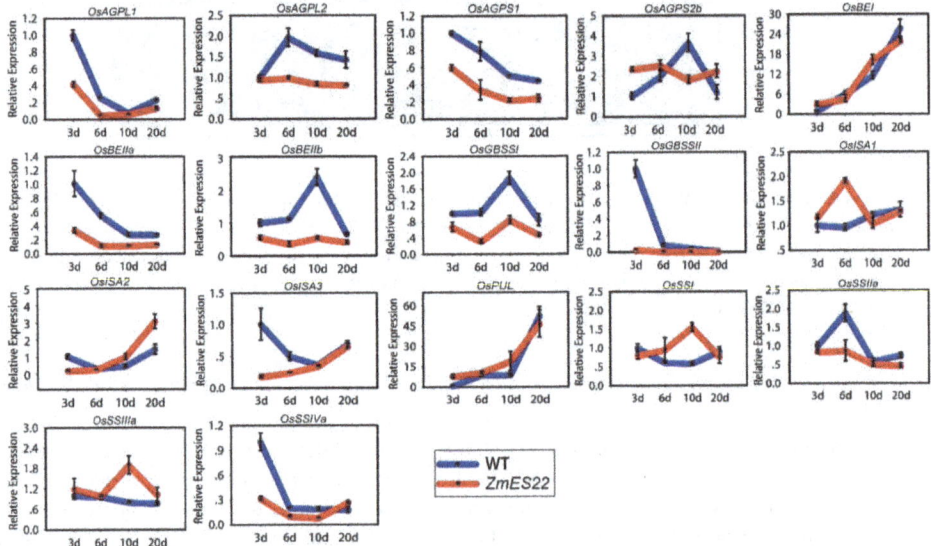

Figure 7. Expression profiles of 17 starch synthesis related genes across diverse developmental stages. Blue line represents wild type (Zhonghua 11), and red line denotes transgenic lines. d is short for days after pollination (DAP). The mRNA expression level of each gene in the three DAP seeds of wild type was used as a control. All data are shown as means ±SD from three biological replicates and two technical replicates. Primers are listed in Supplemental Table S3.

Table 1. Statistics of sequencing data

Sample	Clean Reads	Mapped Reads	Clean Base (Gb)	Mapped Base (Gb)	Mapping Rate (%)	Concordant Pair Rate (%)	Q30 (%)	GC Content (%)
WT-1	66,417,130	61,285,695	6.64	6.13	92.27	85.4	94.03	56.95
WT-2	65,482,702	60,254,372	6.55	6.03	92.02	84.7	94.36	57.31
ZmES22-1	65,615,264	60,672,105	6.56	6.07	92.47	85.7	94.15	57.22
ZmES22-2	65,985,744	60,618,187	6.6	6.06	91.9	84.4	92.45	57.01

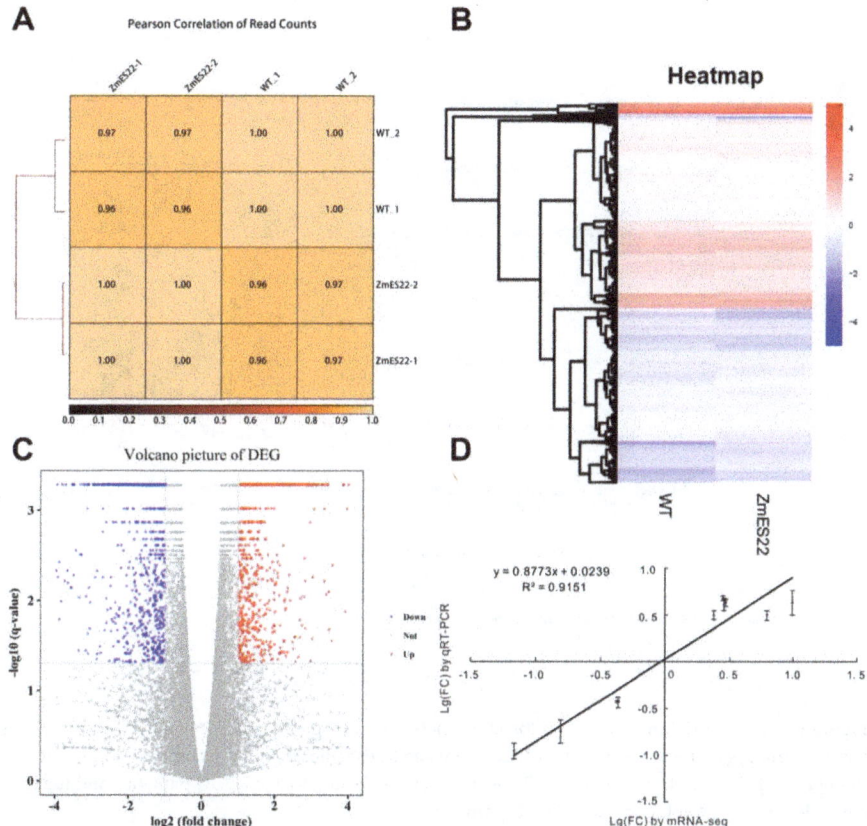

Figure 8. RNA-seq analysis of endosperm at 20 DAP for transgenic rice lines and wild-type plants. (**A**) Pearson correlation of read counts. (**B**) Heat map comparison between Zhonghua 11 and L9. (**C**) A volcano plot of differentially expressed genes (DEGs) about Zhonghua 11 and L9. (**D**) Validation of transcription group data. Expression level changes (log2 (fold change)) of 10 randomly selected DEGs analyzed by RNA-Seq (*x*-axis) were compared with expression data obtained by qRT-PCR (*y*-axis).

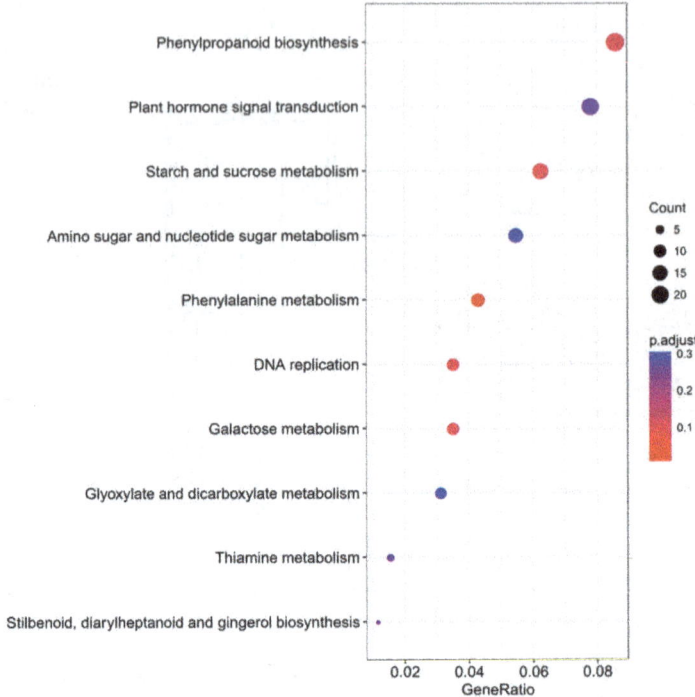

Figure 9. KEGG enrichment analysis for DEGs between transgenic rice lines and wild type plants. KEGG pathways that were enriched for DEGs between transgenic rice lines in comparison to wild type Zhonghua 11. The black circle denotes DEGs that were annotated to one KEGG pathway, and the color panel denoted the *p*-value for each KEGG pathway.

2.6. ZmES22 Could Bind to Promoter GIF1 Gene

To determine if ZmES22 could bind to the promoter of *GIF1*, we firstly extracted 2000 base-pair sequences from upstream of *GIF1* and subjected it web site (http://bioinformatics.psb.ugent.be/webtools/plantcare/html/) to predict if promoter of *GIF1* contained conserved element that MADS type transcription factors could bind to [17]. Surprisingly, a conserved element (CATGT) was located at minus 365 base-pair in upstream of *OsGIF1* gene [18] (Figure S9B). Afterwards, yeast one hybrid assay was explored to determine whether ZmES22 protein could bind to this element. Complete coding sequence of *ZmES22* was inserted into vector containing both activation domain (AD) and was drove by P_{T7}. Two repeated copies of CATGT was synthesized as bait (designated as pGIF1). Simultaneously, a mutant (CAGGT, designated as pmGIF1) was also used as a negative control. As was illustrated in Figure 10, both the growth of the yeast in null and negative control were obviously inhibited in SD/-Ura medium with 900 ng/mL AbA in the yeast one-hybrid assay, however, the yeast co-transformed with P_{T7}-*ZmES22* and pGIF1-AbAi grows well, indicating that ZmES22 binds to the core element of the promoter of *OsGIF1* (Figure 10).

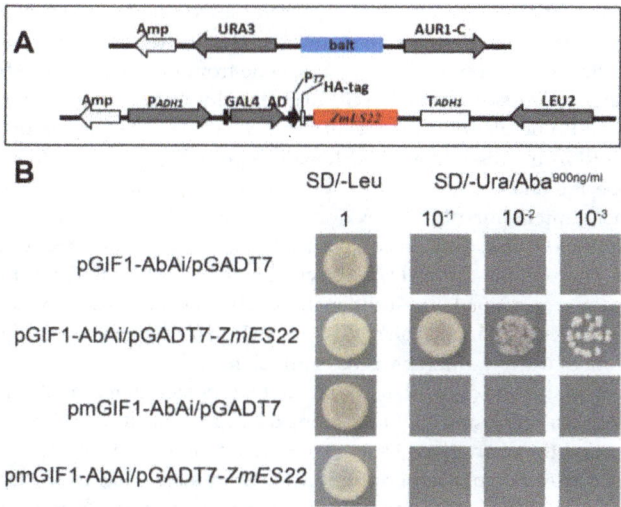

Figure 10. ZmES22 could bind to the core motif of *OsGIF1* via yeast one hybrid assay. (**A**) Schematic structure of yeast expression construct pGAD-*ZmES22* and reporter construct. (**B**) Yeast Y1HGlod was transformed with the vector pGADT7-*ZmES22* and CATGT (*pGIF1*) or mutant tandem repeats (*pmGIF1*) plasmids. The transformants were screened by plating on SD/-Leu/AbA plates to verified the interaction between ZmES22 and the core motif of *GIF1* promoter.

3. Discussion

The endosperm is the tissue that most flowering plants produce in the seeds after fertilization. Endosperm development involves the process of starch synthesis and storage protein accumulation. Recent studies revealed that process of starch synthesis was remarkably conserved ranging from green algae to extant higher plants, suggesting that genes encoding starch biosynthesis related enzymes were functionally conserved across diverse lineages [19,20]. To date, enzymes involved in starch synthesis has been soundly documented in rice. Therefore, rice endosperm is a particularly ideal model to screen and identify key transcription factors that could finely tune the process of starch synthesis in maize [21].

In the present study, transgenic rice that overexpressed one MADS type transcription factor *ZmES22* from maize, exhibited obvious defects with respect to grain characteristics, represented by loosely packed starch granules, reduced 1000-grain weight, and altered apparent amylose and total starch content, suggesting that *ZmES22* might play a key role in regulation of starch synthesis pathway in the transgenic lines. It has been reported that the process of starch biosynthesis was regulated by 17 genes, which mainly encoded multiple subunits or isoforms of four enzymes: ADP-glucose pyrophosphorylase (AGPase), starch synthase (SS), starch branching enzyme (SBE), and starch debranching enzyme (DBE) [2,3]. Interestingly, no significant expression change could be detected for all of these 17 genes in endosperm at 20 DAP between wide type and transgenic rice based on clues from both qRT-PCR and RNA-seq results. However, the mRNA expression levels of majority of starch synthesis related genes, except for *OsISA1*, *OsISA2*, *OsSSI*, and *OsSSIIa*, decreased during the early stages of endosperm development stages compared to that in wild type rice. These results indicated that overexpression of *ZmES22* could negatively affect mRNA expression of the majority of starch synthesis related genes in distinct degree during the early endosperm development stages.

KEGG analysis revealed that DEGs was significantly enriched in starch and sucrose metabolism pathway, in which the mRNA expression of one gene, named *OsGIF1*, decreased as much as 8-fold in transgenic rice. Previous result demonstrated that grain filling rate of *gif1* mutants was slower and accompanied with distinct chalkiness and loosely packed starch granules, while overexpression of *GIF1* driven by its native promoter produces larger grains [6]. The phenotype of *gif1* mutant was consistent to transgenic rice lines that overexpressed *ZmES22* gene from maize. Evidence from qRT-PCR also validated that mRNA expression level of *OsGIF1* continually decreased in overexpression rice lines in comparison to wild type plants. These results indicated that overexpression of *ZmES22* in rice might inhibit the mRNA expression of *OsGIF1*. As is reported that typic MADS type transcription factor are plant specific and often contained four functional domains, the MADS-box conserved domain (MADS-box), intervening (I), K-box domain, which is homologous to keratin (K), and the C terminal domain. MADS-box domain could bind to the promoter and regulate the expression of downstream genes. For example, *ZmMADS47* directly binds the core motif CATGT of promoter of zein genes and activated its expression [18]. Additionally, CATGT element also resided in the upstream of transcription starting sites (−365 base pair), yeast one hybrid assay demonstrated that ZmES22 could bind to the core motif of *OsGIF1* and repress its expression.

The primary results provide evidence that *ZmES22* affect starch synthesis and endosperm development through binding to and downregulating the expression of *OsGIF1*, and in turn influencing carbon distribution and transportation of sucrose on grain filling in rice plant.

In conclusion, a MADS type transcription factor from maize was modestly characterized via overexpression in rice and the molecular mechanism for its anticipant role in regulation of starch synthesis were also explored by RNA-seq and yeast one hybrid assay in the present study. Starch synthesis is a complicated and sophisticated process, which is regulated by numerous transcription factors via protein–protein and protein–DNA interactions. In order to shed light on how *ZmES22* influence the starch synthesis in rice, *ZmES22* mutant are being created via CRISPR/Cas9 system.

4. Materials and Methods

4.1. Plant Materials and Growth Conditions

Ten representative tissues, including root, stem, leaf, tassel, filament, ear, 10 DAP (days after pollination) embryo, 20 DAP embryo, 10 DAP endosperm, and 20 DAP endosperm, were collected from maize inbred line B73 plants, which were grown in a greenhouse with paddy soil at 28 °C under a 14 h light/10 h dark photoperiod. Each tissue was repeatedly sampled from three individual plants as three biological replicates. Both wild type cultivars (*Oryza sativa* L. *japonica* cv. Zhonghua 11) and transgenic rice lines were grown under natural conditions in experimental field plots for Anhui Agricultural University in Anhui Province, China. Rice endosperms at 3, 6, 10, and 20 DAP were harvested for qRT-PCR assay.

4.2. RNA Extraction and Real-Time RT-PCR Analysis

Total RNA of collected samples was extracted using RNAiso Plus Kit (Takara, Kusatsu, Japan) according to manufacturer's instructions. Afterwards, first-strand cDNA was generated using reverse transcription system (Promega, Madison, WI, USA). The qRT-PCR (quantitative real-time PCR) was performed using SYBR Green Master (Roche, Basel, Switzerland) on an ABI 7300 Real Time PCR System (Applied Biosystems, Foster City, CA, USA), and the reactions were performed according to previous report [22]. In detail, the reaction conditions were set as following: 50 °C for 2 min, then 95 °C for 10 min, followed by 40 cycles of 95 °C for 15 s and 60 °C for 1 min. Each cDNA sample was quantified in three replicates. The achieved data were calculated by 2_DDCt method as described previously [23]. Maize *Actin1* gene was used as internal control to normalize the detection threshold for each of three replicates.

4.3. Subcellular Localization

Full-length open reading frame (ORF) of *ZmES22* without the termination codon was amplified and cloned into pCAMBIA1305 vector under the drive of cauliflower mosaic virus (CaMV) 35S promoter. The fusion construct 35S:: *ZmES22*-GFP and empty vector 35S:: GFP were transformed into leaves for 35-day-old tobacco (*Nicotiana bethamiana*) via a syringe without needle, respectively. Infiltrated tobacco plants were transferred into dark condition for 12 h, followed by normal illumination for 48 h. Finally, green fluorescent signals were examined using confocal microscope (Olympus FV1000, Tokyo, Japan).

4.4. Transcriptional Activation Assay

Full-length ORF of *ZmES22* was amplified and inserted into pGBKT7 vector (Clontech, San Deigo, CA, USA), which were fused with the GAL4 DNA-binding domain beforehand. Subsequently, both negative control (null pGBKT7 vector) and positive control (co-transformation of pGBKT7-53 with pGADT7-T vectors), together with fusion construct pGBKT7-*ZmES22* was transformed into yeast strain AH109, respectively. AH109 strain carries HIS3, ADE2 and MEL1 reporter genes. Transformed yeast cells were then cultured on SD/-Trp medium for 3 days at 30 °C and then transferred into SD/-Trp/-His/-Ade/X-α-GAL medium for 3 days at 30 °C.

4.5. Generation of Transgenic Rice Lines

Full-length ORF of *ZmES22* was amplified and inserted into an overexpression vector pCAMBIA1301a under the drive of CaMV 35S promoter and a NOS terminator. Recombination construct pCAMBIA1301a-*ZmES22* also harbored a GUS reporter gene and was transformed into *japonica* rice cultivar Zhonghua 11 via *Agrobacterium* mediated transformation [24]. Both histochemical staining of GUS activity and PCR experiments followed by sanger sequencing were utilized to validate if transgenic rice lines were positive ones.

4.6. Determination of Agronomic Characters and Measurement of Grain Quality

Vernier caliper was adopted to measure the length, width and thickness for 100 uniformly mature seeds at the longest, widest, and thickest point, respectively. 1000-grain weight was determined by counting ten independent repeats of 100-grain samples on an electronic balance. Each measurement was repeated for three times. Embryos and follicles were separated from the embryo and ground into powder. The starch content was measured with total starch determination kit (K-TSTA; Megazyme, Bray, County Wicklow, Ireland) according to manufacturer's protocol. Apparent amylose content (AAC) of the samples was measured using iodine colorimetry (K-AMYL; Megazyme) [15]. Anthrone method was applied to determine soluble sugar content [7].

4.7. Measurement of Starch Blue Value (BV) and Maximum Absorption Wavelength (kmax)

The separation of amylose and amylopectin, and measurement of the blue value (BV) and maximum absorption wavelength (kmax) of starch were referred to the modified alkali impregnation method [25]. Detailedly, 5 mg isolated amylose powder was dissolved in 8 mL 90% dimethyl sulfoxide and then diluted to 100 mL double distilled water. The absorption spectra of the starch-iodine complex were examined ranging from 500 to 800 nm. The BV was A_{600}. While, with respect to amylopectin, 15 mg isolated amylopectin powder was dissolved in 100 mL double distilled water. The absorption spectra were examined ranging from 500 to 700 nm, and the BV was set to A_{680}.

4.8. Observation of Starch Granules by Scanning Electron Microscopy (SEM)

According to the methods in previous report [25], Hitachi S-3000N scanning electron microscope (SEM) (Hitachi, Tokyo, Japan) were used to observe the morphology of starch granules. SEM images were distinguished through cross-sections of mature rice seeds including ventral, central, and dorsal area of mature endosperm.

4.9. RNA-Seq and Data Analysis

Seeds for both overexpression transgenic rice lines and Zhonghua 11 at 20 DAP were collected for RNA-seq, each group were repeated twice as two biological replicates. Subsequently, RNA was isolated and then high throughput sequencing was performed on BGISEQ-500 platform in Beijing Genomics Institute (BGI; Shenzhen, China). After trimming of low-quality and adaptor sequences from raw sequencing reads, clean data were aligned to *Oryza sativa* ssp. *japonica* cv [26]. Nipponbare genome (IRGSP-1.0, http://rapdb.dna.affrc.go.jp/) [27] using TopHat2 software [28]. The resulted BAM alignment files were subject to Cufflinks to calculate gene expression levels [29]. Differentially expressed genes (DEGs) were determined by Cuffdiff with default parameters, based on the following criteria: (1) the minimum fold-change of gene expression was 2.0; (2) the maximum adjusted p value was 0.05 [30]. The RNA-seq data were validated using quantitative real-time PCR analysis for ten randomly selected DEGs. The R package ClusterProfiler were explored to conduct both Gene Ontology (GO) and Kyoto Encyclopedia of Genes and Genomes (KEGG) analysis [31]. The raw sequencing dataset has been submitted to NCBI's Gene Expression Omnibus (GEO; http://www.ncbi.nlm.nih.gov/geo/) under accession number SRP063765.

4.10. Yeast One-Hybrid Assay

Yeast one-hybrid assays were implemented originally according to the Matchmaker® Gold Yeast One-Hybrid Library Screening System User Manual (Clontech). To test the ability of ZmES22 to bind to the core motif CATGT of *GIF1* promoter, CATGT and mutant tandem repeats were cloned and inserted into the *Bam*HI and *Hind*III site of the p53/AbAi vector. Yeast Y1HGlod was transformed with the vector pGADT7-ZmES22 and CATGT or mutant tandem repeats plasmids. To evaluate interaction between ZmES22 and the core motif CATGT of *GIF1* promoter, the transformants were screened by plating on SD /-Leu/AbA plates.

5. Conclusions

Starch is one of the major components of cereal grains, providing sufficient calories for both human diet and animal feed. Therefore, comprehensive understanding molecular basis of starch synthesis process and its regulatory network is of vital importance. In the present study, we identified a gene *ZmES22*, encoding a typical MADS type transcription factor, which were exclusively highly expressed in maize endosperm, indicating its crucial role in endosperm development of maize. When *ZmES22* was overexpressed in rice, the 1000-grain weight, together with total starch content were remarkably reduced, whereas, the soluble sugar content was significantly higher when compared to wild type. Moreover, overexpression *ZmES22* altered the relative fraction of long branched starch and changed the morphology of starch granule from densely packed, polyhedral starch granules into loosely spherical granules with larger spaces. These results demonstrate that *ZmES22* is a negative regulator that could affect the starch biosynthesis process. Moreover, RNA-seq and qRT-PCR results further illustrated that overexpression of *ZmES22* could downregulate mRNA expression level of numerous key genes in starch synthesis pathway, particularly in early developmental stages in transgenic rice lines. Furthermore, ZmES22 could bind to the promoter region of the *OsGIF1* and downregulate its mRNA expression throughout the endosperm developmental stages. Therefore, we proposed that *ZmES22* might affect starch biosynthesis as well as reducing the rate of grain filling by downregulation of *OsGIF1* in rice. Whether knock-down or knockout of the *ZmES22* gene could contribute to increase of yield in maize remains to be demonstrated.

Supplementary Materials: Supplementary materials can be found at http://www.mdpi.com/1422-0067/20/3/483/s1.

Author Contributions: L.G. designed the research; K.Z. conducted the molecular experiments and analyzed the data; H.X. performed the rice transformation; M.G., Z.M., and Y.W. analyzed agronomic characters, detected starch content; L.G. and W.S. drafted the manuscript.

Int. J. Mol. Sci. **2019**, *20*, 483

Funding: This research was supported by grants from Genetically Modified Organisms Breeding Major Projects (2016ZX08003-002) and the National Natural Sciences Foundation of China (31701436).

Conflicts of Interest: The authors declare no conflict of interest.

References

1. Qi, X.; Li, S.; Zhu, Y.; Zhao, Q.; Zhu, D.; Yu, J. ZmDof3, a maize endosperm-specific Dof protein gene, regulates starch accumulation and aleurone development in maize endosperm. *Plant Mol. Biol.* **2017**, *93*, 7–20. [CrossRef] [PubMed]

2. Nakamura, Y. Towards a Better Understanding of the Metabolic System for Amylopectin Biosynthesis in Plants: Rice Endosperm as a Model Tissue. *Plant Cell Physiol.* **2002**, *43*, 718–725. [CrossRef]

3. James, M.G.; Denyer, K.; Myers, A.M. Starch synthesis in the cereal endosperm. *Curr. Opin. Plant Biol.* **2003**, *6*, 215–222. [CrossRef]

4. Li, S.; Wei, X.; Ren, Y.; Qiu, J.; Jiao, G.; Guo, X.; Tang, S.; Wan, J.; Hu, P. OsBT1 encodes an ADP-glucose transporter involved in starch synthesis and compound granule formation in rice endosperm. *Sci. Rep.* **2017**, *7*, 40124. [CrossRef]

5. Lee, S.-K.; Hwang, S.-K.; Han, M.; Eom, J.-S.; Kang, H.-G.; Han, Y.; Choi, S.-B.; Cho, M.-H.; Bhoo, S.H.; An, G.; et al. Identification of the ADP-glucose pyrophosphorylase isoforms essential for starch synthesis in the leaf and seed endosperm of rice (*Oryza sativa* L.). *Plant Mol. Biol.* **2007**, *65*, 531–546. [CrossRef] [PubMed]

6. Wang, E.; Wang, J.; Zhu, X.; Hao, W.; Wang, L.; Li, Q.; Zhang, L.; He, W.; Lu, B.; Lin, H.; et al. Control of rice grain-filling and yield by a gene with a potential signature of domestication. *Nat. Genet.* **2008**, *40*, 1370–1374. [CrossRef]

7. Wang, J.-C.; Xu, H.; Zhu, Y.; Liu, Q.-Q.; Cai, X.-L. OsbZIP58, a basic leucine zipper transcription factor, regulates starch biosynthesis in rice endosperm. *J. Exp. Bot.* **2013**, *64*, 3453–3466. [CrossRef]

8. Sun, C. A Novel WRKY Transcription Factor, SUSIBA2, Participates in Sugar Signaling in Barley by Binding to the Sugar-Responsive Elements of the iso1 Promoter. *Plant Cell Online* **2003**, *15*, 2076–2092. [CrossRef]

9. Schmidt, R.; Schippers, J.H.M.; Mieulet, D.; Watanabe, M.; Hoefgen, R.; Guiderdoni, E.; Mueller-Roeber, B. SALT-RESPONSIVE ERF1 Is a Negative Regulator of Grain Filling and Gibberellin-Mediated Seedling Establishment in Rice. *Mol. Plant* **2014**, *7*, 404–421. [CrossRef]

10. Chen, J.; Yi, Q.; Cao, Y.; Wei, B.; Zheng, L.; Xiao, Q.; Xie, Y.; Gu, Y.; Li, Y.; Huang, H.; et al. ZmbZIP91 regulates expression of starch synthesis-related genes by binding to ACTCAT elements in their promoters. *J. Exp. Bot.* **2016**, *67*, 1327–1338. [CrossRef] [PubMed]

11. Jiang, Y.; Zeng, B.; Zhao, H.; Zhang, M.; Xie, S.; Lai, J. Genome-wide Transcription Factor Gene Prediction and their Expressional Tissue-Specificities in Maize. *J. Integr. Plant Biol.* **2012**, *54*, 616–630. [CrossRef]

12. Cai, H.; Chen, Y.; Zhang, M.; Cai, R.; Cheng, B.; Ma, Q.; Zhao, Y. A novel GRAS transcription factor, ZmGRAS20, regulates starch biosynthesis in rice endosperm. *Physiol. Mol. Biol. Plants* **2017**, *23*, 143–154. [CrossRef] [PubMed]

13. Gramzow, L.; Ritz, M.S.; Theißen, G. On the origin of MADS-domain transcription factors. *Trends Genet.* **2010**, *26*, 149–153. [CrossRef]

14. Ma, H.; Yanofsky, M.F.; Meyerowitz, E.M. AGL1-AGL6, an Arabidopsis gene family with similarity to floral homeotic and transcription factor genes. *Genes Dev.* **1991**, *5*, 484–495. [CrossRef] [PubMed]

15. Takeda, Y.; Hizukuri, S.; Juliano, B.O. Purification and structure of amylose from rice starch. *Carbohydr. Res.* **1986**, *148*, 299–308. [CrossRef]

16. Chen, X.; Guo, L.; Du, X.; Chen, P.; Ji, Y.; Hao, H.; Xu, X. Investigation of glycerol concentration on corn starch morphologies and gelatinization behaviours during heat treatment. *Carbohydr. Polym.* **2017**, *176*, 56–64. [CrossRef]

17. Lescot, M.; Déhais, P.; Thijs, G.; Marchal, K.; Moreau, Y.; de Peer, Y.V.; Rouzé, P.; Rombauts, S. PlantCARE, a database of plant cis-acting regulatory elements and a portal to tools for in silico analysis of promoter sequences. *Nucleic Acids Res.* **2001**, *30*, 325–327. [CrossRef]

18. Qiao, Z.; Qi, W.; Wang, Q.; Feng, Y.; Yang, Q.; Zhang, N.; Wang, S.; Tang, Y.; Song, R. ZmMADS47 Regulates Zein Gene Transcription through Interaction with Opaque2. *PLoS Genet.* **2016**, *12*, e1005991. [CrossRef]

19. Deschamps, P.; Colleoni, C.; Nakamura, Y.; Suzuki, E.; Putaux, J.-L.; Buleon, A.; Haebel, S.; Ritte, G.; Steup, M.; Falcon, L.I.; et al. Metabolic Symbiosis and the Birth of the Plant Kingdom. *Mol. Biol. Evol.* **2008**, *25*, 536–548. [CrossRef]

20. Qu, J.; Xu, S.; Zhang, Z.; Chen, G.; Zhong, Y.; Liu, L.; Zhang, R.; Xue, J.; Guo, D. Evolutionary, structural and expression analysis of core genes involved in starch synthesis. *Sci. Rep.* **2018**, *8*, 12736. [CrossRef]

21. Nelson, O.; Pan, D. Starch Synthesis in Maize Endosperms. *Annu. Rev. Plant Physiol. Plant Mol. Biol.* **1995**, *46*, 475–496. [CrossRef]

22. Zhai, R.; Feng, Y.; Wang, H.; Zhan, X.; Shen, X.; Wu, W.; Zhang, Y.; Chen, D.; Dai, G.; Yang, Z.; et al. Transcriptome analysis of rice root heterosis by RNA-Seq. *BMC Genom.* **2013**, *14*, 19. [CrossRef] [PubMed]

23. Livak, K.J.; Schmittgen, T.D. Analysis of Relative Gene Expression Data Using Real-Time Quantitative PCR and the $2-\Delta\Delta CT$ Method. *Methods* **2001**, *25*, 402–408. [CrossRef] [PubMed]

24. Lin, Y.J.; Zhang, Q. Optimising the tissue culture conditions for high efficiency transformation of indica rice. *Plant Cell Rep.* **2005**, *23*, 540–547. [CrossRef]

25. Fu, F.-F.; Xue, H.-W. Coexpression Analysis Identifies Rice Starch Regulator1, a Rice AP2/EREBP Family Transcription Factor, as a Novel Rice Starch Biosynthesis Regulator. *Plant Physiol.* **2010**, *154*, 927–938. [CrossRef]

26. Bolger, A.M.; Lohse, M.; Usadel, B. Trimmomatic: A flexible trimmer for Illumina sequence data. *Bioinformatics* **2014**, *30*, 2114–2120. [CrossRef]

27. Sakai, H.; Lee, S.S.; Tanaka, T.; Numa, H.; Kim, J.; Kawahara, Y.; Wakimoto, H.; Yang, C.; Iwamoto, M.; Abe, T.; et al. Rice Annotation Project Database (RAP-DB): An Integrative and Interactive Database for Rice Genomics. *Plant Cell Physiol.* **2013**, *54*, e6. [CrossRef]

28. Kim, D.; Pertea, G.; Trapnell, C.; Pimentel, H.; Kelley, R.; Salzberg, S.L. TopHat2: Accurate alignment of transcriptomes in the presence of insertions, deletions and gene fusions. *Genome Biol.* **2013**, *14*, R36. [CrossRef]

29. Trapnell, C.; Hendrickson, D.G.; Sauvageau, M.; Goff, L.; Rinn, J.L.; Pachter, L. Differential analysis of gene regulation at transcript resolution with RNA-seq. *Nat. Biotechnol.* **2013**, *31*, 46–53. [CrossRef]

30. Trapnell, C.; Roberts, A.; Goff, L.; Pertea, G.; Kim, D.; Kelley, D.R.; Pimentel, H.; Salzberg, S.L.; Rinn, J.L.; Pachter, L. Differential gene and transcript expression analysis of RNA-seq experiments with TopHat and Cufflinks. *Nat. Protoc.* **2012**, *7*, 562–578. [CrossRef]

31. Yu, G.; Wang, L.-G.; Han, Y.; He, Q.-Y. clusterProfiler: An R Package for Comparing Biological Themes Among Gene Clusters. *OMICS J. Integr. Biol.* **2012**, *16*, 284–287. [CrossRef] [PubMed]

Article

Effects of *OsCDPK1* on the Structure and Physicochemical Properties of Starch in Developing Rice Seeds

Jian-Zhi Jiang [1], Chun-Hsiang Kuo [1], Bo-Hong Chen [1], Mao-Kei Chen [1], Choun-Sea Lin [2] and Shin-Lon Ho [1,*]

[1]   Department of Agronomy, National Chiayi University, Chiayi 60004, Taiwan; ssdog27@yahoo.com.tw (J.-Z.J.); efab888@yahoo.com.tw (C.-H.K.); chen.po.hung9117@gmail.com (B.-H.C.); fritz29522879@yahoo.com.tw (M.-K.C.)
[2]   Agricultural Biotechnology Research Center, Academia Sinica, Taipei 11529, Taiwan; cslin99@gate.sinica.edu.tw
*   Correspondence: slho@mail.ncyu.edu.tw; Tel.: +886-5-271-7388; Fax: +886-5-271-7386

Received: 2 September 2018; Accepted: 16 October 2018; Published: 19 October 2018

Abstract: Overexpression of a constitutively active truncated form of *OsCDPK1* (*OEtr*) in rice produced smaller seeds, but a double-stranded RNA gene-silenced form of *OsCDPK1* (*Ri*) yielded larger seeds, suggesting that *OsCDPK1* plays a functional role in rice seed development. In the study presented here, we propose a model in which *OsCDPK1* plays key roles in negatively controlling the grain size, amylose content, and endosperm appearance, and also affects the physicochemical properties of the starch. The dehulled transgenic *OEtr* grains were smaller than the dehulled wild-type grains, and the *OEtr* endosperm was opaque and had a low amylose content and numerous small loosely packed polyhedral starch granules. However, the *OEtr* grain sizes and endosperm appearances were not affected by temperature, which ranged from low (22 °C) to high (31 °C) during the grain-filling phase. In contrast, the transgenic *Ri* grains were larger, had higher amylose content, and had more transparent endosperms filled with tightly packed polyhedral starch granules. This demonstrates that *OsCDPK1* plays a novel functional role in starch biosynthesis during seed development and affects the transparent appearance of the endosperm. These results improve our understanding of the molecular mechanisms through which the grain-filling process occurs in rice.

Keywords: rice; *OsCDPK1*; seed development, starch biosynthesis; endosperm appearance

1. Introduction

The quality of rice (*Oryza sativa* L.) grain is defined in terms of several main factors, including (i) eating and cooking qualities, and (ii) milling qualities and appearance [1]. Eating and cooking qualities are determined by the amylose content, amylopectin structure, gelatinization temperature, and pasting viscosity [2], and milling qualities and appearance correlate strongly with the transparency, flouriness, and chalkiness of the endosperm [3]. The filling and accumulation of starch granules in developing rice endosperms can accelerate at high temperatures, causing the starch in the endosperm cells to be packed loosely and the kernel to be chalky. Such grains crack easily during milling, yielding poor eating and cooking qualities [4–6].

It has been shown in many studies that chalky and less-transparent kernels contain more amylopectin and less amylose in the endosperm than do less-chalky and more-transparent kernels [7–9]. Eliminating chalkiness by regulating the amylopectin and amylose content ratios (by affecting biosynthesis) in the endosperm during the grain-filling phase is therefore a key way of improving grain quality. Two enzymes involved in amylose biosynthesis are ADP-glucose pyrophosphorylase and the Waxy gene-encoded granule-bound starch synthase I (GBSSI) [10,11]. The enzymes involved in the biosynthesis and modification of amylopectin are ADP-glucose pyrophosphorylase, soluble starch synthase, the starch-branching enzyme (BE), and the starch-debranching enzyme [10–12]. In higher plants, BE plays an essential role in amylopectin biosynthesis because it is the only enzyme that can add α-1,6-glucosidic linkages to polyglucans [13]. Three BE isoforms—BEI, BEIIa, and BEIIb—have been found in rice [10,14]. The rice mutant *amylose-extender*, which has a null mutation in *BEIIb*, has been found to alter the degree of polymerization (DP) of amylopectin, giving fewer short chains (DP ≤ 17) and more long chains (DP ≥ 18), the changes being related to the dose on the *amylose-extender* locus in the triploid endosperm cells. These results suggest that *BEIIb* might have critical effects on the amylopectin structure and the rheological properties of the starch [12]. Several floury endosperm rice mutants (*flo1–flo7*) have been isolated. Treating fertilized rice egg cells with the chemical mutagen *N*-methyl-*N*-nitrosourea yielded mutants *flo1* and *flo2*, which had floury endosperms [15,16]. The *flo2* mutant has been found, through map-based cloning, to be a member of the tetratricopeptide repeat-motif protein family. The gene mutation decreases the grain size and decreases the starch quality (by decreasing the amylose content) and also changes the fine structure of the amylopectin [17]. The *flo3* mutant was produced through applying gamma-irradiation and ethyl methansulfonate treatment, and had a low 16 kDa globulin content in the endosperm [18]. The white-core floury endosperm mutants *flo4* and *flo5* were produced through T-DNA insertional mutagenesis and were found to have *pyruvate orthophosphate dikinase B* and *starch synthase IIIa* (SSIIIa) gene mutations, respectively [19,20]. The *flo4* mutant endosperm had a low amylose content, suggesting that *pyruvate orthophosphate dikinase B* might play a role in regulating carbon metabolism during the grain filling process [19]. DP analysis of amylopectin in *flo5* was performed, and the amounts of DP 6–8 and DP 16–20 in the mutant endosperm were found to be decreased but the amounts of DP 9–15 and DP 22–29 had increased, indicating that *SSIIIa* strongly affects the chain-length distribution of amylopectin biosynthesized in developing rice grains [20]. The floury endosperm mutant *flo6* had a completely floury white endosperm, but the *flo7* endosperm was floury and white only at the peripheries, and not in the interior [21,22]. Map-based cloning demonstrated that *flo6* was an insertion mutation in the unknown function gene *Os03g0686900* [21] and that *flo7* was a deletion mutation in the unknown function gene *Os10g0463800* [22], suggesting that these genes may play vital roles in starch biosynthesis and granule formation in the endosperm during the grain-filling process.

Calcium ions (Ca^{2+}) are secondary messengers in plant cells, and are used in response to various environmental and developmental stimuli through temporal and spatial fluctuations of the cytosolic Ca^{2+} concentration [23,24]. Calcium-dependent protein kinases (CDPKs) are a major family of calcium sensors that have been characterized in various plant species. CDPKs are Ser/Thr protein kinases that are encoded by multigene families [25,26]. CDPKs have four functional domains: an N-terminal variable domain, a catalytic kinase domain, an autoinhibitory domain, and a calcium-binding EF-hands regulatory domain [27]. Under normal growth conditions (i.e., absence of Ca^{2+} signals), the autoinhibitory domain can interact with the kinase domain thereby inhibits kinase activity [24]. Deletion of the autoinhibitory domain and the Ca^{+2} binding domains from the coding region could bypass Ca^{2+} signal stimulation and resulted in produced a constitutively active enzyme of CDPKs [26,27]. It has been shown in many studies that CDPKs play important physiological roles in response to various environmental stresses and developmental processes [23,24,28–30]. Only in a few studies have CDPKs been shown to play a role in starch biosynthesis during the grain-filling process in rice. The rice *SPK* (which shares 79% of its amino acid sequence with *OsCDPK1*) has been found to encode a sucrose synthase kinase. Expression of antisense *SPK* in transgenic rice produced watery grains because large amounts of sucrose had accumulated in the endosperm due to low sucrose synthase activity, resulting in inefficient sucrose degradation [31]. This indicates that SPK may be a regulator in the starch biosynthesis pathway.

In the study presented here, *OsCDPK1* was found to play pivotal roles in rice-seed development, in the physicochemical properties of the starch produced, and in the appearance of the endosperm.

The pleiotropic effects on various agronomic traits in loss- and gain-of-function of *OsCDPK1* transgenic plants are also characterized.

2. Results

2.1. Phenotypic Changes in Transgenic Rice Plants with Overexpressing or Silenced OsCDPK1

In previous studies, to understand the physiological roles of *OsCDPK1*, the gene-overexpression and gene-silencing approaches were implemented. For the truncated form of *OsCDPK1* (*OEtr*), the coding sequences in which the autoinhibitory region and the calcium-binding domains had been removed were expressed under the control of maize ubiquitin gene promoter to generate the constitutively active form of the *OEtr* transgenic plants [32]. The transgenic plants in *OEtrs* yielded smaller seeds, whereas RNA-interference gene knockdown mutants (*Ris*) yielded larger seeds [32]. In this study, the various agronomic traits in the T4 transgenic lines of *OEtrs* (*OEtr-1, -3* and *-4*) and *Ris* (*Ri-1, -2* and *-3*) were analyzed. The results (Table 1) were consistent with our previous studies [32] and showed that the *Ris* (*Ri-1, -2* and *-3*) lines had, on average, 7.1% and 10.8% increases in plant height and 1000-grain weight, respectively, compared to those of WT (wild type). However, both examined traits decreased in the *OEtrs* (*OEtr-1, -3* and *-4*) lines compared with WT, with 5.8% and 21.1% reductions in plant height and 1000-grain weight, respectively. Furthermore, compared to WT, the heading date and growth duration were shorter in *Ris*, while *OEtrs* showed no significant difference. The average of the heading date of *Ris* (83.2 day) was 6 days shorter than that of WT (89.3 day) and the growth duration of *Ris* (104.8 day) was 10 days shorter than that of WT (115.2 day). The dehulled grain weight, starch content, and amylose content were higher in *Ris* and lower in *OEtrs* than those of WT; for example, the average weights of dehulled grain of *Ris* and *OEtrs* were 111.2% and 84.1%, respectively, of that of WT. The average percentage of starch content versus dehulled grain weight in WT, *OEtrs*, and *Ris* was 73.4%, 64.2%, and 78.1%, respectively. Moreover, the average amylose content in WT, *OEtrs*, and *Ris* was 23.2%, 15.1% and 26.2%, respectively.

Table 1. Comparison of agronomic traits between WT and transgenic lines.

Genotypes	Plant Height (cm)	Heading Day (day)	Growth Duration (day)	1000-Grain (g)	Dehulled Grain (mg/grain)	Starch Content (mg/grain)	Amylose Content (%)
WT	108.2 ± 2.5	89.3 ± 2.4	115.2 ± 3.3	23.2 ± 0.6	21.4 ± 0.3	15.7 ± 0.4 (73.4%) [a]	23.2 ± 0.3
OEtr-1	99.5 ± 3.5 *	92.1 ± 2.3	117.6 ± 2.7	18.2 ± 0.3 **	16.8 ± 0.2 **	10.5 ± 0.1 ** (62.5%) [a]	14.3 ± 0.3 *
OEtr-3	102.6 ± 2.6 *	92.4 ± 3.4 *	119.1 ± 3.8	18.6 ± 0.1 **	17.3 ± 0.4 **	11.8 ± 0.3 ** (68.2%) [a]	15.3 ± 0.2 *
OEtr-4	103. 6 ± 3.1 *	91.1 ± 2.7 *	119.4 ± 3.3	18.1 ± 0.3 **	16.6 ± 0.2 **	10.3 ± 0.1 ** (62.0%) [a]	15.6 ± 0.4 *
Ri-1	115.4 ± 4.6 *	83.6 ± 2.5 **	106.6 ± 3.1 *	25.6 ± 0.3 *	23.9 ± 0.3 *	18.7 ± 0.2 * (78.2%) [a]	25.8 ± 0.1 *
Ri-2	118.7 ± 3.8 *	82.7 ± 2.8 **	103.6 ± 2.1 *	25.3 ± 0.2 *	23.5 ± 0.2 *	18.5 ± 0.2 * (78.7%) [a]	26.6 ± 0.5 *
Ri-3	113.6 ± 4.3 *	83.2 ± 3.4 **	104.3 ± 3.6 *	26.2 ± 0.5 *	24.4 ± 0.2 *	18.9 ± 0.6 * (77.5%) [a]	26.2 ± 0.2 *

Mean values calculated from three independent transgenic lines. All data are presented as mean ± SE. Statistical significance is determined by *t*-test. Values in the same column indicate significant differences between WT and mutant lines at * $p < 0.05$ and ** $p < 0.01$. [a] Percentage of dehulled grain weight.

2.2. Ectopic Overexpression and Silencing of OsCDPK1 in Transgenic Rice Plants Yielded Opaque and Transparent Endosperms, Respectively

We examined the 55-day-old plant phenotypes in the T4 transgenic lines further, and the results were consistent with Table 1. The *Ri-1, -2* and *-3* plants were higher (mean value is 76.6 ± 6.18 cm) and the *OEtr-1, -3 and -4* plants were shorter (52.4 ± 3.58 cm) than the WT plants (63.3 ± 4.63 cm) (Figure 1A,B). At the grain-filling stage, the WT and transgenic plants were transferred to the growth room at an optimal temperature of a cycle of 25 °C for 16 h light and 20 °C for 8 h dark. Fifteen seeds were randomly selected from each individual line and dehulled. We found that the dehulled *Ri-1* grains were longer and that the *OEtr-1* grains were shorter than the WT grains (Figure 1E,F). We also found that the dehulled *OEtr-1* grains were all of the floury-kernel phenotype (Figure 1C,E).

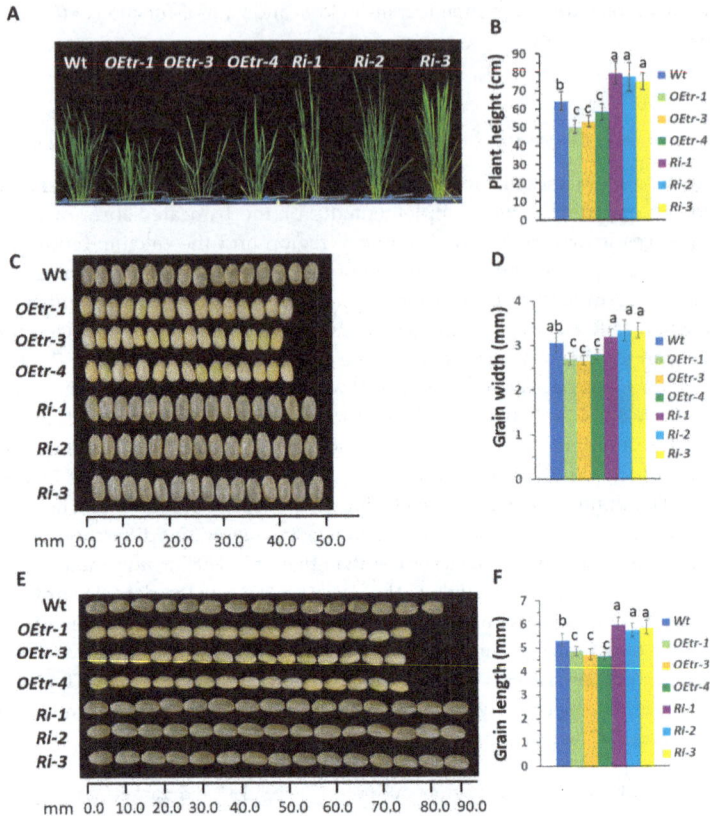

Figure 1. Plant heights and grain morphologies in the WT, the *OEtr-1*, *-3* and *-4* (*OEtrs*), and the *Ri-1*, *-2* and *-3* (*Ris*) plants. After 5 d of flowering, rice plants were transferred to a growth chamber and grown at an optimal temperature (25 °C for 16 h light, 20 °C for 8 h dark). (**A**) Heights of the 55-day-old WT, *OEtrs*, and *Ris* plants. (**B**) Quantification of the plant heights. Each error bar is the standard error for 15 individual plants. (**C**) Endosperm appearances and grain widths for the WT, *OEtrs*, and *Ris* plants. Fifteen grains per line were positioned in a row and measured. (**D**) Quantification of the grain widths. Each error bar is the standard error for 50 individual grains. (**E**) Endosperm appearances and grain lengths for the WT, *OEtrs*, and *Ris* plants. Fifteen grains per line were positioned in a row and measured. (**F**) Quantification of the grain lengths. Each error bar is the standard deviation (*n* = 50). Different letters above the bars indicate significant differences, identified by performing ANOVAs ($p < 0.01$).

We determined whether the temperature affected the endosperm appearance in the transgenic lines by growing rice plants at relatively low temperatures (LT; 22 °C for 16 h light, 20 °C for 8 h dark) and at relatively high temperatures (HT; 31 °C for 16 h light, 28 °C for 8 h dark) during the grain-filling process. The mature seeds were collected from the WT, the *Ri-1*, *Ri-2* and *Ri-3* (*Ris*), and the *OEtr-1*, *OEtr-3* and *OEtr-4* (*OEtrs*) plants. As shown in Figure 2, under LT and HT conditions, all the dehulled *OEtrs* grains had smaller and floury endosperms (100% in the *OEtr-1*, *OEtr-3* and *OEtr-4* lines). In contrast, under LT condition, most of the grains in the WT and *Ris* lines displayed transparency phenotype; the ratios of chalky grains were 16.2% in the WT grains, and 6.0%, 6.7%, and 7.1% in the *Ri-1*, *Ri-2* and *Ri-3* grains, respectively (Figure 2A and Table 2). However, the higher temperature caused a significant increase in the WT and *Ris* lines' endosperms of the chalky phenotype; the ratios of chalky grains were 63.6% in the WT grains, and 50.1%, 51.2% and 45.1% in the *Ri-1*, *Ri-2* and

Ri-3 grains, respectively (Figure 2A and Table 2). Illuminating the kernels with a backlight showed that the *OEtrs* grains all had opaque endosperms regardless of the temperature at which the plants were grown at LT or HT (Figure 2B). However, most of the WT and *Ris* grains displayed transparent endosperms under the LT condition (the *Ris* endosperms being more transparent than those of WT), the proportion of opaque endosperms increased to more than 50% in both the WT and *Ris* lines under HT growth condition (Figure 2B and Figure S1). Cross-sections of the endosperms (Figure 2C) indicated that the kernels in the WT and *Ris* displayed transparent phenotype under LT treatment and showed partial opaque phenotype under HT condition. However, in the *OEtrs*, all endosperms displayed completely floury appearance under both LT and HT conditions, regardless of the different temperature treatments used. These results suggest that *OsCDPK1* affects rice endosperm appearance in a temperature-independent manner.

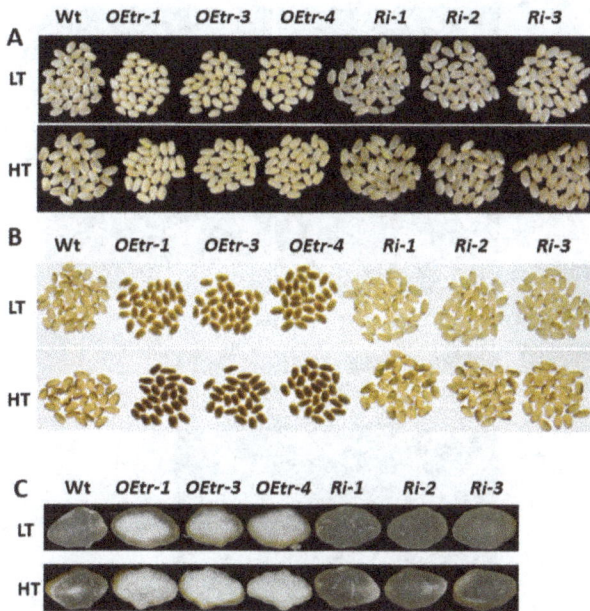

Figure 2. Effects of temperature on the appearances of the WT, *OEtrs*, and *Ris* endosperms. After 5 d of flowering, rice plants were transferred to a growth chamber and grown at a lower temperature (22 °C for 16 h light, 20 °C for 8 h dark) or a higher temperature (31 °C for 16 h light, 28 °C for 8 h dark). (**A,B**) Seeds harvested from the plants grown at the lower and higher temperatures, respectively, illuminated using (**A**) normal lighting and (**B**) backlighting. (**C**) Cross-sections of the endosperms of the seeds from plants grown at either lower or higher temperature. LT: lower temperature; HT: higher temperature.

Table 2. The ratios of chalky grains in wild type (TNG67), *OEtr-1*, and *Ri-1* lines growth under lower temperature (LT; 22 °C for 16 h light, 20 °C for 8 h dark) or higher temperature (HT; 31 °C for 16 h light, 28 °C for 8 h dark) during the rice grain-filling process.

Plant Species	Wild Type	*OEtr-1*	*OEtr-3*	*OEtr-4*	*Ri-1*	*Ri-2*	*Ri-3*
The ratios of chalky grains in LT (%)	16.2 ± 2.19	100	100	100	6.0 ± 0.61	6.7 ± 0.56	7.1 ± 0.61
The ratios of chalky grains in HT (%)	63.6 ± 5.19	100	100	100	50.1 ± 3.61	51.2 ± 4.56	45.1 ± 3.21

Mean values calculated from 100 independent seeds. All data are presented as mean $\pm$ SE. Statistical significance is determined by *t*-test.

2.3. Effect of OsCDPK1 on Starch Granule Morphology in Rice Endosperms

Due to the similar transgenic plant phenotypes and grain morphology (Figures 1 and 2), the *OsCDPK1*-overexpressing line, *OEtr-1*, and the *OsCDPK1*-silencing line, *Ri-1*, were therefore selected for further study. We examined the starch granule morphology in the WT, *Ri-1*, and *OEtr-1* transgenic seeds by analyzing the endosperm cross-sections acquired from scanning electron microscopy. As shown in Figure 3, the three-dimensional structures of the starch granules in the endosperms were irregularly polygonal and polyhedral in all three grains types. The starch granules were large and tightly packed in the WT and *Ri-1* endosperms but small and loosely packed in the *OEtr-1* endosperm (Figure 3B), suggesting that *OsCDPK1* affects the starch granule size and packing density in developing rice seed.

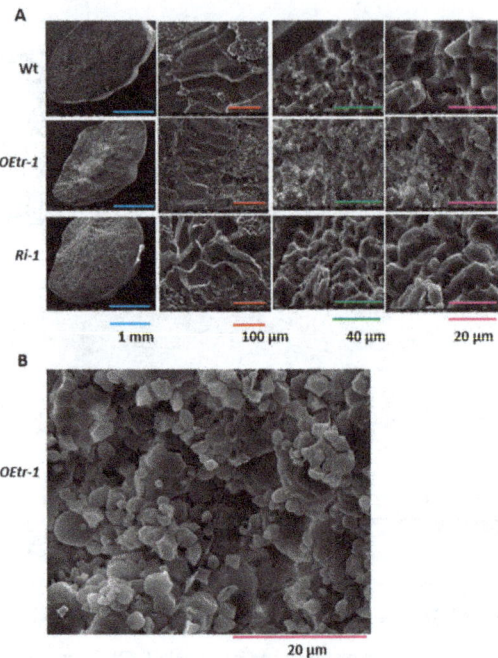

Figure 3. Scanning electron microscopy images of the structures of the starch granules in the rice endosperms. (**A**) The central areas of the cross-sections of mature endosperms from the WT plants (top panel), *OEtr-1* plants (middle panel), and *Ri-1* plants (bottom panel) were acquired using a scanning electron microscope. (**B**) A zoomed-in image of the 20 µm photo of *OEtr-1*. Scale bar colors: blue as 1 mm, red as 100 µm, green as 40 µm, purple as 20 µm.

2.4. Effects of OsCDPK1 on Starch Properties and Gelatinization in the Endosperm

We examined the apparent amylose content in the endosperm by reacting 20 mg of each rice endosperm powder sample with 1 N NaOH to gelatinize the starch. The amylose content was measured using a colorimetric method using an I_2/KI solution [12,33]. The *OEtr-1* samples had less affinity than the other samples for iodine and were light purple, whereas the *Ri-1* and WT samples were dark blue and light blue, respectively (Figure 4A and Figure S2). As described in Table 1, the average amylose content in WT, *OEtrs*, and *Ris* was 23.2%, 15.1% and 26.2%, respectively. Here, the apparent amylose content in WT, *OEtr-1* and *Ri-1* were examined in detail to dissect the roles of *OsCDPK1* in starch biosynthesis. The absorption spectra of the I_2/KI stained solutions and found strong absorbance between 480 and 720 nm for both the *Ri-1* and WT samples, with maxima at 620 nm, but stronger absorbance was found for *Ri-1* than for WT at the same wavelength (Figure 4B). *OEtr-1* absorbance was weaker and decreased as the wavelength increased. We compared the results to the absorbances of potato amylose standards to allow the apparent amylose contents to be determined. The apparent amylose contents of the WT, *OEtr-1*, and *Ri-1* samples were 23.35%, 14.74%, and 26.15%, respectively (Table S2). These results demonstrate that the *Ri-1* seed endosperms had higher amylose content than the WT seed endosperms and that the *OEtr-1* seed endosperms had lower amylose content than the WT seed endosperms. Because the WT, *OEtr-1*, and *Ri-1* amylose contents were different, we investigated starch gelatinization at different urea concentrations (0–9 M). A 20-mg aliquot of a rice endosperm powder was mixed with 1 mL of urea solution, and the mixture was allowed to react for 24 h. The mixture was then centrifuged and the degree of gelatinization determined by measuring the sediment volume. Starch gelatinization started at urea concentrations of 3.0–4.0 M (Figure 4C and Figure S3). In 4.0 M urea, the *Ri-1* sediment volume was 5.3% higher than the WT sediment volume, whereas the *OEtr-1* sediment volume was 36.4% lower than the WT sediment volume (Figure 4D). These results indicate that *OsCDPK1* affects the physicochemical properties of the starch in rice endosperms.

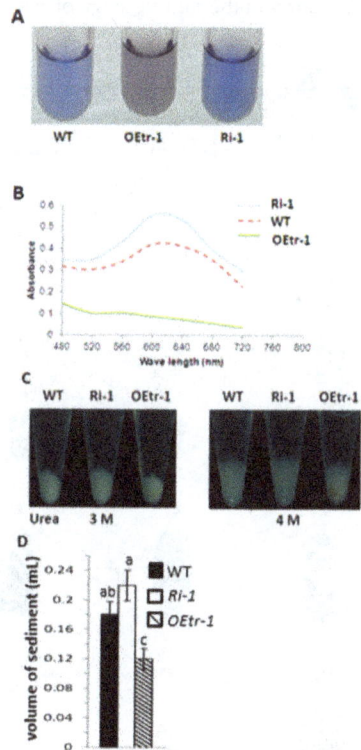

Figure 4. Iodine-staining and gelatinization properties of the starch in the rice endosperms. A 20-mg aliquot of endosperm powder was treated with 1 N NaOH as described in Section 4. (**A**) Supernatants of the iodine-stained WT, *OEtr-1*, and *Ri-1* samples. (**B**) Starch–iodine absorbance spectra of the supernatants. (**C**) Effects of using 3.0 and 4.0 M urea solutions on the gelatinization characteristics of the WT, *OEtr-1*, and *Ri-1* endosperm starch. A 20-mg aliquot of rice powder was mixed in an Eppendorf tube with 1 mL of urea solution and the mixture was shaken for 24 h at 25 °C. The mixture was centrifuged, and the volume of the gelatinized starch sediment was measured. (**D**) Quantification of the gelatinization volume. Different letters above the bars indicate significant differences, identified by performing ANOVAs ($p < 0.05$). Each value is the mean ± SD of three independent measurements.

2.5. OsCDPK1 Expression Profiles in Developing Rice Seeds

Our results have demonstrated that *OsCDPK1* affects rice seed development. It is necessary to track changes in *OsCDPK1* gene expression during rice seed development. The *OsCDPK1::GUS* transgenic line was generated using a *GUS* (*β-glucuronidase*) reporter gene controlled by the *OsCDPK1* promoter (−1706 to +301 bp, i.e., a total of 2007 bp upstream of the translational start site) containing the first intron (607 bp, in the 5′-untranslated region) (Figure 5A). As shown in Figure 5B, strong GUS staining was observed in the ovaries and anthers before flowering, but weaker staining was observed in the styles and lemma. At 1 DAF (days after flowering), strong GUS activity was found only in the ovaries and styles, weaker activity was found in the lemma, and no GUS activity was found in the anthers and stigma. Between 2 and 5 DAF, concentrated GUS staining was found in the rachilla and both ends of the developing seeds and weak staining was found in the lemma. The blue color gradually expanded from both ends toward the central parts of the developing seeds between 6 and 7 DAF, and at 8 DAF the entire seeds were thoroughly stained blue. Staining gradually decreased afterwards, but remained strong between 10 and 14 DAF, then decreased quickly after 14 DAF and had completely gone by 18 DAF. These results suggest that *OsCDPK1* was expressed in a particular temporal and spatial way, predominantly in the middle stage of rice seed development.

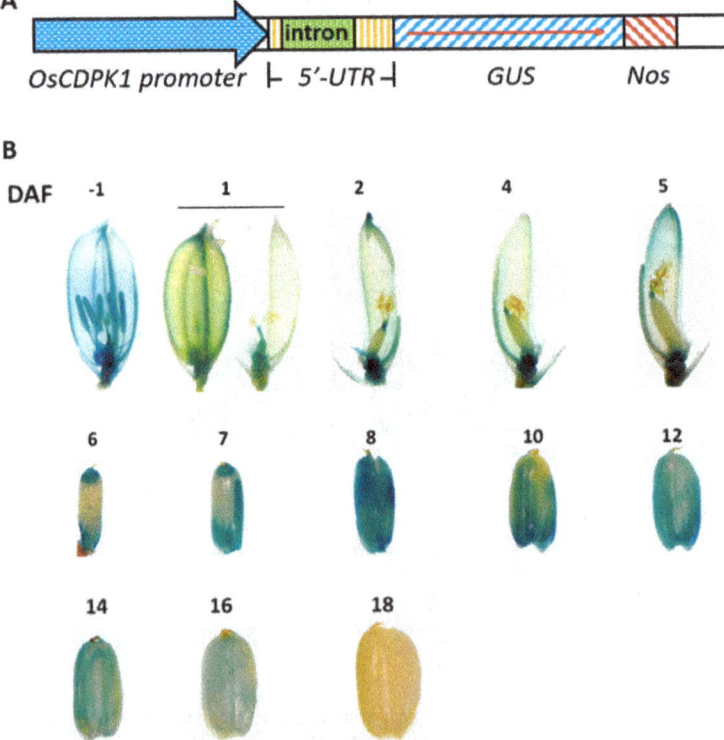

Figure 5. Histochemical GUS activity staining in flowers and developing rice seeds. (**A**) Map of the *OsCDPK1::GUS* expression construct. (**B**) Rice spikelets were collected before flowering and 1–18 days after flowering (DAF). The lemma and palea were partially or completed removed from each spikelet or developing seed before staining. The stained spikelets or immature grains were preserved in 70% ethanol and photographed. −1 DAF means before flowering.

2.6. Effects of OsCDPK1 on the Levels of Starch-Biosynthesis-Related Genes in Developing Rice Seeds

We further investigated the roles of *OsCDPK1* in rice-seed development by analyzing the expression patterns of 12 genes involved in starch biosynthesis. These genes were granule-bound starch synthase (*OsGBSSI*), starch synthase (*OsSSI, OsSSIIa, OsSSIIb, OsSSIIc, OsSSIIIa,* and *OsSSIIIb*), branching enzyme (*OsBEI*), ADP-glucose pyrophosphorylase large subunit (*OsAGPLI, OsAGPLII,* and *OsAGPLIII*), and a small subunit of ADP-glucose pyrophosphorylase (*OsAGPSIIb*). The developing WT, *Ri-1,* and *OEtr-1* rice seeds at 5 and 12 DAF were analyzed. Total RNA was isolated from the dehulled embryo-less half seeds and subjected to quantitative RT-PCR. The relative expression levels of *OsAGPLI, OsAGPSIIb, OsGBSSI, OsSSIIc,* and *OsSSIIIa* were significantly up-regulated in *Ri-1* and down-regulated in *OEtr-1* at 12 DAF, but there was no significant difference at 5 DAF compared with the genes in WT plants (Figure 6). Expression of the seven other genes in the transgenic lines was not significantly different at 5 or 12 DAF from expression in the WT plants. These results suggest that *OsCDPK1* might be involved in regulating starch-biosynthesis-related genes in the mid-development stage of the rice seed.

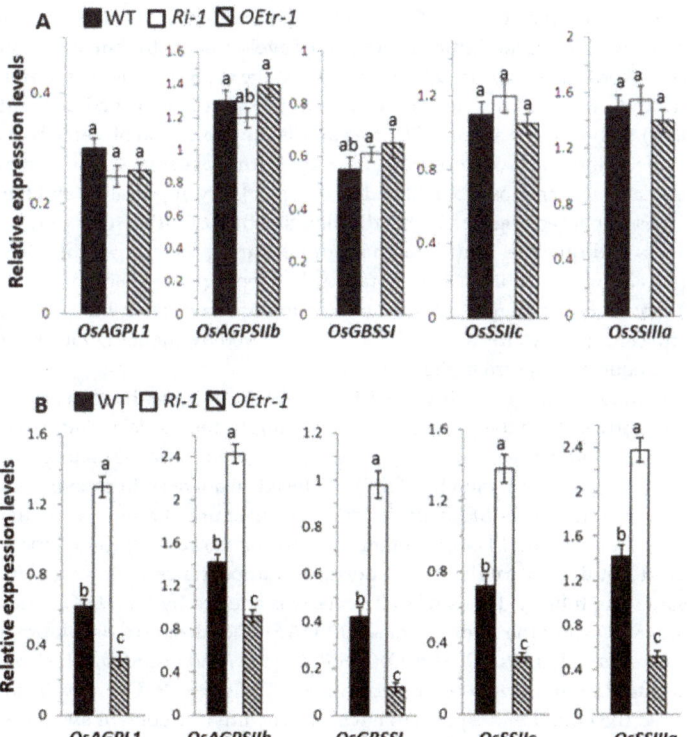

Figure 6. Expression of starch-biosynthesis-related genes during seed development in the WT, *Ri-1,* and *OEtr-1* plants. Total RNA was isolated from developing endosperms (**A**) 5 and (**B**) 12 DAF from the WT, *Ri-1,* and *OEtr-1* plants, and subjected to RT-PCR analysis. The relative expression levels of each gene were normalized to the expression level of the internal control *OsActin.* Different letters above the bars indicate significant differences, identified by performing ANOVAs ($p < 0.01$). Each value is the mean $\pm$ SD of three independent measurements. *OsAGPLI:* ADP-glucose pyrophosphorylase large subunit I; *OsAGPSIIb:* ADP-glucose pyrophosphorylase small subunit IIb; *OsGBSSI:* granule-bound starch synthase I; *OsSSIIc:* starch synthase IIc; *OsSSIIIa:* starch synthase IIIa. Primer sets and gene accession numbers are listed in Table S1.

3. Discussion

The amylose content of endosperms is an important parameter determining the eating quality of rice, which is negatively related with stickiness but positively related to rice grain hardness [34,35]. The key enzymes (genes) involved in the starch (amylopectin and amylose) biosynthesis during rice grain filling are well known. However, the control mechanisms by which the amylose and amylopectin biosynthesis is orchestrated are still not fully understood. Our results indicate for the first time that the protein kinase *OsCDPK1* is functionally negatively correlated with the amylose content, endosperm transparency, and seed size in developing rice seed.

The *OEtr-1* and *Ri-1* seeds had some distinct features compared with the WT seeds. For example, the *OEtr-1* grains were smaller, had lower amylose contents, and had more floury endosperms than the WT grains, and the *Ri-1* grains were larger, had higher amylose contents, and had more transparent endosperms than the WT grains (Figures 1–4). This indicates that the *OsCDPK1* function is closely associated with the rice endosperm starch properties. The *OsCDPK1::GUS* expression profile in the developing rice grains gradually increased immediately after flowering and reached a maximum between 7 and 14 DAF (Figure 5). The *OsCDPK1::GUS* expression timing in the developing rice seeds was similar to that found in a study by Ohdan et al. [36], in which 27 different genes involved in starch biosynthesis were examined during rice-seed development. In that study, all the genes had been differentially expressed before 15 DAF. In this study, the GUS staining was found throughout the endosperm at 12 and 14 DAF of developing seeds, and was more intensified in the interior region at 12 DAF (Figure S4), suggesting that *OsCDPK1* first affects the expression of starch biosynthesis-related genes and later affects the starch composition and endosperm appearance. The starchy endosperm in the *OEtr-1* grains had a low amylose content and an opaque floury appearance, and the starch granules were small and loosely packed (Figures 1–3), indicating that the *OsCDPK1* roles were closely associated with the structures and qualities of the starch granules during the grain filling process. Moreover, *OsCDPK1::GUS* staining was found throughout the developing endosperm at 7–16 DAF (Figure 5B and Figure S4). These results suggest that the expression of some starch-biosynthesis-related genes in the endosperm cells may be affected by *OsCDPK1*, followed by changing the amylose content and resulting in the opaque endosperm in the *OEtr-1* grains.

Several mutations of rice genes involved in starch biosynthesis have been found to alter the structures and properties of the starch produced. For example, the *SSIIIa* mutation (*flo5*) was found to increase the amylose content, alter the amylopectin structure, and cause the endosperm to have a white core [20]. The *waxy* mutant (a mutation in *GBSSI*) produced an amylose-free, floury endosperm [11,12]. The *amylose-extender* mutation in *BEIIb* altered the fine structure of amylopectin and gave a floury endosperm [12]. A mutation in *BEI* (starch-branching enzyme I) caused the amylopectin fine structure to change but did not appear to affect the endosperm appearance [10]. Our data show that the expression of some starch-biosynthesis-related genes was affected by *OsCDPK1*, similar to the results of previous studies. During the middle phase (12 DAF) of endosperm development, examples of genes affected were *OsAGPLI*, *OsAGPSIIb*, *OsGBSSI*, *OsSSIIc*, and *OsSSIIIa*, which were significantly up-regulated in the *Ri-1* and down-regulated in *OEtr-1* (Figure 6). Changes in the expression of these genes caused the *OEtr-1* endosperm to have a lower amylose content and a floury appearance, whereas the *Ri-1* endosperm had a high amylose content and was more transparent (Figures 1, 2 and 4 and Tables 1 and 2). Calcium ions (Ca^{2+}) are secondary messengers in plant cells, with plant CDPK being a sensor to relay calcium signals via binding with calcium via the calcium-binding domains. We therefore suggest that *OsCDPK1* acts as an upstream regulator that is closely associated with starch biosynthesis. Some regulators have been found to regulate the expression of genes encoding key starch-biosynthesis enzymes, and null mutations in these regulators usually alter the amylopectin fine structure, the starch composition, the starch granule morphology and size, and the endosperm appearance. For instance, some starch-biosynthesis-related genes were found to be affected in five independent mutants, *flo2* (a tetratricopeptide repeat motif protein) [17], *flo4* (*pyruvate orthophosphate dikinase B*) [19], *flo7* (an unknown protein) [22], *osbzip58* (a bZIP transcription factor) [37], and *osbt1* (an ADP-glucose transporter) [38]. All these effects decreased the amylose content and changed the amylopectin composition. The endosperm appearance was affected differently in different parts; e.g., *flo2* had endosperm with a floury kernel, *flo4* and *osbzip58* and *osbt1* gave white-core phenotypes, and chalkiness was only found in the peripheral endosperm of *flo7*. Similarly, the low amylose content of the *OEtr-1* endosperm was expressed in a floury morphology (Figures 1 and 4). The results of previous studies and this study together indicate that a low-amylose content of rice endosperm gives a chalky or floury phenotype, suggesting that the amount of amylose present is an important factor affecting the quality and appearance of the starchy endosperm. This raises the question of whether rice can be engineered to have an endosperm with a high amylose content and, therefore, a transparent appearance. Here, we have provided direct evidence that silencing *OsCDPK1* (*Ri-1*) increases the amylose content of the endosperm, making the starchy grain more transparent (Figures 1 and 4 and Tables 1 and 2). These results will be useful in developing rice-breeding strategies aimed at maintaining (or even improving) grain quality in rice to cope with global warming. The transgenic lines *OEtrs* and *Ris* could also be ideal materials for investigating the mechanisms that control rice seed size and starch biosynthesis.

Temperature also strongly affects amylose synthesis during rice grain development. In previous studies, a lower temperature increased the expression of the *waxy* gene and protein and increased the amylose content in developing rice endosperms [39]; a higher temperature had the opposite effects [39,40]. Moreover, during the rice grain-filling process, a temperature higher than the optimum usually causes impaired starch accumulation, resulting in loosely packed starch granules with small air spaces between them, giving high proportions of opaque chalky or floury grains. These effects decrease the market value because the rice will have poor milling qualities (being easily broken), poor cooking and eating qualities, and a poor appearance [41,42]. It has recently been found that a high temperature (33 °C for 12 h light, 28 °C for 12 h dark) also induced the expression of three α-amylase genes—*Amy1A*, *Amy3C*, and *Amy3D*—in developing endosperms, causing the grains to be chalky, probably because of starch degradation and the accumulation of soluble sugars in the endosperm [43]. In contrast, we found that the *Ri-1* endosperm was more transparent at both a low temperature (22 °C for 16 h light, 20 °C for 8 h dark) and a high temperature (31 °C for 16 h light, 28 °C for 8 h dark) than that of the WT plant during the grain-filling process (Figure 2 and Table 2). Our results will be useful in developing rice-breeding strategies aimed at maintaining (or even improving) grain quality in rice selected to cope with global warming. Our results also improve our understanding of the molecular mechanisms involved in amylose biosynthesis.

It has been found in several studies that chalky or floury endosperms might be caused by loosely packed small, round starch granules formed in developing rice seeds [9,19,20,38]. The scanning electron microscopy images indicate that the WT and *Ri-1* endosperms contained closely packed polyhedral starch granules but that the white core of the *OEtr-1* endosperm contained loosely packed small starch granules. The starch granules in the *OEtr-1* endosperm were polyhedral (like in the WT and *Ri-1* endosperm) rather than small and round as in most chalky or floury mutants (Figure 3B). It is, therefore, likely that *OsCDPK1* plays a role in starch biosynthesis and negatively affects the sizes but not the shapes of the starch granules.

We previously found that *OsCDPK1* inhibits the feedback of GA biosynthesis through down-regulating *GA3ox2* and *GA20ox1* [32]. In this study, we put forward a model in which *OsCDPK1* plays key roles in negatively controlling the grain size, amylose content, and endosperm appearance, and also affects the physicochemical properties of the starch (Figure 7). Milled *OsCDPK1*-gene-silenced *Ri-1* grains were larger than WT grains, and there were numerous densely packed polyhedral starch granules accompanied by a high amylose content and a transparent endosperm. In contrast, the *OEtr-1* grains were smaller and contained loosely packed small, but still polyhedral, starch granules. The *OEtr-1* grains had lower amylose content and opaque white-cored endosperms. Notably, the phenotypes of the grain size and the floury endosperm in *OEtr-1*, -2, and -3 were consistent and were unaffected by temperature during the grain-filling process (Figures 1 and 2). *OsCDPK1* therefore plays pleiotropic roles in rice seed development. Our results indicate that *Ri-1* and *OEtr-1* could be ideal materials for investigating the mechanisms controlling rice seed size and starch biosynthesis, and could also be valuable reference samples for rice breeding aimed at simultaneously improving grain yield and quality.

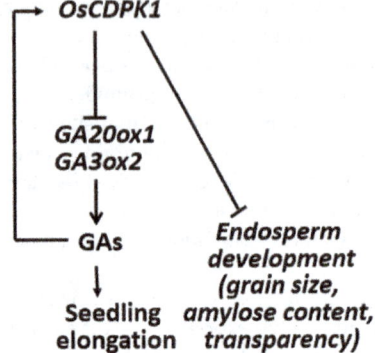

Figure 7. Proposed roles of *OsCDPK1* in the interconnecting GA biosynthesis and signaling pathways and endosperm developmental processes. The model is described in detail in the text. The arrows indicate activation and the blunt ends indicate inhibition.

4. Materials and Methods

4.1. Plant Materials

In our previous studies [32], 8 independent T1 transgenic lines that carried a single transgene in rice were subjected to ectopic overexpression of a constitutively active truncated form of *OsCDPK1* (*OEtr*) in rice seedlings, which all yielded a semi-dwarf phenotype that produced small seeds. By contrast, 5 independent T1 transgenic plants exhibiting a single copy of transgene were subject to *OsCDPK1* gene silencing (*Ri*) by RNA interference, which all gave a slender-like phenotype during seedling development and subsequently produced large seeds. According to the consistency of the phenotypic traits in *OEtr* lines of which all displayed semi-dwarf phenotype and produced small seeds. In *Ri* lines, which all gave a slender-like phenotype and produced large seeds [32]. Therefore, in this study, the homozygous transgenes of the T4 transgenic plants from *OEtr-1*, *OEtr-3*, and *OEtr-4*, and *Ri-1*, *Ri-2*, and *Ri-3* were selected to explore the roles of *OsCDPK1* in rice grain development.

4.2. Callus Induction

Rice *Oryza sativa* L. cv. Tainung 67 was used in the study. Immature seeds were de-hulled, sterilized with 2.4% NaOCl for 30 min, then washed thoroughly with sterile water. The seeds were then incubated on N6 agar medium [44] containing 10 µM 2,4-D to induce calli formation [45]. After about 30 days, the calli derived from the scutella were transferred to the fresh N6 agar medium containing 2,4-D for another 15 days and were subjected to *Agrobacterium*-mediated gene transformation [45].

4.3. Primers

The nucleotide sequences of all the primers used in the real-time polymerase chain reaction (qRT-PCR) amplification are shown in Table S1.

4.4. Construction of OsCDPK1::GUS Expression Vectors

The *OsCDPK1::GUS* expression vector was constructed by amplifying a 2007 bp DNA fragment containing the *OsCDPK1* promoter and its 5′-untranslated region (which contained a 607 bp intron)

(Figure 1A) by PCR using the forward primer *OsCDPK1-5P* (5′-ATACTGCAGTGGTCTTATT AGGTAAGGCC-3′) and the reverse primer *OsCDPK1-3B* (5′-ATAGGATCC TCCAAGAACTCCT TATGCAA-3′). The DNA fragment was cleaved using *Pst*I and *Bam*HI, and then cloned into vector *pBX-2* as described previously [45]. The *OsCDPK1::GUS* construct was linearized by digesting it with *Pst*I, then inserted into the *Pst*I site of the *pSMY1H* binary vector [45], then it was subjected to *Agrobacterium*-mediated gene transformation.

4.5. Plant Transformation

Recombinant binary plasmids were introduced into *Agrobacterium tumefaciens* strain EHA101 by electroporation, and rice calli were transformed as described previously [45].

4.6. Histochemical Staining of GUS Activity in Developing Rice Grains

Rice spikelets collected before flowering and 1–18 DAF were subjected to GUS activity staining to assess *OsCDPK1* gene expression profiles in developing rice grains. The lemma and palea were partially or completely removed from each spikelet (developing seed) before the staining process, and the spikelets were then incubated in a 1-mM 5-bromo-4-chloro-3-indolyl β-D-glucuronide solution (in 100 mM sodium phosphate containing 10 mM EDTA, 0.5 mM potassium ferrocyanide, 0.5 mM potassium ferricyanide, and 0.1% Triton X-100, at pH 7.0) at 37 °C in the dark for 4 h. The stained spikelets or immature grains were then preserved in 70% ethanol and rinsed with water before being photographed.

4.7. Quantitative RT-PCR

Developing seeds were collected from wild-type (WT), *Ri-1*, and *OEtr-1* at 5 and 12 DAF, respectively. Total RNA was isolated from developing endosperms by using TRIzol reagent (Invitrogen, Carlsbad, CA, USA), and DNA contamination was then removed using a TURBO DNA-free kit (Ambion, Foster City, CA, USA). A 5-µg aliquot of the total RNA was used to synthesize first strand cDNA using M-MuLV reverse transcriptase (New England Biolabs, Ipswich, MA, USA) and oligo (dT) primer. Quantitative RT-PCR was performed using an Eco Real-Time PCR System (Illumina, San Diego, CA, USA) following the manufacturer's instructions. Gene-specific primer sets (Table S1) localized at the 3′-untranslated regions for each gene examined were selected to allow assessment of the extent to which the starch-biosynthesis-related genes in WT, *OEtr-1*, and *Ri-1* were expressed. The relative expression levels were normalized to expression in the internal control, *OsActin 1*. All reactions were performed in triplicate.

4.8. Analysis of the Gelatinization Properties of the Starch

A 20-mg aliquot of rice endosperm powder derived from de-embryonic seeds was mixed with 1 mL of urea solution at a concentration of between 0 and 9 M, and the mixture was shaken vigorously for 24 h at room temperature [12]. All the tested mixtures were then centrifuged at 25,000× *g* for 60 min (Eppendorf, model 5427R, North Ryde, Australia) at a same time, and the volume of the gelatinized starch sediment was measured by the volume scale on the Eppendorf tube.

4.9. Apparent Amylose Content Analysis

A 20-mg aliquot of rice endosperm powder was gelatinized by adding 2 mL of 1 N NaOH and incubating the mixture at 25 °C for 24 h. Then, 4 mL of 1 N CH$_3$COOH was added, the mixture was mixed well, and 4 mL H$_2$O was added. A 0.8-mL aliquot of the solution was mixed with 0.2 mL I$_2$/KI (0.2%/2%), then 4 mL H$_2$O was added. The apparent amylose content was measured using the colorimetric method described by [33]. Absorbance at 620 nm was measured, and the apparent amylose content was determined by comparing the absorbance to a calibration curve prepared using potato amylose standards.

4.10. Scanning Electron Microscopy

Dehusked rice grains were cut transversely and analyzed using a scanning electron microscope (Quanta 200; FEI, Hillsboro, OR, USA) following the manufacturer's instructions.

5. Conclusions

Our results demonstrated that *OsCDPK1* plays key roles in negatively controlling the grain size, amylose content, and endosperm appearance, and also affects the physicochemical properties of the starch. Milled *OsCDPK1*-gene-silenced *Ri-1* grains were larger than WT grains, and there were numerous densely packed polyhedral starch granules accompanied by a high amylose content and a transparent endosperm. In contrast, the *OEtr-1* grains were smaller and contained loosely packed small, but still polyhedral, starch granules. The *OEtr-1* grains had lower amylose content and opaque white-cored endosperms. Moreover, the grain phenotypes in *OEtrs* were unaffected by temperature during the grain-filling process. *OsCDPK1* therefore plays pleiotropic roles in rice reproductive developmental processes, in a negative sense. Our results indicate that *Ri-1* and *OEtr-1* could be ideal materials for investigating the mechanisms that control rice seed size and starch biosynthesis, and for rice breeding to improve grain yield and quality.

Supplementary Materials: The following are available online at http://www.mdpi.com/1422-0067/19/10/3247/s1.

Author Contributions: Conceptualization: S.-L.H.; Data curation: J.-Z.J. and S.-L.H.; Investigation: J.-Z.J., C.-H.K., B.-H.C. and M.-K.C.; Methodology: C.-S.L.; Supervision: S.-L.H.; Writing—original draft: S.-L.H.

Funding: This work was supported by grants from the Ministry of Science and Technology of the Republic of China (grant no. MOST 105-2313-B-415-009-, MOST 106-2313-B-415-006- and MOST 107-2313-B-415-006-).

Conflicts of Interest: The authors declare no conflict of interest.

References

1. Rabiei, B.; Valizadeh, M.; Ghareyazie, B.; Moghaddam, M.; Ali, A.J. Identification of QTLs for rice grain size and shape of Iranian cultivars using SSR markers. *Euphytica* **2004**, *137*, 325–332. [CrossRef]
2. Bao, J.; Jin, L.; Xiao, P.; Shen, S.; Sun, M.; Corke, H. Starch physicochemical properties and their associations with microsatellite alleles of starch synthesizing genes in a rice RIL population. *J. Agric. Food Chem.* **2008**, *56*, 1589–1594. [CrossRef] [PubMed]
3. Fitzgerald, M.A.; Lisle, A.J.; Martin, M. Chalky and translucent rice grains differ in starch composition and structure and cooking properties. *Cereal Chem.* **2000**, *77*, 627–632.
4. Jiang, H.; Dian, W.; Wu, P. Effect of high temperature on fine structure of amylopectin in rice endosperm by reducing the activity of the starch branching enzyme. *Phytochemistry* **2003**, *63*, 53–59. [CrossRef]
5. Morita, S.; Shiratsuchi, H.; Takanashi, J.; Fujita, K. Effect of high temperature on grain ripening in rice plants: Analysis of the effects of high night and high day temperatures applied to the panicle and other parts of the plant. *Jpn. J. Crop Sci.* **2004**, *73*, 77–83. [CrossRef]
6. Counce, P.A.; Bryant, R.J.; Bergman, C.J.; Bautista, R.C.; Wang, Y.J.; Siebenmorgen, T.J.; Moldenhauer, K.A.; Meullenet, J.F.C. Rice milling quality, grain dimensions, and starch branching as affected by high night temperatures. *Cereal Chem.* **2005**, *82*, 645–648. [CrossRef]
7. Patindol, J.; Wang, Y.J. Fine structures and physicochemical properties of starches from chalky and translucent rice kernels. *J. Agric. Food Chem.* **2003**, *51*, 2777–2784. [CrossRef] [PubMed]
8. Yamakawa, H.; Hirose, T.; Kuroda, M.; Yamaguchi, T. Comprehensive expression profiling of rice grain filling-related genes under high temperature using DNA microarray. *Plant Physiol.* **2007**, *144*, 258–277. [CrossRef] [PubMed]
9. Fu, F.F.; Xue, H.W. Coexpression analysis identifies rice starch regulator1, a rice AP2/EREBP family transcription factor, as a novel rice starch biosynthesis regulator. *Plant Physiol.* **2010**, *154*, 927–938. [CrossRef] [PubMed]
10. Mizuno, K.; Kimura, K.; Arai, Y.; Kawasaki, T.; Shimada, H.; Baba, T. Starch branching enzymes from immature rice seeds. *J. Biochem.* **1992**, *112*, 643–651. [CrossRef] [PubMed]

11. Smith, A.M.; Denyer, K.; Martin, C. The synthesis of the starch granule. *Annu. Rev. Plant Physiol. Plant Mol. Biol.* **1997**, *48*, 67–87. [CrossRef] [PubMed]

12. Nishi, A.; Nakamura, Y.; Tanaka, N.; Satoh, H. Biochemical and genetic analysis of the effects of amylose-extender mutation in rice endosperm. *Plant Physiol.* **2001**, *127*, 459–472. [CrossRef] [PubMed]

13. Singh, B.K.; Preiss, J. Starch branching enzymes from maize. *Plant Physiol.* **1985**, *79*, 34–40. [CrossRef] [PubMed]

14. Nakamura, Y.; Takeichi, T.; Kawaguchi, K.; Yamanouchi, H. Purification of two forms of starch branching enzyme (Q-enzyme) from developing rice endosperm. *Physiol. Plant.* **1992**, *84*, 329–335. [CrossRef]

15. Satoh, H.; Omura, T. New endosperm mutations induced by chemical mutagens in rice, *Oryza sativa* L. *Jpn. J. Breed.* **1981**, *31*, 316–326. [CrossRef]

16. Kaushik, R.P.; Khush, G.S. Genetic analysis of endosperm mutants in rice *Oryza sativa* L. *Theor. Appl. Genet.* **1991**, *83*, 146–152. [CrossRef] [PubMed]

17. She, K.C.; Kusano, H.; Koizumi, K.; Yamakawa, H.; Hakatae, M.; Imamura, T.; Fukuda, M.; Naito, N.; Tsurumaki, Y.; Yaeshima, M.; et al. A novel factor FLOURY ENDOSPERM 2 is involved in regulation of rice grain size and starch quality. *Plant Cell* **2010**, *22*, 3280–3294. [CrossRef] [PubMed]

18. Nishio, T.; Iida, S. Mutant having a low content of 16-kDa allergenic protein in rice (*Oryza sativa* L.). *Theor. Appl. Genet.* **1993**, *86*, 317–321. [CrossRef] [PubMed]

19. Kang, H.G.; Park, S.; Matsuoka, M.; An, G. White-core endosperm floury endosperm-4 in rice is generated by knockout mutations in the C_4-type pyruvate orthophosphate dikinase gene (*OsPPDKB*). *Plant J.* **2005**, *42*, 901–911. [CrossRef] [PubMed]

20. Ryoo, N.; Yu, C.; Park, C.S.; Baik, M.Y.; Park, I.M.; Cho, M.H.; Bhoo, S.H.; An, G.; Hahn, T.R.; Jeon, J.S. Knockout of a starch synthase gene *OsSSIIIa/Flo5* causes white-core floury endosperm in rice (*Oryza sativa* L.). *Plant Cell Rep.* **2007**, *26*, 1083–1095. [CrossRef] [PubMed]

21. Peng, C.; Wang, Y.H.; Liu, F.; Ren, Y.L.; Zhou, K.N.; Lv, J.; Zheng, M.; Zhao, S.L.; Zhang, L.; Wang, C.M.; et al. *FLOURY ENDOSPERM6* encodes a CBM48 domain-containing protein involved in compound granule formation and starch synthesis in rice endosperm. *Plant J.* **2014**, *77*, 917–930. [CrossRef] [PubMed]

22. Zhang, L.; Ren, Y.L.; Lu, B.Y.; Yang, C.Y.; Feng, Z.M.; Liu, Z.; Chen, J.; Ma, W.W.; Wang, Y.H.; Liu, Y.S.; et al. *FLOURY ENDOSPERM7* encodes a regulator of starch synthesis and amyloplast development essential for peripheral endosperm development in rice. *J. Exp. Bot.* **2016**, *67*, 633–647. [CrossRef] [PubMed]

23. Sanders, D.; Pelloux, J.; Brownlee, C.; Harper, J.F. Calcium at the crossroads of signaling. *Plant Cell* **2002**, *14*, S401–S417. [CrossRef] [PubMed]

24. Harper, J.F.; Breton, G.; Harmon, A. Decoding Ca^{2+} signals through plant protein kinases. *Annu. Rev. Plant Biol.* **2004**, *55*, 263–288. [CrossRef] [PubMed]

25. Mori, I.C.; Murata, Y.; Yang, Y.; Munemasa, S.; Wang, Y.F.; Andreoli, S.; Tiriac, H.; Alonso, J.M.; Harper, J.F.; Ecker, J.R.; et al. CDPKs CPK6 and CPK3 function in ABA regulation of guard cell S-type anion- and Ca^{2+}-permeable channels and stomatal closure. *PLoS Biol.* **2006**, *4*, e327. [CrossRef] [PubMed]

26. Sheen, J. Ca^{2+}-dependent protein kinases and stress signal transduction in plants. *Science* **1996**, *274*, 1900–1902. [CrossRef] [PubMed]

27. Ludwig, A.A.; Romeis, T.; Jones, J.D. CDPK-mediated signalling pathways: Specificity and cross-talk. *J. Exp. Bot.* **2004**, *55*, 181–188. [CrossRef] [PubMed]

28. Romeis, T.; Ludwig, A.A.; Martin, R.; Jones, J.D.G. Calcium-dependent protein kinases play an essential role in a plant defence response. *EMBO J.* **2001**, *20*, 5556–5567. [CrossRef] [PubMed]

29. Abbasi, F.; Onodera, H.; Toki, S.; Tanaka, H.; Komatsu, S. OsCDPK13, a calcium-dependent protein kinase gene from rice, is induced by cold and gibberellin in rice leaf sheath. *Plant Mol. Biol.* **2004**, *55*, 541–552. [CrossRef] [PubMed]

30. Wan, B.; Lin, Y.; Mou, T. Expression of rice Ca^{2+}-dependent protein kinases (CDPKs) genes under different environmental stresses. *FEBS Lett.* **2007**, *581*, 1179–1189. [CrossRef] [PubMed]

31. Asano, T.; Kunieda, N.; Omura, Y.; Ibe, H.; Kawasaki, T.; Takano, M.; Sato, M.; Furuhashi, H.; Mujin, T.; Takaiwa, F.; et al. Rice SPK, a calmodulin-like domain protein kinase, is required for storage product accumulation during seed development: Phosphorylation of sucrose synthase is a possible factor. *Plant Cell* **2002**, *14*, 619–628. [CrossRef] [PubMed]

32. Ho, S.L.; Huang, L.F.; Lu, C.A.; He, S.L.; Wang, C.C.; Yu, S.P.; Chen, J.; Yu, S.M. Sugar starvation- and GA-inducible calcium-dependent protein kinase 1 feedback regulates GA biosynthesis and activates a 14-3-3 protein to confer drought tolerance in rice seedlings. *Plant Mol. Biol.* **2013**, *81*, 347–361. [CrossRef] [PubMed]

33. Juliano, B.O. A simplified assay for milled-rice amylose. *Cereal Sci. Today* **1971**, *16*, 334–340.

34. Cameron, D.K.; Wang, Y.J. A better understanding of factors that affect the hardness and stickiness of long-grain rice. *Cereal Chem.* **2005**, *82*, 113–119. [CrossRef]

35. Li, H.; Prakash, S.; Nicholson, T.M.; Fitzgerald, M.A.; Gilbert, R.G. The importance of amylose and amylopectin fine structure for textural properties of cooked rice grains. *Food Chem.* **2016**, *196*, 702–711. [CrossRef] [PubMed]

36. Ohdan, T.; Francisco, P.B., Jr.; Sawada, T.; Hirose, T.; Terao, T.; Satoh, H.; Nakamura, Y. Expression profiling of genes involved in starch synthesis in sink and source organs of rice. *J. Exp. Bot.* **2005**, *56*, 3229–3244. [CrossRef] [PubMed]

37. Wang, J.C.; Xu, H.; Zhu, Y.; Liu, Q.Q.; Ca, X.L. OsbZIP58, a basic leucine zipper transcription factor, regulates starch biosynthesis in rice endosperm. *J. Exp. Bot.* **2013**, *64*, 3453–3466. [CrossRef] [PubMed]

38. Li, S.; Wei, X.; Ren, Y.; Qiu, J.; Jiao, G.; Guo, X.; Tang, S.; Wan, J.; Hu, P. *OsBT1* encodes an ADP-glucose transporter involved in starch synthesis and compound granule formation in rice endosperm. *Sci. Rep.* **2017**, *7*, 40124. [CrossRef] [PubMed]

39. Sano, Y.; Maekawa, M.; Kikuchi, H. Temperature effects on the Wx protein level and amylose content in the endosperm of rice. *J. Hered.* **1985**, *6*, 221–222. [CrossRef]

40. Asaoka, M.; Okuno, K.; Sugimoto, Y.; Kawakami, J.; Fuwa, H. Effect of environmental temperature during development of rice plants on some properties of endosperm starch. *Starch* **1984**, *36*, 189–193. [CrossRef]

41. Lyman, N.B.; Jagadish, S.V.K.; Nalley, L.L.; Dixon, B.L.; Siebenmorgen, T. Neglecting rice milling yield and quality underestimates economic losses from high-temperature stress. *PLoS ONE* **2013**, *8*, e72157. [CrossRef] [PubMed]

42. Shi, W.; Yin, X.; Struik, P.C.; Solis, C.; Xie, F.; Schmidt, R.C.; Huang, M.; Zou, Y.; Ye, C.; Jagadish, S.V.K. High day- and night-time temperatures affect grain growth dynamics in contrasting rice genotypes. *J. Exp. Bot.* **2017**, *68*, 5233–5245. [CrossRef] [PubMed]

43. Nakata, M.; Fukamatsu, Y.; Miyashita, T.; Hakata, M.; Kimura, R.; Nakata, Y.; Kuroda, M.; Yamaguchi, T.; Yamakawa, H. High temperature-induced expression of rice α-amylases in developing endosperm produces chalky grains. *Front. Plant Sci.* **2017**, *8*, 2089. [CrossRef] [PubMed]

44. Chu, C.C.; Wang, C.C.; Sun, C.S.; Hsu, C.; Yin, K.C.; Chu, C.Y.; Bi, F.Y. Establishment of an efficient medium for another culture of rice through comparative experiments on the nitrogen sources. *Sci. Sin.* **1975**, *5*, 659–668.

45. Ho, S.L.; Tong, W.F.; Yu, S.M. Multiple mode regulation of a cysteine proteinase gene expression in rice. *Plant Physiol.* **2000**, *122*, 57–66. [CrossRef] [PubMed]

Article

OsGPAT3 Plays a Critical Role in Anther Wall Programmed Cell Death and Pollen Development in Rice

Lianping Sun [†], Xiaojiao Xiang [†], Zhengfu Yang, Ping Yu, Xiaoxia Wen, Hong Wang, Adil Abbas, Riaz Muhammad Khan, Yingxin Zhang, Shihua Cheng * and Liyong Cao *

Key Laboratory for Zhejiang Super Rice Research and State Key Laboratory of Rice Biology, China National Rice Research Institute, Hangzhou 310006, China; sunlianping@caas.cn (L.S.); taorangongwangjimm@163.com (X.X.); cnrriyzf@sina.com (Z.Y.); pingping367@163.com (P.Y.); 18883966700@163.com (X.W.); wjiyinh@126.com (H.W.); adilabbasqau@126.com (A.A.); riazkatlang@126.com (R.M.K.); zyxrice@163.com (Y.Z.)
* Correspondence: chengshihua@caas.cn (S.C.); caoliyong@caas.cn (L.C.);
  Tel.: +86-571-6337-0188 (S.C.); +86-571-6337-0329 (L.C.)
† These authors contributed equally to this work.

Received: 13 November 2018; Accepted: 4 December 2018; Published: 12 December 2018

Abstract: In flowering plants, ideal male reproductive development requires the systematic coordination of various processes, in which timely differentiation and degradation of the anther wall, especially the tapetum, is essential for both pollen formation and anther dehiscence. Here, we show that *OsGPAT3*, a conserved glycerol-3-phosphate acyltransferase gene, plays a critical role in regulating anther wall degradation and pollen exine formation. The *gpat3-2* mutant had defective synthesis of Ubisch bodies, delayed programmed cell death (PCD) of the inner three anther layers, and abnormal degradation of micropores/pollen grains, resulting in failure of pollen maturation and complete male sterility. Complementation and clustered regularly interspaced short palindromic repeats (CRISPR)/CRISPR-associated 9 (Cas9) experiments demonstrated that *OsGPAT3* is responsible for the male sterility phenotype. Furthermore, the expression level of tapetal PCD-related and nutrient metabolism-related genes changed significantly in the *gpat3-2* anthers. Based on these genetic and cytological analyses, *OsGPAT3* is proposed to coordinate the differentiation and degradation of the anther wall and pollen grains in addition to regulating lipid biosynthesis. This study provides insights for understanding the function of *GPATs* in regulating rice male reproductive development, and also lays a theoretical basis for hybrid rice breeding.

Keywords: anther wall; tapetum; pollen accumulation; *OsGPAT3*; rice

1. Introduction

Rice is a key gramineous plant that is self-pollinated. Guaranteeing sufficient rice yield requires stable male sterility, which depends on normal development of anther and male gametophyte (pollen) formation [1,2]. Typical rice anthers have four lobes, and the central reproductive microsporocytes (or pollen mother cells) are surrounded by four concentrically organized somatic cell layers in each locule: the epidermis, endothecium, middle layer, and tapetum. Anther development is a multistage process involving localized cellular differentiation and degeneration, cell division and chromosomal behaviors, and synthesis and transportation of nutrients. This process is combined with changes in the structure and external environment to complete anther dehiscence and pollen maturation, and release for pollination and fertilization. All four cell layers possess specific functions and coordinate throughout the whole process to ensure normal anther development and microspore/pollen formation in rice [2–6]. The epidermis is located at the outermost layer of anthers to protect against external environmental stresses to ensure normal development of internal cells at the appropriate time for anther dehiscence and pollen release. The endothecium, with localized secondary thickening, is the second layer and is essential for anther dehiscence and pollen release. Endothecial development is concurrent with pollen maturation and degeneration of the anther tapetum and middle layer. The middle layer, which is situated between the tapetum and endothecium, undergoes programmed degeneration along with the tapetum during the pollen maturation stage [5–8]. The innermost cell layer of the anther wall, the tapetum, directly contacts developing gametophytes. The tapetum undergoes programmed cell death (PCD)-mediated degeneration to supply a series of nutritional components and structural molecules for normal pollen formation and ordinary anther development [9,10]. Proper development and timely degeneration of tapetal cells is essential for providing and supplying nutrients for sporopollenin synthesis and pollen development [1,11–13]. All developmental processes of these cell layers are dominated by specific regulators with extremely precise and systematic molecular mechanisms; mutation of these genes may result in anther development malformations and eventually lead to pollen abortion [14,15].

Programmed cell death (PCD) events, known as apoptosis, are often characterized by nuclear chromatin condensation and degeneration, membrane breakdown, and compactness of cytoplasmic organelles. PCD is usually used as a cytological feature of tapetum degradation in plants [9,12,16]. The PCD process also occurs in the most peripheral layers at the late stages of anther development. Vascular bundle cells, the cavity of dehiscence, the epidermis, and the endothecium are always prepared for apoptosis to provide nutrients for pollen mitosis and maturation [17]. Timely PCD of the anther wall, especially tapetum degradation in vivo, plays a fundamental role in male reproduction. Premature or delayed tapetal PCD and cellular degeneration can cause pollen abortion and male sterility [2,9,12,15]. Recently, multiple genes were identified to play essential roles in the process of rice tapetal PCD and pollen development, including several basic helix–loop–helix (*bHLH*) transcription factors, undeveloped tapetum 1 (*UDT1*, *bHLH164*) [11], tapetum degeneration retardation (*TDR*, *bHLH5*) [12,18], eternal tapetum 1/delayed tapetum degeneration (*EAT1/DTD*, *bHLH141*) [19,20], TDR-interacting protein 2 (*TIP2*, *bHLH142*) [21–24], MYB family transcription factor *GAMYB* [25–27], PHD-finger protein persistent tapetal cell 1 (*PTC1*) [28], and TGA transcription factor *OsTGA10* [29]. These genes were shown to regulate various aspects of anther development, especially tapetal PCD. *GAMYB* probably works upstream of *TDR1* and *PTC1* in parallel with *UDT1* to regulate rice tapetum development and pollen wall formation [27,28]. *TIP2* functions upstream of *TDR* and *EAT1*, but downstream of *UDT1*, and also binds to the promoter of *EAT1* to activate positive effects on regulation of tapetum PCD by promoting aspartic proteases *AP25*, *AP37*, and *OsCP1* [20,22]. *OsTGA10* also regulates tapetum development and pollen formation by interacting with *TIP2* and *TDR* to affect the expression of *AP25* and *MTR* [29]. In addition, some genes that encode enzymes or specific binding proteins are also essential for tapetal degeneration. Loss of function of these genes can result in anther deformity and defects in pollen exine formation. *DTC1* controls tapetal degeneration by modulating the dynamics of reactive oxygen species (ROS) with *OsMT2b* during reproduction of male rice [30]. *DEX1* regulates tapetal cell death and pollen exine formation by binding to Ca^{2+} to modulate cellular Ca^{2+} homeostasis, acting as a component required for tapetal cell death signal transduction [31]. Recessive mutation of the fasciclin glycoprotein *MTR* can dislocate its plasma membrane localization system and cause delayed tapetum PCD and reduced synthesis of Ubisch bodies, which are micron-sized particles on the inner surface of the tapetum in anthers, finally resulting in abortive pollen grains and complete male sterility [32]. The F-box protein *OsADF* is expressed in tapetal cells and microspores and works depending on *TDR* by binding E-box motifs of its promoter [33]. The two aspartic proteases, *AP25* and *AP37*, cysteine protease *OsCP1*, and apoptosis inhibitor *API5* are reported as specific regulators of the tapetal PCD process, and inhibition/mutagenesis of these genes can cause defects in pollen formation and eventually lead to male sterility [12,20,34].

Microspore cells need to undergo a series of processes after they are released from the tetrads to ultimately develop to mature fertile pollen grains, such as pollen wall formation, vacuolation, two rounds of mitosis, and starch enrichment. Normal progress of these processes requires nutritional supply and structural support from aliphatic biopolymers (sporopollenins, Ubisch bodies, epicuticular waxes, and cuticle monomers) and especially the synthesis and transport of sporopollenin. Sporopollenin is made up of polyhydroxylated aliphatic compounds and oxygenated aromatic monomers, such as phenolics, conjugated by ether and ester bonds, and acts as one of the main components of pollen exine, playing a critical role during pollen development by protecting pollen grains from abiotic and biotic stresses. Evidence from previous studies of male-sterile mutants with abnormal pollen wall formation showed that many genes are involved in these biosynthesis/transport mechanisms [1,2,15,35,36]. Two cytochrome P450 family genes, *CYP704B2* and *CYP703A3*, function as catalyzers of ω-hydroxylated fatty acids with 16- and 18-carbon chains and in-chain hydroxylase only for lauric acid to generate 7-hydroxylated lauric acid [37–39]. Defective pollen wall (*DPW*) acts as a fatty acyl-carrier protein reductase and *DPW2* as a fatty-acid acyltransferase to alter the amounts of cutin and waxes, and of lipidic and phenolic compounds, respectively, during anther development and pollen formation [40,41]. The ATP-binding cassette (ABC) transporters *ABCG15/PDA1*, *ABCG26*, and *ABCG3* work collaboratively but perform their own functions to transport different materials for anther development and pollen formation [42–45]. The lipid transfer protein *OsC6* is secreted into anther cuticle, anther locule, and the space between the tapetum and middle layer for pollen exine and orbicule formation downstream of *TDR* and *GAMYB* [46]. In addition, *WDA1* [47], *OsACOS12* [48,49], *OsNP1* [50], and *OsPKS2* [51,52] were also reported to regulate sporopollenin biosynthesis and deposition. Loss of function of these genes caused defects in not only the anther cuticle and pollen wall, but also the number of secretory, lipidic Ubisch bodies. In fact, the functions of these genes often interact; most of the recessive mutants of PCD-induced genes showed apparent defects in Ubisch body patterning and pollen exine formation. In particular, *tdr*, *ptc1*, *eat1-1*, *dex1*, and *dtc1* mutants exhibit obvious microspore collapse and pollen wall degeneration at the microspore stage [12,20,28,30]; *CYP703A3*, *ABCG15*, *ABCG26*, and *OsACOS12* also correspondingly showed high expression levels in tapetal cells and their recessive mutants due to delayed or premature degradation of the tapetum layer [42–44,49].

Acyl coenzyme A (CoA) glycerol-3-phosphate acyltransferases (GPATs), which localize to the endoplasmic reticulum (ER), are generally recognized as important catalyzers for the first step of de novo synthesis of triacylglycerol. GPATs play key roles in regulating cell growth and metabolic processes of membrane lipids, storage lipids, and extracellular lipid polyesters (cutin and suberin) by generating lysophosphatidic acids (LPAs) and acylating glycerol 3-phosphate at the sn-1 or sn-2 hydroxyl with acyl-CoA or acyl-acyl-acyl carrier protein (ACP) to alter glycerolipid triacylglycerol (TAG) biosynthesis [53–55]. The sn-1 GPATs promote acylation to produce lysophosphatidic acid (LPA) for lipid formation, while sn-2 GPATs possess a phosphatase domain to produce sn-2 monoacylglycerol (2-MAG) as the major product for cutin and suberin synthesis in plants. Most GPATs were confirmed to have sn-2 acyl transfer activity [56–60]. In *Arabidopsis*, there are ten GPATs, eight of which (*AtGPAT1* to *AtGPAT8*) belong to the sn-2 family and are divided into three sub-clades. The first clade, *AtGPAT1* to *AtGPA3*, is mainly expressed in flowers and siliques, and show sn-2 acyltransferase during the process of dicarboxylic acyl-CoA substrate utilization but not phosphatase activity. *Arabidopsis gpat1* mutant exhibits less fibrillar material, fewer vesicles in the anther locule, disrupted degeneration of the tapetum, and collapsed pollen grains [61]. *AtGPAT4*, *AtGPA6*, and *AtGPAT8* are unique bifunctional enzymes with both sn-2 acyltransferase and phosphatase activity to produce 2-MAG for cutin synthesis. *AtGPAT5* and *AtGPAt7*, which were identified as part of the suberin-associated clade, function in suberin synthesis in the wounding response for root and seed coat formation [60]. Genetic functions revealed that *AtGPAT1* and *AtGPAT6* play an important role in anther development and pollen formation. Loss of function of these two genes can result in altered ER profiles in tapetal cells, reduced pollen production, and decreased pollination; double mutation of these two genes can cause defective callose degeneration, pollen release, and complete male sterility [60,61]. Homologous sequence alignments and phenotypic analysis on other species reveal that *SlGPAT6*, *OsGPAT3*, and *ZmGPAT3* also have important roles in regulating anther cuticle biosynthesis and pollen exine formation [62–64]. However, studies on the metabolism of GPATs in regulating rice male reproductive development are still limited, particularly in tapetum PCD, and there is a lack of related mutants and phenotype analysis of relevant genes.

In this study, we further characterized *OsGPAT3* in rice male reproductive development, particularly focusing on its function in anther wall degeneration and pollen maturation. The *gpat3-2* mutant exhibited delayed tapetum PCD and Ubisch body formation and abnormal anther wall and pollen degeneration, resulting in complete male sterility. Genetic analysis and map-based cloning revealed that the mutant phenotype was caused by a single-nucleotide polymorphism (SNP) mutation in the first exon of *OsGPAT3*, a land plant sn-2 *GPAT* homolog. Two allelic mutants from our mutant library and three other mutant alleles of *OsGPAT3* generated using clustered regularly interspaced short palindromic repeats (CRISPR)/CRISPR-associated 9 (Cas9) also showed the same male-sterile phenotype. In addition, the expression pattern of many tapetum PCD-induced regulators and nutrition metabolism related genes were significantly altered resulting from recessive mutation of *OsGPAT3*. Thus, our results demonstrate that *OsGPAT3* is essential for anther wall PCD and pollen development in addition to its function during the synthesis of Ubisch bodies for anther cuticle and pollen exine formation. Our study also provides new insights into GPATs on regulating tapetum PCD and pollen maturation during plant reproductive development.

2. Results

2.1. Isolation and Phenotypic Analyses of the gpat3-2 Mutant

The *gpat3-2* mutant was first identified from the ethyl methyl sulfone (EMS)-soaked M1 progeny of an *indica* rice cultivar Zhonghui8015 (Zh8015) in Lingshui, Hainan Province. The *gpat3-2* mutant exhibited complete male sterility and was genetically stable in Hangzhou, Zhejiang Province. When the *gpat3-2* mutant was pollinated with wild-type pollens, all BC_1F_1 plants exhibited normal male fertility. Further identification of anthers in the BC_1F_2 population also indicated that the *gpat3-2* phenotype conformed to the genetic regulation of a single recessive gene ($\chi^2 = 0.17$; $p < 0.05$). Genetic analysis on the F_1 and F_2 progeny generated by the cross *gpat3-2* × 02428, a wide compatibility *japonica* cultivar, also confirmed the monofactorial recessive inheritance of the *gpat3-2* mutant (see Table S1, Supplementary Materials).

Vegetative development, including plant height, heading date, tiller number, other major agronomic traits, and general spikelet morphology of the *gpat3-2* plants did not differ from those of wild-type Zh8015 except for the anthers (Figure 1A–C). The Zh8015 anthers were normal golden yellow and had countless fertile pollen grains in the anther interior (Figure 1C–F); anther dehiscence occurred soon after glumes opened and pollens were deposited for fertilization (Figure 1C,D). Compared with wild-type anthers, the mutant anthers were small and white, without mature pollen grains (Figure 1C–F,H) and failed to dehisce, resulting in completely sterile spikelets (Figure 1C–F,H).

2.2. Defects of Anther Development and Pollen Maturation in gpat3-2 Mutant

To identify defects in the *gpat3-2* mutant, semi-thin transverse sections of anthers at different developmental stages from the wild-type plant and *gpat3-2* mutant were further examined. Rice anther development was delineated into 14 stages, which was consistent with that of *Arabidopsis thaliana* [3], based on the cellular events observed under light microscopy by semi-thin section [65,66]. By stage 7, both wild-type (WT) and *gpat3-2* anthers exhibited an obvious four-layered anther wall from surface to interior, and the wall enwrapped pollen mother cells (PMCs) within the locule. At this stage, the PMCs progressively initiated meiotic division and nestled against the tapetal layer. No obvious defects in the four somatic layers of the anther wall and microsporocytes were detected between the WT and *gpat3-2* mutant until this stage (Figure 2A,B). Subsequently, *gpat3-2* anthers began displaying obvious morphological abnormalities. At late s7 to s8a, the PMCs generated dyed cells, the middle layer became nearly invisible, and the anther wall had three layers in WT anthers. Meanwhile, the tapetal layer began having a weak point of programmed cell death (PCD) (Figure 2C). In the *gpat3-2* anthers, the middle layer was still clearly visible, the anther wall had the four-layer structure, and newly formed dyed cells were misshapen and less darkly stained; this phenomenon was also observed in tapetum cells (Figure 2D). By the end of stage 8b, tetrads were generated, tapetum PCD started, and the middle layer became nearly invisible in the wild-type anthers (Figure 2E). The *gpat3-2* anthers also formed imperfect tetrads, which exhibited unequal cleavage. However, the tapetum layer was electron-dense and expanded, and the middle layer was still apparent and showed no signs of degradation (Figure 2F).

Figure 1. Phenotype comparison between the wild-type Zhonghui8015 (Zh8015) and *gpat3-2* mutant. Plants of the wild-type Zh8015 and *gpat3-2* mutant at the anthesis stage (**A**); Mature spikelets of the wild-type Zh8015and the *gpat3-2* mutant at anthesis (**B**); The stamen morphologies of the wild-type Zh8015and *gpat3-2* mutant; lemmas and paleae were removed for clarity (**C**); Mature anther of the wild-type Zh8015 and *gpat3-2* mutant (**D**); Compressed anthers of wild-type Zh8015 (**E**) and the *gpat3-2* mutant (**F**) after I_2/KI staining (**E,F**); I_2/KI staining of the pollen grains of wild-type Zh8015 (**G**) and the *gpat3-2* mutant (**H**) at stage 13 (**G,H**). Scale bars = 1 mm in (**B–F**), and 100 μm in (**G,H**); ad, anther dehiscence.

At stage 9, microspores were released from the tetrads and tapetal cells, becoming condensed and electron-dense resulting from gradual PCD-induced degradation (Figure 2G). Although microspore cells can be formed after meiosis, most anther walls gradually became distorted and shrank, and the middle layer remained relatively discernible in this period in the mutant anthers. Furthermore, the *gpat3-2* mutant swelled and there were lightly stained tapetal cells, indicating abnormal PCD (Figure 2H). A sharp distinction between the wild-type and *gpat3-2* anthers began appearing at stage 10; normal vacuolated microspores were uniformly attached to the tapetum side as round shapes with dark-stained pollen exine, while the tapetum layer was electron-lucent after gradual PCD (Figure 2I). By comparison, internal cavities of the *gpat3-2* anthers were disordered; the tapetum layer was swollen and lightly stained with obvious degradation characteristics and the middle layer was still visible. In addition, microspores were disrupted and degraded together with the tapetum layer (Figure 2I). From the pollen mitosis stage and the mature pollen stage, wild-type pollen formed a complete double-layer exine and fertile pollen after two mitotic divisions with pollen exine deposition and starch accumulation, while the tapetum layer gradually degenerated and thinned until it almost disappeared at the end of stage 12. During this procedure, the epidermis and endothecium layer further degenerated and anther dehiscence occurred. Mature pollen grains were full of lipids, starch, and other storage nutrients, and were dark-stained with toluidine blue, indicating that the wild-type pollen grains had normal functions and were viable (Figure 2K,M,O). However, degradation of microspores and tapetal cells continued, resulting in linear pollen walls and cell detritus in the *gpat3-2* anther locule. In addition, the outer layers of the anther wall retained the original structure; epidermis, endothecium layer, and middle layer were not degraded until the end of stage 13 in *gpat3-2* anthers (Figure 2L,N,P). These observations suggested that *gpat3-2* carried defects not only in anther wall development and pollen maturation, but also in differentiation and degradation of tapetal cells and anther wall.

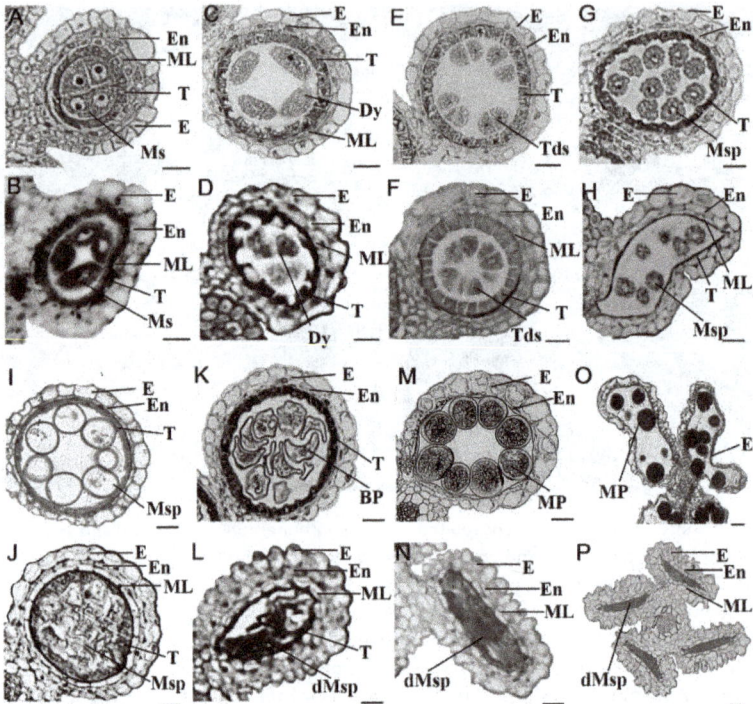

Figure 2. Transverse section analysis of anther development in wild-type Zh8015 and *Osgpat3-2* mutant. Locules from the anther section of Zh8015 (**A,C,E,G,I,K,M,O**) and *Osgpat3-2* (**B,D,F,H,J,L,N,P**) from stage 7 to stage 13 of development (stages 7, 8a, 8b, 9, 10, 11, 12, and 13). BP, bicellular pollen; dMsp, degraded microspores; Dy, dyed cell; E, epidermis; En, endothecium; ML, middle layer; Mp, mature pollen; Ms, microsporocyte; Msp, microspores; T, tapetum; Tds, tetrads. Scale bars = 20 μm.

2.3. Delayed PCD of Osgpat3-2 Tapetal Cells and Anther Wall Cells

The transverse section analysis suggested that *gpat3-2* mutation affected the differentiation and degradation of tapetal cells and anther wall. We, therefore, used a terminal deoxynucleotidyl transferase-mediated deoxyuridine triphosphate (dUTP) nick-end labeling (TUNEL) assay to test the tapetum PCD process during a range of developmental stages in WT and *gpat3-2* anthers.

Microspore mother cells were generated during meiosis and tapetal cells became condensed after meiosis; there was no detectable DNA fragmentation signal in the wild-type or *gpat3-2* tapetal cells. At stage 8a, some TUNEL-positive nuclei were detected in wild-type tapetal cells and the middle layer, indicating that normal PCD started occurring in the wild-type anthers (Figure 3A). At stage

8b, positive signals of DNA fragmentation were much stronger in wild-type tapetal cells. Positive PCD signals were also detected in both the outer layers (endothecium and middle layer) and vascular bundle cells of wild-type anthers at this stage (Figure 3C). However, no visible fragmented DNA signal was observed in the *gpat3-2* mutant anthers at both stage 8a and 8b (Figure 3B,D). At stage 9, when the microspore was released from the tetrad, PCD signals in wild-type tapetal cells became strongest, and the signal in the outer layers, vascular bundle cells, and cavity of dehiscence increased at the same time (Figure 3E). However, PCD signals were still not detected in the *gpat3-2* tapetum; only weak signals of DNA fragmentation were detected in *gpat3-2* outer layers and vascular bundle cells (Figure 3F). At stage 10, positive PCD signals in wild-type tapetal cells and other tissues became much weaker than at stage 9 (Figure 3G), while PCD signals unexpectedly became detectable and very strong in most tissues of *gpat3-2* anthers, including tapetal cells, the outer layers (endothecium and middle layer), vascular bundle cells, and even microspore cells inside the chamber (Figure 3H). At stage 12, PCD signals of wild-type tapetal cells gradually became invisible and only small, positive signals were detected in the outer layers and cavity of dehiscence (Figure 3I). Conversely, PCD signals continued increasing in all parts of *gpat3-2* anthers (Figure 3J). Nevertheless, at the dehiscence stage, none positive PCD signals were detected in wild-type tapetum, but the signals became much stronger in outer layers, vascular bundle cells, and cavity of dehiscence to meet the requirement of anther cracking and pollination (Figure 3K). However, positive PCD signals became strongest in all cell layers of *gpat3-2* anthers at this stage, including the degenerated microspores. These TUNEL assays demonstrate that the PCD of tapetum and peripheral layers were normal and orderly, and the delay and disorder of PCD in tapetum, anther wall, and microspores possibly resulted in the failure of pollen formation in the *gpat3-2* mutant.

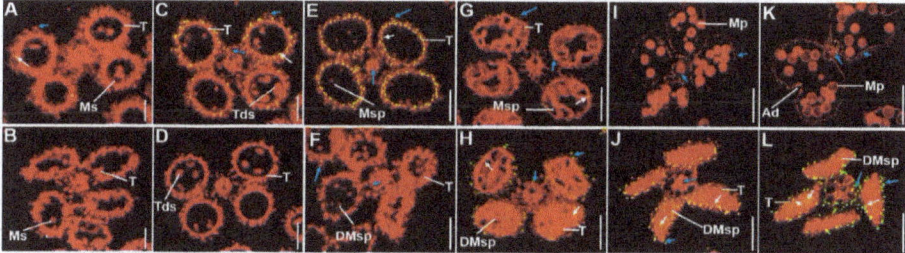

Figure 3. Detection of DNA fragmentation in wild-type and *gpat3-2* anthers using a terminal deoxynucleotidyl transferase-mediated deoxyuridine triphosphate (dUTP) nick-end labeling (TUNEL) assay. The wild-type and *gpat3-2* mutant anthers at stage 8a (**A,B**), stage 8b (**C,D**), stage 9 (**E,F**), late stage 10 (**G,H**), stage 12 (**I,J**), and dehiscence stage (**K,L**). A red signal indicates propidium iodide (PI) staining, while yellow and green fluorescence indicates a TUNEL-positive signal. TUNEL-positive signals detected in the tapetum cells of both wild-type and *gpat3-2* anthers are marked by white arrows, while TUNEL-positive signal observed in the outer cell layers (including the epidermis, endothecium, and middle layer), and vascular bundle cells are marked by blue arrows. Ad, anther dehiscence; DMsp, degenerated microspore; Mp, mature pollen; Ms, microsporocyte; Msp, microspore; Tds, tetrads; T, tapetum. Scale bars = 50 um.

2.4. Osgpat3-2 Exhibits Defects in the Formation of Ubisch Bodies and Pollen Accumulation

For further verification and understanding of abnormalities of the *gpat3-2* mutation, we performed a more detailed scanning electron microscope (SEM) observation of the surfaces of wild-type and *gpat3-2* anthers and pollen grains at different development stages. No obvious differences of the anther morphologies or outer wall structure were observed between wild-type and the *gpat3-2* anthers at stage 9; both showed a smooth epidermis and microspores were released (Figure 4A,B,D-I,D-II). However, further enlargement of the inner surface of locules and pollen exine revealed that the

Ubisch bodies were produced by the tapetum and transported to microspores in wild-type anthers, while the *gpat3-2* anthers still had a smooth inner surface and abnormal pollen exine, indicating the defects of tapetum PCD (Figure 4C,E-I,E-II). From the vacuolated pollen stage (stage 10) to the mature pollen stage (stage 12), the size of wild-type anthers gradually increased to almost double, the anther epidermis constantly thickened and was covered with a three-dimensional spaghetti-like cutin layer, the continuous synthesis and transport of Ubisch bodies resulted in a neat arrangement of inner surface, and the uninucleate microspores enlarged and developed to trinucleate pollen with regular shaped pollen exine, which was formed by secreted tapetum-produced sporopollenin precursors from Ubisch bodies (Figure 4III,V,VII of A–E). By contrast, the size of *gpat3-2* anthers was only slightly increased from stage 9 to 11 and basically no longer increased after this stage (Figure 4A-II,IV,VI,VIII). Anther epidermis was still smooth, which indicated defects in the synthesis of typical cutin and in the synthesis of fatty acids in the anther wall (Figure 4B-II,IV,VI,VIII). The inner wall of anthers gradually became irregular at stages 11 and 12 (Figure 4C-VI,VIII) and was smooth but uneven with randomly distributed flocs of Ubisch bodies and degraded microspores at stages 9 and 10 (Figure 4B-II,IV). Further observation of the microspores from different developmental periods showed that microspores in *gpat3-2* anther locules were gradually collapsed and covered with randomly distributed cutin materials, finally leading to the *gpat3-2* microspores becoming more and more chaotic-like cotton wools (Figure 4E-II,IV,VI,VIII).

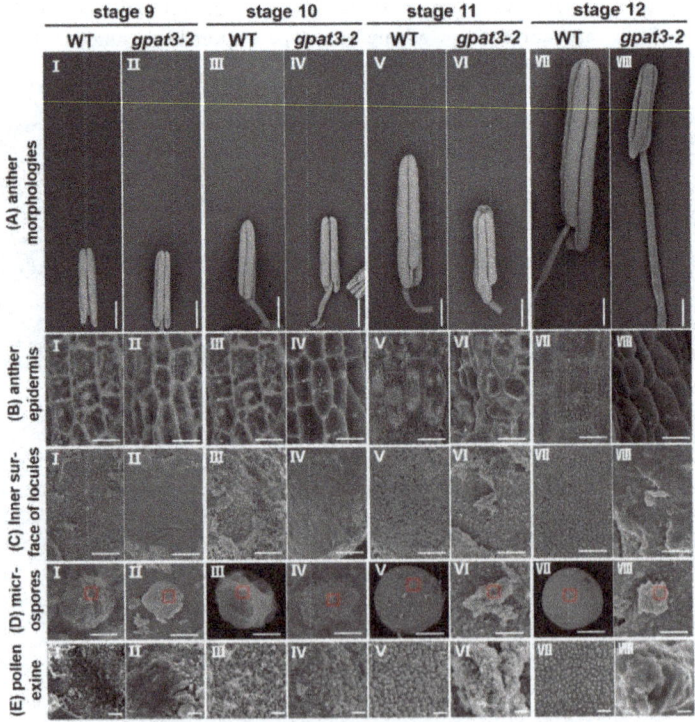

Figure 4. Scanning electron microscopy (SEM) observation of the surface of anther and pollen grains in wild-type (WT) and *gpat3-2* anthers. The anther morphologies (**A**), anther epidermis (**B**), inner surface of anther locules (**C**), microspores (**D**), and pollen exine (**E**) of WT and *gpat3-2* from stages 9 to 12 are shown. (**E**) Macro photograph of the outermost surface of microspores in the areas indicated by red boxes in (**D**). Scale bars = 500 μm in (**A**), 10 μm in (**B,C**), 5 μm in (**D**), and 200 nm in (**E**).

Transmission electron microscopy (TEM) observation verified the defects in *gpat3-2* anthers and pollen grains. Obvious differences were first identified at stage 8a; *gpat3-2* anthers showed less tapetum differentiation but more vacuoles and a thicker middle layer (see Figure S1A–F, Supplementary Materials) and pollen mother cells were agglomerated without nucleus compared with the wild-type anther (see Figure S1G,H, Supplementary Materials). At stage 8b, tapetal cells of wild-type anthers began differentiating and became sparse and loose (see Figure S1I,K,M, Supplementary Materials), while the *gpat3-2* tapetum cells were still electron-dense without any signs of degradation (see Figure S1J,L,N, Supplementary Materials). In addition, tetrads of WT were first surrounded by callose that resembled the pollen primexine structure (see Figure S1O, Supplementary Materials); however, tetrads of *gpat3-2* were still naked without a wrapping layer as before (see Figure S1P, Supplementary Materials). These results were almost the same as previously reported by Men et al. [61]. However, at stage 9, the tapetum gradually turned into a small, loose thread resulting from the vigorous PCD process, and only a very small line of residual cells of the middle layer was observed in wild-type anthers (Figure 5A-1,C-1,E-1). Ubisch bodies, which were believed to secrete tapetum-produced sporopollenin precursors for pollen exine formation, began being released from the tapetum and gradually grew into electron-dense orbicules (Figure 5E,F). Pollen exine of microspores with accumulation of particulate sporopollenin in the wild-type locule was visible at this stage (Figure 5G-1,I-1). Although microspores and pre-Ubisch bodies seemed to be formed from the mutant tapetum at stage 9, the tapetal cells and microspores were still electron-dense without any signs of differentiation (Figure 5B-1,D-1,F-1,H-1). In addition, some accumulation of electron-dense materials occurred on the outer surface of mutant microspores, but these structure were unable to grow into mature Ubisch bodies, and formed an abnormal exine structure (Figure 5J-1). From stage 10 to stage 11, the middle layer disappeared completely, and the endothecium layer constantly degraded and thinned accompanied by the collapse of epidermis (Figure 5A-2,C-2,A-3,C-3); the tapetum layer differentiated and degenerated to a lightly stained and thin layer (Figure 5A-2,C-2,E-2,A-3,C-3,E-3). Microspore exine deposition, which underwent vacuolization and two rounds of mitosis, was completed and eventually formed thick exine with distinctive layers of tectum, bacula, and nexine under normal nutrient supply from continuously generated Ubisch bodies (Figure 5G-2,I-2,G-3,I-3). Although the tapetum of the *gpat3-2* mutant appeared to be degraded, it was still dense and deeply stained (Figure 5B-2,D-2,F-2,B-3,D-3,F-3): the precursor of newly formed pre-Ubisch bodies failed to synthesize complete and mature Ubisch bodies (Figure 5F-2,F-3). In addition, the epidermis and endothecium cells of anthers were deformed, but the middle layer was still obviously visible and thick (Figure 5B-2,D-2,B-3,D-3). The microspores formed previously in *gpat3-2* anthers exhibited irregular pollen exine and were also gradually degraded with tapetal cells, resulting in degraded cell remnants in the anther locule (Figure 5B-2,H-2,J-2,B-3,H-3,J-3). At the mature pollen stage (stage 13), a distorted layer of epidermis was also visible in wild-type anther, and the endothecium cells and tapetum were almost degraded with numerous Ubisch bodies attached to the inner side facing the pollen grains (Figure 5A-4,C-4). At stage 13, wild-type microspores grew to spherical pollen grains full of starch, lipids, and other nutrients surrounded by a normal bilayer exine (Figure 5G-4,I-4, white arrowhead). However, the anther wall of the *gpat3-2* mutant was still a three-layer outer structure with a few lipids deposited on the inner side of the middle layer; the tapetum was prominently electron-lucent and no longer a distinct layer at this time, probably due to serious degradation (Figure 5B-4,D-4,F-4). The *gpat3-2* microspores aborted and collapsed, and the framework of the exine distorted and folded without the internal nexine layers; only remnants of abnormal epidermis in the locule were left, resulting from severe degradation (Figure 5H-4,J-4). Ultimately, all the defective tissue burst out; finally, the anther chamber of the *gpat3-2* mutant was filled with degraded fragments and residual abnormal pollen grains.

Together, all these results indicated that the *gpat3-2* mutation affected the differentiation and degradation of the anther wall, the synthesis and supply of Ubisch bodies, and pollen wall formation. The *gpat3-2* defects were different from the *gpat3* mutant reported previously, because *gpat3* microspores could not be released from the tetrads and were still covered with callose at the young microspore stage [63].

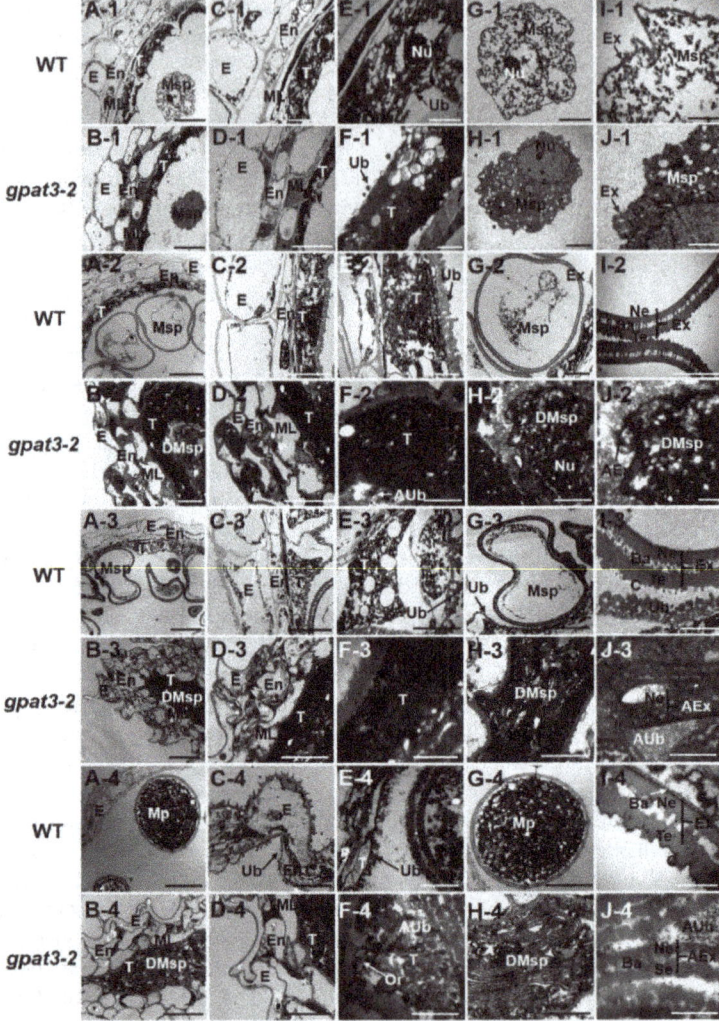

Figure 5. Transmission electron microscopy analysis of anther development in wild-type and *Osgpat3-2* mutant from stages 9–13. The transverse sections of the WT (**A,C,E,G,I**) and *gpat3-2* (**B,D,F,H,J**) anthers at stage 9 (**A1–J1**), stage 10 (**A2–J2**), stage 11 (**A3–J3**), and stage 13 (**A4–J4**) are compared. Anthers of the wild type (**A**) and *Osgpat3-2* (**B**), showing the anther wall with microspores at different development stages. The layers of anther wall in the wild type (**C**) and *Osgpat3-2* (**D**). Higher magnification of the tapetum cells showing Ubisch body in the wild type (**E**) and *Osgpat3-2* (**F**). The development and morphologies of microspores in the wild type (**G**) and *Osgpat3-2* (**H**). The development and structures of pollen exine in the wild type (**I**) and *Osgpat3-2* (**J**). AEx, abnormal exine; AUb, abnormal Ubisch body; Ba, bacula; C, cuticle; DMsp, degenerated microspores; E, epidermis; En, endothecium; Ex, exine; ML, middle layer; Mp, mature pollen; Msp, microspores; Nu, nucleus; Ne, nexine; Or, orbicule; Se, sexine; T, tapetum; Te, tectum; Ub, Ubisch body. Scale bars = 10 μm in (**A,B**), 5 μm in (**C,D**), 0.5 μm in (**E,F**), 2 μm in (**G,H**), and 200 nm in (**I,J**).

2.5. Fine Genetic Mapping and Candidate Gene Analysis of the Osgpat3-2 Mutation

To map the *gpat3-2* gene, map-based cloning was performed using the F_2 population from the cross between *gpat3-2* and 02428. Polymorphisms were confirmed as described previously [39,67]. Seven individuals each of the wild-type and *gpat3-2* phenotypes were chosen for linkage analysis using the bulked segregant analysis (BSA) method and the results revealed that *gpat3-2* is located on the long arm of chromosome 11, flanked by simple sequence repeat (SSR) loci RM27172 and RM27326 (Figure 6A). Further primary mapping was conducted using encryption markup with an insertion/deletion (InDel) marker RD1110 and an SSR marker RM27273 in 176 F_2 recessive individuals; the mutation locus was in a 871.4-kb region between markers RM17273 and RM27326 (Figure 6A). For fine-scale mapping of the *gpat3-2* gene, five significantly polymorphic InDel markers were designed based on polymorphisms between *japonica* Nipponbare and *indica* 9311 [68] (Table S1, Supplementary Materials). Using these newly developed markers along with high-resolution genetic linkage analysis with 1354 recessive individuals from F_2 populations of the cross between *gpat3-2* and 02428, the *GPAT3-2* gene was finally delineated to a 26-kb region between ZH-3 and ZH-6 (Figure 6B).

We sequenced the three open reading frames (ORFs) in this region according to the Rice Genome Annotation Project (http://rice.plantbiology.msu.edu/index.shtml) (Figure 6C; Table S2, Supplementary Materials) of both the WT and *gpat3-2* mutant. We found that the *gpat3-2* mutant carried a single nucleotide mutation (G to A) in the first exon of *Loc_Os11g45400* (Figure 6D–F), which resulted in the corresponding amino acids being mutated directly from tryptophan to a stop codon (Figure 6D,G). Two pairs of calcium-dependent activator protein for secretion 1 (CAPS1) enzyme digestion primers were designed to test and verify this alternative splicing site for this mutation using the endonuclease *NlaIV*, and the enzyme digestion results confirmed our prediction (Figure 6H).

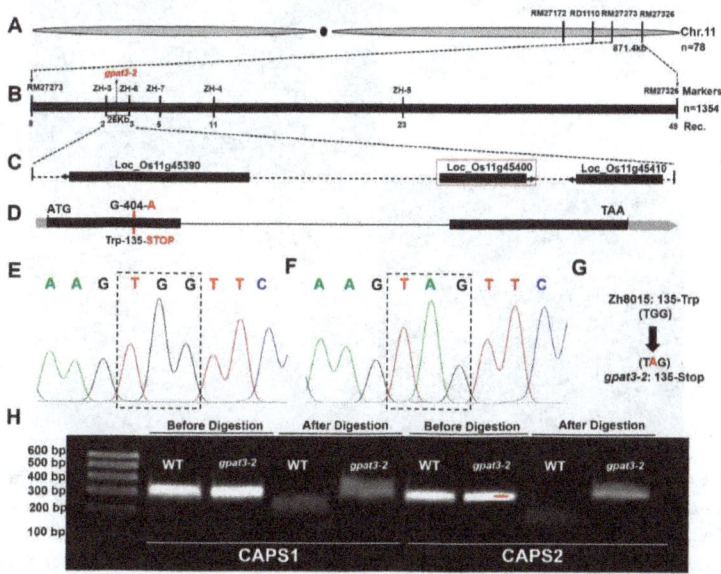

Figure 6. Mapping cloning of the *GPAT3-2* gene. Genetic linkage of the *gpat3-2* locus on chromosome 11 (**A**); Fine mapping of the *GPAT3-2* locus (**B**); The predicted open reading frames (ORFs) in the target region of the rice genome (**C**); Gene structure of the target gene and the *gpat3-2* mutation site (**D**); Sequences of the mutation region in the *Loc_Os11g45400* gene from Zh8015 (**E**) and the *gpat3-2* mutant (**F**); Predicted protein change of the *gpat3-2* mutation (**G**); *NlaIV* electrophoresis before and after enzyme digestion of the mutation site (**H**).

2.6. Function Verification of the OsGPAT3 Gene

To confirm that male sterility was caused by the mutation in *Loc_Os11g45400*, a 7.9-kb *OsGPAT3* genomic fragment (*gOsGPAT3*) was transformed into calli induced from young panicles of homozygous *gpat3-2* mutant plants to rescue the sterile phenotype of *gpat3-2* (Figure 7A). The *gOsGPAT3* fragment included the 3.35-kb upstream sequence, the full-length ORF of *Loc_Os11g45400* from the wild type, and a 1180-bp region downstream from the termination codon sequence. The transgenic positive plants had a normal seed-setting rate, similar to those of wild type (Figure 7B). The complemented lines exhibited a normal seed-setting rate (Figure 7B) and golden yellow anthers (Figure 7C), and pollen grains accumulated abundant starch granules and could be dyed black by 1.2% I_2/KI solution (Figure 7D–F). In addition, anther transverse sections were examined to further verify the anther developmental process in complementary lines. The microspore mother cells secreted normal microspores at stage 9, the middle layer was invisible, and tapetal cells degraded and had fine lines and irregular shapes (Figure 7G). At stage 11, the microspores underwent mitotic divisions and generated binucleate pollen grains, while the endothecium layer and tapetum cells gradually tapered off and epidermis cells were puffy (Figure 7H). Normal developmental morphologies were also visible at stage 13; fertile pollen grains were produced, the epidermis layer was almost all that was left at this stage, and the anther chamber started shrinking and bursting for pollination (Figure 7I). Therefore, the whole developmental process of the anther in complementation lines was restored. These experiments confirmed that the single-nucleotide mutation in *Loc_Os11g45400* was responsible for the no-pollen phenotype of *gpat3-2*.

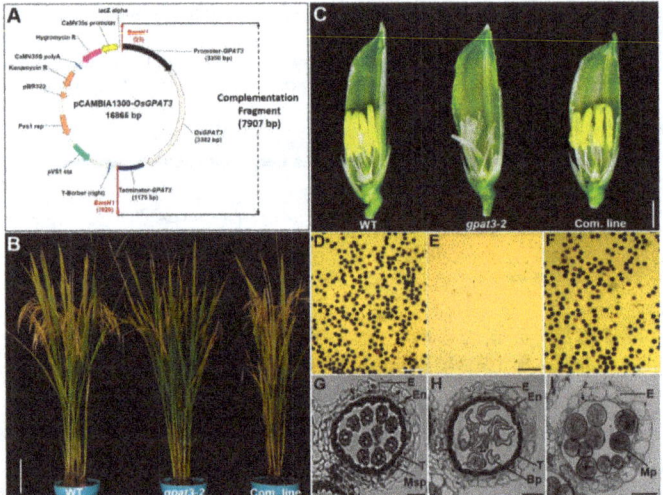

Figure 7. Complementation analysis of the *gpat3-2* mutant using wild-type *OsGPAT3*. The structure of the *gOsGPAT3* plasmid for transformation (**A**). The *gpat3-2* mutant was transformed with the pCAMBIA1300-*OsGPAT3* vector under the 35S promoter for transformation selection and the *OsGPAT3* genomic fragment contained three functional modules, including its native promoter, entire coding sequence, and downstream sequence from the WT for restoration of male fertility. The seed-setting rate of spikelet in wild-type, *gpat3-2*, and complementation plants (**B**). The flower of wild type, *gpat3-2*, and complementation line at stage 12; half of lemma and paleae were removed for clarity (**C**). Pollen grains of wild type (**D**), *gpat3-2* (**E**), and *gOsGPAT3*-complemented mutant (**F**) stained with 1% I_2/KI solution at stage 12. Transverse section analysis of anthers from *gOsGPAT3*-complemented line at stage 9 (**G**), late stage 11 (**H**), and stage 12 (**I**). E, epidermis; En, endothecium; T, tapetum; Msp, microspore; BP, bicellular pollen; Mp, mature pollen. Scale bars = 15 cm in (**B**), 1 mm in (**C**), 5 μm in (**D–F**), and 50 μm in (**G–I**).

To further confirm this result, we designed a target within *Loc_Os11g45400* using CRISPR/Cas9 in the Zh8015 genetic background and obtained three other ideal mutants, *gc-1*, *gc-2*, and *gc-3* (Figure 8A). We also found two allelic mutants, one with a single-base insertion (*gpat3-3*) and the other with a single-nucleotide mutation (*gpat3-4*) (Figure 8A). As expected, these five mutants had small, white anthers without mature pollen grains and with significantly increased transcription levels similar to *gpat3-2* (Figure 8B,C–H,O). A TUNEL assay also uniformly observed disordered anther locule with significantly enhanced PCD signals in almost all layers of anther wall, vascular bundle cells, and cavity of dehiscence compared to wild-type anther (Figure 8I–N). As predicted by the Rice Genome Annotation Project, OsGPAT3 protein includes a typical signal peptide, two transmembrane regions, and a conserved GPAT domain, which contains four acyltransferase motifs, without a phosphatase domain (see Figure S2, Supplementary Materials). Previous studies showed that OsGPAT3 belongs to the first clade of the conserved land plant sn-2 GPAT family that specifically regulates lipid biosynthesis for anther cuticle and pollen exine formation [60–64]. Further multiple comparisons of the *OsGPAT3* sequence found that the *gpat3-2* mutant and all three CRISPR/Cas9-induced mutants had premature stop codons and produced truncated polypeptides, which resulted in destruction of the conserved domains. While the two allelic mutants were somewhat different, the protein structure of *gpat3-3* changed completely after the single-base insertion and *gpat3-4* carried an amino-acid conversion in the most conservative GPAT domain (see Figure S2, Supplementary Materials). The findings indicated that the six mutant lines carried different mutations which disrupted the function of the conserved GPAT domain, transmembrane region, and acyltransferase motifs (Figure 7; Figure S2, Supplementary Materials). Taken together, the results demonstrated the function of *Loc_Os11g45400* in rice anther wall PCD and pollen development.

2.7. Mutation in OsGPAT3 Affects the Expression of Genes Involved in Both Tapetum PCD and Nutrient Metabolism

OsGPAT3 was reported to regulate the biosynthesis of lipid metabolism for anther cuticle and pollen exine formation [63]. Our results indicated that the mutation of *OsGPAT3* caused severely delayed tapetum PCD and anther wall development, failure of Ubisch body formation, and abnormal degradation of microspores. To understand the role of *OsGPAT3* in lipid metabolism, anther wall development, and tapetum PCD, and to explain the phenotype in *gpat3-2*, we compared the expression level of a series of genes regulating tapetum PCD and synthesis/transportation of sporopollenin precursors during male reproductive development between the wild type and *gpat3-2* mutant using qPCR. The expression pattern of *OsGPAT3* is consistent with the defects shown in *gpat3-2* mutant and is exactly like *PTC1* (Figure 9A, Li et al.) [28]. The *gpat3-2* anthers showed higher expression than the wild type at stage 7, lower expression during meiosis, and tetrad formation (stage 8), but significantly higher expression than the wild type after stage 9, when young microspores were released (Figure 7A). Nearly half of the tapetum PCD-related genes were downregulated in the *gpat3-2* mutant, including *GAMYB*, *DEX1*, and three cysteine proteases, *AP25*, *AP37*, and *OsCP1* (Figure 9A). Eight genes, including two homologous genes of *OsGPAT3* in rice (*Loc_Os05g38350* and *Loc_Os10g42720*), four bHLH transcription factors (*UDT1*, *TDR*, *EAT1*, and *TIP2*) that particularly regulate tapetum degeneration, *OsTGA10*, which is the target transcription factor of *OsMADS8*, and *MTR*, showed coincident upregulation in the *gpat3-2* mutant (Figure 9A). This may further prove that the organelles are still active and the metabolic process was still occurring during late stages of the *gpat3-2* mutant.

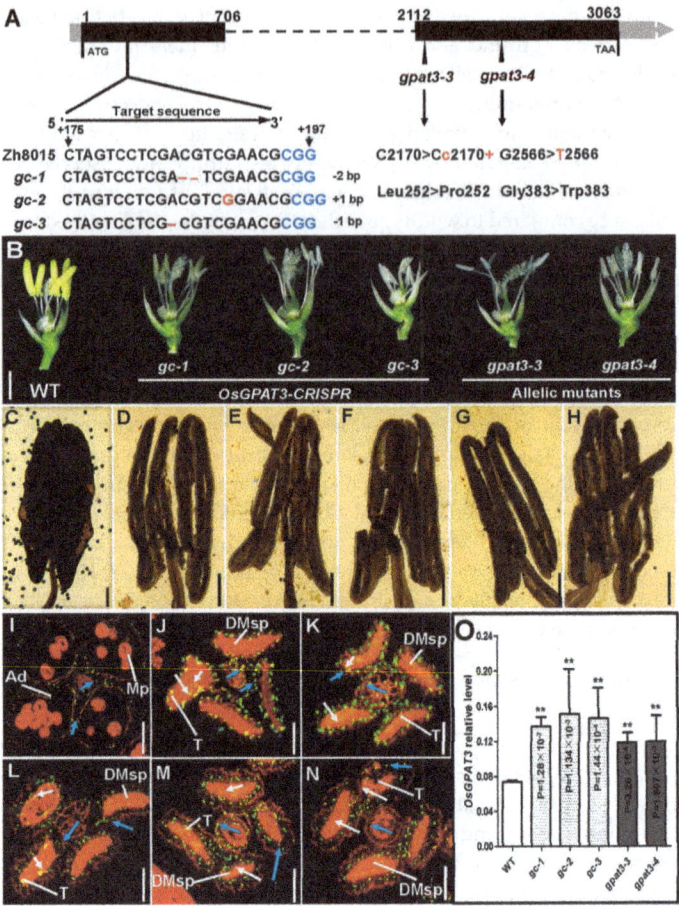

Figure 8. Sequence analysis and phenotypic observation of *OsGPAT3* clustered regularly interspaced short palindromic repeats (CRISPR)/CRISPR-associated 9 (Cas9)-induced mutants and allelic mutants. Gene structure of *OsGPAT3* and mutation analysis of *OsGPAT3* gene in transgenic plants and allelic mutants (**A**). The sequence (5′–CTAGTACTCGACGTCGAAGGCGG–3′) located in the first exon of the *OsGPAT3* gene was selected as the target site of single guide RNA (sgRNA). The black boxes indicate the exons. The blue characters indicate the protospacer adjacent motif (PAM). The red characters indicate the three different types of mutation events generated by CRISPR/Cas9 in the mutants. Phenotypic comparison of the WT and mutant anthers at stage 13; the lemma and paleae were removed for clarity (**B**). Compressed anthers of wild type Zh8015 (**C**) and the mutants (**D**: *gc-1*, **E**: *gc-2*, **F**: *gc-3*, **G**: *gpat3-3*, **H**: *gpat3-4*) after I_2/KI staining. Detection of DNA fragmentation in wild-type (**I**) and mutant (**J**: *gpat3-3*, **K**: *gpat3-4*, **L**: *gc-1*, **M**: *gc-2*, **N**: *gc-3*) anthers using a TUNEL assay at the dehiscence stage (stage 13). White arrows indicate TUNEL-positive signals detected in the tapetum cells of wild-type and *gpat3-2* anthers, while blue arrows indicate TUNEL-positive signal observed in the outer cell layers (including the epidermis, endothecium, and middle layer), and vascular bundle cells. The qPCR analysis of *OsGPAT3* in wild-type, CRISPR/Cas9-induced mutants, and allelic mutants (**O**). Ad, anther dehiscence; DMsp, degenerated microspore; Mp, mature pollen; T, tapetum. Scale bars = 2 mm in (**B**), 500 μm in (**C–H**), and 50 μm in (**I–N**). ** indicates significant differences at $p < 0.01$.

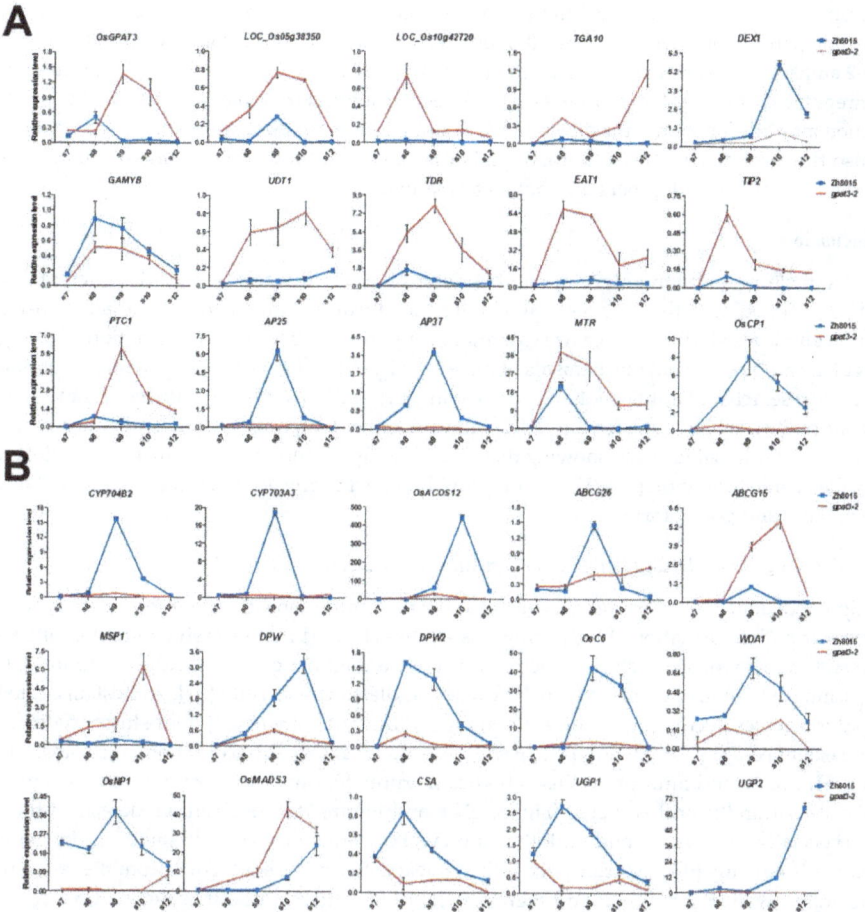

Figure 9. Expression analysis of *GPATs* and male sterility (MS)-involved genes in the WT and *gpat3-2*. The qPCR analysis of genes related to tapetum degeneration retardation in the wild-type and *gpat3-2* anthers at stages 7 to 12 (**A**); The qPCR analysis of genes involved in anther cutin biosynthesis/transport, sugar partitioning, and pollen exine formation in the WT and *gpat3-2* anthers at stages 7 to 12 (**B**). *OsACTIN1* was chosen as a control, and data are shown as means ± SD (*n* = 3).

The expression pattern of genes involved in lipid/carbohydrate metabolism, synthesis/ transportation of sporopollenin precursors and Ubisch bodies, and pollen wall formation also had some interesting features. The majority of genes related to anther cutin biosynthesis and pollen exine formation had significantly reduced expression in the *gpat3-2* mutant, including *CYP704B2*, *CYP703A3*, *OsACOS12*, *DPW*, *DPW2*, *OsC6*, *WDA1*, and *OsNP1*, further confirming the metabolic disorders of these nutrients in the *gpat3-2* mutant (Figure 9B). Three glycometabolism-related regulators, *CSA*, *UGP1*, and *UGP2*, also showed significant downregulation in the *gpat3-2* mutant, indicating that distortion of many other metabolic processes may concomitantly occur in the mutant (Figure 9B). However, the expression levels of two genes were significantly increased in the *gpat3-2* mutant: *MSP1*, a Leu-rich repeat receptor-like protein kinase that affects the number of cells entering into sporogenesis while simultaneously initiating anther wall formation in rice [69], and *OsMADS3*, which regulates late anther development and pollen formation by modulating ROS homeostasis [70]. The expression pattern of two ATP-binding cassette G transporters, *OsABCG15/PDA1* and *OsABCG26*, showed the same upregulation at stage 8 and mature pollen stage (stage 12), but unexpectedly had different expression patterns at stage 8 to stage 10 (Figure 9B). The transcriptional level of *OsABCG15/PDA1* in *gpat3-2* anthers was continuously higher than that of the wild type; however, *OsABCG26* was evidently downregulated at these stages. Together, these results suggest that loss of function of the *OsGPAT3* mutation may affect not only the biosynthesis of lipidic compounds and the anther wall PCD process, but also the formation/transport of Ubisch bodies and sporopollenin precursors, all of which are essential for anther development and pollen formation.

3. Discussion

Rice male reproductive development is determined by a series of specific functional factors, and loss of function of these genes may result in malformed anthers and eventually cause pollen abortion and male sterility, which are essential for hybrid rice breeding. Elucidation of the genetic, molecular, and biochemical mechanisms of male sterility-related genes are of primary importance for both basic research and application of rice breeding [2,15]. In this study, we further characterized the roles of rice *OsGPAT3* in male fertility, PCD-induced anther wall degradation, and pollen maturation on the basis of original research showing that *OsGPAT3* significantly affects anther cuticle biosynthesis and pollen exine formation [63]. This work provides new insights into the role of *OsGPAT3* in anther development and pollen formation.

3.1. OsGPAT3 Controls Ubisch Body Formation and Anther Development in Rice

Sporopollenin precursors are recognized as the most important structural security and nutritional sources for pollen formation [15,71]. Previous studies showed that recessive mutation of *OsGPAT3* can result in metabolism problems such as defective anther cuticle and synthesis failure of Ubisch bodies and pollen exine, which eventually cause complete male sterility [63]. Almost all components of lipid molecules decreased significantly in *osgpat3* mutant anthers [63]. Our study confirmed that *gpat3-2* exhibited a similar phenotype to *gpat3*, i.e., smooth anther surface, abnormal collapse of pollen grains (Figure 4), and amorphic Ubisch bodies (Figure 5F), finally resulting in severe degradation of all cells within the anther locule (Figure 2N,P and Figure 5H). In addition, defects in the anther wall, especially the middle and endothecium layers, were still visible at stage 13. In the *gpat3-2* mutant, this not only blocked transport of lipid substances, such as orbicules from the tapetum layer to the anther wall (Figure 2N,P and Figure 5H), but also hindered anther dehiscence. Expression changes of nutrient metabolism-related genes also confirmed that there are problems in synthesis related pathways. However, the upregulation of *MSP1* and *OsMADS3* may provide a sign that the *gpat3-2* mutant still carried active cell differentiation and degradation for sporogenesis and pollen wall formation. Different expression patterns between *OsABCG26* and OsABCG15/*PDA1* also implied that tapetum-synthesized lipid molecules differ in the process of transport from tapetum cells to anther surface by OsABCG26, and sporopollenin precursor transport to anther locules for anther cuticle formation by OsABCG15/*PDA1* [42–44]. These genes most likely still work to ensure cell activity in the mutant. Thus, our study demonstrated that *OsGPAT3* is required for sporopollenin precursor formation, pollen maturation, and even anther dehiscence in rice. but may functions varied in cell activity and nutrient transportation.

3.2. Loss of Function of OsGPAT3 Causes Abnormal PCD of Both Anther Wall and Pollen Grains

Timely cell differentiation and degradation of the three inner anther somatic layers is of critical importance for the patterning of microspore release, sporopollenin biosynthesis, pollen exine formation, and anther dehiscence. TUNEL assay analysis confirmed that *gpat3-2* exhibited delayed tapetum PCD and abnormal degradation of outer anther layers (Figure 3). The expression pattern of *OsGPAT3* in the *gpat3-2* mutant also suggests a possible feedback regulation of *OsGPAT3* transcription that functions in gradually strengthened degradation of anther wall and microspores in rice. Upregulation of tapetum PCD-related genes such as *bHLH TFs* (*UDT1*, *TDR*, *EAT1*, and *TIP2*) also showed that *gpat3-2* still had vigorous cell activity at stage 13, the appropriate anther dehiscence stage of wild type (Figure 9). However, the expression levels of *OSDEX1*, *GAMYB*, *AP25*, *AP37*, and *OsCP1* decreased significantly resulting from the upregulation of *OsGPAT3*. These results suggest that *OsGPAT3* positively regulates the expression of *OSDEX1*, *GAMYB*, *AP25*, *AP37*, and *OSCP1* in the tapetum, while *bHLH TFS*, *TGA10*, *PTC1*, and *MTR1* may be involved in other parallel regulatory pathways. Unfortunately, the functional network of *OsGPAT3* in rice tapetum development remains unclear; future investigations need to focus on identifying potential upstream genetic regulators using genetic and molecular approaches by expression pattern analysis of *OsGPAT3* in mutants of known PCD or lipid metabolism-associated transcription factors. Researchers can also create more double mutants or even polygene mutations using gene editing.

Our cytological observations showed that the *gpat3-2* mutant can release approximately normal microspores at stage 9 (Figures 1H, 3D and 5), which is significantly different from the *gpat3* mutant reported previously; microspores in *gpat3* were still covered with callose and could not be released from the tetrads at stage 9 [63]. Unfortunately, the newly released micropores in *gpat3-2* mutant gradually shrunk and degraded with the tapetum and other anther tissues at later stages. There were only remnants of abnormal epidermis left until the end in the *gpat3-2* anthers, all CRISPR/Cas9-induced mutants, and allelic mutants (Figure 3H,J,L and Figure 8I–N). This probably resulted from the above defects that hindered synthesis of Ubisch bodies/sporopollenin, but also abnormal PCD of the whole anther wall and pollen grains. In addition, our results also suggested that the middle layer and endothecium layer of the *gpat3-2* mutant had very little degradation (Figures 1 and 5), and the PCD process of all four anther layers were delayed (Figure 3H). Although the anther wall of *gpat3-2* anthers showed accelerated degradation at late development stages, it was too late for anther dehiscence as the pollen grains were already degraded and totally aborted (Figures 2P, 3L and 5). Therefore, our findings demonstrate that *OsGPAT3* is also a key regulator coordinating the differentiation and degradation of different layers for normal pollen exine formation and maturation during late anther development.

3.3. Proposed Functions of OsGPAT3 in Rice Male Reproductive Development

Our results further confirm that *OsGPAT3* affects not only metabolic processes such as lipid and carbohydrate metabolism in the early stages of anther development [63], but also differentiation and degradation of the anther wall in rice anthers to control both anther development and pollen formation. Therefore, a putative summary describing various functions of *OsGPAT3* during the male reproductive development process in rice is given below. *OsGPAT3* directly or indirectly affects various lipid and carbohydrate metabolism-related genes to ensure proper biosynthesis of anther cuticle and glycometabolism/sporopollenin precursors for anther development, pollen exine formation, and pollen maturation. *OsGPAT3* does not, however, directly regulate the transportation pathway of these substances. It also directly or indirectly regulates expression of *GAMYB*, *DEX1*, *AP25*, *AP37*, and *OsCP1* required for tapetum development and degeneration. *GAMYB* showed downregulation probably because it also affects pollen exine/Ubisch body formation and gibberellin-inducible nutrient mobilization [26,72]. Furthermore, *OsGPAT3* directly or indirectly affects differentiation and degradation of the anther wall required for pollen maturation and anther dehiscence. In summary, *OsGPAT3* regulates processes required for proper anther development and pollen maturation in different tissues at different stages of rice anther development. Our study also increased the known novel functions of the *OsGPAT3* gene during male reproductive development in rice.

4. Materials and Methods

4.1. Mutant Material and Plant Growth Conditions

The *gpat3-2* mutant (sp. *indica*), as the pollen acceptor, was crossed with wild-type Zh8015 and 02428 (sp. *japonica*). The heterozygous F_1 plants were then self-pollinated to generate a BC_1F_2 and an F_2 population for genetic analysis and mapping of the *Osgpat3-2* gene. In the F_2 mapping population, male-sterile plants were selected for gene mapping. All plants were grown in paddy fields of the China National Rice Research Institute during the spring of 2016 and 2017 in Lingshui, Hainan Province, China and the summer of 2016 and 2017 in Hangzhou, Zhejiang Province, China.

4.2. Identification of Mutant Anther and Pollen Development

Plant materials were photographed with a Nikon D800 digital camera and a Carl Zeiss SteREO Lumar V12 stereo fluorescence stereomicroscope (Markku Saari, Jena, Germany). For I_2/KI pollen staining, anthers and pollen grains were immersed in 1.2% I_2/KI solution (30 s) and then photographed using a Leica DM2500 microscope [12,67].

For cross-section observation of anther development, materials were collected and fixed on standard plastic sections as described by Zhang et al. and Li et al. [12,66]. Spikelets of wild type and *gpat3-2* mutant at different developmental stages were collected, based on the length of spikelet, and fixed with FAA (5% *v/v* formaldehyde, 5% *v/v* glacial acetic acid, and 50% *v/v* ethanol) overnight at 4 °C, dehydrated in a graded ethanol series (50–100%), embedded in Technovit 7100 resin (Heraeus Kulzer, Hanau, Germany, and polymerized at 60 °C. Transverse sections of 2.5-µm slices were obtained using a Leica RM2265 Fully Automated Rotary Microtome, stained with 0.2% toluidine blue (Chroma, Solms, Germany) and photographed using a Leica DM2000 biological microscope (Chroma, Solms, Germany).

For transmission electron microscopy (TEM), spikelets at various stages of development were collected and fixed in 2.5% glutaraldehyde in phosphate buffer (pH 7.0) for 16–24 h and then washed with phosphate-buffered saline (PBS; pH 7.2) three times, post-fixed with 1% $OsO4$ in phosphate buffer (pH 7.0) for 1 h, and washed three more times in phosphate buffer. Following ethanol dehydration, cutting and staining were performed as described previously [73]. After developmental identification, transverse sections were examined with a Model H-7650 TEM (HITACHI, Tokyo, Japan). For scanning electron microscopy (SEM), anthers at various developmental stages were collected and processed as described previously [67].

4.3. TUNEL Assay

The terminal deoxynucleotidyl transferase-mediated dUTP nick-end labeling (TUNEL) assay was performed as in a previous report [29]. Anthers from wild type and *gpat3-2* at different developmental stages were collected and prepared to generate paraffin sections. The selected paraffin sections were dewaxed in xylene and rehydrated in an ethanol series. TUNEL assay was performed using an In Vitro DeadEnd™ Fluorometric TUNEL System, using fluorescein (Promega, Madison, WI 017959, USA) according to the manufacturer's instructions with some modifications. Signals were observed and imaged under a fluorescence confocal scanner microscope (ZEISS LSM 700, Jena, Germany). All pictures were taken in the same setting.

4.4. Molecular Cloning of Osgpat3-2

For mapping the *Osgpat3-2* locus, total DNA was extracted from fresh leaves using the modified cetyl trimethyl ammonium bromide (CTAB) method [74]. InDel (insertion/deletion) markers within the preliminary-mapping region were developed according to sequence differences between the genome sequence of *japonica* Nipponpare and *indica* 9311 (http://www.gramene.org and http://blast.ncbi.nlm.nih.gov), and polymorphisms between the two parents (*gpat3-2* and 02428) were detected. Potential candidate genes for all open reading frames (ORFs) in the fine-mapping interval were identified by referring to the Rice Genome Annotation Project Database (http://rice.plantbiology.msu.edu/index.shtml). The primers for molecular cloning of *Osgpat3-2* are listed in Table S3 (Supplementary Materials). The PCR products were separated by electrophoresis on 8% non-denaturing polyacrylamide gels and visualized by 0.1% $AgNO_3$ staining and NaOH staining with formaldehyde.

For enzyme digestion analysis of the mutation site, the restriction endonuclease *Nla*IV site of the target gene was identified using Primer Premier 5.0 software. Two pairs of primers were designed to generate 298 bp of DNA fragments amplified by KOD-FX (TOYOBO, Osaka, Japan). Purified PCR products were used for *NlaIV* digestion. A total of 100 μL of this reaction system contained 3 μL of HaeII (20 units per 1 μL), 10 μL of NEBuffer (1×), 60 μL of purified DNA template (3 μg), and 27 μL of distilled deionized H_2O (ddH_2O). Two reaction systems were incubated at 37 °C for 3 h followed by 2.5% agarose gel electrophoresis.

4.5. Complementation of the gpat3-2 Mutant

For complementation, a genomic DNA fragment ~7908 bp containing the entire *Osgpat3-2* coding region, a 3350-bp upstream sequence, and a 1180-bp sequence downstream of the termination codon was amplified from wild-type Zhonghui8015 using the primers listed in Table S1 (Supplementary Materials). The amplified fragment was released by *BamHI* digestion and cloned into *BamHI*-digested binary vector pCAMBIA1300 (CAMBIA, Portland, OR, USA, hygromycin resistance) using an In-Fusion Advantage Cloning kit (catalog no. PT4065; Clontech, San Francisco, CA, USA). Then, calli induced using BC_1F_2 seeds and showing the *gpat3-2* genotype were used for *Agrobacterium tumefaciens*-mediated transformation (all calli were selected by sequencing using the primers listed in Table S3, Supplementary Materials).

4.6. Vector Construction for CRISPR/Cas9-Mediated Mutation

Vector construction of CRISPR/Cas9-mediated mutation was processed essentially as described in Wu et al. (2017) [75]. To create the single guide RNA (sgRNA)/Cas9-induced *OsGPAT3* construct, a 23-bp *OsGPAT3*-specific sgRNA/Cas9 target sequence (red marking sequence) was inserted into the *AarI* site of the pcas9-sgRNA vector, as described in Miao et al. (2013) [76], with some modifications. The primers used are detailed in Table S2 (Supplementary Materials). The aforementioned calli were introduced into wild-type and Zhonghua 11 seeds by *Agrobacterium tumefaciens*-mediated transformation as described previously [77]. The T_0 transgenic mutant plants regenerated from hygromycin-resistant calli were examined for the presence of transgenes using specific Cas-seq primers (Table S3, Supplementary Materials).

4.7. RNA Extraction, First-Strand Synthesis, and qPCR Analysis

Rice anthers at different developmental stages of the *gpat3-2* mutant and wild-type plants were collected for qPCR analysis of gene expression levels. Total RNA was extracted using the TIANGEN RNAprep Pure Plant Kit as described by the supplier. RNA was then reverse-transcribed (RT) from DNase I-treated RNA using Oligo-dT (18) primers in a 20-μL reaction using a SuperScript III Reverse Transcriptase Kit (TOYOBO, Japan). For qPCR, first-strand complementary DNA (cDNA) was diluted three times and then 3 μL of the RT products were used as the template of every PCR reaction using SYBR Premix Ex *Taq* II (TaKaRa) according to the manufacturer's instructions. The qPCR analysis was performed on a Roche LightCycler 480 device using gene-specific primers with the rice *Actin* gene (*Os03g0234200*) as an endogenous control, the relative expression levels were measured using the $2^{-\Delta Ct}$ analysis method, and the results were represented as means ± SD. This analysis examined expression of *OsGPAT3* and two homologs of *OsGPAT3* (Table S4), twelve tapetum PCD-related genes, twelve regulators that participate in lipid metabolism, and three glycometabolism-related regulators [21,31,41,44,50,78].

5. Conclusions

Male reproductive development in rice is important in both the improvement of yield and for an in-depth understanding of the mechanisms of anther development and pollen formation. In this study, we isolated and characterized a candidate recessive gene *OsGPAT3* that regulates anther wall PCD and pollen formation in rice using a typical map cloning method. Complementation analysis and knock-out experiments with the candidate gene further confirmed that the recessive mutation on *OsGPAT3* was responsible for the no-pollen phenotype of *gpat3-2*. Expression patterns of male sterility-related genes also demonstrated that loss of function of *OsGPAT3* caused various alterations in expression levels of both nutrient metabolism-related and tapetum PCD-related regulators, resulting in abnormal anther wall development and pollen formation. Nevertheless, the identification and observation of the *gpat3-2* mutant, together with the five allelic mutants, provided new insights into the function of *OsGPAT3* in regulating anther development and pollen formation in rice.

Supplementary Materials: Supplementary materials can be found at http://www.mdpi.com/1422-0067/19/12/4017/s1.

Author Contributions: L.C., S.C., and L.S. conceived and designed the experiments. X.X. and Z.Y. provided assistance in transection analysis, and TUNEL assay primer design. Y.Z. and P.Y. conducted population construction and phenotypic identification. X.W. and H.W. created the transgenic rice lines and provided assistance in primer design and mapping cloning. A.A. and R.M.K. made significant contributions to manuscript edits. L.S. analyzed the data and drafted the manuscript. Corresponding authors S.C. and L.C. were responsible for the overall concepts and designing the experiments. All authors read and approved the final manuscript.

Funding: This research was funded by grants from the National Key Transgenic Research Project (2016ZX08001-002), the Natural Science Foundation of Innovation Research Group (31521064), the Agricultural Science and Technology Innovation Program of the Chinese Academy of Agricultural Science (CAAS-ASTIP-2013-CNRRI), the Specialized Fund for the Basic Research Operating Expenses Program of Central Public Research Institutes (Y2017CG12), the Natural Science Foundation of Zhejiang Province (LQ17C130003), and the China Postdoctoral Science Foundation (2018M631641).

Acknowledgments: We would like to thank associate professor Jing Yu (Shanghai Jiao Tong University, Shanghai, China) for generously providing many valuable suggestions for the manuscript. In addition, Lianping Sun particularly wishes to thank the god-given arrangements of meeting Siqi Deng last year, and the patience, understanding, and support from her over the past year. Will you marry me and jointly weave our happiness in the rest of our lives?

Conflicts of Interest: The authors declare no conflicts of interest.

References

1. Goldberg, R.B.; Beals, T.P.; Sanders, P.M. Anther development: Basic principles and practical applications. *Plant Cell* **1993**, *5*, 1217–1229. [CrossRef] [PubMed]
2. Zhang, D.B.; Liang, W.Q. In Pushing the boundaries of scientific research: 120 years of addressing global issue. *Science* **2016**, *351*, 1223. [CrossRef]
3. Ma, H. Molecular genetic analyses of microsporogenesis and microgametogenesis in flowering plants. *Annu. Rev. Plant Biol.* **2005**, *56*, 393–434. [CrossRef] [PubMed]
4. Zhao, D.Z. Control of anther cell differentiation: A teamwork of receptor-like kinases. *Sex. Plant Reprod.* **2009**, *22*, 221–228. [CrossRef] [PubMed]
5. Wilson, Z.A.; Song, J.; Taylor, B.; Yang, C. The final split: The regulation of anther dehiscence. *J. Exp. Bot.* **2011**, *62*, 1633–1649. [CrossRef] [PubMed]
6. Walbot, V.; Egger, R.L. Pre-meiotic anther development: Cell fate specification and differentiation. *Annu. Rev. Plant Biol.* **2016**, *67*, 365–395. [CrossRef]
7. Scott, R.J.; Spielman, M.; Dickinson, H.G. Stamen structure and function. *Plant Cell* **2004**, *16*, S46–S60. [CrossRef]
8. Matsui, T.; Omasa, K.; Horie, T. Mechanism of anther dehiscence in rice (*Oryza sativa* L.). *Ann. Bot.* **1999**, *84*, 501–506. [CrossRef]
9. Wu, H.M.; Cheun, A.Y. Programmed cell death in plant reproduction. *Plant Mol. Biol.* **2000**, *44*, 267–281. [CrossRef]
10. Parish, R.W.; Li, S.F. Death of a tapetum: A programme of developmental altruism. *Plant Sci.* **2010**, *178*, 73–89. [CrossRef]
11. Jung, K.H.; Han, M.J.; Lee, Y.S.; Kim, Y.W.; Hwang, I.; Kim, M.J.; Kim, Y.K.; Nahm, B.H.; An, G. Rice *Undeveloped Tapetum1* is a major regulator of early tapetum development. *Plant Cell* **2015**, *17*, 2705–2722. [CrossRef] [PubMed]
12. Li, N.; Zhang, D.S.; Liu, H.S.; Yin, C.S.; Li, X.X.; Liang, W.Q.; Yuan, Z.; Xu, B.; Chu, H.W.; Wang, J.; et al. The rice *tapetum degeneration retardation* gene is required for tapetum degradation and anther development. *Plant Cell* **2006**, *18*, 2999–3014. [CrossRef] [PubMed]

13. Liu, J.; Qu, L.J. Meiotic and mitotic cell cycle mutants involved in gametophyte development in *Arabidopsis*. *Mol. Plant.* **2008**, *1*, 564–574. [CrossRef] [PubMed]

14. Wilson, Z.A.; Zhang, D.B. From Arabidopsis to rice: Pathways in pollen development. *J. Exp. Bot.* **2009**, *60*, 1479–1492. [CrossRef] [PubMed]

15. Shi, J.X.; Cui, M.H.; Yang, L.; Kim, Y.J.; Zhang, D.B. Genetic and biochemical mechanisms of pollen wall development. *Trends Plant Sci.* **2015**, *20*, 741–753. [CrossRef] [PubMed]

16. Gavrieli, Y.; Sherma, Y.; Ben-Sasson, S.A. Identification of programmed cell death in situ via specific labeling of nuclear DNA fragmentation. *J. Cell Biol.* **1992**, *119*, 493–501. [CrossRef] [PubMed]

17. Varnier, A.L.; Mazeyrat-Gourbeyre, F.; Sangwan, R.S.; Clément, C. Programmed cell death progressively models the development of anther sporophytic tissues from the tapetum and is triggered in pollen grains during maturation. *J. Struct Biol.* **2005**, *152*, 118–128. [CrossRef] [PubMed]

18. Zhang, D.S.; Liang, W.Q.; Yuan, Z.; Li, N.; Shi, J.; Wang, J.; Liu, Y.M.; Yu, W.J.; Zhang, D.B. *Tapetum Degeneration Retardation* is critical for aliphatic metabolism and gene regulation during rice pollen development. *Mol. Plant* **2008**, *1*, 599–610. [CrossRef] [PubMed]

19. Ji, C.N.; Li, H.Y.; Chen, L.B.; Xie, M.; Wang, F.P.; Chen, Y.L.; Liu, Y.G. A novel rice bHLH transcription factor, *DTD*, acts coordinately with tdr in controlling tapetum function and pollen development. *Mol. Plant* **2013**, *6*, 1715–1718. [CrossRef] [PubMed]

20. Niu, N.N.; Liang, W.Q.; Yang, X.J.; Jin, W.L.; Wilson, Z.A.; Hu, J.P.; Zhang, D.B. *EAT1* promotes tapetal cell death by regulating aspartic proteases during male reproductive development in rice. *Nat. Commun.* **2013**, *4*, 1445. [CrossRef] [PubMed]

21. Fu, Z.Z.; Yu, J.; Cheng, X.W.; Zong, X.; Xu, J.; Chen, M.J.; Li, Z.Y.; Zhang, D.B.; Liang, W.Q. The rice basic helix-loop-helix transcription factor *TDR INTERACTING PROTEIN2* is a central switch in early anther development. *Plant Cell* **2014**, *26*, 1512–1524. [CrossRef] [PubMed]

22. Ko, S.S.; Li, M.J.; Sun-Ben, K.M.; Ho, Y.C.; Lin, Y.J.; Chuang, M.H.; Hsing, H.X.; Lien, Y.C.; Yang, H.T.; Chang, H.C.; et al. The *bHLH142* transcription factor coordinates with *tdr1* to modulate the expression of *eat1* and regulate pollen development in rice. *Plant Cell* **2014**, *26*, 2486–2504. [CrossRef] [PubMed]

23. Ko, S.S.; Li, M.J.; Lin, Y.J.; Hsing, H.X.; Yang, T.T.; Chen, T.K.; Jhong, C.M.; Ku, M.S. Tightly controlled expression of *bHLH142* is essential for timely tapetal programmed cell death and pollen development in rice. *Front. Plant Sci.* **2017**, *8*, 1258. [CrossRef] [PubMed]

24. Ranjan, R.; Khurana, R.; Malik, N.; Badoni, S.; Parida, S.K.; Kapoor, S.; Tyagi, A.K. *bHLH142* regulates various metabolic pathway-related genes to affect pollen development and anther dehiscence in rice. *Sci. Rep.* **2017**, *7*, 43397. [CrossRef] [PubMed]

25. Kaneko, M.; Inukai, Y.; Ueguchi-Tanaka, M.; Itoh, H.; Izawa, T.; Kobayashi, Y.; Hattori, T.; Miyao, A.; Hirochika, H.; Ashikari, M.; et al. Loss-of-function mutations of the rice *GAMYB* gene impair α-amylase expression in aleurone and flower development. *Plant Cell* **2004**, *16*, 33–44. [CrossRef] [PubMed]

26. Aya, K.; Ueguchi-Tanaka, M.; Kondo, M.; Hamada, K.; Yano, K.; Nishimura, M.; Matsuoka, M. Gibberellin modulates anther development in rice via the transcriptional regulation of *GAMYB*. *Plant Cell* **2009**, *21*, 1453–1472. [CrossRef] [PubMed]

27. Liu, Z.H.; Bao, W.J.; Liang, W.Q.; Yin, J.Y.; Zhang, D.B. Identification of *gamyb-4* and analysis of the regulatory role of *GAMYB* in rice anther development. *J. Integr. Plant Biol.* **2010**, *52*, 670–678. [CrossRef]

28. Li, H.; Yuan, Z.; Vizcay-Barrena, G.; Yang, C.Y.; Liang, W.Q.; Zong, J.; Wilson, Z.A.; Zhang, D.B. *PERSISTENT TAPETAL CELL1* encodes a phd-finger protein that is required for tapetal cell death and pollen development in rice. *Plant Physiol.* **2011**, *156*, 615–630. [CrossRef]

29. Chen, Z.S.; Liu, X.F.; Wang, D.H.; Chen, R.; Zhang, X.L.; Xu, Z.H.; Bai, S.N. Transcription factor *OsTGA10* is a target of the MADS protein OsMADS8 and is required for tapetum development. *Plant Physiol.* **2018**, *176*, 819–835. [CrossRef]

30. Yi, J.; Moon, S.; Lee, Y.S.; Zhu, L.; Liang, W.Q.; Zhang, D.B.; Jung, K.H.; An, G. *Defective Tapetum Cell Death 1 (DTC1)* regulates ROS levels by binding to metallothionein during tapetum degeneration. *Plant Physiol.* **2016**, *170*, 1611–1623. [CrossRef]

31. Yu, J.; Meng, Z.; Liang, W.Q.; Behera, S.; Kudla, J.; Tucker, M.R.; Luo, Z.J.; Chen, M.J.; Xu, D.W.; Zhao, G.C.; et al. A rice Ca^{2+} binding protein is required for tapetum function and pollen formation. *Plant Physiol.* **2016**, *172*, 1772–1786. [CrossRef] [PubMed]

32. Tan, H.X.; Liang, W.Q.; Hu, J.P.; Zhang, D.B. *MTR1* encodes a secretory fasciclin glycoprotein required for male reproductive development in rice. *Dev. Cell* **2012**, *22*, 1127–1137. [CrossRef] [PubMed]

33. Li, L.; Li, Y.X.; Song, S.F.; Deng, H.F.; Li, N.; Fu, X.Q.; Chen, G.H.; Yuan, L.P. An *anther development F-box* (*ADF*) protein regulated by *tapetum degeneration retardation* (*TDR*) controls rice anther development. *Planta* **2015**, *241*, 157–166. [CrossRef] [PubMed]

34. Li, X.W.; Gao, X.Q.; Wei, Y.; Deng, L.; Ouyang, Y.D.; Chen, G.X.; Li, X.H.; Zhang, Q.F.; Wu, C.Y. Rice *APOPTOSIS INHIBITOR5* coupled with two DEAD-Box adenosine 5′-triphosphate-dependent rna helicases regulates tapetum degeneration. *Plant Cell* **2011**, *23*, 1416–1434. [CrossRef] [PubMed]

35. Blackmore, S.; Wortley, A.H.; Skvarla, J.J.; Rowley, J.R. Pollen wall development in flowering plants. *New Phytol.* **2007**, *174*, 483–498. [CrossRef] [PubMed]

36. Zhang, D.B.; Shi, J.X.; Yang, X. Role of lipid metabolism in plant pollen Exine development. *Subcell Biochem.* **2016**, *86*, 315–337. [CrossRef]

37. Li, H.; Pinot, F.; Sauveplane, V.; Werckreichhart, D.; Diehl, P.; Schreiber, L.; Franke, R.; Zhang, P.; Chen, L.; Gao, Y.W.; et al. Cytochrome P450 family member *CYP704B2* catalyzes the v-hydroxylation of fatty acids and is required for anther cutin biosynthesis and pollen exine formation in rice. *Plant Cell* **2010**, *22*, 173–190. [CrossRef]

38. Yang, X.J.; Wu, D.; Shi, J.X.; Yi, H.; Pinot, F.; Grausem, B.; Yin, C.S.; Zhu, L.; Chen, M.J.; Luo, Z.J.; et al. Rice *CYP703A3*, a cytochrome P450 hydroxylase, is essential for development of anther cuticle and pollen exine. *J. Integr. Plant Biol.* **2014**, *56*, 979–994. [CrossRef]

39. Yang, Z.F.; Zhang, Y.X.; Sun, L.P.; Zhang, P.P.; Liu, L.; Yu, P.; Xuan, D.D.; Xiang, X.J.; Wu, W.X.; Cao, L.Y.; et al. Identification of *cyp703a3-3* and analysis of regulatory role of *CYP703A3* in rice anther cuticle and pollen exine development. *Gene* **2018**, *649*, 63–73. [CrossRef]

40. Shi, J.; Tan, H.X.; Yu, X.H.; Liu, Y.Y.; Liang, W.Q.; Ranathunge, K.; Franke, R.B.; Schreiber, L.; Wang, Y.J.; Kai, G.Y.; et al. *Defective pollen wall* is required for anther and microspore development in rice and encodes a fatty acyl carrier protein reductase. *Plant Cell* **2011**, *23*, 2225–2246. [CrossRef]

41. Xu, D.W.; Shi, J.X.; Rautengarten, C.; Yang, L.; Qian, X.L.; Uzair, M.L.; Zhu, L.L.; Luo, Q.L.; An, G.L.; Waßmann, F.L.; et al. *Defective Pollen Wall 2* (*DPW2*) encodes an acyl transferase required for rice pollen development. *Plant Physiol.* **2017**, *173*, 240–255. [CrossRef] [PubMed]

42. Qin, P.; Tu, B.; Wang, Y.P.; Deng, L.C.; Quilichini, T.D.; Li, T.; Wang, H.; Ma, B.T.; Li, S.G. *ABCG15* encodes an ABC transporter protein, and is essential for post-meiotic anther and pollen exine development in rice. *Plant Cell Physiol.* **2013**, *54*, 138–154. [CrossRef] [PubMed]

43. Zhu, L.; Shi, J.X.; Zhao, G.C.; Zhang, D.B.; Liang, W.Q. *Post-meiotic deficient anther1* (*PDA1*) encodes an ABC transporter required for the development of anther cuticle and pollen exine in rice. *J. Plant Biol.* **2013**, *56*, 59–68. [CrossRef]

44. Zhao, G.C.; Shi, J.X.; Liang, W.Q.; Xue, F.Y.; Luo, Q.; Zhu, L.; Qu, G.R.; Chen, M.J.; Schreiber, L.; Zhang, D.B. Two ATP Binding cassette G transporters, rice *ATP binding cassette G26* and *ATP binding cassette G15*, collaboratively regulate rice male reproduction. *Plant Physiol.* **2015**, *169*, 2064–2079. [CrossRef] [PubMed]

45. Chang, Z.Y.; Jin, M.N.; Yan, W.; Chen, H.; Qiu, S.J.; Fu, S.; Xia, J.X.; Liu, Y.C.; Chen, Z.F.; Wu, J.X.; et al. The ATP-binding cassette (ABC) transporter *OsABCG3* is essential for pollen development in rice. *Rice* **2018**, *11*, 58. [CrossRef] [PubMed]

46. Zhang, D.S.; Liang, W.Q.; Yin, C.S.; Zong, J.; Gu, F.W.; Zhang, D.B. *OsC6*, encoding a lipid transfer protein, is required for postmeiotic anther development in rice. *Plant Physiol.* **2010**, *154*, 149–162. [CrossRef]

47. Jung, K.H.; Han, M.J.; Lee, D.Y.; Lee, Y.S.; Schreiber, L.; Franke, R.; Faust, A.; Yephremov, A.; Saedler, H.; Kim, Y.W.; et al. *Wax-deficient anther1* is involved in cuticle and wax production in rice anther walls and is required for pollen development. *Plant Cell* **2006**, *18*, 3015–3032. [CrossRef]

48. Li, Y.L.; Li, D.D.; Guo, Z.L.; Shi, Q.S.; Xiong, S.X.; Zhang, C.; Zhu, J.; Yang, Z.N. *OsACOS12*, an orthologue of Arabidopsis acyl-CoA synthetase5, plays an important role in pollen exine formation and anther development in rice. *BMC Plant Biol.* **2016**, *16*, 256. [CrossRef]

49. Yang, X.J.; Liang, W.Q.; Chen, M.J.; Zhang, D.B.; Zhao, X.X.; Shi, J.X. Rice fatty acyl-CoA synthetase *OsACOS12* is required for tapetum programmed cell death and male fertility. *Planta* **2017**, *246*, 105–122. [CrossRef]

50. Liu, Z.; Lin, S.; Shi, J.X.; Yu, J.; Zhu, L.; Yang, X.J.; Zhang, D.B.; Liang, W.Q. Rice *No Pollen 1* (*NP1*) is required for anther cuticle formation and pollen exine patterning. *Plant J.* **2017**, *91*, 263–277. [CrossRef]

51. Zhu, X.L.; Yu, J.; Shi, J.X.; Tohge, T.; Fernie, A.R.; Meir, S.; Aharoni, A.R.; Xu, D.W.; Zhang, D.B.; Liang, W.Q. The polyketide synthase *OsPKS2* is essential for pollen exine and Ubisch body patterning in rice. *J. Integr. Plant Biol.* **2017**, *59*, 612–628. [CrossRef] [PubMed]

52. Zou, T.; Liu, M.X.; Xiao, Q.; Wang, T.; Chen, D.; Luo, T.; Yuan, G.Q.; Li, Q.; Zhu, J.; Liang, Y.Y.; et al. *OsPKS2* is required for rice male fertility by participating in pollen wall formation. *Plant Cell Rep.* **2018**, *37*, 759–773. [CrossRef] [PubMed]

53. Coleman, R.A.; Lee, D.P. Enzymes of triacylglycerol synthesis and their regulation. *Prog. Lipid Res.* **2004**, *43*, 134–176. [CrossRef]

54. Cao, J.; Li, J.L.; Li, D.; Tobin, J.F.; Gimeno, R.E. Molecular identification of microsomal acyl-CoA:glycerol-3-phosphate acyltransferase, a key enzyme in de novo triacylglycerol synthesis. *Proc. Natl. Acad. Sci. USA* **2006**, *103*, 19695–19700. [CrossRef] [PubMed]

55. Takeuchi, K.; Reue, K. Biochemistry, physiology, and genetics of GPAT, AGPAT, and lipin enzymes in triglyceride synthesis. *Am. J. Physiol. Endocrinol. Metab.* **2009**, *296*, E1195–E1209. [CrossRef] [PubMed]

56. Lung, S.C.; Weselake, R.J. Diacylglycerol acyltransferase: A key mediator of splant triacylglycerol synthesis. *Lipids* **2006**, *41*, 1073–1088. [CrossRef] [PubMed]

57. Li, Y.; Beisson, F.; Koo, A.J.; Molina, I.; Pollard, M.; Ohlrogge, J. Identification of acyltransferases required for cutin biosynthesis and production of cutin with suberin-like monomers. *Proc. Natl. Acad. Sci. USA* **2007**, *104*, 18339–18344. [CrossRef]

58. Pollard, M.; Beisson, F.; Li, Y.; Ohlrogge, J.B. Building lipid barriers: Biosynthesis of cutin and suberin. *Trends Plant Sci.* **2008**, *13*, 236–246. [CrossRef]

59. Beisson, F.; Li-Beisson, Y.; Pollard, M. Solving the puzzles of cutin and suberin polymer biosynthesis. *Curr. Opin. Plant Biol.* **2012**, *15*, 329–337. [CrossRef] [PubMed]

60. Yang, W.; Simpson, J.P.; Li-Beisson, Y.; Beisson, F.; Pollard, M.; Ohlrogge, J.B. A land-plant-specific glycerol-3-phosphate acyltransferase family in Arabidopsis: Substrate specificity, sn-2 preference, and evolution. *Plant Physiol.* **2012**, *160*, 638–652. [CrossRef]

61. Zheng, Z.; Xia, Q.; Dauk, M.; Shen, W.; Selvaraj, G.; Zou, J. Arabidopsis *AtGPAT1*, a member of the membrane-bound glycerol-3-phosphate acyltransferase gene family, is essential for tapetum differentiation and male fertility. *Plant Cell* **2003**, *15*, 1872–1887. [CrossRef] [PubMed]

62. Petit, J.; Bres, C.; Mauxion, J.P.; Tai, F.W.; Martin, L.B.; Fich, E.A.; Joubes, J.; Rose, J.K.; Domergue, F.; Rothan, C. The glycerol-3-phosphate acyltransferase *GPAT6* from tomato plays a central role in fruit cutin biosynthesis. *Plant Physiol.* **2016**, *171*, 894–913. [CrossRef] [PubMed]

63. Men, X.; Shi, J.X.; Liang, W.Q.; Zhang, Q.F.; Lian, G.B.; Quan, S.; Zhu, L.; Luo, Z.J.; Chen, M.J.; Zhang, D.B. *Glycerol-3-Phosphate Acyltransferase 3 (OsGPAT3)* is required for anther development and male fertility in rice. *J. Exp. Bot.* **2017**, *68*, 513–526. [CrossRef] [PubMed]

64. Xie, K.; Wu, S.W.; Li, Z.W.; Zhou, Y.; Zhang, D.F.; Dong, Z.Y.; An, X.L.; Zhu, T.T.; Zhang, S.M.; Liu, S.S.; et al. Map-based cloning and characterization of *Zea mays male sterility33* (*ZmMs33*) gene, encoding a glycerol-3-phosphate acyltransferase. *Theor. Appl. Genet.* **2018**, *131*, 1363–1378. [CrossRef] [PubMed]

65. Zhang, D.B.; Wilson, Z.A. Stamen specification and anther development in rice. *Chin. Sci. Bull.* **2009**, *54*, 2342–2353. [CrossRef]

66. Zhang, D.B.; Luo, X.; Zhu, L. Cytological analysis and genetic control of rice anther development. *J. Genet. Genom.* **2011**, *38*, 379–390. [CrossRef]

67. Sun, L.P.; Zhang, Y.X.; Zhang, P.P.; Yang, Z.F.; Zhou, X.X.; Xuan, D.D.; Li, Z.H.; Wu, W.X.; Zhan, X.D.; Shen, X.H.; et al. Morphogenesis and Gene Mapping of *deformed interior floral organ 1* (*difo1*), a novel mutant associated with floral organ development in rice. *Plant Mol. Biol. Rep.* **2017**, *35*, 330–344. [CrossRef]

68. Shen, Y.J.; Jiang, H.; Jin, J.P.; Zhang, Z.B.; Xi, B.; He, Y.Y.; Wang, G.; Wang, C.; Qian, L.; Li, X.; et al. Development of genome-wide DNA polymorphism database for map-based cloning of rice genes. *Plant Physiol.* **2004**, *135*, 1198–1205. [CrossRef]

69. Nonomura, K.I.; Miyoshi, K.; Eiguchi, M.; Suzuki, T.; Miyao, A.; Hirochika, H.; Kurata, N. The *MSP1* gene is necessary to restrict the number of cells entering into male and female sporogenesis and to initiate anther wall formation in rice. *Plant Cell* **2003**, *15*, 1728–1739. [CrossRef]

70. Hu, L.F.; Liang, W.Q.; Yin, C.S.; Cui, X.; Zong, J.; Wang, X.; Hu, J.P.; Zhang, D.B. Rice *MADS3* regulates ros homeostasis during late anther development. *Plant Cell* **2011**, *23*, 515–533. [CrossRef]

71. Ariizumi, T.; Toriyama, K. Genetic regulation of sporopollenin synthesis and pollen exine development. *Annu. Rev. Plant Biol.* **2011**, *62*, 437–460. [CrossRef] [PubMed]

72. Hong, Y.F.; Ho, T.H.; Wu, C.F.; Ho, S.L.; Yeh, R.H.; Lu, C.A.; Chen, P.W.; Yu, L.C.; Chao, A.; Yu, S.M. Convergent starvation signals and hormone crosstalk in regulating nutrient mobilization upon germination in cereals. *Plant Cell* **2012**, *24*, 2857–2873. [CrossRef] [PubMed]

73. Li, Z.; Zhang, Y.X.; Liu, L.; Liu, Q.E.; Bi, Z.Z.; Yu, N.; Cheng, S.H.; Cao, L.Y. Fine mapping of the *lesion mimic and early senescence 1 (lmes1)* in rice (*Oryza sativa.* L). *Plant Physiol. Biochem.* **2014**, *80*, 300–307. [CrossRef]

74. Chen, D.H.; Ronald, P.C. A rapid DNA minipreparation method suitable for AFLP and other PCR applications. *Plant Mol. Biol. Rep.* **1999**, *17*, 53–57. [CrossRef]

75. Wu, W.X.; Zheng, X.M.; Chen, D.; Zhang, Y.X.; Ma, W.W.; Zhang, H.; Sun, L.P.; Yang, Z.F.; Zhao, C.D.; Zhan, X.D.; et al. *OsCOL16*, encoding a CONSTANS-like protein, represses flowering by up-regulating *Ghd7* expression in rice. *Plant Sci.* **2017**, *260*, 60–69. [CrossRef]

76. Miao, J.; Guo, D.; Zhang, J.; Huang, Q.; Qin, G.; Zhang, X.; Wan, J.; Gu, H.; Qu, L.J. Targeted mutagenesis in rice using CRISPR-Cas system. *Cell Res.* **2013**, *23*, 1233–1236. [CrossRef]

77. Hiei, Y.; Ohta, S.; Komari, T.; Kumashiro, T. Efficient transformation of rice (*Oryza sativa* L.) mediated by Agrobacterium and sequence analysis of the boundaries of the T-DNA. *Plant J.* **1994**, *6*, 271–282. [CrossRef]

78. Zhang, H.; Liang, W.Q.; Yang, X.J.; Luo, X.; Jiang, N.; Ma, H.; Zhang, D.B. *Carbon Starved Anther* encodes a myb domain protein that regulates sugar partitioning required for rice pollen development. *Plant Cell* **2010**, *22*, 672–689. [CrossRef]

International Journal of
Molecular Sciences

Article

WB1, a Regulator of Endosperm Development in Rice, Is Identified by a Modified MutMap Method

Hong Wang [1], Yingxin Zhang [1], Lianping Sun [1], Peng Xu [1], Ranran Tu [1], Shuai Meng [1], Weixun Wu [1], Galal Bakr Anis [1,2], Kashif Hussain [1], Aamiar Riaz [1], Daibo Chen [1], Liyong Cao [1,*], Shihua Cheng [1,*] and Xihong Shen [1,*]

[1] Key Laboratory for Zhejiang Super Rice Research, State Key Laboratory of Rice Biology, China National Rice Research Institute, Hangzhou 311400, Zhejiang, China; wjiyinh@126.com (H.W.); zyxrice@163.com (Y.Z.); slphongjun8868@126.com (L.S.); cnrri_pengxu@163.com (P.X.); 18883948050@163.com (R.T.); mengrice@163.com (S.M.); wuweixun@caas.cn (W.W.); galalanis5@gmail.com (G.B.A.); king3231251@gmail.com (K.S.); aamirriaz33@gmail.com (A.R.); cdb840925@163.com (D.C.)

[2] Rice Research and Training Center, Field Crops Research Institute, Agriculture Research Center, Kafr Elsheikh 33717, Egypt

* Correspondence: caoliyong@caas.cn (L.C.); shcheng@mail.hz.zj.cn (S.C.); xihongshen@126.com (X.S.); Tel.: +86-571-6337-0329 (L.C. & S.C.); +86-571-6337-0233 (X.S.); Fax: +86-571-6337-0265 (L.C. & S.C.); +86-571-6337-0233 (X.S.)

Received: 3 July 2018; Accepted: 19 July 2018; Published: 24 July 2018

Abstract: Abnormally developed endosperm strongly affects rice (*Oryza sativa*) appearance quality and grain weight. Endosperm formation is a complex process, and although many enzymes and related regulators have been identified, many other related factors remain largely unknown. Here, we report the isolation and characterization of a recessive mutation of *White Belly 1* (*WB1*), which regulates rice endosperm development, using a modified MutMap method in the rice mutant *wb1*. The *wb1* mutant develops a white-belly endosperm and abnormal starch granules in the inner portion of white grains. Representative of the white-belly phenotype, grains of *wb1* showed a higher grain chalkiness rate and degree and a lower 1000-grain weight (decreased by ~34%), in comparison with that of Wild Type (WT). The contents of amylose and amylopectin in *wb1* significantly decreased, and its physical properties were also altered. We adopted the modified MutMap method to identify 2.52 Mb candidate regions with a high specificity, where we detected 275 SNPs in chromosome 4. Finally, we identified 19 SNPs at 12 candidate genes. Transcript levels analysis of all candidate genes showed that *WB1* (*Os04t0413500*), encoding a cell-wall invertase, was the most probable cause of white-belly endosperm phenotype. Switching off *WB1* with the CRISPR/cas9 system in Japonica cv. Nipponbare demonstrates that *WB1* regulates endosperm development and that different mutations of *WB1* disrupt its biological function. All of these results taken together suggest that the *wb1* mutant is controlled by the mutation of *WB1*, and that the modified MutMap method is feasible to identify mutant genes, and could promote genetic improvement in rice.

Keywords: *Oryza sativa*; endosperm development; rice quality; *WB1*; the modified MutMap method

1. Introduction

Rice (*Oryza sativa*), one of the most important food crops in the world, provides more than 21% of human caloric needs [1]. With the improvement of living standards, there is increasing demand for high-quality rice, with greater quality of exterior, eating, and processing. Quality of rice appearance and yield are negatively affected by abnormally developed endosperm, which leads to grains with decreased weight and floury endosperm [2–8], shrunken endosperm [9–13], and great chalkiness [14,15].

Therefore, elucidating the mechanisms of endosperm development will be conducive to cultivating rice varieties with better appearance and higher yield.

Previous studies have shown that abnormality of rice endosperm can be caused by disorder of starch biosynthesis in the endosperm. Starch in the endosperm is composed of amylopectin (α-1,6-branched polyglucan) and amylose (α-1,4-polyglucan) [16]. In recent years, many key genes involved in starch biosynthesis have been identified in rice endosperm. The primary substrate of starch biosynthesis in rice endosperm comes from sucrose in the cell during photosynthesis [17], and it must be transported to the endosperm before being converted to glucose and fructose utilized for starch synthesis [18]. Several key genes involved in this process have been identified, including *OsSUT2*, which encodes a sucrose transporter and plays a vital role in transporting sucrose from source to sinks [19], *GIF1*, which encodes cell-wall invertase and is essential for the hydrolysis and uploading of sucrose [14], *OsSWEET4*, which encodes a hexose transporter and enhances sugar import into the endosperm from maternal phloem [20], and some genes (*SUS2*, *SUS3*, *SUS4*) from the sucrose synthase (SUS) genes family, which play an important role in the hydrolysis of sucrose [21]. However, glucose and fructose are not the direct substrate for starch synthesis: both need to be further converted to glucose 1-phosphate (G1P) under the catalysis of a series of enzymes [18]. The reaction of G1P with ATP (Adenosine 5′-triphosphate) produces the activated glucosyl donor ADP (Adenosine diphosphate)-glucose (ADPG), which is catalyzed by the enzyme ADP-glucose pyrophosphoryase (AGPase). In rice, the AGP gene family is made up of six subunit genes: two small subunit genes, *OsAGPS1* and *OsAGPS2* (*OsAGPS2a*, *OsAGPS2b*), and four large subunit genes, *OsAGPL1*, *OsAGPL2*, *OsAGPL3* and *OsAGPL4*. *OsAGPS1*, *OsAGPS2b*, *OsAGPL1* and *OsAGPL2* mainly function in rice endosperm [9,11,22]. In addition, pyruvate orthophosphate dikinase (PPDK) which is encoded by *OsPPDKB*, is involved in activating fructose [23]. In rice, loss-of-function mutants of these genes show abnormally developed endosperm, thus causing negative impacts on rice appearance quality and grain weight.

The activated substrates must cross the membrane of the amyloplasts before amylose and amylopectin are synthesized in the amyloplasts of the endosperm cells. During this transportation from cytoplasm to amyloplast, the major ADP-glucose transporter encoded by the *Brittle1* (*BT1*) imports the ADPG into amyloplasts; mutants with a defect in *BT1* develop shrunken endosperm [12,13]. When the activated substrates have been transported from the cytoplasm to the amyloplast, the gene *Waxy* encodes granule-bound starch synthase I (GBSS I), which primarily controls amylose synthesis [24]. Other genes control amylopectin synthesis in rice endosperm, including *SSI* [25], *SSIIa* [26], and *OsSSIIIa* [27], which encode starch synthase, *ISA1* [28,29] encoding isoamylase-type DBE isoamylase 1, and *OsBEIIb* [30] encoding starch branching enzymeIIb. Loss-of-function mutants of these genes severely disrupt the normal development of the endosperm.

Some regulators involved in starch synthesis during endosperm development have also been identified. *FLO2* mediates a protein-protein interaction, with a mutation of *FLO2* resulting in a floury endosperm [2]. *FLO6* directly interacts with *ISA1*, which affects the formation of starch granules during development of the rice endosperm [4]. *FLO7* encodes a regulator involved in starch synthesis and amylopectin development of the peripheral endosperm [8]. *Rice Starch Regulator1* (*RSR1*), a member of the AP2/EREBP family of transcription factors, negatively regulates starch synthesis [31]. *SUBSTANDARD STARCH GRAIN4* (*SSG4*) regulates the size of starch grains (SGs) in rice endosperm [6].

Deformity of the endosperm can also result from dysregulation of development of the protein bodies and storage proteins in the rice endosperm. *Chalk5*, which encodes a vacuolar H^+-translocating pyrophosphatase with inorganic pyrophosphate hydrolysis and H^+-translocation activity, is a major quantitative trait locus (QTL) which controls grain chalkiness [15]. The rice basic leucine Zipper factor (RISBZ1) and rice prolamin box binding factor (RPBF) are transcriptional activators, which coordinate to regulate the expression of SSP (seed storage protein) genes [32]. Decreased expression of *RISBZ1* (*OsbZIP58*) and *RPBF* in transgenic plants causes opaque endosperm.

Several methods are currently used for gene isolation. The most commonly used one is positional cloning (map-based cloning), by which many rice genes were isolated. However, map-based cloning is more time- and labor-intensive for isolating genes, especially QTLs. Therefore, many researchers have been exploring new methods for genes isolation. MutMap (Figure 1a) [33], based on next-generation sequencing (NGS) [34], is a recently developed method of rapid gene isolation. The MutMap method has been used to isolate some rice genes, including *OsRR22*, a gene responsible for the salinity-tolerant phenotype of *hst1* [35] and *Pii*, a gene enhancing rice blast resistance [36].

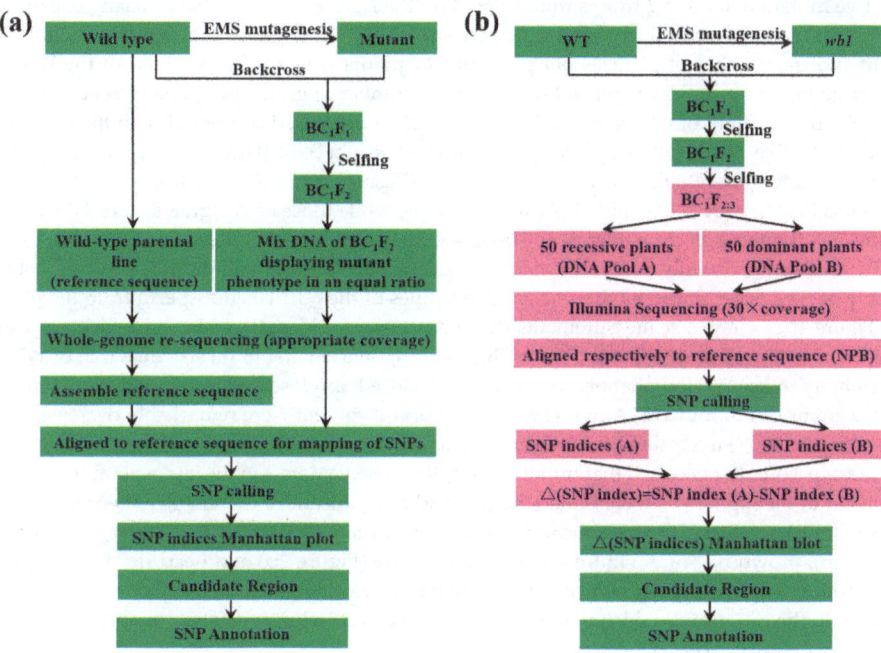

Figure 1. The steps of MutMap method applied to rice. (**a**) Common scheme of MutMap method applied to rice following the protocol described as previously reported [33]; (**b**) The scheme of gene mapping used in this study. The $BC_1F_{2:3}$ progeny formed mapping population. DNA of 50 recessive and 50 dominant plants from mapping population are mixed separately in an equal ratio to form the DNA Pool (A) and Pool (B) followed by the construction of DNA library and Illumina sequencing with 30×coverage, and then the treated sequencing data were aligned with the reference sequence followed by single nucleotide polymorphisms (SNP) calling. The reference sequence is the publicly available Nipponbare rice genome sequence [37]. For each identified SNP, SNP index (A) was obtained from Pool (A) and SNP index (B) corresponds with Pool (B). SNP index (A) minus SNP index (B) is Δ (SNP index) which is used for Manhattan plot, and we can obtain candidate region followed by SNP annotation. The pink color represents the different steps compared to the common scheme of MutMap. EMS: Ethane methyl sulfonate.

Although the understanding of the molecular mechanisms of the formation of rice endosperm has made great progress, rice endosperm is a very complex agronomic trait. Hence, it is still necessary to identify more functional genes and to describe their molecular mechanisms in order to enable systematic and comprehensive understanding of the inheritance of rice endosperm formation. In this study, we isolated *WB1*, which controlled rice endosperm development, via the modified MutMap method, and found that *WB1* played an important role in the regulation of starch synthesis during rice endosperm development. We also verified the target gene by CRISPR/Cas9 system. Our study also played a crucial role in explaining the molecular mechanisms of the formation of rice endosperm and the exploration of new methods for gene mapping.

2. Results

2.1. Phenotypic Characterization and Genetic Analysis of the wb1 Mutant

To identify new regulators of endosperm development, we recovered an endosperm development defective mutant named *wb1* from a mutant pool (in the *Japonica* variety ChangLiGeng background). The *wb1* mutant showed no apparent differences from WT throughout the vegetative stage. Plant height and the number of panicles per plant of *wb1* plants were similar to those of the WT at the mature stage. The number of spikelets per panicle, number of grains per panicle, seed-setting rate and 1000-grain weight of *wb1* were all showed a marked decreased compared with those of the WT (Table S1). Unlike WT, the glume of *wb1* grains showed brown color (Figure 2a–c), and *wb1* displayed markedly more grain chalkiness in the grain belly (Figure 2d,e). Grain chalkiness rate and grain chalkiness degree were 94.8% and 47.6% in *wb1* grains, while those of WT grains were 2.8% and 0.6% (Figure 2i,j). Scanning electron microscope images clearly indicated that the endosperm from grains of *wb1* developed abnormally as a result of loosely packed, spherical starch granules, in contrast to the densely packed, irregularly polyhedral starch granules of the normal endosperm from the grains of WT (Figure 2f,g). Grain size measurements showed that the seed length, width, and thickness were all significantly reduced in *wb1* grains (Figure 2h), resulting in a smaller grain size than that of WT, even occasionally in a shriveled phenotype. We also measured amylose and amylopectin content of the mature grains of *wb1* and WT. Amylose and amylopectin content were remarkably decreased in *wb1* grains (Figure 2k,l), suggesting that the starch accumulation in *wb1* grains was severely disrupted. All these results collectively reveal that mutation of *WB1* caused a defect in the endosperm development, which led to the higher grain chalkiness degree and a significant reduction of 1000-grain weight in *wb1* grains. We also analyzed physical properties of *wb1* and WT grains, including gel consistency (Figure 2n), brown rice rate (Figure 2o), milled rice rate (Figure 2p) and head rice rate (Figure 2q). Each of these was significantly reduced in the mutant, suggesting that dramatic physical changes have occurred in the *wb1* grains, which will further affect rice processing and eating quality.

Figure 2. *Cont.*

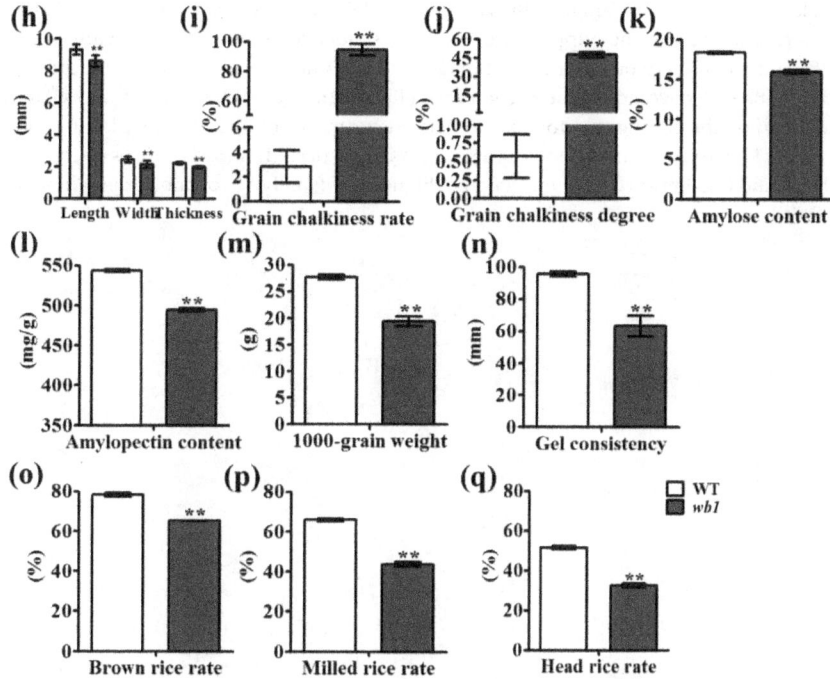

Figure 2. Phenotypic analyses of *wb1*. (**a**) Comparison of representative WT and *wb1* plant panicles; (**b**,**c**) Appearance of WT (**b**) and *wb1* (**c**) grains; (**d**,**e**) Appearance of WT (**d**) and *wb1* (**e**) white grains; (**f**,**g**) Scanning electron microscope images of WT (**f**) and *wb1* (**g**) seed endosperm. Magnification, ×2000; (**h**) Measurements of seed length, width, and thickness of WT and *wb1* grains (*n* = 20); (**i**) Grain chalkiness rate comparison of WT and *wb1* grains (*n* = 6), and grain chalkiness rate is the rate of chalky grains in total grains; (**j**) Grain chalkiness degree comparison of WT and *wb1* grains (*n* = 6), and grain chalkiness degree is the grain chalkiness rate multiplied by grain chalkiness area (the percent area of chalk in a grain); (**k**) Amylose content comparison of WT and *wb1* grains (0.05 g grain powder each, *n* = 3); (**l**) Amylopectin content comparison of WT and *wb1* grains (0.01 g grain powder each, *n* = 3); (**m**) Comparison of 1000-grain weight of WT and *wb1* (*n* = 10); (**n**) Gel consistency comparison of WT and *wb1* grains (*n* = 4); (**o**,**p**,**q**) Comparisons of brown rice, milled rice and head rice rates of WT and *wb1* (25 g paddy each, *n* = 3). Data are given as means ± SD (standard deviation). The asterisks represent statistical significance between WT and *wb1*, determined by a student's *t*-test (** $p \leq 0.01$). Scale bars: (**a**–**e**) 10 mm; (**f**,**g**) 30 μm.

To verify that this locus associated with the *wb1* phenotype was controlled by a single recessive gene, genetic analysis was conducted to examine the phenotype of all plants from BC_1F_1 and F_1 progeny, and of 1087 and 1000 plants from BC_1F_2 and F_2 progeny, respectively. The results showed that BC_1F_1 and F_1 seeds exhibited the wild-type phenotype, while the segregation model of normal to chalky grains fitted well to the expected ratio of a single inheritance, 3:1 (820:267, 745:255), in the BC_1F_2 and F_2 progeny (Table S2).

2.2. Candidate Region of the WB1 Gene Obtained through the Modified MutMap Method

A modified MutMap method (Figure 1b) was applied to isolate the WB1 gene. After re-sequencing for Pool A and Pool B, we obtained 125,252,285 (SRA accession SRP135580) and 120,484,878 (SRA accession SRP135578) cleaned reads for Pool A and Pool B, respectively, corresponding to >20 Gb of total read length with 30× coverage of the rice genome (370 Mb; Table S3). After these cleaned reads were aligned separately to the Nipponbare reference sequence by the BWA software, we obtained 110,119,455 and 105,488,179 unique mapped reads for Pool A and Pool B, respectively, corresponding to 87.55% and 87.92% coverage of the rice genome (Table S4). Then we calculated Δ (SNP indices) or *Fst* value based on the sliding window of the whole genome scan following by plotting the Δ (SNP indices) for all 12 chromosomes of rice (Figure 3b). As we expected, Δ (SNP indices) were distributed randomly around 0 for most parts of the genome (Figure 3b). Finally, we obtained the candidate region of 2.52 Mb (Figure 3b).

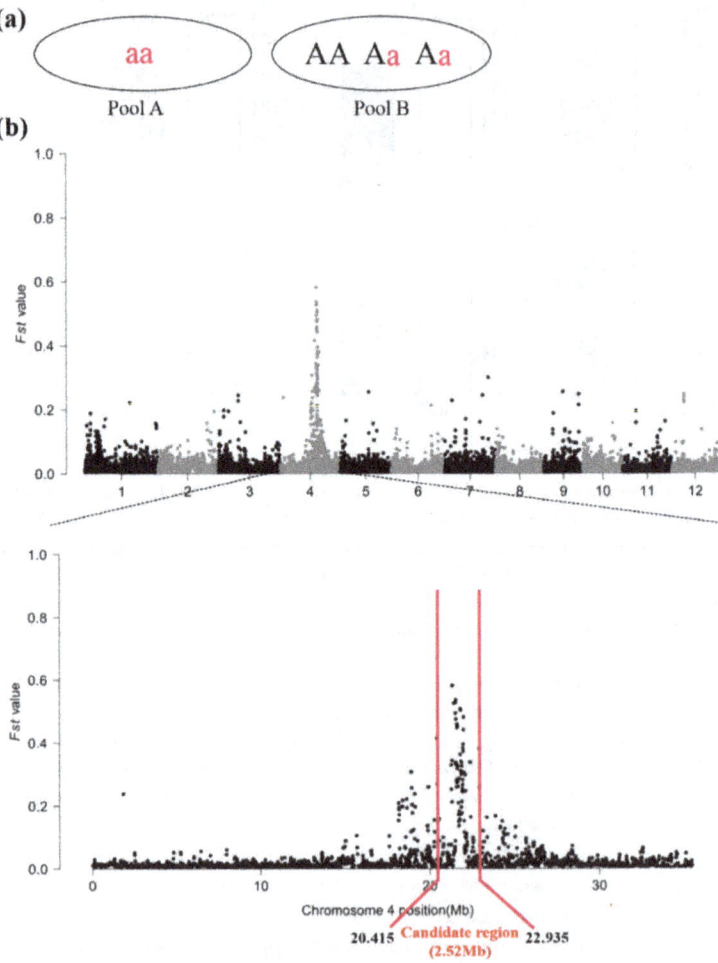

Figure 3. Candidate region of wb1 obtained by the modified MutMap method. (**a**) An example for explanation of Δ (SNP index) for the casual SNP. Theoretically, SNP index (A) would be 1, SNP index (B) 0.333 (1/3), and thus Δ (SNP index) would be 0.667 (1 minus 1/3); (**b**) Δ (SNP indices) Manhattan plot. *Fst* value, defined as the proportion of genetic diversity due to allele frequency differences among populations described by the previous report [38]. Δ (SNP indices) and *Fst* values have the same meaning in this study.

2.3. Screening the SNPs Detected in the Candidate Region

From the Δ (SNP indices) plot (Figure 3b), we obtained the candidate region of 2.52 Mb where we detected 275 SNPs in chromosome 4 followed by gene annotation (Table S5). To identify the true causal SNP, we screened these SNPs with three steps: (i) retaining the SNPs in which Δ (SNP indices) ranged from 0.6 to 0.8; (ii) removing SNPs located in the intergenic region and SNPs which resulted in synonymous substitutions; and (iii) detecting the SNPs between WT and wb1 by sequencing. As a result, we obtained nineteen SNPs, which were located in twelve candidate genes (Table 1).

Table 1. Nineteen SNPs in twelve candidate genes.

Δ (SNP Index)	Accession	Location (bp)	Reference Base (WT)	Altered Base in *wb1*	Type of Mutation	Gene Annotation
0.758		21550665	T	G	Missense (T to P)	
0.754		21550664	G	T	Missense (T to K)	Helicase conserved
0.663	*ORF1*	21550888	C	T	Intron mutation	C-terminal domain
0.655		21550286	T	A	Missense (T to S)	containing protein
0.649		21551279	G	A	Intron mutation	
0.754	*ORF2*	21539737	A	G	3′-UTR mutation	Protein of unknown
0.612		21539457	G	T	Splice region mutation	function DUF668 family protein
0.743	*ORF3*	21331260	G	A	Missense (D to N)	40S ribosomal protein S10
0.734	*ORF4*	21514382	C	T	Intron mutation	Similar to
0.708		21513793	A	G	Intron mutation	H0315E07.10 protein
0.734	*ORF5*	21612944	C	A	Missense (K to N)	CENP-E-like kinetochore
0.663		21610862	C	A	Missense (S to I)	protein
0.733	*ORF6*	20423829	G	A	Missense (A to T)	Glycosyl hydrolases
0.733	*ORF7*	21795109	G	A	Missense (L to F)	Expressed protein
0.672	*ORF8*	21493980	G	A	Intron mutation	Similar to H0315E07.7 protein
0.639	*ORF9*	21897538	C	T	3′-UTR mutation	Nonsense-mediated decay UPF3
0.634	*ORF10*	21970357	C	T	5′-UTR mutation	Peptide transporter PTR2
0.631	*ORF11*	21710470	C	G	Intron mutation	Conserved hypothetical protein
0.61	*ORF12*	21734385	C	T	Nonsense (R to *)	No apical meristem protein

The asterisk indicates the stop codon.

2.4. Identification of the Casual SNP

We detected the expression levels of twelve candidate genes in endosperm tissues at various development stages (5, 10, 15 and 20 DAF) (Figure 4). We successfully detected all genes transcript levels except for that of *ORF8*. *ORF6* maintained relatively high expression level in comparison with other genes and its transcript level changed significantly during the four stages of endosperm development between WT and *wb1* (Figure 4). Although some genes demonstrated higher transcript levels at the DAF15 and DAF20 stages (Figure 4c,d) compared with the DAF5 and DAF10 stages (Figure 4a,b), and the transcript levels of several genes were also significantly altered between WT and *wb1* (Figure 4), all other genes showed low expression levels on the whole in contrast to the transcript levels of *ORF6*. Therefore, we may conclude that the mutation of *ORF6* (*Os04t0413500* or *Os04g33740*) played a major role in the defect of *wb1*.

SNP-20423829 were G to A transitions, presumably caused by EMS mutagenesis [39], and it was located at the site 1659 bp of the third exon of *ORF6* encoding a glycosyl hydrolase. This SNP led to an A159T mutation (codon GCG to ACG; Figure 5a,b). Moreover, results of digestion of restriction endonuclease *Hae* II confirmed this mutant site (Figure 5c). Accordingly, we hypothesized that *wb1* was caused by a missense substitution in *ORF6*. We also found that *ORF6* was a novel allele of *GIF1* (*Os04g33740*) which controlled rice grain filling and yield [14].

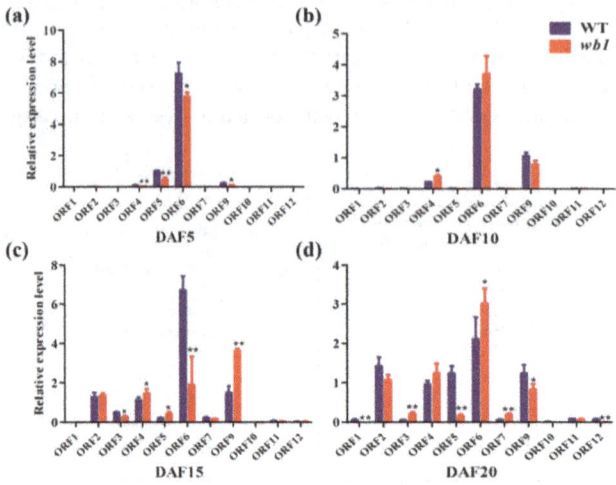

Figure 4. Relative expression analysis of 11 candidate genes based on real-time quantitative PCR (qPCR) at four stages of endosperm development between WT and *wb1*. (**a**) Relative expression analysis of 11 candidate genes at DAF5 stage; (**b**) Relative expression analysis of 11 candidate genes at DAF10 stage; (**c**) Relative expression analysis of 11 candidate genes at DAF15 stage; (**d**) Relative expression analysis of 11 candidate genes at DAF20 stage. All data were compared with transcript levels of WT by Student's *t*-test (* $p \leq 0.05$, ** $p \leq 0.01$). Values were means $\pm$ SD ($n = 3$).

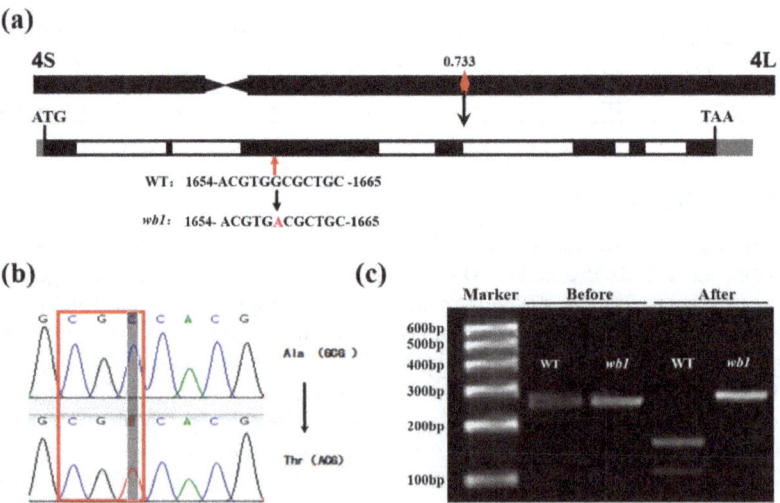

Figure 5. Further verification of causal SNP in *wb1*. (**a**,**b**) Sequencing validation of the causal SNP and the type of mutation; the red arrow indicates the mutant site, and the black arrows indicate the alternations; (**c**) Digestion of restriction endonuclease *Hae*II. "Before" represents the non-treated PCR product and "After" represents the *Hae*II-treated PCR product.

2.5. Function Verification of the WB1 Gene (ORF6) through the CRISPR/Cas9 System in Reverse

To verify our hypothesis, we created six novel alleles of *WB1* through the CRISPR/Cas9 system in *Japonica* cultivar Nipponbare (NPB). We found two target sequences in the third exon of *WB1* corresponding to CRISPR/Cas9 system and obtained six different mutants in T1 lines (Figure 6a). Grains of six mutants displayed brown glumes and grain chalkiness in the grain belly compared with the common grains of NPB (Figure 6b). SEM images distinctly showed that endosperm of six mutants developed abnormally compared to the normal endosperm of NPB (Figure 6b). The phenotypes of six mutants were similar to that of *wb1* (Figure 2b–g). Those results further proved that *WB1* was the target gene responsible for the *wb1* phenotype.

In NPB and six mutant lines, we measured 1000-grain weight (Figure 6h) and the main factors affecting 1000-grain weight, including grain length (Figure 6c), width (Figure 6d), and thickness (Figure 6e), grain chalkiness rate (Figure 6f) and degree (Figure 6g). Duncan's test indicated that the grain chalkiness rate degree were the major factors causing the significant reduction of 1000-grain weight of six mutant lines. The differences in grain length, width, and thickness between six mutant lines and NPB were not similar to the differences between WT and *wb1* (Figure 2h). This discrepancy was probably caused by the longer grain length of WT (~9.3 mm, Figure 2h) compared with that of NPB (~6.6 mm, Figure 6c) and the different mutations of WB1 between wb1 (Figure 5a) and the six mutant lines (Figure 6a).

Interestingly, some differences were also found among the six mutant lines (Figure 6c–h). Those results suggested that the grain chalkiness rate and grain chalkiness degree collectively determined the 1000-grain weight (Figure 6f–h), especially in the mutant line *nc-3*, where grain chalkiness rate and grain chalkiness degree showed significant decreases as compared to the other mutant lines, corresponding to its higher 1000-grain weight. To test whether the differences in the grain chalkiness rate and grain chalkiness degree among the six mutant lines were caused by different mutations of WB1, we performed multiple comparison of WB1 sequences by MEGA 5.0 software (Figure 7). The findings indicated that the six mutant lines showed different mutations which disrupted the substrate binding site and the active site of WB1, except in the mutant line *nc-5* (Figure 7).

2.6. Expression Analysis of Starch Metabolism-Related Genes in Endosperm

We performed qPCR analysis of total RNA extracted from the seed endosperm of WT and *wb1* at various stages (DAF5, DAF10, DAF15, and DAF20) and detected the transcript levels of some genes involved in starch synthesis. As shown in Figure 8, transcript levels of those genes were all altered during development of the rice endosperm. During the critical stages (DAF10 and DAF15) of grain filling, transcript levels of all genes showed a striking contrast between WT and *wb1*. This suggests that altered transcript levels of starch synthesis-related genes are probably involved in the abnormal development of rice endosperm. The higher transcript levels of *WB1*, *OsAPS1*, *OsAPL1*, *OsPPDKB*, and *FLO6* at the mature stage (DAF20) in the *wb1* mutant were probably caused by different maturity of seeds between WT and *wb1*.

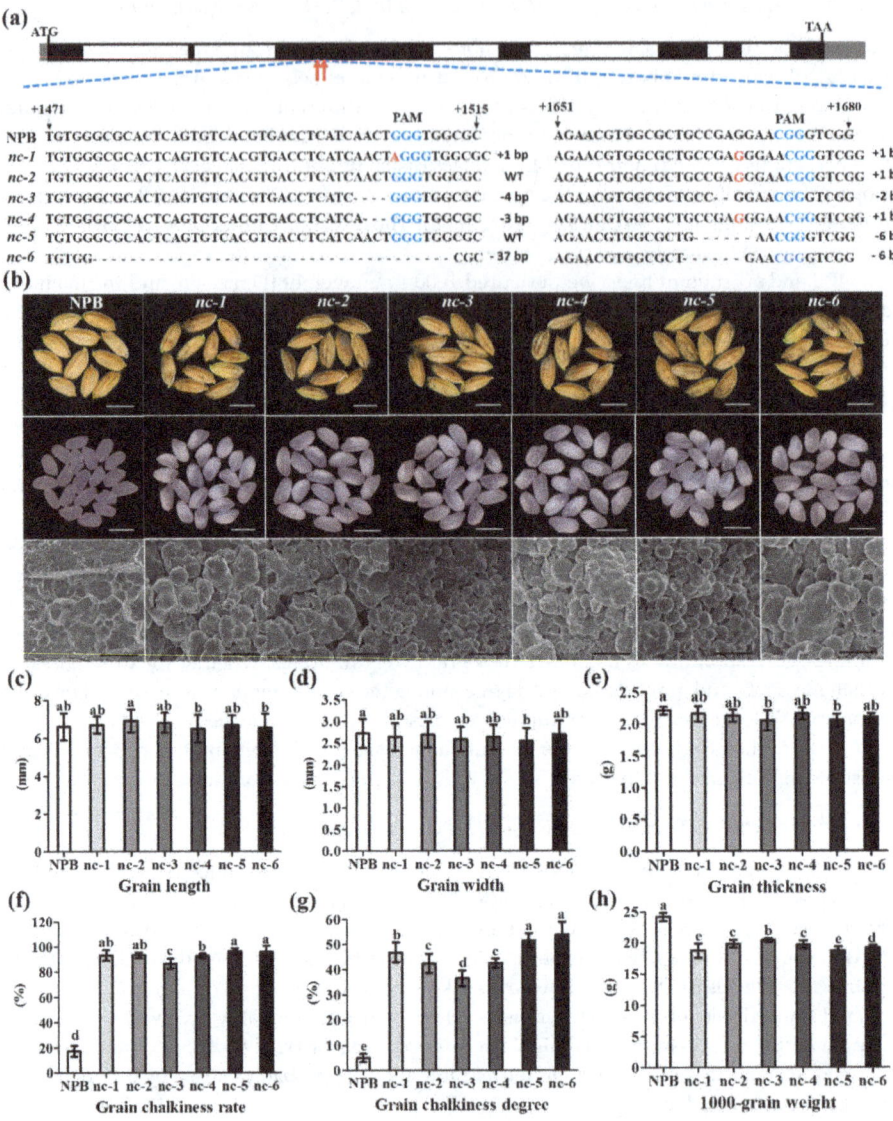

Figure 6. Sequencing validation and phenotypic analyses of six novel allelic mutants. (**a**) Sequencing validation of six novel allelic mutants. Blue color represents the PAM sequence of CRISPR/Cas9 system; red color represents insert bases; "-" represents deletion bases; the red arrows indicate the mutant sites mediated by CRISPR/Cas9 system. (**b**) Appearance and SEM of NPB and mutants grains. Magnification, ×1000; (**c–h**) Measurements of grain length (*n* = 30), width (*n* = 30), thickness (*n* = 10), grain chalkiness rate (*n* = 10), grain chalkiness degree (*n* = 10) and 1000-grain weight (*n* = 10) of NPB and mutant lines. Different letters indicate the statistical difference at $p \leq 0.05$ by Duncan's test. Values were means ± SD. Scale bars: bars of grains figures 5 mm; bars of SEM figures 10 μm.

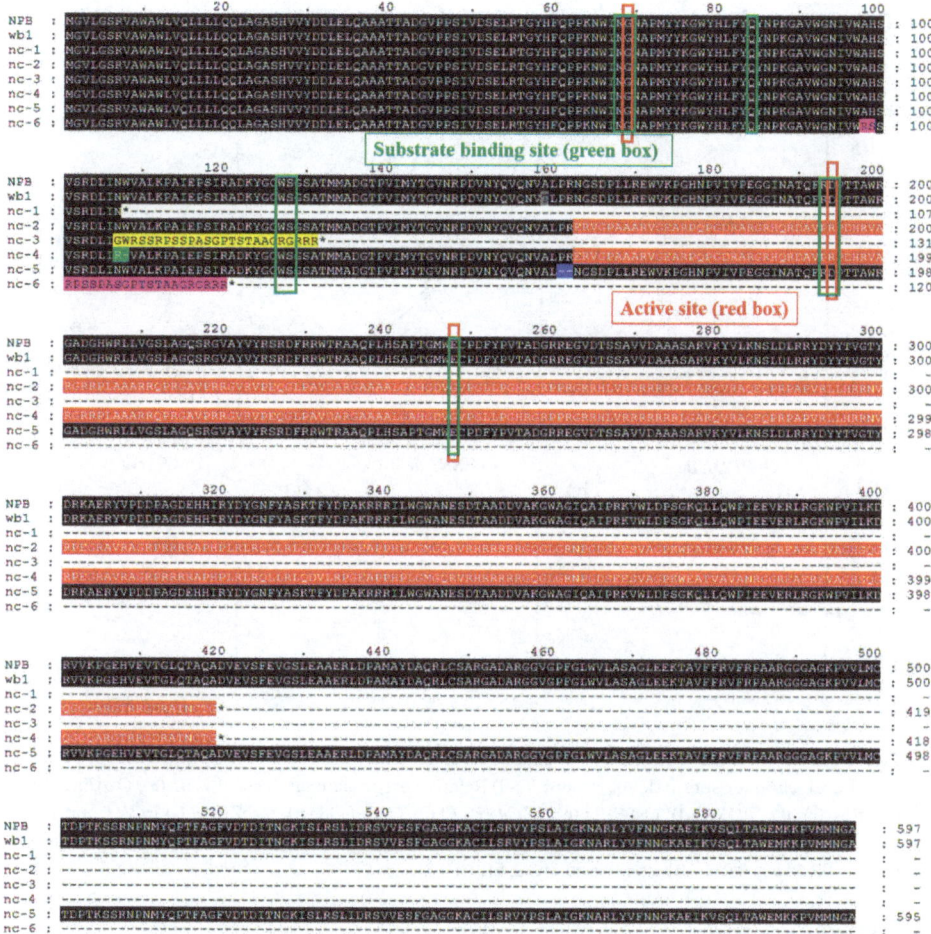

Figure 7. WB1 alignments of NPB (WT) with six mutant lines and *wb1*. Analysis performed with MEGA 5.0 software. The substrate binding site (eight residues, green boxes) and the active site (three residues, red boxes) are indicated by the Blast search program (http://www.ncbi.nlm.nih.gov/BLAST/). Black color indicates a sequence that is consistent with that of WT. Except for the black color, the same color represents the same sequence, and different colors represent the different mutations among the mutant lines.

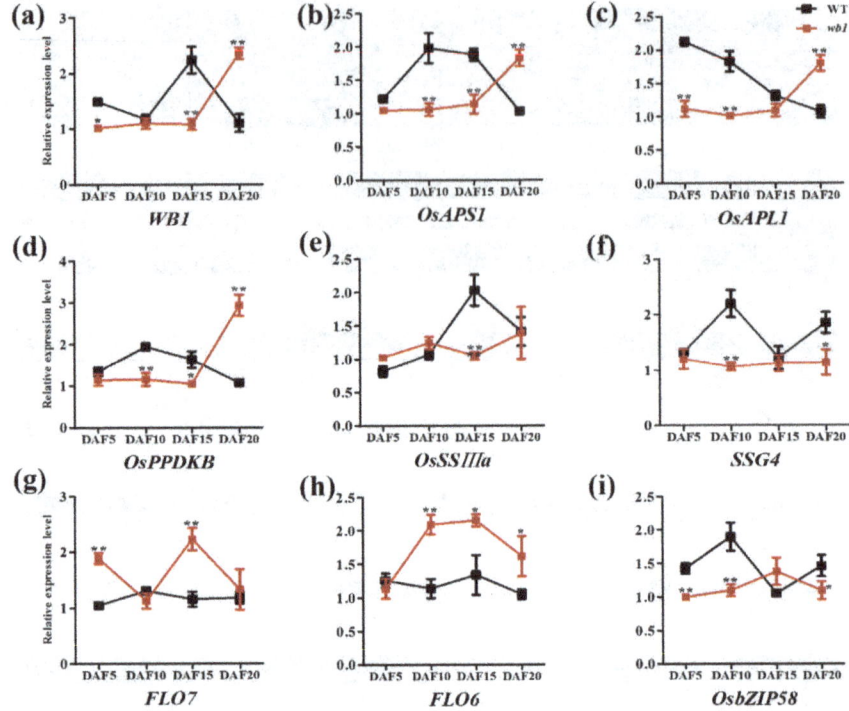

Figure 8. Relative expression analysis of genes associated with starch synthesis by qPCR at four stages of WT and *wb1* endosperm development. (**a–i**) Relative expression analysis of *WB1* (**a**), *OsAPS1* (**b**), *OsAPL1* (**c**), *OsPPDKB* (**d**), *OsSSIIIa* (**e**), *SSG4* (**f**), *FLO7* (**g**), *FLO6* (**h**), *OsbZIP58* (**i**) in WT and the wb1 mutant. All data were compared with the relative expression levels of WT by Student's *t*-test (* $p \leq 0.05$, ** $p \leq 0.01$). Values were means ± SD (*n* = 3).

3. Discussion

3.1. WB1 Controls Rice Endosperm Development

Grain chalkiness is one of the most important factors leading to low grain weight [15] and affecting rice appearance and milling, cooking, and eating quality [40,41]. Grain chalkiness is controlled by complex quantitative trait loci and by climatic conditions during rice grain filling, especially high temperature [42]. *Chalk5*, a major quantitative trait locus controlling grain chalkiness, and affecting head rice rate, is the only one that has been identified and characterized up to now. *Chalk5* is involved in the biogenesis of protein bodies in the endosperm cells [15]. Grain chalkiness can be an indicator of abnormally developed endosperm [2–8]. The major component of rice endosperm is a starch that is mainly composed of amylose and amylopectin. Many genes directly involved in the biosynthesis of amylose and amylopectin in rice endosperm cells have been identified and characterized, such as *Waxy*, *SSI*, *SSIIa*, *OsSSIIIa*, *ISA1*, *OsBEIIb* and *OsPPDKB* [23–28,30]. Loss of function of these genes can result in an abnormally developed endosperm, displaying more grain chalkiness and low grain weight.

Sucrose is produced by the source organ or photosynthetic organ and used as a carbon source for starch biosynthesis in endosperm cells. Sucrose must be transported from source organs into sink organs, which occurs via apoplast and/or sympast. Accordingly, in addition to the genes directly involved in starch biosynthesis of endosperm cells, other genes involved in this process can also affect endosperm development. In the apoplastic pathway, sucrose can be converted by cell-wall invertases into glucose and fructose, which are transported into cells by hexose transporters. Sucrose can also be directly taken by sucrose transporters into sink cells where it is hydrolyzed into glucose and fructose by sucrose hydrolases, including SUS2, SUS3, and SUS4 [21]. *GIF1* encodes a cell-wall invertase that mainly functions in the hydrolysis and uploading of sucrose during early grain-filling. The *gif1* mutant shows slower grain filling, ~24% lower final grain weight, and lower contents of amylose and amylopectin, and markedly more grain chalkiness as a result of abnormally developed and loosely packed starch granules [14]. *OsSWEET4* encodes a hexose transporter that is responsible for transferring hexoses across the BETL (basal endosperm transfer layer) to sustain of rice endosperm, downstream of a cell-wall invertase. The *ossweet4-1* mutant shows incomplete grain filling and significantly decreased grain weight [20]. *OsSUT2* encodes a sucrose transporter that functions in sucrose uptake from the vacuole. The *ossut2* mutant has significantly decreased sugar export ability and 1000-grain weight [19]. In our study, *WB1* encoded the *cell-wall invertase 2* (*OsCIN2*) and was a novel allele of the *GIF1* (*Os04t0413500* or *Os04g33740*) gene which controlled rice grain-filling and thus affected rice endosperm development [14]. Due to the *WB1* gene mutation (Figure 5) which led to grain incomplete filling [14], the physical and chemical properties of grain endosperm of the *wb1* mutant have been altered, like higher grain chalkiness rate (Figure 2i), higher grain chalkiness degree (Figure 2j), markedly more grain chalkiness as a result of loosely packed, spherical granules (Figure 2e,g), lower contents of amylose and amylopectin (Figure 2k,l), and ~30.0% lower 1000-grain weight (Figure 2m). In addition, transcript levels of the starch synthesis-related genes in our study varied during rice endosperm development (Figure 8). All of these observations suggest that the *wb1* mutant exhibits a defect in endosperm development, thus leading to the white-belly endosperm with altered phy-chemical property.

3.2. Different Mutations of WB1 Can Disrupt Its Biological Function

Many genes make up a large regulatory framework that regulates life activity in higher plants. These genes encode active proteins that are responsible for the major functions in this global regulatory framework. The function of each of active protein is determined by its own primary, secondary, and tertiary structure. In rice, mutations of a gene can result in its encoding protein with structural alterations which can affect its biological function followed by phenotypic changes. The *sd1* gene, well known as the genetic basis for the first "green revolution" in rice, encodes a GA 20-oxidase involved in the GA biosynthesis pathway; the *sd1* gene controls the plant height of rice, and mutations (*sd1-d*, *sd1-r*, *sd1-c*, *sd1-j*) in this locus cause the dwarfism of rice to different degrees [43–45].

In our study, six mutants of novel alleles of *WB1* displayed the same phenotype as the *wb1* mutant (Figures 2b–e and 6b), primarily showing higher chalkiness rates (Figure 6f), higher chalkiness degrees (Figure 6g), and lower 1000-grain weight (Figure 6h). The sequence analysis showed that six different mutations have occurred in the WB1 locus (Figure 6a), leading to alterations of the amino acid sequence of the WB1 protein in different types (Figure 7). In the previous research of the *gif1* mutant, the *GIF1* gene revealed a 1-nt deletion in the coding region, causing the premature GIF1 protein which disrupt its biological function (incomplete grain-filling) [14]. The *WB1* gene of *nc-1*, *nc-3* and *nc-6* also caused different premature WB1 proteins with altered substrate binding site and active site (Figure 7). The frame shift mutation of the *WB1* gene of *nc-2* and *nc-4* also disrupted its biological function (Figures 6b and 7). Interestingly, the single amino acid substitution (A159T) of WB1 of *wb1*, and Proline-161 and Arginine-162 deletion of WB1 of *nc-5* led to the dysfunction of WB1 without altered substrate binding site and active site (Figures 2, 6 and 7), suggesting that Alanine-159, Proline-161 and Arginine-162 are required for activity of WB1. Moreover, grain chalkiness degrees and 1000-grain weight showed significant differences among some mutants. However, several mutants showed no differences in grain chalkiness degree and 1000-grain weight (Figure 6g,h); these results suggest that different mutations probably affect the formation of grain chalkiness in different degrees, and further research still needs to be conducted to explain the molecular mechanism. In summary, seven novel alleles (including the *wb1* mutant) had different mutations which disrupted their biological functions.

3.3. The Modified MutMap Method Applied to Isolate WB1 Is Feasible for Gene Mapping

MutMap is a new method used for gene identification [33]. Using MutMap method, researchers can isolate mutant genes and QTLs rapidly, accurately, and conveniently compared to conventional map-based cloning [33,35,36]. Through the MutMap method, researchers only sequence the DNA pool from recessive individuals of F_2 population based on second-generation sequencing, followed by aligning to the assembled whole-genome sequence of wild-type. The population used for the MutMap method is BC_1F_2 population, which can show unequivocal segregation between the mutant and wild-type phenotype. Notably, the MutMap method requires assembling the whole-genome sequence of wild type accurately used as the reference sequence.

Previously, many genes have been identified by the MutMap method in rice. *OsRR22*, responsible for the salinity-tolerance phenotype for the *hst1* mutant, has been identified by a MutMap method: the sequence depth and the average coverage of wild-type are 28.7× and 59.9%; the number of BC_1F_2 individuals is 20 and the sequence depth is 18.4× [35]. Two mutant genes regarding pale green leaf have been identified: the sequence depth and the average coverage of wild-type are >12× and ~95.5%; the number of BC_1F_2 individuals is 20 and the sequence depths are 12.5× and 24.1× [33]. Four mutant genes regarding semi-dwarf phenotype also have been identified: the sequence depth and the average coverage of wild-type are >12× and 82.4%~84.2%; the number of BC_1F_2 individuals is 20 and the sequence depths are 14.2×~16.6× [33].

The MutMap method is subject to high error rates because of multiple factors, including difficulty in determining the number of F_2 progeny showing the mutant phenotype, the average coverage (depth) of genome sequencing, and classification of phenotypes between wild and mutant phenotype [33,46]. Therefore, the greater the number of F_2 progeny showing the mutant phenotype to be bulked, the deeper the average depth of genome to be sequenced, and more accurate classification of phenotypes between wild and mutant type, the lower the rate of false positives [33].

In our study, a modified MutMap method (Figure 1b) was applied to successfully isolate the *WB1* gene related to endosperm development in rice. Compared with the MutMap method, our modified MutMap method has some advantages. Firstly, the individuals of bulked DNA Pools used for sequencing were from $BC_1F_{2:3}$, which has a more stable genetic background than the BC_1F_2 population; because of the reduced effect from other gene mutations on the target phenotype, it was easier to distinguish plants between mutant type and wild type; Secondly, the appropriately elevated number of $BC_1F_{2:3}$ individuals (50, Figure 1b) and sequence depth (30×, Figure 1b) ensured relatively high coverage (87.92% and 87.55%; Table S4); Lastly, we sequenced the DNA pools not only from recessive individuals but also from dominant individuals followed by aligning to the reference sequence Nipponbare, respectively; therefore, it was not necessary to sequence and assemble the whole-genome sequence of wild type used as reference sequence. We directly used the Nipponbare genomic sequence as our reference sequence, so that we greatly reduced the costs required for sequencing and assembling reference sequence. Therefore, the *WB1* gene mapping result showed higher specificity with the single peak in the chromosome 4 (Figure 3b) compared to that by the MutMap method [33]. Besides, the modified MutMap method also has a deficiency that is the significant difference in the whole genome sequence between Nipponbare and Wild-type (reference sequence). Delightedly, several rice genome sequences have been published, like *indica* cultivar 9311 [47], Zhenshan97, Minghui63 [48] and Shuhui498 [49] which can be used as reference sequence directly, expanding the application scope of the modified MutMap method. Overall, the modified MutMap method showed a low error rate, a relatively low cost and a high specificity, and could promote the development of rice genetics.

4. Materials and Methods

4.1. Plant Materials and Growth Condition

The *wb1* (mutant) was initially identified from the ethyl methanesulfonate (EMS)-treated *Japonica* rice variety ChangLiGeng (CLG, Wild Type) M_2 population. The *wb1*, as the pollen acceptor, was crossed with WT and ZhongHui8015 (ZH8015, *Indica*), respectively. The resulting first filial generation (BC_1F_1, F_1) plant was self-pollinated, and the second generation (BC_1F_2, F_2) was used as the genetic analysis population. We collected seeds of 100 individuals from BC_1F_2 population, and then cropped the seeds to obtain 100 pedigrees ($BC_1F_{2:3}$) which were used as the mapping population for the modified MutMap method (Figure 1b). All plants were grown in an experimental paddy field at China National Rice Research Institute (Hangzhou, Zhejiang province and Lingshui, Hainan province, in China) under natural open-air condition.

4.2. Grain Quality Analysis

Scanning electron microscopy was performed as described previously [50]. Measurements of amylose content of mature grains (0.05 g powder) were conducted by HPSEC-MALLS-RI following the method of Fujita et al. (2003) [51]. Quantitative amylopectin content was determined by processing 0.01 g powder of the mature grain, according to a method from a previous report [52]. Each measurement was repeated three times ($n = 3$).

The paddy rice of WT and wb1 were dried to moisture content of 12–14% and were maintained at room temperature at least three months before measuring the brown rice rate, milled rice rate, and head rice rate by grain polisher AH001151 (KETT, Tokyo, Japan), performed as Zhou et al. (2015) [53]. Each measurement of 25 g paddy rice was performed with three replicates ($n = 3$, total 75 g paddy rice). Grain chalkiness rate and grain chalkiness degree were determined using SC-E Rice Quality Inspection and Analysis System (WanShen, Hangzhou, China). The white grains from 12 plants (6 plants from WT and wb1, respectively, $n = 6$) were used for grain chalkiness rate and degree measurements. Gel consistency was measured ($n = 4$) following the protocol described in Li et al. (2014) [15].

4.3. PCR, RNA Isolation and Real-time Quantitative PCR (qPCR)

PCR amplifications of candidate genes were performed using KOD FX DNA Polymerase (TOYOBO). The PCR product of the reaction of restriction enzyme *HaeII* digestion was amplified by KOD-Plus-Neo (TOYOBO). The primer pairs designed for this study are listed in Table S6.

Total RNA was prepared from grains of WT and *wb1* at 5, 10, 15, and 20 DAF (days after flowering) using the TIANGEN RNAprep Pure Plant Kit (Tiangen Biotech, Beijing, China). The first cDNA strand was synthesized from DNase I-treated RNA using Oligo-dT (18) primers in a 20 µL reaction system based on a SuperScriptIII Reverse Transcriptase Kit (TOYOBO). qPCR was performed on a Roche Light Cycle 480 device using THUNDERBIRD SYBR qPCR Mix (TOYOBO). Each reaction was performed with three replicates ($n = 3$). The primers used in this analysis are listed in Table S6.

4.4. DNA Template Preparation, DNA Library Construction, and Re-Sequencing

Genomic DNA was extracted (large scale) from young leaf tissues following the modified hexadecyl trimethylammonium bromide (CTAB) method [54]. Young leaves (total 5 g, 0.1 g per plant) were obtained from 50 plants displaying the mutant or wild phenotype in the $BC_1F_{2:3}$ population and were used to prepare the pooled genomic DNA (Pool A and Pool B, respectively) which was used for illumina sequencing. The DNA concentration was measured by Nanodrop 2000 spectrophotometer. ~1 µg, each for both Pool A and Pool B, of total high-quality pooled DNA samples ($1.8 \leq$ OD260:OD280 ≤ 2.0) was used for re-sequencing library construction. Two libraries with the target insert size of 300 bp were generated by the Illumina Gnomic DNA sample kit according to the manufacturer's instruction. The quality of two libraries was controlled by qPCR. Two libraries were re-sequenced through the Illumina HiSequation 2500 at the BeiJing Berry Genomics Biotechnology Co., Ltd. (Beijing, China) to generate 125 nt paired-end short sequence reads (raw reads) for each pools.

4.5. Re-Sequencing Analysis

The FastQC program was used to evaluate the quality of raw reads (http://www.bioinformatics. babraham.ac.uk/projects/fastqc/).The Illumina paired-end adapters' sequence of raw reads was removed using the FASTX toolkit program (http://hannonlab.cshl.edu/fastx_toolkit/index.html). Removal of low-quality bases (Illumina phred quality score Q < 20) [55] and ≤40 bp of reads was completed using SolexaQA software [56]. The cleaned reads from Pool (A) and Pool (B) have been submitted to the SRA database of NCBI (SRA accessions are SRP135580 and SRP135578, respectively).

The cleaned reads were aligned separately with BWA software (Burrows-Wheeler aligner) [57] to the Nipponbare reference sequence. Alignments were filtered based on the Illumina phred quality score of ≥30, corresponding to 0.1% of the error rate, to obtain the unique mapped reads. Alignment files were converted to SAM files through SAMtools [58], and applied to GATK Pipeline [59] to identify reliable SNPs based on the reference genomic sequence.

4.6. Calculation of Δ (SNP Indices)

Average SNP indices of Pool (A) and Pool (B) were estimated via the sliding window method (sliding window 50 Kb; walking 10 Kb) and the Δ (SNP indices) Manhattan plot was obtained using a custom script written in R version 3.1.1 (https://www.r-project.org/). According to the MutMap method, SNP index (A) would be 1 for the causal SNP or for closely linked SNPs and 0.5 for unlinked loci for each identified SNP in the whole genome sequence, while the SNP index (B) would be 0.333 (1/3) for the causal SNP or closely linked SNPs and 0.5 for unlinked loci. Therefore, Δ (SNP index) would be 0.667 (2/3) for the causal or closely-linked SNPs and 0 for unlinked SNPs.

4.7. Restriction Endonuclease Digestion Analysis

The restriction endonuclease *Hae*II site of the target gene was identified using the primer premier 5.0 software (Premier, Ottawa, ON, Canada). Two pairs of primers were designed (W-H for WT and M-H for wb1, see Table S6) to generate 337 bp of DNA fragments by polymerase chain reaction (PCR). These PCR products were used for restriction endonuclease *Hae*II digestion. A total of 150 μL of this reaction system contained 3 μL *Hae*II (20 units per 1 μL), 15 μL NEBuffer (1×), 50 μL DNA template (2.5 μg), and 82 μL ddH$_2$O. Two reaction systems were incubated at 37 °C for 15 min followed by a 2.0% agarose gel electrophoresis.

4.8. Vector Construction for CRISPR/Cas9-Mediated Mutation

Six novel allelic mutants were created in *Japonica* cv Nipponbare by a CRISPR/Cas-targeted genome editing tool. The pBWA(V)H_cas9i2-CRISPR/Cas9 plasmid (Figure S1) was constructed according the method described in Shan et al. (2013) [60]. To generate pBWA(V)H_cas9i2-CRISPR/Cas9 targeting vector, we used the pBWA(V)H_cas9i2 vector containing codon-optimized Cas9 driven by the 35S promoter, the OsU3 promoter and sgRNA scaffolds, as well as the Cas9 expression backbone vector. The targeting sequence primer pair was ACGTGACCTCATCAACTGGGTGG and AACGTGGCGCTGCCGAGGAACGG. The OsU3 promoter was used to drive the sgRNA expression, and the 35S promoter was used to drive the Cas9 expression. Both the OsU3::gRNA and 35S::Cas9 fragments were cloned into pBWA(V)H_cas9i2 binary vector which was introduced into *Agrobacterium* strain EHA105. Transformed calli were induced from Nipponbare seeds for *Agrobacterium*-mediated transformation as previously described [61]. The T$_0$ transgenic mutant plants regenerated from hygromycin-resistant calli were examined for the presence of transgene using primer pair Cas-seq (Table S6).

5. Conclusions

Breeding of rice with high quality of appearance and high yield is important for rice cultivation. In this study, we isolate and characterize a candidate recessive gene *WB1* that regulates rice endosperm development using a modified MutMap method. The candidate gene *WB1* is further verified by CRISPR/Cas9 system. The *wb1* mutant, as well as six mutants mediated by CRISPR-Cas9 system, all cause a defect in the endosperm development, which lead to the higher grain chalkiness rate and degree and a significant reduction of 1000-grain weight in comparison with that of wild-type plants.

Relative expression analysis of genes associated with starch synthesis by qPCR also suggests that loss of function of *WB1* leads to disorder of starch metabolism-related genes expression, resulting in the abnormal endosperm development. In particular, the modified MutMap method used in this study shows a low error rate, a relatively low cost and a high specificity, and could promote the development of rice genetics. Overall, the gene *WB1* involved in rice endosperm development affects rice quality of appearance and yield, and therefore, it can be used by rice breeders through molecular breeding to improve rice quality of appearance and yield in Green Super Rice.

Supplementary Materials: Supplementary materials can be found at http://www.mdpi.com/1422-0067/19/8/2159/s1.

Author Contributions: H.W. and Y.Z. contributed equally to this work. Conceptualization, H.W.; Data curation, H.W. and Y.Z.; Formal analysis, H.W. and Y.Z.; Funding acquisition, L.C., S.C. and X.S.; Investigation, H.W., P.X., R.T., S.M. and D.C.; Methodology, H.W., Y.Z., L.S., L.C., S.C. and X.S.; Project administration, L.C., S.C. and X.S.; Resources, L.C., S.C. and X.S.; Supervision, L.C., S.C. and X.S.; Validation, Y.Z., L.C., S.C. and X.S.; Visualization, H.W. and R.T.; Writing—original draft, H.W.; Writing—review & editing, Y.Z., L.S., P.X., R.T., W.W., G.B.A., K.H., A.R., L.C., S.C. and X.S.

Funding: This work was funded by grants from the National 863 project (#2014AA10A603), the Key Science and Technology Special Project for Breeding New Cereal Varieties of Zhejiang province (#2016C02050-1), the Agro-Scientific Research in the Public Interest (#201403002), the National Natural Science Foundation of China (#31521064) and the Science and Technology Innovation Project from the Chinese Academy of Agricultural Sciences (CAAS-ASTIP-2013-CNRRI).

Acknowledgments: We would like to thank associate professor Lu Lu (Graduate School of Chinese Academy of Agricultural Sciences, Beijing, China) for critical proofreading, feedback, and editing of the manuscript.

Conflicts of Interest: The authors declare no conflict of interest.

References

1. Fitzgerald, M.A.; McCouch, S.R.; Hall, R.D. Not just a grain of rice: The quest for quality. *Trends Plant Sci.* **2009**, *14*, 133–139. [CrossRef] [PubMed]

2. She, K.C.; Kusano, H.; Koizumi, K.; Yamakawa, H.; Hakata, M.; Imamura, T.; Fukuda, M.; Naito, N.; Tsurumaki, T.; Yaeshima, M.; et al. A novel factor *FLOURY ENDOSPERM2* is involved in regulation of rice grain size and starch quality. *Plant Cell* **2010**, *22*, 3280–3294. [CrossRef] [PubMed]

3. Wang, Y.; Ren, Y.; Liu, X.; Jiang, L.; Chen, L.; Han, X.; Jin, M.; Liu, S.; Liu, F.; Lv, J.; et al. *OsRab5a* regulates endomembrane organization and storage protein trafficking in rice endosperm cells. *Plant J.* **2010**, *64*, 812–824. [CrossRef] [PubMed]

4. Peng, C.; Wang, Y.H.; Liu, F.; Ren, Y.; Zhou, K.; Lv, J.; Zheng, M.; Zhao, S.; Zhang, L.; Wang, C.; et al. *FLOURYE DOSPERM6* encodes a CBM48 domain-containing protein involved in compound granule formation and starch synthesis in rice endosperm. *Plant J.* **2014**, *77*, 917–930. [CrossRef] [PubMed]

5. Ren, Y.; Wang, Y.; Liu, F.; Zhou, K.; Ding, Y.; Zhou, F.; Wang, Y.; Liu, K.; Gan, L.; Ma, W.; et al. *GLUTELIN PRECURSOR ACCUMULATION3* encodes a regulator of post-Golgi vesicular traffic essential for vacuolar protein sorting in rice endosperm. *Plant Cell* **2014**, *26*, 410–425. [CrossRef] [PubMed]

6. Matsushima, R.; Maekawa, M.; Kusano, M.; Kondo, H.; Fujita, N.; Kawagoe, Y.; Sakamoto, W. Amyloplast-localized SUBSTANDARD STARCH GRAIN4 protein influences the size of starch grains in rice endosperm. *Plant Physiol.* **2014**, *164*, 623–636. [CrossRef] [PubMed]

7. Wen, L.; Fukuda, M.; Sunada, M.; Ishino, S.; Ishino, Y.; Okita, T.W.; Ogawa, M.; Ueda, T.; Kumamaru, T. Guanine nucleotide exchange factor 2 for Rab5 proteins coordinated with GLUP6/GEF regulates the intracellular transport of the proglutelin from the Golgi apparatus to the protein storage vacuole in rice endosperm. *J. Exp. Bot.* **2015**, *66*, 6137–6147. [CrossRef] [PubMed]

8. Zhang, L.; Ren, Y.; Lu, B.; Yang, C.; Feng, Z.; Liu, Z.; Chen, J.; Ma, W.; Wang, Y.; Yu, X.; et al. *FLOURY ENDOSPERM 7* encodes a regulator of starch synthesis and amyloplast development essential for peripheral endosperm development in rice. *J. Exp. Bot.* **2016**, *67*, 633–647. [CrossRef] [PubMed]

9. Lee, S.K.; Hwang, S.K.; Han, M.; Eom, J.S.; Kang, H.G.; Han, Y.; Choi, S.B.; Cho, M.H.; Bhoo, S.H.; An, G.; et al. Identification of the ADP-glucose pyrophosphorylase isoforms essential for starch synthesis in the leaf and seed endosperm of rice (*Oryza sativa* L.). *Plant Mol. Biol.* **2007**, *65*, 531–546. [CrossRef] [PubMed]

10. Satoh, H.; Shibahara, K.; Tokunaga, T.; Nishi, A.; Tasaki, M.; Hwang, S.; Okita, T.W.; Kaneko, N.; Fujita, N.; Yoshida, M.; et al. Mutation of the plastidial alpha-glucan phosphorylase gene in rice affects the synthesis and structure of starch in the endosperm. *Plant Cell* **2008**, *20*, 1833–1849. [CrossRef] [PubMed]

11. Tuncel, A.; Kawaguchi, J.; Ihara, Y.; Matsusaka, H.; Nishi, A.; Nakamura, T.; Kuhara, S.; Hirakawa, H.; Nakamura, Y.; Cakir, B.; et al. The rice endosperm ADP-glucose pyrophosphorylase large subunit is essential for optimal catalysis and allosteric regulation of the heterotetrameric enzyme. *Plant Cell Physiol.* **2014**, *55*, 1169–1183. [CrossRef] [PubMed]

12. Cakir, B.; Shiraishi, S.; Tuncel, A.; Matsusaka, H.; Satoh, R.; Singh, S.; Crofts, N.; Hosaka, Y.; Fujita, N.; Hwang, S.K.; et al. Analysis of the rice ADP-glucose transporter (*OsBT1*) indicates the presence of regulatory processes in the amyloplast stroma that control ADP-glucose flux into starch. *Plant Physiol.* **2016**, *170*, 1271–1283. [CrossRef] [PubMed]

13. Li, S.; Wei, X.; Ren, Y.; Qiu, J.; Jiao, G.; Guo, X.; Tang, S.; Wan, J.; Hu, P. *OsBT1* encodes an ADP-glucose transporter involved in starch synthesis and compound granule formation in rice endosperm. *Sci. Rep.* **2017**, *7*, 40124. [CrossRef] [PubMed]

14. Wang, E.; Wang, J.; Zhu, X.; Hao, W.; Wang, L.; Li, Q.; Zhang, L.; He, W.; Lu, B.; Lin, H.; et al. Control of rice grain-filling and yield by a gene with a potential signature of domestication. *Nat. Genet.* **2008**, *40*, 1370–1374. [CrossRef] [PubMed]

15. Li, Y.; Fan, C.; Xing, Y.; Yun, P.; Luo, L.; Yan, B.; Peng, B.; Xie, W.; Wang, G.; Li, X.; et al. Chalk5 encodes a vacuolar H+- translocating pyrophosphatase influencing grain chalkiness in rice. *Nat. Genet.* **2014**, *46*, 398–404. [CrossRef] [PubMed]

16. Hannah, L.C.; James, M. The complexities of starch biosynthesis in cereal endosperms. *Curr. Opin. Biotechnol.* **2008**, *19*, 160–165. [CrossRef] [PubMed]

17. Sauer, N. Molecular physiology of higher plant sucrose transporters. *FEBS Lett.* **2007**, *581*, 2309–2317. [CrossRef] [PubMed]

18. Toyota, K.; Tamura, M.; Ohdan, T.; Nakamura, Y. Expression profiling of starch metabolism-related plastidic translocator genes in rice. *Planta* **2006**, *223*, 248–257. [CrossRef] [PubMed]

19. Eom, J.S.; Cho, J.I.; Reinders, A.; Lee, S.W.; Yoo, Y.; Tuan, P.Q.; Choi, S.B.; Bang, G.; Park, Y.I.; Cho, M.H.; et al. Impaired function of the tonoplast-localized sucrose transporter in rice, OsSUT2, limits the transport of vacuolar reserve sucrose and affects plant growth. *Plant Physiol.* **2011**, *157*, 109–119. [CrossRef] [PubMed]

20. Sosso, D.; Luo, D.; Li, Q.B.; Sasse, J.; Yang, J.; Gendrot, G.; Suzuki, M.; Koch, K.E.; McCarty, D.R.; Chourey, P.S.; et al. Seed filling in domesticated maize and rice depends on SWEET-mediated hexose transport. *Nat. Genet.* **2015**, *47*, 1489–1493. [CrossRef] [PubMed]

21. Tatsuro, H.; Grahamn, S.; Tomio, T. An expression analysis profile for the entire sucrose synthase gene family in rice. *Plant Sci.* **2008**, *174*, 534–543. [CrossRef]

22. Wei, X.J.; Jiao, G.A.; Lin, H.Y.; Sheng, Z.H.; Shao, G.N.; Xie, L.H.; Tang, S.Q.; Xu, Q.; Hu, P.S. *GRAIN INCOMPLETE FILLING 2* regulates grain filling and starch synthesis during rice caryopsis development. *J. Integr. Plant Biol.* **2017**, *59*, 134–153. [CrossRef] [PubMed]

23. Kang, H.; Park, S.; Matsuoka, M.; An, G. White-core endosperm floury endosperm-4 in rice is generated by knockout mutations in the C4-type pyruvate orthophosphate dikinase gene (*OsPPDKB*). *Plant J.* **2005**, *42*, 901–911. [CrossRef] [PubMed]

24. Tian, Z.; Qian, Q.; Liu, Q.; Yan, M.; Liu, X.; Yan, C.; Liu, G.; Gao, Z.; Tang, S.; Zeng, D.; et al. Allelic diversities in rice starch biosynthesis lead to a diverse array of rice eating and cooking qualities. *Proc. Natl. Acad. Sci. USA* **2009**, *106*, 21760–21765. [CrossRef] [PubMed]

25. Fujita, N.; Yoshida, M.; Asakura, N.; Ohdan, T.; Miyao, A.; Hirochika, H.; Nakamura, Y. Function and characterization of *starch synthase I* using mutants in rice. *Plant Physiol.* **2006**, *140*, 1070–1084. [CrossRef] [PubMed]

26. Zhang, G.; Cheng, Z.; Zhang, X.; Gao, X.; Su, N.; Jiang, L.; Mao, L.; Wan, J. Double repression of soluble starch synthase genes *SSIIa* and *SSIIIa* in rice (*Oryza sativa* L.) uncovers interactive effects on the physicochemical properties of starch. *Genome* **2011**, *54*, 448–459. [CrossRef] [PubMed]

27. Fujita, N.; Yoshida, M.; Kondo, T.; Saito, K.; Utsumi, Y.; Tokunaga, T.; Nishi, A.; Satoh, H.; Park, J.H.; Jane, J.L.; et al. Characterization of *SSIIIa*-deficient mutants of rice: The function of *SSIIIa* and pleiotropic effects by *SSIIIa* deficiency in the rice endosperm. *Plant Physiol.* **2007**, *144*, 2009–2023. [CrossRef] [PubMed]

28. Kawagoe, Y.; Kubo, A.; Satoh, H.; Takaiwa, F.; Nakamura, Y. Roles of isoamylase and ADP-glucose pyrophosphorylase in starch granule synthesis in rice endosperm. *Plant J.* **2005**, *42*, 164–174. [CrossRef] [PubMed]

29. Kubo, A.; Fujita, N.; Harada, K.; Matsuda, T.; Satoh, H.; Nakamura, Y. The starch-debranching enzymes isoamylase and pullulanase are both involved in amylopectin biosynthesis in rice endosperm. *Plant Physiol.* **1999**, *121*, 399–410. [CrossRef] [PubMed]

30. Tanaka, N.; Fujita, N.; Nish, A.; Satoh, H.; Hosaka, Y.; Ugaki, M.; Kawasaki, S.; Nakamura, Y. The structure of starch can be manipulated by changing the expression levels of *starch branching enzyme IIb* in rice endosperm. *Plant Biotechnol. J.* **2004**, *2*, 507–516. [CrossRef] [PubMed]

31. Fu, F.F.; Xue, H.W. Coexpression analysis identifies *Rice Starch Regulator1*, a rice AP2/EREBP family transcription factor, as a novel rice starch biosynthesis regulator. *Plant Physiol.* **2010**, *154*, 927–938. [CrossRef] [PubMed]

32. Kawakatsu, T.; Yamamoto, M.P.; Touno, S.M.; Yasuda, H.; Takaiwa, F. Compensation and interaction between RISBZ1 and RPBF during grain filling in rice. *Plant J.* **2009**, *59*, 908–920. [CrossRef] [PubMed]

33. Abe, A.; Kosugi, S.; Yoshida, K.; Natsume, S.; Takagi, H.; Kanzaki, H.; Matsumura, H.; Yoshida, K.; Mitsuoka, C.; Tamiru, M.; et al. Genome sequencing reveals agronomically important loci in rice using MutMap. *Nat. Biotechnol.* **2012**, *30*, 174–178. [CrossRef] [PubMed]

34. Mardis, E.R. Next-Generation DNA Sequencing Method. *Annu. Rev. Genom. Hum. Genet.* **2008**, *9*, 387–402. [CrossRef] [PubMed]

35. Takagi, H.; Tamiru, M.; Abe, A.; Yoshida, K.; Uemura, A.; Yaeqashi, H.; Obara, T.; Oikawa, K.; Utsushi, H.; Kanzaki, E.; et al. MutMap accelerates breeding of a salt-tolerant rice cultivar. *Nat. Biotechnol.* **2015**, *33*, 445–449. [CrossRef] [PubMed]

36. Takagi, H.; Uemura, A.; Yaegashi, H.; Tamiru, M.; Abe, A.; Mitsuoka, C.; Utsushi, H.; Natsume, S.; Kanzaki, H.; Matsumura, H.; et al. MutMap-Gap: Whole-genome resequencing of mutant F_2 progeny bulk combined with de novo assembly of gap regions identifies the rice blast resistance gene *Pii*. *New Phytol.* **2013**, *200*, 276–283. [CrossRef] [PubMed]

37. Matsumoto, T.; Wu, J.; Kanamori, H.; Katayose, Y.; Fujisawa, M.; Namiki, N.; Mizuno, H.; Yamamoto, K.; Antonio, B.A.; Baba, T.; et al. The map-based sequence of the rice genome. *Nature* **2005**, *2436*, 793–800. [CrossRef]

38. Holsinger, K.E.; Weir, B.S. Genetics in geographically structured populations: Defining, estimating and interpreting F(ST). *Nat. Rev. Genet.* **2009**, *10*, 639–650. [CrossRef] [PubMed]

39. Bökel, C. EMS screens: From mutagenesis to screening and mapping. *Methods Mol. Biol.* **2008**, *420*, 119–138. [CrossRef] [PubMed]

40. Singh, N.; Sodhi, N.S.; Kaur, M.; Saxena, S.K. Physico-chemical, morphological, thermal, cooking and textural properties of chalky and translucent rice kernels. *Food Chem.* **2003**, *82*, 433–439. [CrossRef]

41. Cheng, F.M.; Zhong, L.J.; Wang, F.; Zhang, G.P. Differences in cooking and eating properties between chalky and translucent parts in rice grains. *Food Chem.* **2005**, *90*, 39–46. [CrossRef]

42. Lisle, A.J.; Martin, M.; Fitzgerald, M.A. Chalky and translucent rice grains differ in starch composition and structure and cooking properties. *Cereal Chem.* **2000**, *77*, 627–632. [CrossRef]

43. Sasaki, A.; Ashikari, M.; Ueguchi-Tanaka, M.; Itoh, H.; Nishimura, A.; Swapan, D.; Ishiyama, K.; Saito, T.; Kobayashi, M.; Khush, G.S.; et al. Green revolution: A mutant gibberellin-synthesis gene in rice. *Nature* **2002**, *416*, 701–702. [CrossRef] [PubMed]

44. Monna, L.; Kitazawa, N.; Yoshino, R.; Suzuki, J.; Masuda, H.; Maehara, Y.; Tanji, M.; Sato, M.; Nasu, S.; Minobe, Y. Positional cloning of rice semidwarfing gene, *sd-1*: Rice "green revolution gene" encodes a mutant enzyme involved in gibberellin synthesis. *DNA Res.* **2002**, *9*, 11–17. [CrossRef] [PubMed]

45. Spielmeyer, W.; Ellis, M.H.; Chandler, P.M. Semidwarf (*sd-1*), "green revolution" rice, contains a defective gibberellin 20-oxidase gene. *Proc. Natl. Acad. Sci. USA* **2002**, *99*, 9043–9048. [CrossRef] [PubMed]

46. Sims, D.; Sudbery, I.; Ilott, N.E.; Heger, A.; Ponting, C.P. Sequencing depth and coverage: Key considerations in genomic analyses. *Nat. Rev. Genet.* **2014**, *15*, 121–132. [CrossRef] [PubMed]

47. Yu, J.; Hu, S.; Wang, J.; Wong, G.K.; Li, S.; Liu, B.; Deng, Y.; Dai, L.; Zhou, Y.; Zhang, X.; et al. A draft sequence of the rice genome (*Oryza sativa* L. ssp. *indica*). *Science* **2002**, *296*, 79–92. [CrossRef] [PubMed]

48. Zhang, J.W.; Chen, L.L.; Xing, F.; Kudrna, D.A.; Yao, W.; Copetti, D.; Mu, T.; Li, W.; Song, J.M.; Xie, W.; et al. Extensive sequence divergence between the reference genomes of two elite indica rice varieties Zhenshan 97 and Minghui 63. *Proc. Natl. Acad. Sci. USA* **2016**, *113*, E5163–E5171. [CrossRef] [PubMed]
49. Du, H.; Yu, Y.; Ma, Y.; Gao, Q.; Cao, Y.; Chen, Z.; Ma, B.; Qi, M.; Li, Y.; Zhao, X.; et al. Sequencing and de novo assembly of a near complete *indica* rice genome. *Nat. Commun.* **2017**, *8*, 15324. [CrossRef] [PubMed]
50. Zhang, P.; Zhang, Y.; Sun, L.; Sittipun, S.; Yang, Z.; Sun, B.; Xuan, D.; Li, Z.; Yu, P.; Wu, W.; et al. The Rice AAA-ATPase *OsFIGNL1* Is Essential for Male Meiosis. *Fron. Plant Sci.* **2017**, *8*, 1639. [CrossRef] [PubMed]
51. Fujita, N.; Kubo, A.; Suh, S.D.; Wong, K.S.; Jane, J.L.; Ozawa, K.; Takaiwa, F.; Inaba, Y.; Nakamura, Y. Antisense inhibition of isoamylase alters the structure of amylopectin and the physicochemical properties of starch in rice endosperm. *Plant Cell Physiol.* **2003**, *44*, 607–618. [CrossRef] [PubMed]
52. Hovenkamphermelink, J.H.; Devries, J.N.; Adamse, P.; Jacobsen, E.; Witholt, B.; Feenstra, W.J. Rapid estimation of the amylose/amylopectin ratio in small amounts of tuber and leaf tissue of the potato. *Potato Res.* **1988**, *31*, 241–246. [CrossRef]
53. Zhou, L.; Liang, S.; Ponce, K.; Marundon, S.; Ye, G.; Zhao, X. Factors affecting head rice yield and chalkiness in *indica* rice. *Field Crop. Res.* **2015**, *172*, 1–10. [CrossRef]
54. Chen, D.H.; Ronald, P.C. A rapid DNA minipreparation method suitable for AFLP and other PCR applications. *Plant Mol. Biol. Rep.* **1999**, *17*, 53–57. [CrossRef]
55. Ewing, B.; Hillier, L.; Wendl, M.C.; Green, P. Base-calling of automated sequencer traces using phred. I. Accuracy assessment. *Genome Res.* **1988**, *8*, 175–185. [CrossRef]
56. Cox, M.P.; Peterson, D.A.; Biggs, P.J. SolexaQA: At-a-glance quality assessment of Illumina second-generation sequencing data. *BMC Bioinform.* **2010**, *11*, 485. [CrossRef] [PubMed]
57. Li, H.; Durbin, R. Fast and accurate short read alignment with Burrows-Wheeler Transform. *Bioinformatics* **2009**, *25*, 1754–1760. [CrossRef] [PubMed]
58. Li, H.; Handsaker, B.; Wysoker, A.; Fennell, T.; Ruan, J.; Homer, N. The Sequence Alignment/Map (SAM) format and SAM tools. *Bioinformatics* **2009**, *25*, 2078–2079. [CrossRef] [PubMed]
59. McKenna, A.L.; Hanna, M.; Banks, E.; Sivachenko, A.; Cibulskis, K.; Kernytsky, A.; Garimella, K.; Altshuler, D.; Gabriel, S.; Daly, M.; et al. The Genome Analysis Toolkit: A MapReduce framework for analyzing next-generation DNA sequencing data. *Genome Res.* **2010**, *20*, 1297–1303. [CrossRef] [PubMed]
60. Shan, Q.; Wang, Y.; Li, J.; Zhang, Y.; Chen, K.; Liang, Z.; Zhang, K.; Liu, J.; Xi, J.J.; Qiu, J.L.; et al. Targeted genome modification of crop plants using a CRISPR-Cas system. *Nat. Biotechnol.* **2013**, *31*, 686–688. [CrossRef] [PubMed]
61. Hiei, Y.; Ohta, S.; Komari, T.; Kumashiro, T. Efficient transformation of rice (*Oryza sativa* L.) mediated by Agrobacterium and sequence analysis of the boundaries of the T-DNA. *Plant J.* **1994**, *6*, 271–282. [CrossRef] [PubMed]

International Journal of
Molecular Sciences

Article

Isolation and Characterization of a Green-Tissue Promoter from Common Wild Rice (*Oryza rufipogon* Griff.)

Mande Xue [1], Yan Long [1], Zhiqiang Zhao [2], Gege Huang [2], Ke Huang [2], Tianbao Zhang [1], Ying Jiang [3,*], Qianhua Yuan [2,*] and Xinwu Pei [1,*]

[1] MOA Key Laboratory on Safety Assessment (Molecular) of Agri-GMO, Institute of Biotechnology, Chinese Academy of Agricultural Sciences, Beijing 100081, China; mandexue@gmail.com (M.X.); longyan@caas.cn (Y.L.); Zhangtianbao274200@126.com (T.Z.)

[2] College of Tropical Agriculture and Forestry, Hainan University, Haikou 570228, China; zzq19900317@126.com (Z.Z.); Luolengshuang@126.com (G.H.); Huangke199309@163.com (K.H.)

[3] Experimental Center Basic Medical Teaching, Capital Medical University, Beijing 100069, China

* Correspondence: Jiangy@ccmu.edu.cn (Y.J.); qhyuan@163.com (Q.Y.); Peixinwu@caas.cn (X.P.); Tel.: +86-010-83911687 (Y.J.); +86-898-6629-1279 (Q.Y.); +86-010-8210-6119 (X.P.)

Received: 8 June 2018; Accepted: 5 July 2018; Published: 10 July 2018

Abstract: Promoters play a very important role in the initiation and regulation of gene transcription. Green-tissue promoter is of great significance to the development of genetically modified crops. Based on RNA-seq data and RT-PCR expression analysis, this study screened a gene, *OrGSE* (GREEN SPECIAL EXPRESS), which is expressed specifically in green tissues. The study also isolated the promoter of the *OrGSE* gene (OrGSEp), and predicted many cis-acting elements, such as the CAAT-Box and TATA-Box, and light-responding elements, including circadian, G-BOX and GT1 CONSENSUS. Histochemical analysis and quantification of GUS activity in transgenic *Arabidopsis thaliana* plants expressing GUS under the control of OrGSEp revealed that this promoter is not only green tissue-specific, but also light-inducible. The ability of a series of 5′-deletion fragments of OrGSEp to drive GUS expression in *Arabidopsis* was also evaluated. We found that the promoter region from −54 to −114 is critical for the promoter function, and the region from −374 to −114 may contain core cis-elements involved in light response. In transgenic rice expressing GUS under the control of OrGSEp, visualization and quantification of GUS activity showed that GUS was preferentially expressed in green tissues and not in endosperm. OrGSEp is a useful regulatory element for breeding pest-resistant crops.

Keywords: common wild rice; Promoter; Green tissue-specific expression; light-induced

1. Introduction

Promoters are key regulators of transcription and also play critical roles in genetic engineering [1]. Promoters can be divided into three types: constitutive, inducible and tissue-specific. Constitutive promoters are widely used in plant genetic engineering [2]. The cauliflower mosaic virus (CaMV) 35S promoter, which drives the expression of genes in almost all tissues, is an important constitutive promoter in dicotyledonous plants [3]. The *Actin*1 promoter from rice is a classical constitutive promoter [4]. However, inducible promoters greatly increase the transcription level of genes under specific physical or chemical signals [5], this is different from constitutive promoters. At present, a great number of inducible promoters have been isolated, including light-inducible, heat-inducible and trauma-inducible [6–10]. Abiotic stress is a major obstacle for crop production, and the promoters of many genes related to abiotic stress have been cloned and applied in crop biotechnology. Five cold-inducible promoters have been isolated from rice, and these promoters can be applied to engineer plants that are resistant to cold stress [11]. The stress-inducible promoter of *TaSnRK2.8*, which is an important gene for wheat response to abiotic stress, has been isolated and characterized, and this promoter can be used to engineer plants with resistance to various abiotic stresses [12].

Because gene expression is driven in specific tissues and development stages, tissue-specific promoters, also called organ- or cell-specific promoters [13], are different from other promoters [14]. These promoters are significant because they avoid potential negative effects of using constitutive promoters, such as metabolic burden [15–19]. Five non-endosperm tissue-expressed promoters have been isolated from rice, and an exogenous *Cry1Ab* gene (*mCry1Ab*) driven by green tissue-specific promoter was expressed in all tissues except for endosperm [20]. Progress has also been made in identifying promoters that drive expression in roots, which is of interest because crops are faced with root-related pests, pathogens and abiotic stresses. The promoters of the soybean *GmPRP1* and *GmPRP2* genes, which are expressed preferentially in roots [21], were isolated and shown to exhibit root-preferential expression [22].The promoter of the serine/threonine kinase gene *ZmSTK2_USP* was isolated and with the use of a GUS reporter system was found to drive pollen-tissue-specific expression [23]. The promoter of *AtGILT* was shown to drive seed coat-specific expression in *Arabidopsis thaliana*, and further studies demonstrated that in canola this promoter drives expression, specifically in the outer integument of the seed coat, and may be useful for improving canola meal [24].

Green tissue-specific promoters have vast potential in transgenic crop breeding [18], particularly in insect-resistant or herbicide-resistant crops [25]. A growing number of transgenic crops with insect resistance gene expression driven by green tissue-specific promoters have been developed, such as cotton, rice, soybean and maize. Transgenic Bt cotton was developed by driving expression of the *B. thuringiensis* endotoxin (*Cry9C*) genes under the control of the *PNZIP* (*Pharbitis nil* leucine zipper) gene promoter, resulting in preferential expression in plant green tissues as well as lower Bt protein accumulation in transgenic cotton seeds [26]. The *rbcS* promoter is another classical green tissue-specific promoter that has been used to drive expression of the *cry2AX1* gene in rice to confer resistance to leaffolders [27]. The 731-bp 5′ flanking sequence of a potato (*Solanum tuberosum*) gene encoding ribulose-1, 5-bisphosphate carboxylase/oxygenase (rubisco) activase (RCA) was characterized, and GUS reporter gene under the control of StRCAp was expressed throughout the green tissue of light-grown transgenic tobacco seedlings. Further analysis revealed that a 220 bp fragment of StRCAp was sufficient for green-tissue-specific and light-inducible expression [7]. Common wild rice (*Oryza rufipogon* Griff.), which is an ancestor of Asian cultivated rice [28], has abundant genetic diversity and is an important germplasm resource for the improvement of cultivated rice [29]. However, the genes specifically expressed in green tissues in common wild rice have not been reported. Previously, we sequenced the transcriptome of common wild rice and identified root-specific and drought-related genes [30]. In this study, we used this dataset to screen for genes with green tissue-specific expression and cloned the promoter sequence of one of these genes, *OrGSE* (*O. rufipogon* GREEN SPECIAL EXPRESS). The full-length *OrGSE* promoter (OrGSEp) and a series of truncated promoters were fused to the GUS reporter gene to identify putative cis-regulatory elements that confer green tissue-specific and light-inducible expression in *Arabidopsis*. We found that in transgenic rice, GUS driven by the OrGSEp promoter was preferentially expressed in green tissues and not expressed in endosperm and root.

2. Results

2.1. Expression Pattern of OrGSE in Common Wild Rice

RNA-seq data from a previous study [30] were used to screen candidate common wild rice green tissue-expressed genes with higher expression in leaves than in roots (FPKM value in CL >10, FPKM value in CR is <10 and FPKM value in CL 10 times higher than in CR). A total of 1140 unigenes were identified as candidate green tissue-specific expressed genes (Table S3). A series of RT-PCR experiments were performed for confirming the RNA-seq data and combining RNA-seq and RT-PCR results, we selected a novel green tissue-specific expressed gene, comp45689_c0. Blast searches against the rice genome annotation project database (available online: http://rice.plantbiology.msu.edu/) showed that this gene corresponds to *LOC_Os08g02210* and encodes an expressed protein, named OrGSE (*O. rufipogon* GREEN SPECIAL EXPRESS), The BLASTP (Basic Local Alignment Search Tool Protein) analysis revealed that OrGSE shares homology with D27 protein in rice, and prediction of subcellular localization shows that OrGSE is located in chloroplast. Based on RNA-seq FPKM expression values from MSU, *OsGSE* is highly expressed in green tissue; the FPKM value in 20-day-old leaves is >200, but almost 0 in anthers and seeds (Figure S1).

RT-PCR and qPCR experiments confirmed that the *OrGSE* gene is expressed in leaves, stems and spikes, but not in roots and seeds (Figure 1). In addition, preferential expression in green tissue was observed at both the seedling and heading stages (Figure 1).

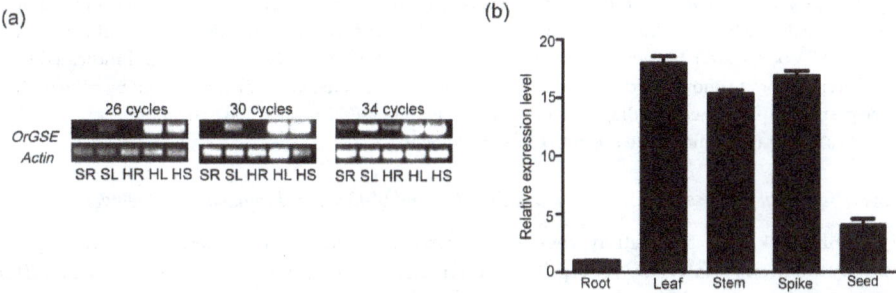

Figure 1. *OrGSE* gene expression in roots and shoots of common wild rice. The mRNA levels of *OrGSE* gene were determined in different tissues and developmental stages by RT-PCR with 26, 30 and 34 cycles of amplification. (**a**) The gel figure of RT-PCR. The rice Actin gene was used as an internal control. SR: Shooting stage Root, SL: Shooting stage Leaf, HR: Heading stage Root, HL: Heading stage Leaf, and HS: Heading stage Stem; (**b**) qPCR analysis of transcript levels of *OrGSE* in different tissues. Data are the means of three replicates, and error bars show the standard error.

2.2. Cloning and Sequence Analysis of the OrGSE Promoter (OrGSEp-374)

A 561 bp fragment including the −374 to +187 region (where the TIS is +1) upstream of *OrGSE* was isolated from common wild rice as the candidate promoter sequence. This promoter fragment, named OrGSEp-374, was submitted to PlantCARE to predict putative cis-elements involved in the regulation of gene expression (Figure 2). Potential regulatory elements were identified within OrGSEp-374 (Table S1), including core elements, such as the TATA-Box, CAAT-Box and GC-Box and cis-elements known to be involved in stress response, such as ABRE (abscisic acid responsive element), CGTCA-motif (response to methyl jasmonate), and LTR (low-temperature response). In addition, a few elements in OrGSEp-374 have been shown to participate in tissue-specific expression, such as the CCGTCC-box (related to meristem-specific activation), the GCN4_motif (endosperm expression) and the Skn-1_motif (required for endosperm expression). Other regulatory elements are involved green tissue-specific expression; most of these elements are light-responsive, such as ACE, G-Box, box II and GT1CONSENSUS. A core element (circadian) involved in circadian control was also identified [31,32].

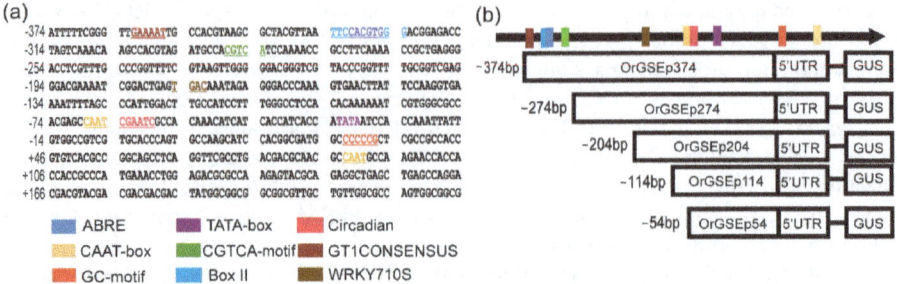

Figure 2. The location of putative cis-acting elements in OrGSEp-374 predicted by the PlantCARE database and schematic diagrams of promoter deletion constructs. (**a**) Putative cis-acting elements in OrGSEp-374, The 5′-region of the *OrGSE* gene containing the 374 bp promoter sequence and 166 bp sequence downstream of the translational start site. The transcription initiation site is defined as +1. The TATA box, CAAT box and other key cis-acting elements are underlined with and indicated by different colors as shown in the legend. The position of each element is also indicated by schematic diagrams; (**b**) The schematic diagrams of the truncated OsGSE-374 constructs. The numbers to the left of these diagrams indicate the position of the 5′-deletion.

2.3. Spatiotemporal Expression Patterns of OrGSEp-374 and 5′-Deletion Fragments in Arabidopsis

OrGSEp-374 and 5′-deletion reporter constructs were transformed into *Arabidopsis* for promoter functional analysis. OrGSEp-374-driven GUS expression was monitored during different developmental stages and in various organs by histochemical staining. GUS expression was detected in the cotyledons and hypocotyls of 3-day-old seedlings (Figure 3). GUS expression was also detected in the leaves of 5-day- and 14-day-old seedlings, and GUS expression level was higher in leaves than in cotyledons (Figure 3). No GUS expression was observed in roots at any stage (Figure 3). During the reproductive stage, GUS expression was observed in leaves, but not in roots or siliques (Figure 4), indicating that OrGSEp-374 drives expression specifically in green tissues.

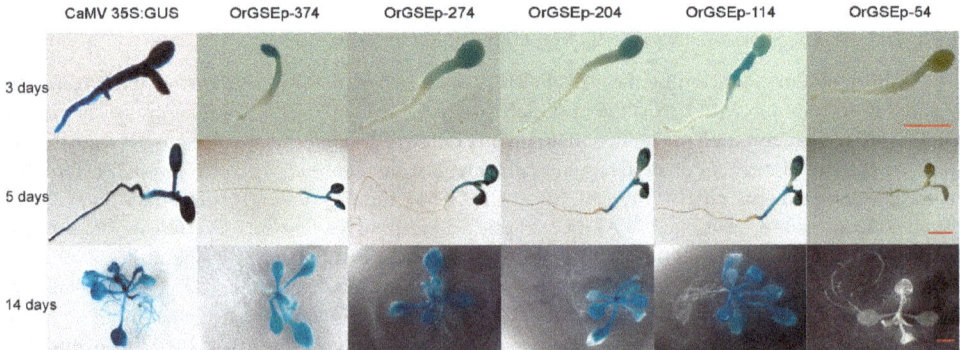

Figure 3. GUS histochemical assays in transgenic *Arabidopsis* T3 seedlings. GUS histochemical assays in transgenic *Arabidopsis* T3 seedlings harboring constructs with GUS expression driven by the CaMV 35S promoter (35S: GUS), OrGSEp-374 (OrGSEp-374) and different 5′-deletion fragments (OrGSEp-274, OrGSEp-204, OrGSEp-114 and OrGSEp-54), during vegetative growth. Photographs were taken 3 days, 5 days and 14 days after seed germination. Bar = 1 cm.

To identify the core elements responsible for green tissue-specific expression, we cloned four different 5′-deletion fragments (Figure 2) into the pBinGlyRed-GUS vector, and transformed these constructs into *Arabidopsis*. Histochemical staining was performed on 3-, 5- and 14-day-old T3 transgenic seedlings. As shown in Figure 3, strong GUS expression was driven by OrGSEp-274, OrGSEp-204 and OrGSEp-114, but no GUS expression was observed in any tissues or stages when GUS was under the control of OrGSEp-54. This result shows that the promoter region from −114 to −54 may contain a key element controlling promoter activity. However, GUS driven by OrGSEp-114 showed weaker expression in leaves than GUS driven by OrGSEp-204, indicating that the promoter region from −204 to −114 may contain an enhancer element (Figure 4). Fluorometric analysis of GUS activity also confirmed that GUS expression driven by OrGSEp-204 was higher than when driven by OrGSEp-114. The results of fluorometric GUS assays were consistent with histochemical staining (Figure 5), supporting the promoter fragment from −54 to −114 playing a critical role in the promoter activity.

Figure 4. GUS staining in siliques, leaves and roots sampled during the reproductive stage from transgenic T3 *Arabidopsis* seedlings carrying OrGSEp-374, OrGSEp-274, OrGSEp-204, OrGSEp-114 and OrGSEp-54.Bar = 1 cm.

2.4. OrGSEp-374 Confers Light-Responsive Expression

OrGSEp-374 sequence analysis showed that the promoter contained many cis-acting elements involved in light responsiveness, such as ACE, Box II, and GT1CONSENSUS (Figure 2). We performed GUS staining and fluorometric assays to determine whether the expression of *OrGSE* was regulated by light. GUS staining showed that *OrGSE* expression was induced by light (Figure 6), and quantitative fluorometric analysis confirmed this result (Figure 6). To find the core elements related to light responsiveness, we analyzed the light-inducible activities of 5′-deletion fragments of OrGSEp-374. GUS expression driven by OrGSEp-374, OrGSEp-274 and OrGSEp-204 was similar under light conditions; however, GUS activity driven by these promoters was significantly reduced in the dark. Interestingly, the pattern of GUS expression in OrGSEp-114-GUS lines was the same under both light and dark conditions (Figure 6). These results demonstrate that the promoter fragment from −374 to −114 contains vital elements involved in light response, but the promoter fragment from −114 to +1 does not. This is consistent with the presence of elements involved in light response in the OrGSEp-374 predicted by PlantCARE (Figure 2; Table S1).

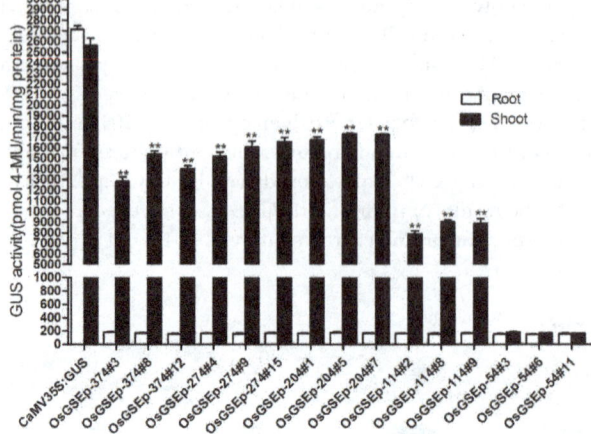

Figure 5. Quantification of GUS activity in transgenic T3 *Arabidopsis* roots and shoots carrying the CaMV 35S promoter, OrGSEp-374 and different 5′-deletion fragments constructs. Gus activity measurements are shown for three positive transgenic lines for each OrGSEp promoter construct. CaMV35S: GUS is an independent transgenic plant carrying the CaMV35S promoter construct. OrGSEp-374#3, OrGSEp-374#8 and OrGSEp-374#12 are three positive transgenic lines carrying the OrGSEp-374 construct. OrGSEp-274#4, OrGSEp-274#9 and OrGSEp-274#15 are three independent transgenic lines carrying the OrGSEp-274 construct. OrGSEp-204#1, OrGSEp-204#5 and OrGSEp-204#7 are three positive lines carrying the OrGSEp-204 construct. OrGSEp-114#2, OrGSEp-114#8 and OrGSEp-114#9 are three lines carrying the OrGSEp-114 construct. OrGSEp-54#3, OrGSEp-54#6 and OrGSEp-54#11 are three transgenic lines carrying the OrGSEp-54 fragment. Data are the means of three replicates, and error bars show the standard error. The "**" indicates that a significant difference ($p < 0.001$) in GUS activity was detected between root and shoots of plants carrying the same promoter construct.

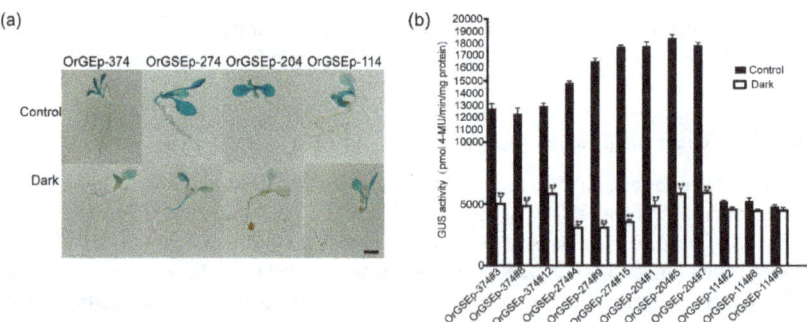

Figure 6. GUS staining and quantification of GUS activity of transgenic *Arabidopsis* seedlings containing different 5′-promoter deletion fragments grown under light and dark conditions. (**a**) Ten-day-old seedlings were stained to observe GUS expression. Bar = 1 cm; (**b**) Quantification of GUS activity in 20-day-old T3 transgenic *Arabidopsis* grown under light and dark conditions. OrGSEp-374#3, OrGSEp-374#8 and OrGSEp-374#12 are three positive transgenic lines carrying the OrGSEp-374 construct. OrGSEp-274#4, OrGSEp-274#9 and OrGSEp-274#15 are three independent transgenic lines carrying the OrGSEp-274 construct. OrGSEp-204#1, OrGSEp-204#5 and OrGSEp-204#7 are three positive lines carrying the OrGSEp-204 construct. OrGSEp-114#2, OrGSEp-114#8 and OrGSEp-114#9 are three lines carrying the OrGSEp-114 construct. Data are the means of three replicates, and standard errors are shown by error bars. The "**" indicates that a significant difference ($p < 0.001$) in GUS activity was detected between roots and shoots of seedlings carrying the same promoter construct.

2.5. The Expression Pattern of OrGSEp-374 in Rice

GUS staining and fluorometric quantification was performed on four independent lines of rice containing the OrGSEp-374 construct. Strong GUS expression was observed in the stem and leaf, weak expression was observed in the anther, ligule, spikelet and embryo, and no expression was observed in the root and endosperm (Figure 7). Further fluorometric analysis of GUS activity showed that the average specific activities in transgenic rice leaves and stems exceeded 15,000 pmol 4-MU min^{-1} mg^{-1} total proteins. However, the GUS activity in the root and panicle was less than 1000 pmol 4-MU min^{-1} mg^{-1} total proteins, which was far lower than the activity level in leaves and stems (Figure 7). GUS activity in the seed was slightly higher than in roots because GUS was weakly expressed in the embryo (Figure 7).

Figure 7. GUS histochemical assays and GUS activity in different tissues of OrGSEp-374 transgenic rice. (**a**) GUS histochemical assays of OrGSEp-374 transgenic rice. A. Root, B. stem, C. leaf, D. ligule, E. spikelet, F. anther, G. seed. H. endosperm. Bar = 1000 μm; (**b**) GUS activity in different tissues of transgenic rice carrying the OrGSEp-374 construct. Boxplots show GUS activity in different tissues of transgenic rice. The lower boundary of each box denotes the 25th percentile, the upper boundary of each box denotes the 75th percentile, and the solid line in the middle of each box denotes the 50th percentile. The two ends of the error bars denote the maximum and minimum values.

3. Discussion

Next-generation sequencing has become a new method for exploring transcriptional regulation and has been used to identify cis-elements involved in gene regulation. To date, many important genes and regulatory elements have been isolated by next-generation sequencing. Using RNA-seq analysis, more than 87 transcription factor genes were identified as being expressed during seed development and fatty acid accumulation. One of these genes, GmDREBL, which belongs to the DREB subfamily of the AP2 family, was shown to participate in fatty acid accumulation based on analysis of *Arabidopsis* plants transformed with the *GmDREBL* gene driven by 35S promoter [33]. These studies illustrate that high-throughput sequencing can be used to identify genes and regulatory elements. In this study, we identified 1140 candidate genes with preferential expression in green tissues by screening an RNA-seq library [30] and performing traditional RT-PCR. We screened a green tissue-specific gene, *OrGSE*, and the promoter of this gene was isolated. GUS reporter gene driven by the promoter was strongly expressed in leaves and stems.

OrGSE shares homology with D27 protein, which is located in chloroplasts. D27 is involved in MAX/RMS/D pathway, in which D27 as a member participates in the biosynthesis of strigolactones, regulating rice tiller bud outgrowth [34]. Although the function of *OrGSE* is unknown, the sequence of *OrGSE* is highly similar with D27. *OrGSE* may share similar function with D27. In *Arabidopsis*, AtD27 is required for the inhibition of secondary bud outgrowth by Strigolactones [35]. According to the sequence similarity between OrGSE and D27, we can infer OrGSE may locate in chloroplasts and participate in rice tiller outgrowth, which can account for OrGSEp drive gene expressed in green tissues.

The tissue-specific expression of transgenes plays a critical role in biotechnology crops because it avoids fitness costs caused by constitutive expression of target genes [9]. However, the information about the core cis-acting elements controlling tissue-specific expression is limited. Two novel cis-elements controlling tissue-specific expression, namely PSE1 (panicle/stem-specific element 1) and LPSRE2 (leaf, panicle/stem and root element 2), were identified in the *DX1* promoter by deletion analysis and gel mobility shift assays [35]. Five tissue-specific expression-related cis-elements in the green tissue-specific promoter PD540 were also characterized. These elements include GEAT, WRKY71OS and TGAC, and are used as references for discovering novel tissue-specific promoters. In this study, we showed that OrGSEp-114 is a short green tissue-specific promoter. However, the above described cis-elements were not found in this novel promoter. The OrGSEp-114 promoter may contain new cis-elements involved in green tissue-specific expression. Our future research will focus on screening cis-acting elements controlling green tissue-specific expression in this promoter by electrophoretic mobility shift assay (EMSA), including at least one novel green tissue-specific element. The OrGSEp-114 has promoter activity and green-specific distinguishing feature, but OrGSEp-54 cannot drive reporter gene expression, which means that 114 bp region of this promoter was sufficient to drive green tissue-specific expression. Cis-elements show that the promoter contains one CAAT-box and GC-motif from −114 to −54. This CAAT-box and GC-motif may be vital to promoter activity (Table S1). Moreover, the promoter from −114 to −54 contains one circadian cis-element, and many studies show that circadian elements participate in green tissue-specific expression. That information will support OrGSEp-114 being a short green tissue-specific promoter.

Light is essential for photosynthesis, and many green tissue-specific promoters contain light-inducible elements [36]. The *IbRbcS* gene from sweet potato shows green tissue-preferential expression and the *IbRbcS1* promoter confers light-responsive expression in transgenic *Arabidopsis* [37]. In this study, results showed that OrGSEp-374, as with the *IbRbcS1* promoter, has both green tissue-specific and light-inducible expression. Based on analysis of the activity of the full-length and truncated promoters in transgenic *Arabidopsis*, we showed that the OrGSEp-374 promoter contains cis-elements involved in light responsiveness, such as ACE, G-box and GT1CONSENSUS, which may regulate green tissue-specific and light-induced expression. Furthermore, analysis of 5′-deletions of this promoter showed that OrGSEp-274 and OrGSEp-204 also have light-induced activity, so the promoter may contain more than one core light-inducible element.

Many insect and fungal diseases influence the normal growth of rice and cause severe yield loss. Sheath blight-resistant rice has been developed by co-expressing the *chitinase* and *oxalate oxidase* 4 genes in green tissues under the control of the green-specific *rbcS* promoter in rice [38]. A transgenic potato resistant to potato tuber moths with 100% tuber moth larval mortality was developed by expressing cry1Ab under the control of the *PEPC* promoter, a green tissue-specific and light-inducible promoter from maize [39]. Cry9C driven by *PNZIP* promoter, a green tissue promoter from *Pharbitis nil*, was used to develop transgenic pest-resistant cotton, and the accumulation of cry9C protein in seeds was 100 times lower than that observed for the seeds of the CaMV 35S:Cry9C line [40]. In our study, we have isolated a novel rice green tissue-specific promoter that does not drive expression in endosperm, and the GUS activity in seeds was much lower than in green tissues. Although the gene driven by this promoter is slightly expressed in embryos, the embryo is linked to bran and will be discarded with the bran during rice bran desquamation, thus alleviating concerns about food safety. OrGSEp is not only not expressed in endosperm but is also not expressed in roots. Thus, the use of this promoter in transgenic crops will ease food safety concerns and reduce the waste of resources that occurs with constitutive expression. This promoter provides a new element for developing insect-resistant and disease-resistant rice and other crops.

4. Materials and Methods

4.1. Plant Materials and Growth Conditions

Common wild rice seeds were collected from Guangdong Province in China. The collection was approved by the supervision department of Guangdong wild rice protection. Plants were grown in a greenhouse under a 12 h light/12 h dark cycle at 28 °C. *Arabidopsis thaliana* (Col-0) seeds were surface-sterilized with 75% (*v*/*v*) ethanol for 10 min and washed for 1 min with 95% (*v*/*v*) ethanol. The sterilized *Arabidopsis* seeds were spread on plates containing 1/2 Murashige and Skoog medium. After stratification at 4 °C for 2 days, the plates were transferred to a plant growth incubator, and seeds were germinated under a 16 h light/8 h dark cycle at 22 °C–24 °C.

4.2. Screening of Green Tissue-Specific Genes and Expression Analysis

Based on RNA-seq data from our previous study [30], genes with higher FPKM (Fragments Per Kilobase per Million) in CL (Control Leaf) than in CR (Control Root) were chosen as candidate green tissue-specific promoter-regulated genes. RT-PCR and Real-time quantitative PCR (qPCR) were used to confirm gene expression patterns. According to expression data, a candidate gene (*OrGSE*) was selected for subsequent analysis. The upstream sequence of *OrGSE* was regarded as a candidate green tissue-specific promoter.

Single-stranded cDNA was synthesized from the total RNA isolated from common wild rice leaves using the 5X All-In-One RT Master Mix (Applied Biological Materials, Vancouver, VAN, Canada). The sequence of *OrGSE* was obtained from a cDNA library. We analyzed sequence conservation by performing a BLAST search against the Michigan State University Rice Genome Annotation Project Database (MSU). We designed a pair of primers (OrGSE-F and OrGSE-R) to amplify the *OrGSE* sequence from leaf cDNA. The PCR amplification with 2× Phanta™ Master Mix was performed according to the manufacturer's instructions (Vazyme, Nanjing, China). The PCR products were purified, cloned into the pEASY-Blunt Cloning Vector (Transgene, Beijing, China) and sequenced. The expression level of *OrGSE* in common wild rice roots, stems and leaves was analyzed using RT-PCR and qPCR. *Actin1(LOC_Os03g61970)* was used as an internal control. Different numbers of cycles were used in RT-PCR amplification for *OrGSE* expression analysis. qRT-PCR was performed using the gene-specific primers listed in Supplemental Table S2 (Actin-F/R and OrGSE-F/R) and a real-time PCR7500 system (Applied Biosystems). Data were collected using the ABI PRISM 7500 sequence detection system. Three biological replicates with independent mRNA isolations were performed, each with three technical repeats, the rice Actin gene was used as an internal control, and the mRNA relative expression level was calculated using the $2^{-\Delta\Delta CT}$ method.

4.3. Cloning of the OrGSE Promoter and Sequence Analysis

Genomic DNA was isolated from common wild rice leaves using the EasyPure Plant Genomic DNA Kit (Transgene Biotech, Beijing, China) and used as the template for amplification of the *OrGSE* promoter. The PCR products were purified and cloned into the plant expression vector pCAMBIA1305.1 using the In-fusion method. This construct was used for sequencing and promoter activity analysis.

Core regulatory elements in the promoter sequences were predicted using the online tool plantCARE (available online: http://bioformatics.psb.ugent.be/webtools/plantcare/html/) [41].

4.4. PCR Amplification of 5'-Deletion Fragments of the OrGSE Promoter

We designed five forward primers with *Bam*H I restriction sites (F-374, F-274, F-204, F-114, F-54) and one reverse primer with an *Eco*R I restriction site (R) to obtain 5'-deletion fragments of the OrGSE promoter. These primers were designed to amplify regions −374 (F-374/R), −274 (F-274/R),

−204 (F-204/R), −114 (F-114/R), and −54 (F-54/R) upstream of the transcription initiation site (TIS), which was designed as +1. The PCR cycling parameters were as follows: 95 °C for 3 min, 35 cycles of 95 °C for 15 s, 60 °C for 15 s, and 72 °C for 1 min followed by 72 °C for 5 min. The full-length promoter and 5'-deletion fragments were cloned into a modified pBinGlyRed vector, which includes GUS plus-his6 that was inserted into the multiple cloning site using the *EcoR* I and *Xma* I restriction sites (Figure S2). The five promoter fragments were named OrGSEp-374, OrGSEp-274, OrGSEp-204, OrGSEp-114 and OrGSEp-54. The constructs were then introduced into *Arabidopsis thaliana* (Col-0), *Agrobacterium*-mediated transformation of *Arabidopsis* was performed through floral dipping [42]. We also amplificatedOrGSEp-374 fragment by 1305GSEp-F/R primers, and it was inserted into pCAMBIA1305.1 vector by *Hind* III and *Nco* I restriction sites, the construct was introduced into Nipponbare (*Oryza sativa* L. ssp. *japonica*) by *Agrobacterium*-mediated transformation.

4.5. Detection of the Expression Pattern of the OrGSEp Promoter and 5'-Deletion Fragments in Different Organs

For GUS histochemical assays, homozygous T3 transgenic *Arabidopsis* seedlings (3-, 5- and 14-day-old) were incubated in GUS staining solution overnight. During the reproductive stage, GUS histochemical assays were performed for transgenic *Arabidopsis* leaves, siliques and root. GUS activity was quantified in the leaves and roots of 3-week-old seedlings. Three independent lines were selected for each promoter deletion construct.

4.6. Inducible Activity Analysis of the OrGSEp Promoter and 5'-Deletion Fragments

To analyze the light inducible activity of the promoter and different 5'-deletion fragments, 3-week-old transgenic *Arabidopsis* plants were grown in the dark for 24 h. Then, the leaves were frozen in liquid nitrogen and stored at −80 °C for GUS fluorometric assay. The control plants were placed under natural growth conditions. For GUS staining, 10-day-old seedlings were placed in the dark for 24 h and transferred to natural condition for 24 h, and control plants were kept under natural conditions.

4.7. GUS Histochemical and Fluorometric Analysis

GUS histochemical and fluorometric analysis was performed as previously described (Wu et al., 2003). Transgenic *Arabidopsis* seedlings at different growth stages and various tissues at the reproductive stage (as described above) were incubated in GUS staining solution (Coolaber, Beijing, China) at 37 °C overnight, and then the samples were cleared with 75% (*v/v*) ethanol. GUS staining was observed under a ZEISS Stemi 508 microscope and photographed with a SONY camera.

Protein was extracted from 100 mg root and leaf tissue from transgenic *Arabidopsis* plants. The protein concentrations were determined using the Bradford method with bovine serum albumin (BSA) as the standard. GUS activity was determined fluorometrically by measuring the amount of 4-methylumbelliferone (4-MU) produced by GUS per milligram of total protein per minute [43]. GUS activity was measured for three lines for each promoter construct, and three replicates were performed for each line. The error bars are reported as "mean ± standard error". The boxplot was drawn by R language gglpot2 package.

Supplementary Materials: Supplementary materials can be found at http://www.mdpi.com/1422-0067/19/7/2009/s1.

Author Contributions: X.P. and Q.Y. conceived and designed the experiments. M.X., Z.Z., K.H., T.Z. and G.H. performed the experiments. M.X. performed the data analysis and wrote the manuscript. Y.L., Y.J. and X.P. revised the manuscript.

Acknowledgments: This research was supported by The National Special Program for Transgenic Research (2016ZX08011-001).

Conflicts of Interest: The authors declare no conflict of interest.

References

1. Li, Y.; Sun, Y.; Yang, Q.C.; Kang, J.M.; Zhang, T.J.; Gruber, M.Y.; Fang, F. Cloning and function analysis of an alfalfa (*Medicago sativa* L.) zinc finger protein promoter MsZPP. *Mol. Biol. Rep.* **2012**, *39*, 8559–8569. [CrossRef] [PubMed]
2. Niu, G.L.; Gou, W.; Han, X.L.; Qin, C.; Zhang, L.X.; Abomohra, A.E.; Ashraf, M. Cloning and functional analysis of phosphoethanolamine methyltransferase promoter from maize (*Zea mays* L.). *Int. J. Mol. Sci.* **2018**, *19*, 191. [CrossRef] [PubMed]
3. Pih, K.T.; Yoo, J.; Fosket, D.E.; Han, I.S. A comparison of the activity of three cauliflower mosaic virus 35s promoters in rice seedlings and tobacco (by-2) protoplasts by analysis of *gus* reporter gene transient expression. *Plant. Sci.* **1996**, *114*, 141–148. [CrossRef]
4. McElroy, D.; Blowers, A.D.; Jenes, B.; Wu, R. Construction of expression vectors based on the rice *actin 1* (*act1*) 5' region for use in monocot transformation. *Mol. Gen. Genet.* **1991**, *231*, 150–160. [CrossRef] [PubMed]
5. Christensen, A.H.; Quail, P.H. Ubiquitin promoter-based vectors for high-level expression of selectable and/or screenable marker genes in monocotyledonous plants. *Transgenic Res.* **1996**, *5*, 213–218. [CrossRef] [PubMed]
6. Eva, C.; Teglas, F.; Zelenyanszki, H.; Tamas, C.; Juhasz, A.; Meszaros, K.; Tamas, L. Cold inducible promoter driven cre-lox system proved to be highly efficient for marker gene excision in transgenic barley. *J. Biotechnol.* **2018**, *265*, 15–24. [CrossRef] [PubMed]
7. Qu, D.; Song, Y.; Li, W.M.; Pei, X.W.; Wang, Z.X.; Jia, S.R.; Zhang, Y.Q. Isolation and characterization of the organ-specific and light-inducible promoter of the gene encoding rubisco activase in potato (*Solanum tuberosum*). *Genet. Mol. Res.* **2011**, *10*, 621–631. [CrossRef] [PubMed]
8. Thomson, B.; Graciet, E.; Wellmer, F. Inducible promoter systems for gene perturbation experiments in *Arabidopsis*. *Methods Mol. Biol.* **2017**, *1629*, 15–25. [CrossRef] [PubMed]
9. Yang, F.; Ding, X.; Chen, J.; Shen, Y.; Kong, L.; Li, N.; Chu, Z. Functional analysis of the *GRMZM2g174449* promoter to identify rhizoctonia solani-inducible cis-elements in maize. *BMC Plant. Biol.* **2017**, *17*, 233. [CrossRef] [PubMed]
10. Yu, S.I.; Lee, B.H. Generation of a stress-inducible luminescent *Arabidopsis* and its use in genetic screening for stress-responsive gene deregulation mutants. *Methods Mol. Biol.* **2017**, *1631*, 109–119. [CrossRef] [PubMed]
11. Li, J.; Qin, R.; Xu, R.; Li, H.; Yang, Y.; Li, L.; Wei, P.; Yang, J. Isolation and identification of five cold-inducible promoters from *Oryza sativa*. *Planta* **2017**, *247*, 99–111. [CrossRef] [PubMed]
12. Zhang, H.; Jing, R.; Mao, X. Functional characterization of *Tasnrk2.8* promoter in response to abiotic stresses by deletion analysis in transgenic *Arabidopsis*. *Front. Plant. Sci.* **2017**, *8*, 1198. [CrossRef] [PubMed]
13. Kakrana, A.; Kumar, A.; Satheesh, V.; Abdin, M.Z.; Subramaniam, K.; Bhattacharya, R.C.; Srinivasan, R.; Sirohi, A.; Jain, P.K. Identification, validation and utilization of novel nematode-responsive root-specific promoters in *Arabidopsis* for inducing host-delivered RNAi mediated root-knot nematode resistance. *Front. Plant. Sci.* **2017**, *8*, 2049. [CrossRef] [PubMed]
14. Nanjareddy, K.; Arthikala, M.K.; Aguirre, A.L.; Gomez, B.M.; Lara, M. Plant promoter analysis: Identification and characterization of root nodule specific promoter in the common bean. *J. Vis. Exp.* **2017**, *23*, 130. [CrossRef] [PubMed]
15. Chavez-Barcenas, A.T.; Valdez-Alarcon, J.J.; Martinez-Trujillo, M.; Chen, L.; Xoconostle-Cazares, B.; Lucas, W.J.; Herrera-Estrella, L. Tissue-specific and developmental pattern of expression of the rice *SPS1* gene. *Plant. Physiol.* **2000**, *124*, 641–654. [CrossRef] [PubMed]
16. Kim, J.; Lee, H.J.; Jung, Y.J.; Kang, K.K.; Tyagi, W.; Kovach, M.; Sweeney, M.; McCouch, S.; Cho, Y.G. Functional properties of an alternative, tissue-specific promoter for rice nadph-dependent dihydroflavonol reductase. *PLoS ONE* **2017**, *12*, e0183722. [CrossRef] [PubMed]
17. Panguluri, S.K.; Sridhar, J.; Jagadish, B.; Sharma, P.C.; Kumar, P.A. Isolation and characterization of a green tissue-specific promoter from pigeonpea [*Cajanus cajan* (L.) millsp.]. *Indian J. Exp. Biol.* **2005**, *43*, 369–372. [PubMed]
18. Xu, W.; Liu, W.; Ye, R.; Mazarei, M.; Huang, D.; Zhang, X.; Stewart, C.N., Jr. A profilin gene promoter from switchgrass (*Panicum virgatum* L.) directs strong and specific transgene expression to vascular bundles in rice. *Plant. Cell. Rep.* **2018**, *37*, 587–597. [CrossRef] [PubMed]

19. Zaidi, M.A.; O'Leary, S.J.; Wu, S.; Chabot, D.; Gleddie, S.; Laroche, A.; Eudes, F.; Robert, L.S. Investigating triticeae anther gene promoter activity in transgenic *brachypodium distachyon*. *Planta* **2017**, *245*, 385–396. [CrossRef] [PubMed]

20. Li, H.; Li, J.; Xu, R.; Qin, R.; Song, F.; Li, L.; Wei, P.; Yang, J. Isolation of five rice non-endosperm tissue-expressed promoters and evaluation of their activities in transgenic rice. *Plant. Biotechnol. J.* **2017**, *16*, 1138–1147. [CrossRef] [PubMed]

21. Suzuki, H.; Fowler, T.J.; Tierney, M.L. Deletion analysis and localization of *SbPRP1*, a soybean cell wall protein gene, in roots of transgenic tobacco and cowpea. *Plant. Mol. Biol.* **1993**, *21*, 109–119. [CrossRef] [PubMed]

22. Chen, L.; Jiang, B.; Wu, C.; Sun, S.; Hou, W.; Han, T. *GmPRP2* promoter drives root-preferential expression in transgenic *Arabidopsis* and soybean hairy roots. *BMC Plant. Biol.* **2014**, *14*, 245. [CrossRef] [PubMed]

23. Wang, H.; Fan, M.; Wang, G.; Zhang, C.; Shi, L.; Wei, Z.; Ma, W.; Chang, J.; Huang, S.; Lin, F. Isolation and characterization of a novel pollen-specific promoter in maize (*Zea mays* L.). *Genome* **2017**, *60*, 485–495. [CrossRef] [PubMed]

24. Wu, L.; El-Mezawy, A.; Shah, S. A seed coat outer integument-specific promoter for *Brassica napus*. *Plant. Cell. Rep.* **2011**, *30*, 75–80. [CrossRef] [PubMed]

25. Shrestha, A.; Khan, A.; Mishra, D.R.; Bhuyan, K.; Sahoo, B.; Maiti, I.B.; Dey, N. WRKY71 and TGA1A physically interact and synergistically regulate the activity of a novel promoter isolated from petunia vein-clearing virus. *Biochim. Biophys. Acta* **2018**, *1861*, 133–146. [CrossRef] [PubMed]

26. Wang, Q.; Zhu, Y.; Sun, L.; Li, L.; Jin, S.; Zhang, X. Transgenic Bt cotton driven by the green tissue-specific promoter shows strong toxicity to lepidopteran pests and lower Bt toxin accumulation in seeds. *Sci. China Life Sci.* **2016**, *59*, 172–182. [CrossRef] [PubMed]

27. Manikandan, R.; Balakrishnan, N.; Sudhakar, D.; Udayasuriyan, V. Development of leaffolder resistant transgenic rice expressing *cry2ax*1 gene driven by green tissue-specific *rbcS* promoter. *World J. Microb. Biot.* **2016**, *32*, 37. [CrossRef] [PubMed]

28. Menguer, P.K.; Sperotto, R.A.; Ricachenevsky, F.K. A walk on the wild side: *Oryza* species as source for rice abiotic stress tolerance. *Genet. Mol. Biol.* **2017**, *40*, 238–252. [CrossRef] [PubMed]

29. Mao, D.; Yu, L.; Chen, D.; Li, L.; Zhu, Y.; Xiao, Y.; Zhang, D.; Chen, C. Multiple cold resistance loci confer the high cold tolerance adaptation of dongxiang wild rice (*Oryza rufipogon*) to its high-latitude habitat. *Theor. Appl. Genet.* **2015**, *128*, 1359–1371. [CrossRef] [PubMed]

30. Tian, X.; Long, Y.; Wang, J.; Zhang, J.; Wang, Y.; Li, W.; Peng, Y.; Yuan, Q.; Pei, X. De novo transcriptome assembly of common wild rice (*Oryza rufipogon* griff.) and discovery of drought-response genes in root tissue based on transcriptomic data. *PLoS ONE* **2015**, *10*, e0131455. [CrossRef] [PubMed]

31. Civan, P.; Svec, M. Genome-wide analysis of rice (*Oryza sativa* L. Subsp. Japonica) tata box and y patch promoter elements. *Genome* **2009**, *52*, 294–297. [CrossRef] [PubMed]

32. Conley, T.R.; Park, S.C.; Kwon, H.B.; Peng, H.P.; Shih, M.C. Characterization of cis-acting elements in light regulation of the nuclear gene encoding a subunit of chloroplast isozymes of glyceraldehyde-3-phosphate dehydrogenase from *Arabidopsis thaliana*. *Mol. Cell. Biol.* **1994**, *14*, 2525–2533. [CrossRef] [PubMed]

33. Zhang, Y.Q.; Lu, X.; Zhao, F.Y.; Li, Q.T.; Niu, S.L.; Wei, W.; Zhang, W.K.; Ma, B.; Chen, S.Y.; Zhang, J.S. Soybean *GmDREBl* increases lipid content in seeds of transgenic *Arabidopsis*. *Sci. Rep.* **2016**, *6*, 34307. [CrossRef] [PubMed]

34. Lin, H.; Wang, R.X.; Qian, Q.; Yan, M.X.; Meng, X.B.; Fu, Z.M.; Yan, C.Y.; Jiang, B.; Su, Z.; Li, J.Y.; et al. DWARF27, an Iron-Containing Protein Required for the Biosynthesis of Strigolactones, Regulates Rice Tiller Bud Outgrowth. *Plant Cell* **2009**, *21*, 1512–1525. [CrossRef] [PubMed]

35. Waters, M.T.; Brewer, P.B.; Bussell, J.D.; SmitSh, S.M.; Beveridge, C.A. The *Arabidopsis* ortholog of rice dwarf27 acts upstream of max1 in the control of plant development by strigolactones. *Plant. Physiol.* **2012**, *159*, 1073–1085. [CrossRef] [PubMed]

36. Ye, R.; Zhou, F.; Lin, Y. Two novel positive cis-regulatory elements involved in green tissue-specific promoter activity in rice (*Oryza sativa* L. ssp.). *Plant. Cell. Rep.* **2012**, *31*, 1159–1172. [CrossRef] [PubMed]

37. Eberhard, S.; Finazzi, G.; Wollman, F.A. The dynamics of photosynthesis. *Annu. Rev. Genet.* **2008**, *42*, 463–515. [CrossRef] [PubMed]

38. Tanabe, N.; Tamoi, M.; Shigeoka, S. The sweet potato *RbcS* gene (*IbRbcS1*) promoter confers high-level and green tissue-specific expression of the gus reporter gene in transgenic *Arabidopsis*. *Gene* **2015**, *567*, 244–250. [CrossRef] [PubMed]

39. Karmakar, S.; Molla, K.A.; Chanda, P.K.; Sarkar, S.N.; Datta, S.K.; Datta, K. Green tissue-specific co-expression of chitinase and oxalate oxidase 4 genes in rice for enhanced resistance against sheath blight. *Planta* **2016**, *243*, 115–130. [CrossRef] [PubMed]

40. Ghasimi Hagh, Z.; Rahnama, H.; Panahandeh, J.; Baghban Kohneh Rouz, B.; Arab Jafari, K.M.; Mahna, N. Green-tissue-specific, c(4)-pepc-promoter-driven expression of *Cry1ab* makes transgenic potato plants resistant to tuber moth (*phthorimaea operculella*, zeller). *Plant. Cell. Rep.* **2009**, *28*, 1869–1879. [CrossRef] [PubMed]

41. Lescot, M.; Dehais, P.; Thijs, G.; Marchal, K.; Moreau, Y.; Van de Peer, Y.; Rouze, P.; Rombauts, S. Plantcare, a database of plant cis-acting regulatory elements and a portal to tools for in silico analysis of promoter sequences. *Nucleic Acids Res.* **2002**, *30*, 325–327. [CrossRef] [PubMed]

42. Clough, S.J.; Bent, A.F. Floral dip: A simplified method for *agrobacterium*-mediated transformation of *Arabidopsis thaliana*. *Plant. J.* **1998**, *16*, 735–743. [CrossRef] [PubMed]

43. Wang, R.; Zhu, M.; Ye, R.; Liu, Z.; Zhou, F.; Chen, H.; Lin, Y. Novel green tissue-specific synthetic promoters and cis-regulatory elements in rice. *Sci. Rep.* **2015**, *5*, 18256. [CrossRef] [PubMed]

Article

Genome-Wide Association Study Reveals Novel Genomic Regions for Grain Yield and Yield-Related Traits in Drought-Stressed Synthetic Hexaploid Wheat

Madhav Bhatta [1], Alexey Morgounov [2], Vikas Belamkar [1] and P. Stephen Baenziger [1,*]

[1] Department of Agronomy and Horticulture, University of Nebraska-Lincoln, Lincoln, NE 68583, USA; madhav.bhatta@huskers.unl.edu (M.B.); vikas.belamkar@unl.edu (V.B.)
[2] International Maize and Wheat Improvement Center (CIMMYT), 06511 Emek, Ankara, Turkey; a.morgounov@cgiar.org
* Correspondence: pbaenziger1@unl.edu; Tel.: +1-402-472-1538

Received: 20 August 2018; Accepted: 29 September 2018; Published: 2 October 2018

Abstract: Synthetic hexaploid wheat (SHW; $2n = 6x = 42$, AABBDD, *Triticum aestivum* L.) is produced from an interspecific cross between durum wheat ($2n = 4x = 28$, AABB, *T. turgidum* L.) and goat grass ($2n = 2x = 14$, DD, *Aegilops tauschii* Coss.) and is reported to have significant novel alleles-controlling biotic and abiotic stresses resistance. A genome-wide association study (GWAS) was conducted to unravel these loci [marker–trait associations (MTAs)] using 35,648 genotyping-by-sequencing-derived single nucleotide polymorphisms in 123 SHWs. We identified 90 novel MTAs (45, 11, and 34 on the A, B, and D genomes, respectively) and haplotype blocks associated with grain yield and yield-related traits including root traits under drought stress. The phenotypic variance explained by the MTAs ranged from 1.1% to 32.3%. Most of the MTAs (120 out of 194) identified were found in genes, and of these 45 MTAs were in genes annotated as having a potential role in drought stress. This result provides further evidence for the reliability of MTAs identified. The large number of MTAs (53) identified especially on the D-genome demonstrate the potential of SHWs for elucidating the genetic architecture of complex traits and provide an opportunity for further improvement of wheat under rapidly changing climatic conditions.

Keywords: marker–trait association; haplotype block; genes; root traits; D-genome; genotyping-by-sequencing; single nucleotide polymorphism; durum wheat; bread wheat; complex traits

1. Introduction

Drought is one of the most important abiotic stresses that reduce crop productivity and is expected to increase with the change in climate [1]. Erratic rainfall patterns caused by climate change may aggravate drought stress and will have a major impact on agriculture [2,3]. The most prominent example of the impact of drought stress on agriculture was the 2012 drought stress in the United States, where moderate to extreme drought stress occurred across the central agricultural states that resulted in crop harvest failure for corn (*Zea mays* L.), sorghum (*Sorghum bicolor* L.), and soybean (*Glycine max* L.), and the agriculture loss due to drought was estimated to be $30 billion [4]. To cope with the challenges of drought stress, plant breeders have been focusing on improving drought tolerance since several decades [2,3,5]. However, the drought tolerance is a complex phenomenon as most of the traits associated with drought tolerance are polygenic in nature, and understating the genetic architecture of drought tolerance is still underway [3] including in wheat (*Triticum sps.*) [2]. Wheat is one of the most important staple cereal crops mainly grown under rainfed conditions [3,6] and is expected to suffer from drought stress [3]. Therefore, breeding for drought tolerance and identifying genomic regions and underlying candidate genes associated with drought tolerance are important for wheat improvement.

Bread wheat (*T. aestivum* L.) has limited genetic and phenotypic diversity available for breeding for drought tolerance [2]. This is mainly due to the genetic bottleneck experienced during its origin and subsequent domestication [7,8]. Diversity can be increased through the production of synthetic hexaploid wheat (SHW) and its utilization in breeding programs [9–11]. Synthetic hexaploid wheat ($2n = 6x = 42$, AABBDD) is produced from an interspecific cross between durum wheat ($2n = 4x = 28$, AABB, *T. turgidum* L.) and goat grass ($2n = 2x = 14$, DD, *Aegilops tauschii* Coss.). The SHWs are reported to have significant genetic variation for biotic [12,13] and abiotic stresses resistance [2,10,14]. However, previous studies focused mainly on biotic stresses including leaf rust (incited by *Puccinia triticina*) [13,15,16], stem rust (incited by *P. graminis*) [15,16], stripe rust (incited by *P. striiformis*) [12,15,16], Fusarium head blight (incited by *Fusarium graminearum*) [13], yellow spot (incited by *Pyrenophora tritici-repentis*) [15,16], septoria nodorum (incited by *Parastagonospora nodorum*) [15,16], Septoria tritici blotch (incited by *Mycosphaerella graminicola*) [13,15], cereal cyst nematode (incited by *Heterodera avenae*) [15], crown rot (incited by *F. pseudograminearum*) [16], and root-lesion nematode (incited by *Pratylenchus thornei* and *P. neglectus*) [15]. Therefore, exploiting genetic variation under abiotic stresses such as drought is needed to further utilize the potential of SHWs.

About 800 quantitative trait loci (QTLs) and marker–trait associations (MTAs) have been reported for drought tolerant traits (agronomic, physiological, root, and yield-related traits) using bi-parental mapping (~691 QTLs) and genome-wide association studies (GWASs; ~109 MTAs) in wheat [3]. However, only 68 QTLs are major QTLs that explain more than 19% of phenotypic variation [3]. This study was conducted to identify novel genomic regions associated with grain yield (GY) and yield-related traits using GWAS performed using 35,648 genotyping-by-sequencing (GBS)-derived single nucleotide polymorphisms (SNPs) in 123 SHWs grown under two drought-stressed growing seasons (2016 and 2017) in Konya, Turkey. Subsequently, the underlying genes for the MTAs identified were investigated for their potential role in drought stress using the functional annotations. To the best of our knowledge, this is the first report on GWAS on GY and yield-related traits under drought stress in SHWs. The results from this study will be a valuable resource for the genetic improvement of GY and yield-related traits in drought stress, introgression of desirable genes from SHWs into elite wheat germplasm, genomic selection, and marker-assisted selection in the breeding program.

2. Results and Discussion

2.1. Weather Conditions

The mean monthly air temperatures were similar at Konya in both growing seasons with 13 °C in 2015–2016 and 12 °C in 2016–2017 compared to the 25-year mean monthly air temperature (11 °C) in Turkey (Table 1). The total rainfall during 2016–2017 (243 mm) was slightly higher than that during 2015–2016 (222.4 mm) growing season. Total rainfalls during the wheat-growing season (September–July) were 48.9% lower in 2015–2016 and 44.2% lower in 2016–2017 compared to the 25-year mean total rainfall (435.1 mm) in Turkey. Although winter wheat water requirements are higher from mid-March to mid-June (from the spring tillering period to the mid-grain filling period), rainfalls were lower in both years compared to the 25-year mean rainfall. The plants were exposed to drought stress from tillering through grain filling. Hence, the results from the present study can be used to understand the effects and genetics of drought in SHW.

Table 1. Mean monthly temperatures and total monthly rainfalls in two growing seasons (2016 and 2017) and 25-year averaged data in Konya, Turkey.

Month	Konya, 2015–2016	Konya, 2016–2017	Turkey, 1991–2015	Konya, 2015–2016	Konya, 2016–2017	Turkey, 1991–2015
	Temperature (temp) ($^\circ$C) [a]	Temp ($^\circ$C)	Temp ($^\circ$C) [b]	Rainfall (mm) [c]	Rainfall (mm)	Rainfall (mm) [d]
September	22.8	17.1	19.0	35.8	11.2	23.1
October	15.3	13.2	13.7	34.4	0.0	48.3
November	7.5	4.9	7.0	5.8	16.6	58.0
December	−0.1	0.5	2.1	8.0	26.8	73.0
January	1.6	0.2	0.1	37.0	9.0	65.6
February	4.9	3.4	1.3	0.4	69.2	60.0
March	8.5	8.2	5.3	37.8	31.0	61.6
April	15.2	12.7	10.4	9.6	33.2	62.7
May	18.4	16.7	15.2	38.4	41.2	54.6
June	23.7	24.4	19.5	15.0	4.8	34.7
July	26.6	27.7	22.9	0.2	0.0	15.1
Total/average	13.1	11.7	10.6	222.4	243.0	435.1

[a] Source: Bahri Dagdas International Agricultural Research Institute. [b] Source: http://sdwebx.worldbank.org/climateportal/index.cfm?page=country_historical_climate&ThisCCode=TUR. [c] Source: Bahri Dagdas International Agricultural Research Institute. [d] Source: http://sdwebx.worldbank.org/climateportal/index.cfm?page=country_historical.

2.2. Phenotypic Variation for Yield and Yield-Related Traits

A combined analysis of variance (ANOVA) across years identified significant cross-over genotype × year interaction for all traits except for flag leaf width (FLW) and stem diameter (STMDIA) (Table S4). Therefore, analysis of variance was computed for both years separately and the results indicated that the SHWs showed significant variation for GY and yield-related traits in each year (Table 2). For instance, GY ranged from 200 g·m^{-2} to 341 g·m^{-2} with an average yield of 259 g·m^{-2} in 2016 and from 241 g·m^{-2} to 392 g·m^{-2} with an average yield of 290 g·m^{-2} in 2017 (Table 2). The large variation among the traits in each year can be attributed to the collection of diverse accessions of SHWs from different countries and different genetic backgrounds [11,14].

Table 2. Phenotypic variation for grain yield and yield-related traits with best linear unbiased predictor values, range, percentage of coefficient of variation (CV), and broad sense heritability (H^2) of 123 synthetic hexaploid wheat grown in two seasons (2016 and 2017) in Konya, Turkey.

Trait	2016				2017			
	Mean	Range	CV	H^2	Mean	Range	CV	H^2
Grain yield (g·m^{-2})	259	200–341	9.7	0.32	290	241–392	9.9	0.56
Harvest index	0.4	0.24–0.66	10.9	0.63	0.34	0.27–0.41	6.3	0.64
Biomass weight (g·m^{-2})	671	537–827	9.1	0.39	865	684–1098	8.9	0.63
Thousand kernel weight (g)	32.1	24–42	10.5	0.75	41	33–50	8	0.90
Grain volume weight (Kg·hL^{-1})	65.6	52–77	7.2	0.91	74	68–77	2.3	0.76
Awn length (cm)	6	2.3–8.6	24.3	0.61	5.6	0.5–8.0	28.3	0.95
Flag leaf length (cm)	22.4	21.8–22.8	0.8	0.91	12	9.9–16.4	7.6	0.53
Flag leaf width (cm)	1	0.96–1.13	2.8	0.67	1	0.9–1.3	6.1	0.49
Flag leaf area (cm^2)	18.9	17.6–19.7	2.2	0.85	10.1	7.7–14	11.6	0.52
Stem diameter (mm)	2.9	2.4–3.5	6.9	0.57	2.9	2.5–4.0	7.4	0.63
Root length (cm)	393	392–395	0.20	0.6	192.2	72–375	20	0.31

Broad-sense heritability (H^2) ranged from low to high (Table 2). Low to moderate H^2 was observed for GY (0.32–0.56), biomass weight (BMWT; 0.39–0.63), FLW (0.49–0.67), and root length (RTLN; 0.31–0.60); moderate H^2 was observed for harvest index (HI; 0.63–0.64) and STMDIA (0.57–0.63); moderate to high H^2 was observed for flag leaf length (FLLN; 0.53–0.91), flag leaf area (FLA; 0.52–0.85) and awn length (AWNLN; 0.61–0.95); and high H^2 was observed for thousand kernel weight (TKW; 0.75–0.90) and grain volume weight (GVWT; 0.76–0.91), indicating the genetic instability of these traits across years under drought stress. Similar H^2 for most of these traits have been observed in previous studies [17–26].

2.3. Principal Component Analysis and Phenotypic Correlation

Principal component (PC) bi-plot analysis showed the association between GY and yield-related traits based on correlation matrix (Figure 1). The first two PCs that explained 43.4% (2016) and 44.9% (2017) of variation better explained the relationship between traits in the two-dimensional space. In the PC biplot, we observed two distinct groupings. The first one comprised of GY, HI, BMWT, TKW, GVWT, AWNLN, and RTLN whereas the second one had FLLN, FLW, and FLA (Figure 1). The traits grouping with GY are the more important traits for improving GY in drought-stressed conditions. The association observed in the PC biplot was supported by the significant correlations of GY with BMWT, HI, TKW, and GVW in both years (Figure S1). Similar correlations for these traits were observed in the previous studies [18,27–32].

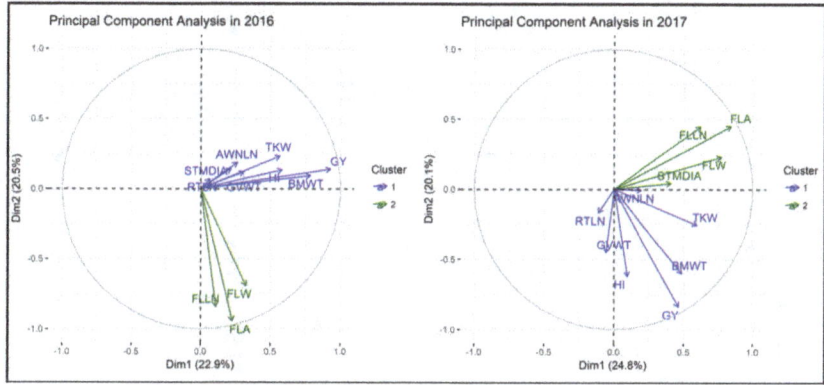

Figure 1. Principal component bi-plot analysis of 123 drought-stressed synthetic hexaploid wheat grown in two seasons (2016 and 2017) in Konya, Turkey. AWLN, awn length; BMWT, biomass weight; FLA, flag leaf area; FLLN, flag leaf length; FLW, flag leaf width; GVWT, grain volume weight; GY, grain yield; HI, harvest index; RTLN, root length; STMDIA, stem diameter; and TKW, thousand kernel weight.

2.4. Population Structure and Genome-Wide Association Study

Population structure analysis of 123 SHWs was performed using 35,648 SNPs [filtered for minor allele frequency (MAF) > 0.05 and missing data < 20%] using the Bayesian clustering algorithm implemented in Structure software and the results showed that these lines were divided into three subgroups (Figure S2 and Table S1). The details of the population structure and genetic diversity of these SHWs have been previously reported in Bhatta et al. [11].

The GWAS identified novel genomic regions for GY and yield-related traits and the MTA explained the high phenotypic variance. The Fixed and random model Circulating Probability Unification algorithm (FarmCPU), with kinship, population structure (Q) or PC, best linear unbiased predictors (BLUPs) for each trait, and 35,648 GBS-derived SNPs was used to identify MTAs. The GBS-derived SNPs were well distributed across each of the chromosome (Figure S3). We identified 194 MTAs distributed across 21 chromosomes for GY and yield-related traits with phenotypic variance explained (PVE) ranging from 1.1% to 32.3% (Figure 2 and Figure S4, Table S5). The highest number of MTAs was observed for GY (29), followed by STMDIA (23), FLA (20), and TKW (20) while the lowest MTAs were observed for HI (10) (Figure 2). Of the 194 MTAs, 75 MTAs were detected on the A genome, with 66 MTAs on the B genome, and 53 MTAs on the D-genome. The highest MTAs were present on chromosome 7A (26 MTAs) and the lowest MTAs were present on chromosome 3D (four MTAs)

(Figure 2). Most of the MTAs identified in the present study were year-specific, suggesting the influence of genotype × environment interaction on the phenotype of the traits measured in two years. However, 120 of the 194 significant SNPs were in 83 genes and 45 of these MTAs were present within genes and their annotations suggested their potential role in drought stress. This result further provided confidence that the MTAs identified in the study are likely reliable MTAs (Table S5).

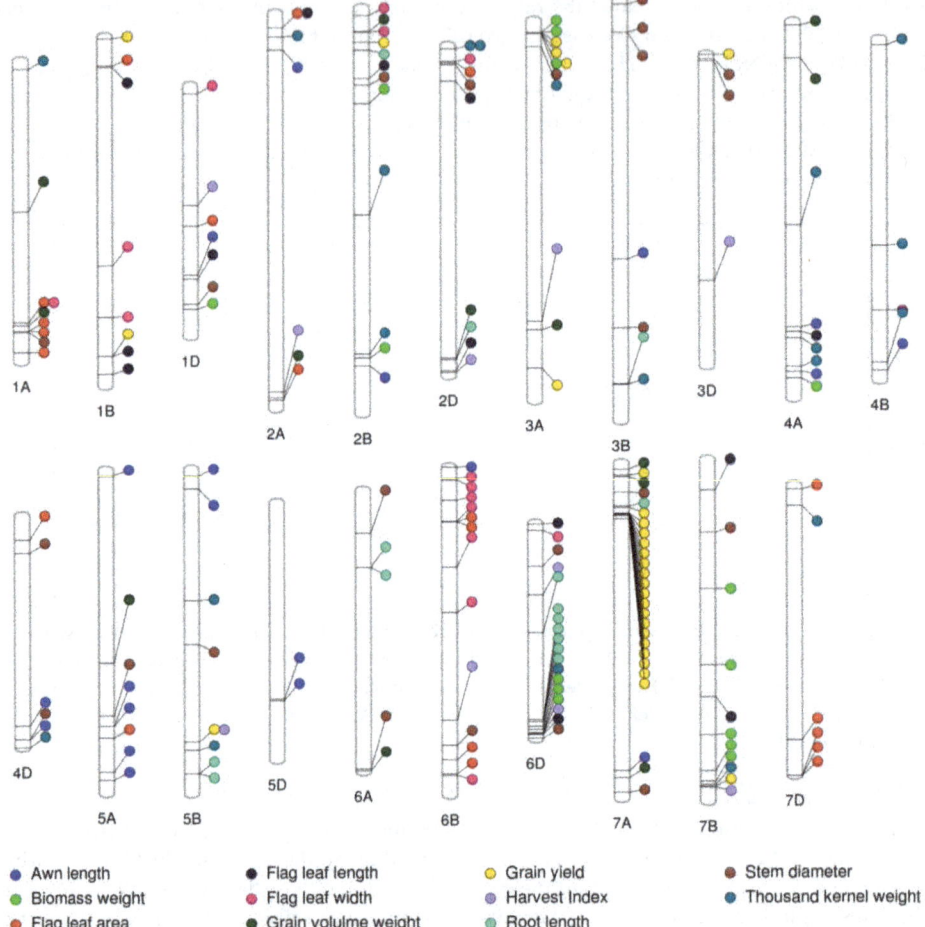

Figure 2. Significant markers trait associations identified on each chromosome for grain yield and yield-related traits obtained from the genome-wide association study of 123 synthetic hexaploid wheats grown in 2016 and 2017 in Konya, Turkey.

2.4.1. Grain Yield

The 29 MTAs for GY were observed in 29 different genomic regions on seven chromosomes including 1B, 2B, 3A, 3D, 5B, 7A, and 7B with PVE ranging from 7.6% to 17.9% (Figure 2, and Table S5). Earlier studies have reported QTLs/MTAs for GY on wheat chromosomes 1B [5,17,19,33], 2B [17,19,29,33,34], 3A [17,30,33,34], 3D [33], 5B [5,28,30,31,33,35], 7A [20,27,30,33], and 7B [17,30,33]. However, it is difficult to align our findings with previous studies due to the use of different marker systems [90K SNP, short sequence repeat (SSR), diversity arrays technology (DART) marker vs. GBS-derived SNP marker], absence of precise location information in published studies, or the use of a different version of the reference wheat genome in previous studies than the International Wheat Genome Sequencing Consortium (IWGSC) RefSeq v1.0. However, identification of several MTAs on the same chromosome as earlier studies provided increased confidence on these associations.

The present study identified four major haplotype blocks (from 19 bp to 433 kb) on chromosome 7A with two to six SNPs associated with GY in 2016 (Figure 3). First haplotype block consisted of six MTAs within the 433 kb range, second haplotype block consisted of four MTAs within the 81 bp range, third haplotype block consisted of two MTAs within the 19 bp range, and fourth haplotype block consisted of three MTAs within the 314 kb range. The PVE on GY by the first, second, third, and fourth haplotype blocks were 17.2%, 24.6%, 21.9%, and 8.2%, respectively.

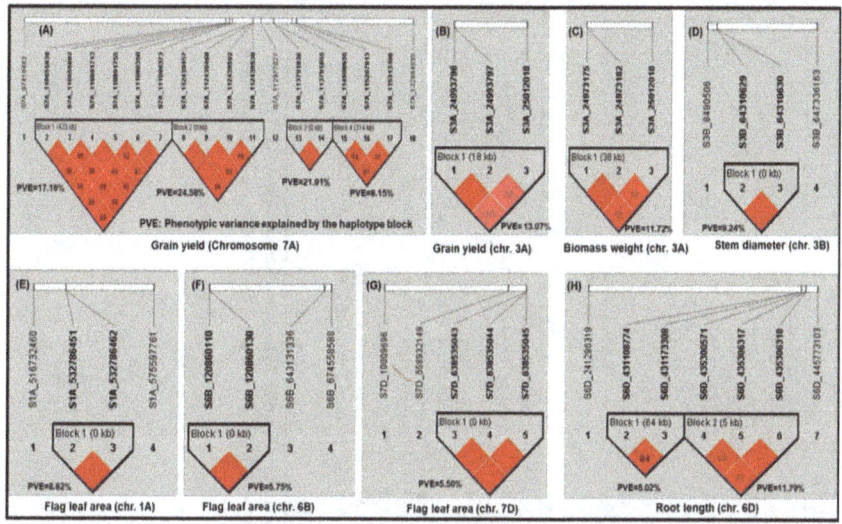

Figure 3. Linkage disequilibrium (LD) values (R^2) and haplotype blocks with significant marker–trait associations (MTAs; $\geq$2) observed (**A**) on chromosome 7A for GY, (**B**) on chromosome 3A for GY, (**C**) on chromosome 3A for BMWT, (**D**) on chromosome 3B for STMDIA, (**E**) on chromosome 1A for FLA, (**F**), on chromosome 6B for FLA, (**G**) on chromosome 7D for FLA, and (**H**) on chromosome 6D for RTLN and phenotypic variance explained (PVE) by each haplotype block. Dark red color represents the strong LD whereas light red color represents the weak LD between pairs of MTAs.

One MTA (S7A_112977027; 112977027 bp) present in-between the second (537 kb away) and third (837 kb) haplotype blocks was within the gene, TraesCS7A01G158200.1, and PVE on GY was 12.8% (Table S5). This gene was annotated as a member of sentrin-specific protease of Ubiquitin-like Protease 1 (*Ulp1*) gene family (Table 3). The *Ulp1* is a small ubiquitin related modifier (SUMO)-specific protease that affects several important biological processes in plants including response to abiotic stress [21]. It has been shown to play a role in drought tolerance in Arabidopsis (*Arabidopsis thaliana*) [36] and rice (*Oryza sativa*) [22,37]. This makes this MTA interesting and a stronger candidate for future functional validation studies.

Table 3. List of significant markers associated with GY and yield-related traits, favorable alleles (underlined), SNP effects, and drought-related putative genes from genome-wide association study of 123 drought stressed synthetic hexaploid wheats grown in 2016 in Konya, Turkey.

Trait	SNP [a]	−log₁₀ (p)	Alleles	SNP Effect	PVE (%)	Gene ID	Annotation
GY	S3A_686179591	4.08	A/G	−14.28	10.7	TraesCS3A01G445100	F-box family protein
GY	S7A_112977027	5.24	A/T	−19.17	12.8	TraesCS7A01G158200.1	Sentrin-specific protease
HI	S3A_593313534	13.56	T/C	0.08	16.0	TraesCS3A01G343700	WRKY transcription factor
HI	S6D_157451060	4.01	A/G	−0.03	6.2	TraesCS6D01G170900.1	Cytochrome P450, putative
HI	S6D_462272376	12.01	G/A	0.02	14.5	TraesCS6D01G382600.1	LOB-domain protein-like
BMWT	S1D_441309135	4.82	C/G	−105.34	14.4	TraesCS1D01G357500.1	Protein DETOXIFICATION
BMWT	S7B_450630784	4.06	A/G	−25.88	10.7	TraesCS7B01G242600.1	F-box family protein
TKW	S2A_47781717	4.52	G/A	0.93	4.2	TraesCS2A01G093500	F-box family protein
TKW	S4A_625466381	4.12	T/G	1.21	15.3	TraesCS4A01G347600	Protein kinase family protein
TKW	S4D_509427923	4.91	C/G	−1.72	10.1	TraesCS4D01G364700	Cytochrome P450 family protein
TKW	S6D_452410667	8.16	A/G	−1.54	17.7	TraesCS6D01G360800	Protein kinase family protein
AWNLN	S4D_461573496	5.71	T/C	0.32	9.0	TraesCS4D01G290700.1	60S ribosomal protein L18a
AWNLN	S5A_562540562	11.67	C/T	−1.71	11.3	TraesCS5A01G361300.1	Guanine nucleotide exchange family protein
FLLN	S1B_667135914	4.38	C/T	−0.16	20.8	TraesCS1B01G447400	Disease resistance protein RPM1
FLW	S6D_16376439	4.85	C/T	−0.02	13.3	TraesCS6D01G040100.1	Mitochondrial transcription termination factor-like
FLA	S1D_278097355	4.74	G/C	0.21	11.5	TraesCS1D01G197200.1	P-loop containing nucleoside triphosphate hydrolases superfamily protein
FLA	S6B_120860110	4.01	G/A	0.17	9.3	TraesCS6B01G125800	Cytochrome P450 family protein, expressed
FLA	S6B_120860130	4.01	A/T	−0.17	9.3	TraesCS6B01G125900	Cytochrome P450 family protein, expressed
STMDIA	S1D_431523575	6.58	A/G	−0.06	10.3	TraesCS1D01G341500	Disease resistance protein (NBS-LRR class) family
STMDIA	S3D_10133372	9.83	G/T	−0.11	8.6	TraesCS3D01G028500.1	Leucine-rich repeat receptor-like protein kinase family protein
STMDIA	S6A_94238211	6.9	T/G	0.06	7.5	TraesCS6A01G122200.1	Protein kinase, putative
RTLN	S5B_669373985	4.62	T/C	0.27	6.9	TraesCS5B01G502200	GRAM domain-containing protein / ABA-responsive
RTLN	S5B_669374027	4.62	T/C	0.27	6.9	TraesCS5B01G502200	GRAM domain-containing protein / ABA-responsive
RTLN	S6D_431108774	4.01	A/G	−0.27	5.8	TraesCS6D01G332800.1	Protein DETOXIFICATION
RTLN	S7A_94404310	4.01	G/A	0.52	7.5	TraesCS7A01G143200.2	Phosphatase 2C family protein

PVE: phenotypic variance explained; GY, grain yield; HI, harvest index; BMWT, biomass weight; TKW, thousand kernel weight; GVWT, grain volume weight; AWNLN, awn length; FLLN, flag leaf length; FLW, flag leaf width; FLA, flag leaf area; STMDIA, stem diameter; RTLN, root length. [a] S+chromosome_chromosome position in bp.

Table 4. List of significant markers associated with GY and yield-related traits, favorable alleles (underlined), SNP effects, and drought-related putative genes obtained from genome-wide association study of 123 drought stressed synthetic hexaploid wheats grown in 2017 in Konya, Turkey.

Trait	SNP [a]	$-\log_{10}(p)$	Alleles	SNP Effect	PVE (%)	Gene-ID	Annotation
GY	S3A_25012018	4.81	A/G	−20.02	12.7	TraesCS3A01G047300	F-box-domain-containing protein
GY	S3D_1203058	4.12	T/G	14.32	12.8	TraesCS3D01G002700	Disease resistance protein RPM1
BMWT	S3A_25012018	6.08	A/G	−59.44	14.4	TraesCS3A01G047300	F-box-domain-containing protein
TKW	S4B_11905230	8.94	C/G	−1.11	3.9	TraesCS4B01G016200.1	LOB domain-containing protein, putative
TKW	S4B_637722874	5.17	T/C	0.86	2.0	TraesCS4B01G344200.1	Zinc finger (C3HC4-type RING finger) family protein
GVWT	S1A_522189599	4.11	A/G	−0.55	2.5	TraesCS1A01G334800	Cytochrome P450
GVWT	S4A_73454791	5.64	C/T	−0.63	5.5	TraesCS4A01G074200.2	Microtubule associated protein family protein, putative, expressed
AWNLN	S5B_43896804	7.31	C/T	−1.13	6.0	TraesCS5B01G038700	F-box family protein
AWNLN	S6B_643657	10.55	C/T	−1.04	5.7	TraesCS6B01G001000	F-box family protein
FLLN	S1B_631203243	5.26	A/G	−0.21	9.8	TraesCS1B01G400600.1	Rp1-like protein
FLLN	S2B_140752747	4.15	G/C	0.29	1.6	TraesCS2B01G167500.1	Cytochrome P450, putative
FLLN	S2D_642055122	4.12	T/C	0.25	5.9	TraesCS2D01G579800	protein kinase family protein
FLLN	S4A_612662321	5.3	C/T	−0.23	5.3	TraesCS4A01G325200	F-box family protein
FLLN	S6D_463762312	5.72	G/A	0.25	6.0	TraesCS6D01G386300	Cytochrome P450, putative
FLW	S1A_516732460	6.99	A/G	−0.03	9.5	TraesCS1A01G326700.1	Citrate-binding protein
FLW	S6B_26200560	7.13	C/A	0.03	12.3	TraesCS6B01G042800	F-box family protein
FLA	S1A_516732460	6.9	A/G	−0.42	8.0	TraesCS1A01G326700.1	Citrate-binding protein
FLA	S2A_764065400	4.18	G/T	−0.19	3.8	TraesCS2A01G563200	NBS-LRR resistance-like protein
STMDIA	S6B_610963076	5.7	T/G	0.06	7.9	TraesCS6B01G346900- TraesCS6B01G347000	NBS-LRR disease resistance protein and F-box protein-like
RTLN	S2D_620326979	4.22	T/C	192.21	9.9	TraesCS2D01G541000.1	Disease resistance protein RPM1

PVE: phenotypic variance explained; GY, grain yield; HI, harvest index; BMWT, biomass weight; TKW, thousand kernel weight; GVWT, grain volume weight; AWNLN, awn length; FLLN, flag leaf length; FLW, flag leaf width; FLA, flag leaf area; STMDIA, stem diameter; RTLN, root length. [a] S+chromosome_chromosome position in bp.

Another major haplotype block (18 kb) of three MTAs was observed on chromosome 3A in 2017 (Figure 3) and the PVE on GY by this haplotype block was 13.1% (Figure 3). The chromosome 3A is known to be an important chromosome that contains useful QTLs for GY and yield-related traits [17–22,30–38] and the haplotype block identified will have a significance in the crop improvement program. All three MTAs present in this haplotype block of chromosome 3A were found in the gene, TraesCS3A01G047300 (Table S5), which was annotated as a member of the F-box gene family (Table 4). These three SNPs were indicated as having a moderate impact on the protein as they resulted in a missense mutation and caused an amino acid change. Such changes may alter the function of the protein [39], which makes this F-box gene a strong candidate for future functional characterization studies under drought tolerance in wheat. The F-box proteins are known to regulate many important biological processes, such as embryogenesis, floral development, plant growth and development, biotic and abiotic stresses, hormonal responses, and senescence [39]. Two other MTAs observed on chromosome 3A and 3D were present within genes (F-box family protein: TraesCS3A01G445100 and disease resistance protein RPM1: TraesCS3D01G002700) that have been previously reported to be involved in drought tolerance [39,40] (Table 4).

The GY haplotype blocks and other MTAs identified in the present study have not been mapped to date and four MTAs were in the genes, of which functional annotations suggested that they are likely involved in drought tolerance. This result implied that haplotype blocks observed on chromosome 3A (3 MTAs) and 7A (16 MTAs), and one MTA on chromosome 3D (1) for GY are novel and may potentially be used in a marker-assisted breeding program, focusing on improving drought tolerance in wheat after validating them in different populations and environments.

2.4.2. Harvest Index

A total of 10 SNPs significantly associated with HI were identified on chromosomes 1D, 2A, 2D, 3A, 3D, 5B, 6B, 6D, and 7B (Figure 2) with PVE ranging from 2.2% to 18.7% (Table S5). Previous studies have reported QTLs/MTAs responsible for HI on chromosomes 2D [27], 3A [27,30], 6B [41], and 7B [30]. To the best of our knowledge, the six MTAs identified for HI on chromosomes 1D, 2A, 3D, 5B, and 6D have not been reported and they are potentially novel MTAs responsible for HI.

Six MTAs for HI detected on chromosomes 2A, 3A, 6B, 6D, and 7B were found in genes and two of these genes have annotations suggesting their involvement in drought stress (Table 3 and Table S5). The two genes are WRKY transcription factor (TraesCS3A01G343700) found on chromosome 3A and cytochrome P450 (TraesCS6D01G170900.1) found on chromosome 6D. The role of WRKY transcription factor is well known in abiotic stresses including drought tolerance [42,43]. The cytochrome P450 genes are a large superfamily of enzymes and are involved in many metabolic pathways including drought tolerance in rice [44] and Arabidopsis [45,46]. The multi-trait marker associated with GY and HI was located on chromosome 5B (S5B_598463062) with PVE ranging from 15.9% to 18.7% (Table S5).

2.4.3. Biomass Weight

The 15 MTAs responsible for BMWT were identified on chromosomes 1D, 2B, 3A, 4A, 6D, and 7B (Figure 2) with PVE ranging from 4.9% to 14.4% (Table S5). Previous studies have reported QTLs/MTAs responsible for BMWT on chromosomes 1D [30,41], 2B [27], 6D [27], and 7B [30]. The four MTAs identified for BMWT on chromosome 3A and 4A have not been reported and they are potentially novel MTAs responsible for BMWT.

A novel haplotype block (38 kb) of three SNPs on chromosome 3A associated with BMWT was identified in 2017 (Figure 3) with PVE by the haplotype block of 11.7%. This MTA (S3A_25012018) was also associated with GY and PVE ranged from 12.7% (GY) to 14.4% (BMWT).

All three MTAs present in this haplotype block were within genes (Table S5) and one of the genes had annotations suggesting its involvement in drought tolerance was an F-box family protein (TraesCS3A01G047300) (Table 4) [39]. Excluding MTAs on haplotype block, eight MTAs for BMWT detected on chromosomes 1D, 2B, 6D, and 7B were found in genes (Table S5) and two of the genes had annotations suggesting its involvement in drought stress (Tables 3 and 4). The genes associated with two MTAs are F-box family protein (TraesCS7B01G242600) [39] and protein DETOXIFICATION containing multi-antimicrobial extrusion protein (MatE) (TraesCS1D01G357500) [23].

2.4.4. Thousand Kernel Weight

A total of 20 MTAs responsible for TKW were detected in 19 different genomic regions on chromosomes 1A, 2A, 2B, 2D, 3A, 3B, 4A, 4B, 4D, 5B, 6D, 7B, and 7D (Figure 2) with PVE ranging from 1.6% to 22.2% (Table S5). Earlier studies have reported QTLs/MTAs for TKW on chromosomes 1A [19,24,29,30], 2A [20], 2B [20,29,30], 2D [19], 3A [24,25,29], 3B [20,26], 4B [5], 5B [24], 7B [30], and 7D [20]. In the present study, only one MTA (S2D_7309581) responsible for TKW was detected on chromosome 2D in both years and assumed to be a stable MTA, because this MTA was detected despite significant genotype x year interaction. The five MTAs identified for TKW on chromosomes 4A, 4D, and 6D have not been previously reported and they are potentially novel MTAs responsible for TKW.

Twelve MTAs responsible for TKW detected on chromosomes 2A, 2B, 3A, 3B, 4A, 4B, 5B, and 6D were found in genes (Table S5) and five of these genes had annotations suggesting their involvement in drought stress (Tables 3 and 4). The genes associated with MTAs involved in drought tolerance are F-box family protein (chromosome 3A; TraesCS3A01G047300) [39], protein kinase family protein (chromosome 4A; TraesCS4A01G347600 and chromosome 6D; TraesCS6D01G360800) [47], cytochrome P450 family protein (chromosome 4D; TraesCS4D01G364700) [44–46], and zinc finger (C3HC4-type RING finger) family protein (chromosome 4B; TraesCS4B01G344200.1) [48–50]. The SNP S4D_509427923 was annotated as a missense variant and thus may have a moderate impact on the protein function (Table S5).

2.4.5. Grain Volume Weight

Thirteen MTAs responsible for GVWT were identified on chromosomes 1A, 2A, 2B, 2D, 3A, 4A, 5A, 6A, and 7A (Figure 2) with PVE ranging from 1.3% to 16.2% (Table S5). Earlier studies have reported QTLs/MTAs for GVWT on chromosomes 1A [33], 2A [33,51], 2B [33,51], 2D [33,51], 5A [33] and 7A [33,51]. Four MTAs identified for GVWT on chromosomes 3A, 4A, and 6A have not been previously reported and they are potentially novel MTAs responsible for GVWT.

Eight MTAs responsible for GVWT detected on chromosomes 1A, 2A, 2B, 4A, 6A, and 7A were found in genes (Table S5) and two of these genes had annotations suggesting their involvement in drought stress (Table 4). The genes associated with two MTAs involved in drought tolerance are cytochrome P450 (TraesCS1A01G334800) [44–46] on chromosome 1A and microtubule-associated protein family protein (TraesCS4A01G074200.2) [52] on chromosome 4A.

2.4.6. Awn Length

Twenty MTAs responsible for AWNLN were observed on chromosomes 1D, 2A, 2B, 3B, 4A, 4B, 4D, 5A, 5B, 5D, 6B, and 7A (Figure 2) with PVE ranging from 1.1% to 20.1% (Table S5). Earlier studies have reported QTLs/MTAs for AWNLN on chromosomes 2A [53,54], 4A [54], 4B [54], 5A [54], and 6B [53,54]. The nine MTAs identified for AWNLN on chromosomes 2B, 3B, 4D, 5B, 5D, and 7A have not been previously reported and they are potentially novel MTAs responsible for AWNLN.

Eleven MTAs responsible for AWNLN detected on chromosomes 1D, 2A, 4D, 5A, 5B, 6B, and 7A were found in genes (Table S5) and four of these genes had annotations suggesting their involvement in drought stress (Tables 3 and 4). The genes associated with four MTAs involved in drought tolerance are 60S ribosomal protein L18a (chromosome 4D; TraesCS4D01G290700) [55], guanine nucleotide exchange family protein (chromosome 5A; TraesCS5A01G361300) [56], and F-box family protein (chromosome 5B; TraesCS5B01G038700 and chromosome 6B; TraesCS6B01G001000) [39]. It has been reported that the putative 60S ribosomal protein L18a is an upregulated transcript in response to drought stress in ears and silks during the flowering stage in maize [55].

2.4.7. Flag Leaf Length

Thirteen MTAs responsible for FLLN were detected on chromosomes 1B, 1D, 2A, 2B, 2D, 4A, 6D, and 7B (Figure 2) with PVE ranging from 1.58% to 32.3% (Table S5). Previous studies have reported QTLs for FLLN on chromosomes 1B [57,58], 2B [57–60], 2D [57,61], 4A [57–59], and 7B [61]. The four MTAs identified for FLLN on chromosomes 1D, 2A, and 6D have not been previously reported and they are potentially novel MTAs responsible for FLLN.

Eleven MTAs responsible for FLLN detected on chromosomes 1B, 1D, 2B, 2D, 4A, 6D, and 7B were found in genes (Table S5) and four of these genes had annotations suggesting their involvement in drought stress (Tables 3 and 4). The genes associated with four MTAs involved in drought stress are F-box family protein (chromosome 4A: TraesCS4A01G325200) [39], cytochrome P450 (chromosome 2B; TraesCS2B01G167500 and chromosome 6D; TraesCS6D01G386300) [44–46], and Rp1-like protein (chromosome 1B; TraesCS1B01G400600) [40].

2.4.8. Flag Leaf Width

Sixteen MTAs responsible for FLW were detected on chromosomes 1A, 1B, 1D, 2B, 2D, 4B, 6B, and 6D (Figure 2) with PVE ranging from 1.6% to 15.2% (Table S5). Previous studies have found QTLs for FLW on chromosomes 1B [57,60,61], 1D [57,59], 2B [57,59], 2D [57,59,61], 4B [59,60], and 6B [57–59]. The two MTAs identified for FLW on chromosomes 1A and 6D have not been previously reported and they are potentially novel MTAs responsible for FLW.

Thirteen MTAs responsible for FLW detected on chromosomes 1A, 1B, 1D, 2D, 4B, 6B, and 6D were found in genes (Table S5) and three of these genes had annotations suggesting their involvement in drought stress (Tables 3 and 4). The genes associated with three MTAs involved in drought stress are citrate-binding protein (chromosome 1A; TraesCS1A01G326700) [62], F-box family protein (chromosome 6B; TraesCS6B01G042800) [39], and mitochondrial transcription termination factor-like (chromosome 6D; TraesCS6D01G040100) [63]. The SNPs S1A_516732460 and S6D_16376439 were annotated as a missense variant and thus may impact the function of the proteins that are annotated as citrate-binding protein and mitochondrial transcription termination factor-like protein, respectively (Table S5).

2.4.9. Flag Leaf Area

Twenty MTAs responsible for FLW were detected on chromosomes 1A, 1B, 1D, 2A, 2D, 4D, 5A, 6B, and 7D (Figure S2) with PVE ranging from 8.1% to 23.1% (Table S5). Previous studies have reported QTLs for FLA on chromosomes 1B [58,59], 1D [57,59,61], 2A [57,59,61], 2D [57,59,61], 4D [58], 5A [57,58,60,61], 6B [58], and 7D [61]. The four MTAs identified for FLA on chromosome 1A have not been previously reported and they are potentially novel MTAs responsible for FLA.

Three novel haplotype blocks were observed for FLA on chromosomes 1A (two MTAs), 6B (two MTAs) and 7D (three MTAs) with PVE by these haplotype block ranging from 5.5% to 8.6% (Figure 3). Fourteen MTAs responsible for FLA detected on chromosomes 1A, 1B, 1D, 2A, 4D, 5A, 6B, and 7D were found in genes (Table S5) and five of these genes had annotations suggesting their involvement in drought stress (Tables 3 and 4). The genes associated with five MTAs involved in drought stress are citrate-binding protein (TraesCS1A01G326700), P-loop containing nucleoside triphosphate hydrolases superfamily protein (TraesCS1D01G197200) [64], cytochrome P450 (TraesCS6B01G125900) [44–46], and NBS-LRR resistance-like protein [40].

The multi-trait marker associated with FLW and FLA was located on chromosome 1A (S1A_516732460) with PVE ranging from 8.0% to 9.5% (Table S4). Another multi-trait marker associated with FLLN and FLA was located on chromosome 2A (S2A_29874199) with PVE ranging from 23.1% to 32.3% (Table S4). The multi-trait MTA indicates that the related candidate gene may affect multiple traits.

2.4.10. Stem Diameter

In the present study, 23 MTAs responsible for STMDIA were identified on chromosomes 1A, 1D, 2B, 2D, 3A, 3B, 3D, 4D, 5A, 5B, 6A, 6B, 6D, 7A, and 7B (Figure 2) with PVE ranging from 2.7% to 28.8% (Table S5). Earlier study has identified one minor QTL (QSd-3B) for STMDIA on chromosome 3B that explained 8.7% of the phenotypic variance [65]. It means that, 19 MTAs detected on chromosomes 1A, 1D, 2B, 2D, 3A, 3D, 4D, 5A, 5B, 6A, 6B, 6D, 7A, and 7B except chromosome 3B in the present study may potentially be a novel MTAs controlling STMDIA under drought stress.

Four MTAs were detected on chromosome 3B and two of them (1 bp apart) were observed in one haplotype block in 2016 with PVE was 9.2% (Figure 3). Fifteen MTAs for STMDIA detected on chromosomes 1D, 2B, 3A, 3B, 6B, 6D, 7A, and 7B were found in genes (Table S5) and four of these genes had annotations suggesting their involvement in drought stress (Tables 3 and 4). The genes associated with four MTAs involved in drought stress are leucine-rich repeat receptor-like protein kinase family protein (TraesCS3D01G028500) [66], protein kinase (TraesCS6A01G122200.1) [47], disease resistance protein (NBS-LRR class) family (TraesCS1D01G341500 and TraesCS6B01G346900) [40], and F-box protein family (TraesCS6B01G347000) [39].

2.4.11. Root Length

RTLN is one of the most important traits under drought stress. We have measured RTLN 3–4 days after anthesis (Zadoks 60 growth stage) under the drought-stressed field condition using WinRhizo® (WinRhizo reg. 2009c, Regent Instruments Inc., Quebec City, QC, Canada). This trait is very unique compared to previous studies where they focused on the roots of seedlings [67–70] rather than direct field-based measurements (labor intensive, time consuming, and expensive). Identification of QTL governing RTLN is very important in wheat, especially for the wheat grown under drought stress. Limited information is available on QTL related to root traits in wheat [67,69–72].

In the present study, 15 MTAs responsible for RTLN were identified on chromosomes 2B, 2D, 3B, 5B, 6A, 6D, and 7A (Figure 2) with PVE ranging from 5.3% to 18.5% (Table S5). Earlier studies have reported QTLs for RTLN on chromosomes 2B [68], 3B [69], 5B [67,69], 6A [68,69], and 6D [67,70] in hexaploid wheat and on chromosomes 2B [71,72], 3B [72], 6A [72], and 7A [72] in tetraploid wheat. The MTA identified for RTLN on chromosome 2D has not been previously reported and it is potentially novel MTAs responsible for RTLN under drought stress. Furthermore, previous studies identified very few QTLs for RTLN on the D-genome of wheat [67,70]. Therefore, the MTAs (eight MTAs) for RTLN detected on the D-genome of SHWs in the present study are potentially novel.

Seven out of eight MTAs responsible for RTLN were present on chromosome 6D. Two haplotype blocks (the haplotype block1 with a size of 64 kb and the haplotype block2 with a size of 5kb) were identified from five out of seven MTAs for RTLN on chromosome 6D with PVE ranging from 5.0% to 11.8% (Figure 3). One SNP (S6D_435300571) present in the haplotype block2 was found in the gene (TraesCS6D01G332800) with PVE of 13.0%. The gene associated with this SNP is protein detoxification gene-containing multi-antimicrobial extrusion protein (MatE) (Table 3) and has been reported to be expressed mainly in the root than shoots under drought stress [73]. For instance, *MatE* family genes such as *HvAACT1* in barley [74] and *TaMate* in wheat [75], encode proteins that are primarily localized to root epidermis cells [74] and required for external resistance [23]. In the present study, this gene was also significantly associated with BMWT on chromosome 1D. This result implied that that this gene plays an important role for RTLN and BMWT in drought-stressed conditions.

Excluding MTAs on haplotype block, eight MTAs for RTLN detected on chromosomes 2D, 3B, 5B, 6A, and 7A were found in genes (Table S5) and four of these genes had annotations suggesting their involvement in drought stress (Tables 3 and 4). The genes associated with four MTAs involved in drought stress are GRAM domain-containing protein/ABA-responsive (TraesCS5B01G502200) [76–81], phosphatase 2C family protein (TraesCS7A01G143200.2) [82], and disease resistance protein RPM1 (TraesCS2D01G541000.1) [40]. The SNPs S7A_94404310 was annotated as a missense variant and thus may have a moderate impact on the protein function (Table S5).

2.5. Potential Candidate Gene Annotations Affecting Yield and Yield-related Traits under Drought Stress

This study identified ~194 MTAs present on different chromosomes and associated with multiple traits. These 62 MTAs were associated with either the same trait in multiple years (MTA stability in different environments) or multiple traits within the same year or across years (suggesting epistasis) (Table S5). Additionally, ~45 of the MTAs were present in genes with annotations relevant to the respective trait under drought stress (Tables 3 and 4). Interestingly, we noticed MTAs associated with the same or related traits were located within genes that had the exact same annotation (Figure 4; and Table S6). For instance, some of the MTAs for GY (2 MTAs), BMWT (2), TKW (1), AWNLN (2), FLLN (1), FLW (1), and STMDIA (1) were located within genes annotated as F-box family protein (Table S6). Similarly, the genes annotated as cytochrome P450 harbored MTAs for HI (1), TKW (1), FLA (2), GVWT (1), and FLLN (2). Additional examples are provided in Figure 4 and Table S6 in Supplementary Materials. This result indicated the likely gene families that are important for GY and yield-related traits under drought stress.

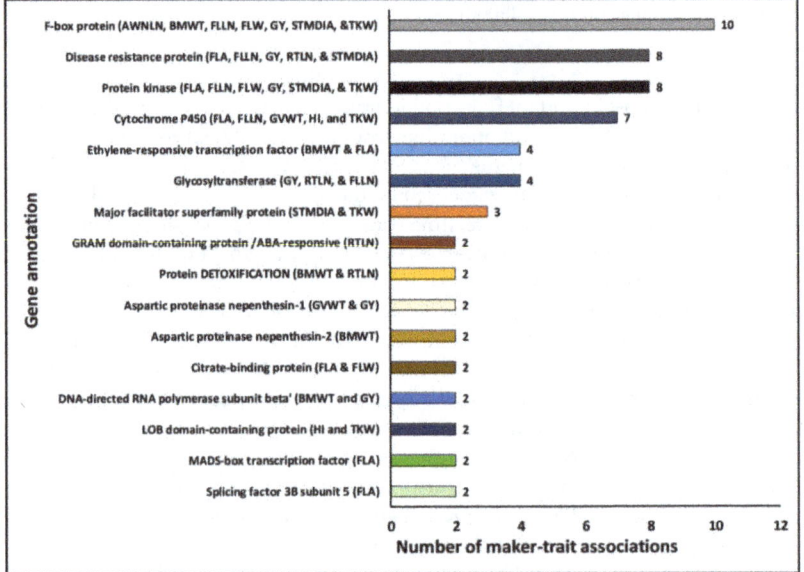

Figure 4. Potential candidate gene functions harboring SNPs affecting yield and yield-related traits under drought stress. The count of marker–trait associations (for either single or multiple traits) located within genes that have the same gene annotation is shown. AWLN, awn length; BMWT, biomass weight; FLA, flag leaf area; FLLN, flag leaf length; FLW, flag leaf width; GVWT, grain volume weight; GY, grain yield; HI, harvest index; RTLN, root length; STMDIA, stem diameter; and TKW, thousand kernel weight.

3. Materials and Methods

3.1. Site Description

A field experiment was conducted during two growing seasons (2016 and 2017) under drought stressed conditions (rainfed) at the research farm located at the Bahri Dagdas International Agricultural Research Institute in Konya, Turkey (37°51'15.894" N, 32°34'3.936" E; Elevation = 1021 m). This site was characterized by a low precipitation (below 300 mm), low humidity, and slightly alkaline clay loam soil [83].

3.2. Plant Materials and Experimental Design

One hundred twenty-three SHWs developed from two introgression programs were used (Table S1). The first group was developed by Kyoto University, Japan from one spring durum ('Langdon') parent crossed with 14 different *Ae. tauschii* accessions resulting in 14 different lines (Table S2). The remaining 109 lines were the second group of synthetics that were developed by International Maize and Wheat Improvement Center (CIMMYT) from the six durum parents crossed with 11 different *Ae. tauschii* accessions mainly from the Caspian Sea Basin area. Initially, 13 crosses among six winter durum wheats were involved in the creation of 13 different winter-type synthetics. However, due to the partial sterility, segregation, and continuous selection in the early generation, 109 lines were selected as unique lines because of their differences in phenotype [14] and their kinship values [11]. The synthetic genotypes used in the present study are unique and have been developed recently and tested for their agronomic traits [14], genetic diversity, and population structure [11]. The detailed information of these SHWs were provided in Bhatta et al. [11].

The experimental design was an augmented design (plot size: 1.2 m × 5 m) with replicated checks ('Gerek' and 'Karahan') in the 2016 growing season and modified alpha-lattice design (plot size: 1.2 m × 5 m) including 123 SHWs and replicated checks ('Gerek' and 'Karahan') with two replications in the 2017 growing season. The SHWs were planted on 20 September in 2015 and harvested on 18 July 2016 for the 2016 growing season, whereas the SHWs were planted on 15 September in 2016 and harvested on 21 July 2017 for the 2017 growing season.

3.3. Trait Measurements

The GY was obtained by harvesting four middle rows of 1.008 m^2 (i.e., 84 m × 120 m) and reported in $g \cdot m^{-2}$. The HI, BMWT, TKW, GVWT, AWNLN, FLLN, FLW, and FLA ($0.8 \times \text{FLLN} \times \text{FLW}$) were measured using previously reported protocols [18,32,59,61]. The STMDIA was measured from five randomly selected plants per plot using a digital Vernier caliper at the second internode from the soil surface at physiological maturity. The RTLN was measured from randomly selected three plants per plot after 3–4 days of flowering (Zadoks 60 growth stage) using WinRhizo software (WinRhizo reg. 2009c, Regent Instruments Inc., Quebec City, QC, Canada).

3.4. Phenotypic Data Analysis

A combined ANOVA was performed using the following model:

$$y_{ijklmn} = \mu + Yr_i + R(Yr)_{ji} + B(R(Yr))_{kji} + C_l + G_{m(kji)} + GXYr_{mi} + e_{ijklmn} \tag{1}$$

where y_{ijklm} is the GY and yield-related trait; μ is overall mean; Yr_i is the effect of ith year; $R(Yr)_{ji}$ is the effect of jth replication within the ith year; $B(R(Yr))_{kji}$ is the effect of kth incomplete block within jth replication of ith environment; C_l is the lth checks; G_{mkji} is the effect of mth genotypes (new variable, where check is coded as 0 and entry is coded is 1 and the genotype is taken as a new variable × entry) within the kth incomplete block of jth replication in the ith year; $GXYr_{mi}$ is the interaction effect of mth genotype and ith year; and e_{ijklmn} is the residual. In the combined ANOVA, year and check were assumed as fixed effects, whereas genotype, genotype × year interaction, replication nested within a year, and incomplete block nested within replications were assumed as random effects.

Individual analyses of variance were performed because most of the traits had highly significant genotype × year interaction and therefore will be discussed hereafter. An augmented design was analyzed using the following model for the estimation of BLUPs in the year 2016:

$$y_{ijkl} = \mu + B_i + C_j + G_{k(i)} + e_{ijkl} \tag{2}$$

where y_{ijkl} is the trait, μ is the overall mean; B_i is the effect of ith incomplete block; C_j is the jth check; $G_{k(i)}$ (new variable, where check is coded as 0 and entry is coded as 1 and genotype is taken as a new variable × entry) is the effect of kth genotypes within the ith block; e_{ijkl} is the residual. In ANOVA calculated for the 2016 datasets, the check was assumed as a fixed effect, whereas genotype and incomplete block were assumed as random effects.

An alpha (α) lattice design with two replications was analyzed using the following model for the estimation of BLUPs in the year 2017:

$$y_{ijkl} = \mu + R_i + B(R)_{ji} + C_k + G_{l(ji)} + e_{ijkl} \tag{3}$$

where y_{ijk} is the trait, μ is the overall mean; R_i is the effect of ith replication; $B(R)_{ji}$ is the effect of jth block within the ith replication; C_k is the kth checks; G_{lji} (new variable, where check is coded as 0 and entry is coded as 1 and the genotype is taken as a new variable × entry) is the effect of kth genotypes within jth incomplete block of ith replication; e_{ijkl} is the residual. In ANOVA calculated for the 2017 datasets, the check was assumed as a fixed effect, whereas genotype, replication, and incomplete block nested within replication were assumed as random effects.

All phenotypic data were analyzed using PROC MIXED in SAS 9.4 (SAS Institute Inc., Cary, NC) [84] using the restricted maximum likelihood (REML) approach unless mentioned otherwise.

Broad-sense heritability for each trait in each year was calculated based on entry mean basis using Equations (4–6) for 2016, 2017, and combine0d experiments, respectively:

$$H^2 = \frac{\sigma^2 g}{\sigma^2 g + \sigma^2 e} \tag{4}$$

$$H^2 = \frac{\sigma^2 g}{\sigma^2 g + \frac{\sigma^2 e}{r}} \tag{5}$$

$$H^2 = \frac{\sigma^2 g}{\sigma^2 g + \frac{\sigma^2 yr}{n} + \frac{\sigma^2 gxyr}{nr}} \tag{6}$$

where σ^2_g, σ^2_{yr}, σ^2_{gxyr}, and σ^2_e are the variance components for genotype, year, genotype × year, and error, respectively, and n and r are the numbers of years and replications, respectively.

Pearson's correlation of GY and yield-related traits was calculated based on BLUPs for each trait in each year using PROC CORR in SAS. The PC biplot analysis (PCA-biplot) was performed based on the correlation matrix to avoid any variation due to the different scales of the measured variables using 'factoextra' package in R software [85].

3.5. Genotyping and SNP Discovery

Genomic DNA was extracted from two to three fresh young leaves of 14-day-old seedlings using BioSprint 96 Plant Kits (Qiagen, Hombrechtikon, Switzerland), as described in Bhatta et al. [11]. The GBS libraries were constructed in 96-plex following digestion with two restriction enzymes, PstI and MspI [86] and pooled libraries were sequenced using Illumina, Inc. (San Diego, CA, USA) next-generation sequencing platforms at the Wheat Genetics Resource Center at Kansas State University (Manhattan, KS, USA). SNP calling was performed using TASSEL v. 5.2.40 GBS v2 Pipeline (available online: https://bitbucket.org/tasseladmin/tassel-5-source/wiki/Tassel5GBSv2Pipeline) [87] with physical alignment to the Chinese Spring genome sequence (RefSeq v1.0) made available by the IWGSC [88] using default settings with the one exception that the number of times for a GBS tag to be present and included for SNP calling was changed from the default value of 1 to 5 to increase the stringency in SNP calling. The identified SNPs with MAF less than 5% and missing data more than 20% were removed from the analysis. All lines had missing data less than 20% and none of them were dropped due to missing percentage-filtering criterion. The GBS-derived SNPs are provided in Table S3 in Supplementary Materials.

3.6. Population Structure and Genome-Wide Association Study Analysis

Population structure of 123 genotypes was assessed using the Bayesian clustering algorithm in the program STRUCTURE v 2.3.4 (available online: https://web.stanford.edu/group/pritchardlab/structure_software/release_versions/v2.3.4/html/structure.html) [89] and principal component (PC) analysis using TASSEL (available online: http://www.maizegenetics.net/tassel) [90], as described in Bhatta et al. [11].

Many GWASs were previously performed using the mixed linear model (MLM), where the population structure (Q) or PC was set as a fixed effect and kinship (K) as a random effect to control false positives [91,92]. However, the MLM may lead to confounding between population structure, kinship, and quantitative trait nucleotides (QTNs) that results in false negatives due to model overfitting [78]. Recently, the multilocus mixed model (MLMM), which tests multiple markers simultaneously by fitting pseudo QTNs, in addition to testing markers in stepwise MLM, has been proposed, which is advantageous over conventional GLM and MLM testing one marker at a time [78]. One of the examples of recently popular GWAS analysis algorithm that is based on MLMM is FarmCPU [78,79]. The FarmCPU uses a fixed effect model (FEM) and a random effect model (REM) iteratively to remove the confounding between testing markers and kinship that results in false negatives, prevents model overfitting, and control false positives simultaneously [78]. Therefore, GWAS was performed on the adjusted BLUPs for each trait in each year to identify SNPs associated with GY and yield-related traits in SHWs using FarmCPU with population structures (Q_1 and Q_2) or first three principal components (PC_1, PC_2, and PC_3) as covariates by looking at the model fit using Quantile-Quantile (Q-Q) plots and FarmCPU-calculated kinship [78] implemented in MVP R software package (available online: https://github.com/XiaoleiLiuBio/MVP). A uniform suggestive genome-wide significance threshold level of p-value = 9.99×10^{-5} ($-\log_{10}p$ = 4.00) was selected for MTAs considering the deviation of the observed test statistics values from the expected test statistics values in the Q-Q plots [28,80] from the two-year results of the present study.

3.7. Haplotype Block Analysis

Haplotype blocks with linkage disequilibrium (LD) values (squared correlation coefficient between locus allele frequency; $R^2 > 0.2$) in adjacent regions (<500 kb) of significant MTAs were visualized and plotted using default parameters (Hardy–Weinberg p-value cut off at 1% and MAF > 0.001) of Haploview software (available online: https://www.broadinstitute.org/haploview/haploview) [81]. PVE by each haplotype block on the trait of interest was calculated using multiple regression analysis that accounted for the population structure by removing the haplotype allele of less than 5% in SAS using PROC REG.

3.8. Putative Candidate Gene Analysis

The genes underlying the MTAs and subsequently their annotations were retrieved using a Perl script and the IWGSC RefSeq v1.0 annotations [88] provided for the Chinese Spring. The underlying genes were further examined for their association with GY and yield-related traits under drought stress using previously published literature. Additionally, the SnpEff program (available online: http://snpeff.sourceforge.net/) was used for SNP annotation and predicting the effects of SNPs on the protein function. The MTAs present within genes or flanked (5 kb) by genes were investigated [93].

4. Conclusions

The present study showed SHWs have large amounts of genetic variation for GY and yield-related traits. The GWAS in 123 SHWs using 35,648 SNPs identified several novel (90 MTAs: 45 MTAs on the A genome, 11 on the B genome, and 34 on the D-genome) genomic regions or haplotype blocks associated with GY and yield-related traits in drought-stressed conditions. Most of the MTAs (120 MTAs) were present in genes ad several of them (45 MTAs) were annotated with functions related to drought stress.

This provided further evidence for the reliability of the MTAs identified. We also identified MTAs on different chromosomes associated with multiple traits but within genes having the same annotation. This resulted in the identification of candidate genes belonging to the same gene family that likely have a major role in affecting GY and yield-related traits under drought stress in SHWs. The large number of MTAs, especially on the D-genome (53 MTAs with 34 MTAs being novel) identified in the present study, demonstrate the potential of SHWs for elucidating the genetic architecture of complex traits and provide an opportunity for further improvement of wheat under rapidly growing drought-stressed environment worldwide.

Supplementary Materials: Supplementary materials can be found at http://www.mdpi.com/1422-0067/19/10/3011/s1.

Author Contributions: Conceptualization, M.B., A.M., and P.S.B.; methodology, M.B.; validation, M.B.; formal analysis, M.B.; investigation, M.B.; resources, M.B., A.M. and P.S.B.; data curation, M.B. and V.B.; writing of the original draft preparation, M.B.; writing of review and editing, M.B., A.M., V.B., and P.S.B.; supervision, A.M. and P.S.B.; project administration, A.M. and P.S.B.

Funding: This research was funded by the Monsanto Beachell-Borlaug International Scholarship Program. This project is a collaborative effort between University of Nebraska–Lincoln, NE, USA, and CIMMYT at Turkey (supported by CRP WHEAT; Ministry of Food, Agriculture and Livestock of Turkey; Bill and Melinda Gates Foundation, and UK Department for International Development (grant number: OPP1133199)), and Omsk State Agrarian University (supported by the Russian Science Foundation Project No. 16-16-10005). Partial funding for P.S. Baenziger is from Hatch project NEB-22-328, AFRI/2011-68002-30029, the USDA National Institute of Food and Agriculture as part of the International Wheat Yield Partnership award number 2017-67007-25939, the CERES Trust Organic Research Initiative, and USDA under Agreement No. 59-0790-4-092 which is a cooperative project with the U.S. Wheat and Barley Scab Initiative. Any opinions, findings, conclusions, or recommendations expressed in this publication are those of the authors and do not necessarily reflect the view of the USDA. Cooperative investigations of the Nebraska Agricultural Research Division, University of Nebraska, and USDA-ARS. The APC was funded by the Hardin Distinguished Graduate Fellowship received by Madhav Bhatta at the University of Nebraska–Lincoln, NE, USA.

Acknowledgments: We would like to thank CIMMYT-Turkey for providing seeds of synthetic hexaploid wheat and Jesse Poland for providing the GBS data. We would like to acknowledge CIMMYT-Turkey staffs including Adem Urglu and Ibrahim Ozturk, Seregey Schepelew from Omsk, Russia, Bahri Dagdas International Agricultural Research Institute staffs including Emel Özer, Fatih Özdemir, and Enes Yakişir.

Conflicts of Interest: The authors declare no conflicts of interest.

Abbreviations

ANOVA	analysis of variance
AWNLN	awn length
BLUP	best linear unbiased predictor
BMWT	biomass weight
FarmCPU	fixed and random model circulating probability unification
FLA	flag leaf area
FLLN	flag leaf length
FLW	flag leaf width
GBS	genotyping-by-sequencing
GVWT	grain volume weight
GWAS	genome wide association study
GY	grain yield
HI	harvest index
IWGC	international wheat genome sequencing consortium
MAF	minor allele frequency
QTN	quantitative trait nucleotide
MLM	mixed linear model
MLMM	multilocus mixed model

MTA	marker trait association
QTL	quantitative trait loci
RTLN	root length
SHW	synthetic hexaploid wheat
SNP	single nucleotide polymorphism
STMDIA	stem diameter
TKW	thousand kernel weight

References

1. Kang, Y.; Khan, S.; Ma, X. Climate change impacts on crop yield, crop water productivity and food security—A review. *Prog. Nat. Sci.* **2009**, *19*, 1665–1674. [CrossRef]
2. Becker, S.R.; Byrne, P.F.; Reid, S.D.; Bauerle, W.L.; McKay, J.K.; Haley, S.D. Root traits contributing to drought tolerance of synthetic hexaploid wheat in a greenhouse study. *Euphytica* **2016**, *207*, 213–224. [CrossRef]
3. Gupta, P.; Balyan, H.; Gahlaut, V.; Gupta, P.K.; Balyan, H.S.; Gahlaut, V. QTL analysis for drought tolerance in wheat: Present status and future possibilities. *Agronomy* **2017**, *7*, 5. [CrossRef]
4. Smith, A.B.; Matthews, J.L. Quantifying uncertainty and variable sensitivity within the US billion-dollar weather and climate disaster cost estimates. *Nat. Hazards* **2015**, *77*, 1829–1851. [CrossRef]
5. Pinto, R.S.; Reynolds, M.P.; Mathews, K.L.; McIntyre, C.L.; Olivares-Villegas, J.-J.; Chapman, S.C. Heat and drought adaptive QTL in a wheat population designed to minimize confounding agronomic effects. *Theor. Appl. Genet.* **2010**, *121*, 1001–1021. [CrossRef] [PubMed]
6. Gummadov, N.; Keser, M.; Akin, B.; Cakmak, M.; Mert, Z.; Taner, S.; Ozturk, I.; Topal, A.; Yazar, S.; Morgounov, A. Genetic gains in wheat in Turkey: Winter wheat for irrigated conditions. *Crop J.* **2015**, *3*, 507–516. [CrossRef]
7. Cavanagh, C.R.; Chao, S.; Wang, S.; Huang, B.E.; Stephen, S.; Kiani, S.; Forrest, K.; Saintenac, C.; Brown-Guedira, G.L.; Akhunova, A.; et al. Genome-wide comparative diversity uncovers multiple targets of selection for improvement in hexaploid wheat landraces and cultivars. *Proc. Natl. Acad. Sci. USA* **2013**, *110*, 8057–8062. [CrossRef] [PubMed]
8. Cox, T. Deepening the wheat gene pool. *J. Crop Prod.* **1997**, *1*, 145–168. [CrossRef]
9. Dreisigacker, S.; Kishii, M.; Lage, J.; Warburton, M. Use of synthetic hexaploid wheat to increase diversity for CIMMYT bread wheat improvement. *Aust. J. Agric. Res.* **2008**, *59*, 413–420. [CrossRef]
10. Ogbonnaya, F. C.; Abdalla, O.; Mujeeb-Kazi, A.; Kazi, A.G.; Xu, S.S.; Gosman, N.; Lagudah, E.S.; Bonnett, D.; Sorrells, M.E.; Tsujimoto, H. Synthetic hexaploids: Harnessing species of the primary gene pool for wheat improvement. *Plant Breed. Rev.* **2013**, *37*, 35–122. [CrossRef]
11. Bhatta, M.; Morgounov, A.; Belamkar, V.; Poland, J.; Baenziger, P.S. Unlocking the novel genetic diversity and population structure of synthetic hexaploid wheat. *BMC Genom.* **2018**, *19*, 591. [CrossRef] [PubMed]
12. Zegeye, H.; Rasheed, A.; Makdis, F.; Badebo, A.; Ogbonnaya, F.C. Genome-wide association mapping for seedling and adult plant resistance to stripe rust in synthetic hexaploid wheat. *PLoS ONE* **2014**, *9*, e105593. [CrossRef] [PubMed]
13. Das, M.K.; Bai, G.; Mujeeb-Kazi, A.; Rajaram, S. Genetic diversity among synthetic hexaploid wheat accessions (*Triticum aestivum*) with resistance to several fungal diseases. *Genet. Resour. Crop Evol.* **2016**, *63*, 1285–1296. [CrossRef]
14. Morgounov, A.; Abugalieva, A.; Akan, K.; Akın, B.; Baenziger, S.; Bhatta, M.; Dababat, A.A.; Demir, L.; Dutbayev, Y.; El Bouhssini, M.; et al. High-yielding winter synthetic hexaploid wheats resistant to multiple diseases and pests. *Plant Genet. Resour.* **2018**, *16*, 273–278. [CrossRef]
15. Ogbonnaya, F.C.; Imtiaz, M.; Bariana, H.S.; McLean, M.; Shankar, M.M.; Hollaway, G.J.; Trethowan, R.M.; Lagudah, E.S.; Van Ginkel, M. Mining synthetic hexaploids for multiple disease resistance to improve bread wheat. *Aust. J. Agric Res.* **2008**, *59*, 421–431. [CrossRef]
16. Jighly, A.; Alagu, M.; Makdis, F.; Singh, M.; Singh, S.; Emebiri, L.C.; Ogbonnaya, F.C. Genomic regions conferring resistance to multiple fungal pathogens in synthetic hexaploid wheat. *Mol. Breed.* **2016**, *36*, 1–19. [CrossRef]
17. Sehgal, D.; Autrique, E.; Singh, R.; Ellis, M.; Singh, S.; Dreisigacker, S. Identification of genomic regions for grain yield and yield stability and their epistatic interactions. *Sci. Rep.* **2017**, *7*. [CrossRef] [PubMed]

18. Bhatta, M.; Eskridge, K.M.; Rose, D.J.; Santra, D.K.; Baenziger, P.S.; Regassa, T. Seeding Rate, genotype, and topdressed nitrogen effects on yield and agronomic characteristics of winter wheat. *Crop Sci.* **2017**, *57*, 951–963. [CrossRef]

19. Ogbonnaya, F.C.; Rasheed, A.; Okechukwu, E.C.; Jighly, A.; Makdis, F.; Wuletaw, T.; Hagras, A.; Uguru, M.I.; Agbo, C.U. Genome-wide association study for agronomic and physiological traits in spring wheat evaluated in a range of heat prone environments. *Theor. Appl. Genet.* **2017**, *130*, 1819–1835. [CrossRef] [PubMed]

20. Sukumaran, S.; Lopes, M.; Dreisigacker, S.; Reynolds, M. Genetic analysis of multi-environmental spring wheat trials identifies genomic regions for locus-specific trade-offs for grain weight and grain number. *Theor. Appl. Genet.* **2018**, *131*, 985–998. [CrossRef] [PubMed]

21. Murtas, G. A Nuclear Protease Required for Flowering-Time Regulation in Arabidopsis Reduces the Abundance of SMALL UBIQUITIN-RELATED MODIFIER Conjugates. *Plant Cell* **2003**, *15*, 2308–2319. [CrossRef] [PubMed]

22. Choudhary, M.K.; Basu, D.; Datta, A.; Chakraborty, N.; Chakraborty, S. Dehydration-responsive nuclear proteome of rice (*Oryza sativa* L.) illustrates protein network, novel regulators of cellular adaptation, and evolutionary perspective. *Mol. Cell. Proteomics* **2009**, *8*, 1579–1598. [CrossRef] [PubMed]

23. Yokosho, K.; Yamaji, N.; Fujii-Kashino, M.; Ma, J.F. Functional Analysis of a MATE gene OsFRDL2 revealed its involvement in Al-induced secretion of citrate, but a lower contribution to Al tolerance in rice. *Plant Cell Physiol.* **2016**, *57*, 976–985. [CrossRef] [PubMed]

24. Sun, C.; Zhang, F.; Yan, X.; Zhang, X.; Dong, Z.; Cui, D.; Chen, F. Genome-wide association study for 13 agronomic traits reveals distribution of superior alleles in bread wheat from the Yellow and Huai Valley of China. *Plant Biotechnol. J.* **2017**, *15*, 953–969. [CrossRef] [PubMed]

25. Zanke, C.D.; Ling, J.; Plieske, J.; Kollers, S.; Ebmeyer, E.; Korzun, V.; Argillier, O.; Stiewe, G.; Hinze, M.; Neumann, K.; et al. Whole genome association mapping of plant height in winter wheat (*Triticum aestivum* L.). *PLoS ONE* **2014**, *9*, e113287. [CrossRef] [PubMed]

26. Golabadi, M.; Arzani, A.; Mirmohammadi Maibody, S.A.M.; Sayed Tabatabaei, B.E.; Mohammadi, S.A. Identification of microsatellite markers linked with yield components under drought stress at terminal growth stages in durum wheat. *Euphytica* **2011**, *177*, 207–221. [CrossRef]

27. Kumar, N.; Kulwal, P.L.; Balyan, H.S.; Gupta, P.K. QTL mapping for yield and yield contributing traits in two mapping populations of bread wheat. *Mol. Breed.* **2007**, *19*, 163–177. [CrossRef]

28. Sukumaran, S.; Dreisigacker, S.; Lopes, M.; Chavez, P.; Reynolds, M.P. Genome-wide association study for grain yield and related traits in an elite spring wheat population grown in temperate irrigated environments. *Theor. Appl. Genet.* **2015**, *128*, 353–363. [CrossRef] [PubMed]

29. Lozada, D.N.; Mason, R.E.; Babar, M.A.; Carver, B.F.; Guedira, G.B.; Merrill, K.; Arguello, M.N.; Acuna, A.; Vieira, L.; Holder, A.; et al. Association mapping reveals loci associated with multiple traits that affect grain yield and adaptation in soft winter wheat. *Euphytica* **2017**, *213*. [CrossRef]

30. Neumann, K.; Kobiljski, B.; Denčić, S.; Varshney, R.K.; Börner, A. Genome-wide association mapping: A case study in bread wheat (*Triticum aestivum* L.). *Mol. Breed.* **2011**, *27*, 37–58. [CrossRef]

31. Edae, E.A.; Byrne, P.F.; Haley, S.D.; Lopes, M.S.; Reynolds, M.P. Genome-wide association mapping of yield and yield components of spring wheat under contrasting moisture regimes. *Theor. Appl. Genet.* **2014**, *127*, 791–807. [CrossRef] [PubMed]

32. Bhatta, M.; Regassa, T.; Rose, D.J.; Baenziger, P.S.; Eskridge, K.M.; Santra, D.K.; Poudel, R. Genotype, environment, seeding rate, and top-dressed nitrogen effects on end-use quality of modern Nebraska winter wheat. *J. Sci. Food Agric.* **2017**, *97*, 5311–5318. [CrossRef] [PubMed]

33. Bordes, J.; Goudemand, E.; Duchalais, L.; Chevarin, L.; Oury, F.X.; Heumez, E.; Lapierre, A.; Perretant, M.R.; Rolland, B.; Beghin, D.; et al. Genome-wide association mapping of three important traits using bread wheat elite breeding populations. *Mol. Breed.* **2014**, *33*, 755–768. [CrossRef]

34. Hoffstetter, A.; Cabrera, A.; Sneller, C. Identifying quantitative trait loci for economic traits in an elite soft red winter wheat population. *Crop Sci.* **2016**, *56*, 547. [CrossRef]

35. Wang, S.-X.; Zhu, Y.-L.; Zhang, D.-X.; Shao, H.; Liu, P.; Hu, J.-B.; Zhang, H.; Zhang, H.-P.; Chang, C.; Lu, J.; et al. Genome-wide association study for grain yield and related traits in elite wheat varieties and advanced lines using SNP markers. *PLOS ONE* **2017**, *12*, e0188662. [CrossRef] [PubMed]

36. Catala, R.; Ouyang, J.; Abreu, I.A.; Hu, Y.; Seo, H.; Zhang, X.; Chua, N.-H. The Arabidopsis E3 SUMO Ligase SIZ1 Regulates Plant Growth and Drought Responses. *Plant Cell* **2007**, *19*, 2952–2966. [CrossRef] [PubMed]

37. Park, H.J.; Kim, W.-Y.; Park, H.C.; Lee, S.Y.; Bohnert, H.J.; Yun, D.-J. SUMO and SUMOylation in plants. *Mol. Cells* **2011**, *32*, 305–316. [CrossRef] [PubMed]

38. Ali, M.L.; Baenziger, P.S.; Ajlouni, Z.A.; Campbell, B.T.; Gill, K.S.; Eskridge, K.M.; Mujeeb-Kazi, A.; Dweikat, I. Mapping QTL for agronomic traits on wheat chromosome 3a and a comparison of recombinant inbred chromosome line populations. *Crop Sci.* **2011**, *51*, 553–566. [CrossRef]

39. Lechner, E.; Achard, P.; Vansiri, A.; Potuschak, T.; Genschik, P. F-box proteins everywhere. *Curr. Opin. Plant Biol.* **2006**, *9*, 631–638. [CrossRef] [PubMed]

40. Lee, H.-A.; Yeom, S.-I. Plant NB-LRR proteins: Tightly regulated sensors in a complex manner. *Brief. Funct. Genomics* **2015**, *14*, 233–242. [CrossRef] [PubMed]

41. Ain, Q.; Rasheed, A.; Anwar, A.; Mahmood, T.; Imtiaz, M.; Mahmood, T.; Xia, X.; He, Z.; Quraishi, U.M. Genome-wide association for grain yield under rainfed conditions in historical wheat cultivars from Pakistan. *Front. Plant Sci.* **2015**, *6*. [CrossRef] [PubMed]

42. He, G.-H.; Xu, J.-Y.; Wang, Y.-X.; Liu, J.-M.; Li, P.-S.; Chen, M.; Ma, Y.-Z.; Xu, Z.-S. Drought-responsive WRKY transcription factor genes TaWRKY1 and TaWRKY33 from wheat confer drought and/or heat resistance in Arabidopsis. *BMC Plant Biol.* **2016**, *16*, 116. [CrossRef] [PubMed]

43. Ning, P.; Liu, C.; Kang, J.; Lv, J. Genome-wide analysis of WRKY transcription factors in wheat (*Triticum aestivum* L.) and differential expression under water deficit condition. *PeerJ* **2017**, *5*, e3232. [CrossRef] [PubMed]

44. Tamiru, M.; Undan, J. R.; Takagi, H.; Abe, A.; Yoshida, K.; Undan, J. Q.; Natsume, S.; Uemura, A.; Saitoh, H.; Matsumura, H.; Urasaki, N.; Yokota, T.; Terauchi, R. A cytochrome P450, OsDSS1, is involved in growth and drought stress responses in rice (*Oryza sativa* L.). *Plant Mol. Biol.* **2015**, *88*, 85–99. [CrossRef] [PubMed]

45. Seki, M.; Narusaka, M.; Ishida, J.; Nanjo, T.; Fujita, M.; Oono, Y.; Kamiya, A.; Nakajima, M.; Enju, A.; Sakurai, T.; et al. Monitoring the expression profiles of 7000 Arabidopsis genes under drought, cold and high-salinity stresses using a full-length cDNA microarray: Expression profiling under abiotic stresses. *Plant J.* **2002**, *31*, 279–292. [CrossRef] [PubMed]

46. Kushiro, T.; Okamoto, M.; Nakabayashi, K.; Yamagishi, K.; Kitamura, S.; Asami, T.; Hirai, N.; Koshiba, T.; Kamiya, Y.; Nambara, E. The Arabidopsis cytochrome P450 CYP707A encodes ABA 8′-hydroxylases: key enzymes in ABA catabolism. *EMBO J.* **2004**, *23*, 1647–1656. [CrossRef] [PubMed]

47. Yan, J.; Su, P.; Wei, Z.; Nevo, E.; Kong, L. Genome-wide identification, classification, evolutionary analysis and gene expression patterns of the protein kinase gene family in wheat and Aegilops tauschii. *Plant Mol. Biol.* **2017**, *95*, 227–242. [CrossRef] [PubMed]

48. Sakamoto, H.; Maruyama, K.; Sakuma, Y.; Meshi, T.; Iwabuchi, M.; Shinozaki, K.; Yamaguchi-Shinozaki, K. Arabidopsis Cys2/His2-Type zinc-finger proteins function as transcription repressors under drought, cold, and high-salinity stress conditions. *Plant Physiol.* **2004**, *136*, 2734–2746. [CrossRef] [PubMed]

49. Ciftci-Yilmaz, S.; Morsy, M.R.; Song, L.; Coutu, A.; Krizek, B.A.; Lewis, M.W.; Warren, D.; Cushman, J.; Connolly, E.L.; Mittler, R. The EAR-motif of the Cys2/His2-type zinc finger protein Zat7 plays a key role in the defense response of Arabidopsis to salinity stress. *J. Biol. Chem.* **2007**, *282*, 9260–9268. [CrossRef] [PubMed]

50. Ma, K.; Xiao, J.; Li, X.; Zhang, Q.; Lian, X. Sequence and expression analysis of the C3HC4-type RING finger gene family in rice. *Gene* **2009**, *444*, 33–45. [CrossRef] [PubMed]

51. Cabral, A.L.; Jordan, M.C.; Larson, G.; Somers, D.J.; Humphreys, D.G.; McCartney, C.A. Relationship between QTL for grain shape, grain weight, test weight, milling yield, and plant height in the spring wheat cross RL4452/'AC Domain'. *PLoS ONE* **2018**, *13*, e0190681. [CrossRef] [PubMed]

52. Zhou, J.; Wang, X.; Jiao, Y.; Qin, Y.; Liu, X.; He, K.; Chen, C.; Ma, L.; Wang, J.; Xiong, L.; et al. Global genome expression analysis of rice in response to drought and high-salinity stresses in shoot, flag leaf, and panicle. *Plant Mol. Biol.* **2007**, *63*, 591–608. [CrossRef] [PubMed]

53. Sourdille, P.; Cadalen, T.; Gay, G.; Gill, B.; Bernard, M. Molecular and physical mapping of genes affecting awning in wheat. *Plant Breed.* **2002**, *121*, 320–324. [CrossRef]

54. Echeverry-Solarte, M.; Kumar, A.; Kianian, S.; Mantovani, E.E.; McClean, P.E.; Deckard, E.L.; Elias, E.; Simsek, S.; Alamri, M.S.; Hegstad, J.; et al. Genome-wide mapping of spike-related and agronomic traits in a common wheat population derived from a supernumerary spikelet parent and an elite parent. *Plant Genome* **2015**, *8*, 0. [CrossRef]

55. Li, H.Y.; Wang, T.Y.; Shi, Y.S.; Fu, J.J.; Song, Y.C.; Wang, G.Y.; Li, Y. Isolation and characterization of induced genes under drought stress at the flowering stage in maize (*Zea mays*): Full Length Research Paper. *DNA Seq* **2007**, *18*, 445–460. [CrossRef] [PubMed]

56. Pandey, S.; Assmann, S.M. The Arabidopsis putative g protein–coupled receptor GCR1 interacts with the G protein α subunit GPA1 and regulates abscisic acid signaling. *Plant Cell* **2004**, *16*, 1616–1632. [CrossRef] [PubMed]

57. Wu, Q.; Chen, Y.; Fu, L.; Zhou, S.; Chen, J.; Zhao, X.; Zhang, D.; Ouyang, S.; Wang, Z.; Li, D.; et al. QTL mapping of flag leaf traits in common wheat using an integrated high-density SSR and SNP genetic linkage map. *Euphytica* **2016**, *208*, 337–351. [CrossRef]

58. Yang, D.; Liu, Y.; Cheng, H.; Chang, L.; Chen, J.; Chai, S.; Li, M. Genetic dissection of flag leaf morphology in wheat (*Triticum aestivum* L.) under diverse water regimes. *BMC Genet* **2016**, *17*. [CrossRef] [PubMed]

59. Fan, X.; Cui, F.; Zhao, C.; Zhang, W.; Yang, L.; Zhao, X.; Han, J.; Su, Q.; Ji, J.; Zhao, Z.; et al. QTLs for flag leaf size and their influence on yield-related traits in wheat (*Triticum aestivum* L.). *Mol. Breed.* **2015**, *35*, 24. [CrossRef]

60. Liu, K.; Xu, H.; Liu, G.; Guan, P.; Zhou, X.; Peng, H.; Yao, Y.; Ni, Z.; Sun, Q.; Du, J. QTL mapping of flag leaf-related traits in wheat (*Triticum aestivum* L.). *Theor. Appl. Genet.* **2018**, *131*, 839–849. [CrossRef] [PubMed]

61. Hussain, W.; Baenziger, P.S.; Belamkar, V.; Guttieri, M.J.; Venegas, J.P.; Easterly, A.; Sallam, A.; Poland, J. Genotyping-by-sequencing derived high-density linkage map and its application to QTL mapping of flag leaf traits in bread wheat. *Sci. Rep.* **2017**, *7*, 16394. [CrossRef] [PubMed]

62. Ros, B.; Thümmler, F.; Wenzel, G. Analysis of differentially expressed genes in a susceptible and moderately resistant potato cultivar upon Phytophthora infestans infection. *Mol. Plant Pathol.* **2004**, *5*, 191–201. [CrossRef] [PubMed]

63. Zhao, Y.; Cai, M.; Zhang, X.; Li, Y.; Zhang, J.; Zhao, H.; Kong, F.; Zheng, Y.; Qiu, F. Genome-Wide identification, evolution and expression analysis of mTERF gene family in maize. *PLoS ONE* **2014**, *9*, e94126. [CrossRef] [PubMed]

64. Cotsaftis, O.; Plett, D.; Johnson, A.A.T.; Walia, H.; Wilson, C.; Ismail, A.M.; Close, T.J.; Tester, M.; Baumann, U. Root-specific transcript profiling of contrasting rice genotypes in response to salinity stress. *Mol. Plant* **2011**, *4*, 25–41. [CrossRef] [PubMed]

65. Hai, L.; Guo, H.; Xiao, S.; Jiang, G.; Zhang, X.; Yan, C.; Xin, Z.; Jia, J. Quantitative trait loci (QTL) of stem strength and related traits in a doubled-haploid population of wheat (*Triticum aestivum* L.). *Euphytica* **2005**, *141*, 1–9. [CrossRef]

66. Shumayla; Sharma, S.; Kumar, R.; Mendu, V.; Singh, K.; Upadhyay, S.K. Genomic dissection and expression profiling revealed functional divergence in Triticum aestivum leucine rich repeat receptor like kinases (TaLRRKs). *Front. Plant Sci.* **2016**, *7*. [CrossRef]

67. Landjeva, S.; Neumann, K.; Lohwasser, U.; Börner, A. Molecular mapping of genomic regions associated with wheat seedling growth under osmotic stress. *Biol. Plant.* **2008**, *52*, 259–266. [CrossRef]

68. Ren, Y.; He, X.; Liu, D.; Li, J.; Zhao, X.; Li, B.; Tong, Y.; Zhang, A.; Li, Z. Major quantitative trait loci for seminal root morphology of wheat seedlings. *Mol. Breed.* **2012**.

69. Bai, C.; Liang, Y.; Hawkesford, M.J. Identification of QTLs associated with seedling root traits and their correlation with plant height in wheat. *J. Exp. Bot.* **2013**, *64*, 1745–1753. [CrossRef] [PubMed]

70. Atkinson, J.A.; Wingen, L.U.; Griffiths, M.; Pound, M.P.; Gaju, O.; Foulkes, M.J.; Le Gouis, J.; Griffiths, S.; Bennett, M.J.; King, J.; et al. Phenotyping pipeline reveals major seedling root growth QTL in hexaploid wheat. *J. Exp. Bot.* **2015**, *66*, 2283–2292. [CrossRef] [PubMed]

71. Canè, M.A.; Maccaferri, M.; Nazemi, G.; Salvi, S.; Francia, R.; Colalongo, C.; Tuberosa, R. Association mapping for root architectural traits in durum wheat seedlings as related to agronomic performance. *Mol. Breed.* **2014**, *34*, 1629–1645. [CrossRef] [PubMed]

72. Maccaferri, M.; El-Feki, W.; Nazemi, G.; Salvi, S.; Canè, M.A.; Colalongo, M.C.; Stefanelli, S.; Tuberosa, R. Prioritizing quantitative trait loci for root system architecture in tetraploid wheat. *J. Exp. Bot.* **2016**, *67*, 1161–1178. [CrossRef] [PubMed]

73. Oono, Y.; Yazawa, T.; Kanamori, H.; Sasaki, H.; Mori, S.; Handa, H.; Matsumoto, T. Genome-wide transcriptome analysis of cadmium stress in rice. *BioMed Res. Int.* **2016**, *2016*. [CrossRef] [PubMed]

74. Furukawa, J.; Yamaji, N.; Wang, H.; Mitani, N.; Murata, Y.; Sato, K.; Katsuhara, M.; Takeda, K.; Ma, J.F. An aluminum-activated citrate transporter in barley. *Plant Cell Physiol.* **2007**, *48*, 1081–1091. [CrossRef] [PubMed]

75. Ryan, P.R.; Raman, H.; Gupta, S.; Horst, W.J.; Delhaize, E. A second mechanism for aluminum resistance in wheat relies on the constitutive efflux of citrate from roots. *Plant Physiol.* **2009**, *149*, 340–351. [CrossRef] [PubMed]

76. Baron, K.N.; Schroeder, D.F.; Stasolla, C. GEm-related 5 (GER5), an ABA and stress-responsive GRAM domain protein regulating seed development and inflorescence architecture. *Plant Sci.* **2014**, *223*, 153–166. [CrossRef] [PubMed]

77. Liu, L.; Li, N.; Yao, C.; Meng, S.; Song, C. Functional analysis of the ABA-responsive protein family in ABA and stress signal transduction in *Arabidopsis*. *Chin. Sci. Bull.* **2013**, *58*, 3721–3730. [CrossRef]

78. Liu, X.; Huang, M.; Fan, B.; Buckler, E.S.; Zhang, Z. Iterative usage of fixed and random effect models for powerful and efficient genome-wide association studies. *PLoS Genet.* **2016**, *12*, e1005767. [CrossRef] [PubMed]

79. Arora, S.; Singh, N.; Kaur, S.; Bains, N.S.; Uauy, C.; Poland, J.; Chhuneja, P. Genome-wide association study of grain architecture in wild wheat Aegilops tauschii. *Front. Plant Sci.* **2017**, *8*. [CrossRef] [PubMed]

80. Sukumaran, S.; Xiang, W.; Bean, S.R.; Pedersen, J.F.; Kresovich, S.; Tuinstra, M.R.; Tesso, T.T.; Hamblin, M.T.; Yu, J. Association mapping for grain quality in a diverse sorghum collection. *Plant Genome J.* **2012**, *5*, 126. [CrossRef]

81. Barrett, J.C.; Fry, B.; Maller, J.; Daly, M.J. Haploview: analysis and visualization of LD and haplotype maps. *Bioinformatics* **2005**, *21*, 263–265. [CrossRef] [PubMed]

82. Singh, A.; Giri, J.; Kapoor, S.; Tyagi, A.K.; Pandey, G.K. Protein phosphatase complement in rice: Genome-wide identification and transcriptional analysis under abiotic stress conditions and reproductive development. *BMC Genom.* **2010**, *11*, 435. [CrossRef] [PubMed]

83. Evaluation of deficit irrigation for efficient sheep production from permanent sown pastures in a dry continental climate. *Agric. Water Manag.* **2013**, *119*, 135–143. [CrossRef]

84. SAS 9.4 Product Documentation. Available online: https://support.sas.com/documentation/94/ (accessed on 16 August 2018).

85. Kassambara, A.; Mundt, F. Factoextra: Extract and visualize the results of multivariate data analyses. 2017. Available online: http://www.sthda.com/english/rpkgs/factoextra/ (accessed on 16 August 2018).

86. Poland, J.A.; Brown, P.J.; Sorrells, M.E.; Jannink, J. Development of high-density genetic maps for barley and wheat using a novel two-enzyme genotyping-by-sequencing approach. *PLoS ONE* **2012**, *7*. [CrossRef] [PubMed]

87. Glaubitz, J.C.; Casstevens, T.M.; Lu, F.; Harriman, J.; Elshire, R.J.; Sun, Q.; Buckler, E.S. TASSEL-GBS: A high capacity genotyping by sequencing analysis pipeline. *PLoS ONE* **2014**, *9*. [CrossRef] [PubMed]

88. International Wheat Genome Sequencing Consortium (IWGSC). Shifting the limits in wheat research and breeding using a fully annotated reference genome. *Science* **2018**, *361*, 7191. [CrossRef] [PubMed]

89. Pritchard, J.K.; Stephens, M.; Donnelly, P. Inference of Population Structure Using Multilocus Genotype Data. *Genetics* **2000**, *155*, 945–959. [PubMed]

90. Bradbury, P.J.; Zhang, Z.; Kroon, D.E.; Casstevens, T.M.; Ramdoss, Y.; Buckler, E.S. TASSEL: Software for association mapping of complex traits in diverse samples. *Bioinformatics* **2007**, *23*, 2633–2635. [CrossRef] [PubMed]

91. Yu, J.; Pressoir, G.; Briggs, W.H.; Vroh Bi, I.; Yamasaki, M.; Doebley, J.F.; McMullen, M.D.; Gaut, B.S.; Nielsen, D.M.; Holland, J.B.; et al. A unified mixed-model method for association mapping that accounts for multiple levels of relatedness. *Nat. Genet.* **2006**, *38*, 203–208. [CrossRef] [PubMed]

92. Zhang, Z.; Ersoz, E.; Lai, C.-Q.; Todhunter, R.J.; Tiwari, H.K.; Gore, M.A.; Bradbury, P.J.; Yu, J.; Arnett, D.K.; Ordovas, J.M.; et al. Mixed linear model approach adapted for genome-wide association studies. *Nature Genetics* **2010**, *42*, 355–360. [CrossRef] [PubMed]
93. Cingolani, P.; Platts, A.; Wang, L.L.; Coon, M.; Nguyen, T.; Wang, L.; Land, S.J.; Lu, X.; Ruden, D.M. A program for annotating and predicting the effects of single nucleotide polymorphisms, SnpEff. *Fly (Austin)* **2012**, *6*, 80–92. [CrossRef] [PubMed]

International Journal of
Molecular Sciences

Article

Transcriptome and Hormone Comparison of Three Cytoplasmic Male Sterile Systems in *Brassica napus*

Bingli Ding, Mengyu Hao, Desheng Mei, Qamar U Zaman, Shifei Sang, Hui Wang, Wenxiang Wang, Li Fu, Hongtao Cheng * and Qiong Hu *

Key Laboratory for Biological Sciences and Genetic Improvement of Oil Crops, Ministry of Agriculture, Oil Crops Research Institute, Chinese Academy of Agricultural Sciences, Wuhan 430062, China; dingbl91@163.com (B.D.); haomengyu@caas.cn (M.H.); deshengmei@caas.cn (D.M.); qamaruzamanch@gmail.com (Q.U.Z.); 15652142445@163.com (S.S.); wanghui06@caas.cn (H.W.); wangwenxiang@caas.cn (W.W.); fuli@caas.cn (L.F.)
* Correspondence: chenghongtao@caas.cn (H.C.); huqiong01@caas.cn (Q.H.); Tel: +86-27-8671-7152 (Q.H.)

Received: 1 November 2018; Accepted: 11 December 2018; Published: 12 December 2018

Abstract: The interaction between plant mitochondria and the nucleus markedly influences stress responses and morphological features, including growth and development. An important example of this interaction is cytoplasmic male sterility (CMS), which results in plants producing non-functional pollen. In current research work, we compared the phenotypic differences in floral buds of different *Brassica napus* CMS (*Polima, Ogura, Nsa*) lines with their corresponding maintainer lines. By comparing anther developmental stages between CMS and maintainer lines, we identified that in the *Nsa* CMS line abnormality occurred at the tetrad stage of pollen development. Phytohormone assays demonstrated that IAA content decreased in sterile lines as compared to maintainer lines, while the total hormone content was increased two-fold in the S_2 stage compared with the S_1 stage. ABA content was higher in the S_1 stage and exhibited a two-fold decreasing trend in S_2 stage. Sterile lines however, had increased ABA content at both stages compared with the corresponding maintainer lines. Through transcriptome sequencing, we compared differentially expressed unigenes in sterile and maintainer lines at both (S_1 and S_2) developmental stages. We also explored the co-expressed genes of the three sterile lines in the two stages and classified these genes by gene function. By analyzing transcriptome data and validating by RT-PCR, it was shown that some transcription factors (TFs) and hormone-related genes were weakly or not expressed in the sterile lines. This research work provides preliminary identification of the pollen abortion stage in *Nsa* CMS line. Our focus on genes specifically expressed in sterile lines may be useful to understand the regulation of CMS.

Keywords: cytoplasmic male sterility (CMS); phytohormones; differentially expressed genes; pollen development; *Brassica napus*

1. Introduction

Oilseed rape is one of most important oil crops worldwide, producing food, biofuel, and industrial compounds, including lubricants and surfactants. Hybrid breeding is a key technique to enhance crop production [1–3], in which cytoplasmic male sterility (CMS) plays an important role in seed production [4]. CMS is a maternally inherited trait and is beneficial for the production of F_1 hybrid seeds by generating infertile pollen without changing vegetative growth and female fertility [5]. CMS systems are not only a useful component for studying pollen development, but also an important way to utilize hybrid vigor [6]. The existence of CMS systems in plants eliminates the laborious and painstaking work of sterilization and manual emasculation in a broad range of crops. CMS can arise spontaneously in breeding lines after wide crosses, interspecific exchange of nuclear or cytoplasmic genomes, and mutagenesis [7]. Initially, it was thought that sterility was caused by mutation within the mitochondrial genome [8], however, further research has revealed that a major cause of CMS is mitochondrial DNA rearrangement, which results in plants unable to generate functional pollen [9]. Mitochondria are important cellular components for energy (ATP, NADH, $FADH_2$)-dependent metabolic pathways, including oxidative phosphorylation, respiratory electron transfer, biosynthesis of amino acids, vitamin cofactors, the Krebs cycle, and programmed cell death [10–12]. Therefore, CMS proteins were hypothesized to cause mitochondrial energy deficiency and failure to meet energy requirements during male reproductive development [13].

Currently, 10 types of CMS systems have been reported in *Brassica napus*, including the natural mutation *pol* CMS [14] and *shan2A* CMS [15], and intergeneric hybridization CMS *nap* CMS [16] and *Nsa* CMS [17]. *Nsa* CMS [17], *Ogu* CMS [18] and *tour* CMS [19] were generated by protoplast fusion of different species, resulting in a source of genetic variation within the cytoplasmic organelles [20]. Both *Pol* CMS and *Ogu* CMS are commonly used as CMS systems for *B. napus* hybrid breeding. CMS is sensitive to harsh environmental factors, including air temperature and exposure time to sunlight [21–23]. However, the *Nsa* CMS system has demonstrated stable male sterility under different environmental conditions, ensuring seed purity during hybrid seed production.

Preliminary work has demonstrated significant differences in plant endogenous hormones between CMS lines and their maintainer lines in different species [24,25]. In sugar beet, it was found the level of endogenous IAA (indole-3-acetic acid), GA3 (gibberellic acid), and ZR (zeatin-riboside), in relation to ABA (abscisic acid), differed at three developmental stages (vegetative, early flowering, and bud development) [26]. It was also demonstrated that pepper CMS line 'Bei-A' and maintainer line 'Bei-B' showed significant hormonal differences [27], with a higher IAA and ABA content and lower ZR_5 and GA_3 content observed within the CMS line [27]. The relationship between phytohormones and CMS has been widely investigated in many species, including *B. napus* [28,29], flax [30], and rice [31]. It has been shown that phytohormones ABA and IAA may be major contributors for CMS. The concentration of ABA and IAA changes at different stages of bud development between male sterile lines and their maintainer lines [29,30]. These studies collectively provide evidence for the importance of determining the endogenous level of ABA and IAA in CMS and maintainer lines when studying cytoplasmic male sterility.

Most recently, attention has focused on the provision of next-generation sequencing (NGS) technology [32–34] and the use of NGS to make studies on expressed genes and genomes in higher plants more feasible [35–37]. Currently, RNA-Seq has been used in higher plants with CMS systems in many species, including tomato [37], rice [38], and *B. napus* [36,39]. A large and growing body of literature has investigated floral buds of CMS and maintainer lines using RNA sequencing and comparative gene expression. In *Pol* CMS, unigenes related to pollen development were analyzed through transcriptome sequencing [36]. These high-throughput results will be useful for understanding the sterility mechanism of *pol* CMS in detail. Another transcriptome study of SaNa-1A CMS was also conducted in *B. napus* [40]. By comparing the sterile line and the maintainer line, many differentially expressed genes (DEGs) involved in metabolic, protein synthesis, and other pathways were identified. These results provide a basis for future research on the CMS mechanism in SaNa-1A. The existence of various CMS lines with different mitochondrial patterns offer new opportunities to explore the genetic regulation of CMS and its associated developmental effects [41].

In the current study, *Pol* CMS, *Ogu* CMS, *Nsa* CMS, and their corresponding maintainer lines (with the same nuclear genome but fertile cytoplasm) were used to carry out transcriptomic and DEG analysis. Simultaneously, we compared the morphological differences in sterile and fertile lines, and analyzed the IAA and ABA contents. We investigated the pollen abortion stage of the *Nsa* CMS line by semi-thin sectioning. This study confirms the stage of pollen abortion in *Nsa* CMS, and illustrates the mode of regulation of the different CMS systems during pollen development at the transcriptomic level.

2. Results

2.1. Phenotypic Characterization of CMS Lines and Maintainer Lines

The flower structure of rapeseed includes four sepals, four petals, six stamens (four long and two short), and one pistil from outwards to inwards. When a flower blooms, mature pollen sticks to the pistil. The pistil is almost the same height as the long stamens, allowing pollination to occur easily. In this study, we obtained three CMS systems (*Nsa* CMS, *Pol* CMS, and *Ogu* CMS) with corresponding maintainer lines. All sterile lines and their maintainer line harbor the same nuclear genome but different cytoplasm. We found that all sterile floral petals were visually wrinkled and smaller than fertile flowers in three CMS systems (Figure 1). Degeneration of stamens and shorter stamen length was observed in sterile lines as compared with the normal fertile flowers. Among the three CMS systems, the stamens of the *pol* CMS sterile line were more seriously degenerated (Figure 1F). However, the pistils of all the sterile floral buds were the same as fertile lines (Figure 1).

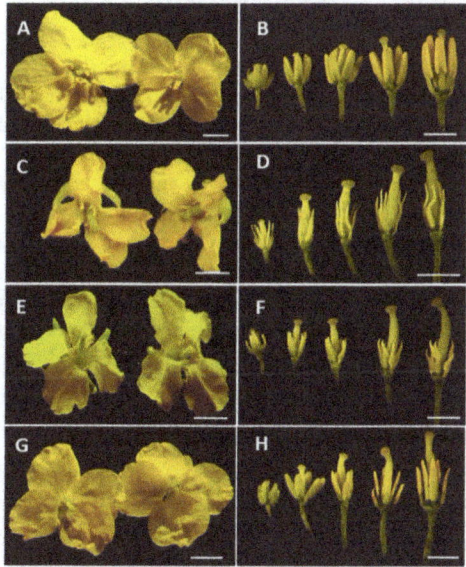

Figure 1. Flower morphology of maintainer and sterile lines of the *Pol*, *Nsa*, and *Ogu* cytoplasmic male sterility (CMS) systems. (**A–B**) Maintainer line; (**C–D**) *Nsa* CMS line; (**E–F**) *Pol* CMS line; (**G–H**) *Ogu* CMS line; Bar = 0.5 cm.

The stage at which pollen abortion occurs within the *Nsa* CMS system has not been determined clearly. For detailed characterization of the developing pollen, ultrathin specimens were observed under a microscope. By observing a semi-thin section of anthers, we conclude that the abortion period of *Nsa* CMS occurred during the tetrad period (Figure 2F). After the tetrad stage, *Nsa* CMS could not produce normal spores at the uni-nuclear stage (Figure 2G). Normal anthers form mature pollen, as shown in Figure 2D. The sterile line did not produce mature pollen but formed a large number of abnormal spores (Figure 2H).

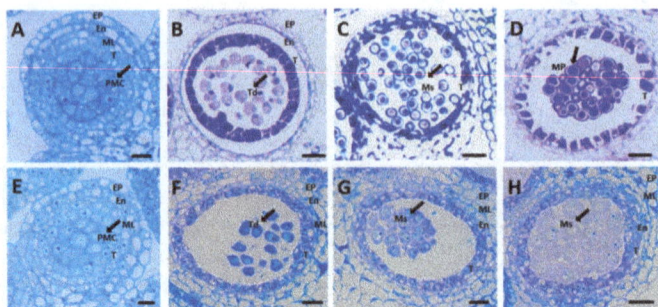

Figure 2. Comparison of maintainer line "*ZS4*" (**A–D**) and sterile line (**E–H**) anthers of *Nsa* CMS with toluidine blue staining. Bar = 10 μm, Ep, epidermis; En, endothecium; ML, middle layer; T, tapetum; Ms, microspore; MP, mature pollen; PMC, primary mother cells.

2.2. IAA and ABA Concentration in CMS and Maintainer Lines

Plant hormones were assessed in CMS and maintainer lines to clarify how plant hormones are altered in the three CMS systems (Figure 3). The ABA and IAA contents in flower buds were detected at S_1 (<2.5 mm size of floral buds) and S_2 stages (>2.5 mm size of floral buds) in CMS and maintainer lines, respectively. We found that ABA levels were significantly higher in all three CMS lines as compared to maintainer lines at both stages. Conversely, IAA content was significantly lower in the *Nsa* CMS line than its maintainer line at the S_1 stage, while *Ogu* and *pol* CMS lines showed no significant difference with their maintainer lines. However, IAA content was significantly lower in all CMS lines as compared to the maintainer lines at the S_2 stage. These results indicate that a significantly higher content of endogenous ABA and lower content of IAA may enhance pollen abortion in sterile lines. ABA content showed increasing and IAA decreasing trends at the S_1 stage compared to the S_2 stage. The ABA content was significantly higher at both stages in CMS lines than in their corresponding maintainer line.

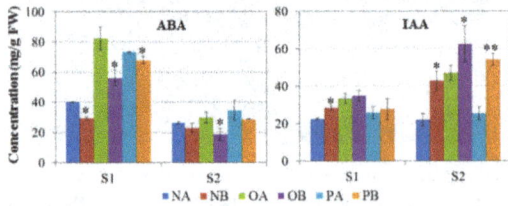

Figure 3. ABA and IAA contents of developing buds in maintainer and male sterile lines of the *Pol*, *Nsa*, and *Ogu* systems. NA, *Nsa* sterile line; NB, *Nsa* maintainer line; OA, *Ogu* sterile line; OB, *Ogu* maintainer line; PA, *Pol* sterile line; PB, *Pol* maintainer line. Asterisks indicate a significant difference was detected between CMS line and maintainer line in S_1 and S_2 stage by t-test at *p<0.05, **p<0.01.

2.3. Differentially Expressed Genes in CMS and Maintainer Lines

Using high-throughput sequencing, differentially expressed genes were detected in the sterile and corresponding maintainer lines. The flower buds used to determine phytohormone levels were also subjected to transcriptome sequence analysis. Three biological replicates were performed with the reproducibility between replicates being ≥90%. In total, 222.15 Gb of clean data were generated (with all samples Q30 ≥ 90%). Differentially expressed genes (DEGs) were identified in Biocloud (Biomarker Technologies). For each CMS system, DEGs were found between the male sterile line and the corresponding maintainer line. DEGs exhibiting a two-fold change or greater were selected according to the *q*-values [39]. At the S_1 stage, we identified 1306, 1262, and 4127 DEGs in the *Nsa*, *Pol*, and *Ogu* systems, respectively. More DEGs (2369, 1690, and 3035) were discovered at the S_2 stage in the three CMS systems. Among the three CMS systems, the largest number of DEGs were observed in the *Ogu* CMS system at the S_1 stage. Among the total 4127 DEGs, 2158 genes were upregulated and 1969 genes were downregulated. The smallest number of DEGs was observed in the *Pol* CMS system, in which 806 genes were upregulated and 456 genes were downregulated at the S_1 stage (Figure 4). Many more upregulated DEGs with high-fold change (>5-fold) were found at the S_2 stage compared to the S_1 stage in all three systems (Figure 4). Furthermore, only the *Ogu* CMS system exhibited more DEGs, including upregulated and downregulated genes, in the S_1 stage than the S_2 stage. More DEGs were observed in the *Pol* CMS, and especially in the *Nsa* CMS system at the S_2 stage than the S_1 stage.

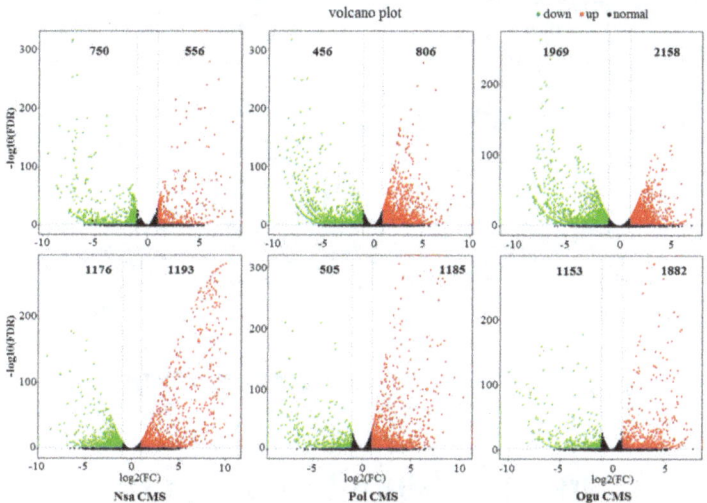

Figure 4. Differentially expressed unigenes and corresponding genes in the sterile and maintainer lines. The genes were selected with "$p \leq 0.01$" and "fold change ≥ 2". The X-axis is the log of 2-fold change in expression between the sterile and maintainer lines at two stages. Y-axis shows the statistical significance of the differences with the value of log10 (FDR). The spots in different colors are representing expression of different genes. Black spots represent genes without significant expression. Red spots mean 2-fold upregulated genes from maintainer lines to sterile lines. Green spots represent significantly 2-fold down-expressed genes from maintainer lines to sterile lines.

2.4. Gene Ontology and Classification of Three CMS Lines

At the S_1 stage, we observed that only 156 unigenes were co-differentially expressed in the three CMS lines, compared to 581 unigenes at the S_2 stage (Figure 5A,B). KEGG classification and functional enrichment was performed for DEGs at both stages (Figure 5C,D). At the S_1 stage, five categories were identified, including environmental information processing, genetic information processing, organismal systems, cellular processes, and metabolism (Figure 5C). At the S_2 stage, genes were divided into four categories, including metabolism, genetic information processing, cellular processes, and environmental information processing (Figure 5D). At the S_1 stage, in the environmental information processing category, 3% of DEGs were associated with plant hormone signal transduction. Only 1% of the DEGs were relative to plant–pathogen interaction. Within the cellular processes category, the highest number of DEGs was related to the peroxisome. Significantly enriched DEGs were identified as being involved in pentose–glucuronate interconversions and starch–sucrose metabolism among the metabolic components category. At the S_2 stage, metabolic components were significantly enriched, including starch–sucrose, arginine–proline, glycerophospholipid, alanine–aspartate–glutamate, and amino–nucleotide sugar metabolism. From these analyses, we can determine starch and sucrose play an important role in metabolism at this stage. Transcriptomic data also revealed that many DEGs were enriched in plant hormonal signal transduction pathways.

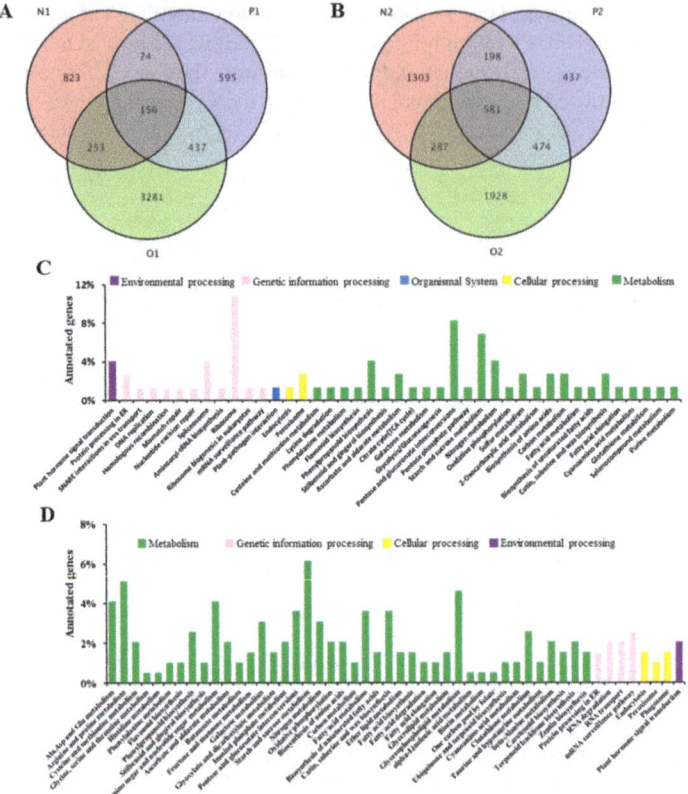

Figure 5. Co-differentially expressed unigenes (**A**, **B**) and GO annotations (**C**, **D**) of differentially expressed genes (DEGs). Venn diagrams of differentially expressed genes at (**A**) the S1 stage and (**B**) the S2 stage in *Nsa* (N), *Ogu* (O), and *Pol* (P) male sterile lines. N1, S1 stage of *Nsa* CMS system; O1, S1 stage of *Ogu* CMS system; P1, S1 stage of *Pol* CMS system; N2, S2 stage of *Nsa* CMS system; O2, S2 stage of *Ogu* CMS system; P2, S2 stage of *Pol* CMS system; (**C**) and (**D**), the X-axis indicates the percentage of genes in each categories, and y-axis showed classification of unigenes.

2.5. Verification of DEGs by RT-PCR

We conducted RT-PCR to validate the results generated by RNA-Seq. To determine whether transcription factors and hormone-related genes were differentially expressed, we quantified expression of these genes in the three CMS lines by semi-quantitative polymerase chain reaction (RT-PCR). Expression of genes encoding transcription factors or involved in ABA or IAA signaling were enriched in maintainer lines compared to male sterile lines (Figure 6). From the RT-PCR results, we found that almost all selected genes were highly expressed in maintainer lines compared to sterile lines. Most of the genes showed higher expression levels at the S_2 stage compared to the S_1 stage in all CMS systems (*Pol*, *Ogu*, and *Nsa*). This result was consistent with the RNA-Seq results.

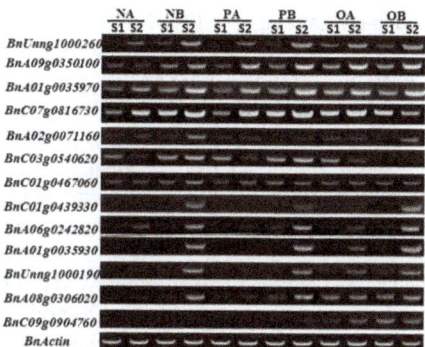

Figure 6. Gene expression difference in three cytoplasmic male sterile materials of S1 and S2 stages. NA, *Nsa* sterile line; NB, *Nsa* maintainer line; OA, *Ogu* sterile line; OB, *Ogu* maintainer line; PA, *Pol* sterile line; PB, *Pol* maintainer line. The *BnActin* gene was used as the control.

3. Discussion

The widespread existence of CMS in plant species may be related to the potential to promote outcrossing and prevent inbreeding depression. Independent CMS lines differ not only in their sequences and origins [42], but also in their phenotype, including changes in microspore development [43] and breakdown pattern of tapetum structure [44]. Pollen development comprises a series of defined physiological events. A large volume of published studies describe the role of energy (ATP, NADH, FADH$_2$) in the pollen abortion process [45]. Our results also show that many energy production or conversion related genes are differentially expressed between male sterile and maintainer lines. In the semi-thin sections of the *Nsa* CMS and maintainer lines (Figure 2), we identified differences in the epidermis, endothecium, tapetum, microspore, and mature pollen. Our results revealed that pollen abortion occurred at the tetrad stage in the *Nsa* CMS line. Following the tetrad stage, *Nsa* CMS plants could not produce normal spores at the uni-nuclear stage (Figure 2H). Pollen abortion occurred due to the breakdown of tapetum and premature or delayed degeneracy [46,47]. In *pol* CMS, abortion was started at stage 4 (pollen development period). Anthers of the sterile line could not differentiate sporogenous cells, with the middle layer, endothecium, and tapetum being indistinguishable. The results in sterile anthers filled with numerous, highly vacuolated cells [36]. Due to abnormal development in the early stage, *pol* CMS lines cannot produce normal tetrads. In *Ogu* CMS, abortion was also identified to start at the tetrad stage by comparing the cell morphology of three central stages (the tetrad, mid-microspore, vacuolated microspore) of pollen development [44]. It was found that the tapetal cells developed a large vacuole at tetrad stage in the *Ogu* CMS line. The anther development of sterile line SaNa-1A, a line with CMS derived from somatic hybrids between *B. napus* and *Sinapis alba*, is also abnormal from the tetrad stage [40]. The abortion phenotype of *Nsa* CMS is the same as the SaNa-1 sterile line in all four stages of pollen development. Similar to SaNa-1 CMS, *Nsa* CMS was derived from somatic hybrids between *B. napus* and *Sinapis arvensis*. Together with *Ogu* CMS, which is derived from intergeneric hybridization between *B. napus* and *Raphanus sativa*, all the alloplasmic CMS systems have the same pollen development abortion stage.

Phytohormones (IAA and ABA) are generally known for their specific role in the induction and promotion of DNA synthesis, and play a role in metabolic pathways [48,49]. It was observed that high and exogenous application of ABA induced pollen abortion by specifically suppressing apoplastic sugar transport in pollen [22,50,51]. The ABA content in younger floral buds was higher than that of elder ones in all three CMS systems, whereas the IAA content showed a converse trend. However, similar to previous studies in other species, sterile lines of the three CMS systems showed some differences in IAA and ABA content in both male sterile and their corresponding maintainer lines [52,53].

Transcriptomic analysis detected a total of 5619 DEGs at the S_1 stage, with 156 co-differentially expressed in all three CMS lines, and 5208 DEGs at the S_2 stage, with 581 co-differentially expressed in all three CMS lines. KEGG analysis divided these co-differentially expressed genes into 42 and 50 categories at the S_1 and S_2 stages, respectively. At both the S_1 and S_2 stages, half of the genes were involved in metabolism. Many genes for mitochondrial energy metabolism and pollen development were also differentially expressed in multiple CMS systems [36,54,55]. It has been shown that the presence of infertility genes affects the transcription of genes involved in the energy metabolism of the mitochondria, resulting in impairment of the normal physiological functions of the mitochondria, which leads to infertility [13].

Previous studies identified a link between plant hormones and cytoplasmic male infertility [22,24]. In addition, transcription factors have also been implicated in pollen infertility [56,57]. Therefore, the expression of some transcription factors and hormone-related genes were selected to verify the results generated by RNA-Seq. RT-PCR results indicated that the expression level of selected genes in maintainer lines was significantly higher than that of the male sterile lines. This provides further support that levels of phytohormone precursor genes and some transcription factors may be correlated with cytoplasmic male sterility. Increased IAA content was detected in maintainer lines compared to male sterile lines in all CMS systems. Coincident with this result, we observed IAA signaling-related genes, including two *IAA19* genes, were significantly enriched in maintainer lines (Figure 6).

4. Materials and Methods

4.1. Plant Materials

Pol CMS, *Ogu* CMS, *Nsa* CMS and their corresponding maintainer lines were used in this study. Materials were cultivated in the field of Oil Crops Research Institute, Chinese Academy of Agricultural Sciences (OCRI-CAAS), Wuhan, China. Anthers at different stages were collected for morphological study, and the abortion stage was studied by semi-thin sectioning of floral buds. Samples of floral buds (<2.5 mm and >2.5 mm) were collected and stored at −70 °C for further RNA-sequencing and hormonal quantification.

4.2. Morphology and Semi-Thin Sections

The floral buds of sterile and fertile lines were examined under the microscope (Olympus: CX31RTSF). At different stages, samples collected from the sterile and fertile line were fixed in FAA solution [38% formaldehyde, 70% ethanol, and 100% acetic acid (1:1:18)]. A vacuum chamber was used to evacuate the air and volatiles from the sample bottles. Fixed floral buds were dehydrated by a graded series of ethanol (70, 85, 95, and 100%) for one hour. Pre-infiltration and penetration by Technovit 7100 resin steps were undertaken to produce semi-thin sections ~3 μm thick. Samples were stained by 1% toluidine blue (Sigma Aldrich, St. Louis, MO, USA) for 3 min, and 5 specimens of each stage were observed to take images under an optical microscope (Olympus: CX31RTSF, Tokyo, Japen).

4.3. Phytohormone (ABA and IAA) Quantification

About 60 floral buds (smaller than 2.5 mm and larger than 2.5 mm) of *Pol* CMS, *Ogu* CMS, *Nsa* CMS, and their corresponding maintainer lines were quantified for ABA and IAA phytohormones. Samples were collected and extracted using methanol compounds [58]. The extraction was carried out by adding 1 mL of MeOH (methyl alcohol) with water (8:2) into each tube containing fresh plant material. Samples were shaken for 30 min before centrifugation at 12000 rpm at 4 °C for 10 min. The supernatant was transferred to a new microcentrifuge tube and dried in a speed vacuum. After drying, 100 μL of MeOH was added to each sample. Each sample was homogenized using a vortex mixer and centrifuged at 12000 RPM at 4 °C for 10 min. Phytohormones within the supernatant were separated by HPLC (Agilent 1200) and analyzed by a hybrid triple quadrupole/linear ion trap mass spectrometry (ABI 4000 Q-Trap, Applied Biosystems, Foster City, CA, USA).

4.4. Illumina Sequencing and Analysis of DEGs

About 60 floral buds were harvested from the plants of each line (*Ogu* CMS, *Nsa* CMS, and their corresponding maintainer lines) at the same time. Samples collected from each line were pooled, frozen in liquid nitrogen, and stored at −70 °C for RNA preparation. Total RNA from two stages of floral buds (<2.5 mm and >2.5 mm) of *pol* CMS, *Ogu* CMS, *Nsa* CMS, and their corresponding maintainer lines were extracted by using RNA kits (Tiangen, Beijing, China) in accordance with the manufacturer's protocol. The integrity of the total RNA was checked by 1% agarose gel electrophoresis. The concentration was detected by Nano-Drop (Thermo Scientific, Madison, WI, USA) and purity of RNA was determined by Agilent 2100 Bio-analyzer (Agilent, Waldbronn, Germany). RNA (10 µL) was sequenced using the Illumina HiSeq 2000 (Illumina, San Diego, CA, USA) and 150 bp of data collected per run. After removing adapters and low-quality data, the resulting clean data was aligned to the *B. napus* reference genome [59]. Potential duplicate molecules were removed from the aligned BAM/SAM format records. FPKM (fragments per kilobase of exon per million fragments mapped) values were used to analyze gene expression by the software Cufflinks [60]. Three biological replicates were performed for each sample.

4.5. Semi-Quantitative (RT-PCR) Analysis of DEGs

The DEG results were confirmed by RT-PCR using the same RNA samples which were used for RNA library construction. Complementary DNA was generated from the RNA template by using the reverse transcription kit (Vazyme, Nanjing, China). Specific primers for differentially expressed genes were designed to amplify 600–750 bp sequences (Table S1). RT-PCR was carried out by using a program of 95 °C for 5 min (initial hot start), 30 cycles of 95 °C for 30 s, 56 °C for 35 s, and 72 °C for 5 min. Three biological replicates were analyzed for each sample.

5. Conclusions

Considerable effort has been taken to identify the pollen abortion stage in *Pol* and *Ogu* CMS lines. Conversely, the pollen abortion stage in the *Nsa* CMS line had not been determined clearly. From this study, we identified the tetrad stage of *Nsa* CMS for pollen abortion by using the semi-thin sectioning of floral buds. This information will help support the application of *Nsa* CMS in plant breeding. Higher content of ABA and lower content of IAA was observed in sterile lines when compared to maintainer lines in all male sterile systems. This result may reveal that ABA and IAA play different roles in fertile pollen development. During the two stages, genes involved in energy production were enriched in maintainer lines in comparison to sterile lines for all CMS systems investigated.

Supplementary Materials: Supplementary materials can be found at http://www.mdpi.com/1422-0067/19/12/4022/s1.

Author Contributions: Data curation, M.H.; Formal analysis, B.D.; Funding acquisition, H.C. and Q.H.; Investigation, S.S.; Project administration, D.M. and Q.H.; Resources, H.W. and L.F.; Software, W.W.; Writing—original draft, B.D., Q.U.Z. and H.C.; Writing—review & editing, Q.H. All authors read and approved the final manuscript.

Funding: This research was funded by the National Key Research and Development program of China (No. 2016YFD0101300), National Natural Science Foundation of China (No. 31600226), Fundamental Research Funds for Central Non-profit Scientific Institution (1610172017005), Hubei Provincial Natural Science Foundation of China (2018CFB248), Science and Technology Innovation Project of the Chinese Academy of Agricultural Sciences (Group No. 118).

Acknowledgments: We thank Rachel Wells in John Innes Center for proof-reading and English correction.

Conflicts of Interest: The authors declare no conflict of interests.

References

1. Wang, H.Z. Review and future development of rapeseed industry in China. *Chin. J. Oil Crop Sci.* **2010**, *2*, 300–302.
2. Fu, T.D. Breeding and utilization of rapeseed hybrid. *Hubei Sci. Technol.* **2000**, 167–169.

3. Allender, C.J.; King, G.J. Origins of the amphiploid species *Brassica napus* L. Investigated by chloroplast and nuclear molecular markers. *BMC Plant Biol.* **2010**, *10*, 54. [CrossRef] [PubMed]
4. Yamagishi, H.; Bhat, S.R. Cytoplasmic male sterility in brassicaceae crops. *Breed Sci.* **2014**, *64*, 38–47. [CrossRef] [PubMed]
5. Shinada, T.; Kikuchi, Y.; Fujimoto, R.; Kishitani, S. An alloplasmic male-sterile line of *Brassica oleracea* harboring the mitochondria from Diplotaxis muralis expresses a novel chimeric open reading frame, *orf72*. *Plant Cell Physiol.* **2006**, *47*, 549–553. [CrossRef] [PubMed]
6. Shiga, T.; Baba, S. Cytoplasmic male sterility in oil seed rape. *Brassica napus* L., and its utilization to breeding. *Japan J. Breed.* **1973**, *23*, 187–197.
7. Hanson, M.R.; Bentolila, S. Interactions of mitochondrial and nuclear genes that affect male gametophyte development. *Plant Cell* **2004**, *16*, 154–169. [CrossRef]
8. Schnable, P.S.; Wise, R.P. The molecular basis of cytoplasmic male sterility and fertility restoration. *Trends Plant Sci.* **1998**, *3*, 175–180. [CrossRef]
9. Horn, R.; Guptac, K.J.; Colombo, N. Mitochondrion role in molecular basis of cytoplasmic male sterility. *Mitochondrion* **2014**, *19*, 198–205. [CrossRef]
10. Siedow, J.N.; Umbach, A.L. Plant mitochondrial electron transfer and molecular biology. *Plant Cell* **1995**, *7*, 821–831. [CrossRef]
11. Logan, D.C. The mitochondrial compartment. *J. Exp. Bot.* **2006**, *57*, 1225–1243. [CrossRef] [PubMed]
12. Youle, R.J.; Karbowski, M. Mitochondrial fission in apoptosis. *Nat. Rev. Mol. Cell Biol.* **2005**, *6*, 657. [CrossRef] [PubMed]
13. Chen, L.; Liu, Y.G. Male sterility and fertility restoration in crops. *Annu. Rev. Plant Biol.* **2014**, *65*, 579–606. [CrossRef] [PubMed]
14. Fu, T.D. Production and research of rapeseed in the People's Republic of China. *Eucarpia. Crucif. News* **1981**, *6*, 6–7.
15. Shen, J.X.; Wang, H.Z.; Fu, T.D.; Tian, B.M. Cytoplasmic male sterility with self-incompatibility, a novel approach to utilizing heterosis in rapeseed (*Brassica napus* L.). *Euphytica* **2008**, *162*, 109–115. [CrossRef]
16. Thompson, K.F. Cytoplasmic male-sterility in oil-seed rape. *Heredity* **1972**, *29*, 253–257. [CrossRef]
17. Hu, Q.; Andersen, S.; Dixelius, C.; Hansen, L. Production of fertile intergeneric somatic hybrids between brassica napus and sinapis arvensis for the enrichment of the rapeseed gene pool. *Plant Cell Rep.* **2002**, *21*, 147–152.
18. Ogura, H. Studies on the new male-sterility in japanese radish, with special reference to the utilization of this sterility towerds the practical raising of hybrid seeds. *Mém. Fac. Agric. Kagoshima Univ.* **1967**, *6*, 1446–1459.
19. Rawat, D.S.; Anand, I.J. Male sterility in indian mustard. *Indian J. Genet. Plant Breed.* **1979**, *39*, 412–414.
20. Kemble, R.J.; Barsby, T.L. Use of protoplast fusion systems to study organelle genetics in a commercially important crop. *Biochem. Cell Biol.* **1988**, *66*, 665–676. [CrossRef]
21. Burns, D.R.; Scarth, R.; McVetty, P.B.E. Temperature and genotypic effects on the expression of *pol* cytoplasmic male sterility in summer rape. *Can. J. Plant Sci.* **1991**, *71*, 655–661. [CrossRef]
22. Zhang, J.K.; Zong, X.F.; Yu, G.D.; Li, J.N.; Zhang, W. Relationship between phytohormones and male sterility in thermo-photo-sensitive genic male sterile (TGMS) wheat. *Euphytica* **2006**, *150*, 241–248. [CrossRef]
23. Zhou, Y.M.; Fu, T.D. Genetic improvement of rapeseed in China. In Proceedings of the 12th International Rapeseed Congress, Wuhan, China, 26–30 March 2007.
24. Sawhney, V.K.; Shukla, A. Male sterility in flowering plants: Are plant growth substances involved? *Am. J. Bot.* **1994**, *81*, 1640–1647. [CrossRef]
25. Tian, C.G.; Zhang, M.Y.; Duan, J. Preliminary study on the changes of phytohormones at different development stage in cytoplasmic male sterility line and its maintainer of rape. *Sci. Agric. Sin.* **1998**, *31*, 20–25.
26. Wang, H.Z.; Wu, Z.D.; Han, Y. Relationships between endogenous hormone contents and cytoplasmic male sterility in sugarbeet. *Sci. Agric. Sin.* **2008**, *41*, 1134–1141.
27. Wu, Z.M.; Hu, K.L.; Fu, J.Q.; Qiao, A.M. Relationships between cytoplasmic male sterility and endogenous hormone content of pepper bud. *J. South China Agric. Univ.* **2010**, *31*, 1–4.
28. Dubas, E.; Janowiak, F.; Krzewska, M.; Hura, T.; Żur, I. Endogenous aba concentration and cytoplasmic membrane fluidity in microspores of oilseed rape (*Brassica napus* L.) genotypes differing in responsiveness to androgenesis induction. *Plant Cell Rep.* **2013**, *32*, 1465–1475. [CrossRef] [PubMed]

29. Shukla, A.; Sawhney, V.K. Abscisic acid: One of the factors affecting male sterility in *Brassica napus*. *Physiol. Plant.* **2010**, *91*, 522–528. [CrossRef]

30. Guan, T.; Dang, Z.; Zhang, J. Studies on the changes of phytohormones during bud development stage in thermo-sensitivity genic male-sterile flax. *Chin. J. Oil Crop Sci.* **2007**, *3*, 248–253.

31. Huang, S.; Zhou, X. Relationship between rice cytoplasmic male sterility and contents of GA (1+4) and IAA. *Acta Agric. Bor-Sin.* **1994**, *9*, 16–20.

32. Wang, Z.; Gerstein, M.; Snyder, M. RNA-seq: A revolutionary tool for transcriptomics. *Nat. Rev. Genet.* **2009**, *10*, 57–63. [CrossRef] [PubMed]

33. Grabherr, M.G.; Haas, B.J.; Yassour, M.; Levin, J.Z.; Thompson, D.A.; Amit, I.; Adiconis, X.; Fan, L.; Raychowdhury, R.; Zeng, Q.; et al. Full-length transcriptome assembly from RNA-Seq data without a reference genome. *Nat. Biotechnol.* **2011**, *29*, 644–652. [CrossRef] [PubMed]

34. Qi, Y.X.; Liu, Y.B.; Rong, W.H. RNA-Seq and its applications: A new technology for transcriptomics. *Hereditas* **2011**, *33*, 1191. [CrossRef] [PubMed]

35. Torti, S.; Fornara, F.; Vincent, C.; Andrés, F.; Nordström, K.; Göbel, U.; Knoll, D.; Schoof, H.; Coupland, G. Analysis of the *Arabidopsis* shoot meristem transcriptome during floral transition identifies distinct regulatory patterns and a leucine-rich repeat protein that promotes flowering. *Plant Cell.* **2012**, *24*, 444–462. [CrossRef] [PubMed]

36. An, H.; Yang, Z.; Yi, B.; Wen, J.; Shen, J.; Tu, J.; Ma, C.; Fu, T. Comparative transcript profiling of the fertile and sterile flower buds of *pol CMS* in *B. napus*. *BMC Genom.* **2014**, *15*, 258. [CrossRef] [PubMed]

37. Jeong, H.J.; Kang, J.H.; Zhao, M.; Kwon, J.K.; Choi, H.S.; Bae, J.H.; Lee, H.A.; Joung, Y.H.; Choi, D.; Kang, B.C. Tomato male sterile 1035 is essential for pollen development and meiosis in anthers. *J. Exp. Bot.* **2014**, *65*, 6693–6709. [CrossRef] [PubMed]

38. Yan, J.; Zhang, H.; Zheng, Y.; Ding, Y. Comparative expression profiling of mirnas between the cytoplasmic male sterile line meixianga and its maintainer line meixiangb during rice anther development. *Planta* **2015**, *241*, 109–123. [CrossRef] [PubMed]

39. Qu, C.; Fu, F.; Liu, M.; Zhao, H.; Liu, C.; Li, J.; Tang, Z.; Xu, X.; Qiu, X.; Wang, R.; et al. Comparative transcriptome analysis of recessive male sterility (RGMS) in sterile and fertile *Brassica napus* lines. *PLoS ONE* **2015**, *10*, e0144118. [CrossRef]

40. Du, K.; Liu, Q.; Wu, X.Y.; Jiang, J.J.; Wu, J.; Fang, Y.J.; Li, A.M.; Wang, Y. Morphological structure and transcriptome comparison of the cytoplasmic male sterility line in *Brassica napus* (SaNa-1A) derived from somatic hybridization and its maintainer line SaNa-1B. *Front. Plant Sci.* **2016**, *7*, 1313. [CrossRef]

41. Leino, M.; Teixeira, R.; Landgren, M.; Glimelius, K. Brassica napus lines with rearranged arabidopsis mitochondria display *CMS* and a range of developmental aberrations. *Theor. Appl. Genet.* **2003**, *106*, 1156–1163. [CrossRef]

42. Mackenzie, S. Male sterility and hybrid seed production. *Plant Biotechnol. Agric.* **2012**, 185–194.

43. Datta, R.; Chamusco, K.C.; Chourey, P.S. Starch biosynthesis during pollen maturation is associated with altered patterns of gene expression in maize. *Plant Physiol.* **2002**, *130*, 1645–1656. [CrossRef] [PubMed]

44. González-Melendi, P.; Uyttewaal, M.; Morcillo, C.N.; Hernández Mora, J.R.; Fajardo, S.; Budar, F.; Lucas, M.M. A light and electron microscopy analysis of the events leading to male sterility in Ogu-INRA *CMS* of rapeseed (*Brassica napus*). *J. Exp. Bot.* **2008**, *59*, 827–838. [CrossRef] [PubMed]

45. Yang, J.H.; Huai, Y.; Zhang, M.F. Mitochondrial atpa gene is altered in a new orf220-type cytoplasmic male-sterile line of stem mustard (*Brassica juncea*). *Mol. Biol. Rep.* **2009**, *36*, 273–280. [CrossRef] [PubMed]

46. Luo, X.D.; Dai, L.F.; Wang, S.B.; Wolukau, J.; Jahn, M.; Chen, J.F. Male gamete development and early tapetal degeneration in cytoplasmic male sterile pepper investigated by meiotic, anatomical and ultrastructural analyses. *Plant Breed.* **2006**, *125*, 395–399. [CrossRef]

47. Li, N.; Zhang, D.S.; Liu, H.S.; Yin, C.S.; Li, X.X.; Liang, W.Q.; Yuan, Z.; Xu, B.; Chu, H.W.; Wang, J. The rice tapetum degeneration retardation gene is required for tapetum degradation and anther development. *Plant Cell* **2006**, *18*, 2999–3014. [CrossRef]

48. Minocha, S.C. The role of auxin and abscisic acid in the induction of cell division in jerusalem artichoke tuber tissue cultured in vitro. *Z. Pflanzenphysiol.* **1979**, *92*, 431–441. [CrossRef]

49. Rook, F.; Hadingham, S.A.; Li, Y.; Bevan, M.W. Sugar and aba response pathways and the control of gene expression. *Plant Cell Environ.* **2006**, *29*, 426–434. [CrossRef]

50. Ji, X.; Dong, B.; Shiran, B.; Talbot, M.J.; Edlington, J.E.; Hughes, T.; White, R.G.; Gubler, F.; Dolferus, R. Control of abscisic acid catabolism and abscisic acid homeostasis is important for reproductive stage stress tolerance in cereals. *Plant Physiol.* **2011**, *156*, 647–662. [CrossRef]

51. De Storme, N.; Geelen, D. The impact of environmental stress on male reproductive development in plants: Biological processes and molecular mechanisms. *Plant Cell Environ.* **2014**, *37*, 1–18.

52. Duca, M. Genetic-phytohormonal interactions in male fertility and male sterility phenotype expression in sunflower (*Helianthus annuus* L.)/interacciones genético-fitohormonales en la expresión fenotípica de la androfertilidad y androesterilidad en girasol (helianthus annuus l.)/interactions génétiques phytohormonales dans l'expression phénotypique d'un male fertile et male sterile du tournesol (helianthus annuus L.). *Helia* **2008**, *31*, 27–38.

53. Singh, S.; Sawhney, V. Abscisic acid in a male sterile tomato mutant and its regulation by low temperature. *J. Exp. Bot.* **1998**, *49*, 199–203. [CrossRef]

54. Yang, J.H.; Zhang, M.F.; Yu, J.Q. Mitochondrial *nad2* gene is co-transcripted with CMS-associated *orfB* gene in cytoplasmic male-sterile stem mustard (*Brassica juncea*). *Mol. Biol. Rep.* **2009**, *36*, 345–351. [CrossRef] [PubMed]

55. Heng, S.; Liu, S.; Xia, C.; Tang, H.; Xie, F.; Fu, T.; Wan, Z. Morphological and genetic characterization of a new cytoplasmic male sterility system (oxa CMS) in stem mustard (*Brassica juncea*). *Theor. Appl. Genet.* **2018**, *131*, 59–66. [CrossRef] [PubMed]

56. Xing, M.; Sun, C.; Li, H.; Hu, S.; Lei, L.; Kang, J. Integrated analysis of transcriptome and proteome changes related to the Ogura cytoplasmic male sterility in cabbage. *PLoS ONE* **2018**, *13*, e0193462. [CrossRef] [PubMed]

57. Li, Y.; Ding, X.; Wang, X.; He, T.; Zhang, H.; Yang, L.; Wang, T.; Chen, L.; Gai, J.; Yang, S. Genome-wide comparative analysis of DNA methylation between soybean cytoplasmic male-sterile line NJCMS5A and its maintainer NJCMS5B. *BMC Genom.* **2017**, *18*, 596. [CrossRef] [PubMed]

58. Wang, H.; Cheng, H.; Wang, W.; Liu, J.; Hao, M.; Mei, D.; Zhou, R.; Fu, L.; Hu, Q. Identification of *BnaYUCCA6* as a candidate gene for branch angle in *Brassica napus* by QTL-seq. *Sci. Rep.* **2016**, *6*, 38493. [CrossRef] [PubMed]

59. Chalhoub, B.; Denoeud, F.; Liu, S.; Parkin, I.A.; Tang, H.; Wang, X.; Chiquet, J.; Belcram, H.; Tong, C.; Samans, B.; et al. Early allopolyploid evolution in the post-Neolithic Brassica napus oilseed genome. *Science* **2014**, *345*, 950–953. [CrossRef] [PubMed]

60. Cheng, H.; Hao, M.; Wang, W.; Mei, D.; Wells, R.; Liu, J.; Wang, H.; Sang, S.; Tang, M.; Zhou, R.; et al. Integrative RNA- and miRNA-Profile Analysis Reveals a Likely Role of BR and Auxin Signaling in Branch Angle Regulation of *B. napus. Int. J. Mol. Sci.* **2017**, *18*, 887. [CrossRef] [PubMed]

International Journal of
Molecular Sciences

Article

BvcZR3 and *BvHs1^{pro-1}* Genes Pyramiding Enhanced Beet Cyst Nematode (*Heterodera schachtii* Schm.) Resistance in Oilseed Rape (*Brassica napus* L.)

Xuanbo Zhong [1], Qizheng Zhou [1], Nan Cui [1], Daguang Cai [2] and Guixiang Tang [1,*]

[1] Zhejiang Provincial Key Laboratory of Crop Genetic Resources, Institute of Crop Science, Zhejiang University, Hangzhou 310058, Zhejiang, China; 21716135@zju.edu.cn (X.Z.); 21716136@zju.edu.cn (Q.Z.); 21816113@zju.edu.cn (N.C.)

[2] Department of Molecular Phytopathology, Christian-Albrechts-University of Kiel, Hermann Rodewald Str. 9, D-24118 Kiel, Germany; dcai@phytomed.uni-kiel.de

* Correspondence: tanggx@zju.edu.cn; Tel.: +86-571-88982243; Fax: +86-571-88982243

Received: 26 February 2019; Accepted: 6 April 2019; Published: 8 April 2019

Abstract: Beet cyst nematode (*Heterodera schachtii* Schm.) is one of the most damaging pests in sugar beet growing areas around the world. The *Hs1^{pro-1}* and *cZR3* genes confer resistance to the beet cyst nematode, and both were cloned from sugar beet translocation line (A906001). The translocation line carried the locus from *B. procumbens* chromosome 1 including *Hs1^{pro-1}* gene and resistance gene analogs (RGA), which confer resistance to *Heterodera schachtii*. In this research, *BvHs1^{pro-1}* and *BvcZR3* genes were transferred into oilseed rape to obtain different transgenic lines by *A. tumefaciens* mediated transformation method. The *cZR3Hs1^{pro-1}* gene was pyramided into the same plants by crossing homozygous *cZR3* and *Hs1^{pro-1}* plants to identify the function and interaction of *cZR3* and *Hs1^{pro-1}* genes. In vitro and in vivo cyst nematode resistance tests showed that *cZR3* and *Hs1^{pro-1}* plants could be infested by beet cyst nematode (BCN) juveniles, however a large fraction of penetrated nematode juveniles was not able to develop normally and stagnated in roots of transgenic plants, consequently resulting in a significant reduction in the number of developed nematode females. A higher efficiency in inhibition of nematode females was observed in plants expressing pyramiding genes than in those only expressing a single gene. Molecular analysis demonstrated that *BvHs1^{pro-1}* and *BvcZR3* gene expressions in oilseed rape constitutively activated transcription of plant-defense related genes such as *NPR1* (non-expresser of *PR1*), *SGT1b* (enhanced disease resistance 1) and *RAR1* (suppressor of the *G2* allele of *skp1*). Transcript of *NPR1* gene in transgenic *cZR3* and *Hs1^{pro-1}* plants were slightly up-regulated, while its expression was considerably enhanced in *cZR3Hs1^{pro-1}* gene pyramiding plants. The expression of *EDS1* gene did not change significantly among transgenic *cZR3*, *Hs1^{pro-1}* and *cZR3Hs1^{pro-1}* gene pyramiding plants and wild type. The expression of *SGT1b* gene was slightly up-regulated in transgenic *cZR3* and *Hs1^{pro-1}* plants compared with the wild type, however, its expression was not changed in *cZR3Hs1^{pro-1}* gene pyramiding plant and had no interaction effect. *RAR1* gene expression was significantly up-regulated in transgenic *cZR3* and *cZR3Hs1^{pro-1}* genes pyramiding plants, but almost no expression was found in *Hs1^{pro-1}* transgenic plants. These results show that nematode resistance genes from sugar beet were functional in oilseed rape and conferred BCN resistance by activation of a CC-NBS-LRR R gene mediated resistance response. The gene pyramiding had enhanced resistance, thus offering a novel approach for the BCN control by preventing the propagation of BCN in oilseed rape. The transgenic oilseed rape could be used as a trap crop to offer an alternative method for beet cyst nematode control.

Keywords: *Hs1^{pro-1}*; *cZR3*; gene pyramiding; *Heterodera schachtii*; resistance

1. Introduction

Beet cyst nematode (*Heterodera schachtii* Schm.) is an important pest of sugar beet that can cause significant reductions in yield. Unlike most cyst nematode, the beet cyst nematode (BCN) has a wide host range and can infect more than 218 [1] plant species, including family Brassicaceae and Chenopodiaceae such as sugar beet (*Beta vulgaris* L.), oilseed rape (*Brassica napus* L.), and spinach (*Spinacia oleiracear*) [2]. In general, nematodes can be controlled by treating with nematicides, growing resistant cultivars, and crop rotation with non-host or trap crops [3]. However, control via nematicides application is difficult and expensive because nematode larvae and eggs are well protected. In addition, the use of nematicides increases the threat of environmental pollution and some are prohibited worldwide. Growing resistant cultivars with a monogenic resistance can induce the emergence of more virulent nematode pathotypes. Oilseed rape (*Brassica napus* L.) is a good host for BCN and always rotates with sugar beet in the agro-farming system. Therefore, breeding for BCN resistant oilseed rape is of great value and importance [4].

A previous investigation on 111 *Brassica* germplasm lines revealed that all lines are susceptible to *H. schachtii* [5]. BCN resistance is found only in a few spices of *Brassiceae* including oil radish (*Raphanussativus* L. *ssp.oleiferus* DC.) and white mustard [6] (*Sinapisalba* L.). Oil radish shows complete resistance and is often used as a trap crop to mitigate the degree of damage in infested fields [2]. Resistant/trap crop could stimulate the hatching of larvae, which invaded the roots, but prevents larvae from fulfilling their life cycle and thus lowers BCN populations. A dominant nematode gene *Hs1^Rph*, which is located in the radish chromosome *d*, confers beet cyst nematode resistance [7]. Efforts have been made to transfer this resistance gene into oilseed rape genome by intergeneric hybridization [4,6,8–10]. However, expression of the resistance gene in a hybrid species remains a great challenge due to the extremely low and unstable inheritance [7]. Cloning of the cyst nematode resistance genes and their subsequent transfer into the rapeseed could be one of the most feasible strategies to induce nematode resistance in oilseed rape.

Hs1^pro-1 gene is the first beet cyst nematode resistance gene cloned from sugar beet translocation line (A906001) by a map-based cloning strategy [11]. The translocation line carried the locus from *B. procumbens* chromosome 1 that confers resistance to *H. schachtii* in sugar beet (*Beta vulgaris* L.) [11]. The *Hs1^pro-1* gene is different from other nematode resistance genes amid the presence of the NBS-LRR structure, such as *Mi* [12], *Gpa2* [13] and *Hero* [14], due to the presence of the NBS-LRR structure. The resistance mechanism of *Hs1^pro-1* gene is based on the gene-for-gene relationship [15]. The transcript of *Hs1^pro-1* gene is upregulated about fourfold after one day of nematode infection. However, no considerable change in the transcript accumulation of *Hs1^pro-1* is recorded in uninfected roots of resistant beet plants [15]. McLean et al. [16] published a complete sequence of the *Hs1^pro-1* protein, which included an additional 176 amino acid at N-terminal extension conferred resistance to soybean (*Glycine max* L.) cyst nematode (*Heterodera glycines*).

Hunger et al. [17] cloned 47 resistance gene analogs (RGAs) from genomic DNA of sugar beet. Most cloned resistance genes belong to nucleotide-binding site leucine-rich repeat (NBS-LRR) gene family [18,19]. Members of this family have been isolated in both dicot and monocot plants, exhibiting resistance to a variety of plant pathogens, including bacteria, fungi, viruses, and nematodes [20]. The cloning of conserved sequence of NBS domain has been successfully applied to isolate resistance gene candidates or resistance gene analogs (RGAs) from plant genomes, using a degenerate primer-based PCR strategy. For example, from sugar beet translocation line (A906001), *cZR3* was cloned using a degenerated primer-based PCR strategy [21,22]. The gene is similar to a subset of CC-NBS-LRR resistance proteins, including nematode resistance genes *Mi* from tomato and *Gpa2* from potato [12,21]. The phylogenetic analysis shows that this gene originates from the ancestral gene from which *I2C1* [23] (vascular wilt disease resistance), *Xa1* [24] (bacterial blight resistance) and *Cre3* [25] (nematode resistance) were also originated. For CC-NBS domains, RGAs are required for initiation of a necrotic hypersensitive response, as well as for cell death on the initial infection sites [26,27]. RGAs are ubiquitous in plant genomes, and often cluster with close linkage to active resistance genes [28,29]. Because no complete resistance could be observed so far by transgenic sugar beet plants (Cai unpublished data), it is proposed that additional genes are required to confer full resistance [30]. We hypothesized that the *cZR3* gene may interact with the *Hs1^{pro-1}* gene to confer resistance against sugar beet nematode in oilseed rape. Therefore, we transferred the *Hs1^{pro-1}* and *cZR3* genes into oilseed rape using hypocotyl explants by *A. tumefaciens* mediated transformation method and pyramided the *cZR3Hs1^{pro-1}* genes by crossing homozygous transgenic *cZR3* and *Hs1* plants. The beet cyst nematode resistance tests in vitro and in vivo were detected in *cZR3*, *Hs1^{pro-1}* and *cZR3 Hs1^{pro-1}* pyramiding plants. Possible resistance mechanisms of *Hs1^{pro-1}*, *cZR3* and *cZR3Hs1^{pro-1}* gene pyramiding were also discussed in the present study.

2. Results

2.1. Generation of Transgenic Plants and Identification

The recombinant plasmid-DNA carrying *Hs1^{pro-1}* and *cZR3* genes (Figure 1A,B) were used for transformation to generate transgenic *Hs1^{pro-1}* and *cZR3* plants. Figure 1C–F shows the process of transgenic *Hs1^{pro-1}* and *cZR3* plants generation. The hypocotyl segments were pre-cultured on the CIM to induce callus formation for two days and then infected with *A. tumefaciens*. After co-cultivation, explants were cultivated on SIM (Figure 1C) supplemented with 500 mg/L Carb to eliminate the extra bacterium and 50 mg/L Kan to select the positive transgenic plants. The adventitious shoots were observed from the cut end of hypocotyl explants 30 days after cultivation in SIM (Figure 1D). The adventitious shoots were cut off and transferred to shoot elongation medium supplemented with 500 mg/L Carb and 50 mg/L Kan. The putative green transgenic shoots were observed 30 after cultivation in SEM (Figure 1E). The green shoots were transferred to rooting medium to develop the roots (Figure 1F). The rooted plantlets were transferred to the soil and the survived putative transgenic plants were assayed by PCR (Figure 1G,H) and Southern blot (Figure 1I). In general, 194 explants were transformed with pAM194-Hs1 vector and four rooting seedlings were obtained, of which three were identified as transgenic seedlings by PCR (Figure 1G) and Southern blot analysis (Figure 1I). A total of 162 explants were transformed with pAM194-cZR3 vector and 12 rooting seedlings were obtained, of which three were identified as transgenic seedlings by PCR (Figure 1H) and Southern blot analysis (Figure 1I).

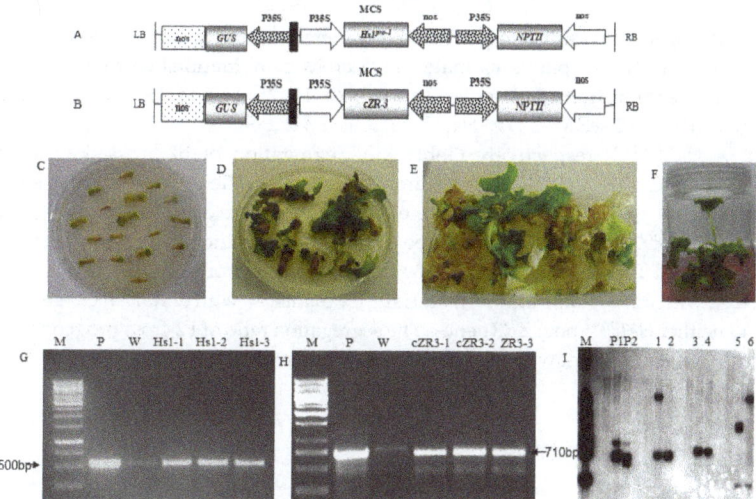

Figure 1. Generation and identification of transgenic *Hs1^{pro-1}* and *cZR3* plants from hypocotyle explants of B. *napus* cv. *Zheshuang 758* by *A. tumefaciens* mediated transformation. (**A,B**) Diagram of

T-DNA region of the binary vector pAM194-*Hs1^pro-1* (**A**) and pAM194-*cZR3* (**B**). LB, T-DNA left border; RB, right border; P35S, Cauliflower mosaic virus (CaMV) 35S promoter; NOS, nopaline synthase terminator; MCS, multi clone site including XhoI restriction site; *Hs1^pro-1*, *Beta vulgaris Hs1^pro-1* gene open read fragment sequence; *cZR3*, *Beta vulgaris* resistance gene sequence; *GUS*, β-glucuronidase report gene; *NPTII*, neomycin phosphotranferase gene for kanamycin resistance. (**C**) Callus induction from hypocotyl explants after co-cultivation with *A. tumefaciens*. (**D**) Shoots regenerated on SIM. (**E**) Shoots elongation on SEM. (**F**) Shoots rooted on RM. (**G,H**) PCR assay for transgenic *Hs1^pro-1* plants (**G**) and *cZR3* plants (**H**) M means 1 kb DNA ladder; P means *Hs1* or *cZR3* plasmid DNA, W means non-transgenic plant DNA; Hs1-1, Hs1-2 and Hs1-3 mean T0 independent transgenic *Hs1^pro-1* plants; cZR3-1, cZR3-2 and cZR3-3 mean T0 independent transgenic *cZR3* plants. (**I**) Southern blot assay for T0 independent transgenic *cZR3* and *Hs1^pro-1* lines, P1P2 means *Hs1* and *cZR3* plasmid DNA, respectively; 1, 3, and 5 mean T0 independent transgenic *Hs1* plants; 2, 4, and 6 mean T0 independent transgenic *cZR3* plants.

2.2. Generation of cZR3Hs1^pro-1 Gene Pyramiding Plants

T1 seeds were obtained by self-pollination of the six T0 independent transgenic *Hs1^pro-1* (*Hs1^pro-1*-1, -2, -3) and *cZR3* (*cZR3*-1, -2, -3) oilseed rape plants. The segregation ratio of T1 progeny was determined by PCR analysis using *Hs1^pro-1* and *cZR3* genes specific primers. The exogenous gene began to segregate in T1 progeny and could be inherited to the next generation (Table 1). The results show the segregation ratio of *Hs1^pro-1*-1, -3 and *cZR3*-1, -2, -3 were nearly 3:1 and the χ^{c2} test was positive.

Table 1. Segregation rates of T1 progeny of three independent transgenic *Hs1^pro-1* and *cZR3* lines.

Lines	Number of Plants			Expected Ratio	*p*-Value
	T1	*Hs1^pro-1* or *cZR3+*	*Hs1^pro-1* or *cZR3−*		
Hs1^pro-1-1	21	19	2	3:1	$p > 0.05$
Hs1^pro-1-2	6	1	5	3:1	$p < 0.05$
Hs1^pro-1-3	18	16	2	3:1	$p > 0.05$
cZR-3-1	8	6	2	3:1	$p > 0.05$
cZR-3-2	10	8	2	3:1	$p > 0.05$
cZR-3-3	37	27	10	3:1	$p > 0.05$

Hs1^pro-1 cZR3 genes pyramiding plants were generated by artificial hybridization using *cZR3* plant as female and *Hs1^pro-1* plants as male. Five cross combinations carried out between *cZR3* and *Hs1^pro-1* independent transgenic lines. In total, 20, 15, 27, 14 and 35 seeds were harvested for F1 *cZR3*-1 × *Hs1^pro-1*-1-2, *cZR3*-2.1 × *Hs1^pro-1*-1-3, *cZR3*-2.2 × *Hs1^pro-1*-1-3, *cZR3*-3 × *Hs1^pro-1*-1-2 and *cZR3*-3 × *Hs1^pro-1*-3, respectively (Table 2). Segregation of F2 crossed progeny occurred and the heterozygous *cZR3Hs1^pro-1* gene pyramiding was identified by PCR analysis using specific gene primers (Figure 2). There were four possible exogenous gene combinations: e.g., Type I: plants, containing both *Hs1^pro-1* and *cZR3* genes, Type II and III: plants including a single exogenous gene *Hs1^pro-1* or *cZR-3* and Type IV: plants including neither *Hs1^pro-1* nor *cZR3*. Of all 111 F2 individuals, we identified 32 with *cZR3*-plants and 19 with *Hs1^pro-1*- plants, 47 with *cZR3Hs1^pro-1* genes pyramiding plants, and 13 neither *Hs1^pro-1* nor *cZR3* genes. The segregation ratio of F2 cross progeny was 47:32:19:13 (9:3:3:1) which were consistent with Mendel's laws of inheritance.

Table 2. Segregation ratios in the F2 generation derived from a cross between homozygous transgenic $Hs1^{pro-1}$ and $cZR3$ plants.

Cross Combination	Number of Plants					Expected Ratio	*p*-Value
	F2	$cZR3Hs1^{pro-1}+$	$cZR3+$	$Hs1^{pro-1}+$	$cZR3Hs1^{pro-1}-$		
$cZR3$-1 $\times Hs1^{pro-1}$-2	20	10	3	4	3	9:3:3:1	$p > 0.05$
$cZR3$-2.1 $\times Hs1^{pro-1}$-3	15	1	12	1	1	9:3:3:1	$p < 0.05$
$cZR3$-2.2 $\times Hs1^{pro-1}$-3	27	16	4	4	3	9:3:3:1	$p > 0.05$
$cZR3$-3 $\times Hs1^{pro-1}$-2	14	6	2	2	4	9:3:3:1	$p < 0.05$
$cZR3$-3 $\times Hs1^{pro-1}$-3	35	15	8	9	3	9:3:3:1	$p > 0.05$

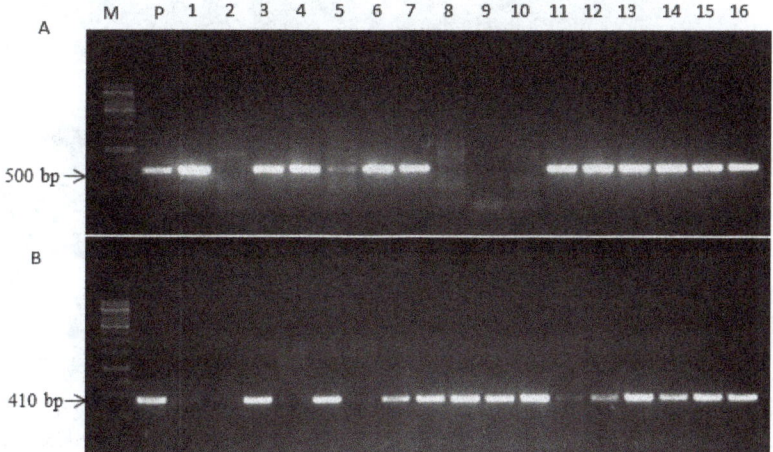

Figure 2. Identification of $cZR3Hs1^{pro-1}$ pyramiding homozygous plants in F2 crossing progeny by PCR assay using specific $cZR3$ and $Hs1^{pro-1}$ gene primers. The amplification fragment length of $Hs1^{pro-1}$ and $cZR3$ gene was 500 bp and 410 bp, respectively. (**A**) Identification of $Hs1^{pro-1}$ gene in F2 crossing progeny. (**B**) Identification of $cZR3$ gene in F2 crossing progeny. M represents a molecular marker (1 kb); P represents plasmid DNA for $Hs1^{pro-1}$ or $cZR3$ binary vector; 1–16 represent different F2 rapeseed plants.

2.3. Beet Cyst Nematode Test In Vitro and In Vivo

To determine the resistance to BCN of the $cZR3$ and $Hs1$ genes and $cZR3Hs1$ gene pyramiding in oilseed rape, BCN resistance tests in vitro and in vivo were performed in T2 $cZR3$, $Hs1^{pro-1}$ and F3 $cZR3Hs1^{pro-1}$ pyramiding seeds. For this, transgenic and gene pyramiding seeds were geminated on agar plates containing 150 mg/L kanamycin for selection of transgenic plants (Figure 3). The surviving seedlings were transferred to six-well plates for nematode infection experiments (Figure 4A), where each plant was inoculated with 200 infective nematode juveniles and was repeated three times for individual plant. The J2 penetration rate was determined one week after nematode inoculation. On average 15–22% of inoculated BCN J2 juveniles penetrated the plants containing exogenous genes as well as the wild type plants. However, a significant difference was observed between the transgenic and the control plants by counting developed females six weeks after nematode inoculation. Most juveniles in transgenic plants were not fully developed and became smaller and translucent, whereas the well-developed nematode females were easily recognizable on wild type plants (Figure 4B). These results demonstrate that $Hs1^{pro-1}$ and $cZR3$ could confer a certain level of resistance to BCN in oilseed rape. There was a significant difference in the reduction of the number of developed females among transgenic generations and $cZR3Hs1^{pro-1}$ gene pyramiding plants. On average, 10.8 ± 0.9 developed females were found in each wild type plant, while 5.3 ± 1.4 developed females in $cZR3$, 7.1 ± 2.4 in $Hs1^{pro-1}$, and 4.1 ± 1.3 in $cZR3{:}Hs1^{pro-1}$ pyramiding plants were counted (Figure 5). Thus, *cZR3Hs1^{pro-1}* gene pyramiding could enhance the BCN resistance by decreasing the number of developed females per plant.

Figure 3. Selection of transgenic rapeseed plants on MS germination containing 150 mg/L kanamycin medium (**A**) and histochemical GUS staining (**B**). The healthy and green plants were used for nematode infestation in vitro and in vivo.

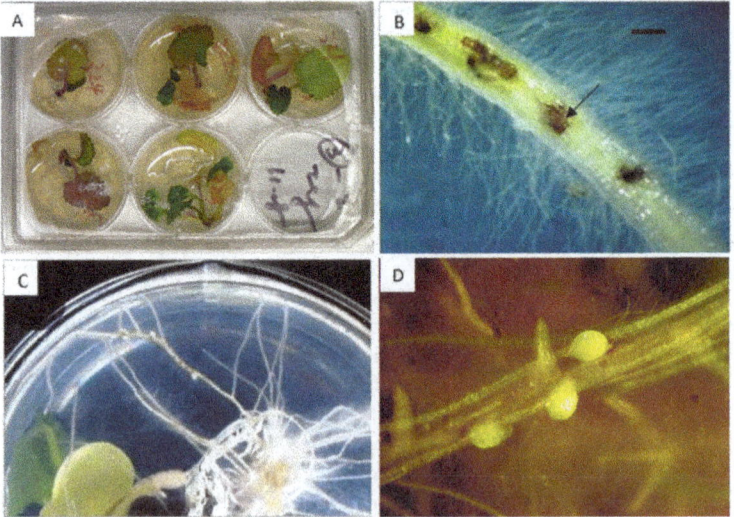

Figure 4. Beet cyst nematode resistance analysis in vitro. (**A**) The kanamycin resistance transgenic plants were planted on the six-well plate with KNOP medium. (**B**) Second stage juveniles (J2s) were inoculated near the plant root and larval penetration could be seen in the oilseed rape roots (dark arrow). The black bar equals 500 μm. (**C**) Female cyst nematode developed on the root surface after 21 days of inoculation with J2. (**D**) The stereo microscope figures of developed female on oilseed rape root.

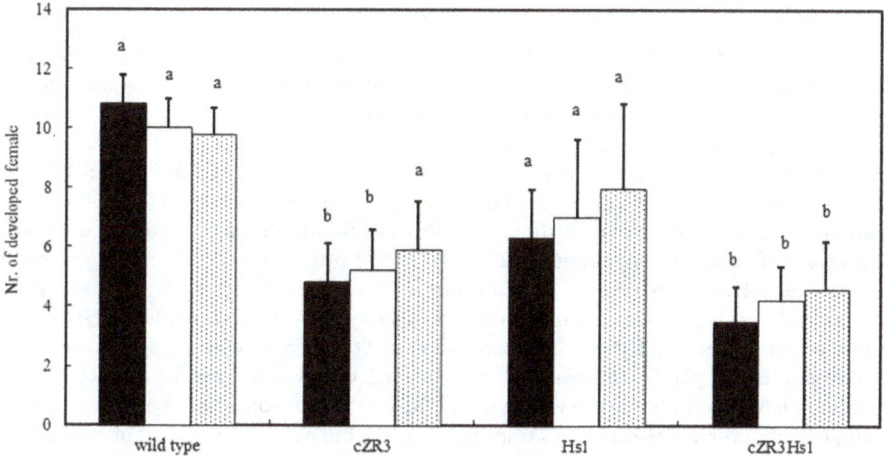

Figure 5. Numbers of developed female per plant on transgenic and wild type plants according to cyst nematode test in vitro. Wild type means non-transgenic oilseed rape (*B. napus* L.); transgenic plants consist of cZR3 and Hs1; and cZR3Hs1 means *cZR3, Hs1^{pro-1}* and *cZR3Hs1^{pro-1}* populations. The black, white, dotted rectangles represent independent wildtype plants and transgenic lines. The averages ± standard errors from three separate replicates are shown. The values with different letters are significantly different at $p \leq 0.05$ as determined by Duncan's test (a, b).

BCN resistance tests in vivo were performed under the greenhouse condition. Six weeks after BCN inoculation, the number of developed females per plant was counted. The number of developed females in wild type was 50 ± 4.7, whereas the number of developed females among *cZR3, Hs1^{pro-1}* and *cZR3Hs1^{pro-1}* was 38 ± 7.4, 33 ± 3.7 and 24 ± 5.2, respectively (Figure 6). There was a significant difference in the number of developed females among wild type, *cZR3* and *Hs1^{pro-1}* transgenic and *cZR3Hs1^{pro-1}* pyramiding plants.

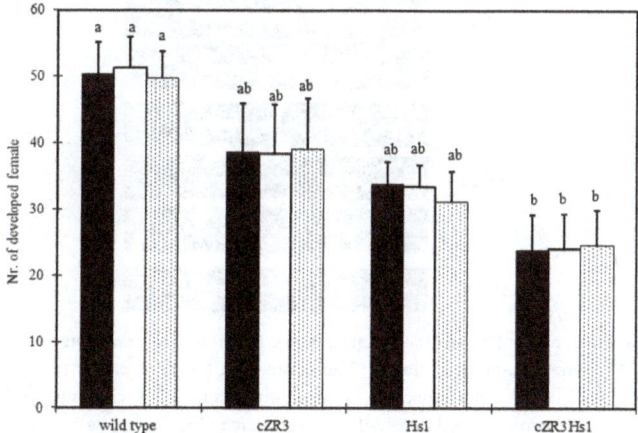

Figure 6. Numbers of developed female per plant on transgenic and wild type plants according to cyst nematode test in vivo. Wild type means non-transgenic oilseed rape (*B. napus* L.); transgenic plants consist of cZR3 and Hs1; and cZR3Hs1 means *cZR3, Hs1^{pro-1}* and *cZR3Hs1^{pro-1}* plants. The black, white, dotted rectangles represent independent wildtype plants and transgenic lines. The averages ± standard errors from three separate replicates are shown. Values with different letters are significantly different at $p \leq 0.05$ as determined by Duncan's test (a, b).

2.4. Determination of RGA-Mediated Signaling Pathways

To clarify whether expression of *cZR3* and *Hs1^pro-1* activates a specific signaling pathway, the transcript levels of four key genes involve in distinct defense pathways were analyzed in transgenic *cZR3*, *Hs1* and *cZR3Hs1* gene pyramiding plants compared with those in wild type by semi-RT-PCR (Figure 7). The gene list included *NPR1* (non-expresser of *PR1*) and *EDS1* (enhanced disease resistance 1), which are key regulators of resistance responses triggered by TIR-NBS-LRR-R-proteins mediated response and *SGT1* (suppressor of the *G2* allele of *skp1*) as well as *RAR1* (required for *Mla12* resistance), which are both involved in the non-TIR-(CC) NBS-LRR-R proteins mediated responses [31]. The expression of *NPR1* gene in transgenic *cZR3* and *Hs1^pro-1* plants seemed to be slightly up-regulated and there was enhanced expression in *cZR3Hs1^pro-1* gene pyramiding plants. The expression of *EDS1* gene did not change significantly among transgenic *cZR3* and *Hs1* plants, *cZR3Hs1^pro-1* gene pyramiding plants and wild type. The expression of *SGT1b* gene was slightly up-regulated in transgenic and *Hs1^pro-1* plants compared with the wild type, however, its response in *cZR3Hs1^pro-1* gene pyramiding plant was similar to the wild type and had no interaction effect. *RAR1* gene expression was significantly up-regulated in *cZR3* transgenic plants, but almost no change in expression was found in *Hs1^pro-1* transgenic plants compared with wild type plants. However, *RAR1* gene expression was also significantly enhanced in *cZR3Hs1^pro-1* gene pyramiding plants. Furthermore, the relative expression of *SGT1b* and *RAR1* genes was also analyzed. Compared with wild type, expression level of *SGT1b* in *cZR3* transgenic plants and *cZR3Hs1^pro-1* pyramiding plants was decreased slightly, while it was increased in *Hs1* transgenic plants (Figure 8A). The expression of *RAR1* gene by qRT-PCR analysis showed that *RAR1* was slightly expressed in wild type and *Hs1^pro-1* transgenic plants. The expression of *RAR1* in *cZR3* transgenic plant was about 50 times higher than the wild type, while the expression of *RAR1* was enhanced by *cZR3Hs1^pro-1* gene pyramiding, which was about 150 times higher than the wild type (Figure 8B).

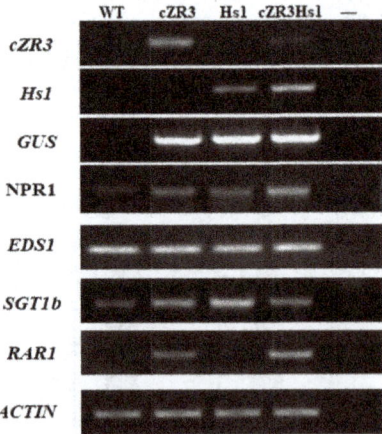

Figure 7. Expression of the key defense related genes in wild-type, transgenic *cZR3* and *Hs1^pro-1*, and *cZR3Hs1^pro-1* gene pyramiding plants. The expression levels of *cZR3*, *Hs1^pro-1*, *GUS*, *NPR1*, *EDS1*, *SGT1* and *RAR1* were determined by semi-quantitative RT-PCR with independent transgenic and pyramiding lines, while oilseed rape wild type *zheshuang 758* served as control. "–" represents the negative control.

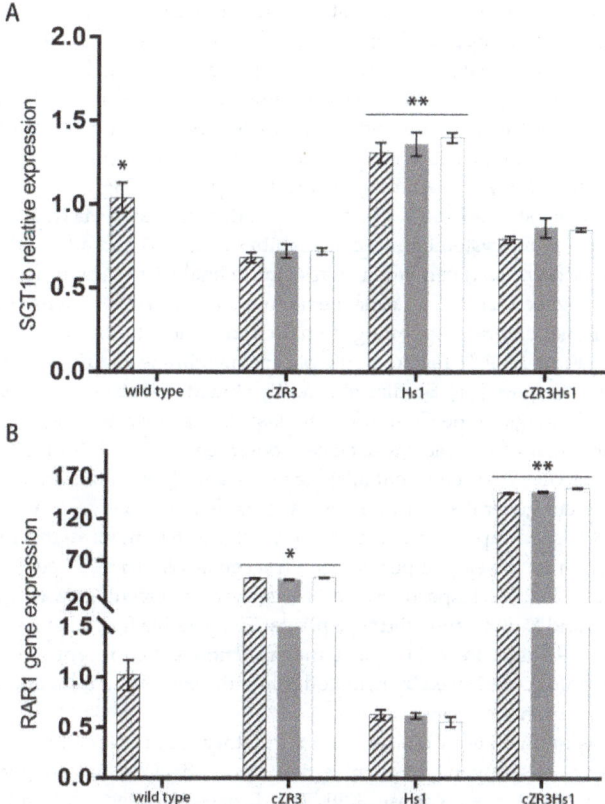

Figure 8. The relative expression of *SGT1b* (**A**) and *RAR1* (**B**) gene among wild type, *cZR3* and *Hs1^pro-1^*, and *cZR3Hs1^pro-1^* pyramiding plants by qRT-PCR. The diagonal stripes, grey, white rectangles represent independent wildtype plants and transgenic lines. The averages ± standard errors from three separate replicates are shown. Significant difference is indicated by different symbols (* and **), as determined by Duncan's test.

3. Discussion

Crops are attacked by nematodes causing considerable economic losses worldwide. The estimated worldwide losses due to plant parasitic nematodes are about $125 billion annually [32]. An integrated strategy, e.g., including trap crop planting, nematicide application and cultivation of resistant sugar beet varieties, could control beet cyst nematode infestation in sugar beets [33]. Effective nematode control could be achieved by cultivating a nematode resistant variety that can reduce nematode populations up to 70% in field conditions [33]. Breeding of resistant cultivars is the most desired and promising alternative because regeneration of whole transgenic sugar beet plants with *A. tumefaciens* mediated transformation is laborious and timing consuming. In the present study, we transferred *BvHs1^pro-1^* and *BvcZR3* resistance genes, into oilseed rape to develop cyst nematode resistance crop. In vitro and in vivo nematode tests showed the developed female reduced 50.93%, 34.26%, 24.0%, and 34.0%, respectively, in *BvcZR3* and *BvHs1^pro-1^* transgenic plants. Our results are consistent with previous studies, where *A. rhizogenes*-mediated transformation was performed to generate transgenic *Hs1^pro-1^* and *cZR3* hairy roots in sugar beet [11,21].

Different R genes often confer resistance to different isolates, races or biotypes of beet cyst nematode. The simultaneous expression of different R genes could broaden the resistance spectrum, as it may provide resistant against various races or isolates [34]. The multiple genes pyramiding strategy has been applied in genetic engineering to achieve durable resistance against phytopathogen [35,36] and nematode resistance [37,38]. Bharathia et al. [36] reported *Allium sativum* (*asal*) and *Galanthus nivalis* (*gna*) lectin genes pyramided into rice lines through sexual crosses between two stable transgenic rice lines, which endowed the pyramided rice lines with enhanced resistance to major sap sucking insects. Urwin et al. [38] transferred cowpea trypsin inhibitor (*CpTI*) and a cystatin gene (*Oc-ID86*) into *Arabidopsis* and the *CpTI Oc-ID86* transgenic plants displayed enhanced resistance against *H. schachtii*. Chan et al. [37] demonstrated via dual gene overexpression system that utilizing a plant cysteine proteinase inhibitor (*CeCPI*) and a fungal chitinase (*PjCHI-1*) in tomato (*Solanum lycopersicum*) can enhance resistance against root-knot nematode (*Meloidogyne incognita*). The *Hs1^{pro-1}* gene encodes a plasma membrane protein with an extensive leucine-rich region, which contains a transmembrane spanning domain and a short hydrophobic C-terminal domain. It could be speculated that *Hs1^{pro-1}* resides in the plasma membrane as a receptor with its N-terminus toward the extracellular space [11]. Similar to recently cloned nematode R-genes, including *Mi* [12] and *Gpa2* [21], *cZR3* belongs to the CC-NBS-LRR class of R-protein lacking a signal sequence [21]. Previous studies show that no complete resistance is observed in *Hs1^{pro-1}* transgenic sugar beet plants and a second gene is proposed to be essential for the resistance [30]. In this study, *cZR3Hs1^{pro-1}* genes pyramiding plants were generated through sexual crosses made between two transgenic *Hs1^{pro-1}* and *cZR3* plants of oilseed rape. Cyst nematode resistance test in vitro and in vivo showed that the number of females that developed per plant on transgenic *cZR3Hs1^{pro-1}* genes pyramiding plants were reduced 52.0% and 62.0%, respectively, and the reduction percentage was significantly different with transgenic *cZR3* and *Hs1^{pro-1}* and wild type plants. The possible functional model for *cZR3Hs1^{pro-1}* mediated response was based on *cZR3*, which may function as a co-receptor with *Hs1^{pro-1}* together recognizing the *Avr* products released by nematodes into the cytoplasm initiating signal transduction that finally leads to resistance response.

Thus far, little is known about the complex regulatory role of cloned nematode R-genes. It is generally believed that these genes recognize nematode effectors and trigger specific signaling pathways that lead to resistance responses [39]. The disease resistance mechanisms in model plant *Arabidopsis thaliana* were extensively studied, and the gene-for-gene hypothesis has been proposed for a long time. R genes interact with the corresponding pathogenic toxic genes, thus causing local reactive oxygen species accumulation, programmed cell death and local allergic reactions [40]. On the one hand, these reactions limit the growth and extend of pathogenic bacteria in infected sites. On the other hand, these reactions release signaling molecules to surrounding cells, further inducing expression of defense genes to boost whole-plant resistance. Some crucial genes mediated by R gene have been found in recent years, such as *RAR1*, *NPR1*, *EDS1* and *SGT1*. Studies have shown that the CC-NBS-LRR R gene mediates disease resistance reaction associated with *RAR1* and *SGT1*, while *NPR1*, *EDS1* and *PAD4* are related with TIR-NBS-LRR R genes [31]. This study proved that cZR-3 and *Hs1^{pro-1}* could independently activate a *RAR1/SGT1* dependent signaling pathway in plants, which is essential for a CC-NBS-LRR R gene mediated resistance response. Hence, the present study provides a new approach to develop BCN resistance in oilseed rape plants based on stacking of *cZR3Hs1^{pro-1}* genes that confer a high level of BCN resistance in transgenic plants.

4. Materials and Methods

4.1. Plant Transformation

Direct shoots regeneration from hypocotyl explants of semi-winter-type oilseed rape cultivar *zheshuang 758* were transformed by *A. tumefaciens*, as described by Tang et al. [41]. The PAM194-*Hs1^{pro-1}* and PAM194-*cZR3* vector contains the complete *Hs1^{pro-1}* ORF (Gene accession number U79733) and *cZR3* (Gene accession number DQ907613), respectively, driven by cauliflower 35S promoter and nopaline synthase terminator, the neomycin phosphotranferase marker gene (*nptII*) and β-glucuronidase (GUS) reporter gene (Figure 1A,B) driven by the cauliflower 35S promoter. Both vectors were presented by Prof. Daguang Cai from Institute of Molecular Phytopathology, Kiel University, Germany. The binary vectors were then transformed into the competent cells of *A. tumefaciens* strain GV3101 by the freeze–thaw method.

In brief, we used about 10 mm hypocotyl segments was obtained by growing seeds of oilseed rape seeds on germination medium [42] (2.22 g/L 1/2MS basal medium from Phytotechnology Laboratories®, Shawnee Mission, Lenexa, KS, USA, 20 g/L sucrose, 8 g/L Agar, pH 5.8) at 25 °C and 16 h light/8 h dark periods for about 4–5 days as explants. The isolated explants were pre-cultured on the callus induction medium (CIM) (MS 4.43 g/L, 1 mg/L 2,4-D, 30 g/L sucrose, 8 g/L Agar, pH 5.6) at 25 °C and 16 h light/8 h dark periods for 3 days. The pre-cultured explants were infected by *A. tumefaciens* suspension (OD600 0.3–0.4) for 10 min, then subsequently blotted on sterile filter paper and placed on CIM at 22 °C under dim light co-cultivation for 3 days. After co-cultivation, the explants were placed on SIM media for shoot generation (SIM) (MS 4.43 g/L, 4 mg/L N-6-benzylaminopurine (BAP), 0.1 mg/L 1-Naphthaleneacetic acid (NAA), 5 mg/L silver nitrate (AgNO$_3$), 500 mg/L carbenicillin (Carb), 50 mg/L kanamycin (Kan), 30 g/L sucrose, 8 g/L Agar, pH 5.8) at 25 °C and 16 h light/8 h dark periods for 2 weeks. The media were replaced every two weeks until multiple shoots were generated. The multiple induced shoots were isolated and transferred to shoot elongation media [43] (SEM) (B5 basal medium from Phytotechnology laboratories® 3.21 g/L, 10 g/L sucrose, 50 mg/L Kan, 9 g/L Agar, pH 5.8). The elongated shoots were transferred to SEM to develop the roots. The rooting plants were first grown in 10 cm diameter pots containing nutrient soil (Pindstrup, Ryomgaard, Denmark). The surviving plants were transferred to greenhouse to vernalize at 4 °C 16 h light/8 h dark periods for 40 days. The vernalized plants were transferred to the greenhouse to grow normally to harvest seeds at 25 °C and 16 h light/8 h dark periods. When the plants began to flower, the inflorescence of each independent putative transformed line was covered with transparent plastic bags and the plants were allowed self-pollination to acquire T1 seeds.

4.2. PCR Analysis

Forty-day-old rapeseed leaves were used to extract genomic DNA from transgenic and wild-type plants using CTAB method. PCR analyses were performed using specific gene primers for *Hs1^{pro-1}* (accession number: U79733.1) (F: 5′-GGCACCATCCAAACTCGG-3′, R: 5′-CGAATAAGTGAGAGGATC-3′), *cZR3* (accession number: DQ907613) (F: 5′-GGCAAAACTGCTCTTGCC-3′ and R: 5′-AGCCCTATCAATAACTCC-3′) and *cZR3* (F: 5′-AGTTATTGATAGGGCTATGG-3′ and R: 5′-ATACTTGAAGCAGTCAGG-3′). And the size of amplifying products are 500-bp [11], 710-bp and 410-bp [22] respectively. The PCR reaction was performed at 94 °C for 3 min, followed by 35 cycles of 94 °C for 1 min, 55 °C (*Hs1^{pro-1}* and *cZR3*) for 1 min and 72 °C for 1 min 20 s with a final extension at 72 °C for 10 min. PCR-amplified products were analyzed on 1% agarose gel, stained with ethidium bromide and fluoresce under ultraviolet light.

4.3. Southern Blot Analysis

Genomic DNA of each independent transgenic *cZR3* and *Hs1^{pro-1}* lines was restricted by Hind III (Takara Bio Inc., Dalian, China) at 37 °C for 5 h, and was separated on 0.7% agarose gel overnight and transferred onto a Hybond N$^+$ nylon membrane (GE Healthcare, RPN 303B, Piscataway, NJ, USA) using the alkaline transfer buffer as recommended by the manufacturer. Southern blots were hybridized using 32P-labeled *cZR-3* and *Hs1^{pro-1}* DNA fragment as probe at 62 °C overnight. The blots were washed twice with 0.5× saline sodium citrate (SSC), 0.2% *w/v* SDS for 30 min and together with the film exposed at −70 °C for 48 h [44].

4.4. Histochemical GUS Assays

The histochemical *GUS* assay was carried out according to Jefferson et al.'s [45] method. Ten milliliters of X-Gluc solution including 50 mM Na$_3$PO$_4$ buffer (pH 7.0), 0.2 mg/mL X-Gluc (5-bromo-4-chloro-3-indolyl β-D-glucuronide) and two drops of Triton-100 were added in the Petri-dish, which contained the putative transgenic leaves and plants. The leaves and plants were covered with X-Gluc solution and incubated at 37 °C for 16 h. Afterwards, the samples were washed using 70% ethanol to remove the chlorophyll. The GUS staining signals were evaluated under a stereomicroscope (Stemi SV 11, Zeiss, Jena, Germany).

4.5. Genes Pyramiding and Progeny Analysis

Hs1^{pro-1} and *cZR3* genes pyramiding were generated by artificial hybridization using T0 putative transgenic *cZR3* line as female and T0 putative transgenic *Hs1^{pro-1}* line as male. F1 seeds obtained from the crossed inflorescence and *cZR3Hs1^{pro-1}* positive F2 plant were screened for the presence of *cZR3Hs1^{pro-1}* genes by PCR.

T1 seeds obtained from self-pollination of 3 T0 independent *Hs1^{pro-1}* and *cZR3* plants were sown in 9 cm Petri-dishes containing germination medium with 150 mg/L kanamycin. The surviving plants were transferred to pots containing peat moss (Pindstrup, Balozi, Lativa) and maintained in green house. All T1 were screened for the presence of *Hs1^{pro-1}* or *cZR3* genes by PCR. Subsequent generations were obtained by self-pollination of transgenic plants and confirmed by PCR.

4.6. Propagation of H. Schactii

In vitro cultured mustard (*Sinapis albacv.* Albatros) was used as host plants for beet cyst nematode (*H. schactii*) stock propagation. The mustard seeds were surface sterilized in 70% ethanol and in 5% NaClO solution, containing 2–3 drops of Tween 20 for 10 and min, then rinsed at least 3 times with sterilized water. The sterilized seeds were sown on half-strength MS media with 0.8% agar on 9 cm Petri-dishes at 25 °C in the dark. After 7-days of germination, the seedlings were transferred to 15 cm Petri-dishes containing 0.2× KNOP medium with 2% (*w/v*) sucrose and 0.8% (*w/v*) Daishin agar (Duchefa, Haarlem, The Netherlands) and placed in dark for 4 weeks at 25 °C [46]. The cysts were picked from the fields and J2 larvae hatched on 50 μm gauze stimulated by incubating cysts in 3 mM ZnCl$_2$ for 7–12 days. The J2 larvae were collected with 10 μm gauze, surface-sterilized with 0.1% HgCl$_2$, washed four times in sterile water and re-suspended in 0.2% (*w/v*) Gelrite (Duchefa). The sterile J2 larvae were directly inoculated to the mustard roots and the cysts were propagated in the mustard roots in vitro. These plates were used as a stock for beet cyst nematode.

4.7. Nematode Resistance Assay In Vitro and In Vivo

T2 transgenic *Hs1^{pro-1}*, *cZR3* and F3 *Hs1^{pro-1}cZR3* gene pyramiding oilseed rape plants were used for nematode infestation analysis, under both in vitro and in vivo. Nematode resistance analysis in vitro was performed according to Sijmons et al. [47]. First, the T1 transgenic *Hs1^{pro-1}* and *cZR3* and F3 *Hs1^{pro-1}cZR3* gene pyramiding seeds were surface sterilized, and then germinated on 15 mm Petri dishes containing half-length MS germination medium supplemented with 150 mg/L kanamycin to select the positive transgenic plants. Ten days after germination, the surviving seedlings were transferred to a six-well plate containing KNOP medium and cultured for further 7 days at 23 ± 1 °C with a fluorescent light illumination regime of 16 h/8 h (day/night, 100 μ·mol m^{-2} s^{-1}) and the relative humidity was 75%. Two hundred sterile infective J2 larvae were inoculated to each plant by a veterinary syringe. The number of females that developed per plant was counted 6 weeks after inoculation under a stereomicroscope (Stemi SV11, Zeiss, Jena, Germany). At least three 6-well plates were performed for each transgenic line and wild type as biological replicates. For nematode resistance test in vivo, the T2 transgenic *Hs1^{pro-1}*, *cZR3* and F3 *Hs1^{pro-1}cZR3* genes pyramiding seeds were germinated on the germination medium with 150 mg/L kanamycin for 10 days to select the positive transgenic ones. The positive transgenic single plants were transplanted in plastic tubes (3 cm × 4 cm × 20 cm) filled with silver sand and moistened with Steiner I nutrient solution [48] and cultured in a greenhouse at 25 °C 16 h/8 h (light/dark) periods for further 2 weeks. Each plant was then inoculated with 2 mL suspension containing approximately 2000 pre-hatched J2 BCN larvae by a veterinary syringe. Six weeks after BCN inoculation, the root system was washed free from sand by high-speed tap water and the number of developed BCN cysts was counted. Each line and wild type were replicated 15 plants at least.

4.8. Semi Real Time PCR Analysis

For real time PCR analysis, the total RNA from leaves and roots was extracted with Trizol (Gibco, BRL life technologies, Grand Island, NY, USA) according to the manufacturer's protocol. First strand cDNA was synthesized with 1 μg of purified total RNA using PrimeScriptTMRT reagent Kit with gDNA Eraser (TaKaRa). Real-time PCR reaction was carried out according to SYBR Premix Ex Taq II (TaKaRa). Specific primers of defensing genes were designed according to sequences of *Arabidopsis* defensing genes in NCBI and summarized in Table 3. The semi-quantitative PCR was performed in 50 μL reactions consisting of 2.5 μL 10 ng/μL cDNA, 5 μL 10× buffer, 0.5 μL 10 mM dNTPs, 5 μL each of 10 pmol/μL primer, 2.5 units of Taq polymerase and 31.5 μL H_2O under the PCR program: 94 °C for 50 s, 51 °C for 1 min and 72 °C for 1 min for 25 cycles, followed by 10 min at 72 °C. Amplicons were separated on a 1% (*w/v*) agarose gel and visualized under UV-light. The levels of gene expression were calculated by comparison of the densities of the PCR products, in which house-keeping *ubiquitin* gene served as an internal control and the mRNA levels for each cDNA probe were normalized to the ubiquitin message RNA level (5'-ACTCTCACCGGAAAGACAATC-3' and 5'-TGACGTTGTCGATGGTGTCAG-3'). Quantitative RT-PCR was carried out using a SYBR Premix Ex22Taq (perfect real time) kit (TaKaRa Biomedicals) on a LightCycler23480 machine (Roche Diagnostics, Rotkreuz, Switzerland) according to the manufacturer's instructions (Roche Diagnostics). The qRT-PCR amplification was performed at 94 °C for 10 s, 58 °C for 10 s and 72 °C for 10 s. All reactions were repeated three times. The relative level of gene expression was calculated using the formula $2^{-\triangle\triangle CP}$ according to Livak and Schmittgen [49].

Table 3. Primers were used in the described gene expression of defense pathway.

Target Gene	Accession Number	Primer Sequence (5' → 3')	References
NPR1	AT1G64280.1	TGAATTGAAGATGACGCTGCT AGGCCTTCTTTAGTGTCTCTTGTA	Wu et al. 2012 [50]
PAD4	AT3G52430	GGTCGACGCTGCCATACTCAAACT AGAGAGATTGGTTTCCGAGCAGAG	Youssef et al. 2013 [51]
RAR1	AT5G51700	CGGCTCCTACTTCATCTCCAG AACATCGCAACATTTCCACCCTCT	Tornero et al. 2002 [52]
SGT1b	AT4G11260	CCCAAACCCAATGTCTCATCAG TCCACTTTCTTAGTCCCAACTTCT	Tör et al. 2002 [53]

4.9. Data Analysis

The segregation rates of the T2 progenies of transgenic oilseed plants, as well as the segregation ratios in the F2 generation from crosses between *cZR3* and *Hs1* homozygous transgenic plants were analyzed using the chi-square test to confirm the expected Mendelian segregation pattern of 3:1 (transgenic: non-transgenic plants) and 9:3:3:1. Data were analyzed using IBMSPSS 23.0 statistical system for windows (SPSS Inc., Armonk, NY, USA). Duncan multiple range tests were performed at the 0.05 level of probability.

Author Contributions: Conceptualization, G.T.; Formal analysis, X.Z. and Q.Z.; Investigation, X.Z.; Resources, G.T. and D.C.; Validation, Q.Z.; Visualization, N.C.; Writing—original draft, G.T.; and Writing—review and editing, X.Z.

Funding: This work was financially supported by the National Natural Science Funds (No. 31071443) and the National Major Special Project for Transgenic Organisms, Ministry of Agriculture in China (No. 2016ZX08004-004-005).

Conflicts of Interest: The authors declare no conflict of interest.

Abbreviations

BCN   beet cyst nematode
RGAs   resistance gene analogs
CIM   callus induction medium
SIM   shoot induced media
SEM   shoot elongation media
BAP   N-6-benzylaminopurine
NAA   Naphthaleneacetic acid
AgNO$_3$ silver nitrate
Carb   carbenicillin
Kan   kanamycin

References

1. Lilley, C.J.; Atkinson, H.J.; Urwin, P.E. Molecular aspects of cysts nematodes. *Mol. Plant Pathol.* **2005**, *6*, 577–588. [CrossRef] [PubMed]
2. Hemayati, S.S.; Akbar, J.E.; Ghaemi, A.R.; Fasahat, P. Efficiency of white mustard and oilseed radish trap plants against sugar beet cyst nematode. *Appl. Soil Ecol.* **2017**, *119*, 192–196. [CrossRef]
3. Jung, C.; Wyss, U. New approaches to control plant parasitic nematodes. *Appl. Microbiol. Biotechnol.* **1999**, *51*, 439–446. [CrossRef]
4. Lelivelt, C.L.C.; Lange, W. Intergeneric crosses for the transfer of resistance to the beet cyst nematode from *Raphanussativus* to *Brassica napus*. *Euphytica* **1993**, *68*, 111–120. [CrossRef]
5. Nielsen, E.L.; Baltenspencer, D.P.; Kerr, E.D.; Rife, C.L. Host suitability of rapeseed for *Heterodera schachtii*. *J. Nematol.* **2003**, *35*, 35–38. [PubMed]
6. Peterka, H.; Budahn, H.; Schrader, O.; Ahne, R.; Schütze, W. Transfer of resistance against the beet cyst nematode from radish (*Raphanussativus*) to rape (*Brassica napus*) by monosomic chromosome addition. *Appl. Genet.* **2004**, *109*, 30–41. [CrossRef] [PubMed]
7. Peterka, H.; Budahn, H.; Zhang, S.S.; Li, J.B. Nematode resistance of rape-radish chromosome addition lines. *Nematology* **2010**, *12*, 269–275.
8. Lelivelt, C.L.C.; Leunissen, E.H.M.; Frederiks, H.J.; Helsper, J.P.F.G.; Krens, F.A. Transfer of resistance to the beet cyst nematode (*Heterodera schachtii* Schm.) from *Sinapis alba* L. (white mustard) to the *Brassica napus* L. gene pool by means of sexual and somatic hybridization. *Appl. Genet.* **1993**, *85*, 688–696. [CrossRef]
9. Brown, J.; Brown, A.P.; Davis, J.B.; Erickson, D. Intergeneric hybridization between *Sinapis alba* and *Brassica napus*. *Euphytica* **1997**, *93*, 163–168. [CrossRef]
10. Zhang, S.S.; Peterka, H.; Budahn, H.; Schrader, O.; Li, C.Y. Chromosomal localization of resistance gene in radish against beet cyst nematode and the stability of additional oil radish chromosomes in rape-radish addition lines. *Sci. Agric. Sin.* **2008**, *4*, 93–101.
11. Cai, D.; Kleine, M.; Kifle, S.; Harloff, H.J.; Sandal, N.N.; Marcker, K.A.; Klein-Lankhorst, R.M.; Salentijn, E.M.; Lange, W.; Stiekema, W.J.; et al. Positional cloning of a gene for nematode resistance in sugar beet. *Science* **1997**, *275*, 832–834. [CrossRef]
12. Milligan, S.B.; Bodeau, J.; Yaghoobi, J.; Kaloshian, I.; Zabel, P.; Williamson, V.M. The root knot nematode resistance gene *Mi* from tomato is a member of the leucine zipper, nucleotide binding, leucine-rich family of plant genes. *Plant Cell* **1998**, *10*, 1307–1319. [CrossRef]
13. Van Der Vossen, E.A.; Van Der Voort, J.N.; Kanyuka, K.; Bendahmane, A.; Sandbrink, H.; Baulcombe, D.C.; Bakker, J.; Stiekema, W.J.; Klein-Lankhorst, R.M. Homologues of a single resistance-gene cluster in potato confer resistance to distinct pathogens: A virus and a nematode. *Plant J.* **2000**, *23*, 567–576. [CrossRef]
14. Ernst, K.; Kumar, A.; Kriseleit, D.; Kloos, D.U.; Phillips, M.S.; Ganal, M.W. The broad-spectrum potato cyst nematode resistance gene (*Hero*) from tomato is the only member of a large gene family of NBS-LRR genes with an annual amino acid repeat in the LRR region. *Plant J.* **2002**, *31*, 127–136. [CrossRef]
15. Thurau, T.; Kifle, S.; Jung, C.; Cai, D. The promoter of the nematode resistance gene *Hs1^{pro-1}* activates a nematode-responsive and feeding site specific gene expression in sugar beet (*Beta vulgaris* L.) and *Arabidopsis thaliana*. *Plant Mol. Biol.* **2003**, *52*, 643–660. [CrossRef]

16. McLean, M.D.; Hoover, G.J.; Bancroft, B.; Makhmoudova, A.; Clark, S.M.; Welacky, T.; Simmonds, D.H.; Shelp, B.J. Identification of the full-length *Hs1pro-1* coding sequence and preliminary evaluation of soybean cyst nematode resistance in soybean transformed with *Hs1pro-1* cDNA. *Can. J. Bot.* **2007**, *85*, 437–441. [CrossRef]

17. Hunger, S.; Gaspero, G.D.; Mohring, S.; Bellin, D.; Schafer-Pregl, R.; Borchardt, D.C.; Durel, C.E.; Werber, M.; Weisshaar, B.; Salamini, F.; et al. Isolation and linkage analysis of expressed disease-resistance gene analogues of sugar beet (*Beta vulgaris* L.). *Genome* **2003**, *46*, 70–82. [CrossRef]

18. Leister, D.; Kurth, J.; Laurie, D.A.; Yano, M.; Sasaki, T.; Devos, K.; Graner, A.; Schulze-Lefert, P. Rapid reorganization of resistance gene homologues in cereal genomes. *Proc. Natl. Acad. Sci. USA* **1998**, *95*, 370–375. [CrossRef]

19. Timmerman-Vaughan, G.M.; Frew, T.J.; Weeden, N. Characterization and linkage mapping of R-gene analogous DNA sequences in pea (*Pisum sativum* L.). *Appl. Genet.* **2000**, *101*, 241–247. [CrossRef]

20. Dangl, J.L.; Jones, J.D. Plant pathogens and integrated defence responses to infection. *Nature* **2001**, *411*, 826–833. [CrossRef]

21. Tian, Y.Y.; Fan, L.J.; Thurau, T.; Jung, C.; Cai, D. The absence of TIR-type resistance gene analogues in the sugar beet (*Beta vulgaris* L.) genome. *Mol. Evol.* **2004**, *58*, 40–53. [CrossRef]

22. Lein, J.C.; Asbach, K.; Tian, Y.; Schulte, D.; Li, C.; Koch, G.; Jung, C.; Cai, D.G. Resistance gene analogues are clustered on chromosome 3 of sugar beet and cosegregate with QTL for rhizomania resistance. *Genome* **2007**, *50*, 61–71. [CrossRef]

23. Ori, N.; Eshed, Y.; Paran, I.; Presting, G.; Aviv, D.; Tanksley, S.; Zamir, D.; Fluhr, R. The I2C family from the wilt disease resistance locus *I2* belongs to the nucleotide binding, leucine-rich repeat superfamily of plant resistance genes. *Plant Cell* **1997**, *9*, 521–532. [CrossRef]

24. Yoshimura, S.; Yamanouchi, U.; Katayose, Y.; Toki, S.; Wang, Z.X.; Kono, I.; Kurata, N.; Yano, M.; Iwata, N.; Sasaki, T. Expression of Xa1, a bacterial blight-resistance gene in rice, is induced by bacterial inoculation. *Proc. Natl. Acad. Sci. USA* **1998**, *95*, 1663–1668. [CrossRef]

25. Lagudah, E.S.; Moullet, O.; Appels, R. Map-based cloning of a gene sequence encoding a nucleotide binding domain and a leucine-rich region at the *Cre3* nematode resistance locus of wheat. *Genome* **1997**, *40*, 659–665. [CrossRef]

26. Whitham, S.; Dinesh-Kumar, S.P.; Choi, D.; Hehl, R.; Corr, C.; Baker, B. The product of the tobacco mosaic virus resistance gene N: Similarity to toll and the interleukin-1 receptor. *Cell* **1994**, *78*, 1101–1115. [CrossRef]

27. Dinesh-Kumar, S.P.; Baker, B.J. Alternatively spliced N resistance gene transcripts: Their possible role in tobacco mosaic virus resistance. *Proc. Natl. Acad. Sci. USA* **2000**, *97*, 1908–1913. [CrossRef]

28. Paal, J.; Henselewski, H.; Muth, J.; Meksem, K.; Menéndez, C.M.; Salamini, F.; Ballvora, A.; Gebhardt, C. Molecular cloning of the potato *Gro1-4* gene conferring resistance to pathotype Ro1 of the root cyst nematode *Globodera rostochiensis*, based on a candidate gene approach. *Plant. J.* **2004**, *38*, 285–297. [CrossRef]

29. Calenge, F.; Van der Linden, C.G.; Van de Weg, E.; Schouten, H.J.; Van Arkel, G.; Denance, C.; Durel, C.E. Resistance gene analogues identified through the NBS-profiling method map close to major genes and QTL for disease resistance in apple. *Appl. Genet.* **2005**, *110*, 660–668. [CrossRef]

30. Schulte, D.; Cai, D.; Kleine, M.; Fan, L.; Wang, S.; Jung, C. A complete physical map of wild beet (*Beta procumbens*) translocation in sugar beet. *Mol. Gen. Genom.* **2006**, *275*, 504–511. [CrossRef]

31. Hammond-Kosack, K.E.; Parker, J.E. Deciphering plant-pathogen communication: Fresh perspectives for molecular resistance breeding. *Curr. Opin. Biotechnol.* **2003**, *14*, 177–193. [CrossRef]

32. Chitwood, D.J. Research on plant-parasitic nematode biology conducted by the United States Department of Agriculture–Agricultural Research Service. *Pest. Manag. Sci.* **2003**, *59*, 748–753. [CrossRef]

33. Hauer, M.; Koch, H.J.; Krüssel, S.; Mittler, S.; Märländer, B. Integrated control of *Heterodera schachtii* Schmidt in Central Europe by trap crop cultivation, sugar beet variety choice and nematicide application. *Appl. Soil Ecol.* **2016**, *99*, 62–69. [CrossRef]

34. Feechan, A.; Kocsis, M.; Riaz, S.; Zhang, W.; Gadoury, D.M.; Walker, M.A.; Dry, I.B.; Reisch, B.; Cadle-Davidson, L. Strategies for RUN1 deployment using RUN2 and REN2 to manage grapevine powdery mildew informed by studies of race specificity. *Phytopathology* **2015**, *105*, 1104–1113. [CrossRef]

35. Abdeen, A.; Virgòs, A.; Olivella, E.; Villanueva, J.; Avilé, X.; Gabarra, R.; Prat, S. Multiple insect resistance in transgenic tomato plants over-expressing two families of plant proteinase inhibitors. *Plant Mol. Biol.* **2005**, *57*, 189–202. [CrossRef]

36. Bharathia, Y.; Kumara, S.V.; Pasalub, I.C.; Balachandranb, S.M.; Reddya, V.D.; Raoa, K.V. Pyramided rice lines harbouring *Allium sativum* (*asal*) and *Galanthus nivalis* (*gna*) lectin genes impart enhanced resistance against major sap-sucking pests. *J. Biotechnol.* **2011**, *152*, 63–71. [CrossRef]

37. Chan, Y.L.; He, Y.; Hsiao, T.T.; Wang, C.J.; Tian, Z.H.; Yeh, K.W. Pyramiding taro cystatin and fungal chitinase genes driven by a synthetic promoter enhances resistance in tomato to root-knot nematode Meloidogyne incognita. *Plant Sci.* **2015**, *231*, 74–81. [CrossRef]

38. Urwin, P.E.; McPherson, M.J.; Atkinson, H.J. Enhanced transgenic plant resistance to nematodes by dual proteinase inhibitor constructs. *Planta* **1998**, *204*, 472–479. [CrossRef]

39. Williamson, M.V.; Kumar, A. Nematode resistance in plants: The battle underground. *Trends Genet.* **2006**, *22*, 396–403. [CrossRef]

40. Glazebrook, J. Contrasting mechanisms of defense against biotrophic and necrotrophic pathogens. *Annu. Rev. Phytopathol.* **2005**, *43*, 205–227. [CrossRef]

41. Tang, G.X.; Kneck, K.; Yang, X.F.; Qin, Y.B.; Zhou, W.J.; Cai, D. A two-step protocol for shoot regeneration from hypocotyl explants of oilseed rape and its application for *Agrobacterium*-mediated transformation. *Biol. Plant.* **2011**, *55*, 21–26. [CrossRef]

42. Murashige, T.; Skoog, F. A revised medium for rapid growth and bioassays in tobacco tissue culture. *Physiol. Plant.* **1962**, *15*, 474–493. [CrossRef]

43. Gamborg, O.L.; Miller, R.A.; Ojima, K. Nutrient requirement of suspensions cultures of soybean root cells. *Exp. Cell Res.* **1968**, *50*, 151. [CrossRef]

44. Rogers, S.O.; Bendich, A.J. Extraction of DNA from milligram amounts of fresh, herbarium and mummified plant tissues. *Plant Mol. Biol.* **1985**, *5*, 69–76. [CrossRef]

45. Jefferson, R.A. Assaying chimeric gene in plants: The GUS gene fusion system. *Plant. Mol. Rep.* **1987**, *5*, 387–405. [CrossRef]

46. Reski, R.; Abel, W.O. Induction of budding on chloronemata and caulonemata of the moss, *Physcomitrella patens*, using isopentenyladenine. *Planta* **1985**, *165*, 354–358. [CrossRef]

47. Steiner, A.A. A universal method for preparing nutrient solutions of a certain desired composition. *Plant Soil* **1961**, *15*, 134–154. [CrossRef]

48. Sijmons, P.C.; Grundler, F.M.W.; Mende, N.; Burrows, P.R. *Arabidopsis thaliana* as a new model host for plant parasitic nematodes. *Plant J.* **2010**, *1*, 245–254. [CrossRef]

49. Livak, K.J.; Schmittgen, T.D. Analysis of relative gene expression data using real-time quantitative PCR and $2^{-\triangle(\triangle CP)}$ method. *Methods* **2001**, *25*, 402–408. [CrossRef]

50. Wu, Y.; Zhang, D.; Chu, J. The Arabidopsis NPR1 protein is a receptor for the plant defense hormone salicylic acid. *Cell Rep.* **2012**, *1*, 639–647. [CrossRef]

51. Youssef, R.M.; Macdonald, M.H.; Brewer, E.P. Ectopic expression of AtPAD4 broadens resistance of soybean to soybean cyst and root-knot nematodes. *BMC Plant Biol.* **2013**, *13*, 67. [CrossRef]

52. Tornero, P. RAR1 and NDR1 contribute quantitatively to disease resistance in *Arabidopsis*, and their relative contributions are dependent on the R gene assayed. *Plant Cell* **2002**, *14*, 1005–1015. [CrossRef]

53. Tör, M.; Gordon, P.; Cuzick, A. *Arabidopsis* SGT1b is required for defense signaling conferred by several downy mildew (*Peronospora parasitica*) resistance genes. *Plant Cell* **2002**, *14*, 993–1003. [CrossRef]

Article

Genome-Wide Characterization and Analysis of Metallothionein Family Genes That Function in Metal Stress Tolerance in *Brassica napus* L.

Yu Pan [1,2,†], Meichen Zhu [2,3,†], Shuxian Wang [2,3], Guoqiang Ma [2,3], Xiaohu Huang [2,3], Cailin Qiao [2,3], Rui Wang [2,3], Xinfu Xu [2,3], Ying Liang [2,3], Kun Lu [2,3], Jiana Li [2,3,*] and Cunmin Qu [2,3,*]

[1] Key Laboratory of Horticulture Science for Southern Mountainous Regions, Ministry of Education, Southwest University, No. 2 Tiansheng Road, Beibei, Chongqing 400715, China; panyu1020@swu.edu.cn

[2] Academy of Agricultural Sciences, Southwest University, Chongqing 400715, China; zmc0809@email.swu.edu.cn (M.Z.); wsxsummer@email.swu.edu.cn (S.W.); mgqdw123456@email.swu.edu.cn (G.M.); hxh9305@email.swu.edu.cn (X.H.); qcl123@email.swu.edu.cn (C.Q.); ruiwang71@163.com (R.W.); xinfuxu@swu.edu.cn (X.X.); yliang@swu.edu.cn (Y.L.); drlukun@swu.edu.cn (K.L.)

[3] Chongqing Rapeseed Engineering Research Center, College of Agronomy and Biotechnology, Southwest University, No. 2 Tiansheng Road, Beibei, Chongqing 400715, China

* Correspondence: ljn1950@swu.edu.cn (J.L.); drqucunmin@swu.edu.cn (C.Q.); Tel.: +86-23-6825-0642 (J.L.); +86-23-6825-0701 (C.Q.)

† These authors contributed equally to this paper.

Received: 25 June 2018; Accepted: 24 July 2018; Published: 26 July 2018

Abstract: *Brassica* plants exhibit both high biomass productivity and high rates of heavy metal absorption. Metallothionein (MT) proteins are low molecular weight, cysteine-rich, metal-binding proteins that play crucial roles in protecting plants from heavy metal toxicity. However, to date, MT proteins have not been systematically characterized in *Brassica*. In this study, we identified 60 MTs from *Arabidopsis thaliana* and five *Brassica* species. All the MT family genes from Brassica are closely related to *Arabidopsis* MTs, encoding putative proteins that share similar functions within the same clades. Genome mapping analysis revealed high levels of synteny throughout the genome due to whole genome duplication and segmental duplication events. We analyzed the expression levels of 16 *Brassica napus* MTs (*BnaMTs*) by RNA-sequencing and real-time RT-PCR (RT-qPCR) analysis in plants under As^{3+} stress. These genes exhibited different expression patterns in various tissues. Our results suggest that *BnaMT3C* plays a key role in the response to As^{3+} stress in *B. napus*. This study provides insight into the phylogeny, origin, and evolution of MT family members in *Brassica*, laying the foundation for further studies of the roles of MT proteins in these important crops.

Keywords: metallothionein; *Brassica*; *Brassica napus*; As^{3+} stress

1. Introduction

Heavy metals are essential micronutrients for various physiological processes in plants. However, excess amounts of essential (copper, zinc) and non-essential metals (cadmium) are toxic to plants, as they inhibit plant growth, impair root development, and decrease chlorophyll contents, resulting in chlorosis [1,2]. Therefore, plants have evolved a suite of mechanisms involving the chelation and sequestration of heavy metals by various amino acids, organic acids, phytochelatins (PCs) and metallothioneins (MTs) [3,4]. These compounds play crucial roles in protecting plants from heavy metal toxicity [5–9].

MTs, low-molecular-weight proteins (7–10 kDa) with a high percentage of cysteine (Cys) residues [10,11], have been widely characterized in various prokaryotic and eukaryotic organisms. Plant MTs are classified into four types according to the arrangement of their Cys residues [12], including the *MT1*, *MT2*, *MT3*, and *MT4* subfamilies [10,13]. MTs play crucial roles in ion homeostasis and tolerance in plants. Seven functional MT genes have been isolated from the model plant *Arabidopsis thaliana*. Of these, *AtMT1a*, *AtMT2a*, *AtMT2b*, and *AtMT3* enhance plant tolerance of Cu ions, especially in leaves [14,15], as well as Cd tolerance in transformed yeast and fava bean (*Vicia faba*) guard cells [16,17]. *AtMT4* modulates Zn homeostasis in seeds and is highly expressed during the late stages of development [18]. Additionally, various *MT* genes show significantly different expression patterns in plants under heavy metal stress. For example, *MT2a* and *MT2b* genes are more highly expressed in the roots of the heavy metal hyperaccumulator *Noccaea caerulescens* than in those of *A. thaliana*, while *MT3* is more highly expressed in shoots of *N. caerulescens* than in the non-hyperaccumulator *Thlaspi arvense* [19,20]. *MT4* mRNAs are primarily expressed in ripening fruits and developing seeds [14]. Therefore, plant MTs are likely involved in many physiological processes, such as seed development and germination [18,21,22], fruit ripening [14], and root development [21,23].

Brassica plants are considered to be highly tolerant to heavy metals (e.g., Cd, Cu, Ni, Zn, Pb, and Se), making them ideal plants for studying metal accumulation in phytoremediation studies [24–27]. Indian mustard (*Brassica juncea*) is a high-biomass-producing crop with the potential to take up and accumulate heavy metals [1,23,27,28]. However, this plant accumulates Cd less effectively than other crops such as maize (*Zea mays*), rice (*Oryza sativa*), and sugar beet (*B. vulgaris*) when it is present at low concentrations in the soil [29]. Rapeseed (*B. napus*) has many advantages for this type of analysis due to its rapid growth, high biomass productivity and efficient heavy metal absorption, and it is therefore also widely used to investigate heavy metal tolerance [26,30–32]. Indeed, while heavy metal tolerance has been well studied in various *Brassica* species, the mechanisms that contribute to the tolerance of these plants to heavy metals remain unclear.

In the present study, we screened the draft genome sequences of *A. thaliana* and various *Brassica* species (*Brassica rapa*, *Brassica oleracea*, *B. napus*, *Brassica juncea* and *Brassica nigra*) for *MT* genes that participate in heavy metal detoxification. We identified 60 *MT* genes and performed a detailed analysis of their duplication patterns, classifications, and chromosomal distribution and motifs, as well as a phylogenetic analysis. Finally, we verified the differential expression profiles of selected rapeseed *MT* genes in different *B. napus* tissues at various developmental stages. We also investigated the expression patterns of *MT* genes in *B. napus* seedlings exposed to heavy metals. Our results provide important information about the origin and evolution of the *MT* gene family in *Brassica* and provide a basis for further studies of the functions of MT family proteins in rapeseed.

2. Results

2.1. Identification and Multiple Sequence Alignment of MT Family Genes

Using the protein sequences of the *MT* family genes from the TAIR10 database (Table 1) as queries, we identified 60 *MT* genes in *A. thaliana* and various *Brassica* plants (*B. rapa*, *B. oleracea*, *B. napus*, *B. juncea* and *B. nigra*). These genes were classified into four subgroups (Figure 1, Table 1). Of these, seven were *MT1* subfamily members, five of which were identified from the corresponding genome databases, except *BolMT1* and *BjuMT1*. *BraMT1* has been reported in *B. rapa* with 45 amino-acid proteins [33]; the difference between these sequences requires further study. In addition, 37 were *MT2* subfamily members, encoding deduced proteins ranging from 56 to 103 amino acids in length; nine were *MT3* subfamily members, encoding proteins from 65 to 69 amino acids in length; and seven were *MT4* subfamily members, encoding proteins from 85 to 120 amino acids in length (Table 1). Of the *MT1* subfamily members, three homologs were identified in *A. thaliana*, while *BnaMT1* and *BraMT1* were identified in *B. napus* and *B. rapa*, respectively. No *MT1* subfamily members were found in *B. oleracea*, *B. juncea*, or *B. nigra*, whereas *BolMT1* and *BjuMT1* have been reported previously [13], indicating that genome gaps may still emerge in *B. oleracea*, and *B. juncea*. Six Cys-X-Cys motifs were equally distributed on the *N*- and *C*-termini of MT1 family proteins, except in the case of BraMT1 and AtMT1B (Figure 2A).

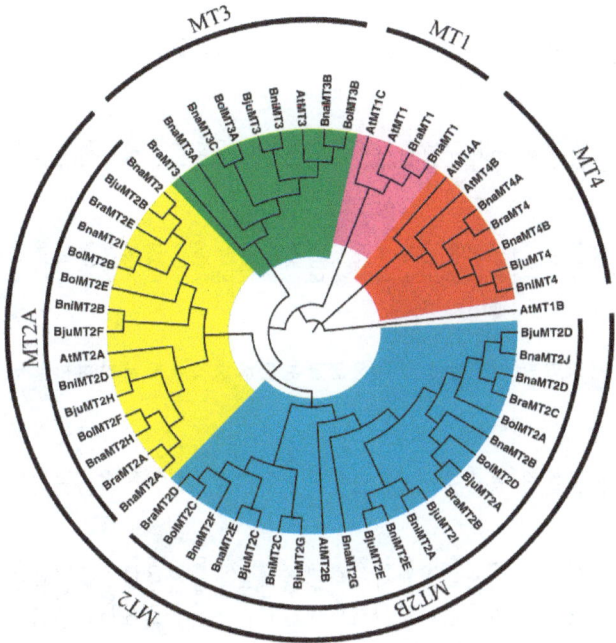

Figure 1. Neighbor-Joining (NJ) phylogenetic tree showed the relationships of *Metallothionein* (*MT*) family genes from *A. thaliana* and various *Brassica* species. The rooted neighbor-joining phylogenetic tree was constructed using MEGA6 and visualized using Figure Tree v1.4.2. The MTs were divided into four subfamilies (MT1–MT4), which are indicated by different colors. Organism name and gene accession numbers are shown in Table 1.

We identified 37 *MT2* subfamily genes in *A. thaliana* (*AtMT2A* and *AtMT2B*) and *Brassica* (five in *B. rapa*, six in *B. oleracea*, ten in *B. napus*, nine in *B. juncea*, and five in *B. nigra*), which were divided into the *MT2A* and *MT2B* subgroups (Figure 2B, Table 1), pointing to the extensive triplication and expansion of these genomes during their evolution in *Brassica* plants. Furthermore, one Cys–Cys and two Cys–X–Cys motifs were almost always present in the N-terminal regions of these proteins, and three Cys–X–Cys were almost always present in their C-terminal regions (Figure 2B). In addition, MT2 subfamily genes encode a deduced protein with the MSCCGGN/S sequence in their N-termini, which is consistent with previous findings [13,34]. Three variant regions were found in the MT2A subgroup and three in the MT2B subgroup (Figure 2B), which might be associated with their roles in metal tolerance.

We identified *MT3* subfamily genes, including one each in *A. thaliana*, *B. rapa*, *B. juncea*, and *B. nigra*, two in *B. oleracea* and three in *B. napus* (Figure 1, Table 1). The alignment of the MT3 amino acid sequences showed a completely conserved sequence, CXXCDCX$_5$C, located in the N-terminus of each protein, and a highly conserved consensus sequence with eight Cys residues at the C-terminus (Figure 2C). In addition, 30–40 amino acids were detected in the Cys-poor linker region between the *N*- and *C*-terminal regions, in accordance with the MT2 subfamily (Figure 2B,C), pointing to a possible evolutionary relationship between the *MT2* and *MT3* family genes.

Table 1. List of *Metallothionein* (*MT*) genes identified in the *A. thaliana* and *Brassica* genomes.

Groups	Name	Gene ID	Chr.	Start (bp)	End (bp)	Length (bp)	Length (aa)	MW (KDa)	pIs	Exon	Intron
MT1	AtMT1	AT1G07600	AtChr1	2338904	2339321	138	45	4.580	4.23	2	1
	AtMT1C	AT1G07610	AtChr1	2341542	2342123	138	45	4.495	4.54	2	1
	AtMT1B	AT5G56795	AtChr5	22972042	22972449	156	51	5.428	10.25	2	1
	BraMT1	Bra015594	BraA10	766670	772706	450	149	16.77	9.32	4	3
	BnaMT1	BnaA10g04950D	BnaA10	2673266	2673770	138	45	4.480	3.92	2	1
	BolMT1	DK501359	UN	UN	UN	138	45	4.412	3.92	UN	UN
	BjuMT1	EF471214	UN	UN	UN	138	45	4.439	3.92	UN	UN
MT2	AtMT2A	AT3G09390	AtChr3	2889486	2890229	246	81	8.163	4.35	2	1
	AtMT2B	AT5G02380	AtChr5	506498	507244	234	77	7.766	4.54	2	1
	BraMT2A	Bra001309	BraA03	15803598	15803933	246	81	8.197	4.17	2	1
	BraMT2B	Bra005720	BraA03	275453	275766	243	80	8.033	4.29	2	1
	BraMT2C	Bra009595	BraA10	16182058	16182453	243	80	8.031	4.29	2	1
	BraMT2D	Bra028875	BraA02	269835	270238	246	81	8.386	4.20	2	1
	BraMT2E	Bra029765	BraA05	23082833	23083178	243	80	8.024	4.35	2	1
	BolMT2A	Bol000591	Scaffold000521	37950	38342	243	80	8.031	4.29	2	1
	BolMT2B	Bol011307	Scaffold000212	445183	445551	243	80	8.054	4.58	2	1
	BolMT2C	Bol012825	BolC02	305134	305536	246	81	8.386	4.20	2	1
	BolMT2D	Bol015273	BolC03	101213	101535	243	80	8.137	4.29	2	1
	BolMT2E	Bol023080	BolC01	37533909	37534079	171	56	5.920	4.15	1	0
	BolMT2F	Bol033925	Scaffold000040	316614	316949	246	81	8.147	4.15	2	1
	BnaMT2A	BnaA03g30680D	BnaA03	14857530	14858170	240	79	7.966	3.81	2	1
	BnaMT2B	BnaA03g54880D	A03_random	44751	45330	243	80	8.077	4.29	2	1
	BnaMT2C	BnaA05g29010D	BnaA05	20416060	20416808	243	80	8.024	4.35	2	1
	BnaMT2D	BnaA10g27170D	BnaA10	17170773	17171571	243	80	8.031	4.29	2	1
	BnaMT2E	BnaAnng00330D	Ann_random	321245	322046	246	81	8.370	4.20	2	1
	BnaMT2F	BnaC02g03550D	BnaC02	1685618	1686330	246	81	8.386	4.20	2	1
	BnaMT2G	BnaC03g00710D	BnaC03	346945	347282	216	71	7.284	4.08	2	1
	BnaMT2H	BnaC03g35960D	BnaC03	21778025	21778740	240	79	7.950	3.79	2	1
	BnaMT2I	BnaC05g43490D	BnaC05	40327486	40328304	243	80	8.028	4.58	2	1
	BnaMT2J	BnaCmng40400D	Cnn_random	38972401	38973067	243	80	8.031	4.29	2	1
	BjuMT2A	BjuA008858	BjuA03	321697	322020	243	80	8.033	4.29	2	1
	BjuMT2B	BjuA020647	BjuA05	29317680	29319366	312	103	10.661	4.24	3	2
	BjuMT2C	BjuA040818	BjuA02	430697	431097	246	81	8.370	4.20	2	1
	BjuMT2D	BjuA044587	BjuA10	19619066	19619458	243	80	8.031	4.29	2	1
	BjuMT2E	BjuB001621	BjuB08	169755	170152	243	80	8.059	4.29	2	1

Table 1. *Cont.*

Groups	Name	Gene ID	Chr.	Start (bp)	End (bp)	Length (bp)	Length (aa)	MW (KDa)	pIs	Exon	Intron
	BjuMT2F	BjuB005939	BjuB01	43381550	43383641	297	98	10.057	3.93	3	2
	BjuMT2G	BjuB012072	BjuB05	20387272	20387699	246	81	8.353	4.11	2	1
	BjuMT2H	BjuB031838	BjuB03	12766757	12767103	222	73	7.596	4.35	2	1
	BjuMT2I	BjuB044439	BjuB02	52757313	52757655	243	80	8.061	4.29	2	1
	BniMT2A	BniB001954-PA	BniB08	30524166	30524483	243	80	8.075	4.29	2	1
	BniMT2B	BniB007929-PA	BniB05	2189985	2190357	243	80	7.978	4.35	2	1
	BniMT2C	BniB023579-PA	BniB02	28213025	28213453	246	81	8.353	4.11	2	1
	BniMT2D	BniB039464-PA	BniB07	32240383	32240733	222	73	7.596	4.35	2	1
	BniMT2E	BniB045064-PA	BniB03	44318189	44318591	225	74	7.569	4.08	2	1
MT3	*AtMT3*	AT3G15353	AtChr3	5180642	5181586	210	69	7.373	4.35	3	2
	BraMT3	Bra027254	BraA05	20683494	20683920	204	67	7.183	4.17	3	2
	BolMT3A	Bol011145	BolC05	28120530	28120940	204	67	7.158	4.15	3	2
	BolMT3B	Bol025753	BolC03	17061982	17062438	198	65	7.016	4.40	3	2
	BnaMT3A	BnaA05g24200D	BnaA05	18177871	18178669	204	67	7.127	4.15	3	2
	BnaMT3B	BnaC03g39060D	BnaC03	24091592	24092154	198	65	7.016	4.40	3	2
	BnaMT3C	BnaC05g38240D	BnaC05	37017131	37017881	204	67	7.158	4.15	3	2
	BjuMT3	BjuB025665	BjuB01	40434253	40434691	204	67	7.187	4.15	3	2
	BniMT3	BniB008959-PA	BniB05	4957466	4957899	204	67	7.187	4.15	3	2
MT4	*AtMT4B*	AT2G23240	AtChr2	9895855	9896325	261	86	8.437	5.58	2	1
	AtMT4A	AT2G42000	AtChr2	17529243	17530443	366	121	12.229	7.62	4	3
	BraMT4	Bra000590	BraA03	11951235	11951571	261	86	8.480	7.37	2	1
	BnaMT4A	BnaA03g23200D	BnaA03	11067719	11068213	261	86	8.480	7.37	2	1
	BnaMT4B	BnaC03g27400D	BnaC03	15895032	15895647	261	86	8.468	6.97	2	1
	BjuMT4	BjuO006263	Contig407_1_341981	122112	122451	261	86	8.500	6.97	2	1
	BniMT4	BniB049568-PA	BniB03	29749524	29749863	261	86	8.472	6.97	2	1

Note, At, *A. thaliana*; Bra, *B. rapa*; Bol, *B. oleracea*; Bni, *B. nigra*; Bna, *B. napus*; and Bju, *B. juncea*; Chr, Chromosome; UN, unknown.

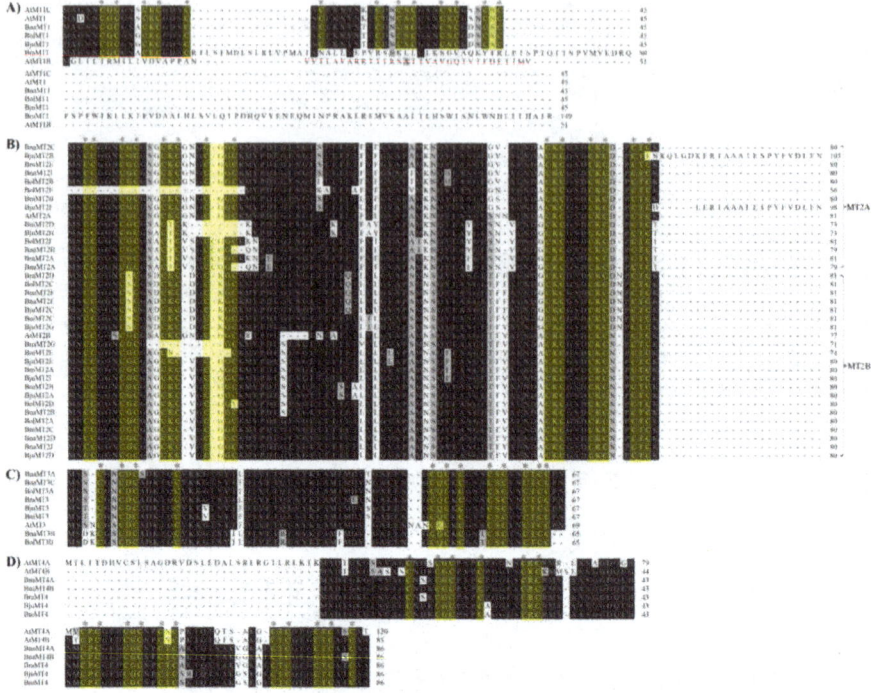

Figure 2. Alignment of MT protein sequences from *A. thaliana* and various *Brassica* species. Black and light gray shading indicate identical and conserved amino acid residues, respectively. (**A**) The MT1 protein sequences; (**B**) the MT2 protein sequences; (**C**) the MT3 protein sequences; (**D**) the MT4 protein sequences. The conserved cysteines regions are highlighted by asterisks and light yellow. The MTs were preliminarily classified by Cobbett and Goldsbrough reported [10]; detailed information is provided in Table 1.

MT4 subfamily genes, which are homologous to plant EC metallothionein-like genes, are different from *MT1–MT3* subfamily members. Two *MT4* subfamily members were found in *A. thaliana* and *B. napus*, one each in *B. rapa*, *B. juncea*, and *B. nigra*, and none in *B. oleracea* (Figure 1, Table 1). These proteins contain three Cys-poor linkers comprising 12–15 amino acids, as well as two Cys-rich regions with a highly conserved consensus sequence among them (Figure 2D).

2.2. Phylogenetic Analysis of MT Family Genes

Based on the multiple sequence alignment of the deduced MT1–MT4 proteins, it was found that Cys-rich regions are widely distributed among MT family proteins. These regions are characterized by conserved consensus sequences, with motifs such as Cys–G–Cys, Cys–K–Cys, and Cys–S–Cys (Figure 2). To investigate the evolutionary relationships among MT family genes from *A. thaliana* and various *Brassica* species, we constructed a NJ phylogenetic tree based on the alignment of MT domains. Based on the phylogenetic tree, the 58 MT domains were classified into four subfamilies (*MT1*, *MT2*, *MT3*, and *MT4*), except for *BolMT1* and *BjuMT1*, which were not annotated in the genome databases, and most genes were grouped with the *AtMTs* (Figure 2). *AtMT1B* represents the outgroup in the phylogenetic tree. In addition, the *MT2* subfamily was classified into two sister groups (Figure 1), *MT2A* and *MT2B*, which is highly consistent with the results of multiple sequence alignments of whole proteins (Figure 2). For example, two sister groups were also identified and found to contain eight and six *MT* family genes, respectively (Figure 2), which also contain highly conserved consensus sequences (Figure 1). These results will be helpful in identifying the functions of *MT* family genes via orthology analysis.

2.3. Genomic Structure and Conserved Motif Analysis of the MT Gene Family

We characterized the gene structures of the *MT* family genes by comparing the full-length CDS and the corresponding genomic DNA sequences using GSDS 2.0 (http://gsds.cbi.pku.edu.cn/index.php). Of the 58 *MT* genes, 44 contain a single intron with a highly conserved structure in each group, i.e., the *MT1*, *MT2* and *MT4* subfamilies. Additionally, *MT3* subfamily members contain two introns, which were also found in *BjuMT2B* and *BjuMT2F* (Figure 3). *BolMT2E* lacks an intron and belongs to the *MT2A* gene family, while *AtMT4A* and *BraMT1* contain three introns with distinct sizes (Figure 3). Most genes in the same subfamily exhibit similar exon–intron structures, but the genomic structures of *BjuMT2B* and *BjuMT2F* are similar to those of the *MT3* subfamily, providing further support for the evolutionary relationship and classification of the *MT* gene family members identified in this study.

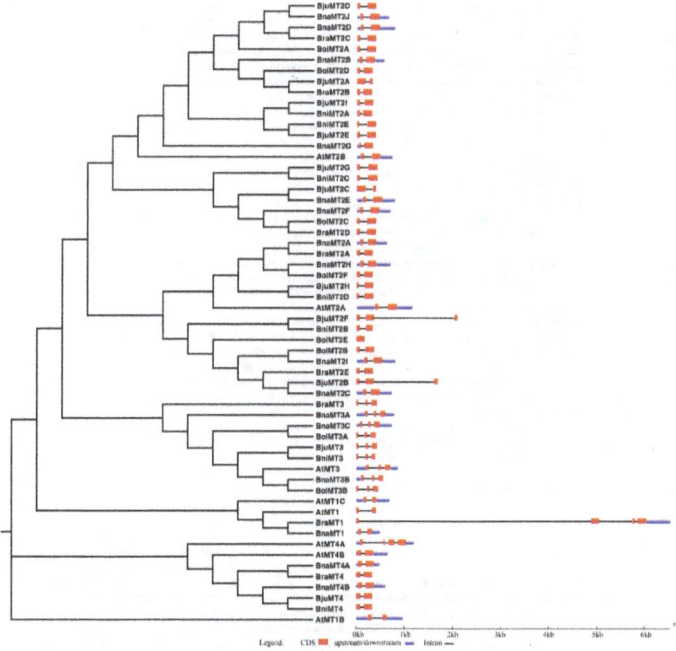

Figure 3. Phylogenetic relationships and genomic structures of the *MT* genes from *A. thaliana* and various *Brassica* species. The red boxes represent exons, solid lines represent introns (connecting two exons), and blue boxes represent untranslated regions (UTRs). The lengths of the *MT* genes are indicated by horizontal lines (kb).

Using MEME v4.12.0 (http://meme-suite.org/tools/meme), six, eight, three, and four conserved motifs were detected in the *MT1*, *MT2*, *MT3*, and *MT4* subfamilies, respectively, in *B. rapa*, *B. oleracea*, *B. napus*, *B. juncea*, and *B. nigra* (Figure 4A–D); the detailed structures of the motifs are shown in Figure S1A–D. All members of the *MT1* subfamily except for *AtMT1B* contain motif 1 (Figure 4A and Figure S1A). All members of the *MT2* subfamily contain motif 1, whereas all members of the *MT2A* subfamily contain motifs 2, 4, and 6, but motif 3 is found only in the *MT2B* subfamily members. The motifs in the *MT2B* subfamily members are more variable than those of the other *MTs* (Figure 4B and Figure S1B), pointing to the triplication and expansion of *Brassica* genomes. All nine genes in the *MT3* subfamily contain motif 1, while the *MT3A* subfamily genes contain motif 2 and the *MT3B* subfamily genes contain motif 3 (Figure 4C and Figure S1C). The *MT4* subfamily genes contain motifs 1, 2, and 3, indicating that these motifs are conserved among these genes (Figure 4D and Figure S1D). In summary, the same conserved motifs are widely found in paralogous/orthologous genes, suggesting that they might have similar functions at the protein level.

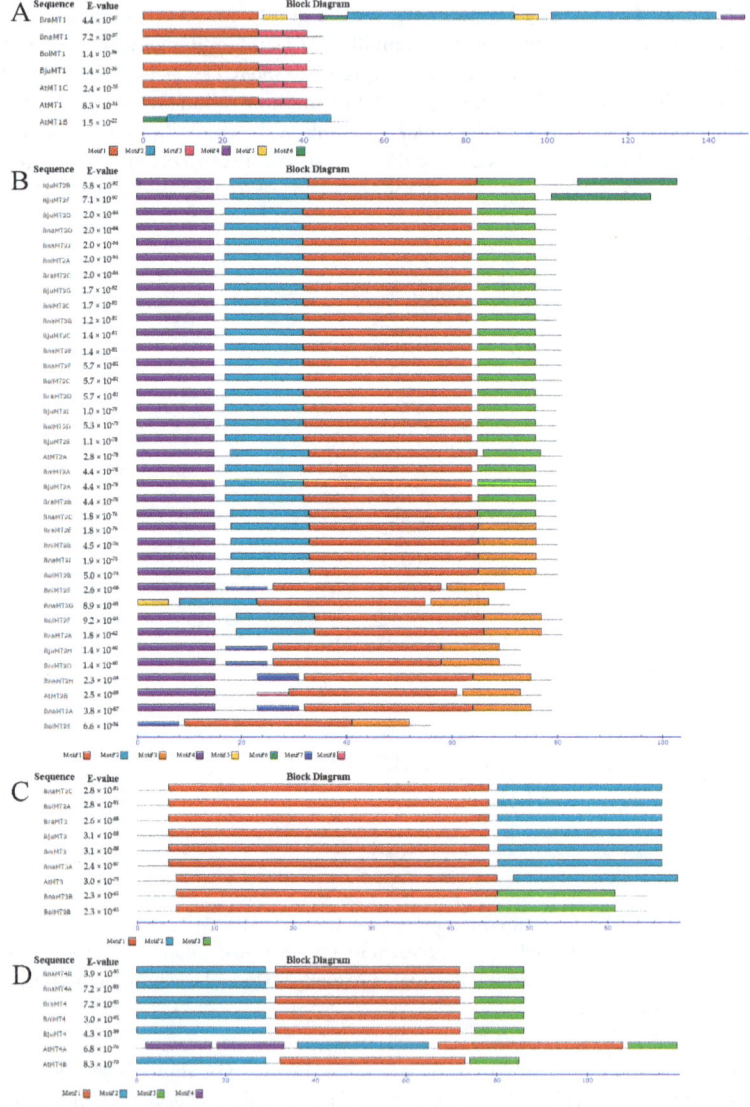

Figure 4. Putative conserved motifs in MT family proteins in various *Brassica* species identified using the MEME search tool. (**A**) the conserved motifs in MT1 family; (**B**) the conserved motifs in MT2 family; (**C**) the conserved motifs in MT3 family; (**D**) the conserved motifs in MT4 family. Different motifs are represented by different colors, and protein names and combined *p* values are shown on left side of this figure. The best possible matched motifs, their functional annotation, and motif width are shown in Figure S1.

2.4. Chromosome Locations and Duplication of MT Genes in Brassica

Brassica includes three diploid species, *B. rapa* (AA, 2*n* = 20), *B. oleracea* (CC, 2*n* = 18), and *B. nigra* (BB, 2*n* = 20) and three allotetraploid species, *B. napus* (AACC, 2*n* = 38), *B. juncea* (AABB, 2*n* = 36), and *B. carinata* (BBCC, 2*n* = 34), and the evolution and relationships between the members of *Brassica* can be well understood according to the U-triangle theory [35]. Five of these species have been completely sequenced, and their sequences are available in the *Brassica* database (BRAD) database. To identify the physical positions of the *MT* genes, we mapped them to the chromosomes in the corresponding *Brassica* species. The 43 *MT* genes are located on 27 chromosomes in the five *Brassica* species with available whole-genome sequences, including four chromosomes (BraA02, BraA03, BraA05, and BraA10) in *B. rapa*, four chromosomes (BolC01, BolC02, BolC03 and BolC05) in *B. oleracea* and five chromosomes in *B. nigra* (BniB02, BniB03, BniB05, BniB07 and BniB08) (Figure 5). Further, we detected high levels of synteny among *MT* family genes in these species. For example *BraMT2D* on chromosome BraA02, *BjuMT2C* on chromosome BjuA02, *BnaMT2F* on chromosome BnaC02, and *BolMT2C* on chromosome BolC02 are located near the top of the chromosomes and are classified into the same subgroups (Figure 2), suggesting that these genes might have undergone whole-genome duplication events during the evolutionary process and might have similar functions. However, some of these genes, e.g., *BnaMT4A* and *BraMT4* on chromosome A03, *BnaMT3A* and *BraMT3* on chromosome A05, and *BnaMT1* and *BraMT1* on chromosome A10 might have undergone segmental duplications (Figure 5). Finally, *BjuMT2H* and *BniMT2E* on chromosome B03 and *BjuMT2E* and *BniMT2A* on chromosome B08 might have undergone gene transposition (Figure 5). Taken together, these results shed light on the evolutionary patterns of these subfamilies among adjacent species.

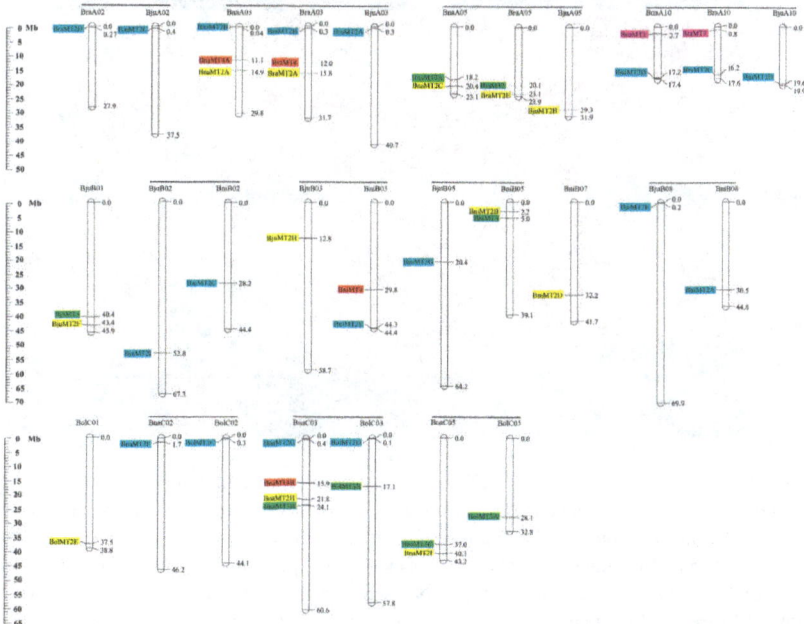

Figure 5. Chromosomal distribution and analysis of duplication events in *MT* family genes among *Brassica* species. Genes from the same subgroups are indicated by the same color, which is consistent with the corresponding family in the phylogenetic tree (Figure 1). The labels on the corresponding chromosomes indicate the names of the source organism and the subgenome. The scales indicate the sizes of various *Brassica* plant genomes (Mb). Bra, *B. rapa*; Bol, *B. oleracea*; Bni, *B. nigra*; Bna, *B. napus*; and Bju, *B. juncea*. The genes located on the scaffold are not shown in the Figure 5.

2.5. Expression Profiles of BnaMT Family Genes in B. napus

Based on the transcriptome sequencing datasets from *B. napus* ZS11 (BioProject ID PRJNA358784), we characterized the expression profiles of the *BnaMT* genes in eight different tissues, covering all stages of rapeseed development (Figure 6, Tables S1 and S2). Among the 16 *BnaMT* genes, *BnaMT1* was more highly expressed in the stems, leaves, and siliques 30 days after pollination than in other tissues (Figure 6). Among *MT2* genes, *BnaMT2A* and *BnaMT2H* were specifically expressed in buds; *BnaMT2C* and *BnaMT2I* were expressed at higher levels in roots, hypocotyls, cotyledons, and buds than in others tissues; *BnaMT2B*, *BnaMT2D*, and *BnaMT2J* were highly expressed throughout plant development, whereas *BnaMT2G* was expressed at low levels; and *BnaMT2E* and *BnaMT2F* were more highly expressed in stems and leaves than in other tissues (Figure 6). *BnaMT3A*, *BnaMT3B*, and *BnaMT3C* were more highly expressed in stems, leaves, and siliques before day 30 than in other tissues (Figure 6). Finally, *BnaMT4A* and *BnaMT4B* were mainly expressed in ripening seeds (Figure 6). The expression patterns of *MT* family genes correspond with the results of the phylogenetic analysis (Figure 2). For example, the expression patterns were similar for *BnaMT2A* and *BnaMT2H*, *BnaMT2C* and *BnaMT2I*, and *BnaMT2B*, *BnaMT2D*, and *BnaMT2J*, which were classified into the same sister groups.

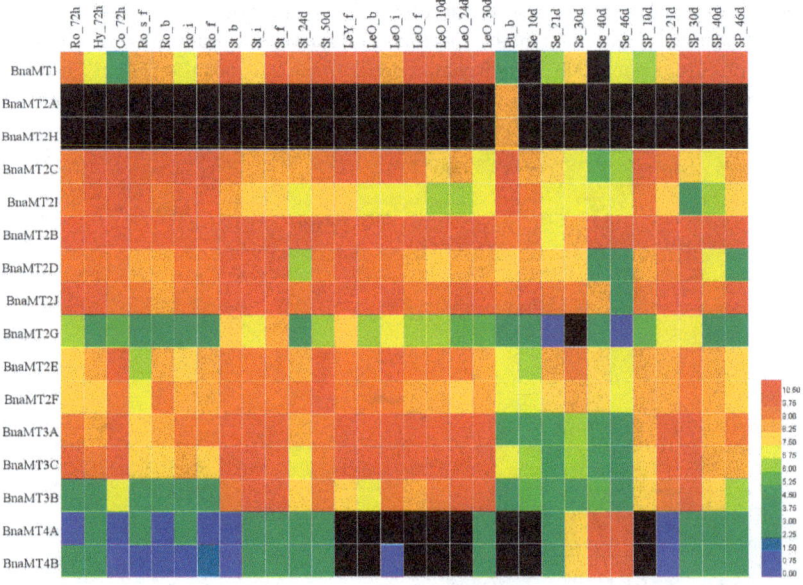

Figure 6. Heatmap of the expression profiles of *BnaMT* family genes in different tissues and organs. The abbreviations above the heatmap indicate the different tissues and organs/developmental stages of *B. napus* ZS11 (listed in Table S1). The expression data was gained from the RNA-seq data and shown as log2, as calculated by fragments per kilo base of exon model permillion (FPKM) values. Black boxes indicate that no expression was detected by RNA-seq analysis. The heatmap was generated using Heatmap Illustrator v1.0 (HemI v1.0, Huazhong University, Wuhan, China; http://hemi.biocuckoo. org/contact.php).

2.6. Expression Analysis of BnaMT Genes in Response to Metal Treatment

MTs are the best-characterized heavy-metal-binding ligands in plants. To analyze the roles of *BnaMTs* in metal tolerance, we compared the expression profiles of *BnaMTs* in the roots, hypocotyls, and cotyledons of *B. napus* plants under As^{3+} stress versus normal conditions via real-time RT-PCR

(RT-qPCR). Under normal conditions, the expression patterns of the *BnaMTs* were similar to the patterns identified by RNA-seq, with different expression profiles detected among different rapeseed varieties (Figure 7, Table S3). For example, *BnaMT2B*, *BnaMT2C*, *BnaMT2D*, and *BnaMT2J* were highly expressed in all tissues; *BnaMT1* and *BnaMT4B* were expressed at lower levels in roots, hypocotyls, and cotyledons; *BnaMT2A*, *BnaMT2F*, *BnaMT2G*, *BnaMT2H*, *BnaMT2I*, *BnaMT3A*, and *BnaMT3B* were expressed at lower levels in roots and hypocotyls than in cotyledons (Figure 7); and *BnaMT2A* and *BnaMT2H* did not exhibit tissue-specific expression in *B. napus* (Figures 6 and 7). After As³⁺ treatment, all *BnaMT* genes were expressed at higher levels in roots than in hypocotyls but were expressed at the highest levels in cotyledons (Figure 7). For example, *BnaMT1* was upregulated by As³⁺ treatment, and *BnaMT2A*, *BnaMT2B*, *BnaMT2F*, *BnaMT2J*, and *BnaMT3B* were more significantly upregulated in cotyledons than in roots and hypocotyls (Figure 7). Importantly, *BnaMT3C* was more highly expressed in varieties B33 and B34 than in B93 and B113 (Figure 7).

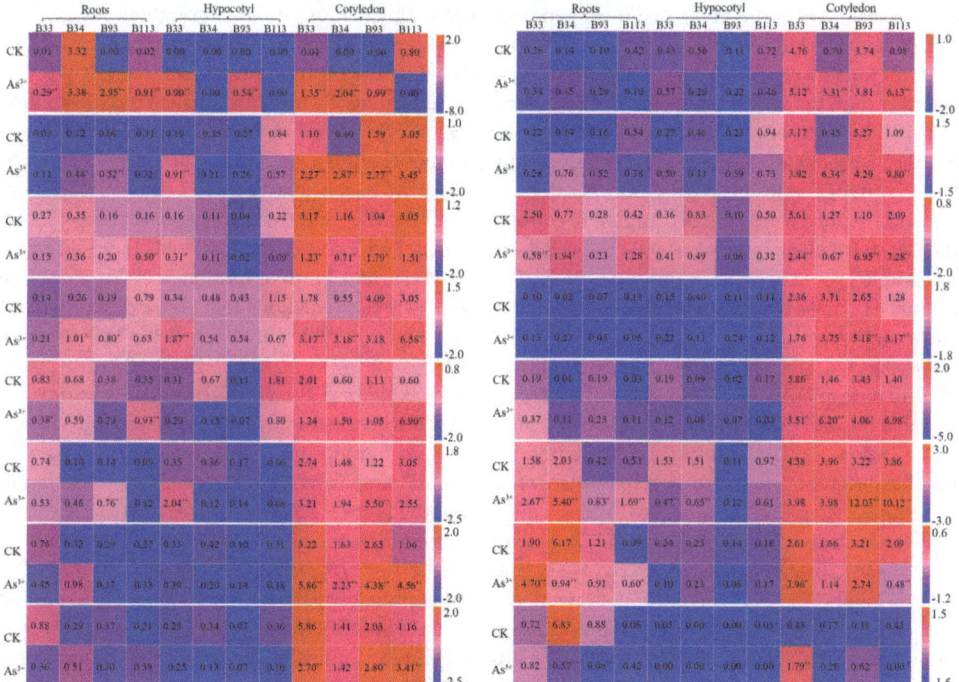

Figure 7. Expression analysis of *BnaMT* family genes in different tissues under control and As³⁺ treatment via real-time RT-PCR (RT-qPCR). Three biological replicates per sample were used for analysis, and three technical replicates were analyzed per biological replicate. Values represent the average of three biological replicates with three technical replicates of each tissue (Table S3). The expression data was gained from the real-time RT-PCR (RT-qPCR) analysis data and shown as log2 as calculated by average values normalized to that of the reference gene *BnACTIN7* (EV116054). *, ** indicates a significance level at 0.05 and 0.01, respectively. The heatmap was generated using Heatmap Illustrator v1.0 (HemI v1.0, Huazhong University, Wuhan, China; http://hemi.biocuckoo. org/contact.php).

3. Discussion

The high-affinity heavy metal chelators, PCs and MTs, play crucial roles in maintaining metal homeostasis during plant development [36–41]. Moreover, the *Brassica* plants had high biomass productivity and high levels of heavy metal absorption, as analyzed in *B. juncea* [1,23], *B. rapa* [33], and *B. napus* [42]. In addition, seven putative *MT* genes have been identified in *Arabidopsis* [15,16], but no comprehensive study of these genes has been reported. *Brassica* species, which were derived from a common ancestor, are ideal model systems for analyzing polyploid evolution and genome duplication [43]. Many analyses have focused on the model plant *A. thaliana* and various *Brassica* species (*B. rapa*, *B. nigra*, *B. oleracea*, *B. napus*, *B. juncea* and *B. carinata*). The whole genome sequences of all species except *B. carinata* are available in BRAD (the *Brassica* Database, http://brassicadb.org/brad/downloadOverview.php). In the present study, we identified 52 *MT* genes from various *Brassica* species based on *A. thaliana MT* gene sequences (Table 1). Phylogenetic analysis revealed that all *MT* family genes are closely associated with *AtMTs* (Figure 2), suggesting that they share similar functions or have undergone gene fusion [44]. Of these, *MT1* subfamily genes from *B. nigra*, *B. oleracea*, and *B. juncea* have not been identified, but *BolMT1* and *BjuMT1* have been identified [13]. The number of *MT2* (10) and *MT3* (3) genes in *B. napus* is nearly equal to the sum of these genes in *B. rapa* (5 *MT2* and *1 MT3*) and *B. oleracea* (6 *MT2* and 2 *MT3*), and most genes showed high levels of synteny throughout the genome, reflecting the fact that whole genome duplications and segmental duplications were a major contributor to the expansion of *MTs* during evolution. However, the deduced protein sequence of *BraMT1* in *B. rapa* is longer than the previously published sequence [33], indicating the need for further study to confirm *BraMT1*. In addition, homologs of *BnaMT4A* and *BraMT4*, *BnaMT3A* and *BraMT3*, and *BnaMT1* and *BraMT1* were not detected in the corresponding genomes (Figure 5, Table 1), in accordance with the finding that gene loss typically occurs after polyploidization in eukaryotes [45–47]. Furthermore, Cys-rich regions were almost completely conserved among MTs, and the distinct spacer sequences in the Cys-poor linkers were also well-conserved, comprising 7 amino acids in *Brassica* MT1s, 40–42 amino acids in *Brassica* MT2s, 32–34 amino acids in *Brassica* MT3s, and 14–15 amino acids in *Brassica* MT4s (Figure 1). These results are in close agreement with previous predictions that Cys-rich regions will show highly conserved *MT* family genes [13,33], and we infer that the variations in the MTs might be associated with their different functions in plants [48].

To date, plant MTs have been widely characterized, exhibiting different tissue-specific expression patterns [8,14,49–52]. For example, *AtMT1A* and *AtMT2B* were predominantly expressed in roots and leaves, while *AtMT2A* and *AtMT3* were highly expressed in roots and young leaves [14]. Likewise, notable differences in expression patterns were also found among *MT1*, *MT2A*, and *MT2B* subfamily genes in *B. napus* (Figure 6). *BnaMT1* was expressed at the highest level in siliques of 30D, except for roots and leaves, whereas *B. napus MT2A* genes (with four members; *BnaMT2A*, *BnaMT2C*, *BnaMT2H* and *BnaMT2I*) and *MT2B* genes (with six members; *BnaMT2B*, *BnaMT2D*, *BnaMT2E*, *BnaMT2F*, *BnaMT2G* and *BnaMT2J*; Figure 2) showed variable expression patterns. For example, *BnaMT2A* and *BnaMT2H* were preferentially expressed in buds, and *BnaMT2G* was expressed at low levels in all organs (Figure 6). These differences may be attributed to concentrations and species differences in future works. However, *B. napus MT3* and *MT4* subfamily genes shared similar expression patterns with *AtMT3* and *AtMT4* [14]. *B. napus MT3s* were mainly expressed in stems, roots, and leaves, and *B. napus MT4s* were primarily expressed in developing seeds (Figure 6). The expression patterns of these *BnaMTs* revealed by RT-qPCR corresponded well with the patterns obtained by transcriptome analysis under normal conditions, although there were differences among *B. napus* varieties (Figure 7). Although no comprehensive heavy metal tolerance mechanisms have been uncovered in *Brassica*, distinctive expression patterns were identified among the *B. napus MT* family members in this work, laying the foundation for investigating the biochemical and physiological functions of MTs in plants.

Heavy metal (Cu, Cd, and As) pollution in agricultural soils has become a critical problem affecting crop production and quality. The absorption of these heavy metals by plants plays an important role in the entry of these metals into the food chain [42]. Further, *MTs* have been shown to play an important role in metal homeostasis and tolerance in plants [4,6,9,10,21,26,33,34]. Strikingly, *Brassica* plants exhibit efficient heavy metal uptake and translocation, as well as a high tolerance to heavy metals [31,42], and several *MT* genes in *Brassica* have been reported, especially in Indian mustard (*Brassica juncea* L.) [23,27,33,53]. Recently, three *B. rapa* metallothionein genes (*BrMT1–3*) displayed differential expression levels under various exogenous stress factors [54], and MT-like, protein-encoding gene transcription was obviously induced in roots and leaves of *B. napus* under As treatment [55]. Here, we investigated the expression profiles of *BnaMT* family genes in *B. napus* under normal conditions and As^{3+} stress. The *BnaMTs* were obviously induced by the As^{3+} treatments, which is in accordance with findings that MTs are involved in the chelation and sequestration of heavy metals [3,4,55]. Like other plant *MT* genes [4,41,54,55], however, these genes also had different expression profiles in different *B. napus* varieties and in different tissues, with high expression levels in cotyledons and low expression levels in hypocotyls, such as *BnaMT1, BnaMT2A, BnaMT2B, BnaMT2C, BnaMT2F, BnaMT2J, BnaMT3B*, and *BnaMT3C* (Figure 7). These results suggest that hypocotyls might merely be involved in the transport of heavy metal ions, but that these ions accumulate in roots and cotyledons. In addition, *BnaMT3C* was obviously increased in the roots and hypocotyls of B33 and B34, but the higher expression levels of B93 and B113 in cotyledons (Figure 7), which comply with the finding that B33 and B34 exhibited better growth than B93 and B113 under As^{3+} treatment (Figure S2), indicate that they play crucial roles in the response to As^{3+} stress in *B. napus*. Our results provide important information for further functional studies of MT family genes in *B. napus*.

4. Materials and Methods

4.1. Identification of MT Family Genes in Brassica

The amino acid sequences of MTs from the Arabidopsis Information Resource (TAIR10) database (ftp://ftp.arabidopsis.org) were used as queries for the BLASTp analysis against the whole genome sequences in the *Brassica* database [56]. The candidate sequences with *E*-values $\leq 1 \times 10^{-20}$ were identified and confirmed using the Hidden Markov model (HMM) searches program (HMMER v3.0, http://hmmer.janelia.org/), and the BLAST analysis of the MTs was performed against a *Brassica* protein database constructed using Geneious v4.8.5 software (http://www.geneious.com/, Biomatters, Auckland, New Zealand). The coding sequences (CDS) of the MTs were identified by BLASTn searches against the *Brassica* genome database. The candidate proteins were named using the species abbreviation of the source organism (italicized), the gene family name, and the positions in the subtribe, e.g., *AtMT1B* and *BnaMT1A*. Physicochemical properties, including the molecular weight (kDa), isoelectric point (pI), and the grand average of hydropathy (GRAVY) value of each deduced protein were determined using the online ExPASy-ProtParam tool (http://web.expasy.org/protparam/).

4.2. Multiple Sequence Alignment and Phylogenetic Analysis of MTs in Brassica

The deduced amino acid sequences of MT proteins from *A. thaliana* and various *Brassica* species, including *B. rapa*, *B. oleracea*, *B. napus*, *B. juncea*, and *B. nigra*, were subjected to multiple protein sequence alignment using the ClustalW software with default settings [57]. To illustrate the evolutionary relationships of MTs in *Brassica*, a neighbor-joining (NJ) phylogenetic tree was generated with the MEGA v6.0 program (Tokyo Metropolitan University, Tokyo, Japan) using the JTT+I+G substitution model and a bootstrap test with 1000 replicates [58]. The phylogenetic trees were visualized using FigTree v1.4.2 (http://tree.bio.ed.ac.uk/software/figtree/).

4.3. Conserved Motif Recognition and Gene Structure Analysis

The CDS of the MTs from the *Brassica* species were retrieved based on their protein sequences, and the corresponding genomic sequences were extracted from the *Brassica* genome sequences. The exon–intron structures of the MTs were analyzed online using the Gene Structure Display Server (GSDS v2.0, http://gsds.cbi.pku.edu.cn/index.php). Conserved motifs were identified using Multipel Expectation Maximization for Motif Elucidation (MEME v4.12.0, http://meme-suite.org/tools/meme) with the following parameters: number of repetitions, any; maximum number of motifs, 15; and optimum width of each motif, between 6 and 300 residues [59]. Each motif with an *E*-value $< 1 \times 10^{-10}$ was retained for motif detection.

4.4. Chromosomal Locations of MT Family Genes in B. napus

The *MT* family genes were mapped to the rapeseed chromosomes according to their physical distances in the GFF genome files, which were downloaded from the *B. napus* genome database (http://www.genoscope.cns.fr/brassicanapus/) [43]. A map of the chromosomal locations of the MTs was constructed using MapChart v2.0 (https://www.wur.nl/en/show/Mapchart.htm) [60].

4.5. Plant Materials and Metal Stress Treatments

B. napus seeds were collected from the Rapeseed Engineering Research Center of Southwest University in Chongqing, China (CERCR). Fifty healthy seeds were selected and soaked in a dish (the diameter was 90 mm) containing deionized water for 24 h. Then morphologically uniform seedlings were selected and plugged into a hydroponic system with a float tray (60 cm × 40 cm × 10 cm) for 7 days. Here, the seedlings were exposed to distilled water and 35 μM As^{3+} solutions, respectively. Meanwhile, they were cultivated under long-day conditions (16 h light/8 h dark, 5000 Lux) at 25 °C. After 7 days, the whole roots, hypocotyls, and cotyledons were sampled to analyze the *MT* gene expression patterns; the tissues were snap frozen in liquid nitrogen and stored at −80 °C prior to total RNA extraction. All experiments were repeated three times.

4.6. Total RNA Extraction and RT-qPCR Analysis

As the *B. napus* cultivars B33 and B34 grow better than B93 and B113 under heavy metal treatment (Figure S2), they were therefore used for expression analysis. Total RNA was isolated from the samples using a DNAaway RNA Mini-Prep Kit (Sangon Biotech, Shanghai, China). For the tissue-specific expression analysis, RNA was extracted from the roots, hypocotyls, and cotyledons and pretreated with gDNA Eraser (Takara, Dalian, China). Subsequently, 1 μg of the total RNA was used to synthesize first-strand cDNA with an RNA PCR Kit (AMV) Ver. 3.0 (Takara, Dalian, China). The cDNA was subjected to RT-qPCR analysis using SYBR Premix Ex Taq II (Takara, Dalian, China) on a Bio-Rad CFX96 Real Time System (Bio-Rad Laboratories, Hercules, CA, USA) as previously described [61]. *BnACTIN7* (EV116054) was employed as a reference gene to normalize MT gene expression levels via the $2^{-\Delta\Delta Ct}$ method [62]. All experiments were performed with three technical replicates, and the values represent the average ± standard error (SE). The specific primer sequences used in this study were obtained from the qPCR Primer Database [63] and are listed in Supplementary Table S4.

4.7. Statistical Analysis

All experiments were repeated three times (three biological replicates). All data were statistically analyzed using the Student's *t*-test with the statistical analysis software package SPSS v15.0 (IBM Corp, Armonk, NJ, USA).

5. Conclusions

In this study, we identified 60 *MTs* from *A. thaliana* and five *Brassica* species. The phylogenetic analysis showed that all *MT* family genes are closely associated with the *AtMTs*. Genome-mapping analysis revealed high levels of synteny throughout the genome due to whole genome duplication and segmental duplication events. In addition, all 16 *BnaMTs* were induced by heavy metal stress, especially in cotyledons versus roots and hypocotyls. Finally, *BnaMT3C* might improve the response to As^{3+} stress in *B. napus*. Our results provide a basis for the further functional analysis of the molecular functions of *MT* family genes in *B. napus*.

Supplementary Materials: Supplementary materials can be found at http://www.mdpi.com/1422-0067/19/8/2181/s1.

Author Contributions: C.Q. (Cunmin Qu) and J.L. conceived and designed the experiments; Y.P. and M.Z. conducted the experiments; S.W., G.M. and X.H. collected and analyzed the data; Y.L., C.Q. (Cailin Qiao), R.W. and X.X. carried out the experiments and performed the software analyses; K.L. and C.Q. (Cunmin Qu) wrote the manuscript; and C.Q. (Cunmin Qu) and J.L. reviewed the manuscript.

Funding: This work was supported by the National Natural Science Foundation of China (31772320, 31571701), The National Key Research and Development Plan (2018YFD0100501), the Department of Agriculture projects of modern agricultural technology system (CARS-12), Chongqing Basic Scientific and Advanced Technology Research (cstc2015jcyjBX0001, cstc2017jcyjAX0321), the 973 Project (2015CB150201), the 111 Project (B12006), the Fundamental Research Funds for the Central Universities (XDJK2016A005, XDJK2016B030).

Acknowledgments: We would also like to thank Kathy Farquharson for critical reading of this manuscript.

Conflicts of Interest: The authors declare no conflict of interest.

References

1. Ebbs, S.; Uchil, S. Cadmium and zinc induced chlorosis in Indian mustard [*Brassica juncea* (L.) czern] involves preferential loss of chlorophyll b. *Photosynthetica* **2008**, *46*, 49–55. [CrossRef]

2. Lingua, G.; Franchin, C.; Todeschini, V.; Castiglione, S.; Biondi, S.; Burlando, B.; Parravicini, V.; Torrigiani, P.; Berta, G. *Arbuscular mycorrhizal* fungi differentially affect the response to high zinc concentrations of two registered poplar clones. *Environ. Pollut.* **2008**, *153*, 137–147. [CrossRef] [PubMed]

3. Hall, J.L. Cellular mechanisms for heavy metal detoxification and tolerance. *J. Exp. Bot.* **2002**, *53*, 1–11. [CrossRef] [PubMed]

4. Lv, Y.; Deng, X.; Quan, L.; Xia, Y.; Shen, Z. Metallothioneins *BcMT1* and *BcMT2* from *Brassica campestris* enhance tolerance to cadmium and copper and decrease production of reactive oxygen species in *Arabidopsis thaliana*. *Plant Soil* **2013**, *367*, 507–519. [CrossRef]

5. Huang, G.Y.; Wang, Y.S. Expression and characterization analysis of type 2 metallothionein from grey mangrove species (*Avicennia marina*) in response to metal stress. *Aquat. Toxicol.* **2010**, *99*, 86–92. [CrossRef] [PubMed]

6. Mir, G.; Domenech, J.; Huguet, G.; Guo, W.J.; Goldsbrough, P.; Atrian, S.; Molinas, M. A plant type 2 metallothionein (MT) from cork tissue responds to oxidative stress. *J. Exp. Bot.* **2004**, *55*, 2483–2493. [CrossRef] [PubMed]

7. Gall, J.E.; Boyd, R.S.; Rajakaruna, N. Transfer of heavy metals through terrestrial food webs: A review. *Environ. Monit. Assess.* **2015**, *187*, 201. [CrossRef] [PubMed]

8. Gu, C.S.; Liu, L.Q.; Deng, Y.M.; Zhu, X.D.; Huang, S.Z.; Lu, X.Q. The heterologous expression of the *Iris lactea* var. chinensis type 2 metallothionein IlMT2b gene enhances copper tolerance in *Arabidopsis thaliana*. *Bull. Environ. Contam. Toxicol.* **2015**, *94*, 247–253. [CrossRef] [PubMed]

9. Hassinen, V.H.; Tuomainen, M.; Peraniemi, S.; Schat, H.; Karenlampi, S.O.; Tervahauta, A.I. Metallothioneins 2 and 3 contribute to the metal-adapted phenotype but are not directly linked to Zn accumulation in the metal hyperaccumulator, *Thlaspi caerulescens*. *J. Exp. Bot.* **2009**, *60*, 187–196. [CrossRef] [PubMed]

10. Cobbett, C.; Goldsbrough, P. Phytochelatins and metallothioneins: Roles in heavy metal detoxification and homeostasis. *Annu. Rev. Plant Biol.* **2002**, *53*, 159–182. [CrossRef] [PubMed]

11. Bourdineaud, J.P.; Baudrimont, M.; Gonzalez, P.; Moreau, J.L. Challenging the model for induction of metallothionein gene expression. *Biochimie* **2006**, *88*, 1787–1792. [CrossRef] [PubMed]

12. Robinson, N.J.; Tommey, A.M.; Kuske, C.; Jackson, P.J. Plant metallothioneins. *Biochem. J.* **1993**, *295*, 1–10. [CrossRef] [PubMed]

13. Leszczyszyn, O.I.; Imam, H.T.; Blindauer, C.A. Diversity and distribution of plant metallothioneins: A review of structure, properties and functions. *Metallomics* **2013**, *5*, 1146–1169. [CrossRef] [PubMed]

14. Guo, W.J.; Bundithya, W.; Goldsbrough, P.B. Characterization of the *Arabidopsis* metallothionein gene family: Tissue-specific expression and induction during senescence and in response to copper. *New Phytol.* **2003**, *159*, 369–381. [CrossRef]

15. Benatti, M.R.; Yookongkaew, N.; Meetam, M.; Guo, W.J.; Punyasuk, N.; Abuqamar, S.; Goldsbrough, P. Metallothionein deficiency impacts copper accumulation and redistribution in leaves and seeds of *Arabidopsis*. *New Phytol.* **2014**, *202*, 940–951. [CrossRef] [PubMed]

16. Guo, W.J.; Meetam, M.; Goldsbrough, P.B. Examining the specific contributions of individual *Arabidopsis* metallothioneins to copper distribution and metal tolerance. *Plant Physiol.* **2008**, *146*, 1697–1706. [CrossRef] [PubMed]

17. Lee, J.; Shim, D.; Song, W.Y.; Hwang, I.; Lee, Y. Arabidopsis metallothioneins 2a and 3 enhance resistance to cadmium when expressed in *Vicia faba* guard cells. *Plant Mol. Biol.* **2004**, *54*, 805–815. [CrossRef] [PubMed]

18. Ren, Y.; Liu, Y.; Chen, H.; Li, G.; Zhang, X.; Zhao, J. Type 4 metallothionein genes are involved in regulating Zn ion accumulation in late embryo and in controlling early seedling growth in *Arabidopsis*. *Plant Cell Environ.* **2012**, *35*, 770–789. [CrossRef] [PubMed]

19. Roosens, N.; Leplae, R.; Bernard, C.; Verbruggen, N. Variations in plant metallothioneins: The heavy metal hyperaccumulator *Thlaspi caerulescens* as a study case. *Planta* **2005**, *222*, 716–729. [CrossRef] [PubMed]

20. Hammond, J.P.; Bowen, H.C.; White, P.J.; Mills, V.; Pyke, K.A.; Baker, A.J.; Whiting, S.N.; May, S.T.; Broadley, M.R. A comparison of the *Thlaspi caerulescens* and *Thlaspi arvense* shoot transcriptomes. *New Phytol.* **2006**, *170*, 239–260. [CrossRef] [PubMed]

21. Yuan, J.; Chen, D.; Ren, Y.; Zhang, X.; Zhao, J. Characteristic and Expression Analysis of a Metallothionein Gene, *OsMT2b*, Down-Regulated by Cytokinin Suggests Functions in Root Development and Seed Embryo Germination of Rice. *Plant Physiol.* **2008**, *146*, 1637–1650. [CrossRef] [PubMed]

22. Zhou, Y.; Chu, P.; Chen, H.; Li, Y.; Liu, J.; Ding, Y.; Tsang, E.W.; Jiang, L.; Wu, K.; Huang, S. Overexpression of *Nelumbo nucifera* metallothioneins 2a and 3 enhances seed germination vigor in *Arabidopsis*. *Planta* **2012**, *235*, 523–537. [CrossRef] [PubMed]

23. An, Z.; Li, C.; Zu, Y.; Du, Y.; Andreas, W.; Gromes, R.; Rausch, T. Expression of *BjMT2*, a metallothionein 2 from *Brassica juncea*, increases copper and cadmium tolerance in *Escherichia coli* and *Arabidopsis thaliana*, but inhibits root elongation in *Arabidopsis thaliana* seedlings. *J. Exp. Bot.* **2006**, *57*, 3575–3582.

24. Kumar, P.B.; Dushenkov, V.; Motto, H.; Raskin, I. Phytoextraction: The use of plants to remove heavy metals from soils. *Environ. Sci. Technol.* **1995**, *29*, 1232–1238. [CrossRef] [PubMed]

25. Salt, D.E.; Blaylock, M.; Kumar, N.P.; Dushenkov, V.; Ensley, B.D.; Chet, I.; Raskin, I. Phytoremediation: A novel strategy for the removal of toxic metals from the environment using plants. *Biotechnology* **1995**, *13*, 468–474. [CrossRef] [PubMed]

26. Cojocaru, P.; Gusiatin, Z.M.; Cretescu, I. Phytoextraction of Cd and Zn as single or mixed pollutants from soil by rape (*Brassica napus*). *Environ. Sci. Pollut. Res.* **2016**, *23*, 10693–10701. [CrossRef] [PubMed]

27. Gasic, K.; Korban, S.S. Expression of *Arabidopsis* phytochelatin synthase in Indian mustard (*Brassica juncea*) plants enhances tolerance for Cd and Zn. *Planta* **2007**, *225*, 1277–1285. [CrossRef] [PubMed]

28. Sridhar, B.B.M.; Diehl, S.V.; Han, F.X.; Monts, D.L.; Su, Y. Anatomical changes due to uptake and accumulation of Zn and Cd in Indian mustard (*Brassica juncea*). *Environ. Exp. Bot.* **2005**, *54*, 131–141. [CrossRef]

29. Ishikawa, S.; Noriharu, A.; Murakami, M.; Wagatsuma, T. Is *Brassica juncea* a suitable plant for phytoremediation of cadmium in soils with moderately low cadmium contamination?—Possibility of using other plant species for Cd-phytoextraction. *Soil Sci. Plant Nutr.* **2006**, *52*, 32–42. [CrossRef]

30. Marchiol, L.; Assolari, S.; Sacco, P.; Zerbi, G. Phytoextraction of heavy metals by canola (*Brassica napus*) and radish (*Raphanus sativus*) grown on multicontaminated soil. *Environ. Pollut.* **2004**, *132*, 21–27. [CrossRef] [PubMed]

31. Solhi, M.; Shareatmadari, H.; Hajabbasi, M.A. Lead and Zinc Extraction Potential of Two Common Crop Plants, *Helianthus Annuus* and *Brassica napus*. *Water Air Soil Pollut.* **2005**, *167*, 59–71. [CrossRef]

32. Touiserkani, T.; Haddad, R. Cadmium-induced stress and antioxidative responses in different *Brassica napus* cultivars. *J. Agric. Sci. Technol.* **2012**, *14*, 929–937.

33. Kim, S.H.; Lee, H.S.; Song, W.Y.; Choi, K.S.; Hur, Y. Chloroplast-targeted BrMT1 (*Brassica rapa* type-1 metallothionein) enhances resistance to cadmium and ROS in transgenicatabidopsis plants. *J. Plant Biol.* **2007**, *50*, 1–7. [CrossRef]

34. Zhang, M.; Takano, T.; Liu, S.; Zhang, X. Abiotic stress response in yeast and metal-binding ability of a type 2 metallothionein-like protein (PutMT2) from *Puccinellia tenuiflora*. *Mol. Biol. Rep.* **2014**, *41*, 5839–5849. [CrossRef] [PubMed]

35. Nagaharu, U. Genome analysis in Brassica with special reference to the experimental formation of *B. napus* and peculiar mode of fertilisation. *J. Jpn. Bot.* **1935**, *7*, 389–452.

36. Von Ruecker, A.A.; Wild, M.; Rao, G.S.; Bidlingmaier, F. Atrial natriuretic peptide protects hepatocytes against damage induced by hypoxia and reactive oxygen. Possible role of intracellular free ionized calcium. *J. Clin. Chem. Clin. Biochem.* **1989**, *27*, 531–537. [PubMed]

37. Palmiter, R.D. The elusive function of metallothioneins. *Proc. Natl. Acad. Sci. USA* **1998**, *95*, 8428–8430. [CrossRef] [PubMed]

38. Tottey, S.; Rondet, S.A.; Borrelly, G.P.; Robinson, P.J.; Rich, P.R.; Robinson, N.J. A copper metallochaperone for photosynthesis and respiration reveals metal-specific targets, interaction with an importer, and alternative sites for copper acquisition. *J. Biol. Chem.* **2002**, *277*, 5490–5497. [CrossRef] [PubMed]

39. Coyle, P.; Philcox, J.; Carey, L.; Rofe, A.M. Metallothionein: The multipurpose protein. *Cell. Mol. Life Sci.* **2002**, *59*, 627–647. [CrossRef] [PubMed]

40. Guo, S.Y.; Xu, D.; Yang, P.; Yu, W.T.; Lv, M.K.; Yuan, D.R.; Yang, Z.H.; Zhang, G.H.; Sun, S.Y.; Wang, X.Q.; et al. A Novel Nonlinear Optical Complex Crystal with an Organic Ligand Coordinated Through an O Atom: Tetrathiocyanatocadmiummercury-Dimethyl Sulfoxide. *Cryst. Res. Technol.* **2001**, *36*, 609–614. [CrossRef]

41. Ansarypour, Z.; Shahpiri, A. Heterologous expression of a rice metallothionein isoform (OsMTI-1b) in *Saccharomyces cerevisiae* enhances cadmium, hydrogen peroxide and ethanol tolerance. *Braz. J. Microbiol.* **2017**, *48*, 537–543. [CrossRef] [PubMed]

42. Benáková, M.; Ahmadi, H.; Dučaiová, Z.; Tylová, E.; Clemens, S.; Tůma, J. Effects of Cd and Zn on physiological and anatomical properties of hydroponically grown *Brassica napus* plants. *Environ. Sci. Pollut. Res.* **2017**, *24*, 20705–20716. [CrossRef] [PubMed]

43. Chalhoub, B.; Denoeud, F.; Liu, S.; Parkin, I.A.; Tang, H.; Wang, X.; Chiquet, J.; Belcram, H.; Tong, C.; Samans, B.; et al. Early allopolyploid evolution in the post-neolithic *Brassica napus* oilseed genome. *Science* **2014**, *345*, 950–953. [CrossRef] [PubMed]

44. Sun, H.; Fan, H.-J.; Ling, H.-Q. Genome-wide identification and characterization of the bHLH gene family in tomato. *BMC Genom.* **2015**, *16*, 9. [CrossRef] [PubMed]

45. Sankoff, D.; Zheng, C.; Zhu, Q. The collapse of gene complement following whole genome duplication. *BMC Genom.* **2010**, *11*, 313. [CrossRef] [PubMed]

46. Wang, X.; Wang, H.; Wang, J.; Sun, R.; Wu, J.; Liu, S.; Bai, Y.; Mun, J.-H.; Bancroft, I.; Cheng, F.; et al. The genome of the mesopolyploid crop species *Brassica Rapa*. *Nat. Genet.* **2011**, *43*, 1035–1039. [CrossRef] [PubMed]

47. Qu, C.; Zhao, H.; Fu, F.; Wang, Z.; Zhang, K.; Zhou, Y.; Wang, X.; Wang, R.; Xu, X.; Tang, Z.; et al. Genome-wide survey of flavonoid biosynthesis genes and gene expression analysis between black- and yellow-seeded *Brassica napus*. *Front. Plant Sci.* **2016**, *7*. [CrossRef] [PubMed]

48. Zimeri, A.M.; Dhankher, O.P.; Mccaig, B.; Meagher, R.B. The Plant MT1 Metallothioneins are Stabilized by Binding Cadmiums and are Required for Cadmium Tolerance and Accumulation. *Plant Mol. Biol.* **2005**, *58*, 839–855. [CrossRef] [PubMed]

49. Framond, A.J.D. A metallothionein-like gene from maize (*Zea mays*) Cloning and characterization. *FEBS Lett.* **1991**, *290*, 103–106. [CrossRef]

50. Zhou, J.; Goldsbrough, P.B. Structure, organization and expression of the metallothionein gene family in *Arabidopsis*. *Mol. Gen. Genet.* **1995**, *248*, 318–328. [CrossRef] [PubMed]

51. Foley, R.C.; Singh, K.B. Isolation of a Vicia faba metallothionein-like gene, expression in foliar trichomes. *Plant Mol. Biol.* **1994**, *26*, 435–444. [CrossRef] [PubMed]

52. Shawber, C.; Nofziger, D.; Hsieh, J.J.; Lindsell, C.; Bögler, O.; Hayward, D.; Weinmaster, G. Notch signaling inhibits muscle cell differentiation through a CBF1-independent pathway. *Development* **1996**, *122*, 3765–3773. [PubMed]

53. Liu, J.X.; Zu, Y.G.; Shi, X.G.; Ai, Y.Z.; Du, Y.J.; Fu, Y.J.; An, Z.G. BjMT2, a metallothionein type-2 from *Brassica juncea*, may effectively remove excess lead from erythrocytes and kidneys of rats. *Environ. Toxicol. Pharmacol.* **2007**, *23*, 168–173. [CrossRef] [PubMed]

54. Ahn, Y.O.; Kim, S.H.; Lee, J.; Kim, H.; Lee, H.S.; Kwak, S.S. Three *Brassica rapa* metallothionein genes are differentially regulated under various stress conditions. *Mol. Biol. Rep.* **2012**, *39*, 2059–2067. [CrossRef] [PubMed]

55. Farooq, M.A.; Gill, R.A.; Ali, B.; Wang, J.; Islam, F.; Ali, S.; Zhou, W. Subcellular distribution, modulation of antioxidant and stress-related genes response to arsenic in *Brassica napus* L. *Ecotoxicology* **2016**, *25*, 350–366. [CrossRef] [PubMed]

56. Altschul, S.F.; Madden, T.L.; Schäffer, A.A.; Zhang, J.; Zhang, Z.; Miller, W.; Lipman, D.J. Gapped BLAST and PSI-BLAST: A new generation of protein database search programs. *Nucleic Acids Res.* **1997**, *25*, 3389–3402. [CrossRef] [PubMed]

57. Larkin, M.A.; Blackshields, G.; Brown, N.; Chenna, R.; McGettigan, P.A.; McWilliam, H.; Valentin, F.; Wallace, I.M.; Wilm, A.; Lopez, R. Clustal W and Clustal X version 2.0. *Bioinformatics* **2007**, *23*, 2947–2948. [CrossRef] [PubMed]

58. Tamura, K.; Stecher, G.; Peterson, D.; Filipski, A.; Kumar, S. MEGA6: Molecular evolutionary genetics analysis version 6.0. *Mol. Biol. Evol.* **2013**, *30*, 2725–2729. [CrossRef] [PubMed]

59. Bailey, T.L.; Boden, M.; Buske, F.A.; Frith, M.; Grant, C.E.; Clementi, L.; Ren, J.; Li, W.W.; Noble, W.S. MEME SUITE: Tools for motif discovery and searching. *Nucleic Acids Res.* **2009**, *37*, W202–W208. [CrossRef] [PubMed]

60. Voorrips, R. MapChart: Software for the graphical presentation of linkage maps and QTLs. *Heredity* **2002**, *93*, 77–78. [CrossRef]

61. Qu, C.; Fu, F.; Lu, K.; Zhang, K.; Wang, R.; Xu, X.; Wang, M.; Lu, J.; Wan, H.; Zhanglin, T.; et al. Differential accumulation of phenolic compounds and expression of related genes in black-and yellow-seeded *Brassica napus. J. Exp. Bot.* **2013**, *64*, 2885–2898. [CrossRef] [PubMed]

62. Wu, G.; Li, Z.; Yuhua, W.; Yinglong, C.; Changming, L. Comparison of Five Endogenous Reference Genes for Specific PCR Detection and Quantification of *Brassica napus. J. Agric. Food Chem.* **2010**, *58*, 2812–2817. [CrossRef] [PubMed]

63. Lu, K.; Li, T.; He, J.; Chang, W.; Zhang, R.; Liu, M.; Yu, M.; Fan, Y.; Ma, J.; Sun, W. qPrimerDB: A thermodynamics-based gene-specific qPCR primer database for 147 organisms. *Nucleic Acids Res.* **2018**, *46*, D1229–D1236. [CrossRef] [PubMed]

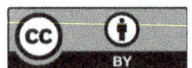

International Journal of
Molecular Sciences

Article

Particle Bombardment of the *cry2A* Gene Cassette Induces Stem Borer Resistance in Sugarcane

Shiwu Gao, Yingying Yang, Liping Xu *, Jinlong Guo, Yachun Su, Qibin Wu, Chunfeng Wang and Youxiong Que *

Key Laboratory of Sugarcane Biology and Genetic Breeding, Ministry of Agriculture and Key Laboratory of Crop Genetics and Breeding and Comprehensive Utilization, College of Crop Science, Fujian Agriculture and Forestry University, Ministry of Education, Fuzhou 350002, China; gaoshiwu2008@126.com (S.G.); yingyingyang13@163.com (Y.Y.); jl.guo@163.com (J.G.); syc2009mail@163.com (Y.S.); wqbaidqq@163.com (Q.W.); 18305999305@163.com (C.W.)
* Correspondence: xlpmail@fafu.edu.cn (L.X.); queyouxiong@fafu.edu.cn (Y.Q.);
  Tel.: +86-0591-8377-2604 (L.X.); +86-0591-8385-2547 (Y.Q.)

Received: 3 May 2018; Accepted: 4 June 2018; Published: 6 June 2018

Abstract: Sugarcane borer is the most common and harmful pest in Chinese sugarcane fields, and can cause damage to the whole plant during the entire growing season. To improve borer resistance in sugarcane, we constructed a plant expression vector pGcry2A0229 with the *bar* gene as the marker and the *cry2A* gene as the target, and introduced it into embryogenic calli of most widely cultivated sugarcane cultivar ROC22 by particle bombardment. After screening with phosphinothricin in vitro and Basta spray, 21 resistance-regenerated plants were obtained, and 10 positive transgenic lines harboring the *cry2A* gene were further confirmed by conventional PCR detection. Real-time quantitative PCR (RT-qPCR) analysis showed that the copy number of the *cry2A* gene varied among different transgenic lines but did not exceed four copies. Quantitative ELISA analysis showed that there was no linear relationship with copy number but negatively correlated with the percentage of borer-infested plants. The analysis of industrial and agronomic traits showed that the theoretical sugar yields of transgenic lines TR-4 and TR-10 were slightly lower than that of the control in both plant cane and ratoon cane; nevertheless, TR-4 and TR-10 lines exhibited markedly lower in frequency of borer-infested plants in plant cane and in the ratoon cane compared to the control. Our results indicate that the introduction of the *cry2A* gene via bombardment produces transgenic lines with obviously increased stem borer resistance and comparable sugar yield, providing a practical value in direct commercial cultivation and crossbreeding for ROC22 has been used as the most popular elite genitor in various breeding programs in China.

Keywords: sugarcane; *cry2A* gene; particle bombardment; stem borer; resistance

---

1. Introduction

Sugarcane is the most important sugar crop, with sucrose accounting for 80% of the total sugar production in the world and accounting for more than 92% of total sugar production in China. As a C_4 crop, sugarcane makes it one of the most important energy crops due to its high biomass, high fiber, and years of ratooning. Nearly 90% of biofuel ethanol is produced by sugarcane in the United States and Brazil [1]. The risk level of transgenic safety for sugarcane is low mainly due to the following three reasons. Firstly, sugarcane is an asexually propagated crop and generally does not bloom during field cultivation in China, indicating little chance of exogenous gene drift by flowering. Secondly, as an industrial raw material for sucrose and fuel ethanol, sugarcane is not a directly circulated food. The processing of sucrose requires as high as 107 °C for crystallization and crystal sucrose belongs to a purified carbohydrate, without protein ingredient. Besides, the fuel ethanol is not edible.

A similar opinion of high food safety level of transgenic sugarcane can be ascribed to the decomposition of protein expressed by the exogenous gene during process of sucrose crystallization [2]. Thirdly, transgenic sugarcane does not affect the microbial community diversity and has no significant effect on enzyme activities in rhizosphere soil, which means better ecological security [3]. To date, two cases involving genetically modified sugarcanes were approved for commercial planting, namely drought resistant transgenic sugarcane in Indonesia and insect-resistant transgenic sugarcane in Brazil.

There are currently five major species of stem borer thriving in Chinese sugarcane fields: *Chilo sacchariphagus* Bjojer, *Scirpophaga nivella* Fabricius, *C. infuscatellus* Snellen, *Argyroploce schistaceana* Snellen, and *Sesamia inferens* Walker. Because several generations occur in one planting season, together with several different species and overlap among generations, stem borer is the most common and harmful pests in the Chinese sugarcane industry. The percentage of dead heart seedlings is normally within the range of 10–20%, but can reach 60% in severely infected sugarcane fields, and the damage incurred during the mid-late stage leads to a significant reduction in sucrose content and the increasing of wind broken stalks [4,5]. Sugarcane, which is heterogeneous polyploid and aneuploid, has a complex genetic background [6,7]. As many as 120 chromosomes are available in modern sugarcane hybrids and there is a lack of stem borer resistance genes in the gene pool [8,9], it is extremely difficult to breed a cultivar with resistance to stem borers in traditional crossbreeding program. Chemical pesticides have long been the main method for preventing and controlling stem borers in Chinese sugarcane, which not only increases production cost but also pollutes the environment.

In 1987, Vaeck et al. successfully introduced the *cry1A(b)* gene into tobacco through *Agrobacterium*-mediated transformation and obtained stem borer-resistant transgenic tobacco [10]. Subsequently, the *Bt* gene was introduced into crops such as cotton [11–14], maize [15], rice [16–19], and tomato [20], resulting in effective improvement in stem borer resistance. In sugarcane, Arencibia et al. first described the transformation of *cry1A* gene to improve stem borer resistance in 1997 [8], followed by numerous reports on the improvement of insect resistance in transgenic sugarcanes such as *cry1A(b)* [21–23], *GNA* [24,25], *cry1Aa3* [26], *cry1Ac* [27–29] and proteinase inhibitor [30,31]. Besides, researchers also attempted to use RNAi technology to control pest damage in sugarcane [32], with success in other crops [33–35]. Our previous study showed that the application of insect-resistant transgenic sugarcane can economically and effectively solve the problem of stem borers in the sugarcane industry [36]. *cry2A*, which has low homology (<45%) with *cry1A*, is another Bt protein. Previous researchers demonstrated that cry2A protein is toxic to several lepidopteran pests, indicating its feasibility to be used as a bio-insecticide [37,38]. However, there is no report about the application of *cry2A* in sugarcane.

Compared to other screening marker genes such as *npt II*, *bar* gene screening can be performed at the early stage of genetic transformation of sugarcane, thereby reducing the workload involving tissue culture and increasing efficiency [39]. Thus, the *bar* gene as a screening marker gene has an obvious advantage in eliminating of pseudotyped transformants during selection of putative transformants after bombardment, as sugarcane is phosphinothricin (PPT)-sensitive. The antibiotics and PPT resistance screening tests of two sugarcane genotypes, FN81-745 (*Saccharum* spp. hybrid) and Badila (*Saccharum officinarum*), showed that the effective concentration of both G418 and Hyg was 30.0 mg/L, while only 0.75 mg/L and 1.0 mg/L for PPT, respectively [39]. In plant genetic transformation, the *bar* gene has been widely used as a screening marker gene [40,41], and its application to sugarcane has also been described in several reports [22,29,42,43].

In the present study, to obtain *cry2A* transgenic sugarcane, a plant expression vector pGcry2A0229 with the *bar* gene as a screening marker and *cry2A* as a target gene was constructed and genetically transformed into sugarcane by particle bombardment. PPT and Basta resistance screening and PCR validation were conducted to confirm the positive *cry2A* gene transgenic sugarcane plants. Then, the copy number of the *cry2A* gene and its protein expression in transgenic lines were determined by Real-time quantitative PCR (RT-qPCR) and quantitative ELISA detection of protein, respectively. Finally, several transgenic lines with better comprehensive traits based on a field survey of industrial and agronomic traits were identified, which provides a transgenic line of potential commercial cultivation and the foundation for crossbreeding of stem borer-resistant traits in sugarcane.

2. Results

2.1. Construction and Verification of the Plant Vector pGcry2A0229

The construction scheme of plant expression vector pGcry2A0229 is depicted in Figure 1. Single-enzyme digestion of pGcry2A0229 using restriction endonuclease *Hin*d III generated the expected single band with a size of 8213 bp. Electrophoresis analysis of double enzyme digestion with *Hin*d III and *Eco*R I showed the expected two bands with sizes of about 4436 and 3777 bp, respectively (Figure 2). The sequencing results confirmed that the pGcry2A0229 was the expected positive recombinant plasmid.

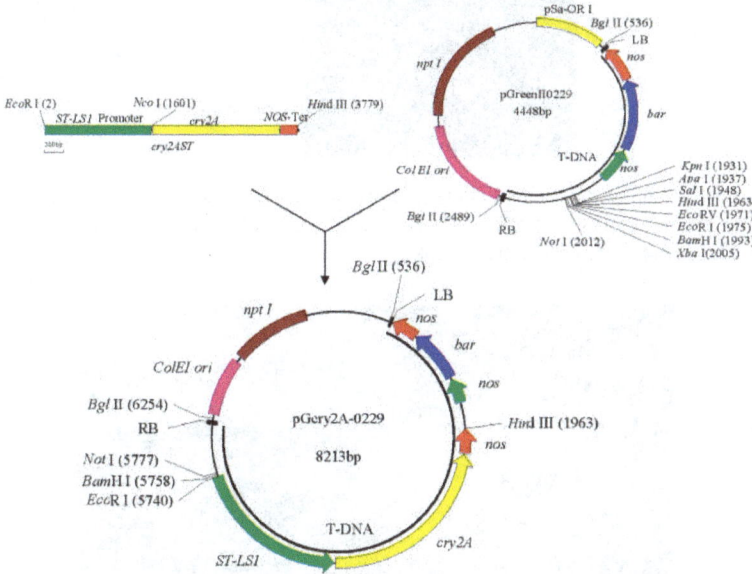

Figure 1. Construction roadmap of the plant expression vector pGcry2A0229.

2.2. Particle Bombardment and Resistance Screening

Micro-bombs were prepared using the tungsten particles and pGcry2A0229 DNA, and bombardment transformation was conducted with the embryogenic calli of sugarcane cultivar ROC22 as the receptor material (Figure 3a). Based on our preliminary experimental results, the embryogenic calli after bombardment was subcultured and differentiated with PPT of 0.8 mg/L. Some tissues gradually differentiated into regenerated plantlets with PPT resistance, whereas most wild-type calli gradually became brown and died (Figure 3b). The negative control died. When the resistant regenerated plantlets grew up to a height of 4–5 cm, they were transferred into the rooting medium without PPT for rooting (Figure 3c). Finally, 95 resistant regenerated plants were obtained.

The resistant regenerated plantlets were transplanted into nutrient pots to ensure their survival. After spray screening with 3.0‰ (*v/v*) Basta, most plants gradually turned yellow, wilted, and died after 15–20 days, and finally only 21 plants survived (Figure 3d).

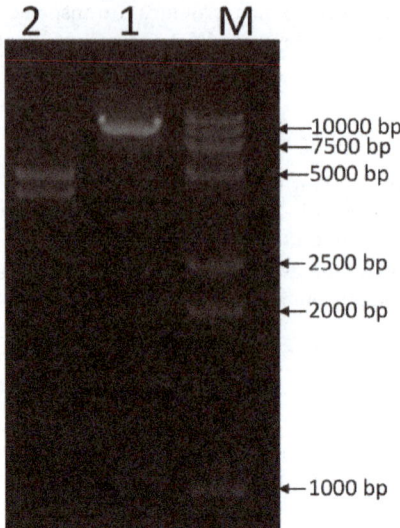

Figure 2. The products of recombinant plasmid pGcry2A0229 digested by restriction enzymes: M, DL15,000 + 2, 000 DNA ladder; 1, The products of pGcry2A0229 digested by *Hin*d III; 2, The products of pGcry2A0229 digested by *Hin*d III and *Eco*R I.

Figure 3. Putative recombinant screening: (**a**) wild-type calli on medium without PPT; (**b**) PPT-resistant plantlets at the differentiation stage on selection medium; (**c**) regenerated plantlets at the stage of rooting culture; and (**d**) spraying screening of resistant plantlets with 3.0‰ Basta.

2.3. PCR Identification of the cry2A Gene in Resistant Regenerated Plants

A total of 21 resistant regenerated plantlets obtained by PPT and Basta screening were verified by PCR amplification of the *cry2A* gene. The results showed that a single band was amplified from 10 samples; the position of the band was consistent with that of the positive plasmid and showed an approximate size of 600 bp. The sequencing results were also consistent with the partial sequence of the *cry2A* gene, whereas no band was amplified from the non-transgenic negative control and ddH$_2$O blank control (Figure 4). Therefore, PCR analysis verified that 10 positive *cry2A* transgenic plants were successfully obtained.

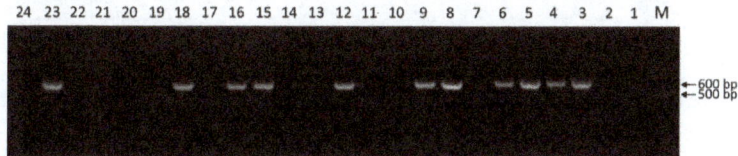

Figure 4. Electrophoretic analysis of PCR amplification products of a putative *cry2A* gene for transgenic sugarcane plants: M, DNA Marker; 1, Blank control of ddH$_2$O; 2, Negative control (non-transgenic sugarcane without bombardment); 3, Positive control (plasmid pGcry2A0229); 4–24, Herbicide Basta-resistant plants.

2.4. RT-qPCR Detection of the cry2A Gene and the Copy Number Estimation in Transgenic Lines

Ten PCR-positive transgenic sugarcane lines were tested by RT-qPCR technique and the copy number of the *cry2A* gene was estimated. The RT-qPCR quantitative standard curve of *cry2A* gene was constructed using the following equation: $y = -3.593x + 43.082$, $R^2 = 0.994$, where the y-axis represents the C_t value, the x-axis represents the logarithm of the initial template copy number. A good correlation between C_t values (18–40) and initial template copy number (10^1–10^8) was observed. According to the linear equation, x, the total copy number of the *cry2A* gene in the sample, was determined, and the number of exogenous *cry2A* copies of a single cell (Table 1) was calculated using the followed formula: Copies/genome = 10^x/[25 ng × 10^{-9} × 6.02 × 10^{23}/(10,000 × 10^6 × 660)]. Table 1 shows that the *cry2A* gene copy number of different transgenic sugarcane lines varied, wherein three lines had two copies, five lines had three copies, and two lines had four copies.

Table 1. Estimated *cry2A* gene copy number of different transgenic lines.

Line	C_t I	C_t II	C_t III	C_t Mean	Copy Number
TR-1	29.96	29.84	30.19	30.00 ± 0.10	1.92
TR-2	29.10	29.39	29.22	29.24 ± 0.08	3.13
TR-3	29.78	30.06	30.20	30.01 ± 0.12	1.90
TR-4	29.95	30.11	30.09	30.05 ± 0.05	1.86
TR-5	29.08	28.97	29.10	29.05 ± 0.04	3.53
TR-6	29.75	29.67	29.63	29.68 ± 0.04	2.35
TR-7	29.70	29.39	29.70	29.60 ± 0.10	2.49
TR-8	29.48	29.51	29.62	29.54 ± 0.04	2.58
TR-9	29.68	29.68	29.78	29.72 ± 0.03	2.30
TR-10	30.00	29.10	29.48	29.53 ± 0.26	2.60
Non-transgenic	37.97	38.22	40.03	38.74 ± 0.65	0.01

2.5. cry2A Protein Expression in the Transgenic Lines

cry2A protein expression in the mature leaves of 10 transgenic sugarcane lines was quantitated by ELISA, and a standard curve was constructed using the Bt protein standard in the kit as the following equation: $y = -1.156x + 3.865$, $R^2 = 0.999$, where the y-axis represents the OD$_{450}$ absorbance, the x-axis represents the concentration of the Bt protein standard. The absorbance value correlated well with the protein standard concentration (R^2 = 0.999). The amount of protein expression in the 10 samples was calculated according to the linear equation (Figure 5), which showed that the *cry2A* protein expression was observed in all 10 transgenic lines at levels within the range of 76.45–90.75 μg/FWg, of which three lines, namely, TR-4, TR-8, and TR-10 had higher protein expression levels (85.86, 82.49 and 90.75 μg/FWg, respectively) and the difference among the three lines was statistically significant.

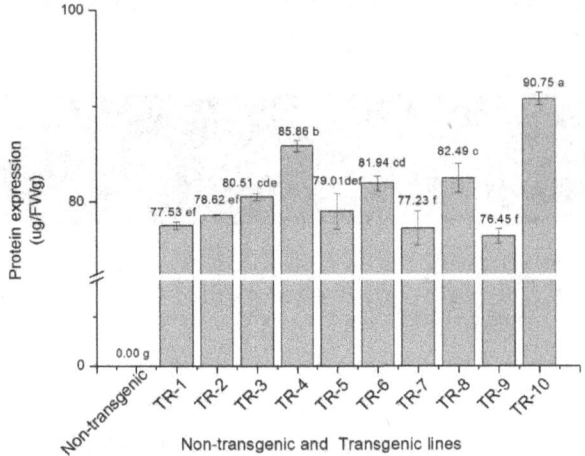

Figure 5. The *cry2A* protein expression in the leaves of non- transgenic and 10 different transgenic sugarcane lines detected by quantitative ELISA. The value is the average of three replicate experiments ± standard deviation (*n* = 3), and the different letters indicate significant difference at 0.05 level.

2.6. Survey of Industrial and Agronomic Traits of the cry2A Transgenic Sugarcane Lines

According to protein expression and field performance, three *cry2A* transgenic sugarcane lines, TR-4, TR-8, and TR-10, were selected for the further field experiment using non-transgenic recipient ROC22 as a control, and plant height, stem diameter, brix, effective stalk number, and other indicators of industrial and agronomic traits at maturity were determined. The results were then subjected to univariate statistical analysis, and the results are shown in Table 2.

In the plant cane, when refer to the plant height of three transgenic lines, both TR-4 and TR-10 were slightly lower than the control, while the significant lower was observed in line TR-8. Although the stem diameters of the three lines were lower than the control, but not statistically significant. The brix (of the three lines) was higher than that of the control, although not statistically significant. The lines TR-4 and TR-8 had slightly lower number of effective stalk per hectare than the control, while line TR-10 is slightly higher than the control, but both had no significant difference. The theoretical sugar yield of TR-4 and TR-10 was comparable to that of the control, whereas TR-8 was significantly lower than the control.

In the ratoon cane, the height and stem diameter of the three transgenic lines were lower than the control, however, only the plant height of TR-8 and the stem diameter of TR-10 were significantly lower than the control. The brix of the three transgenic lines was higher than that of the control, and TR-8 and TR-10 were significantly higher than the control. Although the number of effective stalk per hectare of TR-8 and TR-10 were higher than that of the control, this difference was not statistically significant. Similar to that in the plant cane, the theoretical sugar yield of TR-4 and TR-10 were comparable to that of the control, while significantly lower was observed in TR-8. Compared to the plant cane, the brix of the transgenic lines in the ratoon cane increased, of which TR-8 and TR-10 increased by more than 1.0, while the control ROC22 decreased by 0.43.

Table 2. Industrial and agronomic traits of different *cry2A* transgenic sugarcane lines during the plant and ratoon cane.

Crop Season	Line	Plant Height, H	Stem Diameter, D	Brix, Bx	Effective Stems (stem/ha)	Theoretical Sugar Yield (t/ha)
Plant cane	TR-4	270.23 ± 3.23 [a]	2.33 ± 0.16 [a]	20.74 ± 0.26 [a]	64,989.74 ± 1982.53 [a]	11.02 ± 0.34 [a]
	TR-8	240.27 ± 1.88 [b]	2.30 ± 0.07 [a]	20.39 ± 0.36 [a]	64,989.74 ± 799.89 [a]	9.33 ± 0.13 [b]
	TR-10	258.67 ± 5.06 [a]	2.34 ± 0.12 [a]	20.33 ± 0.32 [a]	69,692.95 ± 709.90 [a]	11.07 ± 0.18 [a]
	Non-transgenic	272.70 ± 4.71 [a]	2.38 ± 0.06 [a]	19.98 ± 0.30 [a]	66,700.00 ± 1385.47 [a]	11.26 ± 0.19 [a]
Ratoon cane	TR-4	271.13 ± 8.05 [a,b]	2.50 ± 0.04 [a,b]	20.91 ± 0.47 [a,b]	54,646.15 ± 1397.39 [a]	10.87 ± 0.28 [a]
	TR-8	248.57 ± 3.48 [b]	2.46 ± 0.02 [a,b]	21.69 ± 0.11 [a]	52,084.61 ± 1597.40 [a]	9.70 ± 0.21 [b]
	TR-10	264.33 ± 5.02 [a,b]	2.36 ± 0.07 [b]	21.85 ± 0.15 [a]	58,061.53 ± 1597.39 [a]	10.73 ± 0.17 [a]
	Non-transgenic	279.43 ± 2.96 [a]	2.66 ± 0.10 [a]	19.85 ± 0.53 [b]	51,230.77 ± 2091.48 [a]	10.96 ± 0.20 [a]

Data followed with the different letters ([a] and [b]) indicate significant difference at 0.05 level, but the same letter ([a] or [b]) indicates that the difference is not significant.

2.7. Insect Resistance Identification of the cry2A Transgenic Sugarcane Lines

Under natural field conditions, the percentage of borer-infested plants of the three lines and the control ROC22 in the plant cane and ratoon cane were investigated. The leaves and stems of the transgenic sugarcane lines showed more pronounced insect resistance compared to that in the non-transgenic control ROC22 (Figure 6). The survey results (Table 3) showed that the percentage of borer-infested plants in the three transgenic lines was lower than the control. In the plant cane, the percentage of borer-infested plants in the TR-10 line was only 26.67%, but as high as 80.0% in the control, and the difference was statistically significant. Although the TR-4 and TR-8 lines compared to the control did not reach significant level, these were only 36.67% and 53.33%, respectively. After one year of ratooning, the percentage of borer-infested plants decreased in the three transgenic lines, but slightly increased in the control, and the percentage of borer-infested plants was 30.0% and 16.67% in TR-4 and TR-10 lines, respectively, which was significantly lower than the 83.33% of the control. Especially, the transgenic lines affected by stem borers only incurred damages in the cane stem cortex (Figure 6b), whereas the control line exhibited more serious damage with injuries in the entire stem (Figure 6c).

Table 3. The percentage of borer-infested plants of different *cry2A* transgenic sugarcane lines in plant and ratoon cane.

Line	Percentage of Borer-Infested Plants (%)	
	Plant Cane	Ratoon Cane
TR-4	36.67 ± 12.02 [a,b]	30.00 ± 11.55 [b]
TR-8	53.33 ± 17.64 [a,b]	50.00 ± 10.00 [a,b]
TR-10	26.67 ± 5.77 [b]	16.67 ± 3.33 [b]
Non-transgenic	80.00 ± 6.67 [a]	83.33 ± 6.67 [a]

Data followed with the different letters ([a] and [b]) indicate significant difference at 0.05 level, but the same letter ([a] or [b]) indicates that the difference is not significant.

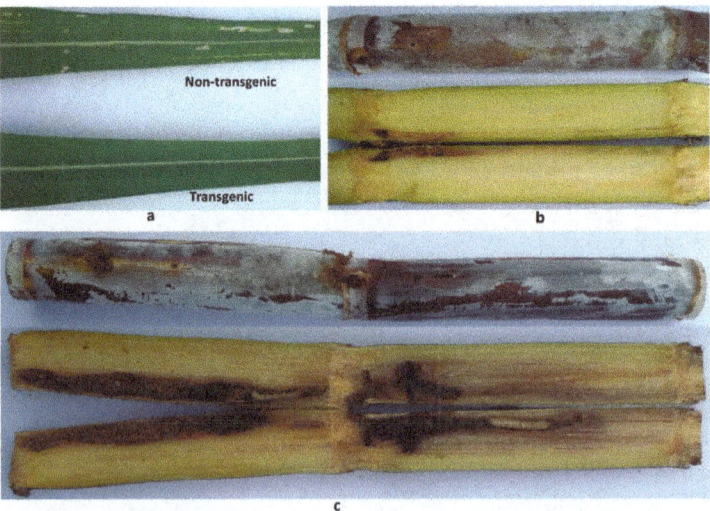

Figure 6. Stem borer damage in sugarcane under natural field conditions: (**a**) symptoms of the transgenic and non-transgenic sugarcane leaf; (**b**) symptoms of the transgenic sugarcane stem (only in cortex); and (**c**) symptoms of non-transgenic sugarcane stem.

3. Discussion and Conclusions

Sugarcane is a perennial crop and, to save costs, ratooning is usually conducted for over three years or even up to five years or more, during which sucrose content reduction and wind broken stalk increases can be caused by stem borer [4,5]. Arencibia et al. (1997) transformed the *cry1A(b)* gene into sugarcane by the cell electroporation, and improved the stem borer resistance [8]. Arvinth et al. (2010) introduced the *cry1Ab* gene into sugarcane, which significantly reduced the percentage of dead heart sugarcane seedlings [22]. The *GNA* gene was integrated into sugarcane genome via *Agrobacterium*-mediated transformation by Zhangsun et al. (2007), and the results showed that transgenic sugarcane plants had a significant resistance to the woolly aphid [25]. Falco et al. (1997) introduced the soybean bowman-birk inhibitor into sugarcane callus using particle bombardment, and it demonstrated that, compared to larvae fed on leaf tissue from untransformed ones, the growth of larvae feeding on leaf tissue from transgenic plants was significantly retarded, however the retardation was not sufficient to prevent the "dead heart" symptom [30]. Weng et al. (2011) [28] and Gao et al. (2016) [29] introduced the *cry1Ac* gene into different sugarcane varieties, and the transgenic sugarcane plants showed much better resistance to stem borer than the non-transgenic ones. ROC22, the most widely cultivated cultivar accounting for more than 60% of Chinese sugarcane acreage in the past 15 years, was used as the receptor in this study. A field test comparing the three transgenic lines with non-transgenic control found that the percentage of borer-infested plants of the transgenic lines in the ratoon cane decreased compared to that in the plant cane, whereas contrarily, slightly increased in the control. In addition, compared to the control, line TR-4, exhibiting markedly lower in frequency of borer-infested plants in the ratoon cane (30.0% vs. 83.3%) and much lower in plant cane (36.67% vs. 80.0%) indicating that the stress of stem borers gradually shifted to non-transgenic control.

Assessment of copy number of transgenic lines is essential to phenotypic studies and investigations on genetic stability. The traditional method for copy number identification is the Southern blot, which is highly cumbersome and strongly operation dependent [44], and various external factors may influence visualization of hybridization bands and thus are often underestimated. Previous research has shown that Southern blotting was not able to accurately determine the number copy numbers of exogenous genes in sugarcane b, whereas RT-qPCR is characterized by high specificity and high sensitivity, and thus more accurate [29,45]. RT-qPCR has been widely used to identify exogenous gene copy number [44–49], even for transgene copy number from 3 to >50 [45]. Sugarcane has a complex genetic background and is a highly heterogeneous polyploid or aneuploid crop, with genome sizes of up to 10 Gb [50]. In the present study, an RT-qPCR assay standard curve for the *cry2A* gene with a slope of −3.593 and a correlation coefficient of 0.994 was established, which indicated that PCR amplification efficiency and C_t values correlate well with the initial template copy number. Based on the standard curve, the *cry2A* gene copy number in the 10 transgenic sugarcane lines was determined, which revealed that the copy number of each transgenic line did not exceed four copies, which is discrepant to the findings of our previous study on *cry1Ac* transgenic sugarcane [29]. These may be related to different exogenous genes introduced and different genotypes of receptor materials.

In the present study, ELISA was used to quantitatively determine the cry2A protein expression levels in the leaves of 10 obtained transgenic lines, which ranged from 76.45 to 90.75 μg/FWg, with significant differences in some lines. However, no clear linear relationship between protein expression and copy number was observed, which was similar to that observed in previous studies [51–53]. The expression of exogenous Bt protein can effectively improve the insect resistance of transgenic plants [18,54]. Weng et al. (2011) introduced a modified *cry1Ac* gene into sugarcane cultivars ROC16 and YT79-177 by particle bombardment, and 17 transgenic plants were positive for Western blot. It also demonstrated that the expression of water-soluble proteins in leaves ranged from 2.2 ng/mg to 50 ng/mg, and when the expression exceeded 9 ng/mg (9 μg/FWg), insect resistance was observed, with the content of cry1Ac protein in transgenic sugarcane positively correlated with its insect resistance [28]. Arvinth et al. (2010) found that the total soluble cry1Ab protein expression in the obtained transgenic sugarcane leaves ranged from 0.007% to 1.73%, and protein expression was negatively correlated to the percentage of dead heart seedlings [22]. In our previous research, the cry1Ac protein expression in *cry1Ac* transgenic sugarcane leaves ranged from 0.85 to 70.9 μg/FWg, and the higher the protein expression, the lower the percentage of borer-infested plants, which exhibited a significant negative correlation [29]. Here, again, we observed that the higher the protein expression, the better the insect-resistant effect, which is consistent with the results of our previous investigation [29] and with other reports [22,28].

Weng et al. (2011) generated a ubiquitin (ubi) initiated *cry1Ac* transgenic sugarcane, and the assessment of industrial and agronomic traits showed that the agronomic traits such as plant height and stem diameter were greatly affected. However, the industrial indicators such as sucrose content and brix exhibited no significant difference compared to the control [28]. Our group previously conducted a field survey on double 35 s initiated *cry1Ac* transgenic sugarcane cultivar FN15, and it showed that only 2 of 14 transgenic lines had slightly greater plant heights than the control, although not statistically significant, whereas the other lines were lower than the control, and all the stem diameters (of the transgenic lines) were lower than that of the control. However, both higher and lower brix than that of the control was observed, and the calculated theoretical sugar yields (of the transgenic lines) were all lower than that of the control though three lines are unobvious [29]. Wang et al. (2017) introduced *cry1Ab* gene into sugarcane cultivar ROC22 by *Agrobacterium*, and investigated the industrial and agronomic traits of five single-copy transgenic lines. The result showed that plant height, stem diameter, brix, effective stalk number of several transgenic lines was only slightly lower than that of the control, while calculated theoretical sugar yield was significantly lower than that of the control [23]. Besides, a three-year field performance trial of transgenic sugarcane with *npt II* gene showed a reduction in growth and cane yield, but, when individual events were analyzed separately, the yields of several transgenic events were comparable to that of no transformants [53]. The present study conducted on plant cane and ratoon cane in the field using the obtained transgenic lines TR-4, TR-8, and TR-10, and their theoretical sugar yields were all lower (9.33–11.07 t/ha for plant cane, and 9.70–10.87 t/ha for ratoon cane) than that of the control (11.26 and 10.96 t/ha). It indicates that the introduction of exogenous *cry2A* gene into sugarcane increased stem borer resistance and reduced the percentage of infested plants, while the expression of the Bt protein consumes energy, thereby resulting in a decrease in sugar yield in generally, though comparable sugar yield of transgenic lines can be obtained, such as TR-4 and TR-10 in this study, which is in line with the results of two previous researches [29,53].

In conclusion, the introduction of the *cry2A* gene via particle bombardment produces the transgenic lines with obviously increased stem borer resistance and comparable sugar yield, providing a practical value in direct commercial cultivation, and crossbreeding for ROC22 has been used as the most popular elite parent in various breeding programs in China.

4. Materials and Methods

4.1. Materials

The plant expression vector pGreenII0229 was obtained from John Innes Center in Norwich, Norfolk, UK, and the clone 2AST1305.1 containing the *cry2A* gene was a gift from Professor Illimar Altosaar of the University of Ottawa in Canada. The pGreen plasmid can help plant genetic transformation because it was a versatile and flexible binary vector [55], and the pGreenII0229 vector contains the *bar* gene as the screening marker gene. The receptor material used for genetic transformation was ROC22, the most widely cultivated sugarcane cultivar in mainland China, which was provided by Key Laboratory of Sugarcane Biology and Genetics and Breeding, Ministry of Agriculture, China.

4.2. Plant Vector Construction of the cry2A Gene

The *cry2A* gene plant expression vector was constructed using the directional cloning strategy. First, plasmid DNA of the 2AST1305.1 cloning vector that harbored the exogenous *cry2A* gene was digested with restriction endonucleases *EcoR* I and *Hind* III. The exogenous gene expression cassette containing the ST-LS1 promoter, the *cry2A* gene, and the *nos* terminator was recovered. Meanwhile, the plasmid DNA of the plant expression vector pGreenII0229 was digested with restriction enzymes *EcoR* I and *Hind* III, and the target fragment containing the *bar* gene as a screening marker gene was recovered. Finally, the two recovered fragments were ligated with T4-DNA ligase to obtain a new *cry2A* gene plant expression vector pGcry2A0229.

4.3. Transformation and Screening

Shoots of ROC22 sugarcane plants that showed robust growth in the field were selected. The leaves were collected from the shoots and disinfected with 75% alcohol, and the outer leaf sheaths were stripped under aseptic conditions. Then, the heart lobe above the growth point was removed and sliced into about 2-mm thick discs, cultured in the dark at 26–28 °C for 2–4 weeks, and then subjected to particle bombardment transformation after callus generation [29]. Before bombardment, the tungsten particles (Bio-Rad, Foster City, CA, USA, 0.7) were coated by the plasmid of pGcry2A0229 DNA as the micro-bombs, with 1.0 μg of DNA each bombardment. The operation was performed according to the protocol of the PDS-1000/He gene gun (Bio-Rad, Hercules, CA, USA). The bombarded and transformed material was restored culture in subculture medium, then subjected to a screening culture using 0.8 mg/L PPT according to our preliminary experiment, until the plantlets had differentiated, which refers to literature for details [29]. Once developing roots, the plants were transplanted into a nutrient pot. Upon reaching a height of about 10 cm and on a sunny day, the plants were sprayed with 3.0‰ Basta solution (v/v) [29]. Calli were inducted on medium consisted of MS, 3.0 mg/L 2,4-D, 30 g/L sucrose, and 6.0 g/L agar powder, at a pH of 5.8. The subculture medium comprised MS, 2 mg/L 2,4-D, 30 g/L sucrose, and 6 g/L agar powder, at a pH of 5.8. The differentiation medium included MS, 1.5 mg/L 6-BA, 1.0 mg/L KT, 0.2 mg/L NAA, 30 g/L sucrose, and 6 g/L agarose, at a pH of 5.8. The rooting medium consisted of $\frac{1}{2}$ MS, 0.2 mg/L 6-BA, 3 mg/L NAA, 60 g/L sucrose, and 5.5 g/L agarose, at a pH of 5.8.

4.4. DNA Extraction and Primer Design

Genomic DNA was extracted from young leaves of resistant plants that survived the PPT and Basta screening and non-transgenic ROC22 negative control plants using a modified CTAB method [56]. Based on the *cry2A* gene sequence, Primer Premier 5 software was used to design PCR and RT-qPCR primers. The PCR primers were as follows: 2ast1178s: 5′-AACAGGCAACAACCCATAGAGG-3′ and 2ast1798r: 5′-AGGGAGCCCACCTTCTTGAG-3′, and the resulting amplified fragment was 620 bp in size. The RT-qPCR primers were as follows: forward primer: 5′-CAACCAGCAGGTGGACAACTT-3′, reverse primer: 5′-AAGAGCTGCTGCATGGTGTTC-3′, and probe: 5′-CTCAACCCGACCCAGAACCCGG-3′.

4.5. PCR Amplification of Putative Transgenic Sugarcane Lines

Using non-transformed ROC22 as the negative control, pGcry2A0229 plasmid DNA containing the *cry2A* gene as the positive control, and ddH$_2$O as a blank control, amplification and identification were performed using an Eppendorf 5331 PCR instrument (Eppendorf, Hamburg, Germany). Each PCR amplification system consisted of the following reagents: 2.5 μL of 10 × PCR buffer (Mg^{2+} Plus), 2.0 μL of a dNTP mixture (2.5 mmol/L each), 1.0 μL of the DNA template (50.0 ng/μL), 1 μL each of the upstream and downstream primers, 0.25 μL of Taq DNA polymerase (5 U/μL), and topped up to 25.0 μL with ddH$_2$O. The reaction conditions were as follows: pre-denaturation at 95 °C for 5 min; followed by 30 cycles of denaturation at 95 °C for 30 s, annealing at 57 °C for 30 s, and extension at 72 °C for 40 s; and a final extension at 72 °C for 10 min. After amplification, the PCR products were electrophoresed on a 1.5% agarose gel and photographed using a gel imaging system.

4.6. Copy Number Calculation in Transgenic Sugarcane Lines by RT-qPCR

The *cry2A* gene was quantitatively detected in the PCR-positive transgenic sugarcane lines using the designed and synthesized RT-qPCR primers. The fluorescence quantitative PCR instrument was an ABI PRISM 7500 Sequence Detection System (Foster City, CA, USA). The total volume of the detection system was 25.0 µL, which contained 12.5 µL of a FastStart Universal Probe Master Mix, 1.0 µL of gDNA (25.0 ng/µL), 1.0 µL (10.0 µmol/L) of the forward primer, 1.0 µL (10.0 µmol/L) of the reverse primer, 0.2 µL (10.0 µmol/L) of probe, and then topped up to a final volume of 25.0 µL with ddH$_2$O. The amplification conditions were as follows: 50 °C for 2 min; 95 °C for 10 min; 40 cycles of 95 °C for 15 s and 60 °C for 1 min; and a final cycle of 95 °C for 15 s, 60 °C for 15 s, and 95 °C for 15 s. Three replicates were used for each sample. At the same time, gradient dilutions of 10^8, 10^7, 10^6, 10^5, 10^4, 10^3, 10^2, and 10^1 copies/µL were prepared using pGcry2A0229 plasmid DNA. Plasmid copy number was calculated using the following equation: Plasmid copy number (copies/µL) = 6.02×10^{23} copies/mol × plasmid concentration (g/µL)/plasmid molecular weight (g/mol)/660 [57]. After the reaction, using log(plasmid copy number) as x-axis and the C_t value as y-axis, a standard curve was generated using the formula $y = kx + b$. Further, based on the C_t value (y) and linear equation, the total copy number (10^x) of the *cry2A* transgenic lines was determined, and then the single cell copy number of each sample was calculated using the following formula: Copies/genome = $10^x/[25 \text{ ng} \times 10^{-9} \times 6.02 \times 10^{23}/(10{,}000 \times 10^6 \times 660)]$ [58].

4.7. Quantitative ELISA of the cry2A Protein in Transgenic Sugarcane Lines

The *cry2A* protein in the leaves of PCR-positive transgenic sugarcane plants was detected using double-antibody sandwich enzyme linked immunosorbent assay (ELISA). Non-transformed ROC22 plants were used as the negative control and ddH$_2$O as a blank control. Gradient dilutions of cry2A protein reference standards in a Qualiplat kit for *cry2A* purchased from Envirologix (Portland, OR, USA) were prepared, with the y-axis representing the OD$_{450}$ absorbance and the x-axis representing the Bt standard protein concentration to construct a standard curve. Quantitative ELISA was conducted according to the protocol provided in the *cry2A* protein assay kit. Three replicates of each sample were prepared.

4.8. Field Trial Design and Assessment of Phenotype Traits of the Transgenic Sugarcane Lines

Three *cry2A* transgenic sugarcane lines with good performance in the field were selected used for further investigation, and non-transgenic ROC22 was used as the control in the field experiment. The experiment followed a randomized block design that consisted of triplicates. The length of the plot was 8.0 m, three rows with a row spacing of 1.3 m were used, the plot area was 31.2 m^2, and 13 buds per meter length. The present study applied common fertilizers at amounts routinely used in the sugarcane field: 345.0 kg/ha of nitrogen fertilizer (N), 240.0 kg/ha of phosphate fertilizer (P$_2$O$_5$), and 360.0 kg/ha of potassium fertilizer (K$_2$O), coupled with normal field management. The industrial and agronomic traits including plant height, stem diameter, brix, effective stalk number, and percentage of borer-infested plants at maturity were investigated, and 20 plants in each plot as the biological repeats in the plot were measured. At the same time, 5 m long and more evenly distributed sections in each plot were selected, and the effective stalk number was counted. The theoretical cane yield per mu and sugar yield per mu were calculated according to the formulae [59]:

Theoretical cane yield = Plant height × Stem diameter^2 × 0.785/1000 × Effective stalk number
Sucrose content (%) = Brix × 1.0825 − 7.703
Theoretical sugar yield = Theoretical cane yield × Sucrose content (%)
DPS analysis software and Tukey method were used for statistical analysis of the collected data.

Author Contributions: Conceived and designed the experiments: S.G., L.X. and Y.Q. Performed the experiments: S.G., Y.Y., J.G., Y.S., Q.W. and C.W. Analyzed the data: S.G., Y.Y., J.G., Y.S., Q.W. and C.W. Wrote the paper: S.G., L.X. and Y.Q. Revised and approved the final version of the paper: L.X., Y.Q. and S.G.

Funding: This work was supported by the earmarked fund for the China Agricultural Research System (CARS-17), Science and Technology Major Project of Fujian Province (2015NZ0002-2), the National Nature Science Foundation of China (31271782), the 948 Program on the Introduction of International Advanced Agricultural Science and Technique of Department of Agriculture (2014-S18).

Acknowledgments: The authors especially thank Illimar Altosaar in University of Ottawa, Canada, for providing the cloning vector 2AST1305.1.

Conflicts of Interest: The authors declare that no competing financial interests exist. The founding sponsors had no role in the design of the study; in the collection, analyses, or interpretation of data; in the writing of the manuscript, and in the decision to publish the results.

Abbreviations

ELISA	Enzyme-linked immuno sorbent assay
CTAB	Cetyltrimethyl ammonium bromide
G418	Geneticin
Hyg	Hygromycin
KT	Kinetin
MS	Murashige and shoog medium
NAA	Naphthalene acetic acid
PPT	Phosphinothricin
2,4-D	2,4-Dichlorophenoxy acetic acid
6-BA	6-Benzyladenine

References

1. Fisher, G.; Teixeira, E.; Hizsnyik, E.T.; Velthuizen, H.V. Land use dynamics and sugarcane production. In *Sugarcane Ethanol: Contributions to Climate Change Mitigation and the Environment*; Zuurbier, P., Vooren, J.V.D., Eds.; Wageningen Academic Publishers: Wageningen, The Netherlands, 2008; pp. 29–62.
2. Lakshmanan, P.; Geijskes, R.J.; Aitken, K.S.; Grof, C.L.P.; Bonnett, G.D.; Smith, G.R. Sugarcane biotechnology: The challenges and opportunities. *In Vitro Cell. Dev. Biol. Plant* **2005**, *41*, 345–363. [CrossRef]
3. Zhou, D.G.; Xu, L.P.; Gao, S.W.; Guo, J.L.; Luo, J.; You, Q.; Que, Y.X. *Cry1Ac* transgenic sugarcane does not affect the diversity of microbial communities and has no significant effect on enzyme activities in rhizosphere soil within one crop season. *Front. Plant Sci.* **2016**, *7*, 265. [CrossRef] [PubMed]
4. Huang, Y.K.; Li, W.F. The effective test of 5% carbosulfan G against *Sesamia inferens* walker and *Chilo infuscatellus* snellen. *Sugar Crops China* **2006**, *4*, 34–35.
5. Wang, Z.P.; Liu, L.; Jiang, H.T.; Zhang, G.M.; Huang, W.H.; Liang, Q.; Duan, W.X.; Li, Y.J.; Wei, J.J.; Qin, Z.Q. Field efficacy test of 22% *Fipronil FS* against sugarcane stem borers and thrips. *Plant Dis. Pests* **2016**, *7*, 23–26.
6. Aitken, K.S.; Jackson, P.A.; McIntyre, C.L. A combination of AFLP and SSR markers provides extensive map coverage and identification of homo(eo)logous linkage groups in a sugarcane cultivar. *Theor. Appl. Genet.* **2005**, *110*, 789–801. [CrossRef] [PubMed]
7. Piperidis, G.; Piperidis, N.; D'Hont, A. Molecular cytogenetic investigation of chromosome composition and transmission in sugarcane. *Mol. Genet. Genom.* **2010**, *284*, 65–73. [CrossRef] [PubMed]
8. Arencibia, A.; Vázquez, R.I.; Prieto, D.; Téllez, P.; Carmona, E.R.; Coego, A.; Hernandez, L.; de la Riva, G.A.; Selman-Housein, G. Transgenic sugarcane plants resistant to stem borer attack. *Mol. Breed.* **1997**, *3*, 247–255. [CrossRef]
9. Zhou, D.G.; Guo, J.L.; Xu, L.P.; Gao, S.W.; Lin, Q.L.; Wu, Q.B.; Wu, L.G.; Que, Y.X. Establishment and application of a loop-mediated isothermal amplification (LAMP) system for detection of *Cry1Ac* transgenic sugarcane. *Sci. Rep.* **2014**, *4*, 4912. [CrossRef] [PubMed]
10. Vaeck, M.; Reynaerts, A.; Höfte, H.; Jansens, S.; De Beuckeleer, M.; Dean, C.; Zabeau, M.; Van Montagu, M.; Leemans, J. Transgenic plants protected from insect attack. *Nature* **1987**, *328*, 33–37. [CrossRef]
11. Perlak, F.J.; Deaton, R.W.; Armstron, T.A.; Fuchs, R.L.; Sims, S.R.; Greenplate, J.T.; Fischhoff, D.A. Insect resistant cotton plants. *BioTechnology* **1990**, *8*, 939–943. [CrossRef] [PubMed]

12. Ribeiro, T.P.; Arraes, F.B.M.; Lourenço-Tessutti, I.T.; Silva, M.S.; Lisei-de-Sá, M.E.; Lucena, W.A.; Macedo, L.L.P.; Lima, J.N.; Amorim, R.M.S.; Artico, S.; et al. Transgenic cotton expressing *Cry10Aa* toxin confers high resistance to the cotton boll weevil. *Plant Biotechnol. J.* **2017**, *15*, 997–1009. [CrossRef] [PubMed]

13. Guo, X.; Huang, C.; Jin, S.X.; Liang, S.; Nie, Y.; Zhang, X.L. *Agrobacterium*-mediated transformation of *Cry1C*, *Cry2A* and *Cry9C* genes into *Gossypium hirsutum* and plant regeneration. *Biol. Plant.* **2007**, *51*, 242–248. [CrossRef]

14. Bakhsh, A.; Rao, A.Q.; Khan, G.A.; Rashid, B.; Shahid, A.A.; Husnain, T. Insect resistance studies of transgenic cotton cultivar harboring *Cry1Ac* and *Cry2A*. *Tabad Tarım Bilim. Arastırma Derg.* **2012**, *5*, 167–171.

15. Koziel, M.G.; Beland, G.L.; Bowman, C.; Carozzi, N.B.; Crenshaw, R.; Crossland, L.; Dawson, J.; Desai, N.; Hill, M.; Kadwell, S.; et al. Field performance of elite transgenic maize plants expressing an insecticidal protein derived from *Bacillus thuringiensis*. *BioTechnology* **1993**, *11*, 194–200. [CrossRef]

16. Fujimoto, H.; Itoh, K.; Yamamoto, M.; Kyozuka, J.; Shimamoto, K. Insect resistant rice generated by introduction of a modified δ-endotoxin gene of *Bacillus thuringiensis*. *BioTechnology* **1993**, *11*, 1151–1155. [CrossRef] [PubMed]

17. Bashir, K.; Husnain, T.; Fatira, T.; Latif, Z.; Mehdi, S.A.; Riazuddin, S. Field evaluation and risk assessment of transgenic indica basmati rice. *Mol. Breed.* **2004**, *13*, 301–312. [CrossRef]

18. Chen, H.; Tang, W.; Xu, C.G.; Li, X.H.; Lin, Y.J.; Zhang, Q.F. Transgenic indica rice plants harboring a synthetic *Cry2A* gene of *Bacillus thuringiensis* exhibit enhanced resistance against lepidopteran rice pests. *Theor. Appl. Genet.* **2005**, *111*, 1330–1337. [CrossRef] [PubMed]

19. Gunasekara, J.M.A.; Jayasekera, G.A.U.; Perera, K.L.N.S.; Wickramasuriya, A.M. Development of a Sri Lankan rice variety Bg 94-1 harbouring *Cry2A* gene of *Bacillus thuringiensis* resistant to rice leaffolder [*Cnaphalocrocis medinalis (Guenée)*]. *J. Natl. Sci. Found. Sri Lanka* **2017**, *45*, 143–157. [CrossRef]

20. Delannay, X.; LaVallee, B.J.; Proksch, R.K.; Fuchs, R.L.; Sims, S.R.; Greenplate, J.T.; Marrone, P.G.; Dodson, R.B.; Augustine, J.J.; Layton, J.G.; et al. Field performance of transgenic tomato plants expressing the *Bacillus thuringinesis* var. kurstaki insect control plant. *BioTechnology* **1989**, *7*, 1265–1269.

21. Arencibia, A.D.; Carmona, E.R.; Cornide, M.T.; Castiglione, S.; O'Relly, J.; Chinea, A.; Oramas, P.; Sala, F. Somaclonal variation in insect-resistant transgenic sugarcane (*Saccharum hybrid*) plants produced by cell electroporation. *Transgenic Res.* **1999**, *8*, 349–360. [CrossRef]

22. Arvinth, S.; Arun, S.; Selvakesavan, R.K.; Srikanth, J.; Mukunthan, N.; Ananda Kumar, P.; Premachandran, M.N.; Subramonian, N. Genetic transformation and pyramiding of aprotinin-expressing sugarcane with *cry1Ab* for shoot borer (*Chilo infuscatellus*) resistance. *Plant Cell Rep.* **2010**, *29*, 383–395. [CrossRef] [PubMed]

23. Wang, W.Z.; Yang, B.P.; Feng, X.Y.; Cao, Z.Y.; Feng, C.L.; Wang, J.G.; Xiong, G.R.; Shen, L.B.; Zeng, J.; Zhao, T.T.; et al. Development and characterization of transgenic sugarcane with insect resistance and herbicide tolerance. *Front. Plant Sci.* **2017**, *8*, 1535. [CrossRef] [PubMed]

24. Sétamou, M.; Bernal, J.S.; Legaspi, J.C.; Mirkov, T.E.; Legaspi, B.C. Evaluation of lectin-expressing transgenic sugarcane against stalkborers (Lepidoptera: Pyralidae): Effects on life history parameters. *J. Econ. Entomol.* **2002**, *95*, 469–477. [CrossRef] [PubMed]

25. Zhangsun, D.T.; Luo, S.L.; Chen, R.K.; Tang, K.X. Improved *agrobacterium*-mediated genetic transformation of GNA transgenic sugarcane. *Biologia* **2007**, *62*, 386–393. [CrossRef]

26. Kalunke, R.M.; Kolge, A.M.; Babu, K.H.; Prasad, D.T. *Agrobacterium* mediated transformation of sugarcane for borer resistance using *Cry1Aa3* gene and one-step regeneration of transgenic plants. *Sugar Technol.* **2009**, *11*, 355–359. [CrossRef]

27. Weng, L.X.; Deng, H.H.; Xu, J.L.; Li, Q.; Wang, L.H.; Jiang, Z.D.; Zhang, H.B.; Li, Q.; Zhang, L.H. Regeneration of sugarcane elite breeding lines and engineering of strong stem borer resistance. *Pest Manag. Sci.* **2006**, *62*, 178–187. [CrossRef] [PubMed]

28. Weng, L.X.; Deng, H.H.; Xu, J.L.; Li, Q.; Zhang, Y.Q.; Jiang, Z.D.; Li, Q.W.; Chen, J.W.; Zhang, L.H. Transgenic sugarcane plants expressing high levels of modified *Cry1Ac* provide effective control against stem borers in field trials. *Transgenic Res.* **2011**, *20*, 759–772. [CrossRef] [PubMed]

29. Gao, S.W.; Yang, Y.Y.; Wang, C.F.; Guo, J.L.; Zhou, D.G.; Wu, Q.B.; Su, Y.C.; Xu, L.P.; Que, Y.X. Transgenic sugarcane with a *Cry1ac* gene exhibited better phenotypic traits and enhanced resistance against sugarcane borer. *PLoS ONE* **2016**, *11*, e0153929. [CrossRef] [PubMed]

30. Falco, M.C.; Silva Filho, M.C. Expression of soybean proteinase inhibitors in transgenic sugarcane plants: Effects on natural defense against Diatraea saccharalis. *Plant Physiol. Biochem.* **2003**, *41*, 761–766. [CrossRef]

31. Nutt, K.A.; Allsopp, P.G.; Geijskes, R.J.; McKeon, M.G.; Smith, G.R.; Hogarth, D.M. Canegrub resistant sugarcane. *Proc. Int. Soc. Suger Cane Technol. Congr.* **2001**, *24*, 582–583.

32. Zhang, Y.L.; Huang, Q.X.; Zhang, S.Z.; Wang, J.H.; Wu, S.R.; Wang, Z.G.; Liu, Z.X. Use of bacterially mediated RNAi technology to study the molt-regulating transcription factor gene *CiHR3* of the sugarcane stem borer, *Chilo infuscatellus. Chin. J. Appl. Entomol.* **2013**, *50*, 1301–1310.

33. Mao, Y.B.; Cai, W.J.; Wang, J.W.; Hong, G.J.; Tao, X.Y.; Wang, L.J.; Huang, Y.P.; Chen, X.Y. Silencing a cotton bollworm P450 monooxygenase gene by plant-mediated RNAi impairs larval tolerance of gossypol. *Nat. Biotechnol.* **2007**, *25*, 1307–1313. [CrossRef] [PubMed]

34. Baum, J.A.; Bogaert, T.; Clinton, W.; Heck, G.R.; Feldmann, P.; Ilagan, O.; Johnson, S.; Plaetinck, G.; Munyikwa, T.; Pleau, M.; et al. Control of coleopteran insect pests through RNA interference. *Nat. Biotechnol.* **2007**, *25*, 1322–1326. [CrossRef] [PubMed]

35. Zha, W.J.; Peng, X.X.; Chen, R.Z.; Du, B.; Zhu, L.L.; He, G.C. Knockdown of midgut genes by ds RNA-transgenic plant-mediated RNA interference in the hemipteran insect *Nilaparvata lugens. PLoS ONE* **2011**, *6*, e20504. [CrossRef] [PubMed]

36. Ye, J.; Yang, Y.Y.; Xu, L.P.; Li, Y.R.; Que, Y.X. Economic Impact of Stem Borer-Resistant Genetically Modified Sugarcane in Guangxi and Yunnan Provinces of China. *Sugar Tech* **2016**, *18*, 1–9. [CrossRef]

37. Karim, S.; Dean, D.H. Toxicity and receptor binding properties of *Bacillus thuringiensis* δ-endotoxins to the midgut brush border membrane vesicles of the rice leaf folders, *Cnaphalocrocis medinalis* and *Marasmia patnalis. Curr. Microbiol.* **2000**, *41*, 276–283. [CrossRef] [PubMed]

38. Alcantara, E.P.; Aguda, R.M.; Curtiss, A.; Dean, D.H.; Cohen, M.B. *Bacillus thuringiensis* δ-endotoxin binding to brush border membrane vesicles of rice stem borers. *Arch. Insect Biochem. Physiol.* **2004**, *55*, 169–177. [CrossRef] [PubMed]

39. Luo, S.L.; Lin, J.Y.; Zhangsun, D.T. Selective test of antibiotics and PPT in different stages of sugarcane tissue culture. *J. Hainan Univ. Nat. Sci.* **2003**, *21*, 259–265.

40. Gordon-Kamm, W.J.; Spencer, T.M.; Mangano, M.L.; Adams, T.R.; Daines, R.J.; Start, W.G.; O'Brien, J.V.; Chambers, S.A.; Adams, W.R., Jr.; Willetts, N.G.; et al. Transformation of maize cells and regeneration of fertile transgenic plants. *Plant Cell* **1990**, *2*, 603–618. [CrossRef] [PubMed]

41. Vasil, V.; Castillo, A.M.; Fromm, M.E.; Vasil, I.K. Herbicide resistant fertile transgenic wheat plants obtained by microprojectile bombardment of regenerable embryogenic callus. *BioTechnology* **1992**, *10*, 667–674. [CrossRef]

42. Gallo-Meagher, M.; Irvine, J.E. Herbicide resistant transgenic sugarcane plants containing the *Bar* gene. *Crop Sci.* **1996**, *36*, 1367–1374. [CrossRef]

43. Manickavasagam, M.; Ganapathi, A.; Anbazhagan, V.R.; Sudhakar, B.; Selvaraj, N.; Vasudevan, A. *Agrobacterium*-mediated genetic transformation and development of herbicide-resistant sugarcane (*Saccharum* species hybrids) using axillary buds. *Plant Cell Rep.* **2004**, *23*, 134–143. [CrossRef] [PubMed]

44. Mason, G.; Provero, P.; Vaira, A.M.; Accotto, G.P. Estimating the number of integrations in transformed plants by quantitative real-time PCR. *BMC Biotechnol.* **2002**, *2*, 20. [CrossRef]

45. Casu, R.E.; Selivanova, A.; Perroux, J.M. High-throughput assessment of transgene copy number in sugarcane using realtime quantitative PCR. *Plant Cell Rep.* **2012**, *31*, 167–177. [CrossRef] [PubMed]

46. Song, P.; Cai, C.Q.; Skokut, M.; Kosegi, B.D.; Petolino, J.F. Quantitative real-time PCR as a screening tool for estimating transgene copy number in WHISKERS™-derived transgenic maize. *Plant Cell Rep.* **2002**, *20*, 948–954.

47. Yang, L.T.; Ding, J.Y.; Zhang, C.M.; Jia, J.W.; Weng, H.B.; Liu, W.X.; Zhang, D.B. Estimating the copy number of transgenes in transformed rice by real-time quantitative PCR. *Plant Cell Rep.* **2005**, *23*, 759–763. [CrossRef] [PubMed]

48. Yi, C.X.; Zhang, J.; Chan, K.M.; Liu, X.K.; Hong, Y. Quantitative realtime PCR assay to detect transgene copy number in cotton (*Gossypium hirsutum*). *Anal. Biochem.* **2008**, *375*, 150–152. [CrossRef] [PubMed]

49. Ji, Z.G.; Gao, X.J.; Ao, J.X.; Zhang, M.H.; Huo, N. Establishment of SYBR Green-base quantitative real-time PCR assay for determining transgene copy number in transgenic soybean. *J. Northeast Agric. Univ.* **2011**, *42*, 11–15.

50. Sun, Y.; Joyce, P.A. Application of droplet digital PCR to determine copy number of endogenous genes and transgenes in sugarcane. *Plant Cell Rep.* **2017**, *36*, 1775–1783. [CrossRef] [PubMed]

51. Dominguez, A.; Guerri, J.; Cambra, M.; Navarro, L.; Morenod, P.; Pena, L. Efficient production of transgenic citrus plants expressing the coat protein gene of citrus tristeza virus. *Plant Cell Rep.* **2000**, *19*, 427–433. [CrossRef]

52. Yang, X.J.; Liu, C.L.; Zhang, X.Y.; Li, F.G. Molecule characterization and expression of T-DNA integration in transformed plants. *Genom. Appl. Biol.* **2010**, *29*, 125–130.

53. Joyce, P.; Hermann, S.; O'Connell, A.; Dinh, Q.; Shumbe, L.; Lakshmanan, P. Field performance of transgenic sugarcane produced using *Agrobacterium* and biolistics methods. *Plant Biotechnol. J.* **2014**, *12*, 411–424. [CrossRef] [PubMed]

54. Fu, J.P.; Wang, B.; Liu, L.J.; Yang, J.Y.; Wang, X.X.; Xing, X.L.; Peng, D.X. Transgenic ramie with *Bt* gene mediated by *Agrobacterium tumefacien* and evaluation of its pest-resistance. *Acta Agron. Sin.* **2009**, *35*, 1771–1777. [CrossRef]

55. Hellens, R.P.; Edwards, E.A.; Leyland, N.R.; Bean, S.; Mullineaux, P.M. pGreen: A versatile and flexible binary Ti vector for Agrobacterium-mediated plant transformation. *Plant Mol. Biol.* **2000**, *42*, 819–832. [CrossRef] [PubMed]

56. Paterson, A.H.; Brubaker, C.L.; Wendel, J.F. A rapid method for extraction of cotton (*Gossypium* spp.) genomic DNA suitable for RFLP or PCR analysis. *Plant. Mol. Biol. Rep.* **1993**, *11*, 122–127. [CrossRef]

57. Li, H.F.; Li, L.; Zhang, L.J.; Xu, Y.L.; Li, W.F. Standard curve generation of *PepT1* gene for absolute quantification using real-time PCR. *J. Shanxi Agric. Univ.* **2010**, *30*, 332–334.

58. Xue, B.T.; Guo, J.L.; Que, Y.X.; Fu, Z.W.; Wu, L.G.; Xu, L.P. Selection of suitable endogenous reference genes for relative copy number detection in sugarcane. *Int. J. Mol. Sci.* **2014**, *15*, 8846–8862. [CrossRef] [PubMed]

59. Luo, J.; Pan, Y.B.; Xu, L.P.; Zhang, Y.Y.; Zhang, H.; Chen, R.K.; Que, Y.X. Photosynthetic and canopy characteristics of different varieties at the early elongation stage and their relationships with the cane yield in sugarcane. *Sci. World J.* **2014**, *2014*, 707095. [CrossRef] [PubMed]

International Journal of
Molecular Sciences

Article

Expression Characteristics and Functional Analysis of the *ScWRKY3* Gene from Sugarcane

Ling Wang [1], Feng Liu [1], Xu Zhang [1], Wenju Wang [1], Tingting Sun [1], Yufeng Chen [1], Mingjian Dai [1], Shengxiao Yu [1], Liping Xu [1,2], Yachun Su [1,2,*] and Youxiong Que [1,2,*]

1 Key Laboratory of Sugarcane Biology and Genetic Breeding, Ministry of Agriculture, Fujian Agriculture and Forestry University, Fuzhou 350002, China; lingw2017@126.com (L.W.); 18359162091@163.com (F.L.); zahngxuqq7@126.com (X.Z.); wwj1470665850@163.com (W.W.); sunting3221@163.com (T.S.); CYF9410@163.com (Y.C.); 18906071567@163.com (M.D.); dbzq666666@163.com (S.Y.); xlpmail@126.com (L.X.)
2 Key Laboratory of Ministry of Education for Genetics, Breeding and Multiple Utilization of Crops, College of Crop Science, Fujian Agriculture and Forestry University, Fuzhou 350002, China
* Correspondence: syc2009mail@163.com (Y.S.); queyouxiong@126.com (Y.Q.); Tel.: +86-591-8385-2547 (Y.S. & Y.Q.)

Received: 26 October 2018; Accepted: 11 December 2018; Published: 14 December 2018

Abstract: The plant-specific WRKY transcriptional regulatory factors have been proven to play vital roles in plant growth, development, and responses to biotic and abiotic stresses. However, there are few studies on the *WRKY* gene family in sugarcane (*Saccharum* spp.). In the present study, the characterization of a new subgroup, IIc WRKY protein ScWRKY3, from a *Saccharum* hybrid cultivar is reported. The ScWRKY3 protein was localized in the nucleus of *Nicotiana benthamiana* leaves and showed no transcriptional activation activity and no toxic effects on the yeast strain Y2HGold. An interaction between ScWRKY3 and a reported sugarcane protein ScWRKY4, was confirmed in the nucleus. The *ScWRKY3* gene had the highest expression level in sugarcane stem pith. The transcript of *ScWRKY3* was stable in the smut-resistant *Saccharum* hybrid cultivar Yacheng05-179, while it was down-regulated in the smut-susceptible *Saccharum* hybrid cultivar ROC22 during inoculation with the smut pathogen (*Sporisorium scitamineum*) at 0–72 h. *ScWRKY3* was remarkably up-regulated by sodium chloride (NaCl), polyethylene glycol (PEG), and plant hormone abscisic acid (ABA), but it was down-regulated by salicylic acid (SA) and methyl jasmonate (MeJA). Moreover, transient overexpression of the *ScWRKY3* gene in *N. benthamiana* indicated a negative regulation during challenges with the fungal pathogen *Fusarium solani* var. *coeruleum* or the bacterial pathogen *Ralstonia solanacearum* in *N. benthamiana*. The findings of the present study should accelerate future research on the identification and functional characterization of the WRKY family in sugarcane.

Keywords: sugarcane; WRKY; subcellular localization; gene expression pattern; protein-protein interaction; transient overexpression

---

1. Introduction

Plant growth and development are vulnerable to several external environmental challenges, such as drought, high salinity, cold, and pathogens. There are complex metabolic regulation mechanisms in plants which enhance their resistance to a wide range of stresses through physiological changes largely controlled at the molecular level [1]. The transcription factors (TFs) in plant cells interact with specific DNA sequences in target gene promoters to activate or inhibit transcription and expression of target genes, thereby regulating the expression of these genes. This modulation causes adaptation to the effects and damage from various stresses [2].

As one of the largest plant-specific families of TFs, WRKY has been proven to be widely implicated in responses to biotic and abiotic stresses [3,4]. WRKY TFs are also a vital part of the signaling pathway network of plants, regulating physiological and biochemical processes [5]. The WRKY proteins were named based on a DNA-binding WRKY domain, which contains approximately 60 amino acid residues. This domain features a WRKYGQK sequence at its N-terminal end together with a $C_{X45}C_{X22-23}H_XH$ (C_2H_2-type) or $C_{X7}C_{X23}H_XC$ (C_2HC-type) zinc finger-like motif at the C-terminal [1,5,6]. Although the DNA-binding domain is highly conserved, the overall structure of the WRKY proteins is highly diverse and can be divided into three groups (I, II, and III) including five subgroups (IIa-IIe) in group II. These groupings are categorized according to the number of WRKY domains and are also based on features of the zinc finger-like motif [6,7]. Previous studies have indicated that WRKYs with similar roles usually have related functions. For example, WRKYs in groups I and III are involved in epidermal development, senescence, and abiotic stress, while WRKYs in group II are related to low phosphorus stress, disease resistance, secondary root formation, and abiotic stress, with a few exceptions observed [8].

Currently, *WRKY* genes have been identified in various plant species. There is a total of 72 *WRKYs* in the model dicot *Arabidopsis thaliana* [5]. In monocots, there are 103 *WRKYs* in *Oryza sativa* [9], 116 *WRKYs* in *Zea mays* [10], 105 *WRKYs* in *Setaria italica* [11], 68 *WRKYs* in *Sorghum bicolor* [12], and 45 *WRKYs* in *Hordeum vulgare* [13]. It has been reported that 30 *WRKY* genes in *A. thaliana* were responsive to salt stress [14]. Also, 58 *WRKY* genes in *Z. mays* and 19 *WRKY* genes in *Phaseolus vulgaris* were related to drought stress [15,16]. Qiu et al. [17] indicated that ten of 13 candidate *WRKY* genes in rice can respond to sodium chloride (NaCl), polyethylene glycol (PEG), low temperature, or high temperature stress. Wu et al. [18] showed that eight of the 15 candidate *WRKY* genes in wheat responded to low temperature, NaCl, or PEG stress. Previous studies showed that 49 *A. thaliana WRKY* genes were induced by *Pseudomonas syringae* or salicylic acid (SA) [19]. Fifteen *WRKYs* were induced by *Magnaporthe grisea*, and 12 of them were simultaneously induced by *Xanthomonas oryzae* pv. *oryzae* [9]. Several other reports also demonstrated that *WRKYs* can positively or negatively regulate the responses of plants to external biotic or abiotic stresses [20–22]. Moreover, numerous studies have reported that *WRKYs* are widely involved in a complicated signal transduction network, which may work together with upstream or downstream components, or may interact with other WRKY proteins during physiological processes or in response to various biotic and abiotic stimuli [1,23,24]. These reports provide the foundation for studying the tolerance mechanism of plant *WRKY* genes to environmental stress.

Sugarcane (*Saccharum* spp.) is not only the foremost sugar-producing crop, but also the one that has potential as a bioenergy resource [25]. The study on signal transduction in sugarcane growth, development, and its responses to the external environment, especially the functional analysis of TFs, is of great significance for sugarcane molecular breeding. As reported, there were 26 WRKY-like proteins discovered in a publicly available sugarcane expressed sequence tag (EST) database via an *in silico* study, and their phylogenetic relationships were determined [26]. Beyond that, only two other sugarcane group IIc WRKY proteins, Sc-WRKY (GenBank Accession No. GQ246458.1) [27] and ScWRKY4 (GenBank Accession No. MG852087.1) [28], have been isolated from *Saccharum* hybrid cultivar FN22 and *Saccharum* hybrid cultivar ROC22 respectively and characterized by molecular techniques. *Sc-WRKY* and *ScWRKY4* were both shown to be related to tolerance enhancement to PEG and NaCl stresses [27,28]. Under biotic treatment, *Sc-WRKY* may play a positive role in response to smut pathogen [27], while *ScWRKY4* may be negatively or probably not involved in this regulation [28], suggesting the functional differentiation of group IIc ScWRKYs in smut pathogen resistance. In this study, a new group IIc *WRKY* gene family member, *ScWRKY3* (GenBank Accession No. MK034706), was screened from our previous sugarcane transcriptome data [29]. The sequence characteristics of ScWRKY3 and its subcellular localization, transcriptional activation activity, and its protein-protein interaction with ScWRKY4 were analyzed. The expression profiles of *ScWRKY3* in sugarcane tissues in response to various stresses were assessed, as well as the effects that occurred in *Nicotiana benthamiana* leaves after challenging with the bacterial pathogen *Ralstonia solanacearum* and the fungal pathogen *Fusarium solani* var. *coeruleum*.

Int. J. Mol. Sci. **2018**, *19*, 4059

2. Results

2.1. Bioinformatics Analysis of ScWRKY3 Gene

There was no nucleic acid sequence difference or amino acid sequence difference in *ScWRKY3* between smut-susceptible *Saccharum* hybrid cultivar ROC22 and smut-resistant *Saccharum* hybrid cultivar Yacheng05-179 (Figure S1). As shown in Figure 1, the *ScWRKY3* gene has a cDNA length of 910 bp containing an open reading frame (ORF) from position 162 to 872, and its encoded amino acid residues contain a conserved WRKY domain from position 166 to 223. Bioinformatics analysis revealed that the ScWRKY3 protein has a molecular weight of 25.98 kDa (Table S1). The theoretical isoelectric point (pI), grand average of hydrophobicity (GRAVY), and instability index (II) of ScWRKY3 were 8.58, -0.49, and 56.08 (Table S1), respectively, suggesting that ScWRKY3 might be an unstable basic hydrophilic protein. Secondary structure prediction showed that ScWRKY3 is mainly composed of random coil (69.07%), alpha-helix (18.22%), and extended strand (12.70%) portions (Figure S2). In addition, the ScWRKY3 protein was predicted to have no signal peptide or transmembrane domain (Figure S3). Euk-mPLoc 2.0 software [30] showed that ScWRKY3 has the highest probability of localization in the nucleus (Figure S4).

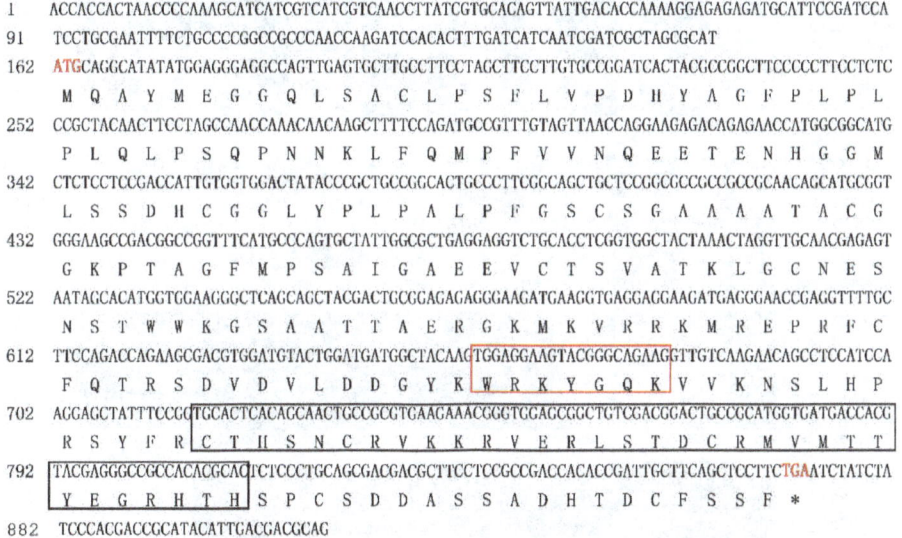

Figure 1. Nucleotide acid sequences and deduced amino acid sequences of the sugarcane *ScWRKY3* gene obtained by PCR amplification. The sequence of the WRKY motif (WRKYGQK) is highlighted in the red box, and that of the C_2H_2 domain ($C_{X4}C_{X23}H_XH$) in the black box. The upstream sequences to start codon ATG (marked in red font) is 5′ untranslated region (UTR) and the downstream sequences to stop codon TGA (marked in red font) is 3′UTR of *ScWRKY3*. *: stop codon.

Amino acid sequence alignment (Figure S5) indicated that the similarity of ScWRKY3 to *S. bicolor* SbWRKY57 (XP_002452824.2), *Miscanthus lutarioriparius* MlWRKY12 (AGQ46321.1), *Z. mays* ZmWRKY51 (XP_020393361.1), *S. italica* SiWRKY12 (XP_004953301.1), *O. sativa* OsWRKY12 (XP_015624962.1) (all these accession numbers in brackets are from GenBank), sugarcane ScWRKY4 and Sc-WRKY were 93%, 93%, 87%, 87%, 66%, 53% and 24%, respectively. A conserved WRKY domain (WRKYGQK) and a conserved zinc-finger motif ($C_{X4}C_{X23}H_XH$) at the C-terminus were found (Figure 2). The phylogenetic tree of sugarcane ScWRKY3, ScWRKY4, Sc-WRKY and WRKYs from other plant species demonstrated that WRKY proteins could be divided into three groups with no obvious distinction between monocots and dicots. ScWRKY3 was classified into group IIc, along with AtWRKY13, OsWRKY22, AtWRKY57, TaWRKY10, Sc-WRKY, and ScWRKY4 (Figure 3). MEME software prediction showed that all WRKYs except TaWRKY46 and OsWRKY46 contained motif 1 (WRKY domain) and motif 2 (zinc-finger domain). In addition, motif 3 (WRKY domain) and motif 4 (unknown domain) were detected in group I WRKYs. Some group IIc WRKYs, for example ScWRKY3, ScWRKY4, AtWRKY13, OsWRKY23 and AtWRKY57, contained motif 4. The WRKYs in group IId had their unique motif 5 (unknown domain) (Figure 3). On the whole, the phylogenetic analysis showed that most WRKYs within the same group generally had a similar structure.

2.2. Subcellular Localization

The recombinant vector pMDC83-*ScWRKY3-GFP* was generated to investigate the subcellular distribution of ScWRKY3. We used 4′,6-diamidino-2-phenylindole (DAPI) staining as a nuclear marker. As shown in Figure 4, the green fluorescence of the control (*35S::GFP*) in *N. benthamiana* was distributed through the whole cell, including the plasma membrane, nucleus, and cytoplasm, while the fusion protein of ScWRKY3::GFP was only found in the nucleus, which was consistent with the software prediction.

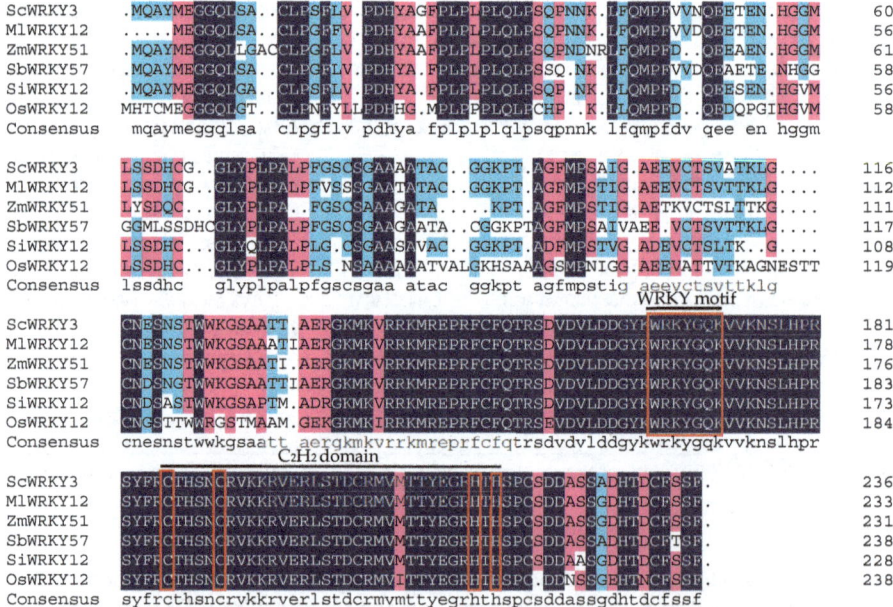

Figure 2. Amino acid sequence alignment of ScWRKY3 and WRKYs from other plant species by DNAMAN (version 6.0.3.99, Lynnon Biosoft) software. The amino acid sequences of *Miscanthus lutarioriparius* MlWRKY12 (AGQ46321.1), *Zea mays* ZmWRKY51 (XP_020393361.1), *Sorghum bicolor* SbWRKY57 (XP_002452824.2), *Setaria italica* SiWRKY12 (XP_004953301.1), and *Oryza sativa* OsWRKY12 (XP_015624962.1) are from GenBank. The black, pink, blue, and white colors indicate the homology level of conservation of the amino acid residues in the alignment at 100, ≥75, ≥50, and <50%, respectively. The sequences of the WRKY motif (WRKYGQK) and the C_2H_2 domain ($C_{X4}C_{X23}H_XH$) are highlighted by the red rectangle.

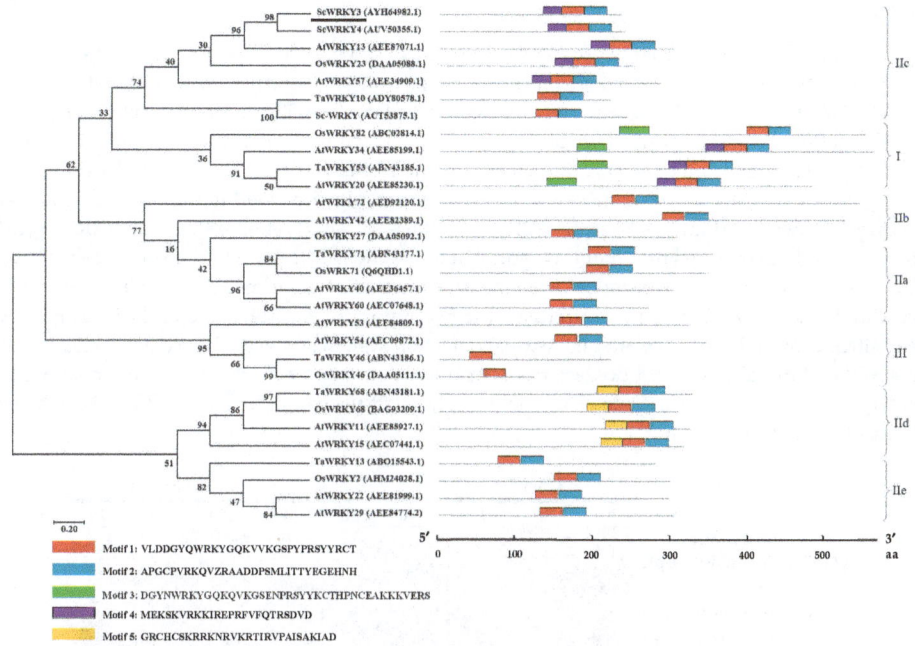

Figure 3. Phylogenetic tree (left) and predicted conserved motifs (right) of ScWRKY3 protein and WRKYs from various plant species. The GenBank accession number of WRKY proteins follows the protein name. Ml, *Miscanthus lutarioriparius*; Sb, *Sorghum bicolor*; Zm, *Zea mays*; Si, *Setaria italic*; Os, *Oryza sativa*; and At, *Arabidopsis thaliana*. The unrooted tree is constructed by the Maximum Likelihood with bootstrapping (1000 iterations) using MEGA7.0 software. ScWRKY3 is underlined. The conserved domains were predicted by MEME Suite 5.0.2 software. The different-colored boxes named at the bottom represent conserved motifs. Gray lines represent the nonconserved sequences, and the position of each WRKY sequence is exhibited proportionally. The motif logo is shown in Figure S6.

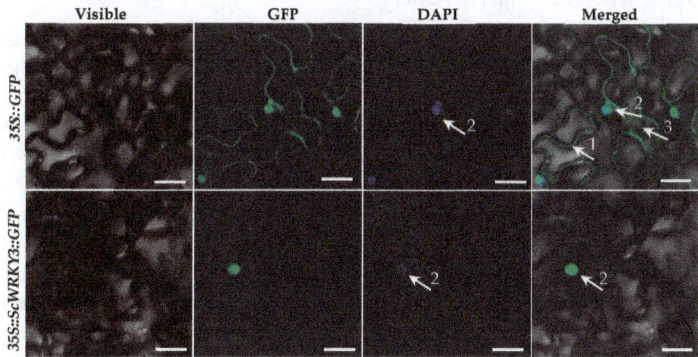

Figure 4. Subcellular localizations of *35S::GFP* and *35S::ScWRKY3::GFP* in *Nicotiana benthamiana* leaves. The epidermal cells of *N. benthamiana* are used for capturing images of visible light, green fluorescence, blue fluorescence, and visible light merged with green and blue fluorescence. White arrows 1, 2, and 3 indicate plasma membrane, nucleus, and cytoplasm, respectively. Scale bar = 50 μm. *35S::GFP*, the *Agrobacterium tumefaciens* strain carrying the empty vector pMDC83-*GFP*. *35S::ScWRKY3::GFP*, the *A. tumefaciens* strain carrying the recombinant vector pMDC83-*ScWRKY3-GFP*. DAPI, 4′,6-diamidino-2-phenylindole.

2.3. Transcription Activation Activity of ScWRKY3

The Y2H Gold-GAL4 yeast two-hybrid system was used to detect the transcriptional activation activity of ScWRKY3. As shown in Figure 5, yeast cells transformed with either the positive control pGADT7+pGBKT7-p53, the negative control pGBKT7-p53, or the recombinant plasmid pGBKT7-*ScWRKY3* all grew well in SDO (SD/-Trp, SD minimal medium without tryptophan) medium plates, while only the positive control turned blue in SDO/X (SD/-Trp/X-α-Gal, SDO plates with X-α-D-galactosidase) medium plates. These results indicated that all the plasmids were successfully transfected into yeast strain Y2HGold. The GAL4-BD combined with ScWRKY3 protein can successfully express tryptophan but cannot activate the *MEL1* gene in the presence of X-α-gal. After aureobasidin A (AbA) resistance screening, the yeast cells transformed with pGBKT7-*ScWRKY3* and with the negative control did not activate the two reporter genes, *AUR1-C* and *MEL1*. However, the positive control did survive, and its X-α-gal detection system showed a blue color, indicating that the ScWRKY3 protein does not possess transcriptional activation activity. This protein showed no toxicity to the yeast strain Y2HGold. This result implies that the bait protein of ScWRKY3 can be used for yeast two-hybrid screening.

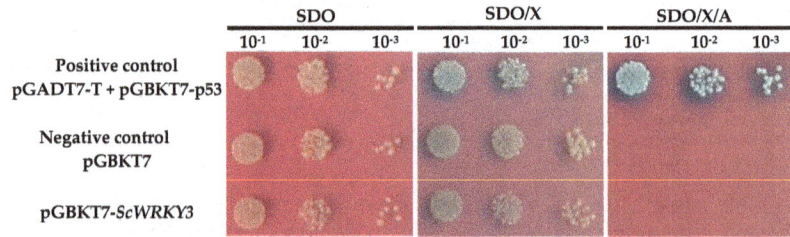

Figure 5. Testing of the ScWRKY3 transactivation activity assay. SDO (SD/-Trp), synthetic dropout medium without tryptophan; SDO/X (SD/-Trp/X-α-Gal), synthetic dropout medium without tryptophan, but plus 5-bromo-4-chloro-3-indoxyl-α-D-galactopyranoside; SDO/X/A (SD/-Trp/X-α-Gal/AbA), synthetic dropout medium without tryptophan, but plus 5-bromo-4-chloro-3-indoxyl-α-D-galactopyranoside and aureobasidin A.

2.4. Interaction Between ScWRKY3 and ScWRKY4

As shown in Figure 6, all plasmid combinations grew normally on DDO (SD/-Leu/-Trp) plates. However, when transferred to QDO (SD/-Ade/-His/-Leu/-Trp) and QDO/X/A (SD/-Ade/-His/-Leu/-Trp/X-α-Gal/AbA) plates, only the AD-ScWRKY4+BD-ScWRKY3 combination and the positive control pGADT7-T+pGBKT7-p53 continued to grow and turned blue with X-α-gal detection. This indicates that ScWRKY4 may function downstream of ScWRKY3. Additionally, we have further proved the above results by bimolecular fluorescence complementation (BiFC). When ScWRKY3 was fused to pUC-SPYNE (this fusion was named ScWRKY3-YFP[N]), and ScWRKY4 was fused to pUC-SPYCE (this fusion was named ScWRKY4-YFP[C]), a fluorescent complex was formed and was visualized in the nucleus of *N. benthamiana* leaf cells. While, when ScWRKY3 was fused to pUC-SPYCE (this fusion was named ScWRKY3-YFP[C]), and ScWRKY4 was fused to pUC-SPYNE (this fusion was named ScWRKY4-YFP[N]), no fluorescent complex was formed in *N. benthamiana* leaf cells. The results were consistent with those of yeast two-hybrid system and showed that there was an interaction between ScWRKY3 and ScWRKY4, and the specific protein complex was located in the nucleus.

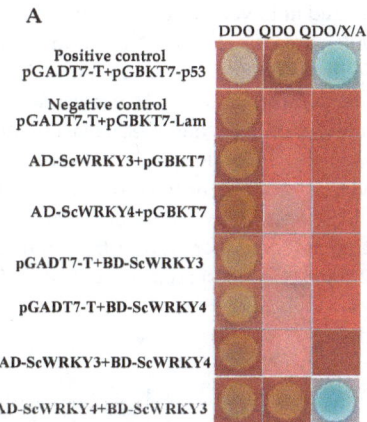

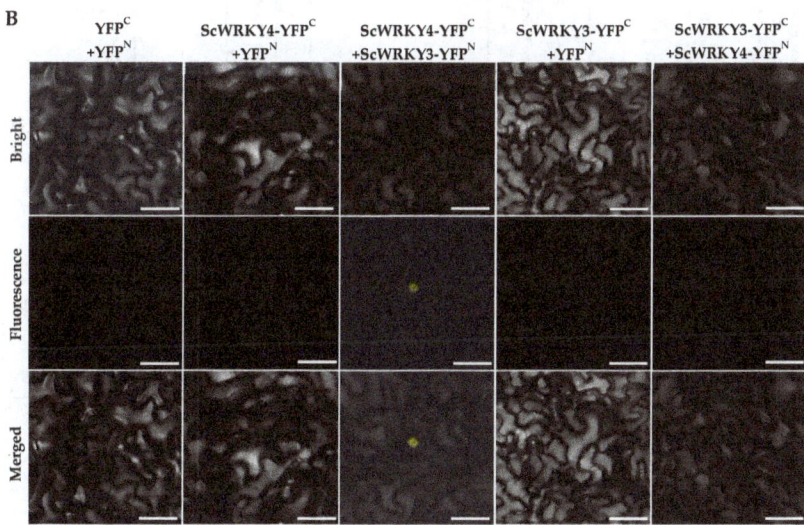

Figure 6. Interaction between ScWRKY3 and ScWRKY4 in yeast and in *Nicotiana benthamiana* leaves. (**A**) The interaction between ScWRKY3 and ScWRKY4 was verified using a yeast two-hybrid system. A variety of BD and AD vectors were combined and transformed into GoldY2H yeast. Left to Right: Transformations were grown and screened on DDO (SD/-Leu/-Trp, SD medium without leucine and tryptophan), QDO (SD/-Ade/-His/-Leu/-Trp, SD medium without adenine, histidine, leucine, or tryptophan), and QDO/X/A (SD/-Ade/-His/-Leu/-Trp/X-α-Gal/AbA, QDO medium with X-α-D-Galactosidase and aureobasidin medium. (**B**) The bimolecular fluorescence complementation (BiFC) assay for the location determination of the interaction between ScWRKY3 and ScWRKY4. Scale bar = 50 μm. YFP^C, YFP^N, ScWRKY4-YFP^C, ScWRKY4-YFP^N, ScWRKY3-YFP^C, and ScWRKY3-YFP^N represent the plasmids pUC-SPYCE and pUC-SPYNE and the recombinant plasmids ScWRKY4-pUC-SPYCE, ScWRKY4-pUC-SPYNE, ScWRKY3-pUC-SPYCE and ScWRKY3-pUC-SPYNE, respectively.

2.5. Gene Expression Patterns of ScWRKY3 in Response to Various Stress Conditions

The expression patterns of the *ScWRKY3* gene in sugarcane tissues and under various stresses were investigated using real-time fluorescent quantitative PCR (qRT-PCR). The results indicated that *ScWRKY3* was constitutively expressed in different sugarcane tissues, with the highest expression level in stem epidermis. It remained at lower expression levels in other tissues (root, bud, leaf, and stem pith) (Figure 7A). During the treatments with NaCl and PEG, the transcript of *ScWRKY3* in ROC22 was remarkably up-regulated by 3.28-fold at 24 h and 38.57-fold at 3 h, respectively, and remained unchanged at other time points (Figure 7B). Moreover, the expression of *ScWRKY3* in ROC22 was markedly down-regulated under both SA and methyl jasmonate (MeJA) treatments, but it was up-regulated under ABA stress with a 1.73-fold higher level than in the control (Figure 7C). These results indicated that the *ScWRKY3* gene might have positive responses to ABA, PEG, and NaCl stimuli but a negative response to SA and MeJA. After infection by the smut pathogen, the expression level of *ScWRKY3* was almost unchanged after 72 h in the smut-resistant cultivar Yacheng05-179 and during the period of 48–72 h in the smut-susceptible cultivar ROC22, while it was significantly down-regulated (0.76-fold) at 24 h in ROC22 (Figure 7D). This suggested that *ScWRKY3* may play a role in the smut pathogen response.

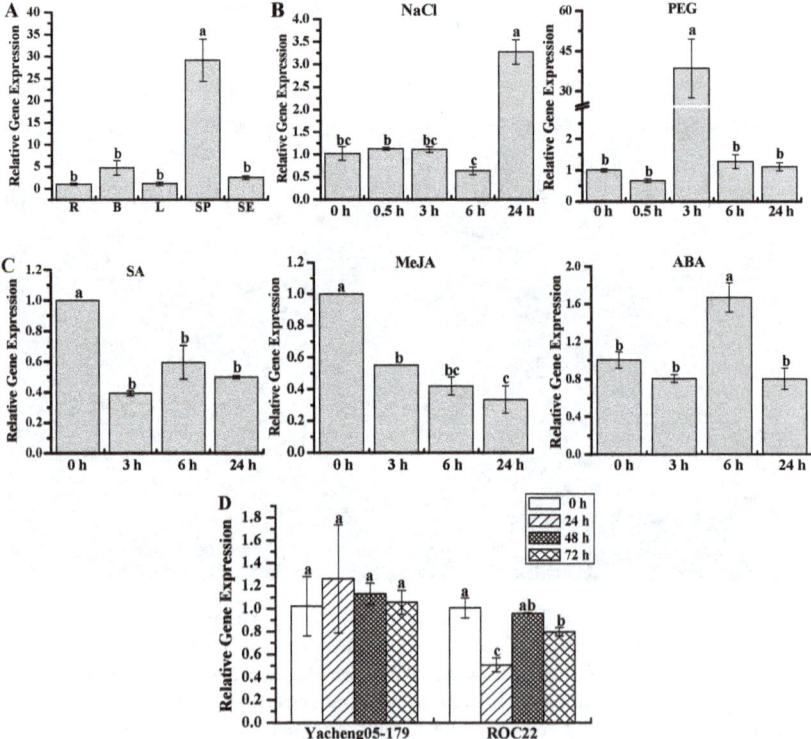

Figure 7. Gene expression assay of *ScWRKY3*. (**A**) Tissue-specific expression analysis of *ScWRKY3* in different 10-month-old ROC22 tissues by qRT-PCR. The tissues (root, bud, leaf, stem pith, and stem epidermis) are represented by R, B, L, SP, and SE, respectively; (**B**) Gene expression patterns of *ScWRKY3* in 4-month-old ROC22 plantlets under abiotic stress. NaCl, sodium chloride (simulating salt stress) (250 mM); PEG, polyethylene glycol (simulating drought treatment) (25.0%); (**C**) Gene expression patterns of *ScWRKY3* in 4-month-old ROC22 plantlets under plant hormone stress. SA, salicylic acid (5 mM); MeJA, methyl jasmonate (25 μM); ABA, abscisic acid (100 μM); (**D**) Gene expression patterns of the *ScWRKY3* gene after infection with smut pathogen. Yacheng05-179 is a smut-resistant *Saccharum* hybrid cultivar, and ROC22 is a smut-susceptible *Saccharum* hybrid cultivar. Data are normalized to the glyceraldehyde-3-phosphate dehydrogenase (*GAPDH*) expression level. All data points are means ± standard error ($n = 3$). Bars superscripted by different lowercase letters indicate significant differences, as determined by Duncan's new multiple range test (p-value < 0.05).

2.6. Transient Overexpression of ScWRKY3 in N. benthamiana Leaves

The *ScWRKY3* gene was inserted into the plant overexpression vector, and was transformed into *N. benthamiana* leaves by the *Agrobacterium tumefaciens*-mediated method to analyze whether the target gene could induce a plant immune response. The transcripts of *ScWRKY3* in *N. benthamiana* leaves were detected using a semi-quantitative PCR technique (Figure 8A). The phenotypic observation after injection for one day was shown in Figure 8B, and no significant difference in superficial characteristics was demonstrated between the experimental group and the control group. However, qRT-PCR results demonstrated that six immunity-associated marker genes, including the hypersensitive response (HR) marker genes, *NtHSR203* and *NtHSR515*, the SA pathway related gene *NtPR1*, the JA pathway associated gene *NtPR3*, and two ethylene synthesis-dependent genes, *NtEFE26* and *NtAccdeaminase*, were all up-regulated with a higher fold change range from 1.47 to 14.16 than the control (Figure 8C). These results suggest that transiently overexpressed *ScWRKY3* may take part in the immune response in *N. benthamiana* leaves.

To detect the effect of *ScWRKY3* in response to pathogen, the *N. benthamiana* leaves, were transformed with the control in the left half blade and with *35S::ScWRKY3* in the right half blade for one day. Then they were inoculated by the bacterial pathogen *R. solanacearum*. As shown in Figure 8D, there was a slight symptomatic difference between the control half-leaves and the *35S::ScWRKY3* half-leaves when injected with *R. solanacearum* for one day and seven days. Moreover, qRT-PCR results revealed that the HR marker genes *NtHSR201* and *NtHSR515* and the SA-related gene *NtNPR1* all showed significantly lower expression in *35S::ScWRKY3*-overexpressing *N. benthamiana* leaves after one day and seven days of *R. solanacearum* inoculation when compared to the control. No remarkable transcript difference or down-regulation of the SA-related gene *NtPR-1a/c* or the JA-associated genes *NtPR2* and *NtPR3* was observed in *35S::ScWRKY3* leaves when compared with controls. Compared to the control, the transcript abundance of *NtHSR203* in *35S::ScWRKY3*-overexpressing leaves was decreased at one day but increased at seven days post-agroinfiltration.

When the leaves of *ScWRKY3*-transiently-overexpressing *N. benthamiana* were inoculated by the fungal pathogen *F. solani* var. *coeruleum* for one day and seven days, a heavier wilting disease symptom was observed in the *N. benthamiana* leaves containing *35S::ScWRKY3* than in the control (Figure 8F). Additionally, in comparison with the control, *NtHSR201*, *NtPR3*, and *NtAccdeaminase* showed significantly higher expression, while *NtHSR515* and *NtNPR1* presented obviously lower expression in *35S::ScWRKY3*-overexpressing leaves at one day or seven days after *F. solani* var. *coeruleum* infection. No statistically significant expression difference in *NtPR-1a/c*, *NtPR2*, or *NtEFE26* was found at one day, while *NtPR-1a/c* was down-regulated, *NtPR2* was up-regulated, and *NtEFE26* was unchanged in *35S::ScWRKY3* leaves after seven days with *F. solani* var. *coeruleum* inoculation. The expression level of *NtHSR203* in *35S::ScWRKY3* was lower at one day but higher at seven days after inoculation than in the control (Figure 8G).

These results demonstrated that in comparison with the transiently overexpressing pEarleyGate 203 vector, *ScWRKY3* transient overexpression in *N. benthamiana* leaves significantly decreased the transcript abundance of *NtHSR515*, *NtPR1*, and *NtPR-1a/c* after *R. solanacearum* or *F. solani* var. *coeruleum* infection. It was anticipated that *ScWRKY3* can negatively regulate the HR marker genes or SA signaling pathway-mediated genes to reduce the tolerance of *N. benthamiana* to pathogens.

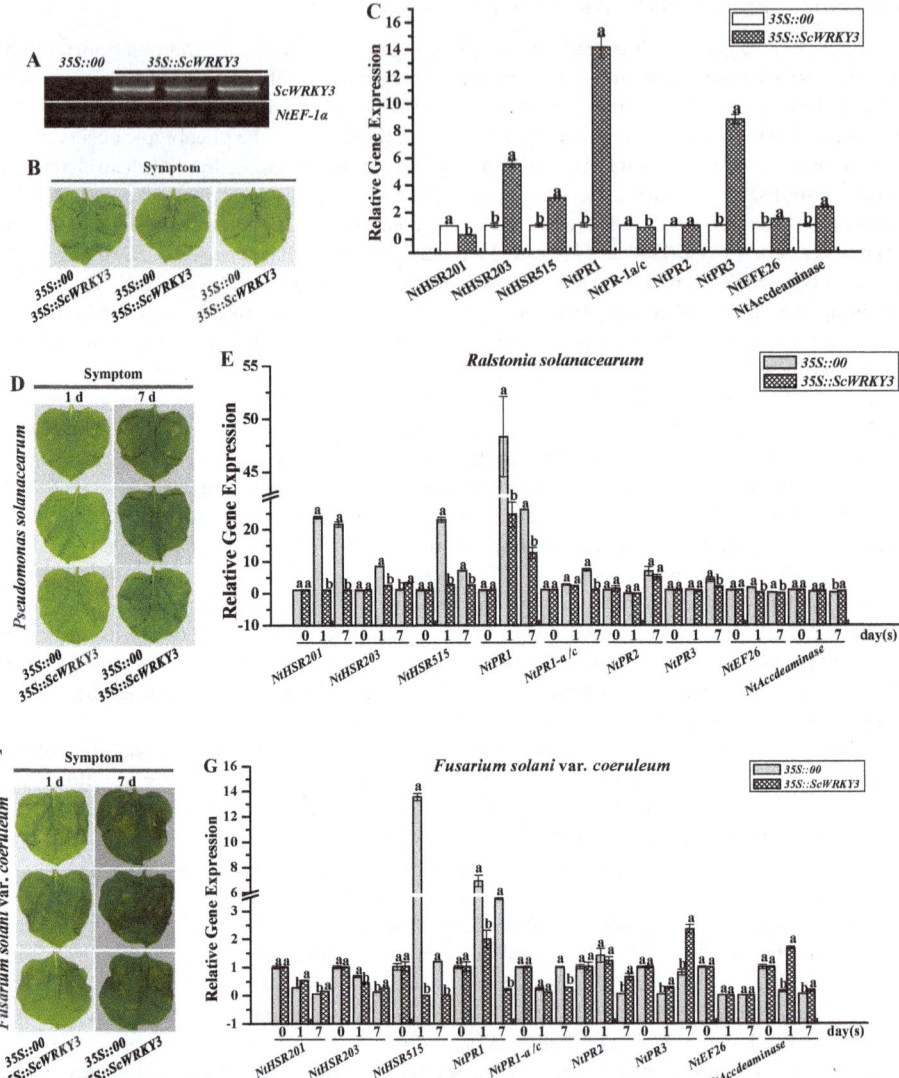

Figure 8. Effects of transient overexpression of *ScWRKY3* in *Nicotiana benthamiana* leaves. (**A**) Semi-quantitative PCR analysis of *ScWRKY3* in *N. benthamiana* leaves after one day of infiltration by *Agrobacterium* strain GV3101 carrying pEarleyGate 203-*ScWRKY3* (*35S::ScWRKY3*) and the empty vector pEarleyGate 203 (*35S::00*). (**B**) Phenotype of *N. benthamiana* leaves after one day of agroinfiltration. (**C**) The transcript level of nine immunity-associated marker genes in the *N. benthamiana* leaves after one day of agroinfiltration. (**D,F**) Disease symptoms of *N. benthamiana* post-inoculation with *Ralstonia solanacearum* and *Fusarium solani* var. *coeruleum* are observed after one day and seven days of agroinfiltration. (**E,G**) The transcripts of nine immunity-associated marker genes in the *N. benthamiana* leaves after inoculation with *R. solanacearum* or *F. solani* var. *coeruleum* for one day and seven days. Data are normalized to the *NtEF-1α* expression level. All data points are means ± standard error (*n* = 3). Bars superscripted by different lowercase letters indicate significant differences, as determined by Duncan's new multiple range test (*p* value < 0.05).

3. Discussion

As one of the largest groups of TFs, the WRKY proteins have been found in a wide range of plant species since the initial *WRKY* cDNA was isolated from sweet potato [31]. Although sugarcane is an important bioenergy and cash crop [25], there are only three reports about WRKYs in sugarcane [26–28]. In this study, a novel sugarcane *ScWRKY3* gene was isolated and identified. As reported, there is a functional similarity of WRKYs in the same or phylogenetically closely related group [2,3]. Phylogenetic tree analysis indicated that ScWRKY3 is a member of the group IIc WRKY proteins, along with Sc-WRKY [27] and ScWRKY4 [28] (Figure 3). This is helpful for further functional comparative studies on the same WRKY family members in sugarcane. The structure of TFs is usually composed of four functional domains, namely, the DNA binding domain, the transcriptional activation or repression domain, the oligomerization sites, and the nuclear localization signals [32]. These four components are the core regions that perform the functions of TFs or interact with the *cis*-acting elements in the promoter regions of various stress related genes [33]. In our study, ScWRKY3 protein had one WRKY domain, which is a DNA binding domain containing 60 amino acids. WRKY domain was mainly composed of motif 1 and motif 2 (Figure 3) and can bind specifically to the DNA sequence motif (T)(T)TGAC(C/T) which is known as the W-box and existed in many promoters of plant defense-related genes [6]. As showed by Wei et al. [10], subgroup IId WRKYs possess two basic amino acid sequences, including a RCHCSK[RK][RK]K[LN]R motif, which may function as a nuclear localization signal, and a KRxIxVPAISxKxAD motif. Similarly, the motif 5 (Figure 3) also contains these amino acid sequences. While further work is required to clarify the function of the other unknown motifs, such as the predicted motif 4 in Figure 3. Previous studies showed that the functions of the two WRKY domains in the group I WRKYs are different. The WRKY domain at the C-terminal can bind to their target DNA, while another WRKY domain at the N-terminal may be as the site where proteins interact with each other [8,12,15].

Subcellular localization analysis is valuable for determining the functions of proteins. The present study showed that the fusion protein of ScWRKY3::GFP was detected in the nucleus of *N. benthamiana* leaf cells (Figure 4), which was consistent with the software prediction results and previous studies on other plant WRKYs [28,34–36]. This indicated that ScWRKY3 may play a role as a nuclear-localized protein to regulate cellular processes.

Transcriptional activity analysis is important for the functional analysis of TFs [37]. Since the GAL4 yeast two-hybrid system was first discovered [38], this method has been increasingly used to study the interactions between WRKY proteins [39]. In this study, the full-length ScWRKY3 cDNA showed no auto-activation (Figure 5), so it could be used as the bait to screen interacting proteins in a yeast two-hybrid system. Post-translational modifications or interactions with cofactors are needed for ScWRKY3 protein to fulfill its function. Screening and identifying WRKY interacting proteins is important to reveal the role of WRKY in plant signal transduction [40,41]. It has been reported that WRKYs have the activities of self-regulation and mutual regulation, and they can form functional homo- or heterodimers among some WRKY proteins or interact with other functional proteins to play roles [4,5]. WRKY6 and WRKY22, which both belong to group II WRKYs in *A. thaliana*, interact with MPK10 and MPK3/MPK6, respectively [42,43]. Previous studies also proved that AtWRKY30, AtWRKY53, AtWRKY54, and AtWRKY70, which all belong to group III of the WRKY proteins, have interactive effects in yeast [39]. Similarly, yeast two-hybrid and BiFC results showed that a fluorescent complex from ScWRKY3 and ScWRKY4 was formed (Figure 6B). This was visualized in the nucleus in *N. benthamiana* leaf cells, which indicated that there may be an interacting relationship between ScWRKY3 and ScWRKY4. In *Arabidopsis*, previous study indicated that WRKYs in the group IIb or group III could interact with themselves and with group IIa WRKYs, while group IId WRKYs could only interact with group IIa WRKY members [41]. Groups IIc ZmWRKY25 and ZmWRKY47 had interactions and may be involved in the response to drought stress by interacting with other WRKYs [15]. Besides, ZmWRKY39 was down-regulated under light drought stress and its phylogenetically closely related protein ZmWRKY106 showed a positive response to this stress, while they had up-regulated co-expression interaction under drought stress [15]. However, the nature of the interaction as well as the biological implications of this process between ScWRKY3 and ScWRKY4 requires further study.

WRKY genes are expressed differentially in different tissues of plants, which demonstrates that *WRKY* genes can be expressed in different physiological conditions and in different types of cells, and may regulate a series of life activities including growth, development, and morphological composition [35,40,44]. Twenty-eight *CiWRKYs* were detected in the roots, stems, and leaves of wild *Caragana intermedia,* and different *CiWRKYs* showed differential expression in various tissues. For example, *CiWRKY69–1* had the highest expression level in roots, while *CiWRKY40–1* and *CiWRKY30* were mainly expressed in leaves [35]. Among the 37 *A. thaliana WRKY* genes reported by Bakshi et al. [45], 12 were specifically expressed in the mature zone of root cells, suggesting that these *WRKY* genes may be involved in the regulation of root cell maturation in *A. thaliana*. *TaWRKY44*, a *WRKY* gene of *Triticum aestivum*, was differentially expressed in all organs examined, including root, stem, leaf, pistil, and stamen, with the highest expression level in the leaves and the lowest expression level in the pistils [40]. It is known that *ScWRKY4* is constitutively expressed in the root, bud, leaf, stem pith, and stem epidermis of sugarcane, with the highest expression level in the stem epidermis [28]. This was similar to the findings on tissue-specific expression of *ScWRKY3* in the present study. Can the coexpression of *ScWRKY3* and *ScWRKY4* in the same tissue be tied together with their BiFC interaction results (Figure 6B)? Can they co-localize in the same organelle in the same tissue? These remain to be validated by future research, for example, only if we obtain the promoters specific to *ScWRKY3* and *ScWRKY4* respectively, can we check that if they co-localize in the same organelle in the same tissue. WRKY TFs are critical for signal transduction, plant growth and stress responses [44]. Salt and drought, the two representative abiotic stresses, adversely affect the growth and development of plants, but the response can be resisted by activating the ABA signal transduction pathway to induce the expression of a series of stress-responsive genes [46]. *AtWRKY1* played a negative role in ABA-mediated drought resistance, and the *AtWRKY1* knockout mutant could enhance the drought tolerance of *A. thaliana* [47]. In the *OsWRKY11* knockout mutant, drought responsive genes were induced to enhance the drought tolerance of rice [36]. The present study showed that the expression of *ScWRKY3* was increased by PEG, NaCl, and exogenous ABA. Previous studies also determined that *Sc-WRKY* and *ScWRKY4* showed up-regulated expression levels under PEG and NaCl treatments [27,28]. Furthermore, under ABA stress, the transcript of *ScWRKY4* was remarkably up-regulated by 1.59-, 2.87-, and 1.26-fold at 0.5 h, 6 h, and 24 h higher than the control, respectively [28]. These results suggest that *ScWRKY3* and *ScWRKY4* may participate in sugarcane resistance to drought and salt stresses which may be mediated through ABA signaling. Similar to the expression characteristics seen in group II *WRKYs* of other plants, the expression level of *AtWRKY57* was remarkably up-regulated in drought conditions. This occurred through the direct activation of the expression of an important functional gene (*AtNCED3*) in the ABA synthesis pathway, which enhanced the tolerance of *A. thaliana* to drought stress [48].

Plants have evolved at least two sets of biochemical defenses to protect themselves against external challenges [49]. One defense response, caused by infection of pathogenic bacteria, initiates a localized hypersensitive reaction to confine the injured site and to prevent further infection of pathogens. This is known as systemic acquired resistance (SAR) [50]. The other response is mainly activating the expression of defense genes through various signal molecules to exhibit resistance in plants. The signal transduction pathways related to this defense response may be mediated by SA, JA, or ABA [51,52]. Previous studies indicated that the *WRKY* genes of group IIc are related to plant immunity. For instance, *AtWRKY28* and *AtWRKY75* can be induced by infection with *Sclerotinia sclerotiorum* in connection with SA- and JA/ET-mediated defense signaling pathways [53]. Overexpression of rice *WRKY89* increased the expression level of SA and enhanced the resistance to rice blast fungus [54]. *AtWRKY57* is negatively associated with resistance to *Botrytis cinerea* in *A. thaliana* by regulating the expression of JA pathway-related genes [49]. *OsWRKY13* enhances the resistance of rice to *X. oryzae* pv. *oryzae* and *M. grisea* [55,56]. In the present study, *ScWRKY3* was down-regulated by smut pathogen in the smut-susceptible cultivar ROC22 at 24 h, while it was almost unchanged in the smut-resistant cultivar Yacheng05-179. Conversely, Wang et al. [28] showed that the expression of *ScWRKY4* was quite stable in ROC22 but was down-regulated in Yacheng05-179 under the stress of smut pathogen. Liu et al. [27] found that the expression of *Sc-WRKY* was remarkably up-regulated at 24 and 60 h after infection by smut pathogen treatment in sugarcane FN22. Moreover, *ScWRKY3* was down-regulated under SA and MeJA treatments, which was opposite to the expression patterns of *ScWRKY4* and *Sc-WRKY* [27,28]. The results revealed that *ScWRKY3* might be a negative regulatory gene in the sugarcane response to smut pathogen. This can be further proved by antimicrobial test in more stable overexpression and knockout plants. In *Arabidopsis*, *AtWRKY25* gene was proved to play a negative regulatory role in the SA-mediated defense response to *Pseudomonas syringae* [57] but a positive regulatory role in response to NaCl stress [58]. Yokotani et al. [59] demonstrated that overexpression of *OsWRKY76* in rice plants suppressed the induction of defense related genes after inoculation with blast fungus (*Magnaporthe oryzae*) but up-regulated the expressions of abiotic stress-associated genes using microarray analysis. It is therefore possible for *WRKY* genes to play opposite roles in biotic resistance and abiotic tolerance. In the future, genetic transformations can be done to further our understanding on the responses of *ScWRKY* genes to biotic and abiotic stresses. In addition, whether there are sequence differences in the promoter regions of the *ScWRKY* genes from different sugarcane cultivars which may cause differences in gene expression patterns for the biotic or abiotic stress need further investigation.

As Liu et al. [60] proved, overexpression of the *Gossypium hirsutum GhWRKY25* in *N. benthamiana* is involved in the regulation of expression of multiple defense-associated marker genes, including the SA-, ET-, and JA-mediated genes, to decrease the resistance to the fungal pathogen *B. cinerea*. *GhWRKY40* has been revealed to be inducible by stress from the bacterial pathogen *R. solanacearum*, and *GhWRKY40* expression was up-regulated by SA, MeJA, and ET [61]. When *GhWRKY40* was transiently overexpressed in *N. benthamiana* leaves, most of the resistance-associated genes, including the SA-, ET-, JA-, and HR-responsive genes, were down-regulated after infection with *R. solanacearum*, which indicated that overexpression of *GhWRKY40* reduces the tolerance to *R. solanacearum* [61]. *A. thaliana* WRKY27 negatively regulated resistance genes during infection with the pathogen *R. solanacearum*, and the symptom development in *R. solanacearum* appeared earlier than in the mutant wrky27-1 plants, which lacks the function of WRKY27 [62]. In this study, most of the immunity-associated marker genes, including the HR-, SA-, JA-, and ET-related genes, were up-regulated when *ScWRKY3* was transiently overexpressed in *N. benthamiana*, suggesting that *ScWRKY3* may play a role in the plant immune response. After infection with the bacterial pathogen *R. solanacearum*, most of the detected immunity-associated marker genes, including the HR marker genes *NtHSR201* and *NtHSR515*, the SA-related genes *NtPR-1a/c* and *NtNPR1*, and the JA-associated gene *NtPR3*, were lower in the *35S::ScWRKY3* overexpressing leaves than in the control, revealing that *ScWRKY3* may play a negative role in the response to the bacterial pathogen *R. solanacearum*. After *F. solani* var. *coeruleum* infection, the wilting disease symptoms were greater in *35S::ScWRKY3* leaves than in the control. The qRT-PCR results showed that the transcript abundance of JA- and ET-related genes was remarkably higher, but the SA-related genes were evidently reduced in *35S::ScWRKY3* leaves compared to their levels in the control. These results indicated that there was a crosstalk between JA-/ET-related genes and SA-related genes in the response to the fungal pathogen *F. solani* var. *coeruleum* when *ScWRKY3* was overexpressed in *N. benthamiana*.

4. Materials and Methods

4.1. Plant Materials and Treatments

In China, ROC22 has been the main sugarcane cultivar grown for the past 20 years and encompasses approximately 60% of the total sugarcane cultivated area. ROC22 is a *Saccharum* hybrid cultivar which is susceptible to smut disease and results in a poor ratoon performance.

Yacheng05-179, an intergeneric hybrid (BC2) with smut resistant properties, is generated from *S. officinarum* × *S. arundinaceum*. In this study, ROC22 and Yacheng05-179 were used as plant materials and collected from the Key Laboratory of Sugarcane Biology and Genetic Breeding, Ministry of Agriculture, Fuzhou, China.

To analyze the tissue-specific expression level of the target gene, nine healthy and uniform 10-month-old ROC22 plants were randomly selected from one field. The white root, bud, +1 leaf, stem pith, and stem epidermis were immediately frozen in liquid nitrogen and kept at -80 °C until extraction of total RNA. Each sample contained three biological replicates.

For biotic treatment, the robust and healthy stems of 10-month-old ROC22 and Yacheng05-179 were harvested and soaked in water for germination at 32 °C. Then the two-bud setts of both sugarcane cultivars were inoculated with 0.5 µL suspensions of 5×10^6 smut spores/mL (plus 0.01% (*v/v*) Tween-20), while the control was inoculated with aseptic water in 0.01% (*v/v*) Tween-20 [63]. The treated samples were cultured at 28 ± 1 °C in a photoperiod of 16-h light and 8-h darkness. Three biological replicates were set, and five buds were randomly chosen at 0 h, 24 h, 48 h, and 72 h for each biological replicate, respectively.

For abiotic and hormone stimuli, healthy and uniform approximately 4-month-old ROC22 plantlets were transferred to water for one week and then treated with six different exogenous stresses. Two groups were separately cultured in aqueous solutions of 250 mM NaCl and 25% PEG 8000, and the leaves were sampled at 0, 0.5, 3, 6, and 24 h, respectively [63–65]. The other three groups were sprayed with 100 µM ABA, 5 mM SA in 0.01% (*v/v*) Tween-20 and 25 µM MeJA for 0, 3, 6, and 24 h, respectively [63–65]. Each treatment was prepared with three biological replicates that contained three plants. All collected samples were immediately frozen in liquid nitrogen and kept at −80 °C until use.

4.2. RNA Extraction and First-strand cDNA Synthesis

The total RNAs of all the samples were extracted with TRIzol® reagent (Invitrogen, Shanghai, China). The RNA quality was determined by 1.0% agarose gel electrophoresis and measured at wavelengths of 260 and 280 nm using a spectrophotometer (NanoVueplus, GE, USA). The residual DNA was removed by DNase I (Promega, Madison, WI, USA). The RevertAid First Strand cDNA Synthesis Kit (Fermentas, Shanghai, China) was used to synthesize the first-strand cDNA from ROC22 and Yacheng05-179 leaves which was treated as templates for cloning the target gene. Prime-Script™ RT Reagent Kit (Perfect Real Time) (TaKaRa Biotechnology, Dalian, China) was used to synthesize the first-strand cDNA of the other samples for expression profile analysis.

4.3. Cloning, Sequencing, and Bioinformatic Analysis of the ScWRKY3 Gene

A gene which codes for a predicted *WRKY* transcriptional regulator named *ScWRKY3* was screened from our previous transcriptome data of sugarcane infected by smut fungus [29]. The specific amplification primers (Table S2) were designed using National Center of Biotechnology Information (NCBI) online software (https://www.ncbi.nlm.nih.gov/tools/primer-blast/). The reverse transcription-polymerase chain reaction (RT-PCR) system contained 1.0 µL cDNA template, 1.0 µL each of the forward and reverse primers (10 µM), 2.5 µL 10× ExTaq buffer (Mg^{2+} plus), 2.0 µL dNTPs (2.5 mM), and 0.125 µL ExTaq enzyme (5.0 U/µL) (TaKaRa Biotechnology, Dalian, China), and 17.375 µL ddH$_2$O. The RT-PCR reaction conditions were as follows: 94 °C for 4 min; 35 cycles of 94 °C for 30 s, 58 °C for 30 s, and 72 °C for 1 min 30 s; and 72 °C for 10 min. The amplified fragment, which had been gel-purifed using a Gel Extraction Kit (Tiangen, Beijing, China), was linked to the pMD19-T vector (TaKaRa Biotechnology, Dalian, China) and transformed into *Escherichia coli* strain DH5α cells. The positive clones were selected for sequencing (Biosune, Fuzhou, China).

The sequence of the *ScWRKY3* gene was analyzed using the ORF Finder (https://www.ncbi.nlm. nih.gov/orffinder/) and a conserved domains program (http://www.ncbi.nlm.nih.gov/Structure/ cdd/wrpsb.cgi) [66]. ProtParam (https://web.expasy.org/protparam/) [67] and NPS@ srever (https: //npsa-prabi.ibcp.fr/cgi-bin/npsa_automat.pl?page=/NPSA/npsa_hnn.html) [68] were used for analyzing the primary structure and secondary structure of the ScWRKY3 protein, respectively. The online programs SignalP 4.1 Server (http://www.cbs.dtu.dk/services/SignalP/) [69,70], TMHMM Server v. 2.0 (http://www.cbs.dtu.dk/services/TMHMM/) [71], and Euk-mPLoc 2.0 Server (http://www.csbio.sjtu.edu.cn/bioinf/euk-multi-2/) [30] were used to predict the signal peptide, the transmembrane domain, and the subcellular localization of the target protein, respectively. The BLASTp program (https://blast.ncbi.nlm.nih.gov/Blast.cgi?PROGRAM=blastp&PAGE_TYPE=BlastSearch&LINK_LOC=blasthome) in NCBI was used to find the homologous amino acid sequences from other plants. The multiple alignment was performed using DNAMAN 6.0.3.99 software. Then the MEGA 7.0 software [72] with the Maximum Likelihood (ML) (1000 BootStrap) method was used to construct the unrooted phylogenetic tree of ScWRKY3 with sugarcane Sc-WRKY, ScWRKY4, and WRKY proteins from *Arabidopsis thaliana* [14] and other plants [13,73]. The online software MEME Suite 5.0.2 (http://meme.sdsc.edu/meme/intro.html) [73] was used to build the logo representations of the conservative domain and the rest of the alignment.

4.4. Subcellular Localization

The complete coding region of *ScWRKY3* without a stop codon was amplified using the primers *ScWRKY3*-Gate-F and *ScWRKY3*-Gate-R (Table S2), which were designed based on the sequences of *ScWRKY3* and the Gateway® donor vector of pDONR221. The gel-purified product was linked into pDONR221 using the Gateway BP Clonase™ II enzyme mix (Invitrogen, Carlsbad, CA, USA) and transformed into DH5α cells and sequenced (Biosune, Fuzhou, China). The Gateway LR Clonase™ II enzyme mix (Invitrogen) was used to ligate pDONR221-*ScWRKY3* into the subcellular localization vector pMDC83-*GFP* [74]. GV3101 cells, carrying the recombinant vector pMDC83-*ScWRKY3-GFP* or the pMDC83-*GFP* vector were inoculated into LB liquid medium supplemented with 35 μg/mL rifampicin and 50 μg/mL kanamycin, and shaken overnight in an incubator at 200 rpm and 28 °C. Subsequently, Murashige and Skoog (MS) liquid medium was used to dilute the cell density of the *Agrobacterium* solutions to an OD_{600} of 0.8. This was supplemented with 200 μM acetosyringone and cultured in the dark for 30 minutes. Then the *Agrobacterium* solutions were injected into the leaves of eight-leaf stage *N. benthamiana* using a 1.0 mL sterilized syringe [75,76]. After two days of infiltration, the treated leaves were collected and stained with 1.0 μg/mL DAPI solution in dark conditions for one h. The subcellular localization result was observed using a Leica Microsystems microscope (model Leica TCS SP8, Mannheim, Germany) with a 10 × lens, a chroma GFP filter set for EGFP (excitation at 488 nm), and a DAPI filter set for chromatin (excitation at 458 nm) [60].

4.5. Analysis of Transcriptional Activation of ScWRKY3 in Yeast Cells

To analyze the transcriptional activation of ScWRKY3, the Y2HGold-GAL4 yeast two hybrid system (containing four reporter genes, including *AUR1-C*, *HIS3*, *ADE2*, and *MEL1*) was used following the manufacturer's instructions for the Matchmaker Gold yeast two-hybrid system [49]. The *ScWRKY3* gene was PCR-amplified from pMD19-T-*ScWRKY3* using primers *ScWRKY3*-BD-F and *ScWRKY3*-BD-R (Table S2). The gel-purified product was double-digested with *Nde* I and *BamH* I enzymes, as was the plasmid pGBKT7. Then the recombinant plasmid pGBKT7-*ScWRKY3* was constructed using T4 DNA ligase (5 U/μL) (Thermo Fisher, Shanghai, China). The pGBKT7 vector, containing the nutritional screening marker gene *TRP1*, was used as a negative control. Plasmids of pGBKT7-53+pGADT7-T have been proven to bind the 53 protein and the T protein in yeast cells. The hybrid vector can activate the reporter gene *AUR1-C* on a plate that contains the AbA antibiotics, so it was used as the positive control. The empty vector plasmid pGBKT7 and plasmids pGBKT7-53+pGADT7-T and pGBKT7-*ScWRKY3* were transformed into yeast strain Y2HGold following the manufacturer's protocol for Y2HGold Chemically Competent Cells (TaKaRa Biotechnology, Dalian, China). The positive colonies were screened from selective medium plates for transferring onto the SDO (SD/-Trp, SD minimal medium without tryptophan), the SDO/X (SD/-Trp/X-α-Gal, SDO plates with X-α-D-Galactosidase), and the SDO/X/A (SD/-Trp/X-α-Gal/AbA, SDO/X plates with aureobasidin A) plates, respectively. Then the transcriptional activation activities were calculated by observing and imaging the growth conditions of the yeast cells after incubating for 2–3 days in a 29 °C incubator.

4.6. Analysis of Interaction Between ScWRKY3 and ScWRKY4

Transcriptional activation analysis in this study and a previous study [28] showed that ScWRKY3 and ScWRKY4 (GenBank Accession No. AUV50355.1) did not possess transcriptional activation activity, and the bait protein has no toxic effect on the yeast strain Y2HGold. Hence, the interacting relationship between ScWRKY3 and ScWRKY4 was identified by a yeast two-hybrid system and BiFC analysis. In the yeast two-hybrid system, AD-ScWRKY3 or AD-ScWRKY4 was used as a prey vector, and BD-ScWRKY4 or BD-ScWRKY3 was used as a bait vector, respectively. A double-enzyme digestion method was used for bait vector and prey vector construction. The specific primers with corresponding restriction enzyme sites are shown in Table S2. pGADT7-T was used as prey control, and pGBKT7-p53 or pGBKT7-Lam was used as the positive or negative bait control, respectively. These combination constructs, including the positive control pGADT7-T + pGBKT7-p53, the negative control pGADT7-T + pGBKT7-Lam, AD-ScWRKY4 + pGBKT7, AD-ScWRKY3 + pGBKT7, pGADT7-T + BD-ScWRKY3, pGADT7-T + BD-ScWRKY4, AD-ScWRKY3 + BD-ScWRKY4, or AD-ScWRKY4 + BD-ScWRKY3, were co-transformed into yeast strain Y2HGold following the manufacturer's protocol for Y2HGold Chemically Competent Cells (TaKaRa Biotechnology, Dalian, China). Subsequently, the transformed yeast cells were selected using yeast selective medium DDO (SD/-Leu/-Trp) to detect whether all the plasmids were successfully transfected into the yeast strain Y2HGold. Then the interaction between ScWRKY3 and ScWRKY4 was detected using QDO (SD/-Ade/-His/-Leu/-Trp) and QDO/X/A (SD/-Ade/-His/-Leu/-Trp/X-α-Gal/AbA) medium [64]. For BiFC vector construction, we used the Gateway method [77]. The coding sequences of *ScWRKY3* and *ScWRKY4* were amplified and linked into the non-fluorescent fragment in the pUC-SPYNE or pUC-SPYCE vector through LR-recombination using Gateway primers (Table S2). The two cooperating plasmids were transformed into the *N. benthamiana* leaves using the *Agrobacterium*-mediated method [78,79]. After five days of infiltration, the presence of fluorescence from yellow fluorescent protein (YFP) was observed using Leica Microsystems (model Leica TCS SP8, Mannheim, Germany) with a 10 × lens and a YFP filter (excitation at 561 nm).

4.7. Expression Patterns of ScWRKY3 in Sugarcane Tissues under Various Stresses

For the expression pattern analysis of *ScWRKY3* in sugarcane tissues (root, bud, leaf, stem pith, and stem epidermis) and in response to various stresses (NaCl, PEG, SA, MeJA, and ABA), the qRT-PCR primers ScWRKY3-QF/R (Table S2) were designed using the Beacon Designer V8.14 software. Glyceraldehyde-3-phosphate dehydrogenase (*GAPDH*) (GenBank Accession Number: CA254672) was used as the reference gene (Table S2). An ABI 7500 Real-Time PCR System (Applied Biosystems, Foster City, CA, USA) and a SYBR Green PCR Master Mix Kit (Roche, Shanghai, China) were used for qRT-PCR analysis. The qRT-PCR reaction system was subjected to 50 °C for 2 min, 95 °C for 10 min, 95 °C for 15 s, and 59 °C for 1 min, for 40 cycles. A melting curve analysis was performed at 95 °C for 15 s, 60 °C for 1 min, 95 °C for 15 s, and 60 °C for 30 s. Each sample was set up for triplicate technical replicates, and sterile water was used as the negative control template. The $2^{-\Delta\Delta C_T}$ method [80], DPS 9.50 software, and Origin 8 software were adopted to calculate the relative expression of the target gene (Tables S3–S6), to analyze the significance level of the experimental data, and to structure the histogram, respectively. To reduce the effects of mechanical injury on the expression of the target gene in the inoculation test with the smut pathogen, the relative expression level of the *ScWRKY3* gene was determined by subtracting the expression of the sterile water at the corresponding time according to Su et al. [81].

4.8. Transient Expression of ScWRKY3 in N. benthamiana

The Gateway LR Clonase^TM II enzyme mix (Invitrogen) was used to ligate pDONR221-*ScWRKY3*, as mentioned above, into the overexpression vector of pEarleyGate 203 [82]. The plasmid of pEarleyGate 203-*ScWRKY3* was transformed from the Gateway LR reaction into *A. tumefaciens* strain GV3101, while the pEarleyGate 203 vector which was transformed into GV3101 alone was used as a control. GV3101 cells were shaken overnight in LB liquid medium supplemented with 35 μg/mL rifampicin and 50 μg/mL kanamycin at 200 rpm and 28 °C. The *Agrobacterium* solutions were collected and resuspended to OD_{600} = 0.8 using the MS liquid medium and supplemented with 200 μM acetosyringone. Then the *Agrobacterium* suspensions were injected into the lower epidermis of the eight-leaf stage of *N. benthamiana* leaves using a 1.0 mL sterilized syringe and cultured at 28 °C with a photoperiod of 16-h light and 8-h darkness [76]. Each group of the injected leaves was collected for RNA extraction to analyze the expression level of *ScWRKY3* in *N. benthamiana* by semi-quantitative PCR with the specific primer *ScWRKY3*-Gate-F/R (Table S2). The *NtEF-1α* (GenBank Accession No. D63396) gene was used as the reference gene. The semi-quantitative PCR program was set as: 94 °C, 4 min; 94 °C, 30 s; 65 °C, 30 s; 72 °C, 1 min plus 30 s; 35 cycles; and 72 °C, 10 min. Two important tobacco pathogens, including the bacterial pathogen of *R. solanacearum* and the fungal pathogen of *F. solani* var. *coeruleum*, were cultured overnight in potato dextrose water (PDW) liquid medium at 200 rpm and 28 °C. Then the two cultured pathogen cells were separately injected into the 1-day overexpressing *N. benthamiana* leaves after being diluted to OD_{600} = 0.6 using 10 mM magnesium chloride ($MgCl_2$) solution. All treated plants were maintained for one week at 28 °C with a photoperiod of 16-h light/8-h darkness to track the changes in leaf symptoms and to analyze the relative transcript level of nine tobacco immunity-associated marker genes, including the HR marker genes *NtHSR201*, *NtHSR203*, and *NtHSR515*; the SA-related genes *NtPR-1a/c* and *NtNPR1*; the JA-associated genes *NtPR2* and *NtPR3*; and the ET synthesis-dependent genes *NtEFE26* and *NtAccdeaminase* (Tables S2 and S7–S9) [83,84]. All the treatments were carried out in three replicates. For representational observation, *Agrobacterium* suspensions carrying the vector pEarleyGate 203 and the recombinant vector pEarleyGate 203-*ScWRKY3* were injected into the left and right side of each selected *N. benthamiana* leaf, respectively.

5. Conclusions

In this study, a novel *ScWRKY3* gene was isolated from sugarcane and functionally characterized. ScWRKY3 belongs to group IIc of the WRKY family as a nucleoprotein, with no auto-activation. It has an interaction with another group IIc sugarcane WRKY protein, ScWRKY4, however the interaction mechanism and its corresponding function need further investigation. ScWRKY3 may participate in sugarcane resistance to drought and salt stimuli, and this resistance may be mediated by ABA signaling pathways. The transcript abundance of *ScWRKY3* was stable in the smut-resistant cultivar Yacheng05-179, while it was down-regulated in the smut-susceptible cultivar ROC22 at 24 h, during inoculation with *S. scitamineum*. In addition, *ScWRKY3* showed a negative regulatory effect on the bacterial pathogen *R. solanacearum* and the fungal disease *F. solani* var. *coeruleum* in *35S::ScWRKY3*-overexpressing *N. benthamiana*. These results may be useful for the functional identification of the WRKY family in sugarcane and good for the interaction analysis of ScWRKY3 with other WRKY proteins or other functional proteins.

Supplementary Materials: Supplementary materials can be found at http://www.mdpi.com/1422-0067/19/12/4059/s1, **Figure S1**: Nucleic acid sequences alignment of *ScWRKY3* in ROC22 and Yacheng05-179; **Figure S2.** Secondary structure prediction of ScWRKY3; **Figure S3.** Signal peptide and transmembrane domain prediction of ScWRKY3; **Figure S4.** Subcellular localization prediction of ScWRKY3; **Figure S5.** Amino acid sequences alignment of ScWRKY3 and other WRKYs; **Figure S6.** The logo of predicted conserved motifs in the WRKYs; **Table S1.** Primary structure analysis of ScWRKY3; **Table S2.** Primers used in this study; **Table S3.** Raw calculations of tissue-specific expression of *ScWRKY3* in different 10-month-old ROC22 tissues by qRT-PCR; **Table S4.** Raw calculations of gene expression patterns of *ScWRKY3* in 4-month-old ROC22 plantlets under abiotic stress; **Table S5.** Raw calculations of gene expression of *ScWRKY3* in 4-month-old ROC22 plantlets under plant hormone stress; **Table S6.** Raw calculations of expression of the *ScWRKY3* gene after infection with smut pathogen; **Table S7.** Raw calculations of the transcript level of nine immunity-associated marker genes in the *Nicotiana benthamiana* leaves after one day of agroinfiltration; **Table S8.** Raw calculations of the transcripts of nine immunity-associated marker genes in the *Nicotiana benthamiana* leaves after inoculation with *Ralstonia solanacearum* for one day and seven days; **Table S9.** Raw calculations of the transcripts of nine immunity-associated marker genes in the *Nicotiana benthamiana* leaves after inoculation with *Fusarium solani* var. *coeruleum* for one day and seven days.

Author Contributions: L.W., Y.S., and Y.Q. conceived and designed the research. L.W. and X.Z. prepared the materials. L.W., F.L., X.Z., W.W., T.S., Y.C., M.D., and S.Y. conducted the experiments. L.W., F.L., and X.Z. analyzed the data. L.W. wrote the manuscript. Y.S., Y.Q., and L.P. helped to revise the manuscript. All authors read and approved the final manuscript.

Funding: This work was funded by the National Natural Science Foundation of China (31101196, 31671752, and 31501363), the Research Funds for Distinguished Young Scientists in Fujian Provincial Department of Education (SYC-2017), the Special Fund for Science and Technology Innovation of Fujian Agriculture and Forestry University (KFA17267A), and the China Agriculture Research System (CARS-17).

Conflicts of Interest: The authors declare no conflict of interest.

References

1. Ulker, B.; Somssich, I.E. WRKY transcription factors: From DNA binding towards biological function. *Curr. Opin. Plant Biol.* **2004**, *7*, 491–498. [CrossRef] [PubMed]
2. Jiang, J.J.; Ma, S.H.; Ye, N.H.; Jiang, M.; Cao, J.S.; Zhang, J.H. WRKY transcription factors in plant responses to stresses. *J. Integr. Plant Biol.* **2017**, *59*, 86–101. [CrossRef] [PubMed]
3. Singh, K.B.; Foley, R.C.; Oñatesánchez, L. Transcription factors in plant defense and stress responses. *Curr. Opin. Plant Biol.* **2002**, *5*, 430–436. [CrossRef]
4. Chen, L.G.; Song, Y.; Li, S.J.; Li, S.J.; Zhang, L.P.; Zou, C.S.; Yu, D.Q. The role of WRKY transcription factors in plant abiotic stresses. *Biochim. Biophys. Acta* **2012**, *1819*, 120–128. [CrossRef] [PubMed]
5. Rushton, P.J.; Somssich, I.E.; Ringler, P.; Shen, Q.J. WRKY transcription factors. *Trends Plant Sci.* **2010**, *15*, 247–258. [CrossRef] [PubMed]
6. Eulgem, T.; Rushton, P.J.; Robatzek, S.; Somssich, I.E. The WRKY superfamily of plant transcription factors. *Trends Plant Sci.* **2000**, *5*, 199–206. [CrossRef]
7. Mohanta, T.K.; Park, Y.H.; Bae, H. Novel genomic and evolutionary insight of WRKY transcription factors in plant lineage. *Sci. Rep.* **2016**, *6*, 37309. [CrossRef] [PubMed]
8. Chen, F.; Hu, Y.; Vannozzi, A.; Wu, K.C.; Cai, H.Y.; Qin, Y.; Mullis, A.; Lin, Z.G.; Zhang, L.S. The WRKY transcription factor family in model plants and crops. *Plant Sci.* **2018**, *36*, 1–25. [CrossRef]
9. Ramamoorthy, R.; Jiang, S.Y.; Kumar, N.; Venkatesh, P.N.; Ramachandran, S. A comprehensive transcriptional profiling of the *WRKY* gene family in rice under various abiotic and phytohormone treatments. *Plant. Cell Physiol.* **2008**, *49*, 865–879. [CrossRef] [PubMed]
10. Wei, K.; Chen, J.; Chen, Y.F.; Wu, L.J.; Xie, D.X. Multiple-strategy analyses of ZmWRKY subgroups and functional exploration of *ZmWRKY* genes in pathogen responses. *Mol. Biosyst.* **2012**, *8*, 1940–1949. [CrossRef] [PubMed]
11. Muthamilarasan, M.; Bonthala, V.S.; Khandelwal, R.; Jaishankar, J.; Shweta, S.; Nawaz, K.; Prasad, M. Global analysis of WRKY transcription factor superfamily in *Setaria* identifies potential candidates involved in abiotic stress signaling. *Front. Plant Sci.* **2015**, *6*, 910. [CrossRef] [PubMed]
12. Pandey, S.P.; Somssich, I.E. The role of WRKY transcription factors in plant immunity. *Plant Physiol.* **2009**, *150*, 1648–1655. [CrossRef] [PubMed]
13. Mangelsen, E.; Kilian, J.; Berendzen, K.W.; Kolukisaoglu, U.H.; Harter, K.; Jansson, C.; Wanke, D. Phylogenetic and comparative gene expression analysis of barley (*Hordeum vulgare*) WRKY transcription factor family reveals putatively retained functions between monocots and dicots. *BMC Genomics* **2008**, *9*, 194. [CrossRef] [PubMed]
14. Wei, X.A.; Yao, W.J.; Jiang, T.B.; Zhou, B.R. Identification of *WRKY* gene in response to abiotic stress from WRKY transcirption factor gene family of *Arabidopsis thaliana*. *J. Northeast. For. Univ.* **2016**, *44*, 45–48.
15. Zhang, T.; Tan, D.; Zhang, L.; Zhang, X.; Han, Z. Phylogenetic analysis and drought-responsive expression profiles of the WRKY transcription factor family in maize. *Agric. Gene* **2017**, *3*, 99–108. [CrossRef]

16. Wu, J.; Chen, J.B.; Wang, L.F.; Wang, S.M. Genome-wide investigation of WRKY transcription factors involved in terminal drought stress response in common bean. *Front. Plant Sci.* **2017**, *8*, 380. [CrossRef] [PubMed]

17. Qiu, Y.Q.; Jing, S.J.; Fu, J.; Li, L.; Yu, D.Q. Cloning and analysis of expression profile of 13 *WRKY* genes in rice. *Chin. Sci. Bull.* **2004**, *49*, 2159–2168. [CrossRef]

18. Wu, H.L.; Ni, Z.F.; Yao, Y.Y.; Guo, G.G.; Sun, Q.X. Cloning and expression profiles of 15 genes encoding WRKY transcription factor in wheat (*Triticum aestivem* L.). *Prog. Nat. Sci. Mater. Int.* **2008**, *18*, 697–705. [CrossRef]

19. Dong, J.; Chen, C.H.; Chen, Z.X. Expression profiles of the *Arabidopsis WRKY* gene superfamily during plant defense response. *Plant Mol. Biol.* **2003**, *51*, 21–37. [CrossRef] [PubMed]

20. Kim, K.C.; Lai, Z.; Fan, B.; Chen, Z. *Arabidopsis* WRKY38 and WRKY62 transcription factors interact with histone deacetylase 19 in basal defense. *Plant Cell* **2008**, *20*, 2357–2371. [CrossRef] [PubMed]

21. Liu, D.L.; Leib, K.; Zhao, P.Y.; Kogel, K.H.; Langen, G. Phylogenetic analysis of barley WRKY proteins and characterization of HvWRKY1 and repressors of the pathogen-inducible gene *HvGER4c*. *Mol. Genet. Genomics* **2014**, *289*, 1331–1345. [CrossRef] [PubMed]

22. Peng, Y.; Bartley, L.E.; Chen, X.W.; Dardick, C.; Chern, M.; Ruan, R.; Canlas, P.E.; Ronald, P.C. *OsWRKY62* is a negative regulator of basal and *Xa21*-mediated defense against *Xanthomonas orozae* pv. *orozae* in rice. *Mol. Plant* **2008**, *1*, 446–458. [CrossRef] [PubMed]

23. Berri, S.; Abbruscato, P.; Faivre-Rampant, O.; Brasileiro, A.C.; Fumasoni, I.; Satoh, K.; Kikuchi, S.; Mizzi, L.; Morandini, P.; Pè, M.E.; et al. Characterization of WRKY co-regulatory networks in rice and *Arabidopsis*. *BMC Plant Biol.* **2009**, *9*, 120. [CrossRef] [PubMed]

24. Banerjee, A.; Roychoudhury, A. WRKY proteins: Signaling and regulation of expression during abiotic stress responses. *Sci. World J.* **2015**, 807560. [CrossRef] [PubMed]

25. Chen, R.K.; Xu, L.P.; Lin, Y.Q. *Modern Sugarcane Genetic Breeding*; China Agriculture Press: Beijing, China, 2011.

26. Lambais, M.R. *In silico* differential display of defense-related expressed sequence tags from sugarcane tissues infected with *Diazotrophic endophytes*. *Genet. Mol. Biol.* **2001**, *24*, 103–111. [CrossRef]

27. Liu, J.X.; Que, Y.X.; Guo, J.L.; Xu, L.P.; Wu, J.Y.; Chen, R.K. Molecular cloning and expression analysis of a WRKY transcription factor in sugarcane. *Afr. J. Biotechnol.* **2012**, *11*, 6434–6444.

28. Wang, L.; Liu, F.; Dai, M.J.; Sun, T.T.; Su, W.H.; Wang, C.F.; Zhang, X.; Mao, H.Y.; Su, Y.C.; Que, Y.X. Cloning and expression characteristic analysis of *ScWRKY4* gene in sugarcane. *Acta Agron. Sin.* **2018**, *44*, 1367–1379.

29. Que, Y.X.; Su, Y.C.; Guo, J.L.; Wu, Q.B.; Xu, L.P. A global view of transcriptome dynamics during *Sporisorium scitamineum* challenge in sugarcane by RNAseq. *PLoS ONE* **2014**, *9*, e106476. [CrossRef] [PubMed]

30. Chou, K.C.; Shen, H.B. A new method for predicting the subcellular localization of eukaryotic proteins with both single and multiple sites: Euk-mPLoc 2.0. *PLoS ONE* **2010**, *5*, e9931. [CrossRef] [PubMed]

31. Ishiguro, S.; Nakamura, K. Characterization of a cDNA encoding a novel DNA-binding protein, *SPF1*, that recognizes SP8 sequences in the 5' upstream regions of genes coding for sporamin and beta-amylase from sweet potato. *Mol. Gen. Genet.* **1994**, *244*, 563–571. [CrossRef] [PubMed]

32. Eulgem, T.; Rushton, P.J.; Schmelzer, E.; Hahlbrock, K.; Somssich, I.E. Early nuclear events in plant defence signalling: Rapid gene activation by WRKY transcription factors. *Embo J.* **1999**, *18*, 4689–4699. [CrossRef] [PubMed]

33. Liu, L.S.; White, M.J.; Macrae, T. Transcription factors and their genes in higher plants. *FEBS J.* **2010**, *262*, 247–257.

34. Fan, Z.Q.; Tan, X.L.; Shan, W.; Kuang, J.F.; Lu, W.J.; Chen, J.Y. BrWRKY65, a WRKY transcription factor, is involved in regulating three leaf senescence-associated genes in Chinese flowering cabbage. *Int. J. Mol. Sci.* **2017**, *18*, 1228.

35. Wan, Y.Q.; Mao, M.Z.; Wan, D.L.; Yang, Q.; Yang, F.Y.; Mandlaa; Li, G.J.; Wang, R.G. Identification of the *WRKY* gene family and functional analysis of two genes in *Caragana intermedia*. *BMC Plant Biol.* **2018**, *18*, 31. [CrossRef] [PubMed]

36. Lee, H.; Cha, J.; Choi, C.; Choi, N.; Ji, H.S.; Park, S.R.; Lee, S.; Hwang, D.J. Rice *WRKY11* plays a role in pathogen defense and drought tolerance. *Rice* **2018**, *11*, 5. [CrossRef] [PubMed]

37. Wang, H.H.; Meng, J.; Peng, X.X.; Tang, X.K.; Zhou, P.L.; Xiang, J.H.; Deng, X.B. Rice *WRKY4* acts as a transcriptional activator mediating defense responses toward *Rhizoctonia solani*, the causing agent of rice sheath blight. *Plant Mol. Biol.* **2015**, *89*, 157–171. [CrossRef] [PubMed]

38. Fields, S.; Song, O. A novel genetic system to detect protein-protein interactions. *Nature* **1989**, *340*, 245. [CrossRef] [PubMed]

39. Sébastien, B.; Li, J.; Palva, E.T. WRKY54 and WRKY70 co-operate as negative regulators of leaf senescence in *Arabidopsis thaliana*. *J. Exp. Bot.* **2012**, *63*, 2667–2679.

40. Wang, X.T.; Zeng, J.; Li, Y.; Rong, X.L.; Sun, J.T.; Sun, T.; Li, M.; Wang, L.Z.; Feng, Y.; Chai, R.H.; et al. Expression of *TaWRKY44*, a wheat *WRKY* gene, in transgenic tobacco confers multiple abiotic stress tolerances. *Front. Plant Sci.* **2015**, *6*, 615. [CrossRef] [PubMed]

41. Chi, Y.C.; Yang, Y.; Zhou, Y.; Zhou, J.; Fan, B.F.; Yu, J.Q.; Chen, Z.X. Protein-protein interactions in the regulation of WRKY transcription factors. *Mol. Plant* **2013**, *6*, 287–300. [CrossRef] [PubMed]

42. Robatzek, S.; Somssich, I.E. Targets of *AtWRKY6* regulation during plant senescence and pathogen defense. *Genes Dev.* **2002**, *16*, 1139–1149. [CrossRef] [PubMed]

43. Popescu, S.C.; Popeseu, G.V.; Bachan, S.; Zhang, Z.; Gerstein, M.; Snyder, M.; Dinesh-Kumar, S.E. MAPK target networks in *Arabidopsis thaliana* revealed using functional protein microarrays. *Genes Dev.* **2009**, *23*, 80–92. [CrossRef] [PubMed]

44. Zhou, L.; Wang, N.N.; Kong, L.; Gong, S.Y.; Li, Y.; Li, X.B. Molecular characterization of 26 cotton *WRKY* genes that are expressed differentially in tissues and are induced in seedlings under high salinity and osmotic stress. *Plant Cell Tissue Org. Cult.* **2014**, *119*, 141–156. [CrossRef]

45. Bakshi, M.; Oelmüller, R. WRKY transcription factors: Jack of many trades in plants. *Plant Signal. Behav.* **2014**, *9*, e27700. [CrossRef] [PubMed]

46. Bartels, D.; Sunkar, R. Drought and salt tolerance in plants. *Crit. Rev. Plant Sci.* **2005**, *24*, 23–58. [CrossRef]

47. Qiao, Z.; Li, C.L.; Zhang, W. WRKY1 regulates stomatal movement in drought-stressed *Arabidopsis thaliana*. *Plant Mol. Biol.* **2016**, *91*, 53–65. [CrossRef] [PubMed]

48. Jiang, Y.J.; Yu, D.Q. *WRKY57* regulates *JAZ* genes transcriptionally to compromise *Botrytis cinerea* resistance in *Arabidopsis thaliana*. *Plant Physiol.* **2016**, *171*, 2771–2782. [CrossRef] [PubMed]

49. Jones, J.D.; Dang, J.L. The plant immune system. *Nature* **2006**, *444*, 323–329. [CrossRef] [PubMed]

50. Métraux, J.P.; Nawrath, C.; Genoud, T. Systemic acquired resistance. *Euphytica* **2002**, *124*, 23–243. [CrossRef]

51. Katagiri, F. A global view of defense gene expression regulation—A highly interconnected signaling network. *Curr. Opin. Plant Biol.* **2004**, *7*, 506–511. [CrossRef] [PubMed]

52. Kunkel, B.N.; Brooks, D.M. Cross talk between signaling pathways in pathogen defense. *Curr. Opin. Plant Biol.* **2002**, *5*, 325–331. [CrossRef]

53. Chen, X.T.; Liu, J.; Lin, G.F.; Wang, A.; Wang, Z.G.; Lu, G.D. Overexpression of *AtWRKY28* and *AtWRKY75* in *Arabidopsis* enhances resistance to oxalic acid and *Sclerotinia sclerotiorum*. *Plant Cell Rep.* **2013**, *32*, 1589–1599. [CrossRef] [PubMed]

54. Wang, H.H.; Hao, J.J.; Chen, X.J.; Hao, Z.N.; Wang, X.; Lou, Y.G.; Peng, Y.L.; Guo, Z.J. Overexpression of rice *WRKY89* enhances ultraviolet B tolerance and disease resistance in rice plants. *Plant Mol. Biol.* **2007**, *65*, 799–815. [CrossRef] [PubMed]

55. Qiu, D.Y.; Xiao, J.; Ding, X.H.; Xiong, M.; Cai, M.; Cao, Y.L.; Li, X.H.; Xu, C.G.; Wang, S.P. *OsWRKY13* mediates rice disease resistance by regulating defense-related genes in salicylate- and jasmonate-dependent signaling. *Mol. Plant-Microbe Interact.* **2007**, *20*, 492–499. [CrossRef] [PubMed]

56. Qiu, D.Y.; Xiao, J.; Xie, W.B.; Liu, H.B.; Li, X.H.; Xiong, L.Z.; Wang, S.P. Rice gene network inferred from expression profiling of plants overexpressing *OsWRKY13*, a positive regulator of disease resistance. *Mol. Plant* **2008**, *1*, 538–551. [PubMed]

57. Zheng, Z.Y.; Mosher, S.L.; Fan, B.F.; Klessig, D.F.; Chen, Z.X. Functional analysis of *Arabidopsis* WRKY25 transcription factor in plant defense against *Pseudomonas syringae*. *BMC Plant Biol.* **2007**, *7*, 2. [CrossRef] [PubMed]

58. Jiang, Y.; Deyholos, M.K. Functional characterization of *Arabidopsis* NaCl-inducible WRKY25 and WRKY33 transcription factors in abiotic stresses. *Plant Mol. Biol.* **2009**, *69*, 91–105. [CrossRef] [PubMed]

59. Yokotani, N.; Sato, Y.; Tanabe, S.; Chujo, T.; Shimizu, T.; Okada, K.; Yamane, K.; Shimono, M.; Sugano, S.; Takatsuji, H.; et al. WRKY76 is a rice transcriptional repressor playing opposite roles in blast disease resistance and cold stress tolerance. *J. Exp. Bot.* **2013**, *64*, 5085–5097. [CrossRef] [PubMed]

60. Liu, X.F.; Song, Y.Z.; Xing, F.Y.; Wang, N.; Wen, F.J. *GhWRKY25*, a group I *WRKY* gene from cotton, confers differential tolerance to abiotic and biotic stresses in transgenic *Nicotiana benthamiana*. *Protoplasma* **2015**, *253*, 1–17. [CrossRef] [PubMed]

61. Wang, X.; Yan, Y.; Li, Y.Z.; Chu, X.Q.; Wu, C.G.; Guo, X.Q. *GhWRKY40*, a multiple stress-responsive cotton *WRKY* gene, plays an important role in the wounding response and enhances susceptibility to *Ralstonia solanacearum* infection in transgenic *Nicotiana benthamiana*. *PLoS ONE* **2014**, *9*, e93577. [CrossRef] [PubMed]

62. Mukhtar, M.S.; Deslandes, L.; Auriac, M.; Marco, Y.; Somssich, I.E. The Arabidopsis transcription factor WRKY27 influences wilt disease symptom development caused by *Ralstonia solanacearum*. *Plant J.* **2008**, *56*, 935. [CrossRef] [PubMed]

63. Su, Y.C.; Guo, J.L.; Ling, H.; Chen, S.S.; Wang, S.S.; Xu, L.P.; Allan, A.C.; Que, Y.X. Isolation of a novel peroxisomal catalase gene from sugarcane, which is responsive to biotic and abiotic stresses. *PLoS ONE* **2014**, *9*, e84426. [CrossRef] [PubMed]

64. Li, H.; Gao, Y.; Xu, H.; Dai, Y.; Deng, D.Q.; Chen, J.M. *ZmWRKY33*, a WRKY maize transcription factor conferring enhanced salt stress tolerances in *Arabidopsis*. *Plant Growth Regul.* **2013**, *70*, 207–216. [CrossRef]

65. Scarpeci, T.E.; Zanor, M.I.; Zanor, M.I.; Mueller-Roeber, B.; Valle, E.M. Overexpression of *AtWRKY30* enhances abiotic stress tolerance during early growth stages in *Arabidopsis thaliana*. *Plant Mol. Biol.* **2013**, *83*, 265–277. [CrossRef] [PubMed]

66. Marchlerbauer, A.; Bo, Y.; Han, L.Y.; He, J.; Lanczycki, C.J.; Lu, S.N.; Chitsaz, F.; Derbyshire, M.K.; Geer, R.C.; Gonzales, N.R.; et al. CDD/SPARCLE: Functional classification of proteins via subfamily domain architectures. *Nucleic Acids Res.* **2017**, *45*, 200–203. [CrossRef] [PubMed]

67. Gasteiger, E.; Hoogland, C.; Gattiker, A.; Duvaud, S.; Wilkins, M.R.; Appel, R.D.; Bairoch, A. Protein identification and analysis tools on the ExPASy server. In *The Proteomics Protocols Handbook*; Humana Press: New York, NY, USA, 2005; pp. 571–607.

68. Combet, C.; Blanchet, C.; Geourjon, C.; Deléage, G. NPS@: Network protein sequence analysis. *Trends Biochem. Sci.* **2000**, *25*, 147–150. [CrossRef]

69. Petersen, T.N.; Brunak, S.; Heijne, G.; Nielsen, H. SignalP 4.0: Discriminating signal peptides from transmembrane regions. *Nat. Methods* **2011**, *8*, 785–786. [CrossRef] [PubMed]

70. Nielsen, H. Predicting secretory proteins with SignalP. *Methods Mol. Biol.* **2017**, *1611*, 59–73. [PubMed]

71. Krogh, A.; Larsson, B.; Heijne, G.V.; Sonnhammer, E.L.L. Predicting transmembrane protein topology with a hidden markov model: Application to complete genomes. *J. Mol. Biol.* **2001**, *305*, 567–580. [CrossRef] [PubMed]

72. Kumar, S.; Stecher, G.; Tamura, K. MEGA7: Molecular evolutionary genetics analysis version 7.0 for bigger datasets. *Mol. Biol. Evol.* **2016**, *33*, 1870–1874. [CrossRef] [PubMed]

73. Wang, C.; Deng, P.; Chen, L.; Wang, X.; Ma, H.; Hu, W.; Yao, N.C.; Feng, Y.; Chai, R.H.; Yang, G.X.; et al. A wheat WRKY transcription factor TaWRKY10 confers tolerance to multiple abiotic stresses in transgenic tobacco. *PLoS ONE* **2013**, *8*, e65120. [CrossRef] [PubMed]

74. Curtis, M.D.; Grossniklaus, U. A gateway cloning vector set for high-throughput functional analysis of genes in planta. *Plant Phyosiol.* **2003**, *133*, 462–469. [CrossRef] [PubMed]

75. Hwang, I.S.; Hwang, B.K. Requirement of the cytosolic interaction between pathogenesis-related protein10 and leucine-rich repeat protein1 for cell death and defense signaling in pepper. *Plant Cell* **2012**, *24*, 1675–1690.

76. Fan, Z.Q.; Kuang, J.F.; Fu, C.C.; Shan, W.; Han, Y.C.; Xiao, Y.Y.; Ye, Y.J.; Lu, W.J.; Lakshmanan, P.; Duan, X.W.; et al. The banana transcriptional repressor *MaDEAR1* negatively regulates cell wall-modifying genes involved in fruit ripening. *Front. Plant Sci.* **2016**, *7*, 1021. [CrossRef] [PubMed]

77. Shen, C.J.; Wang, S.K.; Bai, Y.H.; Wu, Y.R.; Zhang, S.N.; Chen, M.; Guilfoyle, T.J.; Wu, P.; Qi, Y.H. Functional analysis of the structural domain of ARF proteins in rice (*Oryza sativa* L.). *J. Exp. Bot.* **2010**, *61*, 3971–3981. [CrossRef] [PubMed]

78. Schütze, K.; Harter, K.; Chaban, C. Bimolecular fluorescence complementation (BiFC) to study protein-protein interactions in living plant cells. *Methods Mol. Biol.* **2009**, *479*, 189–202. [PubMed]

79. Mersereau, M.; Pazour, G.J.; Das, A. Efficient transformation of *Agrobacterium tumefaciens* by electroporation. *Gene* **1990**, *90*, 149–151. [CrossRef]

80. Livak, K.J.; Schmittgen, T.D. Analysis of relative gene expression data using real-time quantitative PCR and the $2^{-\Delta\Delta C_T}$ method. *Methods* **2001**, *25*, 402–408. [CrossRef] [PubMed]

81. Su, Y.C.; Wang, Z.Q.; Li, Z.; Xu, L.P.; Que, Y.X.; Dai, M.J.; Chen, Y.H. Molecular cloning and functional identification of peroxidase gene *ScPOD02* in sugarcane. *Acta Agron. Sin.* **2017**, *43*, 510–521. [CrossRef]

82. Earley, K.W.; Haag, J.R.; Pontes, O.; Opper, K.; Juehne, S.; Song, K.; Pikaard, C.S. Gateway-compatible vectors for plant functional genomics and proteomics. *Plant J.* **2006**, *45*, 616–629. [CrossRef] [PubMed]

83. Peng, Q.; Su, Y.C.; Ling, H.; Ahmad, W.; Gao, S.W.; Guo, J.L.; Que, Y.X.; Xu, L.P. A sugarcane pathogenesis-related protein, *ScPR10*, plays a positive role in defense responses under *Sporisorium scitamineum*, SrMV, SA, and MeJA stresses. *Plant Cell Rep.* **2017**, *36*, 1427–1440. [CrossRef] [PubMed]

84. Lai, Y.; Dang, F.F.; Lin, J.; Yu, L.; Shi, Y.L.; Xiao, Y.H.; Huang, M.K.; Lin, J.H.; Chen, C.C.; Qi, A.H. Overexpression of a *Chinese cabbage BrERF11*, transcription factor enhances disease resistance to *Ralstonia solanacearum*, in *tobacco. Plant Physiol. Biochem.* **2013**, *62*, 70–78. [CrossRef] [PubMed]

International Journal of
Molecular Sciences

Article

GmBRC1 is a Candidate Gene for Branching in Soybean (*Glycine max* (L.) Merrill)

Sangrea Shim [1,2], Jungmin Ha [1,2], Moon Young Kim [1,2], Man Soo Choi [3], Sung-Taeg Kang [4], Soon-Chun Jeong [5], Jung-Kyung Moon [6] and Suk-Ha Lee [1,2,*]

[1]   Department of Plant Science and Research Institute of Agriculture and Life Sciences,
      Seoul National University, Seoul 08826, Korea; sangreashim@gmail.com (S.S.); saga16@snu.ac.kr (J.H.);
      moonykim@snu.ac.kr (M.Y.K.)
[2]   Plant Genomics and Breeding Institute, Seoul National University, Seoul 08826, Korea
[3]   National Institute of Crop Sciences, Rural Development Administration, Wanju-gun,
      Jeollabuk-do 55365, Korea; mschoi73@korea.kr
[4]   Department of Crop Science & Biotechnology, Dankook University, Cheonan-si,
      Chungcheongnam-do 31116, Korea; kangst@dankook.ac.kr
[5]   Bio-Evaluation Center, Korea Research Institute of Bioscience and Biotechnology, Cheongju-si,
      Chungcheongbuk-do 28116, Korea; scjeong@kribb.re.kr
[6]   National Institute of Agricultural Sciences, Rural Development Administration, Jeonju-si,
      Jeollabuk-do 54874, Korea; moonjk2@korea.kr
*   Correspondence: sukhalee@snu.ac.kr; Tel.: +82-2880-4545; Fax: +82-2877-4550

Received: 15 November 2018; Accepted: 25 December 2018; Published: 1 January 2019

Abstract: Branch number is one of the main factors affecting the yield of soybean (*Glycine max* (L.)). In this study, we conducted a genome-wide association study combined with linkage analysis for the identification of a candidate gene controlling soybean branching. Five quantitative trait nucleotides (QTNs) were associated with branch numbers in a soybean core collection. Among these QTNs, a linkage disequilibrium (LD) block *qtnBR6-1* spanning 20 genes was found to overlap a previously identified major quantitative trait locus *qBR6-1*. To validate and narrow down *qtnBR6-1*, we developed a set of near-isogenic lines (NILs) harboring high-branching (HB) and low-branching (LB) alleles of *qBR6-1*, with 99.96% isogenicity and different branch numbers. A cluster of single nucleotide polymorphisms (SNPs) segregating between NIL-HB and NIL-LB was located within the *qtnBR6-1* LD block. Among the five genes showing differential expression between NIL-HB and NIL-LB, *BRANCHED1* (*BRC1*; Glyma.06G210600) was down-regulated in the shoot apex of NIL-HB, and one missense mutation and two SNPs upstream of *BRC1* were associated with branch numbers in 59 additional soybean accessions. *BRC1* encodes TEOSINTE-BRANCHED1/CYCLOIDEA/PROLIFERATING CELL FACTORS 1 and 2 transcription factor and functions as a regulatory repressor of branching. On the basis of these results, we propose *BRC1* as a candidate gene for branching in soybean.

Keywords: soybean; branching; genome-wide association study (GWAS); near-isogenic line (NIL); *BRANCHED1* (*BRC1*); TCP transcription factor

1. Introduction

Soybean (*Glycine max* (L.)) is a major food crop and a rich source of protein and oil in human diet and animal feed. Two cultivation methods with different planting densities are used to maximize soybean yield. The high-density planting method is mainly practiced in the USA [1], where the yield of soybean increases with planting density until saturation. The low-density planting method is practiced in Asia not only to avoid disease and lodging, but also to reduce seed and labor cost; however, the productivity in low-density planting is lower than that in high-density planting [2]. An important factor of low-density planting is branching plasticity, which offsets yield losses [3]. Branching pattern and branch number are dependent on environmental factors such as planting density and light quality [3–5]. These factors have obstructed the identification of genes regulating branch development. However, variation in branch number among soybean cultivars of diverse origins under low-density planting suggests the presence of genetic differences [6,7]. Thus, breeding for soybean genotypes with optimal plant architecture adapted to a specific planting density is necessary for improving yield and enabling mechanical harvesting.

To date, a dozen quantitative trait loci (QTLs) regulating branch development have been identified using recombinant inbred line (RIL) or F_2 populations in soybean [7–11]. However, the identified QTLs span a large number of plausible genes because of the low resolution of genetic linkage maps and low recombination frequency in mapping populations. Recently, three known QTLs have been narrowed down to identify candidate genes based on a high-density linkage map using the 6K single nucleotide polymorphism (SNP) chip (BARCSoySNP6K), and positive phenotypic correlation has been demonstrated between branch number and total pod number [8]. Among the QTLs, a major one (*qBR6-1*) on chromosome (Chr) 6, with a logarithm of odds (LOD) score of 10.3 and 14.5% of the phenotypic variation in branch numbers, was shown to contain 13 genes [8]. One of these genes was a gene encoding TEOSINTE-BRANCHED1/CYCLOIDEA/PROLIFERATING CELL FACTORS 1 and 2 (TCP) transcription factor, also known as *BRANCHED1* (*BRC1*), which is involved in gene networks of axillary branching via interactions with the auxin hormone network [8]. In soybean, the *BRC1* gene has not been genetically identified.

A genome-wide association study (GWAS) is used to identify associations between genetic loci and traits. It is a powerful method, as it provides high genetic resolution derived from all recombination events that occurred during the evolution of a natural population [12]. GWAS has been used successfully to identify the genetic basis of complex agronomic traits in Arabidopsis (*Arabidopsis thaliana* (L.)), rice (*Oryza sativa* (L.)), and maize (*Zea mays* (L.)) [13–15]. In soybean, GWAS has been performed to identify the association of genetic regions with agronomic traits, including flowering time, maturity date, plant height, and seed oil content [16,17]. However, the detection of false-positive associations with phenotypes is a weakness of GWAS, which is caused by the population structure and kinship relatedness in natural populations [16]. Utilizing linkage analysis in combination with GWAS has been previously used to overcome the limitations of QTL mapping and to validate the results of GWAS [16].

The objective of this study was to narrow down a genomic region of *qBR6-1* to identify a candidate gene responsible for soybean branching. To this end, GWAS was conducted to identify quantitative trait nucleotides (QTNs) associated with branch number in a soybean core collection comprising 400 soybean genotypes grown in three locations. To validate and narrow down a QTN that was found to be overlapped with the previously identified major QTL, *qBR6-1*, we developed and analyzed a set of near-isogenic lines (NILs) carrying high-branching (HB) and low-branching (LB) alleles of *qBR6-1* derived from an F_6 residual heterozygous line (RHL) heterozygous for *qBR6-1*. Using these NILs, we detected differential expression of genes within the linkage disequilibrium (LD) block of the QTN on Chr 6. In addition, we examined nucleotide variations in a selected candidate gene between NIL-HB and NIL-LB, and confirmed their allelic associations with branch numbers in soybean accessions obtained from the United States Department of Agriculture Germplasm Resources Information Network (USDA-GRIN). This study will provide a better understanding of genetic basis underlying the branch development in soybean and valuable information to improve plant architecture for soybean cultivars with high yields.

2. Results

2.1. Variation and Heritability in Branch Number of the Soybean Core Collection

We present variations in branch number of the soybean core collection according to the location of cultivation (Wanju, Cheonan, and Ochang) and best linear unbiased predictor (BLUP) values in Figure S1. The number of branches varied from 2.0–20.7 in Wanju, 1.0–21.3 in Cheonan, and 0.0–14.3 in Ochang (Table S1). The broad-sense heritability (H^2) of branch numbers in the soybean core collection was 57.7% (Table S1). Analysis of variance (ANOVA) showed significant effects of genotype as well as genotype-by-environment (G × E) interaction on branch numbers ($p < 0.0001$). Additionally, a continuous distribution of branch numbers was observed in all three locations, indicating that branch numbers were regulated by multiple genetic factors. To identify reliable QTNs associated with branch numbers, the BLUP values were calculated for each genotype and used in GWAS (Figure S1).

2.2. Population Structure and LD

Covariates from the population structure were stratified using 81,078 SNP markers. The log likelihood (LnP(D)) from STRUCTURE [18] analysis showed a continuous increase with the number of sub-populations (*K*). Therefore, we determined the optimum *K* value using the Δ*K* method. The Δ*K* value classified the core collection into two groups (Figure S2). In a scree plot, the proportion of variance drastically decreased until the number of principal components (PCs) reached two (Figure 1A). The extent of LD estimated by PLINK [19] showed that the average pair-wise squared correlation coefficient (r^2) between alleles was dropped to half of its maximum value at 160–170 kb for the entire genome, with 140–150 kb for euchromatin and 460–470 kb for heterochromatin (Figure 1B).

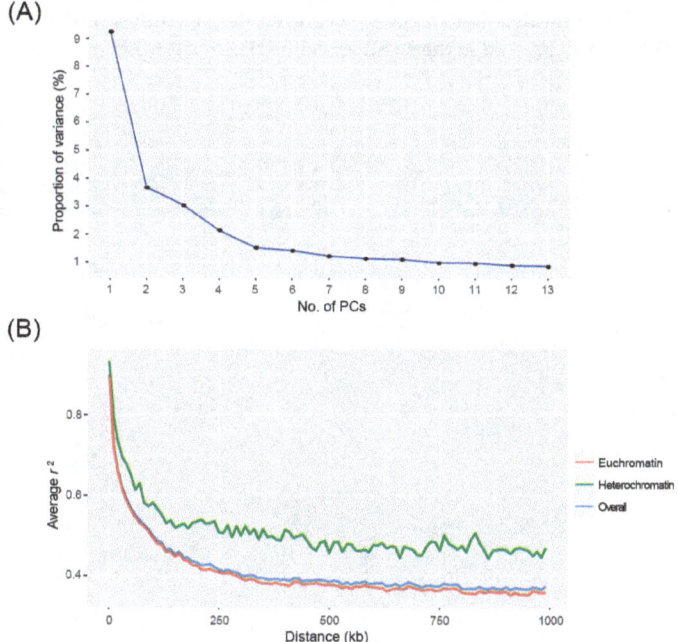

Figure 1. Principal components analysis (PCA) and linkage disequilibrium (LD) decay of the soybean core collection. (**A**) Scree plot showing the proportion of variance explained by principal components (PCs). (**B**) LD decay based on pairwise r^2 values. Blue line represents the overall rate of LD decay. Red and green lines indicate LD decay patterns of euchromatin and heterochromatin, respectively.

2.3. Determination of Genetic Association with Branch Numbers Using GWAS

To determine the best fitted GWAS model for branch numbers, quantile–quantile (QQ) plots generated using the generalized linear model (GLM) + population structure (Q) were compared with those generated using the mixed linear model (MLM) + Q (from principal component analysis; PCA) and kinship matrix (K) (Figure S3). The GLM + Q model showed a strong inflation of *p*-value (blue dots) compared with the MLM + Q (PCA) and K model (red dots), indicating erroneous inflation of false-positive signal. Thus, MLM + Q (PCA) and K model was more appropriate for the identification of QTNs associated with branch numbers in this study.

A total of five significant QTNs showing significant association with branch numbers were identified on Chr 6, 11, 12, and 20 (Table 1 and Figure 2A). The most highly significant QTN, *qtnBR12-1* at 38,057,780 base pair (bp) in the euchromatic region of Chr 12, explained 6.4% of the phenotypic variation in branch numbers. The second most highly significant QTN, *qtnBR6-1* at 20,663,101 bp in heterochromatin of Chr 6, accounted for 5.8% of the phenotypic variation (Figure 2B). Phenotypic variation in branch numbers explained by two other QTNs, *qtnBR11-1* and *qtnBR11-2* at 16,074,992 and 28,613,118 bp, respectively, in heterochromatin of Chr 11 was 5.0% and 5.6%, respectively. The last QTN, *qtnBR20-1*, located at 42,471,316 bp in euchromatin of Chr 20, accounted for 4.9% of the phenotypic variation in branch numbers.

On the basis of the rate of LD decay, we extended the chromosomal regions both upstream and downstream of the QTN positions, up to 140 kb for euchromatin and 460 kb for heterochromatin (Table 1). The LD blocks of *qtnBR6-1*, *qtnBR11-1*, and *qtnBR11-2* were adjacent to a major QTL *qBR6-1* and a minor QTL *qBR11-1* reported previously [8]. The *qtnBR6-1*, spanning 20 protein-coding genes (Figure 2C), overlapped with the QTL *qBR6-1*, which has been shown to play a major role in branch development. Using the publicly available soybean RNA-seq data [20], in silico expression profiling of all 20 genes in *qtnBR6-1* revealed that 13 genes, including the TCP transcription factor gene (*BRC1*; Glyma.06G210600), were expressed in the shoot apical meristem (SAM) (Figure 3A).

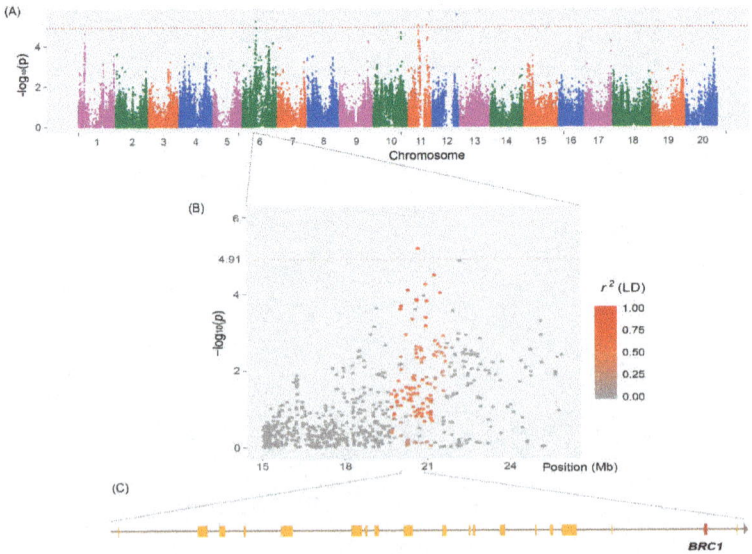

Figure 2. Genome-wide association study (GWAS) of branch numbers in the soybean core collection. (**A**) Genome-wide Manhattan plot of branch numbers. (**B**) LD region harboring *qtnBR6-1* on Chr 6. The pair-wise r^2 values between markers in LD are presented along a color gradient ranging from gray to red. The *p*-value threshold is indicated with red horizontal lines in Manhattan plots. (**C**) Protein-coding genes located in the LD block of *qtnBR6-1*.

Table 1. Quantitative trait nucleotides (QTNs) associated with branch numbers in the soybean core collection identified by genome-wide association study (GWAS).

QTN ID	Marker ID	Chr [a]	Marker Position (bp)	p-Value	Phenotypic R^2 (%)	Chromosomal Location	Linkage Disequilibrium Block	No. of Genes	Known QTL	Reference [b]
qtnBR6-1	AX-90305605	6	20,663,101	6.43×10^{-6}	5.8	Heterochromatin	20,433,101	20,893,101	20	*qBR6-1*	[8]
qtnBR11-1	AX-90512426	11	16,074,992	9.98×10^{-6}	5.0	Heterochromatin	15,844,992	16,304,992	13	*qBR11-1*	[8]
qtnBR11-2	AX-90472718	11	28,613,118	9.51×10^{-6}	5.6	Heterochromatin	28,383,118	28,843,118	23	*qBR11-1*	[8]
qtnBR12-1	AX-90419363	12	38,057,780	2.89×10^{-6}	6.4	Euchromatin	37,987,780	38,127,780	14		
qtnBR20-1	AX-90519199	20	42,471,316	8.88×10^{-6}	4.9	Euchromatin	42,401,316	42,541,316	13		

[a] Chr represents the soybean chromosome. [b] Reference represented reference article for previously reported quantitative trait loci (QTL).

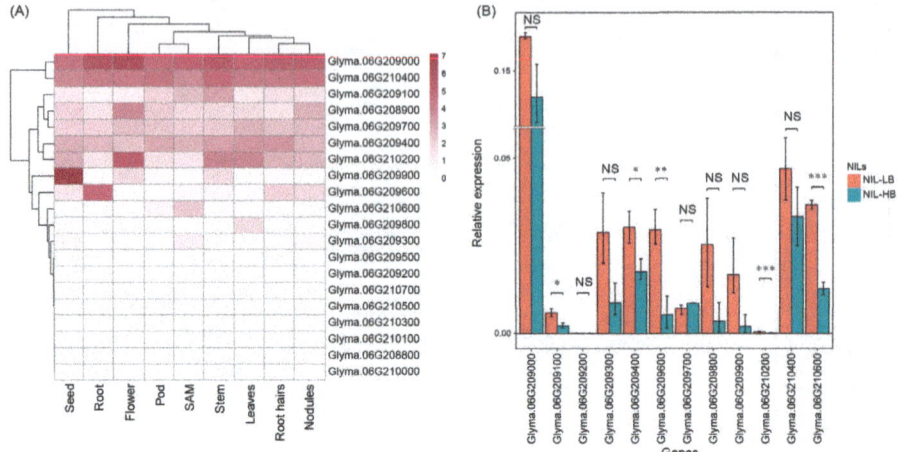

Figure 3. Expression patterns of genes located in the LD block of *qtnBR6-1*. (**A**) Heatmap showing the expression patterns of 20 genes obtained from the publicly available RNA-seq data. Gene expression values of $\log_2$(FPKM+1) were used to generate the heatmap. (**B**) Comparison of expression levels of selected genes in the shoot apical meristem between NILs with high-branching (HB) and low-branching (LB) alleles (NIL-HB and NIL-LB, respectively). The red and green bars indicate mean expression level of genes for nine samples (three biological replicates x three technical replicates) of each NIL (NIL-LB and NIL-HB), respectively. Black variance bars represent standard error of the mean (SEM). Statistically significant differences in relative gene expression between NIL-HB and NIL-LB are indicated with asterisks (*, $p < 0.05$; **, $p < 0.01$; ***, $p < 0.001$). NS, not significant.

2.4. Isogenicity and Phenotypic Differences Between NIL-HB and NIL-LB Associated with qBR6-1

To validate and narrow down *qtnBR6-1*, we explored a set of NILs carrying HB and LB alleles of *qBR6-1*. NIL-HB with more branches contained HB allele derived from paternal genotype SS0404-T5-76. NIL-LB carried LB allele from maternal genotype Jiyu69, showing fewer branches. The NILs and parental genotypes, SS0404-T5-76 and Jiyu69, were resequenced at an average depth of 31.6X (Table S2). On average, 92.4% of the paired-end reads were mapped to the soybean reference genome sequence. Within 895 Mb of consensus genome sequence of the NILs with at least 10X mapping depth, a total of 286,467 nucleotide variants were identified on all chromosomes except Chr 6 (carrier chromosome) between NIL-HB and NIL-LB, resulting in 99.97% isogenicity (Table S3). In addition, the non-QTL region on the carrier Chr 6 showed 99.89% isogenicity. The QTL region of *qBR6-1* on Chr 6 harbored 3798 nucleotide variants between NIL-HB and NIL-LB, of which 96.2% were shared with polymorphisms between parental genotypes SS0404-T5-76 and Jiyu69. Within the LD block of *qtnBR6-1*, SNPs segregating between NIL-HB and NIL-LB clustered in a 343 kb interval (20,555–20,898 kb) harboring 16 protein-coding genes (Figure 4A).

Additionally, we evaluated the effect of planting density on the branching performance of NIL-HB and NIL-LB in the field and greenhouse. The results showed a significant difference in the number of branches between NIL-HB and NIL-LB grown under low planting density (Table 2 and Figure S4). On average, NIL-HB plants had two more branches than NIL-LB plants. High planting density displayed no difference in branch numbers between NIL-HB and NIL-LB.

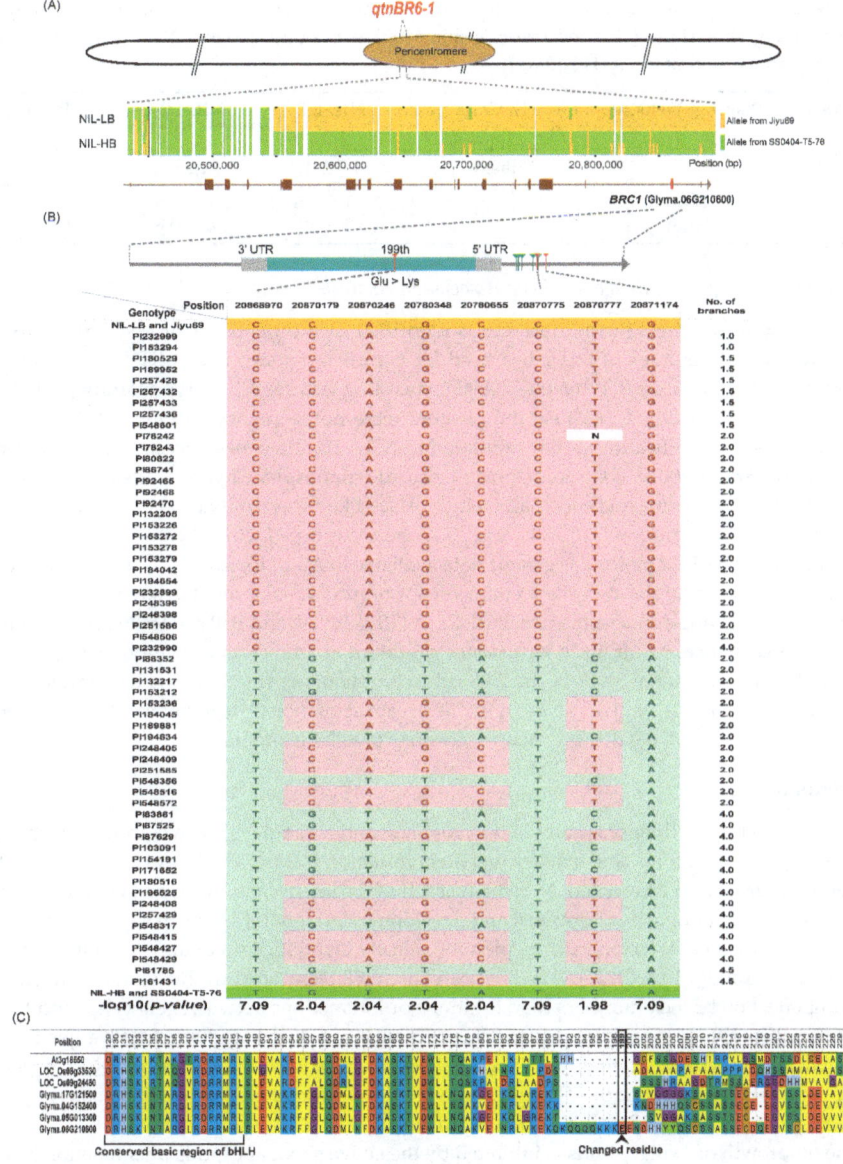

Figure 4. Allelic association of nucleotide variants in *BRC1* with branch numbers and protein sequence alignment of *BRC1*. (**A**) Physical location of *qtnBR6-1* in the pericentromeric region of Chr 6, and a cluster of single nucleotide polymorphism (SNPs) segregating between NIL-HB and NIL-LB. The SNPs between NIL-HB and NIL-LB within the LD block of *qtnBR6-1* are originated from SS0404-T5-76 (high-branching, green) and Jiyu69 (low-branching, orange), respectively. The X-axis indicates the genomic position (bp). (**B**) Allelic association of SNPs in *BRC1* with branch numbers among 59 United States Department of Agriculture (USDA) soybean accessions. Among eight SNPs, one missense mutation and two SNPs upstream of *BRC1*, indicated by red bars, were associated with branch numbers in 59 USDA soybean accessions. (**C**) Protein sequence alignment of *BRC1* orthologues. The missense mutation (glutamate to lysine) at amino acid position 199 is outlined with a black box.

Table 2. Effect of planting density on the branch number of soybean near-isogenic lines (NILs) carrying high-branching (HB) and low-branching (LB) alleles (NIL-HB and NIL-LB, respectively) at the quantitative trait loci (QTL) *qBR6-1*.

Year	Planting Density	Growth Condition	NIL-HB	NIL-LB	*p*-Value
2017	Low	Greenhouse	5.8 ± 1.7	3.4 ± 0.7	0.0001
		Field	14.5 ± 1.1	12.2 ± 2.1	0.019
2018	Low	Field	7.1 ± 1.4	5.1 ± 1.6	0.0003
	High		3.3 ± 1.9	2.6 ± 0.8	0.057

2.5. Candidate Gene Identification and Allelic Association Analysis

To narrow down the 13 candidate genes identified in the LD block of *qtnBR6-1*, cDNA was isolated from the shoot apex of NILs and used for expression analysis by qRT-PCR. Among these 13 genes, a multi-copy gene (Glyma.06G208900) encoding ATPase E1-E2 type family protein could not be analyzed by qRT-PCR because the primers were not sequence-specific. Of the remaining 12 genes, five were significantly down-regulated in NIL-HB; the genes encode MIZU-KUSSEI-like protein (Glyma.06G209100), P-loop containing nucleoside triphosphate hydrolases superfamily protein (Glyma.06G209400), adenine nucleotide alpha hydrolases-like superfamily protein (Glyma.06G209600), unknown protein (Glyma.06G210200), and TCP transcription factor (*BRC1*; Glyma.06G210600) (Figure 3B). On the basis of molecular genetic data available in Arabidopsis [21,22], *BRC1* was identified as the most promising candidate gene responsible for branch development in soybean.

Analysis of the nucleotide sequence of *BRC1* in NILs, Jily69, and SS0404-T5-76 revealed one SNP in the coding sequence, resulting in a missense mutation at amino acid position 199 (glutamate to lysine), and seven additional SNPs in the 2 kb upstream sequence (Figure 4B). Furthermore, the SNP in *BRC1* coding sequence and two SNPs in the upstream sequence were significantly associated with branch number in 59 USDA-GRIN soybean accessions (available online: http://www.ars-grin.gov).

3. Discussion

Shoot branching influences seed yield in soybean and is regulated by a complex mechanism of axillary bud outgrowth following axillary meristem initiation [21,23]. The fate of axillary buds, that is, whether to outgrow into a branch or to remain as a bud, is determined by an orchestrated regulatory process induced by endogenous hormonal and developmental signals [24]. Such a regulatory process is affected by environmental factors such as planting density, shading, light quality, soil nitrogen content, and soil water content [3,4,25–27]. The soybean core collection used in this study showed a more significant effect on branch numbers than locations or interactions between genotypes and locations. We identified five QTNs showing association with branch numbers based on BLUP values from three different locations (Table 1 and Figure 2). The LD block of one of these five QTNs, *qtnBR6-1*, co-localized with a previously reported major QTL, *qBR6-1*, which was validated based on the clustering of SNPs segregating between NIL-HB and NIL-LB (Figure 4A).

The outgrowth of axillary buds is inhibited by the active shoot apex; this phenomenon is referred to as apical dominance. Decapitation abolishes apical dominance and triggers the growth of one or more axillary buds because auxin, which is synthesized in the shoot apex, is mobilized to the lower parts of plants and inhibits branch outgrowth [28]. Considering the relevance of shoot apex to branch development, we compared the expression levels of 13 genes showing transcriptional activity in SAM using publicly available RNA-seq data (Figure 3A). Among the five significantly down-regulated genes in NIL-HB (Figure 3B), we identified *BRC1* (Glyma.06G210600) as the most promising candidate gene responsible for soybean shoot branching because *BRC1* functions as a regulatory hub that integrates hormonal signals and external stimuli for determining the fate of axillary buds [21,22]. In Arabidopsis, *BRC1* is expressed in axillary buds and SAM [21,29]. In pea (*Pisum sativum* L.), *PsBRC1* shows the highest expression level in lateral buds and is also expressed in the shoot apex [30]. In this study,

NIL-LB showed higher transcriptional activity of *BRC1* in the shoot apex than NIL-HB, although the expression of *BRC1* was not examined in axillary buds. Our data are consistent with previous studies showing that *BRC1* orthologues in rice (*Ostb1*) and maize (*tb1*) negatively regulate branch development [21,22,31,32].

The effects of planting density and shading on branch development are well-established in plants [3,5,21,33,34]. High planting density and shading have the same effect on light quality as they both reduce the ratio of red to far red light (R/FR) [24]. Shoot branching increases under high R/FR ratios caused by low planting density, but decreases under low R/FR ratios [3,5,21,33,34]. Similarly, in this study, NILs grown under high planting density produced fewer branches than those grown under low planting density, and showed no difference in branch numbers (Table 2). However, a significant difference in branch numbers was observed between NIL-HB and NIL-LB cultivated at low planting density. This indicates that the gene regulating branch number responds to low planting density. The light signal perceived by phytochrome B (PHYB), a photoreceptor, is transduced to the endogenous signal via *BRC1*, which functions as a molecular mediator [34]. Under shade or high planting density, an elevated FR signal inactivates PHYB by converting the active form of phytochrome (Pr) to the inactive form (Pfr) [35]. The inactive Pfr form of PHYB up-regulates *BRC1*, resulting in the inhibition of branch development [34,35]. These findings link the exogenous light signal with the endogenous molecular regulator in branch development, and suggest *GmBRC1* as the causal gene responsible for branch development in soybean. The function of *BRC1* is highly conserved across other plant species [21,30–32].

The *BRC1* gene and its orthologues are characterized by a highly conserved basic region of basic helix–loop–helix (bHLH) on a functional domain of TCP genes [21,31,32,36], and belong to CYCLODEA/TEOSINTE BRANCHED1 (CYC/TB1)-type TCP [37]. In this study, a missense mutation (glutamate to lysine) in the CYC/TB1-type TCP domain was identified between NIL-HB and NIL-LB, which showed a tight association with branch numbers in 59 USDA soybean accessions (Figure 4B). However, the altered amino acid residue was not located in the highly conserved basic region of bHLH, and was not conserved among other CYC/TB1-type TCP orthologues in Arabidopsis, rice, and soybean (Figure 4C). Thus, it is not clear if the difference in branch numbers between NIL-HB and NIL-LB could be attributed to the amino acid change at position 199. In addition to the SNP in *BRC1* coding sequence, two out of seven SNPs within the 2 kb upstream sequence of *BRC1* showed tight association with branch numbers. As both these SNPs are located in the putative promoter region of *BRC1*, they are predicted to affect *BRC1* expression. This is consistent with previous reports that *BRC1* regulates branching at the transcription level [21,31,32]; *BRC1* expression was down-regulated in NIL-HB (Figure 3B). An example similar to our results is of maize *tb1*; sequence variation in the upstream region of *tb1*, resulting in low expression, is associated with increased branch development in maize [31,38,39].

In conclusion, we propose *BRC1* (Glyma.06G210600) as the candidate gene regulating branch development in soybean. Further functional validation of these results by overexpression or knockout of *GmBRC1* is required for a thorough understanding of the regulatory mechanism of branch development in soybean, which will provide key insights into the complex genetic modules mediating branch development in soybean. Agronomically, soybean cultivars with optimal plant architecture depending on cultivation methods can be developed based on the allelic information of *BRC1* gene. In western countries including USA, soybean cultivars for high yield and mechanical harvesting can be improved by selecting genotypes with alleles contributing low branching phenotype. Besides, introgression of alleles responsible for high branching phenotype to other elite cultivars will enable breeding of high yielding soybean cultivars with high branch number and contribute to labor saving cultivation practice in Asian countries.

4. Materials and Methods

4.1. Plant Materials

A soybean core collection comprising 400 soybean genotypes with diverse origins was obtained from the National Agrobiodiversity Center in the Rural Development Administration (RDA, Jeonju, Korea) for GWAS (Table S4). To validate and narrow down the locus for branching, a set of NILs carrying HB and LB alleles of *qBR6-1* was developed from an RHL selected from the F$_6$ RIL population of Jiyu69 (low-branching) × SS0404-T5-76 (high-branching). Among five individuals in the progeny of RHL, plants showing the highest and lowest number of branches were selected as NILs carrying the HB and LB alleles, respectively, in 2016, and were designated as NIL-HB and NIL-LB, respectively. These NILs were genotyped using simple sequence repeat markers flanking *qBR6-1* (Table S5). We also used 59 USDA-GRIN soybean accessions with known branch numbers to confirm the allelic association of SNPs within a candidate gene between NIL-HB and NIL-LB.

The soybean core collection was grown in three different locations, namely, Wanju (35°50′27.384″ N, 127°2′46.1826″ E), Cheonan (36°49′49.2816″ N, 127°10′1.9122″ E), and Ochang (36°43′14.0982″ N, 127°26′1.1148″ E), in Korea in 2017. NIL-HB and NIL-LB were planted in a greenhouse and experimental field of Seoul National University, Suwon, Korea (37°16′12.094″ N, 126°59′20.756″ E). In the greenhouse, three plants of each line (NIL-HB and NIL-LB) were grown in a rectangular pot (64.3 cm × 23.0 cm × 16.9 cm). The plant-to-plant and row-to-row spacing was 20 and 80 cm, respectively, in low-density planting, and 10 and 40 cm, respectively, in high-density planting. All experiments were performed in triplicate, and field evaluation of branching in NILs was conducted in 2017 and 2018.

4.2. Phenotyping of Branch Number and Statistical Analysis

The number of branches generated on the main stem of the soybean core collection genotypes and NILs was evaluated in three biological replicates. Phenotypic differences between NILs were examined by analysis of variance (ANOVA) using the R software (available online: http://www.R-project.org). To minimize the effect of environmental factors on branch numbers at three different locations in GWAS, the BLUP value was predicted by the lme4 package of R, considering the variation among genotypes and locations [40]. The BLUP values were calculated according to the following equation:

$$Y_{ik} = \mu + G_i + L_k + GL_{ik} + e_{ik}$$

where Y_{ik} represents the phenotypic measurement, μ is the total mean, G_i is the genotypic effect of the ith genotype, L_k is the effect of the kth location, GL_{ik} represents interaction between genotype and location, and e_{ik} is the residual error. The BLUP values of each soybean genotype were calculated with random effect and used as phenotypes for GWAS. Broad-sense heritability (H^2) of branch numbers was calculated using the following equation:

$$H^2 = \frac{\sigma_g^2}{\left(\sigma_g^2 + \frac{\sigma_{gl}^2}{n} + \frac{\sigma_e^2}{nr}\right)}$$

where $\sigma_g{}^2$ represents genotypic variance, $\sigma_{gi}{}^2$ is the variance of interaction between genotype and location, $\sigma_e{}^2$ represents the variance of error components, n represents the number of locations, and r represents the number of replications.

4.3. Population Structure and LD Analysis

SNP genotypic data of the soybean core collection previously produced from the 180K Axiom® SoyaSNP array were explored for GWAS [41]. SNPs with minor allele frequency <0.05 and missing genotype >10% were excluded. The remaining 81,078 SNPs were used in STRUCTURE [18], PCA, kinship analysis, LD analysis, and GWAS. To stratify covariates (Q) from the population structure,

STRUCTURE and PCA analyses were applied. In STRUCTURE, burn-in and Markov chain Monte Carlo (MCMC) values were set at 10,000 and 100,000, respectively. The STRUCTURE analysis was carried out for number of sub-populations (*K*) values ranging from 1 to 13. To evaluate the optimum *K* for this population, the ΔK method was applied. The use of MLM with covariates from STRUCTURE produced erroneous inflation and false-positive signals. Therefore, population stratification was analyzed using PCA, which has been previously proposed as an alternative method for investigating relatedness [42]. Therefore, covariates implemented by PCA analysis in TASSEL v5.2 [43] were adopted for MLM analysis. To determine the optimal number of PCs, a scree plot was generated based on the proportion of variance explained by PCs. The kinship matrix (K) for soybean core collection was analyzed using TASSEL v5.2 [43].

LD was analyzed using PLINK software [19] with LD window length of 1 Mb and an unlimited number of variants within LD window (–r2 –ld-window-kb 1000 –ld-window 99999). Considering the different patterns of LD decay in heterochromatin and euchromatin, genomic regions specified as pericentromeric regions in Soybase (available online: http://soybase.org) were downloaded and used in the LD analysis. The rate of LD decay for the soybean core collection was measured in physical distance, where the average pair-wise r^2 value between alleles dropped to half of its maximum value.

4.4. GWAS

GWAS was conducted using TASSEL v5.2 [43]. Two statistical models, GLM + Q (from STRUCTURE) and MLM + Q (from PCA) and K, were considered. The Q and K were regarded as fixed and random effects in GLM and MLM models, respectively. Quantile–quantile plots of both models were compared for determining the best fit. The threshold *p*-value ($1/n$, where *n* is the number of SNPs (81,078)) was used for the identification of QTNs significantly associated with branch number.

4.5. Expression Patterns of Genes in LD Block of qtnBR6-1

The identified QTNs were extended based on the rate of LD decay, depending on the chromosomal region (euchromatin vs. heterochromatin). To investigate the expression patterns and levels of genes located within the LD block of the *qtnBR6-1*, RNA-seq data (fragments per kilobase of exon model per million mapped reads (FPKM) values) for nine tissues, including flower, leaf, nodule, pod, root, root hair, SAM, seed, and stem, of soybean cv. Williams 82 were obtained from Phytozome v12.0 (available online: https://phytozome.jgi.doe.gov/pz/portal.html) [20]; RNA-seq data for axillary buds were not available in the public database. A heatmap with hierarchical clustering of genes was constructed using the R package pheatmap for visualizing gene expression levels in nine tissues, based on the $\log_2(\text{FPKM} + 1)$ values. Only genes expressed in SAM were selected for further analysis.

4.6. Resequencing of NIL-HB, NIL-LB, and Parental Genotypes

The two NILs (NIL-HB and NIL-LB) and their parental genotypes, SS0404-T5-76 and Jiyu69, were resequenced. Raw sequence reads were mapped to the soybean reference genome (Wm82.a2) downloaded from Phytozome [20] using BWA [44], Samtools [45], and Vcftools [46]. Annotation of SNPs was conducted using SnpEff [47]. Nucleotide positions with more than ten supporting reads per genotype were analyzed further. SNPs segregating between NIL-HB and NIL-LB in the LD block of *qtnBR6-1* were compared with the sequence of SS0404-T5-76 and Jiyu69.

4.7. qRT-PCR Analysis of Candidate Genes in NILs

To determine the expression levels of selected genes in NILs, total RNA was extracted from the shoot apex (<3 mm) of NILs at the R1 stage using Ribospin™ Plant (GeneAll, Seoul, Korea), and cDNA was synthesized using Bio-Rad iScript™ cDNA Synthesis Kit (Hercules, CA, USA). Next, qRT-PCR was performed on a LightCycler® 480 (Roche Diagnostics, Laval, QC, Canada) using Bio-Rad iQ™ SYBR Green Supermix Kit. Primer sequences were designed using PRIMER3plus (available online: http://www.bioinformatics.nl/cgi-bin/primer3plus/primer3plus.cgi) [48]. Appropriate primer pairs that did not amplify orthologues of target genes were selected using a stand-alone version of electronic PCR [49] (Table S5). Each qRT-PCR reaction mixture (20 µL volume) contained 100 ng cDNA template, and 300 µM each of forward and reverse primer. Amplification was performed using the following conditions: initial denaturation at 95 °C for 5 min, followed by 40 cycles of denaturation at 95 °C for 10 s, and annealing and extension at 60 °C for 1 min. Three biological samples of each NIL were analyzed in triplicate to increase statistical power. The *ACTIN11* (*ACT11*) gene was used as a reference for data normalization. Normalized data were analyzed using the method of Livak and Schmittgen [50]. Statistical significance was analyzed using Fisher's least significant difference ($p < 0.05$) in R.

4.8. Analysis of BRC1 SNPs and Amino Acid Sequence

SNPs identified in the candidate gene *BRC1* based on a comparison between NIL-HB and NIL-LB were tested for association with branch number in a collection of 59 soybean accessions obtained from USDA-GRIN. Genomic DNA of each soybean accession was extracted using Exgene™ Plant SV mini DNA extraction kit (GeneAll, Seoul, Korea). On the basis of the identified SNPs between NIL-HB and NIL-LB, primers were designed using PRIMER3Plus (Table S5) [48] and validated using an electronic PCR algorithm [49]. PCR products were sequenced using ABI 3730XL DNA analyzer (Applied Biosystems, Foster, CA, USA). Branch numbers of soybean accessions were downloaded from the GRIN website. Association analysis between branch numbers and allelic variations was performed using ANOVA. Amino acid sequences of *BRC1* orthologues from Arabidopsis, rice, and soybean were aligned using MEGA7 [51].

Supplementary Materials: Supplementary materials can be found at http://www.mdpi.com/1422-0067/20/1/135/s1.

Author Contributions: Conceptualization, S.S., J.H., M.Y.K., and S.-H.L.; Methodology, S.S.; Validation, S.S.; Formal Analysis, S.S.; Investigation, S.S.; Resources, M.S.C., S.-T.K., S.-C.J., and J.-K.M.; Data Curation, S.S.; Writing—Original Draft Preparation, S.S.; Writing—Review and Editing, M.Y.K., J.H., and S.-H.L.; Visualization, S.S.; Supervision, M.Y.K., S.-C.J., and S.-H.L.; Project Administration, S.-H.L.; Funding Acquisition, S.-H.L.

Funding: This work was supported by the Next Generation BioGreen 21 Program (Code No. PJ01322401), Rural Development Administration, Republic of Korea.

Conflicts of Interest: The authors declare that there is no conflict of interest.

Abbreviations

BLUP	Best linear unbiased predictor/prediction
BRC1	BRANCHED1
GLM	Generalized linear model
GWAS	Genome-wide association study
LD	Linkage disequilibrium
MLM	Mixed linear model
NIL	Near-isogenic line
QTL	Quantitative trait locus
QTN	Quantitative trait nucleotide
RHL	Residual heterozygous line
RIL	Recombinant inbred line
SAM	Shoot apical meristem
SNP	Single nucleotide polymorphism
TCP	TEOSINTE BRANCHED1/CYCLODEA/PROLIFERATING CELL FACTOR 1 and 2

References

1. Heatherly, L.G.; Elmore, R.W. Managing Inputs for Peak Production. In *Soybeans: Improvement, Production, and Uses*; Agronomy Monograph 16; American Society of Agronomy, Inc.; Crop Science Society of America, Inc.; Soil Science Society of America, Inc.: Madison, WI, USA, 2004; pp. 451–536. [CrossRef]

2. Cho, Y.; Kim, S. Growth parameters and seed yield compenets by seeding time and seed density of non-/few branching soybean cultivars in drained paddy field. *Asian J. Plant Sci.* **2010**, *9*, 140–145.

3. Agudamu; Yoshihira, T.; Shiraiwa, T. Branch development responses to planting density and yield stability in soybean cultivars. *Plant Prod. Sci.* **2016**, *19*, 331–339. [CrossRef]

4. Board, J. Light Interception Efficiency and Light Quality Affect Yield Compensation of Soybean at Low Plant Populations. *Crop Sci.* **2000**, *40*, 1285–1294. [CrossRef]

5. Cox, W.J.; Cherney, J.H.; Shields, E. Soybeans Compensate at Low Seeding Rates but not at High Thinning Rates. *Agron. J.* **2010**, *102*, 1238–1243. [CrossRef]

6. Board, J.E.; Kahlon, C.S. Morphological Responses to Low Plant Population Differ Between Soybean Genotypes. *Crop Sci.* **2013**, *53*, 1109–1119. [CrossRef]

7. Sayama, T.; Hwang, T.-Y.; Yamazaki, H.; Yamaguchi, N.; Komatsu, K.; Takahashi, M.; Suzuki, C.; Miyoshi, T.; Tanaka, Y.; Xia, Z.; et al. Mapping and comparison of quantitative trait loci for soybean branching phenotype in two locations. *Breeding Sci.* **2010**, *60*, 380–389. [CrossRef]

8. Shim, S.; Kim, M.Y.; Ha, J.; Lee, Y.-H.; Lee, S.-H. Identification of QTLs for branching in soybean (*Glycine max* (L.) Merrill). *Euphytica* **2017**, *213*, 225. [CrossRef]

9. Yao, D.; Liu, Z.Z.; Zhang, J.; Liu, S.Y.; Qu, J.; Guan, S.Y.; Pan, L.D.; Wang, D.; Liu, J.W.; Wang, P.W. Analysis of quantitative trait loci for main plant traits in soybean. *Genet. Mol. Res.* **2015**, *14*, 6101–6109. [CrossRef]

10. Chen, Q.; Zhang, Z.; Liu, C.; Xin, D.; Qiu, H.; Shan, D.; Shan, C.; Hu, G. QTL Analysis of Major Agronomic Traits in Soybean. *Agr. Sci. China* **2007**, *6*, 399–405. [CrossRef]

11. Li, W.; Zheng, D.; Van, K.; Lee, S. QTL Mapping for Major Agronomic Traits across Two Years in Soybean (*Glycine max* L. Merr.). *J. Crop Sci. Biotech.* **2008**, *11*, 171.

12. Rafalski, J.A. Association genetics in crop improvement. *Curr. Opin. Plant Biol.* **2010**, *13*, 174–180. [CrossRef] [PubMed]

13. Zhao, K.; Tung, C.-W.; Eizenga, G.C.; Wright, M.H.; Ali, M.L.; Price, A.H.; Norton, G.J.; Islam, M.R.; Reynolds, A.; Mezey, J.; et al. Genome-wide association mapping reveals a rich genetic architecture of complex traits in *Oryza sativa*. *Nat. Commun.* **2011**, *2*, 467. [CrossRef] [PubMed]

14. Tian, F.; Bradbury, P.J.; Brown, P.J.; Hung, H.; Sun, Q.; Flint-Garcia, S.; Rocheford, T.R.; McMullen, M.D.; Holland, J.B.; Buckler, E.S. Genome-wide association study of leaf architecture in the maize nested association mapping population. *Nat. Genet.* **2011**, *43*, 159–162. [CrossRef] [PubMed]

15. Zhao, K.; Aranzana, M.J.; Kim, S.; Lister, C.; Shindo, C.; Tang, C.; Toomajian, C.; Zheng, H.; Dean, C.; Marjoram, P.; et al. An *Arabidopsis* Example of Association Mapping in Structured Samples. *PLoS Genet.* **2007**, *3*, e4. [CrossRef] [PubMed]

16. Cao, Y.; Li, S.; Wang, Z.; Chang, F.; Kong, J.; Gai, J.; Zhao, T. Identification of Major Quantitative Trait Loci for Seed Oil Content in Soybeans by Combining Linkage and Genome-Wide Association Mapping. *Front. Plant Sci.* **2017**, *8*. [CrossRef]

17. Zhang, J.; Song, Q.; Cregan, P.B.; Nelson, R.L.; Wang, X.; Wu, J.; Jiang, G.-L. Genome-wide association study for flowering time, maturity dates and plant height in early maturing soybean (*Glycine max*) germplasm. *BMC Genomics* **2015**, *16*, 217. [CrossRef] [PubMed]

18. Pritchard, J.K.; Stephens, M.; Donnelly, P. Inference of population structure using multilocus genotype data. *Genetics* **2000**, *155*, 945–959.

19. Purcell, S.; Neale, B.; Todd-Brown, K.; Thomas, L.; Ferreira, M.A.R.; Bender, D.; Maller, J.; Sklar, P.; de Bakker, P.I.W.; Daly, M.J.; et al. PLINK: A Tool Set for Whole-Genome Association and Population-Based Linkage Analyses. *Am. J. Hum. Genet.* **2007**, *81*, 559–575. [CrossRef]

20. Schmutz, J.; Cannon, S.B.; Schlueter, J.; Ma, J.; Mitros, T.; Nelson, W.; Hyten, D.L.; Song, Q.; Thelen, J.J.; Cheng, J.; et al. Genome sequence of the palaeopolyploid soybean. *Nature* **2010**, *463*, 178–183. [CrossRef]

21. Aguilar-Martínez, J.A.; Poza-Carrión, C.; Cubas, P. *Arabidopsis BRANCHED1* Acts as an Integrator of Branching Signals within Axillary Buds. *Plant Cell* **2007**, *19*, 458–472. [CrossRef]

22. Poza-Carrión, C.; Aguilar-Martínez, J.A.; Cubas, P. Role of TCP Gene *BRANCHED1* in the Control of Shoot Branching in Arabidopsis. *Plant Signal Behav.* **2007**, *2*, 551–552. [CrossRef] [PubMed]

23. Ghodrati, G. Study of genetic variation and broad sense heritability for some qualitative and quantitative traits in soybean (*Glycine max* L.) genotypes. *Curr. Opin. Agr.* **2013**, *2*.

24. Rameau, C.; Bertheloot, J.; Leduc, N.; Andrieu, B.; Foucher, F.; Sakr, S. Multiple pathways regulate shoot branching. *Front Plant Sci.* **2015**, *5*. [CrossRef] [PubMed]

25. Board, J.E.; Settimi, J.R. Photoperiod Effect before and after Flowering on Branch Development in Determinate Soybean. *Agron. J.* **1986**, *78*, 995–1002. [CrossRef]

26. Frederick, J.R.; Camp, C.R.; Bauer, P.J. Drought-Stress Effects on Branch and Mainstem Seed Yield and Yield Components of Determinate Soybean. *Crop Sci.* **2001**, *41*, 759–763. [CrossRef]

27. Linkemer, G.; Board, J.E.; Musgrave, M.E. Waterlogging effects on growth and yield components in late-planted soybean. *Crop Sci.* **1998**, *38*, 1576–1584. [CrossRef] [PubMed]

28. Thimann, K.V.; Skoog, F. Studies on the growth hormone of plants the inhibiting action of the growth substance on bud development. *Proc. Natl. Acad. Sci. USA* **1933**, *19*, 714–716. [CrossRef] [PubMed]

29. Niwa, M.; Daimon, Y.; Kurotani, K.; Higo, A.; Pruneda-Paz, J.L.; Breton, G.; Mitsuda, N.; Kay, S.A.; Ohme-Takagi, M.; Endo, M.; et al. BRANCHED1 Interacts with FLOWERING LOCUS T to Repress the Floral Transition of the Axillary Meristems in *Arabidopsis*[C][W][OA]. *Plant Cell* **2013**, *25*, 1228–1242. [CrossRef]

30. Braun, N.; Germain, A. de S.; Pillot, J.-P.; Boutet-Mercey, S.; Dalmais, M.; Antoniadi, I.; Li, X.; Maia-Grondard, A.; Signor, C.L.; Bouteiller, N.; et al. The Pea TCP Transcription Factor PsBRC1 Acts Downstream of Strigolactones to Control Shoot Branching. *Plant Physiol.* **2012**, *158*, 225–238. [CrossRef]

31. Doebley, J.; Stec, A.; Hubbard, L. The evolution of apical dominance in maize. *Nature* **1997**, *386*, 485–488. [CrossRef]

32. Takeda, T.; Suwa, Y.; Suzuki, M.; Kitano, H.; Ueguchi-Tanaka, M.; Ashikari, M.; Matsuoka, M.; Ueguchi, C. The *OsTB1* gene negatively regulates lateral branching in rice. *Plant J.* **2003**, *33*, 513–520. [CrossRef] [PubMed]

33. Finlayson, S.A. Arabidopsis TEOSINTE BRANCHED1-LIKE 1 Regulates Axillary Bud Outgrowth and is Homologous to Monocot TEOSINTE BRANCHED1. *Plant Cell Physiol.* **2007**, *48*, 667–677. [CrossRef] [PubMed]

34. Kebrom, T.H.; Brutnell, T.P.; Finlayson, S.A. Suppression of sorghum axillary bud outgrowth by shade, phyB and defoliation signalling pathways. *Plant Cell Environ.* **2010**, *33*, 48–58. [CrossRef]

35. Minakuchi, K.; Kameoka, H.; Yasuno, N.; Umehara, M.; Luo, L.; Kobayashi, K.; Hanada, A.; Ueno, K.; Asami, T.; Yamaguchi, S.; et al. *FINE CULM1* (*FC1*) Works Downstream of Strigolactones to Inhibit the Outgrowth of Axillary Buds in Rice. *Plant Cell Physiol.* **2010**, *51*, 1127–1135. [CrossRef] [PubMed]

36. Cubas, P.; Lauter, N.; Doebley, J.; Coen, E. The TCP domain: a motif found in proteins regulating plant growth and development. *Plant J.* **1999**, *18*, 215–222. [CrossRef] [PubMed]

37. Sorefan, K.; Booker, J.; Haurogné, K.; Goussot, M.; Bainbridge, K.; Foo, E.; Chatfield, S.; Ward, S.; Beveridge, C.; Rameau, C.; et al. *MAX4* and *RMS1* are orthologous dioxygenase-like genes that regulate shoot branching in *Arabidopsis* and pea. *Genes Dev.* **2003**, *17*, 1469–1474. [CrossRef]

38. Clark, R.M.; Linton, E.; Messing, J.; Doebley, J.F. Pattern of diversity in the genomic region near the maize domestication gene *tb1*. *Proc. Natl. Acad. Sci. USA* **2004**, *101*, 700–707. [CrossRef]

39. Clark, R.M.; Wagler, T.N.; Quijada, P.; Doebley, J. A distant upstream enhancer at the maize domestication gene *tb1* has pleiotropic effects on plant and inflorescent architecture. *Nat. Genet.* **2006**, *38*, 594–597. [CrossRef]

40. Bates, D.; Sarkar, D. lme4: Linear mixed-effects models using S4 classes. 2007.

41. Lee, Y.-G.; Jeong, N.; Kim, J.H.; Lee, K.; Kim, K.H.; Pirani, A.; Ha, B.-K.; Kang, S.-T.; Park, B.-S.; Moon, J.-K.; et al. Development, validation and genetic analysis of a large soybean SNP genotyping array. *Plant J.* **2015**, *81*, 625–636. [CrossRef]

42. Price, A.L.; Patterson, N.J.; Plenge, R.M.; Weinblatt, M.E.; Shadick, N.A.; Reich, D. Principal components analysis corrects for stratification in genome-wide association studies. *Nat. Genet.* **2006**, *38*, 904–909. [CrossRef]

43. Bradbury, P.J.; Zhang, Z.; Kroon, D.E.; Casstevens, T.M.; Ramdoss, Y.; Buckler, E.S. TASSEL: software for association mapping of complex traits in diverse samples. *Bioinformatics* **2007**, *23*, 2633–2635. [CrossRef] [PubMed]

44. Li, H.; Durbin, R. Fast and accurate short read alignment with Burrows-Wheeler transform. *Bioinformatics* **2009**, *25*, 1754–1760. [CrossRef] [PubMed]

45. Li, H.; Handsaker, B.; Wysoker, A.; Fennell, T.; Ruan, J.; Homer, N.; Marth, G.; Abecasis, G.; Durbin, R. 1000 Genome Project Data Processing Subgroup The Sequence Alignment/Map format and SAMtools. *Bioinformatics* **2009**, *25*, 2078–2079. [CrossRef] [PubMed]

46. Danecek, P.; Auton, A.; Abecasis, G.; Albers, C.A.; Banks, E.; DePristo, M.A.; Handsaker, R.E.; Lunter, G.; Marth, G.T.; Sherry, S.T.; et al. The variant call format and VCFtools. *Bioinformatics* **2011**, *27*, 2156–2158. [CrossRef]

47. Cingolani, P.; Platts, A.; Wang, L.L.; Coon, M.; Nguyen, T.; Wang, L.; Land, S.J.; Lu, X.; Ruden, D.M. A program for annotating and predicting the effects of single nucleotide polymorphisms, SnpEff: *SNPs in the genome of Drosophila melanogaster strain w1118; iso-2; iso-3. Fly (Austin)* **2012**, *6*, 80–92. [CrossRef] [PubMed]

48. Untergasser, A.; Nijveen, H.; Rao, X.; Bisseling, T.; Geurts, R.; Leunissen, J.A.M. Primer3Plus, an enhanced web interface to Primer3. *Nucleic Acids Res.* **2007**, *35*, W71–W74. [CrossRef] [PubMed]

49. Rotmistrovsky, K.; Jang, W.; Schuler, G.D. A web server for performing electronic PCR. *Nucleic Acids Res.* **2004**, *32*, W108–W112. [CrossRef]

50. Livak, K.J.; Schmittgen, T.D. Analysis of Relative Gene Expression Data Using Real-Time Quantitative PCR and the 2−ΔΔCT Method. *Methods* **2001**, *25*, 402–408. [CrossRef]

51. Kumar, S.; Stecher, G.; Tamura, K. MEGA7: Molecular Evolutionary Genetics Analysis Version 7.0 for Bigger Datasets. *Mol. Biol. Evol.* **2016**, *33*, 1870–1874. [CrossRef]

Article

Genome-Wide Association Studies of 39 Seed Yield-Related Traits in Sesame (*Sesamum indicum* L.)

Rong Zhou [1,†], Komivi Dossa [1,2,†], Donghua Li [1], Jingyin Yu [1], Jun You [1], Xin Wei [1,3,*] and Xiurong Zhang [1,*]

1   Key Laboratory of Biology and Genetic Improvement of Oil Crops, Oil Crops Research Institute of the Chinese Academy of Agricultural Sciences, Ministry of Agriculture, No. 2 Xudong 2nd Road, Wuhan 430062, China; rongzzzzzz@126.com (R.Z.); dossakomivi@gmail.com (K.D.); ldh360681@163.com (D.L.); yujingyin@caas.cn (J.Y.); youjunbio@163.com (J.Y.)
2   Centre d'Etude Régional Pour l'Amélioration de l'Adaptation à la Sécheresse (CERAAS), Route de Khombole, Thiès, Thiès Escale Thiès BP3320, Senegal
3   College of Life and Environmental Sciences, Shanghai Normal University, Shanghai 200234, China
*   Correspondence: weixin@caas.cn (X.W.); zhangxr@oilcrops.cn (X.Z.);
    Tel.: +86-21-6432-2008 (X.W.); +86-27-8681-1836 (X.Z.)
†   These authors contributed equally to this work.

Received: 23 August 2018; Accepted: 13 September 2018; Published: 17 September 2018

Abstract: Sesame is poised to become a major oilseed crop owing to its high oil quality and adaptation to various ecological areas. However, the seed yield of sesame is very low and the underlying genetic basis is still elusive. Here, we performed genome-wide association studies of 39 seed yield-related traits categorized into five major trait groups, in three different environments, using 705 diverse lines. Extensive variation was observed for the traits with capsule size, capsule number and seed size-related traits, found to be highly correlated with seed yield indexes. In total, 646 loci were significantly associated with the 39 traits ($p < 10^{-7}$) and resolved to 547 quantitative trait loci QTLs. We identified six multi-environment QTLs and 76 pleiotropic QTLs associated with two to five different traits. By analyzing the candidate genes for the assayed traits, we retrieved 48 potential genes containing significant functional loci. Several homologs of these candidate genes in *Arabidopsis* are described to be involved in seed or biomass formation. However, we also identified novel candidate genes, such as *SiLPT3* and *SiACS8*, which may control capsule length and capsule number traits. Altogether, we provided the highly-anticipated basis for research on genetics and functional genomics towards seed yield improvement in sesame.

Keywords: sesame; genome-wide association study; yield; QTL; candidate gene

---

1. Introduction

The use of high-quality oil in human daily food intake is an important part of overall well-being. Sesame (*Sesamum indicum* L.) is a source of an excellent vegetable oil rich in vital minerals, vitamins, phytosterols, polyunsaturated fatty acids, tocopherols and unique classes of lignans such as sesamin and sesamolin, which have been identified as beneficial compounds for human health [1]. Moreover, its seeds have one of the highest oil contents (55%) among major oilseed crops, as well as a high protein content [2]. The world population is growing fast and the demand for vegetable oil in quantity and high-quality is pressing. Vegetable oil consumption is expected to double by 2040 [3]. Therefore, sesame can play a significant role in satisfying this demand.

Sesame is essentially a small-scale farmer crop and its cultivation offers two main advantages: it is a very rewarding crop because of its low production cost and high sale price; and, it is also a very resilient crop, able to provide yield and generate incomes in marginal areas where many other crops cannot grow [4,5]. Over the last decade, the production of sesame seeds has doubled and the growing area has extended to more than 50 countries in the world, showing an ever-increasing interest in this crop [6]. However, sesame has a very low seed yield capacity compared to other oilseed crops [7]. According to the Food and Agriculture Organization, the average seed yield of sesame was only 578 kg/ha in 2016, ranked as the second lowest among the major oil crops [6]. Therefore, understanding the genetic basis of seed yield-related traits and applying that knowledge in sesame breeding programs might be instrumental in developing stable high-yielding sesame varieties.

The yield of any crop is a complex character, which depends upon many independent contributing components. Deep understanding of the relationship between yield and its components is crucial to the selection process and to crop improvement [8]. Sesame seed yield per plant is considered to mainly have three components, namely, the number of capsules per plant, the number of seeds per capsule and seed weight. Some other factors, including plant height, capsule dimensions, the first capsule axis height and the number of internodes, were found to be strongly associated with seed yield in sesame [9,10]. In addition, the plant growth habit, branching type, capsule shattering, management practices, and biotic and environmental factors can significantly affect sesame yield [11]. Beside the variation among cultivars for seed yield components, the within-plant variation is extremely important. For example, some sesame cultivars can have three or more capsules per leaf axil. Mosjidis and Yermanos [12] observed that seed weight from medial capsules is higher than that from lateral capsules. Moreover, Tashiro et al. [13] and later Kumazaki et al. [14] confirmed the significant differences between seed weight between capsules from nodes located at different positions along the main stem within the same plant. Accordingly, dissecting the genetic basis of the seed yield components in sesame may be challenging and will need meticulous analysis of the multiple and complex seed yield components.

Thirteen quantitative trait loci (QTL) were detected for seven seed yield-related traits using the linkage mapping approach in sesame [10]. Genome-wide association study (GWAS) has proven to be advantageous over bi-parental QTL mapping as it captures greater diversity and offers higher resolution for gene and favorable allele discovery in several plant species [15]. Recently, GWAS was also successfully applied to sesame to unravel the genetic basis of the oil production and quality traits, yield related traits, important agronomic traits, as well as salt and drought tolerance [16,17]. The objective of the hereby study was to employ the GWAS approach to comprehensively decipher the genetic basis of 39 seed yield-related traits in sesame and unlock potential alleles and genes for seed yield improvement based on a large and diverse sample phenotyped in three different environments.

2. Results

2.1. Variability and Correlation of the Seed Yield-Related Traits in the Sesame Association Panel

A total of 39 direct and indirect seed yield-related traits were studied and classified into five main trait groups: yield index, seed traits, capsule number, capsule size, and capsule pericarp (Table S1). Ten yield-related traits that were investigated in the previous research of Wei et al. [16] were also included in this study. Descriptive statistics for the traits across the 705 accessions included in this study are listed in the Table S2. Overall, the sesame diversity panel exhibited extensive trait variation across the three environments analyzed (Figure 1 and Figure S1). We selected three contrasting environments for phenotyping (Nanning (NN), Wuhan (WH) and Sanya (SY)) because they represent natural sesame growing areas in China and also cover different geographical regions of China: Central China (WH), South China (SY), Southwest China (NN). The traits appeared to be slightly higher at NN environment compared with WH and SY, but overall the yields are similar among the three locations. Some traits, especially those related to the capsule number and capsule size groups, were stable across environments; however, the traits belonging to the yield index group displayed a high variation. This observation was further confirmed with the broad-sense heritability estimates (Table S2). Generally, a large portion of the phenotypic variance in seed yield components could be attributed to the genotypic effects in sesame.

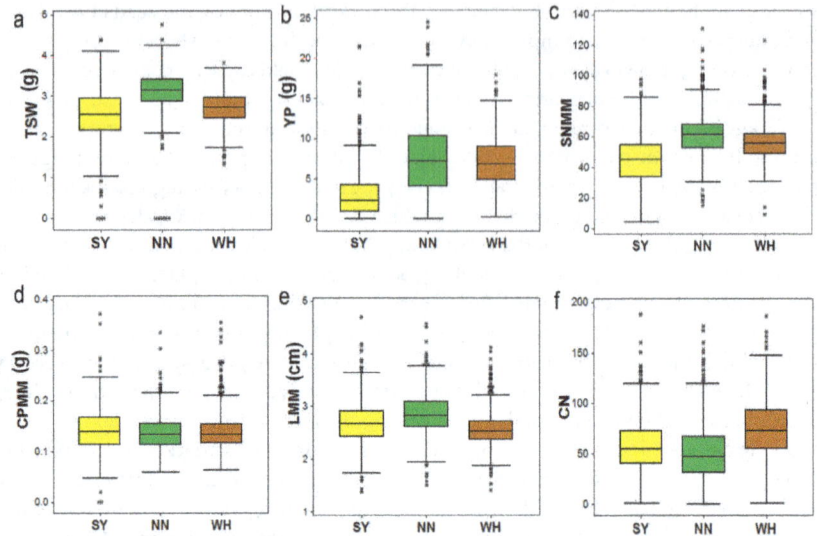

Figure 1. Boxplots displaying variation of six traits across three different environments (SY = Sanya, NN = Nanning and WH = Wuhan). Definition of the labels can be found at the end of this article.

To gain insight into the relationship between the seed yield-related traits, a clustering and correlation analysis was performed (Figure 2). It can be obviously observed that traits from the same group clustered closely, indicating strong correlations with each other. Furthermore, clustering analysis of the phenotype data highlighted three main groups (A, B and C). Group A comprised capsule number (MCNM, CN, MCNB, CNB and LCNB) and yield index (YMB and YB) related traits, which were strongly and positively correlated. This result shows that a high capsule number in a sesame plant leads to a high yield. The second group (B) was composed of mixed traits in relation to yield index, seed traits, and capsule size. From such a cluster, we inferred that accessions with high ratios of seed weight/capsule weight are likely to have a high yield. In addition, we found that high values of seed number and seed weight-traits are favorable for seed yield in sesame. Finally, Group C clustered some capsule pericarp and capsule size-related traits with moderate correlation values. Since no yield index trait was observed in this group, we concluded that it may not directly contribute to seed yield in sesame. More importantly, we found that traits from this group were negatively correlated with traits contributing to a high seed yield in sesame. For example, accessions with high capsule pericarp thickness have lower yield indexes.

2.2. Genetic Variants Associated with Seed Yield-Related Traits in Sesame

To predict significant marker-trait associations for seed yield-related traits, the mixed model was implemented in this study of the phenotype data from each environment. Genome wide association studies (GWAS) revealed 646 statistically significant loci ($p < 10^{-7}$) across the three environments associated with the 39 traits. A total of 6% of the loci were in line with the previous identified yield-related loci [16]. Significant loci were found on all of the 16 linkage groups (LG) of the genome, justifying the complex genetic architecture of the seed yield in sesame. The highest number of significant loci (86) was detected on the LG5, while the LG14 harbored only six significant loci (Table S3, Figure S2). The phenotypic variation explained by the lead loci ranged from 6.01 (SNP2372143) to 17.9% (SNP6737753 and SNP5479753), suggesting a moderate contribution to the traits (Table 1). We defined as a QTL the 88 kb region (corresponding to the linkage disequilibrium (LD) window) surrounding the peak loci and containing at least three significant loci [17]. By combining peak single nucleotide polymorphism (SNP)-trait-environment, a total of 547 QTLs were identified (Figure 3). Furthermore, by comparing peak loci through environments and traits, we uncovered six stable QTLs (detected in different environments for the same trait) and 76 pleiotropic QTLs associated with two to five various traits (Table 1). We compared the detected pleiotropic QTLs between the five groups of traits defined in this study. The results showed that most of the pleiotropic QTLs principally controlled traits from the same group (Figure 4). Few common QTLs could be observed between pairs of trait groups and there was no shared QTL for more than three traits groups. Overall, these results corroborate the phenotypic relationships observed in Figure 2. For example, there is no common QTL for the capsule pericarp and yield index groups; similarly for the capsule size and yield index groups. Conversely, the trait groups related to the yield index and capsule number exhibited the highest number of common QTLs (6), demonstrating that these groups shared similar genetic architectures. The examples presented in Figures 5 and 6, related to the trait-association for the effective capsule number in the main stem (CNM) and length of medial capsule in the main stem (LMM) of the three environments, highlight two stable QTLs detected on LG5 for CNM and LG11 for LMM. Overall, more significant loci were discovered in SY compared to the other environments.

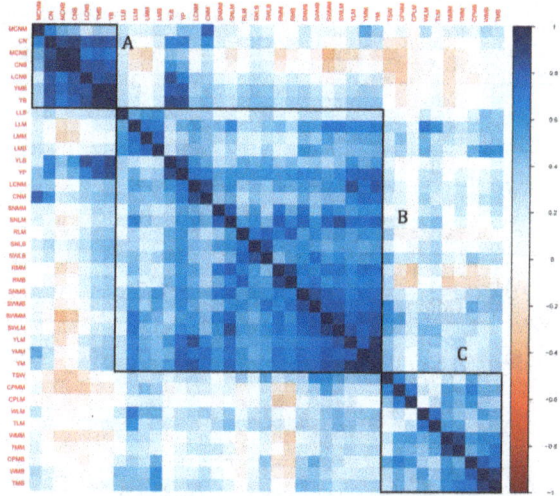

Figure 2. Correlation between all the seed yield-related traits in sesame. Blue color depicts positive correlation while red color means negative correlation. A, B and C correspond to the clusters of traits. Definition of the labels can be found at the end of this article.

Table 1. SNPs stably detected in different environments and for various traits.

LG	Position (bp)	Env.	Traits	PVE (%)	LG	Position (bp)	Env.	Traits	PVE (%)
1	1,700,170	SY	CNB	7.00	5	17,411,684	SY	SNMM	7.29
			CN	7.10				SWMM	6.79
1	1,994,183	SY	YB	8.68				YB	10.65
			YMB	7.41	6	3,404,764	SY	MCNB	7.33
1	4,450,107	SY	YB	7.55				YMB	11.15
			YMB	6.99				TWB	7.13
1	6,149,415	SY	TMM	7.61	6	3,790,583	SY	WMB	8.28
			WMM	6.58				YB	10.91
1	8,185,969	SY	YB	7.03	6	5,995,560	SY	CNB	6.99
			YMB	7.26				YMB	11.32

Table 1. *Cont.*

LG	Position (bp)	Env.	Traits	PVE (%)	LG	Position (bp)	Env.	Traits	PVE (%)
1	9,906,190	SY	YB	6.67	6	9,021,538	SY	YB	8.12
			YMB	6.88				YMB	8.34
1	11,118,941	SY	SNMB	6.32	6	9,971,772	NN	TMM	6.29
			SWMB	6.50				WMM	6.39
1	17,291,730	SY	YB	8.73	6	14,154,329	SY	YB	7.29
			YMB	9.10				YMB	6.59
2	1,201,448	SY	YB	8.76	6	14,581,641	SY	LCNM	10.29
			YMB	9.15				CNM	6.41
2	5,260,400	SY	YB	6.41	6	14,701,957	WH	TMM	7.09
			YB	7.08				WMM	6.96
			YMB	7.87	6	15,551,496	SY	RMM	7.63
2	6,057,670	NN	TMM	9.46				SNMM	6.13
			WMM	11.33	6	21,992,131	SY	YB	10.89
2	7,236,995	SY	YB	6.09				YMB	10.77
			YMB	6.11	7	6,763,527	SY	RMB	6.69
2	8,388,879	NN	TWB	6.32				SWMB	7.65
			TWB	7.24	7	1,702,826	NN	CNM	9.42
2	9,244,103	SY	TMM	6.04			WH	CNM	6.66
		WH	WMM	9.85	8	1,398,196	SY	YB	11.12
			TMM	9.25				YMB	11.99
2	11,245,765	SY	TMM	7.95	8	1,668,572	SY	YB	7.38
			WMM	7.56				SNMM	9.89
2	15,016,082	SY	YB	7.51				YM	6.66
			YMB	7.78				SWMM	8.84
2	17,451,873	SY	SNMB	6.47	8	21,325,953	SY	YB	8.97
			RMM	9.17				YMM	7.68
			SWMM	14.89	9	1,007,867	SY	RMM	8.57
			SWMB	6.83				SNMM	8.58
			SNMM	15.13	9	954,526	WH	SNMM	7.48
3	4,840,197	SY	LCNM	6.12				WMM	8.03
			CNM	6.42				TMM	7.80
3	13,198,513	SY	YB	6.68	10	1,647,805	SY	CNB	12.64
			YMB	7.24				CN	6.14
3	14,990,430	SY	YB	8.68	10	3,823,922	SY	CNB	7.43
			CNB	9.34				MCNB	6.85
			MCNB	9.40	10	7,418,158	NN	TMM	6.24
			YMB	8.81				WMM	6.31
3	16,939,689	SY	YB	6.70	10	8,305,398	SY	YB	7.76
			YMB	6.63				YMB	7.89
3	20,410,997	SY	TMM	6.47	10	10,792,029	SY	YB	7.29
			WMM	6.79				YMB	6.60
3	20,876,555	SY	YB	10.76	10	12,008,065	SY	RMM	6.83
			YMB	11.09				SNMM	6.61

Table 1. *Cont.*

LG	Position (bp)	Env.	Traits	PVE (%)	LG	Position (bp)	Env.	Traits	PVE (%)
3	20,878,243	SY	CNB	8.06	10	14,650,964	SY	YB	6.81
			MCNB	8.44				YMM	7.50
3	24,164,350	SY	TMM	6.84	10	15,097,365	SY	TMM	7.41
			WMM	6.43				WMM	6.90
4	2,505,014	SY	YB	7.98	11	6,996,833	SY	SNMM	6.53
			YMB	7.82				SWMM	7.78
4	6,419,408	SY	WMM	6.99	11	11,923,935	SY	YB	7.10
			CPMM	6.25				YMB	7.77
4	14,211,075	SY	YB	7.47	11	14,876,966	SY	YB	6.33
			YMB	6.10				YMB	6.91
5	202,984	SY	YB	9.50	11	15,137,600	SY	TMM	7.52
			YMM	6.92				WMM	6.55
5	2,854,336	NN	YM	7.85	12	328,609	SY	YM	7.53
			CNM	8.20				YMM	7.78
5	5,479,753	SY	YB	7.61	12	2,356,955	SY	YB	6.24
			CNB	8.89				YMB	6.72
			YMB	7.81	12	4,200,237	SY	YB	6.60
			CNM	13.68				YMB	7.21
		NN	CNM	17.90	12	4,895,688	SY	SNMM	8.67
5	6,737,753	SY	CNB	8.89				SWMM	6.18
			CNM	13.68	13	2,772,629	SY	YB	8.17
		NN	CNM	17.90				YMB	8.49
5	6,738,735	NN	LCNM	7.01	14	194,410	SY	YB	6.20
								YMB	6.70
			CNM	13.30	15	2,174,040	SY	YB	7.20
		SY	LCNM	13.03				YMB	7.22
		WH	CNM	14.79	15	2,372,143	WH	YB	6.29
5	6,757,688	NN	SNMB	8.02				CNM	7.03
			SWMB	10.88				YM	6.01
5	9,869,746	NN	RMB	6.54	15	3,989,016	SY	YB	6.72
			SWMB	6.08				YMB	6.69
5	11,806,702	SY	TWB	6.29	16	555,771	WH	TMM	7.55
			TWB	6.34				WMM	6.40
5	15,855,382	WH	SNMM	6.76	16	1,633,469	SY	YB	7.97
			TMM	9.11				YMB	8.17
			WMM	8.68	16	2,989,809	SY	CNB	9.26
		NN	WMM	6.23				MCNB	8.26
5	17,340,920	SY	CNB	8.64					
			MCNB	10.45					

LG = Linkage group; Env. = Environment; PVE = Phenotypic variance explained; SY = Sanya; NN = Nanning; WH = Wuhan.

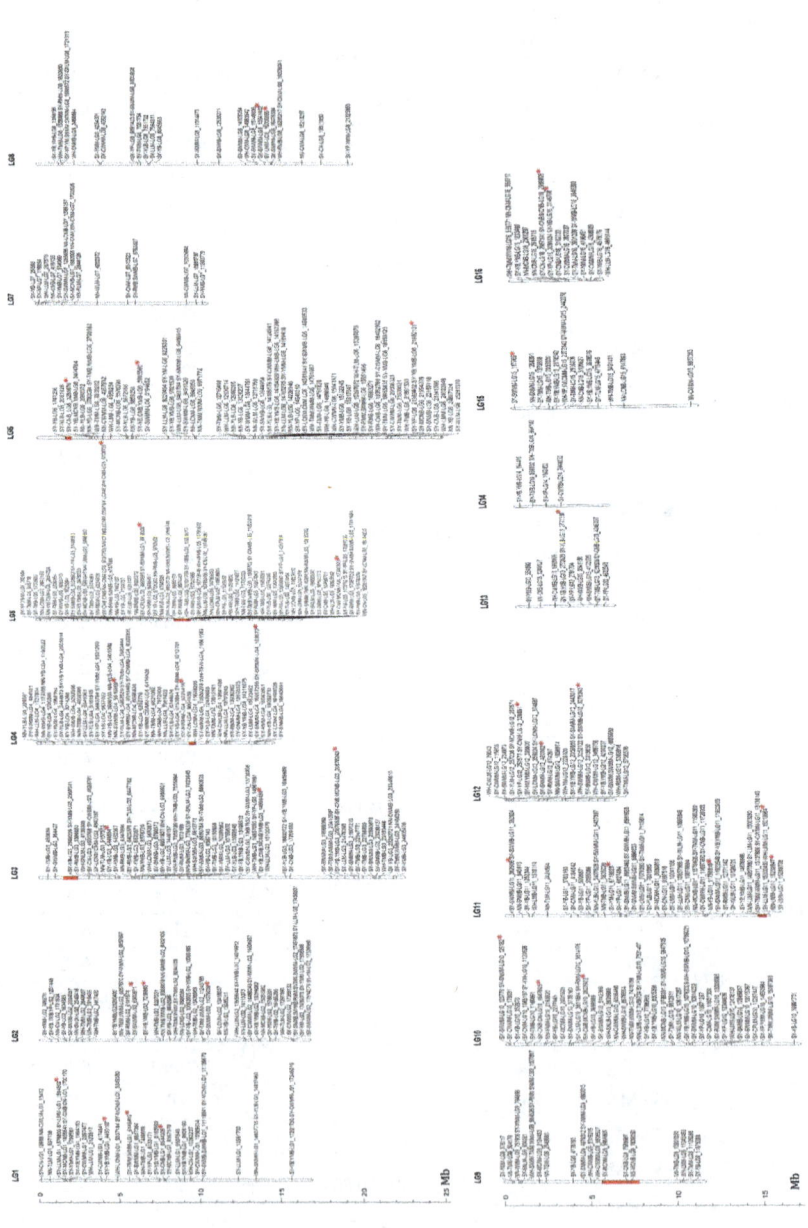

Figure 3. Genomic location of the 547 QTLs identified for seed yield-related traits in sesame. QTLs were named as follow: ENVIRONMENT-TRAIT-LINKAGEGROUP_POSITION. Bars represent the linkage groups of sesame genome. Red portions of the bars represent the previous QTLs detected by Wu et al. [10]. Red stars represent loci previously detected by Wei et al. [16]. Definition of the labels can be found at the end of this article.

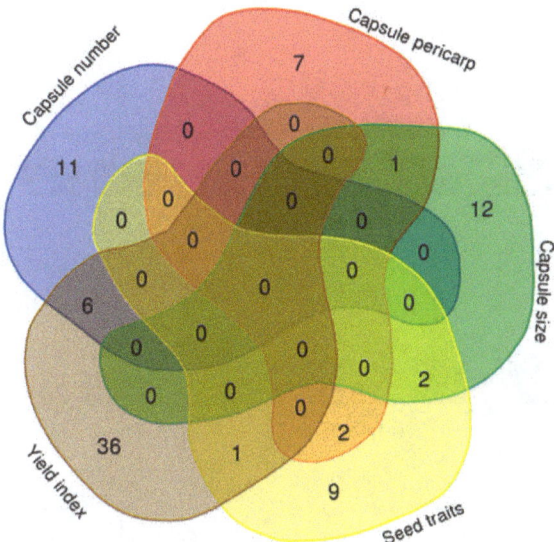

Figure 4. Venn diagram depicting the shared and common QTLs between five groups of seed yield-related traits analyzed in this study.

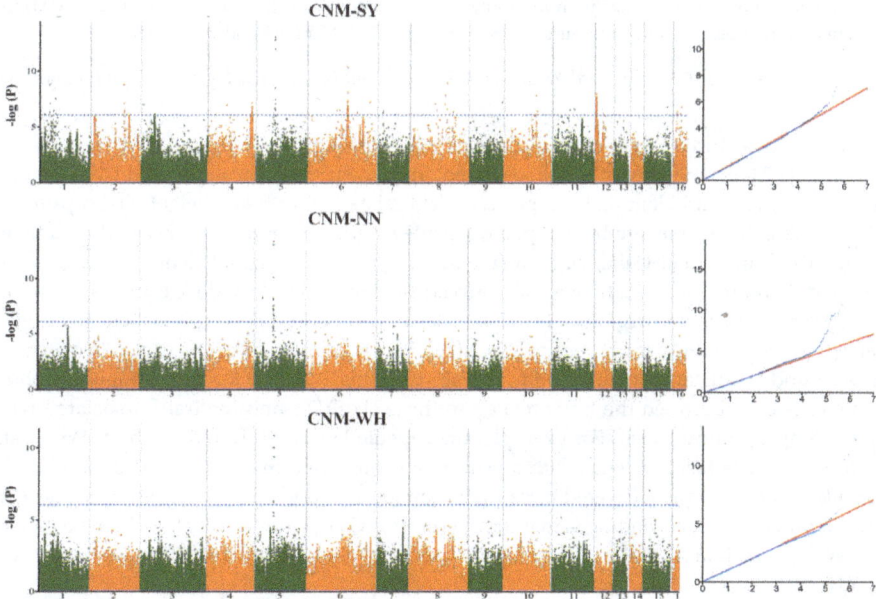

Figure 5. Genome-wide association mapping of effective capsule number in main stem (CNM) in sesame from three different environments (SY = Sanya, NN = Nanning and WH = Wuhan).

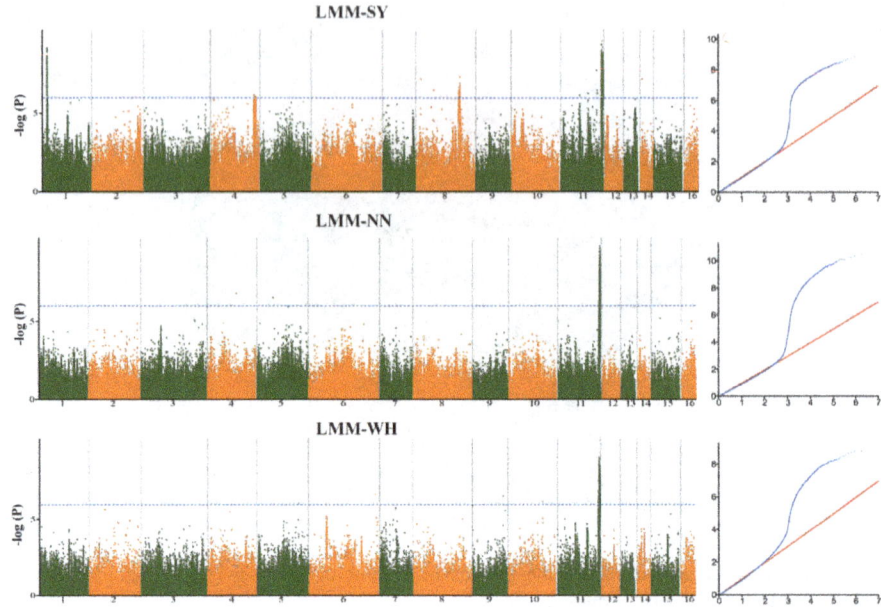

Figure 6. Genome-wide association mapping of length of medial capsule in main stem (LMM) in sesame from three different environments (SY = Sanya, NN = Nanning and WH = Wuhan).

2.3. Comparing Previous QTLs on Seed Yield-Related Traits from Bi-Parental Linkage Mapping with Our GWAS Results

In a previous study, Wu et al. [10] constructed a high-density genetic map of sesame using a population of 224 recombinant inbred lines based on the restriction-site associated DNA sequencing (RAD-seq) approach and identified several seed yield-related QTLs (plant height, first capsule height, capsule axis length, capsule number per plant, capsule length, seed number per capsule and thousand seed weight). Four similar traits, viz., capsule number per plant, capsule length, seed number per capsule and thousand seed weight, were also investigated in our study and we compared both studies to identify common genomic regions. The physical locations of the QTLs were searched on the reference genome [18] following the descriptions of Dossa [19]. Six QTLs detected by Wu et al. [10] matched with regions around significant loci detected in this study (Table 2; Figure 3). Interestingly, we observed a good consistency between the traits related to those six QTLs and the traits associated with the corresponding significant loci. For example, the capsule length QTL (Qcl-12) from Wu et al. [10] corresponded to nine loci associated with capsule size-related traits in our study. Also, the QTL Qcn-11 for capsule number per plant covered three significant loci identified for capsule number based on our GWAS. Another important finding is that the overlapped QTLs from Wu et al. [10] can be pleiotropic since they expanded on several significant loci which were associated with various seed yield traits in our study.

Table 2. Shared genomic regions detected for seed yield-related traits between our GWAS results and previous linkage mapping QTLs.

Traits Linkage Mapping	Code	LG	Start (bp)	End (bp)	Traits GWAS	LG	SNP Position (bp)
Grain number per capsule	Qgn-6	6	1,739,987	2,125,872	YB	6	1,741,236
					YLB	6	2,081,828
Capsule number per plant	Qcn-11	9	6,032,193	8,312,219	MCNM	9	5,988,865
					CNB	9	7,589,997
					MCNB	9	7,839,050
Capsule length	Qcl-3	3	1,566,853	2,593,783	YB	3	2,588,239
					YMB	3	2,588,241
	Qcl-4	5	9,840,981	10,961,395	YLB	5	9,857,730
					RMB	5	9,869,746
					SWMB	5	9,869,746
					TMM	5	9,895,178
					WMB	5	9,974,401
					TMB	5	10,197,769
					TMB	5	10,208,013
					YLB	5	10,705,889
					SWMB	5	10,773,145
					WMB	5	10,781,532
					LLM	5	10,786,506
					CN	5	10,786,597
					LCNM	5	10,790,853
					SWLM	5	10,958,834
	Qcl-8	4	11,220,208	11,670,895	LCNM	4	11,649,295
					WMM	4	11,658,278
					TSW	4	11,661,092
	Qcl-12	11	14,935,946	15,400,039	LLM	11	14,957,580
					LLM	11	15,003,280
					TMM	11	15,137,600
					WMM	11	15,137,600
					CPMM	11	15,138,140
					LLM	11	15,200,435
					LMM	11	15,219,964
					LMM	11	15,239,947
					LMM	11	15,289,738

2.4. Important Candidate Genes Associated with Seed Yield in Sesame

To identify the candidate genes controlling the seed yield-related traits in sesame, all the genes in 88 kb around the peak loci were retrieved [17]. In total, 7149 genes were identified and the number of genes in the LD window ranged between 8 and 42 (Table S4). Within these genes, 48 contained significant loci (Table S5). We particularly focused on these SNP-containing genes as they are more likely to modulate seed yield in sesame. Their homologs in *Arabidopsis* were identified and their functions predicted. Gene ontology analysis of these genes indicated that they are involved in developmental process, DNA and protein metabolism, response to stress, signal transduction, cell organization and biogenesis, transport and transcription (Figure 7a). Several homolog genes in *Arabidopsis* are well known to be directly or indirectly implicated in seed yield and biomass production. For example, the gene *AGL20* (AGAMOUS-like 20) plays an important role in flowering time [20], hence is directly associated with seed yield in *Arabidopsis*. In this study, we detected an intronic SNP located in the gene *SIN_1013997* (homolog of *AGL20*) strongly associated with the branch per plant seed yield and with the medial capsules in branch seed yield. Another important illustration concerns the gene *SIN_1006338* (*SiACS8*), which is located in the pleiotropic QTL associated with four various traits and was detected in all the three environments. A non-synonymous polymorphism (T/C) at the position 6,738,735 bp in this gene modulates the capsule number related traits (LCNM, CNM and CNB). An in-depth analysis suggests that the thymine allele is the favorable allele as it increases the capsule number on the stem and, therefore, leads to a higher yield (Figure 7b). Furthermore, the frequency of the T allele was rapidly increased by recent breeding, from 57% in landraces to 92% in modern cultivars. The gene *SiACS8* was previously identified as being associated with the capsule number per axil, particularly controlling the 1:3 capsules per axil in sesame [16]. These results further support our findings, indicating that *SiACS8* is indeed the causative gene controlling the capsule number trait in sesame. The homolog of *SiACS8* in *Arabidopsis AT4G37770* (*AtACS8*) was reported to be an auxin-induced gene involved in ethylene biosynthesis, suggesting that the number of capsules on sesame stem is under the regulation of plant hormones [21].

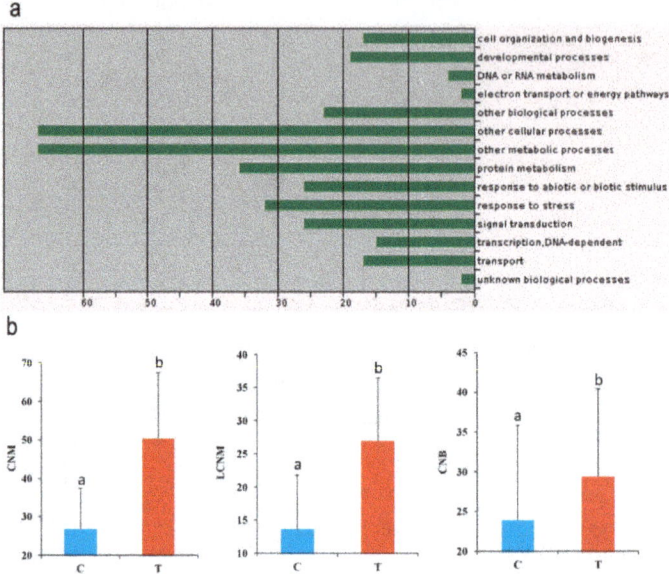

Figure 7. Functional analysis of 48 candidate gene-containing significant SNPs. (**a**) Biological function of the SNP-containing genes. (**b**) Identification of the favorable allele for the gene *SiACS8*. 262 genotypes harboring the C allele and 420 harboring the T allele were used. Different letters above bars represent significant difference ($p < 0.05$) between genotypes. The error bar indicates the standard error of the mean. Definition of the labels can be found at the end of this article.

A total of seven genes (*SIN_1017946*, *SIN_1017109*, *SIN_1021838*, *SIN_1019958*, *SIN_1011780*, *SIN_1019747* and *SIN_1014519*) involved in nutrient assimilation, carbohydrate metabolism, repression of early auxin response and kinase activity contain significant loci strongly associated with the total seed yield per plant (YP). These genes appear to be important in an effective source/sink relationship favorable for a high yield in sesame.

Some strongly associated loci were not located in the genic region; hence, gene expression analysis can give clues to pinpoint the probable candidate genes. As a proof of concept, we focused on the trait LMM and investigated the associated candidate gene. The strongest significant loci (A/G) ($-\log_{10}(p) = 9.06$) for LMM was located on the LG11 at the position 15,219,964 bp. Accessions with the guanine allele have a long capsule size as opposed to accessions with the adenosine allele. Interestingly, the frequency of the G allele in modern cultivars (20%) is comparable with that of landraces (37%), implying that this allele has not yet been intensively selected. Three genes *SIN_1011000*, *SIN_1010995* and *SIN_1010983* were found in the linkage disequilibrium window. Judging from the quantitative real time PCR (qRT-PCR) expression analysis of these genes, only *SIN_1010995* displayed a conspicuous discrepancy between the short and long capsule size accessions at different developmental stages

(Figure 8). The expression level of *SIN_1010995* (*SiLPT3*), a lipid transfer protein, was striking in the short capsule size accession but weakly expressed in the long capsule size accession. LPT3 proteins are described to be involved in cell wall edification, and more precisely in biosynthesis of cutin, which has been proposed to regulate cell adhesion during plant development [22]. The homolog gene of *SiLPT3* in *Arabidopsis AT5G59320.1* (*AtLPT3*) exhibited higher expression in the silique than other organs of *Arabidopsis*, indicating an active role in silique development [23]. Based on these observations, we speculate that *SiLPT3* regulates cell adhesion in the sesame capsule that contributes to the capsule length.

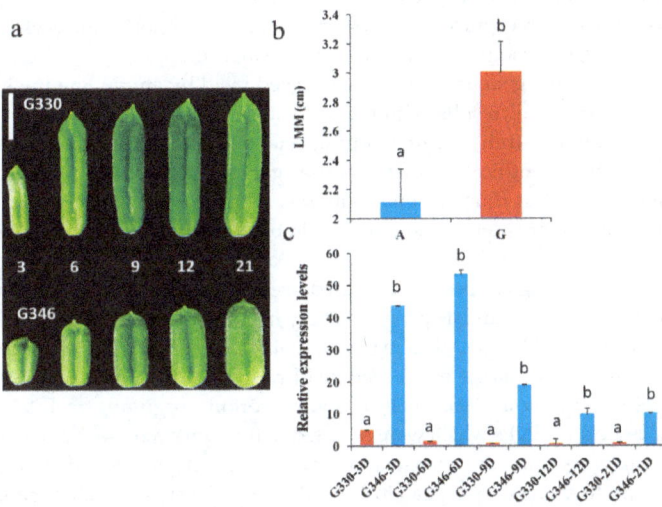

Figure 8. Expression analysis of the candidate gene for LMM trait between two contrasting accessions. (a) Phenotypes of G330 and G346 displaying long and short capsule length, respectively, at 3, 6, 9, 12 and 21 days after pollination. (b) Identification of the favorable allele at the locus 15,219,964 bp on the LG11. A total of 427 genotypes harboring the A allele and 175 harboring the G allele were used. (c) qRT-PCR relative expression level of the gene *SIN_1010995* between G330 and G346 at different days after pollination. Different letters above bars represent significant difference ($p < 0.05$) between genotypes. The error bar indicates the standard error of the mean. The sesame *Actin* gene (*SIN_1006268*) was used as the internal reference and 3 biological replicates and 3 technical replicates were used.

3. Discussion

The seed yield improvement of sesame is a prerequisite for the rapid expansion of the crop. Although sesame has being cultivated for a long time (~5000 years), few efforts have been made for its improvement [5]. In fact, the lack of basic information on the genetics of important agronomical traits, especially the traits complexly inherited, are causing hindrance for the breeders to achieve higher yields [24]. In this study, we observed a high variability for the assayed seed yield related traits, suggesting that our association panel harbors a large diversity necessary for genome wide association studies (GWAS). In a previous comprehensive GWAS for seed quality traits, Wei et al. [16], using the same association panel, found a low population structure, a moderate linkage disequilibrium (LD) decay (88 kb) and recommended that a high marker density, as employed in our study, could give ample power for association analyses. Several authors have studied traits that contribute to the seed yield formation in sesame. Distinctly, the capsule number per plant is a primary determinant for high seed yield in sesame [7,9,10,25,26]. In fact, sesame seeds grow in a capsule; therefore, more capsules on the plant are likely to yield more seeds [4]. Moreover, the number of seeds per capsule and the seed weight are also largely reported as important contributors to seed yield [10,27,28]. Our results match well with those of the literature, as we found that capsule size, capsule number and seed size-related traits are strongly correlated with yield indexes.

Our GWAS results revealed several clusters of significant loci, highlighting important genomic regions associated with seed yield-related traits. Interestingly, many pleiotropic QTLs were identified but an in-depth analysis indicates that very few QTLs were associated with traits from the different groups (Table S1). These results suggest that seed yield component traits from the same group have a similar genetic architecture but traits from different groups may be manipulated independently to increase the seed yield in sesame. Boyles et al. [29] also reported similar observation in sorghum with no overlapping loci for grain yield components.

The GWAS approach is recognized as a powerful tool to reconnect traits back to the underlying genetics and offers higher resolution than classical linkage mapping [30]. Previously, only one study was performed on the genetics of the sesame seed yield by employing the linkage mapping approach [10]. Comparing our results with the previous QTLs, we identified several overlapping loci associated with similar traits. Our study substantially narrows down these QTL regions which will facilitate the identification of the causal genes. In addition, several loci previously identified by Wei et al. [16] in different environments were also detected in this study, implying that these trait-associations are highly stable and could be very useful to accelerate sesame seed yield improvement efforts.

Transcriptome sequencing has been widely used to estimate gene expression changes and enables the efficiency and accuracy of candidate gene discovery in GWAS [31]. In this study, several candidate genes were retrieved from the genomic regions significantly associated with the assayed traits. To effectively pinpoint the causal genes for seed yield-related traits, additional RNA-seq data could be exploited as demonstrated in *Brassica napus*, maize, cotton, sorghum, etc. [31–33]. Nonetheless, genes containing associated SNPs which were detected in this work represent potential candidates for further functional analysis using the transgenic approach [34] and genome-editing technologies using CRISPR/Cas system. Meanwhile, the peak loci could be transformed into allele-specific markers for applications in breeding programs to design sesame varieties with improved seed yield. In fact, Asian, American and European sesame producing countries present higher yields than in Africa [6]. This can be, inter alia, related to the use of elite cultivars. For example, the modern cultivars in our panel have, on average, 70 capsules on the main stem, which is approximately double of the capsule number in landraces, and thus have a higher yield potential. Since several favorable alleles detected in this study have not yet been intensively selected, our GWAS results will undoubtedly assist in incorporating further useful alleles into the elite sesame germplasm for a seed yield increase in the future.

4. Materials and Methods

4.1. Plant Materials

In the present study, 705 cultivated sesame (*Sesamum indicum* L.) accessions were obtained from the germplasm preserved at the China National Gene Bank, Oil Crops Research Institute, Chinese Academy of Agricultural Sciences (Table S6). The panel is composed of 405 traditional landraces and 95 modern cultivars from China, as well as 205 accessions collected from 28 other countries [16]. All the accessions have been self-pollinated for four generations in Sanya, Hainan province, China (109.187° E, 18.38° N, altitude 11 m).

4.2. Field Growth Conditions

Three field trials were set in three environments in China during the years 2013 to 2014 at normal planting seasons [16]. All the accessions were grown at experiment stations in Wuhan (WH), the Hubei province (30.57° N, 114.30° E), Nanning (NN), the Guangxi province (23.17° N, 107.55° E) and Sanya (SY), the Hainan province, (109.187° E, 18.38° N). We recorded ranges of temperature (32–38/25–27 °C, day/night), relative humidity (45–72%) and rainfall (125–210 mm) during the experiment in Wuhan. In Nanning, we recorded ranges of temperature (31–34/25–26 °C, day/night), relative humidity (42–58%) and rainfall (205–235 mm) during our experiment. In Sanya, ranges of temperature (30–33/24–26 °C, day/night), relative humidity (50–75%) and rainfall (159–219 mm) were recorded during our experiment. These data show that Wuhan was the hottest location with the lowest rainfall among the 3 locations. Sanya and Nanning experimental fields have a sandy loam soil while Wuhan experimental field is characterized by a loam soil. The field trials were conducted using a randomized block design with three replications. Each plot had four rows of 2 m long spaced 0.4 m apart. At the four-leaf stage, seedlings were thinned down and eight evenly distributed plants in each row were retained for further analyses. Five uniform plants for each genotype were randomly selected to collect phenotypic data.

4.3. Trait Evaluation

Plants at the two ends of each row were not selected to avoid edge effects. Traits evaluated included (1) weight (g), length (cm), width (cm) and thickness (cm) of the dry capsule pericarp and the seed selected from different parts of the plant: medial or lateral position on the main stem or branch; (2) the seed number was counted in capsules from different parts of the plant: medial or lateral position on the main stem or branch; (3) the seed yield (g) was recorded from different parts of the plant: the capsules at medial or lateral position on the main stem or branch, total yields of the main stem, the branch and the whole plant. Based on the seed and capsule pericarp dry weights recorded from different parts of the plant, the ratio seed weight and pericarp weight were also computed. In total, 39 traits were investigated in this study and categorized into five major trait groups: yield index, seed traits, capsule number, capsule size and capsule pericarp (Table S1).

4.4. Statistical Analysis

All the statistical analyses were performed using R2.3.0 [35]. For each trait, the least square mean and descriptive statistics such as the minimum, maximum, skewness and kurtosis were estimated based on five replicates in each environment. Variation of the different traits in the different environments was represented as boxplot employing the "ggplot2" package [36]. The broad-sense heritability (H^2) was calculated as follow: $H^2 = \sigma^2_a / (\sigma^2_a + \sigma^2_{ae}/E + \sigma^2_\varepsilon/ER)$, where σ^2_a, σ^2_{ay}, and σ^2_ε are estimates of the variances of accession, accession × environment interaction, and error, respectively, estimated by analysis of variance (ANOVA). E represents Environment, and R is the number of replications. Correlation among the seed yield related traits was estimated by Pearson's method at a significance level of $p < 0.05$ using the "corrplot" package [37]. For the correlation analysis, we used the best linear unbiased estimator (BLUE) values of phenotype data from the three environments.

4.5. Genome Wide Association Study Implementation

The association panel used in the present study was previously fully re-sequenced [16]. A total of 1.8 M common single nucleotide polymorphisms (SNPs) covering the whole genome with minor allele frequency >0.03 were retained for the genome wide association studies (GWAS). Phenotype-genotype association was implemented with the EMMAX model [38]. The matrix of pair-wise genetic distance derived from simple matching coefficients was used as the variance–covariance matrix of the random effect. Using the Genetic type 1 Error Calculator, version 0.2 [39], the effective number of independent SNPs were estimated to be 469,175 and the threshold to declare significant associated loci was approximately $p = 10^{-7}$ [16]. Significant associations were also selected on the threshold of $p \leq 0.01$, corrected for multiple comparisons according to the false discovery rate procedure reported by Benjamini and Hochberg [40].

4.6. Candidate Gene Mining

Based on the reference genome [18], all the genes in the 88 kb region corresponding to the average linkage disequilibrium window [16] around the peak associated loci were retrieved. Their homologs in *Arabidopsis thaliana* were predicted and their functions annotated from the database Sinbase 2.0 [18] with a cut off *E*-value of $\leq 1 \times 10^{-40}$. All the genes containing significant associated loci were prioritized. Moreover, for genomic regions where we did not find any associated SNP-containing genes, the putative candidate genes were retained if the homolog genes in *Arabidopsis thaliana* were described to be involved in seed yield or biomass formation. Gene ontology analysis of the candidate genes was performed using the Blast2GO tool v.3.1.3 [41] and plotted with the WEGO tool [42].

4.7. Gene Expression Analysis Based on Quantitative Real-Time PCR

We performed the qRT-PCR expression analysis for all the genes around the strongest associated loci with the capsule length (LMM) trait in order to pinpoint the potential candidate gene. Accession G330 with a long capsule size (~3.65 cm, at maturity stage) and accession G346 with a short capsule size (~1.90 cm, at maturity stage) were selected for this experiment. Capsules from the middle of the main stem were collected from 3 different plants (biological replicates) in Wuhan on 3, 6, 9, 12 and 21 days after pollination. RNA was extracted from fresh capsule tissues and reverse transcribed according to descriptions of Mmadi et al. [43]. In total, three genes were investigated and their gene-specific primers designed using the Primer5.0 tool [44] (Table S7). The qRT-PCR was conducted in triplicate (technical replicates) on a Roche Lightcyler® 480 instrument (Roche Molecular Systems, Inc, Basel, Switzerland) using SYBR Green Master Mix (Vazyme), according to the manufacturer's protocol. Reaction and PCR conditions are the same as the descriptions of Mmadi et al. [43]. The sesame *Actin* gene (*SIN_1006268*) was used as the internal reference and the relative gene expression values were calculated using the $2^{-\Delta Ct}$ method [45].

Supplementary Materials: Supplementary materials can be found at http://www.mdpi.com/1422-0067/19/9/2794/s1. Figure S1. Boxplots displaying variation of 33 traits across three different environments (SY = Sanya, NN = Nanning and WH = Wuhan). Figure S2. Manhattan plots for SNP association of all traits in the three environments (SY = Sanya, NN = Nanning and WH = Wuhan). Table S1. Full name of the 39 assayed traits. Table S2. Summary of descriptive statistics of the 39 traits in three environments. Table S3. List and position of the significant loci detected in this study. Table S4. List and functional annotation of genes around peak loci associated with the assayed traits in this study. Table S5. Candidate gene-containing significant SNPs detected in this study and their homologs in *Arabidopsis thaliana*. Table S6. Full list of the 705 accessions used in this study, their origin and their breeding status. Table S7. Primer sequences for qRT-PCR gene expression analysis.

Author Contributions: R.Z., X.W., K.D., J.Y., J.Y., D.L. participated in data collection and analysis; K.D., R.Z. wrote the manuscript; X.W. and X.Z. conceived and supervised the study. All authors have read and approved the final version of the manuscript.

Funding: This work was funded by the Agriculture Science and Technology Innovation Project of Chinese Academy of Agricultural Sciences (CAAS-ASTIP-2013-OCRI), Project of Crop Germplasm Resources Protection (2018NWB033), China Agriculture Research System (CARS-14).

Conflicts of Interest: The authors declare no conflict of interest. The funders had no role in the design of the study; in the collection, analyses, or interpretation of data; in the writing of the manuscript, and in the decision to publish the results.

Abbreviations

CN	effective capsule number in plant
CNB	effective capsule number in branch
CNM	effective capsule number in main stem
CPLM	dry capsule pericarp weight of lateral capsule in main stem
CPMB	dry capsule pericarp weight of medial capsule in branch
CPMM	dry capsule pericarp weight of medial capsule in main stem
DNA	deoxyribonucleic acid
GWAS	genome wide association study
LCNB	effective lateral capsule number in branch
LCNM	effective lateral capsule number in main stem

LD	linkage disequilibrium
LG	linkage group
LLB	length of lateral capsule in branch
LLM	length of lateral capsule in main stem
LMB	length of medial capsule in branch
LMM	length of medial capsule in main stem
MAF	minor allele frequency
MCNB	effective medial capsule number in branch
MCNM	effective medial capsule number in main stem
NN	Nanning
qRT-PCR	quantitative real-time polymerase chain reaction
QTL	quantitative trait loci
RLM	ratio of seed weight and capsule pericarp weight for lateral capsule in main stem
RMB	ratio of seed weight and capsule pericarp weight for medial capsule in branch
RMM	ratio of seed weight and capsule pericarp weight for medial capsule in main stem
RNA	ribonucleic acid
SNLB	seed number per lateral capsule in branch
SNLM	seed number per lateral capsule in main stem
SNMB	seed number per medial capsule in branch
SNMM	seed number per medial capsule in main stem
SNP	single nucleotide polymorphism
SY	Sanya
SWLB	dry seed weight of per lateral capsule in branch
SWLM	dry seed weight of per lateral capsule in main stem
SWMB	dry seed weight of per medial capsule in branch
SWMM	dry seed weight of per medial capsule in main stem
TLM	thickness of lateral capsule in main stem
TMB	thickness of medial capsule in branch
TMM	thickness of medial capsule in main stem
TSW	thousand seeds weight
WH	Wuhan
WLM	width of lateral capsule in main stem
WMB	width of medial capsule in branch
WMM	width of medial capsule in main stem
YB	yield of branch per plant
YLB	yield of lateral capsules in branch
YLM	yield of lateral capsules in main stem
YM	yield of main stem per plant
YMB	yield of medial capsules in branch
YMM	yield of medial capsules in main stem
YP	yield per plant

References

1. Anilakumar, K.R.; Pal, A.; Khanum, F.; Bawa, A.S. Nutritional, medicinal and industrial uses of sesame (*Sesamum indicum* L.) seeds: An overview. *Agric. Conspec. Sci.* **2010**, *75*, 159–168.
2. Dossa, K.; Wei, X.; Niang, M.; Liu, P.; Zhang, Y.; Wang, L.; Liao, B.; Cissé, N.; Zhang, X.; Diouf, D. Near-infrared reflectance spectroscopy reveals wide variation in major components of sesame seeds from Africa and Asia. *Crop J.* **2018**, *6*, 202–206. [CrossRef]
3. Ingersent, K.A. World agriculture: Towards 2015/2030–An FAO perspective. *J. Agric. Econ.* **2003**, *54*, 513–515.
4. Langham, D.R. Phenology of Sesame. In *Issues in New Crops and New Uses*; Janick, J., Whipley, A., Eds.; ASHS Press: Alexandria, VA, USA, 2007; p. 39.

5. Dossa, K.; Diouf, D.; Wang, L.; Wei, X.; Zhang, Y.; Niang, M.; Fonceka, D.; Yu, J.; Mmadi, M.A.; Yehouessi, L.W.; et al. The emerging oilseed crop *Sesamum indicum* enters the "Omics" era. *Front. Plant Sci.* **2017**, *8*, 1154. [CrossRef] [PubMed]

6. Food and Agriculture Organization Statistical Databases (FAOSTAT). 2017. Available online: http://faostat. fao.org/ (accessed on 19 March 2018).

7. Akhtar, K.P.; Sarwar, G.; Dickinson, M.; Ahmad, M.; Haq, M.A.; Hameed, S.; Iqbal, M.J. Sesame phyllody disease: Its symptomatology, etiology, and transmission in Pakistan. *Turk. J. Agric. For.* **2009**, *33*, 477–486.

8. Yol, E.; Uzun, B. Geographical patterns of sesame accessions grown under Mediterranean environmental conditions, and establishment of a core collection. *Crop Sci.* **2012**, *52*, 2206–2214. [CrossRef]

9. Biabani, A.R.; Pakniyat, H. Evaluation of seed yield-related characters in sesame (*Sesamum indicum* L.) using factor and path analysis. *Pak. J. Biol. Sci.* **2008**, *11*, 1157–1160. [PubMed]

10. Wu, K.; Liu, H.; Yang, M.; Tao, Y.; Ma, H.; Wu, W.; Zuo, Y.; Zhao, Y. High-density genetic map construction and QTLs analysis of grain yield-related traits in sesame (*Sesamum indicum* L.) based on RAD-Seq technology. *BMC Plant. Biol.* **2014**, *14*, 274. [CrossRef] [PubMed]

11. Diouf, M.; Boureima, S.; Diop, T.; Çagirgan, M. Gamma rays-induced mutant spectrum and frequency in sesame. *Turk. J. Field Crops* **2010**, *15*, 99–105.

12. Mosjidis, J.A.; Yermanos, D.M. Plant position effect on seed weight, oil content, and oil composition in sesame. *Euphytica* **1985**, *34*, 193–199. [CrossRef]

13. Tashiro, T.; Fukuda, Y.; Osawa, T. Oil contents of seeds and minor components in the oil of sesame, *Sesamum indicum* L.; as affected by capsule position. *Jpn. J. Crop Sci.* **1991**, *60*, 116–121. [CrossRef]

14. Kumazaki, T.; Yamada, Y.; Karaya, S.; Kawamura, M.; Hirano, T.; Yasumoto, S.; Katsuta, M.; Michiyama, H. Effects of day length and air and soil temperatures on sesamin and sesamolin contents of sesame seed. *Plant Prod. Sci.* **2009**, *12*, 481–491. [CrossRef]

15. Huang, X.; Han, B. Natural variations and genome-wide association studies in crop plants. *Annu. Rev. Plant Biol.* **2014**, *65*, 531–551. [CrossRef] [PubMed]

16. Wei, X.; Liu, K.; Zhang, Y.; Feng, Q.; Wang, L.; Zhao, Y.; Li, D.; Zhao, Q.; Zhu, X.; Zhu, X.; et al. Genetic discovery for oil production and quality in sesame. *Nat. Commun.* **2015**, *6*, 8609. [CrossRef] [PubMed]

17. Li, D.; Dossa, K.; Zhang, Y.; Wei, X.; Wang, L.; Zhang, Y.; Liu, A.; Zhou, R.; Zhang, X. GWAS uncovers differential genetic bases for drought and salt tolerances in sesame at the germination stage. *Genes* **2018**, *9*, 87. [CrossRef] [PubMed]

18. Wang, L.; Yu, S.; Tong, C.; Zhao, Y.; Liu, Y.; Song, C.; Zhang, Y.; Zhang, X.; Wang, Y.; Hua, W.; et al. Genome sequencing of the high oil crop sesame provides insight into oil biosynthesis. *Genome Biol.* **2014**, *15*, R39. [CrossRef] [PubMed]

19. Dossa, K. A physical map of important QTLs, functional markers and genes available for sesame breeding programs. *Physiol. Mol. Biol. Plants* **2016**, *22*, 613–619. [CrossRef] [PubMed]

20. Lee, H.; Suh, S.S.; Park, E.; Cho, E.; Ahn, J.H.; Kim, S.G.; Lee, J.S.; Kwon, Y.M.; Lee, I. The AGAMOUS-LIKE 20 MADS domain protein integrates floral inductive pathways in *Arabidopsis*. *Gene Dev.* **2000**, *14*, 2366–2376. [CrossRef] [PubMed]

21. Tsuchisaka, A.; Theologis, A. Heterodimeric interactions among the 1-amino-cyclopropane-1-carboxylate synthase polypeptides encoded by the *Arabidopsis* gene family. *Proc. Natl. Acad. Sci. USA* **2004**, *101*, 2275–2280. [CrossRef] [PubMed]

22. Shi, J.X.; Malitsky, S.; De Oliveira, S.; Branigan, C.; Franke, R.B.; Schreiber, L.; Aharoni, A. SHINE transcription factors act redundantly to pattern the archetypal surface of Arabidopsis flower organs. *PLoS Genet.* **2011**, *7*, e1001388. [CrossRef] [PubMed]

23. Klepikova, A.V.; Kasianov, A.S.; Gerasimov, E.S.; Logacheva, M.D.; Penin, A.A. A high resolution map of the *Arabidopsis thaliana* developmental transcriptome based on RNA-seq profiling. *Plant J.* **2016**, *88*, 1058–1070. [CrossRef] [PubMed]

24. Rao, P.V.R.; Prasuna, K.; Anuradha, G.; Srividya, A.; Vemireddy, L.R.; Shankar, V.G.; Sridhar, S.; Jayaprada, M.; Reddy, K.R.; Reddy, N.E.; et al. Molecular mapping of important agro-botanic traits in sesame. *Electron. J. Plant Breed.* **2014**, *5*, 475–488.

25. Shim, K.B.; Shin, S.H.; Shon, J.Y.; Kang, S.G.; Yang, W.H.; Heu, S.G. Classification of a collection of sesame germplasm using multivariate analysis. *J. Crop Sci. Biotechnol.* **2006**, *19*, 151–155. [CrossRef]

26. Monpara, B.A.; Khairnar, S.S. Heritability and expected genetic gain from selection in components of crop duration and seed yield in sesame (*Sesamum indicum* L.). *Plant Gene and Trait* **2016**, *7*, 1–5.

27. Emamgholizadeh, S.; Parsaeian, M.; Baradaran, M. Seed yield prediction of sesame using artificial neural network. *Eur. J. Agron.* **2015**, *68*, 89–96. [CrossRef]

28. Ramazani, S.H.R. Surveying the relations among traits affecting seed yield in sesame (*Sesamum indicum* L.). *J. Crop Sci. Biotechnol.* **2016**, *19*, 303–309. [CrossRef]

29. Boyles, R.E.; Cooper, E.A.; Myers, M.T.; Brenton, Z.; Rauh, B.L.; Morris, G.P.; Kresovich, S. Genome-wide association studies of grain yield components in diverse sorghum germplasm. *Plant Genome* **2016**, *9*, 1–17. [CrossRef] [PubMed]

30. Korte, A.; Farlow, A. The advantages and limitations of trait analysis with GWAS: A review. *Plant Meth.* **2013**, *9*, 29. [CrossRef] [PubMed]

31. Lu, K.; Peng, L.; Zhang, C.; Lu, J.; Yang, B.; Xiao, Z.; Liang, Y.; Xu, X.; Qu, C.; Zhang, K.; et al. Genome-wide association and transcriptome analyses reveal candidate genes underlying yield-determining traits in *Brassica napus*. *Front. Plant Sci.* **2017**, *8*, 206. [CrossRef] [PubMed]

32. Mao, H.; Wang, H.; Liu, S.; Li, Z.; Yang, X.; Yan, J.; Li, J.; Tran, L.S.P.; Qin, F. A transposable element in a NAC gene is associated with drought tolerance in maize seedlings. *Nat. Com.* **2015**, *6*, 8326. [CrossRef] [PubMed]

33. Sun, Z.; Wang, X.; Liu, Z.; Gu, Q.; Zhang, Y.; Li, Z.; Ke, H.; Yang, J.; Wu, J.; Wu, L.; et al. Genome-wide association study discovered genetic variation and candidate genes of fibre quality traits in *Gossypium hirsutum* L. *Plant Biotechnol. J.* **2017**, *15*, 982–996. [CrossRef] [PubMed]

34. Chowdhury, S.; Basu, A.; Kundu, S. Overexpression of a new osmotin- like protein gene (*SindOLP*) confers tolerance against biotic and abiotic stresses in sesame. *Front. Plant Sci.* **2017**, *8*, 410. [CrossRef] [PubMed]

35. R Development Core Team. *R: A Language and Environment for Statistical Computing*; R Foundation for Statistical Computing: Vienna, Austria, 2008; ISBN 3-900051-07-0.

36. Wickham, H. *Ggplot2: Elegant Graphics for Data Analysis*; Springer: New York, NY, USA, 2009.

37. Wei, T.; Simko, V. Corrplot: Visualization of a Correlation Matrix, R Package Version 0.77. 2016. Available online: http://CRAN.R-project.org/package=corrplot (accessed on 15 March 2018).

38. Kang, H.; Sul, J.; Service, S.; Zaitlen, N.; Kong, S.; Freimer, N.; Sabatti, C.; Eskin, E. Variance component model to account for sample structure in genome-wide association studies. *Nat. Genet.* **2010**, *42*, 348–354. [CrossRef] [PubMed]

39. Li, M.X.; Yeung, J.M.; Cherny, S.S.; Sham, P.C. Evaluating the effective numbers of independent tests and significant p-value thresholds in commercial genotyping arrays and public imputation reference datasets. *Hum. Genet.* **2012**, *131*, 747–756. [CrossRef] [PubMed]

40. Benjamini, Y.; Hochberg, Y. Controlling the false discovery rate: A practical and powerful approach to multiple testing. *J. R. Stat. Soc. Ser. B* **1995**, *57*, 289–300.

41. Conesa, A.; Götz, S. Blast2GO: A comprehensive suite for functional analysis in plant genomics. *Int. J. Plant Genom.* **2008**, *2008*, 619832. [CrossRef] [PubMed]

42. Ye, J.; Fang, L.; Zheng, H.; Zhang, Y.; Chen, J.; Zhang, Z.; Wang, J.; Li, S.; Li, R.; Bolund, L.; et al. WEGO: A web tool for plotting GO annotations. *Nucleic Acids Res.* **2006**, *34*, 293–297. [CrossRef] [PubMed]

43. Mmadi, M.A.; Dossa, K.; Wang, L.; Zhou, R.; Wang, Y.; Cisse, N.; Sy, M.O.; Zhang, X. Functional characterization of the versatile MYB gene family uncovered their important roles in plant development and responses to drought and waterlogging in sesame. *Genes* **2017**, *8*, 362. [CrossRef] [PubMed]

44. Lalitha, S. Primer premier 5. *Biotechnol. Softw. Int. Rep.* **2000**, *1*, 270–272. [CrossRef]

45. Livak, K.J.; Schmittgen, T.D. Analysis of relative gene expression data using real-time quantitative PCR and the $2^{-\Delta\Delta CT}$ Method. *Methods* **2001**, *25*, 402–408. [CrossRef] [PubMed]

International Journal of
Molecular Sciences

Article

A Conserved Glycine Is Identified to be Essential for Desaturase Activity of IpFAD2s by Analyzing Natural Variants from *Idesia polycarpa*

Pan Wu [1,2], Lingling Zhang [1], Tao Feng [1], Wenying Lu [1,2], Huayan Zhao [3], Jianzhong Li [4] and Shiyou Lü [1,5,*]

[1]   Key Laboratory of Plant Germplasm Enhancement and Specialty Agriculture, Wuhan Botanical Garden, Chinese Academy of Sciences, Wuhan 430074, China; wupanx@126.com (P.W.); zhanglingling@wbgcas.cn (L.Z.); fengtao@wbgcas.cn (T.F.); luwenying16@mails.ucas.ac.cn (W.L.);
[2]   University of Chinese Academy of Sciences, Beijing 100049, China
[3]   Applied Biotechnology Center, Wuhan Institute of Bioengineering, Wuhan 430415, China; huayan_zhao@163.com
[4]   Tianjin Garrison hangu farm, Tianjin 300480, China; lvsy1021@gmail.com
[5]   Sino-Africa Joint Research Center, Chinese Academy of Sciences, Wuhan 430074, China
*   Correspondence: shiyoulu@wbgcas.cn; Tel.: +86-27-87700852

Received: 25 August 2018; Accepted: 5 December 2018; Published: 7 December 2018

Abstract: High amounts of polyunsaturated fatty acids (PUFAs) in vegetable oil are not desirable for biodiesel or food oil due to their lower oxidative stability. The oil from *Idesia polycarpa* fruit contains 65–80% (mol%) linoleic acid (C18:2). Therefore, development of *Idesia polycarpa* cultivars with low PUFAs is highly desirable for *Idesia polycarpa* oil quality. Fatty acid desaturase 2 (FAD2) is the key enzyme converting oleic acid (C18:1) to C18:2. We isolated four FAD2 homologs from the fruit of *Idesia polycarpa*. Yeast transformed with *IpFAD2-1*, *IpFAD2-2* and *IpFAD2-3* can generate appreciable amounts of hexadecadienoic acid (C16:2) and C18:2, which are not present in wild-type yeast cells, revealing that the proteins encoded by these genes have Δ^{12} desaturase activity. Only trace amounts of C18:2 and little C16:2 were detected in yeast cells transformed with *IpFAD2-4*, suggesting IpFAD2-4 displays low activity. We also analyzed the activity of several FAD2 natural variants of *Idesia polycarpa* in yeast and found that a highly conserved Gly376 substitution caused the markedly reduced products catalyzed by IpFAD2-3. This glycine is also essential for the activity of IpFAD2-1 and IpFAD2-2, but its replacement in other plant FAD2 proteins displays different effects on the desaturase activity, suggesting its distinct roles across plant FAD2s proteins.

Keywords: *Idesia polycarpa var*; glycine; FAD2; linoleic acid; oleic acid

---

1. Introduction

Vegetable oils are not only essential resources for nutritional applications, but also for sustainable industrial feedstocks, which are commonly used in paints, lubricants, soaps, biodiesel, etc. [1,2]. The demand for vegetable oils is quickly increasing due to the fast growing population across the world. To meet this demand, many efforts have been made to improve the yields of oil crops or to domesticate wild oilseed plants [3]. *Idesia polycarpa*, a member of *Flacourtiaceae* family, is a local tree species in some Asian countries including Korea, Japan, and China [4]. It is receiving more attention due to the high amount of oil in its fruits, which can potentially be used in the biodiesel industry [4]. In addition, the oil from *Idesia polycarpa* fruit is healthy and edible since it contains 65–80% (mol%) linoleic acid (C18:2). C18:2 is one of the essential polyunsaturated fatty acids in humans and cannot be synthesized by the human body, it can only be obtained from food [5]. However, high amounts of C18:2 in its oil also make it more prone to rancidity and thus decreases its flavor [6]. Previous reports showed that oleic acid (C18:1) had higher oxidative stability than C18:2, and thus the edible oils with higher ratios of C18:1/C18:2 are more desirable [7]. Similarly, an ideal biodiesel composition should also contain more monounsaturated fatty acids and less polyunsaturated fatty acids since high percentages of polyunsaturated fatty acids in biodiesel negatively affect its oxidative stability and cause high rates of nitrogen oxide emission [8]. Hence it would be valuable to breed *Idesia polycarpa* cultivars which produce oils with low C18:2 and high C18:1 contents and it would also be helpful to uncover the desaturation mechanism in woody plants.

The Δ^{12} fatty acid desaturase 2 (FAD2) is the key microsomal enzyme that converts C18:1 to C18:2 [9]. Many efforts have been made to modify the plant oil quality via manipulation of expression levels of FAD2, or through screening natural varieties with altered FAD2 activity. For example, the mutations of GmFAD2-1A and GmFAD2-1B greatly increased the levels of C18:1 in soybean seeds [10]. The *Arachis hypogaea AhFAD2* mutant was used as an introgression line for breeding peanut cultivars with high C18:1 and low C18:2 [11]. Numerous safflower breeding lines with high levels of C18:1 (75–84%) were selected from the natural variations in *FAD2* genes [12]. The mutation of a candidate protein, fatty acid desaturase-2 (FAD2-1D) gene from pima cotton produced less linoleic acid [13,14]. Olive oil extracted from the olive fruit has mostly high level of C18:1 (about 75%) and less C18:2 (about 5.5%), which might be attributed to the suppression of *FAD2* genes by si-RNA [15]. Taken together, the activity of FAD2 is crucial for determining the C18:1/C18:2 ratios in seed or fruit storage lipids and FAD2 is an ideal candidate for improving oil quality of oil crops or trees.

Plant FAD2 proteins belong to a large family of ER localized membrane-bound desaturases [16]. The relationship between structure and function of FAD2 proteins has been extensively studied in the past two decades. FAD2 proteins contain three to six predicated transmembrane domains (PTMDs) and three highly conserved histidine-rich motifs, which are key characteristics of all membrane-bound desaturases. In the conserved histidine-rich motifs, the histidines are proved to be crucial for FAD2 desaturase activity [17]. In addition, four relevant amino acid residues within a distance of five residues from the His boxes of AtFAD2 are responsible for the conversion of this monofunctional desaturase (Δ^{12} desaturase activity) into a bifunctional desaturase/hydroxylase [18]. Besides these His boxes, McCartney et al. [19] found that the deletion of *Arabidopsis thaliana* AtFAD2 C-terminus containing an ER retrieval motif resulted in loss of ER localization and enzyme activity in yeast cells. In addition, Hoffmann et al. [20] showed that a small membrane-peripheral region close to the active center of a monofunctional Δ^{12} desaturase from *Aspergillus nidulans* determines substrate specificity and regioselectivity. Despite much progress in the relationship between FAD2s structure and function, which has been elucidated in the past two decades, some unidentified factors affecting their enzymatic activity remain to be investigated.

C18:2 was the major component of fatty acids in both seed and pericarp of *Idesia ploycarpa*, which respectively accounted for 83.92% and 62.08% of the total fatty acids in the two organs [5]. Our previous study showed that four *IpFAD2* genes exist in *Idesia ploycarpa* [5]. To ascertain if the IpFAD2 proteins are capable of desaturating C18:1 into C18:2, we isolated four *IpFAD2* genes from *Idesia ploycarpa*, and identified their activity in yeast cells. We also assessed the activity of natural FAD2 variants and identified a highly conserved glycine at position 376 of IpFAD2-3, which is critical for normal function of IpFAD2s. To determine if the function of this glycine is conserved across plant FAD2 proteins, we also evaluated the impacts of its substitution on the activity of FAD2 proteins from other plants. Our study provides a clue for genetically modifying the oil quality of *Idesia polycarpa*.

2. Results

2.1. Isolation of FAD2 Orthologs from Idesia ploycarpa

Our previous study showed that four *FAD2* orthologs are present in the fruit of *Idesia polycarpa* [5], which were renamed *IpFAD2-1* (c63420_g2), *IpFAD2-2* (c56614_g1), *IpFAD2-3* (c63420_g1), and *IpFAD2-4* (c50543_g1), respectively, in this study. The entire coding sequence (CDS) of *IpFAD2-1*, *IpFAD2-2*, *IpFAD2-3*, and *IpFAD2-4* were cloned from *Idesia ploycarpa*. The length of the predicted polypeptides encoded by *IpFAD2-1*, *IpFAD2-2*, *IpFAD2-3*, and *IpFAD2-4* CDS are 385, 382, 385, and 380 amino acids respectively. IpFAD2-1, IpFAD2-2, IpFAD2-3 and IpFAD2-4 shared 76.74%, 70.03%, 76.49% and 67.70% identity with AtFAD2 (Figure 1, Supplementary table S1). Four IpFAD2 proteins contain three conserved histidine-rich motifs, which are commonly present in all membrane-bound fatty acid desaturases (Figure 1) [21,22] and also contain an ER-localized motif in C-terminus [19]. The cDNA sequences of *IpFAD2-1*, *IpFAD2-2*, *IpFAD2-3*, and *IpFAD2-4* were submitted to the NCBI Genbank, their accession numbers were MH394208, MH394209, MH394210, and MK105894, respectively.

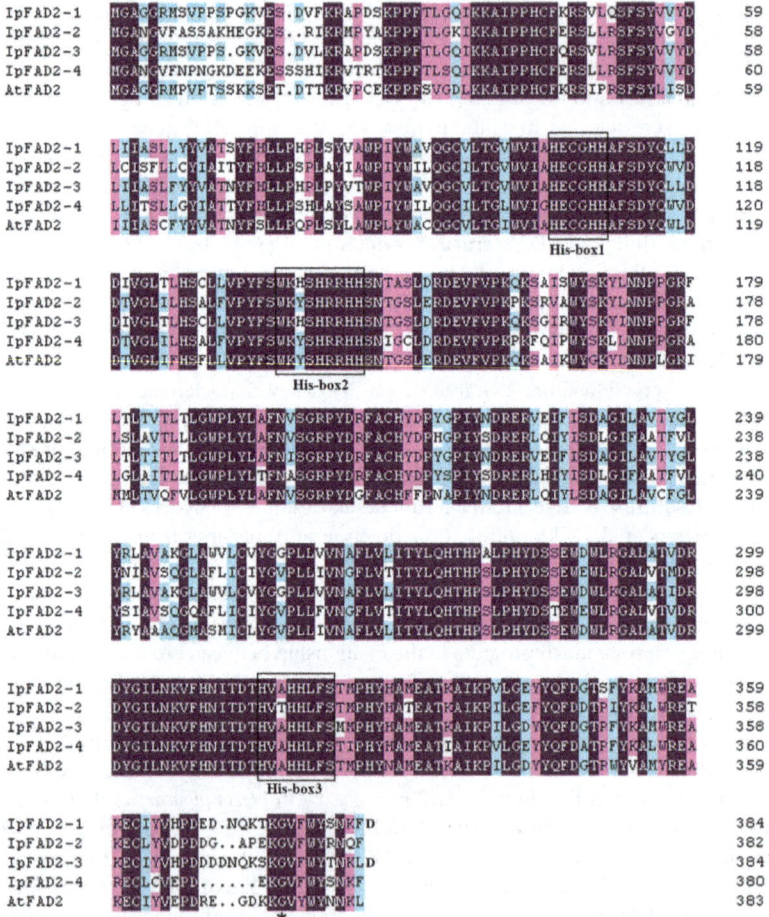

Figure 1. Alignments of predicted amino acid sequences encoded by FAD2 coding sequences from *Idesia polycarpa* and *Arabidopsis thaliana*. The three "histidine-rich motifs" are boxed. The GenBank accession numbers of *IpFAD2-1*, *IpFAD2-2*, *IpFAD2-3*, *IpFAD2-4*, *AtFAD2* are: MH394208, MH394209, MH394210, MK105894, and NP_187819.1 accordingly. The shading colors represent the identity level of amino acids. Black, magenta, and cyan indicate 100%, 80%, and 60% identity, respectively. The asterisk indicates the conserved glycine.

The phylogenetic relationship of the four IpFAD2s with other reported FAD2s was further elucidated. Similar to the previous report, the selected FAD2 proteins were grouped into two major clades, the house-keeping type and seed-type [16] (Figure 2). House-keeping type FAD2s are constitutively and abundantly expressed, while seed-type FAD2s are specifically or highly expressed in developing seeds [16]. As shown in Figure 2, IpFAD2-1 and IpFAD2-3 belong to the house-keeping group, whereas IpFAD2-2 and IpFAD2-4 fall into the seed-type group.

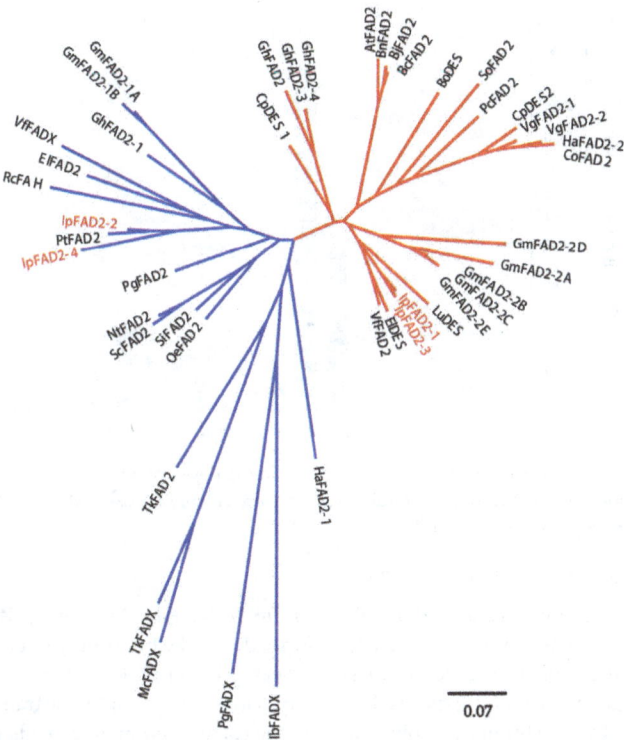

Figure 2. Cladogram of four IpFAD2s and other plant FAD2-related polypeptides. The house-keeping clade and the seed-type clade are labeled in red and blue accordingly. The four FAD2 homologs from *Idesia polycarpa* were designated IpFAD2-1, 2, 3, 4 and were labeled in red. The protein sequences used here were: GmFAD2-1A (Glyma.10G278000.1.p), GmFAD2-1B (Glyma.20G111000.1.p), GmFAD2-2A (Glyma.19G147300.1.p), GmFAD2-2B (Glyma.19G147400.1.p), GmFAD2-2C (Glyma.03G144500.1.p), GmFAD2-2D (Glyma.09G111900.1.p), and GmFAD2-2E (Glyma.15G195200.1.p) from *Glycine max*; CpDES (AAS19533) from *Cucurbita pepo*; IpFAD2-1 (MH394208), IpFAD2-2 (MH394209), IpFAD2-3 (MH394210), and IpFAD2-4 (MK105894) from *Idesia polycarpa*; HaFAD2-1 (AF251842) and HaFAD2-2 (AF251843) from *Helianthus annuus*; IbFADX (AF182520) from *Impatiens balsamina*; PgFADX (AY178446) from *Punica granatum*; McFADX (AF182521) from *Momordica charantia*; TkFADX (AY178444) and TkFAD2 (AY178445) from *Trichosanthes kirilowii*; SiFAD2 (AF192486) from *Sesamum indicum*; ScFAD2 (X92847) from *Solanum commersonii*; RcFAH (EU523112) from *Ricinus communis*; VfFADX (AF525535), and VfFAD2 (AF525534) from *Vernicia fordii*; GhFAD2-1 (X97016), GhFAD2-2 (Y10112), GhFAD2-3 (AF331163), and GhFAD2-4 (AY279315) from *Gossypium hirsutum*; PgFAD2 (AJ437139) from *Punica granatum*; CpDES (AAS19533, and CpDES2 (AAS19533) from *Cucurbita pepo*; AtFAD2 (L26296) from *Arabidopsis thaliana*; BnFAD2 (AF243045) from *Brassica napus*; BjFAD2 (X91139) from *Brassica juncea*; BcFAD2 (AF124360) from *Brassica carinata*; BoDES (AF074324) from *Borago officinalis*; SoFAD2 (AB094415) from *Spinacia oleracea*; PcFAD2 (U86072) from *Petroselinum crispum*; VgFAD2-1 (AF188263) and VgFAD2-2 (AF188264) from *Vernonia galamensis*; CoFAD2 (AF343065) from *Calendula officinalis*; LuDES (ACF49507) from *Linum usitatissimum*.

To identify which gene participated in the production of C18:2 content in fruit, we examined the expression patterns of these genes by RT-PCR analysis in seed and pericarp from fruit 80 days after pollination (DAP) (Figure 3). The expression levels of *IpFAD2-2* far exceeded those of the other *IpFAD2* genes, suggesting *IpFAD2-2* might be mainly responsible for producing high C18:2 content in fruit (Figure 3).

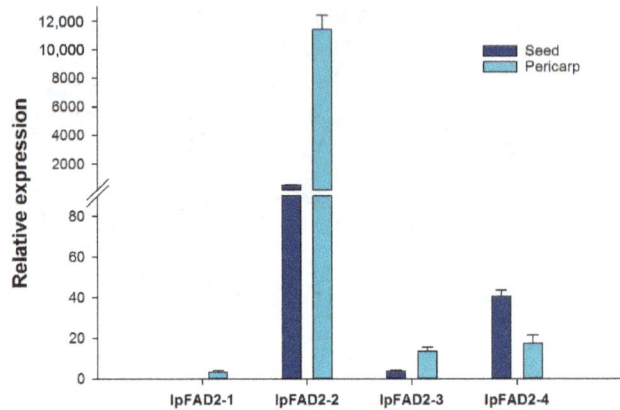

Figure 3. Relative expression levels of IpFAD2 genes in the pericarp and seed of I*desia* polycarpa at 80 days post pollination. Relative expression values were calculated using the $2-\Delta\Delta Ct$ method by using *EF1A* gene as an internal control.

2.2. Three IpFAD2s Possess Desaturase Activity

To examine if the proteins encoded by *IpFAD2-1*, *IpFAD2-2*, *IpFAD2-3*, and *IpFAD2-4* are involved in the desaturation process, four genes were transformed into the budding yeast *S. cerevisiae* INVSc1. Then we checked the expression levels of these transgenes by RT-PCR. As shown in Supplementary Figure S1, all transgenes are highly expressed. The fatty acid compositions in the transformed yeast cells were also analyzed by gas chromatography (GC) and the corresponding fatty acids were confirmed by gas chromatography mass spectrometry (GC-MS) (Supplementary Figure S2). The yeast cells harboring *IpFAD2-1*, *IpFAD2-2* and *IpFAD2-3* produced two novel fatty acids, C16:2 and C18:2, which were not generated in the yeast cells containing blank vector (Figure 4A–D, Supplementary Figure S3), suggesting that the three IpFAD2s have their own catalytic activity. Moreover, the proportion of C18:2 in yeast cells transformed with *IpFAD2-1*, IpFAD2-2, and *IpFAD2-3* is 9.97%, 8.96% and 11.43% of the total fatty acids, much higher than that of C16:2, which only accounts for 2.58%, 4.62%, and 4.45%, respectively. However, only trace amounts of C18:2 (0.66%) and little C16:2 were detected in the yeast cells containing *IpFAD2-4* (Figure 4E). These data indicate that IpFAD2-1, IpFAD2-2, and IpFAD2-3 possess high Δ^{12}-fatty acid desaturation activity using both C16:1 and C18:1 as substrates, furthermore C18:1 is a preferable substrate for all three IpFAD2 proteins, while IpFAD2-4 displays low activity (Figure 4A–E).

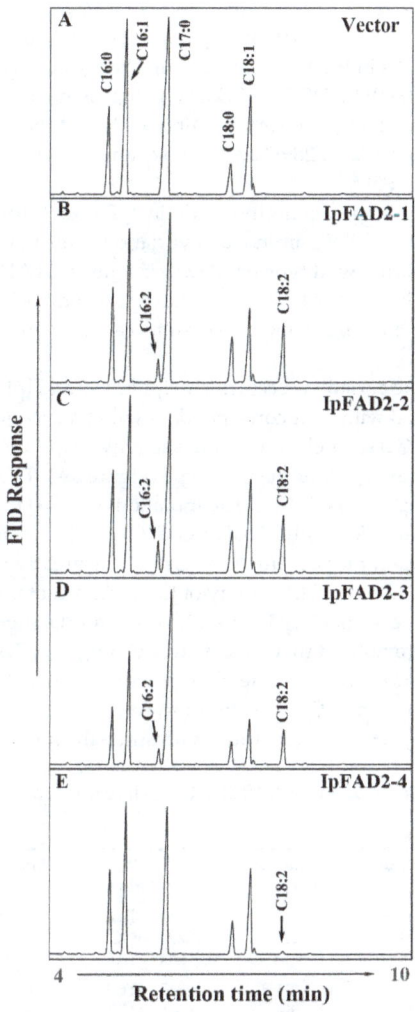

Figure 4. GC analysis of fatty acid methyl esters (FAMEs) isolated from yeast cells expressing IpFAD2s. The FAMEs of total lipid were extracted from yeast transformed with control empty control vector pESC-his (**A**), IpFAD2-1 (**B**), IpFAD2-2 (**C**), IpFAD2-3 (**D**), and IpFAD2-4 (**E**) under induction conditions and analyzed by gas chromatography/flame ionization detector (GC/FID). Major fatty acids peaks are labeled. The newly synthesized fatty acids corresponding to C16:2 Δ9,12 and C18:2 Δ9,12 are indicated by the arrows. Heptadecanoic acid methyl ester (C17:0) is used as the internal standard.

2.3. A Highly Conserved Glycine Residue Identified from the FAD2 Natural Variation is Required for FAD2 Desaturase Activity

Previous reports showed that some natural variations in FAD2 orthologs resulted in an elevated C18:1/C18:2 ratio in oil seed crops with C18:2 as the major fatty acid [10,11,14,23–25]. Thus it is feasible to find some natural FAD2 dysfunction variants in *Idesia ploycarpa*. A small population of five-year-old *Idesia ploycarpa* trees were examined. 32 Single nucleotide polymorphisms (SNPs) were found throughout the CDS region of IpFAD2s. The association between SNPs and the amino acid changes in IpFAD2s was summarized in Table 1. The SNPs caused synonymous mutations in the IpFAD2-1 CDS region, while the SNPs in IpFAD2-2 resulted in changes to five amino acids. The five IpFAD2-2 variants were named IpFAD2-2V1 (Y54S, V164I and V243M), IpFAD2-2V2 (Y54S), IpFAD2-2V3 (A69V), IpFAD2-2V4 (Q115/R and V164I), and IpFAD2-2V5 (V164I and V243M) (Supplementary Figure S4). Two of the five SNPs in IpFAD2-3 resulted in changes to two amino acids. The IpFAD2-3 variant was named as IpFAD2-3V1 (V253I and G376C) (Supplementary Figure S5). Seven of twelve SNPs in IpFAD2-4 resulted in changes to seven amino acids and the IpFAD2-4 variant was named a IpFAD2-4V1 (C151S, F164S, L172F, E289D and C365Y) and IpFAD2-4V2 (P31L, A71V, C151S and F164S) (Supplementary Figure S5).

Since SNPs caused the changes to amino acids in IpFAD2-2, IpFAD2-3, and IpFAD2-4, it is interesting to determine whether these amino acid variations affect desaturase activity of the three proteins. The constructs harboring wild type or FAD2 variants of IpFAD2-2, IpFAD2-3 and IpFAD2-4 were transformed into yeast *S. cerevisiae* INVSc1 and fatty acid composition was examined. The SNPs in IpFAD2-3s brought about the variations of two amino acids (V253I and G376C) in IpFAD2-3V1 (Table 1).

As shown in Figure 5A, the variations occurring in IpFAD2-2 and IpFAD2-4 have little effects on its desaturase activity as compared with their corresponding wild type form, whereas the two amino acid substitutions (V253I and G376C) severely affected the activity of IpFAD2-3V1 since the C18:2 content in IpFAD2-3V1 was only less than 10% of the wild type (Figure 5A). To ascertain which amino acid is responsible for this result, we performed single site mutation on IpFAD2-3 and obtained two *IpFAD2-3* mutants containing a single mutation with V253I or G376C. The G376C mutation caused the dramatic decreasing activity of IpFAD2-3, the percentage of C18:2 was greatly reduced to 8.3% of wild type (Figure 5A, Supplementary Figure S3). The activity of the variant containing V253I mutation was only slightly affected (Figure 5A, Supplementary Figure S3). These results suggested that Gly376 is essential for IpFAD2-3 activity. Here we noticed that Glycine was changed to Cysteine and thus deduced that the redox status might be concerned with the altered activity. To test this possibility, we replaced Gly376 with either alanine or serine. Our results showed that both replacements caused markedly decreased activity (Figure 5A). These results further illustrate the importance of G376 residues.

Table 1. Single nucleotide polymorphisms (SNPs) in the coding region of fatty acid desaturase 2 (*FAD2*) genes from *Idesia ploycarpa*.

FAD2 Gene	SNPSite	Amino Acid Position	SNP Mutation	Amino Acid Mutation	Mutation Type
IpFAD2-1	201	67	TAT→TAC	Tyr	S
(1158 bp)	729	243	GCA→GCG	Ala	S
	765	255	TAT→TAC	Tyr	S
IpFAD2-2	161	54	TAT→TCT	Tyr→Ser	N
(1149 bp)	206	69	GCC→GTC	Ala→Val	N
	344	115	CAG→CGG	Gln→Arg	N
	372	124	ATC→ATT	Ile	S
	399	133	TAC→TAT	Tyr	S
	490	164	AGT→AAT	Ser→Asn	N
	612	204	CGA→CGC	Arg	S
	624	208	CAC→CAT	His	S
	727	243	GTG→ATG	Val→Met	N
	1041	347	GAC→GAT	Asp	S
	1092	364	GTT→GTG	Val	S
	1113	371	CCA→CCC	Pro	S
IpFAD2-3	690	230	GGC→GGT	Gly	S
(1158 bp)	696	232	CTC→CTT	Leu	S
	729	243	GTC→GTA	Val	S
	757	253	GTT→ATT	Val→Ile	N
	1126	376	GGC→TGC	Gly→Cys	N

Table 1. *Cont.*

FAD2 Gene	SNPSite	Amino Acid Position	SNP Mutation	Amino Acid Mutation	Mutation Type
IpFAD2-4	92	31	CCC→CTC	Pro→Leu	N
(1143 bp)	212	71	GCC→GTC	Ala→Val	N
	279	93	CTA→CTC	Leu	S
	451	151	TGC→AGC	Cys→Ser	N
	480	160	CCA→CCG	Pro	S
	491	164	TTC→TCC	Phe→Ser	N
	514	172	CTC→TTC	Leu→Phe	N
	531	177	CCT→CCA	Pro	S
	867	289	GAA→GAT	Glu→Asp	N
	999	333	GCA→GCT	Ala	S
	1002	334	ACT→ACA	Thr	S
	1094	365	TGT→TAT	Cys→Tyr	N

S represents synonymous, N represents nonsynonymous.

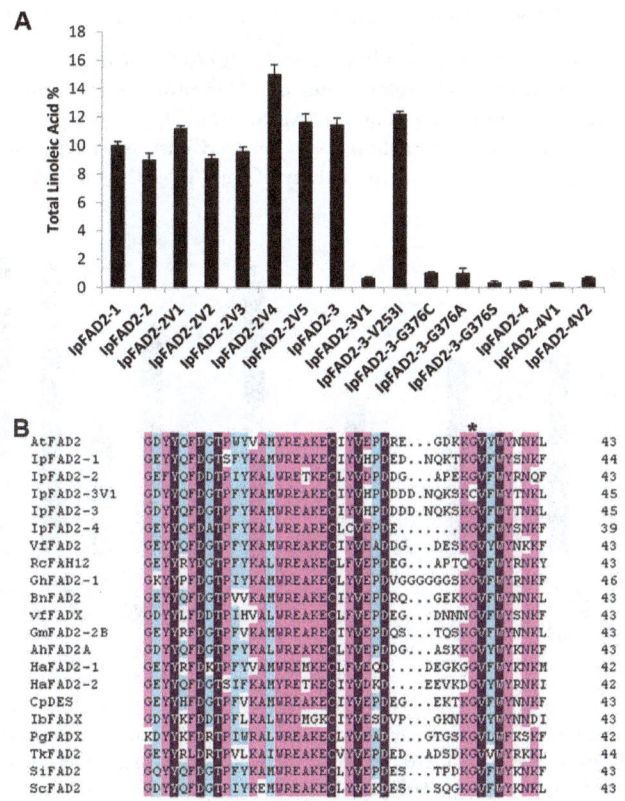

Figure 5. Total C18:2 accumulation in yeast transformed with different FAD2 alleles from *Idesia polycarpa*. (**A**) Total LA accumulation in transgenic yeast expressing natural FAD2 alleles and site-directed mutations from *Idesia polycarpa*. (**B**) Alignment of C-terminal amino acids of 21 FAD2 genes. The GenBank accession numbers of the proteins presented in this figure are shown in Figure 2. The asterisk indicates the conserved Glycine residue changed in IpFA2-3V1. The shading colors represent the identity level of amino acids. Black, magenta, and cyan indicate 100%, ~90%, and ~55% identity, respectively.

To find out the reasons why the substitution of the conserved glycine severely affects the activity of IpFAD2-3, we carefully examined plant FAD2 protein structure. This glycine residue is highly conserved across plant FAD2 proteins (Figure 5B). It is far away from the catalytic center consisting of three histidine-rich motifs [26], but is at -9 position relative to the C-terminus and adjacent to the mini ER retrieval sequence motif (Φ-X-X-K/R/D/E-Φ-COOH, YTNKL in the case of IpFAD2-3) [19]. Thus we hypothesized that the impacts of the glycine residue on IpFAD2-3 might act through disturbing the precise location of IpFAD2-3. To test this hypothesis, we made IpFAD2-3-GFP and IpFAD2-3-G376C-GFP constructs and co-transformed them with an ER membrane marker (CD3-959) into tobacco epidermal cells. Each fluorescent fusion protein was co-localized with the ER membrane marker CD3-959 (Supplementary Figure S6), consistent with the expression pattern of the wild type protein, indicating that G376C does not interrupt the localization of IpFAD2-3 and its impacts on enzyme activity could not act through mis-localizing the protein. The GFP fluorescent signals of tobacco epidermal cells containing either of IpFAD2-3-GFP or IpFAD2-3-G376C-GFP were similar to each other, suggesting that G376C does not affect IpFAD2-3 at protein level (Supplementary Figure S6).

2.4. Gly376 Has Different Effects on the Activity of FAD2 Proteins among Different Species

Since the highly conserved Gly376 is important for IpFAD2-3 activity, we wondered if it has a conserved function in all IpFAD2 proteins and in FAD2 proteins from other species. Firstly, we introduced the mutation into IpFAD2-1 and IpFAD2-2. The mIpFAD2-1 and mIpFAD2-2 variants display markedly reduced desaturase activity as compared with wild type, suggesting that the conserved glycine is also required for enzyme activity of both IpFAD2-1 and IpFAD2-2 (Figure 6).

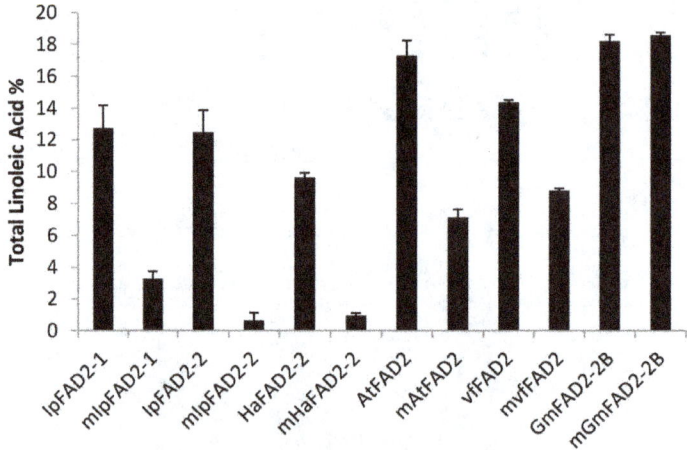

Figure 6. Total C18:2 accumulation in yeast containing FAD2 orthologs from different species. Levels of total C18:2 accumulation in yeast cells expressing plant FAD2s and the corresponding mutated form with the conserved glycine (G376, 375 or 374) substituted by cysteine. Ha, *Helianthus annuus*; At, *Arabidopsis thaliana*; Vf, *Vernicia fordii*; Gm, *Glycine max*. The corresponding mutated residues are: IpFAD2-1, G376C; IpFAD2-2, G374C; HaFAD2-2, G375C; AtFAD2, G375C; VfFAD2, G375C; GmFAD2-2B, G375C. "m" in front of each protein name represents the mutated form of the corresponding FAD2. Mean ± SD; *n* = 3.

To check if the conserved glycine also has similar roles across plant FAD2 proteins, we cloned *FAD2* genes from *Helianthus annuus*, *Arabidopsis thaliana*, *Vernicia fordii* and *Glycine max*, which are *HaFAD2-2, AtFAD2, VfFAD2* and *GmFAD2-2B* accordingly, and also replaced this glycine with cysteine in these FAD2 proteins. The wild type and mutated constructs were then transformed into yeast and FA composition was detected by GC. As shown in Figure 6, the yeast cells containing each wild-type

FAD2s produced a certain amount of C18:2, ranging from 9% to 18% of total fatty acids, indicating that all these selected FAD2s have functional Δ^{12} oleate desaturase activity in heterologous yeast cells. We then checked the C18:2 levels of the yeast cells containing the mutated constructs and found that the effects of the mutation (G376/C) on the activity of FAD2 proteins varied among different proteins. The levels of C18:2 in yeast cells containing the mutated HaFAD2-2 were markedly decreased to less than 10% of wild type (Figure 6). The activity of mutated AtFAD2 and VfFAD2 only showed moderate reduction since the mutants showed about 58% and 38% reductions. The mutated glycine in GmFAD2-2B has no effects on the production of C18:2 (Figure 6). Taken together, these results suggested that the effects of the highly conserved glycine at 376 position of IpFAD2-3 on plant FAD2 activity probably act in a species/protein-specific manner.

3. Discussion

The *FAD2* gene was first identified in Arabidopsis [27], since then many *FAD2* genes have been cloned from different plants [26,28–33]. To date, none of the microsomal Δ^{12} fatty acid desaturases associated with C18:2 have been identified in *Idesia ploycarpa*. Here we isolated four *IpFAD2* genes. Deduced amino acid sequences alignment showed that the four IpFAD2s shared 69.5–78.5% identity with AtFAD2, The *Idesia polycarpa* fruit, consisting of pericarp and seed, produces large amounts of fatty acids, among which about 83.92% of C18:2 is present in pericarp oils, that far exceeds seed oils [5]. To gain insight into the oil accumulation mechanism in pericarp and seed, we studied the expression patterns of four *IpFAD2* paralogues in these two organs by real time-PCR. Each gene shows distinct expression patterns (Figure 3). *IpFAD2-2* is highly expressed in both organs, its transcripts far exceed those of other *IpFAD2* genes, suggesting that *IpFAD2-2* might be a major gene responsible for C18:2 production in the fruit of *Idesia polycarpa*. The expression levels of *IpFAD2-2* varied between two parts, i.e., higher in pericarps and lower in seeds, suggesting its different roles in the two organs. Taken together, *IpFAD2-2* might be a candidate for genetically modifying the ratio of C18:1/C18:2 in *Idesia ploycarpa* fruits in the future. As compared with *IpFAD2-2*, the expression levels of the other three *IpFAD2* genes are very low in fruit (Figure 3). They also exhibit the distinct expression patterns found in pericarps and seeds. The expression of *IpFAD2-3* displays little differences between the two organs while *IpFAD2-1* was preferably expressed in the pericarp. The different expression patterns of these *IpFAD2* genes might cause the different ratios of C18:1/C18:2 in pericarps and seeds.

To identify the activity of IpFAD2s, we cloned these four genes and introduced them into yeast cells. With the exception of IpFAD2-4, the three other IpFAD2 proteins efficiently converted C18:1 to C18:2 in the yeast system (Figure 4). To investigate the reason why IpFAD2-4 displayed low activity, we carefully compared the amino acid sequences of IpFAD2-4 with that of the other proteins and found that IpFAD2-4 shared 85.1% with *IpFAD2-2* (Figures 1 and 2), suggesting that both genes might be derived from duplication events. RT-PCR results showed that it displayed similar patterns to *IpFAD2* though its expression levels are far lower than the latter. But the two proteins showed distinct activity in yeast cells. To precisely compare the activity of these proteins, we need to transform it into plants for functional identification in the near future.

With the rapid advances in biotechnology, genetic engineering has been widely used in identifying gene function or modifying plant quality in the lab due to its simplicity and easy-of-use. But its extensive application in nature is largely restricted since the impacts of genetically modified plants on nature are unpredictable. Natural variants have endured long-term natural selection and natural mutation is thus more reliable, stable, less toxic, and desirable for variety breeding. Natural variations in FAD2 coding region correlated with the C18:1 content have been identified from different plants. D150N and H101D from peanut, S117N and P137R from soybean have been shown to decrease the activity of FAD2 [10,34]. Most substitutions were in or near the His-box, which makes up the catalytic center. Here we checked 32 SNP sites present in FAD2 coding region among 30 *Idesia ploycarpa* natural variants. These SNP sites caused fourteen amino acid substitutions and twelve IpFAD2 alleles. Our data revealed that only the substitution of the highly conserved Gly376 severely affects the IpFAD2-3 activity (Figure 5A), suggesting its important role for IpFAD2-3. Its replacement also severely disrupted the activity of IpFAD2-1 and IpFAD2-2 (Figure 5A). We also expand its mutation to other plant FAD2 proteins and found that this glycine more or less affects the function of most of the tested plant FAD2 proteins except for GmFAD2-2B (Figure 6). Till now it is unknown how the glycine mutation causes the reduced products catalyzed by IpFAD2-3 proteins. Its mutation does not to affect the protein levels since the mutated protein can be normally expressed as shown in Supplementary Figure S6. The effects of this conserved glycine on enzyme activity might be associated with other unknown factors such as protein structure formation, phosphorylation and etc. All in all, the identified G376 in IpFAD2-3 could be a potential site for the manipulation of the desaturase activity of IpFAD2 by genome editing in the near future, and it will be applicable for genetically improving crop quality.

4. Materials and Methods

4.1. Plant Materials

The fresh leaves of 26 five-year-old *Idesia ploycarpa* female trees, which are growing at Huanggang, Hubei province, China, were collected, mixed and quickly frozen in liquid nitrogen for further RNA extraction. Col-0 *Arabidopsis thaliana* are growing in greenhouse condition. Sunflower (*Helianthus annuus*), tung tree (*Vernicia fordi*), and soybean (*Glycine max*) were collected from Wuhan Botanical Garden, Wuhan, China. The 80 days after pollination (DAP) fruits of *Idesia ploycarpa* cultivar 76A were quickly frozen in liquid nitrogen and stored at $-80\,^\circ$C until use.

4.2. Total RNA Extraction and Complementary DNA Synthesis

Total RNA was isolated from 100 mg of frozen leaves and seeds and pericarps of 80 DAP *Idesia ploycarpa* fruit with Trizol reagent (Life Technologies Corporation, Carlsbad, CA, USA) according to the manufacturer's protocol. RNA concentration was determined by NanoDropTM spectrophotometer ND2000 (Thermo Fisher Scientific, Wilmington, DE, USA). Total RNA was then treated with DNase I (Thermo Fisher Scientific, Wilmington, DE, USA) to eliminate residue DNA. About 500 ng DNA-free RNA was used as a template for first-strand complementary DNA (cDNA) synthesis. Reverse-transcription was performed with the M-MLV Reverse Transcriptase (Promega, Madison, WI, USA) and oligo(dT)$_{20}$ primer (Tsingke, Wuhan, China).

4.3. Gene Cloning and Sequence Analysis

Using leaf cDNA as template, we cloned the coding sequences of different *IpFAD2* with corresponding primers listed in Supplementary Table S2. The primers were designed according to the sequence submitted by Li et al. [5]. Due to lack of upstream sequence information, a degenerate primer was designed as the forward primer according to sequence homology used for *IpFAD2-4* cloning. The FAD2 fragments obtained were cloned into the pESC-his vector (Alilent Technologies, Santa Clara, CA, USA) directly and then sequenced. Multiple sequence alignments were performed using DNAman software. For phylogenetic relationship analysis, the protein sequences of IpFAD2 protein sequence and a number of plant FAD2 homologs were aligned with MAFFT v7.154b [35]. Maximum-likelihood (ML) tree was generated using FastTree v2.1.7 [36] and was visualized using FigTree (Available online: http://tree.bio.ed.ac.uk/software/figtree/).

4.4. Real-Time Quantitative PCR

Gene expression analysis was performed by RT-PCR using Applied Biosystems 7500 Fast Real-Time PCR System (Thermo Fisher Scientific, Wilmington, DE, USA). Primers with Tm (melting temperature) 60 $^\circ$C and 18–20 bp in length were designed by Primer 3 (Supplementary Table S2). IpEF1A was selected as the internal reference gene. PCR reaction mix (20 μL per well) contained 10 μL TB Green Premix Ex Taq II (TliRNaseH Plus) (2X) (Takara, Tokyo, Japan), 0.8 μL forward and reverse primers (10 μM) (Tsingke, Wuhan, China), 0.4 μL ROX Reference Dye II (Takara, Tokyo, Japan), 50 ng cDNA and RNA-free water. The two-step thermal cycling conditions were 95 °C for 30 s, followed by 40 cycles of 95 °C for 5 s, 60 °C for 34 s. Corresponding gene expression level was analyzed with the 2-$\Delta\Delta$Ct method. Elongation factor 1-alpha was used as the internal control to normalize the relative amount of mRNAs for all samples.

4.5. Site-Directed Mutagenesis

Mutagenesis was done according to the fast mutagenesis system (Transgen, Beijing, China). In brief, the mutated plasmids were amplified with two primers containing the mutations (Supplementary Table S2) using the TransStart FastPfu DNA polymerase (Transgen, Beijing, China). The PCR conditions were as follows: initial denaturation at 94 °C for 5 min, followed by 25 cycles of 94 °C for 20 s, 55 °C for 20 s, and 72 °C for 3 min, final extension at 72 °C for 10 min. The amplicons were subsequently digested with DMT (Transgen, Beijing, China) enzyme for eliminating the methylated parental plasmid and then purified from agrose gels. The purified products were transformed into DMT competent cells (Transgen, Beijing, China). The mutated clones selected on plates containing antibiotics were verified by sequencing.

4.6. Yeast Transformation and Heterologous Expression of IpFAD2 Variants

Constructs containing the *IpFAD2* genes were transformed into *Saccharomyces cerevisiae* INVSc1 (Invitrogen, Carlsbad, CA, USA) by the LiAc/SS carrier DNA/PEG method. Transformants were incubated in yeast nitrogen base (YNB) liquid medium at 28 °C for 36 h with rotary shaking at 180 rpm and then spread on synthetic defined medium without histidine (SD-his, Clonetech, Mountain View, CA, USA) solidified medium supplemented with glucose. The colonies growing on SD-his medium were then cultured in SD-his liquid medium for another two days and then centrifuged. The pellets were washed with distilled water twice and sub-cultured in SD-his liquid medium containing galactose (2%, *w/v*) for 48 h. Yeast cells were collected for fatty acid analysis.

4.7. Analysis of Fatty Acid Composition in Yeast

The induced yeast cells transformed with *FAD2* cDNA fragments were pelleted and washed with distilled water twice, and total lipids were extracted with hexane and methylated with 5M KOH-methanol. The heptadecanoic acid methyl ester (C17:0) was used as the internal standard. Fatty acid methyl esters were measured by gas chromatography with an Agilent 7820A (Alilent Technologies, Santa Clara, CA, USA). The samples were separated on an Agilent DB-23 capillary column (Alilent Technologies, Santa Clara, CA, USA). The column temperature was programmed with an initial temperature of 180 °C for 1 min, ramping at 3 °C/min to 240 °C, and then holding for 39 min.

4.8. Subcellular Localization Assay

The coding sequences of IpFAD2-3 and IpFAD2-3V1 were amplified with specific primers (Supplemental Table S2). The amplified fragments were cloned into the PMDC83 vector, which generated Pro35S::IpFAD2-3::GFP and Pro35S::IpFAD2-3V1::GFP fusion constructs. The obtained plasmids were transferred into *Agrobacterium tumefaciens* (GV3101) using the freeze-thaw method, and subsequently transformed into leaves of *Nicotiana benthamiana* by infiltration. To precisely localize which compartments the FAD2 proteins reside in, CD3-959 (35S-ER-mCherry), an ER marker [37] was co-transformed with the FAD proteins. The fluorescent signals generated by GFP and mCherry fusion proteins were observed by confocal microscopy.

5. Conclusions

We identified four *FAD2* homologs from fruits of *Idesia polycarpa*. Heterologous expression in yeast showed that three IpFAD2s have strong Δ^{12} fatty acid desaturase activity. Natural variation together with site-directed mutagenesis analysis reveals one natural variation (G376C in IpFAD2-3) that strongly hinders the catalytic activity of IpFAD2-3. Even though this amino acid is highly conserved among plant FAD2 proteins, the effects of its mutation on the Δ^{12} oleate desaturase activity of tested FAD2 proteins are different. Our findings will be helpful to advance understanding the roles of FAD2 proteins in woody plants and also provide a new potential site of IpFAD2s for modifying the ratio of C18:1/C18:2 of *Idesia polycarpa* fruit in the future through genetic engineering. Further characterization of the mechanisms of the effects of G376C substitutions in different FAD2s, either at the enzyme activity level or on other modulatory molecules, is currently under way in our group.

Supplementary Materials: Supplementary materials can be found at http://www.mdpi.com/1422-0067/19/12/3932/s1.

Author Contributions: Conceptualization, P.W. and L.Z.; methodology, T.F. and W.L.; investigation, P.W.; resources, J.L.; writing—original draft preparation, P.W.; writing—review and editing, S.L. and H.Z.; funding acquisition, S.L.

Funding: This research was funded by Biological Resources Service Network from Chinese Academy of Sciences, grant number kfj-brsn-2018-6-007.

Conflicts of Interest: The authors declare no conflict of interest.

References

1. Dyer, J.M.; Stymne, S.; Green, A.G.; Carlsson, A.S. High-value oils from plants. *Plant J.* **2008**, *54*, 640–655. [CrossRef] [PubMed]

2. Lummiss, J.A.M.; Oliveira, K.C.; Pranckevicius, A.M.T.; Santos, A.G.; dos Santos, E.N.; Fogg, D.E. Chemical Plants: High-Value Molecules from Essential Oils. *J. Am. Chem. Soc.* **2012**, *134*, 18889–18891. [CrossRef] [PubMed]

3. Eriksson, D.; Merker, A. Cloning and functional characterization of genes involved in fatty acid biosynthesis in the novel oilseed crop *Lepidium campestre* L. *Plant Breed.* **2011**, *130*, 407–409. [CrossRef]

4. Yang, F.X.; Su, Y.Q.; Li, X.H.; Zhang, Q.; Sun, R.C. Preparation of biodiesel from *Idesia polycarpa var.* vestita fruit oil. *Ind. Crop Prod.* **2009**, *29*, 622–628. [CrossRef]

5. Li, R.J.; Gao, X.; Li, L.M.; Liu, X.L.; Wang, Z.Y.; Lu, S.Y. De novo assembly and characterization of the fruit transcriptome of *idesia polycarpa* reveals candidate genes for lipid biosynthesis. *Front. Plant Sci.* **2016**, *7*, 801. [CrossRef] [PubMed]

6. Pandey, M.K.; Wang, M.L.; Qiao, L.X.; Feng, S.P.; Khera, P.; Wang, H.; Tonnis, B.; Barkley, N.A.; Wang, J.P.; Holbrook, C.C.; et al. Identification of QTLs associated with oil content and mapping *FAD2* genes and their relative contribution to oil quality in peanut (*Arachis hypogaea* L.). *BMC Genet.* **2014**, *15*, 133. [CrossRef] [PubMed]

7. Ge, Y.; Chang, Y.; Xu, W.L.; Cui, C.S.; Qu, S.P. Sequence variations in the *FAD2* gene in seeded pumpkins. *Genet. Mol. Res.* **2015**, *14*, 17482–17488. [CrossRef] [PubMed]

8. Qu, J.; Mao, H.Z.; Chen, W.; Gao, S.Q.; Bai, Y.N.; Sun, Y.W.; Geng, Y.F.; Ye, J. Development of marker-free transgenic *Jatropha* plants with increased levels of seed oleic acid. *Biotechnol. Biofuels* **2012**, *5*, 10. [CrossRef] [PubMed]

9. Miquel, M.; Browse, J. Arabidopsis mutants deficient in polyunsaturated fatty acid synthesis. Biochemical and genetic characterization of a plant oleoyl-phosphatidylcholine desaturase. *J. Biol. Chem.* **1992**, *267*, 1502–1509. [PubMed]

10. Pham, A.T.; Lee, J.D.; Shannon, J.G.; Bilyeu, K.D. Mutant alleles of *FAD2-1A* and *FAD2-1B* combine to produce soybeans with the high oleic acid seed oil trait. *BMC Plant Biol.* **2010**, *10*, 195. [CrossRef]

11. Janila, P.; Pandey, M.K.; Shasidhar, Y.; Variath, M.T.; Sriswathi, M.; Khera, P.; Manohar, S.S.; Nagesh, P.; Vishwakarma, M.K.; Mishra, G.P.; et al. Molecular breeding for introgression of fatty acid desaturase mutant alleles (*ahFAD2A* and *ahFAD2B*) enhances oil quality in high and low oil containing peanut genotypes. *Plant Sci.* **2016**, *242*, 203–213. [CrossRef] [PubMed]

12. Cao, S.; Zhou, X.R.; Wood, C.C.; Green, A.G.; Singh, S.P.; Liu, L.; Liu, Q. A large and functionally diverse family of *Fad2* genes in safflower (*Carthamus tinctorius* L.). *BMC Plant Biol.* **2013**, *13*, 5. [CrossRef] [PubMed]

13. Shockey, J.; Dowd, M.; Mack, B.; Gilbert, M.; Scheffler, B.; Ballard, L.; Frelichowski, J.; Mason, C. Naturally occurring high oleic acid cottonseed oil: Identification and functional analysis of a mutant allele of *Gossypium barbadense fatty acid desaturase-2. Planta* **2017**, *245*, 611–622. [CrossRef] [PubMed]

14. Sturtevant, D.; Horn, P.; Kennedy, C.; Hinze, L.; Percy, R.; Chapman, K. Lipid metabolites in seeds of diverse *Gossypium* accessions: Molecular identification of a high oleic mutant allele. *Planta* **2017**, *245*, 595–610. [CrossRef] [PubMed]

15. Unver, T.; Wu, Z.; Sterck, L.; Turktas, M.; Lohaus, R.; Li, Z.; Yang, M.; He, L.; Deng, T.; Escalante, F.J.; et al. Genome of wild olive and the evolution of oil biosynthesis. *Proc. Natl. Acad. Sci. USA* **2017**, *114*, E9413–E9422. [CrossRef] [PubMed]

16. Dar, A.A.; Choudhury, A.R.; Kancharla, P.K.; Arumugam, N. The *FAD2* gene in plants: Occurrence, regulation, and role. *Front. Plant Sci.* **2017**, *8*, 1789. [CrossRef] [PubMed]

17. Avelange-Macherel, M.H.; Macherel, D.; Wada, H.; Murata, N. Site-directed mutagenesis of histidine residues in the delta 12 acyl-lipid desaturase of Synechocystis. *FEBS Lett.* **1995**, *361*, 111–114. [CrossRef]

18. Broun, P.; Shanklin, J.; Whittle, E.; Somerville, C. Catalytic plasticity of fatty acid modification enzymes underlying chemical diversity of plant lipids. *Science* **1998**, *282*, 1315–1317. [CrossRef] [PubMed]

19. McCartney, A.W.; Dyer, J.M.; Dhanoa, P.K.; Kim, P.K.; Andrews, D.W.; McNew, J.A.; Mullen, R.T. Membrane-bound fatty acid desaturases are inserted co-translationally into the ER and contain different ER retrieval motifs at their carboxy termini. *Plant J.* **2004**, *37*, 156–173. [CrossRef]

20. Hoffmann, M.; Hornung, E.; Busch, S.; Kassner, N.; Ternes, P.; Braus, G.H.; Feussner, I. A small membrane-peripheral region close to the active center determines regioselectivity of membrane-bound fatty acid desaturases from *Aspergillus nidulans. J. Biol. Chem.* **2007**, *282*, 26666–26674. [CrossRef]

21. Los, D.A.; Murata, N. Structure and expression of fatty acid desaturases. *BBA-Lipids Lipid Met.* **1998**, *1394*, 3–15. [CrossRef]

22. Shanklin, J.; Cahoon, E.B. Desaturation and related modifications of fatty acids. *Annu. Rev. Plant Phys.* **1998**, *49*, 611–641. [CrossRef] [PubMed]

23. Jung, S.; Swift, D.; Sengoku, E.; Patel, M.; Teule, F.; Powell, G.; Moore, K.; Abbott, A. The high oleate trait in the cultivated peanut [*Arachis hypogaea* L.]. I. Isolation and characterization of two genes encoding microsomal oleoyl-PC desaturases. *Mol. Gen. Genet.* **2000**, *263*, 796–805. [CrossRef] [PubMed]

24. Thapa, R.; Carrero-Colon, M.; Crowe, M.; Gaskin, E.; Hudson, K. Novel *FAD2-1A* alleles confer an elevated oleic acid phenotype in soybean seeds. *Crop Sci.* **2016**, *56*, 226–231. [CrossRef]

25. Ustun, R.; Uzun, B. Breeding for introgression of *FAD2-1A* and *FAD2-1B* genes to local soybean cultivars of Turkey. *J. Biotechnol.* **2017**, *256*, S103. [CrossRef]

26. Dyer, J.M.; Chapital, D.C.; Kuan, J.C.W.; Mullen, R.T.; Turner, C.; McKeon, T.A.; Pepperman, A.B. Molecular analysis of a bifunctional fatty acid conjugase/desaturase from tung. Implications for the evolution of plant fatty acid diversity. *Plant Physiol.* **2002**, *130*, 2027–2038. [CrossRef] [PubMed]

27. Okuley, J.; Lightner, J.; Feldmann, K.; Yadav, N.; Lark, E.; Browse, J. Arabidopsis *Fad2* gene encodes the enzyme that is essential for polyunsaturated lipid-synthesis. *Plant Cell* **1994**, *6*, 147–158. [CrossRef] [PubMed]

28. Kishore, K.; Sinha, S.K.; Kumar, R.; Gupta, N.C.; Dubey, N.; Sachdev, A. Isolation and characterization of microsomal omega-6-desaturase gene (*fad2-1*) from soybean. *Indian J. Exp. Biol.* **2007**, *45*, 390–397.

29. Li, L.Y.; Wang, X.L.; Gai, J.Y.; Yu, D.Y. Molecular cloning and characterization of a novel microsomal oleate desaturase gene from soybean. *J. Plant Physiol.* **2007**, *164*, 1516–1526. [CrossRef]

30. Pirtle, I.L.; Kongcharoensuntorn, W.; Nampaisansuk, M.; Knesek, J.E.; Chapman, K.D.; Pirtle, R.M. Molecular cloning and functional expression of the gene for a cotton Delta-12 fatty acid desaturase (*FAD-2*). *BBA-Gene Struct. Expr.* **2001**, *1522*, 122–129. [CrossRef]

31. Sakai, H.; Kajiwara, S. Cloning and functional characterization of a Delta 12 fatty acid desaturase gene from the basidiomycete *Lentinula edodes. Mol. Genet. Genomics* **2005**, *273*, 336–341. [CrossRef] [PubMed]

32. Schierholt, A.; Becker, H.C.; Ecke, W. Mapping a high oleic acid mutation in winter oilseed rape (*Brassica napus* L.). *Theor. Appl. Genet.* **2000**, *101*, 897–901. [CrossRef]

33. Tang, G.Q.; Novitzky, W.P.; Griffin, H.C.; Huber, S.C.; Dewey, R.E. Oleate desaturase enzymes of soybean: Evidence of regulation through differential stability and phosphorylation. *Plant J.* **2005**, *44*, 433–446. [CrossRef] [PubMed]

34. Bruner, A.C.; Jung, S.; Abbott, A.G.; Powell, G.L. The naturally occurring high oleate oil character in some peanut varieties results from reduced oleoyl-PC desaturase activity from mutation of aspartate 150 to asparagine. *Crop Sci.* **2001**, *41*, 522–526. [CrossRef]
35. Katoh, K.; Standley, D.M. MAFFT multiple sequence alignment software version 7: Improvements in performance and usability. *Mol. Biol. Evol.* **2013**, *30*, 772–780. [CrossRef]
36. Price, M.N.; Dehal, P.S.; Arkin, A.P. FastTree 2-approximately maximum-likelihood trees for large alignments. *PLoS ONE* **2010**, *5*, e9490.
37. Nelson, B.K.; Cai, X.; Nebenfuhr, A. A multicolored set of in vivo organelle markers for co-localization studies in Arabidopsis and other plants. *Plant J.* **2007**, *51*, 1126–1136. [CrossRef]

 International Journal of
Molecular Sciences

Article

iTRAQ-Based Proteomic Analysis of Ogura-CMS Cabbage and Its Maintainer Line

Fengqing Han [1], Xiaoli Zhang [2], Limei Yang [1], Mu Zhuang [1], Yangyong Zhang [1], Zhansheng Li [1], Zhiyuan Fang [1] and Honghao Lv [1,*]

1   Institute of Vegetables and Flowers, Chinese Academy of Agricultural Sciences, Key Laboratory of Biology and Genetic Improvement of Horticultural Crops, Ministry of Agriculture, Beijing 100081, China; feng857142@163.com (F.H.); yanglimei@163.com (L.Y.); zhuangmu@163.com (M.Z.); zhangyangyong@163.com (Y.Z.); lizhansheng@163.com (Z.L.); fangzhiyuan@163.com (Z.F.)
2   Tianjin Kernel Vegetable Research Institute, The National Key Laboratory of Vegetable Germplasm Innovation, The Enterprise key Laboratory of Tianjin Vegetable Genetics and Breeding, Jinjing Road, Xiqing District, Tianjin 300384, China; zxl19871009@163.com
*   Correspondence: lvhonghao@caas.cn; Tel.: +86-010-621-35629

Received: 21 September 2018; Accepted: 8 October 2018; Published: 15 October 2018

Abstract: Ogura cytoplasmic male sterility (CMS) contributes considerably to hybrid seed production in *Brassica* crops. To detect the key protein species and pathways involved in Ogura-CMS, we analysed the proteome of the cabbage Ogura-CMS line CMS01-20 and its corresponding maintainer line F01-20 using the isobaric tags for the relative and absolute quantitation (iTRAQ) approach. In total, 162 differential abundance protein species (DAPs) were identified between the two lines, of which 92 were down-accumulated and 70 were up-accumulated in CMS01-20. For energy metabolism in the mitochondrion, eight DAPs involved in oxidative phosphorylation were down-accumulated in CMS01-20, whereas in the tricarboxylic acid (TCA) cycle, five DAPs were up-accumulated, which may compensate for the decreased respiration capacity and may be associated with the elevated O_2 consumption rate in Ogura-CMS plants. Other key protein species and pathways involved in pollen wall assembly and programmed cell death (PCD) were also identified as being male-sterility related. Transcriptome profiling revealed 3247 differentially expressed genes between the CMS line and the fertile line. In a conjoint analysis of the proteome and transcriptome data, 30 and 9 protein species/genes showed the same and opposite accumulation patterns, respectively. Nine noteworthy genes involved in sporopollenin synthesis, callose wall degeneration, and oxidative phosphorylation were presumably associated with the processes leading to male sterility, and their expression levels were validated by qRT-PCR analysis. This study will improve our understanding of the protein species involved in pollen development and the molecular mechanisms underlying Ogura-CMS.

Keywords: *Brassica oleracea*; Ogura-CMS; iTRAQ; transcriptome; pollen development

1. Introduction

Cabbage (*Brassica oleracea* L. var. *capitata*) is an important leafy vegetable cultivated worldwide, and it provides substantial amounts of fibre, vitamins, mineral elements and health-promoting nutrients. The Food and Agriculture Organization of the United Nations reported that the global harvested area of vegetables in 2014 was 20,119,000 ha, with cabbage and other cole crops accounting for approximately 12% of the total (2,470,000 ha; see http://faostat3.fao.org/).

Commercially available cabbage mainly consists of hybrid cultivars because of their significant levels of heterosis. Cross-pollination in hybrid seed production is mainly accomplished using male sterility and self-incompatibility. Self-incompatibility has several limitations, such as poor seed purity and high costs of parental reproduction, whereas male sterility is generally more reliable and economically effective [1,2]. Male sterility encompasses genic male sterility (GMS), caused by nuclear genes, and cytoplasmic male sterility (CMS), caused by interactions between mitochondrial and nuclear genes [3]. Currently, CMS represents the most widely used breeding tool in cabbage hybrid seed production [4,5]. Ogura cytoplasmic male sterility (Ogura-CMS) was discovered in radish [6] and has been transferred to several *Brassica* species [5,7–9]. The original Ogura-CMS *B. oleracea* line exhibits poor agronomic traits, which have been improved by protoplast fusion [10]. Ogura-CMS is stable and easy to transfer between species; thus, it has become one of the most important types of CMS in *B. oleracea* [11].

In addition to the crucial breeding role of CMS in harnessing heterosis, it provides important materials for studying gametophyte development and mitochondrial–nuclear interactions, etc. [3]. CMS has been observed in approximately 200 species and is inherited maternally [12]. At least 17 CMS-related genes have been studied at the genetic and molecular levels [13]. These loci share similar characteristics, with CMS primarily caused by either novel mitochondrial open reading frames (ORFs), which generally result from rearrangements or recombination events in the mitochondrial genome [14]. Fertility restoration is mediated by nuclear-encoded fertility restorer (Rf) genes, most of which encode pentatricopeptide repeat (PPR) proteins that counteract the influence of CMS-associated genes [15,16].

Two hypothetical CMS pathways have been proposed: (I) CMS inhibits energy production by disrupting the mitochondrial electron transport chain complex; and (II) CMS impairs the normal growth of cells, at least in *E. coli* and/or yeast, because its products are cytotoxic [13,17]. In *Brassica* and *Raphanus* species, Ogura-CMS is caused by a mitochondrial gene named *orf138*, which encodes a subunit of a large mitochondrial membrane complex ORF138 protein [18–20]. Both plant membrane fractionation and analyses of *E. coli* have suggested that the ORF138 protein forms oligomers in the inner mitochondrial membrane of male-sterile plants, which is similar to another CMS protein, T-URF1321. Although the ORF138 protein severely inhibits bacterial growth, it does not affect respiration [21]. In a later study, Duroc et al. reported that the complex formed by the ORF138 protein in the inner mitochondrial membrane exerted an uncoupling effect because the mitochondria isolated from sterile plants consumed more oxygen than those of fertile plants, and this uncoupling effect was compensated at the cell and tissue levels, especially in vegetative tissues/organs, although the compensatory effects were apparently not efficient in male reproductive organs [17]. Despite these genetic and molecular studies, the mechanisms underlying the interference exerted by CMS genes on male gametophyte development are largely unknown.

High-throughput next-generation sequencing (NGS) has facilitated transcriptome analyses of Ogura-CMS materials (including Chinese cabbage, cabbage, broccoli) and may help provide a comprehensive understanding of the mechanisms underlying Ogura-CMS [22–25]. Xing et al. performed transcriptome and proteome analyses (focused on the transcriptome) and identified gibberellin, and sporopollenin synthesis as important pathways in Ogura-CMS cabbage [25]. Other studies performing proteomic analyses using two-dimensional gel electrophoresis (2-DE) have identified distinct differences in the proteomes of Ogura-CMS and fertile plants [26,27], including the down-accumulated protein species associated with processes that include carbohydrate and energy metabolism, cell wall remodelling, flavonoid synthesis and up-accumulated protein species like protease inhibitors.

Our group developed several elite Ogura-CMS cabbage lines with excellent agronomic performance [5], and they have been successfully used for the hybrid seed production of many elite varieties. F01-20 is an elite cabbage line originally introduced from Canada [28], and its Ogura-CMS line CMS01-20 was bred through crosses with a different Ogura-CMS line using F01-20 as the male parent and subsequent recurrent backcrossing with F01-20 for more than 20 generations. The very similar genetic backgrounds of this CMS line and its maintainer line make them ideal materials for cabbage breeding as well as for studying the molecular mechanisms of Ogura-CMS.

Herein, we describe the isobaric tags for the relative and absolute quantitation (iTRAQ)-based proteome analysis of Ogura-CMS using the cabbage lines CMS01-20 and F01-20. Our goals were to identify essential differential abundance protein species (DAPs) and pathways between male-sterile and male-fertile lines and investigate their potential mechanistic roles in Ogura-CMS.

2. Results

2.1. Morphological and Microscopic Examination

CMS01-20 showed degenerated anthers and no visible pollen compared with its maintainer line F01-20 (Figure 1A,I). We further observed male gametophytes of the two lines at different developmental stages using light microscopy (Figure 1B–H,J–P). No obvious phenotypic differences were observed before the tetrad stage. At the tetrad stage, certain tetrads exhibited irregular shapes. At the late tetrad stage or shortly after the release of microspores, CMS01-20 tapetal cells were swollen and vacuolated and the separation of microspores was delayed, indicating defects in the dissolution of the callose and tetrad walls. At the uninucleate to bicellular microspore stage, the tapetum layers of CMS01-20 showed earlier degradation, and we also observed the aggregation of abnormal and vacuolated immature pollen microspores as described in previous studies [23,29], although these microspores may be held together by the residue of degraded tapetum cells rather than by the tapetal layers. At the mature microspore stage, the aborted microspores were completely degenerated, no pollen was viable in mature locules, and the anthers did not dehisce.

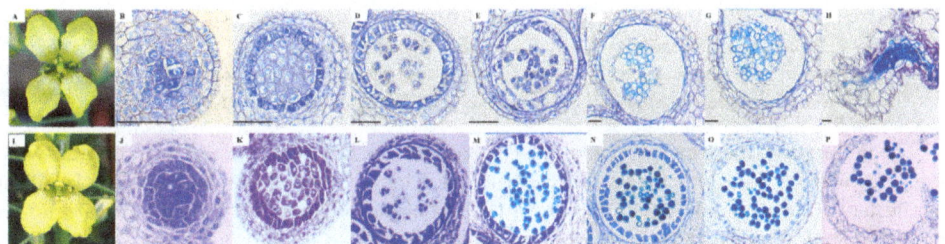

Figure 1. Phenotypes of Ogura cytoplasmic male sterility (Ogura-CMS) line CMS01-20 and its corresponding maintainer line F01-20. (**A–H**) CMS01-20; I-P: F01-20. (**A**) CMS01-20 shows degenerated anthers and no visible pollen; (**I**) F01-20 is fertility; (**B,J**) microsporocyte stage; (**C,K**) meiotic stage; (**D,L**) tetrad stage; (**E,M**) uninucleate stage; (**F,N**) bicellular microspore stage to trinucleate microspores stage. (**G,H,O,P**) mature pollen satge. Bar = 50 μm.

2.2. Overview of the Protein Species Identified Using iTRAQ Data

iTRAQ-based proteomic analysis was employed to assess protein changes between the buds of F01-20 and CMS01-20. A total of 197,216 spectra were generated. After filtering the data with Mascot, 53,777 spectra were matched to known sequences, over half of which (28,865) were unique (Figure 2A). By searching against the cabbage A2 reference genome database, 12,062 unique peptides were identified within 4188 protein species (Figure 2A), which represented 11.8% of all predicted protein-coding loci in the genome. Most of the identified peptides had lengths between 8 and 16 amino acids. More than 60% of the protein species had masses between 30 and 70 kDa (Figure 2B), and approximately 70% of the protein species contained at least two mapped peptides.

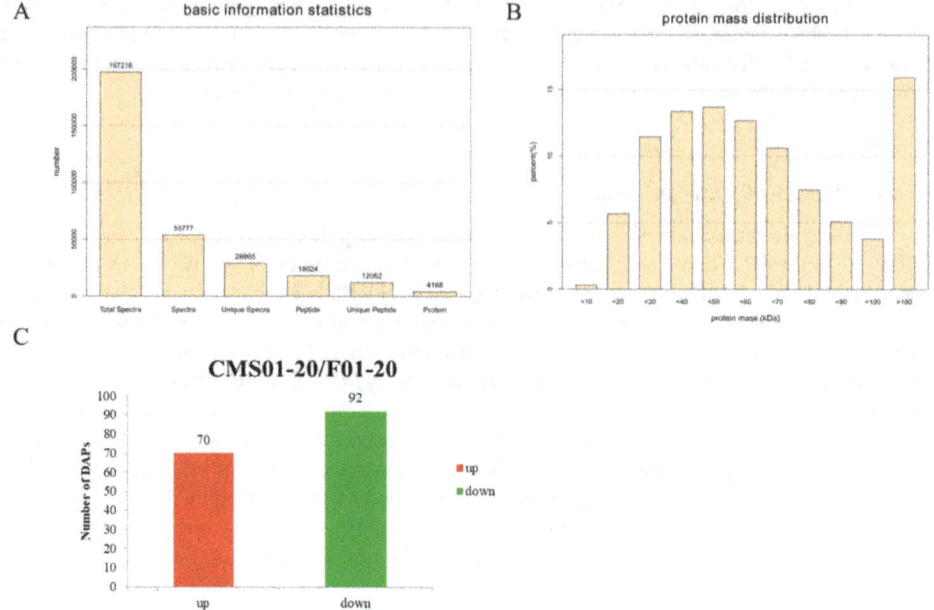

Figure 2. Protein species identification by the isobaric tags for the relative and absolute quantitation (iTRAQ) approach. (**A**) Number of spectra, peptide and protein; (**B**) percentage of protein mass distribution; (**C**) differential abundance protein species (DAPs) between CMS01-20 and F01-20.

To understand the functions of these 4188 protein species in cabbage buds, analyses were conducted with the Gene Ontology (GO), the Kyoto Encyclopaedia of Genes and Genomes (KEGG) and the Clusters of Orthologous Groups of proteins (COG) databases (Figure S1). The GO analysis showed that the biological process terms "metabolic process" (79.0%), "cellular process" (76.3%), "response to stimulus" (79.0%) and "single-organism process" (42.5%) were the most highly overrepresented functional groups; 10.8% of the protein species were involved in "reproductive process/reproduction"; the most overrepresented among the cellular component terms were "cell & cell part" (87.0%), "organelle" (72.6%), "organelle part" (38.6%) and "membrane" (34.9%); and the main for molecular function terms were "binding" (61.7%) and "catalytic activity" (57.4%).

2.3. Overview of the DAPs between CMS01-20 and F01-20

Following the criteria of a fold difference ≥ 1.2 and p value ≤ 0.05, we identified 162 DAPs between CMS01-20 and F01-20, of which 92 were down-accumulated and 70 were up-accumulated in the CMS line (Figure 2C, Table S1).

GO annotations were performed based on the TAIR GO Slim method provided by blast2GO. The GO annotations for 153 (94.4%) DAPs were divided into 35 functional groups, of which 16 were biological process GO terms (the largest category was "metabolic process"); cellular components accounted for 12 GO terms (the largest category was "cell"); and molecular functions accounted for 7 GO terms (the largest category was "catalytic activity") (Figure 3A). Of the DAPs, 112 (70.4%) were assigned to 62 KEGG pathways, and enriched in 14 KEGG pathways (p-value < 0.05) including peroxisome (7.89%), cutin, suberine and wax biosynthesis (3.51%), sulphur metabolism (4.4%), ribosome (16.7%), tryptophan metabolism (5.26%), linoleic acid metabolism (1.75%), alpha-linolenic acid metabolism (3.51%), tropane, piperidine and pyridine alkaloid biosynthesis (3.51%), photosynthesis-antenna proteins (3.51%), carbon metabolism (14.9%), glyoxylate and dicarboxylate metabolism (6.1%), glutathione metabolism (5.3%), photosynthesis (5.3%), citrate cycle (the tricarboxylic acid) (4.4%) as listed in Table S2.

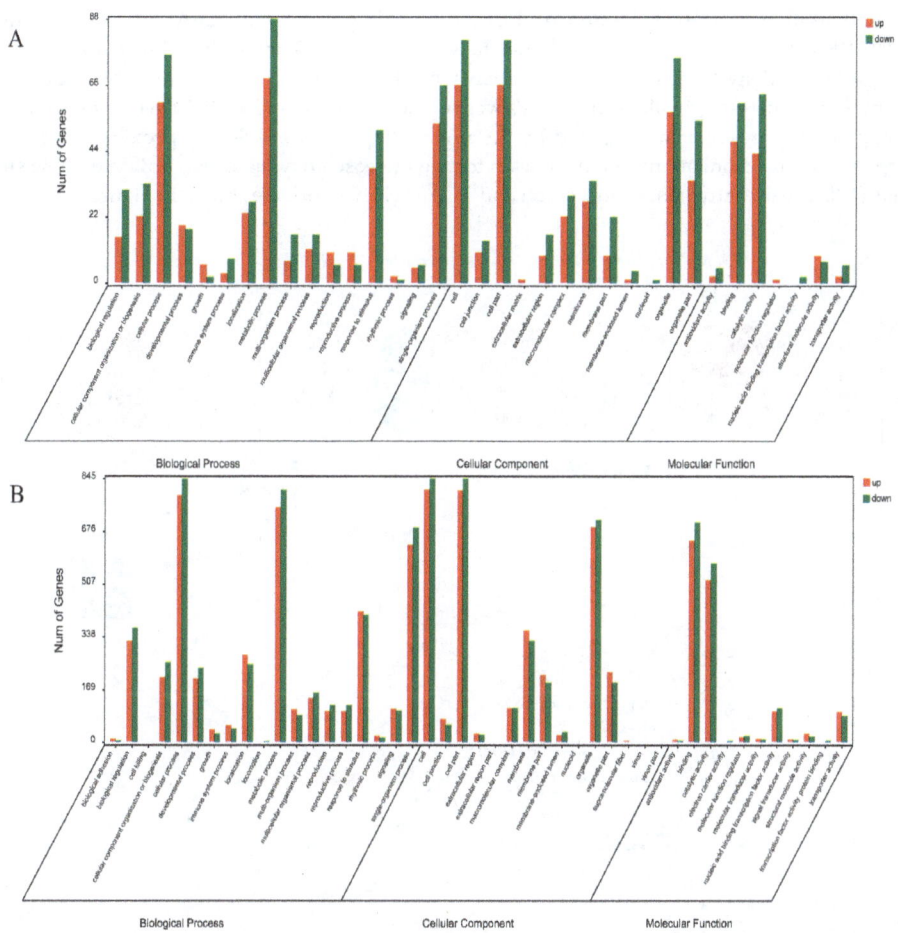

Figure 3. (**A**) Gene ontology categories for differential abundance protein species in the proteome data; (**B**) gene ontology categories for differentially expressed genes in the transcriptome data.

2.4. DAPs Involved in Oxidative Phosphorylation and TCA Cycle

Most CMS are associated with disturbances in the energy metabolism. The ORF138 protein formed a mitochondrial membrane complex that exhibited uncoupling effect, and affected oxygen consumption [17,21]. Thus oxidative phosphorylation and the TCA cycle are important pathways possibly affected by ORF138 protein. We found that all of the eight DAPs involved in oxidative phosphorylation were down-accumulated in CMS01-20 (Table 1), including one ETC complex I protein (Bol015119, NADH-ubiquinone oxidoreductase B18 subunit), one cytochrome *c* protein (Bol012326, cytochrome *c*), one complex IV protein (Bol010838, cytochrome *c* oxidase subunit Vc), and five complex V proteins (Bol009135 and Bol025034, ATP synthase subunit d; Bol017288, Bol025922 and Bol015469, ATP synthase 6 KD subunit). These results suggest CMS02-10 may have a decreased energy-generation capacity.

The citrate cycle (TCA cycle) provides NADH, FADH$_2$ and H$^+$ for oxidative phosphorylation. Five DAPs were annotated to be involved in TCA cycle: pyruvate dehydrogenase (Bol022522 and Bol008536) catalyses the synthesis of acetyl-CoA; citrate synthase 4 (Bol02950) catalyses the condensation of acetyl-CoA and oxaloacetate yielding citrate and CoA; aconitate hydratase (Bol029048) catalyses the conversion of citrate to isocitrate; 2-oxoglutarate dehydrogenase (Bol008657, 2-oxoglutarate dehydrogenase E1 component) catalyses the conversion of 2-oxoglutarate to succinyl-CoA. Interestingly, all the five DAPs were up-accumulated in CMS01-20, which may represent a compensatory mechanism triggered by the uncoupling effect of ORF138 protein [17], and this compensatory mechanism may be the reason for the increased oxygen consumption of male sterile plants [17]. The oxidative phosphorylation and TCA cycle network are shown in Figure 4.

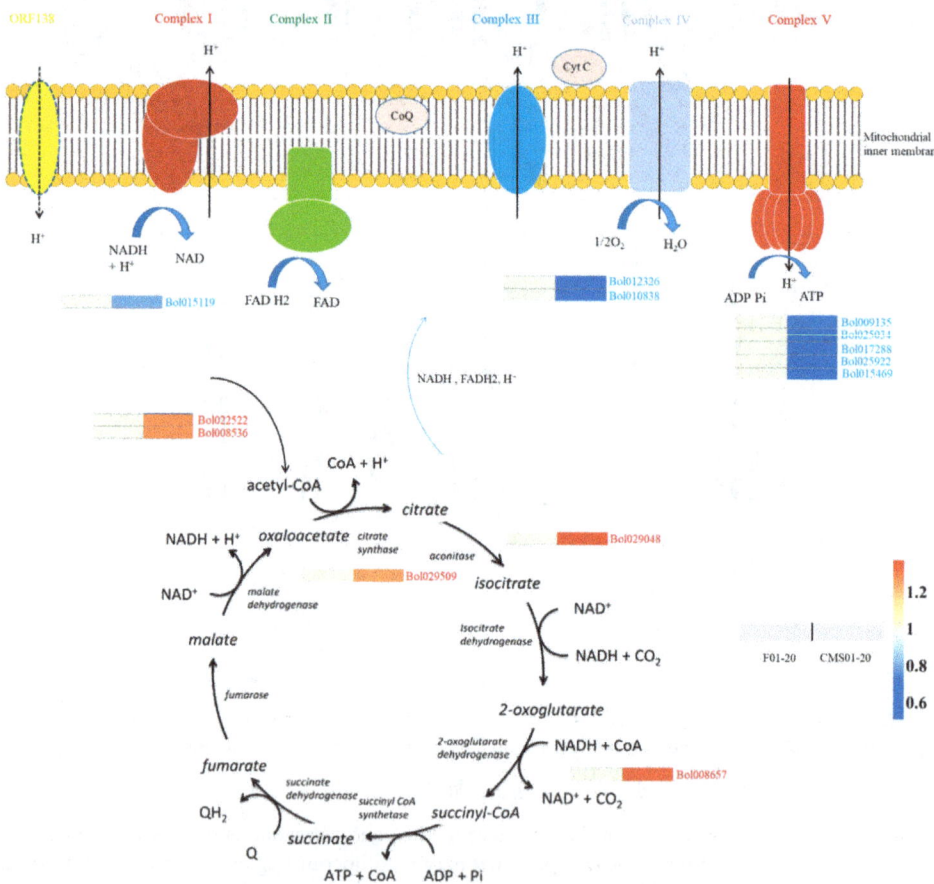

Figure 4. Differential abundance protein species involved in the oxidative phosphorylation and the tricarboxylic acid cycle. The possible uncoupling role of the ORF138 protein was also indicated on mitochondrial inner membrane. The fold changes of differential abundance protein species are indicated by the colour filled in the squares.

Table 1. Differential abundance protein species involved in oxidative phosphorylation, the tricarboxylic acid cycle, pollen wall, tetrad wall and programmed cell death.

	ID	Description	Up/Down in CMS Line
Oxidative phosphorylation	Bol015119	NADH-ubiquinone oxidoreductase B18 subunit	down
	Bol009135	ATP synthase subunit d, mitochondrial-like	down
	Bol025034	ATP synthase subunit d, mitochondrial-like	down
	Bol017288	mitochondrial ATP synthase 6 KD subunit	down
	Bol025922	mitochondrial ATP synthase 6 KD subunit	down
	Bol015469	ATP synthase 6 kDa subunit	down
	Bol012326	cytochrome c	down
	Bol010838	cytochrome c oxidase subunit Vc	down
TCA cycle	Bol022522	pyruvate dehydrogenase E1 component subunit beta-2	up
	Bol008536	pyruvate dehydrogenase E1 component subunit beta-2	up
	Bol008657	2-oxoglutarate dehydrogenase	up
	Bol029048	aconitate hydratase 1	up
	Bol029509	citrate synthase 4	up
pollen wall	Bol013698	LAP5; Chalcone and stilbene synthase family protein	down
	Bol025267	LAP6; Chalcone and stilbene synthase family protein	down
	Bol007277	MS2; fatty acyl-CoA reductase	down
	Bol034656	LAP5; Chalcone and stilbene synthase family protein	down
	Bol040704	cytochrome P450 703A2	down
	Bol010336	MS2; fatty acyl-CoA reductase	down
tetrad wall	Bol009974	probable glucan endo-1,3-beta-glucosidase A6	down
	Bol037314	O-Glycosyl hydrolases family 17 protein;	down
	Bol033052	beta-D-xylosidase 1	up
	Bol030909	beta-glucosidase 43 isoform X2	down
PCD	Bol006999	catalase-3	down
	Bol026973	catalase-3	down
	Bol035942	allene oxide synthase	down
	Bol037061	peroxisomal	down
	Bol005496	stromal ascorbate peroxidase	down
	Bol004624	glutathione S-transferase F9	down
	Bol033376	glutathione S-transferase F9	down

2.5. Other Ogura-CMS Related DAPs and Pathways

Notably, terms related to cell wall assembly were significantly enriched among the DAPs. Fifty-five protein species were involved in cellular component organization or biogenesis, among which 33 were down-accumulated and 22 were up-accumulated in CMS01-20. We focused on protein species involved in the assembly (or degeneration) of pollen exine and the tetrad wall. We identified six DAPs involved in the synthesis of pollen exine. In the flavonoid biosynthesis pathway, LAP5 (Bol013698, Bol034656), LAP6 (Bol025267) and CYP703A/CYP703A2 (Bol040704) were down-accumulated in CMS01-20. In unsaturated fatty acid and fatty acid elongation pathways, MS2 (Bol010336, Bol007277) was down-accumulation in CMS01-20 (Table 1, Figure 5). These genes are vital for the development of viable pollen; therefore, they may be partially responsible for the phenotype of Ogura-CMS plants, such as the underdeveloped exine and the aberrant/aborted microspores observed here and in previous studies [23,29].

The degeneration of the callose wall and the outer wall (or pollen mother cell wall) is required for the release of microspores. Pectinase, endo-β-1,3-glucanases, exo-β-1,3-glucanase, and endo-β-1,4-glucanase enzymes are likely involved in this degeneration process [30]. We identified four DAPs that may be involved in the degradation of callose walls (Table 1, Figure 5). Bol009974 (predicted probable glucan endo-1,3-beta-glucosidase A6), Bol037314 (O-Glycosyl hydrolases family 17 protein) and Bol030909 (predicted beta-glucosidase) were down-accumulated in CMS01-20. Bol033052 (predicted beta-D-xylosidase, glycosyl hydrolase family 3) was up-accumulated in CMS01-20. They may be responsible for the pollen separation defects of CMS01-20 (shown in Figure 1). However, we did not identify any DAPs associated with pectin degradation.

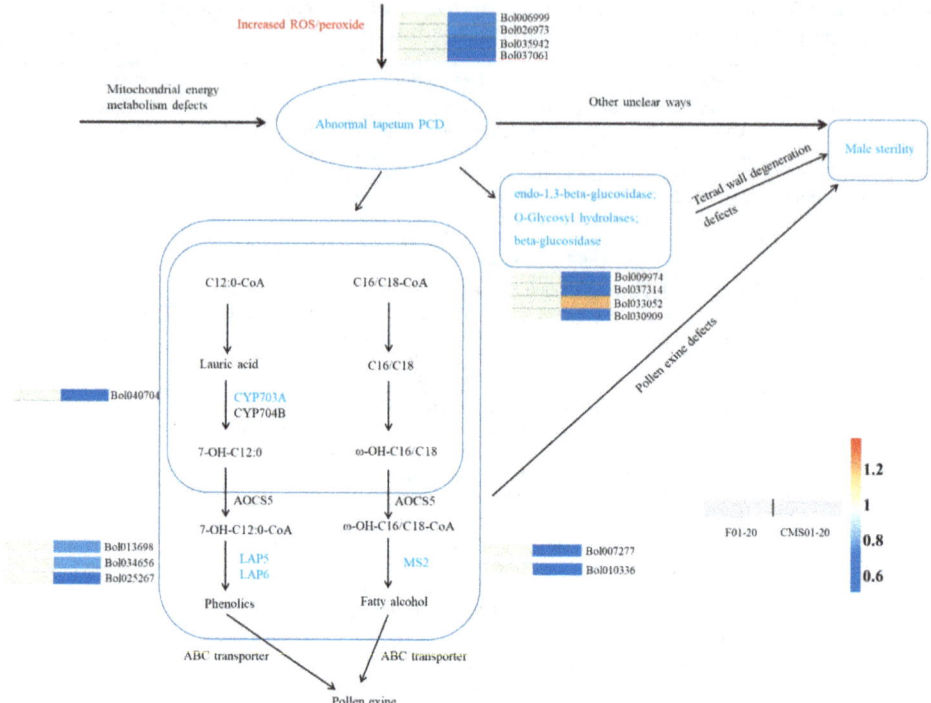

Figure 5. A possible network according to which abnormal tapetal programmed cell death is triggered by reactive oxygen species (ROS), resulting in male sterility. The differential abundance protein species involved in pollen exine formation and tetrad wall degeneration are also shown. The fold changes of differential abundance protein species are indicated by the colour filled in the squares.

Several previous studies have proposed a hypothetical mechanism in which CMS proteins trigger abnormal programmed cell death (PCD), which is usually associated with an increase in reactive oxygen species (ROS) and the release of cytochrome *c* in male organs, such as the tapetum [31–33]. Additionally, tapetal PCD often depends on the generation of ROS that can be detoxified by antioxidative enzymes including catalases [34,35]. In this study, we identified nine DAPs associated with PCD (Table 1, Figure 5). In the glyoxylate and dicarboxylate metabolism pathway, two catalase-3 proteins (Bol006999 and Bol026973) and one glycolate oxidase (Bol037061) were down-accumulated in CMS01-20. In the alpha-linolenic acid metabolism/linolenic acid metabolism pathway, four DAPs were down-accumulated in CMS01-20, among which allene oxide synthase (Bol035942) was a key enzyme catalysing the dehydration of the hydroperoxide to an unstable allene oxide. In the glutathione metabolism pathway, five down-accumulated DAPs and one up-accumulated DAP were identified. Among these five down-accumulated DAPs, Bol005496, Bol004624 and Bol033376 showed peroxidase activity. ROS may burst due to the down-accumulation of these enzymes and may trigger the abnormal PCD of the tapetum. Additionally, the disruption of allene oxide synthase DDE2 (the homologue of Bol035942) in *Arabidopsis* resulted in male sterility. Two oxygen-evolving enhancer protein species (Bol023353, Bol041074) were also associated with PCD and down-accumulated in CMS01-20, but they were predicted to be involved in the photosynthesis pathway and thus may not be CMS-related protein species.

We also identified DAPs and pathways that were similarly found in previous proteomic and/or transcriptomic analysis cases for male-sterile plants [25,36,37]. For example, the ribosome pathway (12 up-accumulated and 7 down-accumulated) and protein processing in the endoplasmic reticulum pathway (4 up-accumulated and 0 down-accumulated) was also identified in male-sterile cabbage and soybean [25,36,37], but their roles are largely unknown. For all the remaining DAPs and pathways, we did not find clues as to their possible roles in Ogura-CMS.

2.6. Joint Proteome–Transcriptome Analysis

To better understand the mechanisms underlying Ogura-CMS in cabbage, we analysed the transcriptomes of the CMS line CMS01-20 and its maintainer F01-20 via NGS. The RNA-seq libraries for F01-20 and CMS01-20 produced 91,952,648 and 98,470,304 clean reads, respectively. In total, 32,687 transcripts were identified for F01-20, and 32,680 transcripts for CMS01-20. A total of 3247 differentially expressed genes (DEGs) ($p < 0.05$) were identified, including 1525 up-accumulated and 1722 down-accumulated genes (Table S3). The GO analysis classified these DEGs into 45 GO categories, which showed a similar pattern to that of the DAPs (Figure 3B).

Integrative analyses comparing proteome and transcriptome data were performed between CMS01-20 and F01-20. The genes fell into nine groups based on the calculated $\log_2$ accumulation ratios of their protein species and transcripts (Figure 6A,B). These results showed poor correlation between the mRNA and protein species accumulation patterns as has been described in many previous studies [36,38,39]; this may be related to a combination of translational regulation, protein localization, protein modification, degradation, and other factors.

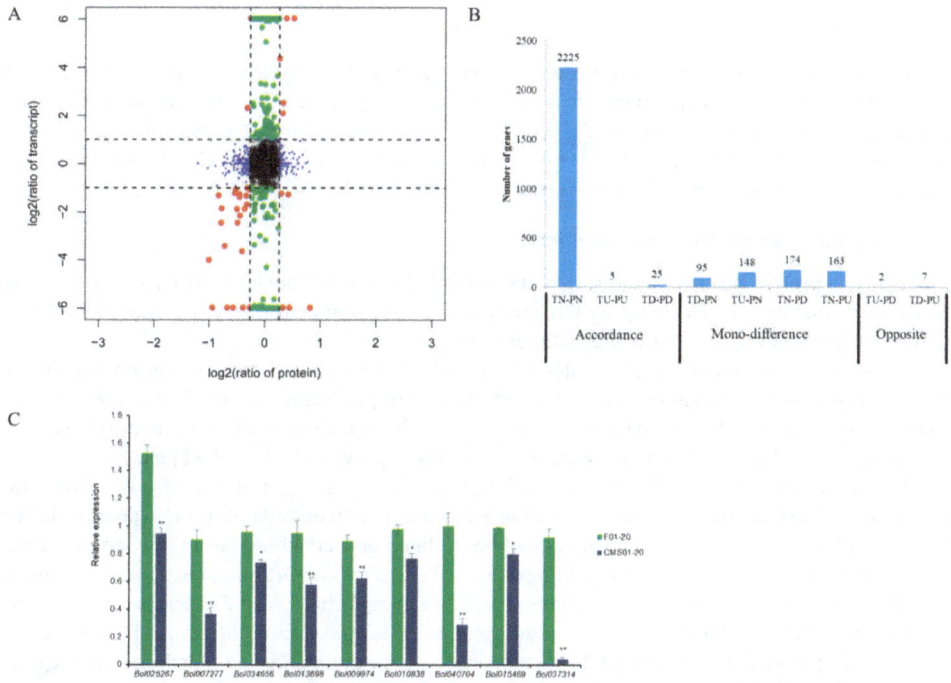

Figure 6. Integrative analyses comparing proteome and transcriptome data. (**A**) Genes were divided into nine groups according to $\log_2$ ratios of the protein species (*y*-axis) and transcripts (*x*-axis); (**B**) Number of genes among the nine groups in (**A**). T: transcript; P: protein species; N, no difference; U: up-accumulation; D, down-accumulation. (**C**) Expression validation for nine key genes by qRT-PCR. * $p < 0.05$, ** $p < 0.01$.

Although this conjoint analysis showed little overlap between the gene accumulation at the transcript and protein levels, certain noteworthy genes involved in sporopollenin synthesis (Bol013698,

Bol025267, Bol034656, Bol040704, Bol007277), callose wall degeneration (Bol009974, Bol037314), and oxidative phosphorylation (Bol015469, Bol010838) showed accordance patterns. The expression levels of these nine genes were validated by qRT-PCR analysis, which revealed that all of the genes showed expression patterns consistent with the RNA-seq data (Figure 6C).

3. Materials and Methods

3.1. Plant Materials and Sample Preparation

Cabbage CMS line 01-20 (CMS01-20) and its maintainer line F01-20 were used in this study. These lines were sown on 20 August, transferred to a cold frame for vernalisation on 20 November and finally transplanted to a greenhouse on 3 March 2016 for bolting and flowering. All plant materials were obtained from the Institute of Flowers and Vegetables of the Chinese Academy of Agriculture Sciences (IVFCAAS, Beijing, China). During the flowering stage, flower buds with different lengths were sampled to observe a range of microspore developmental stages and identify differences between CMS01-20 and F01-20 using an Olympus CX31 optical microscope (Olympus Japan Co., Tokyo, Japan). Based on the microscopic examination results, flower buds before the bicellular microspore stage ($\leq$3.5 mm) were collected for the transcriptome and proteome analyses. All collected buds were immediately frozen in liquid nitrogen and stored at -80 °C. Three biological replicate were performed for all experiments.

3.2. Microscopy

Flower buds with different lengths were fixed in formalin-aceto-alcohol (FAA), dehydrated in an ethanol series, embedded in paraffin, sectioned into 3–5 μm transverse slices using a microtome and stained with 1% toluidine blue as described by Lou et al. [40]. Then the anther transverse sections were observed with an Olympus CX31 optical microscope (Olympus Japan Co., Tokyo, Japan) and photographed with a Nikon 550D camera (Canon, Tokyo, Japan).

3.3. iTRAQ Analysis and Protein Species Annotation

The total protein species was extracted and subjected to iTRAQ labelling, strong cation exchange (SCX) separation and LC-electrospray ionization tandem mass spectrometry (LC-MS/MS) analysis using the same method as described by Chu et al. [41].

After converting them into MGF files, the raw iTRAQ data files were used for protein species identification and quantification. Database searches were performed using Mascot version 2.3.02 (Matrix Science, Boston, MA, USA) against a cabbage database, including 35,400 sequences from the *B. oleracea* genome A2 [42]. The search parameters were set as previously described [38].

Protein species with a fold change $\geq$1.2 (CMS01-20 vs. F01-20) and a false discovery rate (FDR) < 0.05 in at least two replicates were defined as differential abundance protein species (DAPs). All protein species identified were functionally annotated and classified based on Gene Ontology (GO) annotations (http://www.geneontology.org/), the Clusters of Orthologous Groups of proteins (COG) database (http://www.ncbi.nlm.nih.gov/COG/) and the Kyoto Encyclopaedia of Genes and Genomes (KEGG) database (http://www.genome.jp/kegg/pathway.html). DAPs were further analysed using the GO and KEGG databases to identify significantly enriched functional subcategories and metabolic pathways.

3.4. RNA-Seq Analysis and Conjoint Analysis with Proteome Data

Total RNA was extracted using an RNAprep pure Plant Kit (TIANGEN, Beijing, China) following the manufacturer's instructions. High-quality RNA from each sample was used for cDNA library construction and RNA sequencing on an Illumina HiSeq 2500TM platform (Gene Denovo Biotechnology Co., Guangzhou, China). To obtain clean high-quality reads, adapter sequences, low-quality reads (>50% bases with Q-value $\leq$ 20) and unknown bases (>10% N bases) were removed from the raw reads. The short read alignment tool Bowtie2 [43] was used to map reads to a ribosomal RNA (rRNA) database. After removing the rRNA mapped reads, each sample read was then mapped to the reference genome (ftp://brassicadb.org/Brassica_oleracea/) with TopHat2 (version 2.0.3.12) [44]. Gene expression levels were normalized using the FPKM (fragments per kilobase of transcript per million mapped reads) method [45]. The edgeR package (http://www.rproject.org/) was used to identify differentially expressed genes (DEGs) between two samples. We defined genes with a fold change ≥ 2 and an FDR < 0.05 as significant DEGs. DEGs were then analysed for the enrichment of GO functions and KEGG pathways. GO terms or pathways with FDR ≤ 0.05 were defined as significantly enriched in DEGs.

For the conjoint analysis of DAPs and DEGs, the transcriptome and proteome data were combined using the same *B. oleracea* genome A2 database. Thresholds of "FDR ≤ 0.05, $|\log 2 FC| \geq 1$" and "$p \leq 0.05$, $|FC| \geq 1.5$" were set to select DEGs and DAPs, respectively. The correlation between the expression levels of the DAPs and their corresponding mRNAs were analysed by Pearson correlation tests.

3.5. Quantitative RT-PCR Analysis

Quantitative real-time RT-PCR (qRT-PCR) analyses were performed to validate the results from the DEGs. Total RNA was extracted from the buds of CMS01-20 and F01-20 plants using an RNAprep pure Plant Kit (TIANGEN, Beijing, China) according to the manufacturer's instructions. RNA was treated with RNase-free DNase I (Fermentas, Harrington, QC, Canada) to remove genomic DNA. First-strand cDNA was synthesized using a PrimeScript 1st Strand cDNA Synthesis Kit (Takara, Kyoto, Japan). qRT-PCR reactions were conducted using SYBR Premix Ex Taq II (Tli RNase H Plus; Takara, Dalian, China) with a CFX96 Touch Real-Time PCR Detection System (Bio-Rad, Hercules, CA, USA). Three biological replicates (with three technical replicates for each biological replicate) were analysed for each gene. The relative expression level of each gene was estimated by the $2^{-\Delta\Delta Ct}$ method [46]. The *B. oleracea actin* gene (GenBank accession number AF044573.1) [23] was used as an internal control.

4. Discussion

As an important type of CMS in *Brassica* and *Raphanus* species, Ogura-CMS contributes significantly to hybrid seed production. Researchers have long been interested in Ogura-CMS, although the mechanisms underlying its ability to interfere with pollen development remain unclear. Previous transcriptome and proteome analyses of Ogura-CMS provided basic knowledge of the DAPs between Ogura-CMS plants and their maintainer lines [22–27]. iTRAQ is also an efficient and reliable approach for the relative and absolute quantification of protein species, and it has been applied in proteome analyses of CMS and GMS in several plants, including cabbage [25,37] soybean [36], cotton [38], and cybrid pummelo [47]. Herein, we reported the iTRAQ analysis of an elite cabbage Ogura-CMS line and its corresponding maintainer line. Many more protein species were identified using this method compared to the traditional 2-DE technique [26,27]. In total, 162 DAPs were identified, with 92 protein species down-accumulated and 70 up-accumulated in CMS01-20. These DAPs are mainly involved in carbon metabolism, energy metabolism, and cell wall assembly, etc.

Many CMS-related proteins are associated with deficiencies in the ETC and oxidative phosphorylation [3]. In sunflower, a chimeric mitochondrial ORF522 protein has been described that likely decreases ATP hydrolysis via mitochondrial ATP synthase [48]. In a CMS tobacco, the ATP/ADP ratio is significantly decreased in the floral buds of male-sterile plants [49]. In wild beet, mitochondrial gene *G* alters the molecular weight of a respiratory chain complex subunit, and the male-sterile *G* cytoplasm plants exhibit severely reduced cytochrome *c* oxidase activity [50]. In the HL CMS line of rice, ORFH79 disrupts the F_0F_1-ATPase, and reduced protein quantity and enzyme activity are observed in sterile plants [51]. In a more recent study, ORFH79 was confirmed to decrease the enzymatic activity of the mitochondrial ETC complex III by interacting with P61, a subunit of the ETC complex III, which resulted in a deficiency in ATP production and an increase in reactive oxygen species

(ROS) content [52]. In the present study, we identified eight predicted ETC components differentially accumulated between the floral buds of CMS01-20 and F01-20. Five of these genes are predicted to be components of the mitochondrial ATP synthase complex and were down-accumulated in the CMS line, similar to a previous report in rapeseed [27]. Additionally, two cytochrome *c* (oxidase) genes and one NADH-ubiquinone oxidoreductase were significantly down-accumulated in sterile plants. These results suggest CMS02-10 may have a decreased respiration capacity, although Duroc et al. reported that ORF138 protein does not impair the capacities of electron transport chain complexes I, II, IV, or ATP synthase [17]. Interestingly, all DAPs involved in the TCA cycle were up-accumulated in CMS01-20. The TCA cycle provides precursors for many biochemical pathways and produces energy in the form of ATP largely via oxidative phosphorylation. The up-accumulation of TCA genes may compensate for the decreased respiration capacity and may be associated with the elevated O_2 consumption rate observed in mitochondria from Ogura-CMS plants [17].

Pollen grains are covered by an exine wall that consists of sporopollenin, which provides essential protection from the environment and is involved in interactions with female stigma cells. Although the mechanisms of exine formation are not well understood, dozens of genes regulating sporopollenin biosynthesis have been characterized [53]. Six genes involved in sporopollenin synthesis were identified in the present study. Except for *Bol010336*, these genes were all down-accumulated at both the transcript and protein levels in CMS01-20, which was confirmed by the qRT-PCR analysis, whereas *Bol010336* was down-accumulated at only the protein level. *LAP5* (homologue to *Bol013698*, *Bol034656*) encodes an anther-specific chalcone and stilbene synthase (CHS) family protein, but does not present CHS activity in vitro, and it may act as a multifunctional enzyme or could be involved in a novel pathway for sporopollenin synthesis [54,55]. The mutation of this gene results in abnormal exine patterning. *LAP6* (homologue to *Bol025267*) is similar to *LAP5*, and double mutants of *LAP5* and *LAP6* exhibit strong male sterility because of a lack of exine on the surface of the pollen grains. *CYP703A/CYP703A2* (homologue to *Bol040704*) is specifically expressed in the anthers of land plants, and is involved in catalysing medium-chain saturated fatty acids and thus is essential for sporopollenin synthesis; moreover, *cyp703a* mutants produces pollen grains without exine, and they display a partial male-sterile phenotype [56]. *Male sterility 2* (homologue to *Bol010336*) is the first gene identified through a genetic approach using mutants with exine defects [57], and it encodes a fatty acid reductase that is responsible for the accumulation of C16 and C18 fatty alcohols, which are essential for pollen exine wall biosynthesis [58]. The homologue of this gene in moss shows a conserved function, suggesting that *MS2* is a core component of the sporopollenin biosynthetic pathway [59]. However, in Ogura-CMS plants, the down-accumulation of these protein species/genes may not explain the vacuolated and early degenerated tapetum phenotypes. Conversely, these genes may be down-accumulated because of the abnormal tapetum development (possibly caused by abnormal programmed cell death) because the tapetum supplies necessary metabolites, nutrients, and sporopollenin precursors for the normal development of the male gamete [60]. Indeed, Ogura-CMS tapetal cells showed reduced secretory activity [29]. Thus, other genes are likely involved in the impaired function of Ogura-CMS tapetal cells.

At the late tetrad stage, some key enzymes are secreted from tapetal cells to dissolve the callose wall and the outer wall. Three *quartet* (*QRT*) genes have been identified in *A. thaliana*, and they are involved in the pectin degradation of the pollen mother wall [61,62]. Mutants of these three *QRT* genes produce microspore tetrads that fail to separate, although these adhered microspores are viable [61,62]. Tratt reported other enzymes that are likely involved in dissolving the pollen tetrad walls, including endo-β-1,3-glucanases, exo-β-1,3-glucanase, and endo-β-1,4-glucanase [30]. In the present study, a light microscopy examination suggested a delayed separation of CMS01-20 microspores, which may be caused by these tetrad wall degradation-related genes. We also identified four down-accumulated genes that may be involved in the degradation of tetrad walls. In Ogura-CMS *Brassica napus*, Sheoran et al. also reported the down-accumulation of protein species associated with cell wall remodelling, including β-1,3-glucanase and pectinesterase using a 2-DE approach [27]. We also considered the down-regulation of these protein species/genes as a result of abnormal tapetal cells.

Premature PCD in the tapetum was observed in PET1-CMS cytoplasm sunflowers as indicated by cell condensation, oligonucleosomal cleavage of nuclear DNA, chromatin separation into delineated masses, and initial mitochondrial persistence [31]. This early PCD may be caused by the release of cytochrome *c* from the mitochondria into the cytosol of tapetal cells [31]. HL CMS rice showed a PCD phenotype in microspores accompanied by inner mitochondrial membrane disruption [32] that was triggered by chronic oxidative stress caused by increased ROS levels and reduced superoxide dismutase (SOD), ascorbate peroxidase (APX) and catalase activity in mitochondria. In wild abortive CMS (CMS-WA) rice, Luo et al. reported that WA352 interacts with nuclear cytochrome *c* oxidase 11 (COX11) to inhibit its function in peroxide metabolism, and this interaction was demonstrated to be responsible for premature PCD in the tapetum and male sterility [33]. Although PCD in the tapetum is a feature of normal development, premature or delayed tapetum PCD usually results in male sterility due to tapetum's crucial role in pollen development [13,33]. Similar to several previous studies of Ogura-CMS plants, we observed a premature PCD phenotype in the tapetum [23,29], which may be responsible for the down-accumulation of many pollen wall assembly and pollen development-related genes, ultimately leading to male sterility. Therefore, identifying the key genes involved in tapetum PCD is crucial. In this study, we identified nine DAPs associated with PCD, especially, four protein species that are highly related to PCD: the two predicted catalase-3 proteins (Bol006999, and Bol026973, which catalyses the breakdown of hydrogen peroxide into water and oxygen), one allene oxide synthase (Bol035942, which catalyses dehydration of the hydroperoxide to an unstable allene oxide in the JA biosynthetic pathway), and one glycolate oxidase (Bol037061, which encodes a glycolate oxidase that modulates reactive oxygen species-mediated signal transduction). Decreased accumulation of these protein species, especially catalase and allene oxide synthase, may trigger PCD in the tapetum mediated by hydrogen peroxide or other ROS [13,63,64]. In addition, the homologue of Bol035942 in *Arabidopsis* is DDE2 (AT5G42650), which is an enzyme involved in jasmonic acid biosynthesis, and *dde2-2* mutants show male sterility and exhibit filament elongation and defects in anther dehiscence [65]. Thus, *Bol037061* may be associated with the small indehiscent anther phenotype of Ogura-CMS.

5. Conclusions

The present study provided an iTRAQ-based proteome analysis of Ogura-CMS using the cabbage Ogura-CMS line CMS01-20 and its isogenic maintainer line F01-20. A total of 4188 proteins were identified, and 162 were designated as DAPs. Key pathways and DAPs involved in processes including energy metabolism in mitochondrion, assembly/degeneration of pollen exine and the tetrad wall, and programmed cell death were found to be closely related to male sterility. Transcriptome profiling revealed 3247 differentially expressed genes between the CMS line and the fertile line. Additionally, the integrative analyses of the transcriptome and proteome data revealed nine Ogura-CMS-related genes showing accordance accumulation patterns at the transcript and protein levels, and the expression levels of these nine genes were validated by qRT-PCR. This study improves our understanding of the genes associated with pollen development and the molecular mechanisms of Ogura-CMS.

Supplementary Materials: The following are available online at http://www.mdpi.com/1422-0067/19/10/3180/s1, Figure S1: Gene Ontology (GO) and Clusters of Orthologous Groups of proteins (COG) classification of total identified proteins. Table S1: Information of the DAPs between F01-20 and CMS01-20. Table S2: Kyoto Encyclopaedia of Genes and Genomes (KEGG) analysis of DAPs between F01-20 and CMS01-20. Table S3: Total number and bioinformatic analysis of differentially expressed genes (DEGs) between F01-20 and CMS01-20.

Author Contributions: Data curation, F.H.; Formal analysis, F.H.; Funding acquisition, Z.F.; Project administration, Z.F.; Resources, L.Y.; Supervision, Z.F. and H.L.; Validation, L.Y., M.Z., Y.Z., Z.L. and H.L.; Writing—original draft, F.H. and X.Z.; Writing—review & editing, F.H.

Funding: This work was supported by grants from the National Key Research and Development Program of China (2017YFD0101804), the Central Public-interest Scientific Institution Basal Research Fund (No. 1610102017013), the Science and Technology Innovation Program of the Chinese Academy of Agricultural Sciences

(CAAS-ASTIP-IVFCAAS), and the earmarked fund for the Modern Agro-Industry Technology Research System, China (nycytx-35-gw01).

Acknowledgments: The work reported here was performed in the Key Laboratory of Biology and Genetic Improvement of Horticultural Crops, Ministry of Agriculture, Beijing 100081, China.

Conflicts of Interest: The authors declare no conflict of interest.

References

1. Fang, Z.Y.; Liu, Y.M.; Yang, L.M.; Wang, X.W.; Zhuang, M.; Zhang, Y.Y.; Sun, P.T. A survey of research in genetic breedings of cabbage in China. *Acta Hort. Sin.* **2002**, *29*, S657–S663.
2. Prakash, C.; Verma, T. Heterosis in cytoplasmic male sterile lines of cabbage. *Crucif. Newslett.* **2004**, *25*, 49–50.
3. Chen, L.; Liu, Y.G. Male sterility and fertility restoration in crops. *Annu. Rev. Plant Biol.* **2014**, *65*, 579–606. [CrossRef] [PubMed]
4. Fang, Z.Y.; Sun, P.T.; Liu, Y.M.; Yang, L.M.; Wang, X.W.; Hou, A.F.; Bian, C.S. A male sterile line with dominant gene (*MS*) in cabbage (*Brassica oleracea* var. *capitata*) and its utilization for hybrid seed production. *Euphytica* **1997**, *97*, 265–268.
5. Fang, Z.Y.; Sun, P.T.; Liu, Y.M. Investigation of different types of male sterility and application of dominant male sterility in cabbage. *China Veg.* **2001**, *1*, 6–10.
6. Ogura, H. Studies of a new male-sterility in Japanese radish, with special reference to the utilization of this sterility towards the practical raising of hybrid seeds. *Mem. Fac. Agric. Kagoshima Univ.* **1968**, *6*, 39–78.
7. Bannerrot, H.; Boulidard, L.; Couderon, Y.; Temple, J. Transfer of cytoplasmic male sterility from *Raphanus sativus* to *Brassica oleracea. Proc. Eucarpia Meet. Crucif.* **1974**, *25*, 52–54.
8. Sigareva, M.; Earle, E. Direct transfer of a cold-tolerant Ogura male-sterile cytoplasm into cabbage (*Brassica oleracea* ssp. *capitata*) via protoplast fusion. *Theor. Appl. Genet.* **1997**, *94*, 213–220. [CrossRef]
9. Walters, W.T.; Mutschler, A.M.; Earle, D.E. Protoplast fusion-derived Ogura male-sterile cauliflower with cold tolerance. *Plant Cell Rep.* **1992**, *10*, 624–628. [CrossRef] [PubMed]
10. Pelletier, G.; Primard, C.; Vedel, F.; Chetrit, P.; Remy, R.; Renard, M. Intergeneric cytoplasmic hybridization in Cruciferae by protoplast fusion. *Mol. Genet. Genomics* **1983**, *191*, 244–250. [CrossRef]
11. Wang, Q.B.; Zhang, Y.Y.; Fang, Z.Y.; Liu, Y.M.; Yang, L.M.; Zhuang, M. Chloroplast and mitochondrial SSR help to distinguish allo-cytoplasmic male sterile types in cabbage (*Brassica oleracea* L. var. *capitata). Mol. Breed.* **2012**, *30*, 709–716. [CrossRef]
12. Chase, C.D. Cytoplasmic male sterility: A window to the world of plant mitochondrial-nuclear interactions. *Trends Genet.* **2007**, *23*, 81–90. [CrossRef] [PubMed]
13. Hu, J.; Huang, W.; Huang, Q.; Qin, X.; Yu, C.; Wang, L.; Li, S.; Zhu, R.; Zhu, Y. Mitochondria and cytoplasmic male sterility in plants. *Mitochondrion* **2014**, *19*, 282–288. [CrossRef] [PubMed]
14. Tuteja, R.; Saxena, R.K.; Davila, J.; Shah, T.; Chen, W.; Xiao, Y.; Fan, G.; Saxena, K.B.; Alverson, A.J.; Spillane, C.; et al. Cytoplasmic male sterility-associated chimeric open reading frames identified by mitochondrial genome sequencing of four *Cajanus* genotypes. *DNA Res.* **2013**, *20*, 485–495. [CrossRef] [PubMed]
15. Carlsson, J.; Glimelius, K. Cytoplasmic male-sterility and nuclear encoded fertility restoration. *Plant Mitochond.* **2011**, *1*, 469–491.
16. Woodson, J.D.; Chory, J. Coordination of gene expression between organellar and nuclear genomes. *Nat. Rev. Genet.* **2008**, *9*, 383–395. [CrossRef] [PubMed]
17. Duroc, Y.; Hiard, S.; Vrielynck, N.; Ragu, S.; Budar, F. The Ogura sterility-inducing protein forms a large complex without interfering with the oxidative phosphorylation components in rapeseed mitochondria. *Plant Mol. Biol.* **2009**, *70*, 123–137. [CrossRef] [PubMed]
18. Bonhomme, S.; Budar, F.; Lancelin, D.; Small, I.; Defrance, M.C.; Pelletier, G. Sequence and transcript analysis of the *Nco2.5* Ogura specific fragment correlated with cytoplasmic male sterility in *Brassica* cybrids. *Mol. Gen. Genet.* **1992**, *235*, 340–348. [CrossRef] [PubMed]
19. Grelon, M.; Budar, F.; Bonhomme, S.; Pelletier, G. Ogura cytoplasmic male-sterility (CMS)-associated *orf138* is translated into a mitochondrial membrane polypeptide in male-sterile *Brassica* cybrids. *Mol. Gen. Genet.* **1994**, *243*, 540–547. [CrossRef] [PubMed]

20. Krishnasamy, S.; Makaroff, C.A. Characterization of the radish mitochondrial *orfB* locus: Possible relationship with male sterility in Ogura radish. *Curr. Genet.* **1993**, *24*, 156–163. [CrossRef] [PubMed]

21. Duroc, Y.; Gaillard, C.; Hiard, S.; Defrance, M.C.; Pelletier, G.; Budar, F. Biochemical and functional characterization of ORF138, a mitochondrial protein responsible for Ogura cytoplasmic male sterility in Brassiceae. *Biochimie* **2005**, *87*, 1089–1100. [CrossRef] [PubMed]

22. Dong, X.; Kim, W.K.; Lim, Y.P.; Kim, Y.K.; Hur, Y. Ogura-CMS in Chinese cabbage (*Brassica rapa* ssp. *pekinensis*) causes delayed expression of many nuclear genes. *Plant Sci.* **2013**, *199*, 7–17. [PubMed]

23. Wang, S.; Wang, C.; Zhang, X.X.; Chen, X.; Liu, J.J.; Jia, X.F.; Jia, S.Q. Transcriptome de novo assembly and analysis of differentially expressed genes related to cytoplasmic male sterility in cabbage. *Plant Physiol. Biochem.* **2016**, *105*, 224–232. [CrossRef] [PubMed]

24. Shu, J.; Zhang, L.; Liu, Y.; Li, Z.; Fang, Z.; Yang, L.; Zhuang, M.; Zhang, Y.; Lv, H. Normal and abortive buds transcriptomic profiling of broccoli ogu cytoplasmic male sterile line and its maintainer. *Int. J. Mol. Sci.* **2018**, *19*, 2501. [CrossRef] [PubMed]

25. Xing, M.; Sun, C.; Li, H.; Hu, S.; Lei, L.; Kang, J. Integrated analysis of transcriptome and proteome changes related to the Ogura cytoplasmic male sterility in cabbage. *PLoS ONE* **2018**, *13*, e0193462. [CrossRef] [PubMed]

26. Mihr, C.; Baumgärtner, M.; Dieterich, J.H.; Schmitz, U.K.; Braun, H.P. Proteomic approach for investigation of cytoplasmic male sterility (CMS) in *Brassica. J. Plant Physiol.* **2001**, *158*, 787–794. [CrossRef]

27. Sheoran, I.S.; Sawhney, V.K. Proteome analysis of the normal and Ogura (ogu) CMS anthers of *Brassica napus* to identify proteins associated with male sterility. *Botany* **2010**, *88*, 217–230. [CrossRef]

28. Liu, X.; Han, F.; Kong, C.; Fang, Z.; Yang, L.; Zhang, Y.; Zhuang, M.; Liu, Y.; Li, Z.; Lv, H. Rapid introgression of the Fusarium wilt resistance gene into an elite cabbage line through the combined application of a microspore culture, genome background analysis, and disease resistance-specific marker assisted foreground selection. *Front. Plant Sci.* **2017**, *8*, 354. [CrossRef] [PubMed]

29. González-Melendi, P.; Uyttewaal, M.; Morcillo, C.N.; Mora, J.R.H.; Fajardo, S.; Budar, F.; Lucas, M.M. A light and electron microscopy analysis of the events leading to male sterility in Ogu-INRA CMS of rapeseed (*Brassica napus*). *J. Exp. Bot.* **2008**, *59*, 827–838. [CrossRef] [PubMed]

30. Tratt, J. Identifying the Wall-Degrading Enzymes Responsible for Microspore Release from the Pollen Tetrad. Ph.D. Thesis, University of Bath, Bath, UK, 2016.

31. Balk, J.; Leaver, C.J. The PET1-CMS mitochondrial mutation in sunflower is associated with premature programmed cell death and cytochrome *c* release. *Plant Cell* **2001**, *13*, 1803–1818. [CrossRef] [PubMed]

32. Li, S.; Wan, C.; Kong, J.; Zhang, Z.; Li, Y.; Zhu, Y. Programmed cell death during microgenesis in a Honglian CMS line of rice is correlated with oxidative stress in mitochondria. *Funct. Plant Biol.* **2004**, *31*, 369–376. [CrossRef]

33. Luo, D.P.; Xu, H.; Liu, Z.L.; Guo, J.X.; Li, H.Y.; Chen, L.T.; Fang, C.; Zhang, Q.Y.; Bai, M.; Yao, N.; et al. A detrimental mitochondrial-nuclear interaction causes cytoplasmic male sterility in rice. *Nat. Genet.* **2013**, *45*, 573–577. [CrossRef] [PubMed]

34. Yi, J.; Moon, S.; Lee, Y.S.; Zhu, L.; Liang, W.; Zhang, D.; Jung, K.H.; An, G. Defective Tapetum Cell Death 1 (DTC1) regulates ROS levels by binding to metallothionein during tapetum degeneration. *Plant Physiol.* **2016**, *170*, 1611–1623. [CrossRef] [PubMed]

35. Min, L.; Zhu, L.; Tu, L.; Deng, F.; Yuan, D.; Zhang, X. Cotton *GhCKI* disrupts normal male reproduction by delaying tapetum programmed cell death via inactivating starch synthase. *Plant J.* **2013**, *75*, 823–835. [CrossRef] [PubMed]

36. Li, J.; Ding, X.; Han, S.; He, T.; Zhang, H.; Yang, L.; Yang, S.; Gai, J. Differential proteomics analysis to identify proteins and pathways associated with male sterility of soybean using iTRAQ-based strategy. *J. Proteomics* **2016**, *138*, 72–82. [CrossRef] [PubMed]

37. Ji, J.; Yang, L.; Fang, Z.; Zhuang, M.; Zhang, Y.; Lv, H.; Liu, Y.; Li, Z. Complementary transcriptome and proteome profiling in cabbage buds of a recessive male sterile mutant provides new insights into male reproductive development. *J. Proteomics* **2018**, *179*, 80–91. [CrossRef] [PubMed]

38. Liu, J.; Pang, C.; Wei, H.; Song, M.; Meng, Y.; Ma, J.; Fan, S.; Yu, S. iTRAQ-facilitated proteomic profiling of anthers from a photosensitive male sterile mutant and wild-type cotton (*Gossypium hirsutum* L.). *J. Proteomics* **2015**, *126*, 68–81. [CrossRef] [PubMed]

39. Pan, Z.; Zeng, Y.; An, J.; Ye, J.; Xu, Q.; Deng, X. An integrative analysis of transcriptome and proteome provides new insights into carotenoid biosynthesis and regulation in sweet orange fruits. *J. Proteomics* **2012**, *75*, 2670–2684. [CrossRef] [PubMed]

40. Lou, P.; Kang, J.; Zhang, G.; Bonnema, G.; Fang, Z.; Wang, X. Transcript profiling of a dominant male sterile mutant (*Ms-cd1*) in cabbage during flower bud development. *Plant Sci.* **2007**, *172*, 111–119. [CrossRef]

41. Chu, P.; Yan, G.; Yang, Q.; Zhai, L.; Zhang, C.; Zhang, F.; Guan, R. iTRAQ-based quantitative proteomics analysis of *Brassica napus*, leaves reveals pathways associated with chlorophyll deficiency. *J. Proteomics* **2015**, *113*, 244–259. [CrossRef] [PubMed]

42. Liu, S.; Liu, Y.; Yang, X.; Tong, C.; Edwards, D.; Parkin, I.A.; Zhao, M.; Ma, J.; Yu, J.; Huang, S.; et al. The *Brassica oleracea* genome reveals the asymmetrical evolution of polyploid genomes. *Nat. Commun.* **2014**, *5*, 3930. [CrossRef] [PubMed]

43. Langmead, B.; Salzberg, S.L. Fast gapped-read alignment with Bowtie 2. *Nat. Methods* **2012**, *9*, 357–359. [CrossRef] [PubMed]

44. Kim, D.; Pertea, G.; Trapnell, C.; Pimentel, H.; Kelley, R.; Salzberg, S. TopHat2: Accurate alignment of transcriptomes in the presence of insertions, deletions and gene fusions. *Genome Biol.* **2013**, *14*, R36. [CrossRef] [PubMed]

45. Trapnell, C.; Williams, B.A.; Pertea, G.; Mortazavi, A.; Kwan, G.; van Baren, M.J.; Salzberg, S.L.; Wold, B.J.; Pachter, L. Transcript assembly and quantification by RNA-Seq reveals unannotated transcripts and isoform switching during cell differentiation. *Nat. Biotechnol.* **2010**, *28*, 511–515. [CrossRef] [PubMed]

46. Livak, K.J.; Schmittgen, T.D. Analysis of relative gene expression data using real-time quantitative PCR and the $2^{-\Delta\Delta Ct}$ method. *Methods* **2001**, *25*, 402–408. [CrossRef] [PubMed]

47. Zheng, B.B.; Fang, Y.N.; Pan, Z.Y.; Sun, L.; Deng, X.X.; Grosser, J.W.; Guo, W.W. iTRAQ-based quantitative proteomics analysis revealed alterations of carbohydrate metabolism pathways and mitochondrial proteins in a male sterile *Cybrid pummelo*. *J. Proteome Res.* **2014**, *13*, 2998–3015. [CrossRef] [PubMed]

48. Sabar, M.; Gagliardi, D.; Balk, J.; Leaver, C.J. ORFB is a subunit of F1F(O)-ATP synthase: Insight into the basis of cytoplasmic male sterility in sunflower. *EMBO Rep.* **2003**, *4*, 381–386. [CrossRef] [PubMed]

49. Bergman, P.; Edqvist, J.; Farbos, I.; Glimelius, K. Male-sterile tobacco displays abnormal mitochondrial atp1 transcript accumulation and reduced floral ATP/ADP ratio. *Plant Mol. Biol.* **2000**, *42*, 531–544. [CrossRef] [PubMed]

50. Ducos, E.; Touzet, P.; Boutry, M. The male sterile *G*, cytoplasm of wild beet displays modified mitochondrial respiratory complexes. *Plant J.* **2001**, *26*, 171–180. [CrossRef] [PubMed]

51. Zhang, H.; Li, S.; Yi, P.; Wan, C.; Chen, Z.; Zhu, Y. A Honglian CMS line of rice displays aberrant F_0 of F_0 F_1-ATPase. *Plant Cell Rep.* **2007**, *26*, 1065–1071. [CrossRef] [PubMed]

52. Wang, K.; Gao, F.; Ji, Y.; Liu, Y.; Dan, Z.; Yang, P.; Zhu, Y.; Li, S. ORFH79 impairs mitochondrial function via interaction with a subunit of electron transport chain complex III in Honglian cytoplasmic male sterile rice. *New Phytol.* **2013**, *198*, 408–418. [CrossRef] [PubMed]

53. Ariizumi, T.; Toriyama, K. Genetic regulation of sporopollenin synthesis and pollen exine development. *Annu. Rev. Plant Biol.* **2011**, *62*, 437–460. [CrossRef] [PubMed]

54. Shirley, B.W.; Kubasek, W.L.; Storz, G.; Bruggemann, E.; Koornneef, M.; Ausubel, F.M.; Goodman, H.M. Analysis of *Arabidopsis* mutants deficient in flavonoid biosynthesis. *Plant J.* **1995**, *8*, 659–671. [CrossRef] [PubMed]

55. Dobritsa, A.A.; Lei, Z.; Nishikawa, S.I.; Urbanczyk-Wochniak, E.; Huhman, D.V.; Preuss, D.; Sumner, L.W. *LAP5* and *LAP6* encode anther-specific proteins with similarity to chalcone synthase essential for pollen exine development in *Arabidopsis thaliana*. *Plant Physiol.* **2010**, *110*, 157446. [CrossRef]

56. Morant, M.; Jørgensen, K.; Schaller, H.; Pinot, F.; Møller, B.L.; Werck-Reichhart, D.; Bak, S. CYP703 is an ancient cytochrome P450 in land plants catalyzing in-chain hydroxylation of lauric acid to provide building blocks for sporopollenin synthesis in pollen. *Plant Cell* **2007**, *19*, 1473–1487. [CrossRef] [PubMed]

57. Aarts, M.G.M.; Hodge, R.; Kalantidis, K.; Florack, D.; Wilson, Z.A.; Mulligan, B.J.; Stiekema, W.J.; Scott, R.; Pereira, A. The *Arabidopsis MALE STERILITY 2* protein shares similarity with reductases in elongation/condensation complexes. *Plant J.* **1997**, *12*, 615–623. [CrossRef] [PubMed]

58. Chen, W.; Yu, X.H.; Zhang, K.; Shi, J.; De Oliveira, S.; Schreiber, L.; Shanklin, J.; Zhang, D. *Male Sterile2* encodes a plastid-localized fatty acyl carrier protein reductase required for pollen exine development in *Arabidopsis*. *Plant Physiol.* **2011**, *157*, 842–853. [CrossRef] [PubMed]

59. Wallace, S.; Chater, C.C.; Kamisugi, Y.; Cuming, A.C.; Wellman, C.H.; Beerling, D.J.; Fleming, A.J. Conservation of *Male Sterility 2* function during spore and pollen wall development supports an evolutionarily early recruitment of a core component in the sporopollenin biosynthetic pathway. *New Phytol.* **2015**, *205*, 390–401. [CrossRef] [PubMed]

60. Piffanelli, P.; Ross, J.H.E.; Murphy, D.J. Biogenesis and function of the lipidic structures of pollen grains. *Sexual Plant Reprod.* **1998**, *11*, 65–80. [CrossRef]

61. Preuss, D.; Rhee, S.Y.; Davis, R.W. Tetrad analysis possible in *Arabidopsis* with mutation of the *QUARTET* (*QRT*) genes. *Science* **1994**, *264*, 1458–1460. [CrossRef] [PubMed]

62. Rhee, S.Y.; Osborne, E.; Poindexter, P.D.; Somerville, C.R. Microspore separation in the *quartet 3* mutants of *Arabidopsis* is impaired by a defect in a developmentally regulated polygalacturonase required for pollen mother cell wall degradation. *Plant Physiol.* **2003**, *133*, 1170–1180. [CrossRef] [PubMed]

63. Jiang, P.; Zhang, X.; Zhu, Y.; Zhu, W.; Xie, H.; Wang, X. Metabolism of reactive oxygen species in cotton cytoplasmic male sterility and its restoration. *Plant Cell Rep.* **2007**, *26*, 1627–1634. [CrossRef] [PubMed]

64. Gechev, T.S.; Hille, J. Hydrogen peroxide as a signal controlling plant programmed cell death. *J. Cell Biol.* **2005**, *168*, 17–20. [CrossRef] [PubMed]

65. Von Malek, B.; van der Graaff, E.; Schneitz, K.; Keller, B. The *Arabidopsis* male-sterile mutant *dde2-2* is defective in the *ALLENE OXIDE SYNTHASE* gene encoding one of the key enzymes of the jasmonic acid biosynthesis pathway. *Planta* **2002**, *216*, 187–192. [CrossRef] [PubMed]

International Journal of
Molecular Sciences

Article

Differentially Expressed Genes Associated with the Cabbage Yellow-Green-Leaf Mutant in the *ygl-1* Mapping Interval with Recombination Suppression

Xiaoping Liu [†], Hailong Yu [†], Fengqing Han, Zhiyuan Li, Zhiyuan Fang, Limei Yang, Mu Zhuang, Honghao Lv, Yumei Liu, Zhansheng Li, Xing Li and Yangyong Zhang *

Institute of Vegetables and Flowers, Chinese Academy of Agricultural Sciences, Key Laboratory of Biology and Genetic Improvement of Horticultural Crops, Ministry of Agriculture, Beijing 100081, China; 82101181120@caas.cn (X.L.); yuhailong@caas.cn (H.Y.); feng857142@163.com (F.H.); 82101181071@caas.cn (Z.L.); fangzhiyuan@caas.cn (Z.F.); yanglimei@caas.cn (L.Y.); zhuangmu@caas.cn (M.Z.); lvhonghao@caas.cn (H.L.); liuyumei@caas.cn (Y.L.); lizhansheng@caas.cn (Z.L.); xzdlixing@126.com (X.L.)
* Correspondence: zhangyangyong@caas.cn; Tel.: +86-010-8210-8756
† These authors contributed equally to this work.

Received: 12 August 2018; Accepted: 20 September 2018; Published: 27 September 2018

Abstract: Although the genetics and preliminary mapping of the cabbage yellow-green-leaf mutant YL-1 has been extensively studied, transcriptome profiling associated with the yellow-green-leaf mutant of YL-1 has not been discovered. Positional mapping with two populations showed that the yellow-green-leaf gene *ygl-1* is located in a recombination-suppressed genomic region. Then, a bulk segregant RNA-seq (BSR) was applied to identify differentially expressed genes (DEGs) using an F_3 population (YL-1 × 11-192) and a BC_2 population (YL-1 × 01-20). Among the 37,286 unique genes, 5730 and 4118 DEGs were detected between the yellow-leaf and normal-leaf pools from the F_3 and BC_2 populations. BSR analysis with four pools greatly reduced the number of common DEGs from 4924 to 1112. In the *ygl-1* gene mapping region with suppressed recombination, 43 common DEGs were identified. Five of the DEGs were related to chloroplasts, including the down-regulated *Bo1g087310*, *Bo1g094360*, and *Bo1g098630* and the up-regulated *Bo1g059170* and *Bo1g098440*. The *Bo1g098440* and *Bo1g098630* genes were excluded by qRT-PCR. Hence, we inferred that these three DEGs (*Bo1g094360*, *Bo1g087310*, and *Bo1g059170*) in the mapping interval may be tightly associated with the development of the yellow-green-leaf mutant phenotype.

Keywords: cabbage; yellow-green-leaf mutant; recombination-suppressed region; bulk segregant RNA-seq; differentially expressed genes

1. Introduction

Yellow-green-leaf mutants have been extensively studied in many species, including *Arabidopsis thaliana* [1], barley [2], *Brassica napus* [3], rice [4–6], cabbage [7], and muskmelon [8]. Leaf color mutants are an ideal model for studying mechanisms of photosynthesis and light morphogenesis, since yellow-green-leaf mutants are commonly related to chlorophyll synthesis or degradation [9,10].

Chlorophyll is the most important pigment related to photosynthesis. In *Arabidopsis*, 27 genes involved in 15 steps in the pathway from glutamyl-tRNA to chlorophylls a and b have been identified. Leaf color mutants commonly result from blocking a portion of the chlorophyll synthesis pathway, such as the synthesis of 5-aminolevulinic acid (ALA) [11]. Runge et al. [12] isolated and classified some chlorophyll-deficient xantha mutants of *Arabidopsis thaliana* and found that some of the mutants were blocked at various steps of the chlorophyll pathway between ALA and protochlorophyllide (Pchlide), and the latter did not accumulate in the dark.

Bulked segregant analysis (BSA) is a powerful strategy that is commonly used in gene mapping [13]. Futschik and Schlötterer showed that sequencing of pools of samples from individuals are often more effective for Single Nucleotide Polymorphisms (SNP) discovery and provide more accurate allele frequency estimates [14]. Typically, two populations are used for BSA: a backcross (BC) population [15,16] and an F_2 population [17,18]. Mackay and Caligari [19] found that quantitative trait loci (QTLs) are more easily detected in BC populations than in F_2 populations.

In recent years, transcriptome analysis based on deep RNA sequencing (RNA-seq) has been used for the estimation of genome-wide gene expression levels [20,21]. Transcriptome sequencing encompasses mRNA transcript expression analysis. Combined RNA-seq analysis can be used for purposes such as novel transcript prediction, gene structure refinement, alternative splicing analysis, and SNP/InDel analysis [22]. Bulk segregant RNA-seq (BSR) has been applied to identify differentially expressed genes (DEGs) and trait-associated SNPs [23,24].

A yellow-green-leaf mutant (YL-1) was discovered in cabbage [10], and measurements of photosynthetic pigment contents, chloroplast ultrastructure, and chlorophyll fluorescence parameters indicated that YL-1 was deficient in its total chlorophyll content [10]. In a previous study, we mapped *ygl-1*, which controls the yellow-green-leaf phenotype, to chromosome C01 [7]. The linkage distance of the mapping interval was only 0.75 cM, but the physical distance in the reference genome TO1000 was ~10 Mb, indicating that recombination suppression existed in this interval. In this study, the recombination-suppressed region was identified by gene mapping. Two runs of BSR were performed using BC and F_3 populations, with the aim of obtaining DEGs associated with the yellow-green-leaf mutant.

2. Materials and Methods

2.1. Plant Materials

Group I: The F_2, BC_1P_1, and F_3 populations were constructed using as parents the yellow-green-leaf cabbage mutant YL-1 (P_1) and the normal green leaf cabbage inbred line 01-20 (P_2). The F_2, BC_1P_1 population was employed for *ygl-1* mapping.

Group II: The BC_1P_1 and BC_2P_1 populations were constructed using as parents the mutant YL-1 (P_1) and the normal green leaf Chinese kale inbred line 11–192 (P_3) (Supplementary Figure S1). The BC_2P_1 population was employed for *ygl-1* mapping.

The F_3 population in group I and the BC_2 population in group II were used for RNA-seq analysis. All plant materials came from the Cabbage and Broccoli Research Group, the Institute of Vegetables and Flowers (IVF), and the Chinese Academy of Agricultural Sciences (CAAS).

2.2. Identification of Recombination Suppression in the ygl-1 Gene-Mapping Interval

The sequences of 24 markers from the 02-12 reference genome (Supplementary Table S1) were aligned to chromosome C01 and the scaffold of the TO1000 reference genome [25] (Figure 1). Based on this alignment, we propose that possible assembly errors might exist in the 02-12 reference genome. Hence, InDel primers designed based on the TO1000 reference genome were applied for further mapping. The rates of recombination in the two populations were compared with the normal level in the cabbage genome (~600 kb/cM) to analyze the recombination-suppressed region.

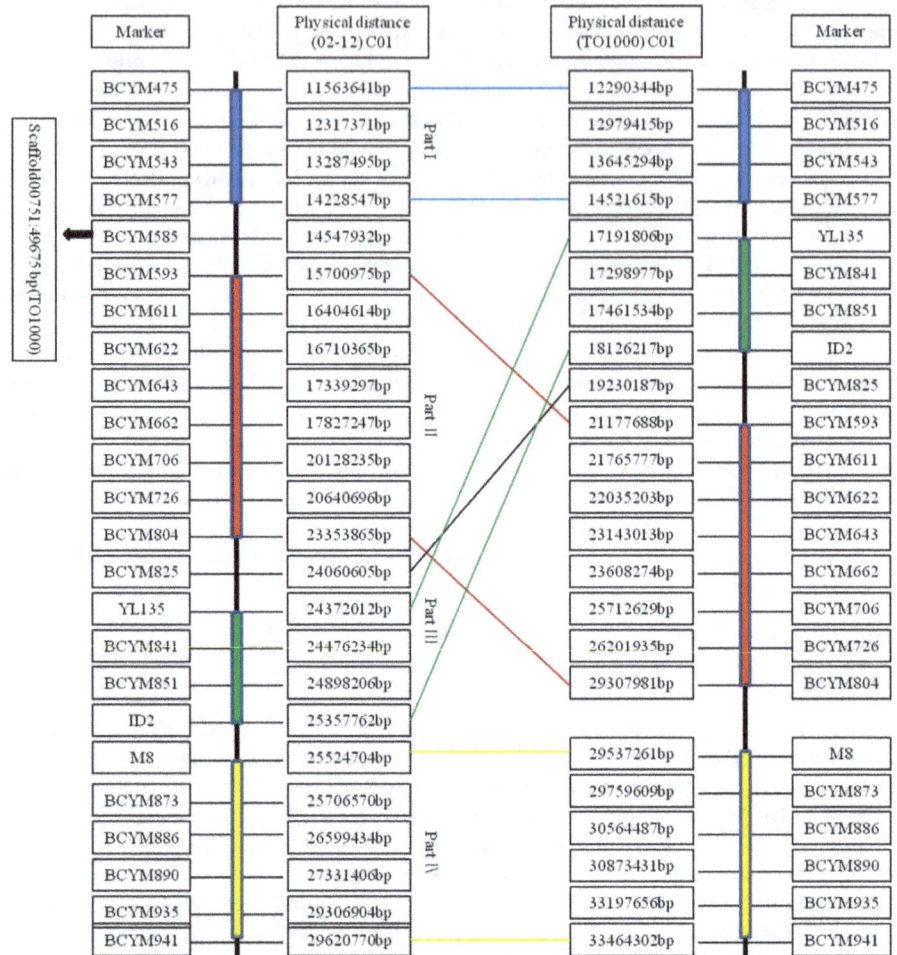

Figure 1. The physical distances of 24 InDel markers in the two reference genomes (02-12 and TO1000).

2.3. BSA, RNA Isolation, and Library Construction

Before RNA isolation, leaf samples from the two populations (the F_3 population in group I and the BC_2 population in group II) were harvested to prepare four bulk groups: Bulk F_yellow (consisting of equal amounts of leaf tissues from 20 yellow-green-leaf F_3 individuals), Bulk F_normal (20 normal-green-leaf F_3 individuals), BC_yellow (20 yellow-green-leaf BC_2 individuals), and BC_normal (20 normal-green-leaf BC_2 individuals).

Total RNA extraction was performed according to instructions of the manufacturer of the TIANGEN kit employed for extraction (Invitrogen, Carlsbad, CA, USA). RNA purity was determined using a NanoDrop spectrophotometer (Thermo Fisher Scientific Inc., Wilmington, DE, USA), 1% formaldehyde gel electrophoresis, and a 2100 Bioanalyzer (Agilent Technologies, Santa Clara, CA, USA).

A total amount of 1 μg of RNA per sample was employed for RNA sample preparation. Sequencing libraries were generated using the NEBNext® Ultra™ RNA Library Prep Kit for Illumina®

(Illumina, CA, USA) following the manufacturer's recommendations. The cDNA library products were sequenced in a paired-end flow cell using an Illumina HiSeq™ 2000 system.

3. Data Analysis

Reads containing adaptor sequences, low-quality reads (bases with more than 50% of quality scores ≤5), and unknown bases (>5% N bases) were removed from each dataset to obtain more reliable results, because such data negatively affect bioinformatics analyses. The sequencing reads were then aligned to the reference database for the *B. oleracea* genome (TO1000) (http://plants.ensembl. org/Brassica_oleracea/Info/Index) (accessed on 5 May 2017) [25] using HISAT [26]. Differential expression analysis to identify DEGs was performed using DESeq [27], with a threshold q value (or false discovery rate [FDR]) < 0.01 & |log$_2$(fold change)| > 1 for significant differential expression. DEGs were displayed using Circos v0.66 [28]. GO (http://www.geneontology.org/) (accessed on 7 May 2017) [29] enrichment analysis of the DEGs was implemented using GOseq, in which gene length bias was corrected. GO functional analysis provides GO functional classifications and annotations for DEGs. Various genes usually cooperate with each other to exercise their biological functions. A pathway-related database was therefore obtained based on Kyoto Encyclopedia of Genes and Genome (KEGG) results (http://www.genome.jp/kegg/) (accessed on 11 May 2017) [30].

Gene Expression Validation

DEGs associated with the yellow-green-leaf mutant were subjected to quantitative real-time RT-PCR (qRT-PCR) analysis. The primers designed according to the gene CDS sequences using DNAMAN are listed in Supplementary Table S6. Three technical replicates were performed for each gene. First-strand cDNA was synthesized using the PrimeScript™ RT reagent Kit (TAKARA BIO, Inc., Shiga, Japan). qRT-PCR was performed with the SYBR Premix Ex Taq™ Kit (Takara, Dalian, China) with the following cycling parameters: 95 °C for five min, followed by 40 cycles of 95 °C for 10 s and 55 °C for 30 s, with a final cycle of 95 °C for 15 s, 55 °C for 60 s, and 95 °C for 15 s. Relative transcription levels were analyzed using the $2^{-\Delta\Delta Ct}$ method [31]. qRT-PCR was performed in a BIO-RAD CFX96 system (Bio-Rad, Hercules, CA, USA), and the actin gene was employed as the internal control [32].

4. Results

4.1. Identification of the Recombination-Suppressed Region

In a previous study [7], we mapped *ygl-1*, which controls the yellow-green-leaf phenotype, to chromosome C01 using a population derived from YL-1 and 01-20. The *ygl-1* gene is flanked by the InDel markers ID2 and M8, and the interval between these two markers is 167 kb (C01: 25,357,762–25,524,704 bp) in the 02-12 reference genome.

However, these two markers are anchored to the TO1000 reference genome, in which the interval between ID2 (C01: 18,126,217 bp) and M8 (C01: 29,537,261 bp) is 11.41 Mb, which is approximately 680 times greater than the distance (167 kb) in the 02-12 reference genome. Then, 24 markers from the 02-12 reference genome (Supplementary Table S1) were aligned to chromosome C01 and the scaffold of the TO1000 reference genome (Figure 1). In the 02-12 reference genome, the physical interval between BCYM475 (11,563,641 bp) and BCYM941 (29,620,770 bp) could be divided into four parts [Part I: BCYM475 (11,563,641 bp) to BCYM577 (14,228,547 bp); Part II: BCYM593 (15,700,975 bp) to BCYM804 (23,353,865 bp); Part III: YL135 (24,372,012 bp) to ID2 (25,357,762 bp); and Part IV: BCYM873 (25,706,570 bp) to BCYM941 (29,620,770 bp)]. The physical locations of Part I and Part IV in the two reference genomes were parallel. However, the physical locations of Part II and Part III were opposite. The makers' order of linkage map was consistent with the physical map order of TO1000 reference genome but not 02-12 reference genome. Therefore, we proposed that an assembly error might exist in the 02-12 reference genome.

InDel primers designed based on the TO1000 reference genome were then applied for further mapping of the *ygl-1* gene. A total of 43 of the 62 pairs of InDel primers designed based on the TO1000 reference genome exhibited polymorphisms according to the F$_3$ population. The genetic distances of the 16 InDel markers are shown in Table 1 (the sequences of these 16 markers are provided in Supplementary Table S2). The *ygl-1* gene was flanked by the InDel markers T1-36 (18,069,792 kb) and T1-58 (29,537,314 kb), with genetic distances of 0.42 cM and 0.42 cM, respectively. The interval distance between the two markers was 11.47 Mb based on the TO1000 reference genome. In the mapping region, spanning a physical distance of 11.47 Mb with a genetic difference of only 0.84 cM, the recombination rate was almost twenty times lower than the normal level for the cabbage genome (~600 kb/cM), suggesting that recombination suppression existed in this region.

Table 1. Genetic distances of the InDel primers to the *ygl-1* in the two mapping populations.

| YL-1 × 01-20 | | YL-1 × 11-192 | |
Primers	Genetic Distance (cM)	Primers	Genetic Distance (cM)
T2-3	9.21	T2-3	13.3
T2-5	6.90	-	-
T1-1	6.28	-	-
T1-14	4.39	T1-14	6.5
T1-18	3.97	T1-18	4.4
T1-26	2.51	T1-26	2.3
T1-28	1.46	T1-28	1.5
T1-30	1.05	T1-30	1.3
T1-34	0.63	T1-34	0.3
T1-36	0.42	T1-36	0.00
T1-58	0.42	T1-58	0.7
T2-6	0.42	T2-6	1.04
T2-10	0.63	T2-10	1.04
T2-14	0.63	T2-14	1.04
T2-16	3.14	T2-16	2.61
T2-18	5.02	T2-18	6.02

Another BC$_2$P$_1$ population, constructed with YL-1 and 11–192, was used to further identify recombination suppression. The *ygl-1* gene was flanked by InDel markers T1-34 (17,301,717 kb) and T1-58 (29,537,314 kb), with genetic distances of 0.3 cM and 0.7 cM, respectively. This result further demonstrated the existence of a recombination-suppressed region in the *ygl-1* mapping interval.

In a previous study [7], we showed that the region between markers the BCYM585 (14,547,932 bp) and BCYM825 (24,060,605 bp) exhibits recombination suppression. In Figure 1, the sequence of BCYM585 was aligned to an unanchored scaffold (Scaffold00751), and the sequence of BCYM825 was aligned to a physical distance of 19,230,187 bp based on the TO1000 reference genome. Part II was aligned between 21,177,688 bp and 29,307,981 bp based on the TO1000 reference genome. These results showed that the recombination-suppressed region observed between the markers T1-36 (18,069,792 kb) and T1-58 (29,537,314 kb) in this study was consistent with the recombination-suppressed region between the markers BCYM585 and BCYM825 identified in our previous study [7].

4.2. BSR Analysis, DEGs between the Yellow-Green-Leaf and Normal-Leaf Pools

BSR was applied to obtain DEGs using the F$_3$ segregated population constructed with YL-1 and 01-20 and the BC$_2$ population constructed with YL-1 and 11-192. A total of 339,481,468 reads were generated from the four cDNA libraries. Among these reads, 82,143,852 were obtained from BC_normal, 91,405,984 from BC_yellow, 86,447,180 from F_normal, and 79,484,452 from F_yellow. The GC contents of the sequences of the four libraries were all approximately 47%, and all Q30% scores (reads with average quality scores >30) were >90%, indicating that the accuracy and quality of the sequencing data were sufficient for further analysis. The sequenced reads were aligned to the *B. oleracea* genome reference (TO1000) (http://plants.ensembl.org/Brassica_oleracea/Info/Index) (accessed on 5 May 2017). An overview of the sequencing process is shown in Supplementary Table S3. The density distribution and boxplot of all the genes exhibited similar patterns among the four samples, indicating that no bias occurred in the construction of the cDNA libraries (Supplementary Figure S2).

The number of DEGs identified between the yellow-green-leaf and normal-leaf samples is shown in Table 2 (Supplementary Figure S3). In the yellow-green-leaf pools, there were approximately 20% fewer down-regulated genes than up-regulated genes. In total, 5730 and 4118 (4924 on average) DEGs were detected between the yellow-green-leaf and normal-leaf pools for the F_3 and BC_2 populations. As shown in the Venn diagram presented in Figure 2, 1884 common DEGs were shared between the DEGs identified in BC_normal vs. BC_yellow and the DEGs identified in F_normal vs. F_yellow, representing approximately half of the total number of DEGs in either population. Cross-comparison showed that only 1112 DEGs (Supplementary Table S4) were common between yellow-leaf and normal-leaf bulks. Thus, BSR analysis using four pools greatly reduced the number of DEGs from 4924 to 1112.

Table 2. Numbers of DEGs between the yellow-leaf and normal-leaf samples.

	No. of DEGs	No. of Up-Regulated DEGs	Percentage (%)	No. of Down-Regulated DEGs	Percentage (%)
BC_normal vs. BC_yellow	4118	2384	58	1734	42
BC_normal vs. F_yellow	8009	4894	60	3315	40
F_normal vs. F_yellow	5730	3226	56	2504	44
F_normal vs. BC_yellow	5405	2844	53	2561	47

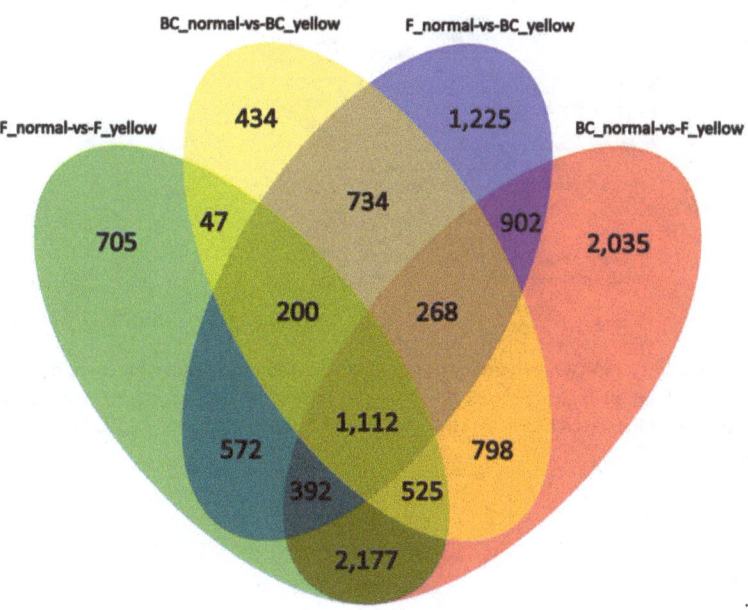

Figure 2. Venn diagram showing the numbers of overlapping and nonoverlapping DEGs (FDR < 0.01 and fold change > 2.0 or < −2.0) in the indicated segments from normal-leaf samples and yellow-leaf samples.

These 1112 DEGs were assigned into three Gene Ontology (GO) classes: biological process, cellular component, and molecular function. Thirty of the most significantly enriched of GO terms are shown in Figure 3, including "carbohydrate binding", "sequence-specific DNA binding transcription factor activity", "receptor activity", "brassinosteroid sulfotransferase activity", "unfolded protein binding" and "protein phosphatase inhibitor activity" under GO molecular functions and "endoplasmic reticulum lumen", "plant-type cell wall", "cytoplasm", "vacuolar membrane", "apoplast", and "nucleus" under GO cellular components. Seventeen biological function or functional groups were enriched in the GO biological process category. In certain biological functions, genes play roles by interacting with each other, and KEGG pathway analysis helps provide an in-depth understanding of the biological functions of genes. A total of 1112 DEGs were annotated in the KEGG database, and 117 KEGG pathways were assigned. These 117 pathways were divided into three levels. Level one included "genetic information processing", "metabolism", "cellular processes", "organismal systems", and "environmental information processing." The nineteen terms in level two are shown in Figure 4.

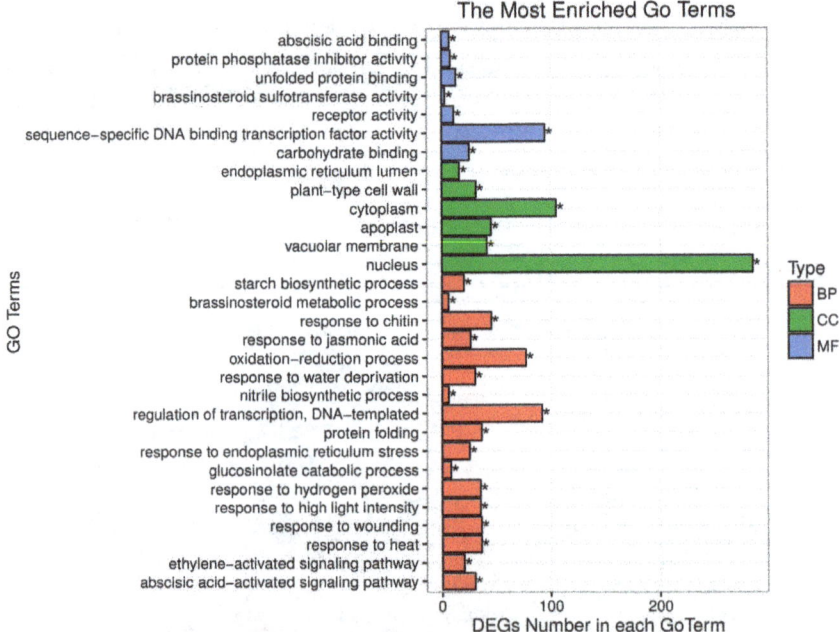

Figure 3. The thirty top GO assignments of 1112 DEGs. Blue: molecular function, green: cellular component, and red: biological process. The Y-axis represents the GO Term; the X-axis represents the number of DEGs for each GO Term. "*" indicates significant enrichment of the GO Term.

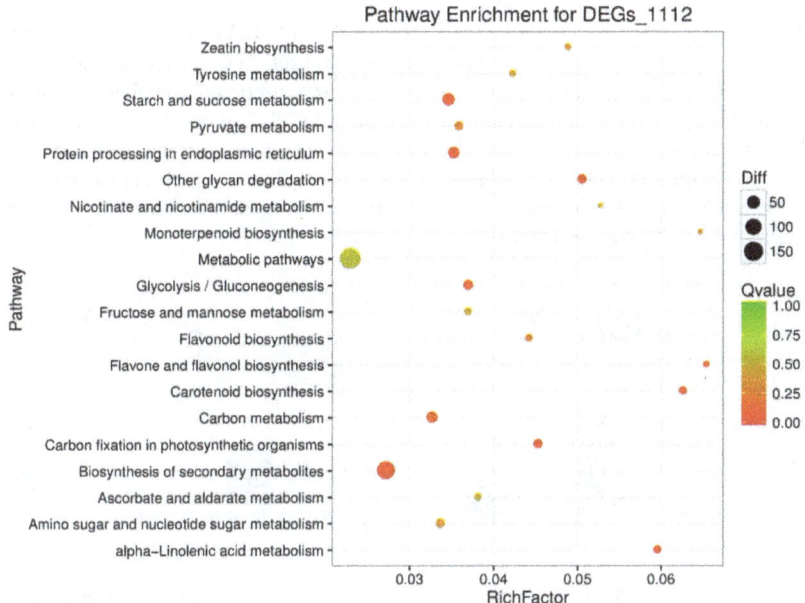

Figure 4. The top-20 enriched KEGG pathways of the 1112 DEGs. The Y-axis represents the pathway term; the X-axis represents the rich factor. The sizes of the points represent different DEG numbers, such that the bigger the point, the greater the DEG number. The colors represent different Q-values.

4.3. DEGs Involved in B. oleracea Chlorophyll Synthesis

The chlorophyll a, chlorophyll b, and total chlorophyll contents of the yellow-green-leaf mutant YL-1 were significantly lower than those of wild-type plants over the entire growth period [10]. Among the 1112 identified DEGs, 18 DEGs related to chlorophyll were clustered, which are shown in Supplementary Figure S4, including nine down- and nine up-regulated DEGs. These 18 DEGs were distributed among different chromosomes. Among the nine chromosomes, there were more DEGs on C01, C03, and C06 than on the other chromosomes (Supplementary Figure S5). In the 11.47 Mb recombination suppression region, two genes *Bo1g088040* (homologous gene *AT1G58290*, *HEMA1*) and *Bo1g098190* (homologous gene *AT1G61520*, *LCA3*) were related to chlorophyll according to the annotations, but there were not DEGs among these four pools by transcriptomics analysis and semi-quantitative PCR. Besides, no sequence variation was detected in the CDS region of these two genes of YL-1, compared with the sequences of 01-20, 11-192, and reference genome TO1000.

DEGs located in the *ygl-1* mapping interval with recombination suppression were selected for further analysis. In the BC_normal vs. BC_yellow comparison, 82 DEGs were found in the 11.47 Mb genomic region, with 45 DEGs being down-regulated and 37 being up-regulated. In the F_normal vs. F_yellow comparison, 105 DEGs were found in the 11.47 Mb genomic region, with 47 DEGs being down-regulated and 58 being up-regulated. Among these four pools, 43 common DEGs were present, with 20 DEGs being down-regulated and 23 being up-regulated (Supplementary Table S5). According to the annotations, five of these genes were related to chloroplasts (Table 3), including the down-regulated genes *Bo1g087310*, *Bo1g094360*, and *Bo1g098630* and the up-regulated genes *Bo1g059170* and *Bo1g098440*. These five genes were applied in qRT-PCR and RT-PCR analyses of the three parents (01-20, YL-1, 11-192). The relative normalized expression of these five genes is shown in Figure 5. The primers of qRT-PCR were supplied on Supplementary Table S6. Based on the relative normalized expression, it can be observed that the expression of *Bo1g059170*, *Bo1g087310*, and *Bo1g094360* genes was consistent with the results of BSR, whereas the relative expression of the *Bo1g098440* and *Bo1g098630* genes differed from the results of BSR. We inferred that these two genes' transcription levels were irrelevant to the yellow-green-leaf trait. In the other three genes that related to chloroplasts, *Bo1g087310* (homologous gene *AT1G56340*, Calreticulins-1) plays important roles in calciumion binding, plant growth, and plant height [33]. *Bo1g059170* (homologous gene *AT3G51420*) is involved in strictosidine synthase activity and plant defense [34], and *Bo1g094360* (homologous gene *AT3G08840*) functions in D-alanine-D-alanine ligase activity (Table 3) [35]. Hence, we inferred that these three candidate genes (*Bo1g094360, Bo1g087310,* and *Bo1g059170*) may be responsible for the development of the yellow-green-leaf mutant phenotype.

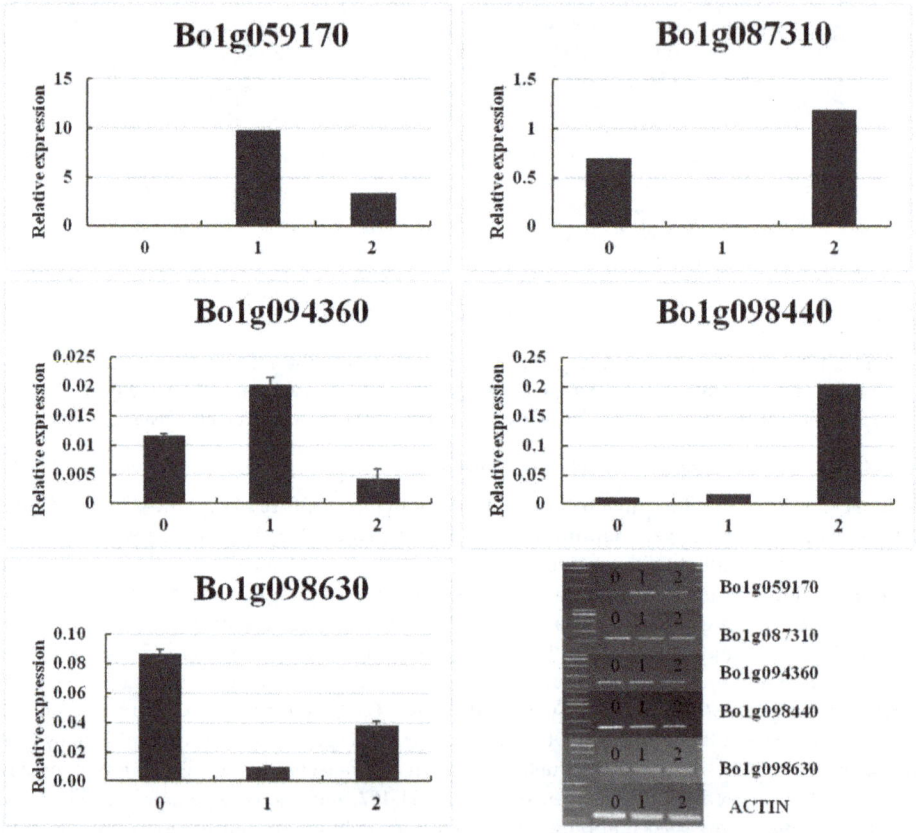

Figure 5. qRT-PCR and RT-PCR validation of transcripts of five DEGs associated with the yellow-green-leaf mutant. 0: the parent 01-20, 1: Mutant YL-1, 2: the parent 11-192.

Table 3. DEGs related to chloroplasts in the recombination-suppressed region.

Gene ID [a]	Physical Distance (TO1000)	F Normal [b]	F_Yellow [b]	BC_Normal [b]	BC_Yellow [b]	Diff [c]	A.T. Annotation [d]
Bo1g087310	C1:25381300-25383803	1837.98	156.85	1920.64	287.42	Down	Calreticulins-1, response to oxidative stress, response to cadmiumion, response to salt stress, calciumion homeostasis;
Bo1g094360	C1:27829353-27834745	48.65	10.53	29.70	2.04	Down	D-alanine-D-alanine ligase activity
Bo1g098630	C1:29261755-29263303	4002.89	475.81	1119.81	125.36	Down	GPT2: glucose-6-phosphate/phosphate translocator 2
Bo1g059170	C1:18110687-18112080	167.35	828.45	277.80	858.19	Up	SSL4: strictosidine synthase-like 4
Bo1g098440	C1:29037892-29038492	129.41	285.27	120.70	427.05	Up	Protein of unknown function, DUF538

[a] Five *B. oleracea* DEGs related to chloroplasts (reference genome TO1000). [b] Expression levels in the four samples. [c] Differential regulation: up-regulation and down-regulation. [d] GO annotations for seven Bo to AT best-hit genes obtained from The Arabidopsis Information Resource (TAIR).

5. Discussion

5.1. Efficiency of BSR in DEG Detection

BSA (an efficient method for rapidly identifying markers linked to mutant phenotypes) combined with RNA-seq has been performed to map important agronomic traits at the transcription level in some species, such as catfish [23], onion [36] maize [37], Chinese cabbage [38], Chinese wheat cultivar [39], polyploid wheat [40], etc. Using BSR, Kim et al. [35] identified the candidate gene, AcPMS1, which is involved in DNA mismatch repair, for the fertility restoration of cytoplasmic male sterility in onions. Ramirez-Gonzalez et al. [24] mapped *Yr15* to a 0.77-cM interval in hexaploid wheat using a segregated F_2 population through BSR. In the present study, RNA-seq analysis of four bulks detected only 1112 common DEGs between the four pools (4924 on average), which can reduce the number of genes related to the phenotype. Therefore, BSR was further demonstrated to be an efficient method for analyzing the genes associated with the yellow-green-leaf mutant phenotype.

5.2. DEGs Analysis Associated with the Yellow-Green-Leaf in a Recombination-Suppressed Region via RNA-Seq

In recent years, the fine mapping of important agronomic traits in *Brassica* has developed rapidly [41–43]. Some yellow leaf color genes have been mapped in *Brassica* crops. A mutation responsible for chlorophyll deficiency in *Brassica juncea* was mapped between amplified fragment length polymorphism (AFLP) markers EA4TG4 and EA7MC1, with genetic distances of 33.6 and 21.5 cM, respectively [44]. In *B. napus*, Wang et al. [45] mapped the *CDE1* locus to a 0.9 cM interval of chromosome C08, and Zhu et al. [3] mapped a chlorophyll-deficient mutant between the markers BnY5 and CB10534, which are closely linked to the chlorophyll deficiency gene *BnaC.YGL*, with genetic distances of 3.0 and 3.2 cM on C06, respectively. Gene mapping for the above leaf color mutant was based on normal recombination in the segregated population. Recombination suppression was reported in many species, such as tomato [46], barley [47], petunia [48], *Populus* [49], hexaploid wheat [50], and buffelgrass [51]. In this study, we identified a large recombination suppression region spanning ~11 Mb on C01. However, recombination rate of *Brassica oleracea* C01 in previous studies seemed to be normal. The genetic map was constructed based on *Brassica oleracea* re-sequencing data; the C01 linkage groups spanned 97.59 cM, with an average distance of 1.15 cM between neighboring loci; and no recombination suppression was found [52]. Lv et al. (2016) [53] constructed a high-density genetic map while describing a comprehensive QTL analysis of key agronomic traits of cabbage. On C01, twelve markers existed between the markers Indel481 (17,365,179 bp) and

Indel14 (28,513,070 bp), which showed recombination was observed to be normal at the 17.3–28.5 Mb. In the present study, recombination suppression was observed at C01: 18,069,792–29,537,314 bp in the mapping of *ygl-1* gene using the population constructed from YL-1 and 01-20. Moreover, a recombination-suppressed region was identified in the same area while mapping *ygl-1* using another population constructed from YL-1 and 11-192. These two populations have one same parent YL-1. Therefore, we speculated that the suppression of recombination may be due to the YL-1 mutant.

In the recombination-suppressed region, it is difficult to identify candidate genes using fine mapping. Some research has revealed genes related to the phenotype by RNA-seq, such as *Fhb1* in wheat [54] and *BPH15* in rice [55]. In the *ygl-1* gene-mapping interval, a total of 10478 SNPs and Indels, with 455 genes, were identified in the recombination-suppressed region, including 78 genes related to chloroplasts. Comparison of the two bulk RNA-seq groups showed that only 43 genes were common DEGs, only five of which were related to chloroplasts. Furthermore, three of these five genes' expression by qRT-PCR were consistent with the results of BSR. Therefore, BSA combined with RNA-seq was able to greatly reduce the number of DEGs, demonstrating that this method is an effective alternative for identifying candidate genes in a recombination-suppressed region.

5.3. Assembly Error in the Reference Genome

Brassica oleracea reference genome sequencing was completed in 2014 [25,56]. However, the 02-12 reference genome assemblies have been woefully incomplete, and some assembly errors have been identified in recent studies. For example, Lee et al. [47] revised 27 v-blocks, 10 s-blocks, and several other blocks in the 02-12 reference genome assembly during the mapping of clubroot resistance QTLs through genotyping-by-sequencing. The purple leaf gene (*BoPr*) in the ornamental kale was mapped on an unanchored scaffold by Liu et al. (2017) [57]. In a previous study [7], we identified possible assembly errors in the 02-12 reference genome. According to the comparison of marker positions with the TO1000 reference, the physical locations of Part II and Part III in the 02-12 reference genome likely represent assembly errors (Figure 1). The makers' order of linkage map was consistent with the physical map order of TO1000 reference genome. All the results showed that the TO1000 reference genome is reliable. These results will contribute to the improvement of the cabbage genome.

6. Conclusions

In conclusion, we mapped the yellow-green-leaf gene *ygl-1* on a recombination-suppressed genomic region by two populations. Bulk segregant RNA-seq (BSR) was applied to identify differentially expressed genes using two segregate populations. BSR analysis with four pools greatly reduced the number of common DEGs from 4924 to 1112. Eighteen DEGs related to chlorophyll were clustered. In the *ygl-1* gene mapping region with suppressed recombination, 43 common DEGs were identified. Five of the genes were related to chloroplasts; the *Bo1g098440* and *Bo1g098630* genes were excluded by qRT-PCR. Hence, *Bo1g059170*, *Bo1g087310*, and *Bo1g094360* in the mapping interval may be tightly associated with the development of the yellow-green-leaf mutant phenotype. Further studies on these genes may reveal the molecular mechanism of yellow-green-leaf formation in *B. oleracea*.

Supplementary Materials: Supplementary materials can be found at http://www.mdpi.com/1422-0067/19/10/2936/s1.

Author Contributions: X.L. and H.Y. developed the F$_2$ and BC. populations and wrote and revised the manuscript. H.Y., F.H., Z.L., and X.L. isolated the samples, performed the marker assays, and analyzed the marker data. F.H., Z.F., L.Y., M.Z., H.L., Y.L., Z.L., and Y.Z. conceived the study and critically reviewed the manuscript. All authors read and approved the final manuscript.

Funding: This work was financially supported by grants from the National Natural Science Foundation of China (31572141), the Major State Research Development Program (2016YFD0101702), the Science and Technology Innovation Program of Chinese Academy of Agricultural Sciences (CAAS-ASTIP-IVFCAAS), and the earmarked fund for the Modern Agro-Industry Technology Research System, China (nycytx-35-gw01).

Acknowledgments: This work was performed in the Key Laboratory of Biology and Genetic Improvement of Horticultural Crops, Ministry of Agriculture, Beijing 100081, People's Republic of China. The work reported here

Int. J. Mol. Sci. **2018**, *19*, 2936

was performed in the Key Laboratory of Biology and Genetic Improvement of Horticultural Crops, Ministry of Agriculture, Beijing 100081, China.

Conflicts of Interest: The authors declare no conflict of interest.

References

1. Carol, P.; Stevenson, D. Mutations in the *Arabidopsis* gene IMMUTANS cause a variegated phenotype by inactivating a chloroplast terminal oxidase assoeiated with phytoene desaturation. *Plant Cell.* **1999**, *11*, 57–68. [CrossRef] [PubMed]
2. Svensson, J.T.; Crosatti, C. Transcriptome analysis of cold acclimation in barley albina and xantha mutants. *Plant Physiol.* **2006**, *141*, 257–270. [CrossRef] [PubMed]
3. Zhu, L.; Zeng, X. Genetic characterisation and fine mapping of a chlorophyll-deficient mutant (*BnaC.ygl*) in Brassica napus. *Mol. Breed.* **2014**, *34*, 603–614. [CrossRef]
4. Chen, H.; Cheng, Z. A knockdown mutation of YELLOW—GREEN LEAF2, blocks chlorophyll biosynthesis in rice. *Plant Cell Rep.* **2013**, *32*, 1855–1867. [CrossRef] [PubMed]
5. Li, C.; Hu, Y. Mutation of FdC2 gene encoding a ferredoxin-like protein with C-terminal extension causes yellow-green-leaf phenotype in rice. *Plant Sci.* **2015**, *238*, 127–134. [CrossRef] [PubMed]
6. Ma, X.; Sun, X. Map-based cloning and characterization of the novel yellow-green-leaf gene *ys83* in rice (*Oryza sativa*). *Plant Physiol. Biochem.* **2016**, *111*, 1–9. [CrossRef] [PubMed]
7. Liu, X.; Yang, C. Genetics and fine mapping of a yellow-green-leaf gene (*ygl-1*) in cabbage (*Brassica oleracea* var. capitata L.). *Mol. Breed.* **2016**, *36*, 1–8. [CrossRef]
8. Whitaker, T.W. Genetic and Chlorophyll Studies of a Yellow-Green Mutant in Muskmelon. *Plant Physiol.* **1952**, *27*, 263–268. [CrossRef] [PubMed]
9. Zhong, X.M.; Sun, S.F. Research on photosynthetic physiology of a yellow-green mutant line in maize. *Photosynthetica* **2015**, *53*, 1–8. [CrossRef]
10. Yang, C.; Zhang, Y.Y. Photosynthetic Physiological Characteristics and Chloroplast Ultrastructure of Yellow Leaf Mutant YL-1 in Cabbage. *Acta Hortic. Sin.* **2014**, *41*, 1133–1144.
11. Ladygin, V.G. Spectral features and structure of chloroplasts under an early block of chlorophyll synthesis. *Biophysics* **2006**, *51*, 635–644. [CrossRef]
12. Runge, S.; Cleve, B.V.; Lebedev, N.; Armstrong, G.; Apel, K. Isolation and classification of chlorophyll-deficient xantha mutants of *Arabidopsis thaliana*. *Planta* **1995**, *197*, 490–500. [CrossRef] [PubMed]
13. Chantret, N.; Sourdille, P. Location and mapping of the powdery mildew resistance gene MIRE and detection of a resistance QTL by bulked segregant analysis (BSA) with microsatellites in wheat. *Theor. Appl. Genet.* **2000**, *100*, 1217–1224. [CrossRef]
14. Futschik, A.; Schlötterer, C. The next generation of molecular markers from massively parallel sequencing of pooled DNA samples. *Genetics* **2010**, *186*, 207–218. [CrossRef] [PubMed]
15. Zeng, F.; Yi, B.; Tu, J.; Fu, T. Identification of AFLP and SCAR markers linked to the male fertility restorer gene of pol, CMS (*Brassica napus* L.). *Euphytica* **2009**, *165*, 363–369. [CrossRef]
16. Wang, Y.; Thomas, C.E.; Dean, R.A. Genetic mapping of a Fusarium wilt resistance gene (*Fom-2*) in melon (*Cucumis melo* L.). *Mol. Breed.* **2000**, *6*, 379–389. [CrossRef]
17. Subudhi, P.K.; Borkakati, R.P.; Virmani, S.S.; Huang, N. Molecular mapping of a thermosensitive genetic male sterility gene in rice using bulked segregant analysis. *Genome* **1997**, *40*, 188–194. [CrossRef] [PubMed]
18. Cheema, K.K.; Grewal, N.K. A novel bacterial blight resistance gene from Oryza nivara mapped to 38 kb region on chromosome 4L and transferred to *Oryza sativa* L. *Gen. Res.* **2008**, *90*, 397–407. [CrossRef] [PubMed]
19. Mackay, I.J.; Caligari, P.D. Efficiencies of F_2 and backcross generations for bulked segregant analysis using dominant markers. *Crop Sci.* **2000**, *40*, 626–630. [CrossRef]
20. Zhang, G.; Guo, G.W. Deep RNA sequencing at single base-pair resolution reveals high complexity of the rice transcriptome. *Genome Res.* **2010**, *20*, 646–654. [CrossRef] [PubMed]
21. Song, H.K.; Hong, S.E. Deep RNA Sequencing Reveals Novel Cardiac Transcriptomic Signatures for Physiological and Pathological Hypertrophy. *PLoS ONE* **2012**, *7*, e35552. [CrossRef] [PubMed]
22. Jarvie, T.; Harkins, T. Transcriptome sequencing with the Genome Sequencer FLX system. *Nat. Methods* **2008**, *5*. [CrossRef]

23. Wang, R.; Sun, L. Bulk segregant RNA-seq reveals expression and positional candidate genes and allele-specific expression for disease resistance against enteric septicemia of catfish. *BMC Genomics* **2013**, *14*, 929–939. [CrossRef] [PubMed]

24. Ramirez-Gonzalez, R.H.; Segovia, V. RNA-Seq bulked segregant analysis enables the identification of high-resolution genetic markers for breeding in hexaploid wheat. *Plant Biotech. J.* **2014**, *13*, 613–624. [CrossRef] [PubMed]

25. Parkin, I.A.; Koh, C. Transcriptome and methylome profiling reveals relics of genome dominance in the mesopolyploid *Brassica oleracea*. *Genome Biol.* **2014**, *15*, R77. [CrossRef] [PubMed]

26. Kim, D.; Langmead, B.; Salzberg, S.L. HISAT: A fast spliced aligner with low memory requirements. *Nat. Methods* **2015**, *12*, 357–359. [CrossRef] [PubMed]

27. Anders, S.; Huber, W. Differential expression analysis for sequence count data. *Genome Biol.* **2010**, *11*, R106. [CrossRef] [PubMed]

28. Krzywinski, M.; Schein, J. Circos: An information aesthetic for comparative genomics. *Genome Res.* **2009**, *19*, 1639–1645. [CrossRef] [PubMed]

29. Ashburner, M.; Ashburner, M. Gene ontology: Tool for the unification of biology. *Nat. Genet.* **2000**, *25*, 25–29. [CrossRef] [PubMed]

30. Kanehisa, M.; Goto, S. The KEGG resource for deciphering the genome. *Nucleic Acids Res.* **2004**, *32*, D277–D280. [CrossRef] [PubMed]

31. Livak, K.J.; Schmittgen, T.D. Analysis of relative gene expression data using real-time quantitative PCR and the $2^{-\Delta\Delta CT}$ method. *Methods* **2001**, *25*, 402–408. [CrossRef] [PubMed]

32. Guo, J.; Zhang, Y. Transcriptome sequencing and de novo analysis of a recessive genic male sterile line in cabbage *(Brassica oleracea* L. var. capitata). *Mol. Breed.* **2016**, *36*, 117–119. [CrossRef]

33. Piippo, M.; Allahverdiyeva, Y. Chloroplast-mediated regulation of nuclear genes in Arabidopsis thaliana in the absence of light stress. *Physiol. Gen.* **2006**, *25*, 142–152. [CrossRef] [PubMed]

34. Sohani, M.M.; Schenk, P.M.; Schultz, C.J.; Schmidt, O. Phylogenetic and transcriptional analysis of a strictosidine synthase-like gene family in *Arabidopsis thaliana*, reveals involvement in plant defence responses. *Plant Biol.* **2009**, *11*, 105–117. [CrossRef] [PubMed]

35. Jyothi, T.; Duan, H.; Liu, L.; Schuler, M.A. Bicistronic and fused monocistronic transcripts are derived from adjacent loci in the Arabidopsis genome. *RNA* **2005**, *11*, 128–138.

36. Kim, S.; Kim, C.W.; Park, M.; Choi, D. Identification of candidate genes associated with fertility restoration of cytoplasmic male-sterility in onion (*Allium cepa* L.) using a combination of bulked segregant analysis and RNA-seq. *Theor. Appl. Genet.* **2015**, *128*, 2289–2299. [CrossRef] [PubMed]

37. Liu, C.; Zhou, Q.; Dong, L.; Wang, H.; Liu, F.; Weng, J.; Li, X.; Xie, C. Genetic architecture of the maize kernel row number revealed by combining QTL mapping using a high-density genetic map and bulked segregant RNA sequencing. *BMC Genomics* **2016**, *17*, 915. [CrossRef] [PubMed]

38. Huang, Z.; Peng, G.; Liu, X.; Deora, A.; Falk, K.C.; Gossen, B.D.; McDonald, M.R.; Yu, F. Fine Mapping of a Clubroot Resistance Gene in Chinese Cabbage Using SNP Markers Identified from Bulked Segregant RNA Sequencing. *Front. Plant Sci.* **2017**, *8*, 1448–1459. [CrossRef] [PubMed]

39. Wang, Y.; Xie, J.; Zhang, H. Mapping stripe rust resistance gene YrZH22, in Chinese wheat cultivar Zhoumai 22 by bulked segregant RNA-Seq (BSR-Seq) and comparative genomics analyses. *Theor. Appl. Genet.* **2017**, *130*, 2191–2201. [CrossRef] [PubMed]

40. Trick, M.; Adamski, N.M.; Mugford, S.G.; Jiang, C.C.; Febrer, M.; Uauy, C. Combining SNP discovery from next-generation sequencing data with bulked segregant analysis (BSA) to fine-map genes in polyploid wheat. *BMC Plant Biol.* **2012**, *12*, 14–19. [CrossRef] [PubMed]

41. Lei, S.; Yao, X. Towards map-based cloning: Fine mapping of a recessive genic male-sterile gene (*BnMs2*) in *Brassica napus* L. and syntenic region identification based on the *Arabidopsis thaliana* genome sequences. *Theor. Appl. Genet.* **2007**, *115*, 643–651. [CrossRef] [PubMed]

42. Shimizu, M.; Pu, Z.J. Map-based cloning of a candidate gene conferring Fusarium yellows resistance in *Brassica oleracea*. *Theor. Appl. Genet.* **2015**, *128*, 119–130. [CrossRef] [PubMed]

43. Liang, J.; Ma, Y. Map-based cloning of the dominant genic male sterile *Ms-cd1* gene in cabbage (*Brassica oleracea*). *Theor. Appl. Genet.* **2016**, *5*, 1–9. [CrossRef] [PubMed]

44. Tian, Y.; Huang, Q. Inheritance of chlorophyll-deficient mutant *L638-y* in *Brassica juncea* L. and molecular markers for chlorophyll deficient gene *gr1*. *J. Northwest A F Univ.* **2012**, *12*, 17–19.

45. Wang, Y.; He, Y. Fine mapping of a dominant gene conferring chlorophyll-deficiency in *Brassica napus*. *Sci. Rep.* **2016**, *6*, 314–319. [CrossRef] [PubMed]

46. Sherman, J.D.; Stack, S.M. Two-dimensional spreads of synaptonemal complexes from Solanaceous plants. VI. Highresolution recombination nodule map for tomato (*Lycopersicon esculentum*). *Genetics* **1995**, *141*, 683–708. [PubMed]

47. Wei, F.; Gobelman-Werner, K. The Mla (powdery mildew) resistance cluster is associated with three NBS-LRR gene families and suppressed recombination within a 240-kb DNA interval on chromosome 5S (1HS) of barley. *Genetics* **1999**, *153*, 1929–1948. [PubMed]

48. Ten, H.R.; Robbins, T.P. Localization of T-DNA Insertions in Petunia by Fluorescence in Situ Hybridization: Physical Evidence for Suppression of Recombination. *Plant Cell* **1996**, *8*, 823–830.

49. Stirling, B.; Newcombe, G. Suppressed recombination around the MXC3 locus, a major gene for resistance to poplar leaf rust. *Theor. Appl. Genet.* **2001**, *103*, 1129–1137. [CrossRef]

50. Neu, C.; Stein, N.; Keller, B. Genetic mapping of the Lr20-Pm1 resistance locus reveals suppressed recombination on chromosome arm 7AL in hexaploid wheat. *Genome* **2002**, *45*, 737–744. [CrossRef] [PubMed]

51. Jessup, R.W.; Burson, B.L. Disomic Inheritance, Suppressed Recombination, and Allelic Interactions Govern Apospory in Buffelgrass as Revealed by Genome Mapping. *Crop Sci.* **2002**, *42*, 1688–1694. [CrossRef]

52. Lee, J.; Izzah, N.K. Genotyping-by-sequencing map permits identification of clubroot resistance QTLs and revision of the reference genome assembly in cabbage (*Brassica oleracea* L.). *DNA Res.* **2015**, *14*, S113. [CrossRef] [PubMed]

53. Lv, H.; Wang, Q. Whole-Genome Mapping Reveals Novel QTL Clusters Associated with Main Agronomic Traits of Cabbage (*Brassica oleracea* var. *capitata* L.). *Front. Plant Sci.* **2016**, *7*, 989–999. [CrossRef] [PubMed]

54. Schweiger, W.; Schweiger, W.; Steiner, B.; Vautrin, S.; Nussbaumer, T.; Siegwart, G.; Zamini, M.; Jungreithmeier, F.; Gratl, V.; Lemmens, M.; et al. Suppressed recombination and unique candidate genes in the divergent haplotype encoding *Fhb1*, a major Fusarium head blight resistance locus in wheat. *Theor. Appl. Genet.* **2016**, *129*, 1607–1623. [CrossRef] [PubMed]

55. Lv, W.; Du, B. BAC and RNA sequencing reveal the brown planthopper resistance gene *BPH15*, in a recombination cold spot that mediates a unique defense mechanism. *BMC Genomics* **2014**, *15*, 674–679. [CrossRef] [PubMed]

56. Liu, S.; Liu, Y.; Yang, X.; Tong, C.; Edwards, D.; Parkin, I.A. The *Brassica oleracea* genome reveals the asymmetrical evolution of polyploid genomes. *Nat. Commun.* **2014**. [CrossRef] [PubMed]

57. Liu, X.P.; Gao, B.Z.; Han, F.Q.; Fang, Z.Y.; Yang, L.M.; Zhuang, M.; Lv, H.H.; Liu, Y.M.; Li, Z.S.; Cai, C.C.; et al. Genetics and fine mapping of a purple leaf gene, BoPr, in ornamental kale (*Brassica oleracea* L. var. acephala). *BMC Genomics* **2017**, *18*, 230–239. [CrossRef] [PubMed]

International Journal of
Molecular Sciences

Article

Identification and Characterization of *EI* (*Elongated Internode*) Gene in Tomato (*Solanum lycopersicum*)

Xiaorong Sun [†], Jinshuai Shu [†], Ali Mohamed Ali Mohamed, Xuebin Deng, Xiaona Zhi, Jinrui Bai, Yanan Cui, Xiaoxiao Lu, Yongchen Du, Xiaoxuan Wang, Zejun Huang, Yanmei Guo, Lei Liu * and Junming Li *

Institute of Vegetables and Flowers, Chinese Academy of Agricultural Sciences, Key Laboratory of Biology and Genetic Improvement of Horticultural Crops, Ministry of Agriculture, 12 Zhongguancun Nandajie Street, Beijing 100081, China; sunxiaorong@caas.cn (X.S.); shujinshuai@caas.cn (J.S.); almuttaried@hotmail.com (A.M.A.M.); dengxuebin@caas.cn (X.D.); 18763825752@163.com (X.Z.); baijinrui@caas.cn (J.B.); 15621567802@163.com (Y.C.); luxiaoxiao914@163.com (X.L.); duyongchen@caas.cn (Y.D.); wangxiaoxuan@caas.cn (X.W.); huangzejun@caas.cn (Z.H.); guoyanmei@caas.cn (Y.G.)
* Correspondence: liulei02@caas.cn (L.L.); lijunming@caas.cn (J.L.); +86-10-82109530 (L.L. & J.L.)
† These authors equally contributed to this work.

Received: 19 April 2019; Accepted: 3 May 2019; Published: 5 May 2019

Abstract: Internode length is an important agronomic trait affecting plant architecture and crop yield. However, few genes for internode elongation have been identified in tomato. In this study, we characterized an elongated internode inbred line P502, which is a natural mutant of the tomato cultivar 05T606. The mutant P502 exhibits longer internode and higher bioactive GA concentration compared with wild-type 05T606. Genetic analysis suggested that the elongated internode trait is controlled by quantitative trait loci (QTL). Then, we identified a major QTL on chromosome 2 based on molecular markers and bulked segregant analysis (BSA). The locus was designated as *EI* (*Elongated Internode*), which explained 73.6% genetic variance. The *EI* was further mapped to a 75.8-kb region containing 10 genes in the reference Heinz 1706 genome. One single nucleotide polymorphism (SNP) in the coding region of *solyc02g080120.1* was identified, which encodes gibberellin 2-beta-dioxygenase 7 (SlGA2ox7). SlGA2ox7, orthologous to AtGA2ox7 and AtGA2ox8, is involved in the regulation of GA degradation. Overexpression of the wild *EI* gene in mutant P502 caused a dwarf phenotype with a shortened internode. The difference of *EI* expression levels was not significant in the P502 and wild-type, but the expression levels of GA biosynthetic genes including *CPS, KO, KAO, GA20ox1, GA20ox2, GA20ox4, GA3ox1, GA2ox1, GA2ox2, GA2ox4,* and *GA2ox5,* were upregulated in mutant P502. Our results may provide a better understanding of the genetics underlying the internode elongation and valuable information to improve plant architecture of the tomato.

Keywords: tomato; *Elongated Internode* (*EI*); QTL; GA2ox7

1. Introduction

Plant height is an important component of plant architecture, and is highly correlated with the yield [1]. One of the decisive factors affecting plant height is internode length. The reduced plant height or internode length of semi-dwarf varieties has improved the harvest index and biomass production. The introduction of the "green revolution" semi-dwarf gene *SD1* in rice and *Rht* in wheat resulted in substantial increases in grain yields and helped to avert predicted food shortages in Asia during the 1960s and 1970s [2–5]. To explore the genetic potential, several genes or quantitative trait loci (QTL) controlling the internode length in rice [6–8], maize [9–11], wheat [12,13], and sorghum [14,15] have been identified. In a recent report, the semi-dwarf gene *SBI* was cloned, which could shorten the basal internode of rice. Moreover, the *SBI* allele-introduced varieties have great potential for improving lodging resistance and yield [16].

The molecular genetic analysis of dwarf and slender mutants revealed that most of the mutations are related to the biosynthesis pathways and signal transduction of phytohormones, mostly gibberellins (GAs) [2,9,17–19]. GAs are a group of tetracyclic diterpenoid that affect plant developmental processes such as stem elongation [20]. Nowadays, more than 130 GAs have been identified, but relatively few GAs (e.g., GA_1, GA_3, GA_4, and GA_7) have intrinsic biological activity [21]. The rice *SD1* encodes GA20ox2 and catalyzes the conversion of GA_{12}/GA_{53} to bioactive GA precursors GA_9/GA_{20}. The recessive semi-dwarf mutant *sd1* can be restored by exogenous GA_3 [22,23]. DELLA protein is a negative regulatory factor in the GA signaling pathway, which inhibits plant growth. The semi-dominant mutations that occurred in Arabidopsis *GAI*, maize *D8*, wheat *Rht*, and rice *GAI* were caused by the gain of function of DELLA protein, which led to the dwarfism [3,24–28]. In contrast, the loss-of-function of DELLA protein in barley *SLN1*, rice *SLR1*, and tomato *PRO* increased growth and caused a GA-constitutive response phenotype [29–31]. In addition, other phytohormones including brassinosteroid (BR), indole-3-acetic acid (IAA), and strigolactones (SLs) have also been proven to be involved in the regulation of internode length or plant height [7,14,32–34].

Tomato (*Solanum lycopersicum*) is the second most consumed vegetable crop and is widely grown around the world [35]. The tomato internode length not only affects the plant architecture, but also plays an important role in mechanized harvesting. However, there are few reports on the genetic regulation of internode length in tomato. Evidence has shown that the well-known dwarf cultivar Micro-Tom, which has the characteristics of extreme dwarfing, dark and wrinkled leaves, has at least two mutations (*d* and *mnt*) affecting internode length [36]. However, Micro-Tom was produced for ornamental purposes, and as a conventional model system for research due to its small size, rapid growth, and easy transformation. Due to its characteristics of extreme dwarfing, it seems very limited in a practical breeding program. Moreover, the tomato *br* locus contributes to a shorter internode and could be useful source for tomato short internode breeding. However, the *br* locus was only mapped to a 763.1-kb region on chromosome 1, and the gene has not been cloned yet [37]. Therefore, to clone new gene or locus that controls internode elongation is of great theoretical and practical significance. It is helpful to clarify the regulatory mechanism of tomato internode elongation and to provide the possibility of establishing a breeding approach. Furthermore, with the completion of tomato genome sequencing as well as the rapid development of sequencing, marker development has become easier, which has accelerated the speed of gene cloning [38].

In this study, we performed a phenotypic characterization of the mutant P502, which shows a significantly elongated internode compared with wild-type 05T606. Then, we report on the molecular identification of the *EI* (*Elongated Internode*) gene by map-based cloning. *EI*, which is expressed in the root, stem, and leaf, encodes the GA2ox7 enzyme involved in the GA metabolism pathway. Overexpressing of *EI* can cause a dwarf phenotype with short internode. Our results indicate that *EI* plays an important role in controlling the tomato internode elongation.

2. Results

2.1. Characterization of Elongated Internode Inbred Line P502

To compare the dynamic difference in the internode lengths of P502 and wild-type 05T606, we recorded the mean internode lengths of 20-day-old, 25-day-old, 30-day-old, 35-day-old, and 40-day-old seedlings. The results showed that P502 had a longer internode than the wild-type across the seedling stages (Figure 1a,b). Additionally, we compared individual internode lengths including the first, second, third, fourth, and fifth internodes of 40-day-old seedlings. The results indicated that each internode of P502 was significantly greater than the corresponding internode of the wild-type (Figure 1c,d). An analysis of the longitudinal sections of the third internode with a scanning electron microscope revealed that the cells were much longer in P502 than in the wild-type (Figure 1e,f). These results suggest that the mutant P502 phenotype is characterized by the elongated internode.

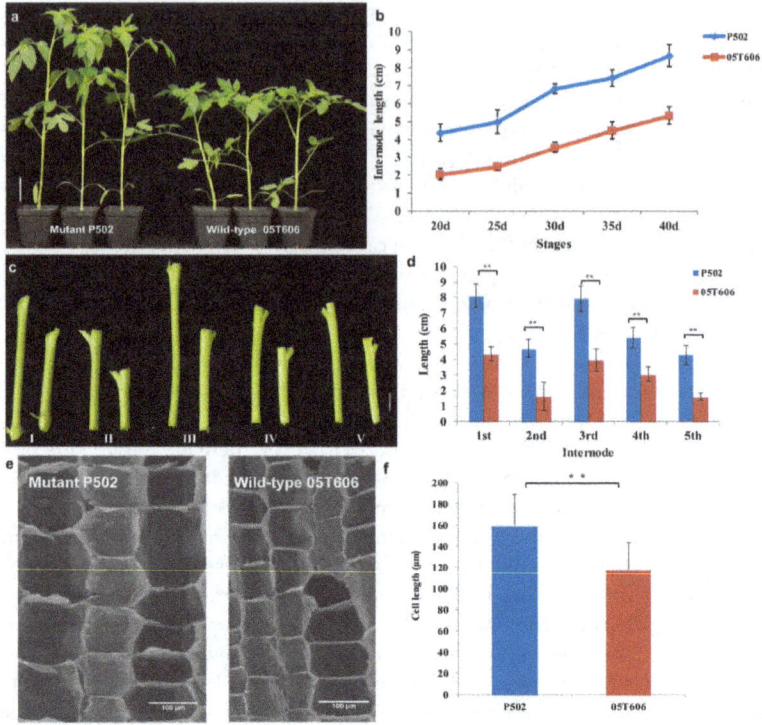

Figure 1. The phenotypic characterization of the mutant P502 and wild-type 05T606 at 40-day-old seedlings. Morphological phenotypes (**a**) and the statistical data of internode lengths (**b**) of the mutant P502 and wild-type 05T606. (**c**) Morphological phenotypes of the first five internode lengths of the mutant P502 (left) and wild-type (right). (**d**) Statistical data of internode length in (**c**). Longitudinal sections (**e**) and the statistical data (**f**) of the third internode pith cell length of P502 and wild-type. Scale bar is 5 cm in (**a**), 1 cm in (**c**), 100 μm in (**e**). A Student's *t* test indicated a significant difference ((**b,d**), $n = 30$ plants; (**f**) $n = 90$ cells) in (**d,f**). ** $p < 0.01$. All data are given as mean ± SD.

2.2. Elongated Internode Mutation Is Related to the GA Metabolic Pathway

The GAs stimulate cell elongation, and are effective internode elongation regulators. Paclobutrazol (PAC) inhibits the oxidation of *ent*-kaurene, an early step in GA biosynthesis, and can reduce endogenous GA level [39]. To examine the responses to GAs, we sprayed 20-day-old wild-type 05T606 and mutant P502 seedlings with exogenous GA_3 and PAC. For the 40-day-old seedlings, the plant height of 05T606 and P502 increased by 58.2% and 20.5%, respectively, after GA_3 treatment. However, the PAC treatment decreased the height of 05T606 by 57.6%, and decreased the height of P502 by 61.0% (Figure 2a–d). Next, we measured the endogenous GA concentration of the first five internodes in 30-day-old seedlings of P502 and 05T606 plants. There was an increase in the bioactive GA_1 in P502, and bioactive GA_4 was only detected in mutant P502 (Figure 2e). These results indicate that the elongated internode of mutant P502 is related to the GA metabolic pathway and was caused by a higher level of bioactive GAs.

Figure 2. The line P502 is a GA-sensitive mutant. (**a,b**) The morphological phenotypes of 30-day-old wild-type and P502 seedlings after treatment with GA_3 (10^{-5} M) and PAC (10^{-7} M, GA biosynthesis inhibitor), respectively. (**c,d**) The statistical data of wild-type and P502 plant height in different stages, respectively. (**e**) Concentration of endogenous bioactive GAs in the first five internodes of 30-day-old mutant P502 and wild-type seedlings. The water treatment was used as control and the scale bar is 5 cm in (**a,b**). Data for (**c,d**) are based on three replicates of eight plants per group. N.D. represents not detectable. Data for (**e**) are based on three independent biological replicates, and the asterisk indicates a statistically significant difference (Student's *t*-test, ** $p < 0.01$). All data are given as mean ± SD.

2.3. Genetic Analysis of the Elongated Internode Trait

To determine whether the elongated internode trait is controlled by a single gene or QTL, the first five average internode lengths of the P_1 (Heinz 1706), P_2 (P502), F_1 (Heinz 1706 × P502), and F_2 population were recorded in the spring of 2016, 2017, and 2018. The internode length frequency of the F_2 population exhibited a continuous and skewed distribution in different years (Figure 3), and the F_1 was biased toward the parent Heinz 1706. Moreover, the differences of P_1, P_2, and F_1 were significant (Table 1). These results indicate that the elongated internode length is a quantitative trait and that this population was ideal for elongated internode QTL analysis.

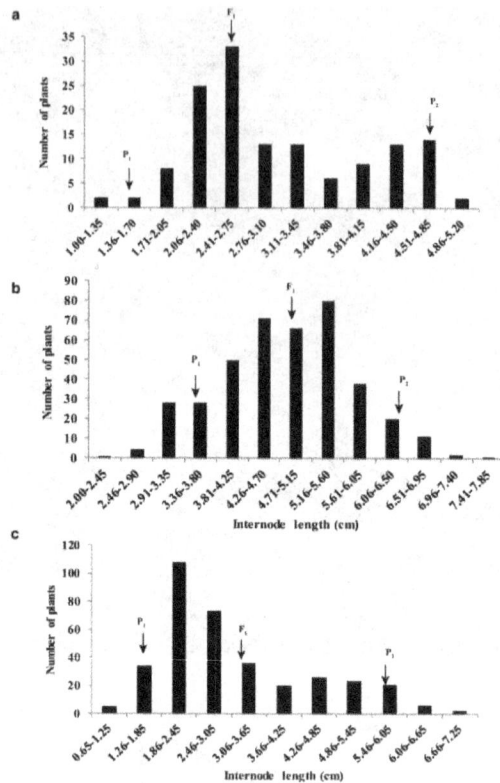

Figure 3. Internode length frequency distribution of F_2 individuals at 40-day-old seedlings. (**a**) 2016 (*n* = 140). (**b**) 2017 (*n* = 400). (**c**) 2018 (*n* = 354). The mean internode lengths were recorded from the first internode to the fifth internode (starting from the cotyledons).

Table 1. Internode lengths of P_1, P_2, and F_1 population plants.

Materials	2016 (cm) [a]	2017 (cm) [a]	2018 (cm) [a]
P_1 (Heinz 1706)	1.52 ± 0.35 a	1.48 ± 0.63 a	1.85 ± 0.30 a
P_2 (P502)	4.62 ± 0.19 c	4.66 ± 0.60 c	5.54 ± 0.48 c
F_1 (Heinz 1706 × P502)	2.74 ± 0.28 b	2.53 ± 0.47 b	3.10 ± 0.24 b

[a] Values followed by different letters (a, b, and c) are significantly different (*p* < 0.01). *n* = 15 plants.

2.4. Map-Based Cloning of the EI Gene

A total of 372 InDel markers distributed on 12 chromosomes were screened, and 90 were polymorphic between the parents. Of these 90 markers, four markers (D55, D57, D64, and D67) located on chromosome 2 were polymorphic between the E and N bulks. The bands of the E pool were consistent with those of the parent P502, whereas the bands of the N pool were heterozygous. These results suggest that the locus is present on chromosome 2.

The 354 F_2 individuals, derived from Heinz 1706 × P502, were used for QTL analysis. The results showed that there was only a single peak, with a logarithm of odds score of 102.4 explaining 73.6% phenotypic variance (Figure 4a). Therefore, we concluded that the locus named *elongated internode* (*ei*) was located between markers D64 and HP2509. Another 956 F_2 individuals were screened for recombinants with the flanking markers D64 and HP2509. The detected recombinants were analyzed with another seven CAPS and InDel markers between the flanking markers (Figure 4b). According to the genotypes of the recombinants and the phenotypes of F_3 individuals, we narrowed the *ei* locus to an interval between CAPS17 and InDel6 (Table 2), corresponding to a 75.8-kb region on chromosome 2 of the reference Heinz 1706 genome.

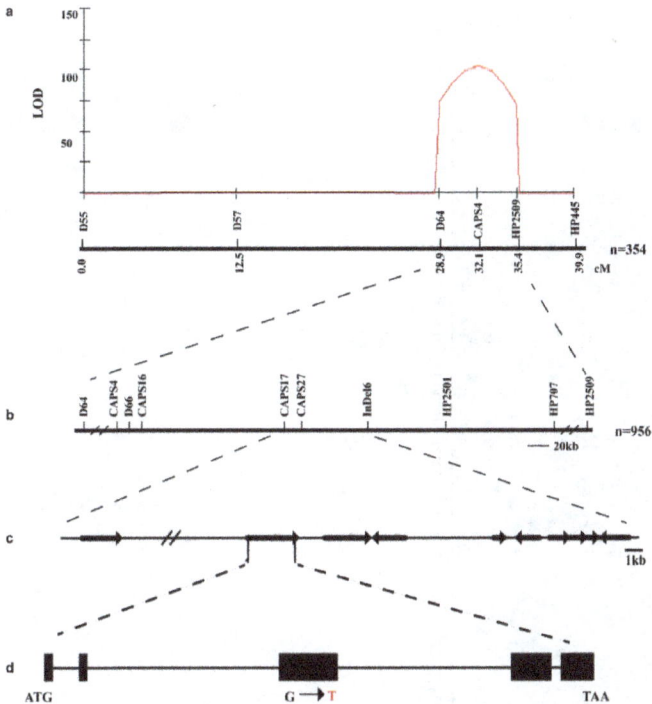

Figure 4. Map-based cloning of *ei* locus. (**a**) The genetic map of *ei* on chromosome 2, mapped using 354 F_2 individuals and six polymorphic markers. (**b**) The *ei* locus was fine-mapped to the interval between markers CAPS17 and InDel6. (**c**) Predicted genes in the region encompassing the *ei* locus. The arrows indicate the direction of transcription. (**d**) Gene structure and the mutation site. The black rectangles and black line indicate exons and introns, respectively. The red letter represents the mutation base in mutant P502.

Table 2. Genotypes of F$_2$ recombinants and phenotypes of F$_{2:3}$ individuals.

Recombinants	Genotype (F$_2$) [a]										Phenotype(F$_3$)	
	D64	CAPS4	D66	CAPS16	CAPS17	CAPS27	InDel6	HP2501	HP707	HP2509	N [b]	E [b]
6–11	b	h	h	h	h	h	h	h	h	h	-	-
13–5	b	b	b	h	h	h	h	h	h	h	44	15
7–48	b	b	b	b	h	h	h	h	h	h	45	15
7–71	b	b	b	b	h	h	h	h	h	h	43	14
15–543	b	b	b	b	b	h	h	h	h	h	45	14
17–18	b	b	b	b	b	h	h	h	h	h	45	15
15–741	h	h	h	h	h	h	b	b	b	b	46	14
1–2	h	h	h	h	h	h	b	b	b	b	45	15
1–72	h	h	h	h	h	h	b	b	b	b	44	16
6–21	h	h	h	h	h	h	h	b	h	b	-	-

[a] b in green backgroud: homozygous like P502; h in gray background: heterozygous; [b] N: the number of normal internode plants; E: the number of elongated internode plants; -: no data.

The Solanaceae Genomics Network website (SGN; http://solgenomics.net) searches [40] indicated that there were 10 genes in this region (Figure 4c, Table 3). By analyzing the sequenced P502 genome, we determined that the DNA sequence had no mutations in the other nine genes, whereas *solyc02g080120.1* contained a SNP (G–T) in the exon region (Figure 4d). Thus, *solyc02g080120.1* was amplified using genomic DNA extracted from 05T606 and P502 plants and eight primer pairs (A1, A2, A3, A4, A5, A6, A7, and A8; Table S1). The amplification results revealed that the *EI* gene consists of 4626 bp (with five exons and four introns). Moreover, the *EI* gene in mutant P502 includes a SNP mutation in the third exon (G2152T) (Figure S1). To confirm this mutation site, the *solyc02g080120.1* coding sequence (CDS) was amplified by RT-PCR (CDS-F and CDS-R primers) (Table S1). Sequences of the CDS further confirmed the presence of a SNP mutation in the coding region, which resulted in an amino acid mutation in mutant P502.

Table 3. Ten predicted genes in the 75.8-kb fine mapping interval according to the reference genome.

Gene ID [a]	Position	Functional Annotation
solyc02g080110.2	SL2.50ch02: 44414686..44417879 (+)	Unknown Protein (AHRD V1)
solyc02g080120.1	SL2.50ch02: 44432042..44436667 (+)	Gibberellin 2-beta-dioxygenase 7
solyc02g080130.2	SL2.50ch02: 44439349..44443883 (+)	Chaperone dnaj-like protein
solyc02g080140.2	SL2.50ch02: 44444670..44447584 (-)	cysteine-rich PDZ-binding protein
solyc02g080150.1	SL2.50ch02: 44458406..44458810 (+)	uncharacterized LOC101262168
solyc02g080160.2	SL2.50ch02: 44460578..44462486 (-)	probable xyloglucan endotransglucosylase/hydrolase protein 8
solyc02g080170.1	SL2.50ch02: 44471643..44473397 (+)	pentatricopeptide repeat-containing protein At4g21170
solyc02g080180.1	SL2.50ch02: 44474353..44475381 (+)	Probable dolichyl-diphosphooligosaccharide—protein glycosyltransferase subunit 3B
solyc02g080190.2	SL2.50ch02: 44477570..44478612 (+)	Nuclear transport factor 2
solyc02g080200.2	SL2.50ch02: 44478656..44480593 (-)	pectinesterase-like

[a] Genes were identified based on the tomato model (ITAG release 2.40, SL2.50) in SGN (https://solgenomics.net/) (Access on 21 June 2014).

To determine whether the G-to-T transition was directly associated with the elongated internode phenotype, a co-segregation analysis was conducted with a functional marker (KASP) developed from this SNP. The KASP marker (S-A1: GAAGGTGACCAAGTTCATGCTCACAAGCTTCACAAGAATGGGG; S-A2: GAAGGTCGGAGTCAACGGATTGCACAAGCTTCACAAGAATGGGT; S-C: GTGATACTCCATGGTTTACAACTTGGAA) was used to validate the genotypes of 354 F_2 individuals derived from the cross of Heinz 1706 × P502 hybridization. The results showed that this KASP marker was co-segregated with internode length (Figure S2), with a 100% accuracy rate. It further confirmed that SNP mutation is associated with the elongated internode.

2.5. Protein Sequence Alignment and Phylogenetic Analysis of SlGA2ox7

According to the gene annotation, *EI* encodes the SlGA2ox7 enzyme, which comprises 380 amino acids. The sequence alignment and phylogenetic analysis revealed that GA2ox7 and GA2ox8 are clustered in one group, which is separate from GA20oxs and GA3oxs (Figure 5a), indicating that GA2ox7 and GA2ox8 are conserved in tobacco, Arabidopsis, maize, and grape. Moreover, amino acid position 112 of wild-type SlGA2ox7 is a hydrophilic glycine, whereas it is a hydrophobic valine in P502.

The SlGA2ox7 at this position is located within a conserved region of the PcbC superfamily (Figure 5b), indicating that the altered hydrophobicity of the amino acid may affect the function of SlGA2ox7.

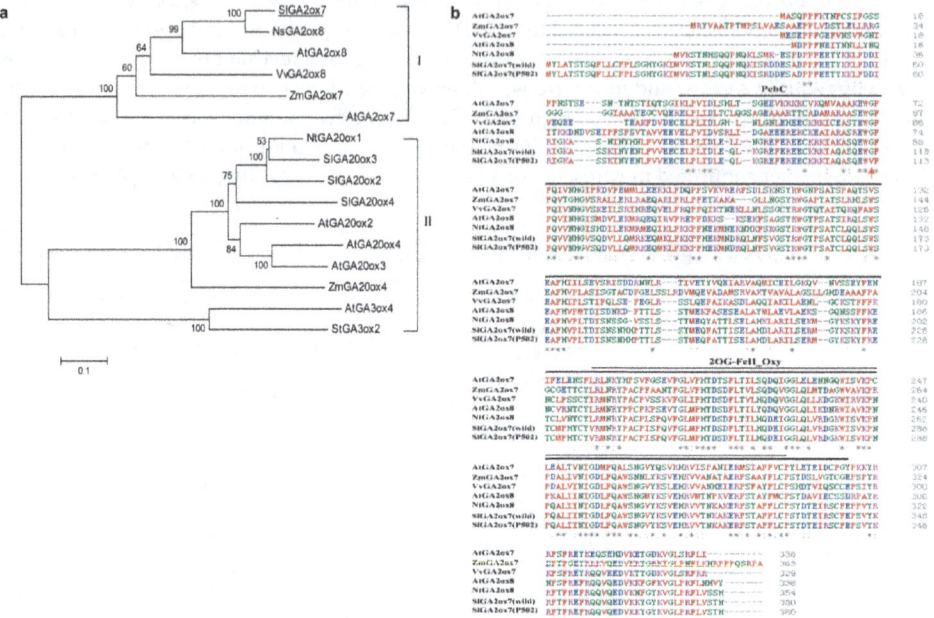

Figure 5. Phylogenetic analysis and sequence alignment of SlGA2ox7 with various species. (**a**) Phylogenetic analysis of SlGA2ox7. The phylogenetic tree was generated using the neighbor-joining method built in MEGA6.0, and the inferred phylogeny was tested by bootstrap analysis with 1000 replicate datasets. Numbers shown at the tree forks indicate the frequency of occurrence among all bootstrap iterations performed. (**b**) Alignment of the SlGA2ox7 sequences. The proteins were aligned with the ClustalW program. The bold black line and the thin black line indicate the PcbC domain (81–343) and 2OG-FeII_Oxy domain (237–332), respectively. The red arrow represents the amino acid at position 112 of SlGA2ox7.

2.6. Overexpression of Wild-Type EI in P502 Resulted in Dwarfism

To confirm the function of *EI*, the recombinant plasmid 35S: *EI* was introduced into the mutant P502. Transgenic plants were obtained after screening for regenerated shoots on selection medium containing kanamycin. The transgenic plants were analyzed further by PCR with primers NPTII-F and NPTII-R, and two positive transgenic plants (T_0-1 and T_0-2) were obtained. The *EI*-overexpressing transgenic T_1 homozygous lines (OE-1 and OE-2) exhibited dwarfism with shortened internodes (CK: 36.00 ± 1.32 cm; OE-1: 24.7 ± 1.14 cm; OE-2: 8.53 ± 0.91 cm) (Figure 6a). The *EI* expression level was 1.44-fold and 14-fold higher in the OE-1 and OE-2 plants, respectively, than in the P502 control plants (Figure 6b). These results indicated that overexpression of *EI* could result in dwarf phenotype with shortened internodes, and the degree of shortness is related to the expression level of *EI*.

Figure 6. Overexpression of wild-type *EI* gene reduced the internode length of mutant P502. (a) Phenotype of P502 (CK) and T_1 transgenic plants (OE-1 and OE-2) at 40-day-old seedlings. (b) Expression level of *EI* in T_1 lines (OE-1 and OE-2). Total RNA was isolated from internode of P502 and transgenic T_1 plants at 40-day-old seedlings. Data represent mean $\pm$ SD based on three independent biological and three technical experiments. Scale bar is 5 cm in (a). Statistical significances were calculated based on two-tailed, two-sample Student's *t*-test at * $p < 0.05$ and ** $p < 0.01$.

2.7. The Expression Analysis of GA Metabolic Pathway-Related Genes

To study the spatiotemporal expression patterns of *EI*, total RNA was extracted from the leaves, stem (the third internode), and roots of 40-day-old wild-type and mutant P502 seedlings. The results of a qRT-PCR assay indicated that *EI* was expressed in the leaves, stem, and roots (Figure 7a), and the expression level in the leaves was more than 18-fold higher than that in the stem and roots. Interestingly, the differences of expression levels were not significant ($p > 0.05$) in the wild-type and mutant P502 (Figure 7a), indicating the G-to-T mutation does not alter the expression level of the *EI*.

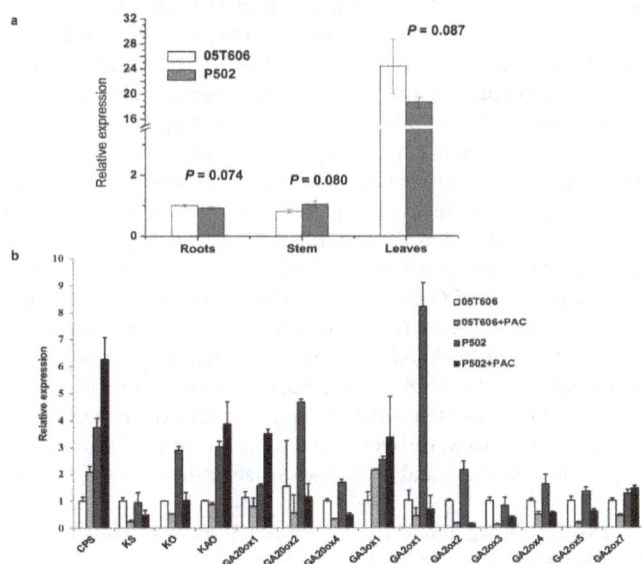

Figure 7. Expression analysis in the qRT-PCR assay. (a) Expression patterns of *EI* in the roots, stem, and leaves of P502 and the wild-type. Total RNA was isolated from P502 and wild-type at 40-day-old seedlings. A Student's *t* test was used for statistical analysis. (b) Expression levels of the GA biosynthetic genes after 10^{-7} M paclobutrazol (PAC, GA biosynthesis inhibitor) treatment. Data represent mean $\pm$ SD based on three independent biological and three technical experiments.

Many genes are involved in the GA biosynthetic pathway. CPS, KS, and KO are each encoded by a single gene in most plant species examined. However, the cytosol-localized GA20ox, GA3ox, and GA2ox each is encoded by a small gene family [41]. The qRT-PCR assay indicated that *CPS*, *KO*, *KAO*, *GA20ox1*, *GA20ox2*, *GA20ox4*, *GA3ox1*, *GA2ox1*, *GA2ox2*, *GA2ox4*, and *GA2ox5*, were more highly expressed in P502 than in the wild-type 05T606. PAC treatment decreased the expression levels of *KS*, *KO*, *GA20ox2*, *GA20ox4*, *GA2ox1*, *GA2ox2*, *GA2ox3*, *GA2ox4*, and *GA2ox5*, but increased the expression level of *CPS* and *GA3ox1*. However, the expression of *EI* (*GA2ox7*) in P502 was not significantly changed after PAC treatment (Figure 7b).

3. Discussion

The GA-related mutants have been categorized into GA-deficient (GA-sensitive) mutants and GA-insensitive mutants according to their response to exogenous GAs [42]. In GA-deficient dwarfs, the normal phenotype can be restored by the application of exogenous GAs and the mutations are usually due to a deficiency in the GA metabolic pathway [43]. In GA-insensitive types, the mutants are deficient in GA signaling and exhibit altered GA responses or the constitutive activation of GA responses [29–31,44]. In our study, the plant height of wild-type and the mutant P502 increased by 58.2% and 20.5%, respectively, after GA_3 treatment, whereas PAC treatment decreased the height of the wild-type by 57.6%, and the mutant P502 by 61.0% (Figure 2a–d). These results indicated that the mutant P502 was responsive to GA_3 and PAC, and the mutation was related to the GA metabolic pathway. In addition, it was found that mutant P502 was far less sensitive to GA_3 and more sensitive to PAC compared with the wild-type. This may be caused by a higher GA concentration of mutant P502 (Figure 2e), which reduced its sensitivity to GA_3 and increased its sensitivity to PAC.

GA2ox members are thought to disable GA functions by hydroxylating the C-2 position of active GAs or their precursors. The genes encoding 2-oxidases have been isolated from Arabidopsis, rice, spinach, and pea [45–48]. However, few GA2-oxidase genes have been isolated from tomato. In our study, we isolated the tomato *EI* gene by map-based cloning. The *EI* gene encodes SlGA2ox7, which belongs to GA2-oxidase. A point mutation in the exon region of *EI* gene resulted in the amino acid mutation (glycine to valine) of SlGA2ox7. GA2ox7 or GA2ox8 is conserved in various species (Figure 5a), and amino acid mutation occurs in the conserved domain of the PcbC superfamily (Figure 5b). Overexpression of *EI* inhibited the internode elongation of mutant P502, leading to dwarfism with a shortened internode (Figure 6a). These results are consistent with the research reported by Schomburg et al. [45], who revealed that the overexpression of *AtGA2ox7* and *AtGA2ox8* induced a dwarf phenotype of Arabidopsis and tobacco. Similar results have been obtained in transgenic plants overexpressing GA 2-oxidases from rice (*O. sativa*) [49]. Taken together, the elongated internode is caused by the loss-of-function of SlGA2ox7.

So far, three different kinds of GA deactivation have been identified. One type of GA2oxs hydroxylates the C-2 of active C_{19}-GAs (GA_1 and GA_4) or C_{19}-GA precursors (GA_{20} and GA_9) to produce biologically inactive GAs (GA_8, GA_{34}, GA_{29}, and GA_{51}, respectively) [18]. Another type of GA2oxs including AtGA2ox7, AtGA2ox8, OsGA2ox5, OsGA2ox6, and SoGA2ox3 (*Spinacia oleracea*) accept C_{20}-GAs (GA_{12} and GA_{53}) as their substrates to produce GA_{110} and GA_{97}, respectively [45,50]. In addition, the recombinant SoGA2ox1 can work on both C_{19}-GA and C_{20}-GA substrates [47]. Although SlGA2ox7 is orthologous to NsGA2ox8 and AtGA2ox8, whether they catalyze the same substrate needs further research.

Bioactive GA (GA_1, GA_3, GA_4, and GA_7) concentrations are maintained mainly by the balanced activities of GA 3-oxidases (GA3oxs) and GA 20-oxidases (GA20oxs), essential enzymes regulating GA biosynthesis, and GA 2-oxidases (GA2oxs) necessary for GA inactivation [51]. In our study, the expression of *EI* (*GA2ox7*) did not significantly change in the mutant P502 compared with the wild-type 05T606 (Figure 7a), indicating that the mutation of *EI* did not affect its transcript level. However, the expression of GA biosynthesis pathway genes including *CPS*, *KO*, *KAO*, *GA20ox1*, *GA20ox2*, *GA20ox4*, *GA3ox1*, *GA2ox1*, *GA2ox2*, *GA2ox4* and *GA2ox5* were increased (Figure 7b).

Similarly, the *dw* mutant of the soybean had lower expression levels of *CPS* and *GA20oxs* than the wild-type [19]. These results indicated that the mutations of genes related to GA synthesis may regulate the expression of other genes involved in the GA biosynthesis pathway by changing the GA concentrations. After exogenous PAC treatment, the growths of the mutant P502 and wild-type 05T606 were blocked. The expression levels of *KS*, *KO*, *GA20ox2*, *GA20ox4*, *GA2ox1*, *GA2ox2*, *GA2ox3*, *GA2ox4*, and *GA2ox5*, were downregulated (Figure 7b), revealing that these genes may be involved in the homeostatic maintenance of bioactive GA levels. Interestingly, the expression of the *EI* (*GA2ox7*) gene in the mutant P502 was not significantly changed after PAC treatment. We speculated that the mutation of *EI* may result in more complex regulation of GA homeostasis.

Tomato cultivars with determinate growth habit, compact, and short internode have been developed for commercial use [35]. The *sp* gene controlling determinate growth habit was cloned and the introduction of the *sp* allele into tomato cultivars has transformed the industry by creating a major modification in plant architecture [52]. However, few genes controlling internode elongation have been cloned. In our study, the *EI* gene for internode elongation was identified by map-based cloning and this gene encodes SlGA2ox7. Increased expression of *EI* caused different degrees of dwarfism, which may provide a useful resource for improving the plant architecture in tomato. Meanwhile, the co-segregated KASP marker developed in our study might be useful for high throughput maker assistant selection (MAS) in short internode breeding programs.

4. Materials and Methods

4.1. Plant Materials

Three determinate tomato inbred lines, P502, 05T606, and Heinz 1706, were used in this study. The mutant P502 shows an elongated internode, derived from a natural mutant of tomato cultivar 05T606. These were generated at the Institute of Vegetables and Flowers, Chinese Academy of Agricultural Sciences. Heinz 1706 was obtained from the Tomato Genetics Resource Center and displays a normal internode length, which is significantly shorter than the mutant P502 [53].

The mutant P502 (as the male parent) and Heinz 1706 (as the female parent) were hybridized to obtain an F_1 generation, and F_1 plants were self-pollinated to generate the F_2 population, which were used for inheritance analysis and fine mapping. All of the plant materials were grown in a greenhouse in Beijing, China. The average day and night temperatures were set at 25 °C and 20 °C, respectively.

4.2. Scanning Electron Microscopy Observation

To measure the cell lengths, the third internode was collected from 40-day-old 05T606 and P502 seedlings. The internode was cut into 5-mm segments and fixed in 3.5% glutaraldehyde for 24 h at room temperature. After washing in 0.1 M phosphate buffer (pH 7.4), the samples were fixed in 1% osmic acid for 2 h, dehydrated in a graded ethanol series, and dried in a Leica-EM CPD 300 desiccator (Leica, Frankfurt, Germany). Longitudinal sections were prepared by cutting the middle of the internode, which was then coated with a gold film. The pith cells at approximately the center of the stem were visualized and photographed with the Hitachi SU 8010 scanning election microscope (Hitachi, Tokyo, Japan). The cell length was measured with IMAGEJ software [54].

4.3. Exogenous GA_3 Treatments and Endogenous GA Quantification

To assess the response of P502 and 05T606 to GAs, the aerial parts of the 20-day-old seedlings were separately sprayed with 10^{-5} M GA_3 (Sigma, St. Louis Missouri, USA) and 10^{-7} M paclobutrazol (PAC, GA biosynthesis inhibitor; Biotopped, Beijing, China) [55] solutions containing 0.02% Tween-20 at an interval of one day. Control plants were sprayed with water. We sprayed ten times, and stopped at 40-day-old seedlings. The effects of GA_3 and PAC on stem expansion (from the cotyledons to the uppermost internode) were evaluated every four days. Each treatment was completed with three replicates (each with eight plants).

To determine the concentration of endogenous GAs, the first five internodes from the 30-day-old seedlings of P502 and wild-type 05T606 were collected into three biological replicates. Each biological replicate contained 1 g of tissue fresh weight. Tissue was immediately frozen in liquid nitrogen, and then was stored at −80 °C. The phytohormone extraction and quantitative profiling of GAs (GA_1, GA_4, GA_9, GA_{19}, and GA_{20}) were performed by HPLC-MS/MS [56].

4.4. DNA Extraction and Molecular Marker Development

A single young leaf was collected from each plant at 2-week-old seedlings. Genomic DNA was extracted according to a modified CTAB method [57] and then diluted to a concentration of 100–150 ng/µL in RNase (10 mg/mL) H_2O (1:100). To develop new markers, the elongated internode line P502 was sequenced with the Illumina HiSeq PE150 system, with a 50× genome coverage (Sequence Read Archive accession number: PRJNA540748). According to differences with the reference genome sequence (Heinz 1706), insertion and deletion (InDel) and competitive allele specific PCR (KASP) markers were designed using Primer Premier 5.0 software [58], and cleaved amplified polymorphic sequence (CAPS) markers were designed by dCAPS Finder 2.0 [59].

4.5. Mapping Strategy

The mean internode lengths from the first internode to the fifth internode (starting from the cotyledons) of 40-day-old seedlings were recorded for phenotypic analysis [36,53]. The bulked segregant analysis (BSA) strategy was used for the quick identification of molecular markers linked with the target locus [60]. Two DNA bulks, the elongated internode bulk (E bulk) and the normal internode bulk (N bulk), were generated by pooling equal amounts of DNA from ten elongated internode and ten normal F_2 plants, respectively.

To screen for polymorphic makers, the two parents were genotyped with 372 InDel markers across 12 tomato chromosomes (unpublished). All polymorphic markers were used to analyze the two bulked DNA samples. The target chromosome was identified based on the BSA results. Subsequently, QTL mapping was conducted according to the internode lengths of 354 F_2 individuals and genotypes of ideal markers on the target chromosome by using JoinMap 4.0 and MapQTL 6.0 [61,62]. After flanking markers were identified, another 956 F_2 individuals derived from the same cross were used for selecting recombinants. The F_3 individuals from eight F_2 recombinants were obtained for the confirmation of the progeny phenotype. Each $F_{2:3}$ contained 60 plants, and the internode lengths were evaluated on 40-day-old seedlings. Details regarding the polymorphic markers are provided in Table 4.

Table 4. Markers used for mapping of *EI* gene.

Primer Name	Forward Primer (5′-3′)	Reverse primer (5′-3′)	Type	Enzyme
D55	AATGACTTACCTACTGGAAAGC	GATTGATCACCCTTTGGATA	InDel	
D57	GAGACATCACTTTGCCTTTC	AAAAGTCTCTCCGCCTATGT	InDel	
D64	TTGTTACCGCTTACTTTGGT	CACAGCTGTTGATTTCTTCA	InDel	
CAPS4	GCATTGCAACCTATTCTCAC	TCTGTAGTTTCCGTCTTCTT	CAPS	*Hae*III
D66	CGTTGTCTAGGTCAATAGCC	AGGTGTTACACTTTCTACGTCT	InDel	
CAPS16	AGAGAAGGAGGATTCGGGTT	ATAGGGGCATTATCAAAAGG	CAPS	*Bsr*DI
CAPS17	TAAGTTAGCCATATAAAAC	AAATGACACAGCGAGACA	CAPS	*Mbo*II
CAPS27	GAGAAAATTATTTGGGATAC	ATTAAAACTTTGATGCCTAC	CAPS	*Mfe*I
InDel6	ACAATCCCAGTTTATGTGAT	ATATTTGGTGTTTTCTGTTT	InDel	
HP2501	CTTTTCACAAAACTAACACAGG	TGACAATATAAGCATTTGTCGC	InDel	
HP707	TCCGATGTAACATCACGCAA	GTTGATCACCTTCAGACAGC	InDel	
D67	AGCTTTTATAGCACGTACCG	CCATACTCTACTTATGCTGCAA	InDel	
HP2509	ACCTCGACACTGGTTCACTC	GTGACTCATATACACCCTTACCTA	InDel	
HP445	GAGAACATCTGTACCAGCCT	CAAGTATCTATATGCCTGACAAC	InDel	

4.6. RNA Extraction and qRT-PCR

Total RNA was extracted using an RNA pure kit (Aidlab, Beijing, China) following the user manual. First-strand cDNA was synthesized using TransScript One-Step gDNA Removal and cDNA Synthesis SuperMix (Transgene, Beijing, China). A quantitative real-time polymerase chain reaction (qRT-PCR) assay was conducted using a SYBR Green reagent (Yeasen, Shanghai, China) and the LightCycler 480 Real-Time detection system (Roche, Basel, Switzerland). Housekeeping gene *actin* (*solyc03g078400*) was used as an internal control to normalize the data. Details regarding the qRT-PCR primers arc listed in Table S1 [63]. The qRT-PCR data for each sample were validated with three biological and three technical replicates. The relative expression levels were quantified according to the $2^{-\Delta\Delta Ct}$ method [64].

4.7. Protein Sequence Alignment and Phylogenetic Analysis

The sequence was retrieved from the National Center for Biotechnology Information database (NCBI). The BLASTP program [65] was used for homology searches in GenBank. A multiple protein sequence alignment was performed by the ClustalW program, and the phylogenetic tree was constructed according to the neighbor-joining method of the MEGA 6.0 program with 1000 bootstrap replicates [66].

4.8. Plasmid Construction and Transformation

The genomic clone including the whole *EI* coding region was amplified from full-length cDNA of wild-type 05T606 with primers OE-F (5'-CACGGGGGACTCTAGAATGTACTTAGCCACCTCCA-3') and OE-R (5'-GATCGGGGAAATTCGAGCTCTTAGTGAGTTGAGACAAGAAAC-3'). The amplified fragment was cloned into the *Xba*I and *Sac*I sites of the binary vector *pBI*121 by using an In-Fusion HD Cloning Kit (Takara, Dalian, China) to generate an *EI* transformation plasmid under the control of the CaMV35S promoter. The plasmid mediated by *Agrobacterium tumefaciens* strain GV3101 was transformed into the mutant P502 as described by the method of Sharma et al. [67]. After screening for regenerated shoots on the selection medium, the transgenic plants were further confirmed by PCR using NPTII-F (5'-GACAATCGGCTGCTCTGA-3') and NPTII-R (5'-AACTCCAGCATGAGATCC-3') primers. The positive transgenic plants were selected and the T_1 generation was obtained for phenotypic observation and gene expression analysis.

Accession numbers: SlGA2ox7 (XP_004232746), NsGA2ox8 (NP_001289506.1), AtGA2ox8 (NP_193852.2), AtGA2ox7 (NP_175509.1), VvGA2ox8 (NP_001268435.1), ZmGA2ox7 (NP_001148252.2), NtGA20ox1 (NP_001313089.1), SlGA20ox3 (NP_001234579.1), SlGA20ox2 (NP_001234628.2), SlGA20ox4 (NP_001234363.1), AtGA20ox2 (NP_199994.1), AtGA20ox4 (NP_176294.1), AtGA20ox3 (NP_196337.1), ZmGA20ox4 (NP_001308615.1) StGA3ox2 (NP_001275412.1), and AtGA3ox4 (NP_178149.1).

Supplementary Materials: Supplementary materials can be found at http://www.mdpi.com/1422-0067/20/9/2204/s1.

Author Contributions: J.L., L.L., J.S., and X.S. conceived and designed the experiments; X.S., J.S., A.M.A.M., and X.D. performed the experiments; X.S. analyzed the data and prepared the manuscript; J.S. improved the manuscript; X.Z., J.B., Y.C., and X.L. assisted in the field experiment; J.L., L.L., J.S., Y.D., X.W., Z.H., and Y.G. contributed reagents and materials; J.L. provided guidance on the whole study and revised the manuscript. All authors read and approved the final manuscript.

Funding: This work was supported by the National Natural Science Foundation of China (Grant No. 31872103), the National Key Research and Development Program of China (Grant No. 2016YFD0101703), the China Postdoctoral Science Foundation (2017M623289, 2018T111123), the Fundamental Research Funds for Central Non-profit Scientific Institution (IVF-BRF2018007), the Key Laboratory of Biology and Genetic Improvement of Horticultural Crops, Ministry of Agriculture, China, and the Science and Technology Innovation Program of the Chinese Academy of Agricultural Sciences (Grant No. CAAS-ASTIPIVFCAAS).

Conflicts of Interest: The authors declare that they have no conflicts of interest.

References

1. Salas Fernandez, G.S.; Becraft, P.W.; Yin, Y.; Lübberstedt, T. From dwarves to giants? Plant height manipulation for biomass yield. *Trends Plant Sci.* **2009**, *14*, 454–461. [CrossRef] [PubMed]

2. Itoh, H.; Tatsumi, T.; Sakamoto, T.; Otomo, K.; Toyomasu, T.; Kitano, H.; Ashikari, M.; Ichihara, S.; Matsuoka, M. A rice semi-dwarf gene, *Tan-Ginbozu* (*D35*), encodes the gibberellin biosynthesis enzyme, *ent*-kaurene oxidase. *Plant Mol. Biol.* **2004**, *54*, 533–547. [CrossRef]

3. Peng, J.; Richards, D.E.; Hartley, N.M.; Murphy, G.P.; Devos, K.M.; Flintham, J.E.; Beales, J.; Fish, L.J.; Worland, A.J.; Pelica, F.; et al. Green revolution' genes encode mutant gibberellin response modulators. *Nature* **1999**, *400*, 256–261. [CrossRef]

4. Hedden, P. The genes of the Green Revolution. *Trends Genet.* **2003**, *19*, 5–9. [CrossRef]

5. Khush, G.S. Green revolution: The way forward. *Nat. Rev. Genet.* **2001**, *2*, 815–822. [CrossRef] [PubMed]

6. Md. Babul, A.; Piao, R.; Reflinur; Md. Lutfor, R.; Yunjoo, L.; Jeonghwan, S.; Backki, K.; Hee-Jong, K. Characterization and mapping of *d13*, a dwarfing mutant gene, in rice. *Genes Genom.* **2015**, *37*, 893–903.

7. Hong, Z.; Ueguchi-Tanaka, M.; Umemura, K.; Uozu, S.; Fujioka, S.; Takatsuto, S.; Yoshida, S.; Ashikari, M.; Kitano, H.; Matsuoka, M. A rice brassinosteroid-deficient mutant, *ebisu dwarf* (*d2*), is caused by a loss of function of a new member of cytochrome P450. *Plant Cell* **2003**, *15*, 2900–2910. [CrossRef]

8. Zhou, H.; He, S.; Cao, Y.; Chen, T.; Du, B.; Chu, C.; Zhang, J.; Chen, S. OsGLU1, a putative membrane-bound endo-1,4-ß-D-glucanase from rice, affects plant internode elongation. *Plant Mol. Biol.* **2006**, *60*, 137–151. [CrossRef]

9. Chen, Y.; Hou, M.; Liu, L.; Wu, S.; Shen, Y.; Ishiyama, K.; Kobayashi, M.; McCarty, D.R.; Tan, B.C. The maize *DWARF1* encodes a gibberellin 3-oxidase and is dual localized to the nucleus and cytosol. *Plant Physiol.* **2014**, *166*, 2028–2039. [CrossRef]

10. Li, X.; Zhou, Z.; Ding, J.; Wu, Y.; Zhou, B.; Wang, R.; Ma, J.; Wang, S.; Zhang, X.; Xia, Z.; et al. Combined linkage and association mapping reveals QTL and candidate genes for plant and ear height in maize. *Front Plant Sci.* **2016**, *7*, 833. [CrossRef]

11. Teng, F.; Zhai, L.; Liu, R.; Bai, W.; Wang, L.; Huo, D.; Tao, Y.; Zheng, Y.; Zhang, Z. *ZmGA3ox2*, a candidate gene for a major QTL, *qPH3.1*, for plant height in maize. *Plant J.* **2013**, *73*, 405–416. [CrossRef] [PubMed]

12. Tian, X.; Wen, W.; Xie, L.; Fu, L.; Xu, D.; Fu, C.; Wang, D.; Chen, X.; Xia, X.; Chen, Q.; et al. Molecular mapping of reduced plant height gene *Rht24* in Bread Wheat. *Front. Plant Sci.* **2017**, *8*, 1379. [CrossRef]

13. Zhai, H.; Feng, Z.; Li, J.; Liu, X.; Xiao, S.; Ni, Z.; Sun, Q. QTL analysis of spike morphological traits and plant height in Winter Wheat (*Triticum aestivum* L.) using a high-density SNP and SSR-based linkage map. *Front Plant Sci.* **2016**, *7*, 1617. [CrossRef] [PubMed]

14. Multani, D.S.; Briggs, S.P.; Chamberlin, M.A.; Blakeslee, J.J.; Murphy, A.S.; Johal, G.S. Loss of an MDR transporter in compact stalks of maize *br2* and sorghum *dw3* mutants. *Science* **2003**, *302*, 81–84. [CrossRef]

15. Yamaguchi, M.; Fujimoto, H.; Hirano, K.; Araki-Nakamura, S.; Ohmae-Shinohara, K.; Fujii, A.; Tsunashima, M.; Song, X.J.; Ito, Y.; Nagae, R.; et al. Sorghum *Dw1*, an agronomically important gene for lodging resistance, encodes a novel protein involved in cell proliferation. *Sci. Rep.* **2016**, *6*, 28366. [CrossRef] [PubMed]

16. Liu, C.; Zheng, S.; Gui, J.; Fu, C.; Yu, H.; Song, D.; Shen, J.; Qin, P.; Liu, X.; Han, B.; et al. *Shortened Basal Internodes* encodes a gibberellin 2-oxidase and contributes to lodging resistance in rice. *Mol. Plant* **2018**, *11*, 288–299. [CrossRef] [PubMed]

17. Helliwell, C.A.; Sheldon, C.C.; Olive, M.R.; Walker, A.R.; Zeevaart, J.A.; Peacock, W.J.; Dennis, E.S. Cloning of the Arabidopsis *ent*-kaurene oxidase gene GA₃. *Proc. Natl. Acad. Sci. USA* **1998**, *95*, 9019–9024. [CrossRef]

18. Sakamoto, T.; Miura, K.; Itoh, H.; Tatsumi, T.; Ueguchi-Tanaka, M.; Ishiyama, K.; Kobayashi, M.; Agrawal, G.K.; Takeda, S.; Abe, K.; et al. An overview of gibberellin metabolism enzyme genes and their related mutants in rice. *Plant Physiol.* **2004**, *135*, 1642–1653. [CrossRef] [PubMed]

19. Li, Z.; Guo, Y.; Ou, L.; Hong, H.; Wang, J.; Liu, Z.; Guo, B.; Zhang, L.; Qiu, L. Identification of the dwarf gene *GmDW1* in soybean (*Glycine max* L.) by combining mapping-by-sequencing and linkage analysis. *Theor. Appl. Genet.* **2018**, *131*, 1001–1016. [CrossRef]

20. Richards, D.E.; King, K.E.; Ait-Ali, T.; Harberd, N.P. How gibberellin regulates plant growth and development: A molecular genetic analysis of gibberellin signaling. *Annu. Rev. Plant Physiol. Plant Mol. Biol.* **2001**, *52*, 67–88. [CrossRef]

21. Hedden, P.; Thomas, S.G. Gibberellin biosynthesis and its regulation. *Biochem. J.* **2012**, *444*, 11–25. [CrossRef]

22. Sasaki, A.; Ashikari, M.; Ueguchi-Tanaka, M.; Itoh, H.; Nishimura, A.; Swapan, D.; Ishiyama, K.; Saito, T.; Kobayashi, M.; Khush, G.S.; et al. Green revolution: A mutant gibberellin-synthesis gene in rice. *Nature* **2002**, *416*, 701–702. [CrossRef] [PubMed]

23. Spielmeyer, W.; Ellism, M.H.; Chandler, P.M. *Semidwarf* (*sd-1*), 'green revolution' rice, contains a defective gibberellin 20-oxidase gene. *Proc. Natl. Acad. Sci. USA* **2002**, *99*, 9043–9048. [CrossRef] [PubMed]

24. Koorneef, M.; Elgersma, A.; Hanhart, C.J.; Van Loenen Martinet, E.P.; Van Rijn, L.; Zeevaart, J.A. A gibberellin insensitive mutant of *Arabidopsis thaliana*. *Physiol Plant.* **2010**, *65*, 33–39. [CrossRef]

25. Peng, J.; Carol, P.; Richards, D.E.; King, K.E.; Cowling, R.J.; Murphy, G.P.; Harberd, N.P. The Arabidopsis *GAI* gene defines a signaling pathway that negatively regulates gibberellin responses. *Gene Dev.* **1997**, *11*, 3194–3205. [CrossRef] [PubMed]

26. Harberd, N.P.; Freeling, M. Genetics of dominant gibberellin-insensitive dwarfism in maize. *Genetics* **1989**, *121*, 827–838.

27. Winkler, R.G.; Freeling, M. Physiological genetics of the dominant gibberellin-nonresponsive maize dwarfs, *Dwarf 8* and *Dwarf 9*. *Planta* **1994**, *193*, 341–348. [CrossRef]

28. Ogawa, M.; Kusano, T.; Katsumi, M.; Sano, H. Rice gibberellin-insensitive gene homolog, *OsGAI*, encodes a nuclear-localized protein capable of gene activation at transcriptional level. *Gene* **2000**, *245*, 21–29. [CrossRef]

29. Fu, X.D.; Richards, D.E.; Ait-Ali, T.; Hynes, L.W.; Ougham, H.; Peng, J.; Harberd, N.P. Gibberellin-mediated proteasome-dependent degradation of the barley DELLA protein SLN1 repressor. *Plant Cell* **2002**, *14*, 3191–3200. [CrossRef]

30. Ikeda, A.; Ueguchi-Tanaka, M.; Sonoda, Y.; Kitano, H.; Koshioka, M.; Futsuhara, Y.; Matsuoka, M.; Yamaguchi, J. *Slender* rice, a constitutive gibberellin response mutant, is caused by a null mutation of the *SLR1* gene, an ortholog of the height-regulating gene *GAI/RGA/RHT/D8*. *Plant Cell* **2001**, *13*, 999–1010. [CrossRef]

31. Carrera, E.; Ruiz-Rivero, O.; Peres, L.E.; Atares, A.; Garcia-Martinez, J.L. Characterization of the *procera* tomato mutant shows novel functions of the SlDELLA protein in the control of flower morphology, cell division and expansion, and the auxin-signaling pathway during fruit-set and development. *Plant Physiol.* **2012**, *160*, 1581–1596. [CrossRef]

32. Yamamuro, C.; Ihara, Y.; Wu, X.; Noguchi, T.; Fujioka, S.; Takatsuto, S.; Ashikari, M.; Kitano, H.; Matsuoka, M. Loss of function of a rice *brassinosteroid insensitive1* homolog prevents internode elongation and bending of the lamina joint. *Plant Cell* **2000**, *12*, 1591–1606. [CrossRef]

33. Lin, H.; Wang, R.; Qian, Q.; Yan, M.; Meng, X.; Fu, Z.; Yan, C.; Jiang, B.; Su, Z.; Li, J.; et al. DWARF27, an iron-containing protein required for the biosynthesis of strigolactones, regulates rice tiller bud outgrowth. *Plant Cell* **2009**, *21*, 1512–1525. [CrossRef]

34. Jiang, L.; Liu, X.; Xiong, G.; Liu, H.; Chen, F.; Wang, L.; Meng, X.; Liu, G.; Yu, H.; Yuan, Y.; et al. *DWARF 53* acts as a repressor of strigolactone signalling in rice. *Nature* **2013**, *504*, 401–405. [CrossRef]

35. Foolad, M.R. Genome mapping and molecular breeding of tomato. *Int. J. Plant Genom.* **2007**, *2007*, 64358. [CrossRef] [PubMed]

36. Martí, E.; Gisbert, C.; Bishop, G.J.; Dixon, M.S.; Garcíamartínez, J.L. Genetic and physiological characterization of tomato cv. Micro-Tom. *J. Exp. Bot.* **2006**, *57*, 2037–2047. [CrossRef]

37. Barabaschi, D.; Tondelli, A.; Desiderio, F.; Volante, A.; Vaccino, P.; Valè, G.; Cattivelli, L. Next generation breeding. *Plant Sci.* **2016**, *242*, 3–13. [CrossRef]

38. Tong, G.L.; Samuel, F.H. Fine mapping of the *brachytic* locus on the tomato genome. *J. Amer. Soc. Hort. Sci.* **2018**, *143*, 239–247.

39. Cowling, R.J.; Kamiya, Y.; Seto, H.; Harberd, N.P. Gibberellin dose-response regulation of GA$_4$ gene transcript levels in Arabidopsis. *Plant Physiol.* **1998**, *117*, 1195–1203. [CrossRef] [PubMed]

40. Mueller, L.A.; Solow, T.H.; Taylor, N.; Skwarecki, B.; Buels, R.; Binns, J.; Lin, C.; Wright, M.H.; Ahrens, R.; Wang, Y.; Herbst, E.V.; Keyder, E.R.; Menda, N.; Zamir, D.; Tanksley, S.D. The SOL Genomics Network: A comparative resource for Solanaceae biology and beyond. *Plant Physiol.* **2005**, *138*, 1310–1317. [CrossRef]

41. Olszewski, N.; Sun, T.P.; Gubler, F. Gibberellin signaling: Biosynthesis, catabolism, and response pathways. *Plant Cell* **2002**, *14*, S61–S80. [CrossRef]

42. Mitsunaga, S.; Tashiro, T.; Yamaguchi, J. Identification and characterization of gibberellin-insensitive mutants selected from among dwarf mutants of rice. *Theor. Appl. Genet.* **1994**, *87*, 705–712. [PubMed]

43. Hedden, P.; Phillips, A.L. Gibberellin metabolism: New insights revealed by the genes. *Trends. Plant Sci.* **2000**, *5*, 523–530. [CrossRef]

44. Gomi, K.; Matsuoka, M. Gibberellin signalling pathway. *Curr. Opin. Plant Biol.* **2003**, *6*, 489–493. [CrossRef]

45. Schomburg, F.M.; Bizzell, C.M.; Lee, D.J.; Zeevaart, J.A.; Amasino, R.M. Overexpression of a novel class of gibberellin 2-oxidases decreases gibberellin levels and creates dwarf plants. *Plant Cell* **2003**, *15*, 151–163. [CrossRef]

46. Lo, S.F.; Yang, S.Y.; Chen, K.T.; Hsing, Y.I.; Zeevaart, J.A.; Chen, L.J.; Yu, S.M. A novel class of gibberellin 2-oxidases control semidwarfism, tillering, and root development in rice. *Plant Cell* **2008**, *20*, 2603–2618. [CrossRef] [PubMed]

47. Lee, D.J.; Zeevaart, J.A. Differential Regulation of RNA Levels of Gibberellin Dioxygenases by Photoperiod in Spinach. *Plant Physiol.* **2002**, *130*, 2085–2094. [CrossRef] [PubMed]

48. Martin, D.N.; Proebsting, W.M.; Hedden, P. The *SLENDER* gene of pea encodes a gibberellin 2-oxidase. *Plant Physiol.* **1999**, *121*, 775–781. [CrossRef] [PubMed]

49. Huang, J.; Tang, D.; Shen, Y.; Qin, B.; Hong, L.; You, A.; Li, M.; Wang, X.; Yu, H.; Gu, M.; et al. Activation of gibberellin 2-oxidase 6 decreases active gibberellin levels and creates a dominant semi-dwarf phenotype in rice *(Oryza sativa* L.). *J. Genet. Genom.* **2010**, *37*, 23–36. [CrossRef]

50. Lee, D.J.; Zeevaart, J.A. Molecular Cloning of GA 2-Oxidase3 from Spinach and Its Ectopic Expression in Nicotiana sylvestris. *Plant Physiol.* **2005**, *138*, 243–254. [CrossRef]

51. Sun, T.P. Gibberellin metabolism, perception and signaling pathways in Arabidopsis. *Arabidopsis Book* **2008**, *6*, e0103. [CrossRef]

52. Pnueli, L.; Carmel-Goren, L.; Hareven, D.; Gutfinger, T.; Alvarez, J.; Ganal, M.; Zamir, D.; Lifschitz, E. The *SELF-PRUNING* gene of tomato regulates vegetative to reproductive switching of sympodial meristems and is the ortholog of CEN and TFL1. *Development* **1998**, *125*, 1979–1989.

53. Sun, X.R.; Liu, L.; Zhi, X.N.; Bai, J.R.; Cui, Y.N.; Shu, J.S.; Li, J.M. Genetic analysis of tomato internode length via mixed major gene plus polygene inheritance model. *Sci. Hortic.* **2019**, *246*, 759–764. [CrossRef]

54. Collins, T.J. ImageJ for microscopy. *BioTechniques* **2007**, *43*, 25–30. [CrossRef] [PubMed]

55. García-Hurtado, N.; Carrera, E.; Ruiz-Rivero, O.; López-Gresa, M.P.; Hedden, P.; Gong, F.; García-Martínez, J.L. The characterization of transgenic tomato overexpressing gibberellin 20-oxidase reveals induction of parthenocarpic fruit growth, higher yield, and alteration of the gibberellin biosynthetic pathway. *J. Exp. Bot.* **2012**, *63*, 5803–5813. [CrossRef]

56. Dong, C.F.; Gu, H.R.; Ding, C.L.; Xu, N.X.; Zhang, W.J. Effects of gibberellic acid on forage quality of rice *(Oryza sativa)* straw. *Acta Pratacult. Sin.* **2016**, *25*, 94–102.

57. Saghaimaroof, M.A.; Soliman, K.M.; Jorgensen, R.A.; Allard, R.W. Ribosomal DNA spacer-length polymorphisms in barley: Mendelian inheritance, chromosomal location, and population dynamics. *Proc. Natl. Acad. Sci. USA* **1984**, *81*, 8014–8018. [CrossRef]

58. Lalitha, S. Primer Premier 5. *Biotech Softw. Internet Report.* **2000**, *1*, 270–272. [CrossRef]

59. Neff, M.M.; Turk, E.; Kalishman, M. Web-based primer design for single nucleotide polymorphism analysis. *Trends Genet.* **2002**, *18*, 613–615. [CrossRef]

60. Michelmore, R.W.; Paran, I.; Kesseli, R.V. Identification of markers linked to disease-resistance genes by bulked segregant analysis: A rapid method to detect markers in specific genomic regions by using segregating populations. *Proc. Natl. Acad. Sci. USA* **1991**, *88*, 9828–9832. [CrossRef]

61. Van Ooijen, J.W. *JoinMap*® *4.0. Software for the Calculation of Genetic Linkage Maps in Experimental Populations*; Kyazma: Wageningen, The Netherlands, 2006.

62. Van Ooijen, J.W. *MapQTL*® *6.0, Software for the Mapping of Quantitative trait Loci in Experimental Populations*; Kyazma: Wageningen, The Netherlands, 2009.

63. Li, J.; Sima, W.; Ouyang, B.; Wang, T.; Ziaf, K.; Luo, Z.; Liu, L.; Li, H.; Chen, M.; Huang, Y.; et al. Tomato *SlDREB* gene restricts leaf expansion and internode elongation by downregulating key genes for gibberellin biosynthesis. *J. Exp. Bot.* **2012**, *63*, 6407–6420. [CrossRef]

64. Livak, K.J.; Schmittgen, T.D. Analysis of relative gene expression data using real-time quantitative PCR and the $2^{-\Delta\Delta CT}$ method. *Methods* **2001**, *25*, 402–408. [CrossRef] [PubMed]

65. Altschul, S.; Madden, T.; Schäffer, A. Gapped BLAST and PSI-BLAST: A new generation of protein database search programs. *Nucleic Acids Res.* **1997**, *25*, 3389–3402. [CrossRef]

66. Tamura, K.; Stecher, G.; Peterson, D.; Filipski, A.; Kumar, S. MEGA6: Molecular evolutionary genetics analysis version 6.0. *Mol. Biol. Evol.* **2013**, *30*, 2725–2729. [CrossRef]

67. Sharma, M.K.; Solanke, A.U.; Jani, D.; Singh, Y.; Sharma, A.K. A simple and efficient *Agrobacterium*-mediated procedure for transformation of tomato. *J. Biosci.* **2009**, *34*, 423. [CrossRef] [PubMed]

Article

Normal and Abortive Buds Transcriptomic Profiling of *Broccoli ogu* Cytoplasmic Male Sterile Line and Its Maintainer

Jinshuai Shu, Lili Zhang, Yumei Liu *, Zhansheng Li, Zhiyuan Fang, Limei Yang, Mu Zhuang, Yangyong Zhang and Honghao Lv

Institute of Vegetables and Flowers, Chinese Academy of Agricultural Sciences, Key Laboratory of Biology and Genetic Improvement of Horticultural Crops, Ministry of Agriculture, 12 Zhongguancun Nandajie Street, Beijing 100081, China; shujinshuai@caas.cn (J.S.); zhanglilina@126.com (L.Z.); lizhansheng@caas.cn (Z.L.); fangzhiyuan@caas.cn (Z.F.); yanglimei@caas.cn (L.Y.); zhuangmu@caas.cn (M.Z.); zhangyangyong@caas.cn (Y.Z.); lvhonghao@caas.cn (H.L.)
* Correspondence: liuyumei@caas.cn or liuyumeicaas@126.com; Tel.: +86-10-8210-8756; Fax: +86-10-6217-4123

Received: 30 June 2018; Accepted: 14 August 2018; Published: 24 August 2018

Abstract: Bud abortion is the main factor affecting hybrid seeds' yield during broccoli cross breeding when using *ogura* cytoplasmic male sterile (*ogu* CMS) lines. However, the genes associated with bud abortion are poorly understood. We applied RNA sequencing to analyze the transcriptomes of normal and abortive buds of broccoli maintainer and *ogu* CMS lines. Functional analysis showed that among the 54,753 annotated unigenes obtained, 74 and 21 differentially expressed genes in common were upregulated and downregulated in *ogu* CMS abortive buds compared with *ogu* CMS normal buds, maintainer normal, and abortive buds, respectively. Nineteen of the common differentially expressed genes were enriched by GO terms associated with glycosyl hydrolases, reactive oxygen species scavenging, inhibitor, and protein degradation. Ethylene-responsive transcription factor 115 and transcriptional factor basic helix-loop-helix 137 were significantly upregulated; transcription factors DUO1 and PosF21/RF2a/BZIP34 were downregulated in *ogu* CMS abortive buds compared with the other groups. Genes related to polygalacturonase metabolism, glycosyl hydrolases, oxidation reduction process, phenylalanine metabolism, and phenylpropanoid biosynthesis were significantly changed in *ogu* CMS abortive buds. Our results increase our understanding of bud abortion, provide a valuable resource for further functional characterization of *ogu* CMS during bud abortion, and will aid in future cross breeding of *Brassica* crops.

Keywords: broccoli; cytoplasmic male sterile; bud abortion; gene expression; transcriptome; RNA-Seq

---

1. Introduction

Bud abortion is a very common biological phenomenon in *Brassica* species. During bud abortion, the buds stop growing, convert from green into yellow progressively from the base to the top of the bud, and eventually wither off before flowering [1,2], which is detrimental for hybrid seeds yield during cross breeding. In recent years, bud abortion has attracted considerable attention in *Brassica* species cross breeding [3–5], as it reduces the hybrid seed yield production of many *Brassica* crops and also affects the efficiency of *Brassica* crops cross breeding. In the past decade, to solve the bud abortion problem, many studies have investigated the factors that cause bud abortion [1–8]. However, most of the studies focused on morphology and anatomy, and few studies have reported the factors contributing to *Brassica* bud abortion at the molecular level. Therefore, the fundamental mechanism of bud abortion is poorly understood.

High-throughput sequencing technologies have played important roles in revealing the molecular mechanisms of various organismal biological processes. RNA sequencing (RNA-Seq) technology is an important high throughput sequencing technology that produces functional genomic data for non-model plants that lack genomic sequence data. RNA-Seq technology has been widely applied to assist in determining differentially expressed genes (DEGs) involved in different biological processes in many species and may be a promising method to address the genes associated with bud abortion in *Brassica* species. Surprisingly, no studies on bud abortion using RNA-Seq technology have been documented so far.

Brassica species display obvious heterosis, and using cytoplasmic male sterile (CMS) lines to produce hybrid seeds is an important approach that uses this heterosis. Broccoli (*Brassica oleracea* var. *italica*) is an important vegetable crop and plays a vital role in the vegetable production industry worldwide. More importantly, broccoli is also used to produce health products and develop anti-cancer drugs [9–11]. Thus, broccoli is the most important economical *Brassica* vegetable crop. However, a large problem exists in broccoli cross breeding: the hybrid seed purity cannot reach 100% when using inbred lines or self-incompatible lines to produce hybrid seeds. To improve the purity of broccoli hybrid seeds, many breeders execute wide crosses between broccoli cultivars and other *Brassica* CMS materials to transfer CMS genes. Our group used multiple broccoli inbred lines to with the cabbage CMS material *ogura* CMSR$_3$629 (*ogu* CMS), which was introduced by the Asgrow Seed Co. (USA), which has been used to breed a plurality of broccoli CMS lines with a 100% sterile rate. However, the obtained CMS lines often show serious bud abortion, which leads to delayed flowering time and sharply reduced amounts of flowers, some serious bud abortion CMS lines cannot flower [5]. These problems limit the use of the CMS lines considerably and have reduced their breeding efficiency dramatically. So far, several studies have focused on the causes of bud abortion, e.g., the *RsVPE1* gene encodes a vacuolar processing enzyme that is involved in radish floral bud abortion under heat stress [1]. The expression of stress response, energy metabolism, amino acid synthesizing and processing, signal transduction, disease resistance and senescence, transcription and translation, and transmembrane transport-related genes were different between normal and abortive buds, as detected by cDNA-amplified length polymorphism technology [6–8]. However, as far as we know, no study has investigated the molecular events occurring in bud abortion based on high throughput sequencing during broccoli cross breeding, or any other crop, thus little information is available concerning the genes involved in bud abortion. Thus, it is necessary to determine the elements involved in broccoli bud abortion at the molecular level, which will increase our understanding of the molecular events involved in bud abortion of broccoli and other crops. The objectives of this study were to identify the functional genes involved in bud abortion and to determine the genes expression characteristics related to broccoli bud abortion, based on RNA-Seq technology.

2. Results

2.1. High-Throughput Transcriptome Sequencing and Unigene Assembly

To explore gene expression and gene networks that control bud abortion of *ogu* CMS lines in broccoli, we performed RNA-Seq analysis of normal and abortive buds from *ogu* CMS line CMS93219 and its maintainer line ML93219. The cytoplasmic male sterility original of CMS93219 was cabbage CMS material *ogu* CMSR$_3$629 and the backcross paternal was broccoli ML93219. CMS93219 was stabilized via sixteen generations of backcrossing before this study. There were no visible differences between CMS93219 and ML93219 in terms of the plants and the heading morphologies of the harvested material, except that CMS93219 showed significant bud abortion and ML93219 showed only slight bud abortion before or at the early anthesis stage (Figure 1A–D). The average number of abortive buds per branch of CMS93219 was 35.73 throughout the whole growth phase, accounting for 83.88% of the total bud number, which is extremely significantly higher than the 19.60 per branch (33.33%) in ML93219 (*t*-test, $p < 0.01$) (Figure 1E,F). Neither the abortive nor the normal buds of CMS93219 contained pollen, while the aborted buds of ML93219 contained less pollen than normal buds. Therefore, bud abortion and pollen abortion character were closely connected between ML93219 and CMS93219, and pollen abortion may promote bud abortion. Gene expression changes are usually associated with morphologic changes; therefore, we chose the maximum buds stage (around 3 days before anthesis) for transcriptome analyses (Figure 1B-I,B-II,D-I,D-II).

Figure 1. Morphological characterization of two broccoli lines with different degrees of bud abortion. (**A,B**) ML93219, (**C,D**) CMS93219. B-I and B-II represent the normal and abortive bud of ML93219, respectively. D-I and D-II represent the normal and abortive bud of CMS93219, respectively. The bars in (**E,F**) represent the standard deviation ($n = 15$). Asterisks indicate that the average number of abortive buds and the abortive buds rate per branch are very significantly different between ML93219 and CMS93219 (unpaired t test, $p < 0.01$).

Twelve cDNA libraries from normal buds and abortive buds of the ML93219 and CMS93219, containing three biological replicates for each sample, were subjected to Illumina sequencing. After filtering invalid reads and data cleaning, 663,701,918 clean reads containing 99,555,287,700 nucleotides (99.6 Gb), with a mean length of 150 bp, were obtained. The Q20 and GC percentage were 92.93–94.33% and 46.46–47.12%, respectively (Table 1).

Table 1. Summary of broccoli bud transcriptome sequencing data.

Sample Name	Raw Reads	Clean Reads (Clean/Raw)	Clean Bases (Gb)	Q20 (%)	GC Content (%)	Mapped Reads (Mapped/Clean)
ML_NB1	57,767,246	55,194,896 (95.55%)	8.28	94.30	46.93	41,252,694 (74.74%)
ML_NB2	44,577,570	43,849,450 (98.37%)	6.58	92.93	46.46	32,614,176 (74.38%)
ML_NB3	61,251,852	59,738,162 (97.53%)	8.96	94.03	46.96	44,511,290 (74.51%)
ML_AB1	61,588,740	60,493,582 (98.22%)	9.08	94.32	46.73	45,965,016 (75.98%)
ML_AB2	54,823,836	53,530,342 (97.64%)	8.02	93.90	46.74	40,115,268 (74.94%)
ML_AB3	58,528,852	56,475,844 (96.49%)	8.48	94.33	46.81	42,816,816 (75.81%)
CMS_NB1	54,455,258	53,170,054 (97.64%)	7.98	94.23	47.02	39,776,946 (74.81%)
CMS_NB2	53,265,514	52,284,140 (98.16%)	7.84	93.97	47.12	38,965,364 (74.53%)
CMS_NB3	51,816,268	50,884,086 (98.20%)	7.64	93.79	46.91	37,800,172 (74.29%)
CMS_AB1	64,340,176	63,146,694 (98.15%)	9.48	94.16	47.04	47,846,088 (75.77%)
CMS_AB2	56,566,280	55,530,456 (98.17%)	8.32	94.27	46.78	42,582,890 (76.68%)
CMS_AB3	60,701,770	59,404,212 (97.86%)	8.92	94.09	46.70	44,549,934 (74.99%)

Note: ML_NB and ML_AB represent the normal and abortive bud samples of ML93219, respectively. CMS_NB and CMS_AB represent the normal and abortive bud samples of CMS93219, respectively.

After assembling the clean reads, 97,347 transcripts (91,137,323 nucleotides) and 66,050 unigenes (51,896,834 nucleotides) were obtained. The average length of the transcripts and unigenes were 936 and 786 bp, respectively (Table 2).

Table 2. Summary of de novo transcriptome length distribution.

Length	200–500 bp	500–1 kbp	1 k–2 kbp	>2 kbp	Total	Min (bp)	Mean (bp)	Median (bp)	Max (bp)	N50	Total Nucleotides
No. of transcripts	44,030	21,348	21,467	10,502	97,347	201	936	578	16,361	1510	91,137,323
No. of Unigenes	37,344	12,545	10,691	5470	66,050	201	786	424	16,361	1363	51,896,834

2.2. Gene Annotation and Functional Classification

A total of 54,753 unigenes (82.90%) were annotated in at least one database and 6070 unigenes (9.19%) were annotated in seven databases. In the NCBI non-redundant protein sequences (Nr), NCBI non-redundant nucleotide sequences (Nt), KEGG Ortholog (KO), a manually annotated and reviewed protein sequence database (SwissProt), protein family (Pfam), Gene Ontology (GO), eukaryotic ortholog groups (KOG) databases, 44,294 (67.06%), 50,157 (75.93%), 12,403 (18.77%), 29,861 (45.2%), 24,660 (37.33%), 30,285 (45.85%) and 12,492 (18.91%) unigenes were annotated, respectively (Table 3).

Table 3. Annotation of unigene sequences in broccoli buds.

Sequence Database	Number of Annotated Unigenes	Percentage of Annotated Unigene Sequences (%)	Software and Parameters
Annotated in Nr	44,294	67.06	NCBI blast 2.2.28+, *e*-value = 1×10^{-5}
Annotated in Nt	50,157	75.93	NCBI blast 2.2.28+, *e*-value = 1×10^{-5}
Annotated in KO	12,403	18.77	KAAS, KEGG Automatic Annotation Server, *e*-value = 1×10^{-10}
Annotated in SwissProt	29,861	45.2	NCBI blast 2.2.28+, *e*-value = 1×10^{-5}
Annotated in PFAM	24,660	37.33	HMMER 3.0 package, hmmscan, *e*-value = 0.01
Annotated in GO	30,285	45.85	Blast2GO v2.5 [12] and self-write the script, *e*-value = 1×10^{-6}
Annotated in KOG	12,492	18.91	NCBI blast 2.2.28+, *e*-value = 1×10^{-3}
Annotated in all Databases	6070	9.19	-
Annotated in at least one Database	54,753	82.89	-
Total Unigenes	66,050	100	-

We used the KOG functional annotation system for the assembled unigenes to obtain clues to the functions of the unigenes. As a result, 12,492 unigenes have defined, specific protein functions, accounting for 22.82% of the total annotated unigenes and involving 26 KOG functional classes. The five largest categories were "General function prediction only", "Posttranslational modification, protein turnover, chaperones", "Signal transduction mechanisms", "Transcription", "Intracellular trafficking, secretion, and vesicular transport". Two categories, "Unnamed protein" and "Cell motility", accounted for the lowest percentages (Figure 2).

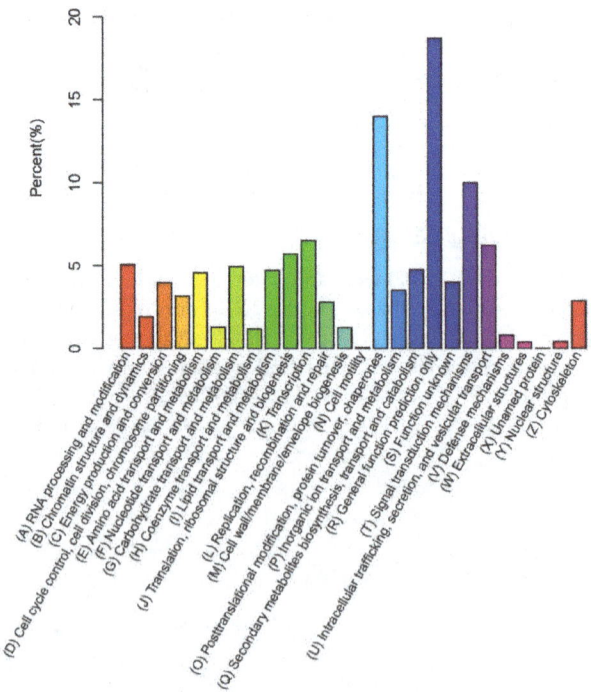

Figure 2. Clusters of eukaryotic orthologous groups (KOG) functional classification of the broccoli bud transcriptome.

To classify the functions of the unigenes, GO assignments were used based on the annotations from the Nr and Pfam databases. The results showed that 30,285 unigenes could be categorized into 57 functional groups, which were separated into three main categories: biological process (25 subcategories), cellular component (18 subcategories), and molecular function (14 subcategories) (Figure 3). The "cellular process," "metabolic process," and "single-organism process" were the major subcategories of the biological processes category; "cell" and "cell part," "binding," and "catalytic activity" were remarkable in the cellular component and molecular function categories, respectively. The classification result indicated that biological processes play a notable role during broccoli bud development, while the terms "biological phase," "cell aggregation," "synapse," "synapse part," "receptor regulator activity," and "metallochaperone activity" were rare.

To predict the activated biochemical pathways in broccoli buds, the unigenes were annotated against the KEGG database (e-value = 1×10^{-10}). A total of 264 KEGG pathways represented by 12,403 unigenes were obtained (Table S1). The main pathways were "Biosynthesis of amino acids [ko01230] (454 unigenes)," "Plant hormone signal transduction [ko04075]" (447 unigenes), "Carbon metabolism [ko01200]" (433 unigenes), and "Starch and sucrose metabolism [ko00500]" (366 unigenes). Moreover, the pathways such as "Endocytosis [ko04144]," "RNA degradation [ko03018]," "ubiquitin mediated proteolysis [ko04120]," "Fatty acid degradation [ko00071]," "Valine, leucine and isoleucine degradation [ko00280]," "Proteasome [ko03050]," "Lysine degradation [ko00310]," and "Apoptosis [ko04210]," which contained 235, 187, 184, 82, 79, 77, 49, and 31 unigenes, respectively, which indicated that protein degradation and cell death were associated with bud abortion.

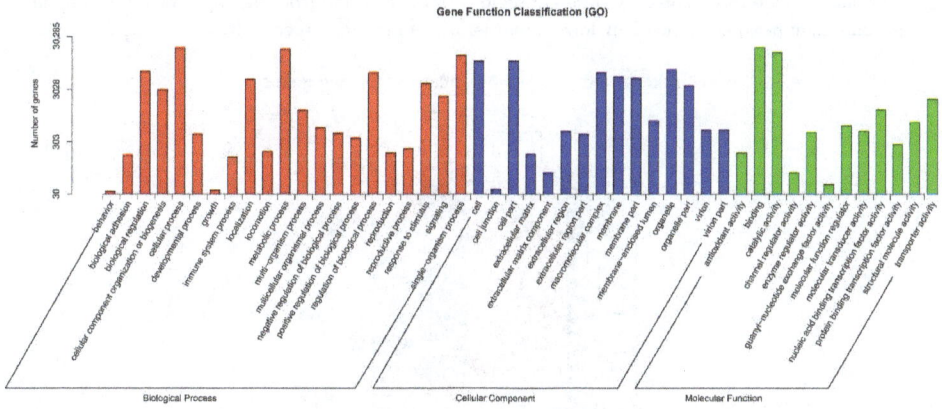

Figure 3. Gene Ontology (GO) classifications of the broccoli bud transcriptome.

2.3. Genes Express Differences and DEGs Clustering

The 97,347 assembly unigenes sequences obtained from de novo assemble of a merged set of 663,701,918 clean reads were used as the reference. We acquired the number of readcounts that aligned to each unigene and calculated the expected number of Fragments Per Kilobase of transcript sequence per Millions base pairs sequenced (FPKM) value to estimate the expression levels of the unigenes. The FPKM distribution and heatmap indicated the presence of many DEGs among the four samples, and the expression patterns of aborted buds of ML93219 (ML_AB) and aborted buds of *ogu* CMS93219 (CMS_AB), and normal buds of ML93219 (ML_NB) and normal buds of *ogu* CMS93219 (CMS_NB) were more similar, respectively (Figures 4 and 5). Changes in the DEGs indicated that the abundances of transcripts were different at the same stage of normal and abortive buds development between ML93219 and CMS93219.

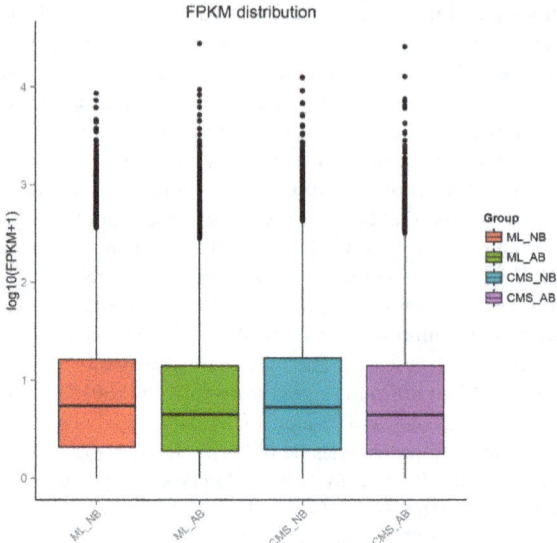

Figure 4. Boxplot of Fragments per kb per million fragments (FPKM) distribution for the four samples. Five statistics are represented by different regions of the Boxplot; from the top down they are the maximum, upper quartile, median, lower quartile, and minimum, respectively.

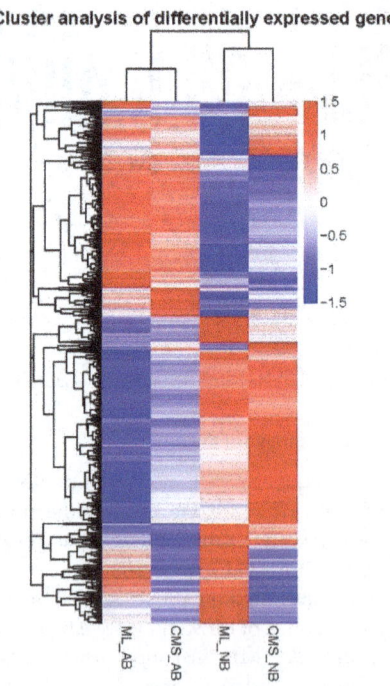

Figure 5. Cluster analysis of differentially expressed genes among the four samples. Heatmap of differentially expressed genes among the four samples. Red indicates high expression, and blue indicates low expression. Color from red to blue represents descending log10 (FPKM + 1).

2.4. DEGs in Normal and Abortive Buds of ogu CMS

To explore the reference sample CMS_AB gene expression levels, the gene expression variations were determined using three comparisons, between CMS_AB and CMS_NB, between CMS_AB and ML_NB and between CMS_AB and ML_AB. Compared with CMS_NB, there were 6575 and 5482 genes up- and downregulated in CMS_AB, respectively. Compared with ML_NB, in CMS_AB, there were 6192 and 6321 genes that were up- and downregulated, respectively. Compared with ML_AB, there were 182 and 825 genes that were up- and downregulated in CMS_AB, respectively (Figure 6a–c).

To determine the key genes involved in *ogu* CMS line bud abortion process, we compared the DEGs of normal and abortive buds from ML93219 and CMS93219 to define the differentially expressed genes in common. In CMS_AB, 74 and 21 genes were upregulated and downregulated, respectively, compared with CMS_NB, ML_NB and ML_AB (Figure 7; Tables S2–S5), which are most likely associated with *ogu* CMS bud abortion in broccoli.

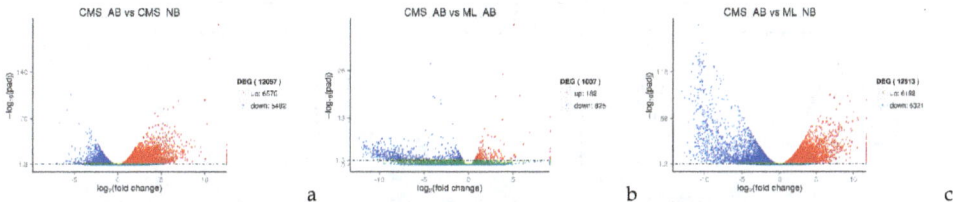

Figure 6. Differentially expressed genes (DEGs) between normal and abortive buds from ML93219 and CMS93219. (**a**) DEGs between CMS AB and CMS NB. (**b**) DEGs between CMS AB and ML AB. (**c**) DEGs between CMS AB and ML NB. The *x*-axis indicates the log$_2$ (fold change) between the two samples. The *y*-axis indicates $-$log$_{10}$ (padj) (*p*-adjusted). The scatter points in the figure represent individual genes, the green dots indicate genes with no significant differences, the red dots indicate up-regulated genes with significant differences, and the blue dots indicate down-regulated genes with significant differences. The screening condition for DEGs is padj < 0.05.

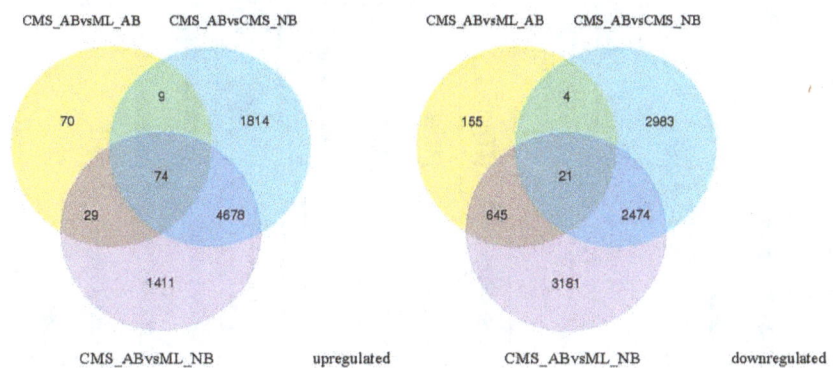

Figure 7. DEGs between normal and abortive buds from broccoli maintainer and *ogu* cytoplasmic male sterile (CMS) lines.

2.5. Validation of RNA-Seq Data by Quantitative Real-Time Reverse Transcription PCR (qRT-PCR)

We performed qRT-PCR to verify the DEGs identified by RNA-Seq, using the same samples that were used for the RNA-Seq analysis. Among the 21 randomly selected DEGs based on the expression level fold-change and differences in expression, 11 genes displayed higher expression quantification and 10 genes showed lower expression quantification in CMS_AB, compared with CMS_NB, ML_NB, and ML_AB. All 21 genes showed the same expression patterns in qRT-PCR as they did in the RNA-Seq databases experiment, indicating the high reliability of our RNA-Seq data (Figure 8).

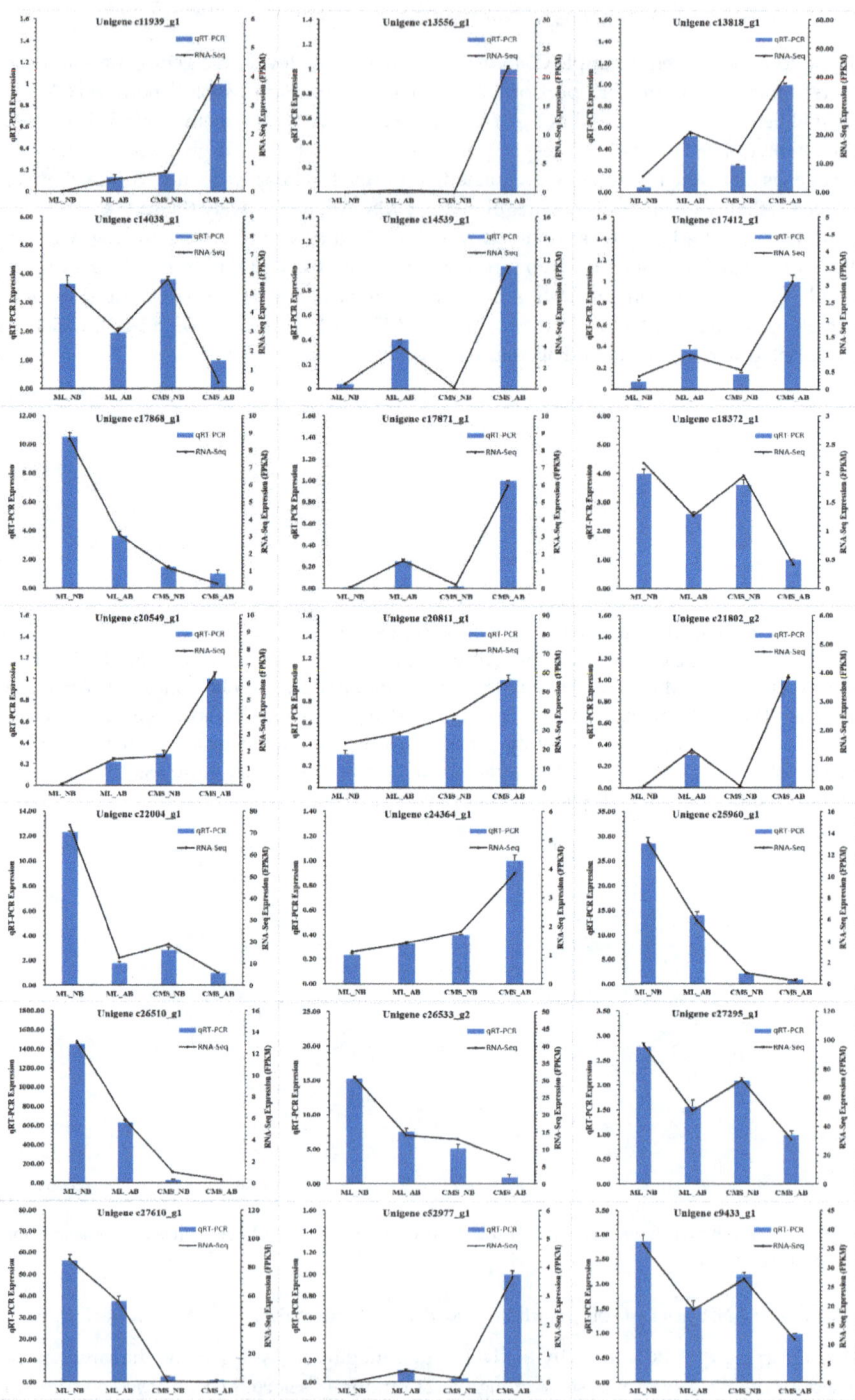

Figure 8. Verification of the DEGs by qRT-PCR. Eleven DEGs with higher expression and ten DEGs with lower expression in CMS_AB were selected for qRT-PCR validation. The relative expression level of each gene was expressed as the FPKM among four samples in the RNA-Seq data (black line) and qRT-PCR data (blue bar). To normalize the expression data, the broccoli *β-actin* gene was used as the internal control. The bars represent the standard deviation.

2.6. Ogu CMS Bud Abortion-Related Genes in Broccoli

To further understand the function and biological process of the DEGs, GO term (corrected $p < 0.05$) and KEGG pathway (corrected $p < 0.05$) enrichment were performed to analyse the 95 differentially coexpressed genes. Nineteen were enriched and the most significantly enriched GO terms were "cell wall organization" and "external encapsulating structure organization" (corrected $p = 3.26 \times 10^{-5}$) in the biological process (BP) group, "extracellular region" (corrected $p = 0.003937$) in the cellular component (CC) group and "polygalacturonase activity" (corrected $p = 0.015471$) in the molecular function (MF) group (Table 4). The GO terms of "cell wall organization or biogenesis" and "extracellular region" contained 10 and 13 genes, respectively, and were the biggest categories in the biological process and cellular component groups (Table 4), respectively. Only "polygalacturonase activity" GO terms in molecular function group was enriched and contained 4 genes (Table 4). Furthermore, there were 8 same genes were enriched in "sucrose metabolic process," "starch metabolic process," "disaccharide metabolic process," "cellular glucan metabolic process," "glucan metabolic process" and "oligosaccharide metabolic process" (Table 4); 5 same genes were enriched between "external encapsulating structure" and "cell wall" (Table 4). "Phenylalanine metabolism" and "Phenylpropanoid biosynthesis" (K00430) contained the same three genes (Gene ID: c20440_g1, c20661_g1, c9433_g1) that were significantly enriched in KEGG pathways; meanwhile, the three genes were enriched by GO enrichment (Tables 4 and 5).

Based on the enrichment results and functional annotation of the 19 enriched differential expressed genes in common by GO term, we found that "Glycosyl hydrolases" and "Reactive oxygen species (ROS) scavenger" related genes accounted for a high proportion (Figure 9). In addition, "Inhibitor," "Plant defense," and "Cell division and expansion" related genes were also significantly upregulated in CMS_AB and "Transporter" related gene Kinesin-4 downregulated in CMS_AB (Figure 9). Furthermore, five of the differentially expressed genes in common (GI: c11939_g1, c14539_g1, c25960_g1, c52977_g1, c9433_g1) enriched by GO terms were verified by qRT-PCR, and had significantly different expression between the abortive buds and normal buds of CMS93219 (Figure 8). These results suggested that most of the differentially expressed genes in common found in our study are required for *ogu* CMS bud abortion in broccoli.

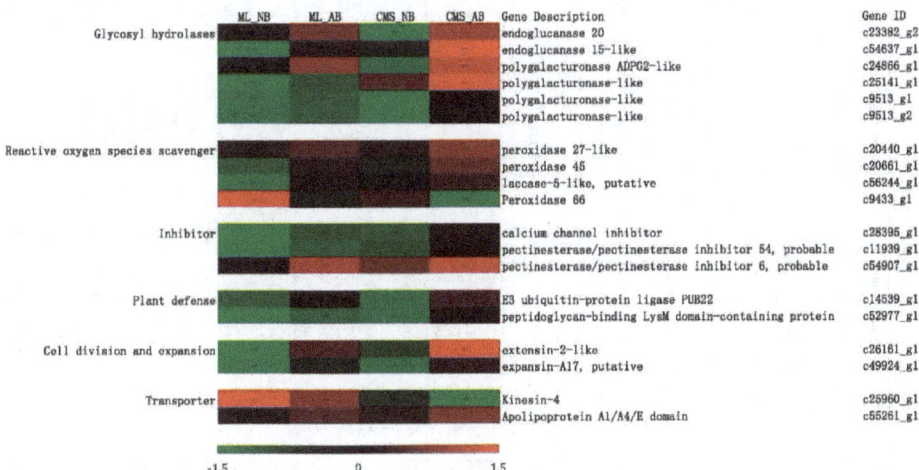

Figure 9. Heatmap of the 19 enriched common differentially expressed genes in normal and abortive broccoli buds. The bar represents the expression levels for each gene ($\log_{10}$ (FPKM + 1)) in the ML_NB, ML_AB, CMS_NB, and CMS_AB groups, as indicated by red or green rectangles. Red means upregulation of genes and green means downregulation.

Table 4. Results of DEGs enriched by Gene Ontology (GO) term.

GO Accession	Description	Term Type	p Value	Corrected p Value	DEG Item	Gene Names
GO:0071555	cell wall organization	Biological process	9.34×10^{-10}	3.25×10^{-6}	9	c23382_g2, c9513_g2, c9513_g1, c24866_g1, c25141_g1, c11939_g1, c54907_g1, c49924_g1, c26161_g1
GO:0045229	external encapsulating structure organization	Biological process	1.34×10^{-9}	3.25×10^{-6}	9	c54907_g1, c26161_g1, c49924_g1, c23382_g2, c24866_g1, c25141_g1, c11939_g1, c9513_g1, c9513_g2
GO:0071554	cell wall organization or biogenesis	Biological process	1.66×10^{-9}	3.25×10^{-6}	10	c49924_g1, c26161_g1, c52977_g1, c54907_g1, c9513_g1, c9513_g2, c24866_g1, c25141_g1, c11939_g1, c23382_g2
GO:0005985	sucrose metabolic process	Biological process	6.13×10^{-6}	0.005404	8	c54907_g1, c54637_g1, c23382_g2, c25141_g1, c11939_g1, c24866_g1, c9513_g2, c9513_g1
GO:0005982	starch metabolic process	Biological process	6.46×10^{-6}	0.005404	8	c23382_g2, c25141_g1, c11939_g1, c24866_g1, c9513_g2, c9513_g1, c54907_g1, c54637_g1
GO:0005984	disaccharide metabolic process	Biological process	8.99×10^{-6}	0.005849	8	c54907_g1, c54637_g1, c23382_g2, c9513_g2, c9513_g1, c11939_g1, c25141_g1, c24866_g1
GO:0006073	cellular glucan metabolic process	Biological process	1.2×10^{-5}	0.006395	8	c24866_g1, c11939_g1, c25141_g1, c9513_g1, c9513_g2, c23382_g2, c54637_g1, c54907_g1
GO:0044042	glucan metabolic process	Biological process	1.2×10^{-5}	0.006395	8	c23382_g2, c9513_g1, c9513_g2, c25141_g1, c11939_g1, c24866_g1, c54907_g1, c54637_g1
GO:0009311	oligosaccharide metabolic process	Biological process	1.66×10^{-5}	0.008084	8	c23382_g2, c25141_g1, c11939_g1, c24866_g1, c9513_g2, c9513_g1, c54907_g1, c54637_g1
GO:0044264	cellular polysaccharide metabolic process	Biological process	3.33×10^{-5}	0.013006	9	c54637_g1, c24364_g1, c54907_g1, c24866_g1, c11939_g1, c25141_g1, c9513_g1, c9513_g2, c23382_g2
GO:0005976	polysaccharide metabolic process	Biological process	5.27×10^{-5}	0.017134	9	c25141_g1, c11939_g1, c24866_g1, c9513_g2, c9513_g1, c23382_g2, c54637_g1, c24364_g1, c54907_g1
GO:0044723	single-organism carbohydrate metabolic process	Biological process	7.34×10^{-5}	0.022634	11	c11939_g1, c25141_g1, c24866_g1, c9513_g1, c9513_g2, c62_g1, c23382_g2, c17871_g1, c54637_g1, c54907_g1, c24364_g1
GO:0044262	cellular carbohydrate metabolic process	Biological process	0.00015	0.043887	9	c23382_g2, c9513_g1, c9513_g2, c25141_g1, c11939_g1, c24866_g1, c24364_g1, c54907_g1, c54637_g1
GO:0030312	external encapsulating structure	Cellular component	3.87×10^{-6}	0.005404	6	c23382_g2, c54907_g1, c49924_g1, c11939_g1, c26161_g1, c24866_g1
GO:0005576	extracellular region	Cellular component	7.52×10^{-6}	0.005507	11	c56244_g1, c28895_g1, c55261_g1, c14539_g1, c49924_g1, c20661_g1, c24866_g1, c25141_g1, c20440_g1, c9513_g2, c9513_g1
GO:0005618	cell wall	Cellular component	2.31×10^{-5}	0.010397	5	c11939_g1, c26161_g1, c24866_g1, c49924_g1, c54907_g1
GO:0071944	cell periphery	Cellular component	4.68×10^{-5}	0.016257	8	c49924_g1, c50518_g1, c26161_g1, c54907_g1, c11939_g1, c24866_g1, c23382_g2, c22601_g2
GO:0004650	polygalacturonase activity	Molecular function	6.12×10^{-6}	0.005404	4	c9513_g1, c9513_g2, c25141_g1, c24866_g1
GO:0004553	hydrolase activity, hydrolyzing O-glycosyl compounds	Molecular function	2.96×10^{-5}	0.012366	8	c54637_g1, c25141_g1, c24866_g1, c9513_g1, c9513_g2, c23382_g2, c562_g1, c17871_g1
GO:0016798	hydrolase activity, acting on glycosyl bonds	Molecular function	4.72×10^{-5}	0.016257	8	c17871_g1, c562_g1, c23382_g2, c9513_g1, c9513_g2, c25141_g1, c24866_g1, c54637_g1

Table 5. Results of DEGs enriched by Kyoto Encyclopedia of Genes and Genomes (KEGG) pathway.

#Term	ID	Input Number	p Value	Corrected p-Value	Input
Phenylalanine metabolism	ko00360	3	0.000337	0.00639837	c9433_g1, c20661_g1, c20440_g1
Phenylpropanoid biosynthesis	ko00940	3	0.001398	0.013279387	c9433_g1, c20661_g1, c20440_g1
Propanoate metabolism	ko00640	1	0.042347	0.179678515	c17871_g1
Cutin, suberine and wax biosynthesis	ko00073	1	0.048213	0.179678515	c24364_g1
Fatty acid elongation	ko00062	1	0.051033	0.179678515	c21267_g1
Endocrine and other factor-regulated calcium reabsorption	ko04961	1	0.060379	0.179678515	c22601_g2
alpha-Linolenic acid metabolism	ko00592	1	0.083359	0.179678515	c27465_g3
Photosynthesis	ko00195	1	0.08789	0.179678515	c28693_g1
Synaptic vesicle cycle	ko04721	1	0.090598	0.179678515	c22601_g2
Parkinson's disease	ko05012	1	0.095095	0.179678515	c25708_g1
Glycerolipid metabolism	ko00561	1	0.104024	0.179678515	c57011_g1
Peroxisome	ko04146	1	0.132037	0.187274571	c24364_g1
Pyruvate metabolism	ko00620	1	0.145727	0.187274571	c17871_g1
Cysteine and methionine metabolism	ko00270	1	0.149117	0.187274571	c17871_g1
Huntington's disease	ko05016	1	0.161716	0.187274571	c22601_g2
Glycerophospholipid metabolism	ko00564	1	0.164214	0.187274571	c57011_g1
Oxidative phosphorylation	ko00190	1	0.174958	0.187274571	c25708_g1
Glycolysis/Gluconeogenesis	ko00010	1	0.177418	0.187274571	c17871_g1
Endocytosis	ko04144	1	0.208772	0.208772251	c22601_g2

2.7. Transcription Factors Are Involved in Broccoli ogu CMS Bud Abortion Control

Among the 95 common differentially expressed genes, four transcription factors were identified, including predicted transcription factor basic helix-loop-helix (bHLH) 137 (GI: c13818_g1), ethylene-responsive transcription factor (ERF) 115 (GI: c21802_g2), transcription factor DUO1 (GI: c18372_g1), and PosF21/RF2a/BZIP34 (GI: c14038_g1), which were distributed in four different gene families. Compared with CMS_NB, ML_NB, and ML_AB, transcription factors bHLH137 ERF115 were upregulated in CMS_AB, suggesting that these transcription factors may function as positive regulators in *ogu* CMS bud abortion in broccoli. Conversely, transcription factors DUO1 and PosF21/RF2a/BZIP34 were downregulated in CMS_AB, suggesting that they are negative regulators. Moreover, the expression quantifications of the four transcription factors were confirmed by qRT-PCR (Figure 8), suggesting that the four transcription factors play key roles in *ogu* CMS bud abortion in broccoli.

3. Discussion

The bud abortion phenomenon is a very complex bioprocess and demands many molecular events during bud development. In this study, we used RNA-Seq technology to explore the genes involved in *ogu* CMS bud abortion process and to provide a comprehensive analysis of the genes involved in *ogu* CMS bud abortion control in broccoli. Compared with CMS_NB, ML_NB, ML_AB, there were 6575, 182, and 6192 genes that were upregulated and 5482, 825 and 6321 genes that were downregulated in CMS_AB, respectively (Figure 6a–c), among which 74 genes were significantly upregulated and 21 genes were significantly downregulated equally in CMS_AB with serious bud abortion (Figure 7; Tables S2–S5). qRT-PCR proved that our RNA-Seq data was highly reliable (Figure 8). Functional categories of the common differentially expressed genes by GO term enrichment analysis showed that gene associated with cell wall composition and metabolism, such as cell wall organization, cell wall organization or biogenesis; extracellular region, such as external encapsulating structure organization; sugar metabolism, such as sucrose metabolic process, starch metabolic process, disaccharide metabolic process, cellular glucan metabolic process, glucan metabolic process, oligosaccharide metabolic process related genes were strongly induced in the buds abortion of broccoli (Table 4). These results indicated that bud abortion may be closely related to cell wall organization, external encapsulating structures and sugar metabolism related genes play an important role in bud abortion in broccoli. Our results were consistent with the differential expression genes obtained by cDNA-AFLP technique in radish and cabbage between normal bud and dead bud [6–8].

3.1. Genes Related to Programmed Cell Death Are Involved in Bud Abortion

Programmed cell death (PCD) is an important physiological process in single cells and penetrates the whole plant life cycle, which can help plants to control and organize the destruction of non-functional or redundant damaged cells [13–15]. Although PCD is a natural result of ageing, it may be switched on by environmental stress or irregular development in plants [15]. Increases in caspase-like proteases [16] and ROS [17,18] activities of metacaspase gene family related genes [19] are the important features of PCD. In this study, several genes associated with PCD were significantly changed in *ogu* CMS abortive buds compared with normal buds and the maintainer abortive buds (CMS_NB, ML_NB and ML_AB). Among the 8833 differential expression genes between ML_AB vs. ML_NB and CMS_AB vs. CMS_NB, there was 11.75% participate in the redox process, 11.08% with redox enzyme activity and 55.38% with catalytic activity, indicating that redox process involved in bud abortion in broccoli. ROS scavengers related genes, such as: peroxidase 27-like and peroxidase 45, L-lactate dehydrogenase, laccase-5-like and fatty acyl-CoA reductase 7 (Table S5), which were all upregulated in abortive buds compared with normal buds, suggesting that higher levels of ROS were produced in abortive buds and ROS scavenger-related genes were closely related with broccoli bud abortion. Moreover, 13 caspase-like and metacaspase activity genes involved in cell apoptosis were also discovered, such as metacaspase-1, metacaspase-3, metacaspase-5, metacaspase-6, metacaspase-7, metacaspase-9, pentapeptide repeats protein, caspase recruitment domain, BTB/POZ domain-containing protein POB1. The expression of these genes in abortion buds was significantly higher than that of normal buds suggesting that bud abortion was related with PCD.

3.2. Glycosyl Hydrolases, Inhibitors and Plant Defence Related Genes Are Implicated in Bud Abortion

In this study, several glycosyl hydrolase-related genes were determined as significantly upregulated in *ogu* CMS abortive buds, such as endoglucanase 20, endoglucanase 15-like, polygalacturonase ADPG2-like and polygalacturonase-like. However, polygalacturonase plays an important role during the life cycle of cell separation, being involved in cell wall modification, abscission and dehiscence in *Arabidopsis thaliana* [20,21], and endoglucanase is involved in cell wall biogenesis or degradation, cellulose degradation and polysaccharide degradation [22], which suggested that glycosyl hydrolase-related genes were required for *ogu* CMS bud abortion in broccoli. Pectinesterase/pectinesterase inhibitors modify cell walls via demethylesterification of cell wall pectin, which negatively regulates catalytic activity. As a voltage-gated inward-rectifying Ca^{2+} channel (VDCC) across the vacuole membrane, the calcium channel is an essential components of the slow vacuolar (SV) channel and is the major ROS-responsive Ca^{2+} channel, which is the possible target of Al-dependent inhibition and is involved in the regulation of stomatal movement [23,24]. In our study, pectinesterase/pectinesterase inhibitor 6 and 54 probable and calcium channel inhibitor-related genes were highly expressed in *ogu* CMS abortive buds, indicating that these genes may be important in controlling *ogu* CMS bud abortion. Moreover, E3 ubiquitin-protein ligase and peptidoglycan-binding LysM domain-containing protein both play important roles in the plant defence response [23,25–30] and nicotianamine synthase is involved in the cellular response to ethylene stimulus [31]: these genes were distinguished significantly upregulated in *ogu* CMS abortive buds, suggesting their likely involvement in *ogu* CMS bud abortion.

3.3. Transcription Factors Associated with Bud Development

Transcription factors are critical to regulate gene expression during plant development and in response to biotic and abiotic stresses [32–38]. In the present study, predicted transcription factors bHLH137, ERF115, DUO1 and PosF21/RF2a/BZIP34 were obviously changed in abortive broccoli buds compared with normal buds. Interestingly, the genes encoding stress-responsive transcription factors ERF115 and gibberellin-responsive transcription factors bHLH137 showed similar expression patterns: both were significantly upregulated in abortive buds (ML_AB and CMS_AB) compared with normal buds (ML_NB and CMS_NB), and in *ogu* CMS abortive buds (CMS_AB), they were significantly upregulated compared with maintainer abortive buds (ML_AB) (Figure 8). Conversely, the genes encoding transcription factors DUO1 and PosF21/RF2a/BZIP34, which are positive regulators of transcription showed similar expression patterns: both were significantly downregulated in abortive buds, and in *ogu* CMS abortive buds they were significantly downregulated compared with maintainer abortive buds (Figure 8). ERF 115 as a transcriptional activator of the phytosulfokine PSK5 peptide hormone family that binds to the GCC-box pathogenesis-related promoter element and limit quiescent center cell division activity when surrounding stem cells are damaged and is also a proteolytic target of the APC/C-FZR1 complex [39]. bHLH transcription factors belong to a family of transcriptional regulators and have a range of different roles in plant cells and tissue development, as well as in plant metabolism [40]. Transcription factor DUO1 could be involved in pollen sperm cell differentiation [41]. PosF21/RF2a/BZIP34 is a transcription factor with an activatory role [42,43], which might be involved in the sporophytic control of cell wall patterning and gametophytic control of pollen development, and play a role in the control of metabolic pathways regulating cellular transport and lipid metabolism [44,45]. Therefore, transcription factors play a key role in the complex regulatory networks of *ogu* CMS bud abortion.

3.4. Molecular Mechanisms Associated with CMS

Previous studies have shown that the genes involved in reactive oxygen species (ROS) homeostasis or antioxidative system balance may be important factors contributing to pollen abortion in cotton [46] and wheat [47]. Carbohydrate and energy metabolisms, oxidation-reduction system and phenylpropanoid metabolism pathways related genes may be important factor for CMS in soybean [48], rapeseed [49], cabbage [50], onion [51] and wheat [52]. Male sterility might be related to energy metabolism turbulence, excessive ethylene synthesis, and suffocation of starch synthesis in pepper [53]. In addition, pentatricopeptide repeat proteins, heat shock proteins, stress proteins, MYB, bHLH and heat shock transcription factors and anther development related genes may be the candidates for pollen abortion in *Brassica* crops [49,50,54]. In this study, we found that genes related to polygalacturonase metabolism, glycosyl hydrolases, oxidation reduction process, phenylalanine metabolism, phenylpropanoid biosynthesis were significantly changed in *ogu* CMS abortive buds compared with the other groups. Ethylene-responsive transcription factor 115 and transcriptional factor basic helix-loop-helix 137 were both significantly upregulated in *ogu* CMS abortive buds. Therefore, our results were basically consistent with the results of previous studies [46–54], and the genes discovered related to energy metabolism, oxidation reduction process and phenylpropanoid biosynthesis, ethylene-responsive transcription factor 115 and transcriptional factor basic helix-loop-helix 137 may be important factors contributing to *Broccoli ogu* CMS pollen abortion and bud abortion. Further experiments are needed to elucidate the molecular mechanisms of these genes that lead to broccoli CMS and bud abortion.

4. Materials and Methods

4.1. Plant Materials

Broccoli (*Brassica oleracea* var. *italica*) maintainer ML93219 (showing slight bud abortion) and *ogu* CMS93219 (serious levels of bud abortion) were bred by the Institute of Vegetables and Flowers, Chinese Academy of Agricultural Sciences. The backcross paternal line of *ogu* CMS93219 was ML93219 and the number of backcross generations was sixteen. In the spring of 2015, the plants were grown in an experimental greenhouse at the Institute of Vegetable and Flowers, Chinese Academy of Agricultural Sciences, Changping (Beijing, China). We handled the main bouquet using the approach proposed by Shu et al. [55]. When the plants began to flower, four kinds of bud samples, ML_NB, ML_AB, CMS_NB and CMS_AB, were collected and labelled with three biological replicates. To ensure the integrity of the sample RNA, isolated buds were immediately frozen in liquid nitrogen and stored at −80 °C before RNA extraction.

4.2. RNA Extraction and Quality Testing

Total RNA was extracted using an EASYspin Plus Plant RNA-38 Kit, according to the manufacturer's instructions (Juhuatech Co., Ltd., Beijing, China). The integrity and purity of the RNA samples were determined by 1% agarose gels electrophoresis, and the RNA concentration was measured by Qubit® RNA Assay Kit in Qubit® 2.0 Flurometer (Life Technologies Corporation, Carlsbad, CA, USA) and the integrity of the RNA was assessed by an RNA Nano 6000 Assay Kit of the Agilent Bioanalyzer 2100 system (Agilent Technologies Inc., Santa Clara, CA, USA).

4.3. RNA-Seq Library Construction and Illumina Sequencing

Twelve strand-specific RNA-Seq libraries were constructed with cDNA fragments of 250–300 bp in length. An Illumina TruSeq PE Cluster Kit v3-cBot-HS on a cBot Cluster Generation System was then used to cluster the samples, according to the manufacturer's instructions. After cluster generation, the libraries were sequenced on an Illumina Hiseq™ 4000 system and reads were generated.

4.4. RNA-Seq Data Quality Control and Transcriptome de Novo Assembly

We obtained clean reads by removing the adaptor reads, unknown sequences "N" (reads containing more than 10% unknown nucleotides), low quality reads (reads containing more than 50% bases with Q-value $\leq$ 5) from the raw data. The Q20, Q30 and GC-content were then calculated based on the clean reads. The high quality clean reads were used for downstream analyses. Transcriptome *de novo* assembly was executed using Trinity [56] with min_kmer_cov set to 2 by default and other parameters set at their defaults. After assembly, the longest transcripts of each gene were selected as the unigenes.

4.5. Unigene Function Annotation

We annotated the unigenes based on seven databases, NCBI blast (2.2.28+) was used to search against the Nr (E-value = 1×10^{-5}), Nt (E-value = 1×10^{-5}), Swiss-Prot (E-value = 1×10^{-5}) and KOG databases (E-value = 1×10^{-3}). The unigenes were divided into 26 groups and their participation in different metabolic pathways based on KOG annotation was assessed. Pfam annotated was determined using the HMMER 3.0 package [57], hmmscan (e-value = 0.01). GO annotations for the unigenes were determined by Blast2GO v2.5 [12] with the self-write script (e-value = 1×10^{-6}) based on the annotation result of Nr and Pfam, which has three ontologies: molecular function, cellular component and biological process [58]. KEGG [59] related annotations were identified by the KAAS and KEGG Automatic Annotation Server [60] (E-value = 1×10^{-10}) to determine the metabolic pathway of unigenes.

4.6. Analysis of DEGs

Alignment results of bowtie were counted by RSEM [61]. FPKM [62] values were used to calculate the gene expression levels of the four groups of normal and abortive buds from the maintainer and *ogu* CMS lines. FPKM has become the most commonly used method to estimate the level of gene expression and takes into account the effects of sequencing depth and gene length on the calculation of gene expression [62]. There were three biological replicates; therefore, the calculated gene expression could be used directly to compare the gene expression between samples. Referring to the statistical method of Storey and Tibshirani [63], |log$_2$Fold change| > 1 and p-adjusted < 0.05 were set as the threshold for significantly differential expression. p-Values were adjusted to control the false discovery rate, referring to Benjamini and Hochberg's approach [64]. Then, based on the Wallenius non-central hyper-geometric distribution [65], GO and KEGG functional enrichment analysis of the DEGs was executed by the GOseq [66] and KOBAS software [60], respectively.

4.7. qRT-PCR Validation

qRT-PCR analyses with the three biological replicates samples used for RNA-Seq were performed to verify the DGEs results. Twenty-one common differentially expressed genes were randomly selected that accounted for about 22.1% of the 95 common differentially expressed genes. Specific primers were designed using the Primer-BLAST tool (available online: http://www.ncbi.nlm.nih.gov/tools/primer-blast/index.cgi?LINK_LOC=BlastHome) in NCBI and synthesized by Sangon Biotech Co., Ltd. (Shanghai, China). cDNAs were reverse transcribed from total RNA using a PrimeScript RT reagent Kit (Takara, Dalian, China). qRT-PCR was carried out according to the SYBR PrimeScript RT-PCR Kit manufacturer specifications (Takara) on an ABI Prism®7900 Real-Time PCR System (Applied Biosystems, Foster City, CA, USA). To normalize the gene expression data, we used the broccoli *β-actin* gene as an internal standard [67]. The $2^{-\Delta\Delta Ct}$ method [68] was used to determine the relative expression of genes. The standard deviation was calculated based on the three biological replicates. The specific primers sequences are listed in Table S6.

5. Conclusions

In this study, we found that buds abortion was related with polygalacturonase metabolism, glycosyl hydrolases, oxidation reduction process, phenylalanine metabolism, and phenylpropanoid biosynthesis. Moreover, 19 common differentially expressed genes associated glycosyl hydrolases, reactive oxygen species scavenging, inhibitor, plant defense, cell division and expansion, transporter, and four transcriptional factors—ethylene-responsive transcription factor 115, transcriptional factor basic helix-loop-helix 137, transcription factors DUO1, and PosF21/RF2a/BZIP34—may be associated with *ogu* CMS abortive buds. In conclusion, our results not only increased our understanding of *ogu* CMS bud abortive mechanisms and provided a valuable resource for further functional characterization of *ogu* CMS bud abortion, but also laid the foundation for molecular breeding to overcoming bud abortion in broccoli, as well as other *Brassica* crops in the future.

Supplementary Materials: Supplementary materials can be found at http://www.mdpi.com/1422-0067/19/9/2501/s1. Table S1: Sequences of the specific primers; Table S2: KEGG classification of the unigenes; Table S3: Unigenes expression fold change of CMS_ABvsCMS_NB; Table S4: Unigenes expression fold change of CMS_ABvsML_AB; Table S5: Unigenes expression fold change of CMS_ABvsML_NB; Table S6: FPKM and annotation of the 95 common differentially expressed genes.

Author Contributions: Conceptualization, J.S. and Y.L.; Data curation, J.S.; Formal analysis, J.S., L.Z., Z.L. and H.L.; Funding acquisition, Y.L.; Investigation, J.S.; Methodology, J.S.; Resources, Y.L., Z.F., L.Y., M.Z. and Y.Z.; Supervision, Y.L.; Writing—original draft, J.S. and L.Z.; Writing—review & editing, J.S. and Y.L. All authors read and approved the final manuscript.

Acknowledgments: This work was supported by the National Natural Science Foundation of China (Grant No. 31372067), the China Agriculture Research System (Grant No. CARS-25-A), Key Projects in the National Science and Technology Pillar Program of China (Grant No. 2013BAD01B04), the Key Laboratory of Biology and Genetic Improvement of Horticultural Crops, Ministry of Agriculture, China, and the Science and Technology Innovation Program of the Chinese Academy of Agricultural Sciences (Grant No. CAAS-ASTIP-IVFCAAS). We thank Liwen Bianji, Edanz Group China (available online: www.liwenbianji.cn/ac), for editing the English text of a draft of this manuscript.

Conflicts of Interest: The authors declare no conflict of interest.

References

1. Zhang, J.; Li, Q.F.; Huang, W.W.; Xu, X.Y.; Zhang, X.L.; Hui, M.X.; Zhang, M.K.; Zhang, L.G. A vacuolar processing enzyme *RsVPE1* gene of radish is involved in floral bud abortion under heat stress. *Int. J. Mol. Sci.* **2013**, *14*, 13346–13359. [CrossRef] [PubMed]

2. Zhang, J.; Sun, X.L.; Zhang, L.G.; Hui, M.X.; Zhang, M.K. Analysis of differential gene expression during floral bud abortion in radish (*Raphanus sativus* L.). *Genet. Mol. Res.* **2013**, *12*, 2507–2516. [CrossRef] [PubMed]

3. Liu, J.; Zhang, L.G.; Wang, F.M.; Hui, M.X.; Zhang, M.K. Observation of histocytological feature on radish flower bud during aborting. *Acta Agric. Bor. Occid. Sin.* **2008**, *5*, 272–276.

4. Wang, Q.B.; Zhang, Y.Y.; Liu, Y.M.; Yang, L.M.; Zhuang, M.; Sun, P.T. The floral and seed setting characteristics in two types of male sterile lines of cabbage (*Brassica oleracea* L. var. *capitata*). *Acta Hortic. Sin.* **2011**, *38*, 61–68.

5. Shu, J.S.; Liu, Y.M.; Li, Z.S.; Zhang, L.L.; Fang, Z.Y.; Yang, L.M.; Zhuang, M.; Zhang, Y.Y.; Li, Z.S.; Sun, P.T. Study on the floral characteristics and structure in two types of male sterile lines of broccoli (*Brassica oleracea* var. *italica*). *J. Plant. Genet. Resour.* **2014**, *15*, 113–119.

6. Jia, J.; Zhang, L.G. mRNA differential display and EST sequence analysis of aborted bud and normal bud in radish (*Raphanus sativus*). *Acta Agric. Nucl. Sin.* **2008**, *22*, 426–431.

7. Zhang, L.G.; Jia, J.; Zhang, S.L.; Zhang, Y. cDNA-AFLP differential expression analysis of genes related with aborting bud in Chinese cabbage. *J. Agric. Biotechnol.* **2010**, *18*, 489–492.

8. Wang, Q.; Liu, Z.; Feng, H. cDNA-AFLP differential expression analysis of genes about bud aborting in genetic male sterile line of *Brassica campestris* L. ssp. *chinensis* (L.) *Makino* var. *rosularis* Tsen *et* Lee. *Mol. Plant. Breed.* **2014**, *12*, 118–126.

9.   Kensler, T.W.; Chen, J.G.; Egner, P.A.; Fahey, J.W.; Jacobson, L.P.; Stephenson, K.K.; Ye, L.X.; Coady, J.L.;
      Wang, J.B.; Wu, Y.; et al. Effects of glucosinolate-rich broccoli sprouts on urinary levels of aflatoxin-DNA
      adducts and phenanthrene tetraols in a randomized clinical trial in He Zuo township, Qidong, People's
      Republic of China. *Cancer Epidemiol. Biomark. Prev.* **2005**, *14*, 2605–2613. [CrossRef] [PubMed]
10.   Canene-Adams, K.; Lindshield, B.L.; Wang, S.; Jeffery, E.H.; Clinton, S.K.; Erdman, J.W., Jr. Combinations of
      tomato and broccoli enhance antitumor activity in dunning r3327-h prostate adenocarcinomas. *Cancer Res.*
      2007, *67*, 836–843. [CrossRef] [PubMed]
11.   Jeffery, E.H.; Araya, M. Physiological effects of broccoli consumption. *Phytochem. Rev.* **2009**, *8*, 283–298.
      [CrossRef]
12.   Götz, S.; García-Gómez, J.M.; Terol, J.; Williams, T.D.; Nagaraj, S.H.; Nueda, M.J.; Robles, M.; Talon, M.;
      Dopazo, J.; Conesa, A. High-throughput functional annotation and data mining with the Blast2GO suite.
      Nucleic Acids Res. **2008**, *36*, 3420–3435.
13.   Thomas, S.G.; Franklin-Tong, V.E. Self-incompatibility triggers programmed cell death in Papaver pollen.
      Nature **2004**, *429*, 305–309. [CrossRef] [PubMed]
14.   Reape, T.J.; Molony, E.M.; McCabe, P.F. Programmed cell death in plants: Distinguishing between different
      modes. *J. Exp. Bot.* **2008**, *59*, 435–444. [CrossRef] [PubMed]
15.   Rybaczek, D.; Musiałek, M.W.; Balcerczyk, A. Caffeine-induced premature chromosome condensation results
      in the apoptosis-like programmed cell death in root meristems of *Vicia faba*. *PLoS ONE* **2015**, *10*, e0142307.
      [CrossRef] [PubMed]
16.   Bosch, M.; Franklin-Tong, V.E. Temporal and spatial activation of caspase-like enzymes induced by
      self-incompatibility in Papaver pollen. *Proc. Natl. Acad. Sci. USA* **2007**, *104*, 18327–18332. [CrossRef]
      [PubMed]
17.   Wilkins, K.A.; Bancroft, J.; Bosch, M.; Ings, J.; Smirnoff, N.; Franklin-Tong, V.E. Reactive oxygen species and
      nitric oxide mediate actin reorganization and programmed cell death in the self-incompatibility response of
      papaver. *Plant Physiol.* **2011**, *156*, 404–416. [CrossRef] [PubMed]
18.   Zhang, F.J.; Wang, Z.Q.; Dong, W.; Sun, C.Q.; Wang, H.B.; Song, A.P.; He, L.Z.; Fang, W.M.; Chen, F.D.;
      Teng, N.J. Transcriptomic and proteomic analysis reveals mechanisms of embryo abortion during
      chrysanthemum cross breeding. *Sci. Rep.* **2014**, *4*, 6536. [CrossRef] [PubMed]
19.   Fagundes, D.; Bohn, B.; Cabreira, C.; Leipelt, F.; Dias, N.; Bodanese-Zanettini, M.H.; Cagliari, A. Caspases in
      plants: Metacaspase gene family in plant stress responses. *Funct. Integr. Genom.* **2015**, *15*, 639–649. [CrossRef]
      [PubMed]
20.   Gonzalez-Carranza, Z.H.; Elliott, K.A.; Roberts, J.A. Expression of polygalacturonases and evidence to
      support their role during cell separation processes in *Arabidopsis thaliana*. *J. Exp. Bot.* **2007**, *58*, 3719–3730.
      [CrossRef] [PubMed]
21.   Ogawa, M.; Kay, P.; Wilson, S.; Swain, S.M. ARABIDOPSIS DEHISCENCE ZONE POLYGALACTURONASE1
      (ADPG1), ADPG2, and QUARTET2 are polygalacturonases required for cell separation during reproductive
      development in *Arabidopsis*. *Plant Cell* **2009**, *21*, 216–233. [CrossRef] [PubMed]
22.   Libertini, E.; Li, Y.; McQueen-Mason, S.J. Phylogenetic analysis of the plant endo-β-1, 4-glucanase gene
      family. *J. Mol. Evol.* **2004**, *58*, 506–515. [CrossRef] [PubMed]
23.   Kawano, T.; Kawano, T.; Kadono, T.; Fumoto, K.; Lapeyrie, F.; Kuse, M.; Isobe, M.; Furuichi, T.; Muto, S.
      Aluminum as a specific inhibitor of plant TPC1 Ca^{2+} channels. *Biochem. Biophys. Res. Commun.* **2004**,
      324, 40–45. [CrossRef] [PubMed]
24.   Peiter, E.; Maathuis, F.J.; Mills, L.N.; Knight, H.; Pelloux, J.; Hetherington, A.M.; Sanders, D. The vacuolar
      Ca^{2+}-activated channel TPC1 regulates germination and stomatal movement. *Nature* **2005**, *434*, 404–408.
      [CrossRef] [PubMed]
25.   Zhao, J.W.; Wang, J.L.; An, L.L.; Doerge, R.W.; Chen, Z.J.; Grau, C.R.; Meng, J.L.; Osborn, T.C. Analysis of
      gene expression profiles in response to Sclerotinia sclerotiorum in *Brassica napus*. *Planta* **2007**, *227*, 13–24.
      [CrossRef] [PubMed]
26.   Libault, M.; Wan, J.; Czechowski, T.; Udvardi, M.; Stacey, G. Identification of 118 *Arabidopsis* transcription
      factor and 30 ubiquitin-ligase genes responding to chitin, a plant-defense elicitor. *Mol. Plant Microbe Interact.*
      2007, *20*, 900–911. [CrossRef] [PubMed]

27. Seo, D.H.; Ryu, M.Y.; Jammes, F.; Hwang, J.H.; Turek, M.; Kang, B.G.; Kwak, J.M.; Kim, W.T. Roles of four *Arabidopsis* U-Box E3 ubiquitin ligases in negative regulation of abscisic acid-mediated drought stress responses. *Plant Physiol.* **2012**, *160*, 556–568. [CrossRef] [PubMed]

28. Wan, J.R.; Zhang, X.C.; Neece, D.; Ramonell, K.M.; Clough, S.; Kim, S.Y.; Stacey, M.G.; Stacey, G. A LysM receptor-like kinase plays a critical role in chitin signaling and fungal resistance in *Arabidopsis*. *Plant Cell* **2008**, *20*, 471–481. [CrossRef] [PubMed]

29. Brotmana, Y.; Landau, U.; Pnini, S.; Lisec, J.; Balazadeh, S.; Mueller-Roeber, B.; Zilberstein, A.; Willmitzer, L.; Chet, I.; Viterbo, A. The LysM receptor-like kinase LysM RLK1 is required to activate defense and abiotic-stress responses induced by overexpression of fungal chitinases in *Arabidopsis* plants. *Mol. Plant.* **2012**, *5*, 1113–1124. [CrossRef] [PubMed]

30. Tanaka, K.; Nguyen, C.T.; Liang, Y.; Cao, Y.; Stacey, G. Role of LysM receptors in chitin-triggered plant innate immunity. *Plant. Signal. Behav.* **2013**, *8*, e22598. [CrossRef] [PubMed]

31. Garcia, M.J.; Lucena, C.; Romera, F.J.; Alcantara, E.; Perez-Vicente, R. Ethylene and nitric oxide involvement in the up-regulation of key genes related to iron acquisition and homeostasis in *Arabidopsis*. *J. Exp. Bot.* **2010**, *61*, 3885–3899. [CrossRef] [PubMed]

32. Xu, Z.S.; Chen, M.; Li, L.C.; Ma, Y.Z. Functions and application of the AP2/ERF transcription factor family in crop improvement. *J. Integr. Plant Biol.* **2011**, *53*, 570–585. [CrossRef] [PubMed]

33. Chen, M.K.; Shpak, E.D. *ERECTA* family genes regulate development of cotyledons during embryogenesis. *FEBS Lett.* **2014**, *588*, 3912–3917. [CrossRef] [PubMed]

34. Nakano, T.; Fujisawa, M.; Shima, Y.; Ito, Y. The AP2/ERF transcription factor SlERF52 functions in flower pedicel abscission in tomato. *J. Exp. Bot.* **2014**, *65*, 3111–3119. [CrossRef] [PubMed]

35. Ito, Y.; Nakanot, T. Development and regulation of pedicel abscission in tomato. *Front. Plant Sci.* **2015**, *6*, 442. [CrossRef] [PubMed]

36. Jisha, V.; Dampanaboina, L.; Vadassery, J.; Mithöfer, A.; Kappara, S.; Ramanan, R. Overexpression of an AP2/ERF type transcription factor *OsEREBP1* confers biotic and abiotic stress tolerance in rice. *PLoS ONE* **2015**, *10*, e0127831. [CrossRef] [PubMed]

37. Mishra, S.; Phukan, U.J.; Tripathi, V.; Singh, D.K.; Luqman, S.; Shukla, R.K. PsAP2 an AP2/ERF family transcription factor from *Papaver somniferum* enhances abiotic and biotic stress tolerance in transgenic tobacco. *Plant Mol. Biol.* **2015**, *89*, 173–186. [CrossRef] [PubMed]

38. Babitha, K.C.; Vemanna, R.S.; Nataraja, K.N.; Udayakumar, M. Overexpression of *EcbHLH57* transcription factor from *Eleusine coracana* L. in tobacco confers tolerance to salt, oxidative and drought stress. *PLoS ONE* **2015**, *10*, e0137098. [CrossRef] [PubMed]

39. Heyman, J.; Cools, T.; Vandenbussche, F.; Heyndrickx, K.S.; Van Leene, J.; Vercauteren, I.; Vanderauwera, S.; Vandepoele, K.; De Jaeger, G.; Van Der, S.D.; et al. ERF115 controls root quiescent center cell division and stem cell replenishment. *Science* **2013**, *342*, 860–863. [CrossRef] [PubMed]

40. Heim, M.A.; Jakoby, M.; Werber, M.; Martin, C.; Weisshaar, B.; Bailey, P.C. The basic helix–loop–helix transcription factor family in plants: A genome-wide study of protein structure and functional diversity. *Mol. Biol. Evol.* **2003**, *20*, 735–747. [CrossRef] [PubMed]

41. Borg, M.; Brownfield, L.; Khatab, H.; Sidorova, A.; Lingaya, M.; Twel, D. The R2R3 MYB transcription factor DUO1 activates a male germline-specific regulon essential for sperm cell differentiation in *Arabidopsis*. *Plant Cell* **2011**, *23*, 534–549. [CrossRef] [PubMed]

42. Aeschbacher, R.A.; Schrott, M.; Potrykus, I.; Saul, M.W. Isolation and molecular characterization of PosF21, an *Arabidopsis thaliana* gene which shows characteristics of a b-Zip class transcription factor. *Plant J.* **1991**, *1*, 303–316. [CrossRef] [PubMed]

43. Yin, Y.H.; Zhu, Q.; Dai, S.H.; Lamb, C.; Beachy, R.N. RF2a, a bZIP transcriptional activator of the phloem-specific rice tungro bacilliform virus promoter, functions in vascular development. *EMBO J.* **1997**, *16*, 5247–5259. [CrossRef] [PubMed]

44. Shen, H.; Cao, K.; Wang, X. A conserved proline residue in the leucine zipper region of AtbZIP34 and AtbZIP61 in *Arabidopsis thaliana* interferes with the formation of homodimer. *Biochem. Biophys. Res. Commun.* **2007**, *362*, 425–430. [CrossRef] [PubMed]

45. Gibalová, A.; Renák, D.; Matczuk, K.; Dupl'áková, N.; Cháb, D.; Twell, D.; Hongys, D. AtbZIP34 is required for *Arabidopsis* pollen wall patterning and the control of several metabolic pathways in developing pollen. *Plant Mol. Biol.* **2009**, *70*, 581–601.

46. Yang, P.; Han, J.; Huang, J. Transcriptome sequencing and de novo analysis of cytoplasmic male sterility and maintenance in JA-CMS cotton. *PLoS ONE* **2014**, *9*, e112320. [CrossRef] [PubMed]

47. Liu, Z.; Shi, X.; Li, S.; Hu, G.; Zhang, L.; Song, X. Tapetal-Delayed programmed cell death (PCD) and oxidative stress-induced male sterility of *Aegilops uniaristata* cytoplasm in wheat. *Int. J. Mol. Sci.* **2018**, *19*, 1708. [CrossRef] [PubMed]

48. Li, J.; Han, S.; Ding, X.; He, T.; Dai, J.; Yang, S.; Gai, J. Comparative transcriptome analysis between the cytoplasmic male sterile line NJCMS1A and its maintainer NJCMS1B in soybean (*Glycine max* (L.) Merr.). *PLoS ONE* **2015**, *10*, e0126771. [CrossRef] [PubMed]

49. Du, K.; Liu, Q.; Wu, X.; Jiang, J.; Wu, J.; Fang, Y.; Li, A.; Wang, Y. Morphological structure and transcriptome comparison of the cytoplasmic male sterility line in Brassica napus (SaNa-1A) derived from somatic hybridization and its maintainer line SaNa-1B. *Front. Plant Sci.* **2016**, *7*, 1313. [CrossRef] [PubMed]

50. Wang, S.; Wang, C.; Zhang, X.X.; Chen, X.; Liu, J.J.; Jia, X.F.; Jia, S.Q. Transcriptome de novo assembly and analysis of differentially expressed genes related to cytoplasmic male sterility in cabbage. *Plant Physiol. Biochem.* **2016**, *105*, 224–232. [CrossRef] [PubMed]

51. Liu, Q.; Lan, Y.; Wen, C.; Zhao, H.; Wang, J.; Wang, Y. Transcriptome sequencing analyses between the cytoplasmic male sterile line and its maintainer line in Welsh onion (*Allium fistulosum* L.). *Int. J. Mol. Sci.* **2016**, *17*, 1058. [CrossRef] [PubMed]

52. Zhang, G.; Ye, J.; Jia, Y.; Zhang, L.; Song, X. ITRAQ-based proteomics analyses of sterile/fertile anthers from a thermo-sensitive cytoplasmic male-sterile wheat with *Aegilops kotschyi* cytoplasm. *Int. J. Mol. Sci.* **2018**, *19*, 1344. [CrossRef] [PubMed]

53. Wu, Z.; Cheng, J.; Qin, C.; Hu, Z.; Yin, C.; Hu, K. Differential proteomic analysis of anthers between cytoplasmic male sterile and maintainer lines in *Capsicum annuum* L. *Int. J. Mol. Sci.* **2013**, *14*, 22982–22996. [CrossRef] [PubMed]

54. Mei, S.; Liu, T.; Wang, Z. Comparative transcriptome profile of the cytoplasmic male sterile and fertile floral buds of radish (*Raphanus sativus* L.). *Int. J. Mol. Sci.* **2016**, *17*, 42. [CrossRef] [PubMed]

55. Shu, J.; Liu, Y.; Li, Z.; Zhang, L.; Fang, Z.; Yang, L.; Zhuang, M.; Zhang, Y.; Lv, H. Organelle simple sequence repeat markers help to distinguish carpelloid stamen and normal cytoplasmic male sterile sources in broccoli. *PLoS ONE* **2015**, *10*, e0138750. [CrossRef] [PubMed]

56. Grabherr, M.G.; Haas, B.J.; Yassour, M.; Levin, J.Z.; Thompson, D.A.; Amit, I.; Adiconis, X.; Fan, L.; Raychowdhury, R.; Zeng, Q.; et al. Full-length transcriptome assembly from RNA-Seq data without a reference genome. *Nat. Biotechnol.* **2011**, *29*, 644–652. [CrossRef] [PubMed]

57. Zhang, Z.; Wood, W.I. A profile hidden Markov model for signal peptides generated by HMMER. *Bioinformatics* **2003**, *19*, 307–308. [CrossRef] [PubMed]

58. Ashburner, M.; Ball, C.A.; Blake, J.A.; Botstein, D.; Butler, H.; Cherry, J.M.; Davis, A.P.; Dolinski, K.; Dwight, S.S.; Eppig, J.T.; et al. Gene Ontology: Tool for the unification of biology. *Nat. Genet.* **2000**, *25*, 25–29. [CrossRef] [PubMed]

59. Kanehisa, M.; Araki, M.; Goto, S.; Hattori, M.; Hirakawa, M.; Itoh, M.; Katayama, T.; Kawashima, S.; Okuda, S.; Tokimatsu, T.; et al. KEGG for linking genomes to life and the environment. *Nucleic Acids Res.* **2008**, *36*, D480–D484. [CrossRef] [PubMed]

60. Mao, X.; Cai, T.; Olyarchuk, J.G.; Wei, L. Automated genome annotation and pathway identification using the KEGG Orthology (KO) as a controlled vocabulary. *Bioinformatics* **2005**, *21*, 3787–3793. [CrossRef] [PubMed]

61. Li, B.; Dewey, C. RSEM: Accurate transcript quantification from RNA-Seq data with or without a reference genome. *BMC Bioinform.* **2011**, *12*, 323. [CrossRef] [PubMed]

62. Trapnell, C.; Williams, B.A.; Pertea, G.; Mortazavi, A.; Kwan, G.; van Baren, M.J.; Salzberg, S.L.; Wold, B.J.; Pachter, L. Transcript assembly and quantification by RNA-Seq reveals unannotated transcripts and isoform switching during cell differentiation. *Nat. Biotechnol.* **2010**, *28*, 511–515. [CrossRef] [PubMed]

63. Storey, J.D.; Tibshirani, R. Statistical significance for genomewide studies. *Proc. Nat. Acad. Sci. USA* **2003**, *100*, 9440–9445. [CrossRef] [PubMed]

64. Benjamini, Y.; Hochberg, Y. Controlling the false discovery rate: A practical and powerful approach to multiple testing. *J. R. Stat. Soc. Ser. B Methodol.* **1995**, *57*, 289–300.

65. Wallenius, K.T. Biased Sampling: The Non-Central Hypegeometric Probability Distribution. Ph.D. Thesis, Stanford University, Stanford, CA, USA, 1963.

66. Young, M.D.; Wakefield, M.J.; Smyth, G.K.; Oshlack, A. Gene ontology analysis for RNA-Seq: Accounting for selection bias. *Genome Biol.* **2010**, *11*, R14. [CrossRef] [PubMed]
67. Gómez-Lobato, M.E.; Hasperuéb, J.H.; Civelloa, P.M.; Chaves, A.R.; Martínez, G.A. Effect of 1-MCP on the expression of chlorophyll degrading genes during senescence of broccoli (*Brassica oleracea* L.). *Sci. Hortic* **2012**, *144*, 208–211.
68. Livak, K.J.; Schmittgen, T.D. Analysis of relative gene expression data using real-time quantitative PCR and the $2^{-\Delta\Delta CT}$ Method. *Methods* **2001**, *25*, 402–408. [CrossRef] [PubMed]

International Journal of
Molecular Sciences

Article

Molecular Mapping of QTLs for Heat Tolerance in Chickpea

Pronob J. Paul [1,2], Srinivasan Samineni [1], Mahendar Thudi [1], Sobhan B. Sajja [1],
Abhishek Rathore [1], Roma R. Das [1], Aamir W. Khan [1], Sushil K. Chaturvedi [3],
Gera Roopa Lavanya [2], Rajeev. K. Varshney [1] and Pooran M. Gaur [1,4,*]

[1] International Crops Research Institute for the Semi-Arid Tropics (ICRISAT), Patancheru Hyderabad 502324,
    India; pronobjpaul@gmail.com (P.J.P.); s.srinivasan@cgiar.org (S.S.); t.mahendar@cgiar.org (M.T.);
    S.Sobhan@cgiar.org (S.B.S.); a.rathore@cgiar.org (A.R.); r.das@cgiar.org (R.R.D.); A.khan@cgiar.org (A.W.K.);
    r.k.varshney@cgiar.org (R.K.V.)
[2] Department of Genetics and Plant Breeding, Sam Higginbottom University of Agriculture,
    Technology and Sciences (SHUATS), Allahabad 211007, India; lavanya.roopa@gmail.com
[3] ICAR-Indian Institute of Pulses Research (ICAR-IIPR), Kanpur 208024, India; sushilk.chaturvedi@gmail.com
[4] The UWA Institute of Agriculture, University of Western Australia, Perth, WA 6009, Australia
* Correspondence: p.gaur@cgiar.org; Tel.: +91-40-3071-3356

Received: 19 June 2018; Accepted: 19 July 2018; Published: 25 July 2018

Abstract: Chickpea (*Cicer arietinum* L.), a cool-season legume, is increasingly affected by
heat-stress at reproductive stage due to changes in global climatic conditions and cropping
systems. Identifying quantitative trait loci (QTLs) for heat tolerance may facilitate breeding
for heat tolerant varieties. The present study was aimed at identifying QTLs associated with
heat tolerance in chickpea using 292 F_{8-9} recombinant inbred lines (RILs) developed from the
cross ICC 4567 (heat sensitive) × ICC 15614 (heat tolerant). Phenotyping of RILs was undertaken
for two heat-stress (late sown) and one non-stress (normal sown) environments. A genetic map
spanning 529.11 cM and comprising 271 genotyping by sequencing (GBS) based single nucleotide
polymorphism (SNP) markers was constructed. Composite interval mapping (CIM) analysis revealed
two consistent genomic regions harbouring four QTLs each on CaLG05 and CaLG06. Four major
QTLs for number of filled pods per plot (FPod), total number of seeds per plot (TS), grain yield per
plot (GY) and % pod setting (%PodSet), located in the CaLG05 genomic region, were found to have
cumulative phenotypic variation of above 50%. Nineteen pairs of epistatic QTLs showed significant
epistatic effect, and non-significant QTL × environment interaction effect, except for harvest index
(HI) and biomass (BM). A total of 25 putative candidate genes for heat-stress were identified in the
two major genomic regions. This is the first report on QTLs for heat-stress response in chickpea.
The markers linked to the above mentioned four major QTLs can facilitate marker-assisted breeding
for heat tolerance in chickpea.

Keywords: abiotic stress; *Cicer arietinum*; candidate genes; genetics; heat-stress; molecular breeding

1. Introduction

In recent years, the adverse impact of climate change on agriculture is well recognized all over
the globe. The ever-increasing day and night temperature is going to affect the production of crops,
especially those grown in the winter [1]. In this context, heat-stress due to rise in temperatures remains
a challenge in developing crop varieties that are adaptive to changing climatic conditions.

Chickpea is a nutrient-rich grain legume crop cultivated in arid and semi-arid regions.
The chickpea grain is an excellent source of proteins along with a wide range of essential amino
acids and vitamins. In the fight against hidden hunger all over the globe, the role of legumes

such as chickpea is indispensable. Grown in over 60 countries and traded in over 190 countries, chickpea is the second most consumed pulse crop in the world after common bean [2]. Due to global warming, several noticeable changes occurred in the cropping system and intensity in the recent past. These are delaying the cultivation of chickpea to relatively hot conditions [1]. Generally, the crop faces heat-stress during reproductive phase under late sown condition in the tropical and semi-arid regions [3]. Reports state that the exposure to temperature, 35 °C and above, even for a few days, during reproductive phase has a negative impact on optimum yield in chickpea [4,5]. Unlike drought and other abiotic stresses, until recently, the importance of breeding for heat-stress conditions in chickpea has not been realized [1].

Grain yield under heat-stress is considered to be one of the important criteria for assessing heat tolerance in chickpea [3–5]. However, chickpea yield is known to be highly influenced by environments [6]. Due to genotype by environment (G × E) interaction, breeding for heat tolerance through conventional breeding approaches based on yield parameter sometimes limits selection for heat-stress tolerance in chickpea.

In recent years, progress has been made in genomics-enabled trait dissection in several crop plants, including chickpea. Several studies have been carried out earlier to identify the quantitative trait loci (QTLs) for tolerance to various biotic stresses [7,8], and abiotic stresses like drought tolerance [9], and salinity tolerance [10–12] in chickpea. Moreover, genomic regions associated with heat tolerance have been reported in several crops, including wheat, rice, maize, barley, potato, tomato, cowpea, azuki bean, brassica [13]. Pod setting (seed set) and grain yield have been used as proxy traits to detect QTLs for heat tolerance in different crops [14–18]. Similarly, in chickpea, the number of filled pods, total number of seeds, biomass, and harvest index were found to be significantly associated with heat tolerance [3,19]. However, to date, QTLs for heat tolerance have not been reported in chickpea.

In this study, genotyping by sequencing (GBS)-based single nucleotide polymorphism markers were used to identify key genomic regions responsible for heat tolerance. In addition, putative candidate genes for heat tolerance in these genomic regions were identified using the available chickpea genome sequence information [20].

2. Results

2.1. Response of Parents and Recombinant Inbred Lines (RILs) under Heat-Stress and Non-Stress Environments

The descriptive analysis of parents and RILs are presented in Table 1. Predicted means for all the traits in parents differed significantly in both heat-stress environments, except biomass per plot (BM). In the non-stress environment, predicted means for grain yield per plot (GY), BM, harvest index (HI) and %PodSet were non-significant between parents, while filled pods per plot (FPod) and total number of seeds per plot (TS) were significant. The range of variation in all the traits was high in stress environments (Table 1). The combined analysis of variance (ANOVA) for both the stress environments revealed that significant variation existed in RILs for all the traits measured, except BM, whereas under non-stress environment relatively low genetic variability was observed. Transgressive segregants in both directions were observed for several traits in the RIL population (Figure 1a,b).

The potential use of a trait in a breeding program relies on the heritability of that trait. Under both the heat-stress conditions, the heritability of all the traits was high (72.0–90.7%), except BM in summer 2014 (49.8%). Whereas, under non-stress environment the heritability of the traits was moderate (47.6–66.0%) (Table 1).

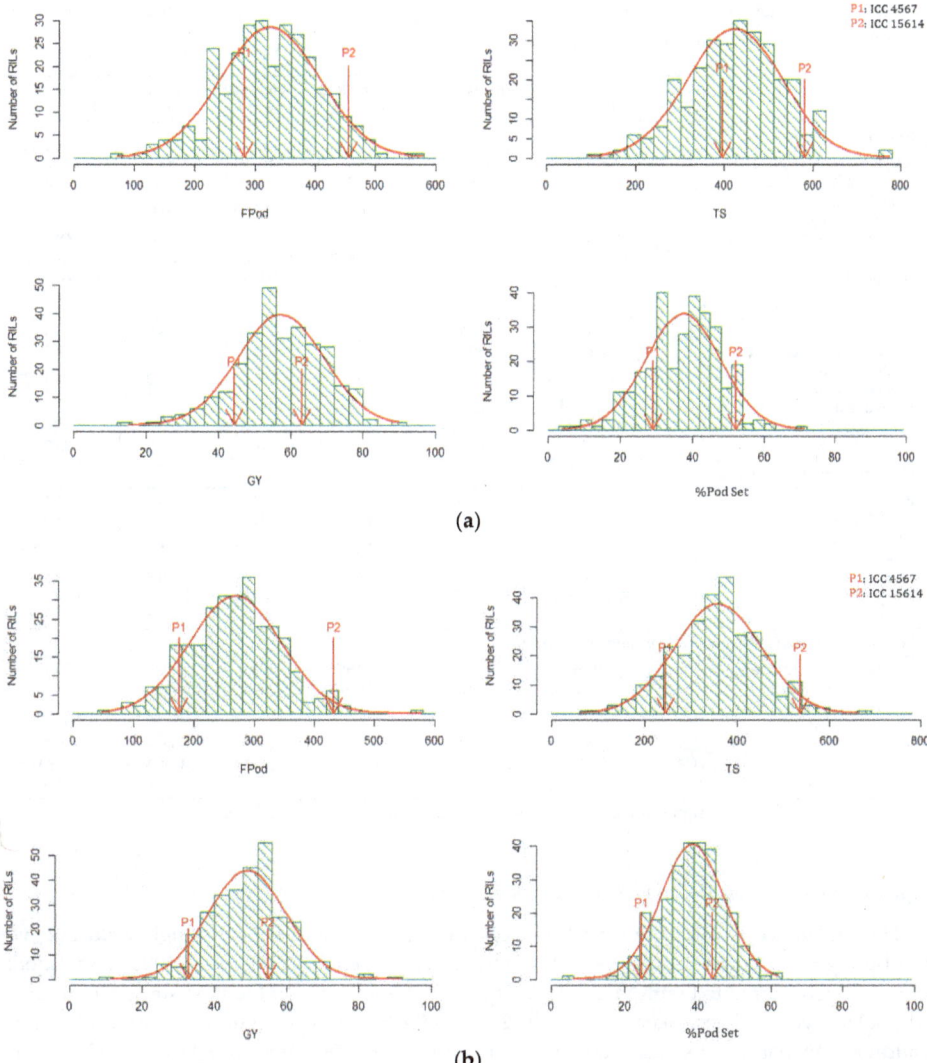

Figure 1. (**a**) Frequency distribution of Number of Filled Pods per Plot (FPod), Total Number of Seeds per Plot (TS), Grain Yield per Plot (GY, g), and Percent Pod Setting (%PodSet) in RIL population (ICC 4567 × ICC 15614). P1 is heat sensitive parent ICC 4567 and P2 is heat tolerant parent ICC 15614. The left portion of the P1 on the *X*-axis indicates the negative transgressive segregants, conversely, the right portion of the P2 on the *X*-axis indicates the positive transgressive segregants in heat-stress environment, 2013; (**b**) Frequency distribution of Number of Filled Pods per Plot (FPod), Total Number of Seeds per Plot (TS), Grain Yield per Plot (GY, g), and Percent Pod Setting (%PodSet) in RIL population (ICC 4567 × ICC 15614). P1 is heat sensitive parent ICC 4567 and P2 is heat tolerant parent ICC 15614. The left portion of the P1 on the *X*-axis indicates the negative transgressive segregants, conversely, the right portion of the P2 on the *X*-axis indicates the positive transgressive segregants in heat-stress environment, 2014.

Table 1. Summary statistics and heritability (H^2) values for the measured traits of 292 RILs in non-stress and heat-stress environments.

Trait	Visual Score	Filled Pods Plot^{-1}	Total No. of Seeds Plot^{-1}	Grain Yield Plot^{-1} (g)	Biomass Plot^{-1} (g)	Harvest Index	Percent PodSet (%)
Non-stress Environment, 2013							
ICC 4567 (heat sensitive)	-	406.8	429.2	76.0	144.8	52.1	67.7
ICC 15614 (heat tolerant)	-	538.7	553.0	70.2	132.3	53.9	75.6
Contrast analysis between parents	-	−131.9 *	−123.9 *	5.8 ^{ns}	12.5 ^{ns}	−1.9 ^{ns}	−7.9 ^{ns}
Mean of RILs	-	459.0	486.3	73.5	139.7	53.0	68.8
Range of RILs	-	360.8–580.1	378.3–604.7	57.6–93.3	118.1–165.2	45.5–59.2	48.1–84.2
Heritability (%)	-	62.1	60.5	57.6	47.6	63.4	66.0
Heat-stress environment, 2013							
ICC 4567 (heat sensitive)	2	281.3	395.1	44.3	147.6	34.2	28.8
ICC 15614 (heat tolerant)	5	455.6	580.7	62.9	125.9	50.6	52.0
Contrast analysis between parents	−0.5 *	−174.3 *	−185.6 *	−18.6 *	21.7 ^{ns}	−16.3 *	−23.1 *
Mean of RILs	3.0	323.9	421.3	57.1	114.6	50.6	37.3
Range of RILs	(1–5)	70.5–578.3	91.9–772.4	14.9–89.8	32.9–185.6	34.5–69.1	3.7–71.3
Heritability (%)	79.8	86.9	86.3	82.2	83.2	72.0	90.7
Heat-stress environment, 2014							
ICC 4567 (heat sensitive)	2	175.3	242.0	32.6	123.2	23.9	24.4
ICC 15614 (heat tolerant)	5	431.2	534.9	54.8	111.6	52.0	43.9
Contrast analysis between parents	−0.6 *	−255.9 *	−292.9 *	−22.1 *	11.7 ^{ns}	−28.2 *	−19.6 *
Mean of RILs	3.0	268.0	355.7	49.0	119.7	40.9	38.4
Range of RILs	(1–5)	46.9–576.8	61.8–665.8	11.0–91.6	65.4–142.4	12.8–63.4	5.8–61.6
Heritability (%)	86.5	86.8	86.6	80.9	49.8	91.3	84.7
Pooled environments (Heat-stress environments, 2013 and 2014)							
ICC 4567 (heat sensitive)	2	201.6	278.1	37.5	134.8	28.6	26.1
ICC 15614 (heat tolerant)	5	453.6	570.3	59.6	116.4	51.2	48.7
Contrast analysis between parents	−0.6 *	−252 *	−292.2 *	−22 *	18.4 ^{ns}	−22.6 *	−22.6 *
Mean of RILs	3.0	296.0	388.5	53.0	117.2	45.8	37.9
Range of RILs	(1–5)	42.2–516	54.9–672.5	9.01–82.3	37.14–157.5	24.13–58.8	2.61–63.9
Heritability (%)	72.2	81.6	82.3	73.1	19.2	NA	81.6

* significant at *p* = 0.05, ns = Not significant, NA = Not available.

2.2. Relationship between Yield and Yield Determining Traits

Heat tolerance is a complex trait and can be estimated indirectly through yield and yield contributing traits under heat-stress. All the traits- visual score (VS), FPod, TS, BM and %PodSet were positively correlated with yield (*r* = 0.51 **–0.90 **) under both the heat-stress environments and pooled over analysis except HI (*r* = 0.32 **) under heat-stress environment of 2013 (Table 2). In addition, VS had positive association with FPod (*r* = 0.68 **–0.80 **) and TS (*r* = 0.67 **–0.79 **). Likewise, %PodSet was found to have a strong positive correlation with FPod (*r* = 0.59 **–0.77 **) and TS (*r* = 0.60 **–0.78 **) under both the heat-stress environments as well as in pooled analysis (Table 2). In contrast, under non-stress environment, the correlation with yield was low for %PodSet (*r* = 0.17 **) and HI (*r* = 0.33 **), and high for other traits (*r* = 0.63 **–0.91 **) (Table 2). Regression analysis between the traits and yield revealed that all the traits exhibited medium to high variation for yield (25% to 81%) in both stress environments as well as pooled over years (Figure S3a–c). In non-stress environment, %PodSet had low contribution (3%) whereas BM was found to have high variation for yield (82%) (Figure S3d). A significant correlation between the yield and yield contributing traits under heat-stress environment indicated that these traits can be used in direct or indirect selection for improving heat tolerance in chickpea.

Table 2. Correlation among the different traits evaluated in RIL population in two heat-stress environments, non-stress environment and pooled over years.

Environments	Traits	VS	FPod	TS	BM	HI	%PodSet	GY
HSE-2013	VS	1						
HSE-2014	VS	1						
Pooled years	VS	1						
HSE-2013	FPod	0.68 **	1					
HSE-2014	FPod	0.78 **	1					
Pooled years	FPod	0.80 **	1					
HSE-2013	TS	0.67 **	0.97 **	1				
HSE-2014	TS	0.78 **	0.96 **	1				
Pooled years	TS	0.79 **	0.97 **	1				
HSE-2013	BM	0.69 **	0.70 **	0.68 **	1			
HSE-2014	BM	0.15 **	0.40 **	0.38 **	1			
Pooled years	BM	0.61 **	0.67 **	0.65 **	1			
HSE-2013	HI	−0.04 ns	0.22 **	0.25 **	−0.35 **	1		
HSE-2014	HI	0.83 **	0.84 **	0.84 **	0.08 ns	1		
Pooled years	HI	0.62 **	0.70 **	0.72 **	0.24 **	1		
HSE-2013	%PodSet	0.63 **	0.72 **	0.73 **	0.62 **	0.00	1	
HSE-2014	%PodSet	0.61 **	0.59 **	0.60 **	0.05 **	0.62 **	1	
Pooled years	%PodSet	0.71 **	0.77 **	0.78 **	0.50 **	0.59 **	1	
HSE-2013	GY	0.66 **	0.88 **	0.89 **	0.74 **	0.32 **	0.63 **	1
HSE-2014	GY	0.73 **	0.90 **	0.89 **	0.57 **	0.84 **	0.50 **	1
Pooled years	GY	0.79 **	0.89 **	0.88 **	0.78 **	0.76 **	0.69 **	1
	Traits	FPod	TS	BM	HI	%PodSet	GY	
NSE-2013	FPod	1						
NSE-2013	TS	0.94 **	1					
NSE-2013	BM	0.60 **	0.63 **	1				
NSE-2013	HI	0.15 **	0.22 **	−0.07 ns	1			
NSE-2013	%PodSet	0.23 **	0.27 **	0.17 **	0.05 ns	1		
NSE-2013	GY	0.63 **	0.69 **	0.91 **	0.33 **	0.17 **	1	

** Significant at $p < 0.01$, respectively. ns: Non-significant. HSE-2013: Heat-stress environment—2013; HSE-2014: Heat-stress environment-2014; NSE-2013: Non-stress environment-2013; Pooled years: Pooled over HSE-2013 and HSE-2014; VS, Visual Score; FPod, Number of Filled Pods per Plot; TS, Total Number of Seeds Per Plot; BM, Biomass; HI, Harvest Index; %PodSet, Percentage Pod Setting; GY, Grain Yield per Plot.

2.3. Sequencing Data and SNP Discovery

The parents of the mapping population (ICC 4567 × ICC 15614) were sequenced at higher depth (5× coverage), and a total of 19.63 million reads containing 1.70 Gb for ICC 4567, and 15.79 million reads containing 1.37 Gb for ICC 15614, were generated. In addition, 3333.41 million reads containing 289.70 Gb were generated from 292 RILs. The number of reads generated varied from 6.86 million (RIL099) to 20.66 million (RIL112) with an average of 11.42 million per line. The single nucleotide polymorphisms (SNPs), identified using the software SOAP, were analyzed to remove heterozygous SNPs in the parents, and a set of 396 SNPs were identified across 292 RILs. The sequence details of all SNPs have been provided in Table S1a,b.

2.4. Genetic Linkage Map and Marker Distribution

The 396 polymorphic SNPs obtained from GBS were used for genetic map construction. The genetic linkage map covered 529.11 cM of the chickpea genome with an average interval of 1.95 cM between markers (Table S2 and Figure S1). The highest number of markers was in CaLG04 (57), while the lowest number of markers was in CaLG08 (10) (Figure S1). CaLG08 showed the highest marker density with 1.78 markers per cM on average. The lowest marker density was observed for CaLG02, which had 0.29 markers per cM on average. Overall, the map had on average 0.51 markers per cM (Table S2).

2.5. QTL Analysis

2.5.1. Genomic Region on CaLG05

A promising genomic region harbouring major QTLs for four traits—FPod, TS, GY, and %PodSet flanked by markers Ca5_44667768 and Ca5_46955940—was identified on CaLG05 (Table 3). The four QTLs—*qfpod02_5*, *qts02_5*, *qgy02_5*, and *q%podset06_5*—were found in both the stress environments spanning 6.9 cM (corresponding to ~2.28 Mb on physical map) (Figure 2a). The phenotypic variation for GY-QTL (*qgy02_5*) was 16.04% (LOD 11.69) and 16.56% (LOD 12.00) in heat-stress environments I (2013) and II (2014), respectively. QTLs for FPod—*qfpod02_5* in this genomic region demonstrated phenotypic variation of 11.57% (LOD 8.37) and 12.03% (LOD 7.79), respectively, in the consecutive stress environments (Table 3). Similarly, QTLs for the TS *qts02_5* in heat-stress environments I (2013) and II (2014) explained phenotypic variation of 12.0% (LOD 8.54) and 10.0% (LOD 7.30). The QTL for %PodSet (*q% podset06_5*), which has been considered as an important selection criterion for heat tolerance in chickpea, had a phenotypic variation of 11.51% (LOD 8.04) and 13.30% (LOD 9.20) in the heat-stress environments of 2013 and 2014, respectively (Table 3).

All the major QTLs present in the genomic region of CaLG05 were found to exist in the pooled analysis for the two stress environments (Table 3). In CaLG05, two major QTLs for VS and HI were found explaining 15.1% (LOD 11.1) and 18.5% (LOD 13.0) of phenotypic variation, under the heat-stress environment (2014), respectively (Table S3). In contrast, during the stress environment in 2013, one major QTL for VS was found close to the genomic region on CaLG05 with a phenotypic variation of 13.88% (LOD 12.05) (Table S3). Through single marker analysis (SMA), Ca5_44667768 was co-segregated with the four major QTLs in this genomic region.

Table 3. Identification of QTLs associated with heat tolerance in ICC 4567 × ICC 15614 derived RIL population.

LG	Marker Interval	Trait	QTL Name	Heat-Stress Environment, 2013				Heat-Stress Environment, 2014				Pooled Environments			
				Position (cM)	%PVE	LOD	Add	Position (cM)	%PVE	LOD	Add	Position (cM)	%PVE	LOD	Add
CaLG05	Ca5_44667768-Ca5_46955940	FPod	qfpod02_5	4.41	11.57	8.37	27.93	5.41	12.03	7.79	27.31	5.41	12.03	9.41	28.83
		TS	qts02_5	5.41	12.00	8.54	36.14	5.41	10.00	7.30	31.27	5.41	10.00	9.07	35.27
		GY	qgy02_5	4.41	16.04	11.69	4.72	4.41	16.56	12.00	4.61	4.41	16.56	13.17	4.64
		%PodSet	q%podset06_5	6.41	11.51	8.04	3.47	6.41	13.30	9.20	3.40	6.41	13.30	9.48	3.47
CaLG06	Ca6_7846335-	VS	qvs05_6	62.41	11.07	9.79	0.05	61.51	9.04	7.26	0.06	61.51	9.04	9.54	0.06
		FPod	qfpod03_6	62.41	6.56	5.10	20.88	63.40	5.92	4.10	19.01	62.41	5.92	5.22	19.91
	Ca6_14353624	GY	qgy03_6	62.41	4.43	3.68	2.48	62.41	3.92	3.21	2.24	62.41	3.92	3.58	2.24
		%PodSet	q%podset08_6	63.41	8.44	6.22	3.00	65.41	6.96	4.61	2.46	64.41	6.96	5.97	2.77

VS, Visual Score; FPod, Number of Filled Pods per Plot; TS, Total Number of Seeds per Plot; %PodSet, Percentage Pod Setting; GY, Grain Yield per Plot; %PVE, Percentage of Phenotypic Variance Explained; Add, additive effect, where a positive value indicates that ICC 15614 allele was favorable, and a negative value ICC 4567 allele was favorable; LOD, likelihood of Odds Ratio; LG, Linkage Group.

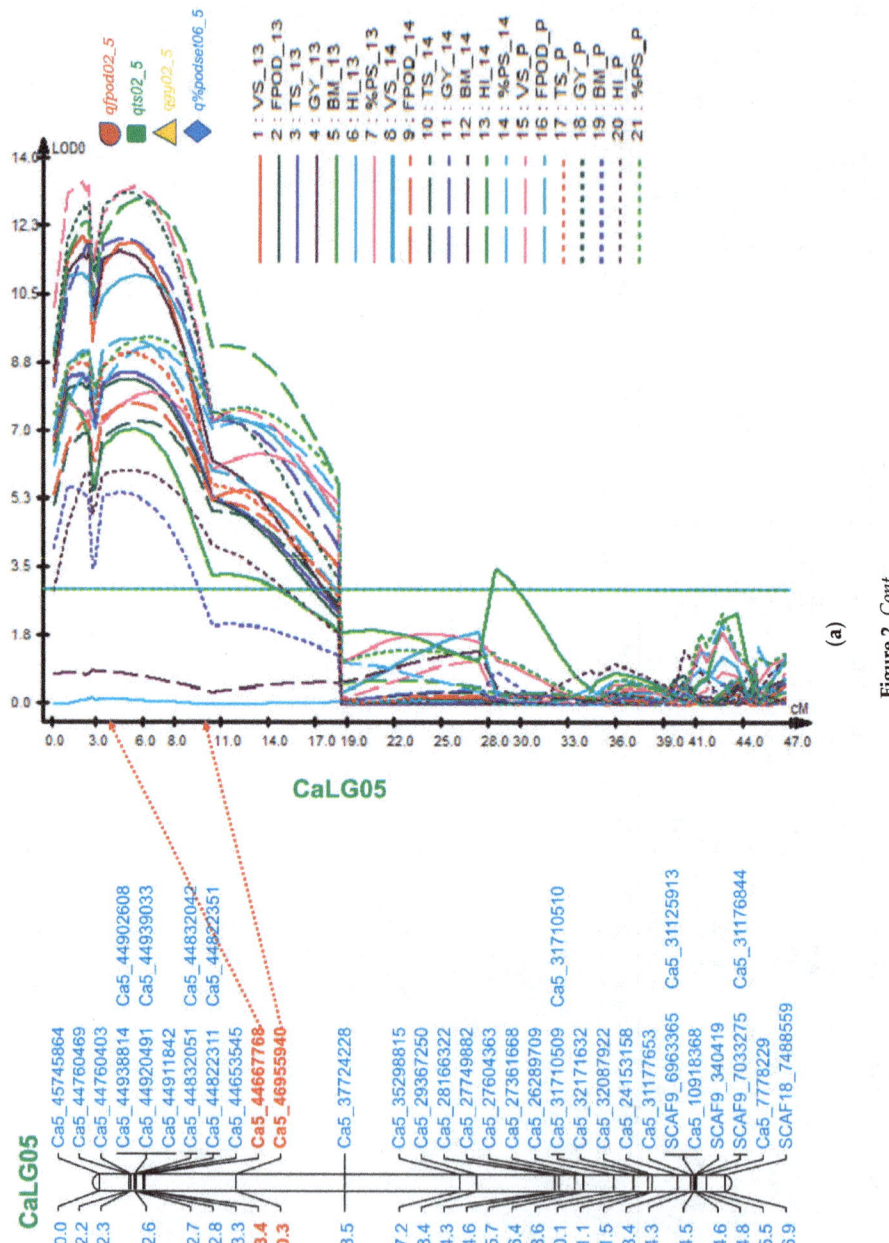

(a)

Figure 2. *Cont.*

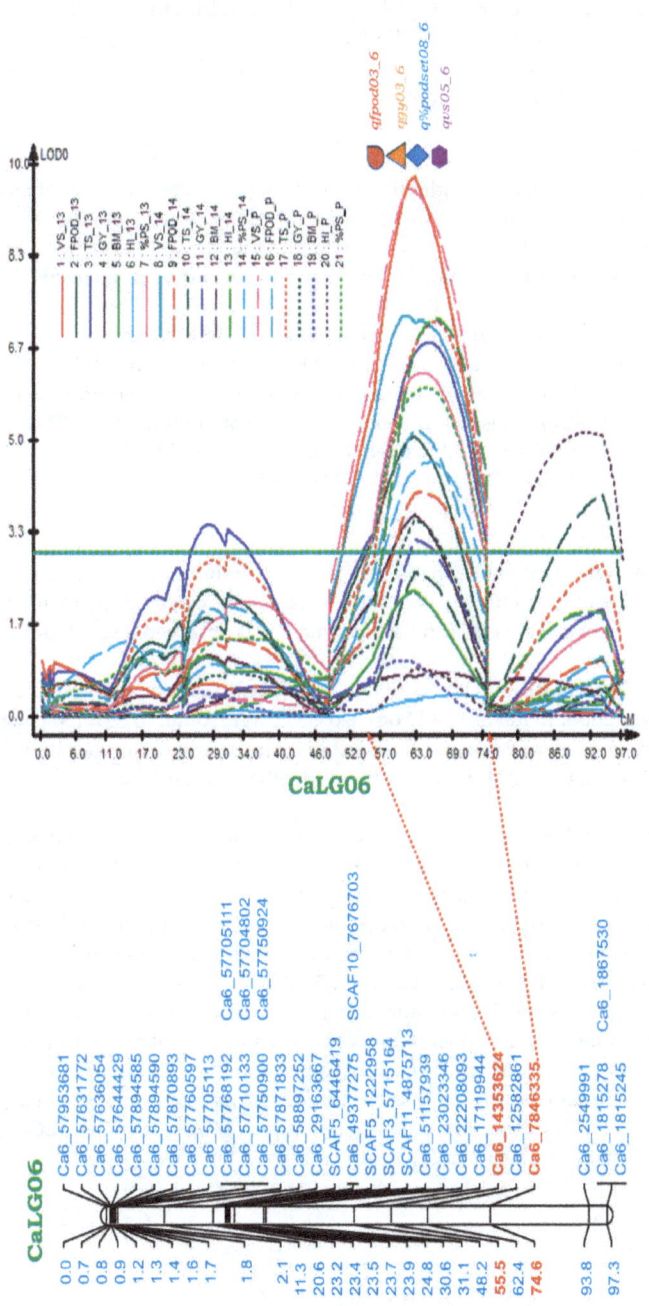

Figure 2. (**a**) Likelihood of odds ratio (LOD) curves obtained by composite interval mapping for quantitative trait loci (QTL) mapped over two heat-stress environments, 2013, 2014 and their pooled years together. Four major QTLs–*qfpod02_5*, *qts02_5*, *qgy02_5*, *q%podset06_5* of the four traits–Number of Filled Pods per Plot (FPod), Total Number of Seeds per Plot (TS), Grain Yield per Plot (GY) and Percent Pod Setting (%PodSet) in the genomic region on CaLG05 flanked by markers Ca5_44667768 and Ca5_46955940. The vertical lines indicate the threshold LOD value (2.5) determining significant QTL; (**b**) Likelihood of odds ratio (LOD) curves obtained by composite interval mapping for quantitative trait loci (QTL) mapped over two heat-stress environments, 2013, 2014 and their pooled years together. Four QTLs, *qfpod03_6*, *qgy03_6*, *q% podset08_6*, *qos05_6* for the traits Number of Filled Pods per Plot (FPod), Grain Yield per Plot (GY), Percent Pod Setting (%PodSet) and visual score on podding behaviour (VS) in the genomic region on CaLG06 with the marker interval Ca6_14353624-Ca6_7846335, in the RIL mapping population of ICC 4567 × ICC 15614. The vertical lines indicating the threshold LOD value (2.5) determining significant QTL.

2.5.2. Genomic Region on CaLG06

A second genomic region, harbouring QTLs for four important traits in this study, was identified having the marker interval Ca6_14353624—Ca6_7846335 (Table 3 and Figure 2b). The QTLs for FPod (*qfpod03_6*), GY (*qgy03_6*), %PodSet (*q% podset08_6*), and VS (*qvs05_6*) spanned a genetic length of 19.14 cM (~6.50 Mb on physical map) in CaLG06. The range of phenotypic variation shown by various traits in this genomic region was from 3.92 to 11.07% (Table 3).

2.5.3. QTLs Identified on Other LGs

In the present work, a total of 13 QTLs were identified consistently across two heat-stress environments showing both major and minor effects for various traits measured. Apart from the QTLs identified in CaLG05 and CaLG06, a QTL for GY (*qgy01_1*) was found in the same position (40.0 cM) demonstrating 7.33% and 10% of phenotypic variation in the first and second year, respectively, on CaLG01 (Table S4).

On CaLG02, QTL for FPod (*qfpod01_2*) occurred at the same position (65.81 cM) in consecutive years with a phenotypic variation of 4.9% (LOD 3.38) and 5.8% (LOD 4.0). Similarly, QTL for TS (*qts01_2*) was found explaining 5.6% and 8.1% phenotypic variation under heat-stress environments (2013 and 2014), respectively. A major QTL (*q%podset03_4*) with phenotypic variation 12.5% (LOD 4.72) for %PodSet in 2013 was also observed in 2014 with 7.8% phenotypic variation and LOD value of 3.6 with same marker interval (Ca4_13699195-Ca4_7818876) on CaLG04 (Table S4).

2.5.4. Mapping of Epistatic QTLs (E-QTLs)

Epistatic interaction analysis revealed that 19 QTL pairs were involved in the epistatic interactions covering seven LGs (Table 4). A significant effect was observed for all the epistatic interactions. However, no significant interaction between epistasis and environment was observed, except for the trait biomass (BM).

Two epistatic QTL pairs for VS were found to have loci distributed on four different LGs accounting for 3.43% phenotypic variation. In the case of FPod, two QTLs were found to be interacting in the same LG, CaLG02. Another QTL pair was found for FPod to interact with each other in two different LGs (Table 4). These two epistatic QTL pairs for FPod together explained a phenotypic variation of 2.94%.

The highest number of epistatic QTL pairs (nine pairs) were detected for TS in this population and have contributed up to 12.38%. The epistatic interaction for TS was found in all the linkage groups, except CaLG03 and CaLG07. One QTL interaction pair was detected for GY interacting from CaLG01 to the locus on CaLG02 with a phenotypic variation 0.83% (Table 4 and Figure S2). Similarly, in the case of %PodSet, four epistatic QTL pairs were found to interact with each other in three linkage groups CaLG01, CaLG03, and CaLG04 showing a phenotypic variation of 5.79%.

In addition, an interaction between non-QTL, and additive and additive × environment-QTL was found in the case of BM, which showed 1.22% phenotypic variation. Concurrently, five loci (loci located at 10.1 cM and 26.4 cM in CaLG01, 2.2 cM and 75.6 cM in CaLG04, and at 44.5 cM in CaLG05) were observed to have interaction simultaneously with several other loci affecting the expression of the particular trait. Two loci (*eqts2_1/eqpodset2_1* in CaLG01 and *neqfpod4_5/neqts9_5* in CaLG05) controlling two or three different traits were also interacted with other loci (Table 4).

Table 4. Epistatic effect, and epistatic × environment interaction QTL found in RIL population (ICC 4567 × ICC 15614) in two heat-stress environments, 2013 and 2014.

SL. No.	Trait	QTL_i	LG	Marker Interval (QTL.i)	Position (QTL_i)	QTL_j	LG	Marker Interval (QTL.j)	Position (QTL_j)	AA	h² (%) (AA)	h² (%) (AAE)
1	VS	eqts1_1	1	Ca1_1732919-Ca1_4429044	48.5	eqts4_7	7	Ca7_3634430-Ca7_6584610	4.6	−0.02 ***	1.02	0.12
2	VS	neqts2_4	4	Ca4_48498166-Ca4_48498181	2.6	neqts3_5	5	Ca5_29367250-Ca5_28166322	30.4	0.03 ***	2.41	0.17
3	FPod	eqfpod1_2	2	Ca2_24709295-Ca2_30876552	30.7	eqfpod2_2	2	Ca2_34481663-Ca2_35860429	64.8	−8.85 ***	0.73	0.01
4	FPod	neqfpod3_4	4	Ca4_48497765-Ca4_48458381	2.2	neqfpod4_5/neqts9_5	5	SCAF9_6963365-Ca5_31125913	44.5	13.10 ***	2.21	0.01
5	TS	eqts1_1	1	Ca1_11321839-Ca1_11411540	10.8	eqts11_6	6	Ca6_51157939-Ca6_23023346	27.8	13.15 ***	0.42	0.01
6	TS	eqts2_1/eqpodset2_1	1	Ca1_39746426-Ca1_34727065	26.4	eqts14_8	8	Ca8_14753681-Ca8_14587797	5.6	9.78 ***	0.46	0.02
7	TS	eqts4_2	2	Ca2_34481663-Ca2_35860429	65.8	eqts12_6	6	Ca6_12582861-Ca6_7846335	62.4	−9.79 ***	0.38	0.05
8	TS	eqts4_2	2	Ca2_34481663-Ca2_35860429	65.8	eqts14_8	8	Ca8_14753681-Ca8_14587797	5.6	16.97 ***	0.96	0.01
9	TS	eqts7_5	5	Ca5_45745864-Ca5_44760469	2	eqts13_6	6	Ca6_2549991-Ca6_1815278	93.8	−8.86 ***	0.6	0.00
10	TS	eqts2_1/eqpodset2_1	1	Ca1_39746426-Ca1_34727065	26.4	neqts10_6	6	Ca6_58897252-Ca6_29163667	14.4	17.68 ***	2.22	0.03
11	TS	neqts3_2	2	Ca2_32483185-Ca2_32979328	47.7	neqts6_4	4	Ca4_47243660-Ca4_44753224	22.3	13.47 ***	2.12	0.01
12	TS	neqts5_4	4	Ca4_48458381-Ca4_48475589	2.2	neqts8_5	5	Ca5_27604363-Ca5_273361668	35.7	10.76 ***	2.52	0.03
13	TS	neqts5_4	4	Ca4_48458381-Ca4_48475589	2.2	neqts9_5/neqfpod4_5	5	SCAF9_6963365-Ca5_31125913	44.5	12.02 ***	2.7	0.00
14	GY	eqgy1_1	1	Ca1_1732919-Ca1_4429044	45.5	eqgy2_2	2	Ca2_34481663-Ca2_35860429	63.8	1.41 ***	0.83	0.01
15	BM	aacqbm1_1	1	Ca1_11685790-Ca1_11372972	9.1	neqbm1_3	3	Ca3_24194574-Ca3_22539683	52.9	−2.09 ***	1.22	0.21
16	%PodSet	eqpodset1_1	1	Ca1_11685790-Ca1_11372972	10.1	eqpodset6_4	4	Ca4_13699195-Ca4_7818876	75.6	−1.33 ***	0.83	0.01
17	%PodSet	eqpodset2_1/eqts2_1	1	Ca1_39746426-Ca1_34727065	26.4	eqpodset6_4	4	Ca4_13699195-Ca4_7818876	75.6	1.89 ***	0.99	0.03
18	%PodSet	eqpodset1_1	1	Ca1_11685790-Ca1_11372972	10.1	neqpodset4_4	4	Ca4_48478303-Ca4_48475461	2.5	−1.38 ***	2.13	0.02
19	%PodSet	neqpodset3_3	3	Ca3_9400875-SCAF14_6484051	63.2	neqpodset5_4	4	Ca4_48269138-Ca4_47243656	11	−1.44 ***	1.84	0.00

VS, Visual Score; FPod, Number of Filled Pods per Plot; TS, Total Number of Seeds per Plot; GY, Grain Yield per Plot; BM, Biomass; %PodSet, Percentage Pod Setting. QTL_i and QTL_j, the two QTL/non-QTL involved in epistatic interaction; AA, additive × additive effect interactions; AAE, epistatic × environment effect interactions; h² (AA): the contribution rate of additive × additive effect interactions; h² (AAE): the contribution rate of epistatic × environment effect interactions. *** Significant at the 0.001 probability level. The underlined QTLs denotes those with an additive effect. *eqpodset2_1/eqts2_1* or *eqts2_1/eqpodset2_1* and *neqts9_5/neqfpod4_5* or *neqfpod4_5/neqts9_5* indicates co-localized loci.

3. Discussion

3.1. Phenotypic Evaluation of RILs and Parents in Field Condition

Sowing during the month of February proved to be an ideal condition to expose chickpea crop to heat-stress and selecting heat tolerance lines in earlier studies under field conditions at ICRISAT, Patancheru, India [19,21]. A recent study on chickpea reported 34 °C as the threshold temperature for pod setting and also observed that at 35 °C, pod set was reduced by 50% in chickpea genotypes [19]. The average maximum temperatures (37.5 °C and 36.7 °C in summer 2013 and summer 2014, respectively) in both the heat-stress environments found were ideal for phenotyping RIL population. An average maximum temperature of 29.4 °C was recorded in non-stress environment, which was considered as control for this study. This temperature was ideal for sowing in the non-stress environment for the timely sown crop [22].

The frequency distribution of measured traits showed the characteristics of continuous variation (Figure 1a,b). Paliwal et al. (2012) [23] in RILs of wheat and Buu et al. (2014) [24] in BC$_2$F$_2$ population in rice, reported several transgressive segregants for heat tolerance. Similarly, in this present study, transgressive segregants in both directions were observed, indicating that both parents have contributed alleles for heat tolerance in the RILs (Figure 1a,b). A significant variation found among the RILs for all the traits indicate the presence of genetic diversity in the selected parents for the selected traits under heat-stress condition. Parents differed significantly for all the traits in both the heat-stress environments, except biomass (BM).

High heritability (H^2) values were observed for all the traits measured under both the heat-stress environments, except for biomass in summer 2014, which indicates that there is a high probability of achieving the same kind of results if the trial is repeated under similar growing conditions.

Yield under high temperatures is the key objective for heat tolerance breeding in chickpea. Traits such as FPod, %PodSet and TS contributing to increased yield under high-temperature stress can be treated as a proxy for heat tolerance. The presence of significant correlations between yield and other traits in heat-stress environments indicated that these traits can be used as selection criteria for heat tolerance.

FPod and TS had a strong correlation with yield (88 to 90%) under both the stress environments. Such high correlation of these traits toward yield was reported earlier in chickpea under abiotic stress [10,11]. In addition, VS and %PodSet was also found to have good correlation (50 to 79%) with yield. However, BM and HI showed large difference in correlation with yield in both the heat-stress conditions. Positive and strong association of the four traits-FPod, TS, VS and %PodSet with grain yield revealed the importance of these traits in determining yield under heat-stress environment. Hence, detecting QTLs of these traits under stress would be helpful in heat tolerance programme.

3.2. QTL Mapping for Heat Tolerance

The genomic region in CaLG05 harbours QTLs for FPod, TS, GY, and %PodSet, which were reportedly associated with heat tolerance in chickpea [3,19]. Interestingly, the positions of the QTLs (*qts02_5, qgy02_5, q% podset06_5*) for TS, GY, and %PodSet were identified in the same position over the years, which strongly confirm the QTLs in these positions.

The presence of four major co-localized QTLs (*qfpod02_5, qts02_5, qgy02_5, and q% podset06_5*) suggests tight linkage or the phenomenon of pleiotropy and the phenotypic correlations between these traits were highly significant in both the stress environments. Moreover, the tolerant parent ICC 15614 is contributing the desirable alleles for all the QTLs found in the two genomic regions in CaLG05 and CaLG06.

Identification of QTLs at the same positions in both the heat-stress environments indicate their possible practical utility in breeding for heat-stress tolerance in subsequent studies [25]. Several co-localized QTLs for various traits were found which could possibly due to pleiotropy or tightly linked QTLs. Fine mapping of the target genomic region will further help in resolving the issues of pleiotropy and tight linkage. The incorporation of a higher number of markers into the existing genetic map can further narrow down the genomic regions identified.

QTLs for traits such as FPod, TS, and GY were not expressed under non-stress condition, confirming the fact that these QTLs were only expressed under high-temperature condition. Two major QTLs for HI were identified in CaLG01 and CaLG04 explaining the phenotypic variation of 12.03% (LOD 8.8) and 12.53% (LOD 7.9), respectively. In addition, three minor QTLs including one for HI and two for %PodSet were found in different LGs. The fewer number of detected QTLs and their unique positions in the non-stress environment is a strong evidence that there is no correspondence between QTLs found in non-stress with the QTLs found in heat-stress environment. This phenomenon proves the fact that those QTLs identified in heat-stress condition were independent and exclusive for heat tolerance.

3.3. Epistatic QTLs for Heat Tolerance

Epistatic interaction is one of the key factors controlling the expression of a complex trait. The epistatic interaction analysis of QTLs provides a more comprehensive knowledge of the QTLs and their genetic behaviour underlying the trait [26,27].

In the current study, 19 pairs of digenic epistatic QTLs were found to be associated with the six traits: VS, FPod, TS, GY, BM, and %PodSet. Maximum number epistatic QTLs loci were observed for TS (nine), followed by %PodSet (four). In this study, some loci such as *eqts2_1/eqpodset2_1*, *eqts2_1/eqpodset2_1*, *eqpodset2_1/eqts2_1*, *neqts9_5/neqfpod4_5*, *neqfpod4_5/neqts9_5* were simultaneously controlling more than one trait indicating the pleiotropy nature of the traits.

Four categories of epistatic interaction were found in this study such as, additive × additive, additive × non-QTL, non-QTL × non-QTL, and additive × (additive-environment) × non-QTL interaction. FPod and VS showed two epistatic interactions each. Out of two epistatic interactions, one additive × additive epistatic interaction was found for both FPod and VS.

For GY, one additive × additive QTL epistatic interaction was found. For TS, five additive × additive QTL epistatic interactions, three non-QTL × non-QTL interaction and one additive × non-QTL interactions were observed. Similarly, two additive × additive QTL interactions, one non-QTL × non-QTL interaction and one additive × non-QTL interaction were observed for %PodSet. All the epistatic interactions were found to be significant.

The additive effects were found in both directions for all the traits. Nine interactions had negative additive effects, meaning that recombinant allele combinations could increase the particular trait value. Similarly, ten epistatic QTL interactions having positive additive effects, indicating parental allele combinations, would help to improve the trait [28].

Presence of epistatic interactions for a given trait will make the selection difficult. Interestingly, all major QTLs had no epistatic interaction and this will increase the heritability of the trait and make the selection easy.

3.4. Putative Candidate Genes for Heat Tolerance

Recent progress in functional genomics facilitates the elucidation of the important role of candidate genes for expression of tolerance against abiotic stress in plants [29–31]. In the present study, mining of the candidate genes for heat tolerance revealed 236 genes in 2.28 Mb (44.6–46.9 Mb) region in CaLG05 and 550 genes in 6.50 Mb (7.85–14.35 Mb) in CaLG06 (Tables S5 and S6). Based on functional categorization, many genes were found to be associated with biological processes (168 genes in CaLG05 and 365 genes in CaLG06) in the two genomic regions.

Gene ontology classification revealed a total of 25 putative candidate genes (11 in CaLG05 and 14 in CaLG06) known to function, directly or indirectly, as heat-stress response genes in several plant species (Table S9a,b). Of the 25 candidate genes, five genes encode protein like farnesylated protein 6 (AtFP6), ethylene-responsive transcription factor ERF114, ethylene-responsive transcription factor CRF4, F-box protein SKP2B, and ethylene-responsive transcription factor RAP2-11. These genes were identified to have key roles in heat acclimation and growth of plants under severe heat-stress condition. Many transcription factors, enzyme, and stress responsive element binding factors responsible for heat tolerance in various plant species were reported earlier [32]. Furthermore, various heat shock proteins (HSPs), ethylene forming enzymes (EFEs), and ethylene-responsive element factors (ERFs) were found to be candidate genes for heat tolerance in soybean and cowpea, two of the plant species closest to chickpea [32].

The role of various heat shock proteins and heat-stress transcription factors has been widely accepted and reported in different crops [33]. The role of HSP90 transcription factors under heat-stress conditions was also reported in chickpea [34]. Five putative genes were identified in the two examined genomic regions, encoding for either heat shock proteins or heat shock transcription factors contributing for thermo-tolerance.

Oxidative stress can occur in parallel with heat-stress through the formation of reactive oxygen species (ROS) [35]. Three putative candidate genes were also observed in this study to have a role in defying oxidative stress and recovering plants from heat-stress damage. These genes encode different types of proteins like protein tansparent testa glabra 1, peroxidase 52, and zinc finger protein CONSTANS-LIKE 5. In addition, certain signalling molecules like ethylene, abscisic acid (ABA), and salicylic acid are among a few to have a significant role in the development of heat tolerance [36]. In this study, a few genes—MYB44, AKH3, and RAN1—were found to involve with these signalling molecules through upregulation process to mitigate the heat-stress. Being a preliminary study, evaluation of these putative candidate gene-functions in chickpea through fine mapping and gene expression study is necessary to use them for further study.

4. Materials and Methods

4.1. Plant Material and Treatment Condition

A mapping population of 292 RILs developed from a cross between a heat sensitive parent ICC 4567 and a heat tolerant parent ICC 15614 was used for the study. Field experiments were carried out at ICRISAT, Patancheru, India (17°30′ N; 78°16′ E; altitude 549 m) on a vertisol soil. The F8-9 RIL population was evaluated under two heat-stress environments (in summer, February–May 2013 and February–May 2014) and in one non-heat-stress environment (in winter, November–February 2013).

In all the environments, the field was solarized using polythene mulch during the preceding summer to sanitize the field, especially to avoid incidence of root diseases. Sowing was done on the ridges using ridge and furrow method with inter- and intra-row spacing of 60 × 10 cm. Each plot consisted of a 2 m long row. Need-based insecticide sprays were provided to control pod borer (*Helicoverpa armigera*) and the experimental plots were kept weed-free through manual weeding. Before sowing, seeds were treated with the mixture of fungicides 0.5% Benlate® (E.I. DuPont India Ltd., Gurgaon, India) + Thiram® (Sudhama Chemicals Pvt., Ltd., Gujarat, India).

The experimental design was laid out in a 15 × 20 alpha lattice design with three replications. The sowing for the non-stress environment was done on the residual moisture in the last week of November 2013 and provided with essential irrigation. The planting was done in the first week of February for stress environments to expose the reproductive phase of RILs to heat-stress (>35 °C). The stress experiments were provided with irrigation to avoid the confounding effect of moisture stress during the heat screening.

In chickpea, a temperature higher than 35 °C during reproductive phase adversely affects growth, development, and yield [1,19]. The parents used for developing RIL population for this study showed significant variations at this temperature (35 °C and above) in an earlier study [19] (Devasirvatham et al., 2013). The mean daily day/night temperatures during the reproductive phase of RILs in heat-stress environment 2013 and heat-stress environment 2014 were 37.5/22.5 °C and 36.7/22.9 °C, respectively (Figure 3). Whereas under normal season (non-stress environment), the mean daily temperatures were 29.6/15.5 °C.

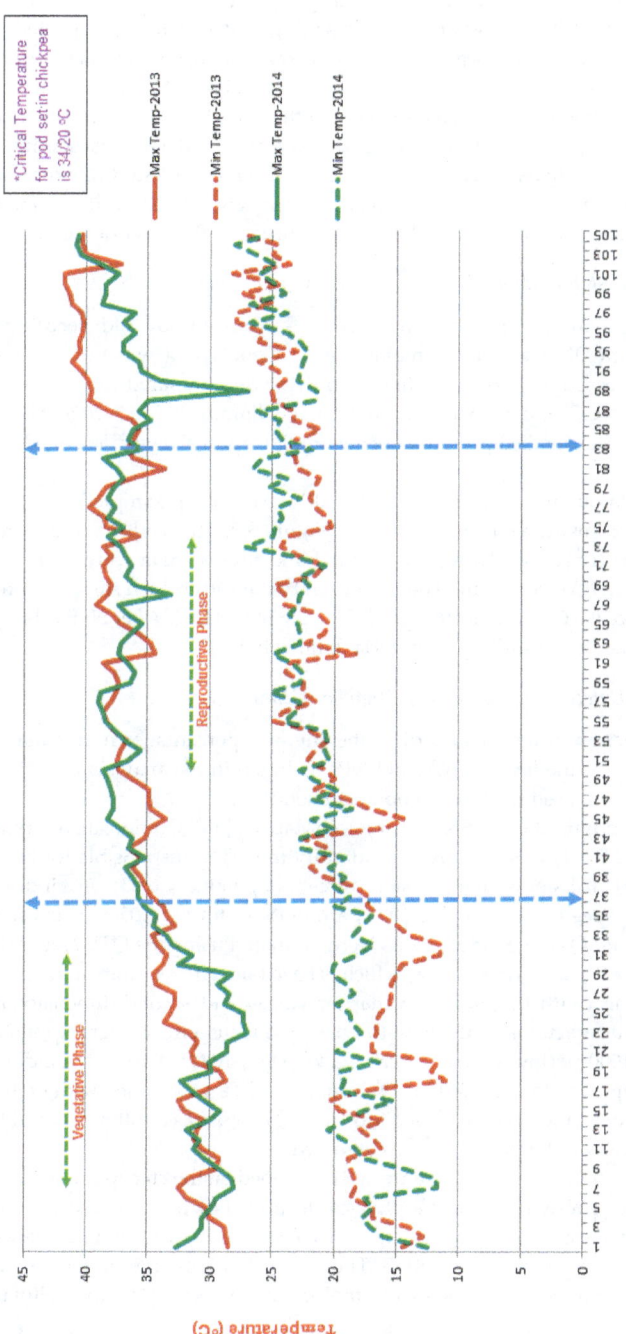

Figure 3. Daily maximum and minimum temperatures (°C) during the late sown crop growing period (stress season) in 2013 and 2014 (34/19 °C is the threshold temperature for the maximum and minimum temperatures for chickpea yield, respectively. The maximum day temperatures were 39.8 °C and 39.0 °C, and maximum night temperatures were 24.9 °C and 27.2 °C in heat-stress environments 2013, and 2014, respectively. Crop growing period was 2nd week of February to 3rd week of May).

4.2. Variables Measured

Number of filled pods per plot (FPod), total number of seeds per plot (TS), grain yield per plot (GY, g), harvest index (HI, %), biomass (BM, g) and percent pod setting (%PodSet), were reportedly found to be associated with heat tolerance in chickpea [3,19]. These six traits along with visual score on podding behaviour (VS) were recorded in the RIL population. The data for FPod, TS, GY, BM, and HI were recorded from a half-meter (0.5 m) long continuous patch out of the 2-m plot. VS at maturity and %PodSet were recorded from the entire plot. For visual scoring, score-1 was considered most sensitive (least number of pod-bearing ability), whereas, score-5 was taken as the most tolerant (maximum number of pod-bearing ability) under heat-stress. In the non-stress environment, all RILs were assumed to behave more or less the same. Hence, no visual score data were recorded in this environment.

4.3. DNA Extraction, Genotyping, and SNP Calling

DNA from 292 RILs, along with the parents, was isolated from 15-day old seedlings following the high-throughput mini-DNA extraction method [37]. Genotyping was done using GBS approach [38]. The GBS libraries from the parental lines and RILs were prepared using ApeKI endonuclease (recognition site: G/CWCG) and were sequenced using the Illumina HiSeq 2000 platform (Illumina Inc, San Diego, CA, USA). The detailed procedure of genotyping approach was described by Jaganathan et al. (2015) [25].

For SNP calling the raw reads obtained were first de-bimultiplexed using sample barcodes, and adapter sequences were removed using a custom Perl script (Figure S5). The reads having more than 50% of low-quality base pairs (Phred < 5%) were discarded and filtered data were used for calling SNPs after due quality check (Q score > 20). The high-quality data from each sample were aligned to the draft genome sequence (CaGAv1.0) of chickpea [20] using SOAP [39]. After SNP calling, the polymorphic loci were determined by following the criteria defined in [25].

4.4. Linkage Map Construction, QTL Detection and Mining of Candidate Genes

By adopting a stringent selection criterion including the missing percentage, minor allele frequency, and percent heterozygosity, the final number of SNPs included in the analysis were 396. The selected panel of robust SNPs were used for construction of genetic maps.

A linkage map was constructed with the 396 SNPs using JoinMap 4.1 [40]. Composite interval mapping in QTL Cartographer-V 2.5 [41] was employed to identify the QTLs responsible for heat tolerance with a forward and backward stepwise regression (threshold *p*-value < 0.05). A window size of 10 cM, along with a walking speed of 1.0 cM, and 1000 permutations for *p* < 0.05 were chosen for the QTL analysis. QTL × QTL and QTL × E interactions were estimated using the QTL Network version 2.0 (http://ibi.zju.edu.cn/software/qtlnetwork/) which is based on a mixed linear model.

First-dimensional genome scan (with the option to map epistasis) and second-dimensional genome scan (to detect epistatic interactions with or without single-locus effect) were applied. A significance level of 0.05 with 1000 permutations, 1.0 cM walk speed, 10.0 cM testing window and filtration window size were employed for the epistatic QTL analysis. QTL was named with prefix "q" for main-effect QTL, "eq" for epistatic QTL and "neq" for non-QTL epistasis followed by the abbreviated trait name and the identity of the linkage group involved.

The identified markers along with the flanking sequences were mapped on the chickpea reference genome CaGAv1.0 [20]. The genes present within the physical locations of these markers were extracted from the genome features file and were searched against TrEMBL and Swiss-Prot databases. Further functional annotation was done using UniProtKB. The Gene Ontology annotations were categorized into three categories: biological processes (BP), molecular function (MF) and cellular components (CC).

4.5. Statistical Analyses

Analysis of Variance, Predicted Means (BLUP), Heritability, and Correlations

The analysis of variance (ANOVA) for the RIL population was performed using GenStat (17th Edition), for individual environments using mixed model analysis. For each trait and environment, the analysis was performed considering entry and block (nested within replication) as random effects and replication as fixed effect.

To pool the data across environments, and to make the error variances homogeneous, individual variances were estimated and modelled for the error distribution using residual maximum likelihood (ReML) procedure. *Z* value and *F* value were calculated for random effects and fixed effects, respectively. For single and multi-environment, QTL mapping was performed using predicted means (BLUP-Best Linear Unbiased Prediction) [42].

Broad-sense heritability was estimated by following Falconer et al., 1996 [43] as

$$H^2 = Vg/(Vg + Ve/nr);$$

and pooled broad-sense heritability was estimated by following Hill et al., 2012 [44] as

$$H^2 = Vg/\{(Vg) + (Vge/ne + Ve/(ne \times nr))\}$$

Whereas, H^2 is broad-sense heritability, Vg is genotypic variance, Vge is G × E interaction variance, Ve is residual variance, ne is number of environments, and nr is number of replications. Pearson correlation analysis and linear regressions were fitted using Microsoft Excel 2016 (Microsoft Corp., 1985, Redmond, WA, USA).

5. Conclusions

The present study identified two potential genomic regions harbouring major QTLs for several heat responsive traits that are directly related to heat tolerance in chickpea. The two regions consistently appeared at the same map position across two years. Epistatic effects were not observed for major QTLs and no QTL × E interaction in the CaLG05 region. The results laid a foundation in understanding heat tolerance and increases the confidence of breeders to proceed with early generation selection for heat tolerance through marker-assisted breeding. In addition, the candidate genes identified in the two genomic regions further help to understand the mechanism of heat tolerance.

Supplementary Materials: The following are available online at http://www.mdpi.com/1422-0067/19/8/2166/s1. Figure S1: Intra-specific genetic map of chickpea RIL population (ICC 4567 × ICC 15614) with 271 GBS-based SNPs covering 529.11 cM. Genetic distances (cM) were shown on the left side and the markers were shown on the right side of the bars. The map was constructed using JoinMap 4.1 and Kosambi function, Figure S2: The epistatic QTLs on linkage groups detected by QTLNetwork v 2.0 in the RIL population (ICC 4567 × ICC 15614). Lines joining two QTLs represents the epistatic interaction between them, Figure S3: Relationship of visual score on podding behaviour (VS), Number of Filled Pods per Plot (FPod), Total Number of Seeds per Plot (TS), Biomass (BM), Harvest Index (HI) and Percent Pod Setting (%PodSet) with Grain Yield per Plot (GY) (**a**) during heat-stress environment of 2013 (**b**) during heat-stress environment of 2014 (**c**) of pooled environments (heat-stressed environments, 2013 and 2014) (**d**) during non-stress environment of 2013 (Due to non-availability of VS data, no relationship of VS with GY is presented in non-stress environment, 2013). X-axis represents yield components traits e.g., VS, FPod, TS, BM HI and %PodSet; *Y*-axis represents GY; (No. of RILs-292), Figure S4: Likelihood of odds ratio (LOD) curves obtained by composite interval mapping for quantitative trait loci (QTL) mapped for the traits-visual score on podding behaviour (VS), Number of Filled Pods per Plot(FPod), Total Number of Seeds per Plot (TS), Grain Yield per Plot (GY), Biomass (BM), Harvest Index (HI), and Percent Pod Setting (%PodSet) in RIL population (ICC 4567 × ICC 15614) (**a**) in the heat-stress environment-2013 (**b**) in the heat-stress environment-2014 (**c**) in the pooled environments (heat-stress environments, 2013, and 2014) (**d**) in the non-stress environment-2013 (Due to non-availability of VS data, VS was not mapped in non-stress environment, 2013). The vertical lines indicating the threshold LOD value (2.5) determining significant QTL, Figure S5: Pipeline of Bioinformatics analysis: GBS data processing and SNP calling, Table S1: (**a**) Summary sequence data generated genotyping-on 292 RILs and two parents (ICC 4567 and ICC 15614) using GBS approach; (**b**) Summary of called SNPs on 292 RILs and two parents (ICC 4567 and ICC 15614) using GBS approach, Table S2: Features of intra-specific genetic map developed using 271 SNPs and RIL population ICC 4567 × ICC 15614, Table S3: Summary of QTLs identified in two heat-stress environments, pooled environments and non-stressed environment in RIL population (ICC 4567 × ICC 15614), Table S4: Consistent QTLs found across heat-stress environments (2013 and 2014) in RIL population (ICC 4567 × ICC 15614), Table S5: Gene ontology classification for CaLG05, Table S6: Gene ontology classification for CaLG06, Table S7: Gene ontology categorization of 236 genes identified on the genomic region flanked by markers Ca5_44667768-Ca5_46955940 on CaLG05, Table S8: Gene ontology categorization of 550 genes identified on the genomic region flanked by markers Ca6_7846335-Ca6_14353624 on CaLG06, Table S9: (**a**) List of putative candidate genes found to be associated with heat stress on CaLG05 in chickpea, (**b**) List of putative candidate genes found to be associated with heat stress on CaLG06 in chickpea.

Author Contributions: P.M.G. conceived the idea and coordinated this project. P.M.G. and S.S. were involved in developing the mapping population. P.M.G., S.S., S.K.C., S.B.S. and G.R.L. provided guidance to P.J.P. in conducting field experiments and phenotyping. A.R., R.R.D. and S.S. helped P.J.P. in statistical data analysis. P.J.P. and M.T. were involved in genotyping of the mapping population, construction of linkage maps and QTL analysis. P.J.P., A.W.K. and M.T. were involved in bioinformatics work. P.J.P., P.M.G., MT, S.B.S. and S.S. contributed to writing of the manuscript, and R.K.V. and G.R.L. provided their inputs. All the authors reviewed and approved the final manuscript.

Funding: National Food Security Mission (NFSM), Govt. of India; and Tropical Legumes II (TL II) project of Bill and Melinda Gates Foundation (BMGF) for financial support and Department of Science and Technology (DST), Govt. of India, for a fellowship to P.J.P.

Conflicts of Interest: The authors declare no conflicts of interest

Abbreviations

%PodSet	Pod Setting Percentage
ANOVA	Analysis of Variance
BLUP	Best Linear Unbiased Prediction
BM	Biomass
CaLG	*Cicer arietinum* Linkage Group
CIM	Composite Interval Mapping
cM	Centimorgan
FPod	Number of Filled Pods Per Plot
GY	Grain Yield Per Plot
HI	Harvest Index
ICRISAT	International Crops Research Institute for the Semi-Arid Tropics
LG	Linkage Group
QTL	Quantitative Trait Loci
ReML	Residual Maximum Likelihood
RIL	Recombinant Inbred Line
TS	Total Number of Seeds Per Plot
VS	Visual Scoring

References

1. Gaur, P.M.; Jukanti, A.K.; Samineni, S.; Chaturvedi, S.K.; Basu, P.S.; Babbar, A.; Jayalakshmi, V.; Nayyar, H.; Devasirvatham, V.; Mallikarjuna, N.; et al. *Climate Change and Heat Stress Tolerance in Chickpea. Climate Change and Plant Abiotic Stress Tolerance*; Wiley-VCH Verlag GmbH & Co. KGaA: Weinheim, Germany, 2014; pp. 837–856.

2. Food and Agriculture Organization (FAO). Food and Agricultural Organization of the United Nation, FAO Statistical Database. 2015. Available online: http://faostat3.fao.org/download/Q/QC/E (accessed on 8 February 2018).

3. Krishnamurthy, L.; Gaur, P.M.; Basu, P.S.; Chaturvedi, S.K.; Tripathi, S.; Vadez, V.; Rathore, A.; Varshney, R.K.; Gowda, C.L.L. Large genetic variation for heat tolerance in the reference collection of chickpea (*Cicer arietinum* L.) germplasm. *Plant Genet. Resour.* **2011**, *9*, 59–69. [CrossRef]

4. Devasirvatham, V.; Gaur, P.M.; Mallikarjuna, N.; Tokachichu, R.N.; Trethowan, R.M.; Tan, D.K.Y. Effect of high temperature on the reproductive development of chickpea genotypes under controlled environments. *Funct. Plant. Biol.* **2012**, *39*, 1009–1018. [CrossRef]

5.   Wang, J.; Gan, Y.T.; Clarke, F.; McDonald, C.L. Response of chickpea yield to high temperature stress during reproductive development. *Crop Sci.* **2006**, *46*, 2171–2178. [CrossRef]

6.   Dehghani, H.; Sabaghpour, S.H.; Ebadi, A. Study of genotype × environment interaction for chickpea yieldin Iran. *Agron. J.* **2010**, *102*, 1–8. [CrossRef]

7.   Gaur, P.M.; Thudi, M.; Samineni, S.; Varshney, R.K. Advances in chickpea genomics. In *Legumes in the Omic Era*; Springer: New York, NY, USA, 2014; pp. 73–94.

8.   Sabbavarapu, M.M.; Sharma, M.; Chamarthi, S.K.; Swapna, N.; Rathore, A.; Thudi, M.; Gaur, P.M.; Pande, S.; Singh, S.; Kaur, L.; et al. Molecular mapping of QTLs for resistance to Fusarium wilt (race 1) and Ascochyta blight in chickpea (*Cicer arietinum* L.). *Euphytica* **2013**, *193*, 121–133. [CrossRef]

9.   Varshney, R.K.; Thudi, M.; Nayak, S.N.; Gaur, P.M.; Kashiwagi, J.; Krishnamurthy, L.; Jaganathan, D.; Koppolu, J.; Bohra, A.; Tripathi, S.; et al. Genetic dissection of drought tolerance in chickpea (*Cicer arietinum* L.). *Theor. Appl. Genet.* **2014**, *127*, 445–462. [CrossRef] [PubMed]

10. Pushpavalli, R.; Krishnamurthy, L.; Thudi, M.; Gaur, P.M.; Rao, M.V.; Siddique, K.H.; Colmer, T.D.; Turner, N.C.; Varshney, R.K.; Vadez, V. Two key genomic regions harbour QTLs for salinity tolerance in ICCV 2× JG 11 derived chickpea (*Cicer arietinum* L.) recombinant inbred lines. *BMC Plant Biol.* **2015**, *15*, 124. [CrossRef] [PubMed]

11. Vadez, V.; Krishnamurthy, L.; Thudi, M.; Anuradha, C.; Colmer, T.D.; Turner, N.C.; Siddique, K.H.; Gaur, P.M.; Varshney, R.K. Assessment of ICCV 2 × JG 62 chickpea progenies shows sensitivity of reproduction to salt stress and reveals QTL for seed yield and yield components. *Mol. Breed.* **2012**, *30*, 9–21. [CrossRef]

12. Samineni, S. Physiology, Genetics and QTL Mapping of Salt Tolerance in Chickpea (*Cicer arietinum* L.). Ph.D. Thesis, The University of Western Australia, Perth, Australia, 2011.

13. Jha, U.C.; Bohra, A.; Singh, N.P. Heat stress in crop plants: Its nature, impacts and integrated breeding strategies to improve heat tolerance. *Plant Breed.* **2014**, *133*, 679–701. [CrossRef]

14. Jagadish, S.V.K.; Craufurd, P.Q.; Wheeler, T.R. Phenotyping parents of mapping populations of rice for heat tolerance during anthesis. *Crop Sci.* **2008**, *48*, 1140–1146. [CrossRef]

15. Ye, C.; Argayoso, M.A.; Redoña, E.D.; Sierra, S.N.; Laza, M.A.; Dilla, C.J.; Mo, Y.; Thomson, M.J.; Chin, J.; Delaviña, C.B.; et al. Mapping QTL for heat tolerance at flowering stage in rice using SNP markers. *Plant Breed.* **2012**, *131*, 33–41. [CrossRef]

16. Xiao, Y.; Pan, Y.; Luo, L.; Zhang, G.; Deng, H.; Dai, L.; Liu, X.; Tang, W.; Chen, L.; Wang, G.L. Quantitative trait loci associated with seed set under high temperature stress at the flowering stage in rice (*Oryza sativa* L.). *Euphytica* **2011**, *178*, 331–338. [CrossRef]

17. Pinto, R.S.; Reynolds, M.P.; Mathews, K.L.; McIntyre, C.L.; Olivares-Villegas, J.J.; Chapman, S.C. Heat and drought adaptive QTL in a wheat population designed to minimize confounding agronomic effects. *Theor. Appl. Genet.* **2010**, *121*, 1001–1021. [CrossRef] [PubMed]

18. Zhang, G.L.; Chen, L.Y.; Xiao, G.Y.; Xiao, Y.H.; Chen, X.B.; Zhang, S.T. Bulked segregant analysis to detect QTL related to heat tolerance in rice (*Oryza sativa* L.) using SSR markers. *Agric. Sci. China* **2009**, *8*, 482–487. [CrossRef]

19. Devasirvatham, V.; Gaur, P.M.; Mallikarjuna, N.; Raju, T.N.; Trethowan, R.M.; Tan, D.K.Y. Reproductive biology of chickpea response to heat stress in the field is associated with the performance in controlled environments. *Field Crop Res.* **2013**, *142*, 9–19. [CrossRef]

20. Varshney, R.K.; Song, C.; Saxena, R.K.; Azam, S.; Yu, S.; Sharpe, A.G.; Cannon, S.; Baek, J.; Rosen, B.D.; Tar'an, B.; et al. Draft genome sequence of chickpea (*Cicer arietinum*) provides a resource for trait improvement. *Nat. Biotechnol.* **2013**, *31*, 240–246. [CrossRef] [PubMed]

21. Gaur, P.M.; Srinivasan, S.; Gowda, C.L.L.; Rao, B.V. Rapid generation advancement in chickpea. *J. SAT Agric. Res.* **2007**, *3*, 3.

22. Berger, J.D.; Milroy, S.P.; Turner, N.C.; Siddique, K.H.M.; Imtiaz, M.; Malhotra, R. Chickpea evolution has selected for contrasting phenological mechanisms among different habitats. *Euphytica* **2011**, *180*, 1–15. [CrossRef]

23. Paliwal, R.; Röder, M.S.; Kumar, U.; Srivastava, J.P.; Joshi, A.K. QTL mapping of terminal heat tolerance in hexaploid wheat (*T. aestivum* L.). *Theor. Appl. Genet.* **2012**, *125*, 561–575. [CrossRef] [PubMed]

24. Buu, B.C.; Ha, P.T.T.; Tam, B.P.; Nhien, T.T.; Van Hieu, N.; Phuoc, N.T.; Giang, L.H.; Lang, N.T. Quantitative trait loci associated with heat tolerance in rice (*Oryza sativa* L.). *Plant Breed. Biotechnol.* **2014**, *2*, 14–24. [CrossRef]

25. Jaganathan, D.; Thudi, M.; Kale, S.; Azam, S.; Roorkiwal, M.; Gaur, P.M.; Kishor, P.K.; Nguyen, H.; Sutton, T.; Varshney, R.K. Genotyping-by-sequencing based intra-specific genetic map refines a "QTL-hotspot" region for drought tolerance in chickpea. *Mol. Genet. Genom.* **2015**, *290*, 559–571. [CrossRef] [PubMed]

26. Bocianowski, J. Epistasis interaction of QTL effects as a genetic parameter influencing estimation of the genetic additive effect. *Genet. Mol. Biol.* **2013**, *36*, 093–100. [CrossRef] [PubMed]

27. Gowda, S.J.M.; Radhika, P.; Mhase, L.B.; Jamadagni, B.M.; Gupta, V.S.; Kadoo, N.Y. Mapping of QTLs governing agronomic and yield traits in chickpea. *J. Appl. Genet.* **2011**, *52*, 9–21. [CrossRef] [PubMed]

28. Qi, L.; Mao, L.; Sun, C.; Pu, Y.; Fu, T.; Ma, C.; Shen, J.; Tu, J.; Yi, B.; Wen, J. Interpreting the genetic basis of silique traits in *Brassica napus* using a joint QTL network. *Plant Breed.* **2014**, *133*, 52–60. [CrossRef]

29. Urano, K.; Kurihara, Y.; Seki, M.; Shinozaki, K. 'Omics' analyses of regulatory networks in plant abiotic stress responses. *Curr. Opin. Plant Biol.* **2010**, *13*, 132–138. [CrossRef] [PubMed]

30. Sreenivasulu, N.; Sopory, S.K.; Kishor, P.K. Deciphering the regulatory mechanisms of abiotic stress tolerance in plants by genomic approaches. *Gene* **2007**, *388*, 1–13. [CrossRef] [PubMed]

31. Vij, S.; Tyagi, A.K. Emerging trends in the functional genomics of the abiotic stress response in crop plants. *Plant Biotechnol. J.* **2007**, *5*, 361–380. [CrossRef] [PubMed]

32. Pottorff, M.; Roberts, P.A.; Close, T.J.; Lonardi, S.; Wanamaker, S.; Ehlers, J.D. Identification of candidate genes and molecular markers for heat-induced brown discoloration of seed coats in (*Vigna unguiculata* (L.) Walp). *BMC Genom.* **2014**, *15*, 328. [CrossRef] [PubMed]

33. Maestri, E.; Klueva, N.; Perrotta, C.; Gulli, M.; Nguyen, H.T.; Marmiroli, N. Molecular genetics of heat tolerance and heat shock proteins in cereals. *Plant Mol. Biol.* **2002**, *48*, 667–681. [CrossRef] [PubMed]

34. Agarwal, G.; Garg, V.; Kudapa, H.; Doddamani, D.; Pazhamala, L.T.; Khan, A.W.; Thudi, M.; Lee, S.H.; Varshney, R.K. Genome-wide dissection of AP2/ERF and HSP90 gene families in five legumes and expression profiles in chickpea and pigeonpea. *Plant Biotechnol. J.* **2016**, *14*, 1563–1577. [CrossRef] [PubMed]

35. Wahid, A.; Gelani, S.; Ashraf, M.; Foolad, M.R. Heat tolerance in plants: An overview. *Environ. Exp. Bot.* **2007**, *61*, 199–223. [CrossRef]

36. Larkindale, J.; Huang, B. Effects of abscisic acid, salicylic acid, ethylene and hydrogen peroxide in thermotolerance and recovery for creeping bentgrass. *Plant Growth Regul.* **2005**, *47*, 17–28. [CrossRef]

37. Cuc, L.M.; Mace, E.S.; Crouch, J.H.; Quang, V.D.; Long, T.D.; Varshney, R.K. Isolation and characterization of novel microsatellite markers and their application for diversity assessment in cultivated groundnut (*Arachis hypogaea*). *BMC Plant Biol.* **2008**, *8*, 55. [CrossRef] [PubMed]

38. Elshire, R.J.; Glaubitz, J.C.; Sun, Q.; Poland, J.A.; Kawamoto, K.; Buckler, E.S.; Mitchell, S.E. A robust, simple genotyping-by-sequencing (GBS) approach for high diversity species. *PLoS ONE* **2011**, *6*, e19379. [CrossRef] [PubMed]

39. Li, R.; Yu, C.; Li, Y.; Lam, T.W.; Yiu, S.M.; Kristiansen, K.; Wang, J. SOAP2: An improved ultrafast tool for short read alignment. *Bioinformatics* **2009**, *25*, 1966–1967. [CrossRef] [PubMed]

40. Van Ooijen, J.J. *JoinMap®4.1, Software for the Calculation of Genetic Linkage Maps in Experimental Populations*; Kyazma BV: Wageningen, The Netherlands, 2006.

41. Wang, S.; Basten, C.J.; Zeng, Z.B. *Windows QTL Cartographer 2.5*; Department of Statistics, North Carolina State University: Raleigh, NC, USA, 2012.

42. Searle, S. *Linear Models*; John Wiley & Sons, Inc.: New York, NY, USA, 1971.

43. Falconer, D.S.; Mackay, T.F.; Frankham, R. *Introduction to Quantitative Genetics.* *Trends in Genetics*; Longman Frankel: Harlow, UK, 1996; Volume 12, p. 280.

44. Hill, J.; Becker, H.C.; Tigerstedt, P.M. *Quantitative and Ecological Aspects of Plant Breeding*; Springer Science & Business Media: Berlin, Germany, 2012.

Article

Transcriptome Analyses in Different Cucumber Cultivars Provide Novel Insights into Drought Stress Responses

Min Wang [1,2,†], Biao Jiang [1,2,†], Qingwu Peng [1], Wenrui Liu [1], Xiaoming He [1], Zhaojun Liang [1] and Yu'e Lin [1,*]

[1] Vegetable Research Institute, Guangdong Academy of Agricultural Sciences, Guangzhou 510640, China; w.jsun@163.com (M.W.); jiangbiao198354@163.com (B.J.); pengqingwu@gdaas.cn (Q.P.); liuwr10@126.com (W.L.); xiaominghe626@163.com (X.H.); liangzhaojun@gdaas.cn (Z.L.)
[2] Guangdong Key Laboratory for New Technology Research of Vegetables, Guangzhou 510640, China
*   Correspondence: cucumber200@163.com; Tel.: +86-020-3846-9441
†   These authors contributed equally to this work.

Received: 21 May 2018; Accepted: 10 July 2018; Published: 16 July 2018

Abstract: Drought stress is one of the most serious threats to cucumber quality and yield. To gain a good understanding of the molecular mechanism upon water deficiency, we compared and analyzed the RNA sequencing-based transcriptomic responses of two contrasting cucumber genotypes, L-9 (drought-tolerant) and A-16 (drought-sensitive). In our present study, combining the analysis of phenotype, twelve samples of cucumber were carried out a transcriptomic profile by RNA-Seq under normal and water-deficiency conditions, respectively. A total of 1008 transcripts were differentially expressed under normal conditions (466 up-regulated and 542 down-regulated) and 2265 transcripts under drought stress (979 up-regulated and 1286 down-regulated). The significant positive correlation between RNA sequencing data and a qRT-PCR analysis supported the results found. Differentially expressed genes (DEGs) involved in metabolic pathway and biosynthesis of secondary metabolism were significantly changed after drought stress. Several genes, which were related to sucrose biosynthesis (*Csa3G784370* and *Csa3G149890*) and abscisic acid (ABA) signal transduction (*Csa4M361820* and *Csa6M382950*), were specifically induced after 4 days of drought stress. DEGs between the two contrasting cultivars identified in our study provide a novel insight into isolating helpful candidate genes for drought tolerance in cucumber.

Keywords: *Cucumis sativus* L.; RNA-Seq; DEGs; sucrose; ABA; drought stress

1. Introduction

Drought stress generally occurs when soil water is deficient, leading to a continuous loss of water by transpiration or evaporation [1]. Water deficiency, a key limiting factor in plant growth and development, impacts plant elongation and expansion growth [2,3]. In order to survive under drought stress, plants have to make corresponding adjustments by regulating gene expression of stress-related and signal transduction pathways [4–6], such as reactive oxygen species (ROS)-related genes [7], transcription factors (TFs) [8], and the abscisic acid (ABA) signal transduction pathway [9,10].

Cucumber (*Cucumis sativus* L.), one of the most important vegetable crops in Cucurbitaceae, is originally from the southern Himalayas and shows a preference for warm and moist environment [11]. Previous studies about cucumber resistance on drought have been carried out in different aspects [12–15]. Carbon monoxide (CO) is involved in hydrogen gas (H_2)-induced adventitious root development under stimulated drought stress and alleviates oxidative damage by altering relative physiological index [12]. *CsCER1* is involved in the fruit cuticle synthesis, and overexpressing the gene has been shown to improve the drought tolerance under water-deficiency conditions [13]. Exogenously applied hydrogen peroxide could considerably enhance the cucumber drought resistance by increasing the plant's antioxidative defense system and its capacity for osmotic adjustment [14]. Tobacco PR-2d promoter/*uidA* (GUS) gene is induced in transgenic cucumber and improves the response to biotic and abiotic stimuli [15].

Comparing transcriptome by RNA-seq of various genotypes in different species is one of the most suitable techniques for exploring resistant genes under abiotic stress and elucidating the role of various biological pathways, as well as mechanisms for influencing tolerance to adverse environments [16,17]. When compared with microarray and expressed sequence tag, advantages of RNA-seq showed determination of alternative splicing (AS) events, novel transcripts and digital gene expression at the isoform level [18,19]. In cucumber, the RNA-seq method has been widely employed for performing crucial agricultural functions such as fruit development [20], parthenocarpy [21], flower sex expression [22], and other plant responses to abiotic stresses [23–25]. A transcriptome profiling reveals a mechanism of fruit trichome formation, which plays an important role in plant defense against biotic and abiotic stresses [23]. A total of 121 genes were significantly induced under melatonin treatment, which promoted the cucumber lateral root formation under salt stress [24]. Zhao et al. [25] examined over 23,000 transcripts in cucumber leaves, and found that 364 genes were differentially expressed in response to nitrogen deficiency, providing novel insights into the responses of cucumber to N starvation at the global transcriptome level [25]. However, to the best of our knowledge, no research has been performed on the drought stress in cucumber using compared transcriptome.

In this study, we carried out RNA-sequencing analysis in cucumber to explore the transcriptional variations between a drought-tolerant cultivar L-9 and a drought-sensitive cultivar A-16 under normal and drought conditions. Different drought stress-responsive novel transcript isoforms were identified between L-9 and A-16. Furthermore, we analyzed the differential gene expression patterns in response to drought stresses. Functional categorization of differentially expressed transcripts was carried out to reveal various metabolic pathways involved in drought responses. Overall, this study provides a theoretical basis for further study of the regulatory mechanism of drought tolerance in cucumber.

2. Results

2.1. A-16 Cultivar Is Sensitive to Drought Stress

Ten-day seedlings of L-9 and A-16 (120 plants for three biological replicates, respectively) grown under normal condition (Figure 1A) were treated with water deficiency for 7 days and recovered for 3 days (Figure 1B). Both L-9 and A-16 showed vigorous development before drought; however, A-16 began to exhibit wilting at the top of the growth point after drought stress, and its leaves turned chlorotic and yellow (Figure 1A,B). Approximately 13% of the drought treated A-16 plants survived after the subsequent 3-day recovery, compared with 77% of L-9 plants (Figure 1C). There were no difference of malondialdehyde (MDA) and the enzyme superoxide dismutase (SOD) between L-9 and A-16 before drought, while A-16 presented a prominent increase of MDA and significant decrease of SOD at the 4th day after drought treatment (Figure 1D,E).

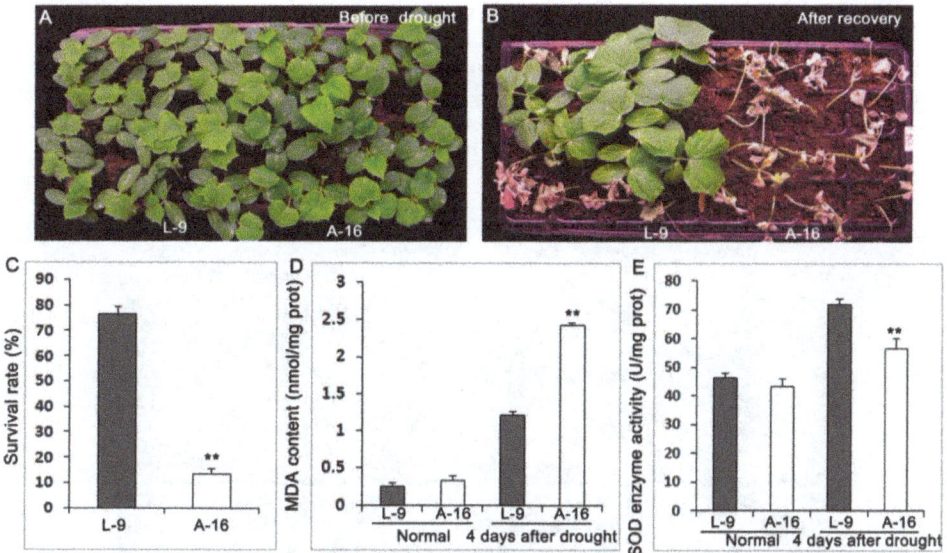

Figure 1. Phenotypes of L-9 and A-16 before drought and after recovery of drought stress. (**A**) L-9 and A-16 plants were grown under normal conditions for 14 days. (**B**) After 7 days drought treatment, seedlings recovered for 3 days. (**C**) Survival rate of plants following the 7-day drought treatment. (**D,E**) Measurement of MDA content (**D**) and SOD enzyme activity (**E**) under normal conditions and 4 days after drought. Data is presented as the mean ± standard deviation ($n = 9$). ** $p < 0.01$; Student's *t*-test.

Before drought, there was no significant difference in chlorophyll content between L-9 and A-16 (Figure 2A). However, the relative content of chlorophyll a decreased to ~34% in L-9 vs. ~52% in A-16, and the chlorophyll b decreased to ~14% and ~33% in L-9 and A-16 after drought treatment, respectively (Figure 2B). These above results indicated that L-9 showed more significant drought tolerance than A-16. In order to compare the ultrastructure of chloroplasts between L-9 and A-16, we used the transmission electron microscopy to observe the leaves at seedling stage. The leaf cells of L-9 contained normal chloroplasts, which showed well-organized lamellar structures with normally stacked grana and thylakoid membranes (Figure 2C–E). However, most cells of A-16 were heteroplastidic, with many more starch grains (Figure 2F–H). These observations implied that the sensitivity to drought stress of A-16 might be related to the abnormal development of chloroplasts in leaves at the early seedling stage.

Additionally, we investigated whether stomatal numbers of A-16 was different from L-9 using scanning electron microscopy (SEM). The result showed that the number of stomas in L-9 (Figure 3A,B) was much less than A-16 (Figure 3C,D) in the same field size, indicating that L-9 lost water more easily when encountering drought stress.

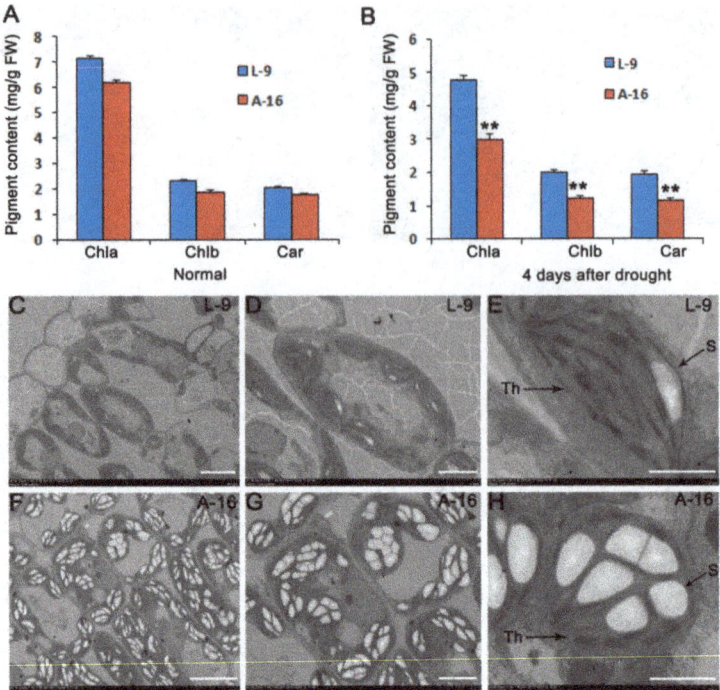

Figure 2. TEM observation of L-9 and A-16 leaves at seedling stage. (**A**) Chlorophyll content of L-9 and A-16 before drought. (**B**) Chlorophyll content of L-9 and A-16 during drought. Data is presented as the mean ± standard deviation (*n* = 9). ** *p* < 0.01; Student's *t*-test. (**C–H**) Transmission electron microscopic photos of cells from L-9 and A-16. (**C–E**) Mesophyll cells in L-9 plants showed normal, well-ordered chloroplasts. (**F–H**) Cells in A-16 plants displayed some abnormalities and accumulated starch grains. Th: thylakoid, S: starch granule. Bar in (**C,F**): 100 μm. Bar in (**D,G**): 50 μm. Bar in (**E,H**) : 20 μm.

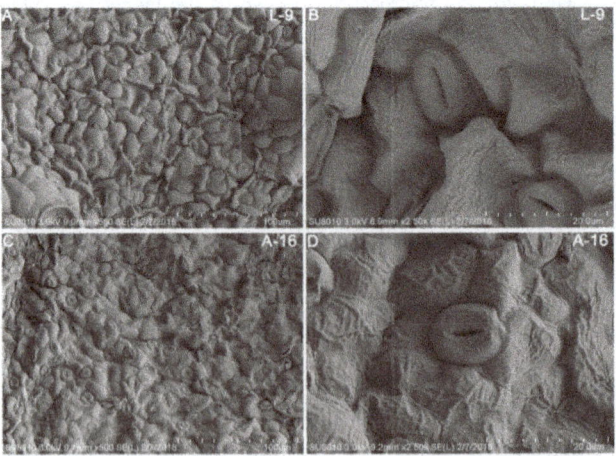

Figure 3. SEM observation of L-9 and A-16 leaves at seedling stage. (**A,B**) Scanning electron microscopy (SEM) images of leaves in L-9. (**C,D**) Scanning electron microscopy (SEM) images of leaves in A-16.

2.2. Drought Stress Results in Extensive Transcriptomic Reprogramming

In order to explore the transcriptional variations between L-9 and A-16 under normal and drought conditions, respectively, we carried out RNA-sequencing. A total of about 23 million clean reads were obtained per sample (Table 1) after removing the low-quality and adaptor-containing reads. At least 1.14 Gb clean data were acquired for each sample (Table 1). In total, the expression of 21,019 genes was detected. Approximately 96% of the clean reads were mapped to the reference cucumber genome [26], with more than 68% among them being uniquely mapped (Table 1). Finally, we identified 1008 (Table S1) and 2265 (Table S2) differentially expressed genes (DEGs) in the comparison of L-9 vs. A-16 under normal conditions and drought stress, respectively. Among them, under normal conditions, 466 genes were up-regulated and 542 down-regulated (gene expression in A-16 compared with L-9) (Figure 4A). Additionally, 979 up-regulated and 1286 down-regulated genes were identified during drought stress (Figure 4B). Next, in order to validate the RNA-seq results, we randomly selected 16 DEGs and conducted qRT-PCR analysis. The results showed that there was a strong positive correlation (two tailed, R^2 = 0.973) between the RNA-seq and qRT-PCR result (Figure 5), which indicated the accuracy of the RNA-seq data.

Table 1. Mapping results of RNA sequencing reads of the cucumber between L-9 and A-16 under normal condition (C) and 4 days after drought (D).

Sample	Total Clean Reads	Total Clean Bases (Gb)	Total Mapping Ratio %	Uniquely Mapping Ratio %
A-16_C1	23,001,330	1.15	0.9664	0.8976
A-16_C2	22,799,582	1.14	0.9662	0.8986
A-16_C3	23,275,316	1.16	0.966	0.9013
A-16_D1	23,202,361	1.16	0.963	0.8943
A-16_D2	23,239,914	1.16	0.9625	0.8981
A-16_D3	23,127,940	1.16	0.959	0.8924
L-9_C1	23,343,741	1.17	0.966	0.8971
L-9_C2	23,065,366	1.15	0.9617	0.8901
L-9_C3	23,055,733	1.15	0.9634	0.8923
L-9_D1	22,973,680	1.15	0.9639	0.8927
L-9_D2	23,183,117	1.16	0.963	0.897
L-9_D3	23,037,373	1.15	0.956	0.8932

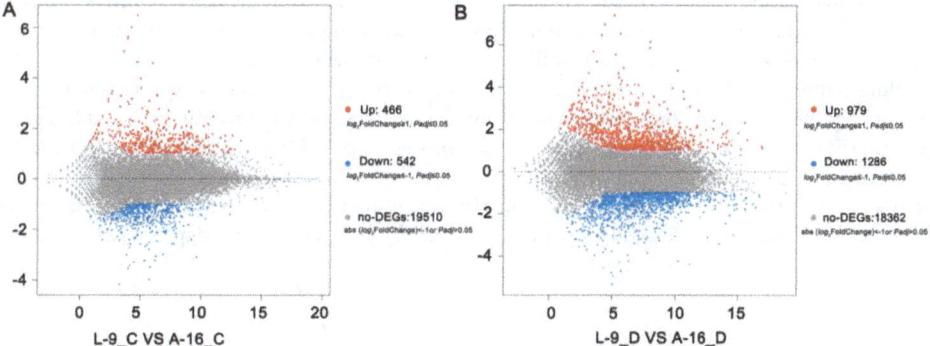

Figure 4. Comparison of different genes expression (DEGs) in leaves between L-9 and A-16 under normal conditions (**A**) and 4 days after drought (**B**). *x*- and *y*-axes represent log2 values of gene expression. Red, brown, and blue correspond to up-regulated, unaltered, and down-regulated gene expression, respectively. If a gene was expressed in just one sample, its expression value in another sample was replaced by the minimum value of all expressed genes in normal and drought samples. The screening threshold is given at the top of the figure.

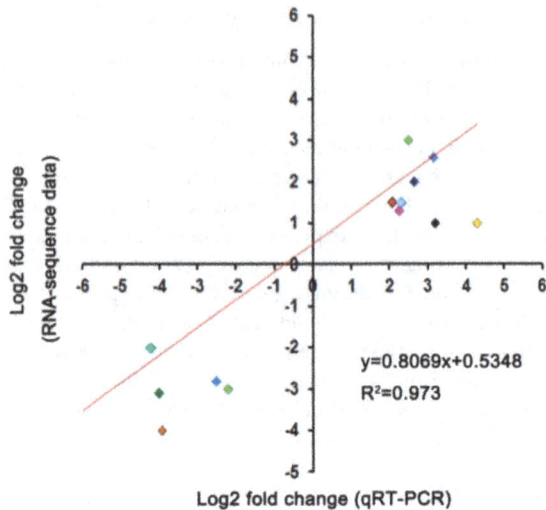

Figure 5. qRT-PCR validation of differentially expressed genes under drought stress. Correlation between the fold change analyzed by RNA-seq (*x*-axis) and data obtained using qRT-PCR. The different colors represent different genes expression.

2.3. Functional Classification of Drought-Responsive Genes

The gene ontology (GO) standardized classification system for gene function was used to analyze DEGs and understand the molecular events involved in drought response. Three categories, including "biological process," "molecular function", and "cellular components", were classified under normal conditions (Figure S1A and Table S3) and drought stress (Figure S1B and Table S4), respectively. The number of the three category genes was prominently increased at 4 days after drought treatment, especially in the metabolic process, membrane, and catalytic activity, followed by subcategories such as cellular process, cell, and binding (Figure S1B).

Next, to examine DEG-associated pathways, they were searched in the KEGG pathway database. The top 20 enriched pathways are shown in Figure 6. The main pathways under normal conditions were "biosynthesis of secondary metabolites", "plant hormone signal transduction", and "MAPK signaling pathway" (Figure 6A and Table S5). When exposed to drought stress, genes related to "metabolic pathways" and "biosynthesis of secondary metabolites" were mostly enriched (Figure 6B and Table S6), indicating that these pathways and processes possibly participated in plant drought resistance. In addition, the category of "starch and sucrose metabolism" was only detected under stress conditions, suggesting these changed genes might contribute to the increased resistance of drought. Under water deficiency, we found that some genes were responsive to water deprivation (Table 2).

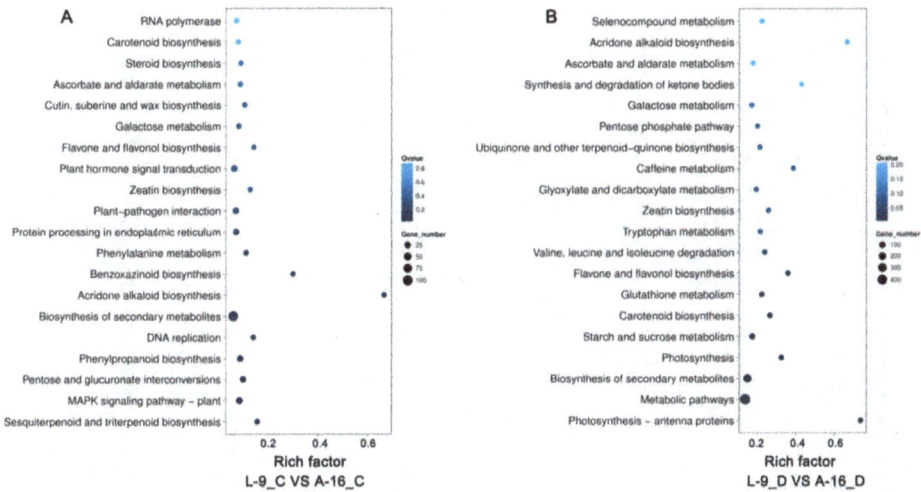

Figure 6. KEGG enrichment of annotated DEGs under three comparisons of normal conditions (**A**) and drought stress (**B**). The *y*-axis indicates the KEGG pathway and the *x*-axis indicates the enrichment factor. A high q-value is represented by light blue, and a low q-value is represented by dark blue.

Table 2. Genes related to sucrose biosynthesis and response to water deprivation.

Gene ID	L-9 Expression	A-16 Expression	Regulation	*p*-Value	Annotation
Csa2G401440	2237.5	1091.6	Down	0.00434039	Sucrose-phosphate synthase
Csa3G784370	3412.9	1102.8	Down	1.20×10^{-6}	Sucrose phosphatase
Csa3G149890	10,202.0	3362.5	Down	3.87×10^{-12}	Glucose-1-phosphate adenylyltransferase
Csa4G001950	2345.5	7726.7	Up	6.84×10^{-6}	Sucrose synthase
Csa4G420150	492.7	239.1	Down	1.17×10^{-5}	4-α-Glucanotransferase
Csa5G568310	4872.5	2423.0	Down	3.78×10^{-6}	Phosphoglucomutase
Csa2G004720	1255.1	2945.9	Up	1.42×10^{-6}	Multiprotein-bridging factor
Csa5G207960	11,815.2	4338.2	Down	1.77×10^{-9}	Omega-3 fatty acid desaturase
Csa3G808370	47.5	102.9	Up	0.00031088	Seed maturation protein LEA 4

2.4. Expression of Genes Involved in Sucrose Biosynthesis and Response to Water Deprivation

Based on the results of GO and KEGG analysis, we chose several DEGs, which were involved in the starch and sucrose synthesis and response to drought stress. A total of 9 transcripts were selected, including 6 genes with sucrose or starch and 3 genes with response to water deprivation (Table 2 and Table S7). The qRT-PCR assay was employed to validate A-16 and L-9 of RNA-seq results under normal and drought stress, respectively. The results showed that no significant changes were detected between these two cultivars before treatment. However, when treated with drought stress for 4 days, six genes were significantly down regulated in A-16, especially genes involved in the sucrose metabolic process, sucrose biosynthetic process, and starch biosynthetic process. The expression of the remaining three genes including genes related to sucrose synthase activity and response to water deprivation, increased significantly in A-16 when compared with L-9 (Figure 7). These results of qRT-PCR were consistent with the RNA-sequencing data.

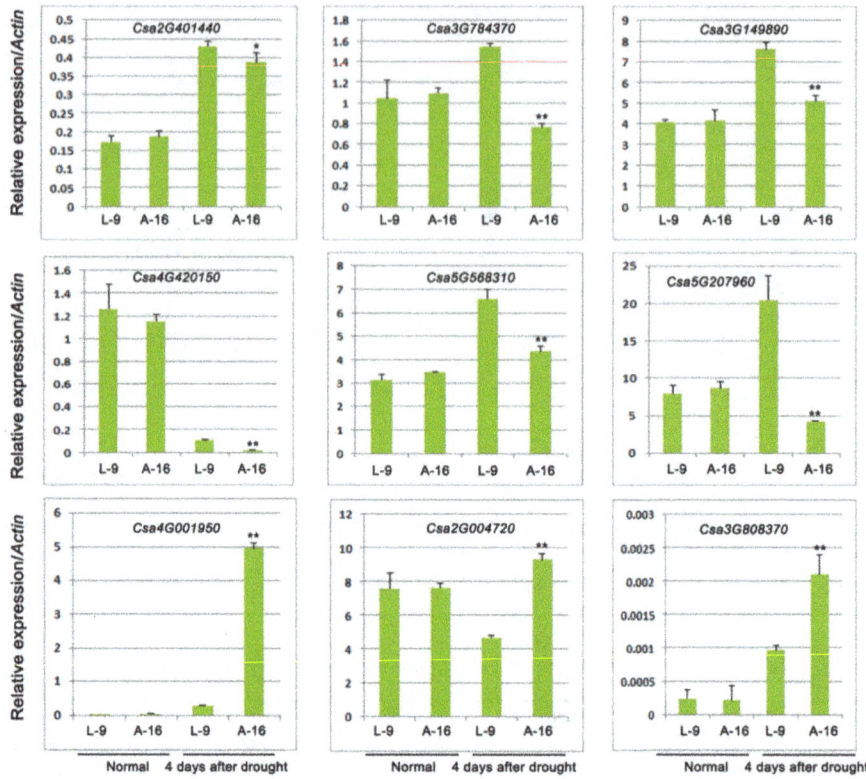

Figure 7. Relative expression of genes related to sucrose biosynthesis and response to water deprivation. Data is presented as the mean ± standard deviation ($n = 9$). * $0.01 \leq p \leq 0.05$, ** $p \leq 0.01$, Student's *t* test.

2.5. Analysis of Abscisic Acid (ABA)-Related Genes

Previous studies have reported that plant hormone, especially ABA, plays crucial roles in the regulation of the developmental process and signaling network involved in plant responses to drought stress [27]. Therefore, we selected the ABA-related genes among DEGs of drought stress from RNA-sequencing data (Table 3 and Table S7). In the present study, six genes related to the ABA signaling pathway were verified. The result showed that four genes were up-regulated and two down-regulated prominently (Figure 8), which was consistent with the RNA sequencing results.

Table 3. Genes involved in ABA signaling pathway.

Gene ID	L-9 Expression	A-16 Expression	Regulation	*p*-Value	Annotation
Csa3G135070	89.6	20.4	Down	3.05×10^{-9}	Calcium-dependent protein kinase
Csa3G133140	463.1	1758.6	Up	1.72×10^{-7}	3-Ketoacyl-CoA thiolase 1
Csa4G361820	1298.2	4273.3	Up	1.04×10^{-17}	NAC domain-containing protein
Csa4G430830	276.8	50.0	Down	1.29×10^{-18}	Calcium-dependent protein kinase-like protein
Csa6G382950	125.3	405.7	Up	4.08×10^{-8}	NAC domain-containing protein
Csa6G408800	47.5	102.9	Up	0.0003109	Circadian clock coupling factor,

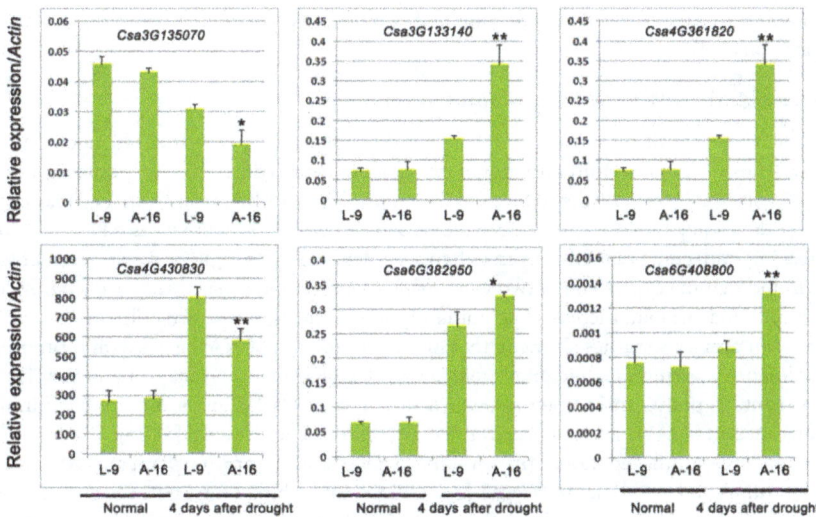

Figure 8. Relative expression of genes involved in ABA signaling pathway. Data is presented as the mean $\pm$ standard deviation ($n = 9$). * $0.01 \leq p \leq 0.05$, ** $p \leq 0.01$, Student's t test.

3. Discussion

The analysis and availability of diverse genetic resources could offer important information for understanding the molecular basis of variability in their response to drought stress [16]. In the study, we characterized two cucumber genotypes for their significantly different response to drought (L-9 and A-16) stress. A-16 exerted drought sensibility under water deficiency with increased MDA content and decreased SOD enzyme activity and chlorophyll content. Through the analysis of the transcript level by RNA-seq, we found that the number of DEGs increased significantly at the 4th day after drought treatment. Among them, several DEGs related to the sucrose synthesis and ABA signaling pathway were possibly involved in the drought response tolerance with prominent expression changes between the two cultivars.

3.1. A-16 Has Less Stomata in the Leaf Than L-9

Previous studies have reported that the regulation of stomatal opening and closure is crucial to the normal transpiration and plays an important role in the resistance of drought stress [28]. In rice, *am1* mutant showed drought resistance and highly percentage of completely closed stomata when compared with the wild type [29]. Drought-tolerant variety *dca1* has a lower number of stomata and more completely closed stomata than the control [30]. In our present study, we found that the number of stomata in L-9 was less than in A-16 in the same field size, indicating that L-9 could enhance its tolerance to drought stress by regulating the number of stomata.

3.2. Analysis of Sucrose and Starch Biosynthetic Process in Drought Stress

Sugar metabolism and starch biosynthesis are involved in the plant tolerance under drought stress [31]. Soluble sugar content is identified as a good marker in selecting the durum with drought tolerance [32]. The accumulation of soluble sugars in plant different tissues is reinforced when faced with different environmental stresses [33]. Under water deficiency, the soluble sugar was significantly accumulated in *Arabidopsis* leaves, resulting in its resistance to drought [34]. In our study, we found that most of genes involved in the sucrose and starch biosynthetic process were significantly up-regulated in the drought tolerant cultivar L-9, indicating that more sucrose and starch content might attribute to its resistance on drought stress.

3.3. Analysis of ABA Signal under Drought Stress

ABA plays essential role in the plant drought resistance because it could not only regulate the stomatal closure but also influence genes expression involved in stress-response and metabolic changes [35,36]. NAC transcriptional factors, which respond to ABA, could enhance plant tolerance under water deficiency [37–39]. In rice, both *OsNAC45* and *OsNAC52* were induced by ABA and their overexpressing transgenic plants showed enhanced tolerance to drought and salt treatments [38,40]. Here, we found that the expression of two genes (*Csa4M361820* and *Csa6M382950*), encoding the NAC domain-containing protein, increased prominently in the drought tolerant cultivar L-9, which was consistent with previous studies showing that higher expression of NAC genes could promote plant drought tolerance. Calcium-dependent protein kinase (CDPK), an important group of Ser/Thr protein kinases presents in plants and some protozoans that decode Ca^{2+} signals, are involved in the ABA signal transduction [41,42] and function in the plant response to drought [43,44]. Overexpression of *ZmCK3* (a maize calcium-dependent protein kinase gene) could improve plant survival rates under drought conditions in transgenic *Arabidopsis* [44]. CPK10, interacting with HSP1 (heat shock protein 1), plays important roles in ABA and Ca^{2+} mediated regulation of stomatal movements, leading to different tolerance to water deficiency [43]. *VfCPK1* of *Vicia faba* and *AtCPK11* of *Arabidopsis* are specifically induced by drought and ABA, respectively [45,46]. In this study, the drought-sensitive cultivar A-16 showed significantly decreased expression of *CDPK* genes (*Csa3M135070* and *Csa4M430830*) when compared with L-9, implying that high expression of *CDPK* might contribute to the drought tolerance in L-9.

3.4. Analysis of Cuticular Waxes Biosynthesis under Drought Stress

In addition, we also found that the "Cutin, suberin, and wax biosynthesis" pathway appeared under normal condition. The aerial surfaces of vascular plants are covered with a cuticle layer, including two major types of lipids, cutin and waxes [47]. Cuticular waxes play important roles in ensuring that plants grow and survive under various different biotic and abiotic stresses, which could help plants prevent non-stomatal water loss, and protect them against UV radiation and bacterial and fungal pathogens [48–50]. In the present study, several DEGs were enriched in the cutin pathway involved in lipid mechanism and were significantly down-regulated in A-16 when compared with L-9, suggesting that the decreased expression of related genes in cutin, suberin, and wax biosynthesis might be responsible for A-16's sensitivity to drought stress.

Overall, we firstly carried out RNA-Seq to analyze the regulation mechanism under water deficiency in cucumber. Several crucial genes involved in sucrose biosynthesis and ABA signal transduction were changed during drought stress. Our study not only provided a foundation for the further understanding of the regulation molecular on drought tolerance, but also explored valuable genes involved in drought tolerant, which will contribute to the improvement of drought resistant varieties in cucumber.

4. Materials and Methods

4.1. Plant Materials and Drought Treatment

Two cucumber cultivars, namely L-9 (South China type cucumber variety) and A-16 (North China type cucumber variety), were used in the study. Seeds were germinated overnight on wet filter in a culture dish at 28 °C in a dark environment. After that, the seedlings were grown in a feeding block under 14/10 h with 28/18 °C in day/night, respectively, in a culture room (5500 lux). When plants were grown to the two true leaves stage, they were subjected to lack of water for 7 days. After that, seedlings recovered for 3 days to normal condition. L-9 and A-16 seedlings were 120 for three biological replicates, respectively. Ten normal leaves were sampled from 10 plants before drought treatment, while drought-treated leaves were randomly sampled at the 4th day after drought treatment. Each biological replicate had a total of 10 leaves from 10 plants randomly selected. The samples were immediately frozen in liquid nitrogen and consistently stored at −80 °C until further analysis. In addition, leaf samples of three randomly selected biological replicates were then collected from both L-9 and A-16 plants (twelve samples in total).

4.2. Quantitative Analysis of Chlorophyll Content

Chlorophyll content was measured based on the procedure [51]. In detail, 0.2 g freshly-sampled leaves were homogenized in 5 mL solution with acetone and 0.1 M NH$_4$OH at a ratio of 9:1 and then centrifuged at 3000× *g* for 20 min. The obtained supernatants were then washed three times using hexane and finally the pigment content was measured by spectrophotometer at the absorption wavelengths of 663 and 645 nm (Beckman Coulter DU-800, Brea, CA, USA). According to the two formulas (Ca = 13.95 × D$_{665}$ − 6.88 × D$_{649}$ × 6 (mg/L); Cb = 24.96 × D$_{649}$ − 7.32 × D$_{665}$ × 5 (mg/L)), the concentrations of chlorophyll a and chlorophyll b were finally calculated, respectively.

4.3. Analysis of Malondialdehyde (MDA) Content by TBA Method

MDA content was measured according to the following procedures. Briefly, 0.5 g freshly-sampled leaves were dipped into 0.5% trichloroacetic acid (TCA) and ground into powder, then centrifuged at 3000× *g* for 20 min. A total of 2 mL supernatant was added to 2 mL 0.5% thibabituric acid (TBA) 0.5% TCA, after that, the mixture was boiled at 100 °C for 30 min. Then, absorption wavelengths of supernatants on 450 nm, 532 nm, 600 nm were recorded. According to the given formula (CMDA = 6.45 × (A$_{532}$ − A$_{600}$) − 0.56 × A$_{450}$ (μmol/L)), the MDA content was finally calculated.

4.4. Scanning Electron Microscopy (SEM)

Leaves of L-9 and A-16 seedlings under normal conditions were air-dried. The leaf abaxial epidermis was visualized under a HITACHI SU8020 variable pressure scanning electron microscope (SEM) (Hitachi, Tokyo, Japan) and imaged with an H-7500 transmission electron microscope (Hitachi).

4.5. Transmission Electron Microscopy (TEM)

Leaves of L-9 and A-16 seedlings under normal conditions were fixed overnight in 2.5% glutaraldehyde in 0.1 M phosphate buffer (pH 7.4) at 4 °C, then post-fixed in 2% (*v/v*) OsO4 in phosphate buffer. A series of 80 nm sections was cut using a Reichert OM2 ultramicrotome (Reichert, Deprew, New York, NY, USA), stained in 2% uranylacetate and 10 mM lead citrate (pH 12), before observation in a HitachiH-7650 (Hitachi) transmission electron microscope.

4.6. BGISEQ-500 Library Construction

A total of twelve samples (three biological replicates each of L-9 and A-16 at normal and drought stress, respectively) were used for RNA extraction with TRIZOL reagent according to the manufacturer's protocol (TaKaRa, Shiga, Japan). Each biological replicate had a total of 10 leaves from 10 plants, selected randomly. After extraction, RNA was then purified (using DNAse) and concentrated using an RNeasyMinElute clean up kit (Qiagen, Duesseldorf, Germany). Then, 2.5 μg RNA of each sample was prepared for constructing BGISEQ-500 library according to the protocol of previous study [52]. Library quality was tested using the Agilent Bioanalyzer (Life Technologies, Carlsbad, CA, USA) 2100 system and the genome reference was the cucumber 9930 genome (http://cucurbitgenomics.org/, Two years).

4.7. Screening and Significant Test for Differentially Expressed Genes (DEGs)

Gene expression level was calculated by quantifying the reads according to the RPKM (reads per kilobase per million reads) method [53]. Then the NOISeq was used to identify DEGs, which existed in the normal and drought stress transcriptome libraries according to the following criteria: fold change ≥ 2 and divergence probability ≥ 0.8. GO enrichment for these DEGs was performed using WEGO software [54]. To further obtain knowledge of DEG biological functions, pathway enrichment analysis was carried out according to the KEGG database [55], the major public pathway-related database.

4.8. Quantitative Real-Time PCR (qRT-PCR) Identification

Quantitative real-time PCR analysis was performed using the total RNA from seedling leaves of both the normal and drought stress treatment. Twenty μL cDNA was obtained using the QuantiTect Reverse Transcription Kit (Qiagen, Duesseldorf, Germany). Quantitative qRT-PCR (20 μL reaction volume) was carried out with 0.5 μL of cDNA, 0.2 μM of primer mix and SYBR Premix Ex Taq Kit (TaKaRa,Shiga, Japan). In an ABI PRISM 7900HT system (Life Technologies, Carlsbad, CA, USA), cucumber α-TUBULIN (*TUA*) gene was used as normal. qRT-PCR was carried out on an ABI 7500 Real-Time PCR System (Applied Biosystems, USA). In addition, all qRT-PCR primers were listed in the Table S8.

4.9. Statistical Analysis

The linux rhel6.7 x64 R-3.4.2 and MEGA6 were used to perform the heat-map and cluster analysis. Significant differences were detected by IBM SPSS Statistics 20 (by Student's *t* test). Relative gene expressions were calculated using the $2^{-\Delta\Delta Ct}$ method [56]. In addition, GraphPad Prism 5 was used for chart preparation.

Supplementary Materials: The following are available online at http://www.mdpi.com/1422-0067/19/7/2067/s1.

Author Contributions: M.W. and Y.L. designed the experiment. M.W. and B.J. performed most of the experiments. Q.P., W.L., X.H. and Z.L. performed part of the experiment. M.W. wrote the paper. Y.L. edited the manuscript.

Funding: This work was supported by the National Key Research and Development Program of China: 2016YFD0100204-16, Special Fund for Agro-scientific Research in the Public Interest: 201503110-07, Science and Technology Planning Project of Guangdong Province, China: 2016B020201008, 2015B020231004, The presidential foundation of Guangdong Academy of Agricultural Sciences: 201813, Guangdong special project youth top-notch talent project, 2016TQ03N529, the Science and Technology Program of Guangdong, 2015B020231004.

Conflicts of Interest: The authors declare no conflict of interest.

Abbreviations

RNA seq	RNA sequencing
DEGs	Differently expressed genes
ABA	Abscisic acid
MDA	Malondialdehyde
SOD	Enzyme activity of superoxide dismutase
SEM	Scanning electron microscopy
TEM	Transmission electron microscopy
GO	Gene ontology
qRT-PCR	Quantitative Real-Time PCR

References

1. Jaleel, C.A.; Manivannan, P.A.; Wahid, A.; Farooq, M.; Al-Juburi, H.J.; Somasundaram, R.A.; Panneerselvam, R. Drought stress in plants: A review on morphological characteristics and pigments composition. *Int. J. Agric. Biol.* **2009**, *11*, 100–105.

2. Kusaka, M.; Ohta, M.; Fujimura, T. Contribution of inorganic components to osmotic adjustment and leaf folding for drought tolerance in pearl millet. *Phys. Plant.* **2005**, *125*, 474–489. [CrossRef]

3. Shao, H.B.; Chu, L.Y.; Shao, M.A.; Jaleel, C.A.; Hong, M.M. Higher plant antioxidants and redox signaling under environmental stresses. *C. R. Biol.* **2008**, *331*, 433–441. [CrossRef] [PubMed]

4. Bohnert, H.J.; Nelson, D.E.; Jensen, R.G. Adaptations to environmental stresses. *Plant Cell* **1995**, *7*, 1099–1111. [CrossRef] [PubMed]

5. Xiong, L.; Zhu, J.K. Molecular and genetic aspects of plant responses to osmotic stress. *Plant Cell Environ.* **2002**, *25*, 131–139. [CrossRef] [PubMed]

6. Shinozaki, K.; Yamaguchi-Shinozaki, K. Gene networks involved in drought stress response and tolerance. *J. Exp. Bot.* **2007**, *58*, 221–227. [CrossRef] [PubMed]

7. Kwak, J.M.; Mori, I.C.; Pei, Z.M.; Leonhardt, N.; Torres, M.A.; Dangl, J.L.; Bloom, R.E.; Bodde, S.; Jone, J.D.G.; Schroeder, J.I. NADPH oxidase AtrbohD and AtrbohF genes function in ROS-dependent ABA signaling in Arabidopsis. *EMBO J.* **2003**, *22*, 2623–2633. [CrossRef] [PubMed]

8. Fujita, Y.; Fujita, M.; Shinozaki, K.; Yamaguchi-Shinozaki, K. ABA-mediated transcriptional regulation in response to osmotic stress in plants. *J. Plant Res.* **2011**, *124*, 509–525. [CrossRef] [PubMed]

9. Ma, Y.; Szostkiewicz, I.; Korte, A.; Moes, D.; Yang, Y.; Christmann, A.; Grill, E. Regulators of PP2C phosphatase activity function as abscisic acid sensors. *Science* **2009**, *324*, 1064–1068. [CrossRef] [PubMed]

10. Park, S.Y.; Fung, P.; Nishimura, N.; Jensen, D.R.; Fujii, H.; Zhao, Y.; Lumba, S.; Santiago, J.; Rodrigues, A.; Tsz-fung, F.C.; et al. Abscisic acid inhibits type 2c protein phosphatases via the pyr/pyl family of start proteins. *Science* **2009**, *324*, 1068–1071. [CrossRef] [PubMed]

11. Malepszy, S. Cucumber (*Cucumis Sativus* L.). In *Crops II*; Bajaj, Y.P.S., Ed.; Springer: Berlin/Heidelberg, Germany, 1988.

12. Chen, Y.; Wang, M.; Hu, L.; Liao, W.; Dawuda, M.M.; Li, C. Carbon monoxide is involved in hydrogen gas-induced adventitious root development in cucumber under simulated drought stress. *Front. Plant Sci.* **2017**, *8*, 128. [CrossRef] [PubMed]

13. Wang, W.; Zhang, Y.; Xu, C.; Ren, J.; Liu, X.; Black, K.; Gai, X.; Wang, Q.; Ren, H. Cucumber *ECERIFERUM1*, (*CsCER1*), which influences the cuticle properties and drought tolerance of cucumber, plays a key role in VLC alkanes biosynthesis. *Plant Mol. Biol.* **2015**, *87*, 219–233. [CrossRef] [PubMed]

14. Sun, Y.; Wang, H.; Liu, S.; Peng, X. Exogenous application of hydrogen peroxide alleviates drought stress in cucumber seedlings. *S. Afr. J. Bot.* **2016**, *106*, 23–28. [CrossRef]

15. Yin, Z.; Hennig, J.; Szwacka, M.; Malepszy, S. Tobacco PR-2d promoter is induced in transgenic cucumber in response to biotic and abiotic stimuli. *J. Plant Physiol.* **2004**, *161*, 621–629. [CrossRef] [PubMed]

16. Garg, R.; Shankar, R.; Thakkar, B.; Kudapa, H.; Krishnamurthy, L.; Mantri, N.; Varshney, R.K.; Bhatia, S.; Jain, M. Transcriptome analyses reveal genotype-and developmental stage-specific molecular responses to drought and salinity stresses in chickpea. *Sci. Rep.* **2016**, *6*, 19228. [CrossRef] [PubMed]

17. Zhou, Y.; Yang, P.; Cui, F.; Zhang, F.; Luo, X.; Xie, J. Transcriptome Analysis of Salt Stress Responsiveness in the Seedlings of Dongxiang Wild Rice (*Oryza rufipogon* Griff.). *PLoS ONE* **2016**, *11*, e0146242. [CrossRef] [PubMed]

18. Cui, J.Y.; Gunewardena, S.S.; Yoo, B.; Liu, J.; Renaud, H.J.; Lu, H.; Zhong, X.B.; Klaassen, C.D. RNA-Seqreveals different mRNA abundance of transporters and their alternative transcript isoforms during liver development. *Toxicol. Sci.* **2012**, *127*, 592–608. [CrossRef] [PubMed]

19. Zhao, S.; Fung-Leung, W.P.; Bittner, A.; Ngo, K.; Liu, X. Comparison of RNA-Seq and microarray in transcriptomeprofiling of activated T cells. *PLoS ONE* **2014**, *9*, e78644.

20. Ando, K.; Grumet, R. Transcriptional profiling of rapidly growing cucumber fruit by 454-pyrosequencing analysis. *J. Am. Soc. Hortic. Sci.* **2010**, *135*, 291–302.

21. Li, J.; Wu, Z.; Cui, L.; Zhang, T.; Guo, Q.; Xu, J.; Jia, L.; Lou, Q.; Huang, S.; Li, Z.; et al. Transcriptome comparison of global distinctive features between pollination and parthenocarpic fruit set reveals transcriptional phytohormone cross-talk in cucumber (*Cucumis sativus* L.). *Plant Cell Physiol.* **2014**, *55*, 1325–1342. [CrossRef] [PubMed]

22. Guo, S.; Zheng, Y.; Joung, J.G.; Liu, S.; Zhang, Z.; Crasta, O.R.; Sobral, B.W.; Xu, Y.; Huang, S.; Fei, Z. Transcriptome sequencing and comparative analysis of cucumber flowers with different sex types. *BMC Genom.* **2010**, *11*, 384. [CrossRef] [PubMed]

23. Chen, C.; Liu, M.; Jiang, L.; Liu, X.; Zhao, J.; Yan, S.; Yang, S.; Ren, H.; Liu, R.; Zhang, X. Transcriptome profiling reveals roles of meristem regulators and polarity genes during fruit trichome development in cucumber (*Cucumis sativus* L.). *J. Exp. Bot.* **2014**, *65*, 4943–4958. [CrossRef] [PubMed]

24. Zhang, N.; Zhang, H.J.; Zhao, B.; Sun, Q.Q.; Cao, Y.Y.; Li, R.; Wu, X.X.; Weeda, S.; Li, L.; Ren, S.; et al. The RNA-seq approach to discriminate gene expression profiles in response to melatonin on cucumber lateral root formation. *J. Pineal Res.* **2014**, *56*, 39–50. [CrossRef] [PubMed]

25. Zhao, W.; Yang, X.; Yu, H.; Jiang, W.; Sun, N.; Liu, X.; Liu, X.; Zhang, X.; Wang, Y.; Gu, X. RNA-Seq-based transcriptome profiling of early nitrogen deficiency response in cucumber seedlings provides new insight into the putative nitrogen regulatory network. *Plant Cell Physiol.* **2015**, *56*, 455–467. [CrossRef] [PubMed]

26. Huang, S.; Li, R.; Zhang, Z.; Li, L.; Gu, X.; Fan, W.; Lucas, W.J.; Wang, X.; Xie, B.; Ni, P.; et al. The genome of the cucumber, *Cucumis sativus* L. *Nat. Genet.* **2009**, *41*, 1275–1281. [CrossRef] [PubMed]

27. Bari, R.; Jones, J.D. Role of plant hormones in plant defence responses. *Plant Mol. Biol.* **2009**, *69*, 473–488. [CrossRef] [PubMed]

28. Lim, C.W.; Baek, W.; Jung, J.; Kim, J.H.; Lee, S.C. Function of ABA in Stomatal Defense against Biotic and Drought Stresses. *Int. J. Mol. Sci.* **2015**, *16*, 15251–15270. [CrossRef] [PubMed]

29. Sheng, P.; Tan, J.; Jin, M.; Wu, F.; Zhou, K.; Ma, W.; Heng, Y.; Wang, J.; Guo, X.; Zhang, X.; et al. Albino midrib 1, encoding a putative potassium efflux antiporter, affects chloroplast development and drought tolerance in rice. *Plant Cell Rep.* **2014**, *33*, 1581–1594. [CrossRef] [PubMed]

30. Cui, L.G.; Shan, J.X.; Shi, M.; Gao, J.P.; Lin, H.X. DCA1 acts as a transcriptional co-activator of DST and contributes to drought and salt tolerance in rice. *PLoS Genet.* **2015**, *11*, e1005617. [CrossRef] [PubMed]

31. Mohammadkhani, N.; Heidari, R. Drought-induced Accumulation of Soluble Sugars and Proline in Two Maize Varieties. *World Appl. Sci. J.* **2008**, *3*, 448–453.

32. Al Hakimi, A.; Monneveux, P.; Galiba, G. Soluble sugars, proline and relative water content (RCW) as traits for improving drought tolerance and divergent selection for RCW from triticumpolonicum into triticum durum. *J. Genet. Breed.* **1995**, *49*, 237–244.

33. Prado, F.E.; Boero, C.; Gallardo, M.; González, J.A. Effect of nacl on germination, growth, and soluble sugar content in chenopodium quinoa willd. seeds. *Bot. Bull. Acad. Sin.* **2000**, *41*, 27–34.

34. Sperdouli, I.; Moustakas, M. Interaction of proline, sugars, and anthocyanins during photosynthetic acclimation of Arabidopsis thaliana to drought stress. *J. Plant Physiol.* **2012**, *169*, 577–585. [CrossRef] [PubMed]

35. Seki, M.; Umezawa, T.; Urano, K.; Shinozaki, K. Regulatory metabolic networks in drought stress responses. *Curr. Opin. Plant Biol.* **2007**, *10*, 296–302. [CrossRef] [PubMed]

36. Danquah, A.; de Zelicourt, A.; Colcombet, J.; Hirt, H. The role of ABA and MAPK signaling pathways in plant abiotic stress responses. *Biotechnol. Adv.* **2014**, *32*, 40–52. [CrossRef] [PubMed]

37. Hu, H.; Dai, M.; Yao, J.; Xiao, B.; Li, X.; Zhang, Q.; Xiong, L. Overexpressing a NAM, ATAF, and CUC (NAC) transcription factor enhances drought resistance and salt tolerance in rice. *Proc. Natl. Acad. Sci. USA* **2006**, *103*, 12987–12992. [CrossRef] [PubMed]

38. Zheng, X.; Chen, B.; Lu, G.; Han, B. Overexpression of a NAC transcription factor enhances rice drought and salt tolerance. *Biochem. Biophys. Res. Commun.* **2009**, *379*, 985–989. [CrossRef] [PubMed]

39. Liu, G.; Li, X.; Jin, S.; Liu, X.; Zhu, L.; Nie, Y.; Zhang, X. Overexpression of rice NAC gene SNAC1 improves drought and salt tolerance by enhancing root development and reducing transpiration rate in transgenic cotton. *PLoS ONE* **2014**, *9*, e86895. [CrossRef] [PubMed]

40. Gao, F.; Xiong, A.; Peng, R.; Jin, X.; Xu, J.; Zhu, B.; Chen, J.; Yao, Q. OsNAC52, a rice NAC transcription factor, potentially responds to ABA and confers drought tolerance in transgenic plants. *Plant Cell Tissue Organ Cult.* **2010**, *100*, 255–262. [CrossRef]

41. Zhou, X.; Zhang, H. Roles of calcium-dependent protein kinases in ABA-regulation of stomatal moment in Poplar. *Sci. Technol. Eng.* **2004**, *4*, 80–83. (In Chinese)

42. Zhu, S.Y.; Yu, X.C.; Wang, X.J.; Zhao, R.; Li, Y.; Fan, R.C.; Shang, Y.; Du, S.Y.; Wang, X.F.; Wu, F.Q.; et al. Two Calcium-Dependent Protein Kinases, CPK$_4$ and CPK$_{11}$, Regulate Abscisic Acid Signal Transduction in *Arabidopsis*. *Plant Cell* **2007**, *19*, 3019–3036. [CrossRef] [PubMed]

43. Zou, J.J.; Wei, F.J.; Wang, C.; Wu, J.J.; Ratnasekera, D.; Liu, W.X.; Wu, W.H. Arabidopsis calcium-dependent protein kinase CPK10 functions in abscisic acid- and Ca^{2+}-mediated stomatal regulation in response to drought stress. *Plant Physiol.* **2010**, *154*, 1232–1243. [CrossRef] [PubMed]

44. Wang, C.T.; Song, W. *ZmCK$_3$*, a maize calcium-dependent protein kinase gene, endows tolerance to drought and drought stresses in transgenic *Arabidopsis*. *J. Plant Biochem. Biotechnol.* **2014**, *23*, 249–256. [CrossRef]

45. Liu, G.; Chen, J.; Wang, X. *VfCPK1*, a gene encoding calcium-dependent protein kinase from *Vicia faba*, is induced by drought and abscisic acid. *Plant Cell Environ.* **2006**, *29*, 2091–2099. [CrossRef] [PubMed]

46. Huang, K.; Peng, L.; Liu, Y.; Yao, R.; Liu, Z.; Li, X.; Yang, Y.; Wang, J. Arabidopsis calcium-dependent protein kinase AtCPK1 plays a positive role in salt/drought-stress response. *Biochem. Biophys. Res. Commun.* **2017**, *49*. [CrossRef] [PubMed]

47. Bernard, A.; Joubès, J. Arabidopsis cuticular waxes: Advances in synthesis, export and regulation. *Prog. Lipid Res.* **2013**, *52*, 110–129. [CrossRef] [PubMed]

48. España, L.; Heredia-Guerrero, J.A.; Reina-Pinto, J.J.; Fernández-Muñoz, R.; Heredia, A.; Domínguez, E. Transient silencing of CHALCONE SYNTHASE during fruit ripening modifies tomato epidermal cells and cuticle properties. *Plant Physiol.* **2014**, *166*, 1371–1386. [CrossRef] [PubMed]

49. Reisige, K.; Gorzelanny, C.; Daniels, U.; Moerschbacher, B.M. The C28 aldehyde octacosanal is a morphogenetically active component involved in host plant recognition and infection structure differentiation in the wheat stem rust fungus. *Physiol. Mol. Plant Pathol.* **2006**, *68*, 33–40. [CrossRef]

50. Shepherd, T.; Wynne, G.D. The effects of stress on plant cuticular waxes. *New Phytol.* **2006**, *171*, 469–499. [CrossRef] [PubMed]

51. Suzuki, Y.; Makino, A. Availability of rubisco small subunit up-regulates the transcript levels of large subunit for stoichiometric assembly of its holoenzyme in rice. *Plant Physiol.* **2012**, *160*, 533–540. [CrossRef] [PubMed]

52. Fehlmann, T.; Reinheimer, S.; Geng, C.; Su, X.; Drmanac, S.; Alexeev, A.; Zhang, C.; Backes, C.; Ludwig, N.; Hart, M.; et al. cPAS-based sequencing on the BGISEQ-500 to explore small non-coding RNAs. *Clin. Epigenet.* **2016**, *8*, 123. [CrossRef] [PubMed]

53. Li, B.; Dewey, C.N. RSEM: Accurate transcript quantification from RNA-Seq data with or without a reference genome. *BMC Bioinform.* **2011**, *12*, 323. [CrossRef] [PubMed]

54. Ye, J.; Fang, L.; Zheng, H.; Zhang, Y.; Chen, J.; Zhang, Z.; Wang, J.; Li, S.; Li, R.; Bolund, L.; et al. WEGO: A web tool for plotting GO annotations. *Nucleic Acids Res.* **2006**, *34*, W293. [CrossRef] [PubMed]

55. Kanehisa, M.; Araki, M.; Goto, S.; Hattori, M.; Hirakawa, M.; Itoh, M.; Katayama, T.; Kawashima, S.; Okuda, S.; Tokimatsu, T.; et al. KEGG for linking genomes to life and the environment. *Nucleic Acids Res.* **2008**, *36*, 480–484. [CrossRef] [PubMed]

56. Vandesompele, J.; de Preter, K.; Pattyn, F.; Poppe, B.; van Roy, N.; de Paepe, A.; Speleman, F. Accuratenormalization of real-time quantitative RT-PCR data by geometric averaging of multiple internal controlgenes. *Genome Biol.* **2002**, *3*, research0034.1. [CrossRef] [PubMed]

Article

Interactions between WUSCHEL- and CYC2-like Transcription Factors in Regulating the Development of Reproductive Organs in *Chrysanthemum morifolium*

Yi Yang [1], Ming Sun [1], Cunquan Yuan [1], Yu Han [1], Tangchun Zheng [1], Tangren Cheng [1], Jia Wang [1] and Qixiang Zhang [1,2,*]

[1] Beijing Key Laboratory of Ornamental Plants Germplasm Innovation & Molecular Breeding, National Engineering Research Center for Floriculture, Beijing Laboratory of Urban and Rural Ecological Environment, Engineering Research Center of Landscape Environment of Ministry of Education, Key Laboratory of Genetics and Breeding in Forest Trees and Ornamental Plants of Ministry of Education, School of Landscape Architecture, Beijing Forestry University, Beijing 100083, China; yiyang921124@126.com (Y.Y.); 13683295193@163.com (M.S.); yuancunquan@163.com (C.Y.); hanyu19880514@126.com (Y.H.); zhengtangchun@126.com (T.Z.); chengtangren@163.com (T.C.); wangjia8248@163.com (J.W.)

[2] Beijing Advanced Innovation Center for Tree Breeding by Molecular Design, Beijing Forestry University, Beijing 100083, China

* Correspondence: zqxbjfu@126.com; Tel.: +86-10-62338347; Fax: +86-10-62336321

Received: 17 February 2019; Accepted: 11 March 2019; Published: 14 March 2019

Abstract: *Chrysanthemum morifolium* is a gynomonoecious plant that bears both female zygomorphic ray florets and bisexual actinomorphic disc florets in the inflorescence. This sexual system is quite prevalent in Asteraceae, but poorly understood. CYCLOIDEA (CYC) 2 subclade transcription factors, key regulators of flower symmetry and floret identity in Asteraceae, have also been speculated to function in reproductive organs and could be an entry point for studying gynomonoecy. However, the molecular mechanism is still unclear. On the other hand, the *Arabidopsis* WUSCHEL (WUS) transcription factor has been proven to play a vital role in the development of reproductive organs. Here, a *WUS* homologue (*CmWUS*) in *C. morifolium* was isolated and characterized. Overexpression of *CmWUS* in *A. thaliana* led to shorter siliques and fewer stamens, which was similar to *CYC2*-like genes reported before. In addition, both *CmWUS* and *CmCYC2* were highly expressed in flower buds during floral organ differentiation and in the reproductive organs at later development stages, indicating their involvement in the development of reproductive organs. Moreover, CmWUS could directly interact with CmCYC2d. Thus, our data suggest a collaboration between CmWUS and CmCYC2 in the regulation of reproductive organ development in chrysanthemum and will contribute to a further understanding of the gynomonoecious sexual system in Asteraceae.

Keywords: *Chrysanthemum morifolium*; WUS; CYC2; gynomonoecy; reproductive organ; flower symmetry

---

1. Introduction

The inflorescence of *C. morifolium* (Asteraceae) is always comprised of two kinds of florets: the bilaterally symmetric female ray florets and radially symmetric bisexual disc florets [1]. Different sex expression and flower symmetry in ray and disc florets are significant features of chrysanthemum inflorescence. This gynomonoecious sexual system is quite prevailing in Asteraceae and has been considered to play a pivotal role in reducing herbivore damage and pollen-pistil interference, as well as in attracting pollinators [2,3]. However, the genetic mechanism of gynomonoecy is poorly understood.

Interestingly, the connection between shifts in flower symmetry and the development of reproductive organs has been discovered and recorded in many species, including Asteraceae members [4–6], and the flower symmetry genes have been speculated to be involved in breeding system [6–8].

CYCLOIDEA (CYC) 2 subclade transcription factors, which belong to ECE-CYC/TB1 clade of plant-specific TCP family [9,10], have been proven to be essential for the regulation of flower symmetry in angiosperms [11] and inflorescence architecture in Asteraceae [6,12]. *CYC* of *Antirrhinum* was the first gene isolated in this subclade and is expressed in the dorsal domain of floral meristem from initiation and maintained throughout the differentiation of petals and stamens [13]. *CYC* promotes the growth of dorsal petals and arrests the development of dorsal stamen to form a staminode [13,14]. Gaudin et al. [15] speculated that *CYC* could directly or indirectly suppress the expression of *cyclinD3b* and other cell cycle genes in the staminode. Studies in *Opithandra* further indicated the negative effects of *OpdCYC* on *OpdcyclinD3* genes and the correlation between the expressions of *OpdCYC* and the abortion of both dorsal and ventral stamen [16]. On the other hand, Preston et al. [17] found that expression patterns of *CYC2*-like genes were not corelated with patterns of stamen arrest in *Veronica montana* and *Gratiola officinalis*. In contrast, in *Papaveracea*, *CYC2*-like genes promote stamen initiation and growth [18].

Previous studies have shown that *CYC2*-like genes in gerbera (*Gerbera hybrida*) are functionally redundant in regulating ray floret identity by promoting ligule growth and suppressing stamen development [19–21]. In addition to stamens, *CYC2*-like genes have also been speculated to have late functions in the development of ovaries and carpels in Asteraceae [6,12]. Both in gerbera and sunflower (*Helianthus annuus*), *CYC2*-like genes are highly expressed in ovary, stigma and style tissues [22]. Expression levels of *AcCYC2a* and *AcCYC2d* are also increased in the developing ovules of *Anacyclus clavatus* [23]. Moreover, constitutive expression of all the gerbera *CYC2*-like genes, except *GhCYC2*, in *A. thaliana* leads to shorter siliques with fewer seeds. In addition, stamen development is also severely disrupted in the transgenic lines ectopically expressing *GhCYC4* and *GhCYC7* [20]. Also, different from the empty achenes in the zygomorphic ray florets of wild type plants, actinomorphic ray florets can produce filled achenes through hand pollination in the *turf* mutant [24], which is caused by insertion of TEs in the TCP domain of *HaCYC2c* in sunflower [25–27]. Still, the molecular relationship between *CYC2*-like genes and the development of reproductive organs awaits more research to elaborate.

In addition to *CYC2*-like genes, our previous comparative transcriptome analysis between ray and disc florets in *C. morifolium* [28] has predicted other candidate transcription factor genes during inflorescence development and organ determination for further studies. Among them, a *WUSCHEL*-like gene, which was highly expressed in the central disc florets, has attracted our attention. *WUSCHEL* (*WUS*) is a member of the WUSCHEL-RELATED HOMEOBOX (WOX) family [29] and takes part in several regulatory networks in shoot and floral meristems [30–32]. In *A. thaliana*, *WUS* is expressed in anther stomium cells during early stages and is required for anther development [33]. In ovules, *WUS* is confined to be expressed in the nucellus and is essential for the initiation of integument [34,35]. Reduced stamens and disappeared carpel in *wus* mutants of *Arabidopsis* also suggest crucial functions of *WUS* in the development of reproductive organs [36]. In *Cucumis sativus* (cucumber), CsWUS directly interacts with CsSPL, a vital factor in male and female fertility, and takes part in the regulatory network that controls the development of reproductive organs [37].

We have identified six *CmCYC2* genes in *C. morifolium* before, and they were also found to be strongly expressed in ray florets [38]. In contrast to *CmCYC2*, *CmWUS* was highly expressed in disc florets [28]. These two distinct expression patterns between ray and disc florets suggested their involvements in the development of inflorescence. In particular, whether *CmWUS* and *CmCYC2* are connected in reproductive organ development of chrysanthemum is an interesting problem worthy of study. Here, we isolated *CmWUS* and expressed it ectopically in *A. thaliana* for functional analysis. Additionally, expression patterns of *CmWUS* and *CmCYC2* during inflorescence development were compared. Furthermore, we performed yeast two-hybrid (Y2H) and bimolecular fluorescence complementation (BiFC) assays to determine protein-protein interactions between CmWUS and

CmCYC2. These results show a direct interaction between CmWUS and CmCYC2 and may help to understand the genetic and molecular mechanisms of reproductive organ development in Asteraceae.

2. Results

2.1. Identification and Phylogenetic Analysis of CmWUS

To identify the function of *CmWUS* in reproductive organ development, we isolated the ORF of *CmWUS* (912 bp) from inflorescences of *C. morifolium* 'Fen Ditan'. The encoded CmWUS protein (303 amino acids) was aligned with WUS-like sequences from other species. As shown in Supplementary Figure S1A, the WOX domain [29] was highly conserved. The signature motifs WUS-box and EAR-like motif [39] were also identified in CmWUS at the carboxyl terminus. A neighbor joining phylogenetic tree (Supplementary Figure S1B) was constructed based on the full length of amino acid sequences of 15 WOX family members from *A. thaliana* and WUS from other species. As described by Graaff, Laux and Rensing [29], these WOX members could be divided into three clades: the ancient clade, the intermediate clade and the WUS clade. The phylogenetic analysis confirmed that CmWUS belongs to WUS clade of WOX family and is closely related to WUS-like from other species of Asterceae: *H. annuus, L. sativa* and *C. cardunculus*.

2.2. Overexpression of CmWUS in A. thaliana Inhibits the Development of Reproductive Organs and Affects Flower Symmetry

The *CmWUS* ORF was overexpressed in *A. thaliana* (Columbia) for functional analysis during floral development. The transgenic lines in which *CmWUS* was highly expressed were confirmed by qPCR assay. Three *35S::CmWUS* lines (line 6, 8,13) with higher and consistent expression levels were selected for detailed analysis. The wild type *Arabidopsis* flowers are polysymmetric with four sepals, four petals, four medial and two lateral stamens and two fused carpels (Figure 1A). Meanwhile, in our transgenic lines, the flowers were changed into monosymmetric with one symmetry plane. The petals on both sides of the lateral stamens were arranged close to each other and the development of the lateral stamens was also inhibited (Figure 1B–E). As listed in Figure 1K, the number of stamens in three *35S::CmWUS* lines were reduced to 4 to 5. In addition, they produced shorter siliques than the wild type. In addition to these three transgenic lines, line 1 showed a stronger phenotype, with flower meristems that were ectopically initiated on the surface of inflorescence stems (Figure 1H,I), which was consistent with the phenotype of *Arabidopsis* overexpressing *AtWUS* [40,41]. Furthermore, petals were slightly curled at the edges (Figure 1F) and siliques were much shorter than wild type (Figure 1G) in line 1.

Figure 1. Ectopic expression of *CmWUS* in *A. thaliana* (Columbia). (**A**) Flower of wild type *A. thaliana* (Columbia). (**B–E**) Flower phenotypes in transgenic line 6, 8 and 13. Petals on both sides of the two lateral stamens were arranged close to each other and the development of the lateral stamens was also inhibited. The number of stamens were reduced to 4 (**B,C**) and 5 (**D,E**). Stamens are marked with white arrows. (**F**) Slightly curled petals at the edges of the flowers in transgenic line 1. (**G**) Siliques of transgenic line 1 (left) were much shorter than wild type (right). (**H,I**) Ectopic initiated flower buds on the surface of inflorescence stems in transgenic line 1. (**J**) qPCR detection of *CmWUS* transcripts in wild type (WT) and transgenic lines of *A. thaliana*. The endogenous *Arabidopsis ACTIN* was chosen as a housekeeper gene. (**K**) Statistics of silique length and stamen number in wild type and transgenic lines of *Arabidopsis*. Statistically significant differences are indicated with lowercase letters (Fisher's LSD, $p < 0.05$). Bars = 1 mm.

2.3. High Expression of CmWUS and CmCYC2 in the Reproductive Organs of C. morifolium

Three developmental phases of flower buds—initiation of floral primordia (I), differentiation of floral organs (II) and growth of floral organs (III) [38]—were selected (Figure 2A–C) to analyze the expression of *CmWUS* at early stages of inflorescence development in *C. morifolium*. As shown in Figure 2D, the expression level of *CmWUS* increased from stage I to stage II and then decreased to the lowest at stage III, which was similar to *CmCYC2* genes reported previously [38] and indicates their involvement in floral organ differentiation.

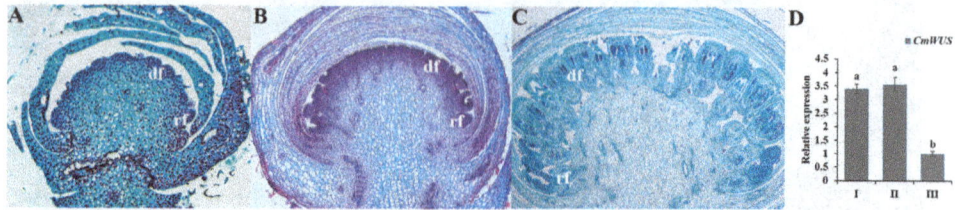

Figure 2. Expression patterns of *CmWUS* in flower buds of *C. morifolium* 'Fen Ditan' at early stages of inflorescence development. (**A–C**) Morphological characteristics of flower buds at three stages: I, initiation of floral primordia (**A**); II, differentiation of floral organs (**B**); and III, growth of floral organs (**C**) were analyzed at a histological level. Abbreviations: rf = ray florets, df = disc florets. (**D**) Expression levels of *CmWUS* in flower buds at stage I, II and III of inflorescence development. The expression levels are relative to the flower buds at stage III. Expression levels of *PP2Acs* are utilized for normalization. Error bars show the standard deviation of three biological replicates. Statistically significant differences are indicated with different lowercase letters (Fisher's LSD, $p < 0.05$).

qPCR assays were also performed to compare the expression patterns of *CmWUS* and *CmCYC2* at later stages of inflorescence development (Figure 3) between ray and disc florets. As shown in Figure 4A, *CmWUS* was expressed extremely highly in disc florets, especially at stage 1. The expression of *CmWUS* in ray florets was also detected, but was pretty weak compared to disc florets. Unlike *CmWUS*, *CmCYC2* genes, especially *CmCYC2c* and *CmCYC2d*, were expressed at relatively higher levels in ray florets than disc florets. To further explore the possible roles of *CmWUS* and *CmCYC2* genes, their expression levels in different tissues of *C. morifolium* 'Fen Ditan' at late development stages were studied. As shown in Figure 4B, *CmCYC2* and *CmWUS* were primarily expressed in floral organs and were strongly expressed in pistils (including ovary, style and stigma). *CmWUS* was also expressed in stamens, but the expression level was not as high as in pistils like *CmCYC2d*. *CmCYC2* genes were also expressed at high levels in petals, especially in ray petals, while *CmWUS* was not, which may explain the differences in expression levels of *CmWUS* and *CmCYC2* between ray and disc florets. Thus, we speculate that *CmWUS* and *CmCYC2* genes are all involved in the regulation of reproductive organ (especially the pistils) development.

Figure 3. Inflorescence morphology of *C. morifolium* 'Fen Ditan' and five later stages of inflorescence development.

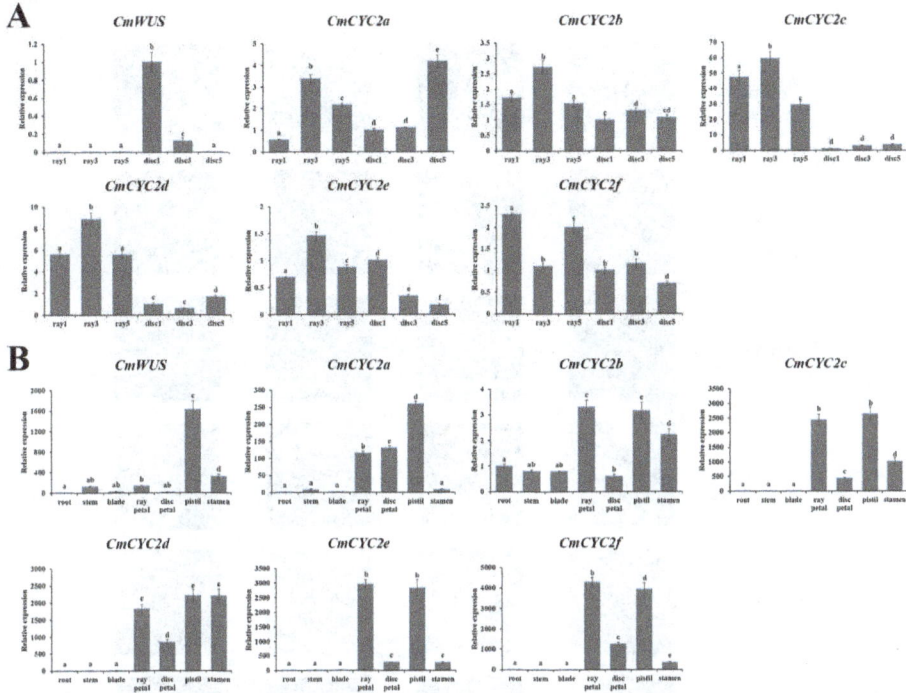

Figure 4. Comparative expression analysis of *CmWUS* and *CmCYC2* genes in *C. morifolium* 'Fen Ditan' at later stages of inflorescence development. (**A**) Gene expression patterns between ray and disc florets at later stages (stage 1, 3 and 5) of inflorescence development. The expression levels are relative to the disc florets at stage 1. (**B**) Relative expression levels of *CmWUS* and *CmCYC2* genes in different tissues of *C. morifolium* 'Fen Ditan'. Tissues analyzed including: root, stem, blade, ray petal, disc petal, pistil (including stigma, style and ovary) and stamen. The expression levels are relative to the root sample. Expression levels of *PP2Acs* are utilized for normalization. Error bars show the standard deviation of three biological replicates. Statistically significant differences are indicated with different lowercase letters (Fisher's LSD, $p < 0.05$).

2.4. Protein-Protein Interactions between CmWUS and CmCYC2

Since *CmWUS* and *CmCYC2* were both highly expressed in the reproductive organs, we further examined the interactions between CmWUS and CmCYC2 to reveal their relationship. The GFP and DAPI fluorescence indicated that CmWUS and CmCYC2 were mainly localized to the cell nucleus (Figure 5). In yeast two-hybrid (Y2H) assays, CmWUS had no autoactivation activity and was used as a bait. The results are shown in Figure 6. CmWUS could not form a homodimer, which was the opposite to the results in *Arabidopsis*, and this may be caused by the differences in the homodimerization interacting amino acids at the central part of the CmWUS sequence (Supplementary Figure S1A) [42]. Furthermore, CmWUS could dimerize with CmCYC2b and CmCYC2d, and the interactions with CmCYC2c, CmCYC2e and CmCYC2f were quite weak. Bimolecular fluorescence complementation (BiFC) assays were performed to provide further evidence for the interactions. There was no interaction in YFP^N/YFP^C, CmCYC2-YFP^N/YFP^C, CmWUS-YFP^N/YFP^C or YFP^N/CmWUS-YFP^C combinations. As shown in Figure 7, only in the combination of CmCYC2d-YFP^N/CmWUS-YFP^C, YFP fluorescence was detected. Taken together, CmWUS could directly interact with CmCYC2d, and the CmWUS-CmCYC2d complex is localized to the cell nucleus.

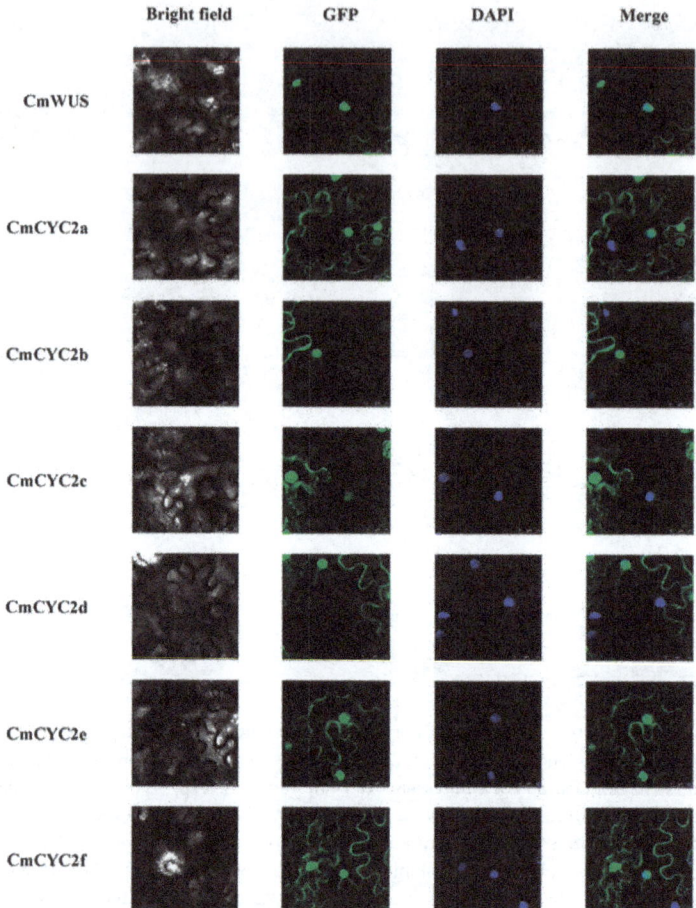

Figure 5. Subcellular localization of CmWUS and CmCYC2. pSuper1300-CmWUS and pSuper1300-CmCYC2 constructs were transiently transformed into the leaves of *Nicotiana benthamiana*. The fusion proteins (CmWUS-GFP and CmCYC2-GFP) were observed under the confocal laser scanning microscope. The merge pictures were made up of the GFP and DAPI pictures. The green and blue fluorescence show the position of proteins and nuclei, respectively. Bars = 25 μm.

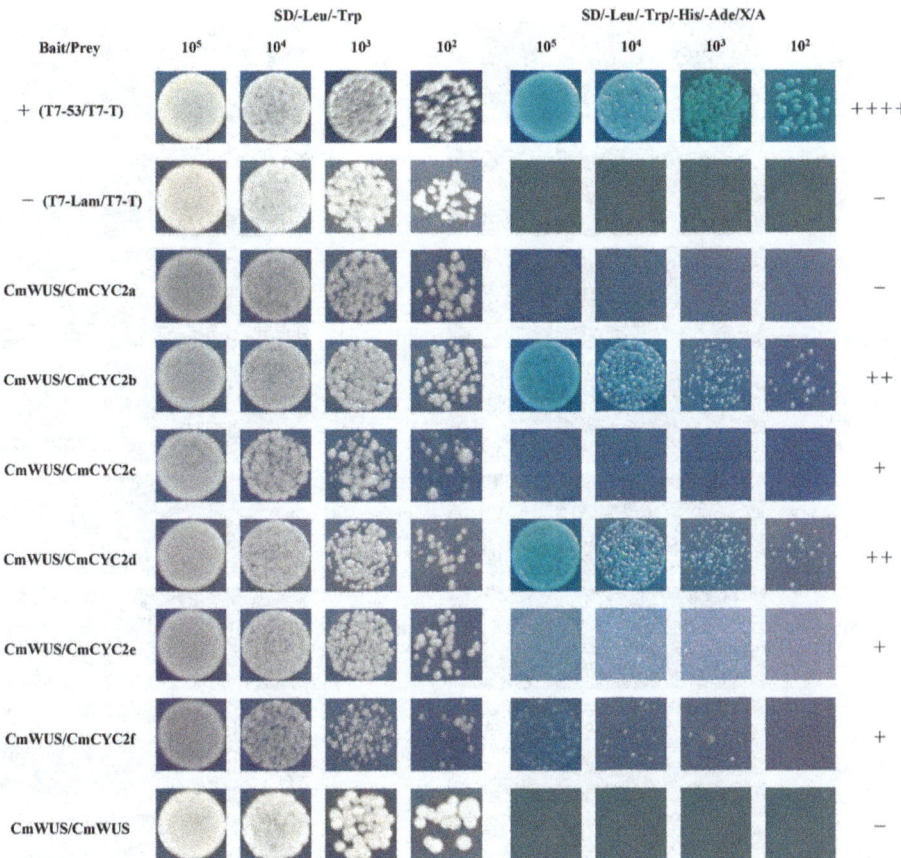

Figure 6. Yeast two-hybrid (Y2H) analysis of protein-protein interactions between CmWUS and CmCYC2. Clones containing each combination of bait and prey vectors were cultured on both nonselective media (SD/-Trp/-Leu) and selective media (SD/-Leu/-Trp/-His/-Ade/X/A). T7-53/T7-T and T7-Lam/T7-T are the positive and negative control. "+" represents the intensity of the interaction and "-" means no interaction.

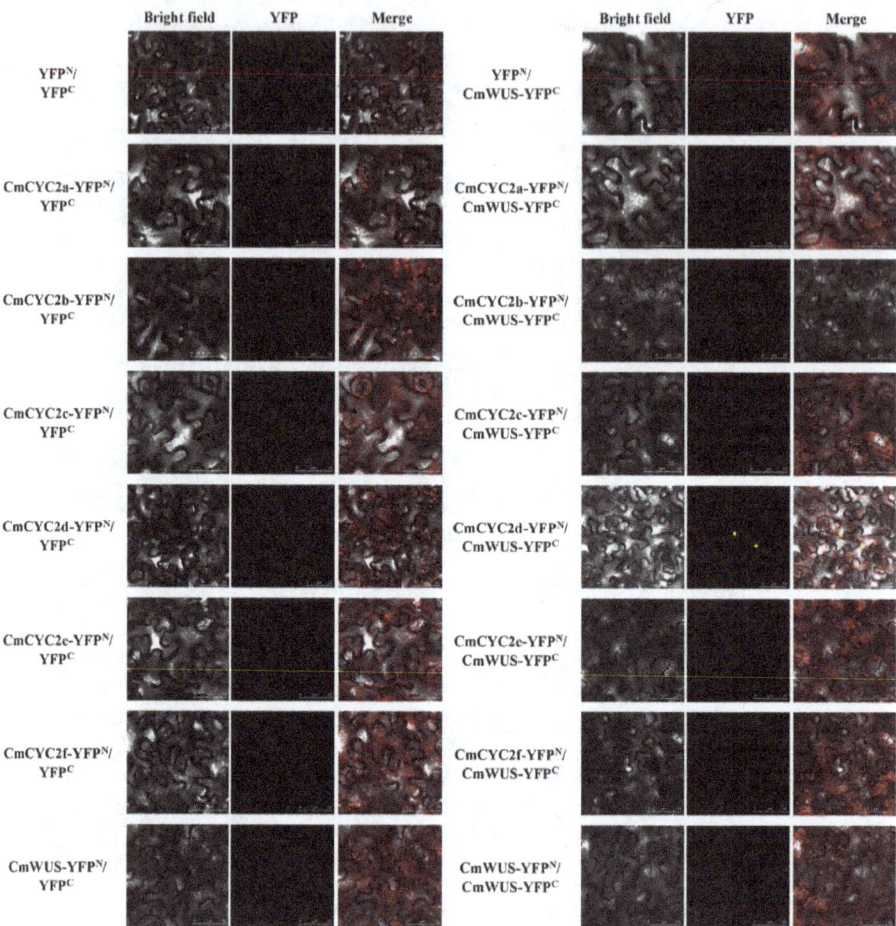

Figure 7. Bimolecular fluorescence complementation (BiFC) analysis of the interactions between CmWUS and CmCYC2 proteins in the epidermal cells of *N. benthamiana* leaves. CmCYC2 and CmWUS were fused to the N-terminal and C-terminal fragment of pCambia1300-YFP respectively and then co-transformed into *N. benthamiana* leaf cells. The confocal laser scanning microscope was used for visualizing. The yellow fluorescence shows the position of protein. Bars = 50 μm.

3. Discussion

3.1. Ectopic Expression of CmWUS in A. thaliana Indicates Possible Conserved Functions in Floral Meristems

Bifunctional transcription factor WUS plays a vital role in the stem cell maintenance of shoot and floral meristems and has been proven to be sufficient for the meristem reestablishment in the inflorescence stem [39–41]. To elucidate the functions of *CmWUS*, we first analyzed the sequence in detail. The WUS-box motif, which was elementary for WUS function in both shoot and floral meristems [39], was highly conserved. Also, the transcriptional repression related EAR motif [43] was identified at the carboxyl terminus. We further explored the function of *CmWUS* during flower development through overexpression in *A. thaliana*. In our transgenic line 1, clustered flower buds were ectopically initiated on the inflorescence stems. This phenotype was consistent with *sef*, a gain-of-function mutant caused by the overexpression of endogenous *WUS* [41]. Therefore, we speculated that *CmWUS* may retain conserved functions in floral meristems. In *sef* mutant, the floral identity gene *LFAFY (LFY)* was also activated [41] and it could cooperate with *WUS* to activate *AGAMOUS (AG)*, a MADS-box gene which specifies the identity of carpel and stamen [30,39,44]. This *WUS/LFY-AG* regulatory loop could be a possible explanation of the ectopic floral buds [40,41].

Another noteworthy phenotype in transgenic line 1 was the curled petals, indicating more active cell proliferation in abaxial side. *WOX1* and *WOX3*, which belong to WUS clade of WOX family [29], have been reported to regulate leaf and floral organ development and affect the abaxial-adaxial balance [45,46]. Thus, *CmWUS* may also be involved in petal morphogenesis through the regulation of abaxial-adaxial patterning. However, this still requires more research to elucidate.

3.2. Proposed Interaction between CmWUS and CmCYC2 in Regulating Reproductive Organ Development

Changes in the number of stamens always come after the shifts in flower symmetry, and it has been reported in Asteraceae that mutations of floret symmetry could affect the development of stamens and carpels [6,47]. *CYC2*-like genes, key factors of flower symmetry, are vital in determining floret identity and regulating floral organ development in Asteraceae [12]. In the transgenic *Arabidopsis* lines with constitutive expression of gerbera *CYC2*-like genes, the siliques were shorter than wild type. Moreover, overexpression of *GhCYC4* and *GhCYC7* could disrupt the development of petals and stamens and carpels were unable to produce normal siliques [20]. In this study, *35S::CmWUS* lines also produced shorter siliques and fewer stamens with variations in flower symmetry. In addition, the transcriptional level of *CmWUS* and *CmCYC2* genes during inflorescence development were compared in chrysanthemum. All the genes were highly expressed at the early stages of flower bud differentiation [38] and may be involved in floral organ development. At later stages, tissue-specific expression analysis revealed that they were all highly expressed in reproductive organs. In general, based on the transgenic *Arabidopsis* phenotypes and gene expression patterns, we conclude that *CmWUS* and *CmCYC2* genes may play an important role in the development of reproductive organs in chrysanthemum. Furthermore, Y2H and BiFC analyses indicated that CmWUS directly interact with CmCYC2d, an ortholog of GhCYC3 that has been proven to suppress stamen development in gerbera [20,38]. Hence, CmWUS and CmCYC2d may act together to affect the development of reproductive organs. This may further explain the mechanism of *CYC2*-like genes in the regulation of reproductive organ development. In addition, previous studies of CYC2-like proteins in gerbera and sunflower have shown redundant functions and higher capacity to form dimers within CYC2 subclade [12,19,20,22,26]. Thus, CmCYC2d could be the mediator between CmWUS and CmCYC2 and a complex regulatory network involving CmWUS and CmCYC2 subclade may exist in regulating reproductive organ development in chrysanthemum.

3.3. WUS Can Be a Bridge to Connect MADS-box and ECE (CYC/TB1)

It has been speculated that the flower morphology-related ECE and MADS-box genes may be closely linked [6,9,12,48]. In *Antirrhinum*, B-class MADS-box gene *DEF* and C-class gene *PLENA* are suggested to be required in the maintenance of *CYC* in whorl 2 and whorl 3, respectively [49]. *CYC2*-like genes are also indicated to be involved in regulating sepal identity by suppressing B-class genes in *Cysticapnos* [18]. In the *mtaga mtagb* double mutant of *Medicago truncatula*, the abnormal petals are related to the upregulation of *CYC2*-like genes [50]. Also, *GhSOC1* is thought to function upstream of CYC2 subclade genes in *Gerbera* [12,51]. However, the regulatory connections between MADS-box and ECE genes still remain to be illustrated. On the other hand, WUS acts as an activator in regulating the expression of C-class MADS-box gene *AG* in floral patterning and *AG* represses *WUS* directly or indirectly through activation of *KNUCKLES* at later stages of floral development in turn [39,44,52]. Furthermore, an A-class gene, *APETALA2 (AP2)*, antagonizes *AG* through promoting the expression of *WUS* in the floral meristem [53]. In this study, *CmWUS* and *CmCYC2* were found to be highly expressed in the reproductive organs of chrysanthemum and CmWUS could directly interact with CmCYC2d. A connection between WUS and ECE was established. Taken together, WUS, ECE and

MADS-box may be linked together during floral development and WUS acts as the adaptor to connect MADS-box and ECE.

In conclusion, this study characterized a *WUS*-like gene, *CmWUS*, in *C. morifolium* and revealed a remarkable link between CmWUS and CmCYC2 subclade. Since the significant function of CmWUS in reproductive organ development, our findings will help fill in the missing link of CmCYC2 in regulating the development of reproductive organs, especially in pistils, and contribute to a further understanding of the molecular mechanisms of gynomonoecy in Asteraceae.

4. Materials and Methods

4.1. Plant Materials and Growth Condition

C. morifolium 'Fen ditan' (Figure 3) and *A. thaliana* were cultivated in a greenhouse of Beijing Forestry University, China. They were grown under photoperiods of 8 h light (24 °C)/16 h dark (20 °C) and 16 h light (22 °C)/8 h dark (19 °C), respectively.

4.2. Gene Cloning

Total RNA was extracted from the inflorescences of *C. morifolium* 'Fen Ditan' with Plant RNA Kit (Omega, Norcross, GA, USA), and then used as template to synthesize first strand of cDNA with TransScript One-Step gDNA Removal and cDNA Synthesis SuperMix (Transgen, Beijing, China). Partial sequence of *CmWUS* in chrysanthemum was retrieved from our previously published RNAseq data [28]. SMARTerTM RACE 5′/3′ Kit (Clontech, Mountain View, CA, USA) was used for 5′ and 3′ RACE. 5′-GSP and 3′-GSP (Supplementary Table S1), gene-specific primers for RACE, were designed according to the instructions. Based on the 5′- and 3′-ends, *CmWUS*-F1 and *CmWUS*-R1 (Supplementary Table S1) were designed to amplify the open reading frame (ORF) sequence of *CmWUS*. Six *CmCYC2* genes (GenBank ID: *CmCYC2a*, KU595430.1; *CmCYC2b*, KU595431.1; *CmCYC2c*, KU595428.1; *CmCYC2d*, KU595426.1; *CmCYC2e*, KU595427.1; *CmCYC2f*, KU595429.1) were amplified with primers reported before [38]. All the PCR products were cloned into pCloneEZ-Blunt TOPO vectors (Taihe, Beijing, China), transformed into *Escherichia coli* DH5α cells (Tiangen, Beijing, China) and sequenced by Taihe (Beijing, China). The coding sequence of *CmWUS* (GenBank accession number: MK124768) has been uploaded to the NCBI database.

4.3. Bioinformatics Analysis

ClustalX software was used to perform alignment of multiple sequences, including CmWUS and WUS-like sequences from other species. GeneDoc software was used to edit the alignment. A phylogenetic tree was constructed by MEGA 7 based on the neighbor-joining method with 1000 bootstrap replicates, using the full length of the amino acid sequences of WUS homologs from various species and 15 WOX family members from *A. thaliana*. The accession numbers of sequences used here were as follows: AtWUS, *A. thaliana*, NM_127349.4; AtWOX1, AY251394.1; AtWOX2, NM_125325.3; AtWOX3, NM_128422.3; AtWOX4, FJ440850.1; AtWOX5, AY251398.1; AtWOX6, AY251399.2; AtWOX7, NM_120659.2; AtWOX8, AY251400.1; AtWOX9, AY251401.1; AtWOX10, NM_101923.1; AtWOX11, AY251402.1; AtWOX12, AY251403.1; AtWOX13, AY251404.1; AtWOX14, NM_101922.3; AmWUS, *Antirrhinum majus*, AAO23113.1; BnWUS, *Brassica napus*, XM_013803833.2; CcWUS, *Cynara cardunculus* var. *scolymus*, XM_025106474.1; CsWUS, *Citrus sinensis*, NM_001288918.1; GmWUS, *Glycine max*, XP_003517180.2; HaWUS, *Helianthus annuus*, HE616565.1; LsWUS, *Lactuca sativa*, XM_023909093.1; MtWUS, *Medicago truncatula*, XP_003612158.1; NtWUS, *Nicotiana tabacum*, XM_016619508.1; SlWUS, *Solanum lycopersicum*, ADZ13564.1; StWUS, *Solanum tuberosum*, XP_006340731.1; VvWUS, *Vitis vinifera*, XM_002266287.3.

4.4. Overexpression of CmWUS in A. thaliana

CmWUS was amplified using primers *CmWUS*-F2 and *CmWUS*-R2 (Supplementary Table S1) and subcloned into *NcoI/BstEII*-cleaved pCambia1304 vector under the CaMV35S promoter using In-Fusion® HD Cloning Kit System (Clontech, Mountain View, CA, USA). The resulting pCambia1304-*CmWUS* vector was transformed into *A. thaliana* (Columbia) via *Agrobacteriaum tumefaciens* GV3101with the floral dip method [54]. The seeds were selected on MS medium containing hygromycin B (50 mg/L; Roche, Basel, Switzerland). qRT-PCR was performed using young leaves to confirm positive lines with primers *CmWUS*-F3/R3 and *AtACTIN*-F/R (Supplementary Table S1). Three independent homozygous T_3 lines with higher and consistent expression levels were selected for floral phenotype analysis. Forty flowers were analyzed and the significant differences were determined according to Fisher's LSD ($p < 0.05$) with SPSS 20.0.

4.5. Microscope Observations

The floral buds of *C. morifolium* 'Fen Ditan' at different stages were fixed in FAA (50% ethanol: acetic acid: formaldehyde = 90:5:5, v/v), dehydrated with a graded ethanol series (50%–100%) and then transferred into xylene (100%). All the samples were embedded in paraffin and cut into 8 μm sections using a microtome (Leica, Wetzlar, Germany). After that, paraffin was removed from the sections with xylene, and then safranin (1%) and fast green (0.5%) were used for histological staining. All the sections were examined and photographed under a light microscope (Zeiss, Jena, Germany) after sealed with neutral gum.

4.6. Gene Expression Analysis in C. morifolium

Floral buds of *C. morifolium* 'Fen Ditan' at different stages were collected for analysis of gene expression patterns. To compare expression patterns of *CmWUS* and *CmCYC2* genes at later stages of inflorescence development between ray and disc florets, samples were pooled from the flower heads of *C. morifolium* 'Fen Ditan' at different stages (Figure 4). To analyze tissue-specific expression of *CmWUS* and *CmCYC2* genes, vegetative and reproductive tissues were collected from the inflorescences of *C. morifolium* 'Fen Ditan' at stage 4 and 5 of inflorescence development (Figure 4). Particularly, pistil samples were dissected from both ray and disc florets, while stamen samples were pooled from disc florets only. Total RNA was extracted as described above and PrimeScriptTM RT reagent Kit (Perfect Real Time; TaKaRa, Shiga, Japan) was used to synthesize the first strand of cDNA. Quantitative real-time PCR experiments were performed using the PikoReal real-time PCR system (Thermo Fisher Scientific, Waltham, MA, USA) with a 10 μL mix of SYBR Premix ExTaq II (5 μL; Takara, Shiga, Japan), forward and reverse primers (10 μM, 0.5 μL each), cDNA (2 μL) and sterile distilled water (2 μL). The qPCR primers of *CmCYC2* genes and the reference gene *PP2Acs* were reported before [28,38,55]. *CmWUS*-F3/R3 (Supplementary Table S1) was used as qPCR primer of *CmWUS*. Three biological replicates were conducted with three technical replicates each. $2^{-\Delta\Delta Ct}$ method [56] was used to calculate the relative expression levels.

4.7. Subcellular Localization

CmWUS and *CmCYC2* genes were amplified and subcloned into *SalI/SpeI*- cleaved pSuper1300-*GFP* vectors to generate the transformation plasmids *35S::CmWUS::GFP* and *35S::CmCYC2::GFP*. The plasmids were transformed into *A. tumefaciens* and injected into the leaves of *Nicotiana benthamiana* following the procedure reported before [57]. TCS SP8 (Leica, Wetzlar, Germany) confocal laser scanning microscope was used to assess subcellular localization at 488 and 408 nm for GFP and DAPI fluorescence, respectively. Primers used for subcellular localization are listed in Supplementary Table S1.

4.8. Y2H Assay

Matchmaker Gold Yeast Two-Hybrid System (Clontech, Mountain View, CA, USA) was used to carry out Y2H assays. *CmCYC2* and *CmWUS* were amplified and subcloned into the pGADT7 (prey) and pGBKT7 (bait) vectors. The reconstructed pGADT7 and pGBKT7 vectors were transformed into Y187 and Y2H gold yeast strains and cultured on SD/-Leu and SD/-Trp plates, respectively. If the colonies containing bait vector are significantly smaller than colonies containing the empty pGBKT7 vector on SD/-Trp plates, then the bait is toxic to the yeast cells. To test the bait for autoactivation, Y2H gold yeast cells containing pGBKT7-*CmWUS* vector were cultured on SD/-Trp, SD/-Trp/X-α-Gal SD/-Trp/X-α-Gal/Aureobasidin A (AbA) and SD–Trp/–His/–Ade plates. If the colonies grow on both SD/-Trp and SD/-Trp/X-α-Gal plates, but not on SD/-Trp/X-α-Gal/AbA and SD–Trp/–His/–Ade plates, then the bait cannot autoactivate the AbA^r and His3/Ade2 reporter. After the testing of toxicity and autoactivation, diploid mating was conducted as described previously [58], and the transformed colonies were cultured on SD/-Trp/-Leu and SD/-Leu/-Trp/-His/-Ade/X-α-Gal/AbA (SD/-Leu/-Trp/-His/-Ade/X/A) plates to test for possible interactions. Y2H screenings were performed in triplicate. Primers used for Y2H assays are listed in Supplementary Table S1.

4.9. BiFC Assay

CmCYC2 and *CmWUS* genes were amplified and subcloned into the pCambia1300-YFP^N and pCambia1300-YFP^C vectors. Co-expression was conducted in the leaves of tobacco (*N. benthamiana*) as described in Subcellular Localization. TCS SP8 (Leica, Wetzlar, Germany) confocal laser scanning microscope was used to detect YFP fluorescence at 514 nm. Primers used for BiFC assays are listed in Supplemental Table S1.

Supplementary Materials: Supplementary materials can be found at http://www.mdpi.com/1422-0067/20/6/1276/s1. Figure S1: Alignment and phylogenetic analysis of CmWUS; Table S1: Primers used.

Author Contributions: Conceptualization, Y.Y. and Q.Z.; Formal analysis, Y.Y.; Investigation, Y.Y.; Methodology, Y.Y., Y.H. and T.Z.; Project administration, Q.Z.; Resources, M.S., C.Y., T.C. and J.W.; Supervision, C.Y. and Q.Z.; Validation, Y.Y.; Writing—original draft, Y.Y.; Writing—review & editing, C.Y., Y.H. and T.Z.

Funding: This work was supported by the Fundamental Research Funds for the Central Universities (NO.BLX2015-03) and Special Fund for Beijing Common Construction Project.

Acknowledgments: Conceptualization, Y.Y. and Q.Z.; Data curation, Y.Y.; Formal analysis, Y.Y.; Investigation, Y.Y.; Methodology, Y.Y. and Y.H.; Resources, M.S., C.Y., T.Z., T.C., J.W. and Q.Z.; Software, T.C. and J.W.; Supervision, M.S., C.Y. and Q.Z; Writing—original draft, Y.Y.; Writing—review & editing, C.Y., Y.H. and T.Z.

Conflicts of Interest: The authors declare that this research is carried on the absence of any financial or commercial relationships that could be interpreted to a potential conflict of interest.

References

1. Gillies, A.C.M.; Cubas, P.; Coen, E.S.; Abbott, R.J. Making rays in the Asteraceae: Genetics and evolution of radiate versus discoid flower heads. In *Developmental Genetics and Plant Evolution*; Quentin, C.B., Cronk, R.M.B., Julie, A.H., Eds.; Taylor & Francis: London, UK, 2002; Volume 65, pp. 233–246.

2. Bertin, R.I.; Kerwin, M.A. Floral sex ratios and gynomonoecy in Aster (Asteraceae). *Am. J. Bot.* **1998**, *85*, 235–244. [CrossRef] [PubMed]

3. Bertin, R.I.; Connors, D.B.; Kleinman, H.M. Differential herbivory on disk and ray flowers of gynomonoecious asters and goldenrods (Asteraceae). *Biol. J. Linn. Soc.* **2010**, *101*, 544–552. [CrossRef]

4. Sun, M.; Ganders, F.R. Outcrossing rates and allozyme variation in rayed and rayless morphs of *Bidens pilosa. Heredity* **1990**, *64*, 139–143. [CrossRef]

5. Andersson, S. Pollinator and nonpollinator selection on ray morphology in *Leucanthemum vulgare* (oxeye daisy, Asteraceae). *Am. J. Bot.* **2008**, *95*, 1072–1078. [CrossRef]

6. Fambrini, M.; Pugliesi, C. CYCLOIDEA 2 clade genes: Key players in the control of floral symmetry, inflorescence architecture, and reproductive organ development. *Plant Mol. Biol. Rep.* **2016**, *35*, 20–36. [CrossRef]

7. Kalisz, S.; Ree, R.H.; Sargent, R.D. Linking floral symmetry genes to breeding system evolution. *Trends Plant Sci.* **2006**, *11*, 568–573. [CrossRef]

8. Hileman, L.C.; Cubas, P. An expanded evolutionary role for flower symmetry genes. *J. Biol.* **2009**, *8*, 90. [CrossRef] [PubMed]

9. Howarth, D.G.; Donoghue, M.J. Phylogenetic analysis of the "ECE" (CYC/TB1) clade reveals duplications predating the core eudicots. *Proc. Natl. Acad. Sci. USA* **2006**, *103*, 9101–9106. [CrossRef]

10. Martin-Trillo, M.; Cubas, P. TCP genes: A family snapshot ten years later. *Trends Plant Sci.* **2010**, *15*, 31–39. [CrossRef]

11. Hileman, L.C. Bilateral flower symmetry—How, when and why? *Curr. Opin. Plant Biol.* **2014**, *17*, 146–152. [CrossRef]

12. Broholm, S.K.; Teeri, T.H.; Elomaa, P. Molecular control of inflorescence development in Asteraceae. In *Advances in Botanical Research*; Fornara, F., Ed.; Academic Press: Oxford, UK, 2014; Volume 72, pp. 297–333.

13. Luo, D.; Carpenter, R.; Vincent, C.; Copsey, L.; Coen, E. Origin of floral asymmetry in *Antirrhinum*. *Nature* **1996**, *383*, 794–799. [CrossRef]

14. Luo, D.; Carpenter, R.; Copsey, L.; Vincent, C.; Clark, J.; Coen, E. Control of organ asymmetry in flowers of *Antirrhinum*. *Cell* **1999**, *99*, 367–376. [CrossRef]

15. Gaudin, V.; Lunness, P.A.; Fobert, P.R.; Towers, M.; Rioukhamlichi, C.; Murray, J.A.H.; Coen, E.; Doonan, J.H. The Expression of *D-Cyclin* Genes Defines Distinct Developmental Zones in Snapdragon Apical Meristems and Is Locally Regulated by the *Cycloidea* Gene. *Plant Physiol.* **2000**, *122*, 1137–1148. [CrossRef]

16. Song, C.F.; Lin, Q.B.; Liang, R.H.; Wang, Y.Z. Expressions of ECE-CYC2 clade genes relating to abortion of both dorsal and ventral stamens in *Opithandra* (Gesneriaceae). *BMC Evol. Biol.* **2009**, *9*, 244. [CrossRef]

17. Preston, J.C.; Kost, M.A.; Hileman, L.C. Conservation and diversification of the symmetry developmental program among close relatives of snapdragon with divergent floral morphologies. *New Phytol.* **2009**, *182*, 751–762. [CrossRef]

18. Zhao, Y.; Pfannebecker, K.; Dommes, A.B.; Hidalgo, O.; Becker, A.; Elomaa, P. Evolutionary diversification of CYC/TB1-like TCP homologs and their recruitment for the control of branching and floral morphology in *Papaveraceae* (basal eudicots). *New Phytol.* **2018**, *220*, 317–331. [CrossRef]

19. Broholm, S.K.; Tahtiharju, S.; Laitinen, R.A.; Albert, V.A.; Teeri, T.H.; Elomaa, P. A TCP domain transcription factor controls flower type specification along the radial axis of the Gerbera (Asteraceae) inflorescence. *Proc. Natl. Acad. Sci. USA* **2008**, *105*, 9117–9122. [CrossRef]

20. Juntheikki-Palovaara, I.; Tahtiharju, S.; Lan, T.; Broholm, S.K.; Rijpkema, A.S.; Ruonala, R.; Kale, L.; Albert, V.A.; Teeri, T.H.; Elomaa, P. Functional diversification of duplicated CYC2 clade genes in regulation of inflorescence development in *Gerbera hybrida* (Asteraceae). *Plant J.* **2014**, *79*, 783–796. [CrossRef]

21. Elomaa, P.; Zhao, Y.; Zhang, T. Flower heads in Asteraceae—Recruitment of conserved developmental regulators to control the flower-like inflorescence architecture. *Hortic. Res.* **2018**, *5*, 36. [CrossRef]

22. Tahtiharju, S.; Rijpkema, A.S.; Vetterli, A.; Albert, V.A.; Teeri, T.H.; Elomaa, P. Evolution and diversification of the CYC/TB1 gene family in Asteraceae—A comparative study in Gerbera (Mutisieae) and sunflower (Heliantheae). *Mol. Biol. Evol.* **2012**, *29*, 1155–1166. [CrossRef]

23. Bello, M.A.; Cubas, P.; Alvarez, I.; Sanjuanbenito, G.; Fuertes-Aguilar, J. Evolution and expression patterns of CYC/TB1 genes in *Anacyclus*: Phylogenetic insights for floral symmetry genes in Asteraceae. *Front. Plant Sci.* **2017**, *8*, 589. [CrossRef]

24. Mizzotti, C.; Fambrini, M.; Caporali, E.; Masiero, S.; Pugliesi, C. A *CYCLOIDEA*-like gene mutation in sunflower determines an unusual floret type able to produce filled achenes at the periphery of the pseudanthium. *Botany* **2015**, *93*, 171–181. [CrossRef]

25. Fambrini, M.; Salvini, M.; Pugliesi, C. A transposon-mediate inactivation of a *CYCLOIDEA*-like gene originates polysymmetric and androgynous ray flowers in *Helianthus Annuus*. *Genetics* **2011**, *139*, 1521–1529. [CrossRef]

26. Chapman, M.A.; Tang, S.; Draeger, D.; Nambeesan, S.; Shaffer, H.; Barb, J.G.; Knapp, S.J.; Burke, J.M. Genetic analysis of floral symmetry in Van Gogh's sunflowers reveals independent recruitment of *CYCLOIDEA* genes in the Asteraceae. *PLoS Genet.* **2012**, *8*, e1002628. [CrossRef]

27. Fambrini, M.; Basile, A.; Salvini, M.; Pugliesi, C. Excisions of a defective transposable CACTA element (*Tetu1*) generate new alleles of a *CYCLOIDEA*-like gene of *Helianthus annuus*. *Gene* **2014**, *549*, 198–207. [CrossRef]

28. Liu, H.; Sun, M.; Du, D.; Pan, H.; Cheng, T.; Wang, J.; Zhang, Q.; Gao, Y. Whole-transcriptome analysis of differentially expressed genes in the ray florets and disc florets of *Chrysanthemum morifolium*. *BMC Genom.* **2016**, *17*, 398. [CrossRef]

29. Graaff, E.V.D.; Laux, T.; Rensing, S.A. The WUS homeobox-containing (WOX) protein family. *Genom. Biol.* **2009**, *10*, 248. [CrossRef]

30. Lohmann, J.U.; Hong, R.L.; Hobe, M.; Busch, M.A.; Parcy, F.; Simon, R.; Weigel, D. A Molecular link between stem cell regulation and floral patterning in *Arabidopsis*. *Cell* **2001**, *105*, 793–803. [CrossRef]

31. Yadav, R.K.; Perales, M.; Gruel, J.; Girke, T.; Jönsson, H.; Reddy, G.V. WUSCHEL protein movement mediates stem cell homeostasis in the *Arabidopsis* shoot apex. *Genes Dev.* **2012**, *25*, 2025–2030. [CrossRef]

32. Zhou, Y.; Yan, A.; Han, H.; Li, T.; Geng, Y.; Liu, X.; Meyerowitz, E.M. HAIRY MERISTEM with WUSCHEL confines CLAVATA3 expression to the outer apical meristem layers. *Science* **2018**, *361*, 502–506. [CrossRef]

33. Deyhle, F.; Sarkar, A.K.; Tucker, E.J.; Laux, T. *WUSCHEL* regulates cell differentiation during anther development. *Dev. Biol.* **2007**, *302*, 154–159. [CrossRef]

34. Doerks, T.; Copley, R.R.; Schultz, J.; Ponting, C.P.; Bork, P. Systematic identification of novel protein domain families associated with nuclear functions. *Genom. Res.* **2002**, *12*, 47–56. [CrossRef]

35. Yamada, T.; Sasaki, Y.; Hashimoto, K.; Nakajima, K.; Gasser, C.S. *CORONA*, *PHABULOSA* and *PHAVOLUTA* collaborate with BELL1 to confine WUSCHEL expression to the nucellus in *Arabidopsis* ovules. *Development* **2016**, *143*, 422–426. [CrossRef]

36. Laux, T.; Mayer, K.F.; Berger, J.; Jürgens, G. The *WUSCHEL* gene is required for shoot and floral meristem integrity in *Arabidopsis*. *Development* **1996**, *122*, 87–96.

37. Liu, X.; Ning, K.; Che, G.; Yan, S.; Han, L.; Gu, R.; Li, Z.; Weng, Y.; Zhang, X. CsSPL functions as an adaptor between HD-ZIP III and CsWUS transcription factors regulating anther and ovule development in *Cucumis sativus* (cucumber). *Plant J.* **2018**, *94*, 535–547. [CrossRef]

38. Huang, D.; Li, X.; Sun, M.; Zhang, T.; Pan, H.; Cheng, T.; Wang, J.; Zhang, Q. Identification and Characterization of *CYC*-Like Genes in Regulation of Ray Floret Development in *Chrysanthemum morifolium*. *Front. Plant Sci.* **2016**, *7*, 1633. [CrossRef]

39. Ikeda, M.; Mitsuda, N.; Ohme-Takagi, M. *Arabidopsis* WUSCHEL is a bifunctional transcription factor that acts as a repressor in stem cell regulation and as an activator in floral patterning. *Plant Cell* **2009**, *21*, 3493–3505. [CrossRef]

40. Gallois, J.L.; Nora, F.R.; Mizukami, Y.; Sablowski, R. *WUSCHEL* induces shoot stem cell activity and developmental plasticity in the root meristem. *Genes Dev.* **2004**, *18*, 375–380. [CrossRef]

41. Xu, Y.Y.; Wang, X.M.; Li, J.; Li, J.H.; Wu, J.S.; Walker, J.C.; Xu, Z.H.; Chong, K. Activation of the *WUS* gene induces ectopic initiation of floral meristems on mature stem surface in *Arabidopsis thaliana*. *Plant Mol. Biol.* **2005**, *57*, 773–784. [CrossRef]

42. Rodriguez, K.; Perales, M.; Snipes, S.; Yadav, R.K.; Diaz-Mendoza, M.; Reddy, G.V. DNA-dependent homodimerization, sub-cellular partitioning, and protein destabilization control *WUSCHEL* levels and spatial patterning. *Proc. Natl. Acad. Sci. USA* **2016**, *113*, 6307–6315. [CrossRef]

43. Paponov, I.A.; Teale, W.; Lang, D.; Paponov, M.; Reski, R.; Rensing, S.A.; Palme, K. The evolution of nuclear auxin signalling. *BMC Evol. Biol.* **2009**, *9*, 126. [CrossRef]

44. Lenhard, M.; Bohnert, A.; Jurgens, G.; Laux, T. Termination of stem cell maintenance in *Arabidopsis* floral meristems by interactions between *WUSCHEL* and *AGAMOUS*. *Cell* **2001**, *105*, 805–814. [CrossRef]

45. Vandenbussche, M.; Horstman, A.; Zethof, J.; Koes, R.; Rijpkema, A.S.; Gerats, T. Differential recruitment of WOX transcription factors for lateral development and organ fusion in Petunia and *Arabidopsis*. *Plant Cell* **2009**, *21*, 2269–2283. [CrossRef]

46. Honda, E.; Yew, C.L.; Yoshikawa, T.; Sato, Y.; Hibara, K.I.; Itoh, J.I. *LEAF LATERAL SYMMETRY1*, a member of the *WUSCHEL-RELATED HOMEOBOX3* gene family, regulates lateral organ development differentially from other paralogs, *NARROW LEAF2* and *NARROW LEAF3* in Rice. *Plant Cell Physiol.* **2018**, *59*, 376–391. [CrossRef]

47. Berti, F.; Fambrini, M.; Turi, M.; Bertini, D.; Pugliesi, C. Mutations of corolla symmetry affect carpel and stamen development in *Helianthus annuus*. *Can. J. Bot.* **2005**, *83*, 1065–1072. [CrossRef]

48. Preston, J.C.; Hileman, L.C. Parallel evolution of TCP and B-class genes in Commelinaceae flower bilateral symmetry. *Evodevo* **2012**, *3*, 6. [CrossRef]

49. Clark, J.I.; Coen, E.S. The *cycloidea* gene can respond to a common dorsoventral prepattern in *Antirrhinum*. *Plant J.* **2002**, *30*, 639–648. [CrossRef]

50. Zhu, B.; Li, H.; Wen, J.; Mysore, K.S.; Wang, X.; Pei, Y.; Niu, L.; Lin, H. Functional specialization of duplicated *AGAMOUS* homologs in regulating floral organ development of *Medicago truncatula*. *Front. Plant Sci.* **2018**, *9*, 854. [CrossRef]

51. Ruokolainen, S.; Ng, Y.P.; Albert, V.A.; Elomaa, P.; Teeri, T.H. Over-expression of the *Gerbera hybrida At-SOC1-like1* gene *Gh-SOC1* leads to floral organ identity deterioration. *Ann. Bot.* **2011**, *107*, 1491–1499. [CrossRef]

52. Liu, X.; Kim, Y.J.; Muller, R.; Yumul, R.E.; Liu, C.; Pan, Y.; Cao, X.; Goodrich, J.; Chen, X. *AGAMOUS* terminates floral stem cell maintenance in *Arabidopsis* by directly repressing *WUSCHEL* through recruitment of Polycomb Group proteins. *Plant Cell* **2011**, *23*, 3654–3670. [CrossRef]

53. Huang, Z.; Shi, T.; Zheng, B.; Yumul, R.E.; Liu, X.; You, C.; Gao, Z.; Xiao, L.; Chen, X. *APETALA2* antagonizes the transcriptional activity of *AGAMOUS* in regulating floral stem cells in *Arabidopsis thaliana*. *New Phytol.* **2017**, *215*, 1197–1209. [CrossRef]

54. Clough, S.J.; Bent, A.F. Floral dip: A simplified method for *Agrobacterium*-mediated transformation of *Arabidopsis thaliana*. *Plant J.* **1998**, *16*, 735–743. [CrossRef]

55. Gu, C.; Chen, S.; Liu, Z.; Shan, H.; Luo, H.; Guan, Z.; Chen, F. Reference gene selection for quantitative real-time PCR in Chrysanthemum subjected to biotic and abiotic stress. *Mol. Biotechnol.* **2011**, *49*, 192–197. [CrossRef]

56. Schmittgen, T.D.; Livak, K.J. Analyzing real-time PCR data by the comparative C(T) method. *Nat. Protoc.* **2008**, *3*, 1101–1108. [CrossRef]

57. Zhao, K.; Zhou, Y.; Ahmad, S.; Xu, Z.; Li, Y.; Yang, W.; Cheng, T.; Wang, J.; Zhang, Q. Comprehensive cloning of *Prunus mume* dormancy associated MADS-Box genes and their response in flower bud development and dormancy. *Front. Plant Sci.* **2018**, *9*, 17. [CrossRef]

58. Zhou, Y.; Xu, Z.; Yong, X.; Ahmad, S.; Yang, W.; Cheng, T.; Wang, J.; Zhang, Q. SEP-class genes in *Prunus mume* and their likely role in floral organ development. *BMC Plant Biol.* **2017**, *17*, 10. [CrossRef]

Article

Chrysanthemum *DgWRKY2* Gene Enhances Tolerance to Salt Stress in Transgenic Chrysanthemum

Ling He, Yin-Huan Wu, Qian Zhao, Bei Wang, Qing-Lin Liu * and Lei Zhang

Department of Ornamental Horticulture, Sichuan Agricultural University, 211 Huimin Road, Wenjiang District, Chengdu 611130, Sichuan, China; heling@stu.sicau.edu.cn (L.H.); s20141825@sicau.edu.cn (Y.-H.W.); s20167109@stu.sicau.edu.cn (Q.Z.); s20167108@stu.sicau.edu.cn (B.W.); 14069@sicau.edu.cn (L.Z.)
* Correspondence: 13854@sicau.edu.cn; Tel./Fax: +86-28-8629-0881

Received: 22 May 2018; Accepted: 10 July 2018; Published: 16 July 2018

Abstract: WRKY transcription factors (TFs) play a vital part in coping with different stresses. In this study, *DgWRKY2* was isolated from *Dendranthema grandiflorum*. The gene encodes a 325 amino acid protein, belonging to the group II WRKY family, and contains one typical WRKY domain (WRKYGQK) and a zinc finger motif (C-X4-5-C-X22-23-H-X1-H). Overexpression of *DgWRKY2* in chrysanthemum enhanced tolerance to high-salt stress compared to the wild type (WT). In addition, the activities of antioxidant enzymes (superoxide dismutase (SOD), peroxidase (POD), catalase (*CAT*)), proline content, soluble sugar content, soluble protein content, and chlorophyll content of transgenic chrysanthemum, as well as the survival rate of the transgenic lines, were on average higher than that of the WT. On the contrary, hydrogen peroxide (H_2O_2), superoxide anion ($O_2{}^-$), and malondialdehyde (MDA) accumulation decreased compared to WT. Expression of the stress-related genes *DgCAT*, *DgAPX*, *DgZnSOD*, *DgP5CS*, *DgDREB1A*, and *DgDREB2A* was increased in the *DgWRKY2* transgenic chrysanthemum compared with their expression in the WT. In conclusion, our results indicate that *DgWRKY2* confers salt tolerance to transgenic chrysanthemum by enhancing antioxidant and osmotic adjustment. Therefore, this study suggests that *DgWRKY2* could be used as a reserve gene for salt-tolerant plant breeding.

Keywords: transgenic chrysanthemum; WRKY transcription factor; salt stress; gene expression; *DgWRKY2*

---

1. Introduction

High-salt stress is one of the most important factors that seriously affects and inhibits the growth and yield of plants [1]. Environmental stresses affect plant growth, causing plants to evolve mechanisms to face these challenges [2]. Under salt stress, transcription factors (TFs) can regulate the expression of multiple stress-related genes, which enhance tolerance to salt compared with the activity of a functional gene [3]. These genes are involved in the salt stress response in plants, forming a complex regulatory network [4]. Therefore, by using transcription factors, the plants' resistance can be improved.

WRKYs are a massive TF family, dominating the genetic transcription of plants. WRKY was named after the highly conserved sequence motif WRKYGQK. The WRKY proteins are divided into 3 types: class I contains two conserved WRKY domains and a zinc finger structure C-X4-5-C-X22-23-H-X1-H; class II contains one conserved WRKY domain and the same zinc finger structure, and most of the WRKY proteins found to date are this type; there is one conserved domain in class III, the zinc finger structure C-X7-C-X22-23-H-X1-C [5]. Overexpression of genes is a commonly used method to study gene function. Many studies had shown that WRKY TFs played a vital role in the physiological processes of plants [6–9]. It has also been proved that overexpression of some WRKY genes successfully increase plant tolerance to abiotic stress. *TaWRKY93* may increase salinity tolerance by enhancing osmotic adjustment, maintaining membrane stability, and increasing transcription of stress-related genes [10]. During salt treatment, *NbWRKY79* enhanced the tolerance of the transgenic plants to oxidant stress. Therefore, it increased the salt tolerance of *Nicotiana benthamiana* [11]. *RtWRKY1* conferred tolerance to salt stress in transgenic *Arabidopsis* by regulating plant growth, osmotic balance, Na^+/K^+ homeostasis, and the antioxidant system [12]. *VvWRKY30* increased salt resistance by regulating reactive oxygen species (ROS)-scavenging activity and the accumulation of osmoticum [13].

Physiological traits are important indicative indexes of botanical abiotic resistance. Plants produce ROS in the body under environmental pressures, including accumulation of superoxide anions (O_2^-), hydroxyl ions (OH^-), hydroxyl radicals (-OH), hydrogen peroxide (H_2O_2), and other types. These species not only lead to membrane lipid peroxidation of plant cells, affecting the redox state of the protein, but also cause oxidative damage to nucleic acids [14]. The plant antioxidant defense system consists of a variety of enzymes (superoxide dismutase (SOD), peroxidase (POD), catalase (*CAT*), ascorbate peroxidase (*APX*), etc.), which act as active oxygen scavengers in plants [15]. Much evidence has shown that the production and removal of ROS are closely related to the mechanism of salt tolerance [16,17]. As penetrating agents, soluble sugar, soluble protein, and proline maintain osmotic balance together. Under salt stress, transcription factors may participate in the regulation of the expression of many salt tolerance-related functional genes, so as to obtain stronger stress resistance than can be imparted by functional genes. The key genes encoding antioxidant enzymes (*Cu/ZnSOD*, *CAT*, *APX*, etc.) can increase the efficiency of ROS elimination in plant cells, so much so that the plant's tolerance to abiotic stresses is improved [18–20]. The proline synthase gene (*P5CS*) can effectively increase the tolerance of transgenic plants to osmotic stress [21]. The DREB (dehydration-responsive element binding proteins) transcription factor can specifically bind to the DRE *cis*-acting element or the core sequence with the DRE element (CCGAC), regulate the expression of stress-related genes, and mediate the transmission of abiotic stress signals [22–25].

Chrysanthemums are cut flower with high economic benefits and appreciable value, but it is sensitive to salinity, which can cause slow growth, plant chlorosis, and even death [26]. In a previous study, we obtained a database of the chrysanthemum transcriptome in response to salinity conditions by using high-throughput sequencing [27]. A large number of salt-induced transcripts were found in the data, especially from the WRKY family. Previously, we identified four WRKY genes (*DgWRKY1*, *DgWRKY3*, *DgWRKY4*, and *DgWRKY5*) and demonstrated that they can increase the salt tolerance of tobacco or chrysanthemum [28–31]. In order to analyze the WRKY family in chrysanthemum from multiple angles to complement our information, the salt stress-related gene *DgWRKY2* was isolated from chrysanthemum. This study investigated the importance of *DgWRKY2* as a transcription regulator under salt stress.

2. Results

2.1. Isolation and Characterization of DgWRKY2

DgWRKY2 contained a complete open reading frame of 1107 bp, which encoded a protein of 325 amino acids with a calculated molecular mass of 36.55 kDa. The theoretical isoelectric point is PI = 6.66 (Figure 1). Multi-sequence alignment analysis of the amino acid sequences of *DgWRKY2* and eight other genes showed that *DgWRKY2* contains a WRKY domain and a zinc finger structure (C-X4-5-C-X22-23-H-X1-H). It was further confirmed that the cloned cDNA sequence was a WRKY transcription factor II family member (Figure 2). Phylogenetic analysis showed that *DgWRKY2* is most closely related to *AtWRKY28* from *Arabidopsis thaliana* (Figure 3).

```
1     GCCAAAGCATAAATAAAGACTCCTCACCTATTATTTGTCAAACTTCTCCTCTCTTCTATC
61    TCTTTGTGTTAATTTGTGTTTCTTCATATTGAACAACCATGTCTCAAGACCAAAGAGATC
1                                            M  S  Q  D  Q  R  D

121   TATACTATCATGATCCATTTCATGATGATCAAAGAACTAGTGAGACTCTTTTTTCATTCT
8     L  Y  Y  H  D  P  F  H  D  D  Q  R  T  S  E  T  L  F  S  F

181   TTGGCCCGAATTCTGGCATTAGAGACGAGTCTTCGCCTCCAAATCATCAAAGATTTCAAG
28    F  G  P  N  S  G  I  R  D  E  S  S  P  P  N  H  Q  R  F  Q

241   ACTACATGGGCTTGACTCACTTTTTTAATAATGGATCTAGTGATTATAACTCACCAGCTA
48    D  Y  M  G  L  T  H  F  F  N  N  G  S  S  D  Y  N  S  P  A

301   CCACTTTTGGTTACACATCTTCTTCATCACAACAAGTGTTTGCTCTTCAAGATGACCAAA
68    T  T  F  G  Y  T  S  S  S  Q  Q  V  F  A  L  Q  D  D  Q

361   AGCCAATTATGGATCATGGAAACTTGGTTGGAGCTAGTGAGATGACTCCTGTTACTCCCA
88    K  P  I  M  D  H  G  N  L  V  G  A  S  E  M  T  P  V  T  P

421   ACTCTTCTTCTATCCTTTCGTCGTCTACTGAGGCTCCTGATGACGAGCCCGAGCTTAACG
108   N  S  S  S  I  L  S  S  S  T  E  A  P  D  D  E  P  E  L  N

481   AAGGTAAGAAGGAGAATCAACAAAAAGGTGTATTGGAAGATGGAGGAGAAAGCTCTAAGA
128   E  G  K  K  E  N  Q  Q  K  G  V  L  E  D  G  G  E  S  S  K

541   AAGTGGTCAAGCAAAAGAAGAAAGACGAAAAGAAGCCAAGAGAGCCACGGTTCGCCTTTA
148   K  V  V  K  Q  K  K  K  D  E  K  K  P  R  E  P  R  F  A  F

601   TGACAAAGAGCGATATTGATCATCTTGAAGACGGATATCGCTGGAGGAAGTATGGACAGA
168   M  T  K  S  D  I  D  H  L  E  D  G  Y  R  W  R  K  Y  G  Q

661   AAGCAGTCAAGAACAGCCCTTATCCGAGAAGCTACTATCGATGTACGACTCAAAAGTGCA
188   K  A  V  K  N  S  P  Y  P  R  S  Y  Y  R  [C] T  T  Q  K  [C]

721   CCGTGAAGAAGCGAGTAGAGAGATCCTATCAAGATCCATCGACCGTGATCACTACGTATG
208   T  V  K  K  R  V  E  R  S  Y  Q  D  P  S  T  V  I  T  T  Y

781   AAGGACAACACAACCACCACTTGCCAGCAACACTTAGAGGAAATGTTGGTGGAATGTTGT
228   E  G  Q  [H] N  [H] H  L  P  A  T  L  R  G  N  V  G  G  M  L

841   ACCCACAGTCTATGTTAGCGGCGCAAAGTGCAATGATGGCTAGTGGTGGCTCAAGCTTCT
248   Y  P  Q  S  M  L  A  A  Q  S  A  M  M  A  S  G  G  S  S  F

901   CACACGAGTTTCTATCTCAAATACCTCATGGCTTCTACAACCCTAATGGCGGTGCAAGCA
268   S  H  E  F  L  S  Q  I  P  H  G  F  Y  N  P  N  G  G  A  S

961   GTGGTTCGATCTACCAACAAACCTCTCTAGCACCAATGCATCAGCAACTTCAGATTCCTG
288   S  G  S  I  Y  Q  Q  T  S  L  A  P  M  H  Q  Q  L  Q  I  P

1021  ATTATGGGCTTTTACAAGATATGGTCCCCTCCATGACTTTTAAACAAGAACCCTAATCAA
308   D  Y  G  L  L  Q  D  M  V  P  S  M  T  F  K  Q  E  P  *

1081  TATCCACCCCCACCCCCACCCCACCCT
```

Figure 1. The nucleotide sequence and the deduced amino acid sequence of *DgWRKY2*. The WRKY domain is underlined. The cysteine and histidine in the zinc-finger motifs are boxed.

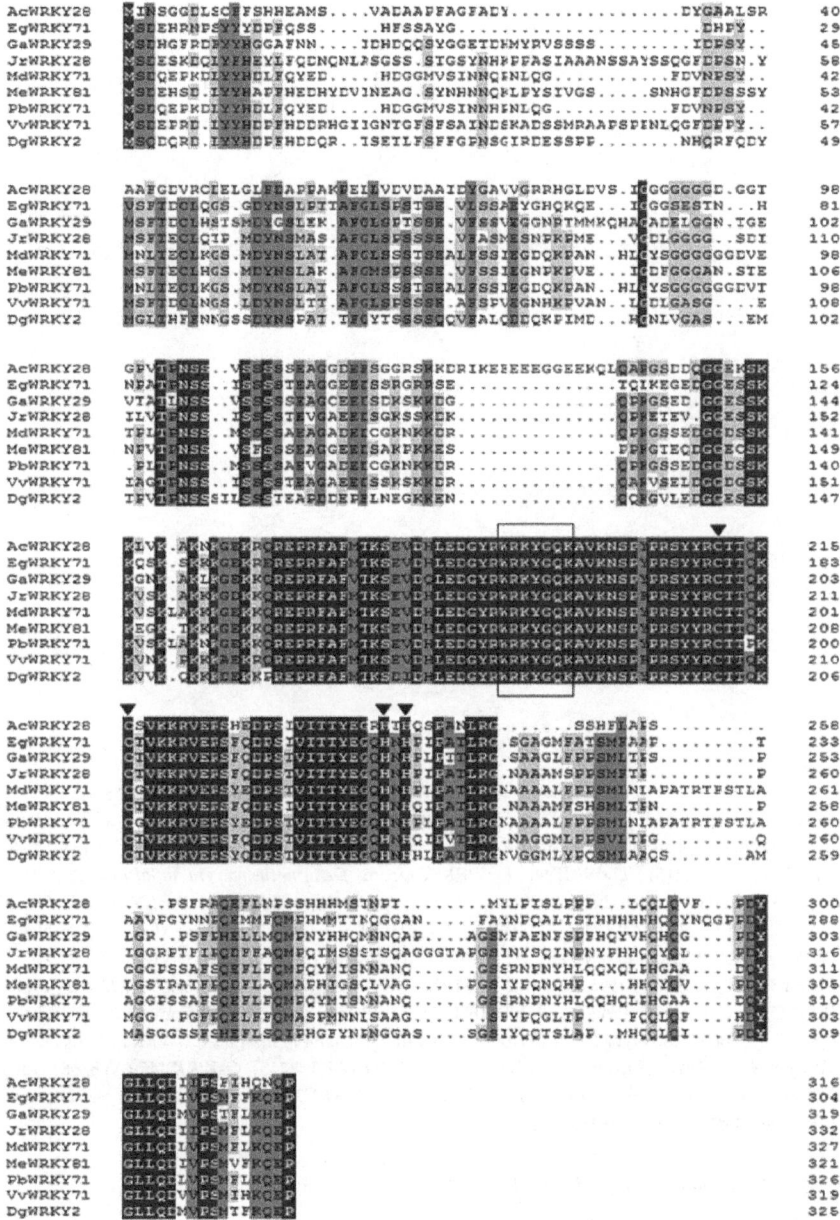

Figure 2. Comparison between the amino acid sequences deduced for the *DgWRKY2* gene. Amino acid residues conserved in all sequences are shaded in black, and those conserved in four sequences are shaded in light gray. The completely conserved WRKYGQK amino acids are boxed. The cysteine and histidine in zinc finger motifs are indicated by arrowheads (▼).

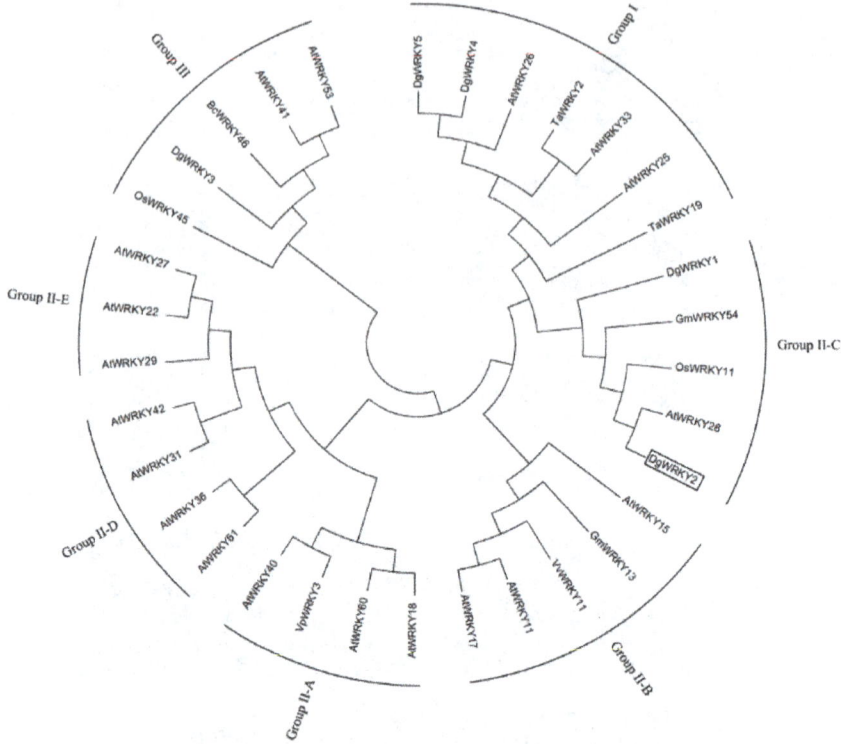

Figure 3. Phylogenetic tree analysis of the WRKY protein in different plants. The phylogenetic tree was drawn using the MEGA 5.0 program with the neighbor-joining method. *DgWRKY2* is boxed. The plant WRKY proteins used for the phylogenetic tree are as follows: *DgWRKY1* (KC153303), *DgWRKY3* (KC292215), *DgWRKY4*, *DgWRKY5* from *Dendranthema grandiflorum*; *AtWRKY11* (NP_849559), *AtWRKY15* (NP_179913.1), *AtWRKY17* (NP_565574.1), *AtWRKY18* (NP_567882), *AtWRKY22* (AEE81999), *AtWRKY25* (NP_180584), *AtWRKY26* (AAK28309), *AtWRKY27* (NP_568777), *AtWRKY28* (NP_193551), *AtWRKY29* (AEE84774), *AtWRKY31* (NP_567644), *AtWRKY33* (NP_181381), *AtWRKY36* (NP_564976), *AtWRKY40* (NP_178199), *AtWRKY41* (NP_192845), *AtWRKY42* (NP_192354), *AtWRKY53* (NP_194112), *AtWRKY60* (NP_180072), *AtWRKY61* (NP_173320) from *Arabidopsis thaliana*. *TaWRKY2* (EU665425), *TaWRKY19* (EU665430) from *Triticicum aestivum*. *GmWRKY13* (DQ322694), *GmWRKY54* (DQ322698) from *Glycine max*. *OsWRKY11* (AK108745), *OsWRKY45* (AY870611) from *Oryza sativa*. *VvWRKY11* (EC935078) from *Vitis vinifera*. *VpWRKY3* (JF500755) from *Vitis pseudoreticulata*. *BcWRKY46* (HM585284) from *Brassica campestris*.

2.2. Salt-Tolerance Analysis of DgWRKY2 Transgenic Chrysanthemum

To determine whether *DgWRKY2* overexpression enhanced salt tolerance, chrysanthemum transgenic lines with overexpressed *DgWRKY2* were produced by *Agrobacterium*-mediated transformation. *DgWRKY2* transcription levels in up in five transgenic lines (OE-3, OE-11, OE-17, OE-21 and OE-24) were detected by qRT-PCR (Figure 4A). We compared the salt stress tolerance between OE-17 and OE-21 transgenic chrysanthemum and the WT. Under normal growth conditions, the phenotypic differences were not significant. The growth rate was consistent. By contrast, under salt stress, wilting and yellowing of leaves of the WT plants were evident (Figure 4B). After the recovery period (2 weeks), the survival rate in the WT was 40.23%, while the survival rates in transgenic lines

OE-17 and OE-21 were 79.07% and 82.60%, respectively (Figure 4C). The survival rate of transgenic chrysanthemums was significantly higher than that of the WT.

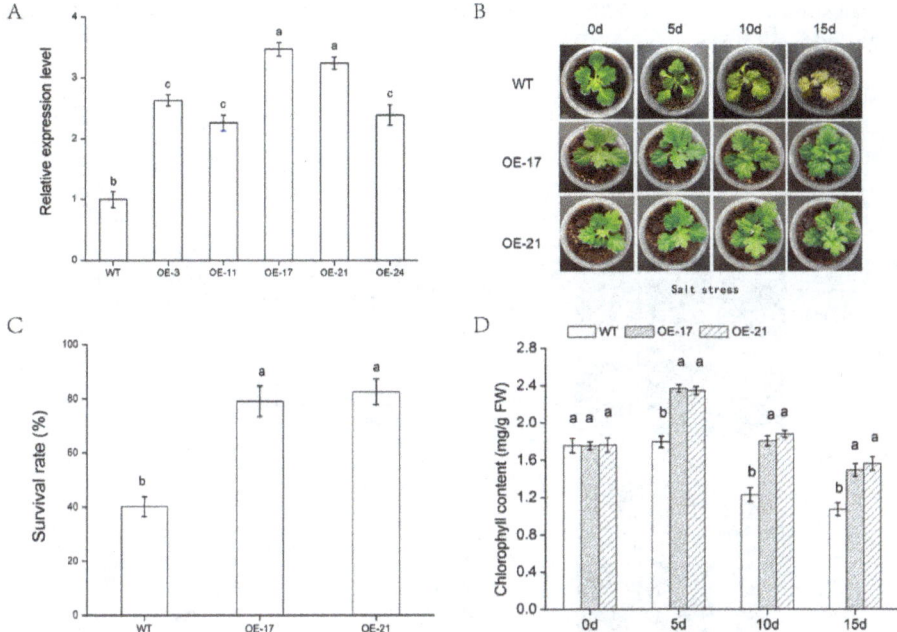

Figure 4. Expression of chrysanthemum in different strains under salt stress. (**A**) Relative expression level of *DgWRKY2* in transgenic chrysanthemums. The different normal letters indicate a significant difference at the 0.05 level among different strain lines, the same below; (**B**) comparison of transgenic plants and wild type plants after different periods under salt stress; (**C**) chrysanthemum survival statistics after recovery; (**D**) chlorophyll contents of chrysanthemum leave under salt stress.

2.3. Analysis of Chlorophyll Content and under Salt Stress

Salt stress significantly inhibited plant photosynthesis [32]. The content of chlorophyll in the leaves of the WT decreased obviously at the 10th day, while reaching the minimum value at the 15th day. However, the chlorophyll content from the transgenic chrysanthemum lines OE-17 and OE-21 increased significantly, by 35% and 33% at the 5th day, and decreased gradually later on. In general, the decrease of chlorophyll content in transgenic chrysanthemum is lower than that of the WT (Figure 4D).

2.4. Accumulation of H_2O_2, O_2^-, and MDA in DgWRKY2 Transgenic Chrysanthemum under Salt Stress

Reactive oxygen species in plant cells have a strong toxic effect. In order to study the effect of transgenic lines on the scavenging of reactive oxygen species, H_2O_2 and O_2^- in different lines were investigated with DAB and NBT staining. Under normal circumstances, there was no significant difference in H_2O_2 and O_2^- between the WT and two transgenic lines. After treatment with salt stress, the H_2O_2 content in each line increased significantly (Figure 5A,B). The contents of O_2^- showed an upward trend with the increase of stress time (Figure 5C,D), but it was not as obvious as that of H_2O_2. Under salt stress, despite the rising trend, the accumulation of H_2O_2 and O_2^- in the transgenic lines was much lower than that of the WT. These results indicate that the overexpression of *DgWRKY2* might regulate the activity of antioxidant protective enzymes, conferring greater tolerance to salt stress in transgenic plants. Similarly, under salt stress, the MDA accumulation level of overexpressed lines was apparently lower than that of the WT (Figure 5E). In all, these results provided strong evidence that the accumulation of ROS in *DgWRKY2* overexpression chrysanthemum was lower than that of WT under salt stress. Thus, *DgWRKY2* overexpression reduced the ROS level and alleviated the oxidant damage under salt stress.

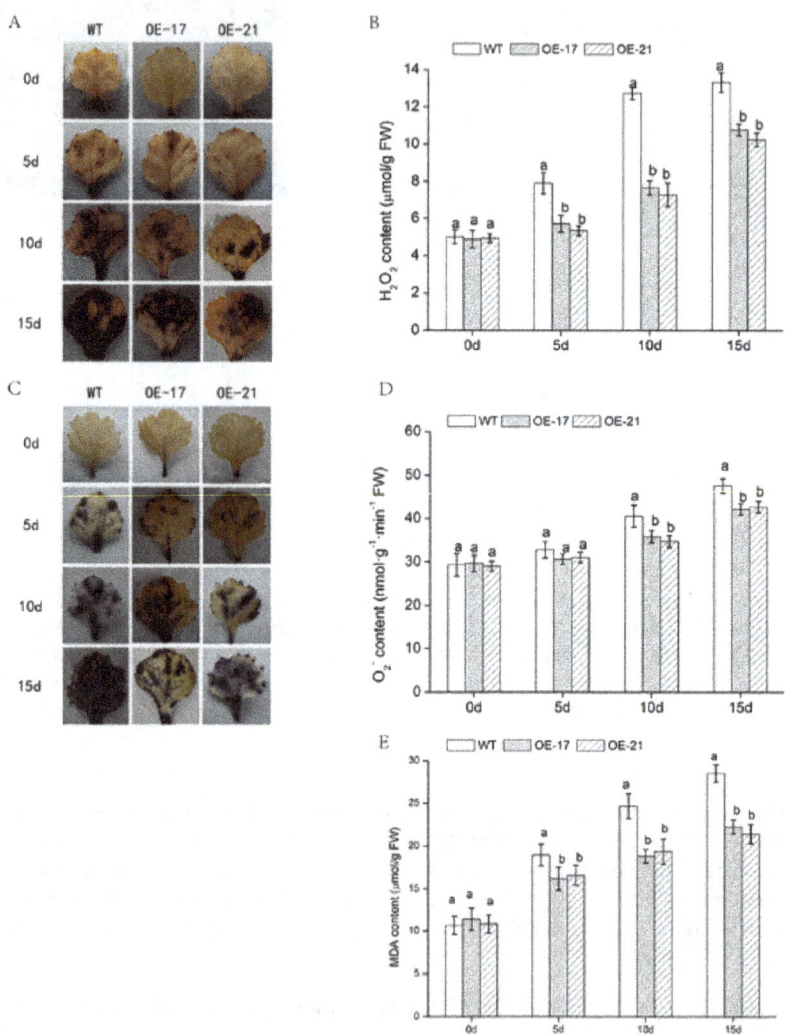

Figure 5. The levels of oxidative damage in WT and *DgWRKY2* overexpression lines of chrysanthemum were analyzed. (**A**) Diaminobenzidine (DAB) staining of chrysanthemum leaves during salt stress treatment; (**B**) changes in H_2O_2 content under salt stress; (**C**) nitroblue tetrazolium (NBT) staining of chrysanthemum leaves during salt stress treatment; (**D**) changes in O_2^- content under salt stress; (**E**) changes in malondialdehyde (MDA) content of chrysanthemum leaves under salt stress. Data represent means and standard errors of three replicates. The different letters above the columns indicate significant differences ($p < 0.05$) according to Duncan's multiple range test.

2.5. Physiological Changes in DgWRKY2 Transgenic Chrysanthemum

Antioxidant enzymes play an important part in botanical stress tolerance. We observed activities of SOD, POD, and *CAT* in the leaves of *DgWRKY2* lines and WT plants at different stages of treatment. Under normal growth conditions, the activities of these three enzymes had no obvious differences in any of the lines. Under salt treatment conditions, there was an increase in the WT and overexpressed lines. Moreover, compared with WT, these increases were extraordinarily greater in the overexpressed lines (Figure 6A–C). As a result, overexpression of *DgWRKY2* increases the antioxidant enzyme activity of transgenic chrysanthemum to counteract injury from ROS. Thus, this reduced oxidative damage.

Osmotic adjustment is one of the most basic characteristics of plant salt tolerance, while proline is the most widely distributed compatible penetrant [33,34]. Under salt stress, we measured the proline content of transgenic lines and the WT in order to understand the osmoregulation ability of transgenic plants (Figure 7A). There was little difference in proline content between transgenic lines and WT under normal circumstances. By contrast, under salt stress, there was a remarkable increase in proline content for both. Nevertheless, the accumulation of proline in the transgenic lines was significantly higher than that of the WT under salt stress. These results indicate that *DgWRKY2* upregulated the accumulation of proline in the transgenic lines under salt stress.

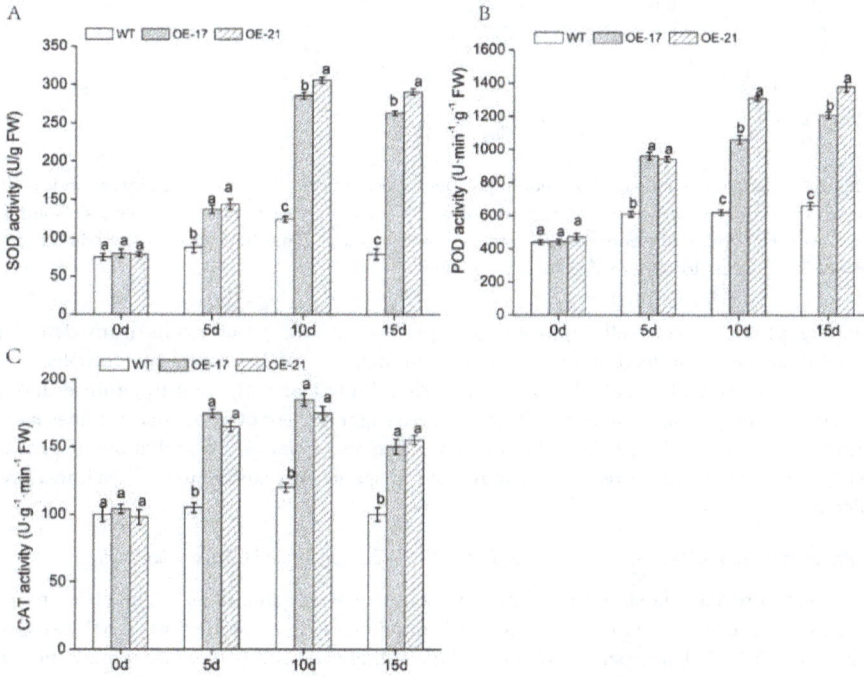

Figure 6. Changes in antioxidant enzyme activities of chrysanthemum leaves under salt stress. (**A**) Superoxide dismutase (SOD) activity under salt stress; (**B**) peroxidase (POD) activity under salt stress; (**C**) catalase (*CAT*) activity under salt stress. Data represent means and standard errors of three replicates. The different letters above the columns indicate significant differences ($p < 0.05$) according to Duncan's multiple range test.

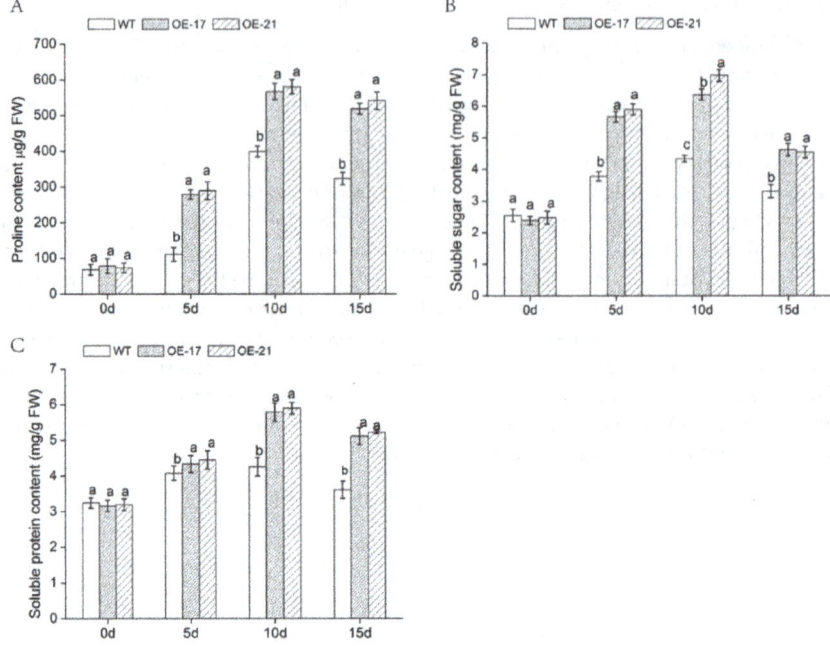

Figure 7. Changes in contents of osmotic adjustment substances of chrysanthemum leaves under salt stress. (**A**) Proline content under salt stress; (**B**) soluble sugar content under salt stress; (**C**) soluble protein content under salt stress. The different letters above the columns indicate significant differences (*p* < 0.05) according to Duncan's multiple range test.

Soluble proteins keep cells appropriately permeable and protect cells from dehydration, while stabilizing and protecting the structure and function of biological macromolecules [35]. We observed the content of soluble protein and of soluble sugar of these three lines under salt stress. In this environment, soluble protein and soluble sugar content of overexpressed lines increased significantly compared with the WT. (Figure 7B,C). The above data suggest that overexpression of *DgWRKY2* enhanced the osmoregulation ability of transgenic chrysanthemum while it increased its salt tolerance.

2.6. Expression of Abiotic Stress-Related Genes in DgWRKY2 Transformed Chrysanthemum

In order to reveal the signal regulatory network of transgenic lines in the stress resistance process, we measured the expression of several functional genes involved in signal transduction pathways by qRT-PCR. Under standard circumstances, there was little difference in the expression of abiotic stress-response genes. When exposed to salt stress, the expression level of the gene encoding ROS-scavenging enzymes (*CAT*, *APX*, and *Cu/ZnSOD*) in the transgenic lines was much higher than in the WT (Figure 8A–C). Additionally, *P5CS*, a gene related to proline synthase, showed an expression level with a similar trend (Figure 8D). Furthermore, other genes, such as *DREB1A* and *DREB2A*, that are closely related to plant responses to environmental stresses, were all significantly upregulated in the overexpressed lines compared to the WT under salinity conditions (Figure 8E,F). Our data suggest that *DgWRKY2* overexpression could reduce osmotic pressure by clearing excess ROS and accumulating proline, thereby promoting salt tolerance.

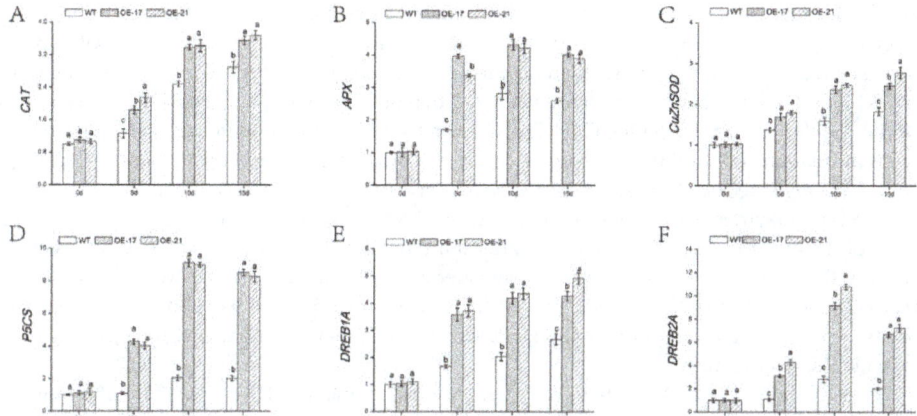

Figure 8. Expression of stress-related genes in wild type (WT) and overexpressed lines. (**A**) Expression analysis of *Cu/ZnSOD* under salt stress; (**B**) expression analysis of *CAT* under salt stress; (**C**) expression analysis of ascorbate peroxidase (*APX*) under salt stress; (**D**) expression analysis of *P5CS* in chrysanthemum under salt stress; (**E**) expression analysis of *DREB1A* under salt stress; (**F**) expression analysis of *DREB2A* under salt stress. Data represent means and standard errors of three replicates. The different letters above the columns indicate significant differences ($p < 0.05$) according to Duncan's multiple range test.

3. Discussion

To date, the WRKY gene has been cloned from *Arabidopsis thaliana* [36], wheat [37,38], rice [39], soybean [40], chrysanthemum [28,29], birch [41], and other plants. It was confirmed that the WRKY gene is related to plant stress resistance. We isolated a new WRKY transcription factor—*DgWRKY2*—from chrysanthemum, and found it to be induced by salt stress. The deduced amino acid sequence of the *DgWRKY2* gene from this study contains one WRKY domain (WRKYGQK) and a zinc finger structure (C-X4-5-C-X22-23-H-X1-H), which could be considered part of the group II WRKY family.

The same group of WRKY proteins might have similar capabilities. Previous studies have shown that *GmWRKY54* might improve the salt and cold tolerance of plants through the regulation of *DREB2A* and STZ/Zat10 [40]. *OsWRKY11* overexpression increased rice drought tolerance [42]. The expression of *AtWRKY28* changed significantly under NaCl stress, indicating that *AtWRKY28* had much to do with adaptation to environmental stress [43]. In our previous study, an overexpression *DgWRKY1* tobacco line was more tolerant to salt stress than the WT [28]. *DgWRKY2* belongs to group II with *GmWRKY54*, *OsWRKY11*, *AtWRKY28*, and *DgWRKY1*, thus, we hypothesized that *DgWRKY2* has a positive effect on salt tolerance. In addition, the previous studies demonstrated that *DgWRKY3*, *DgWRKY4*, and *DgWRKY5* also played a positive regulatory role on salt stress [29–31]. *DgWRKY1* and *DgWRKY3* were only studied for their role in salt tolerance in tobacco, and the salt tolerance in chrysanthemum has yet to be studied. Previous studies have confirmed that *DgWRKY4* and *DgWRKY5* belong to the group III, and *DgWRKY2* in this study belongs to group II. The results of these studies showed that *DgWRKY4* and *DgWRKY5* imparted stronger salt tolerance than *DgWRKY2*. This is partly due to different groups playing different roles in the stress regulatory network. Additional work is needed to understand the mechanisms.

In this study, the *DgWRKY2* overexpression transgenic chrysanthemum was compared with the WT from physiological and biochemical aspects, and the function of *DgWRKY2* overexpression was verified. Chlorophyll content in chrysanthemum leaves continued to decrease in the late stage of salt stress. We speculated that ROS inhibited the photosynthesis of chrysanthemum [44]. However, chlorophyll content in the overexpressed lines was higher than that of the WT at respectively different salt stress stages. Increased ROS activity causes a great deal of physiological and metabolic changes in plants, enabling them to cope with environmental stress. In *CmWRKY17*-overexpressing plants, *CmWRKY17* altered the salinity sensitivity via regulation of ROS levels [45]. *NbWRKY79* was involved with the regulation of SOD, POD, *CAT*, and *APX* activities, which resulted in the suppression of ROS accumulation so that the plant could endure less oxidative damage under salt conditions [11]. *MsWRKY11* might reduce ROS levels and thus increase salt tolerance in soybean [46]. The activity of antioxidant enzymes SOD, POD, and *CAT* in *DgWRKY2* overexpression lines increased, and the activity of the enzymes was significantly higher than that of the WT at each stage of salt treatment. Moreover, the content of H_2O_2 and O_2^- in transgenic chrysanthemum leaves was also lower than that of WT. The above results indicate that *DgWRKY2* overexpression could enhance plant antioxidant capacity by increasing the activities of SOD, POD, and *CAT*, thereby enhancing the salt tolerance of transgenic chrysanthemum.

Accumulation of MDA content can lead to membrane lipid peroxidation of plant cells, causing changes in the cell membrane structure and permeability, reducing cell function [47]. In contrast, proline prevents membrane lipid peroxidation, maintains normal cellular structure, and maintains a stable cell osmotic pressure [48]. Under salt treatment, compared with the WT, MDA content of the overexpressed lines was lower, but proline content was higher. The contents of soluble sugar and soluble protein in *DgWRKY2* overexpression lines were higher than those of WT. The results suggest that *DgWRKY2* might increase its salt tolerance by regulating the osmotic pressure of plant cells.

The expression of antioxidant genes (*Cu/ZnSOD*, *CAT*, and *APX*) was upregulated under salinity, which is consistent with physiological results. Under salt stress, the expression of antioxidant enzyme genes was significantly higher in *RtWRKY1*-overexpressed *Arabidopsis* than in the wild type [12]. The *P5CS* gene is associated with a proline-synthesizing enzyme in plants. When the expression of the *P5CS* gene was induced by environmental stress, the proline content in plants increased. Under salt stress, the expression of genes related to proline biosynthesis was upregulated in *VvWRKY30* transgenic lines compared with their expression in the WT [13]. These results show that transgenic plants exhibited increased expression levels of *P5CS* under stress conditions. The DREB gene belongs to the *AP2/EREBP* transcription factor family. These TFs are closely related to the response of plants to the environment [49,50]. In this study, *DREB1A* was upregulated to a greater extent in overexpressed lines than in WT, and *DREB2A* first increased and later decreased. Previous studies indicated that *OsDREB2A* might participate in abiotic stress by directly binding the DREB element to regulate the expression of downstream genes. Overexpression of *OsDREB2A* in soybean might be used to improve its tolerance to salt stress [51]. Cong found that overexpression of the *OjDREB* gene improved salt tolerance in tobacco plant [52]. These results suggest that enhanced salt tolerance was associated with the induction of downstream stress-related gene expression in *DgWRKY2* transgenic plants.

4. Materials and Methods

4.1. Plant Materials

The experimental material used for treatment is a wild-type chrysanthemum: *Dendranthema grandiforum*—'Jinba'. All plant materials were provided by Sichuan Agricultural University, Chengdu, China. Chrysanthemum seedlings grew on MS culture medium (200 μL m^{-2} s^{-1}, 16 h photoperiod, 25 °C/22 °C day/night temperature, and 70% relative humidity) for 20 days. Then, 20-day-old seedlings were planted in basins filled with a 1:1 mixture of peat and perlite, incubated for 3 days, and watered once daily (70% of field capacity). Seedlings at the six-leaf stage were harvested, frozen in liquid nitrogen immediately, and stored at −80 °C for RNA extraction.

4.2. Cloning of DgWRKY2 and Sequence Analysis

The RNA extraction of chrysanthemum leaves was performed by TRNzol reagent (Mylab, Beijing, China). The full-length cDNA of the *DgWRKY2* sequence was obtained by PCR (polymerase chain reaction) utilizing gene-specific primers (Table 1). The RACE reactions were carried out according to the manufacturer's protocol (Invitrogen RACE cDNA amplification kit, Clontech, Mountain View, CA, USA). The fragment generated was cloned into pEASY-T1 Cloning Kit (Transgene Biotech, Beijing, China) and sequenced.

Table 1. Primers and their sequences in experiment.

Primer	Sequence (5′-3′)
DgWRKY2	F: ATTTGTCAAACTTCTCCTCTCTTCT
	R: GTGGGGGTGGGGGTGGATA
EF1a	F: TTTTGGTATCTGGTCCTGGAG
	R: CCATTCAAGCGACAGACTCA
Cu/Zn SOD	F: CCATTGTTGACAAGCAGATTCCACTCA
	R: ATCATCAGGATCAGCATGGACGACTAC
CAT	F: TACAAGCAACGCCCTTCAA
	R: GACCTCTGTTCCCAACAGTCA
APX	F: GTTGGCTGGTGTTGTTGCT
	R: GATGGTCGTTTCCCTTAGTTG
P5CS	F: TTGGAGCAGAGGTTGGAAT
	R: GCAGGTCTTTGTGGGTGTAG
DREB1A	F: CGGTTTTGGCTATGAGGGGT
	R: TTCTTCTGCCAGCGTCACAT
DREB2A	F: GATCGTGGCTGAGAGACTCG
	R: TACCCCACGTTCTTTGCCTC

The sequence of *DgWRKY2* was analyzed by the National Center for Biotechnology Information (NCBI, http://www.ncbi.nlm.nih.gov/gorf/gorf.html) to obtain its open reading frame (ORF). Identification of protein domains and significant sites was performed with Motifscan (http://myhits. isb-sib.ch/cgi-bin/motif_scan). The phylogenetic tree was drawn with the MEGA 5.0 program (Sudhir Kumar, Arizona State University, Tempe, AZ, USA) using the neighbor-joining method.

4.3. Generation of Transgenic Chrysanthemum

The pEASY-WRKY2 cloning vector was constructed by TA cloning technology (The complementarity between the vector 3′-T overhangs and PCR product 3′-A overhangs allows direct ligation of Taq-amplified PCR products into the T-vector). The plasmid containing the pEASY-WRKY2 and pBI121 expression vectors were double digested with *Sac*I and *Xba*I to construct the *pBI121-DgWRKY2* expression vector. The fused construction of *pBI121-DgWRKY2* was transformed into the leaf disk of chrysanthemum by *Agrobacterium tumefaciens* (strain LBA4404) [53]. Callus induction from chrysanthemum was used to form seedlings [54]. The obtained *DgWRKY2* transgenic chrysanthemum lines (OE-17 and OE-21) were employed in subsequent experiments. The transgenic lines OE-17 and OE-21 were expanded for subsequent replication experiments.

4.4. Expression of DgWRKY2 under Salt Treatment

The method of RNA extraction is the same as above. Then RNA was used for first-strand cDNA synthesis with reverse transcriptase (TransScript II All-in-one First-Strand cDNA Synthesis SuperMix for PCR, Transgene, Beijing, China) according to the manufacturer's protocol. Quantitative real-time PCR (qRT-PCR) was performed by SsoFast EvaGreen supermix (Bio-Rad, Hercules, CA, USA) and Bio-Rad CFX96TM detection system. The gene elongation factor 1α (*EF1α*) was used as a reference for quantitative expression analysis. A final 20 μL qPCR reaction mixture contained: 10 μL SsoFast

EvaGreen supermix, 2 μL diluted cDNA sample, and 300 nM primers. Then, the reactions were incubated following the standard process: 1 cycle of 95 °C for 30 s, 40 cycles of 15 s at 95 °C and 30 s at 60 °C, and a single melting cycle from 65 to 95 °C. To avoid experimental errors, each reaction was repeated at least three times. To avoid variables and statistic error, a negative control group was set up, in which water supplanted the above solution. Relative expression levels were calculated by the $2^{-\Delta\Delta Ct}$ method [55].

4.5. Salt Treatment of Transgenic Chrysanthemum and Stress Tolerance Assays

Two overexpressed lines (OE-17 and OE-21) and the WT of chrysanthemum, all 20 days old, were sown into a 1:1 mixture of peat and perlite, then cultured in a light incubator (200 μL m^{-2} s^{-1}, 16 h photoperiod, 25 °C/22 °C day/night temperature, and 70% relative humidity). Soil-grown chrysanthemum seedlings at the six-leaf stage were irrigated with an increasing concentration of NaCl solution: 100 mm for 1–5 days (d), 200 mm for 6–10 days, and 400 mm for 11–15 days, using Chen as a reference [56]. Under salt stress, leaves 4–5 were harvested at 0, 5, 10, and 15 days for both physiological and molecular experiments. After a 2-week recovery, the surviving plants were collected to calculate the survival rate.

4.6. Determination of Physiological Indexes of Transgenic Chrysanthemum under Salt Stress

Activities of superoxide dismutase (SOD), peroxidase (POD), and catalase (*CAT*) were measured according to Li [57]. Malondialdehyde (MDA) content in chrysanthemum was measured according to Zhang [58]. Accumulation of proline, soluble sugar, and soluble protein was measured according to Sun [59]. The chlorophyll content was detected according to Jin [60].

4.7. Histochemical Detection of Reactive Oxygen Species (ROS)

Nitroblue tetrazolium (NBT) and diaminobenzidine (DAB) staining was measured according to Shi [61]. The standard steps were as follows: chrysanthemum leaves were completely immersed in 10 mm phosphate buffer (pH = 7.8) containing 1 mg/mL NBT or DAB at room temperature. The leaves were not placed in 95% ethanol for decolorization until the spots appeared. After that, the sample was observed, and photos of the sample were taken. Finally, H_2O_2 and O_2^- concentration were determined by detection kits (Nanjing Jiancheng Bioengineering Institute, Nanjing, China).

4.8. Expression of Salt Stress Response Genes in Dgwrky2 Transgenic Chrysanthemum

To evaluate the expression of abiotic stress-related genes, RNA from the WT and transgenic lines was extracted for reverse transcription. Transgenic chrysanthemum stress-responsive gene expression was detected by qRT-PCR. The abiotic stress-response genes monitored were *Cu/ZnSOD*, *CAT*, *APX*, *P5CS*, *DREB1A*, and *DREB2A*. All relevant primers used in the study are listed in Table 1.

4.9. Statistical Analysis

All experiments were performed three times for biological repetition to avoid all types of error. All data were analyzed by SPSS version 24.0 (International Business Machines Corporation, Armonk, NY, USA). A one-way analysis of variance, Tukey's multiple range test ($p < 0.05$), was employed to identify the treatment means to avoid static errors.

5. Conclusions

In summary, this study demonstrated that *DgWRKY2* could positively regulate salt stress tolerance. To alleviate the damage of salt stress to plants, *DgWRKY2* overexpression improved expression of stress-related genes, resulting in relatively enhanced photosynthetic capacity, greatly increased activities of antioxidant enzymes, and high accumulation of proline, soluble sugar, and soluble protein. This indicates that *DgWRKY2* may enhance the sensitivity to salinity by enabling antioxidant and osmotic adjustment capabilities. Overall, this study identified *DgWRKY2* as a potential genetic resource for plant salt tolerance. Not only did *DgWRKY2* play an important role in supplementing and perfecting chrysanthemum-tolerant germplasm resources, but it could also be used as a reserved gene for salt-tolerant plant breeding.

Author Contributions: L.H., Y.-H.W. and Q.-L.L. conceived and designed the experiments. L.H., Y.-H.W., B.W., Q.-L.L. and Q.Z. performed the experiments. L.Z. and L.H. analyzed the data. L.H. wrote the paper. All authors read and approved the manuscript.

Funding: This research was funded by National Natural Science Foundation of China grant number 31770742.

Acknowledgments: This study was supported by Sichuan Agricultural University Ornamental Horticulture lab. We would like to acknowledge the contribution of Qing-Lin Liu for the provision of experimental materials and instruments.

Conflicts of Interest: The authors declare that the research was conducted in the absence of any commercial or financial relationships that could be construed as a potential conflict of interest.

References

1. Jamil, A.; Riaz, S.; Ashraf, M.; Foolad, M.R. Gene Expression Profiling of Plants under Salt Stress. *Crit. Rev. Plant Sci.* **2011**, *30*, 435–458. [CrossRef]
2. Munns, R. Comparative physiology of salt and water stress. *Plant Cell Environ.* **2002**, *25*, 239–250. [CrossRef] [PubMed]
3. Bing, L.; Zhao, B.C.; Shen, Y.Z.; Huang, Z.J.; Ge, R.C. Progress of Study on Salt Tolerance and Salt Tolerant Related Genes in Plant. *J. Hebei Norm. Univ.* **2008**, *2*, 243–248.
4. Tuteja, N. Mechanisms of high salinity tolerance in plants. *Methods Enzymol.* **2007**, *428*, 419–438. [CrossRef] [PubMed]
5. Eulgem, T.; Rushton, P.J.; Robatzek, S.; Somssich, I.E. The WRKY superfamily of plant transcription factors. *Trends Plant Sci.* **2000**, *5*, 199–206. [CrossRef]
6. Zhang, Y.; Yu, H.; Yang, X.; Li, Q.; Ling, J.; Wang, H.; Gu, X.; Huang, S.; Jiang, W. CsWRKY46, a WRKY transcription factor from cucumber, confers cold resistance in transgenic-plant by regulating a set of cold-stress responsive genes in an ABA-dependent manner. *Plant Physiol. Biochem.* **2016**, *108*, 478. [CrossRef] [PubMed]
7. Bakshi, M.; Oelmüller, R. WRKY transcription factors: Jack of many trades in plants. *Plant Signal. Behav.* **2014**, *9*, e27700. [CrossRef] [PubMed]
8. Tripathi, P.; Rabara, R.C.; Rushton, P.J. A systems biology perspective on the role of WRKY transcription factors in drought responses in plants. *Planta* **2014**, *239*, 255–266. [CrossRef] [PubMed]
9. Guo, Y.; Cai, Z.; Gan, S. Transcriptome of Arabidopsis leaf senescence. *Plant Cell Environ.* **2004**, *27*, 521–549. [CrossRef]
10. Qin, Y.; Tian, Y.; Liu, X. A wheat salinity-induced WRKY transcription factor TaWRKY93 confers multiple abiotic stress tolerance in Arabidopsis thaliana. *Biochem. Biophys. Res. Commun.* **2015**, *464*, 428. [CrossRef] [PubMed]
11. Nam, T.N.; Le, H.T.; Mai, D.S.; Tuan, N.V. Overexpression of NbWRKY79, enhances salt stress tolerance in Nicotiana benthamiana. *Acta Physiol. Plant.* **2017**, *39*, 121. [CrossRef]
12. Du, C.; Zhao, P.; Zhang, H.; Li, N.; Zheng, L.; Wang, Y. The Reaumuria trigyna transcription factor RtWRKY1 confers tolerance to salt stress in transgenic Arabidopsis. *J. Plant Physiol.* **2017**, *215*, 48–58. [CrossRef] [PubMed]
13. Zhu, D.; Hou, L.; Xiao, P.; Guo, Y.; Deyholos, M.K.; Liu, X. VvWRKY30, a grape WRKY transcription factor, plays a positive regulatory role under salinity stress. *Plant Sci.* **2018**. [CrossRef]
14. Powers, S.K.; Lennon, S.L.; Quindry, J.; Mehta, J.L. Exercise and cardioprotection. *Curr. Opin. Cardiol.* **2002**, *17*, 495–502. [CrossRef] [PubMed]
15. Jiang, M.; Zhang, J. Water stress-induced abscisic acid accumulation triggers the increased generation of reactive oxygen species and up-regulates the activities of antioxidant enzymes in maize leaves. *J. Exp. Bot.* **2002**, *53*, 2401–2410. [CrossRef] [PubMed]

16. Meloni, D.A.; Oliva, M.A.; Martinez, C.A.; Cambraia, J. Photosynthesis and activity of superoxide dismutase, peroxidase and glutathione reductase in cotton under salt stress. *Environ. Exp. Bot.* **2003**, *49*, 69–76. [CrossRef]

17. Moradi, F.; Ismail, A.M. Responses of photosynthesis, chlorophyll fluorescence and ROS-scavenging systems to salt stress during seedling and reproductive stages in rice. *Ann. Bot.* **2007**, *99*, 1161–1173. [CrossRef] [PubMed]

18. Negi, N.P.; Sharma, V.; Sarin, N.B. Pyramiding of Two Antioxidant Enzymes CuZnSOD and cAPX from Salt Tolerant Cell Lines of Arachis hypogeae Confers Drought Stress Tolerance in Nicotiana tabacum. *Indian J. Agric. Biochem.* **2017**, *30*, 141. [CrossRef]

19. Hui, Y.; Qiang, L.; Park, S.C.; Wang, X.; Liu, Y.J.; Zhang, Y.G.; Tang, W.; Kou, M.; Ma, D.F. Overexpression of CuZnSOD, and APX, enhance salt stress tolerance in sweet potato. *Plant Physiol. Biochem.* **2016**, *109*, 20–27. [CrossRef]

20. Yang, Z.; Zhou, Y.; Ge, L.; Li, G.; Liu, Q.; Xu, Y.; Jiang, L.; Yang, Y.; School of Agriculture, Jiangxi Agricultural University; School of Sciences, Jiangxi Agricultural University. Expression of Cucumber CsCAT3 Gene under Stress and Its Salt Tolerance in Transgenic Arabidopsis thaliana. *Mol. Plant Breed.* **2018**.

21. Guerzoni, J.T.S.; Belintani, N.G.; Moreira, R.M.P.; Hoshino, A.A.; Domingues, D.S.; Filho, J.C.B.; Vieira, L.G.E. Stress-induced Δ1-pyrroline-5-carboxylate synthetase (P5CS) gene confers tolerance to salt stress in transgenic sugarcane. *Acta Physiol. Plant.* **2014**, *36*, 2309–2319. [CrossRef]

22. Wang, W.; Vinocur, B.; Altman, A. Plant responses to drought, salinity and extreme temperatures: Towards genetic engineering for stress tolerance. *Planta* **2003**, *218*, 1–14. [CrossRef] [PubMed]

23. Qin, F.; Kakimoto, M.; Sakuma, Y.; Maruyama, K.; Osakabe, Y.; Tran, L.S.; Shinozaki, K.; Yamaguchi-Shinozaki, K. Regulation and functional analysis of ZmDREB2A in response to drought and heat stresses in *Zea mays* L. *Plant J.* **2010**, *50*, 54–69. [CrossRef] [PubMed]

24. Zhou, M.L.; Ma, J.T.; Zhao, Y.M.; Wei, Y.H.; Tang, Y.X.; Wu, Y.M. Improvement of drought and salt tolerance in Arabidopsis and Lotus corniculatus by overexpression of a novel DREB transcription factor from Populus euphratica. *Gene* **2012**, *506*, 10–17. [CrossRef] [PubMed]

25. Ma, J.T.; Yin, C.C.; Guo, Q.Q.; Zhou, M.L.; Wang, Z.L.; Wu, Y.M. A novel DREB transcription factor from Halimodendron halodendron, leads to enhance drought and salt tolerance in Arabidopsis. *Biol. Plant.* **2014**, *59*, 74–82. [CrossRef]

26. Akça, Y.; Samsunlu, E. The effect of salt stress on growth, chlorophyll content, proline and nutrient accumulation, and k/na ratio in walnut. *Am. Bank.* **2012**, *1999*, 1513–1520.

27. Wu, Y.H.; Wang, T.; Wang, K.; Liang, Q.Y.; Bai, Z.Y.; Liu, Q.L.; Jiang, B.B.; Zhang, L. Comparative Analysis of the Chrysanthemum Leaf Transcript Profiling in Response to Salt Stress. *PLoS ONE* **2016**, *11*, e0159721. [CrossRef] [PubMed]

28. Liu, Q.L.; Xu, K.D.; Pan, Y.Z.; Jiang, B.B.; Liu, G.L.; Jia, Y.; Zhang, H.Q. Functional Analysis of a Novel Chrysanthemum WRKY Transcription Factor Gene Involved in Salt Tolerance. *Plant Mol. Biol. Rep.* **2014**, *32*, 282–289. [CrossRef]

29. Liu, Q.L.; Zhong, M.; Li, S.; Pan, Y.Z.; Jiang, B.B.; Jia, Y.; Zhang, H.Q. Overexpression of a chrysanthemum transcription factor gene, DgWRKY3, intobacco enhances tolerance to salt stress. *Plant Physiol. Biochem.* **2013**, *69*, 27–33. [CrossRef] [PubMed]

30. Wang, K.; Wu, Y.H.; Tian, X.Q.; Bai, Z.Y.; Liang, Q.Y.; Liu, Q.L.; Pan, Y.Z.; Zhang, L.; Jiang, B.B. Overexpression of DgWRKY4 Enhances Salt Tolerance in Chrysanthemum Seedlings. *Front. Plant Sci.* **2017**, *8*, 1592. [CrossRef] [PubMed]

31. Liang, Q.Y.; Wu, Y.H.; Wang, K.; Bai, Z.Y.; Liu, Q.L.; Pan, Y.Z.; Zhang, L.; Jiang, B.B. Chrysanthemum WRKY gene DgWRKY5 enhances tolerance to salt stress in transgenic chrysanthemum. *Sci. Rep.* **2017**, *7*, 4799. [CrossRef] [PubMed]

32. Diao, M.; Ma, L.; Wang, J.; Cui, J.; Fu, A.; Liu, H. Selenium Promotes the Growth and Photosynthesis of Tomato Seedlings Under Salt Stress by Enhancing Chloroplast Antioxidant Defense System. *J. Plant Growth Regul.* **2014**, *33*, 671–682. [CrossRef]

33. Ben, K.R.; Abdelly, C.; Savouré, A. Proline, a multifunctional amino-acid involved in plant adaptation to environmental constraints. *Biol. Aujourdhui* **2012**, *206*, 291. [CrossRef]

34. Chaleff, R.S. Further characterization of picloram tolerant mutance of *Nicotinana tabacum*. *Theor. Appl. Genet.* **1980**, *58*, 91–95. [CrossRef]

35. Wang, F.; Liu, P.; Zhu, J. Effect of magnesium (Mg) on contents of free proline, soluble sugar and protein in soybean leaves. *J. Henan Agric. Sci.* **2004**, *6*, 35–38.

36. Fu, Q.T.; Yu, D.Q. Expression profiles of AtWRKY25, AtWRKY26 and AtWRKY33 under abiotic stresses. *Hereditas* **2010**, *32*, 848–856. [CrossRef] [PubMed]

37. Qin, Y.X. Salt-Tolerant Drought-Tolerant Wheat Gene TaWRKY79 and Application Thereof. CN 102703465A, 3 October 2012.

38. Tian, Y.C.; Qin, Y.X. Wheat Salt-Tolerant and Drought-Resistant Gene TaWRKY80 and Application Thereof. CN 102703466B, 3 July 2013.

39. Wang, H.; Hao, J.; Chen, X.; Hao, Z.; Wang, X.; Lou, Y.; Peng, Y.; Guo, Z. Overexpression of rice WRKY89 enhances ultraviolet B tolerance and disease resistance in rice plants. *Plant Mol. Biol.* **2007**, *65*, 799–815. [CrossRef] [PubMed]

40. Zhou, Q.Y.; Tian, A.G.; Zou, H.F.; Xie, Z.M.; Lei, G.; Huang, J.; Wang, C.M.; Wang, H.W.; Zhang, J.S.; Chen, S.Y. Soybean WRKY-type transcription factor genes, GmWRKY13, GmWRKY21, and GmWRKY54, confer differential tolerance to abiotic stresses in transgenic Arabidopsis plants. *Plant Biotechnol. J.* **2008**, *6*, 486–503. [CrossRef] [PubMed]

41. Wang, F.; Hou, X.; Tang, J.; Wang, Z.; Wang, S.; Jiang, F.; Li, Y. A novel cold-inducible gene from Pak-choi (*Brassica campestris*, ssp. *chinensis*), BcWRKY46, enhances the cold, salt and dehydration stress tolerance in transgenic tobacco. *Mol. Biol. Rep.* **2012**, *39*, 4553. [CrossRef]

42. Song, Y.; Jing, S.J.; Yu, D.Q. Overexpression of the stress-induced OsWRKY08 improves osmotic stress tolerance in Arabidopsis. *Chin. Sci. Bull.* **2009**, *54*, 4671–4678. [CrossRef]

43. Zhong, G.M.; Wu, L.T.; Wang, J.M.; Yang, Y.; Li, X.F. Subcellular localization and expression analysis of transcription factor AtWRKY28 under biotic stresses. *J. Agric. Sci. Technol.* **2012**, *14*, 57–63.

44. Zhao, Y.; Zhou, Y.; Jiang, H.; Li, X.; Gan, D.; Peng, X.; Zhu, S.; Cheng, B. Systematic Analysis of Sequences and Expression Patterns of Drought-Responsive Members of the HD-Zip Gene Family in Maize. *PLoS ONE* **2011**, *6*, e28488. [CrossRef] [PubMed]

45. Raghavendra, A.S.; Padmasree, K.; Saradadevi, K. Interdependence of photosynthesis and respiration in plant cells: Interactions between chloroplasts and mitochondria. *Plant Sci.* **1994**, *97*, 1–14. [CrossRef]

46. Li, P.; Song, A.; Gao, C.; Wang, L.; Wang, Y.; Sun, J.; Jiang, J.; Chen, F.; Chen, S. Chrysanthemum WRKY gene CmWRKY17, negatively regulates salt stress tolerance in transgenic chrysanthemum and Arabidopsis plants. *Plant Cell Rep.* **2015**, *34*, 1365–1378. [CrossRef] [PubMed]

47. Wang, Y.; Jiang, L.; Chen, J.; Tao, L.; An, Y.; Cai, H.; Guo, C. Overexpression of the alfalfa WRKY11 gene enhances salt tolerance in soybean. *PLoS ONE* **2018**, *13*, e0192382. [CrossRef] [PubMed]

48. Skórzyńskapolit, E. Lipid peroxidation in plant cells, its physiological role and changes under heavy metal stress. *Acta Soc. Bot. Pol.* **2007**, *76*, 49–54. [CrossRef]

49. Jain, M.; Mathur, G.; Koul, S.; Sarin, N. Ameliorative effects of proline on salt stress-induced lipid peroxidation in cell lines of groundnut (*Arachis hypogaea* L.). *Plant Cell Rep.* **2001**, *20*, 463–468. [CrossRef]

50. Tang, M.; Liu, X.; Deng, H.; Shen, S. Over-expression of JcDREB, a putative AP2/EREBP domain-containing transcription factor gene in woody biodiesel plant Jatropha curcas, enhances salt and freezing tolerance in transgenic Arabidopsis thaliana. *Plant Sci. Int. J. Exp. Plant Biol.* **2011**, *181*, 623. [CrossRef] [PubMed]

51. Sakuma, Y.; Maruyama, K.; Osakabe, Y.; Qin, F.; Seki, M.; Shinozaki, K. Functional analysis of an Arabidopsis transcription factor, DREB2A, involved in drought-responsive gene expression. *Plant Cell* **2006**, *18*, 1292–1309. [CrossRef] [PubMed]

52. Mallikarjuna, G.; Mallikarjuna, K.; Reddy, M.K.; Kaul, T. Expression of OsDREB2A, transcription factor confers enhanced dehydration and salt stress tolerance in rice (*Oryza sativa*, L.). *Biotechnol. Lett.* **2011**, *33*, 1689–1697. [CrossRef] [PubMed]

53. An, G.; Watson, B.D.; Chiang, C.C. Transformation of Tobacco, Tomato, Potato, and Arabidopsis thaliana Using a Binary Ti Vector System. *Plant Physiol.* **1986**, *81*, 301–305. [CrossRef] [PubMed]

54. Xue, J.P.; Yu, M.; Zhang, A.M. Studies on callus induced from leaves and plantlets regeneration of the traditional Chinese medicine Chrysanthemum morifolium. *China J. Chin. Mater. Med.* **2003**, *28*, 213–216.

55. Schmittgen, T.D. Analysis of relative gene expression data using real-time quantitative PCR and the 2(-Delta Delta C(T)) Method. *Methods* **2001**, *25*, 402–408. [CrossRef]

56. Chen, L.; Chen, Y.; Jiang, J.; Chen, S.; Chen, F.; Guan, Z.; Fang, W. The constitutive expression of Chrysanthemum dichrum ICE1 in Chrysanthemum grandiflorum improves the level of low temperature, salinity and drought tolerance. *Plant Cell Rep.* **2012**, *31*, 1747–1758. [CrossRef] [PubMed]

57. Li, H.S. *Principles and Techniques of Plant Physiological and Biochemical Experiment*, 3rd ed.; Higher Education Press: Beijing, China, 2015; pp. 182–184.

58. Zhang, L.; Tian, L.H.; Zhao, J.F.; Song, Y.; Zhang, C.J.; Guo, Y. Identification of an apoplastic protein involved in the initial phase of salt stress response in rice root by two-dimensional electrophoresis. *Plant Physiol. Plant Signal. Behav.* **2009**, *149*, 916–928. [CrossRef] [PubMed]

59. Sun, H.J.; Wang, S.F.; Chen, Y.T. Effects of salt stress on growth and physiological index of 6 tree species. *For. Res.* **2009**, *22*, 315–324.

60. Jin, Y.; Donglin, L.I.; Ding, Y.; Wang, L. Effects of salt stress on photosynthetic characteristics and chlorophyll content of Sapium sebiferum seedlings. *J. Nanjing For. Univ.* **2011**, *35*, 29–33.

61. Shi, J.; Fu, X.Z.; Peng, T.; Huang, X.S.; Fan, Q.J.; Liu, J.H. Spermine pretreatment confers dehydration tolerance of citrus in vitro plants via modulation of antioxidative capacity and stomatal response. *Tree Physiol.* **2010**, *30*, 914–922. [CrossRef] [PubMed]

International Journal of
Molecular Sciences

Article

Identification of Two Novel R2R3-MYB Transcription factors, *PsMYB114L* and *PsMYB12L*, Related to Anthocyanin Biosynthesis in *Paeonia suffruticosa*

Xinpeng Zhang [1,†], Zongda Xu [2,†], Xiaoyan Yu [2,*], Lanyong Zhao [2,*], Mingyuan Zhao [2], Xu Han [2] and Shuai Qi [2]

[1]   State Key Laboratory of Crop Biology, College of Horticulture Science and Engineering, Shandong Agricultural University, Taian 271018, China; zhxpnd123@163.com
[2]   College of Forestry, Shandong Agricultural University, Taian 271018, China; xuzoda@163.com (Z.X.); zhaomingy9@163.com (M.Z.); hanxusdau@163.com (X.H.); shuaiqi@sdau.edu.cn (S.Q.)
*   Correspondence: yxyxst20040214@163.com (X.Y.); sdzly369@163.com (L.Z.); Tel.: +86-0538-824-2216 (X.Y. & L.Z.)
†   These authors contributed equally to this work.

Received: 30 January 2019; Accepted: 22 February 2019; Published: 28 February 2019

Abstract: Flower color is a charming phenotype with very important ornamental and commercial values. Anthocyanins play a critical role in determining flower color pattern formation, and their biosynthesis is typically regulated by R2R3-MYB transcription factors (TFs). *Paeonia suffruticosa* is a famous ornamental plant with colorful flowers. However, little is known about the R2R3-MYB TFs that regulate anthocyanin accumulation in *P. suffruticosa*. In the present study, two R2R3-MYB TFs, namely, *PsMYB114L* and *PsMYB12L*, were isolated from the petals of *P. suffruticosa* 'Shima Nishiki' and functionally characterized. Sequence analysis suggested that *PsMYB114L* contained a bHLH-interaction motif, whereas *PsMYB12L* contained two flavonol-specific motifs (SG7 and SG7-2). Subsequently, the in vivo function of *PsMYB114L* and *PsMYB12L* was investigated by their heterologous expression in *Arabidopsis thaliana* and apple calli. In transgenic *Arabidopsis* plants, overexpression of *PsMYB114L* and of *PsMYB12L* caused a significantly higher accumulation of anthocyanins, resulting in purple-red leaves. Transgenic apple calli overexpressing *PsMYB114L* and *PsMYB12L* also significantly enhanced the anthocyanins content and resulted in a change in the callus color to red. Meanwhile, gene expression analysis in *A. thaliana* and apple calli suggested that the expression levels of the flavonol synthase (*MdFLS*) and anthocyanidin reductase (*MdANR*) genes were significantly downregulated and the dihydroflavonol 4-reductase (*AtDFR*) and anthocyanin synthase (*AtANS*) genes were significantly upregulated in transgenic lines of *PsMYB114L*. Moreover, the expression level of the *FLS* gene (*MdFLS*) was significantly downregulated and the *DFR* (*AtDFR/MdDFR*) and *ANS* (*AtANS/MdANS*) genes were all significantly upregulated in transgenic lines plants of *PsMYB12L*. These results indicate that *PsMYB114L* and *PsMYB12L* both enhance anthocyanin accumulation by specifically regulating the expression of some anthocyanin biosynthesis-related genes in different plant species. Together, these results provide a valuable resource with which to further study the regulatory mechanism of anthocyanin biosynthesis in *P. suffruticosa* and for the breeding of tree peony cultivars with novel and charming flower colors.

Keywords: *P. suffruticosa*; R2R3-MYB; overexpression; anthocyanin; transcriptional regulation

---

1. Introduction

Paeonia suffruticosa is a very popular ornamental flowering plant that was first cultivated more than 1600 years ago in China and is currently distributed worldwide. This species is in the Paeoniaceae family and has been named 'the king of flowers' for its showy and colorful flowers [1]. Approximately 1500 cultivars of *P. suffruticosa* with a variety of flower colors have been produced by breeders worldwide [2]. Among the many flower colors of this species, most fit into two clusters: monochrome color (red, pink, white, purple, black, blue, green, and yellow) and double color. Cultivars with a double-color phenotype are rarer and more sought after, and thus have great ornamental and commercial value [3]. Among them, *P. suffruticosa* 'Shima Nishiki', a well-known chimeric cultivar, was selected from the bud mutation of *P. suffruticosa* 'Taiyoh'. 'Shima Nishiki' usually has red and pink petals on the same flower, and this aesthetically pleasing double-color phenotype can be stably inherited [4]. Therefore, the 'Shima Nishiki' cultivar is regarded as an important experimental material with which to study the molecular regulatory mechanism of flower color and in the breeding of new cultivars [5].

Anthocyanins are important soluble flavonoid compounds that are widely distributed in the leaves, flowers, fruits, seeds and other tissues of many plants [6]. Anthocyanin composition and concentration are usually closely related to flower color intensity [7,8]. The anthocyanin biosynthetic pathway is well known to be highly conserved in many ornamental plants [9–14]. Anthocyanin biosynthesis and accumulation are usually regulated by a series of structural genes and regulatory genes [15,16]. The structural genes encode enzymes associated with anthocyanin biosynthesis, including chalcone synthase (CHS), chalcone isomerase (CHI), flavanone 3-hydroxylase (F3H), flavonoid 3'-hydroxylase (F3'H), dihydroflavonol 4-reductase (DFR) anthocyanin synthase (ANS), Flavonol synthase (FLS), and anthocyanidin reductase (ANR) [17–19] (Figure 1). Among them, FLS is a dedicated enzyme involved in flavonol biosynthesis, and ANR is a key enzyme for proanthocyanidin biosynthesis. The regulatory genes can be divided into three families R2R3-MYB, bHLH, and WD40 [20–22] and they usually form a regulatory complex to activate the expression of anthocyanin biosynthetic genes [23–25].

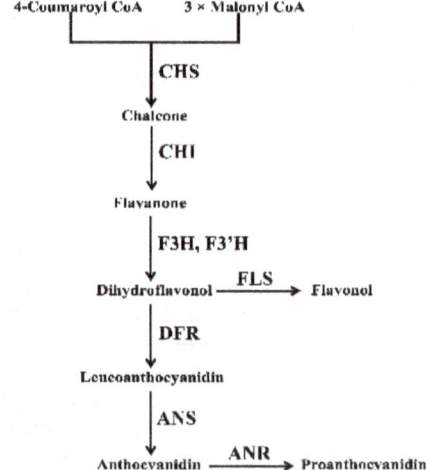

Figure 1. A general schematic diagram of the metabolic pathway related to anthocyanin biosynthesis. CHS, chalcone synthase; CHI, chalcone isomerase; F3H, flavanone 3-hydroxylase; F3'H, flavonoid 3'-hydroxylase; DFR, dihydroflavonol 4-reductase; ANS, anthocyanidin synthase; FLS, flavonol synthase; ANR, anthocyanidin reductase.

Many structural genes have been characterized and cloned in *P. suffruticosa* [12,26–28]. In the MYB-bHLH-WDR (MBW) complex, R2R3-MYB transcription factors (TFs) usually play critical roles in anthocyanin biosynthesis and accumulation [29,30]. Many R2R3-MYB TFs involved in anthocyanin biosynthesis have been isolated and characterized from various plants, including

Arabidopsis thaliana [24], *Zea mays* [31], *Vitis vinifera* [32], *Malus* crabapple [33], *Petunia hybrida* [34], *Antirrhinum majus* [29], *Dendranthema morifolium* [35] and *Phalaenopsis aphrodite* [14]. In *P. suffruticosa*, most previous studies were focused primarily on the preliminary investigation of R2R3-MYB TFs based on transcriptome sequencing and qRT-PCR analyses [12,36–38], whereas whether and how R2R3-MYB TFs control anthocyanin biosynthesis and accumulation in *P. suffruticosa* are almost unknown.

In the present study, two novel R2R3-MYB TFs, namely, *PsMYB114L* and *PsMYB12L*, were cloned in *P. suffruticosa*. Subsequently, the expression patterns of *PsMYB114L* and *PsMYB12L* were determined at five developmental stages in *P. suffruticosa* 'Shima Nishiki'. Furthermore, the function of these two TFs was further verified by heterologous expression in *Arabidopsis* and apple calli. These results will provide valuable insights into understanding the putative roles of *PsMYB114L* and *PsMYB12L* in regulating anthocyanin biosynthesis in *P. suffruticosa*.

2. Results

2.1. Cloning and Analysis of the PsMYB114L and PsMYB12L Genes

Based on the functional annotation and gene expression analysis of transcriptome sequencing data in *P. suffruticosa* 'Shima Nishiki' [39], we filtered two MYB unigenes exhibiting relatively high expression differences between the red and pink petals as the targeted genes of this study.

The full-length cDNA sequences of the two novel MYB genes were obtained with PCR amplification. By conducting GenBank BLAST searches of the amino acid sequences of these two genes, we found that these genes have the highest homology with transcription factor *MYB114*-like [*Quercus suber*] and transcription factor *MYB12*-like [*Juglans regia*], respectively. Therefore, we named these genes *PsMYB114L* and *PsMYB12L*. Sequencing results revealed that *PsMYB114L* (Figure S1A,B) and *PsMYB12L* (Figure S1C,D) contained an open reading frame (ORF) of 600 and 1140 bp encoding 199 and 379 amino acids and that their predicted proteins had a molecular mass of 22.81 and 42.61 kDa and a theoretical isoelectric point (pI) of 8.53 and 4.86, respectively.

Multiple sequence alignment of amino acids revealed that *PsMYB114L* and *PsMYB12L*, belonging to the SANT superfamily (which typically consists of tandem repeats of three alpha-helices arranged in a helix-turn-helix motif, with each alpha helix containing a bulky aromatic residue), and other known R2R3-MYB TFs related to anthocyanin biosynthesis contained a highly conserved R2R3 DNA-binding domain. The presence of this conserved domain means that *PsMYB114L* and *PsMYB12L* are also R2R3-MYB TFs and may perform similar functions in regulating anthocyanin biosynthesis. Furthermore, *PsMYB114L* had a bHLH-interaction motif ([D/E]Lx2[R/K]x3Lx6Lx3R) in the R3 domain at the N terminus and did not have any conserved motifs at the C terminus (Figure 2A). Moreover, *PsMYB12L* did not have any bHLH-interaction motifs at the N terminus, whereas it contained two flavonol-specific motifs [40], namely, SG7 ([K/R][R/x][R/K]xGRT[S/x][R/G]xx[M/x]K) and SG7-2 ([W/x][L/x]LS), at the C terminus (Figure 2B).

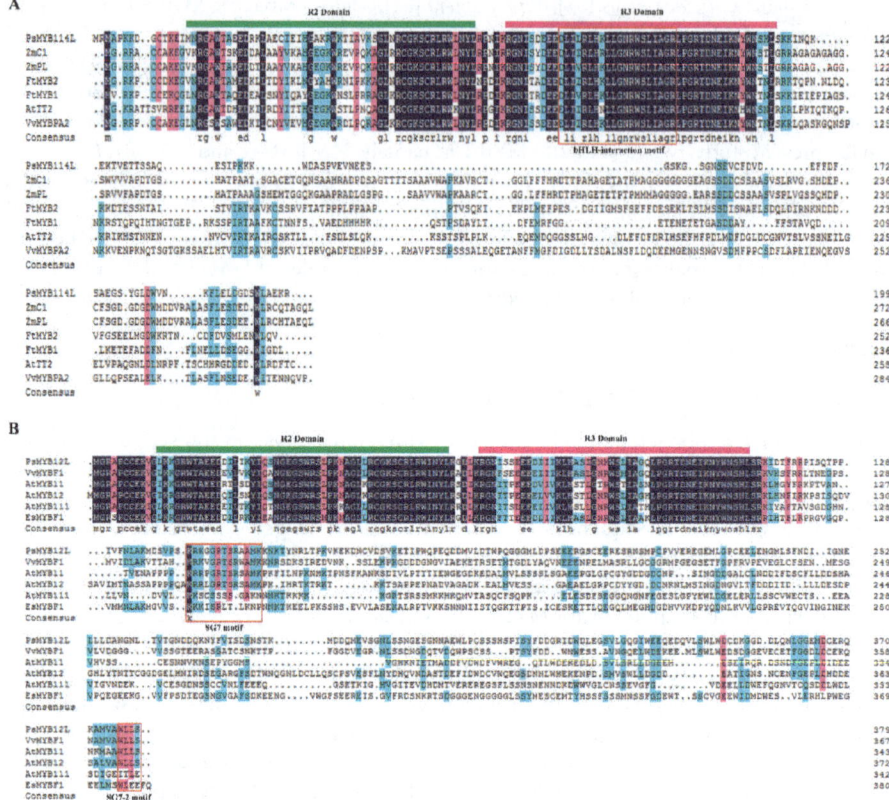

Figure 2. Amino acid sequence alignment analysis of the *PsMYB114L* (**A**) and *PsMYB12L* (**B**) genes with other known R2R3-MYB TFs. The green and pink long lines indicate the R2 and R3 domain, respectively. The red boxes show the conserved bHLH-interaction, SG7 and SG7-2 motifs. The NCBI GenBank accession numbers of these sequences are as follows: *ZmC1, Zea mays,* AF320613.3; *ZmPL, Zea mays,* NM_001112415.1; *FtMYB2, Fagopyrum tataricum,* JF313346.1; *FtMYB1, Fagopyrum tataricum,* JF313344.1; *AtTT2, Arabidopsis thaliana,* NM_122946.3; *VvMYBPA2, Vitis vinifera,* NM_001281024.1; *VvMYBF1, Vitis vinifera,* FJ948477.2; *AtMYB11, Arabidopsis thaliana,* NM_116126.3; *AtMYB12, Arabidopsis thaliana,* NM_130314.4; *AtMYB111, Arabidopsis thaliana,* NM_124310.3; *EsMYBF1, Epimedium sagittatum,* KU365320.1.

To better evaluate the phylogenetic relationships of *PsMYB114L, PsMYB12L* and 16 other known MYB TFs related to the regulation of anthocyanin biosynthesis, a phylogenetic tree was constructed based on the amino acid sequences of these 18 MYB TFs from different species using the neighbor-joining method. The phylogenetic analysis indicated that these 18 MYB TFs were classified into four groups (Flavonol, Anthocyanin, Anthocyanin/Proanthocyanidin and Proanthocyanidin) based on their specific roles in the flavonoid biosynthesis pathway. Among them, *PsMYB114L* had the closest phylogenetic relationship with *ZmC1* and *ZmPL*, which are involved in regulating anthocyanin biosynthesis, whereas *PsMYB12L* belongs to a subgroup of MYB proteins that includes VvMYBF1, EsMYBF1, AtMYB11, AtMYB12 and AtMYB111, which regulate flavonol synthesis and had the closest phylogenetic relationship with *VvMYBF1* (Figure 3).

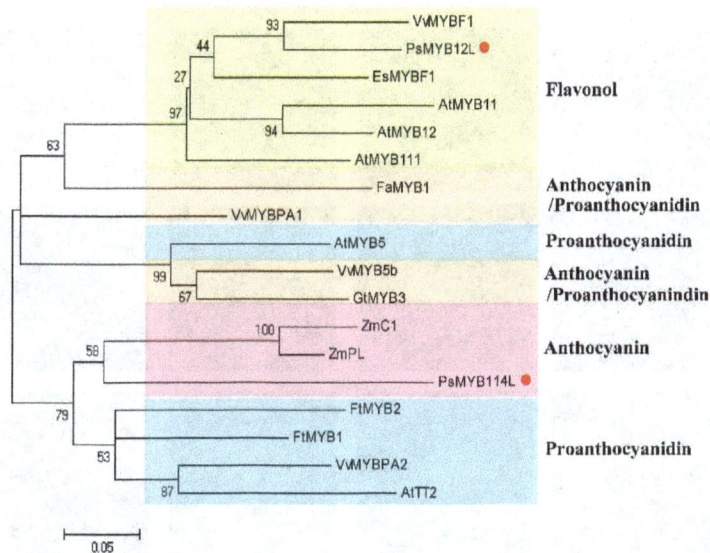

Figure 3. Phylogenetic analysis of the *PsMYB114L* and *PsMYB12L* genes with R2R3-MYB TFs from other species. The NCBI GenBank accession numbers of these sequences are as follows: *FaMYB1, Fragaria x ananassa,* AF401220.1; *VvMYBPA1, Vitis vinifera,* NM_001281231.1; *VvMYB5b, Vitis vinifera,* NM_001280925.1; *AtMYB5, Arabidopsis thaliana,* AF401220.1; *GtMYB3, Gentiana triflora,* AB289445.1; *ZmC1, Zea mays,* AF320613.3; *ZmPL, Zea mays,* NM_001112415.1; *FtMYB2, Fagopyrum tataricum,* JF313346.1; *FtMYB1, Fagopyrum tataricum,* JF313344.1; *AtTT2, Arabidopsis thaliana,* NM_122946.3; *VvMYBPA2, Vitis vinifera,* NM_001281024.1; *VvMYBF1, Vitis vinifera,* FJ948477.2; *AtMYB11, Arabidopsis thaliana,* NM_116126.3; *AtMYB12, Arabidopsis thaliana,* NM_130314.4; *AtMYB111, Arabidopsis thaliana,* NM_124310.3; *EsMYBF1, Epimedium sagittatum,* KU365320.1.

2.2. Subcellular Localization of PsMYB114L and PsMYB12L

To examine the subcellular localization of *PsMYB114L* and *PsMYB12L*, the recombinant vector (*PsMYB114L*-GFP/*PsMYB12L*-GFP) and the control vector (pCAMBIA1301-GFP) were introduced into the tobacco leaves. Our results were basically consistent with those of previous studies [41,42]. The green fluorescent protein (GFP) fluorescence of the control vector was clearly distributed throughout the entire cell (Figure 4A), and the *PsMYB114L*-GFP/*PsMYB12L*-GFP vector displayed a strong fluorescence signal in the nucleus and cytoplasm of tobacco cells (Figure 4B,C). Therefore, we speculated that the two R2R3-MYB TFs (*PsMYB114L*/*PsMYB12L*) were simultaneously localized and functioned in the nucleus and cytoplasm.

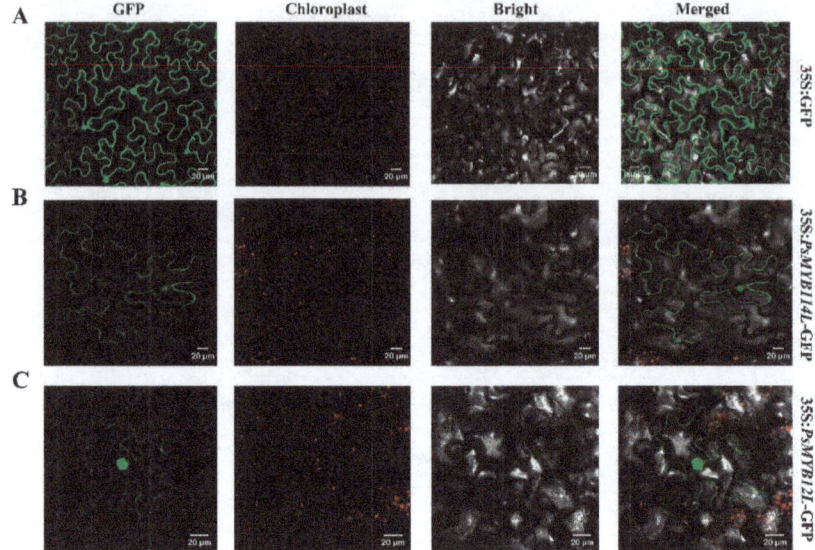

Figure 4. Subcellular localization analysis of the *PsMYB114L* and *PsMYB12L* genes. (**A**) Control vector (pCAMBIA1301-GFP) expressed in epidermal cells of tobacco leaves. (**B**) Recombinant vector (*PsMYB114L*-GFP) expressed in epidermal cells of tobacco leaves. (**C**) Recombinant vector (*PsMYB12L*-GFP) expressed in epidermal cells of tobacco leaves. White lines at the bottom right of the picture represent 20 μm in the respective pixel. GFP, GFP fluorescence; Chloroplast, Chloroplast fluorescence; Bright, Bright field; Merged, Superposition of bright field and fluorescence.

2.3. Expression Patterns of PsMYB114L and PsMYB12L in P. suffruticosa 'Shima Nishiki'

qRT-PCR analysis was conducted to survey the expression patterns of *PsMYB114L* and *PsMYB12L* in *P. suffruticosa* 'Shima Nishiki' (Figure 5). Petal samples of this cultivar were collected at five developmental stages (Figure S2). The expression levels of the *PsMYB114L* gene peaked at S3 and then decreased from S3 to S5, whereas the *PsMYB12L* gene exhibited the highest expression at S4. Furthermore, the expression levels of the eight anthocyanin biosynthesis-related genes (*PsCHS*, *PsCHI*, *PsF3H*, *PsF3'H*, *PsDFR*, *PsANS*, *PsFLS*, and *PsANR*) were analyzed. Among these genes, *PsF3'H*, *PsDFR*, and *PsANS* showed a trend similar to that of *PsMYB12L*, whereas *PsFLS* and *PsANR* showed a trend similar to that of *PsMYB114L*.

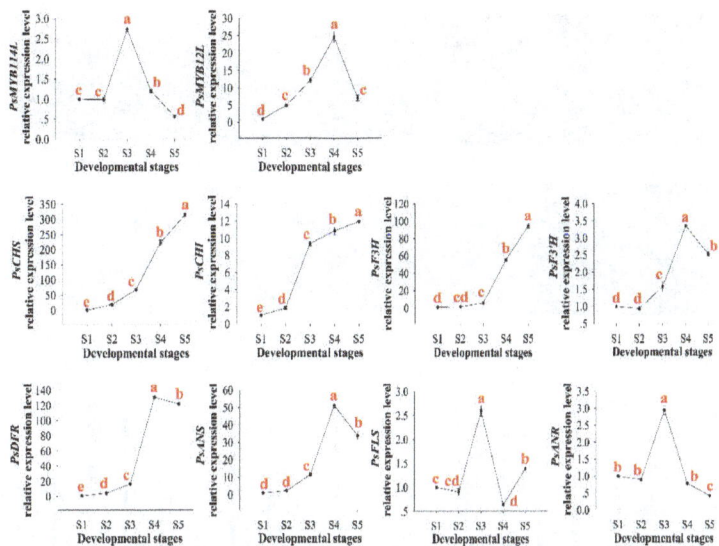

Figure 5. The expression patterns of the *PsMYB114L* gene, *PsMYB12L* gene and anthocyanin biosynthesis-related structural genes in *P. suffruticosa* 'Shima Nishiki'. S1, flower bud emerging stage; S2 small bell-like flower-bud stage; S3, large bell-like flower-bud stage; S4, bell-like flower-bud extending stage; S5, color exposing stage. Different lowercase letters indicate significant differences at $p < 0.05$.

2.4. Overexpression of PsMYB114L and PsMYB12L in Arabidopsis

To characterize the functions of *PsMYB114L* and *PsMYB12L*, these two genes under the expression of the 35S promoter were genetically transformed into *Arabidopsis*. Phenotypic investigations of the transgenic lines of *PsMYB114L* and *PsMYB12L* revealed that their leaves were much deeper in color than those of Col-0 and showed a purple-red color (Figure 6A). Meanwhile, these transgenic lines of the two genes were confirmed by PCR analysis (Figure 6B). Furthermore, the total anthocyanin content results indicated that the transgenic lines of *PsMYB114L* and *PsMYB12L* produced much more anthocyanin than Col-0 (Figure 6C,D).

Additionally, the expression levels of anthocyanin biosynthesis-related genes (*AtCHS*, *AtCHI*, *AtF3H*, *AtF3'H*, *AtDFR*, *AtANS*, *AtFLS*, and *AtANR*) in the Col-0 and the transgenic *Arabidopsis* plants of *PsMYB114L* and *PsMYB12L* were analyzed with qRT-PCR experiments. Compared with the Col-0, overexpression of *PsMYB114L* upregulated the expression of most of the genes (*AtCHS*, *AtCHI*, *AtF3H*, *AtF3'H*, *AtDFR*, and *AtANS*) in transgenic *PsMYB114L* plants; among them, both of the *AtDFR/AtANS* genes showed a relatively high difference between the Col-0 and transgenic plants, whereas *AtFLS* and *AtANR* were downregulated in transgenic *PsMYB114L* plants (Figure 6E).

For *PsMYB12L* overexpression in *Arabidopsis*, the expression levels of all eight genes were upregulated in transgenic *PsMYB12L* plants. Among them, the four genes (*AtCHS*, *AtCHI*, *AtDFR*, and *AtANS*) all showed a relatively high difference between the Col-0 and transgenic plants (Figure 6F).

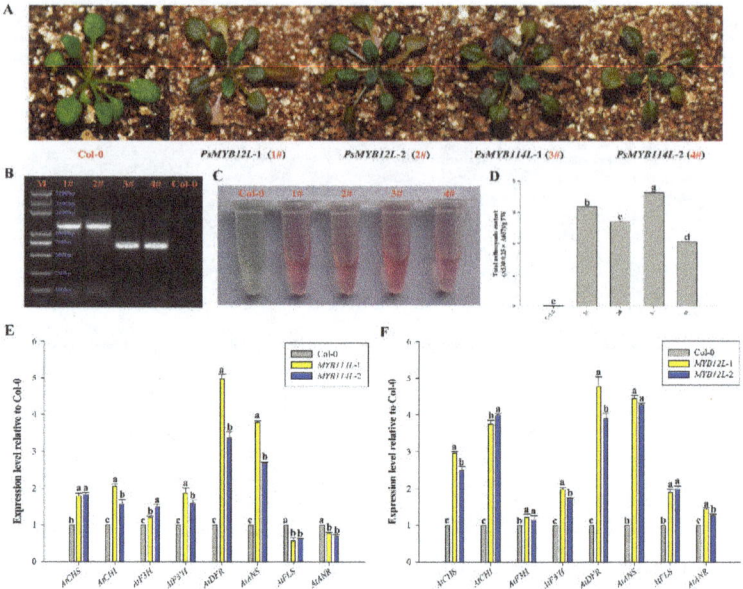

Figure 6. Overexpression analysis of the *PsMYB114L* and *PsMYB12L* genes in *Arabidopsis*. (**A**) Comparison of leaf colors in transgenic *Arabidopsis* plants and Col-0. (**B**) Results of positive PCR detection in transgenic *Arabidopsis* plants. (**C**) Anthocyanin extraction solutions for transgenic *Arabidopsis* plants and Col-0. (**D**) Total anthocyanin content in transgenic *Arabidopsis* plants and Col-0. (**E**) Expression analysis of anthocyanin biosynthesis-related genes in transgenic *Arabidopsis* plants of *PsMYB114L* and Col-0. (**F**) Expression analysis of anthocyanin biosynthesis-related genes in transgenic *Arabidopsis* plants of *PsMYB12L* and Col-0. Col-0, *Arabidopsis thaliana* ecotype Columbia; 1# and 2#, two transgenic lines of the *PsMYB12L* gene; 3# and 4#, two transgenic lines of the *PsMYB114L* gene. Different lowercase letters indicate significant differences at *p* < 0.05.

2.5. Overexpression of PsMYB114L and PsMYB12L in Apple Calli

For further functional validation, the two genes (*PsMYB114L* and *PsMYB12L*) were ectopically expressed in the calli of 'Orin' apple. Interestingly, after light and low-temperature treatments, the WT had almost no phenotypic changes, but an especially obvious color change was observed in the transgenic lines of *PsMYB114L* and *PsMYB12L* (Figures 7A and 8A). PCR amplification confirmed that these transgenic apple calli carry *PsMYB114L* and *PsMYB12L* (Figures 7B and 8B). With regard to the total anthocyanin content, the transgenic lines of *PsMYB114L* and *PsMYB12L* all accumulated markedly higher amounts of anthocyanins than did the WT (Figures 7C,D and 8C,D).

Additionally, the expression levels of anthocyanin biosynthesis-related genes (*MdCHS*, *MdCHI*, *MdF3H*, *MdF3'H*, *MdDFR*, *MdANS*, *MdFLS*, and *MdANR*) in the WT and the transgenic lines of *PsMYB114L* and *PsMYB12L* were analyzed by qRT-PCR. Compared with the WT, overexpression of *PsMYB114L* downregulated the expression of most of the genes, specifically, *MdCHS*, *MdCHI*, *MdF3H*, *MdF3'H*, *MdFLS*, and *MdANR*, and upregulated the expression of *MdDFR* and *MdANS* in transgenic *PsMYB114L* calli (Figure 7E).

For *PsMYB12L* overexpression, the expression levels of most genes, including *MdCHS*, *MdF3H*, *MdF3'H*, *MdDFR*, *MdANS*, and *MdANR*, were upregulated, but those of *MdCHI* and *MdFLS* were downregulated in transgenic *PsMYB12L* calli (Figure 8E).

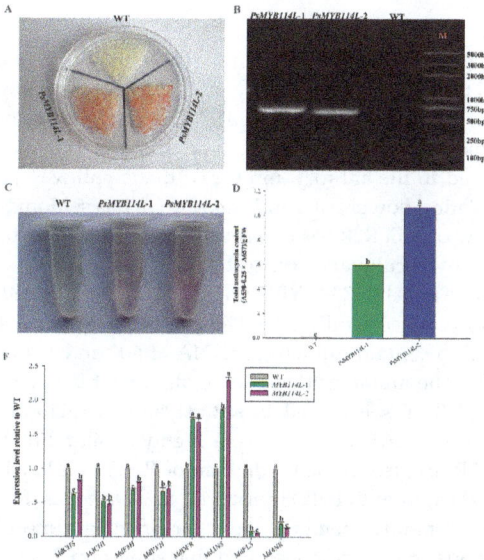

Figure 7. Overexpression analysis of the *PsMYB114L* gene in apple calli. (**A**) Colors observed in transgenic apple calli and the WT. (**B**) Results of positive PCR detection in transgenic apple calli. (**C**) Anthocyanin extraction solutions for transgenic apple calli and the WT. (**D**) Total anthocyanin content in transgenic apple calli and the WT. (**E**) Expression analysis of anthocyanin biosynthesis-related genes in transgenic apple calli and the WT. WT, Wild-type 'Orin' apple calli; *PsMYB114L*-1 and *PsMYB114L*-2, two transgenic lines of the *PsMYB114L* gene. Different lowercase letters indicate significant differences at *p* < 0.05.

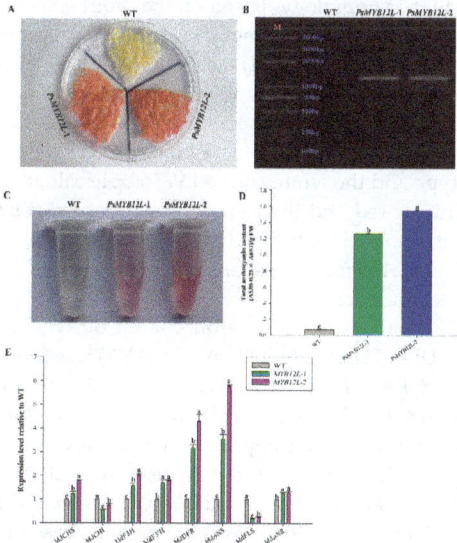

Figure 8. Overexpression analysis of the *PsMYB12L* gene in apple calli. (**A**) Colors observed in transgenic apple calli and the WT. (**B**) Results of positive PCR detection in transgenic apple calli. (**C**) Anthocyanin extraction solutions for transgenic apple calli and the WT. (**D**) Total anthocyanin content in transgenic apple calli and the WT. (**E**) Expression analysis of anthocyanin biosynthesis-related genes in transgenic apple calli and the WT. *PsMYB12L*-1 and *PsMYB12L*-2, two transgenic lines of the *PsMYB12L* gene. Different lowercase letters indicate significant differences at *p* < 0.05.

3. Discussion

Flower color is a very important trait in many ornamental plants and has a close association with their ornamental and commercial value. Many prior studies have shown that anthocyanins are a key factor influencing flower color [43–45]. R2R3-MYB TFs comprise one of the largest gene families in plants and play key roles in regulating anthocyanin accumulation by activating the expression of structural genes involved in the anthocyanin biosynthetic pathway [46,47]. However, the role of R2R3-MYB TFs in regulating flower color in *P. suffruticosa* has seldom been functionally verified. Therefore, determining how certain R2R3-MYB TFs regulate anthocyanin production in *P. suffruticosa* would aid in breeding improved cultivars with desirable flower colors.

In the present study, two novel R2R3-MYB TFs (*PsMYB114L* and *PsMYB12L*) possibly involved in anthocyanin biosynthesis were successfully cloned and characterized from the petals of *P. suffruticosa* 'Shima Nishiki' and found to contain full-length cDNA of 600 and 1140 bp encoding 199 and 379 amino acids, respectively. The amino acid sequence alignment between *PsMYB114L*/*PsMYB12L* and other known R2R3-MYB TFs involved in anthocyanin regulation indicated that the R2R3 domain distributions of these R2R3-MYB TFs were highly similar, but a bHLH-interaction motif ([D/E]Lx2[R/K]x3Lx6Lx3R) existed in the R3 domain of *PsMYB114L*, whereas *PsMYB12L* did not contain this motif for interaction with bHLH proteins. In *Arabidopsis*, based on a similar function, 125 TFs of R2R3-MYB gene-family members were classified into more than 25 subgroups [48]. Furthermore, many previous studies demonstrated that subgroup 7 [49,50], characterized by both the SG7 ([K/R][R/x][R/K]xGRT[S/x][R/G]xx[M/x]K) and SG7-2 ([W/x][L/x]LS) motifs, specifically regulated flavonol biosynthesis. *PsMYB12L* contained these two motifs (SG7 and SG7-2) at the C terminus of the protein, but *PsMYB114L* lacked these two motifs.

Phylogenetic analysis indicated that *PsMYB12L* and 5 flavonol-regulating R2R3-MYB TFs (*VvMYBF1*, *EsMYBF1*, and *AtMYB11/12/111*) belonging to subgroup 7 [30,51] were clustered together, and *PsMYB114L* and certain R2R3-MYB TFs belonging to subgroup 5 (*AtTT2*, *ZmC1*, *VvMYBPA2*, etc.) [52–55] had relatively higher homology. Based on the motif analysis of amino acid sequences and phylogenetic analysis, *PsMYB114L* might regulate anthocyanin production by combinatorially interacting with a basic helix-loop-helix (bHLH) factor [25,56,57]. *PsMYB12L* might independently regulate the expression of anthocyanin biosynthesis-related genes without the MBW complex [58].

In addition, we conducted further ectopic transgenic studies by overexpressing *PsMYB114L*/*PsMYB12L* in *Arabidopsis* and apple calli. In contrast to the green-colored leaves of the Col-0 *A. thaliana* ecotype and the white-colored WT apple calli, the leaves of these transgenic *Arabidopsis* plants turned purple-red and the transgenic calli of *PsMYB114L* and *PsMYB12L* were red, which was in agreement with their remarkably higher anthocyanin content. The color and total anthocyanin content analyses of *Arabidopsis* and apple calli indicated that these two R2R3-MYB TFs contribute to anthocyanin accumulation in transgenic lines.

Subsequently, qRT-PCR analysis of seven anthocyanin biosynthesis-related genes (*MdCHS*, *MdCHI*, *MdF3H*, *MdF3'H*, *MdDFR*, *MdANS*, *MdFLS*, and *MdANR*) was further performed in *Arabidopsis* and apple calli. In terms of *PsMYB114L*, the qRT-PCR results in *Arabidopsis* showed that the expression levels of *AtDFR* and *AtANS* were significantly upregulated, whereas *AtFLS* and *AtANR* were downregulated to a certain extent compared with the levels in the Col-0. Furthermore, the qRT-PCR results in apple calli showed that the expression levels of *MdDFR* and *MdANS* were upregulated to a certain extent, whereas *MdFLS* and *MdANR* (especially *MdFLS*) were significantly downregulated compared with the levels in the WT. Meanwhile, based on the results of expression patterns of *PsMYB114L* in *P. suffruticosa* 'Shima Nishiki', we have known that *PsMYB114L* have a positive correlation with *PsFLS* and *PsANR*. By comparing these three qRT-PCR results in *Arabidopsis*, apple calli, and *P. suffruticosa*, we found differences in the expression patterns of some anthocyanin biosynthesis-related genes. Previous studies have showed that many R2R3-MYB TFs usually regulate flavonoid biosynthesis by interacting with the promoter of the targeted structural genes [55,58]. For promoter region, in general, the sequence of the same structural gene in different plant species also differs greatly. Therefore, it is possible that the same MYB TFs performed different regulatory mechanisms of flavonoid biosynthesis in different species [59]. Dihydroflavonol is the direct substrate for two key genes (*FLS* and *DFR*) in the flavonoid biosynthetic pathways, and these two genes usually show a competitive interaction in producing colored anthocyanidin and colorless flavonols [60]. In this study, the strong upregulation of *AtDFR* and *AtANS* may have played key roles in activating the branch of the anthocyanin biosynthesis, resulting in purple-red leaves in the transgenic *Arabidopsis* plants of *PsMYB114L*, whereas the strong downregulation of *MdFLS* would inhibit the branch of the flavonol biosynthesis, resulting in the production of anthocyanins and a red-colored phenotype in the transgenic calli of *PsMYB114L*. Furthermore, because *PsMYB114L* has a bHLH-interaction motif, it may form an MBW complex and contribute to anthocyanin accumulation by regulating the expression of these key genes (*AtDFR*, *AtANS*, *MdFLS*, and *MdANR*) in *Arabidopsis* and apple calli.

With regard to *PsMYB12L*, the qRT-PCR results showed that the expression levels of the four genes (*AtCHS*, *AtCHI*, *AtDFR* and *AtANS*) were all significantly upregulated in the transgenic *Arabidopsis* plants. Moreover, the expression levels of *MdDFR* and *MdANS* were significantly upregulated in the transgenic calli of *PsMYB12L*, but *MdFLS* was significantly downregulated. Meanwhile, based on the results of expression patterns of *PsMYB12L* in *P. suffruticosa* 'Shima Nishiki', we have known that *PsMYB12L* have a positive correlation with *PsDFR*, *PsANS*, and *PsF3'H*. By comparing these three qRT-PCR results in *Arabidopsis*, apple calli, and *P. suffruticosa*, we can found that the two key anthocyanin biosynthesis-related genes (*DFR* and *ANS*) showed a very similar expression pattern. We considered that *PsMYB12L* should be a specific transcriptional regulator on *DFR* and *ANS* genes in these three species. Furthermore, we also found differences in the expression patterns of the *FLS* gene in *Arabidopsis* and apple calli, and considered that the expression difference of the *FLS* gene is likely caused by the promoter sequence specificity of this gene in these two species [61]. Based on the motif analysis of *PsMYB12L*, we speculated that the TF may be a flavonol-specific MYB regulator. Many flavonol-specific MYB TFs have been isolated and functionally verified in various plants, such as *A. thaliana*, *Vitis vinifera*, and *Epimedium sagittatum* [50,62,63]. Furthermore, many flavonol-specific MYB TFs negatively regulate anthocyanin accumulation by inducing higher expression of the *FLS* gene. By overexpressing *AtMYB12* in tobacco, the expression of *NtCHS*, *NtCHI*, and *NtFLS* was specifically activated; moreover, the flowers of the transgenic plants were paler in color than their wild-type counterparts [64]. Ectopic expression analysis of *EsMYBF1* in transgenic tobacco indicated that *NtCHS*, *NtCHI*, *NtF3H*, and *NtFLS* were upregulated but *NtDFR* and *NtANS* were significantly downregulated, and the accumulation of anthocyanins in transgenic tobacco flowers was also remarkably decreased [63]. A study on the overexpression of *PpMYB15* in tobacco showed that it can significantly activate the expression of *NtCHS*, *NtCHI*, *NtF3H*, and *NtFLS*, while it had no effects on the expression of *NtDFR* and *NtANS*, resulting in pale-pink or pure white flowers in transgenic tobacco plants [40]. Compared with the expression of anthocyanin biosynthesis-related genes documented in the abovementioned studies, in this study *AtCHS/MdCHS*, *AtCHI*, *AtF3H/MdF3H*, and *AtFLS* had a somewhat similar expression pattern and *MdFLS*, *AtDFR/MdDFR* and *AtANS/MdANS* exhibited the opposite pattern. However, the lower expression of the *MdFLS* gene and the higher expression of *AtDFR/MdDFR* and *AtANS/MdANS* were consistent with the significantly higher anthocyanin accumulation in transgenic lines of *PsMYB12L*. Beacuse *PsMYB12L* has the flavonol-specific motif and lacks the bHLH-interaction motif, it alone enhances anthocyanin production by regulating the expression of these key genes (*AtDFR/MdDFR*, *AtANS/MdANS*, and *MdFLS*) independently of bHLH cofactors in *Arabidopsis* and apple calli.

4. Materials and Methods

4.1. Plant Materials

The tree peony cultivar *P. suffruticosa* 'Shima Nishiki' was grown in the experimental nursery of Forestry College, Shandong Agricultural University, Tai'an, Shandong, China. Flower samples were collected at five early flower-bud developmental stages (flower bud emerging stage (S1), small bell-like flower-bud stage (S2), large bell-like flower-bud stage (S3), bell-like flower-bud extending stage (S4), and color exposing stage (S5)) (Figure S2) [65]. All these samples were immediately frozen in liquid nitrogen and then stored at –80 °C for further experiments.

The *A. thaliana* ecotype Columbia (Col-0) was used for genetic transformation and phenotypic analysis in the present study. The plants were grown under a 16 h light/ 8 h dark photoperiod at 23 °C/21 °C

Furthermore, calli of the wild type (WT) of 'Orin' apple were subcultured on Murashige and Skoog (MS) medium with 1.5 mg L^{-1} 6-benzyl adenine (6-BA) and 0.5 mg L^{-1} 2,4-dichlorophenoxyacetic acid (2,4-D) at room temperature (24 °C) in a continuous dark environment at 15-day intervals [66]. Subsequently, the calli were used for genetic transformation and phenotypic analysis.

4.2. Total RNA Extraction and cDNA Synthesis

Total RNA was extracted from all samples according to instructions of the EASY Spin Plant RNA Rapid Extraction Kit (Aidlab Biotech, Beijing, China). The purity and concentration of all RNA samples were assessed using a Nanodrop 2000C spectrophotometer (Thermo Fisher Scientific, Wilmington, Delaware, DE, USA), and RNA quality was detected using 1 % agarose gels. Furthermore, cDNA was synthesized with 1 μg of total RNA using 5× All-In-One RT MasterMix (with an AccuRT Genomic DNA Removal Kit) (ABM, Vancouver, BC, Canada).

4.3. Cloning of the PsMYB114L and PsMYB12L Genes in P. suffruticosa

In this study, based on the transcriptome sequencing data of *P. suffruticosa* 'Shima Nishiki' in our laboratory, two R2R3-MYB transcription factors were filtered by analyzing the functional annotations of MYB unigenes and performing gene expression analysis.

The cDNA of the 'Shima Nishiki' cultivar's petals was used as the template. The full-length coding sequence (CDS) of the *PsMYB114L* (MK518073) and *PsMYB12L* (MK518074) genes was amplified using PCR. The complete 5' CDS of the *PsMYB114L* and *PsMYB12L* genes was identified from the transcriptome sequencing data of *P. suffruticosa* 'Shima Nishiki'. The cDNA 3' end sequence of these candidate genes was obtained using nested PCR technology using *PsMYB114L*-1-F/*PsMYB114L*-2-F and *PsMYB12L*-1-F/*PsMYB12L*-2-F as forward primers (Table S1), respectively, and B26 was used as the common reverse primer. The full-length cDNA of the *PsMYB114L* and *PsMYB12L* genes was amplified with the forward primers *PsMYB114L*-F1/*PsMYB12L*-F1 and the reverse primers *PsMYB114L*-R1/*PsMYB12L*-R1 (Table S1). The PCR program of gene amplification was as follows: initial denaturation at 95 °C for 1 min, followed by 30 cycles of 98 °C for 10 s, 60 °C for 15 s and 68 °C for 60 s. The PCR products were purified and cloned into the pTOPO-Blunt Simple vector for sequencing.

4.4. Subcellular Localization

The full-length cDNA without the termination codon of *PsMYB114L*/*PsMYB12L* was amplified with special primers (*PsMYB114L*-GFPF/*PsMYB12L*-GFPF and *PsMYB114L*-GFPR/*PsMYB12L*-GFPR) (Table S1) with restriction sites (*Xba* I and *Kpn* I) and subcloned into the pCAMBIA1301-GFP vector between the *Xba* I and *Kpn* I sites to create the *PsMYB114L*-GFP/*PsMYB12L*-GFP fusion construct. The recombinant vectors (*PsMYB114L*-GFP/*PsMYB12L*-GFP) and control vector (pCAMBIA1301-GFP) were then introduced into tobacco leaves by agroinfiltration. These infiltrated plants were grown for over 72 h in a growth chamber, and the GFP fluorescence of samples was observed under a Nikon C2-ER confocal laser scanning microscope (Nikon, Tokyo, Japan) [67].

4.5. Overexpression Vector Construction

The full-length cDNA of the *PsMYB114L* and *PsMYB12L* genes from the petals of *P. suffruticosa* 'Shima Nishiki' was amplified using recombinant primers (*PsMYB114L*-F2/*PsMYB12L*-F2 and

PsMYB114L-R2/*PsMYB12L*-R2) (Table S1) with restriction sites (*Spe* I and *BstE* II). Based on the predesigned vector construction procedure, the pCAMBIA1304 empty vector and the pTOPO-Blunt Simple vector containing the target genes (*PsMYB114L* and *PsMYB12L*) with restriction sites were double digested separately between the *Spe* I and *BstE* II sites and then recombined (Figure S3A–C). Subsequently, the two recombinant vectors pCAMBIA1304-*PsMYB114L* (Figure S3D) and pCAMBIA1304-*PsMYB12L* (Figure S3E) were verified successfully by PCR and sequencing with the forward vector validation primer 1304Ve-F and the reverse primers *PsMYB114L*-R2/*PsMYB12L*-R2 (Table S1). These two overexpression constructs were also introduced into *Agrobacterium tumefaciens* strain GV3101 using the freeze-thaw method.

4.6. Stable Transformation of Arabidopsis

The transformation of *Arabidopsis* was performed using the floral dip transformation method [68]. An *A. tumefaciens* infection solution (OD600 = 0.8–1.2) containing 5 % sucrose and 0.01 % Silwet L-77 was prepared to infect inflorescences, and the infection time per inflorescence was 15 s. Subsequently, these plants were transferred to a dark treatment for 24 h. These steps were repeated twice more according to the growth state of the plant. Mature T1 seeds were harvested, surface sterilized, and then sown on MS medium with 30 mg L^{-1} hygromycin B to screen for positive transformants. The resistant seedlings were transplanted into soil and then placed in a light incubator (16 h light/8 h dark at 23 °C/21 °C). When these transgenic *Arabidopsis* plants had grown to a certain size, they were further verified with gene-specific primers by PCR.

4.7. Stable Transformation of Apple Calli

To transform apple calli, 15-day-old WT apple calli were incubated with *A. tumefaciens* infection solution that carried pCAMBIA1304-PsMYB114L/pCAMBIA1304-PsMYB12L for 20 min, and the apple calli were then cocultured on MS medium supplemented with 0.5 mg L^{-1} 2,4-D and 1.5 mg L^{-1} 6-BA for 2 days at 24 °C in the dark. Subsequently, the apple calli were washed three times with sterile water and transferred to a selective medium that contained 15 mg L^{-1} hygromycin B for transgene selection. The transgenic apple calli were cocultured in the selective medium containing appropriate concentrations of an antibiotic and transferred to a light incubator with constant light (photon flux density of ~100 μmol s^{-1} m^{-2}) and low-temperature (15 °C) treatments for phenotypic observation [69,70].

4.8. Measurement of Total Anthocyanin Content

Total anthocyanin were extracted from the rosette leaves of 25-day-old *Arabidopsis* plants and apple calli cultured for 7 days. Anthocyanin extraction was performed using a methanol–HCl method [71]. Approximately 0.1 g of each sample was incubated in 5 mL of 0.1 % acidic methanol solution (CH$_3$OH:HCl:H$_2$O = 70:0.1:29.9, $v/v/v$) overnight in the dark at 4 °C. The absorbance of each extract was measured at 530 and 657 nm with a UV-1600 spectrophotometer (SHIMADZU, Kyoto, Japan). The total anthocyanin content was calculated using the following equation: Q$_{Total\ Anthocyanin}$ = (A530 − 0.25 × A657) × FM^{-1}. There were three biological replicates for each sample.

4.9. Quantitative Real-Time PCR (qRT-PCR) Analysis

qRT-PCR was performed to analyze the expression levels of anthocyanin biosynthesis-related genes in all plant materials in this study. The qRT-PCR experiments were conducted using SYBR® Premix Ex Taq™ (Tli RNaseH Plus) (TaKaRa, Kyoto, Japan) on a Bio-Rad CFX96™ Real-Time system (Bio-Rad, Hercules, CA, USA) with three biological replicates according to the manufacturer's instructions. The PCR conditions were as follows: 95 °C for 30 s, 40 cycles of 95 °C for 5 s and 60 °C for 30 s and then a dissociation stage at 95 °C for 10 s, 65 °C for 5 s and 95 °C for 5 s. The *Psubiquitin* gene, *AtActin2* gene and *MdActin* gene were used as internal controls to normalize the expression levels in *P. suffruticosa*, *A. thaliana* and *Malus domestica*, respectively. All gene-specific primers used in this study are shown in Table S1 [39,66]. The relative expression levels of genes were calculated using the $2^{-\Delta\Delta Ct}$ method [72].

4.10. Sequence and Statistical Analysis

Multiple sequence alignment was performed using DNAMAN 8.0 software (Lynnon Biosoft, San Ramon, CA, USA). Homology search of sequences was carried out using the GenBank BLAST. Phylogenetic tree construction of sequences was performed using MEGA 5.0 software (Arizona State University, Tempe, AZ, USA) with the bootstrap values from 1000 replicates. Primers were designed using Primer Premier 5.0 software (PREMIER Biosoft International, Palo Alto, CA, USA). All experiments were repeated three times, and the data are expressed as the mean ± standard error. Variance analyses were performed using SPSS software ver. 17.0 (SPSS Inc., Chicago, IL, USA). *p*-values of < 0.05 were considered statistically significant.

5. Conclusions

In conclusion, two novel R2R3-MYB TFs, namely *PsMYB114L* and *PsMYB12L*, were successfully cloned from the petals of *P. suffruticosa* 'Shima Nishiki' and functionally characterized by heterologous expression in *Arabidopsis* and apple calli. Based on the above results, we preliminarily demonstrated the potential functional roles of *PsMYB114L* and *PsMYB12L* in regulating anthocyanin biosynthesis. These results provide a valuable resource for further understanding the molecular regulatory mechanisms of anthocyanin biosynthesis and accumulation in *P. suffruticosa* and breeding improved cultivars of *P. suffruticosa* with desirable flower colors in the future.

Supplementary Materials: Supplementary materials can be found at http://www.mdpi.com/1422-0067/20/5/1055/s1. Table S1. Primers used in this study. Figure S1. Full-length cDNA amplification of the *PsMYB114L* and *PsMYB12L* genes. Figure S2. Flowers of *P. suffruticosa* 'Shima Nishiki' at five developmental stages. Figure S3. Construction of the recombinant expression vectors of pCAMBIA1304-*PsMYB114L* and pCAMBIA1304-*PsMYB12L*.

Author Contributions: X.Y. and L.Z. conceived and designed the research. X.Z. and Z.X. participated in the specific design of the study. X.Z. and Z.X. performed the experiments and the data analysis, and drafted the manuscript. M.Z., X.H., and S.Q. contributed analysis tools and helped analyze the data. All authors contributed to manuscript revision and approved the final version.

Funding: This research was funded by Forestry Science and Technology Innovation Projects of Shandong Province (Grant Number LYCX06-2018-32).

Conflicts of Interest: The authors declare no conflict of interest.

References

1. Zhang, J.J.; Shu, Q.Y.; Liu, Z.A.; Ren, H.X.; Wang, L.S.; De Keyser, E. Two EST-derived marker systems for cultivar identification in tree peony. *Plant Cell Rep.* **2012**, *31*, 299–310. [CrossRef] [PubMed]
2. Shi, Q.Q.; Zhou, L.; Wang, Y.; Li, K.; Zheng, B.Q.; Miao, K. Transcriptomic analysis of *Paeonia delavayi* wild population flowers to identify differentially expressed genes involved in purple-red and yellow petal pigmentation. *PLoS ONE* **2015**, *10*, e0135038. [CrossRef] [PubMed]
3. Zhao, D.Q.; Tao, J. Recent advances on the development and regulation of flower color in ornamental plants. *Front. Plant Sci.* **2015**, *6*, 261. [CrossRef] [PubMed]
4. Zhang, X.P.; Zhao, M.Y.; Guo, J.; Zhao, L.Y.; Xu, Z.D. Anatomical and biochemical analyses reveal the mechanism of double-color formation in *Paeonia suffruticosa* 'Shima Nishiki'. *3 Biotech* **2018**, *8*, 420. [CrossRef] [PubMed]
5. Noman, A.; Aqeel, M.; Deng, J.M.; Khalid, N.; Sanaullah, T.; He, S.H. Biotechnological advancements for improving loral attributes in ornamental plants. *Front. Plant Sci.* **2017**, *8*, 530. [CrossRef] [PubMed]
6. Li, C.H.; Qiu, J.; Yang, G.S.; Huang, S.R.; Yin, J.M. Isolation and characterization of a R2R3-MYB transcription factor gene related to anthocyanin biosynthesis in the spathes of *Anthurium andraeanum* (Hort.). *Plant Cell Rep.* **2016**, *35*, 2151–2165. [CrossRef] [PubMed]
7. Grotewold, E. The genetics and biochemistry of floral pigments. *Annu. Rev. Plant Biol.* **2006**, *57*, 761–780. [CrossRef] [PubMed]

8. Miyagawa, N.; Miyahara, T.; Okamoto, M.; Hirose, Y.; Sakaguchi, K.; Hatano, S.; Ozeki, Y. Dihydroflavonol 4-reductase activity is associated with the intensity of flower colors in delphinium. *Plant Biotechnol.* **2015**, *32*, 249–255. [CrossRef]

9. Zhang, J.L.; Pan, D.R.; Zhou, Y.F.; Wang, Z.C.; Hua, S.M.; Hou, L.L.; Sui, F.F. Cloning and expression of genes involved in anthocyanins synthesis in ornamental sunflower. *Acta Hortic. Sin.* **2009**, *36*, 73–80.

10. Chen, S.M.; Li, C.H.; Zhu, X.R.; Deng, Y.M.; Sun, W.; Wang, L.S.; Chen, F.D.; Zhang, Z. The identification of flavonoids and the expression of genes of anthocyanin biosynthesis in the chrysanthemum flowers. *Biol. Plant.* **2012**, *56*, 458–464. [CrossRef]

11. Zhao, D.Q.; Tao, J.; Han, C.X.; Ge, J.T. Flower color diversity revealed by differential expression of flavonoid biosynthetic genes and flavonoid accumulation in herbaceous peony (*Paeonia lactiflora* Pall.). *Mol. Biol. Rep.* **2012**, *39*, 11263–11275. [CrossRef] [PubMed]

12. Zhang, C.; Wang, W.N.; Wang, Y.J.; Gao, S.L.; Du, D.N.; Fu, J.X.; Dong, L. Anthocyanin biosynthesis and accumulation in developing flowers of tree peony (*Paeonia suffruticosa*) 'Luoyang Hong'. *Postharvest Biol. Technol.* **2014**, *97*, 11–22. [CrossRef]

13. Shi, S.G.; Yang, M.; Zhang, M.; Wang, P.; Kang, Y.X.; Liu, J.J. Genome-wide transcriptome analysis of genes involved in flavonoid biosynthesis between red and white strains of *Magnolia sprengeri* pamp. *BMC Genom.* **2014**, *15*, 706. [CrossRef] [PubMed]

14. Hsu, C.C.; Chen, Y.Y.; Tsai, W.C.; Chen, W.H.; Chen, H.H. Three R2R3-MYB transcription factors regulate distinct floral pigmentation patterning in *Phalaenopsis* spp. *Plant Physiol.* **2015**, *168*, 175–191. [CrossRef] [PubMed]

15. Nakatsuka, T.; Nishihara, M.; Mishiba, K.; Yamamura, S. Temporal expression of flavonoid biosynthesis-related genes regulates flower pigmentation in gentian plants. *Plant Sci.* **2005**, *168*, 1309–1318. [CrossRef]

16. Jaakola, L. New insights into the regulation of anthocyanin biosynthesis in fruits. *Trends Plant Sci.* **2013**, *18*, 477–483. [CrossRef] [PubMed]

17. Tanaka, Y.; Nakamura, N.; Togami, J. Altering flower color in transgenic plants by RNAi-mediated engineering of flavonoid biosynthetic pathway. *Methods Mol. Biol.* **2008**, *442*, 245–257. [PubMed]

18. Petroni, K.; Tonelli, C. Recent advances on the regulation of anthocyanin synthesis in reproductive organs. *Plant Sci.* **2011**, *181*, 219–229. [CrossRef] [PubMed]

19. Tian, J.; Chen, M.C.; Zhang, J.; Li, K.T.; Song, T.T.; Zhang, X.; Yao, Y.C. Characteristics of dihydrofavonol 4-reductase gene promoters from different leaf colored Malus crabapple cultivars. *Hortic. Res.* **2017**, *4*, 17070. [CrossRef] [PubMed]

20. Hichri, I.; Barrieu, F.; Bogs, J.; Kappel, C.; Delrot, S.; Lauvergeat, V. Recent advances in the transcriptional regulation of the flavonoid biosynthetic pathway. *J. Exp. Bot.* **2011**, *62*, 2465–2483. [CrossRef] [PubMed]

21. Huang, Y.J.; Song, S.; Allan, A.C.; Liu, X.F.; Yin, X.R.; Xu, C.J.; Chen, K.S. Differential activation of anthocyanin biosynthesis in *Arabidopsis* and tobacco over-expressing an R2R3 MYB from Chinese bayberry. *Plant Cell Tiss. Org. Cult.* **2013**, *1113*, 491–499. [CrossRef]

22. Schaart, J.G.; Dubos, C.; Romero De La Fuente, I.; van Houwelingen, A.M.M.L.; de Vos, R.C.H.; Jonker, H.H.; Xu, W.J.; Routaboul, J.M.; Lipinec, L.; Bovy, A.G. Identification and characterization of MYB-bHLH-WD40 regulatory complexes controlling proanthocyanidin biosynthesis in strawberry (*Fragaria* × *ananassa*) fruits. *New Phytol.* **2013**, *197*, 454–467. [CrossRef] [PubMed]

23. Koes, R.; Verweij, W.; Quattrocchio, F. Flavonoids: A colorful model for the regulation and evolution of biochemical pathways. *Trends Plant Sci.* **2005**, *10*, 236–242. [CrossRef] [PubMed]

24. Gonzalez, A.; Zhao, M.Z.; Leavitt, J.M.; Lloyd, A.M. Regulation of the anthocyanin biosynthetic pathway by the TTG1/bHLH/Myb transcriptional complex in *Arabidopsis* seedlings. *Plant J.* **2008**, *53*, 814–827. [CrossRef] [PubMed]

25. Albert, N.W.; Davies, K.M.; Lewis, D.H.; Zhang, H.B.; Montefiori, M.; Brendolise, C.; Boase, M.R.; Ngo, H.; Jameson, P.E.; Schwinn, K.E. A conserved network of transcriptional activators and repressors regulates anthocyanin pigmentation in eudicots. *Plant Cell* **2014**, *24*, 962–980. [CrossRef] [PubMed]

26. Zhou, L.; Wang, Y.; Peng, Z. Molecular characterization and expression analysis of chalcone synthase gene during flower development in tree peony (*Paeonia suffruticosa*). *Afr. J. Biotechnol.* **2011**, *10*, 1275–1284.

27. Zhou, L.; Wang, Y.; Peng, Z. Cloning and expression analysis of dihydroflavonol 4-reductase gene *PsDFR1* from tree peony (*Paeonia suffruticosa*). *Plant Physiol. J.* **2011**, *47*, 885–892.

28. Zhou, L.; Wang, Y.; Ren, L.; Shi, Q.Q.; Zheng, B.Q.; Miao, K.; Guo, X. Overexpression of *PsCHI1*, a homologue of the chalcone isomerase gene from tree peony (*Paeonia suffruticosa*), reduces the intensity of flower pigmentation in transgenic tobacco. *Plant Cell Tiss. Org. Cult.* **2014**, *116*, 285–295. [CrossRef]

29. Schwinn, K.; Venail, J.; Shang, Y.J.; Mackay, S.; Alm, V.; Butelli, E.; Oyama, R.; Bailey, P.; Davies, K.; Martin, C. A small family of *MYB*-regulatory genes controls foral pigmentation intensity and patterning in the Genus *Antirrhinum*. *Plant Cell* **2006**, *18*, 831–851. [CrossRef] [PubMed]

30. Dubos, C.; Stracke, R.; Grotewold, E.; Weisshaar, B.; Martin, C.; Lepiniec, L. MYB transcription factors in *Arabidopsis*. *Trends Plant Sci.* **2010**, *15*, 573–581. [CrossRef] [PubMed]

31. Carey, C.C.; Strahle, J.T.; Selinger, D.A.; Chandler, V.L. Mutations in the *pale aleurone color1* regulatory gene of the *Zea mays* anthocyanin pathway have distinct phenotypes relative to the functionally similar *TRANSPARENT TESTA GLABRA1* gene in *Arabidopsis thaliana*. *Plant Cell* **2004**, *16*, 450–464. [CrossRef] [PubMed]

32. Walker, A.R.; Lee, E.; Bogs, J.; McDavid, D.A.J.; Thomas, M.R.; Robinson, S.P. White grapes arose through the mutation of two similar and adjacent regulatory genes. *Plant J.* **2007**, *49*, 772–785. [CrossRef] [PubMed]

33. Tian, J.; Peng, Z.; Zhang, J.; Song, T.T.; Wan, H.H.; Zhang, M.L.; Yao, Y.C. McMYB10 regulates coloration via activating *McF3'H* and later structural genes in ever-red leaf crabapple. *Plant Biotechnol. J.* **2015**, *13*, 948–961. [CrossRef] [PubMed]

34. Albert, N.W.; Lewis, D.H.; Zhang, H.B.; Schwinn, K.E.; Jameson, P.E.; Davies, K.M. Members of an R2R3-MYB transcription factor family in *Petunia* are developmentally and environmentally regulated to control complex floral and vegetative pigmentation patterning. *Plant J.* **2011**, *65*, 771–784. [CrossRef] [PubMed]

35. Zhu, L.; Shan, H.; Chen, S.M.; Jiang, J.F.; Gu, C.S.; Zhou, G.Q.; Chen, Y.; Song, A.P.; Chen, F.D. The heterologous expression of the chrysanthemum R2R3-MYB transcription factor *CmMYB1* alters lignin composition and represses flavonoid synthesis in *Arabidopsis thaliana*. *PLoS ONE* **2013**, *8*, e65680. [CrossRef] [PubMed]

36. Zhang, Y.X.; Zhang, L.; Gai, S.P.; Liu, C.Y.; Lu, S. Cloning and expression analysis of the R2R3-*PsMYB1* gene associated with bud dormancy during chilling treatment in the tree peony (*Paeonia suffruticosa*). *Plant Growth Regul.* **2015**, *75*, 667–676. [CrossRef]

37. Gao, L.X.; Yang, H.X.; Liu, H.F.; Yang, J.; Hu, Y.H. Extensive transcriptome changes underlying the flower color intensity variation in *Paeonia ostii*. *Front. Plant Sci.* **2016**, *6*, 1205. [CrossRef] [PubMed]

38. Shi, Q.Q.; Li, L.; Zhang, X.X.; Luo, J.R.; Li, X.; Zhai, L.J.; He, L.X.; Zhang, Y.L. Biochemical and comparative transcriptomic analyses identify candidate genes related to variegation formation in *Paeonia rockii*. *Molecules* **2017**, *22*, 1364. [CrossRef] [PubMed]

39. Zhang, X.P.; Zhao, L.Y.; Xu, Z.D.; Yu, X.Y. Transcriptome sequencing of Paeonia suffruticosa 'Shima Nishiki' to identify differentially expressed genes mediating double-color formation. *Plant Physiol. Biochem.* **2018**, *123*, 114–124. [CrossRef] [PubMed]

40. Cao, Y.L.; Xie, L.F.; Ma, Y.Y.; Ren, C.H.; Xing, M.Y.; Fu, Z.S.; Wu, X.Y.; Yin, X.R.; Xu, C.J.; Li, X. *PpMYB15* and *PpMYBF1* Transcription factors are involved in regulating flavonol biosynthesis in peach fruit. *J. Agric. Food Chem.* **2019**, *67*, 644–652. [CrossRef] [PubMed]

41. Gong, B.H.; Yi, J.; Sui, J.J.; Wu, J.; Wu, Z.; Cheng, Y.H.; Wu, C.Y.; Liu, C.; Yi, M.F. Cloning and expression analysis of LlHsfA1 from *Lilium longiforum*. *Acta Hortic. Sin.* **2014**, *41*, 1400–1408.

42. Ding, K.; Ma, P.D.; Jia, Y.Y.; Pei, T.L.; Bai, Z.Q.; Liang, Z.S. Subcellular localization and transactivation analysis of three R2R3-MYB in *Salvia miltiorrhiza* Bunge. *Acta Agric. Boreali-Occidentalis Sin.* **2018**, *27*, 586–594.

43. Moreau, C.; Ambrose, M.J.; Turner, L.; Hill, L.; Noel Ellis, T.H.; Hofer, J.M.I. The *b* gene of pea encodes a defective flavonoid 3',5'-hydroxylase and confers pink flower color. *Plant Physiol.* **2012**, *159*, 759–768. [CrossRef] [PubMed]

44. Davies, K.M.; Albert, N.W.; Schwinn, K.E. From landing lights to mimicry: The molecular regulation of flower colouration and mechanisms for pigmentation patterning. *Funct. Plant Biol.* **2012**, *39*, 619–638. [CrossRef]

45. Li, X.; Lu, M.; Tang, D.Q.; Shi, Y.M. Composition of carotenoids and flavonoids in narcissus cultivars and their relationship with flower color. *PLoS ONE* **2015**, *10*, e0142074. [CrossRef] [PubMed]

46. Zhang, Q.; Hao, R.J.; Xu, Z.D.; Yang, W.R.; Wang, J.; Cheng, T.R.; Pan, H.T.; Zhang, Q.X. Isolation and functional characterization of a R2R3-MYB regulator of *Prunus mume* anthocyanin biosynthetic pathway. *Plant Cell. Tiss. Org. Cult.* **2017**, *131*, 417–429. [CrossRef]

47. Feng, K.; Xu, Z.S.; Que, F.; Liu, J.X.; Wang, F.; Xiong, A.S. An R2R3-MYB transcription factor, OjMYB1, functions in anthocyanin biosynthesis in *Oenanthe javanica*. *Planta* **2018**, *247*, 301–315. [CrossRef] [PubMed]

48. Stracke, R.; Werber, M.; Weisshaar, B. The *R2R3-MYB* gene family in *Arabidopsis thaliana*. *Curr. Opin. Plant Biol.* **2001**, *4*, 447–456. [CrossRef]

49. Stracke, R.; Ishihara, H.; Huep, G.; Barsch, A.; Mehrtens, F.; Niehaus, K.; Weisshaar, B. Differential regulation of closely related R2R3-MYB transcription factors controls flavonol accumulation in different parts of the *Arabidopsis thaliana* seedling. *Plant J.* **2007**, *50*, 660–677. [CrossRef] [PubMed]

50. Czemmel, S.; Stracke, R.; Weisshaar, B.; Cordon, N.; Harris, N.N.; Walker, A.R.; Robinson, S.P.; Bogs, J. The grapevine R2R3-MYB transcription factor VvMYBF1 regulates flavonol synthesis in developing grape berries. *Plant Physiol.* **2009**, *151*, 1513–1530. [CrossRef] [PubMed]

51. Schwinn, K.E.; Ngo, H.; Kenel, F.; Brummell, D.A.; Albert, N.W.; McCallum, J.A.; Pither-Joyce, M.; Crowhurst, R.N.; Eady, C.; Davies, K.M. The onion (*Allium cepa* L.) *R2R3-MYB* gene *MYB1* regulates anthocyanin biosynthesis. *Front. Plant Sci.* **2016**, *7*, 1865. [CrossRef] [PubMed]

52. Paz-Ares, J.; Wienand, U.; Peterson, P.A.; Saedler, H. Molecular cloning of the *c* locus of *Zea mays*: A locus regulating the anthocyanin pathway. *EMBO J.* **1986**, *5*, 829–834. [CrossRef] [PubMed]

53. Nesi, N.; Jond, C.; Debeaujon, I.; Caboche, M.; Lepiniec, L. The Arabidopsis *TT2* gene encodes an R2R3 MYB domain protein that acts as a key determinant for proanthocyanidin accumulation in developing seed. *Plant Cell* **2001**, *13*, 2099–2114. [PubMed]

54. Terrier, N.; Torregrosa, L.; Ageorges, A.; Vialet, S.; Verriès, C.; Cheynier, V.; Romieu, C. Ectopic expression of VvMybPA$_2$ promotes proanthocyanidin biosynthesis in grapevine and suggests additional targets in the pathway. *Plant Physiol.* **2009**, *149*, 1028–1041. [CrossRef] [PubMed]

55. Gu, Z.Y.; Zhu, J.; Hao, Q.; Yuan, Y.W.; Duan, Y.W.; Men, S.Q.; Wang, Q.Y.; Hou, Q.Z.; Liu, Z.A.; Shu, Q.Y.; et al. A novel R2R3-MYB transcription factor contributes to petal blotch formation by regulating organ-specific expression of *PsCHS* in tree peony (*Paeonia suffruticosa*). *Plant Cell Physiol.* **2018**, pcy232. [CrossRef] [PubMed]

56. Goff, S.A.; Cone, K.C.; Chandler, V.L. Functional analysis of the transcriptional activator encoded by the maize B-gene: Evidence for a direct functional interaction between two classes of regulatory proteins. *Genes Dev.* **1992**, *6*, 864–875. [CrossRef] [PubMed]

57. Zimmermann, I.M.; Heim, M.A.; Weisshaar, B.; Uhrig, J.F. Comprehensive identifcation of Arabidopsis thaliana MYB transcription factors interacting with R/B-like BHLH proteins. *Plant J.* **2004**, *40*, 22–34. [CrossRef] [PubMed]

58. Wang, N.; Xu, H.F.; Jiang, S.H.; Zhang, Z.Y.; Lu, N.L.; Qiu, H.R.; Qu, C.Z.; Wang, Y.C.; Wu, S.J.; Chen, X.S. MYB12 and MYB22 play essential roles in proanthocyanidin and flavonol synthesis in red-fleshed apple (*Malus sieversii* f. *niedzwetzkyana*). *Plant J.* **2017**, *90*, 276–292. [CrossRef] [PubMed]

59. Liu, C.Y.; Long, J.M.; Zhu, K.J.; Liu, L.L.; Yang, W.; Zhang, H.Y.; Li, L.; Xu, Q.; Deng, X.X. Characterization of a Citrus R2R3-MYB transcription factor that regulates the flavonol and hydroxycinnamic acid biosynthesis. *Sci. Rep.* **2016**, *6*, 25352. [CrossRef] [PubMed]

60. Davies, K.M.; Schwinn, K.E.; Deroles, S.C.; Manson, D.G.; Lewis, D.H.; Bloor, S.J.; Bradley, J.M. Enhancing anthocyanin production by altering competition for substrate between flavonol synthase and dihydroflavonol 4-reductase. *Euphytica* **2003**, *131*, 259–268. [CrossRef]

61. Wang, F.B.; Kong, W.L.; Wong, G.; Fu, L.F.; Peng, R.H.; Li, Z.J.; Yao, Q.H. AtMYB12 regulates flavonoids accumulation and abiotic stress tolerance in transgenic *Arabidopsis thaliana*. *Mol. Genet. Genom.* **2016**, *291*, 1545–1559. [CrossRef] [PubMed]

62. Mehrtens, F.; Kranz, H.; Bednarek, P.; Weisshaar, B. The Arabidopsis transcription factor MYB12 is a flavonol-specific regulator of phenylpropanoid biosynthesis. *Plant Physiol.* **2005**, *138*, 1083–1096. [CrossRef] [PubMed]

63. Huang, W.J.; Khaldun, A.B.M.; Chen, J.J.; Zhang, C.J.; Lv, H.Y.; Yuan, L.; Wang, Y. A R2R3-MYB transcription factor regulates the flavonol biosynthetic pathway in a traditional Chinese medicinal plant, *Epimedium sagittatum*. *Front. Plant Sci.* **2016**, *7*, 1089. [CrossRef] [PubMed]

64. Luo, J.; Butelli, E.; Hill, L.; Parr, A.; Niggeweg, R.; Bailey, P.; Weisshaar, B.; Martin, C. AtMYB12 regulates caffeoyl quinic acid and flavonol synthesis in tomato: Expression in fruit results in very high levels of both types of polyphenol. *Plant J.* **2008**, *56*, 316–326. [CrossRef] [PubMed]

65. Ren, X.X.; Wang, S.L.; Xue, J.Q.; Zhu, F.Y.; Liu, C.J.; Zhang, X.X. Molecular cloning and expression analysis of cryptochrome gene *PsCRY2* in tree peony. *Acta Hortic. Sin.* **2015**, *42*, 2229–2236. [CrossRef]

66. An, J.P.; An, X.H.; Yao, J.F.; Wang, X.N.; You, C.X.; Wang, X.F.; Hao, Y.J. BTB protein MdBT2 inhibits anthocyanin and proanthocyanidin biosynthesis by triggering MdMYB9 degradation in apple. *Tree Physiol.* **2018**, *38*, 1578–1587. [CrossRef] [PubMed]

67. Verma, A.; Lee, C.; Morriss, S.; Odu, F.; Kenning, C.; Rizzo, N.; Spollen, W.G.; Lin, M.; McRae, A.G.; Givan, S.A.; et al. The novel cyst nematode effector protein 30D08 targets host nuclear functions to alter gene expression in feeding sites. *New Phytol.* **2018**, *219*, 697–713. [CrossRef] [PubMed]

68. Clough, S.J.; Bent, A.F. Floral dip: A simplified method for *Agrobacterium*-mediated transformation of *Arabidopsis thaliana*. *Plant J.* **1998**, *16*, 735–743. [CrossRef] [PubMed]

69. An, J.P.; Liu, X.; Li, H.H.; You, C.X.; Wang, X.F.; Hao, Y.J. Apple RING E3 ligase MdMIEL1 inhibits anthocyanin accumulation by ubiquitinating and degrading MdMYB1 protein. *Plant Cell Physiol.* **2017**, *58*, 1953–1962. [CrossRef] [PubMed]

70. An, J.P.; Qu, F.J.; Yao, J.F.; Wang, X.N.; You, C.X.; Wang, X.F.; Hao, Y.J. The bZIP transcription factor MdHY5 regulates anthocyanin accumulation and nitrate assimilation in apple. *Hortic. Res.* **2017**, *4*, 17023. [CrossRef] [PubMed]

71. Zhao, D.Q.; Jiang, Y.; Ning, C.L.; Meng, J.S.; Lin, S.S.; Ding, W.; Tao, J. Transcriptome sequencing of a chimaera reveals coordinated expression of anthocyanin biosynthetic genes mediating yellow formation in herbaceous peony (*Paeonia lactilora* Pall.). *BMC Genom.* **2014**, *15*, 689. [CrossRef] [PubMed]

72. Schmittgen, T.D.; Livak, K.J. Analyzing real-time PCR data by the comparative CT method. *Nat. Protoc.* **2008**, *3*, 1101–1108. [CrossRef] [PubMed]

International Journal of
Molecular Sciences

Article

RrGT2, A Key Gene Associated with Anthocyanin Biosynthesis in *Rosa rugosa*, Was Identified Via Virus-Induced Gene Silencing and Overexpression

Xiaoming Sui [†], Mingyuan Zhao [†], Zongda Xu, Lanyong Zhao * and Xu Han *

Flower Research Laboratory, College of Forestry, Shandong Agricultural University, Taian 271018, China;
suixiaomingjiayou@163.com (X.S.); zhaomingy9@163.com (M.Z.); xuzoda@163.com (Z.X.)
* Correspondence: sdzly369@163.com (L.Z.); hanxusdau@163.com (X.H.); Tel.: +86-0538-824-2216 (L.Z. & X.H.)
† These authors contributed equally.

Received: 21 November 2018; Accepted: 12 December 2018; Published: 14 December 2018

Abstract: In this study, a gene with a full-length cDNA of 1422 bp encoding 473 amino acids, designated *RrGT2*, was isolated from *R. rugosa* 'Zizhi' and then functionally characterized. *RrGT2* transcripts were detected in various tissues and were proved that their expression patterns corresponded with anthocyanins accumulation. Functional verification of *RrGT2* in *R. rugosa* was performed via VIGS. When *RrGT2* was silenced, the *Rosa* plants displayed a pale petal color phenotype. The detection results showed that the expression of *RrGT2* was significantly downregulated, which was consistent with the decrease of all anthocyanins; while the expression of six key upstream structural genes was normal. Additionally, the in vivo function of *RrGT2* was investigated via its overexpression in tobacco. In transgenic tobacco plants expressing *RrGT2*, anthocyanin accumulation was induced in the flowers, indicating that *RrGT2* could encode a functional GT protein for anthocyanin biosynthesis and could function in other species. The application of VIGS in transgenic tobacco resulted in the treated tobacco plants presenting flowers whose phenotypes were lighter in color than those of normal plants. These results also validated and affirmed previous conclusions. Therefore, we speculated that glycosylation of *RrGT2* plays a crucial role in anthocyanin biosynthesis in *R. rugosa*.

Keywords: *Rosa rugosa*; *RrGT2* gene; Clone; VIGS; Overexpression; Tobacco; Flower color; Anthocyanin

1. Introduction

Rosa rugosa is an important ornamental plant species that belongs to the genus *Rosa* in the family *Rosaceae*. This species is native to China and is widely distributed worldwide. Because of its unique fragrance, color, cold resistance and drought resistance, there is great potential for the development of this species for use in garden applications. Many varieties of roses exist but most of them are traditional colors such as pink and purple. A few varieties are white and some lack yellow, bright red, orange and compound colors and so forth. [1]. Therefore, the development of innovative rose colors has become the main goal of breeders. At present, scientific research on *R. rugosa* in China and abroad has focused mainly on the development and protection of wild *R. rugosa* resources [2], the analysis of the genetic diversity of *R. rugosa* [3], the optimization of *R. rugosa* essential oil extraction [4], the nutritional value of *R. rugosa* [5], the cultivation and propagation technology of *R. rugosa* [6,7] and strategies for attaining high *R. rugosa* yields. Innovating new *R. rugosa* flower colors has occurred mainly by improving cultivation and management techniques or by trying to cross different varieties during *R. rugosa* breeding; relatively less molecular biology technology has been used in the innovation of new flower colors. However, due to the decrease in wild *R. rugosa* resources and the lack of natural variation in recent years, the existing *R. rugosa* varieties can no longer meet the various needs of gardening applications [1]. However, there is still much room for development in the breeding of new *R. rugosa* varieties via molecular biology. Therefore, studying the mechanism of *R. rugosa* flower color formation and enriching flower color during breeding are urgently needed. Analysis of the pigment composition of rose varieties and the study of the expression characteristics of the key genes encoding enzymes that catalyze the synthesis of rose pigments are important prerequisites for the molecular breeding of rose color traits [8]. Anthocyanins determine the color of higher plant organs. Structural genes (*CHS*, *CHI*, *F3H*, *F3'H*, *DFR*, *ANS*, *3GT*, etc.) and regulatory genes (*MYB*, mostly those of the *R2R3-MYB*, *BHLH* and *WD40* classes) [9,10] related to the anthocyanin biosynthesis pathway have been cloned and sequenced and related protein functional studies have been performed in many plant species, such as petunia, maize, snapdragon and so on [11,12]. However, less anthocyanin-related research has been conducted in rose than in those species.

Flower coloration is caused by the accumulation of pigments comprising mainly flavonoids, carotenoids and betalains. Among them, flavonoids, which comprise chalcones, flavones, flavonols, flavandiols, anthocyanins and proanthocyanidins, are the most important pigments [13,14]. Anthocyanins, which are derived from the anthocyanin biosynthesis pathway, are the largest group of water-soluble plant flavonoids found in the organs of plants, including crop species [15–19]. Anthocyanins are unstable in plants; they exist mainly in the form of glycosides within the vacuole [20]. The flavonoid 3-*O*-glycosyltransferase (*3GT*) gene lies downstream in the anthocyanin synthesis pathway. The enzyme encoded by this gene can catalyze the glycosylation of UDP-glucose to replace the 3 hydroxyl groups of anthocyanin molecules and cause anthocyanin glycosylation to produce colored and stable anthocyanins. Glycosylation can alter the hydrophilicity, biochemical activity and subcellular localization of anthocyanins, which is beneficial to their transport and storage in cells and organisms [21–23].

For a long time, GT genes had failed to be cloned and their functions in plant secondary metabolism were unclear. However, since the first cDNA sequence of a GTase was isolated by transposon tagging in maize, an increasing number of GT genes have been cloned and identified [24]. Studies have shown that the *3GT* gene is expressed only in red grape (*Vitis vinifera*) varieties and not in white grape varieties. When a *3GT* transgene was transformed into a colorless embryo, a pale-red bud was produced [25,26]. Studies by Afifi et al. on the expression of five key genes involved in anthocyanin synthesis in grape cell tissue indicate that the presence of the cytotoxic factor eutypine inhibits the expression of *3GT* and thus reduces anthocyanin contents [27]. This finding indicates that *3GT* is the key gene involved in grape skin color (from white to red) and is also a key gene in the anthocyanin biosynthesis pathway [28]. In *Gentiana triflora*, *3GT* expression occurs mostly in blue petals and rarely in white flowers [29]. Expression of the *3GT* gene is very important for anthocyanin accumulation in many plant species and its expression intensity is positively correlated with anthocyanin synthesis [30].

In this study, we cloned and identified the *RrGT2* gene from the petals of *R. rugosa* for the first time. We carried out detailed bioinformatic and homology analyses of the *RrGT2* gene. VIGS results in perennial *Rosa* plants under field conditions suggested that *RrGT2* is related to the biosynthesis of anthocyanins in *R. rugosa*. Stable transformation of the *RrGT2* gene in tobacco showed that its overexpression was positively correlated with the accumulation of anthocyanins. The results of VIGS in transgenic tobacco also confirmed this conclusion. We verified the functions of the *RrGT2* gene in anthocyanin metabolism in both the positive and negative directions to provide useful information for subsequent color-improvement projects in *R. rugosa*.

2. Results

2.1. Cloning of RrGT2 and Sequence Analysis

In the early stages, we screened the *RrGT2* gene by comparing the differentially expressed genes in the anthocyanin pathway within the *R. rugosa* transcriptome data of our laboratory. The full-length CDS of *RrGT2* (MK034141) was cloned and then confirmed by sequencing [31]. The complete open reading

frame from the ATG start codon to the TAA termination codon encodes a 473 amino acid protein
(Figure 1A). Multiple sequence alignment (Figure 2) revealed that the *RrGT2* protein, which belongs to
the GTB superfamily, displays strong species specificity in the N-terminal region and PSPG conserved
domains that consist of 44 amino acid residues in the C-terminal region. A phylogenetic tree (Figure 1B)
was constructed from the amino acid sequences of 21 plants, including the sequence of *RrGT2*, using
MEGA 5.0 software. The results showed that the *RrGT2* gene shared the highest homology percentages
with *FaUGT* (*Fragaria* × *ananassa*) and *FvGT* (*Fragaria vesca* subsp.), both of which were 89%.

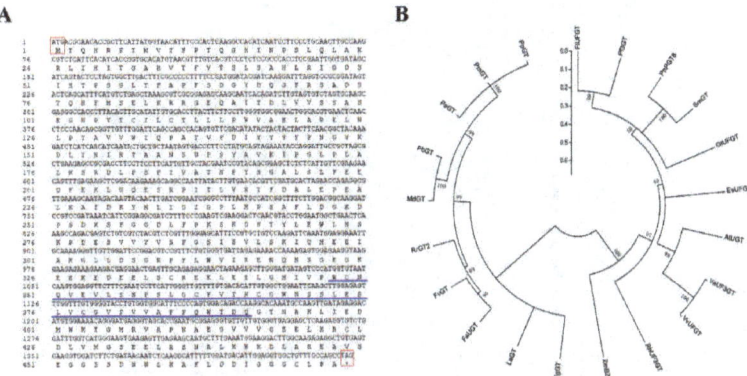

Figure 1. cDNA sequence analysis and phylogenetic tree analysis of the *RrGT2* gene. (**A**) cDNA
sequence of *RrGT2* and its deduced amino acids. The red box shows the start codon and the termination
codon as well as the amino acids they encode. The PSPG domains are underlined by the blue line.
(**B**) Phylogenetic tree of amino acid sequences of *RrGT2* and *GT* members from other plant species.
The tree was constructed by the neighbor-joining method using MEGA 5.0 software. The branch
numbers represent the percentage of bootstrap values from 1000 sampling replicates and the scale
indicates the branch lengths. The gene names from various plant species and the NCBI GenBank
accession numbers for the sequences are as follows: *RhUF3GT* (AB599928.1) from a *Rosa* hybrid cultivar;
AtUGT (UGTNM-121711) from *Arabidopsis thaliana*; *EsUFGT* (KJ648620) from *Epimedium sagittatum*;
FaUGT (KP337600.1) from *Fragaria* × *ananassa*; *FiUFGT* (AF127218.1) from *Forsythia* × *intermedia*; *FvGT*
(XM_004298174.2) from *Fragaria vesca* subspecies; *GtUFGT* (D85186.1) from *Gentiana triflora*; *LaGT*
(XM_019560329.1) from *Lupinus angustifolius*; *MdGT* (XM_008350196.2) from *Malus* × *domestica*; *PaGT*
(XM_021963118.1) from *Prunus avium*; *PbGT* (XM_009339472.2) from *Pyrus* × *bretschneideri*; *Pf3GT*
(AB002818) from *Perilla frutescens*; *PhPGT8* (AB027454) from *Petunia* × *hybrida*; *PmGT* (XM_008224513.2)
from *Prunus mume*; *PpGT* (XM_007221257.2) from *Prunus persica*; *SmGT* (X77369.1) from *Solanum
melongena*; *VaUF3GT* (FJ169463.1) from *Vitis amurensis*; *VvUFGT* (AF000371) from *Vitis vinifera*; *ZmBZ1*
(NM_001112416.1) from *Zea mays*; and *EgGT* (XM_012988948.1) from *Erythranthe guttatus*.

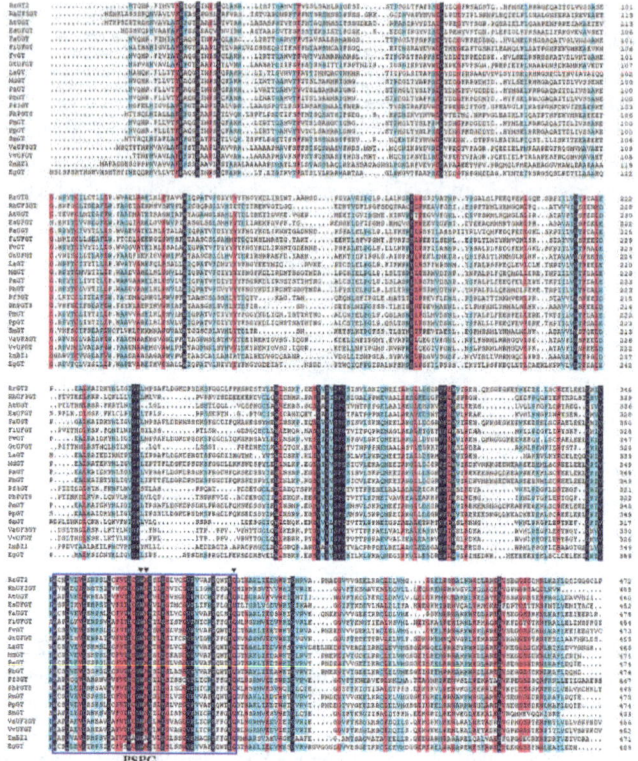

Figure 2. Amino acid sequence homology analysis of the *RrGT2* gene and GTs from other species. Alignments were performed using DNAMAN (version 6.0). The blue box shows the PSPG domains. The black triangles (from left to right) indicate the 22nd, 23rd and 44th amino acids in the PSPG box. The accession numbers are the same as those in Figure 1B.

2.2. Temporal and Spatial Expression Patterns of RrGT2 in Rosa

Before analyzing the expression patterns of *RrGT2*, we cloned a gene from the cDNA of *R. davurica* with the full-length primers of *RrGT2* and sequenced the gene sequence which was identical to that of *RrGT2*. So, we named it *RdGT2*.

The expression levels of the *RrGT2* and *RdGT2* gene, which significantly differed, were assessed during five flowering stages. In *R. rugosa* 'Zizhi' (Figure 3A), the highest expression level of *RrGT2* was observed during the full opening stage and the lowest was observed during the budding stage. In *Rosa davurica* (Figure 3B), the expression level of *RdGT2* was also highest during the full opening stage but lowest during the half opening stage. The expression patterns of the *RrGT2* gene in *R. rugosa* 'Zizhi' and *R. davurica* exhibited approximately the same trend.

The expression levels of the *RrGT2* and *RdGT2* gene, which also significantly differed, were assessed in seven different tissue types. The expression level in the leaves, stems and sepals was relatively high but was relatively low in the other tissues in both *R. rugosa* 'Zizhi' (Figure 3C) and *R. davurica* (Figure 3D).

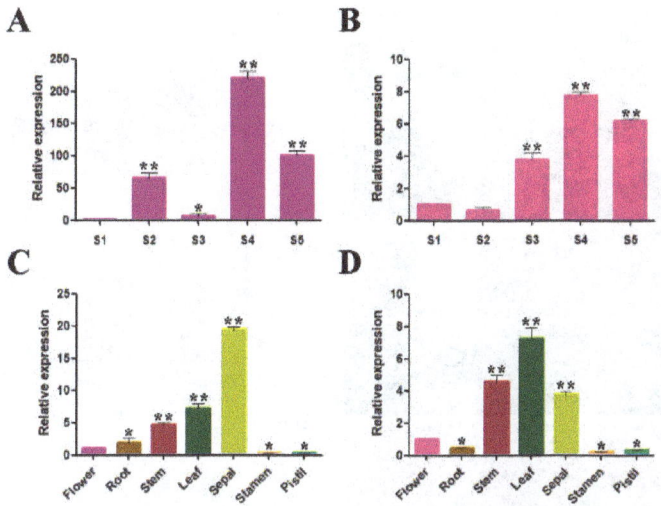

Figure 3. Temporal and spatial expression patterns of *RrGT2*. Relative expression of the *RrGT2* and *RdGT2* gene during the five flowering stages of *R. rugosa* 'Zizhi' (**A**) and *R. davurica* (**B**). S1, budding stage; S2, initial opening stage; S3, half opening stage; S4, full opening stage; S5, wilting stage. Relative expression of the *RrGT2* and *RdGT2* gene in seven different tissues of *R. rugosa* 'Zizhi' (**C**) and *R. davurica* (**D**). The error bars represent the SDs of triplicate reactions. The experiment was repeated three times and each yielded similar results. * $p < 0.05$ and ** $p < 0.01$ indicate significant differences between different flowering stages and between different tissue types.

2.3. VIGS of RrGT2 Reduced the Transcript Abundance of the Endogenous RrGT2 Gene

At 14 days after infection, GFP detection was performed on the newly grown leaves of the infected plants (TRV-GFP and TRV-GFP-*RrGT2*) and on the untreated leaves of both *Rosa* species. GFP imaging (Figures 4A and 5A) showed that the leaves treated with VIGS (TRV-GFP and TRV-GFP-*RrGT2*) showed green fluorescence under longwave ultraviolet light, while the untreated leaves in the control group showed red fluorescence. The corresponding leaves were collected for qRT-PCR detection and the *RrGAPDH* gene was used as an internal control [32] to confirm the efficiency of VIGS. The results (Figures 4D and 5D) showed that the abundance of the *RrGT2* transcript significantly decreased only in the leaves treated with TRV-GFP-*RrGT2* but was expressed normally in the leaves of the plants in the control group and TRV-GFP group.

At 30–35 days after infection, the flowers of *R. rugosa* 'Zizhi' changed from the budding stage to the initial opening stage, by which time notable differences in flower color could be observed with the naked eye. The petals in the control group and TRV-GFP group showed no definitive changes in color but the petals in the TRV-GFP-*RrGT2* group were clearly lighter in color (Figure 4B).

At 35–40 days after infection, GFP imaging and qRT-PCR were performed on the blossoming petals of untreated plants and infected plants (TRV-GFP and TRV-GFP-*RrGT2*) of both *Rosa* species. The results were as expected: the petals treated with VIGS (TRV-GFP and TRV-GFP-*RrGT2*) showed green fluorescence under longwave ultraviolet light, while the petals in the control group showed red fluorescence (Figures 4C and 5B). The results of qRT-PCR (Figures 4E and 5E) were consistent with those of the above detection. In both *Rosa* species, the relative expression trends of the *RrGT2* gene were also essentially consistent: the relative expression of the *RrGT2* gene in the TRV-GFP group was essentially the same as that in the control group. However, the transcript abundance of the endogenous *RrGT2* gene in the petals treated with TRV-GFP-*RrGT2* was significantly downregulated.

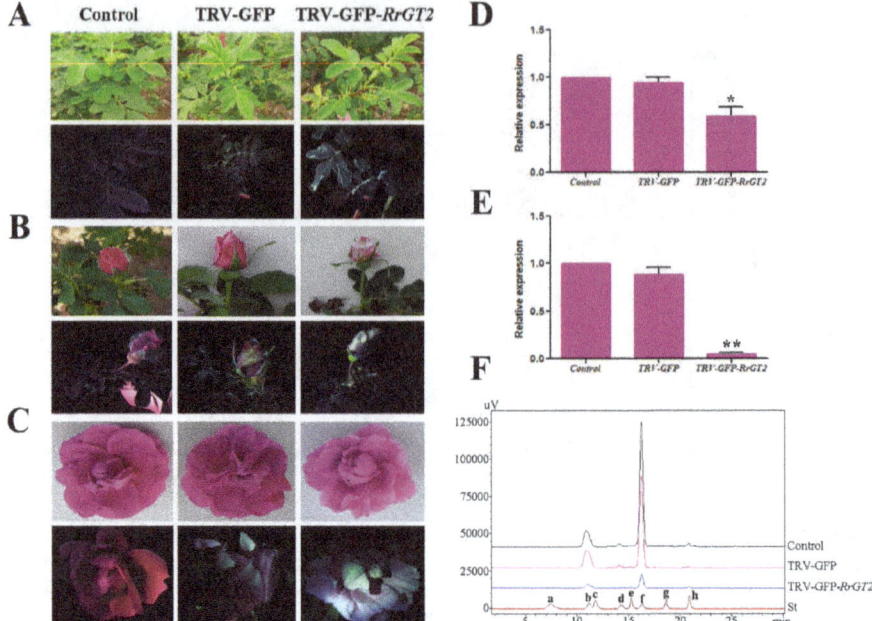

Figure 4. Validation of VIGS in *R. rugosa* 'Zizhi.' Comparisons between the control group and VIGS-treated groups (TRV-GFP and TRV-GFP-*RrGT2*) with respect to leaves (**A**), flowers between budding stage and initial opening stage (**B**) and flowers at the full opening stage (**C**). The plants were imaged under normal light and ultraviolet illumination. qRT-PCR detection of leaves (**D**) and flowers at the full opening stage (**E**) in the control and VIGS-treated groups. *RrGAPDH* was used as an internal control. The error bars represent the SDs of triplicate reactions. The experiment was repeated three times and each yielded similar results. * and ** indicate a significant difference from the control at $p < 0.05$ and $p < 0.01$, respectively, according to Student's *t*-test. (**F**) HPLC-derived chromatograms of flowers at the full opening stage. Eight kinds of anthocyanin standards (St) were used for detection: (a) Dp3G5G; (b) Cy3G5G; (c) Dp3G; (d) Pg3G5G; (e) Cy3G; (f) Pn3G5G; (g) Pg3G; and (h) Pn3G.

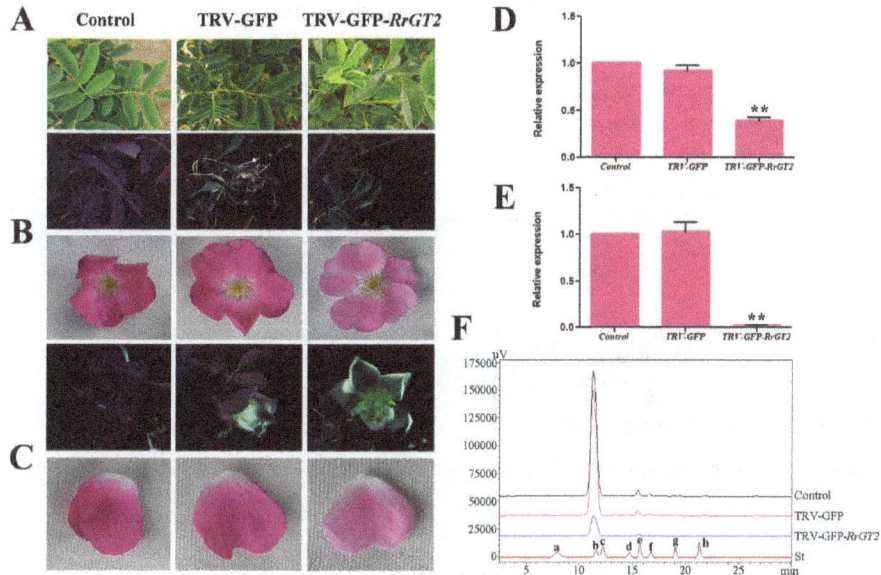

Figure 5. Validation of VIGS in *R. davurica*. Comparisons between the control group and VIGS-treated groups (TRV-GFP and TRV-GFP-*RrGT2*) with respect to leaves (**A**) and flowers at the full opening stage (**B**). The plants were imaged under normal light and ultraviolet illumination. (**C**) The color contrast of a single petal. qRT-PCR detection of leaves (**D**) and flowers at the full opening stage (**E**) in the control and VIGS-treated groups. *RrGAPDH* was used as an internal control. The error bars represent the SDs of triplicate reactions. The experiment was repeated three times and each yielded similar results. ** indicates a significant difference from the control at $p < 0.01$ according to Student's *t*-test. (**F**) HPLC-derived chromatograms of flowers at the full opening stage. Eight kinds of anthocyanin standards (St) were used for detection: (a) Dp3G5G; (b) Cy3G5G; (c) Dp3G; (d) Pg3G5G; (e) Cy3G; (f) Pn3G5G; (g) Pg3G; and (h) Pn3G.

2.4. Reduced Anthocyanin Accumulation in Rosa Petals Was Related to the VIGS of RrGT2

To clarify the role of the posttranscriptional silencing of the *RrGT2* gene in the outcome of the current color change, we compared the relative expression of the six key structural genes, *RrCHS* (KT809351), *RrCHI* (KT809352), *RrF3H* (KT809354), *RrF3'H* (MG735186), *RrDFR* (KT809350) and *RrANS* (KT809353), upstream of the *RrGT2* gene in the anthocyanin pathway (Figure 6). The results showed that the relative expression of the six genes in the control group, TRV-GFP group and TRV-GFP-*RrGT2* group of *R. rugosa* 'Zizhi' (Figure 7A) and *R. davurica* (Figure 7B) did not clearly change. Therefore, the effects of these upstream structural genes can be excluded. It is speculated that the change in flower color after VIGS treatment was related to the posttranscriptional silencing of the *RrGT2* gene.

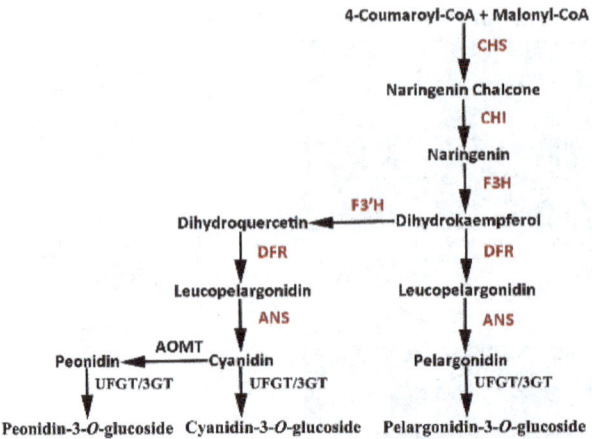

Figure 6. The metabolic pathway analysis of structural genes involved in anthocyanin biosynthesis of *R. rugosa*.

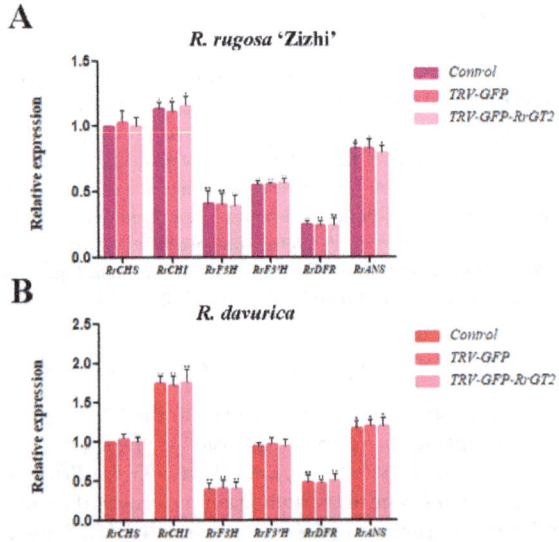

Figure 7. Relative expression levels of six key structural genes (*RrCHS*, *RrCHI*, *RrF3H*, *RrF3'H*, *RrDFR* and *RrANS*) upstream of *RrGT2* in the anthocyanin pathway of *R. rugosa* 'Zizhi' (**A**) and *R. davurica* (**B**) at the full opening stage. *RrGAPDH* was used as the internal control. The error bars represent the SDs of triplicate reactions. The experiment was repeated three times and each yielded similar results. * and ** indicate a significant difference from the relative expression levels of *RrCHS* in the control group at $p < 0.05$ and $p < 0.01$, respectively, according to Student's *t*-test.

2.5. HPLC Analysis of Rosa

The anthocyanin HPLC-generated chromatograms for *R. rugosa* 'Zizhi' (Figure 4F) and *R. davurica* (Figure 5F) showed that the components were well separated. Comparisons with standards allowed the contents of different substances to be calculated by their peak area (Table S2). In 'Zizhi,' six kinds of anthocyanins were detected: Cy3G5G, Pg3G5G, Cy3G, Pn3G5G, Pg3G and Pn3G. Pn3G5G had the highest content, while Cy3G5G had the second highest; the contents of the other four anthocyanins were relatively low. Compared with those in the control group and TRV-GFP group, the contents of several anthocyanins in response to the VIGS treatment were clearly reduced. Pn3G5G exhibited the greatest decrease in content, followed by Cy3G5G; the content of Pg3G was no longer detectable (Figure 8A). In *R. davurica*, the six anthocyanins listed above were also detected. However, Cy3G5G had the highest content and Cy3G had the second highest content; the contents of the other four anthocyanins were relatively low. Compared with those in the control group and TRV-GFP group, the contents of the six anthocyanins in response to the VIGS treatment were clearly reduced. Cy3G5G exhibited the greatest decrease in content, followed by Cy3G; no detection of Pn3G was observed (Figure 8B).

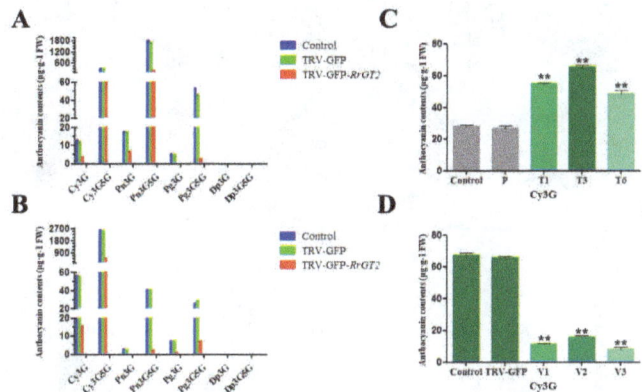

Figure 8. Comparative analysis of anthocyanin components and contents. Anthocyanin components and contents in the flowers of *R. rugosa* 'Zizhi' (**A**) and *R. davurica* (**B**) subjected to different VIGS treatments. (**C**) Anthocyanin component and contents in the flowers of transgenic tobacco. (**D**) Anthocyanin component and contents in the flowers of transgenic tobacco subjected to different VIGS treatments. ** indicates a significant difference from the control at *p* < 0.01 according to Student's *t*-test.

2.6. Overexpression of RrGT2 Increased the Anthocyanin Accumulation in Tobacco

The *RrGT2* gene was ectopically expressed in tobacco using the binary vector pCAMBIA1304-*RrGT2*. Six independent transgenic tobacco lines that overexpressed the *RrGT2* gene were obtained from Hyg-resistance selection and were cultured under the same conditions. PCR analysis confirmed the presence of the transformed *RrGT2* gene in all the transgenic lines as well as the absence of endogenous *RrGT2* in the tobacco plants of the control group (wild type) and empty vector group (the plants were transformed with an empty pCAMBIA1304 vector) (Figure 9A). qRT-PCR analysis revealed that the expression level of *RrGT2* was significantly higher in the transgenic plants, especially T1, T3 and T6, than in the plants of the control group and empty vector group (Figure 9B). Therefore, those three lines were used for further experiments.

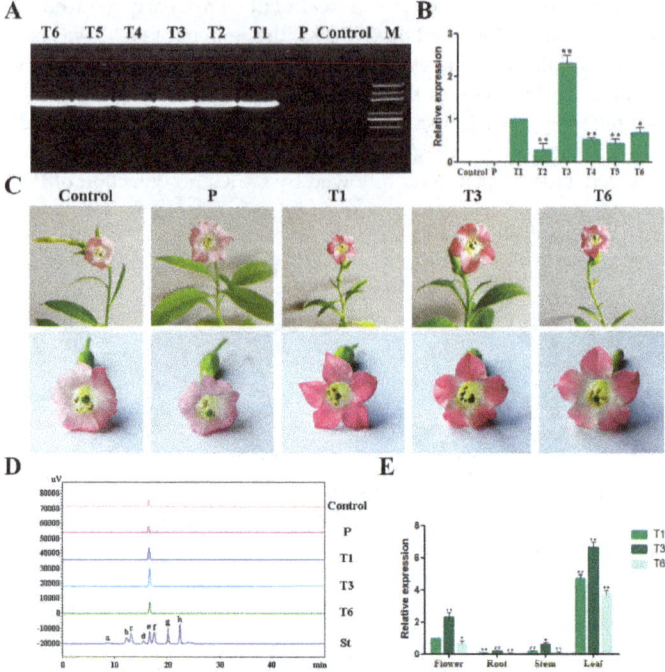

Figure 9. Analysis of tobacco lines overexpressing the *RrGT2* gene. (**A**) The results of positive PCR detection in transgenic tobacco lines. (**B**) The results of qRT-PCR detection in transgenic tobacco lines. (**C**) Phenotypic comparison of flowers between the transgenic tobacco group and the control group and empty vector group. (**D**) HPLC-derived chromatograms of the flowers of the control group, empty vector group and three transgenic tobacco lines. Eight kinds of anthocyanin standards (St) were used for detection: (a) Dp3G5G; (b) Cy3G5G; (c) Dp3G; (d) Pg3G5G; (e) Cy3G; (f) Pn3G5G; (g) Pg3G; and (h) Pn3G. (**E**) Temporal and spatial expression patterns of *RrGT2* in the three transgenic tobacco lines. The error bars represent the SDs of triplicate reactions. The experiment was repeated three times and each yielded similar results. * and ** indicate a significant difference in the relative expression levels at $p < 0.05$ and $p < 0.01$, respectively, according to Student's *t*-test.

Interestingly, there was no substantial difference in morphology between the transgenic plants and wild-type plants. However, the flower color of the transgenic tobacco lines harboring *RrGT2* was affected; compared with that of the tobacco plants in the control group and empty vector group, the petal pigmentation of the transgenic tobacco plants harboring *RrGT2* was markedly deeper (Figure 9C). This change in corolla color of the transgenic tobacco was already visible prior to anthesis. To confirm that the deeper flower color was attributed to increased pigment levels synthesized from the anthocyanin pathway, the total anthocyanins were determined qualitatively and quantitatively via HPLC. Previous studies have reported that cyanidin-3-*O*-rutinoside mainly exists in the petals of wild-type tobacco [33–35]. Even in our study, Cy3G was detected mainly in the flowers of OE-*RrGT2*. Compared with those of the plants in the control group and empty vector group, the contents of anthocyanins in the flowers of T1, T3 and T6 were significantly greater (Figures 8C and 9D and Table S3). The anthocyanin contents in the three transgenic lines were basically consistent with the trends of the relative expression of *RrGT2*. The expression of *RrGT2* in the transgenic tobacco was clearly correlated with the increased pigmentation observed in the petals.

2.7. Expression Patterns of RrGT2 in Transgenic Tobacco

The expression patterns of the *RrGT2* gene in three transgenic tobacco lines were analyzed by qRT-PCR (Figure 9E). The expression levels of *RrGT2*, which significantly differed, were assessed in four different tissue types. The expression levels in the leaves and flowers were relatively high but were relatively low in the stems and roots. Of all the tissue types of the three transgenic lines, the highest gene expression occurred in T3, followed by T1 and the lowest was in T6.

2.8. VIGS of RrGT2 Reduced the Anthocyanin Accumulation in Transgenic Tobacco

Previous tests showed that the *RrGT2* transgenic tobacco line T3 had the highest level of gene expression, so it was used as the experimental object of VIGS. At 14 days after infection, GFP detection was performed on the newly grown leaves of tobacco plants in the control group, TRV-GFP group and TRV-GFP-*RrGT2* group. GFP imaging (Figure 10A) showed that the leaves treated with VIGS (TRV-GFP and TRV-GFP-*RrGT2*) showed green fluorescence under longwave ultraviolet light, while the untreated leaves in the control group showed red fluorescence. The corresponding leaves were collected for qRT-PCR detection and the *NtACT* gene was used as an internal control. The results (Figure 10C) showed that the abundance of the *RrGT2* transcript significantly decreased only in the leaves treated with TRV-GFP-*RrGT2* but was expressed normally in the control group and TRV-GFP group. In addition, interestingly, compared with those of the control group and TRV-GFP group, the phenotypes of the leaves of the tobacco plants in the group treated with TRV-GFP-*RrGT2* were similar to those in response to photobleaching.

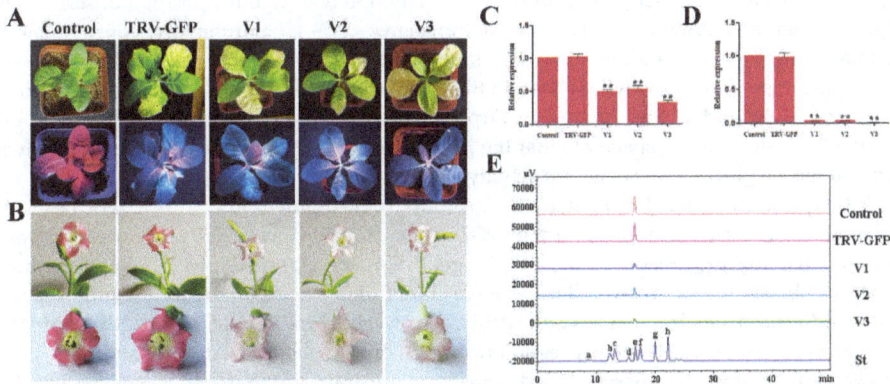

Figure 10. Analysis of OE-*RrGT2* tobacco plants treated with VIGS. (**A**) Comparisons between the control group and the VIGS-treated groups (TRV-GFP and TRV-GFP-*RrGT2*) with respect to leaves at 14 days after infection. The plants were imaged under normal light and ultraviolet illumination. V1, VIGS-*RrGT2*-1; V2, VIGS-*RrGT2*-2; V3, VIGS-*RrGT2*-3. (**B**) Phenotypic comparison of the flowers between the OE-*RrGT2*-VIGS tobacco group and the control group and empty vector group. qRT-PCR detection in the leaves (**C**) and flowers (**D**) of the control and VIGS-treated groups. The error bars represent the SDs of triplicate reactions. The experiment was repeated three times and each yielded similar results. ** indicates a significant difference from the control at $p < 0.01$ according to Student's *t*-test. (**E**) HPLC-derived chromatograms of the flowers in the control group and VIGS-treated groups. Eight kinds of anthocyanin standards (St) were used for detection: (a) Dp3G5G; (b) Cy3G5G; (c) Dp3G; (d) Pg3G5G; (e) Cy3G; (f) Pn3G5G; (g) Pg3G; and (h) Pn3G.

At 60–75 days after infection, notable differences in flower color could be observed with the naked eye. The flowers in the control group and TRV-GFP group showed no definitive changes in color but the flowers in the TRV-GFP-*RrGT2* group were clearly lighter in color (Figure 10B). The corresponding flowers were collected for qRT-PCR detection and the results (Figure 10D) showed that the abundance of the *RrGT2* transcript significantly decreased only in the flowers treated with TRV-GFP-*RrGT2* but was expressed normally in the control group and TRV-GFP group. To confirm that the lighter flower color was attributed to decreased levels of pigments synthesized from the anthocyanin pathway, the total anthocyanins were determined qualitatively and quantitatively via HPLC. Compared with those in the flowers of the control group and TRV-GFP group, the anthocyanin contents in the VIGS-*RrGT2*-1, -2 and -3 flowers were significantly lower (Figures 8D and 10E and Table S4), which was essentially consistent with the qRT-PCR results.

3. Discussion

At present, research on the flower color of *R. rugosa* requires very innovative and practical studies. Although many genes have been reported to regulate the formation of flower color, few reports on downstream structural genes such as GTs exist. The final formation of anthocyanins depends on the glycosylation of GTs, so it is very important to determine the function and influence of the *RrGT2* gene in *Rosa* color formation. In this study, we successfully cloned the *RrGT2* gene, which had a full-length cDNA of 1422 bp and encoded 473 amino acids, from the petals of *R. rugosa* 'Zizhi.'

The amino acid sequence alignment between *RrGT2* and GTs from 21 other species indicated that *RrGT2* has a common PSPG motif of the GT superfamily (Figure 2). Previous studies have shown that the conserved PSPG region is related to the substrate recognition and catalytic activity of protein-based enzymes [36–42]. If the 44 amino acids of the PSPG domain were numbered, those at positions 22, 23 and 44 would play an important role in the selection of enzyme proteoglycan donors. The twenty-second position of tryptophan (Trp, W) can correctly bind UDP-glucose, while arginine (Arg, R) can make UDP-glucuronic acid bind correctly; the twenty-third position of serine (Ser, S) is highly conserved among UDP-glucuronosyltransferases [43–45] and the forty-fourth position of glutamine (Gln, Q) and histidine (His, H) is strongly conserved among glucosyltransferases and galactotransferases, respectively [38]. Within the PSPG domain of the *RrGT2* gene, the amino acids at positions 22, 23 and 44 are tryptophan (Trp, W), asparagine (Asn, N) and glutamine (Gln, Q), respectively. Therefore, we speculate that the *RrGT2* gene uses UDP-glucose as the main glycosyl donor and has no glucuronyltransferase activity [45].

The expression of the *RrGT2* gene during floral development and in different tissues was investigated. The expression trends of the *RrGT2* gene differed during different flowering periods, indicating that the expression of the *RrGT2* gene was developmentally regulated during the anthocyanin biosynthesis process. Studies have shown that the accumulation of anthocyanins in red-skinned sand pear, strawberries and litchi is positively correlated with the activity of *UF3GT*. Boss et al. [28] also detected the expression of *UF3GT* in the peels of red grape that accumulated anthocyanin but not in other tissues of red grape or white grape without anthocyanin accumulation. The tissue-specific anthocyanin expression was similar to that of *F3GT* genes in peach, in which expression levels were greatest in tissues with pigment accumulation but relatively low in unpigmented organs [46]. Notably, the stems of both *Rosa* species were purple, which is consistent with the relatively high expression level of the *RrGT2* gene in that tissue. Interestingly, *R. davurica* is one of the parents of *R. rugosa* 'Zizhi,' so we speculate that this reason might explain the similar expression patterns between both *Rosa* species. However, with respect to the relatively low expression levels in flowers, this did not mean that the *RrGT2* gene had no effect on flower color formation. We believe that this was the result of using flowers at the budding stage as one of the tissue types. During the budding stage, the expression of the *RrGT2* gene was lowest but during the other stages, it was very high. In addition, *RrGT2* was highly expressed in the leaves and sepals of both *Rosa* species, so we infer that *RrGT2* is also involved in the glycosylation of secondary metabolites in leaves and sepals and plays an important role.

To clarify the role of *RrGT2* in the formation of *R. rugosa* flower color, the VIGS technique was used to specifically silence the *RrGT2* gene in both *Rosa* species as well as to detect and analyze the phenotypes of the flowers. The VIGS system, which involves TRV1 and TRV2, is a powerful tool for use in the functional characterization of genes in vivo [47]. At present, few reports exist about the use of the VIGS system in plant floral organs and most of the tested species thus far have been members of the Solanaceae family. For example, VIGS technology was used to study the genes controlling floral fragrance in *Petunia hybrida* [48] and the roles of the *SlMADSI*, *NbMADS4-1* and *NbMADS4-2* genes in tobacco flowers were also determined via VIGS [49]. Furthermore, the TRV recombinant virus vector was successfully used to induce the silencing of the *CHS* and *GLO1* genes in *Gerbera jamesonii* [50]. In the present study, we developed a VIGS system for use with perennial *Rosa* plants grown naturally in the field as experimental materials for the first time and we used this system to study key genes of *Rosa* color and obtained a preliminary result of gene silencing efficiency (Table S5) and other relevant results. Compared with the control conditions, the conditions resulting from the established optimal VIGS system resulted in clearly lighter petal color of both *Rosa* species, which was consistent with the significantly downregulated transcript abundance of the endogenous *RrGT2* gene. The relative expression of the six key upstream structural genes (*RrCHS*, *RrCHI*, *RrF3H*, *RrF3'H*, *RrDFR* and *RrANS*) remained unchanged. In the anthocyanin biosynthesis pathway, the upstream genes are precursors for anthocyanin biosynthesis [29]. Therefore, silencing of the *RrGT2* gene might lead to such a change.

The contents of eight anthocyanins, Cy3G, Cy3G5G, Pg3G, Pg3G5G, Pn3G, Pn3G5G, Dp3G and Dp3G5G, were analyzed qualitatively and quantitatively via HPLC. The results showed that the most abundant anthocyanin in the petals of *R. rugosa* 'Zizhi' was Pn3G5G, which is consistent with the results of Zhang et al. [51]. The content of Cy3G5G was the second highest and the other anthocyanin contents were relatively low; no presence of Dp3G or Dp3G5G was detected. With respect to *R. davurica*, this is the first time different kinds and contents of anthocyanins were detected in the petals. The Cy3G5G content was predominant; that is, the coloration of the *R. davurica* petals was affected mainly by Cy3G5G, while the other anthocyanins contributed little to flower color. After performing the VIGS treatment, we again analyzed both *Rosa* species via HPLC. The results showed that the contents of all the different kinds of anthocyanins decreased to some extent and that the decrease in the contents of several major anthocyanins was clear in both *Rosa* species. These results are in agreement with the lighter flower color phenotypes and the relatively downregulated expression level of the endogenous *RrGT2* gene in response to VIGS treatment. Therefore, it can be inferred that *RrGT2* is a key structural gene that directly affects the formation of anthocyanins in *R. rugosa*.

In plant secondary metabolism, numerous glycosides have already been isolated as biologically active compounds and some of them have been widely used as important medicines. GTs usually act in the final stages of plant secondary metabolism and are used for stabilizing and solubilizing various low-molecular-mass compounds, such as flower pigments [52,53] and for regulating the action of functional compounds such as plant hormones [54–56]. To date, most studies on the characteristics of GT enzymes have been derived from recombinant proteins produced in bacterial cells and characterized in vitro. However, very few published studies exist on the characterization of GTs in vivo. To investigate the function of the *RrGT2* gene in anthocyanin biosynthesis in vivo, *RrGT2* was first transferred into tobacco, which enhanced the flower coloration of the transgenic tobacco plants. In addition, after analyzing the tissue-specific expression of various transgenic tobacco lines, we found that the *RrGT2* gene was highly expressed not only in the flowers but also in the leaves. This phenomenon was consistent with the results of tissue-specific expression analysis in *Rosa*. Therefore, we speculated that the *RrGT2* gene may play an important role in the regulation of the growth and development of leaves, even whole plants, in addition to the anthocyanin biosynthesis pathway. However, whether in *R. rugosa* or in tobacco, the transcriptional regulatory mechanisms related to *RrGT2* expression patterns are still unclear, which requires further study in the future. Anyway, the results of the transgenic experiments proved that the exogenous *RrGT2* enzymes could also affect the synthesis of anthocyanins in different species. In other words, the function of *RrGT2* in anthocyanin biosynthesis and other aspects can be exchanged among plant species.

In this study, we used VIGS technology twice to explore the function of the *RrGT2* gene. Notably, we have applied VIGS technology to transgenic tobacco. Compared with the results of the transgenic experiments, the results of these VIGS experiments successfully verified the function of *RrGT2* in tobacco in the reverse direction. As far as we know, no such reports currently exist, so this study remains novel. In our experiments, we observed that the tobacco flowers were significantly pale in color when the *RrGT2* gene was successfully silenced. As determined via HPLC detection, this phenomenon was mostly due to decreased levels of Cy3G, which is synthesized from the anthocyanin pathway. In addition, we also observed that the phenotypes of the leaves of tobacco plants whose *RrGT2* gene was silenced were similar to those in response to photobleaching. This phenomenon was consistent with the above results of tissue-specific expression analysis of the *RrGT2* gene in *Rosa* and transgenic tobacco. This undoubtedly confirmed our previous inference about the function of the *RrGT2* gene in leaves.

4. Materials and Methods

4.1. Plant Materials

With respect to *Rosa*, *R. rugosa* 'Zizhi' and *R. davurica* plants cultivated in the *Rosa* germplasm nursery of Shandong Agricultural University were used as test materials. We collected petals at the budding stage, initial opening stage, half opening stage, full opening stage and wilting stage as well as seven different tissue samples (roots, stems, leaves, sepals, stamens, pistils and petals at the budding stage) in the mornings of sunny days from 20 April to 10 May 2017. After they were flash frozen in liquid nitrogen, all samples, which were collected in triplicate, were put into a −80 °C refrigerator for storage.

With respect to tobacco, wild-type plants were used as transgenic materials. After the tobacco seeds were disinfected by soaking in 70–75% ethanol for 2 min, rinsing with aseptic water once, soaking with 3.5% NaClO for 10–15 min and then rinsing with aseptic water 5 times, they were sown in Murashige and Skoog (MS) solid medium (without antibiotics). After 3 days of vernalization at 4 °C in darkness, the seeds were placed in a growth chamber (25 °C, 16 h/23 °C, 8 h day/night, 60% relative humidity) for approximately 30 days. The germless tobacco seedlings that grew well were selected as follow-up experimental materials.

4.2. Extraction of Total RNA and Synthesis of First-Strand cDNA

Total RNA was extracted via an EASY Spin Plant RNA Rapid Extraction Kit (Aidlab Biotech, Beijing, China) in accordance with the manufacturer's specifications. The integrity of the RNA was measured by gel electrophoresis with 1.0% nondenatured agarose, the purity and concentration of the RNA were detected by a Nanodrop 2000C ultra-microspectrophotometer (Thermo Fisher Scientific, Wilmington, DE, USA) and the qualified RNA was preserved at −80 °C. First-strand cDNA was synthesized via a 5× All-In-One RT MasterMix Reverse Transcription Kit (ABM Company, Vancouver, Canada) in accordance with both the manufacturer's protocol and the requirements of RT-PCR and qRT-PCR.

4.3. Cloning of the Full-Length CDS of RrGT2

We identified the *RrGT2* gene that contained the complete 5′ CDS from the *R. rugosa* transcriptome data in our laboratory. The cDNA 3′ terminal sequence of the target gene was then amplified by 3′-RACE technology. *RrGT2*-F and *RrGT2*-R primers (Table S1) were designed and amplified according to the full-length cDNA sequence of the *RrGT2* gene [31].

4.4. Tobacco Stable Transformation

The plasmids of *pCAMBIA1304* vectors and the *RrGT2* gene with restriction sites (*SpeI* and *BstEII*) were extracted and digested by two enzymes. The digestion products were then ligated with DNA ligase (Figure S2A) and transformed into *Agrobacterium tumefaciens*.

A. tumefaciens-mediated leaf disc transformation [57] was used to transform tobacco. First, the leaves of the cultured tobacco sterile seedlings were pruned to the appropriate size. *A. tumefaciens* infection was carried out after 2 days of preculture. Acetosyringone (AS) was added to the infective liquid and an empty *pCAMBIA1304* vector was used as a control. After infection, the plants were subjected to a dark treatment for 2 days, after which they were cultured in a growth chamber (25 °C, 16 h/23 °C, 8 h day/night, 60% relative humidity). Differentiation culture and rooting culture were carried out on MS media supplemented with relevant hormones and antibiotics. Hygromycin (Hyg) was used for screening resistant seedlings. The plantlets exhibiting good growth potential were transplanted into small flowerpots that contained substrate after seedling refining and the original growth environment was maintained.

4.5. VIGS in Rosa

On the basis of a modified TRV-GFP vector, a TRV-GFP-*RrGT2* recombinant viral vector was constructed. pTRV1 and pTRV2-GFP are two RNA strands of the TRV-GFP virus vector and the multiple cloning sites are mainly within pTRV2-GFP. For silencing *RrGT2* specifically in *R. rugosa*, a 406 bp fragment of the *RrGT2* gene was amplified and cloned into pTRV2-GFP (Figure S1A).

The pTRV1, pTRV2-GFP and pTRV2-GFP-*RrGT2* plasmids were transformed into *A. tumefaciens*, which was then cultured in YEB media that contained kanamycin, rifampicin and AS at 28 °C for 14–16 h until an OD_{600} = 1.5 was reached. Before infection, pTRV1 was added to the infection liquid that contained pTRV2-GFP and pTRV2-GFP-*RrGT2* in equal volume; the solution was subsequently mixed, forming a complete TRV-GFP and TRV-GFP-*RrGT2* virus carrier. The mixed bacterial solution was kept at room temperature in darkness for 4 h [58–63].

Perennial *Rosa* plants that grew naturally in the field were used as experimental materials and the experimental treatment time (from mid-March to mid-April) was approximately one month before *R. rugosa* flowering. And before setting an inoculation date, we will look into the weather for at least a week to avoid bad weather. In addition, we set the specific inoculation time between 14:00 and 16:00 in the afternoon, because during this time, the environment temperature is relatively high, which is more in line with the inoculation operation of VIGS. Another reason is that inoculation in the afternoon leads to a faster transition to night, making dark processing more real. Because it was difficult to inject the leaves and twigs with syringes and because vacuum infiltration could not be used in the field, we used the method that involved first scratching the leaves and twigs and then infecting them with *A. tumefaciens*. To improve the infection efficiency, 0.01% Silwet L-77 was added to the infection liquid and the plants were subjected to darkness for 24 h after infection for 10 min.

4.6. VIGS in Transgenic Tobacco

The final concentration of the virus infective fluid needed to reach OD_{600} = 1.0. In addition, the other aspects of the preparation of the virus infection solution and other preliminary preparation works were consistent with the above methods. First, we selected one of several *RrGT2* transgenic tobacco lines as the experimental object of VIGS. Tobacco infiltration was then performed as described by Liu et al. [62]. The *A. tumefaciens* cultures containing pTRV1 and pTRV2 or their derivatives (1:1, *v/v*) were injected into the lowest leaf of four-leaf stage plants by using a 1 mL needleless syringe. After infiltration, the tobacco plants were subjected to a dark treatment for 12 h, after which they were cultured in a growth chamber (25 °C, 16 h/23 °C, 8 h day/night, 60% relative humidity).

4.7. GFP Imaging

The detection and imaging of the visualized GFP in *Rosa* and transgenic tobacco plants after VIGS treatment were performed at night with a handheld high-intensity ultraviolet lamp (Model SB-100P/F; Spectronics Corporation, Westbury, NY, USA) and a Nikon D90 camera, respectively.

4.8. qRT-PCR Detection

We analyzed the gene expression by qRT-PCR on a Bio-Rad CFX96™ Real-Time PCR instrument (Bio-Rad, Inc., Philadelphia, PA, USA). The qRT-PCR mixture (total volume of 20 µL) contained 10 µL of SYBR® Premix Ex Taq™ (TaKaRa, Inc., Kusatsu, Japan), 8.2 µL of ddH$_2$O, 0.4 µL of each primer and 1 µL of cDNA. The PCR program consisted of an initial step of 95 °C for 30 s; 40 cycles of 95 °C for 5 s and 60 °C for 30 s; and then a dissociation stage of 95 °C for 10 s, 65 °C for 5 s and 95 °C for 5 s. Each gene was assessed via three biological replicates. The relative expression levels of the genes were calculated by the $2^{-\Delta\Delta Ct}$ method [64].

4.9. Total Anthocyanin Extraction and HPLC Analysis

All samples (0.1 g fresh weight) were homogenized in liquid nitrogen, after which they were extracted with 5 mL of an acidic methanol solution (70:0.1:29.9, $v/v/v$; CH$_3$OH:HCl:H$_2$O) at 4 °C in darkness for 24 h and then sonicated for 30 min [65]. After centrifugation, each extract was passed through a membrane filter (0.22 mm).

Qualitative and quantitative analyses of anthocyanins were performed via HPLC. The chromatographic analysis was conducted using a Prominence LC-20AT series HPLC system (Shimadzu, Inc., Kyoto, Japan) with a detection wavelength of 530 nm and the column (TC-C18 column, 5 µm, 4.6 mm × 250 mm) was maintained at 30 °C. The eluent consisted of an aqueous solution A (0.1% formic acid in water) and organic solvent B (acetonitrile). The gradient elution program was modified as described previously [66]: 0 min, 10% B; 15 min, 17% B; 20 min, 23% B; 25 min, 23% B; and 30 min, 10% B. Moreover, the eluent flow rate was 1.0 mL/min, with a 10 µL injection volume. Cy3G, Cy3G5G, Pg3G, Pg3G5G, Pn3G, Pn3G5G, Dp3G and Dp3G5G (EXTRASYNTHESE Trading Company, Lille-Lezennes, France) were used as references for anthocyanin analysis. Three independent biological replicates were measured for each sample.

4.10. Statistical Analyses

Three independent biological replicates were measured for each sample and the data presented as the mean ± standard error (SE). Where applicable, data were analyzed by Student's *t* test in a two-tailed analysis. Values of $p < 0.05$ or <0.01 were considered to be statistically significant.

5. Conclusions

In conclusion, the *RrGT2* gene from *R. rugosa* was successfully cloned and characterized. Our results demonstrated that *RrGT2* has all the conserved amino acid residues that are typical of the GT enzyme. Transcript analysis revealed that *RrGT2* was expressed in specific tissues and was developmentally regulated, suggesting that *RrGT2* might act as a modified enzyme in the anthocyanin biosynthesis pathway. The functional verification of *RrGT2* in *Rosa* via VIGS revealed that *RrGT2* is a key structural gene that directly affects the formation of anthocyanins in *R. rugosa*. By overexpressing *RrGT2* in tobacco, we found an increase in anthocyanins in flowers, indicating that *RrGT2* encodes a functional GT protein for anthocyanin glucosylation and could function in other species. Furthermore, VIGS of the *RrGT2* gene in transgenic tobacco resulted in a decrease in the total content of anthocyanins that accumulated in the flowers, which further confirmed that *RrGT2* is involved in the modification of flower color.

Supplementary Materials: Supplementary materials can be found at http://www.mdpi.com/1422-0067/19/12/4057/s1. Table S1. Primers used in the present study. Table S2. Anthocyanin contents in the flowers of *R. rugosa* 'Zizhi' and *R. davurica* subjected to different VIGS treatments (µg·g^{-1} FW). Table S3. Anthocyanin contents in the flowers of transgenic tobacco (µg·g^{-1} FW). Table S4. Anthocyanin contents in the flowers of transgenic tobacco subjected to different VIGS treatments (µg·g^{-1} FW). Table S5. Silencing efficiency of VIGS in *R. rugosa* 'Zizhi' and *R. davurica*. Figure S1. Construction and validation of the recombinant virus vector TRV-GFP-*RrGT2*. Figure S2. Construction and validation of the recombinant expression vector pCAMBIA1304-*RrGT2*.

Author Contributions: Conceptualization, X.S. and L.Z.; Data curation, X.S., M.Z. and X.H.; Formal analysis, M.Z. and X.H.; Funding acquisition, Z.X. and L.Z.; Investigation, X.S.; Methodology, X.S. and Z.X.; Project administration, Z.X., L.Z. and X.H.; Resources, L.Z.; Software, X.S. and M.Z.; Supervision, Z.X., L.Z. and X.H.; Validation, X.S., M.Z. and X.H.; Visualization, X.S. and M.Z.; Writing—original draft, X.S.; Writing—review & editing, Z.X., L.Z. and X.H.

Funding: This research received no external funding.

Acknowledgments: This project was supported by the Agricultural Seed Project of Shandong Province ([2014] No. 96) and National Science Foundation of China (NSFC) (31700622).

Conflicts of Interest: The authors declare no conflict of interest.

Abbreviations

cDNA	complementary DNA
RACE	rapid amplification of cDNA ends
CDS	coding DNA sequence
PSPG	plant secondary product glycosyltransferases
PCR	polymerase chain reaction
qRT-PCR	quantitative real-time polymerase chain reaction
VIGS	virus-induced gene silencing
TRV	tobacco rattle virus
GFP	green fluorescent protein
TRV-GFP	a modified TRV virus with GFP gene inserted into TRV virus
HPLC	high-performance liquid chromatography

References

1. Feng, L.G.; Shao, D.W.; Sheng, L.X.; Zhao, L.Y.; Yu, X.Y. Study on investigation and morphological variation of wild *Rosa rugosa* in China. *J. Shandong Agric. Univ. (Nat. Sci. Ed.)* **2009**, *40*, 484–488.
2. Yang, M.; Zhao, L.Y. Research and classification on the germplasm resources of the Pingyin *Rosa rugosa* in Shandong province. *Landsc. Archit. J. Chin.* **2003**, *7*, 61–63.
3. Yu, S.C.; Feng, Z.; Zhao, L.Y. Research on quantitative taxonomy of cultivars in Pingyin rose. *Acta Hortic. Sin.* **2005**, *32*, 327–330.
4. Feng, L.G.; Sheng, L.X.; Tao, J.; Zhao, L.Y.; Shao, D.W. Comparative studies on aroma components and contents of wild *Rosa rugosa* in China. *J. Yangzhou Univ. (Agric. Life Sci. Ed.)* **2009**, *30*, 90–94.
5. Chen, Z.J.; Zang, F.S.; Dai, Y.Q.; Yuan, X.Y.; Li, S.; Wu, Y.M.; Ni, Y.Y. Nutrient components analysis of rose fruit. *Food Res. Dev.* **2012**, *33*, 194–198.
6. Wang, Y. Natural hybridization and speciation. *Biodivers. Sci.* **2017**, *25*, 565–576. [CrossRef]
7. Henderson, I.R.; Salt, D.E. Natural genetic variation and hybridization in plants. *J. Clin. Exp. Dent.* **2017**, *9*, e1212–e1217. [CrossRef] [PubMed]
8. Chen, S.M.; Zhu, X.R.; Chen, F.D.; Luo, H.L.; Lv, G.S.; Fang, W.M.; Zhang, Z. Expression characteristics of anthocyanin structural genes in different flower color chrysanthemum cultivars. *J. Northwest Plants* **2010**, *30*, 453–458.
9. Grotewold, E. The genetics and biochemistry of floral pigments. *Annu. Rev. Plant Biol.* **2006**, *57*, 761–780. [CrossRef] [PubMed]
10. Koes, R.; Verweij, W.; Quattrocchio, F. Flavonoids: A colorful model for the regulation and evolution of biochemical pathways. *Trends Plant Sci.* **2005**, *10*, 236–242. [CrossRef] [PubMed]
11. Quattrocchio, F.; Wing, J.; van der Woude, K.; Souer, E.; de Vetten, N.; Mol, J.; Koes, R. Molecular analysis of the anthocyanin 2 gene of petunia and its role in the evolution of flower color. *Plant Cell* **1999**, *11*, 1433–1444. [CrossRef] [PubMed]
12. Paz-Ares, J.; Ghosal, D.; Wienand, U.; Peterson, P.; Saedler, H. The regulatory *c1* locus of *Zea mays* encodes a protein with homology to *myb* proto-oncogene products and with structural similarities to transcriptional activators. *EMBO J.* **1987**, *6*, 3553–3558. [CrossRef] [PubMed]
13. Winkelshirley, B. Flavonoid Biosynthesis. A Colorful Model for Genetics, Biochemistry, Cell Biology and Biotechnology. *Plant Physiol.* **2001**, *126*, 485–493. [CrossRef]

14. Lu, Y.F.; Zhang, M.L.; Meng, X.N.; Wan, H.H.; Zhang, J.; Tian, J.; Hao, S.X.; Jin, K.N.; Yao, Y.C. Photoperiod and shading regulate coloration and anthocyanin accumulation in the leaves of Malus crabapples. *Plant Cell Tissue Organ Cult.* **2015**, *121*, 619–632. [CrossRef]

15. Holton, T.A.; Cornish, E.C. Genetics and biochemistry of anthocyanin biosynthesis. *Plant Cell* **1995**, *7*, 1071. [CrossRef] [PubMed]

16. Kim, S.H.; Lee, J.R.; Hong, S.T.; Yoo, Y.K.; An, G.; Kim, S.R. Molecular cloning and analysis of anthocyanin biosynthesis genes preferentially expressed in apple skin. *Plant Sci.* **2003**, *165*, 403–413. [CrossRef]

17. Charles, B.; Imin, N.; Djordjevic, M.A. Flavonoids: New roles for old molecules. *J. Integr. Plant Biol.* **2010**, *52*, 98–111.

18. Martens, S.; Preuß, A.; Matern, U. Multifunctional flavonoid dioxygenases: Flavonol and anthocyanin biosynthesis in *Arabidopsis thaliana* L. *Phytochemistry* **2010**, *71*, 1040–1049. [CrossRef] [PubMed]

19. Zhang, Q.; Hao, R.J.; Xu, Z.D.; Yang, W.R.; Wang, J.; Cheng, T.R.; Pan, H.T.; Zhang, Q.X. Isolation and functional characterization of a R2R3-MYB regulator of *Prunus mume* anthocyanin biosynthetic pathway. *Plant Cell Tissue Organ Cult.* **2017**, *131*, 417–429. [CrossRef]

20. Ge, C.L.; Huang, C.H.; Xu, X.B. Research on anthocyanins biosynthesis in fruit. *Acta Hortic. Sin.* **2012**, *39*, 1655–1664.

21. Vogt, T.; Jones, P. Glycosyltransferases in plant natural product synthesis: Characterization of a supergene family. *Trends Plant Sci.* **2000**, *5*, 380–386. [CrossRef]

22. Lim, E.K.; Bowles, D.J. A class of plant glycosyltransferases involved in cellular homeostasis. *EMBO J.* **2004**, *23*, 2915–2922. [CrossRef] [PubMed]

23. Brazier-Hicks, M.; Edwards, R. Functional importance of the family 1 glucosyltransferase UGT72B1 in the metabolism of xenobiotics in *Arabidopsis thaliana*. *Plant J.* **2005**, *42*, 556–566. [CrossRef] [PubMed]

24. Fedoroff, N.V.; Furtek, D.B.; Nelson, O.E. Cloning of the bronze locus in maize by a simple and generalizable procedure using the transposable controlling element Activator (Ac). *Proc. Natl. Acad. Sci. USA* **1984**, *81*, 3825–3829. [CrossRef] [PubMed]

25. Kobayashi, S.; Ishimaru, M.; Ding, C.K.; Yakushiji, H.; Goto, N. Comparison of UDP-glucose: Flavonoid 3-O-glucosyltransferase (UFGT) gene sequences between white grapes (*Vitis vinifera*) and theirsports with red skin. *J. Plant Sci.* **2001**, *160*, 543–550. [CrossRef]

26. Li, X.G.; Yu, Z.Y. Research progress of anthocyanin. *J. North. Hortic.* **2003**, *4*, 6–8.

27. Afifi, M.; El-Kereamy, A.; Legrand, V.; Chervin, C.; Monje, M.-C.; Nepveu, F.; Roustan, J.-P. Control of anthocyanin biosynthesis pathway gene expression by eutypine, atoxin from Eutypa lata in grape cell tiussue cultures. *J. Plant Physiol.* **2003**, *160*, 971–975. [CrossRef] [PubMed]

28. Boss, P.K.; Davies, C.; Robinson, S.P. Anthocyanin composition and anthocyanin pathway gene expression in grapewine sports differing in berry skin colour. *Aust. J. Grape Wine Res.* **1996**, *2*, 163–170. [CrossRef]

29. Nakatsuka, T.; Nishihara, M.; Mishiba, K.; Yamamura, S. Two different mutations are involved in the formation of white-flowered gentian plants. *J. Plant Sci.* **2005**, *169*, 949–958. [CrossRef]

30. Ju, Z.G. Activities of chalcone and UDP Gal: Flavonoid-3-O-glycosyhransferase in relation to anthocyanin synthesis in apple. *J. Sci. Hortic.* **1995**, *63*, 175–185. [CrossRef]

31. Sui, X.M.; Wang, Y.; Zhao, M.Y.; Han, X.; Zhao, L.Y.; Xu, Z.D. Cloning and Expression Analysis of *RrGT2* Gene Related to Anthocyanin Biosynthesis in *Rosa rugosa*. *Am. J. Plant Sci.* **2018**, *9*, 2008–2019. [CrossRef]

32. Qi, Y. Cloning and Expression Analysis of Anthocyanin Biosynthesis Related Gene in *Rosa rugosa*. Master's Thesis, Shandong Agricultural University, Tai'an, China, 2016.

33. Aharoni, A.; De Vos, C.R.; Wein, M.; Sun, Z.; Greco, R.; Kroon, A.; Mol, J.N.; O'Connell, A.P. The strawberry FaMYB1 transcription factor suppresses anthocyanin and flavonol accumulation in transgenic tobacco. *Plant J.* **2001**, *28*, 319–332. [CrossRef] [PubMed]

34. Deluc, L.; Barrieu, F.; Marchive, C.; Lauvergeat, V.; Decendit, A.; Richard, T.; Carde, J.P.; Mérillon, J.M.; Hamdi, S. Characterization of a grapevine R2R3-MYB transcription factor that regulates the phenylpropanoid pathway. *Plant Physiol.* **2006**, *140*, 499–511. [CrossRef] [PubMed]

35. Nakatsuka, T.; Sato, K.; Takahashi, H.; Yamamura, S.; Nishihara, M. Cloning and characterization of the UDP-glucose: Anthocyanin 5-O-glucosyltransferase gene from blue-flowered gentian. *J. Exp. Bot.* **2008**, *59*, 1241–1252. [CrossRef] [PubMed]

36. Wang, C.X.; Yang, M.Z.; Pan, N.S.; Chen, Z.L. Resistance to potato virus X infection in transgenic tobacco plants with coat protein gene of virus. *J. Integr. Plant Biol.* **1993**, *35*, 819–824.

37. Zhang, X.Y.; Xue, Q.Z. Introduction of soybean glycinin gene into rice (*Oryza sativa* L.) with Agrobacterium-mediated transformation. *J. Zhejiang Agric. Univ. (Agric. Life Sci.)* **2001**, *27*, 495–499.

38. Kubo, A.; Arai, Y.; Nagashima, S.; Yoshikawa, T. Alteration of sugar donor specificities of plant glycosyltransferases by a single point mutation. *J. Arch. Biochem. Biophys.* **2004**, *429*, 198–203. [CrossRef] [PubMed]

39. Broothaerts, W.; Mitchell, H.J.; Weir, B.; Kaines, S.; Smith, L.M.; Yang, W.; Mayer, J.; Roa-Rodríguez, C.; Jefferson, R. Genetransfer to plants by diverse species of bacteria. *J. Nat.* **2005**, *433*, 629–633. [CrossRef] [PubMed]

40. Herrera-Estrella, L.; Simpson, J.; Martinez-Trujillo, M. Transgenic plants: An historical perspective. *J. Methods Mol. Biol.* **2005**, *286*, 3–32.

41. Wang, X.Q. Structure, mechanism and engineering of plant natural product glycosyltransferases. *J. FEBS Lett.* **2009**, *583*, 3303–3309. [CrossRef] [PubMed]

42. Yonekura-Sakakibara, K.; Hanada, K. An evolutionary view of functional diversity in family 1 glycosyltransferases. *J. Plant J.* **2011**, *66*, 182–193. [CrossRef] [PubMed]

43. Shao, H.; He, X.; Achnine, L.; Blount, J.W.; Dixon, R.A.; Wang, X. Crystal structures of a multifunctional triterpene/flavonoid glycosyltransferase from *Medicago truncatula*. *J. Plant Cell* **2005**, *17*, 3141–3154. [CrossRef] [PubMed]

44. Li, L.; Modolo, L.V.; Escamilla-Trevino, L.L.; Achnine, L.; Dixon, R.A.; Wang, X. Crystal structure of Medicago truncatula UGT85H2-insights into the trusctural basis of a multifunctional (iso) flavonoid glycosyltransferase. *J. Mol. Biol.* **2007**, *370*, 951–963. [CrossRef] [PubMed]

45. Modolo, L.V.; Li, L.N.; Pan, H.Y.; Blount, J.W.; Dixon, R.A.; Wang, X.Q. Crystal structures of glycosyltransferase UGT/8G1 reveal the molecular basis for glycosylation and deglycosylation of (iso) flavonoids. *J. Mol. Biol.* **2009**, *392*, 1292–1302. [CrossRef] [PubMed]

46. Cheng, J.; Wei, G.; Zhou, H.; Gu, C.; Vimolmangkang, S.; Liao, L.; Han, Y. Unraveling the mechanism underlying the glycosylation and methylation of anthocyanins in peach. *Plant Physiol.* **2014**, *166*, 1044–1058. [CrossRef] [PubMed]

47. Sun, D.; Nandety, R.S.; Zhang, Y.; Reid, M.S.; Niu, L.; Jiang, C.Z. A petunia ethylene-responsive element binding factor, PhERF2, plays an important role in antiviral RNA silencing. *J. Exp. Bot.* **2016**, *67*, 3353–3365. [CrossRef] [PubMed]

48. Spitzer, B.; Zvi, M.M.B.; Ovadis, M.; Marhevka, E.; Barkai, O.; Edelbaum, O.; Marton, I.; Masci, T.; Alon, M.; Morin, S.; et al. Reverse Genetics of floral scent: Application of Tobacco rattle virus-based gene silencing in Petunia. *Plant Physiol.* **2007**, *145*, 1241–1250. [CrossRef] [PubMed]

49. Dong, Y.Y.; Burch-Smith, T.M.; Liu, Y.L.; Mamillapalli, P.; Dinesh-Kumar, S.P. A Ligation-independent cloning Tobacco rattle virus vector for high-throughput virus-induced gene silencing identifies roles for NbMADS4-1 and -2 in floral development. *Plant Physiol.* **2007**, *145*, 1161–1170. [CrossRef] [PubMed]

50. Deng, X.B.; Elomaa, P.; Nguyen, C.X.; Hytönen, T.; Valkonen, J.P.T.; Teeri, T.H. Virus-induced gene silencing for *Asteraceae*-a reverse genetics approach for functional genomics in *Gerbera hybrida*. *Plant Biotechnol. J.* **2012**, *10*, 970–978. [CrossRef] [PubMed]

51. Zhang, L.; Xu, Z.D.; Tang, T.F.; Zhang, H.; Zhao, L.Y. Analysis of anthocyanin related compounds and metabolic pathways in the flowering process of *Rosa rugosa* 'Zizhi'. *Sci. Agric. Sin.* **2015**, *48*, 235–236.

52. Brugliera, F.; Holton, T.A.; Stevenson, T.W.; Farcy, E.; Lu, C.Y.; Cornish, E.C. Isolation and characterization of a cDNA clone corresponding to the Rt locus of *Petunia hybrida*. *Plant J.* **1994**, *5*, 81–92. [CrossRef] [PubMed]

53. Kroon, J.; Souer, E.; de Graaff, A.; Xue, Y.; Mol, J.; Koes, R. Cloning and structural analysis of the anthocyanin pigmentation locus Rt of Petunia hybrida: Characterization of insertion sequences in two mutant alleles. *Plant J.* **1994**, *5*, 69–80. [CrossRef] [PubMed]

54. Szerszen, J.B.; Szczyglowski, K.; Bandurski, R.S. A gene from Zea mays involved in conjugation of growth hormone indole-3-acetic acid. *Science* **1994**, *265*, 1699–1701. [CrossRef] [PubMed]

55. Martin, R.C.; Mok, M.C.; Mok, D.W.S. Isolation of a cytokinin gene, ZOG1, encoding zeatin *O*-glucosyltransferase from *Phaseolus lunatus*. *Proc. Natl. Acad. Sci. USA* **1999**, *96*, 284–289. [CrossRef] [PubMed]

56. Martin, R.C.; Mok, M.C.; Habben, J.E.; Mok, D.W.S. A maize cytokinin gene encoding an *O*-glucosyltransferase specific to cis-zeatin. *Proc. Natl. Acad. Sci. USA* **2001**, *98*, 5922–5926. [CrossRef] [PubMed]

57. Horsch, R.; Fry, J.E.; Hoffmann, N.L.; Eichholtz, D.Z.; Rogers, S.G.; Fraley, R.T. A simple and general method for transferring genes into plants. *Science* **1985**, *227*, 1229–1232.

58. Bachan, S.; Dinesh-Kumar, S.P. Tobacco rattle virus (TRV) based virus-induced gene silencing. *Antivir. Resist. Plants* **2012**, *894*, 83–92.

59. Burch-Smith, T.M.; Schiff, M.; Liu, Y.; Dinesh-Kumar, S.P. Efficient virus-induced gene silencing in Arabidopsis. *Plant Physiol.* **2006**, *142*, 21–27. [CrossRef] [PubMed]

60. Chen, J.C.; Jiang, C.Z.; Gookin, T.E.; Hunter, D.A.; Clark, D.G.; Reid, M.S. Chalcone synthase as a reporter in virus-induced gene silencing studies of flower senescence. *Plant Mol. Biol.* **2004**, *55*, 521–530. [CrossRef] [PubMed]

61. Jiang, C.Z.; Chen, J.C.; Reid, M. Virus-induced gene silencing in ornamental plants. *Methods Mol. Biol.* **2011**, *744*, 81–96. [PubMed]

62. Liu, Y.; Schiff, M.; Dinesh-Kumar, S.P. Virus-induced gene silencing in tomato. *Plant J.* **2002**, *31*, 777–786. [CrossRef] [PubMed]

63. Quadrana, L.; Rodriguez, M.C.; López, M.; Bermúdez, L.; Nunes-Nesi, A.; Fernie, A.R.; Descalzo, A.; Asis, R.; Rossi, M.; Asurmendi, S.; et al. Coupling virus-induced gene silencing to exogenous green fluorescence protein expression provides a highly efficient system for functional genomics in Arabidopsis and across all stages of tomato fruit development. *Plant Physiol.* **2011**, *156*, 1278–1291. [CrossRef] [PubMed]

64. Schmittgen, T.D.; Livak, K.J. Analyzing real-time PCR data by the comparative C (T) method. *Nat. Protoc.* **2008**, *3*, 1101. [CrossRef] [PubMed]

65. Yang, Q.; Yuan, T.; Sun, X.B. Preliminary studies on the changes of flower color during the flowering period in two tree peony cultivars. *Acta Hortic. Sin.* **2015**, *42*, 930–938.

66. Xie, D.Y.; Sharma, S.B.; Wright, E.; Wang, Z.Y.; Dixon, R.A. Metabolic engineering of proanthocyanidins through co-expression of anthocyanidin reductase and the PAP1 MYB transcription factor. *Plant J.* **2006**, *45*, 895–907. [CrossRef] [PubMed]

International Journal of
Molecular Sciences

Article

Constitutive Expression of *Aechmea fasciata SPL14* (*AfSPL14*) Accelerates Flowering and Changes the Plant Architecture in *Arabidopsis*

Ming Lei [1,2,3,4], Zhi-ying Li [1,2,3,4], Jia-bin Wang [1,2,3,4], Yun-liu Fu [1,2,3,4], Meng-fei Ao [1,2,3,4] and Li Xu [1,2,3,4,*]

[1] Institute of Tropical Crops Genetic Resources, Chinese Academy of Tropical Agricultural Sciences, Danzhou 571737, China; leiming_catas@126.com (M.L.); xllizhiying@vip.163.com (Z.-y.L.); jiabinwangfuhu@sina.com (J.-b.W.); fyljj_2007@126.com (Y.-l.F.); 17889981612@163.com (M.-f.A.)
[2] Key Laboratory of Crop Gene Resources and Germplasm Enhancement in Southern China, Danzhou 571737, China
[3] Key Laboratory of Tropical Crops Germplasm Resources Genetic Improvement and Innovation, Danzhou 571737, China
[4] Mid Tropical Crop Gene Bank of National Crop Resources, Danzhou 571737, China
* Correspondence: xllzy@263.net; Tel.: +86-898-2330-0284

Received: 21 June 2018; Accepted: 14 July 2018; Published: 18 July 2018

Abstract: Variations in flowering time and plant architecture have a crucial impact on crop biomass and yield, as well as the aesthetic value of ornamental plants. *Aechmea fasciata*, a member of the Bromeliaceae family, is a bromeliad variety that is commonly cultivated worldwide. Here, we report the characterization of *AfSPL14*, a squamosa promoter binding protein-like gene in *A. fasciata*. *AfSPL14* was predominantly expressed in the young vegetative organs of adult plants. The expression of *AfSPL14* could be upregulated within 1 h by exogenous ethephon treatment. The constitutive expression of *AfSPL14* in *Arabidopsis thaliana* caused early flowering and variations in plant architecture, including smaller rosette leaves and thicker and increased numbers of main inflorescences. Our findings suggest that AfSPL14 may help facilitate the molecular breeding of *A. fasciata*, other ornamental and edible bromeliads (e.g., pineapple), and even cereal crops.

Keywords: *Aechmea fasciata*; squamosa promoter binding protein-like; flowering time; plant architecture; bromeliad

1. Introduction

The squamosa promoter binding protein (SBP)-like (SPL) proteins are plant-specific transcription factors (TFs) that play essential roles in the regulation networks of plant growth and development [1]. The genes encoding SPL proteins were first identified in snapdragon (*Antirrhinum majus*), and were then found in almost all other green plants [2–5]. All SPL proteins contain a highly conserved DNA-binding domain termed the SBP domain, which consists of approximately 76 amino acid residues and features two zinc-binding sites and a bipartite nuclear localization signal (NLS) [6]. Many studies of various species have revealed the diverse functions of SPLs, which are involved in a broad range of important biological processes including the leaf development [7–10], embryonic development [11], fertility controlling [12,13], copper homeostasis [14,15], as well as the biosynthesis of phenylpropanoids and sesquiterpene. In addition to affecting these developmental aspects, several SPL factors which can be regulated by miR156, an evolutionary highly conserved microRNA (miRNA), also play crucial roles in the control of flowering time. The overexpression of *AtSPL3*, *AtSPL4*, *AtSPL5*, *AtSPL9*, *AtSPL15*, and *OsSPL16* can significantly promote flowering [3,9,16–19]. AtSPL9, together with AtSPL3 and the AtSPL2/10/11 group promote the floral meristem identity by directly regulating the same or different target genes [9,19]. Interestingly, compared with the positive regulation of accelerated flowering by the SPLs described above, AtSPL14 appears to be a negative regulator of vegetative-phase changes and floral transitions [20]. Another function of miR156-regulated SPL factors is in plant architecture formation and yield. *Teosinte Glume Architecture (TGA1)*, an *SPL* gene, is responsible for the liberation of the kernel during domestication and evolution in *Z. mays* [21]. In *Triticum aestivum*, TaSPL3/17 play important roles in reducing the number of tillers and the outgrowth rate of axillary buds [22]. Two SPL homologs, TaSPL20 and TaSPL21, together reduce plant height and increase the thousand-grain weight [23]. In switchgrass (*Panicum virgatum*), miR156-regulated SPL4 suppresses the formation of both aerial and basal buds and controls the shoot architecture [24]. In *O. sativa*, higher expression of *OsSPL14* can reduce the tiller number, increase the lodging resistance, promote panicle branching and enhance grain yields [25,26]. The multifaceted functions of SPLs demonstrate complex and interesting regulation networks underlying plant lifestyles.

Bromeliaceae is one of the most morphologically diverse families and is widely distributed in tropical and subtropical areas [27]. Although certain cultivated species of bromeliads are appreciated for their edible fruits (e.g., pineapple: *Ananas comosus*) or medicinal properties (e.g., *Bromelia antiacantha*), the vast majority are appreciated for their ornamental value [28]. However, the unsynchronized natural flowering time of cultivated bromeliads always results in increased cultivation and harvesting costs and decreased economic value of fruits and ornamental flowers [29]. To date, several efforts were made to uncover the mechanism of flowering of bromeliads induced by age, photoperiod, autonomous and exogenous ethylene, or ethephon [29–33], but the precise molecular mechanism remained unknown.

Here, we characterized the *SPL* gene *AfSPL14*, from *Aechmea fasciata*, a popular ornamental flowering bromeliad. Phylogenetic analyses showed that AfSPL14 is closely related to OsSPL17, OsSPL14, AtSPL9, and AtSPL15. Furthermore, the expression of *AfSPL14* transcripts responded to plant age and exogenous ethephon treatment. The constitutive expression of *AfSPL14* in *Arabidopsis* promotes branching and accelerates flowering under long-day (LD) conditions. These results suggested that AfSPL14, a TF of the SPL family, might be involved in the process of flowering and in plant architecture variations of *A. fasciata*.

2. Results

2.1. Isolation and Sequence Analysis of AfSPL14 in A. fasciata

The *SPL* cDNA was isolated using the rapid amplification of cDNA ends (RACE) technique, and then named *AfSPL14*. The cDNA of *AfSPL14* was 1504-bp long and presented a 123-bp 5′ untranslated region (UTR), a 331-bp 3′ UTR, and a 1050-bp open reading frame (ORF), which was predicted to encode a 349-amino acid protein with a molecular weight (MW) and an isoelectric point (pI) of 37.53 kDa and 9.08, respectively.

To investigate the evolutionary relationships between the AfSPL14 and SPL proteins of other species, a phylogenetic tree was constructed using the neighbor-joining method with 1000 bootstrap replicates with 13 SBPs of *Physcomitrella patens*, 16 SPLs of *Arabidopsis*, and 19 SPLs of *O. sativa* (Table S1). Because the alignment of the full-length protein sequences showed no consensus sequences except for SBP domains (data not shown), only the highly-conserved SBP domains were used for the phylogenetic analysis. The unrooted phylogenetic tree classified all SBP domains into seven groups (I-VII), and AfSPL14 was clustered into group III, with AtSPL9, AtSPL15, OsSPL7, OsSPL14, and OsSPL17; however, this group did not contain SBP domains of *P. patens* (Figure 1), which was similar to the results obtained by others [34]. The fact that AfSPL14 has been classified with three SPL genes in *O. sativa* and two in *Arabidopsis* suggests that the SBP domains of these six SPLs might have undergone species-specific evolutionary processes after speciation.

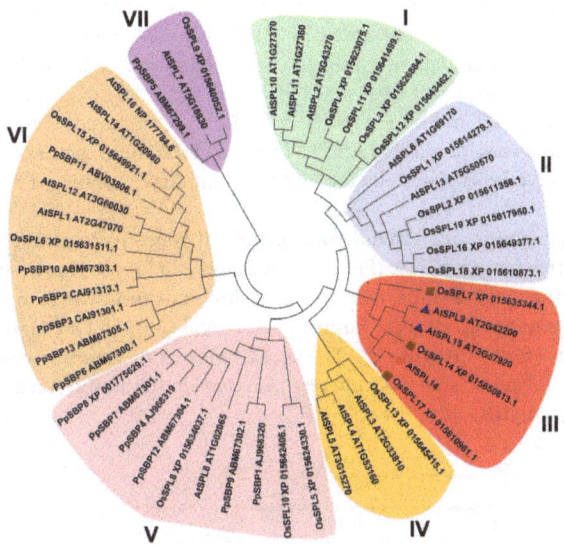

Figure 1. Phylogenetic analysis of AfSPL14 and SPLs of *Arabidopsis*, *O. sativa* and *P. patens* based on the conserved SBP domains. The unrooted tree was created using the neighbor-joining method with 1000 bootstrap replicates with 13 SBPs of *P. patens*, 16 SPLs of *Arabidopsis*, and 19 SPLs of *O. sativa*. The sequences of all these SBP domains are listed in Table S1.

The multiple sequence alignment of the AfSPL14 protein with SPL homologs of other species indicated that the SBP domain was highly conserved among the species (Figure 2a). All SBP domains could be divided into four motifs, which were zinc finger-like structure 1 (Zn1), Zn2, joint peptide (Jp) of Zn1 and Zn2, and NLS. The first zinc finger was C4H, and the second zinc finger-like structure was C2HC. The Jp plays a crucial role in modifying the protein-DNA interaction process [6], and was also highly conserved in all aligned sequences (Figure 2a). In addition, the bipartite NLS motif, which partially overlapped Zn2, was highly conserved (Figure 2a,b).

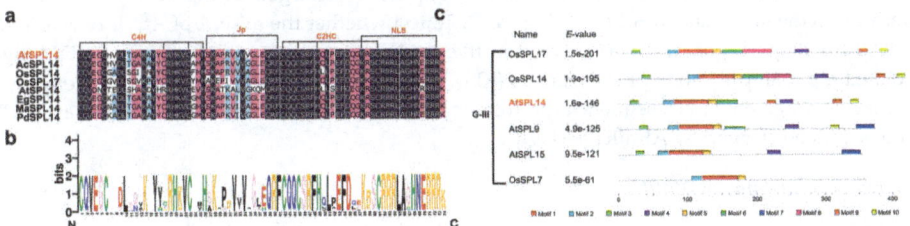

Figure 2. Sequence alignment and logo view of variable SBP domains, and putative motifs of variable SPLs in group III. (**a**) Multiple alignment of the SBP domains using DNAMAN software. The four conserved motifs, which include two zinc finger-like structures (C4H, C2HC), Jp and NLS, are indicated; (**b**) Sequence logo view of the consensus SBP domains. The overall height of the stack and the height of each letter represent the sequence conservation at that position and the relative frequency of the corresponding amino acid at that position, respectively; (**c**) putative motifs of variable SPLs in group III identified by MEME software online (http://meme-suite.org/tools/meme). The color boxes represent different putative motifs for which the sequences are listed in Table S2 online. G-III indicates group III from Figure 1.

To further examine conserved sequences other than the SBP domain, the online Multiple EM for Motif Elicitation (MEME) tool was used to identify putative motifs in the SPL proteins in group III [35]. As shown in Figure 2c, all members of group III contained motifs 1, 2, and 5, and these motifs had similar distributions. Actually, motif 2 belongs to C4H, and motif 5 belongs to NLS; motif 1 contains Jp, C2HC, and partials of C4H and NLS (Figure 2c, Table S2). Compared with OsSPL14 and OsSPL17, which contained all motifs except motif 7, AfSPL14 had a similar motif distribution but lacked motifs 7 and 8 (Figure 2c). These results suggested that AfSPL14 might have conserved functions with OsSPL14 and OsSPL17.

Exon-intron organization of all members of group III genes were generated based on genome sequences and the corresponding CDSs (Figure 3). As shown in Figure 3, each member of these genes had two introns and three exons; thus, they shared a similar exon-intron composition. All members of rice and *Arabidopsis* in group III were targets of miR156 [9,36], and a putative miR156 target site was also observed in *AfSPL14* (Figure 3). The consistency of the motif investigation, exon-intron organization, and phylogenetic analysis indicates putative similarities in functional regions and sites among the genes in group III.

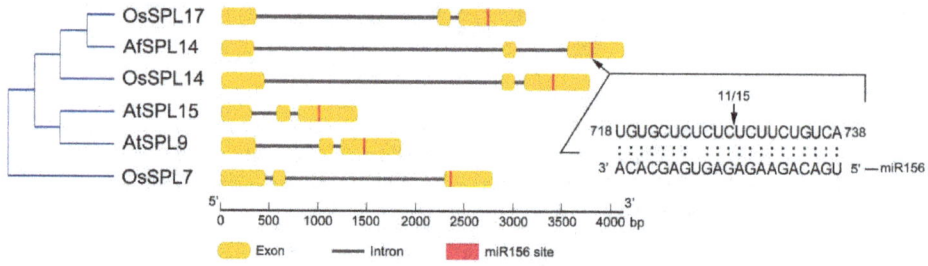

Figure 3. Exon-intron structures of *SPL* genes in group III from Figure 2 and AfmiR156 cleavage site in *AfSPL14* determined by 5′ RLM-RACE. For the determination of the AfmiR156 cleavage site in *AfSPL14*, 15 clones were selected randomly for sequencing, and 11 of them were cleaved in the position indicated by the arrow towards the base interval of *AfSPL14* RNA sequence.

2.2. AfSPL14 Was a Target of miR156 of A. fasciata (AfmiR156)

As all members of rice and *Arabidopsis* in group III were targets of miR156, there was also a putative miR156 target site in *AfSPL14* (Figure 3). To test whether the mRNA of *AfSPL14* was indeed targeted for degradation and was cleaved at the predicted position by AfmiR156, 5′ RNA ligase mediated rapid amplification of cDNA ends (RLM-RACE) was carried out to map the 5′ terminus of the cleavage fragment. DNA sequencing results of the amplified product demonstrated that *AfSPL14* could be indeed cleaved by AfmiR156 (Figure 3).

2.3. Transcript Profiling of AfSPL14 in A. fasciata

To gain insights into the role of AfSPL14 in *A. fasciata*, we determined the gene's expression profiles in various organs at different developmental stages via reverse transcription followed by quantitative real-time PCR (RT-qPCR). The transcripts of *AfSPL14* could be detected in almost all tested tissues except the roots of the adult plant prior to flower bud differentiation (Figure 4a). *AfSPL14* mRNA was more abundant in the central leaves and stems regardless of the developmental stage (Figure 4a,b). The accumulation of *AfSPL14* transcripts in the central leaves and stems showed significant changes during development, with the highest level observed in adult plants prior to flower bud differentiation, a relatively lower level observed in juvenile plants, and the lowest level observed in 39-day-after-flowering (DAF) adult plants (Figure 4a,b), suggesting that AfSPL14 might be involved in phase transitions.

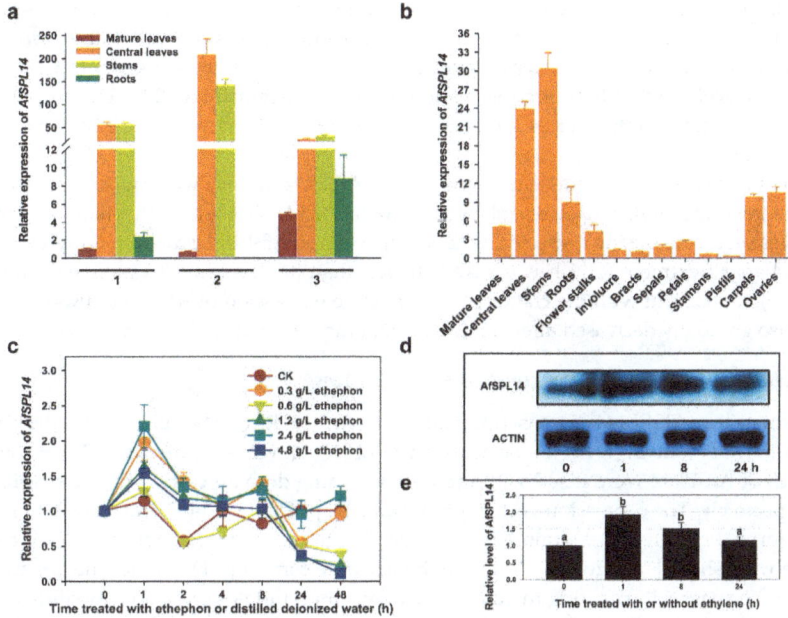

Figure 4. Expression of *AfSPL14* transcripts in various tissues of *A. fasciata* and immunoblot analysis of AfSPL14 in central leaves treated with or without ethephon. (**a**) Expression level of *AfSPL14* transcripts in various tissues of juvenile and adult plants. (1) juvenile plants; (2) adult plants prior to flower bud differentiation; (3) 39-DAF flowering adult plants. Samples were collected at 10:00 am. (**b**) Expression level of *AfSPL14* transcripts in the vegetative and reproductive organs of 39-DAF flowering adult plants. Samples were collected at 10:00 am. (**c**) Expression level of *AfSPL14* transcripts in the central leaves of *A. fasciata* in response to exogenous ethephon treatment at different concentrations for different time. In the panels, 0, 1, 2, 4, and 8 h represents the samples collected at 10:00, 11:00, 12:00, 14:00, and 18:00, respectively; 24 h and 48 h represent the treated samples collected at 10:00 am at the next day and the next two days, respectively. For CK, 10 mL of distilled deionized H_2O was poured into the cylinder shapes of *A. fasciata*. 0 h represents the samples treated without ethephon or distilled deionized H_2O. (**d**) Immunoblot analysis of the AfSPL14 protein level in the central leaves of *A. fasciata* treated with 10 mL of 0.6 $g \cdot L^{-1}$ exogenous ethephon for 1, 8, and 24 h, or without ethephon (0 h). The total proteins were separated using SDS-PAGE, and the transferred proteins were then probed with a rabbit polyclonal AfSPL14 antibody or a rabbit polyclonal Actin antibody, respectively. (**e**) Relative level of AfSPL14 protein in the central leaves of *A. fasciata* treated with 10 mL of 0.6 $g \cdot L^{-1}$ exogenous ethephon for 1, 8 and 24 h, or without ethephon (0 h). Three independent experiments were performed, the values are shown as the means and error bars indicate the standard deviation (*n* = 3). ANOVA was conducted, and means were separated by DNMRT.

2.4. Response to Exogenous Ethephon Treatment

To induce bromeliad flowering, ethylene or ethephon is widely used [29]. In fact, flowering induction by ethylene or ethephon is age-dependent. Plants of *A. comosus* which were somewhat less than about 1.0 kg fresh weight in subtropical regions respond only minimally to ethylene or ethephon [29]. Similar to 'Smooth Cayenne' and other variations of *A. comosus*, adult plants (but not juveniles) of *A. fasciata* could be induced by ethephon. Our previous investigation also showed that above 96% of 12-month-old adult plants could be induced to flower by 10 mL of exogenous ethephon treatment at 0.6 $g \cdot L^{-1}$ within two weeks, but that none of 6-month-old juvenile plants flowered under the same condition [37]. Here, we investigated the possible response of *AfSPL14* in adult plants of *A. fasciata* to exogenous ethephon treatment at different concentrations. As shown in Figure 4c, the expression of *AfSPL14* transcripts in the central leaves of adult plants prior to flower bud differentiation increased transiently after treatment for 1 h. Interestingly, a rapid decrease of the expression level of *AfSPL14* transcripts was observed after treatment for 2 h, almost reaching lower levels than in control plants after 24 h (Figure 4c), suggesting the remarkable effect of ethylene on the expression of *AfSPL14*.

To investigate the effect of ethylene on the level of AfSPL14 protein, we extracted the total proteins from the central leaves of *A. fasciata* treated with or without 10 mL of $0.6\ \mathrm{g\cdot L^{-1}}$ ethephon. An immunoblot analysis was performed using a specific antibody against the AfSPL14 protein (Figure 4d). The level of AfSPL14 after treatment for 1 h was ~200% higher than the level in the untreated central leaves (Figure 4d,e). Consistent with the changes in the relative expression of *AfSPL14* mRNA, the level of AfSPL14 also gradually decreased after continuous treatment for 8 and 24 h (Figure 4d,e).

2.5. AfSPL14 Does Not Exhibit Transactivation Activity in Yeast

To test whether AfSPL14 is a transcriptional activator, the ORF (1-349 amino acids), the N terminus containing the SBP domain (1–141 amino acids) (AfSPL14N), and the C terminus (142–349 amino acids) (AfSPL14C) of AfSPL14 were fused with the GAL4 binding domain carried by the pGBKT7 (pBD) vector, respectively. The expression vectors pBD-AfSPL14, pBD-AfSPL14N, and pBD-AfSPL14C were then transformed into the yeast strain Y2HGold carrying the dual reporter genes AUR1-C and MEL1, respectively. As shown in Figure 5, similar to the negative control pBD, but not the positive control pGAL4, all the yeast cells carrying the three tested vectors could not grow on a medium containing SD/−Trp/+AbA/+X-α-Gal, indicating that AfSPL14 could not activate the transcription of the dual reporter genes in yeast.

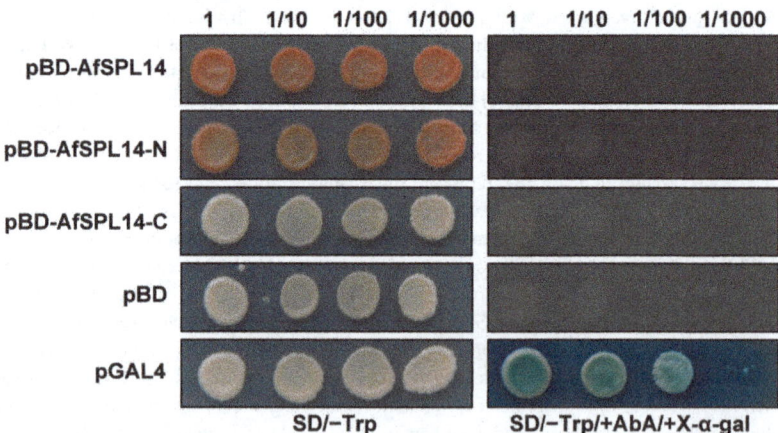

Figure 5. Transactivation activity assay of AfSPL14 in yeast cells. The pGBKT7 (pBD) vectors were fused with the full-length of AfSPL14 (pBD-AfSPL14), the N terminus of AfSPL14 (pBD-AfSPL14-N) and the C terminus of AfSPL14 (pBD-AfSPL14-C), respectively. Each kind of these constructs was then transformed into Y2HGold cells which contained the reporter genes *AUR1-C* and *MEL1*. pBD and pGAL4 plasmids were transformed into Y2HGold cells and used as negative and positive controls, respectively. Yeast clones containing the right constructs grew on SD/−Trp medium at dilutions of 1, 1/10, 1, 100, and 1/1000 for three to five days, and were then transferred onto SD/−Trp/+AbA/+X-α-Gal medium for continuous growth for three further days to test their transactivation activities. SD: synthetic dropout; AbA: Aureobasidin A; SD/−Trp: SD medium without Trp; SD/−Trp/+AbA/+X-α-gal: SD medium without Trp, but with 40 mg/L X-α-gal and 200 µg/L AbA.

2.6. Constitutive Expression of AfSPL14 in Arabidopsis

To assess the function of AfSPL14 in flowering, we induced the ectopic expression of *AfSPL14* with the 35S CaMV promoter (*Pro35S::AfSPL14*) in *Arabidopsis* ecotype Columbia (Col-0) (WT) (Figure S1). Under LD conditions, the flowering time of *Pro35S::AfSPL14* transgenic plants was significantly earlier ($p = 6.26 \times 10^{-8}$) than that of the WT and the WT transformed with the empty vector (Vector) (Figure 6a–c). Although the difference of the number of rosette leaves was minor between the *Pro35S::AfSPL14* transgenic plants and WT, the statistical analysis indicated that the number was significantly lower ($p = 4.69 \times 10^{-5}$) in the *Pro35S::AfSPL14* transgenic plants (Figure 6c). The *Pro35S::AfSPL14* transgenic lines were also smaller than the WT (Figure 6a–c). In addition, the *Pro35S::AfSPL14* transgenic plants showed morphological changes in the reproductive phase. A comparison between the WT and Vector, which only has one main inflorescence per plant, showed that a majority of the transformants of *Pro35S::AfSPL14* developed two main inflorescences (Figure 6b,d). Interestingly, a second inflorescence could be developed from the base of the main inflorescence, and it could also develop from the node of the main inflorescence or even be divided randomly from the non-node position of the main inflorescence (Figure 6b,d). Another change in the reproductive phase was the thickening of the main inflorescence in the *Pro35S::AfSPL14* transgenic plants compared with that of the WT (Figure 6e,f).

To further confirm whether the expression of *AfSPL14* in the *Pro35S::AfSPL14* transgenic plants altered the expression of downstream flowering genes, RT-qPCR analysis was performed with the *Arabidopsis* shoot apices grown under LD conditions as materials. The *Arabidopsis* shoot apices were harvested from the central parts of *Arabidopsis* seedlings, and contain the youngest rosette leaves. As expected, compared to WT, the expression level of the genes *SUPPRESSOR OF OVEREXPRESSION OF CONSTANS 1 (SOC1)*, *FRUITFULL (FUL)* and *APETALA1 (AP1)*, which encode floral inductive factors, was substantially upregulated at the shoot apex of *Pro35S::AfSPL14* transgenic plants (Figure 7a,b,e). However, the expression level of another gene encoding plant-specific transcription factor LEAFY (LFY), which is also a positive regulator inducing flowering at the shoot apex, showed no clear difference between WT and *Pro35S::AfSPL14* transgenic plants (Figure 7c). The expression level of *Flowering Locus T (FT)*, an integrator of flowering pathways and defined as a florigen, was also considerably upregulated at the shoot apex of *Pro35S::AfSPL14* transgenic plants (Figure 7d). In addition, the expression of floral organ identity genes, such as *AtAP2* and *AtAP3*, was also upregulated (Figure 7f,g).

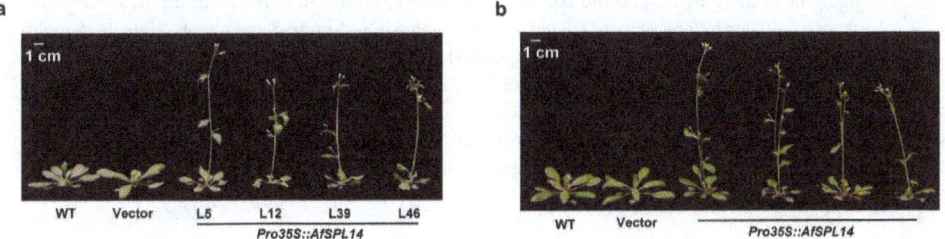

Figure 6. *Cont.*

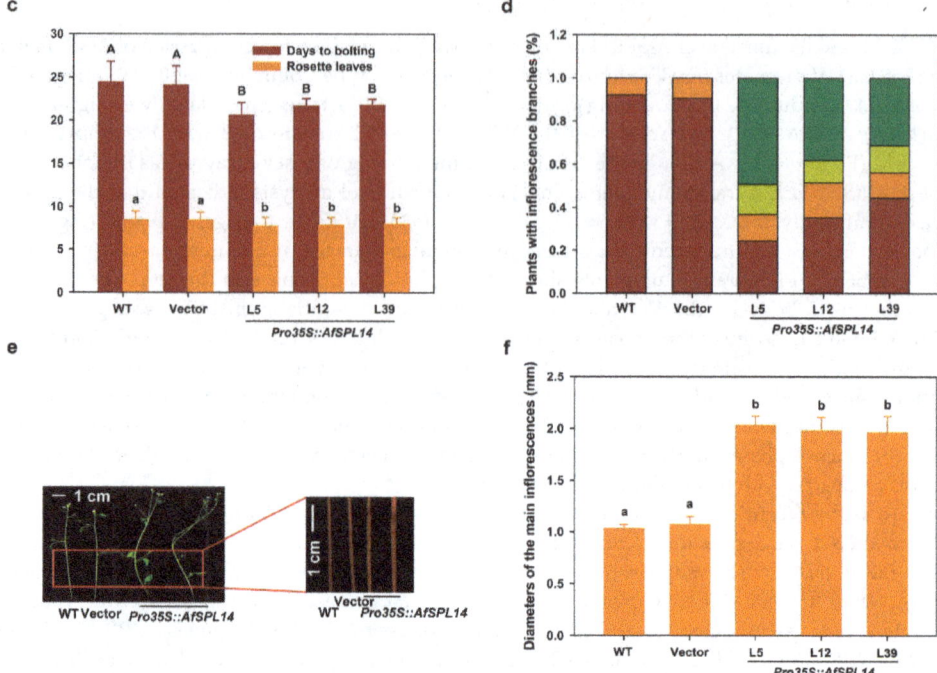

Figure 6. Phenotype analysis of *Pro35S::AfSPL14* transgenic plants. (**a**) Flowering *Pro35S::AfSPL14* transgenic plants shown next to WT and WT transformed with the empty vector (Vector) under LD conditions. L5, L12, L39, and L46 indicate the different lines. (**b**) Flowering *Pro35S::AfSPL14* transgenic plants that had two main inflorescences under LD conditions. (**c**) Days and number of rosette leaves to bolting of the WT, Vector and *Pro35S::AfSPL14* transgenic plants grown under LD conditions. Values are the means ± standard deviation. Seventy-nine plants were scored for each line. Difference letters indicate statistical differences. (**d**) Percentages of plants which have two main inflorescences grown under LD conditions. The number of plants with the second main stem developed from the base (orange), the node (ginger), and the non-node (dark green) position of the main inflorescences and plants with only one inflorescence (dark red) was calculated. One hundred and twenty-eight 38-day-old, long-day-grown plants and ninety-six 55-day-old, short-day-grown plants were counted for each line. (**e**) Bending and thicker main inflorescences of *Pro35S::AfSPL14* transgenic plants under LD conditions. (**f**) The diameter of main inflorescences of WT, Vector and transgenic plants. Forty-eight 38-day-old, long-day-grown plants were counted for each line. The diameter of the positions which were 5 cm distance from the basal of the main inflorescences was measured. ANOVA was conducted, and means were separated by DNMRT.

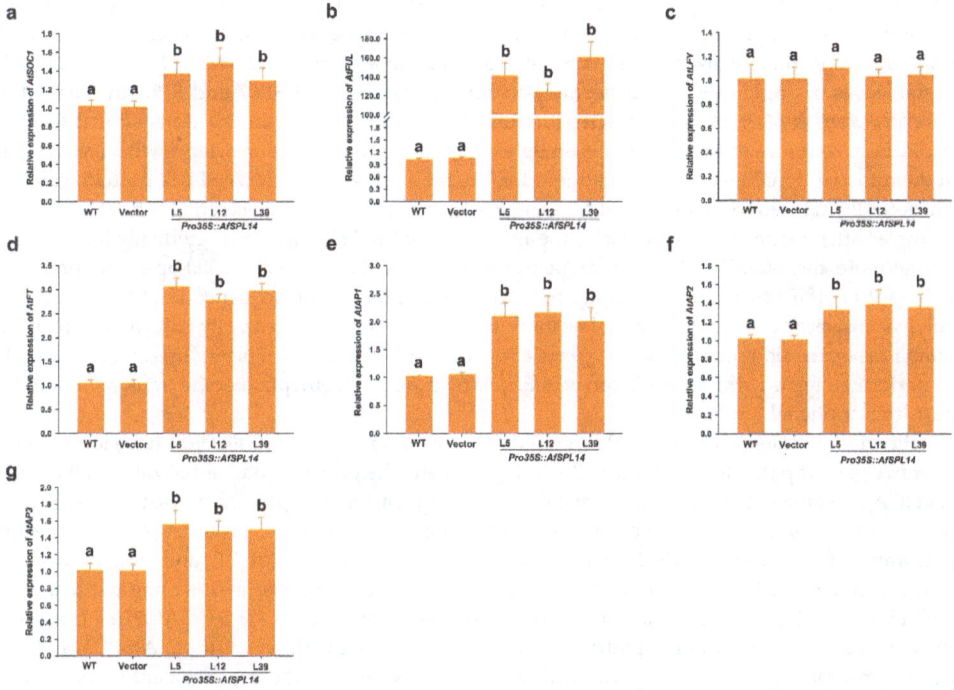

Figure 7. RT-qPCR analysis of flowering related genes at the shoot apex of WT, Vector and *Pro35S::AfSPL14* transgenic plants. Relative expression of three flowering promoting genes, *SUPPRESSOR OF OVEREXPRESSION OF CONSTANS 1 (SOC1)* (**a**), *FRUITFULL (FUL)* (**b**), *LEAFY* (**c**), and one florigen *Flowering Locus T (FT)* (**d**), and three flowering organ identify genes, *APETALA1 (AP1)* (**e**), *AP2* (**f**) and *AP3* (**g**) was performed. Fourteen-day-old long-day-grown seedlings were used. Three biological replicates and three technical replicates were performed. Transcript levels were normalized using *AtACTB* gene as a reference. All primers used here are listed in Table S3 online. ANOVA was conducted, and means were separated by DNMRT.

3. Discussion

SPL proteins are plant-specific TFs, and have been reported in many plants, including *Antirrhinum majus* [2], *Arabidopsis thaliana* [3], *Chlamydomonas* [14], *O. sativa* [36], *P. patens* [5], tomato [38], *Triticum aestivum* [39], *Castor Bean* [40], *Prunus mume* [41], *Citrus* [42], pepper [43], Petunia [44], *Brassica napus* [45] and *Chrysanthemum* [46]. In the present study, we identified the *SPL* gene *AfSPL14* in *A. fasciata*, an economically valuable, short-day ornamental plant that exhibits crassulacean acid metabolism (CAM). We discussed the correlation between this gene and the plant hormone ethylene, which has been widely used to induce flowering of members in numbers of the Bromeliaceae family. We also suggested that AfSPL14 was a putative flowering inducer and ideal plant architecture generator, based on its heterologous constitutive expression in *Arabidopsis*.

Compared with many other plant species in which the role of ethylene in the regulation of flowering appears complicated, in a majority of bromeliads including pineapple and *A. fasciata*, flowering can be triggered by a small burst of ethylene production in the meristem in response to exogenous ethylene or ethephon treatment [30,33]. In previous studies, several ethylene biosynthesis, signaling and responsive genes were identified and characterized [30–33,47]. Here, we found that exogenous ethephon induced the expression of *AfSPL14* transcripts rapidly and dramatically within

1 h (Figure 4c). Interestingly, the expression level of *AfSPL14* transcripts gradually declined after continuous treatment for 8 h (Figure 4c). In fact, several *SPLs* in some other species also could be transiently upregulated and then downregulated by ethylene, for example, *MdSBP20* and *MdSBP27* in the leaves of apple (*Malus × domestica* Borkh.) cv. 'Fuji'34, and *SPL7* and *SPL9* in the fruit of *Cavendish banana* [48]. A more precise identification of the changes in the translational level of AfSPL14 in response to the exogenous ethephon treatment showed a consistence with the changes at the transcriptional level. After treatment for 1 h, the expression of the AfSPL14 protein was also dramatically induced to a higher level compared with that in the untreated central leaves (Figure 4d,e). However, after treatment for 8 h and 24 h, the amount of AfSPL14 also decreased gradually (Figure 4d,e). Furthermore, three 5′-ATGTA-3′ core sequences were enclosed in the nearly 3000-bp-length promoter sequence of *AfSPL14* promoter (Figure S2). The 5′-ATGTA-3′ core sequence might interact with ethylene insensitive 3 (EIN3), a crucial factor in the ethylene signaling pathway that could activate or inhibit the expression of downstream genes at the transcriptional level. Further investigation should be performed regarding the regulation of *AfSPL14* by exogenous ethephon at the transcriptional and post-transcriptional levels.

Previous studies of the molecular regulation of the model species *Arabidopsis* have identified at least five genetic pathways relevant to flowering, namely: the photoperiod, vernalization, gibberellic acid (GA), and the autonomous and aging pathways [49]. During this process, at least 180 genes were involved [50]. *SPLs* are indispensable among these genes, and are involved in several signaling pathways. For example, AtSPL3 and AtASPL9 act independently of *FT*, and directly activate flower-promoting *MADS box* genes, thus defining a separate endogenous flowering pathway [18]. AtSPL9 could also acts upstream of *FT* and promotes *FT* expression [51,52]. AtSPL15 integrates the GA pathway and the aging pathway to promote flowering [53]. In addition, AtSPL3/4/5 link developmental aging and photoperiodic flowering [54]. Moreover, the enhancement of the miR156 site-mutated *OsSPL14* gene could also accelerate flowering [55]. Phylogenetic and motif analyses of AfSPL14 and the SPLs of *Arabidopsis* and OsSPL14 showed that the former was similar to AtASPL9, AtSPL15, and OsSPL14 (Figures 1–3), implying a putative conserved function, such as, flowering promotion. Recently, an age-dependent flowering pathway was identified by the regulation of CmNF-YB8, a nuclear factor, through directly triggering miR156-SPL-regulated processes in the short day plant chrysanthemum (*Chrysanthemum morifolium*) [56]. The flowering of pineapple and *A. fasciata* is also age dependent, and the juvenile plants cannot flower naturally, even when treated with exogenous ethylene [30,33]. The expression of *AfSPL14* transcripts was higher in the central leaves and stems of adult plants prior to flower bud differentiation compared with that of the juvenile plants (Figure 4a). This fact is similar to the increasing pattern of accumulation of *AtSPL9* and *AtSPL15* in the meristem with age [9,17,53], and inconsistent with the expression profile of *AfAP2-1*, a putative flowering TF encoding gene identified in *A. fasciata* [33]. These results suggest that AfSPL14 might act positively in the juvenile-to-vegetative phase transition and flowering pathway regulated by developmental age.

The constitutive expression of *AfSPL14* in *Arabidopsis* significantly promoted flowering under LD conditions (Figure 6a–c), which was inconsistent with the flowering-delayed phenotype caused by the constitutive expression of *AfAP2-1* in *Arabidopsis* [33], thus suggesting that AfSPL14 is an activator of flowering integrator and floral inductive genes such as *AtFT*, *AtAP1*, *AtSOC1*, and *AtFUL* (Figure 7a,b,d,e). A previous study demonstrated that the overexpression of AtSPL3 could strongly induce *AtFUL*, but has a weaker effect, or no effect at all, on *AtSOC1* in the shoot apex in *Arabidopsis* [18]. Interestingly, compared with that of WT, the expression level of *AtFUL* at the shoot apex of *Pro35S::AfSPL14* transgenic plants was upregulated dramatically, while *AtSOC1* was slightly induced (Figure 7a,b), suggesting that similar to AtSPL3, AfSPL14 might also induce flowering via an endogenous pathway.

Phylogenetic and motif analyses of AfSPL14 with variable SPLs suggested that it was closer and more similar to OsSPL14 than to AtSPL9 and AtSPL15 (Figures 2 and 3); this is consistent with the evolutionary distances among *A. fasciata*, rice and *Arabidopsis*. In addition, the genes have similar exon-intron structures (Figure 3). Higher expression of *OsSPL14* could reduce the tiller number, increase the lodging resistance, promote panicle branching, and enhance the grain yield [25,26]. Importantly, in addition to the acceleration of flowering, the constitutive expression of *AfSPL14* in *Arabidopsis* also promotes the number of main inflorescences and produces thicker and sturdier culms (Figure 6a–f). However, we did not find the transactivator activity of AfSPL14 in yeast cells (Figure 5), in opposition to the results reported on OsSPL14 [26]. These results suggested functional conservation and diversification in AfSPL14 and OsSPL14. Interestingly, the repression of *AtSPL10* caused reduced apical dominance, and increased the number of main inflorescences [10]. A loss-of-function mutation of *AtSPL9* and *AtSPL15* resulted in altered main stem architecture and enhanced branching [17]. The main inflorescence-changed phenotypes of constitutive expressed *AfSPL14* in *Arabidopsis* and loss-of-function *AtSPL9*, *AtSPL10*, and *AtSPL15* mutants appeared to be similar.

Certain SPLs can be regulated by miR156, two miRNAs that can regulate the expression of SPL proteins at the post-transcriptional level [57]. Similar to *OsSPL14*, *AfSPL14* also had a miR156 cleavage site in its CDS sequence (Figure 3). Because of a point mutation in the OsmiR156-directed site of *OsSPL14*, grain yield was enhanced [25,26]. Many SPLs positively regulate grain yield [23,58–60]. The thicker main inflorescence phenotype in *AfSPL14*-constitutive expressed *Arabidopsis* implied that this gene might act positively in the regulation of flower stalk diameter in *A. fasciata*. Further investigation should focus on the morphological changes of flowers in *AfSPL14*-overexpressed and/or *AfSPL14*-silenced *A. fasciata*, the morphological changes of flowers in *AfSPL14*-overexpressed and/or *AfSPL14*-silenced pineapple, and the cloning and functional characterization of possible homologs of *AfSPL14* in pineapple.

4. Materials and Methods

4.1. Plant Materials and Sample Preparation

The *A. fasciata* specimens used in this study were planted in a greenhouse (ambient temperature of 30–32 °C) located in the experimental area of the Institute of Tropical Crop Genetic Resources, Chinese Academy of Tropical Agricultural Sciences (CATAS). For the tissue-specific expression and western blot analyses, different tissue samples, including mature leaves, central leaves, stems, roots, and various flower organs, were collected.

The wild-type (WT) and transgenic plants of *Arabidopsis* used in this study were of the Columbia ecotype (Col-0). Seeds were surface sterilized in 0.1% $HgCl_2$ for 10 min and then washed with sterilized distilled water five times. The washed seeds were then plated on MS medium containing sugar (2%) and agar (0.8%) and incubated in the dark at 4 °C for 2 days. The plates were then moved to a chamber at 23 °C under LD (16 h light) conditions, with a photon flux density (120 μmol m^{-2} s^{-1}) for continuous growth.

4.2. Isolation and Sequencing of the AfSPL14 Gene

Total RNA was extracted from the central leaves of *A. fasciata* using the hexadecyl trimethyl ammonium bromide (CTAB) method [33], and then used for the RACE at the 5′ and 3′ ends according to the manufacturer's instructions for the SMARTer™ RACE cDNA Amplification Kit (Clontech, Tokyo, Japan). The specific 5′ and 3′ fragments were cloned into pEASY-blunt vectors (Transgen, Beijing, China), and then sequenced by Thermo Fisher Scientific (Guangzhou, China). The gene accession number of *AfSPL14* is MF114304. The primers used here are listed in Table S3 online.

4.3. Bioinformatic Analysis

The ORF of *AfSPL14* was predicted using the ORF Finder (https://www.ncbi.nlm.nih.gov/orffinder/). The sequence logo was generated by the online WebLogo 3 platform (http://weblogo.threeplusone.com/). A phylogenetic tree was constructed with MEGA version 6.0 using the neighbor-joining method with 1000 bootstrap replications [61]. The scheme of exon-intron structures was generated by Gene Structure Display Server 2.0 (http://gsds.cbi.pku.edu.cn/index.php). Putative motifs of variable SPLs were identified by MEME software online with default settings (http://meme-suite.org/tools/meme).

4.4. 5′ RLM-RACE

5′ RLM-RACE was performed according to the manufacturer's instructions of FirstChoice® RLM-RACE Kit (Thermo Fisher Scientific, New York, NY, USA). For the next amplification, 10 μg of total RNA, which was isolated from central leaves of 12-month-old *A. fasciata* plants using the CTAB method [33], was used. The gene specific primers of *AfSPL14* for the first and second PCR products are *AfSPL14*-5outer and *AfSPL14*-5inner, respectively. The second PCR products were gel purified and subcloned into pEASY-T3 Vector (Transgen, Beijing, China) for sequencing. Primers used for 5′ modified RACE are listed in Table S3 online.

4.5. RT-qPCR

First-strand cDNA was synthesized using the TransScript One-Step gDNA Removal and cDNA Synthesis SuperMix (Transgen, Beijing, China) according to the manufacturer's instructions. Quantitative real-time PCR (qPCR) was conducted on a Therma PikoReal 96™ Real-Time PCR System (Thermo Fisher Scientific, Waltham, MA, USA) using the TransStart Tip Green qPCR SuperMix Kit (Transgen, Beijing, China). The total reactions (20 μL) described in this protocol converted total RNA (500 ng ~ 5 μg) into the first-strand cDNA. The first-strand reaction products (20 μL) were diluted with sterilized distilled H_2O 5 times, and diluted products (1 μL) were used for total qPCR reactions (10 μL). Three biological replicates and three technical replicates were performed. The relative expression levels of specific genes were calculated using the $2^{-\Delta\Delta Ct}$ method with the *β-actin* gene (*ACTB*) of *A. fasciata* or *Arabidopsis* as the internal control [62]. All primers used for qPCR are listed in Table S3 online.

4.6. Transgenic Plants

For the transgenic constructs, the coding sequence (CDS) of *AfSPL14* was cloned into the KpnI-SalI sites of the binary vector Cam35S-gfp under the control of the cauliflower mosaic virus (CaMV) 35S promoter. The constructs were then delivered into *Agrobacterium tumefaciens* strain EHA105 by the freeze-thaw method [63]. Col-0 background *Arabidopsis* was transformed using the floral dipping method [64]. For the selection of transgenic plants, the seeds were planted on MS agar medium supplemented with hygromycin (25 mg/L). Seedlings conferring resistance to hygromycin were then transplanted in a chamber at 23 °C under LD conditions. Transgenic plants were verified by genomic PCR and RT-PCR using primers *AfSPL14-OX* F and *AfSPL14-OX* R, which were listed in Table S3 online. T3 transgenic plants were used for next experiments.

4.7. Transactivation Analysis of AfSPL14 in Yeast Cells

The yeast strain Y2HGold was transformed with plasmids containing the pGBKT7 (pBD) vector with the ORF or fragments of AfSPL14 fused in frame with GAL4 DNA binding domain. The primers used are listed in Table S3 online. pBD and pGAL4 were used as negative and positive controls, respectively. Transformants were selected on synthetic dropout (SD) medium lacking tryptophan (SD/−Trp) (Clontech, Tokyo, Japan) and then dripped onto SD/−Trp/+AbA/+X-α-gal to determine the transactivation activity.

4.8. Exogenous Ethephon Treatment of A. fasciata

To test the response to ethylene, adult (12-month-old) *A. fasciata* plants which were grown in pots in our greenhouse were treated with ethephon (10 mL) at 0.3 g·L^{-1}, 0.6 g·L^{-1}, 1.2 g·L^{-1}, 2.4 g·L^{-1}, 4.8 g·L^{-1} for 1, 2, 4, 8, 24, or 48 h. All treatments were applied by pouring the specific concentration of ethephon solution into the leaf whorl of each plant, with the same quantity of water as control. The central leaves were then physically isolated and immediately frozen in liquid nitrogen for further research.

4.9. SDS-PAGE and Immunoblot Analysis

Total proteins were extracted from the central leaves of adult *A. fasciata* plants. The physically isolated and immediately frozen central leaves (0.5 g) were homogenized with extraction buffer (1 mL) (Tris (1 mol/L, pH 6.8); DL-dithiothreitol (0.2 mol/L); sodium dodecyl sulfate (4% (g/mL)); glycerol (20%)) by using pestle and mortar. After centrifugation at 12,000 rotation per minute (rpm) for 15 min at 4 °C, the supernatants were transferred into new tubes and 4 times volume of acetones were added. After being vortexed for 2 min and then placed on ice for 1 h, the mixture was centrifuged again, as above. The supernatants were discarded and 4 times volume of acetone was added. The mixture was then vortexed and centrifuged again; the extracted proteins were diluted with $0.5\times$ extraction buffer, boiled at 100 °C for 10 min, and then centrifuged at 13,000 rpm for 5 min. The supernatants (total central leaf proteins) were separated using 15% sodium dodecyl sulfate polyacrylamide gel electrophoresis (SDS-PAGE) containing urea (6 mol/L). After electrophoresis, the proteins were transferred onto nitrocellulose membranes (Amersham Biosciences, Pittsburgh, PA, USA) and probed with a rabbit polyclonal AfSPL14 antibody (Jiaxuan Biotech, Beijing, China) or a rabbit polyclonal Actin antibody (Agrisera, Vännäs, Sweden). After incubation with horseradish peroxidase conjugated goat anti-rabbit IgG (Jiaxuan Biotech, Beijing, China), the signals were detected by enhanced chemiluminescence (Jiaxuan Biotech, Beijing, China). X-ray films were scanned and analyzed using ImageMaster™ 2D Platinum software (GE Healthcare, Pittsburgh, PA, USA). Protein concentration of each extract was determined by using a protein assay kit (Bio-Rad, Hercules, CA, USA) with BSA as the standard.

4.10. Data Analysis

Values represent means ± standard deviation of two or three biological replicates. ANOVA was conducted, and the means were separated by Duncan's New Multiple Range Test (DNMRT).

Supplementary Materials: Supplementary materials can be found at http://www.mdpi.com/1422-0067/19/7/2085/s1.

Author Contributions: Conceptualization, L.X. and M.L.; Methodology, L.X.; Software, J.-b.W.; Validation, M.L., and L.X.; Formal Analysis, J.-b.W.; Investigation, M.L. and M.-f.A.; Resources, Z.-y.L., Y.-l.F. and L.X.; Data Curation, M.L., J.-b.W. and M.-f.A.; Writing-Original Draft Preparation, M.L.; Writing-Review & Editing, L.X.

Acknowledgments: This work was supported by grants from the National Natural Science Foundation of China (31372106, 31601793), the National Science Foundation of Hainan Province (20163127) and the Fundamental Scientific Research Funds for CATAS-TCGRI (1630032016006).

Conflicts of Interest: The authors declare no conflict of interest.

Abbreviations

SBP	SQUAMOSA PROMOTER BINDING PROTEIN
SPL	SBP-LIKE
miRNA	microRNA
TGA1	*TEOSINTE GLUME ARCHITECTURE*
LD	long day
RACE	Rapid amplification of cDNA ends
UTR	Untranslated region
ORF	Open reading frame
MW	Molecular weight
pI	Isoelectric point
Jp	Joint peptide
NLS	Nuclear localization signal

RT-qPCR	Reverse transcription followed by quantitative real-time PCR
RLM-RACE	RNA ligase mediated rapid amplification of cDNA ends
DAF	Day after flowering
WT	Wild Type
SOC1	*SUPPRESSOR OF OVEREXPRESSION OF CONSTANS 1*
FUL	*FRUITFULL*
AP1	*APETALA1*
LFY	*LEAFY*
FT	*Flowering Locus T*
AP2	*APETALA2*
AP3	*APETALA3*
CAM	Crassulacean acid metabolism
EIN3	ETHYLENE INSENSITIVE 3
GA	Gibberellic acid
CTAB	Hexadecyl trimethyl ammonium bromide
CDS	the coding sequence
SDS-PAGE	Sodium dodecyl sulfate polyacrylamide gel electrophoresis

References

1. Preston, J.C.; Hileman, L.C. Functional evolution in the plant *SQUAMOSA-PROMOTER BINDING PROTEIN-LIKE (SPL)* gene family. *Front. Plant Sci.* **2013**, *4*, 80. [CrossRef] [PubMed]
2. Klein, J.; Saedler, H.; Huijser, P. A new family of DNA binding proteins includes putative transcriptional regulators of the *Antirrhinum majus* floral meristem identity gene *SQUAMOSA*. *Mol. Gen. Genet.* **1996**, *250*, 7–16. [PubMed]
3. Cardon, G.H.; Hohmann, S.; Nettesheim, K.; Saedler, H.; Huijser, P. Functional analysis of the *Arabidopsis thaliana SBP-box* gene *SPL3*: A novel gene involved in the floral transition. *Plant J.* **1997**, *12*, 367–377. [CrossRef] [PubMed]
4. Arazi, T.; Talmor-Neiman, M.; Stav, R.; Riese, M.; Huijser, P.; Baulcombe, D.C. Cloning and characterization of micro-RNAs from moss. *Plant J.* **2005**, *43*, 837–848. [CrossRef] [PubMed]
5. Riese, M.; Höhmann, S.; Saedler, H.; Münster, T.; Huijser, P. Comparative analysis of the *SBP-box* gene families in *P. patens* and seed plants. *Gene* **2007**, *401*, 28–37. [CrossRef] [PubMed]
6. Yamasaki, K.; Kigawa, T.; Inoue, M.; Tateno, M.; Yamasaki, T.; Yabuki, T.; Aoki, M.; Seki, E.; Matsuda, T.; Nunokawa, E.; Ishizuka, Y.; et al. A novel zinc-binding motif revealed by solution structures of DNA-binding domains of *Arabidopsis* SBP-family transcription factors. *J. Mol. Biol.* **2004**, *337*, 49–63. [CrossRef] [PubMed]
7. Moreno, M.A.; Harper, L.C.; Krueger, R.W.; Dellaporta, S.L.; Freeling, M. *Liguleless1* encodes a nuclear-localized protein required for induction of ligules and auricles during maize leaf organogenesis. *Genes Dev.* **1997**, *11*, 616–628. [CrossRef] [PubMed]
8. Lee, J.; Park, J.J.; Kim, S.L.; Yim, J.; An, G. Mutations in the rice *liguleless* gene result in a complete loss of the auricle, ligule, and laminar joint. *Plant Mol. Biol.* **2007**, *65*, 487–499. [CrossRef] [PubMed]
9. Wang, J.W.; Schwab, R.; Czech, B.; Mica, E.; Weigel, D. Dual effects of miR156-targeted *SPL* genes and *CYP78A5/KLUH* on plastochron length and organ size in *Arabidopsis thaliana*. *Plant Cell* **2008**, *20*, 1231–1243. [CrossRef] [PubMed]
10. Shikata, M.; Koyama, T.; Mitsuda, N.; Ohme-Takagi, M. *Arabidopsis SBP-box* genes *SPL10*, *SPL11* and *SPL2* control morphological change in association with shoot maturation in the reproductive phase. *Plant Cell Physiol.* **2009**, *50*, 2133–2145. [CrossRef] [PubMed]
11. Nodine, M.D.; Bartel, D.P. MicroRNAs prevent precocious gene expression and enable pattern formation during plant embryogenesis. *Genes Dev.* **2010**, *24*, 2678–2692. [CrossRef] [PubMed]
12. Xing, S.; Salinas, M.; Hohmann, S.; Berndtgen, R.; Huijser, P. miR156-targeted and nontargeted SBP-box transcription factors act in concert to secure male fertility in *Arabidopsis*. *Plant Cell* **2010**, *22*, 3935–3950. [CrossRef] [PubMed]

13. Xing, S.; Salinas, M.; Garcia-Molina, A.; Hohmann, S.; Berndtgen, R.; Huijser, P. *SPL8* and miR156-targeted *SPL* genes redundantly regulate *Arabidopsis* gynoecium differential patterning. *Plant J.* **2013**, *75*, 566–577. [CrossRef] [PubMed]

14. Kropat, J.; Tottey, S.; Birkenbihl, R.P.; Depege, N.; Huijser, P.; Merchant, S. A regulator of nutritional copper signaling in *Chlamydomonas* is an SBP domain protein that recognizes the GTAC core of copper response element. *Proc. Natl. Acad. Sci. USA* **2005**, *102*, 18730–18735. [CrossRef] [PubMed]

15. Yamasaki, H.; Hayashi, M.; Fukazawa, M.; Kobayashi, Y.; Shikanai, T. SQUAMOSA promoter binding protein-like7 is a central regulator for copper homeostasis in *Arabidopsis*. *Plant Cell* **2009**, *21*, 347–361. [CrossRef] [PubMed]

16. Wu, G.; Poethig, R.S. Temporal regulation of shoot development in *Arabidopsis thaliana* by miR156 and its target *SPL3*. *Development* **2006**, *133*, 3539–3547. [CrossRef] [PubMed]

17. Schwarz, S.; Grande, A.V.; Bujdoso, N.; Saedler, H.; Huijser, P. The microRNA regulated *SBP-box* genes *SPL9* and *SPL15* control shoot maturation in *Arabidopsis*. *Plant Mol. Biol.* **2008**, *67*, 183–195. [CrossRef] [PubMed]

18. Wang, J.W.; Czech, B.; Weigel, D. MiR156-regulated SPL transcription factors define an endogenous flowering pathway in *Arabidopsis thaliana*. *Cell* **2009**, *138*, 738–749. [CrossRef] [PubMed]

19. Yamaguchi, A.; Wu, M.F.; Yang, L.; Wu, G.; Poethig, R.S.; Wagner, D. The microRNA-regulated SBP-box transcription factor SPL3 is a direct upstream activator of *LEAFY*, *FRUITFULL*, and *APETALA1*. *Dev. Cell* **2009**, *17*, 268–278. [CrossRef] [PubMed]

20. Stone, J.M.; Liang, X.; Nekl, E.R.; Stiers, J.J. *Arabidopsis* AtSPL14, a plant-specific SBP-domain transcription factor, participates in plant development and sensitivity to fumonisin B1. *Plant J.* **2005**, *41*, 744–754. [CrossRef] [PubMed]

21. Wang, H.; Nussbaum-Wagler, T.; Li, B.; Zhao, Q.; Vigouroux, Y.; Faller, M.; Bomblies, K.; Lukens, L.; Doebley, J.F. The origin of the naked grains of maize. *Nature* **2005**, *436*, 714–719. [CrossRef] [PubMed]

22. Liu, J.; Cheng, X.; Liu, P.; Sun, J. miR156-regulated TaSPLs interact with TaD53 to regulate *TaTB1* and *TaBA1* expression in bread wheat. *Plant Physiol.* **2017**, *174*, 1931–1948. [CrossRef] [PubMed]

23. Zhang, B.; Xu, W.; Liu, X.; Mao, X.; Li, A.; Wang, J.; Chang, X.; Zhang, X.; Jing, R. Functional conservation and divergence among homoeologs of *TaSPL20* and *TaSPL21*, two *SBP-box* genes governing yield-related traits in hexaploid wheat. *Plant Physiol.* **2017**, *174*, 1177–1191. [CrossRef] [PubMed]

24. Gou, J.; Fu, C.; Liu, S.; Tang, C.; Debnath, S.; Flanagan, A.; Ge, Y.; Tang, Y.; Jiang, Q.; Larson, P.R.; Wen, J.; Wang, Z.Y. The miR156-SPL4 module predominantly regulates aerial axillary bud formation and controls shoot architecture. *New Phytol.* **2017**, *216*, 829–840. [CrossRef] [PubMed]

25. Jiao, Y.; Wang, Y.; Xue, D.; Wang, J.; Yan, M.; Liu, G.; Dong, G.; Zeng, D.; Lu, Z.; Zhu, X.; et al. Regulation of *OsSPL14* by OsmiR156 defines ideal plant architecture in rice. *Nat. Genet.* **2010**, *42*, 541–544. [CrossRef] [PubMed]

26. Miura, K.; Ikeda, M.; Matsubara, A.; Song, X.J.; Ito, M.; Asano, K.; Matsuoka, M.; Kitano, H.; Ashikari, M. OsSPL14 promotes panicle branching and higher grain productivity in rice. *Nat. Genet.* **2010**, *42*, 545–549. [CrossRef] [PubMed]

27. Givnish, T.J.; Barfuss, M.H.; Van Ee, B.; Riina, R.; Schulte, K.; Horres, R.; Gonsiska, P.A.; Jabaily, R.S.; Crayn, D.M.; Smith, J.A.; et al. Phylogeny, adaptive radiation, and historical biogeography in Bromeliaceae: Insights from an eight-locus plastid phylogeny. *Am. J. Bot.* **2011**, *98*, 872–895. [CrossRef] [PubMed]

28. Zanella, C.M.; Janke, A.; Palma-Silva, C.; Kaltchuk-Santos, E.; Pinheiro, F.G.; Paggi, G.M.; Soares, L.E.; Goetze, M.; Buttow, M.V.; Bered, F. Genetics, evolution and conservation of Bromeliaceae. *Genet. Mol. Biol.* **2012**, *35*, 1020–1026. [CrossRef] [PubMed]

29. Bartholomew, D.P.; Paull, R.E.; Rohrbach, K.G. *The Pineapple-Botany, Production and Uses*; CABI Publishing: Wallingford, UK, 2003; p. 176.

30. Trusov, Y.; Botella, J.R. Silencing of the *ACC synthase* gene *ACACS2* causes delayed flowering in pineapple [*Ananas comosus* (L.) Merr.]. *J. Exp. Bot.* **2006**, *57*, 3953–3960. [CrossRef] [PubMed]

31. Lv, L.; Duan, J.; Xie, J.; Wei, C.; Liu, Y.; Liu, S.; Sun, G. Isolation and characterization of a *FLOWERING LOCUS T* homolog from pineapple (*Ananas comosus* (L.) Merr). *Gene* **2012**, *505*, 368–373. [CrossRef] [PubMed]

32. Lv, L.L.; Duan, J.; Xie, J.H.; Liu, Y.G.; Wei, C.B.; Liu, S.H.; Zhang, J.X.; Sun, G.M. Cloning and expression analysis of a *PISTILLATA* homologous gene from pineapple (*Ananas comosus* L. Merr). *Int. J. Mol. Sci.* **2012**, *13*, 1039–1053. [CrossRef] [PubMed]

33. Lei, M.; Li, Z.Y.; Wang, J.B.; Fu, Y.L.; Ao, M.F.; Xu, L. *AfAP2-1*, An age-dependent gene of *Aechmea fasciata*, responds to exogenous ethylene treatment. *Int. J. Mol. Sci.* **2016**, *17*, 303. [CrossRef] [PubMed]

34. Li, J.; Hou, H.; Li, X.; Xiang, J.; Yin, X.; Gao, H.; Zheng, Y.; Bassett, C.L.; Wang, X. Genome-wide identification and analysis of the *SBP-box* family genes in apple (*Malus x domestica* Borkh.). *Plant Physiol. Biochem.* **2013**, *70*, 100–114. [CrossRef] [PubMed]

35. Bailey, T.L.; Elkan, C. Fitting a mixture model by expectation maximization to discover motifs in biopolymers. *Int. Conf. Intell. Syst. Mol. Biol.* **1994**, *2*, 28–36.

36. Xie, K.; Wu, C.; Xiong, L. Genomic organization, differential expression, and interaction of SQUAMOSA promoter-binding-like transcription factors and microRNA156 in rice. *Plant Physiol.* **2006**, *142*, 280–293. [CrossRef] [PubMed]

37. Li, Z.; Wang, J.; Zhang, X.; Lei, M.; Fu, Y.; Zhang, J.; Wang, Z.; Xu, L. Transcriptome sequencing determined flowering pathway genes in *Aechmea fasciata* treated with ethylene. *J. Plant Growth Regul.* **2016**, *35*, 316–329. [CrossRef]

38. Salinas, M.; Xing, S.; Hohmann, S.; Berndtgen, R.; Huijser, P. Genomic organization, phylogenetic comparison and differential expression of the SBP-box family of transcription factors in tomato. *Planta* **2012**, *235*, 1171–1184. [CrossRef] [PubMed]

39. Zhang, B.; Liu, X.; Zhao, G.; Mao, X.; Li, A.; Jing, R. Molecular characterization and expression analysis of *Triticum aestivum squamosa-promoter binding protein-box* genes involved in ear development. *J. Integr. Plant Biol.* **2013**, *56*, 571–581. [CrossRef] [PubMed]

40. Zhang, S.; Ling, L. Genome-wide identification and evolutionary analysis of the *SBP-box* gene family in castor bean. *PLoS ONE* **2014**, *9*, e86688. [CrossRef] [PubMed]

41. Xu, Z.; Sun, L.; Zhou, Y.; Yang, W.; Cheng, T.; Wang, J.; Zhang, Q. Identifiation and expression analysis of the *SQUAMOSA promoter-binding protein (SBP)-box* gene family in *Prunus mume*. *Mol. Genet. Genom.* **2015**, *290*, 1701–1715. [CrossRef] [PubMed]

42. Shalom, L.; Shlizerman, L.; Zur, N.; Doron-Faigenboim, A.; Blumwald, E.; Sadka, A. Molecular characterization of *SQUAMOSA PROMOTER BINDING PROTEIN-LIKE (SPL)* gene family from *Citrus* and the effect of fruit load on their expression. *Front. Plant Sci.* **2015**, *6*, 389. [CrossRef] [PubMed]

43. Zhang, H.X.; Jin, J.H.; He, Y.M.; Lu, B.Y.; Li, D.W.; Chai, W.G.; Khan, A.; Gong, Z.H. Genome-wide identification and analysis of the SBP-box family genes under *Phytophthora capsici* stress in pepper (*Capsicum annuum* L.). *Front. Plant Sci.* **2016**, *7*, 504. [CrossRef] [PubMed]

44. Preston, J.C.; Jorgensen, S.A.; Orozco, R.; Hileman, L.C. Paralogous *SQUAMOSA PROMOTER BINDING PROTEIN-LIKE (SPL)* genes differentially regulate leaf initiation and reproductive phase change in petunia. *Planta* **2016**, *243*, 429–440. [CrossRef] [PubMed]

45. Cheng, H.; Hao, M.; Wang, W.; Mei, D.; Tong, C.; Wang, H.; Liu, J.; Fu, L.; Hu, Q. Genomic identification, characterization and differential expression analysis of *SBP-box* gene family in *Brassica napus*. *BMC Plant Biol.* **2016**, *16*, 196. [CrossRef] [PubMed]

46. Song, A.; Gao, T.; Wu, D.; Xin, J.; Chen, S.; Guan, Z.; Wang, H.; Jin, L.; Chen, F. Transcriptome-wide identification and expression analysis of *chrysanthemum* SBP-like transcription factors. *Plant Physiol. Biochem.* **2016**, *102*, 10–16. [CrossRef] [PubMed]

47. Li, Y.H.; Wu, Q.S.; Huang, X.; Liu, S.H.; Zhang, H.N.; Zhang, Z.; Sun, G.M. Molecular cloning and characterization of four genes encoding ethylene receptors associated with pineapple (*Ananas comosus* L.) flowering. *Front. Plant Sci.* **2016**, *7*, 710. [CrossRef] [PubMed]

48. Bi, F.; Meng, X.; Ma, C.; Yi, G. Identification of miRNAs involved in fruit ripening in Cavendish bananas by deep sequencing. *BMC Genom.* **2015**, *16*, 776. [CrossRef] [PubMed]

49. Srikanth, A.; Schmid, M. Regulation of flowering time: All roads lead to Rome. *Cell. Mol. Life Sci.* **2011**, *68*, 2013–2037. [CrossRef] [PubMed]

50. Fornara, F.; de Montaigu, A.; Coupland, G. SnapShot: Control of flowering in *Arabidopsis*. *Cell* **2010**, *141*, 550. [CrossRef] [PubMed]

51. Jung, J.H.; Seo, Y.H.; Seo, P.J.; Reyes, J.L.; Yun, J.; Chua, N.H.; Park, C.M. The GIGANTEA-regulated microRNA172 mediates photoperiodic flowering independent of CONSTANS in *Arabidopsis*. *Plant Cell* **2007**, *19*, 2736–2748. [CrossRef] [PubMed]

52. Wu, G.; Park, M.Y.; Conway, S.R.; Wang, J.W.; Weigel, D.; Poethig, R.S. The sequential action of miR156 and miR172 regulates developmental timing in *Arabidopsis*. *Cell* **2009**, *138*, 750–759. [CrossRef] [PubMed]

53. Hyun, Y.; Richter, R.; Vincent, C.; Martinez-Gallegos, R.; Porri, A.; Coupland, G. Multi-layered regulation of *SPL15* and cooperation with *SOC1* integrate endogenous flowering pathways at the *Arabidopsis* shoot meristem. *Dev. Cell* **2016**, *37*, 254–266. [CrossRef] [PubMed]
54. Jung, J.H.; Lee, H.J.; Ryu, J.Y.; Park, C.M. SPL3/4/5 integrate developmental aging and photoperiodic signals into the FT-FD module in *Arabidopsis* flowering. *Mol. Plant* **2016**, *9*, 1647–1659. [CrossRef] [PubMed]
55. Luo, L.; Li, W.; Miura, K.; Ashikari, M.; Kyozuka, J. Control of tiller growth of rice by *OsSPL14* and Strigolactones, which work in two independent pathways. *Plant Cell Physiol.* **2012**, *53*, 1793–1801. [CrossRef] [PubMed]
56. Wei, Q.; Ma, C.; Xu, Y.; Wang, T.; Chen, Y.; Lu, J.; Zhang, L.; Jiang, C.Z.; Hong, B.; Gao, J. Control of *chrysanthemum* flowering through integration with an aging pathway. *Nat. Commun.* **2017**, *8*, 829. [CrossRef] [PubMed]
57. Ling, L.Z.; Zhang, S.D. Exploring the evolutionary differences of *SBP-box* genes targeted by miR156 and miR529 in plants. *Genetica* **2012**, *140*, 317–324. [CrossRef] [PubMed]
58. Wang, S.; Wu, K.; Yuan, Q.; Liu, X.; Liu, Z.; Lin, X.; Zeng, R.; Zhu, H.; Dong, G.; Qian, Q.; et al. Control of grain size, shape and quality by OsSPL16 in rice. *Nat. Genet.* **2012**, *44*, 950–954. [CrossRef] [PubMed]
59. Chuck, G.S.; Brown, P.J.; Meeley, R.; Hake, S. Maize SBP-box transcription factors unbranched2 and unbranched3 affect yield traits by regulating the rate of lateral primordia initiation. *Proc. Natl. Acad. Sci. USA* **2014**, *111*, 18775–18780. [CrossRef] [PubMed]
60. Si, L.; Chen, J.; Huang, X.; Gong, H.; Luo, J.; Hou, Q.; Zhou, T.; Lu, T.; Zhu, J.; Shangguan, Y.; et al. OsSPL13 controls grain size in cultivated rice. *Nat. Genet.* **2016**, *48*, 447–456. [CrossRef] [PubMed]
61. Tamura, K.; Stecher, G.; Peterson, D.; Filipski, A.; Kumar, S. MEGA6: Molecular Evolutionary Genetics Analysis version 6.0. *Mol. Biol. Evol.* **2013**, *30*, 2725–2729. [CrossRef] [PubMed]
62. Livak, K.J.; Schmittgen, T.D. Analysis of relative gene expression data using real-time quantitative PCR and the $2^{-\Delta\Delta CT}$ method. *Methods* **2001**, *25*, 402–408. [CrossRef] [PubMed]
63. Holsters, M.; de Waele, D.; Depicker, A.; Messens, E.; van Montagu, M.; Schell, J. Transfection and transformation of *Agrobacterium tumefaciens*. *Mol. Genet. Genom.* **1978**, *163*, 181–187. [CrossRef]
64. Clough, S.J.; Bent, A.F. Floral dip: A simplified method for *Agrobacterium*-mediated transformation of *Arabidopsis thaliana*. *Plant J.* **1998**, *16*, 735–743. [CrossRef] [PubMed]

International Journal of
Molecular Sciences

Article

AdRAP2.3, a Novel Ethylene Response Factor VII from *Actinidia deliciosa*, Enhances Waterlogging Resistance in Transgenic Tobacco through Improving Expression Levels of *PDC* and *ADH* Genes

De-Lin Pan [†], Gang Wang [†], Tao Wang, Zhan-Hui Jia, Zhong-Ren Guo and Ji-Yu Zhang *

Institute of Botany, Jiangsu Province and Chinese Academy of Sciences, Nanjing 210014, China;
PPxsperfect@163.com (D.-L.P.); wg20092011@163.com (G.W.); immmorer@163.com (T.W.);
13915954315@163.com (Z.-H.J.); zhongrenguo@cnbg.net (Z.-R.G.)
* Correspondence: maxzhangjy@163.com; Tel.: +86-025-8434-7033
† These authors contributed equally to this work.

Received: 25 January 2019; Accepted: 5 March 2019; Published: 8 March 2019

Abstract: APETALA2/ethylene-responsive factor superfamily (AP2/ERF) is a transcription factor involved in abiotic stresses, for instance, cold, drought, and low oxygen. In this study, a novel ethylene-responsive transcription factor named *AdRAP2.3* was isolated from *Actinidia deliciosa* 'Jinkui'. *AdRAP2.3* transcription levels in other reproductive organs except for the pistil were higher than those in the vegetative organs (root, stem, and leaf) in kiwi fruit. Plant hormones (Salicylic acid (SA), Methyl-jasmonate acid (MeJA), 1-Aminocyclopropanecarboxylic Acid (ACC), Abscisic acid (ABA)), abiotic stresses (waterlogging, heat, 4 °C and NaCl) and biotic stress (*Pseudomonas Syringae* pv. *Actinidiae*, *Psa*) could induce the expression of *AdRAP2.3* gene in kiwi fruit. Overexpression of the *AdRAP2.3* gene conferred waterlogging stress tolerance in transgenic tobacco plants. When completely submerged, the survival rate, fresh weight, and dry weight of transgenic tobacco lines were significantly higher than those of wile type (WT). Upon the roots being submerged, transgenic tobacco lines grew aerial roots earlier. Overexpression of *AdRAP2.3* in transgenic tobacco improved the pyruvate decarboxylase (PDC) and alcohol dehydrogenase (ADH) enzyme activities, and improved the expression levels of waterlogging mark genes *NtPDC*, *NtADH*, *NtHB1*, *NtHB2*, *NtPCO1*, and *NtPCO2* in roots under waterlogging treatment. Overall, these results demonstrated that *AdRAP2.3* might play an important role in resistance to waterlogging through regulation of *PDC* and *ADH* genes in kiwi fruit.

Keywords: ethylene-responsive factor; *Actinidia deliciosa*; *AdRAP2.3*; gene expression; waterlogging stress; regulation

1. Introduction

The plant APETALA2/ethylene-responsive factor (AP2/ERF) superfamily is one of the largest plant transcription factor families participating in plant development and resistance to biotic and abiotic stresses [1–4]. Transcription factors are known as regulating plant stress responses through binding to *cis*-acting elements in the promoters of stress-related genes or interacting with other transcription factors [5]. AP2/ERF superfamily members contain 60-70 highly conserved amino acids [6]. The AP2/ERF superfamily has been characterized based on either one or two AP2 domains: the AP2 subfamily has two AP2/ERF domains, the RAV subfamily has one AP2 domain and a B3 domain, C-repeat binding transcription factor/dehydrate responsive element binding factor (CBF/DREB) and the ERF subfamily have one AP2 domain [6]. ERF subfamily has conserved 14th alanine acid (A) and 19th aspartic acids (D) in the AP2/ERF domain. ERF subfamily transcription factors have been reported to be involved in biotic and abiotic stress [5]. The ERF subfamily was further divided into six small subgroups (V, VI, VII, VIII, IX, and X) based on the similarity of the amino acid sequences of their DNA-binding domain [1,7].

Members of the ERF subgroup VII have been identified to be involved in the hypoxic stress in *Arabidopsis thaliana*, including *RAP2.12*, *RAP2.2*, *RAP2.3*, *HRE1*, and *HRE2* [8,9]. The ERF VIIs were regulated by proteasome-mediated proteolysis via the oxygen-dependent branch of the N-end rule pathway [10,11]. Previous research showed that *RAP2.12* regulates central metabolic processes, for example, respiration, tricarboxylic acid (TCA) cycle, and amino acid metabolism [12]. *RAP2.2* has been proven to be induced by darkness, and overexpression of *RAP2.2* resulted in improving plant survival rate, increasing *ADH1* and *PDC1* expression levels, and increasing ADH and PDC enzyme activities [13]. In addition to *A. thaliana*, members of ERF subgroup VII have been found in other plants, for example, *Oryza sativa* [14] and *Petunia hybrida* [15].

The kiwi fruit is one of the most recently domesticated fruit crops. However, the majority of current kiwi fruit cultivars are susceptible to waterlogging stress. In many regions of China, especially in Eastern China, kiwi fruit plants suffer from excess rainfall during the summer rainy season. Waterlogging affects fruit yield severely and trees even die, which restricts the development of the kiwi fruit industry [16]. Kiwi fruit 'Jinkui' is tolerant to waterlogging, and understanding its waterlogging stress responses will be important for improving the tolerance of other kiwi fruit varieties. The transcriptome sequencing analysis in 'Jinkui' under waterlogging stress showed that the family of AP2/ERF transcription factors contained 14 upregulated and 14 downregulated members in the treatment samples. The transcription comp67160_0_seq1, which was a ERF VIIs member named *AdRAP2.3*, was strongly upregulated in roots of *Actinidia deliciosa* during the first 96 h after waterlogging treatment [17]. In order to analyze the function of the *AdRAP2.3* gene, the complete coding sequence (CDS) was cloned from the roots of 'Jinkui' under waterlogging stress according to the sequence comp67160_0_seq1 in this study. The expression patterns of *AdRAP2.3* in response to adverse stresses were investigated. The function of *AdRAP2.3* was further investigated by overexpression of *AdRAP2.3* in transgenic tobacco.

2. Results

2.1. Cloning and Sequence Analysis of AdRAP2.3

The *AdRAP2.3* gene was isolated from *A. deliciosa* roots according to the sequence comp67160_0_seq1 [17]. The gene *AdRAP2.3* contains an 837 bp complete open reading frame (ORF), encoding 278 amino acids (Figure S1) with a predicted molecular weight (MW) of 31.27 kDa, a theoretical isoelectric point (pI) of 5.32, an instability index of 37.29, and an average hydrophilic coefficient of −0.831. The protein is stable and hydrophilic. The predicted *AdRAP2.3* has a typical conserved DNA-binding domain (AP2/ERF domain) of 58 amino acids. Moreover, AdRAP2.3 has conserved alanine (A) and aspartic acid (D) residues at the 14th and 19th positions in the AP2/ERF domain, as is characteristic of the ERF subfamily (Figure S1).

The analysis of the secondary structure of the amino acid sequence showed that the amino acid sequence encoded by *AdRAP2.3* contains the AP2 domain, which is position from 60 aa to 120 aa (Figure S2). The position 60 to 90 is DNA binding sites according to Figure S3A. The three-dimensional structure of the protein encoded by the *AdRAP2.3* gene consists of one α-helix and three β-sheets according to Figure S3.

The 278 amino acid sequence predicted by the *AdRAP2.3* sequence was aligned with the amino acid sequences of other ERF VII subgroup proteins from *Arabidopsis*, *Oryza*, *Capsicum*, *Lycopersicon*, and *Actinidia*. The phylogenetic trees showed that RAP2.2 and RAP2.12 are closely related; SUB1C, SUB1B, and SUB1A-1 are closely related. PSR94738.1 is an ERF RAP2-3 like from *A. chinensis*, which is closely related to *AdRAP2.3* (Figure S4).

2.2. Expression Patterns of AdRAP2.3 in A. deliciosa

QRT-PCR assays were performed to investigate the expression pattern of *AdRAP2.3* in different organs and tissues of *A. deliciosa* (Figure 1). The expression level of *AdRAP2.3* is the highest in young fruit. *AdRAP2.3* transcription levels in other reproductive organs (except for pistils) were higher than those in the vegetative organs (root, stem, and leaf).

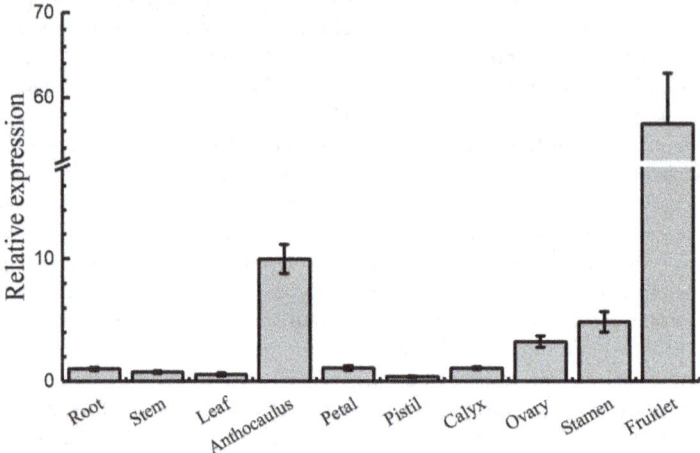

Figure 1. Expression analysis of gene *AdRAP2.3* in different kiwi fruit organs. *AdActin* transcription levels were used to normalize the samples. The mean and standard deviation were obtained from three independent experiments.

The biotic stresses were considered to be correlated with the hormone signals [18,19]. Therefore, the expression patterns of *AdRAP2.3* under the treatment of SA, MeJA, ACC and ABA were analyzed by real-time PCR. As shown in Figure 2A, after SA treatment, *AdRAP2.3* mRNA accumulation increases with time and reaches a maximum at 12 h. Upon treatment with MeJA, the transcription level is induced obviously and reaches a peak at 4 h, then reduces at 48 h (Figure 2B). The transcription levels were induced, with an induction peak at 4 h, and then declines over time until 48 h in kiwi fruit after treatment with ACC (Figure 3C). *AdRAP2.3* mRNA accumulation increases with time and reaches a maximum at 12 h and then decreases in kiwi fruit after ABA treatment (Figure 2D). These results indicated that plant hormones (SA, MeJA, ACC, and ABA) could induce the expression of *AdRAP2.3* gene in kiwi fruit.

Changes of *AdRAP2.3* transcription levels in *A. deliciosa* in response to various abiotic stresses were also analyzed by real-time PCR (Figure 3). As shown in Figure 3A, after 48 °C heat treatment, *AdRAP2.3* transcription level increases obviously at 2 h, reaching more than 30 times that in the control (0 h). The transcription level then decreases at 4 h (0 h). After 6 h recovery, the transcription level increases obviously. As shown in Figure 3B,C, after 4 °C and 0.2 M NaCl treatment, *AdRAP2.3* transcription level keeps going up during the first 48 h. *AdRAP2.3* mRNA accumulation increases about 400 times at 24 h and 600 times at 96 h after treatment with waterlogging (Figure 3D). Drought does not induce the expression of *AdRAP2.3* (Figure 3E). These results indicated that abiotic stresses including waterlogging, heat, low temperature (4 °C), and NaCl could induce the expression of *AdRAP2.3* gene in kiwi fruit. Compared with other environmental stresses, *AdRAP2.3* is strongly induced by waterlogging.

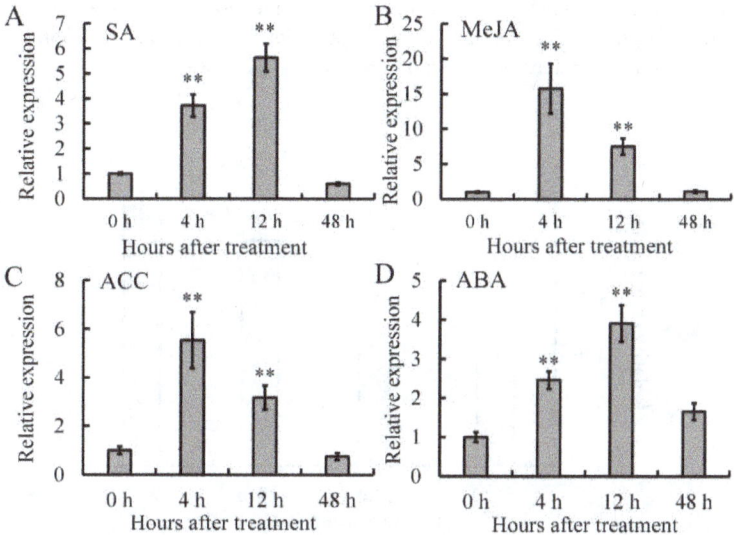

Figure 2. Expression analysis of *AdRAP2.3* in *A. deliciosa* 'Jinkui' leaves under different hormone treatments. (**A**) Salicylic acid (SA); (**B**) Methyl-jasmonate acid (MeJA); (**C**) 1-Aminocyclopropanecarboxylic Acid (ACC); (**D**) Abscisic acid (ABA). *AdActin* transcription levels were used to normalize the samples. The mean value and standard deviation were obtained from three independent experiments. The data represent averages ±Standard error (SE) of three biological repeats with three measurements per sample. ** indicates significant differences in comparison with the control at *p* < 0.01.

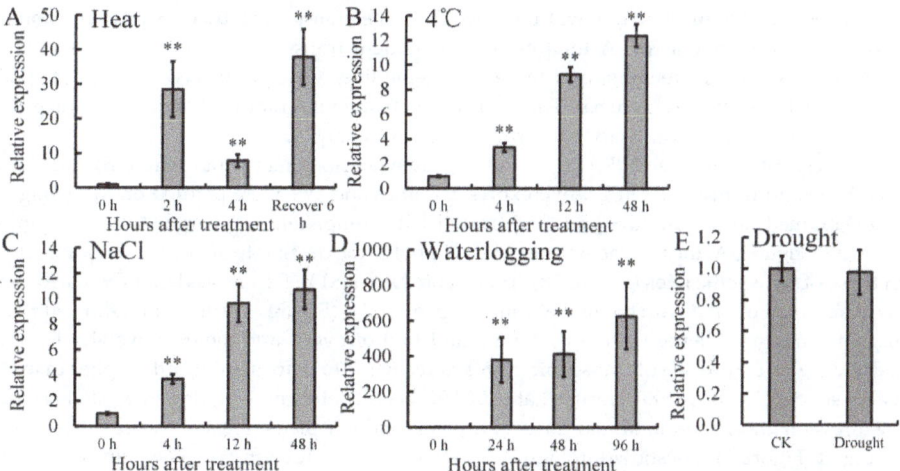

Figure 3. Expression analysis of *AdRAP2.3* gene under different abiotic stresses in kiwifruit. (**A**) Heat; (**B**) 4 °C; (**C**) NaCl; (**D**) waterlogging; (**E**) drought. *AdActin* transcription levels were used to normalize the samples. The mean value and standard deviation were obtained from three independent experiments. The data represent averages ± SE of three biological repeats with three measurements per sample. ** indicates significant differences in comparison with the control at *p* < 0.01.

Pseudomonas Syringae pv. *actinidiae* (*Psa*) is known as a serious disease to kiwi fruit; it causes cankers, cracks, and a reddish bacterial ooze on trunks [20]. So we investigated the *AdRAP2.3* expression in response to *Psa* infection, and the expression levels were analyzed in phloem treated with *Psa* infection (Figure 4). The *AdRAP2.3* transcription level increased significantly at 24 and 96 h, indicating that the *AdRAP2.3* gene is involved in responses to *Psa* stress.

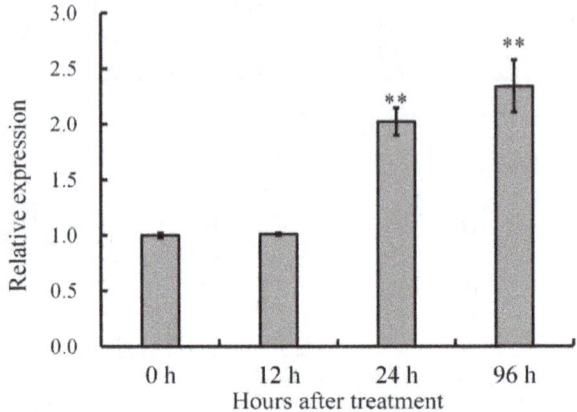

Figure 4. Expression analysis of *AdRAP2.3* gene under *Pseudomonas syringae* pv. *actinidiae* infection in kiwi fruit. *AdActin* transcription levels were used to normalize the samples. The mean value and standard deviation were obtained from three independent experiments. The data represent averages ± SE of three biological repeats with three measurements per sample. ** indicates significant differences in comparison with the control at $p < 0.01$.

2.3. Overexpression of AdRAP2.3 Enhanced Waterlogging Tolerance in Transgenic Plants

To investigate the function of kiwi fruit *AdRAP2.3* gene, transgenic tobacco plants overexpression of *AdRAP2.3* were generated. A total of 40 independent transgenic lines (T_0) were selected by hygromycin-resistance screening, and these transgenic lines were confirmed by β-glucuronidase (GUS) and PCR detection (Figure S5). The seeds from three representative *AdRAP2.3*-overexpressing lines (#1, #14 and #26) were selected for further functional analysis.

To investigated whether *AdRAP2.3* could increase waterlogging tolerance in transgenic tobacco plants. Seeds from three *AdRAP2.3*-overexpression lines and WT were planted on Murashige and Skoog (MS) medium for 10 days and then treated with completely submerged for 48 h (Figure 5). As shown in Figure 5A, the transgenic lines grow euphylla, the death rate of WT is significantly higher than those of transgenic lines (Figure 5B). Transgenic lines and WT were seeded on MS medium for 20 days and then treated with completely submerged for 7 d (Figure 6). The results showed that the transgenic lines grow better than WT after 5 d and 14 d recovery, and the fresh weight (Figure 6B) and dry weight (Figure 6C) of transgenic plants at recovery 14 d are significantly higher than those of WT respectively. To further verity that *AdRAP2.3* can confer the waterlogging tolerance of the overexpression lines, two-month-old transgenic plants and WT planted in soil were treated with root submerged (Figure 7). As shown in Figure 7B, after 7 d of treatment, transgenic lines grow aerial roots, but WT do not. These results showed that overexpression of *AdRAP2.3* could enhance the waterlogging tolerance in transgenic tobacco.

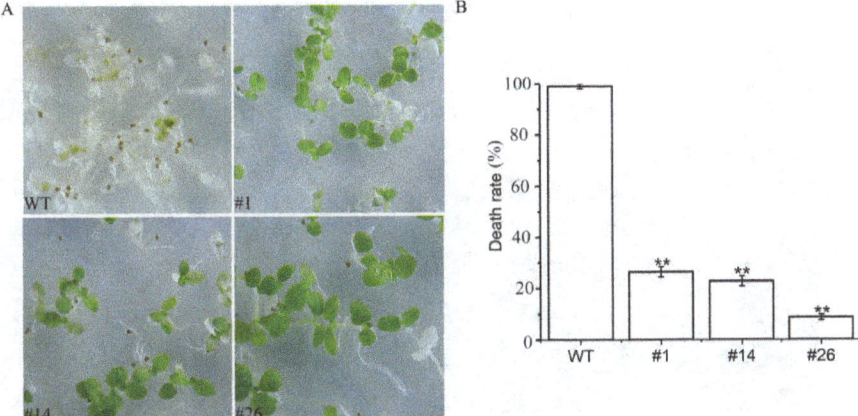

Figure 5. Phenotype (**A**) and death rate (**B**) in 10-day-old WT and transgenic tobacco line seedlings 48 h after they were completely submerged. WT: Wild type; #1,#14,#26: Transgenic tobacco lines. The data represent averages ± SE of three biological repeats with three measurements per sample. ** indicates significant differences in comparison with the WT at $p < 0.01$.

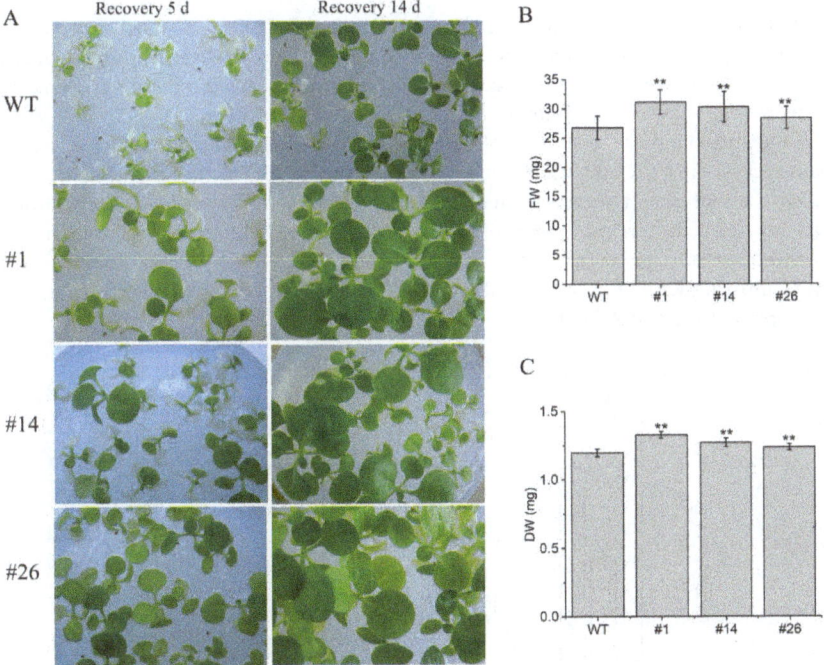

Figure 6. Phenotype, fresh weight (FW), and dry weight (DW) 7 d after completely submerged treatment for 20-day-old WT and transgenic tobacco lines seedlings. (**A**) Phenotype analysis; (**B**) fresh weight; (**C**) dry weight. WT: Wild type; #1, #14, #26: Transgenic tobacco lines. The data represent averages ± SE of three biological repeats with three measurements per sample. ** indicates significant differences in comparison with the WT at $p < 0.01$.

Figure 7. Morphological adaptations of two-month-old transgenic lines and WT plants were observed subjected to root submerged. (**A**) No waterlogging damage; (**B**) waterlogging for seven days. WT: Wild type; #1, #14, #26: Transgenic tobacco lines. Blue arrows indicated the aerial roots

2.4. Physiological Changes in Transgenic Plants under Waterlogging Stress

Alcoholic fermentation through the coupled activity of PDC and ADH enzyme is of great importance in a plant's ability to tolerate waterlogging [21–23]. All three transgenic lines showed improved resistance to waterlogging stress, and had similar results. The result of transgenic line #14 to improve waterlogging was a medium for all three transgenic lines. Transgenic line #14 was selected to explore the physiological mechanism of the waterlogging stress tolerance conferred by *AdRAP2.3* gene overexpression. PDC and ADH enzyme activities in root of WT and transgenic plants were measured in the control and in roots submerged for 24 d treatment (Figure 8). In the control, the PDC and ADH enzyme activities in the roots of transgenic lines are significantly higher than those of the WT. In terms of waterlogging stress, PDC and ADH enzyme activities in roots of transgenic lines are also significantly higher than those of WT. It can be concluded that overexpression of the *AdRAP2.3* gene can improve PDC and ADH enzyme activities.

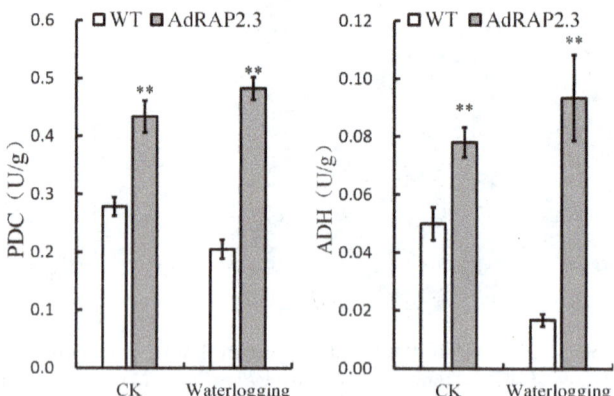

Figure 8. PDC and ADH enzyme activity measurements in roots of wild-type and AdRAP2.3 overexpressing lines #14. CK: normal growth conditions; WT: Wild type. The data represent averages ± SE of three biological repeats with three measurements per sample. ** indicates significant differences in comparison with the WT at $p < 0.01$.

2.5. Waterlogging-Related Genes Changes in Transgenic Plants under Waterlogging Stress

ADH, PDC, HB1, HB2, PCO1, and *PCO2* have been proved as marker genes in response to low oxygen stress [10,12]. There are no significantly difference between WT and transgenic lines on transcription levels of *NtPDC, NtADH, NtHB1, NtHB2,* and *NtPCO1* in roots under control condition (Figure 9). However, under waterlogging stress, the expression levels of *NtPDC, NtADH, NtHB1, NtHB2,* and *NtPCO1* in root of transgenic lines are significantly higher than those of WT (Figure 9). *NtPCO2* expression levels in roots are higher than those of WT under normal condition and waterlogging stress. Taken together, these results indicated overexpression *AdRAP2.3* can upregulate the expression levels of *NtPDC, NtADH, NtHB1, NtHB2, NtPCO1,* and *NtPCO2* under waterlogging stress.

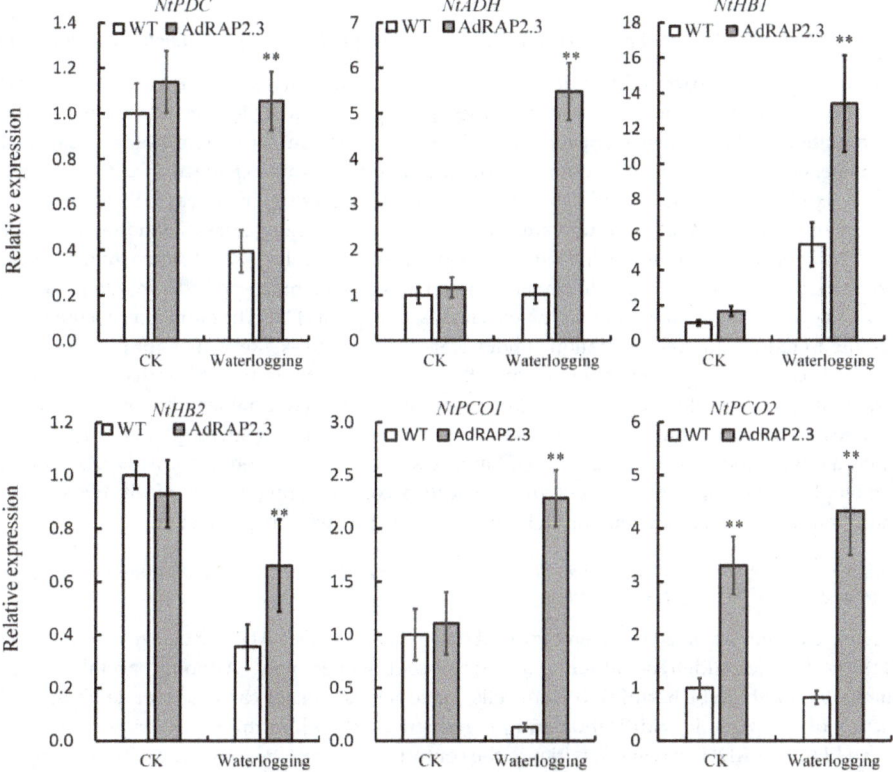

Figure 9. QRT-PCR analysis of the expression of the genes *NtPDC, NtADH, NtHB1, NtHB2, NtPCO1,* and *NtPCO2* in the root line of transgenic tobacco and the wild type during 24 d waterlogging stress. CK: Normal growth conditions; WT: Wild type. The data represent averages ± SE of three biological repeats with three measurements per sample. ** indicates significant differences in comparison with the WT at $p < 0.01$.

3. Discussion

3.1. Kiwi Fruit AdRAP2.3 Plays a Key Role in Resistance to Waterlogging Stress

Waterlogging or submergence caused O_2 deprivation in the soil [24,25]. ERF members are important regulators of low oxygen tolerance extensively studied in many plants [8,10,11]. The ERF family plays a crucial role in the determination of survival of *Arabidopsis* and rice, which could reduce oxygen availability [11,26]. Previous studies showed that there are five ERF VIIs genes in *Arabidopsis*; two ERF VIIs genes (*HYPOXIA RESPONSIVE ERF1/2*) were greatly enhanced at the transcriptional and translational levels by O_2 deprivation at multiple developmental stages, and the other three ERF VIIs genes were constitutively expressed (*RAP2.12, RAP2.2*, and *RAP2.3*) and further upregulated by darkness or ethylene in *A. thaliana* [11,27,28]. Transgenic *Arabidopsis* plants overexpressing *HRE1* showed an improved tolerance of anoxia [11]. The transcriptome sequencing analysis showed that there are 28 AP2/ERF transcription factors regulated by waterlogging in kiwi fruit. In this study, *A. deliciosa AdRAP2.3* gene was induced significantly by waterlogging, and overexpression of the *AdRAP2.3* gene conferred waterlogging stress tolerance in transgenic tobacco plants. When completely submerged, transgenic tobacco lines had a significantly higher survival rate, fresh weight, and dry weight compared to the WT. These results suggested that kiwi fruit *AdRAP2.3* plays a key role in resistance to waterlogging.

3.2. AdRAP2.3 Could Enhance Resistance to Waterlogging through Promoting Pneumatophore Production

Waterlogging may trigger different molecular and physiological disorders in plants. These include significant deterioration of plant water statues [29], decreases in leaf gas exchange variables, photoinhibition of photosystems (PSI and PSII) [30], and decreases in root system biomass. Using a light microscope, we see that the root anatomical structures of plants exhibit changes in root diameter, stele diameter, epidermal thickness, and xylem thickness under waterlogging stress [31]. In addition, the formation of aerenchym tissue is found after treated with waterlogging stress, including an increasing number of aerenchym cells and increasing length of aerenchym cells [31,32]. Aerenchyma formation under waterlogging stress is one of the most effective mechanisms to provide an adequate oxygen supply and overcome the stress-induced hypoxia imposed on plants [33]. The numerous pneumatophores contribute to morphological adaptations under waterlogging stress. Regardless of pneumatophores, stressed oil palm seedlings were able to adjust their leaf water status and gas exchange to cope with waterlogging [34]. Waterlogging subjects plant roots to an anoxic environment, limiting mitochondrial aerobic respiration and causing energy loss. In response to hypoxia stress, plants can temporarily compensate with anaerobic respiration [35]. In this study, the overexpression of *AdRAP2.3* induces pneumatophores under waterlogging in transgenic tobacco, suggesting that kiwi fruit *AdRAP2.3* could enhance resistance to waterlogging through promoting pneumatophore production.

3.3. Kiwi Fruit AdRAP2.3 Enhances Waterlogging Resistance in Transgenic Tobacco through Improving Expression Levels of PDC and ADH Genes

Three enzymes involved in anaerobic metabolic pathways, PDC, ADH, and lactate dehydrogenase (LDH), produce acetaldehyde, ethanol, and lactic acid, respectively. Although acetaldehyde and ethanol are thought to be harmful to plant cells, lactic acid is a major cause of root death through its induction of cytoplasmic acidification and pH reduction [36–38]. In this study, transgenic lines can improve PDC and ADH enzyme activities in the control. In response to long-term waterlogging stress, the PDC and ADH enzyme activities of WT decrease, whereas those of transgenic lines increase to sustain substrate-level adenosine triphosphate (ATP) production and promote hypoxia acclimation. These results indicate that *AdRAP2.3* promotes PDC and ADH enzyme activities.

Previously, a large number of high-throughput sequence (Tag-seq) analyses based on the Solexa Genome Analyzer platform were performed to analyze the gene expression profiling of plants under waterlogging stress. Differentially expressed genes (DEGs) are obtained, mainly linked to carbon metabolism, photosynthesis, reactive oxygen species generation/scavenging, and hormone synthesis/signaling [17,39–42]. Some waterlogging-responsive genes were isolated and identified, such as *RAP2.12, RAP2.2* [13], *HRE1, HRE2, AdPDC1* [43], *AdPDC2* [44], *AdADH1* [45], and *AdADH2* [45]. *RAP2.12* mRNA was described as sufficient for activating the anaerobic response in *Arabidopsis* [11]. The closest RAP2.12 homologue, RAP2.2, has been suggested to be functionally redundant in the induction of the anaerobic gene expression [13,46]. *HRE1* and *HRE2* are expressed at low levels under aerobic conditions and strongly upregulated by hypoxia [11]. Overexpression of *AdPDC1*, *AdPDC2*, *AdADH1*, and *AdADH2* in *Arabidopsis* enhanced waterlogging tolerance [43–45]. RAP2.12 was re-localized from the plasma membrane to the nucleus as O_2 concentrations declined, with increased accumulation of hypoxia-responsive mRNAs, including PDC1 and hypoxia-responsive ERF1/2 (HRE1/2) [10]. In this study, we further analyzed the expression of the hypoxia marker genes *NtADH*, *NtPDC*, *NtHB1*, *NtHB2*, *NtPCO1*, and *NtPCO2* in the roots of WT and a transgenic line under submerged root stress, and the results showed that *AdRAP2.3* regulates these hypoxia marker genes mRNA levels under waterlogging conditions. Thus, kiwi fruit *AdRAP2.3* enhances waterlogging resistance in transgenic tobacco by improving expression levels of *PDC* and *ADH* genes.

In summary, the results showed that the increase of *AdRAP2.3* expression during waterlogging stress was much higher than that during other environmental stresses and that the kiwi fruit *AdRAP2.3* gene is required during waterlogging stress. Upon roots being submerged, transgenic tobacco lines grew aerial roots earlier than the WT. Overexpression of the *AdRAP2.3* gene in transgenic tobacco improved the activities of the PDC and ADH enzymes, and the expression levels of waterlogging mark genes *NtPDC* and *NtADH* in roots under waterlogging treatment. Overall, these results suggested that *AdRAP2.3* might play an important role in resistance to waterlogging through regulation of *PDC* and *ADH* genes in kiwi fruit. In future studies we will concentrate on introducing the *AdRAP2.3* gene into kiwi fruit varieties with poor waterlogging resistance, such as 'Hongyang', to determine whether this can enhance resistance to waterlogging.

4. Materials and Methods

4.1. Plant Materials and Growth Conditions

The kiwi fruit cultivar 'Jinkui' was obtained from the Institute of Botany (Nanjing), Jiangsu Province and Chinese Academy of Sciences, China. The cutting seedlings from 'Jinkui' were grown in pots containing a 7:2:1 mixture of peat mold, vermiculite, and perlite in the greenhouse. Tissue culture seedlings of 'Jinkui' grew in a rooting medium (1/2 MS medium containing 0.6 mg/mL 1-naphthlcetic acid (NAA)) for one month. Shoots of 'Jinkui' were selected, surface sterilized, and grown in MS medium for one week. The conditions were: average temperature of 25 °C, relative air humidity of 60%, photoperiod of 16 h/8 h (light/dark), and quantum irradiance of 160 μmol m^{-2} s^{-1}.

4.2. Treatments

To analyze the tissue-specific gene expression, different organs and tissues of 'Jinkui', including the root, stem, leaf, anthocaulus, petal, pistil, calyx, ovary, stamen, and fruitlet (20 days after full blossom, DAFB), were collected. To analyze gene expression patterns in response to hormones, 0.1 mM SA, 0.05 mM MeJA, 0.01 mM ACC, and 0.01 mM ABA were sprayed on the surface of the culture tissue seedlings leaves, and the leaves were collected at 0, 4, 12, and 48 h to analyze the gene expression pattern in response to stresses including heat, cold, salt, waterlogging, and drought. For heat and cold stress, the tissue seedlings were cultured at 48 °C for 0, 2, and 4 h, had a 23 °C recovery for 6 h, and then were exposed to 4 °C for 0, 4, 12, and 48 h, after which the leaves were collected. For salt stress, the seedlings were cultured in 28 cm × 14 cm × 14 cm containers with 0.2 M NaCl, and the leaves were collected at 0, 4, 12, and 48 h. For waterlogging, the seedlings were waterlogged in a 28 cm × 14 cm × 14 cm container filled with tap water to 2.5 cm above the level of the soil surface, and roots were sampled at 0, 24, 48, and 96 h. For drought, the cutting seedlings were cultured without watering for 14 d (the control was irrigated normally) and the leaves were sampled [45]. To analyze the gene expression pattern in response to *Psa*, bacterial cells were suspended in distilled water and adjusted to an OD_{600} = 0.2, then injected into the 'Jinkui' shoots, which were carved with a knife for 0, 24, 48, or 96 h, and then phloem from the shoots was collected [47]. Different samples were snap frozen in liquid nitrogen and stored at −80 °C for later experiments.

4.3. RNA Extraction and cDNA Synthesis

Total RNA was extracted from samples according to a method reported previously [48]. The purity and content of total RNA were detected by a spectrophotometer (Bruker BioSpin GmbH, Rheinstetten, Germany) and 1.0% agarose electrophoresis. The cDNA was achieved with a PrimeScript^{TM} RT reagent kit with gDNA Eraser (Perfect Real Time, TaKaRa, Cat. #RR047Q, Dalian, China), which could eliminate the residual DNA. The cDNA samples were diluted 1:10 with sterile double distilled water and stored at −20 °C before being used.

4.4. Gene Clone and Sequence Analysis

The cDNA sample from the 'Jinkui' root treatment with waterlogging for four days was used to amplify the complete CDS of *AdRAP2.3*. Gene-specific primers F1 and F2 were designed according to the comp67160_0_seq1 sequence (Table S1) [17]. The open reading frame (ORF) was predicted by DNAstar 7.1.0(http://korwin-mikke.pl). Multiple alignments of the deduced amino acid sequence were performed using the the BioEdit software (v 7. 0. 5, Ibis Therapeutics, Carlsbad, CA, USA), and a phylogenetic tree was constructed by 1000 Bootstrap statistical tests with the Neighbor Joining (NJ) model by Mega 7.0 (https://www.megasoftware.net/). Molecular weight and isoelectric point were analyzed by Exspay (https://web.expasy.org/protparam/). The three-dimensional structure of *AdRAP2.3* was predicted by Swiss-plot (http://swissmodel.expasy.org/).

4.5. Gene Expression Analysis Using Quantitative Real-Time PCR

The qRT-PCR was carried out on an Applied Biosystems 7300 Real Time PCR System (Applied Biosystems, Waltham, MA, USA) using TaKaRa Company SYBR Premix Ex TaqTM II(Perfect Real Time, TaKaRa, code: DRR041A, Dalian, China). *AdActin* was used as internal reference gene to monitor cDNA abundance [49]. The quantitative PCR reaction system with specific primer contains 1 μL cDNA template, 10 μL 2 × SYBR Premix Ex Taq^{TM} II, 0.3μL (10 pm) of each primer (Table S1), and 8.4 μL ddH$_2$O. The primer sequences used are listed in Table S1. The reaction procedure is as follows: denaturation at 95 °C for 1 min, 95 °C for 20 s, 57 °C for 20 s, and 72 °C for 20 s; 45 cycles in total. Each sample set is repeated three times. After the reaction, the $2^{-\triangle\triangle Ct}$ method was used to analyze the gene expression level. The statistical significance was assessed using SPSS 17.0 (SPSS Corp., Chicago, IL, USA).

4.6. Construction Binary Vector and Transformation of Tobacco

The coding sequence of *AdRAP2.3* was amplified by PCR using a specific primer pair (F1 and F2) modified to include 5′BamH I and 3′Sac I restriction sites. The fragment was inserted into the binary vector pCAMBIA 1301 under the control of the *Cauliflower mosaic virus* (CaMV35S) promoter. Then, the plasmid was introduced into *Agrobacterium tumefaciens*. Agrobacterium-mediated transformation of tobacco was performed by the leaf disc method [50]. The transgenic plants were detected by GUS staining and PCR analysis. The seeds of transgenic lines were harvested at the same stage and stored for subsequent analysis.

4.7. Phenotype Analysis of Transgenic Tobacco under Waterlogging Resistance

Seeds from T$_1$ progeny transgenic lines (#1, #14, and #26) and WT were surface sterilized and sown on the MS medium for 10 days; the death rate was measured after they were completely submerged for 48 h. Seedlings of transgenic lines and WT grew on MS medium for 20 d, the waterlogging treatment was performed for 7 d, and then seedlings were returned to normal growth conditions for 14 d. After 14 d recovery, the fresh weight and dry weight were measured. *Nicotiana tabacum* seedlings (#1, #14, #26, and WT) were transplanted into pots from the medium and grown in a greenhouse for two months under a 16/8-h light/dark cycle at 22/25 °C and 60% relative humidity. The pots were flooded in 28 cm × 14 cm × 14 cm containers filled with tap water to 2.5 cm above the level of the soil surface. The seedlings' phenotypic changes during waterlogging stress assays were observed and photographed. The roots of transgenic lines and WT were collected after 24 d waterlogging and were later used for the measurement of enzyme activities and expression levels of downstream genes.

4.8. Measurement of Anaerobic Respiration and ADH and PDC Activities

The enzyme ADH and PDC activities in roots of transgenic lines #14 and WT after 24 d waterlogging were measured spectrophotometrically by monitoring the oxidation of Nicotinamide adenine dinucleotide (NADH) at 340 nm [21]. The PDC assay reaction was carried out for 60 s at 25 °C. The ADH assay reaction was carried out for 10 min at 37 °C. One unit of PDC or ADH was defined as the amount of enzyme required to decompose 1 μmol of NADH per minute per gram fresh weight; at least 10 independent plants were evaluated in each test, and all tests were repeated three times.

4.9. Expression Analysis of Downstream Genes

The expression levels of downstream genes in roots of transgenic line #14 and the WT after 24 d waterlogging were analyzed by qRT-PCR. The downstream genes *NtPDC*, *NtADH*, *NtHB1*, *NtHB2*, *NtPCO1*, and *NtPCO2* were acquired from NCBI (https://www.ncbi.nlm.nih.gov/). *NtTub* was used as an internal reference gene to monitor cDNA abundance [50]. The primers are listed in Table S1. Statistically significant differences were calculated with SPSS 17.0 (SPSS Corp., Chicago, IL, USA).

Supplementary Materials: The following are available online at http://www.mdpi.com/1422-0067/20/5/1189/s1. **Figure S1**: Nucleotide sequence and deduced amino acid of *AdRAP2.3* gene from *A. deliciosa*. **Figure S2**: Protein sequence alignment of AdRAP2.3 with the PSR94738.1, HRE1, RAP2.3 proteins. **Figure S3**: Secondary and three-dimensional structure of main part of AdRAP2.3. **Figure S4**: Neighbor-joining phylogenetic analysis of AdRAP2.3 with other ERF VII subgroup proteins from Arabidopsis (RAP2.2, RAP2.12, RAP2.3, HRE2, HRE1), Oryza (SUB1B, SUB1A-1, SUB1C, SK1, SK2), Capsicum (CaPF1), Lycopersicon (JERF3) and Actinidia (PSR94738.1). **Figure S5**: GUS and PCR detection of transgenic tobacco. **Table S1**: Sequence of primers.

Author Contributions: J.-Y.Z. designed and initiated this study. G.W. and Z.-H.J. carried out the bioinformatics analyses. D.-L.P. performed the experiments. D.-L.P. and J.-Y.Z. wrote the manuscript. T.W. and Z.-R.G. helped in discussions of the manuscript. All authors read and approved the final manuscript.

Acknowledgments: This research is supported by the Natural Science Foundation of Jiangsu Province of China (General Program, BK20171328).

Conflicts of Interest: The authors declare no conflict of interest.

References

1. Nakano, T.; Suzuki, K.; Fujimura, T.; Shinshi, H. Genome-wide analysis of the ERF gene family in *Arabidopsis* and rice. *Plant Physiol.* **2006**, *140*, 411–432. [CrossRef] [PubMed]

2. Li, Z.J.; Tian, Y.S.; Xu, J.; Fu, X.Y.; Gao, J.J.; Wang, B.; Han, H.J.; Wang, L.J.; Peng, R.H.; Yao, Q.H. A tomato ERF transcription factor, SlERF84, confers enhanced tolerance to drought and salt stress but negatively regulates immunity against *Pseudomonas syringae* pv. tomato *DC3000*. *Plant Physiol. Biochem.* **2018**, *132*, 683–695. [CrossRef] [PubMed]

3. Riechmann, J.L.; Meyerowitz, E.M. The AP2/EREBP family of plant transcription factors. *Biol. Chem.* **1998**, *379*, 633–646. [PubMed]

4. Liang, Y.Q.; Li, X.S.; Zhang, D.Y.; Gao, B.; Yang, H.L.; Wang, Y.C.; Guan, K.Y.; Wood, A.J. *ScDREB8*, a novel A-5 type of DREB gene in the desert moss *Syntrichia caninervis*, confers salt tolerance to *Arabidopsis*. *Plant Physiol. Biochem.* **2017**, *120*, 242–251. [CrossRef] [PubMed]

5. Wu, D.Y.; Ji, J.; Wang, G.; Guan, C.F.; Jin, C. LchERF, a novel ethylene-responsive transcription factor from *Lycium chinense*, confers salt tolerance in transgenic tobacco. *Plant Cell Rep.* **2014**, *33*, 2033–2045. [CrossRef] [PubMed]

6. Li, X.P.; Zhu, X.Y.; Mao, J.; Zou, Y.; Fu, D.W.; Chen, W.X.; Lu, W.J. Isolation and characterization of ethylene response factor family genes during development, ethylene regulation and stress treatments in papaya fruit. *Plant Physiol. Biochem.* **2013**, *70*, 81–92. [CrossRef] [PubMed]

7. Sakuma, Y.; Liu, Q.; Dubouzet, J.G.; Abe, H.; Shinozaki, K.; Yamaguchi-Shinozaki, K. DNA-Binding specificity of the ERF/AP2 domain of Arabidopsis DREBs, transcription factors involved in dehydration- and cold-inducible gene expression. *Biochem. Biophys. Res. Commun.* **2002**, *290*, 998–1009. [CrossRef]

8. Gibbs, D.J.; Lee, S.C.; Isa, N.M.; Gramuglia, S.; Fukao, T.; Bassel, G.W.; Correia, C.S.; Corbineau, F.; Theodoulou, F.L.; Bailey-Serres, J.; et al. Homeostatic response to hypoxia is regulated by the N-end rule pathway in plants. *Nature* **2011**, *479*, 415–418. [CrossRef]

9. Gasch, P.; Fundinger, M.; Müller, J.T.; Lee, T.; Bailey-serres, J.; Mustroph, A. Redundant ERF-VII transcription factors bind an evolutionarily-conserved cis-motif to regulate hypoxia-responsive gene expression in *Arabidopsis. Plant Cell* **2015**, *28*, 160–180. [CrossRef]

10. Voesenek, L.A.C.J.; Bailey-Serres, J. Flood adaptive traits and processes: An overview. *New Phytol.* **2015**, *206*, 57–73. [CrossRef]

11. Licausi, F.; Kosmacz, M.; Weits, D.A.; Giuntoli, B.; Giorgi, F.; Voesenek, L.A.C.J.; Perata, P.; Dongen, J.T. Oxygen sensing in plants is mediated by an N-end rule pathway for protein destabilization. *Nature* **2011**, *479*, 419–422. [CrossRef] [PubMed]

12. Paul, M.V.; Iyer, S.; Amerhauser, C.; Lehmann, M.; Dongen, J.T.; Geigenberger, P. Oxygen sensing via the ethylene response transcription factor RAP2.12 affects plant metabolism and performance under both normoxia and hypoxia. *Plant Physiol.* **2016**, *172*, 141–153. [CrossRef] [PubMed]

13. Hinz, M.; Wilson, I.W.; Yang, J.; Buerstenbinder, K.; Llewellyn, D.; Dennis, E.S.; Sauter, M.; Dolferus, R. *Arabidopsis* RAP2.2: An ethylene response transcription factor that is important for hypoxia survival. *Plant Physiol.* **2010**, *153*, 757–772.

14. Perata, P.; Voesenek, L.A.C.J. Submergence tolerance in rice requires *Sub1A*, an ethylene-response-factor-like gene. *Trends Plant Sci.* **2007**, *12*, 43–46. [CrossRef] [PubMed]

15. Liu, J.X.; Li, J.Y.; Wang, H.N.; Fu, Z.D.; Liu, J.; Yu, Y.X. Identification and expression analysis of ERF transcription factor genes in petunia during flower senescence and in response to hormone treatments. *J. Exp. Bot.* **2011**, *62*, 825–840. [CrossRef] [PubMed]

16. Zhang, J.Y.; Pan, D.L.; Wang, G.; Xuan, J.P.; Wang, T.; Guo, Z.R. Genome-wide analysis and expression pattern of the AP2/ERF gene family in kiwifruit under waterlogging stress treatment. *Int. J. Environ. Agric. Res.* **2017**, *3*, 83–90.

17. Zhang, J.Y.; Huang, S.N.; Mo, Z.H.; Xuan, J.P.; Jia, X.D.; Wang, G.; Guo, Z.R. De novo transcriptome sequencing and comparative analysis of differentially expressed genes in kiwifruit under waterlogging stress. *Mol. Breed.* **2015**, *35*, 208. [CrossRef]

18. Vidhyasekaran, P. Abscisic acid signaling system in plant innate immunity. In *Plant Hormone Signaling Systems in Plant Innate Immunity. Signaling and Communication in Plants*; Springer: Dordrecht, The Netherlands, 2015; Volume 2, pp. 245–309.

19. Pál, M.; Szalai, G.; Kovács, V.; Gondor, O.K.; Janada, T. Salicylic acid-mediated abiotic stress tolerance. In *Salicylic Acid*; Hayat, S., Ahmad, A., Alyemeni, M., Eds.; Springer: Dordrecht, The Netherlands, 2013; pp. 183–247.

20. Janse, J.D.; Scortichini, M. Characterization of *Pseudomonas Syringae* pv. *actinidiae*, the causal agent of bacterial canker of kiwifruit by whole cell protein electrophoresis and fatty acid analysis. In *Pseudomonas Syringae Pathovars and Related Pathogens. Developments in Plant Pathology*; Rudolph, K., Burr, T.J., Mansfield, J.W., Stead, D., Vivian, A., von Kietzell, J., Eds.; Springer: Dordrecht, The Netherlands, 1997; Volume 9, p. 499.

21. Yin, D.M.; Chen, S.M.; Chen, F.D.; Guan, Z.Y.; Fang, W.M. Morpho-anatomical and physiological responses of two Dendranthema species to waterlogging. *Environ. Exp. Bot.* **2010**, *68*, 122–130. [CrossRef]

22. Ricard, B.; Couee, I.; Raymond, P.; Saglio, P.H.; Saintges, V.; Pradet, A. Plant metabolism under hypoxia and anoxia. *Plant Physiol. Biochem.* **1994**, *32*, 1–10.

23. Tadege, M.; Dupuis, I.; Kuhlemeier, C. Ethanolic fermentation: New functions for an old pathway. *Trends Plant Sci.* **1999**, *4*, 320–325. [CrossRef]

24. Bailey-Serres, J.; Voesenek, L.A. Flooding stress: Acclimations and genetic diversity. *Ann. Rev. Plant Boil.* **2008**, *59*, 313–339. [CrossRef] [PubMed]

25. Yamauchi, T.; Watanabe, K.; Fukazawa, A.; Mori, H.; Abe, F.; Kawaguchi, K.; Oyanagi, A.; Nakazono, M. Ethylene and reactive oxygen species are involved in root aerenchyma formation and adaptation of wheat seedlings to oxygen-deficient conditions. *J. Exp. Bot.* **2014**, *65*, 261–273. [CrossRef] [PubMed]

26. Licausi, F.; Van Dongen, J.T.; Giuntoli, B.; Novi, G.; Santaniello, A.; Geigenberger, P.; Perata, P. HRE1 and HRE2, two hypoxia-inducible ethylene response factors, affect anaerobic responses in *Arabidopsis thaliana*. *Plant J.* **2010**, *62*, 302–315. [CrossRef] [PubMed]

27. Gibbs, D.J.; Bacardit, J.; Bachmair, A.; Holdsworth, M.J. The eukaryotic N-end rule pathway: Conserved mechanisms and diverse functions. *Trends Cell Biol.* **2014**, *24*, 603. [CrossRef] [PubMed]

28. Licausi, F.; Pucciariello, C.; Perata, P. New role for an old rule: N-end rule-mediated degradation of ethylene responsive factor proteins governs low oxygen response in plants. *J. Integr. Plant Boil.* **2013**, *55*, 31–39. [CrossRef] [PubMed]

29. Rasheed-Depardieu, C.; Parelle, J.; Tatin-Froux, F.; Parent, C.; Capelli, N. Short-term response to waterlogging in Quercus petraea and Quercus robur: A study of the root hydraulic responses and the transcriptional pattern of aquaporins. *Plant Physiol. Biochem.* **2015**, *97*, 323–330. [CrossRef] [PubMed]

30. Yan, K.; Zhao, S.J.; Cui, M.X.; Han, G.X.; Wen, P. Vulnerability of photosynthesis and photosystem I in Jerusalem artichoke (*Helianthus tuberosus* L.) exposed to waterlogging. *Plant Physiol. Biochem.* **2018**, *125*, 239–246. [CrossRef] [PubMed]

31. Purnobasuki, H.; Nurhidayati, T.; Hariyanto, S.; Jadid, N. Data of root anatomical responses to periodic waterlogging stress of tobacco (*Nicotiana tabacum*) varieties. *Data Brief* **2018**, *20*, 2012–2016. [CrossRef] [PubMed]

32. Yin, D.M.; Chen, S.M.; Chen, F.D.; Jiang, J.F. Ethylene promotes induction of aerenchyma formation and ethanolic; fermentation in waterlogged roots of *Dendranthema* spp. *Mol. Biol. Rep.* **2013**, *40*, 4581–4590. [CrossRef]

33. Zhang, X.; Fan, Y.; Shabala, S.; Koutoulis, A.; Shabala, L.; Johnson, P.; Hu, H.L.; Zhou, M.X. A new major-effect QTL for waterlogging tolerance in wild barley (*H. spontaneum*). *Theor. Appl. Genet.* **2017**, *130*, 1559–1568. [CrossRef]

34. Ponte, N.H.T.; Santos, R.I.; Filho, W.R.; Cunha, R.L.; Magalhaes, M.M.; Pinheiro, H.A. Morphological assessments evidence that higher number of pneumatophores improves tolerance to long-term waterlogging in oil palm (*Elaeis guineensis*) seedlings. *Flora* **2019**, *250*, 52–58. [CrossRef]

35. Wang, W.Q.; Zhang, F.S.; Zheng, Y.Z.; Mei, H.X. Comparison of morphology, physiology and mineral element contents among genotypes of sesame with different tolerance to waterlogging under anaerobic condition. *Chin. J. Appl. Ecol.* **2002**, *13*, 421–424.

36. Zhou, C.P.; Bai, T.; Wang, Y.; Wu, T.; Zhang, X.Z.; Xu, X.F.; Han, Z.H. Morpholoical and enzymatic responses to waterlogging in three *Prunus* species. *Sci. Hortic.* **2017**, *221*, 62–67. [CrossRef]

37. An, Y.Y.; Qi, L.; Wang, L.J. ALA pretreatment improves waterlogging tolerance of fig plants. *PLoS ONE* **2016**, *11*, e0147202. [CrossRef] [PubMed]

38. Bahmaniar, M.A. The influence of continuous rice cultivation and different waterlogging periods on the morphology, clay mineralogy, Eh, pH and K in paddy soils. *Eurasian Soil Sci.* **2008**, *41*, 87–92. [CrossRef]

39. Qi, X.H.; Xu, X.W.; Lin, X.J.; Zhang, W.J.; Chen, X.H. Identification of differentially expressed genes in cucumber (*Cucumis sativus* L.) root under waterlogging stress by digital gene expression profile. *Genomics* **2012**, *99*, 160–168. [CrossRef] [PubMed]

40. Qi, B.Y.; Yang, Y.; Yin, Y.L.; Xu, M.; Li, H.G. De novo sequencing, assembly, and analysis of the Taxodium 'Zhongshansa' roots and shoots transcriptome in response to short-term waterlogging. *BMC Plant Biol.* **2014**, *14*, 201. [CrossRef] [PubMed]

41. Xu, X.W.; Chen, M.Y.; Ji, J.; Xu, J.; Qi, X.H.; Chen, X.H. Comparative RNA-seq based transcriptome profiling of waterlogging response in cucumber hypocotyls reveals novel insights into the de novo adventitious root primordia initiation. *BMC Plant Biol.* **2017**, *17*, 129. [CrossRef]

42. Zhang, P.; Lyu, D.G.; Jia, L.T.; He, J.L.; Qin, S.J. Physiological and de novo transcriptome analysis of the fermentation mechanism of *Cerasus sachalinensis* roots in response to short-term waterlogging. *BMC Genom.* **2017**, *18*, 649. [CrossRef]

43. Zhang, J.Y.; Huang, S.N.; Wang, G.; Xuan, J.P.; Guo, Z.R. Overexpression of *Actinidia deliciosa pyruvate decarboxylase* 1 gene enhances waterlogging stress in transgenic *Arabidopsis thaliana*. *Plant Physiol. Biochem.* **2016**, *106*, 244–252. [CrossRef]

44. Luo, H.T.; Zhang, J.Y.; Wang, G.; Jia, Z.H.; Huang, S.N.; Wang, T.; Guo, Z.R. Functional characterization of waterlogging and heat stresses tolerance gene *Pyruvate decarboxylase* 2 from *Actinidia deliciosa*. *Int. J. Mol. Sci.* **2017**, *18*, 2377. [CrossRef] [PubMed]

45. Zhang, J.Y.; Huang, S.N.; Chen, Y.H.; Wang, G.; Guo, Z.R. Identification and characterization of two waterlogging responsive alcohol dehydrogenase genes (*AdADH1* and *AdADH2*) in *Actinidia deliciosa*. *Mol. Breed.* **2017**, *37*, 52. [CrossRef]

46. Weits, D.A.; Giuntoli, B.; Kosmacz, M.; Parlanti, S.; Hubberten, H.M.; Riegler, H.; Hoefgen, R.; Perata, P.; Dongen, J.T.; Licausi, F. Plant cysteine oxidases control the oxygen-dependent branch of the N-end-rule pathway. *Nat. Commun.* **2014**, *5*, 3425. [CrossRef] [PubMed]

47. Wang, T.; Wang, G.; Jia, Z.H.; Pan, D.L.; Zhang, J.Y.; Guo, Z.R. Transcriptome analysis of kiwifruit in response to *Pseudomonas syringae* pv. actinidiae infection. *Int. J. Mol. Sci.* **2018**, *19*, 373. [CrossRef] [PubMed]

48. Cai, B.H.; Zhang, J.Y.; Gao, Z.H.; Qu, S.C.; Tong, Z.G.; Mi, L.; Qiao, Y.S.; Zhang, Z. An improved method for isolation of total RNA from the leaves of *Fragaria* spp. *Jiangsu J. Agric. Sci.* **2008**, *24*, 875–877.

49. Yin, X.R.; Allan, A.C.; Xu, Q.; Burdon, J.; Dejnoprat, S.; Chen, K.S.; Fergusom, L.B. Differential expression of kiwifruit genes in response to postharvest abiotic stress. *Postharvest Biol. Technol.* **2012**, *66*, 1–7. [CrossRef]

50. Horsch, R.B. A simple and general method for transferring genes into plants. *Science* **1985**, *227*, 1229–1231.

International Journal of
Molecular Sciences

Article

DNA Methylation Analysis of Dormancy Release in Almond (*Prunus dulcis*) Flower Buds Using Epi-Genotyping by Sequencing

Ángela S. Prudencio [1,†], Olaf Werner [2,†], Pedro J. Martínez-García [1], Federico Dicenta [1], Rosa M. Ros [2] and Pedro Martínez-Gómez [1,*]

1   Department of Plant Breeding, CEBAS-CSIC, P.O. Box 164, Espinardo, 30100 Murcia, Spain;
    asanchez@cebas.csic.es (A.S.P.); pjmartinez@cebas.csic.es (P.J.M.-G.); fdicenta@cebas.csic.es (F.D.)
2   Department of Plant Biology, Faculty of Biology, University of Murcia, Espinardo, 30100 Murcia, Spain;
    werner@um.es (O.W.); rmros@um.es (R.M.R.)
*   Correspondence: pmartinez@cebas.csic.es; Tel.: +34-968-396-200
†   These authors contributed equally to this work.

Received: 24 September 2018; Accepted: 7 November 2018; Published: 10 November 2018

Abstract: DNA methylation and histone post-translational modifications have been described as epigenetic regulation mechanisms involved in developmental transitions in plants, including seasonal changes in fruit trees. In species like almond (*Prunus dulcis* (Mill.) D.A: Webb), prolonged exposure to cold temperatures is required for dormancy release and flowering. Aiming to identify genomic regions with differential methylation states in response to chill accumulation, we carried out Illumina reduced-representation genome sequencing on bisulfite-treated DNA from floral buds. To do this, we analyzed almond genotypes with different chilling requirements and flowering times both before and after dormancy release for two consecutive years. The study was performed using epi-Genotyping by Sequencing (epi-GBS). A total of 7317 fragments were sequenced and the samples compared. Out of these fragments, 677 were identified as differentially methylated between the almond genotypes. Mapping these fragments using the *Prunus persica* (L.) Batsch v.2 genome as reference provided information about coding regions linked to early and late flowering methylation markers. Additionally, the methylation state of ten gene-coding sequences was found to be linked to the dormancy release process.

Keywords: *Prunus*; flowering; bisulfite sequencing; genomics; epigenetics; breeding

---

1. Introduction

The almond tree (*Prunus dulcis* (Mill.) D.A. Webb), like the rest of the *Prunus* species, is a deciduous fruit tree that undergoes a cyclical process of flowering, sprouting, development, and winter rest, called dormancy. The dormancy state protects the plant from potential damage from cold weather during the winter [1,2]. The dormancy period is overcome when the tree accumulates sufficient chilling hours (the chilling requirement). After dormancy release, the tree is able to sprout and flower under favorable climatic conditions [3,4]. The study of molecular mechanisms leading to dormancy release and flowering is of great interest for almond breeding programs aiming to adapt new cultivars to specific growing areas [5,6]. The dormancy release process involves sensing environmental cues (such as temperature), signal transduction, and gene expression regulation to establish a suitable response according to the stimuli received [7,8]. Transcription reprogramming leading to dormancy release may thus be mediated by epigenetic mechanisms [9,10].

Epigenetics are chemical modifications affecting DNA or structural proteins (histones) within the chromatin. Two types of epigenetic modifications have been described: DNA methylation

(5′ Methylated Cytosine, 5mC) in plants and histone Post-Translational Modifications (PTMs), which include the acetylation and methylation of histones [11,12].

Epigenetic changes are part of the transcriptional regulation machinery of genomes. The dynamic but heritable character of such modifications makes them interesting regulators mediating adaptive responses to environmental changes, such as seasonal cycles, and in the long term, climate change [13]. DNA methylation is associated with cell status stability and regulation of expression. DNA methylation occurs in three sequence contexts: CG and CHG, which are found in promoter and coding regions, and CHH (where H = A, C or T), found in non-coding regions and transposable elements (TEs) [14].

Several works have described the role of epigenetics in the regulation of dormancy in deciduous plant species. Santamaría et al. [15], for instance, described a methylation decrease concomitant with H4 deacetylation and the progress of dormancy release in *Castanea sativa* Mill. In peach (*Prunus persica*), de la Fuente et al. [16] identified a genome-wide pattern of the PTM Trymethylation of Histone 3 on Lis (K) residue 27 (H3K27me3) during bud dormancy release, and Lloret et al. [17] found a relationship between gene expression, PTMs, and sorbitol synthesis during bud dormancy progression and release. Rothkegel et al. [18] showed that DNA methylation is one of the mechanisms participating in the regulation of MADS-box (MCM1-AGAMOUS-DEFICIENS-SRF) genes controlling bud dormancy in sweet cherry (*Prunus avium* L.). In apple (*Malus domestica* (Suckow) Borkh.), genome methylation patterns have been linked to chilling acquisition during dormancy [19]. In the case of almond, preliminary results from the transcriptome sequencing of non-dormant and dormant flower buds showed differential expression in a DNA methyltransferase gene and in the *S-ADENOSYL METHIONINE SYNTHETASE* gene responsible for the synthesis of the molecule SAM (S-adenosyl methionine), which donates the methyl group to the DNA molecule [20]. In addition, DNA methylation phenomena have also been associated with floral self-incompatibility [21] and with bud falling phenomena [22] in this species.

Genome-wide analysis of DNA methylation can be done by bisulfite sequencing, which uses Next Generation Sequencing (NGS) to analyze digested and bisulfite-treated DNA samples. The epi-Genotyping by Sequencing (epiGBS) technique was developed to represent a small part of the genome for cost-effective exploration and comparative analysis of DNA methylation and genetic variation in hundreds of de novo samples. Furthermore, this method makes it possible to genotype samples without a prior reference genome [23].

The objective of this work was to analyze the DNA methylation status of almond flower buds using epi-GBS for the first time. For this purpose, we evaluated dormant and non-dormant flower buds from two almond genotypes with different chilling requirements and flowering times using the epi-GBS protocol.

2. Results

2.1. Evaluation of the Quality of the Epi-GBS Analysis

We sequenced 9518 fragments (about a 1244 kb size) and identified 7317 methylated or unmethylated fragments. Furthermore, we were able to reconstruct the original sequence of 4377 fragments. The total length of the "mock genome" obtained by merging the reconstructed fragments was 662,458 bp. Regarding the quality of this epi-GBS analysis, the absence of a secondary peak towards the right of the read coverage histograms shows that the data do not suffer from PCR duplication bias in either year (Figure 1; data from the flower buds sampled in 2015–2016). These read coverage results show the uniformity of the reads and correct PCR amplification (good quality) around the whole genome in both contexts and seasons of study.

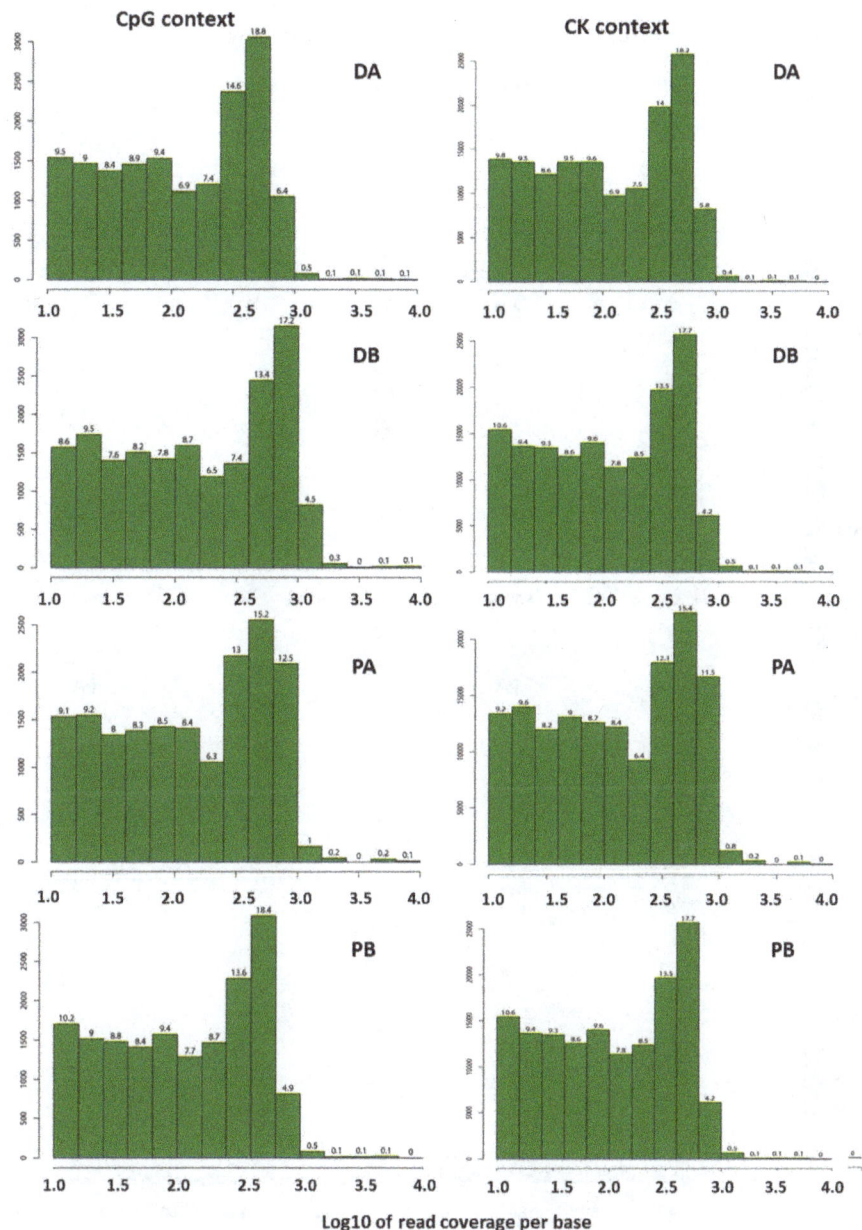

Figure 1. Read coverage of the samples tested in the CpG and CK (including CHG and CHH) contexts during the 2015–2016 season. "D" = 'Desmayo Largueta', "P" = 'Penta', "A" = dormant bud stage, "B" = non-dormant bud stage.

The histograms of CpG methylation showed that roughly 70% to 75% of the cytosine positions in the CpG context of the mock genome were unmethylated, around 10% of the positions were completely methylated and the remaining positions were partially methylated to varying degrees in both seasons of study (Figure 2; data from 2015–2016 season).

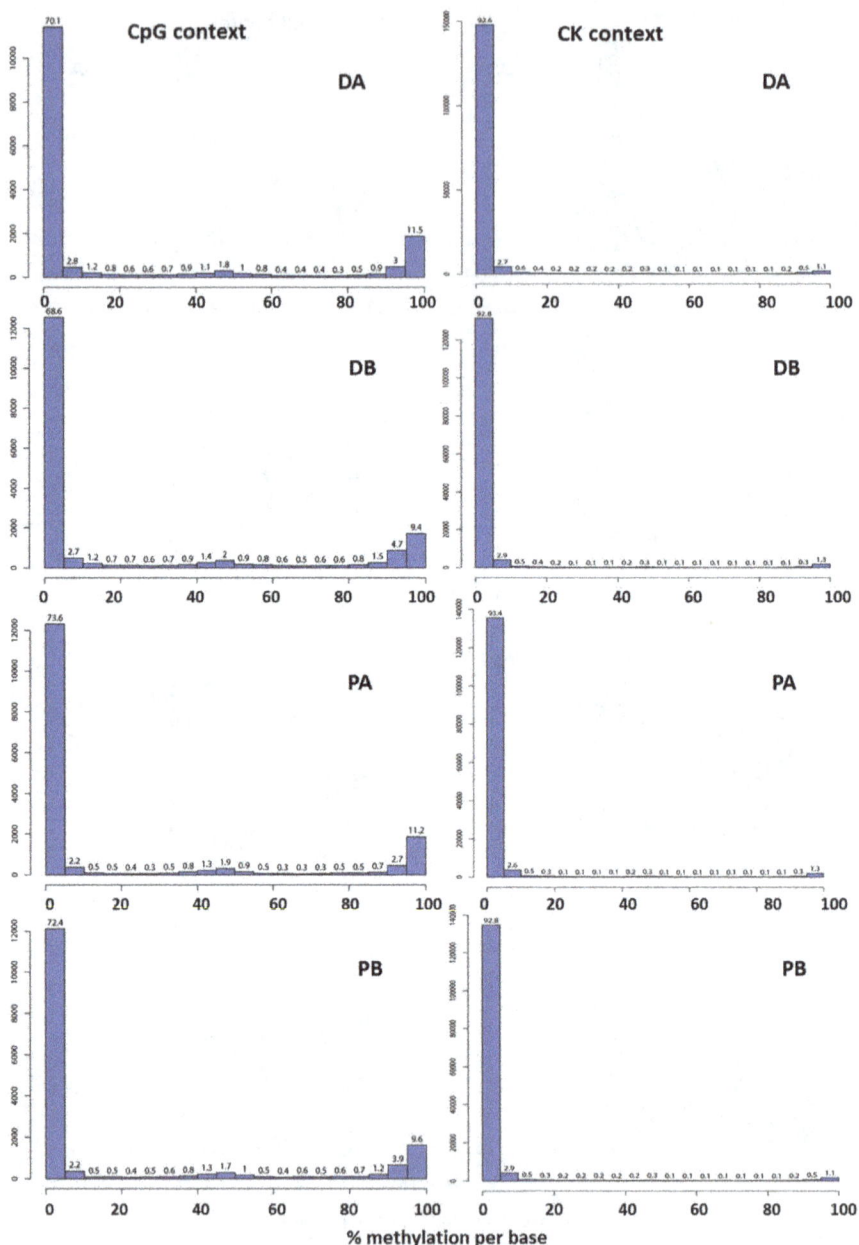

Figure 2. Percentage of DNA methylation of the samples tested in the CpG and CK (including CHG and CHH) contexts during the 2015–2016 season. "D" = 'Desmayo Largueta', "P" = 'Penta', "A" = dormant bud stage, "B" = non-dormant bud stage.

Furthermore, the correlation analyses clearly show that samples of the same variety cluster together independently of the developmental stage. The Pearson's correlation coefficient was constantly 0.99 in comparisons within each variety and in the range of 0.84 to 0.85 in comparisons between samples of different varieties (Figure 3A). These results were also corroborated by clustering analysis, in which samples belonging to the same variety were close together while the two varieties were separated by long branches (Figure 3B). The DNA methylation pattern is generally variety-dependent rather than dormancy-dependent.

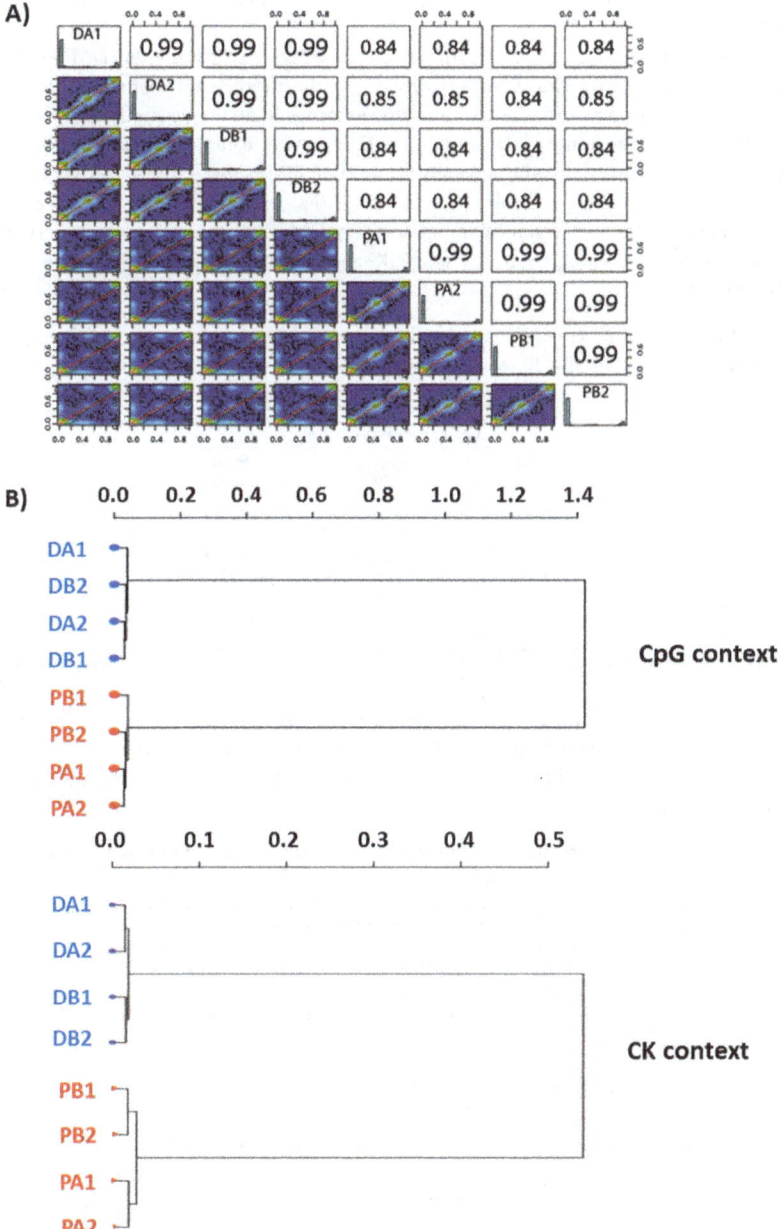

Figure 3. (**A**) Almond correlations and (**B**) clustering analysis of the methylated fragments in both CpG and CK contexts. "D" = 'Desmayo Largueta', "P" = 'Penta', "A" = dormant bud stage, "B"=non-dormant bud stage. "1" samples are from the 2015–2016 season and "2" samples are from the 2016–2017 season.

2.2. Differentyally Methylated Genes Detected

Quantitative analysis showed that 7317 different fragments were methylated in at least one sample: 5109 'Cs' were methylated in 'Desmayo Largueta' A samples; 5089 'Cs' were methylated in 'Desmayo Largueta' B samples; 4955 'Cs' were methylated in 'Penta' A samples; and 5003 'Cs' were methylated in 'Penta' B samples.

Table 1 shows that the number of differentially methylated fragments (DMFs) detected was variable depending on the comparison performed. Furthermore, a total of 677 DMFs were found between 'Desmayo Largueta' and 'Penta' genotype samples in all stages analyzed. However, when comparing dormancy state samples (A and B), 23 DMFs were found between 'Desmayo Largueta' stage A and 'Desmayo Largueta' stage B samples and 48 DMFs were found between 'Penta' stage A samples and 'Penta' stage B samples. Of those DMFs, ten were common between 'Desmayo Largueta' and 'Penta' in the A to B stage comparison. The DMFs were divided into hypermethylated or hypomethylated categories using 'Penta' or 'stage B' samples as the reference. The fragment sequences are included in Table S1.

Table 1. Number of differentially methylated fragments (DMFs) detected by epi-Genotyping by Sequencing (epi-GBS) according to sample comparisons.

Differentially Methylated Fragments		
D–P genotype comparison		
Hypo (<)	Hyper (>)	Stages
307	370	A and B
	677 DMFs	
A–B stage comparison		
Hypo (<)	Hyper (>)	Genotype
3	20	D
21	27	P
3	7	D and P
	10 DMFs	

"D" = 'Desmayo Largueta', "P" = 'Penta'; "A" = dormant bud stage, "B" = non-dormant bud stage.

More than 99% of the identified DMFs were mapped on the *Prunus persica* v2.1 genome (Table S2), and those located between 2 kb upstream and 1 kb downstream from the gene coding sequences were selected for subsequent annotation analysis. The number of differentially methylated genes (DMGs) thus identified is shown in Tables 2 and 3.

Table 2. The number of differentially methylated genes (DMGs) identified from sequenced fragments mapping onto the *Prunus persica* genome (v2.1).

Differentially Methylated Genes			
Methylation State	**Gene Position**	**Gene Hits**	**Genes Identified**
Hypermethylated	Upstream	36	
	Inside	291	
	Downstream	134	
Total		461	423
Hypomethylated	Upstream	19	
	Inside	201	
	Downstream	80	
Total		300	281
Equally-methylated	Upstream	6	
	Inside	41	
	Downstream	8	
Total		55	27
Total DMGs		816	731

The methylation state refers to the number of 5' Methylated Cytosines (5mCs) in 'Desmayo Largueta' samples compared to 'Penta' samples. The category "equally methylated" refers to genes whose number of 5mCs was the same between samples but in which the 5mCs were located in different fragment positions. The gene position is based on gene orientation with respect to the fragment-mapping region ("upstream", "inside", and "downstream").

DMGs were classified according to their position with respect to the fragment-mapping region. The most frequently mapped fragments were those within gene regions ("inside" DMGs) followed by the 5′regulatory regions ("downstream" DMGs) and, finally, in the 3′ regions ("upstream" DMGs).

Data shown in Table 2 indicate that DMGs were found as hypermethylated in 'Desmayo Largueta' samples (in both the A and B stages) to a greater extent than in 'Penta' samples (423).

Table 3. The number of differentially methylated genes between the A and B dormancy states of flower buds, identified from sequenced fragments mapped to the *P. persica* genome (v2.1).

Methylation State	Gene Position	Differentially Methylated Genes		
		'Desmayo Largueta'	'Penta'	Common
Hyper-methylated	Upstream	7	3	1
	Inside	5	2	2
	Downstream	5	9	4
Total		17	14	7
Hypo-methylated	Upstream	1	-	-
Total		18	14	7

The methylation state refers to the number of 5mCs in stage A (dormant buds) samples with respect to the number found in stage B (non-dormant buds) samples. The gene position is based on gene orientation with respect to the fragment-mapping region.

We found enriched hypermethylated genes in 'Desmayo Largueta' flower bud samples in the following processes related to primary metabolism in the "biological function" GO (Gene Ontology) category: amino-acid and carbohydrate synthesis and protein phosphorylation (Figure S1). ATP binding and protein kinase and phosphatase activity were the two main "molecular function" GO terms found (Figure 4).

In a gene-level analysis, the following candidate genes appeared as hypermethylated in 'Desmayo Largueta' flower bud samples: genes related to transcription regulation, including transcription factors (Prupe.1G395600, Prupe.5G088700, Prupe.6G343100); genes linked to RNA-mediated silencing (Prupe.7G221200); genes linked to chromatin remodelling (Prupe.8G221300, and LATE ELONGATED HYPOCOTYL (LHY), encoded by Prupe.2G200400); and, especially, genes involved in the auxin response (Prupe.1G000200, Prupe.1G067400, Prupe.7G048400) and AUXIN RESPONSE FACTOR (ARF) signal transduction (Prupe.3G010900, Prupe.5G217700, Prupe.7G228800). We also identified DNA repair proteins, such as those encoded by Prupe.1G510000, Prupe.2G013900, Prupe.3G029600, Prupe.3G16000 and Prupe.5G066100. Finally, proteins participating in oxidoreduction processes, such as LATE EMBRYOGENESIS ABUNDANT (LEA) proteins encoded by Prupe.4G026900 and Prupe.4G02700, also appeared as hypermethylated in 'Desmayo Largueta' flower bud samples (Table S3).

Regarding the hypomethylated genes in the 'Desmayo Largueta' samples, we found cellular protein localization within the "biological function" GO category (Figure S2). ATP-coupled transmembrane transport and ATP-binding activity, on the other hand, appeared in the "molecular function" GO category (Figure 5).

We were able to identify a wide range of DNA-binding proteins encoded by the hypomethylated genes in the 'Desmayo Largueta' samples: histone methyltransferases (encoded by Prupe.1G050800 and Prupe.7G271600); the transcription factor NUCLEAR FACTOR-Y (NF-Y) (encoded by Prupe.2G47600); DNA topoisomerases (encoded by Prupe.1G173400 and Prupe.1G173500); and FAR-RED IMPAIRED RESPONSE 1 (FAR1) (Prupe.1G196400). Interestingly, a single gene coding for a HYDROPHOBIC SEED PROTEIN (HSP) also appeared as hypomethylated in the 'Desmayo Largueta' samples.

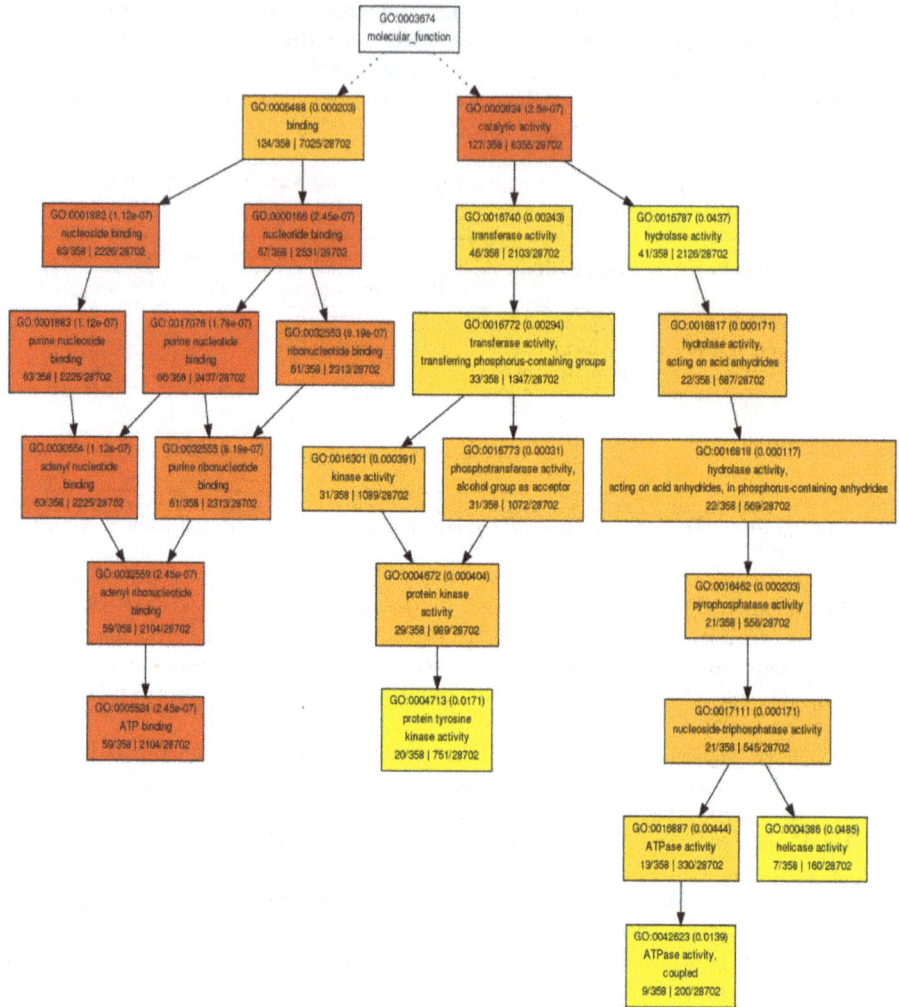

Figure 4. GO terms of the "Molecular function" category represented in genes identified as hypermethylated in 'Desmayo Largueta' flower buds in both the "A" = dormant bud stage and "B" = non dormant bud stage.

2.3. Differentyally Methylated Genes Related to Bud Dormancy

More identified DMGs were found as hypermethylated in stage A samples (dormant buds) than in stage B samples (non-dormant buds) (Table 3). Furthermore, just one hypomethylated gene could be functionally annotated, and it was mapped in the 3' regulatory region of the gene Prupe.4G277200, which encodes for a REGULATION OF CHROMOSOME CONDENSATION (RCC1) protein (Table 4).

Common stage A hypermethylated genes coded for a MITOGEN-ACTIVATED PROTEIN (MAP)-kinase and a phosphatase (Prupe.4G270800 and Prupe.1G287200); an LEUCINE RICH REPEAT-TOLL INTERLEUKIN 1 RECEPTOR (LRR-TIR) apoptotic ATPase associated with disease resistance (Prupe.3G130700); a GDSL (Gly, Asp, Ser and Leu motif) lipase (Prupe.6G307900); an Nt-C2 family protein (Prupe.2G074400); and a Glycerophosphatidylinositol (GPI) anchor synthase (Prupe.2G019300). The Prupe.1G125600 gene was annotated, but its protein function is unknown

(Table 4). Moreover, genes coding for VACUOLAR PROTEIN SORTING 1 (VPS1) proteins were detected as hypermethylated in stage A in both 'Desmayo Largueta' and 'Penta'. VPS1 genes corresponded to Prupe.3G026400 in 'Desmayo Largueta' samples and Prupe.2G029500 in 'Penta' samples.

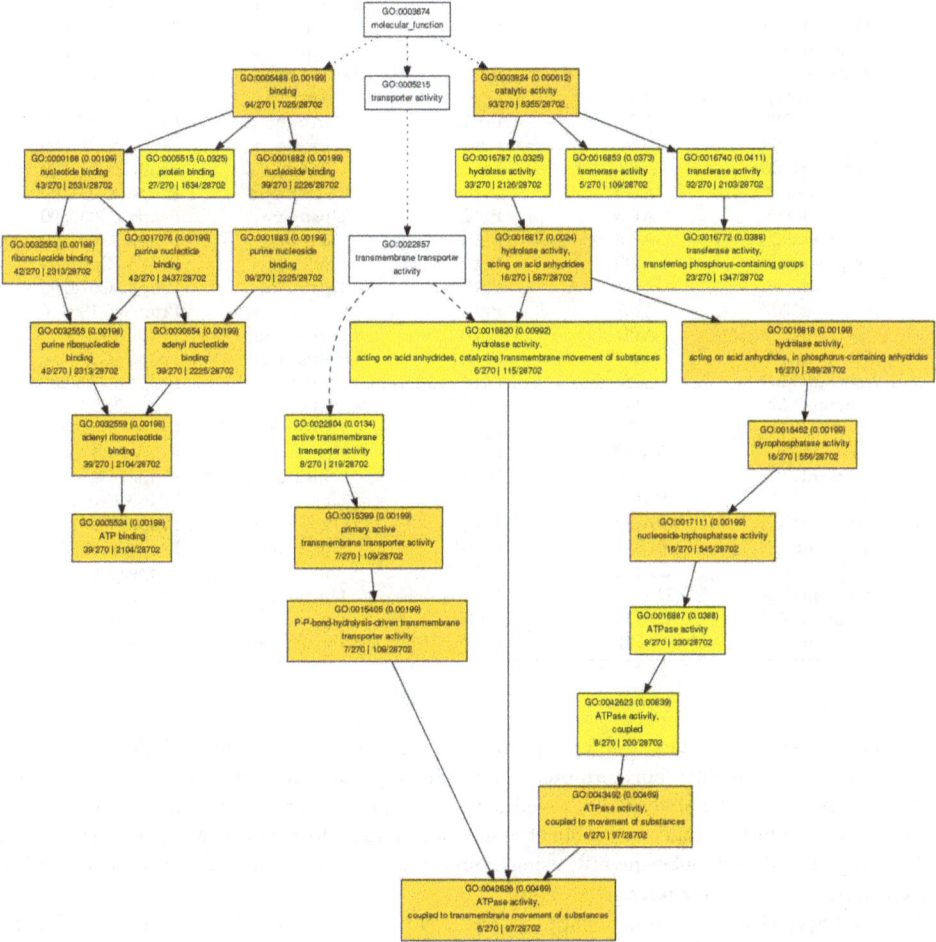

Figure 5. "Molecular function" GO terms represented in genes identified as hypomethylated in 'Desmayo Largueta' flower buds in both the "A" = dormant bud stage and "B" = non dormant bud stage.

Three additional LRR-TIR apoptotic ATPases were identified as encoded by hypermethylated genes in 'Desmayo Largueta' stage A samples (Table 4). Other genes that were found coded for defense proteins, a phosphatase, a cornichon protein associated with cell polarity and a Cytochrome P-450 (CytP450) protein member. On the other hand, hypermethylated genes in 'Penta' stage A samples coded for the ASSEMBLY PROTEIN 180 (AP180) clathrin assembly protein, the Tryptophan-Aspartic acid-Sterile Alpha Motif domain containing protein (WDSAM1) ubiquitination protein, lipases, a glycosyl hydrolase and a FCF2 rRNA processing protein recently described in yeast (Table 4).

Table 4. Hypermethylated genes identified in 'Desmayo Largueta' (D) and 'Penta' (P) stage A flower buds compared to those found in stage B flower buds (Chromosome, Fragment ID, Prupe.ID, Functional annotation).

FragmentID	Comparison	Chromosome	Gene Position	Prupe.Gene Code
fragment1289	AB	Pp01	Downstream	Prupe.1G287200
fragment1294	AB	Pp01	Upstream	Prupe.1G125600
fragment3735	PAPB	Pp01	Downstream	Prupe.1G099900
fragment32	AB	Pp03	Downstream	Prupe.3G130700
fragment341	AB	Pp06	Downstream	Prupe.6G307900
fragment797	AB	Pp02	Inside	Prupe.2G074400
fragment1708	DADB	Pp02	Inside	Prupe.2G019300
fragment4206	PAPB	Pp02	Inside	Prupe.2G019300
fragment341	DADB	Pp02	Downstream	Prupe.2G031100
fragment255	DADB	Pp02	Upstream	Prupe.2G053300
fragment92	DADB	Pp02	Inside	Prupe.2G057100
fragment92	DADB	Pp02	Inside	Prupe.2G057800
fragment797	DADB	Pp02	Upstream	Prupe.2G074300
fragment3707	PAPB	Pp04	Downstream	Prupe.4G186400
fragment4154	PAPB	Pp04	Downstream	Prupe.4G253800
fragment238	AB	Pp04	Downstream	Prupe.4G270800
fragment3299	PAPB	Pp06	Downstream	Prupe.6G307900
fragment1289	DADB	Pp02	Inside	Prupe.2G146000
fragment341	DADB	Pp03	Upstream	Prupe.3G026400
fragment797	DADB	Pp03	Downstream	Prupe.3G130700
fragment2263	DADB	Pp05	Inside	Prupe.5G036900
fragment157	PAPB	Pp06	Upstream	Prupe.6G014500
fragment1289	DADB	Pp05	Downstream	Prupe.5G038500
fragment483	PAPB	Pp02	Upstream	Prupe.2G039500
fragment507	DADB	Pp06	Inside	Prupe.6G097800
fragment1708	DADB	Pp06	Upstream	Prupe.6G331300
fragment2727	DADB	Pp06	Upstream	Prupe.6G331300
fragment849	PAPB	Pp01	Downstream	Prupe.1G105700

3. Discussion

Using our epiGBS variant as a first approach is a less expensive technique than complete GBS with highly accurate results. Furthermore, without the sequenced genome of the species, it is easier to perform bioinformatic analysis with well-defined fragments obtained by epiGBS. In this work, a conversion with bisulfite and a subsequent sequencing were performed to evaluate the 5mC variants of the samples analyzed. Subsequently, using bioinformatic analysis, these differentially methylated regions were mapped in the reference genome.

Applying epi-GBS to 'Desmayo Largueta' and 'Penta' flower bud gDNA samples provided data about methylation (5mC) variants depending on the genotype and dormancy state of the flower buds. Quality evaluation of the analysis showed that more than 90% of all cytosine positions were completely unmethylated and that only 1.0% to 1.3% of the positions were completely methylated, with higher methylation in the CpG context. These results agree with previous results in different plant species indicating that CG methylation is the typical genomic region for DNA methylation with less methylation abundance in the CHG and CHH contexts [23]. As a result of the specificity of the methylome of plants with respect to that of animals, we adapted bisulfite conversion methods to allow for correct analysis in plants for all cytosine contexts [24].

DNA methylomes have now been analyzed in many plants species, including Arabidopsis, rice, maize, and tomato. DNA methylation results in these species indicate that the distribution of methylation marks across the genomes is generally conserved, although variations can be observed between species depending on several factors, including transposon abundance and genome size [25]. In agreement with our results, polymorphism (5mC variants) can be observed by comparing different genotypes or even accessions within the same species. In addition, recent results evaluating methylomes from 1,227 different accessions of *Arabidopsis* distributed worldwide have shown important polymorphisms between accessions [26].

It is interesting to note the high degree of differential methylation that seems to be fixed between the two almond genotypes analyzed. This fact is of practical importance in cultivar improvement for developing epigenetic markers based on methylation variants and taking into account the high flexibility of methylation patterns in relation to external signals in order to identify markers based on methylation polymorphisms. In contrast to standard sequencing, bisulfite sequencing makes it possible to obtain information that conditions the phenotype. As a consequence, knowing the methylation state might help us understand the genetic determinism of important agronomic traits more deeply. Although the methylation patterns are highly variable in response to different external factors, the markers that we have detected in our almond genotypes are conserved in different stages of development and in different years and can therefore be considered as stable and conserved epigenetic marks.

In this study, data showed important differences between genotypes, which displayed different phenotypes in terms of breeding traits (chilling requirements for dormancy release, flowering and ripening times, almond production and almond characteristics) (Tables 1 and 2). It is remarkable that more hypermethylated than hypomethylated fragments were identified in stage A (dormant flower buds) almond samples in both genotypes (Table 3). This is concordant with the general decrease in 5mC during dormancy progression in *C. sativa* [15]. On the other hand, the most frequently mapped fragments were within the gene regions ("inside" DMGs), followed by the 5′ regulatory regions ("downstream" DMGs) and, in last place, in the 3′ regions ("upstream" DMGs). According to Vining et al. [27], 5mC in promoters and gene body parts is related to a repressed state of chromatin, a condition that inhibits the accessibility of the transcriptional machinery.

Among the genes found as hypermethylated in 'Desmayo Largueta' with respect to 'Penta' flower buds (Table S3), ARF transcription factors were highly represented. Its known that the expression of genes like ARFs involved in the auxin response are subjected to epigenetic regulation [9,28], and ARF transcriptional regulation is required for developmental processes like germination [29]. Accordingly, Zhang et al. [30] observed a flowering delay in Arabidopsis when *ARF6* and *ARF8* were repressed. Nonetheless, the reason underlying the hypermethylated state of genes participating in the auxin response pathway in both dormant and non-dormant flower buds of the early flowering genotype 'Desmayo Largueta' has yet to be unraveled.

Another hymermethylated gene in 'Desmayo Largueta' flower bud samples was a member of the LEA family (Table S3). LEA proteins are involved in osmoprotection, which is activated in response to low temperatures [31]. When hypermethylated, this gene showed a repressed state of expression, although in low chilling requirement cultivars like 'Desmayo Largueta', osmoprotection would not be so necessary or may be regulated in a different way. The LEA gene family has been characterized by Du et al. [32] in *Prunus mume* (Siebold) Siebold & Zucc., and differential expression has been identified during bud dormancy in this species [33].

The LHY protein, on the other hand, is a well-described flowering time regulator in response to the photoperiod [34,35], and the gene network controlling this trait has been studied [36]. 'Desmayo Largueta' is a low-chill cultivar whose dormancy period takes place under short photoperiod conditions such as the experimental conditions of this work. It would be interesting to study LHY behavior during dormancy progression in different almond cultivars.

The methylation variants observed may be associated with evolutionary changes related to each genotype's features [37]. It will be interesting to distinguish which variants are related to traits of agronomic interest in order to explore adaptive mechanisms to the environment [38]. Recently, for instance, Garg et al. [39] identified conserved methylation polymorphisms distributed throughout rice varieties with different responses to drought resistance.

We found other hypermethylated genes in dormant (A) flower buds with respect to dormancy released (B) flower buds, including a MAP kinase (MAPK) and a phosphatase (Prupe.1G287200). MAP kinases and phosphatases have been found to participate in the initial response to cold induced by an increase in Ca2+ [31]. Furthermore, MAPK3 has been shown to be a central regulator of seed dormancy in barley [40]. Regarding the other genes hypermethylated in the A state, LRR-TIR apoptotic ATPases may be activated in a type of programmed cell death (PCD) called developmental cell death (DCD), leading to a differentiation of cells after dormancy release, as occurs in floral morphogenesis or in the pollen tube [41,42]. Nt-C2 and VSP1 proteins, on the other hand, are involved in vesicular trafficking from the cell membrane, and this process has been linked to cell wall differentiation and appears to be important in the dormancy release process [43,44]. Finally, GPI anchoring (a post-translational modification of proteins consisting of glycosylation) proteins are involved in intercellular signaling, as occurs in flowering transition as shown in *Populus* genus by Rinne et al. [45].

4. Materials and Methods

4.1. Plant Material and Experimental Design

We used flower buds from 'Desmayo Largueta', a traditional almond cultivar with very low chilling requirements and an extra-early flowering time, and 'Penta', a cultivar released from the CEBAS-CSIC Almond Breeding Program (Murcia, South-East Spain) with high chilling requirements and an extra-late flowering time. The plant material consisted of flower buds at stages A (dormancy phase) and B (after dormancy release) that were referenced to the phenological stages described by Felipe [46] (Figure 6). Dormancy release evaluation was performed by the forcing method according to Prudencio et al. [6]. Almond flower buds were picked from the experimental field of CEBAS-CSIC during two seasons of study: 2015–2016 and 2016–2017.

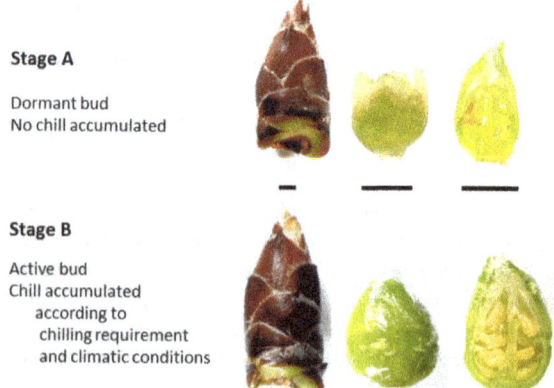

Figure 6. Plant material from 'Desmayo Largueta' and 'Penta' almond cultivars. Flower buds in the dormant state (**A**) and after dormancy release (**B**) according to Felipe [41]. Scale bars represent 1 mm in each case.

4.2. Epi-GBS Protocol

Every sample ('Desmayo Largueta' stage A, 'Penta' stage A, 'Desmayo Largueta' stage B, and 'Penta' stage B from the first and second season of study) consisted of a pool of ten flower buds. Genomic DNA was extracted from each sample following the method described by Doyle and Doyle [47]. The DNA samples were quantified using Qubit (Thermo Fisher Scientific, Alcobendas, Spain) and diluted to 1 µg in 100 µL. A total of 20 µL was digested using *PstI* restriction enzyme. Adaptors consisting of barcoded oligos were ligated to every sample (Table S4). Non-phosphorylated hemimethylated adapters were used to reduce costs. Fragmented samples (libraries generated by restriction) were pooled and purified and subsequently subjected to nick translation with C-dNTPs (Zymo Research, Irvine, CA, USA) and 7.5 µL of DNA PolI (NEB, Ipswich, MA, USA) in NEB buffer 2. An EZ DNA Methylation-Lightning kit (Zymo Research) was used for bisulfite treatment, and fragments were selected by size with a Thermo Scientific Size Selection kit (Thermo Fisher Scientific). Libraries were amplified using the Kapa HiFi HotStart Uracil+ ReadyMix (Roche, Barcelona, Spain) and purified with Magjet NGS Cleanup (Thermo Fisher Scientific). Paired-end Illumina 2500 reads (2 × 100 bases) were generated by Macrogen (Seoul, Korea) [23].

4.3. Bioinformatic Analysis of DNA Methylation

The process_radtags program of the Stacks 1.48 pipeline [48]. Checking the integrity of the restriction site was disabled with the "-disable_rad_check" option and quality filtering with the default settings was disabled with the exception the rad_check. This was necessary because the bisulfite treatment changes unmethylated cytosines in the recognition sequence of *PstI*, and, as a result, checking the restriction cut site would filter out all fragments. The ustacks program of the pipeline was used to align the fragments into perfectly matching stacks. The default settings were used with the exception of -M, which was set to 4 in order to increase the maximum distance (in nucleotides) between stacks. Finally, cstacks was used to build a catalog of consensus loci. A custom C program was used for the reconstruction of the original sequences of the fragments by comparing the reads with origins in the "Watson" and "Crick" strands of the genomic DNA. The reconstructed DNA fragments were merged by another custom C program to produce one continuous "mock genome". Bismark_v0.19.0 [49] was used to align the original fragments to the mock genome and to extract the methylation information. The Bismark coverage reports were used as input for the methylKit R package [50]. A methyl kit was used to elaborate histograms of C-methylation and coverage and to assess sample similarity and correlation using the default settings. For the hierarchical clustering of the samples, dist was set to "correlation" and method to "ward". Finally, we used the calculate DiffMeth function of a MethylKit to search for differentially methylated cytosines with the settings difference = 25, qvalue = 0.01. We looked for both hypermethylated and hypomethylated bases setting type = hyper and = hypo, respectively. The positions of the differentially methylated cytosines were extracted from the MethylKit files. Another custom made C-program was used to identify the original fragments where these differentially methylated cytosines were located.

4.4. Gene Finding and Annotation

The sequence of each fragment was mapped against the *P. persica* reference genome (v2.0) [51] with Gmap [52]. Two different output files, in the gff3 and SAM format, were obtained. The gff3 ouput files were processed to extract the boundary coordinates (start and end positions) of each hit using command line tools. After that, the boundary coordinates were used by a second custom python script to retrieve three different categories of annotations based on gene locations on the *P. persica* reference genome: upstream and downstream genes (in a size window of 10,000 bp) and "inside genes" (fragments within gene sequence). Finally, SAM format files were processed using a custom python script to extract the alignment information (number of exons, percentage of coverage, percentage of identity, and amino acid changes). Functional annotation of genes selected by distance to the mapped fragment was carried out using AgriGO software using Singular Enrichment Analysis (SEA), and Fisher's test [53] (Figure 7).

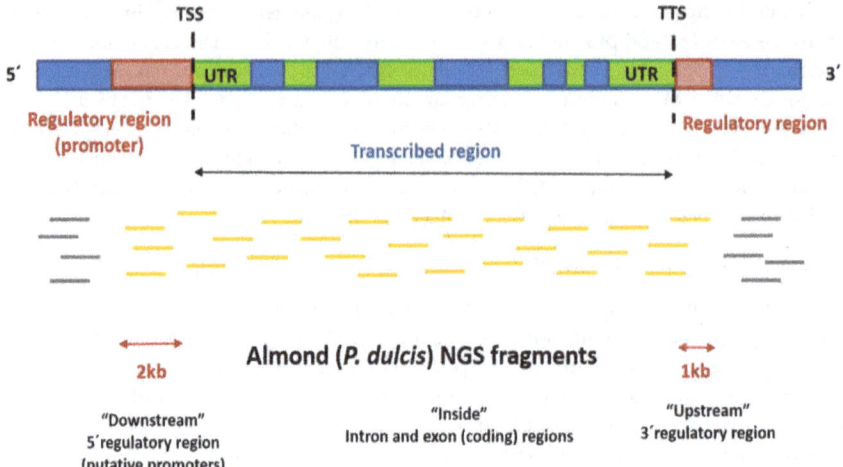

Figure 7. Schematic representation of a plant gene and classification of almond (*Prunus dulcis*) DMFs mapped in the *P. persica* genome. DMFs mapping from 2 kb upstream TSS to 1 kb downstream TTS were selected for functional annotation (fragments colored in orange). DMFs were classified according to the gene position—downstream, inside, or upstream—with respect to the fragment-mapping region. Fragments mapped in intergenic regions (colored in grey) were discarded as putative gene regulatory regions. **TSS:** Transcriptional Start Site, **UTR:** Untranslated Transcription Region, **TTS:** Transcriptional Terminal Site. Exons and introns within the transcribed region are colored in blue and green, respectively.

5. Conclusions

In this study, we applied the epi-GBS protocol to almond (*P. dulcis*) DNA samples for the first time. The technical potential is evident in the discovery of epigenetic variants, based on 5mC, that are genotype-dependent. According to the results obtained, the DNA methylation (5mC) pattern is generally genotype-dependent rather than dormancy state-dependent. Comparative DNA methylation studies of both almond varieties released from breeding programs and traditional varieties will surely contribute to our knowledge of methylation variants and provide candidate epialleles linked to agronomic traits. Such polymorphisms can be screened in large populations using NGS to confirm the locus methylation state associated with a given character of interest. In spite of coverage limitation, we were able to identify genes whose DNA methylation state changed between the dormant and active state of the flower buds. This was possible in both the traditional early-flowering genotype 'Desmayo Largueta' and the extra-late-flowering genotype 'Penta' from the CEBAS-CSIC Almond Breeding Program. Furthermore, common genes arose from the analysis. In the future, it would be interesting to improve the technique coverage to obtain a greater representation of the genome. Ultimately, the results will be an essential complement to RNAseq experiments in bud dormancy progression.

Supplementary Materials: The supplementary materials can be found at http://www.mdpi.com/1422-0067/19/11/3542/s1. Figure S1. GO terms of the "Biological Function" category represented in genes identified as hypermethylated in 'Desmayo Largueta' flower buds, in both the A and B dormancy states, compared to 'Penta' flower buds. Figure S2. GO terms of the "Biological Function" category represented in genes identified as hypomethylated in 'Desmayo Largueta' flower buds, in the both A and B dormancy states, compared to 'Penta' flower buds. Table S1. Total NGS fragments identified by Fragment ID. Table S2. Percentage of DMFs mapped in the *Prunus persica* genome (v2.1), Mapping region, Coverage, Percent identity, Number of exons, Amino acid changes, Comparison and Category. Table S3. Hypermethylated, Hypomethylated and Equally-methylated genes detected in 'Desmayo Largueta' flower buds with respect to 'Penta' flower buds in both the A and B stages (Chromosome, Fragment ID, Prupe.ID, Functional annotation). Gene list for GO annotation. Table S4. Sample barcodes. "D" = 'Desmayo Largueta', "P" = 'Penta', "A" = dormant bud stage; "B" = non-dormant bud stage.

Int. J. Mol. Sci. **2018**, *19*, 3542

DNA samples with number "1" correspond to the 2015–2016 season, and those with number "2" correspond to the 2016–2017 season.

Author Contributions: A.S.P. and O.W. participated in the design and coordination of the study. A.S.P. and O.W. carried out the epiGBS protocol. A.S.P., O.W. and P.J.M.-G. carried out data analysis. A.S.P., O.W., P.J.M.-G., R.M.R., P.M.-G. and F.D. participated in the manuscript elaboration and discussion.

Funding: This study has been supported by Grants 19308/PI/14 and 19879/GERM/15 of the Seneca Foundation of the Region of Murcia (Spain) and the Almond Breeding project of the Spanish Ministry of Economy and Competiveness.

Conflicts of Interest: The authors declare no conflict of interest. The funding sponsors had no role in the design of the study; in the collection, analyses, or interpretation of data; in the writing of the manuscript; or in the decision to publish the results.

Abbreviations

5mC	5′ Methylated Cytosine
AP180	ASSEMBLY PROTEIN 180
ARF	AUXIN RESPONSE FACTOR
CytP450	Cytochrome P-450
DMF	Differentially Methylated Fragment
DMG	Differentially Methylated Gene
epiGBS	Epi-Genotyping By Sequencing
FAR1	FAR-RED IMPAIRED RESPONSE 1
GDSL	Gly, Asp, Ser, and Leu motif
GO	Gene Ontology
GPI	Glycerophosphatidylinositol
H3K27me3	Trymethylation of Histone 3 on Lis (K) residue 27
HSP	HYDROPHOBIC SEED PROTEIN
LEA	LATE EMBRYOGENESIS ABUNDANT
LHY	LATE ELONGATED HYPOCOTYL
LRR-TIR	LEUCINE RICH REPEAT-TOLL INTERLEUKIN 1 RECEPTOR
MADS	MCM1-AGAMOUS-DEFICIENS-SRF
MAPK	MITOGEN-ACTIVATED PROTEIN KINASE
NGS	Next Generation Sequencing
NF-Y	NUCLEAR FACTOR-Y
PTMs	Post-Translational Modifications
SAM	S-ADENOSYL METHIONINE
VPS1	VACUOLAR PROTEIN SORTING 1
WDSAM1	Tryptophan-Aspartic acid-Sterile Alpha Motif 1

References

1. Vitasse, Y.; Lenz, A.; Körner, C. The interaction between freezing tolerance and phenology in temperate deciduous trees. *Front. Plant Sci.* **2014**, *5*. [CrossRef] [PubMed]
2. Beauvieux, R.; Wenden, B.; Dirlewanger, E. Bud Dormancy in Perennial Fruit Tree Species: A Pivotal Role for Oxidative Cues. *Front. Plant Sci.* **2018**, *9*, 657. [CrossRef] [PubMed]
3. Lang, B.A.; Early, J.D.; Martin, G.C.; Darnell, R.L. Endo-, para-, and eco-dormancy physiological terminology and classification for dormancy research. *HortScience* **1987**, *22*, 371–377.
4. Egea, J.; Ortega, E.; Martínez-Gómez, P.; Dicenta, F. Chilling and heat requirements of almond cultivars for flowering. *Environ. Exp. Bot.* **2003**, *50*, 79–85. [CrossRef]
5. Martínez-Gómez, P.; Prudencio, A.S.; Gradziel, T.M.; Dicenta, F. The delay of flowering time in almond: A review of the combined effect of adaptation, mutation and breeding. *Euphytica* **2017**, *213*, 197. [CrossRef]
6. Prudencio, A.S.; Martínez-Gómez, P.; Dicenta, F. Evaluation of breaking dormancy, flowering and productivity of extra-late and ultra-late flowering almond cultivars during cold and warm seasons in South-East of Spain. *Sci. Hort.* **2018**, *235*, 39–46. [CrossRef]

7. Cooke, J.E.K.; Eriksson, M.E.; Juntilla, O. The dynamic nature of bud dormancy in trees: Environmental control and molecular mechanisms. *Plant Cell Environ.* **2012**, *35*, 1707–1728. [CrossRef] [PubMed]

8. Abbott, A.G.; Zhebentyayeva, T.; Barakat, A.; Liu, Z. The genetic control of bud-break in trees. *Adv. Bot. Res.* **2015**, *74*, 201–228.

9. Yaish, M.W.; Colasanti, J.; Rothstein, S.J. The role of epigenetics processes in controlling flowering time exposed to stress. *J. Exp. Bot.* **2011**, *62*, 3727–3735. [CrossRef] [PubMed]

10. Ríos, G.; Leida, C.; Conejero, C.; Badenes, M.L. Epigenetic regulation of bud dormancy events in perennial plants. *Front. Plant Sci.* **2014**, *5*, 247. [PubMed]

11. Saze, H. Epigenetic memory transmission through mitosis and meiosis in plants. *Semin. Cell Dev. Biol.* **2008**, *19*, 527–536. [CrossRef] [PubMed]

12. Feng, S.; Jacobsen, S.E. Epigenetic modifications in plants: An evolutionary perspective. *Curr. Opin. Plant Biol.* **2011**, *14*, 179–186. [CrossRef] [PubMed]

13. Lämke, J.; Bäurle, I. Epigenetic and chromatin-based mechanisms in environmental stress adaptation and stress memory in plants. *Genome Biol.* **2017**, *18*, 124. [CrossRef] [PubMed]

14. Pascual, J.; Cañal, M.J.; Correia, B.; Escandon, M.; Hasbún, R.; Meijón, M.; Pinto, G.; Valledor, L. Can Epigenetics Help Forest Plants to Adapt to Climate Change? In *Epigenetics in Plants of Agronomic Importance: Fundamentals and Applications: Transcriptional Regulation and Chromatin Remodelling in Plants*; Alvarez-Venegas, R., De la Peña, C., Casas-Mollano, J.A., Eds.; Springer: Cham, Switzerland, 2014; pp. 125–146.

15. Santamaría, M.; Hasbún, R.; Valera, M.; Meijón, M.; Valledor, L.; Rodríguez, J.L.; Rodríguez, R. Acetylated H4 histone and genomic DNA methylation patterns during bud set and bud burst in *Castanea sativa*. *J. Plant Physiol.* **2009**, *166*, 1360–1369. [CrossRef] [PubMed]

16. de la Fuente, L.; Conesa, A.; Lloret, A.; Badenes, M.L.; Ríos, G. Genome-wide changes in histone H3 lysine 27 trimethylation associated with bud dormancy release in peach. *Tree Genet. Gen.* **2015**, *11*, 45. [CrossRef]

17. Lloret, A.; Martínez-Fuentes, A.; Agustí, M.; Badenes, M.L.; Ríos, G. Chromatin-associated regulation of sorbitol synthesis in flower buds of peach. *Plant Mol. Biol.* **2017**, *95*, 507–517. [CrossRef] [PubMed]

18. Rothkegel, K.; Sánchez, E.; Montes, C.; Greve, M.; Tapia, S.; Bravo, S.; Almeida, A.M. DNA methylation and small interference RNAs participate in the regulation of MADS-box genes involved in dormancy in sweet cherry (*Prunus avium* L.). *Tree Physiol.* **2017**, *37*, 1739–1751. [CrossRef] [PubMed]

19. Kumar, G.; Rattan, U.K.; Singh, A.K. Chilling-Mediated DNA Methylation Changes during Dormancy and Its Release Reveal the Importance of Epigenetic Regulation during Winter Dormancy in Apple (*Malus × domestica* Borkh.). *PLoS ONE* **2016**, *11*, e0149934. [CrossRef] [PubMed]

20. Prudencio, A.S.; Dicenta, F.; Martínez-Gómez, P. Gene expression analysis of flower bud dormancy breaking in almond using RNA-Seq. In Proceedings of the VII International Symposium on Almonds & Pistachios, Adelaida, Australia, 5–9 November 2017.

21. Fernández i Martí, A.; Gradziel, T.M.; Socias i Company, R. Methylation of the Sf locus in almond is associated with S-RNase loss of function. *Plant Mol. Biol.* **2014**, *86*, 681–689. [CrossRef] [PubMed]

22. Fresnedo-Ramírez, J.; Chan, H.M.; Parfitt, D.E.; Crisosto, C.H.; Gradziel, T.M. Genome-wide DNA-(de) methylation is associated with Noninfectious Bud-failure exhibition in Almond (*Prunus dulcis* [Mill.] D.A.Webb). *Sci. Rep.* **2017**, *7*, 42686.

23. van Gurp, T.P.; Wagemaker, N.V.; Wouters, B.; Vergeer, P.; Ouborg, J.N.; Werhoeben, K.J.V. epiGBS: Reference-free reduced representation bisulfite sequencing. *Nat. Meth.* **2016**, *13*, 322–329. [CrossRef] [PubMed]

24. Niederhuth, C.E.; Bewick, A.J.; Ji, L. Widespread natural variation of DNA methylation within angiosperms. *Gen. Biol.* **2016**, *17*, 194. [CrossRef] [PubMed]

25. How-Kit, A.; Emeline, T.; Deleuze, J.F.; Gallusci, P. Locus-Specific DNA Methylation Analysis and Applications to Plants. In *Plant Epigenetics*; Rajewsky, N., Jurga, S., Barciszewski, J., Eds.; Springer: Berlin, Germany, 2017; pp. 303–328.

26. Kawakatsu, T.; Huang, S.S.; Jupe, F. Epigenomic diversity in a global collection of *Arabidopsis thaliana* accesions. *Cell* **2016**, *166*, 492–505. [CrossRef] [PubMed]

27. Vining, K.J.; Pomraning, K.R.; Wilhelm, L.J.; Priest, H.D.; Pellegrini, M.; Mockler, T.C. Dynamic DNA cytosine methylation in the *Populus trichocarpa* genome: Tissue-level variation and relationship to gene expression. *BMC Gen.* **2012**, *13*, 27. [CrossRef] [PubMed]

28. Xiao, W.; Custard, K.D.; Brown, R.C.; Lemmon, B.E.; Harada, J.J.; Goldberg, R.B.; Fischer, R.L. DNA methylation is critical for Arabidopsis embryogenesis and seed viability. *Plant Cell* 2006, *18*, 805–814. [CrossRef] [PubMed]

29. Liu, P.P.; Montgomery, T.A.; Fahlgren, N.; Kasschau, K.D.; Nonogaki, H.; Carrington, J.C. Repression of AUXIN RESPONSE FACTOR10 by microRNA160 is critical for seed germination and post-germination stages. *Plant J.* 2007, *52*, 133–146. [CrossRef] [PubMed]

30. Zhang, G.-Z.; Jin, S.-H.; Li, P.; Jiang, X.-Y.; Li, Y.-J.; Hou, B.K. Ectopic expression of UGT84A2 delayed flowering by indole-3-butyric acid-mediated transcriptional repression of ARF6 and ARF8 genes in *Arabidopsis*. *Plant Cell Rep.* 2017, *36*, 1995–2006. [CrossRef] [PubMed]

31. Bañuelos, M.L.G.; Moreno, L.V.; Winzerling, J.; Orozco, J.A.; Gardea, A.A. Winter metabolism in deciduous trees: Mechanisms, genes and associated proteins. *Rev. Fitotec. Mex.* 2008, *31*, 295–308.

32. Du, D.; Zhang, Q.; Cheng, T.; Pan, H.; Yang, W.; Sun, L. Genome-wide identification and analysis of late embryogenesis abundant (LEA) genes in *Prunus mume*. *Mol. Biol. Rep.* 2013, *40*, 1937–1946. [CrossRef] [PubMed]

33. Yamane, H.; Kashiwa, Y.; Kakehi, E.; Yonemori, K.; Mori, H.; Hayashi, K.; Iwamoto, K.; Tao, R.; Kataoka, I. Differential expression of dehydrin in flower buds of two Japanese apricot cultivars requiring different chilling requirements for bud break. *Tree Physiol.* 2006, *26*, 1559–1563. [CrossRef] [PubMed]

34. Fujiwara, S.; Oda, A.; Yoshida, R.; Niinuma, K.; Miyata, K.; Tomozoe, Y.; Tajima, T.; Nakagawa, M.; Hayashi, K.; Coupland, G.; et al. Circadian Clock Proteins LHY and CCA1 Regulate SVP Protein Accumulation to Control Flowering in Arabidopsis. *Plant Cell* 2008, *20*, 2960–2971. [CrossRef] [PubMed]

35. Park, B.S.; Eo, H.J.; Jang, I.-C.; Kang, H.-G.; Song, J.T.; Seo, H.S. Ubiquitination of LHY by SINAT5 regulates flowering time and is inhibited by DET1. *Biochem. Biophys. Res. Commun.* 2010, *398*, 242–246. [CrossRef] [PubMed]

36. Park, M.-J.; Kwon, Y.-J.; Gil, K.-E.; Park, C.-M. LATE ELONGATED HYPOCOTYL regulates photoperiodic flowering via the circadian clock in Arabidopsis. *BMC Plant Biol.* 2016, *16*, 114. [CrossRef] [PubMed]

37. Varriale, A. DNA Methylation in Plants and Its Implications in development, Hybrid Vigour, and Evolution. In *Plant Epigenetics*; Rajewsky, N., Jurga, S., Barciszewski, J., Eds.; Springer: Berlin, Germany, 2017; pp. 263–280.

38. Viggiano, L.; de Pinto, M.C. Dynamic DNA Methylation Patterns in Stress Response. In *Plant Epigenetics*; Rajewsky, N., Jurga, S., Barciszewski, J., Eds.; Springer: Berlin, Germany, 2017; pp. 281–302.

39. Garg, R.; Narayana Chevala, V.V.S.; Shankar, R.; Jain, M. Divergent DNA methylation patterns associated with gene expression in rice cultivars with contrasting drought and salinity stress response. *Sci. Rep.* 2015, *5*, 14922. [CrossRef] [PubMed]

40. Nakamura, S.; Pourkheirandish, M.; Morishige, H.; Kubo, Y.; Nakamura, M.; Ichimura, K.; Seo, S.; Kanamori, H.; Wu, J.; Ando, T.; et al. Mitogen-Activated Protein Kinase Kinase 3 Regulates Seed Dormancy in Barley. *Curr. Biol.* 2016, *26*, 775–781. [CrossRef] [PubMed]

41. Koonin, E.V.; Aravind, L. Origin and evolution of eukaryotic apoptosis: The bacterial connection. *Cell Death Differ.* 2002, *9*, 394. [CrossRef] [PubMed]

42. Del Duca, S.; Serafini-Fracassini, D.; Cai, G. Senescence and programmed cell death in plants: Polyamine action mediated by transglutaminase. *Front. Plant Sci.* 2014, *5*, 120. [CrossRef] [PubMed]

43. Ebine, K.; Ueda, T. Roles of membrane trafficking in plant cell wall dynamics. *Front. Plant Sci.* 2015, *6*, 878. [CrossRef] [PubMed]

44. Kim, S.J.; Brandizzi, F. The plant secretory pathway: An essential factory for building the plant cell wall. *Plant Cell Physiol.* 2014, *55*, 687–693. [CrossRef] [PubMed]

45. Rinne, P.L.H.; Welling, A.; Vahala, J.; Ripel, L.; Ruonala, R.; Kangasjärvi, J.; van der Schoot, C. Chilling of Dormant Buds Hyperinduces FLOWERING LOCUS T and Recruits GA-Inducible 1,3-β-Glucanases to Reopen Signal Conduits and Release Dormancy in Populus. *Plant Cell* 2011, *23*, 130–146. [CrossRef] [PubMed]

46. Felipe, A.J. Phenological states of almond. In Proceedings of the Third GREMPA Colloquium, Bari, Italy, 3–7 October 1977; pp. 101–103. (In Italian)

47. Doyle, J.J.; Doyle, M. A rapid DNA isolation procedure for small quantities of fresh leaf tissue. *Phytochem. Bull.* 1987, *19*, 11–15.

48. Catchen, J.; Hohenlohe, P.A.; Bassham, S.; Amores, A.; Cresko, W.A. Stacks: An analysis tool set for population genomics. *Mol. Ecol.* **2013**, *22*, 3124–3140. [CrossRef] [PubMed]

49. Krueger, F.; Andrews, S.R. Bismark: A flexible aligner and methylation caller for Bisulfite-Seq applications. *Bioinformatics* **2011**, *27*, 1571–1572. [CrossRef] [PubMed]

50. Akalin, A.; Kormaksson, M.; Li, S.; Garrett-Bakelman, F.E.; Figueroa, M.E.; Melnick, A.; Mason, C.E. methylKit: A comprehensive R package for the analysis of genome-wide DNA methylation profiles. *Genome Biol.* **2012**, *13*, R87. [CrossRef] [PubMed]

51. Verde, I.; Jenkins, J.; Dondini, L.; Micali, S.; Pagliarani, G.; Vendramin, E. The Peach v2.0 release: High-resolution linkage mapping and deep resequencing improve chromosome-scale assembly and contiguity. *BMC Gen.* **2017**, *18*, 225. [CrossRef] [PubMed]

52. Wu, T.D.; Watanabe, C.K. GMAP: A genomic mapping and alignment program for mRNA and EST sequences. *Bioinformatics* **2005**, *21*, 1859–1875. [CrossRef] [PubMed]

53. Tian, T.; Yue, L.; Hengyu, Y.; Qi, Y.; Xin, Y.; Zhou, D.; Wenying, X.; Zhen, S. agriGO v2.0: A GO analysis toolkit for the agricultural community, 2017 update. *Nucleic Acids Res.* **2017**, *45*, W122–W129. [CrossRef] [PubMed]

International Journal of
Molecular Sciences

Article

Identification, Classification, and Functional Analysis of *AP2/ERF* Family Genes in the Desert Moss *Bryum argenteum*

Xiaoshuang Li [1,†], Bei Gao [2,†], Daoyuan Zhang [1,*], Yuqing Liang [1,3], Xiaojie Liu [1,3], Jinyi Zhao [4], Jianhua Zhang [5] and Andrew J. Wood [6]

1   Key Laboratory of Biogeography and Bioresource in Arid Land, Xinjiang Institute of Ecology and Geography, Chinese Academy of Sciences, Urumqi 830011, China; lixs@ms.xjb.ac.cn (X.L.); liangyuqing14@mails.ucas.ac.cn (Y.L.); liuxiaojie215@mails.ucas.ac.cn (X.L.)
2   School of Life Sciences and State Key Laboratory of Agrobiotechnology, The Chinese University of Hong Kong, Hong Kong, China; gaobei@link.cuhk.edu.hk
3   University of Chinese Academy of Sciences, Beijing 100049, China
4   School of Life Science, University of Liverpool, Liverpool L169 3BX, UK; j.zhao46@student.liverpool.ac.cn
5   Department of Biology, Hong Kong Baptist University, Hong Kong, China; jzhang@hkbu.edu.hk
6   Department of Plant Biology, Southern Illinois University, Carbondale, IL 62901-6899, USA; wood@plant.siu.edu
*   Correspondence: zhangdy@ms.xjb.ac.cn; Tel.: +86-991-7823109
†   These authors contributed equally to this paper.

Received: 8 October 2018; Accepted: 13 November 2018; Published: 19 November 2018

Abstract: *Bryum argenteum* is a desert moss which shows tolerance to the desert environment and is emerging as a good plant material for identification of stress-related genes. *AP2/ERF* transcription factor family plays important roles in plant responses to biotic and abiotic stresses. *AP2/ERF* genes have been identified and extensively studied in many plants, while they are rarely studied in moss. In the present study, we identified 83 *AP2/ERF* genes based on the comprehensive dehydrationrehydration transcriptomic atlas of *B. argenteum*. BaAP2/ERF genes can be classified into five families, including 11 AP2s, 43 DREBs, 26 ERFs, 1 RAV, and 2 Soloists. RNA-seq data showed that 83 *BaAP2/ERFs* exhibited elevated transcript abundances during dehydration–rehydration process. We used RT-qPCR to validate the expression profiles of 12 representative *BaAP2/ERFs* and confirmed the expression trends using RNA-seq data. Eight out of 12 BaAP2/ERFs demonstrated transactivation activities. Seven BaAP2/ERFs enhanced salt and osmotic stress tolerances of yeast. This is the first study to provide detailed information on the identification, classification, and functional analysis of the *AP2/ERFs* in *B. argenteum*. This study will lay the foundation for the further functional analysis of these genes in plants, as well as provide greater insights into the molecular mechanisms of abiotic stress tolerance of *B. argenteum*.

Keywords: *AP2/ERF* genes; *Bryum argenteum*; transcriptome; gene expression; stress tolerance

---

1. Introduction

Bryum argenteum is an important component of the desert biological soil crusts in the Gurbantunggut and Tengger Deserts of northwestern China [1,2]. *B. argenteum* has gained increasing attention as a model organism due to its comprehensive tolerances to the desert environment, such as frequent desiccation–rehydration events and high UV radiation [3,4]. Wood et al. (2007) reported that *B. argenteum* is among the most desiccation tolerant (DT) moss species and is classified as category "A" [5]. Studies on *B. argenteum* have focused on the ecological aspects of vegetative desiccation tolerance, including morphological, structural, and physiological responses to adapt to the desert environment [3,4,6,7]. *B. argenteum* is emerging as a model moss for studying the molecular mechanisms of DT and as a source of stress-related genes [8].

APETALA2/Ethylene Responsive Factor (AP2/ERF) is one of the largest transcription factor (TF) families of plants, and the family members have been demonstrated to play important roles in plant metabolism, development, and stresses response [9]. *AP2/ERF* genes have been identified and studied extensively in the context of plant stress tolerance in many plants [10–12]. The *AP2/ERF* gene family has been rarely studied in moss species, however, the largest TFs families found in the plant transcription factor databases (TFDB) are *AP2/ERF* genes annotated in the mosses *Physcomitrella patens* and *Sphagnum fallax* [13,14]. Moreover, *AP2/ERFs* were demonstrated to be regulated in response to multiple stresses, such as salinity and UV in *P. patens* [15], and *PpDBF1* gene was reported to confer drought, salt, and cold tolerances in transgenic tobacco [16]. Additionally, AP2/ERFs also demonstrated to be the most abundant TFs in the DT moss *Syntrichia caninervis* [17]. The majority of *DREB* (Dehydration-Responsive Element-Binding Protein) genes in *S. caninervis* responded to dehydration and/or rehydration treatments [18,19], indicating that AP2/ERF transcriptional factors also play a central regulatory role during stress responses in moss species.

AP2/ERF classification employs a well-established method based on Arabidopsis and rice [20] and this method has been widely used to classify the *AP2/ERF* family genes in many plant species [11,21]. Two classic classification methods have been proposed for the plant AP2/ERF superfamily based upon the number of AP2 domains and sequence similarities [22]. Sakuma et al. classified the AP2/ERF superfamily into five families: AP2, RAV (Related to ABI3/VP1), DREB ERF and Soloists [22]. Furthermore, *DREBs* are further classified into A1–A6 groups and *ERFs* are divided into the groups B1–B6 [22]. Nakano et al. (2006) classified AP2/ERF proteins into three major families: AP2, ERF (include both DREBs and ERFs) and RAV [20]. The ERF family is then further sub-divided into twelve groups in Arabidopsis and fifteen groups in rice according to the structure and similarity of the AP2 domain [20].

High-throughput sequencing has been an effective tool to identify stress-related genes, and transcriptome-based identification and selection of *AP2/ERF* superfamily genes have been widely used in many non-model plants, such as *Hevea brasiliensis* [23], tea [24], as well as the desiccation tolerant moss *S. caninervis* [19]. Previously, we generated a de novo transcriptome for *B. argenteum* and established a desiccation–rehydration transcriptomic atlas which covered five different hydration stages [8,25]. Using this combined dataset, we established a fully comprehensive desiccation–rehydration transcriptome dataset containing 76,206 high-quality *B. argenteum* transcripts. We found that AP2/ERF was the second largest TF family in *B. argentum*. Moreover, *AP2/ERF* genes were also the most abundant differentially expressed TFs during the desiccation and rehydration process (Figure S1), indicating that the *AP2/ERF* family genes play key roles in *B. argentum* response to moss specific desiccation–rehydration process. Hence, in this study, we aimed to identify and classify the *AP2/ERF* gene family in *B. argentum* based on a comprehensive transcriptome dataset. Additionally, we investigated the gene expression patterns of *BaAP2/ERF* genes under desiccation–rehydration treatment based on RNA-seq data as well as real-time quantitative PCR (RT-qPCR) assay, and their transactivation activities were also analyzed. We further evaluated the stress tolerance ability of eight representative *BaAP2/ERF* genes in yeast system. This is the first report on identification, classification, characterization, and evaluation the stress tolerance functions of the *AP2/ERF* gene family in desiccation tolerant moss *B. argentums*. This study will provide candidate genes for molecular breeding to improve crop stress tolerance, and could be helpful for understanding the molecular mechanisms of the stress responses in *B. argentum*.

2. Results

2.1. Identification of the AP2/ERF Family Genes in B. argenteum

Based on Hidden Markov model (HMM) profiles and BLAST search, 83 AP2/ERF predicted proteins were identified from the *B. argenteum* transcriptome. The unigenes ranged from 559 to 8076 bp in length, and the corresponding deduced polypeptide sequences ranged from 145 to 1983 aa. The sequences of 83 *AP2/ERFs* were submitted to Genbank under accession numbers MK170284-MK170366. Among these 83 unigenes, 73 genes have an intact open reading frame (ORF) (87%), ranging from 163 to 1983 aa. AP2 domain analysis demonstrated that all 83 predicted proteins had full-length AP2 domains (ca. 60 aa) (100%). Seven out of 83 genes had two AP2 domains, one gene had both an AP2 and a B3 domain, while the other 75 genes had a single AP2 domain. Based on the number of AP2 domains, BaAP2/ERFs can be preliminarily classified as follows: 7 genes having two AP2 domains were classified as members of the AP2 family, 1 gene with both an AP2 and B3 domain was classified as a member of RAV family, and the 75 remaining genes with only one AP2 domain were considered as ERF family members (including both DREBs and ERFs).

2.2. Classification of the AP2/ERF Genes in B. argenteum

Family classification was further confirmed by constructing the phylogenetic tree using the AP2 domain of AP2/ERF in *B. argenteum*, the model plant Arabidopsis and the model moss *P. patens*. We first generated the gene tree of AP2/ERFs with *B. argenteum* and Arabidopsis. The tree showed that *BaAP2/ERF* genes were divided into five families according to Arabidopsis' classification method, including AP2, DREB, ERF, RAV and Soloist (Figure 1). Almost half of the *B. argenteum BaAP2/ERF* genes grouped into the DREB family (with 43 members). The second largest group was the ERF family (27 genes), followed by the AP2 family (10 genes), RAV (1 gene) and Soloists (2 genes). However, the number of the AP2 family genes classified by gene tree was inconsistent with the classification result based on the AP2 domain counting. There were two inconsistencies: one was the TR14027|c0_g1_i1 gene which had two AP2 domains, therefore should be classified as AP2 family gene, while it was grouped together with ERF family in the gene tree, and the other was five genes had single AP2 domain but were clustered together with AP2 family in Arabidopsis. To further confirm their classification, we generated another two gene trees using only *BaAP2/ERF* genes (Figure S2), and with *AP2/ERF* genes both in *B. argenteum* and the model moss *P. patens* (Figure S3). We conclude that there were 11 AP2s, 43 DREBs, 26 ERFs, 1 RAV and 2 Soloists in *B. argenteum* based on the available transcriptome data.

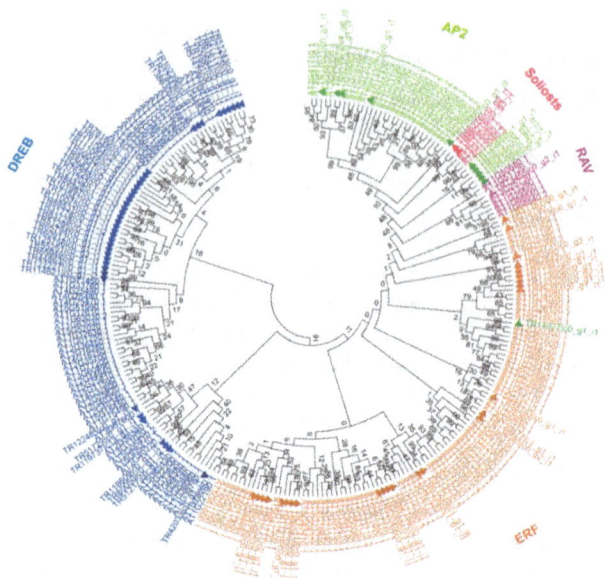

Figure 1. Phylogenetic analysis of AP2/ERF superfamily genes in *B. argenteum* and Arabidopsis. The gene tree was constructed using the neighbor-joining method using 83 BaAP2/ERFs and 176 AtAP2/ERFs, the evolutionary distances were computed using the Poisson correction method with pairwise deletion. Bootstrap values from 1000 replicates were used to assess the robustness of the tree. Different subfamilies were marked with various colors, the BaAP2/ERFs were labeled with rectangles to distinguish from AtAP2/ERF.

DREB and ERF are the two major families in the plant AP2/ERF superfamily, and which are demonstrated to play important roles in abiotic and biotic stress response. DREB and ERF can be subdivided into twelve subfamilies, namely A1–A6 and B1–B6 [22], or ten groups named Groups I–X [20], and each group/subfamily have different functions [11]. To detail classify the *DREB* and *ERF* family genes, we constructed a phylogenetic tree using 69 BaERFs (including 43 DREBs and 26 ERFs) and 31 ERFs in Arabidopsis representative of each subfamily/group. We identified 10 out of 12 DREB/ERF subfamilies in *B. argenteum* based upon the classification method of Sakamu et al. [22], A2 (3 genes), A3 (1 gene), A4 (1 gene), A5 (10 genes), A6 (3 genes), B1 (8 genes), B2 (4 genes), B3 (8 genes), B4 (4 genes), and B6 (2 genes), while no members of the A1 and B5 subfamilies were found in BaAP2/ERFs (Figure 2). Accordingly, based on Nakano et al.'s classification method [20], *BaAP2/ERF* superfamily genes contained 9 out of 10 groups (lacking Group VI), including Groups I–X, and the members for each group were Group I (3 members), Group II (10 members), Group III (one member), Group IV (4 members), Group V (2 members), Group VII (4 members), Group VIII (8 members), Group IX (8 members) and Group X (4 members) (Figure 2). Moreover, we found that 26 *ERFs* can be clearly classified into specific subfamilies/groups, while more than half of *DREBs* (25/43 genes) cannot be classified to any exist subfamilies, which was clustered into one separate clade and was named as *Bryum* group (Ba-clade) (Figure 2). To know whether these Ba-clade genes are specific to Bryum or moss species, we first performed BLASTp search in NCBI database using the full-length sequence of 25 Ba-clade DREBs/ERF and found that all 25 Ba-clade DREBs have homologies in other plants, with the highest sequence identities to model moss *P. patens*. We then performed BLASTp search in ONE KP database with mosses, liverworts, and hornworts transcriptome data. The result also show that all Ba-clade genes have homologies in mosses, liverworts, and hornworts. Moreover, half of Ba-clade DREBs shared highest sequence identities to *Funaria* (Table S1).

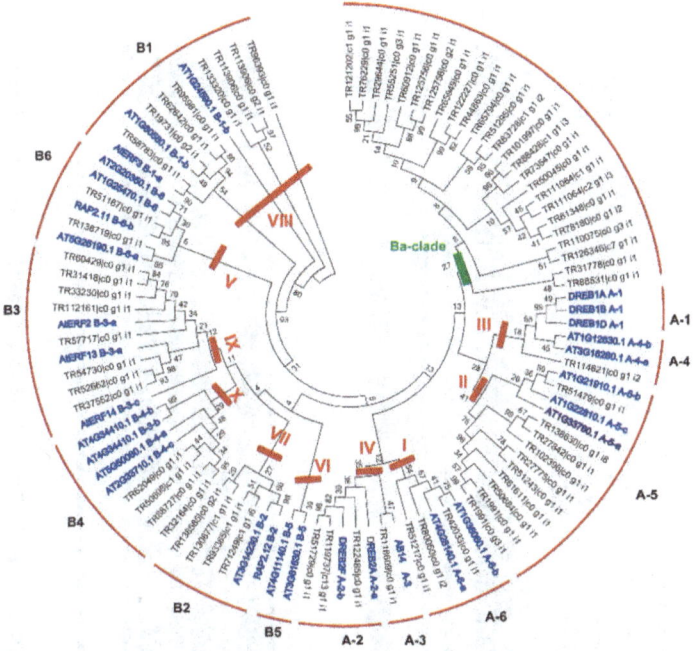

Figure 2. Phylogenetic analysis of ERF family genes in *B. argenteum* and Arabidopsis. The gene tree was constructed using 69 BaERFs and 31 AtERFs representative of each subfamily or group of ERF family genes in Arabidopsis. The evolutionary distances were computed using the neighbor-joining method and Poisson model with pairwise deletion. Bootstrap values from 1000 replicates were used to assess the robustness of the tree. To distinguish ERFs from *B. argenteum* and Arabidopsis, AtERFs and BaERFs were marked in blue and dark, respectively. Previously reported subfamily names (A1–A6 and B1–B6) and group names (Group I to Xb–L) were employed [20,22]. The Bryum-unique clade (Ba-unique) was labeled in green, and other groups were labeled in red.

2.3. Conserved Amino Acids and Motifs Analysis of BaERF Genes

To analyze the amino acids conservation of the AP2 domains, 69 BaERF deduced polypeptide sequences were aligned with 31 AtERFs representative of each gene subfamily in Arabidopsis. Multiple sequence alignment showed that BaERF sequences shared significant amino acid similarity with AtERFs except Ba-clade DREBs (Figure 3). Ba-clade DREBs did not belong to either subfamily/group based on existing classification. Ba-clade DREBs had more diverse amino acids composition in the AP2 domain, especially in the region between two β-sheets and the α-helix (marked with pink boxes). For example, a consensus sequence "TAE" in the C-terminal of α-helix was very conserved among AtERFs and BaERFs, however, in Ba-clade DREBs, TPE and TEE/Q/I patterns also existed (Figure 3). The motif composition analyses also supported this phenomenon. Eight motifs were detected in 69 BaERFs in total; among them, motifs 1–3 represented the typical AP2 domains in Arabidopsis as well as most of the well-classified BaERFs, of which motif 1 contained the β3 sheet and α-helix of AP2 domain, motif 2 corresponded to β1 and β2 sheet, and motif 3 was located in the very C-terminal of the AP2 domain (Figure 4). Motif 4 was similar to motif 1 which corresponded to the β3 sheet and α-helix; in the same way, motif 5 was similar to motif 2 which represent β1 and β2 sheet, and motif 8 was similar to motif 3, but their amino acids compositions differed from the classic AP2 domains in Arabidopsis. Interestingly, motif 4, 5 and 8 constituted the unique Ba-clade DREB AP2 domain. Additionally, we identified motif 7 as an A-5 DREB specific amino acids pattern.

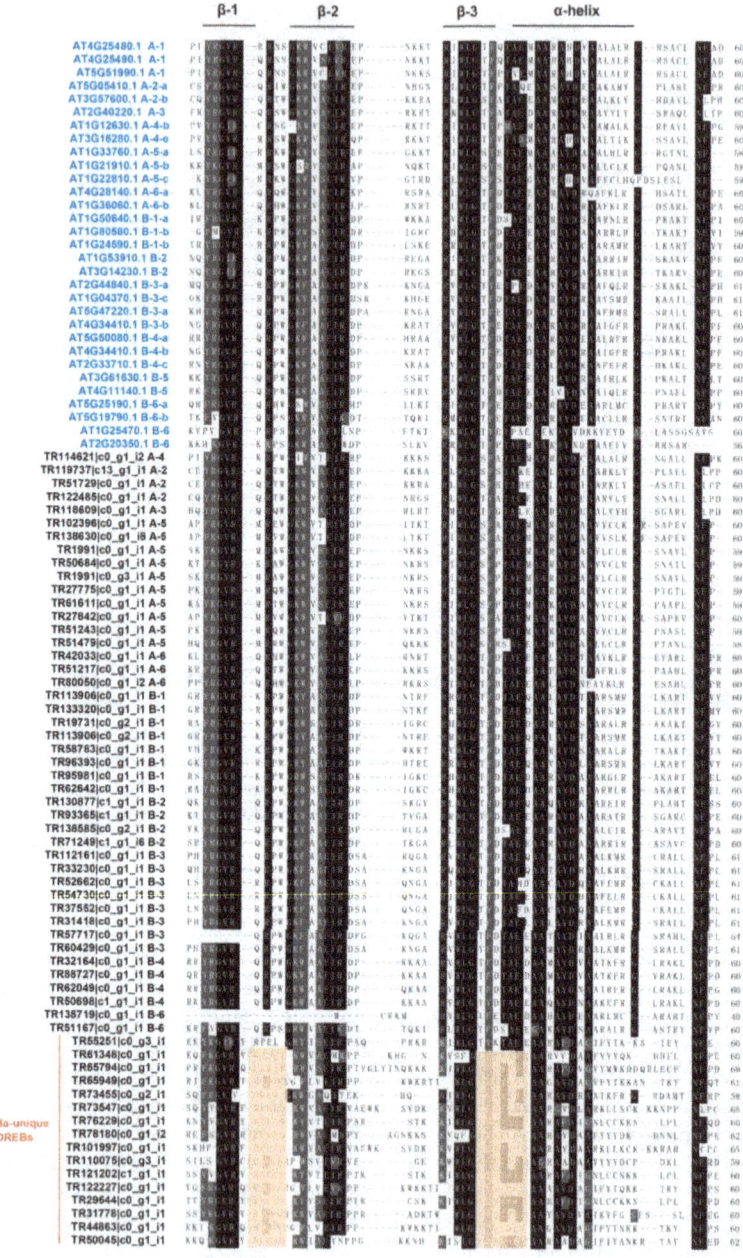

Figure 3. Sequence alignments of AP2 domains of representative ERF proteins in *B. argenteum* and Arabidopsis. Thirty-one AtERF genes representative of each ERF subfamily/group were aligned with 69 BaERFs. The subfamily for each ERF is depicted on the right. The locus names of AtERFs and BaERFs are marked in blue and black, respectively. The identical and conserved amino acid residues are indicated with black and light gray shading, respectively. The black bars represent three β sheets and α helix regions. The Ba-unique DREBs are grouped in pink bar, and two regions representing diverse amino acids compositions are marked with pink boxes.

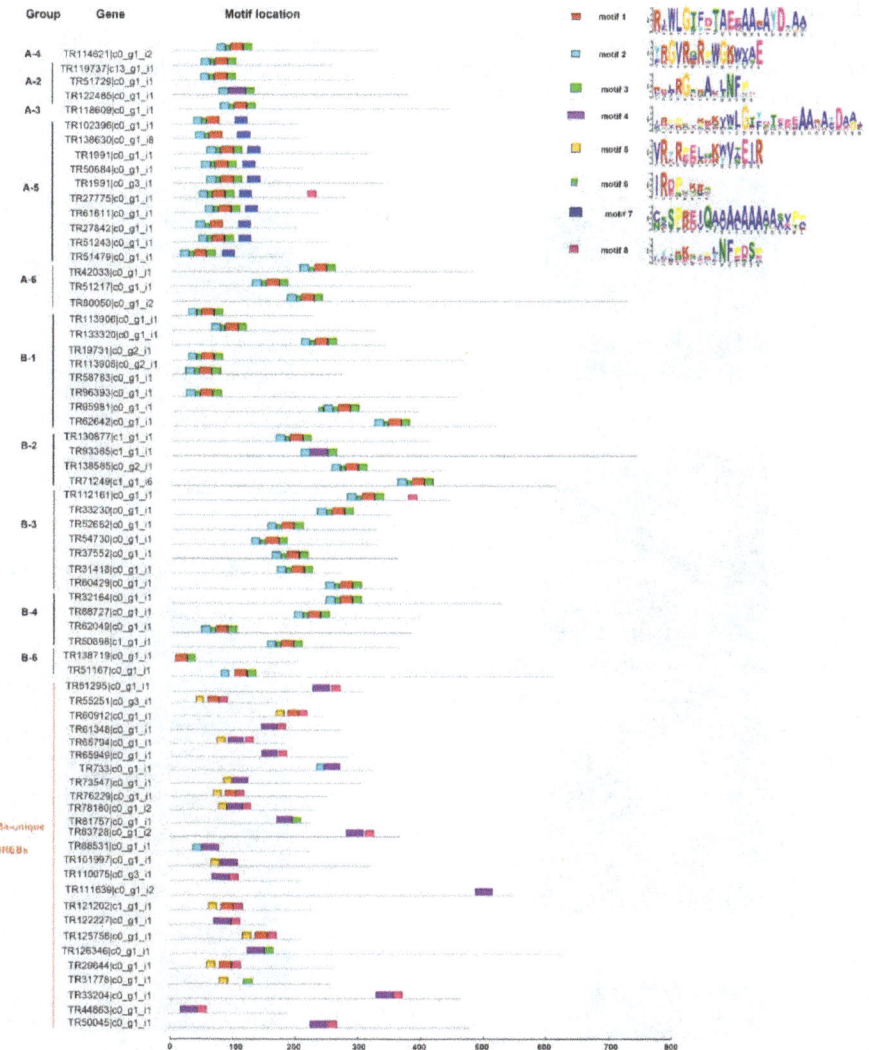

Figure 4. Motif analyses of BaERFs with intact ORFs using MEME online software. BaAP2/ERF proteins with complete ORFs were used for motif prediction. Parameters are as follows: any number of repetitions per sequence, motif width ranges of 6–50 amino acids, and 8 as the maximum number of motifs. Each of the sequence has an E-value less than 10. Motif composition and deduced amino acid sequence of each motif are presented.

2.4. Gene Expression Analysis of all the BaAP2/ERFs during Moss Specific Dehydration–Rehydration Process Using RNA-seq Data

To evaluate the potential function of 83 *BaAP2/ERFs* genes under dehydration–rehydration stress treatment, we investigated the gene expression pattern based on the RNA-seq datasets (H0, D2, D24, R2, and R48). The results show that all 83 *BaAP2/ERFs* belonged to AP2, DREB, ERF, RAV and Soloists families exhibited elevated transcript amounts during dehydration–rehydration process (Figure 5). The majority of *BaAP2/ERF* transcripts were more abundant in both dehydration (D2 and D24) and early-rehydration (R2) stages. For example, 6 out of 11 *AP2* family transcripts and 13 out of 17

DREB transcripts were more abundant in dehydration–rehydration stages. Some transcripts also showed different expression patterns of accumulation. TR88531 | c0_g1_i1 and TR76229 | c0_g1_i1 (Bryum-unique *DREB* genes) transcripts were more abundant in D2 and D24, while TR110575 | c0_g3_i1 (Bryum-unique *DREB* gene) and TR130877 | c0_g1_i1 (*ERF* gene) were more abundant in R2 and R48.

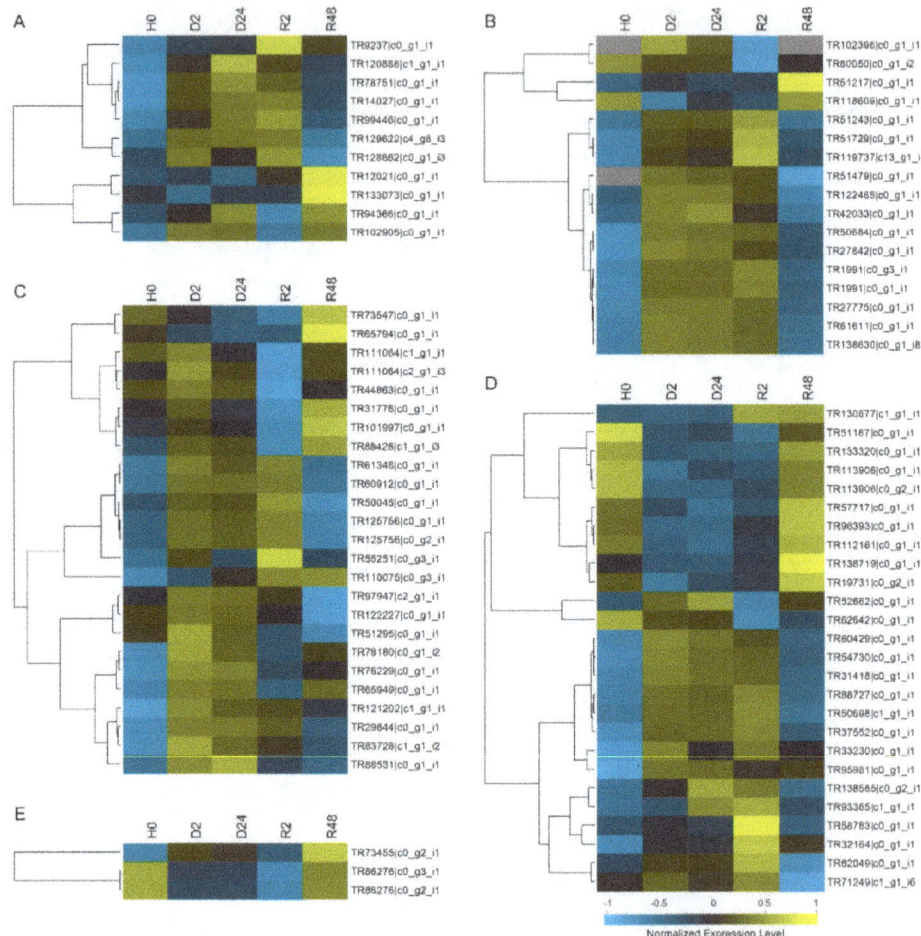

Figure 5. Heat map of the relative expression levels of all identified BaAP2/EFR genes during dehydration–rehydration process of *B. argenteum*. Color scores were normalized by the log2 transformed counts of RPKM values. Yellow represents high expression, while blue represents low expression. Expression differences in the transcripts were clustered by the hierarchical complete linkage clustering method using an uncentered correlation similarity matrix. Prior to the clustering analysis, expression data in unit of FPKM were pretreated using the standardization tools in Cluster 3.0. The heat maps were drawn using the Java Treeview package. Expression profiles (in log2 based values) of the: (**A**) *AP2*; (**B**) *DREB*; (**C**) unclassified *DREB* group (Ba-unique clade); (**D**) *ERF*; and (**E**) *RAV* and *Soloist* genes in *B. argenteum* response to dehydration–rehydration treatment.

2.5. Diverse Gene Expression Patterns of Twelve BaAP2/ERFs during Dehydration–Rehydration Process Using RT-qPCR

The expression profiles of 12 transcripts, representative of different family members of *BaAP2/ERF* (one *AP2*, four *DREBs*, two Ba-unique *DREBs*, three *ERFs* and two *Soloists*) that demonstrated a diverse pattern of induced gene expression were validated by RT-qPCR (Figure 6). RT-qPCR results confirmed the expression trends observed with RNA-seq data. RT-qPCR demonstrated that 9 out of 12 genes increased, reached a peak and then decreased, while the other three genes changed slightly and rapidly increased to the maximum fold at rehydration (R48) stage. Most genes reached an expression peak at rehydration stage (R2 or R48), except TR27842 | c0_g1_i1 and TR42033 | c0_g1_i1, which peaked at D2 and D24, respectively (Figure 6). The expression pattern can be divided into four types: (1) transcripts that accumulate in response to desiccation stress (e.g., TR27842 | c0_g1_i1 which was strongly induced by desiccation treatment (almost 20-fold compared to H0) and rapidly reduced after rehydration); (2) transcripts which modestly accumulate in response to both desiccation and rehydration (e.g.,TR42033 | c0_g1_i1 gene, the gene expression level of which changed within two-fold during desiccation and rehydration process); (3) transcripts which accumulate in response to desiccation and remain elevated upon rehydration (e.g., TR29644 | c0_g1_i1, which was 10-fold increased after desiccation treatment, and then reached a peak (more than 40-fold) at R2 stage); and (4) transcripts which accumulate in response to rehydration (e.g., TR138719 | c0_g1_i1, which was highly induced by rehydration (R48)).

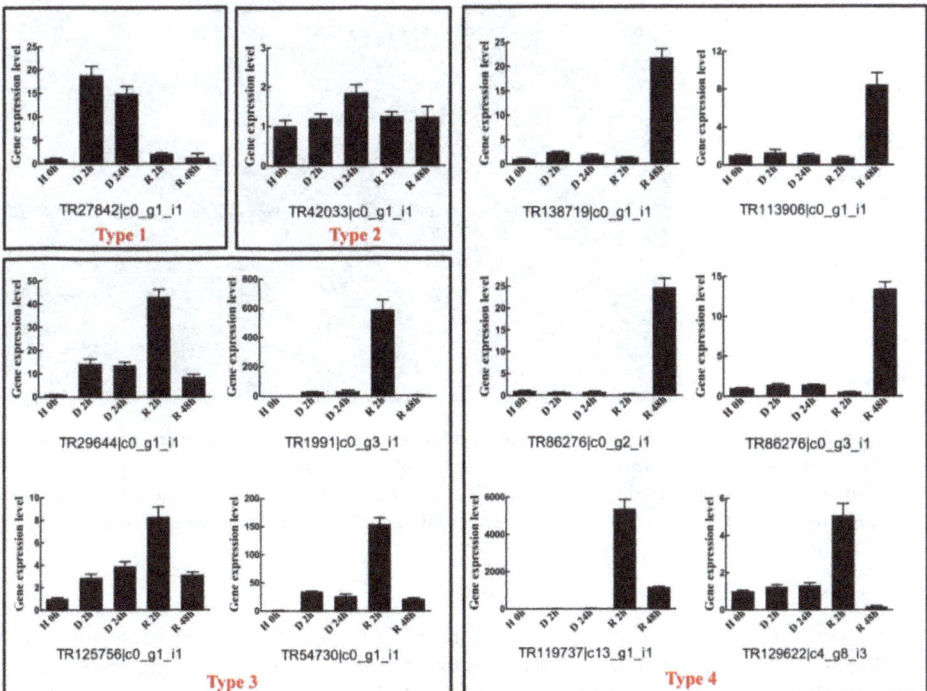

Figure 6. RT-qPCR validation of gene expression patterns of 12 representative *BaAP2/ERF* genes during *B. argenteum* dehydration–rehydration process. RT-qPCR quantitative gene expression data are shown as the mean ± SE. The relative gene expression levels were calculated relative to 0 h and using the $2^{-\Delta\Delta Ct}$ method.

2.6. Transactivation Activity Analyses of Twelve BaAP2/ERFs

We further investigated the transactivation activity of the above 12 *BaAP2/ERFs* using a yeast-based transcriptional activity assay. The results show that 8 out of 12 BaAP2/ERFs proteins can grow well on SD-Trp, SD-Trp-His medium and exhibit α-galactosidase activity on SD-Trp-His medium containing x-α-gal (Figure 7). This indicates that these AP2/ERF proteins (one AP2 (TR129622 | c4_g8_i3), four DREBs (TR119737 | c13_g1_i1, TR27842 | c0_g1_i1, TR1991 | c0_g3_i1, and TR42033 | c0_g1_i1), one Ba-unique DREB (TR29644 | c0_g1_i1) and two ERFs (TR54730 | c0_g1_i1 and TR138719 | c0_g1_i1)) demonstrate transactivation activities. Four proteins (one Ba-unique DREB (TR125756 | c0_g1_i1), one ERF (TR113906 | c0_g1_i1) and two Soloists (TR86276 | c0_g2_i1 and TR86276 | c0_g3_i1)) grew similarly to the negative control indicating that these proteins might not function as transcriptional activators in this yeast heterologous system.

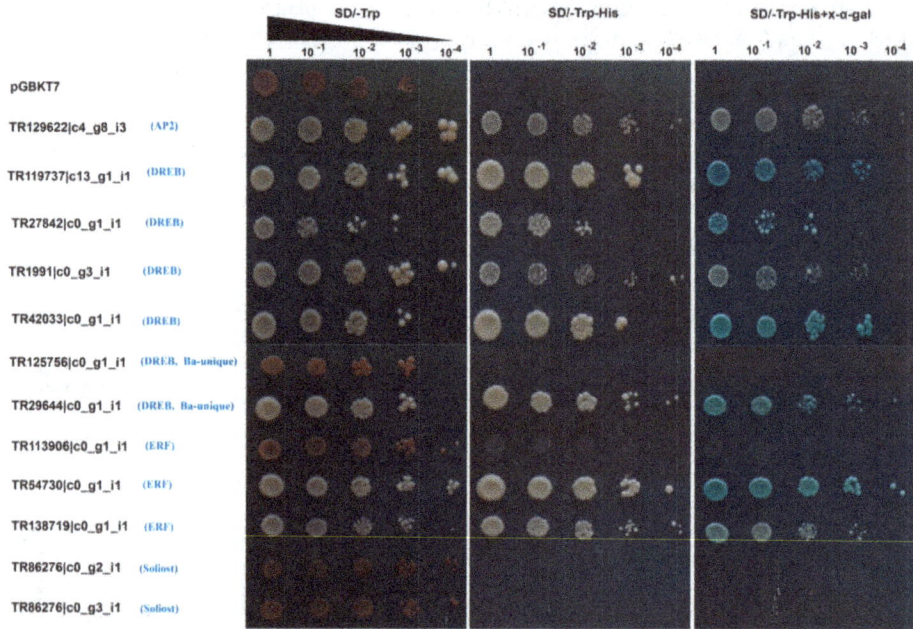

Figure 7. Transactivation activities of 12 BaAP2/ERF proteins in yeast. Yeast cells Y2H expressing the fusion proteins were cultured and adjusted to an OD600 of 2.0, then series diluted and dropped with 2 μL on nutritional selective medium SD/−Trp, SD/−Trp−His and SD/−Trp−His+x-α-gal. Yeast cells expressing the empty vector pGBKT7 was used as negative control. Photos were taken after incubating at 30 °C for 2–4 days.

2.7. Stress Tolerance Ability Evaluation in Transgenic Yeast

To investigate the ability of BaAP2/ERF proteins to enhance abiotic stress tolerance in heterologous expression system, eight representative *BaAP2/ERFs*, driven by a galactose-inducible promoter (pYES2), were introduced into *S. cerevisiae* (INVSc1). After 45 h of exposure to 5 M NaCl and 3 M Sorbitol, the growth patterns of all BaAP2/ERF transformed *S. cerevisiae* (pYES2-*BaAP2/ERF*) were similar to the empty vector (pYES2) under non-stress conditions (Figure 8). Seven out of eight *BaAP2/ERF* (except TR86276 | c0_g3_i1) transformed *S. cerevisiae* survived better than the empty vector under salt and osmotic stresses, especially under salt stress, indicating that these seven BaAP2/ERF proteins (TR119737 | c13_g1_i1, TR27842 | c0_g1_i1, TR1991 | c0_g3_i1, TR29644 | c0_g1_i1,

Int. J. Mol. Sci. **2018**, *19*, 3637

TR54730 | c0_g1_i1, TR138719 | c0_g1_i1, and TR86276 | c0_g2_i1) were functional in yeast cells and improved the yeast tolerance to salt and osmotic stresses.

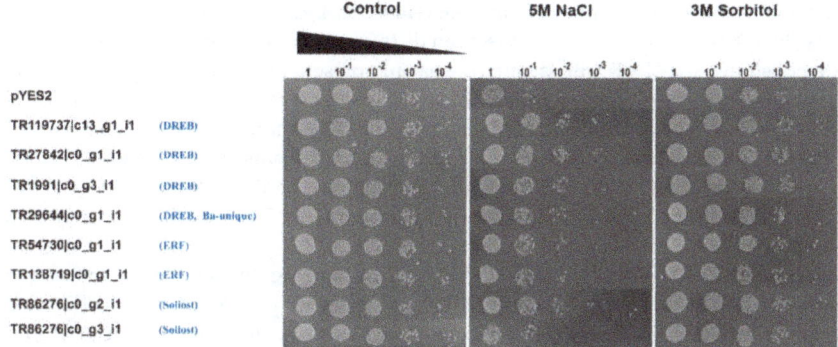

Figure 8. Growth of *S. cerevisiae* yeast cells transformed with the pYES2-BaAP2/ERFs under salt and osmotic stress conditions. To test the salt and osmotic tolerances, the same quantity of yeast culture sample was re-suspended in 5 M NaCl and 3 M Sorbitol, respectively, at 30 °C for 45 h. For non-stress control, an equivalent number of yeast cells was re-suspended in 200 μL of sterile water and incubated at 30 °C for 45 h. Serial dilutions of 1:10 transformed yeast cells were grown on SC-ura medium for two days.

3. Discussion

AP2/ERF genes play central roles during plant stress responses and have been widely identified in both dicotyledonous and monocotyledonous plants using genomic or transcriptomic data [20,22], however, little is known about the functions of *AP2/ERF* genes in moss species. Several recent studies have demonstrated that AP2/ERF transcription factors play an important role in the stress responses of bryophytes [15,16,19]. *B. argenteum* is extremely tolerant to desiccation stress and is a promising model for the identification of stress related genes [8], however, no *AP2/ERF* gene in *B. argenteum* has been reported until now.

The AP2/ERF family of TFs in Arabidopsis comprises five subfamilies of TFs, classified based on sequence similarity, number of AP2 domains, and the presence of other characteristic domains [26]. It is reported that different gene families have different functions. The *AP2* gene family is associated with plant flower development, while *ERF* and *DREB* family genes accumulate in response to biotic and abiotic stress, respectively [9]. Genes annotated to a specific group within the same gene family are also reported to have different functions. A-1 type *DREB* transcripts accumulate in response to cold stress, while A-2 type *DREB* transcripts accumulate in response to osmotic and heat stresses [27]. Hence, a precise and detailed classification of *AP2/ERF* genes within a genome is an important tool for predicting gene expression and function.

Classification of the AP2/ERF superfamily is based on the number of AP2 domains and by constructing a phylogenetic tree comparing the moss AP2 domains to Arabidopsis or rice. The resulting gene tree construction has been a classic and reliable method of annotation which is widely employed in many non-model plants [24,28–31]. However, classification by gene tree should be used cautiously as the results can be inconsistent with an AP2 domain counting based classification. *Soloist* genes with a single AP2 domain always grouped together with *AP2* family genes which have two domains (as reported in *Hevea brasiliensis* and *Vitis vinifera* [23,28]). In this study, we found that *BaSoloists* clustered with *AtSoloists* and were mixed together with *AP2* family genes. In addition, five genes which contained a single AP2 domain also clustered with *AP2* family genes in Arabidopsis. To confirm the classification as AP2 family members, we constructed two more phylogenetic trees: one using only *AP2/ERF* genes in *B. argenteum* and another one using *AP2/ERF* genes from the model moss

P. patens. Finally, these five genes were classified into *AP2* family given their greater homology with the *AP2* family genes.

The AP2 domain of *AP2/ERF* genes is conserved in plants, however amino acid variations within the AP2 domain have been documented and can refine classification of the gene [23,29]. For example, the motifs "HLG" and "WLG" in the β3 sheet can distinguish a *Soloist* gene from an *ERF* gene. Li et al. (2017) demonstrated the "EVR" motif pattern was only present in the A-1 group of *DREB* genes and "ERK" was specific to the B-6 subfamily of *ERF* genes in the β2 sheet of AP2 domain [19]. Based on this A-1 DREB-specific amino acid, in the present study, we finally confirmed the TR11462 | c0_g1_i2 gene was A-4 type of DREB rather than A-1 type. Specific motif elements are also helpful for robust gene classification. In the DREB family, ERF-associated amphiphilic repression (EAR) motif was specifically present in the A-5a group genes, which contained (L/F) DLN (L/F) xP residues and may be essential for repression function [32–34].

Moss species have unique genes which are challenging to annotate compared to other organisms [19]. In *S. caninervis*, the majority of *ScERF* genes can be classified while few *ScERF* genes are not clustered with any Arabidopsis group, and clustered as a unique clade [19]. Similarly, in this study, half of *DREBs* cannot be classified relative to other plant genes and clustered as a Ba-unique clade. The amino acids compositions of AP2 domains were also supported that Ba-unique clade genes have diverse amino acids composition and showed more diverse motif patterns compared with other *DREB* genes. Furthermore, BLASTp search in One KP and NCBI database showed that, although Ba-unique *DREBs* have homolog genes in angiosperm, they shared very low amino acid identities. Some Ba-unique *DREBs* had very high sequence identities with other moss genes, while these moss genes were rarely characterized and no functional analysis were reported until now. Our results extend the idea that *AP2/ERF* genes in moss species can be different from angiosperm genes, and moss-unique genes may have novel and/or altered functions. It is necessary to explore their functions in future work.

Gene expression pattern was considered to be directly connected with the gene function [35], and, in this study, the expression of all 83 *BaAP2/ERFs* genes were induced during dehydration–rehydration process. Moreover, the 12 representative *BaAP2/ERF* genes in different families exhibited differential expression in response to dehydration and rehydration treatment. Within the same family, the gene expression patterns were different suggesting a functional diversity of *AP2/ERF* genes in response to dehydration and rehydration stress in *B. argenteum*. AP2/ERF proteins are important transcriptional factors which can activate many down-stream genes, thus improving the overall stress tolerance of plants [10]. In the present study, 8 out of 12 BaAP2/ERFs proteins demonstrated transactivation activity in the yeast system. Based upon patterns of gene expression and transactivation activity analysis, we selected eight representative *BaAP2/ERF* genes for further functional test in yeast and the result showed that seven of them improved the yeast tolerance to salt and osmotic stresses. Our results demonstrated that *BaAP2/ERFs* genes play crucial roles in *B. argenteum* response to stresses.

4. Materials and Methods

4.1. Identification of the AP2/ERF Protein Family in B. argenteum

B. argenteum transcripts (76,206) were obtained from a hydration–dehydration–rehydration transcriptome [25] (data were deposited at NCBI-SRA with accession SRP077772, https://www.ncbi.nlm.nih.gov/sra/?term=SRP077772) and served as the source for the *AP2/ERF* gene identification and presented in this study. Two methods were used together to identify the putative *AP2/ERF* genes from *B. argenteum*. Firstly, 176 Arabidopsis AP2 predicted amino acid sequences and 171 *P. patens* AP2 predicted amino acid sequences were downloaded from the plant transcription factor database (PlantTFDB v3.0) (http://planttfdb.cbi.edu.cn/) [14], and used as queries to search against the *B. argenteum* transcriptome database using tBLASTn program (E value of 1×10^{-3}. Second, the HMM profiles PF00847 (AP2 domain) and PF02362 (B3 domain) were downloaded from Pfam database v27.0 (http://pfam.sanger.ac.uk/) [36], and the profiles were queried using hmm search command included in the HMMER (v3.0) software (E value cutoff at 1×10^{-3}). All candidate *BaAP2/ERF* genes identified through these two methods were confirmed with Conserved Domain Database (CDD http://www.ncbi.nlm.nih.gov/cdd/) [37] and SMART (http://smart.embl-heidelberg.de/) [38] searches to ensure the presence of an AP2 domain. An AP2 domain, length of approximately 60 amino acids was considered to be a full-length AP2 domain [23]. All the predicted peptide sequences were filtered with a minimum length of 80 amino acids. Sequences which shared >98% matches were considered redundant.

4.2. Sequence Analysis and Classification of BaAP2/ERF Genes Using Phylogenetic Tree

ORFs were predicted with the ORF Finder at NCBI (http://www.ncbi.nlm.nih.gov/gorf/gorf. html). Protein sequence motif detection was performed with MEME program (http://meme-suite.org/index.html) [39] using the parameters: zero or one repetition per sequence, motif width ranges of 6–40 amino acids, and 8 as the maximum number of motifs. Multiple sequence alignment was performed with ClustalW [40], phylogenetic trees were constructed by the neighbor-joining method (with 1000 bootstrap replicates) using MEGA 6.06 the evolutionary distances were computed using the Poisson correction method with pairwise deletion. Sequence similarity was analyzed using BLASTp search with NCBI and ONE KP (https://db.cngb.org/onekp/) database. All the *BaAP2/ERF* sequences were submitted to the GenBank database using *Bank*It (http://www.ncbi.nlm.nih.gov/BankIt/).

4.3. Gene Expression Analysis of BaAP2/ERF Genes Using RNA-seq Data

Expression differences in the transcripts under dehydration–rehydration condition were clustered by the hierarchical complete linkage clustering method using an uncentered correlation similarity matrix. Prior to the clustering analysis, expression data in unit of Fragments Per Kilobase of transcript per Million fragments mapped (FPKM) were pretreated using the standardization tools in Cluster 3.0: (a) log transform data; (b) center genes (mean); and (c) normalize genes [41]. The heat maps were drawn by using the Java Treeview package [42,43].

4.4. Gene Expression Pattern Analysis of BaAP2/ERF Genes Using RT-qPCR Assay

B. argenteum gametophytes were cultured in solid Knop medium at 25 °C with 16 h/8 h photoperiod in a climate chamber as described previously [8]. For desiccation–rehydration treatment, the well-hydrated gametophytes in Knop solid medium were transferred to 90 cm open Petri dish and air-dried for 2 h (D2) and 24 h (D24), and the desiccated gametophytes (D24) samples were subsequently rehydrated with deionized water for 2 h (R2) and 48 h (R48). All treatments were performed at 25 °C with RH ≈ 25–27%, and the well-hydrated gametophores in Knop medium without any treatment was served as the control (H0). Three biological replicates were collected for each of the time point of different treatments.

Total RNAs of *B. argenteum* gametophytes were extracted using MiniBEST plant RNA kit (Takara, Japan). Gel electrophoresis and a NanoDrop 2000 spectrophotometer (Thermo Fisher Scientific, Waltham, MA, USA) were used for RNA quality test and quantitative analysis. High quality RNA samples were used for subsequent reverse transcription. First strand cDNA was synthesized using PrimeScript^TM RT reagent kit (Takara, Shiga Prefecture, Japan).

Twelve *BaAP2/ERF* genes representative different groups/subfamilies were selected to verify the gene expression pattern obtained from transcriptome data under desiccation and rehydration condition. RT-qPCR primers were designed with Primer Premier 5.0 and the primer specificities were tested by running BLAST search against the local *B. argenteum* transcriptional data. Each primer pair was further assessed using melting-curve analysis after RT-qPCR. All primer information for RT-qPCR is shown in Table S2. RT-qPCR experiments were carried out using CFX96 Real-Time PCR Detection System (Bio-Rad, Hercules, CA, USA) with SYBR *Premix Ex Taq*^TM kit (Takara, Shiga Prefecture, Japan). The PCR reaction mixture consisted of 2 µL cDNA sample (1:5 diluted), 0.4 µL each of the forward and reverse primers (10 μM), 10 μL master mix and 7.2 μL PCR-grade water in a final volume of 20 μL. Three biological replicates and three technical replicates of each biological replicate were used for all samples. The RT-qPCR program was as follows: initial denaturation step of 30 s at 95 °C and 40 cycles of PCR (94 °C for 5 s and 60–62 °C for 30 s). The gene relative expression levels were calculated relative to the H0 samples using the $2^{-\Delta\Delta Ct}$ method. The *ACT* gene was used to normalize the RT-qPCR data [8]. Figures were generated using Sigmaplot 12.0.

4.5. Gene Cloning, Vector Construction and Transcriptional Activation Analysis in Yeast Cells

To further evaluation of transactivation activity of 12 *BaAP2/ERF* genes, we cloned these 12 genes into the pMD18-T clone vector. After sequence analysis, the PCR products of these genes were cloned separately into the pGBKT7 vector using the in-fusion PCR cloning system (Clontech, Mountain View, CA, USA). Positive plasmids containing different *BaAP2/ERF* genes were transformed into the Y2H yeast strain (Clontech, Mountain View, CA, USA). All primers used for cloning and vector construction are listed in Tables S3 and S4. The cell concentration of yeast positive transformants were adjusted to an OD600 of 2.0, the yeast cells were then diluted serially (1, 10^{-1}, 10^{-2}, 10^{-3}, and 10^{-4}) and dropped with 2 μL on synthetic dropout (SD) medium without tryptophan (SD/−Trp), without tryptophan and histidine (SD/−Trp−His), and with SD/−Trp−His plates containing x-α-gal with the final concentration of 40 mg/L (SD/−Trp−His+x-α-gal). Yeast cells expressing the empty vector pGBKT7 was used as negative control. The plates were incubated at 30 °C for 2–4 days before photographing. Adobe Illustrator CS5 was used for image processing.

4.6. Stress Tolerance Studies in Yeast

Eight representative *BaAP2/ERF* genes including four *DREBs* (TR119737|c13_g1_i1, TR27842|c0_g1_i1, TR1991|c0_g3_i1 and TR29644|c0_g1_i1), two *ERFs* (TR54730|c0_g1_i1 and TR138719|c0_g1_i1) and two *Soloists* (TR86276|c0_g2_i1 and TR86276|c0_g3_i1) were selected to study the stress tolerance ability under salt and osmotic stress conditions in yeast. The ORF of TR1991|c0_g3_i1, TR29644|c0_g1_i1, TR54730|c0_g1_i1, TR138719|c0_g1_i1, TR86276|c0_g2_i1 and TR86276|c0_g3_i1 were amplified from pGBKT7-*BaAP2/ERF* plasmids of transcriptional activation assay, using primers shown in Table S5, and inserted into the yeast expression vector pYES2 using the in-fusion PCR cloning system. The ORF of two *DREB* genes TR119737|c13_g1_i1 and TR27842|c0_g1_i1 were obtained from pGBKT7-*BaAP2/ERF* plasmids using *Not*I and *Eco*RI restriction enzymes digestion and inserted into the *Not*I and *Eco*RI sites of the yeast expression vector pYES2, which contains a URA3 selection marker driven by the GAL1 promoter. Subsequently, eight pYES2-*BaAP2/ERF* plasmids and the empty pYES2 control plasmids were introduced into yeast strain INVSc1 (Invitrogen, Carlsbad, CA, USA) using a lithium acetate procedure, according to the pYES2 vector kit instructions (Invitrogen, Carlsbad, CA, USA). The transformants were screened by growth on a uracil-deficient synthetic complete (SC-ura) medium with 2% (*w/v*) glucose at 30 °C for 2 days.

For the stress assay, yeast cells harboring both pYES2-BaAP2/ERFs and the empty pYES2 vector (control) were incubated in SC-ura liquid medium containing 2% glucose at 30 °C for approximately 20 h with shaking (180 rpm). After incubation, the optical densities of the yeast cell were determined at OD600. The culture samples were adjusted to contain an equal OD600 of 0.4 as a starting concentration in 10 mL of induction SC-ura medium (supplemented with 2% *w/v* galactose). After incubation for approximately 24 h, the yeast cell densities were recalculated and adjusted to contain an equal number of cells (OD600 = 2) in 200 μL solutions with 5 M NaCl or 3M Sorbitol for the salt or osmotic stress, and the same quantity of yeast cells was re-suspended in 200 μL of sterile water was served as the control. After incubating at 30 °C for 45 h, the cells were 10-fold serially diluted with sterile water, and 2 μL aliquots of each dilution were spread on SC-ura medium containing 2% (*w/v*) glucose and growth performance was compared after growing at 30 °C for 2 days [18,44].

Int. J. Mol. Sci. **2018**, *19*, 3637

5. Conclusions

This is the first report on identification, classification, characterization, and functional evaluation of the *AP2/ERF* gene family in the desiccation tolerant moss *B. argentums*. Eighty-three AP2/ERF predicted proteins were identified from the *B. argenteum* transcriptome and classified within the *AP2/ERF* gene family. The gene expression pattern was analyzed in response to a well characterized dehydration–rehydration based upon RT-qPCR and RNA-seq data. We verified the transactivation activities of 12 representative *BaAP2/ERF* genes. Furthermore, eight *BaAP2/ERF* genes were tested for stress tolerance functions in yeast. We conclude that TR29644|c0_g1_i1 (*DREB*-Ba-unique), TR119737|c13_g1_i1 (*DREB*), TR54730|c0_g1_i1 (*ERF*), TR27842|c0_g1_i1 (*DREB*) and TR86276|c0_g2_i1 (*Soloist*) genes strongly respond to environmental stress and that these genes are correlated with enhanced salt- and osmotic-stress tolerance in transgenic yeast. These genes are promising candidate genes for further functional analysis and demonstrate great potential in plant molecular breeding.

Supplementary Materials: Supplementary Materials can be found at http://www.mdpi.com/1422-0067/19/11/3637/s1, Figure S1: The 20 most abundant predicted transcription factor families in the *B. argenteum* transcripotome datasets; Figure S2: Phylogenetic analysis of *AP2/ERF* family genes in *B. argenteum*; Figure S3: Phylogenetic analysis of *AP2/ERF* family genes in *B. argenteum* and *P. patens*; Table S1: Sequence alignment results of 25 Ba-clade DREBs using IKP BLAST; Table S2: Primer information of 12 *BaAP2/ERF* genes for RT-qPCR analysis; Table S3: Primer information of 12 *BaAP2/ERF* genes for gene cloning; Table S4: Primer information of 12 *BaAP2/ERF* genes for fusing to pGBKT7 vector; Table S5: Primer information of *BaAP2/ERF* genes for infusing to pYES2 vector.

Author Contributions: Formal analysis, B.G.; Funding acquisition, D.Z.; Investigation, X.L. (Xiaoshuang Li), Y.L., X.L. (Xiaojie Liu) and J.Z.; Supervision, D.Z.; Validation, A.J.W.; Visualization, B.G.; Writing—original draft, X.L. (Xiaoshuang Li) and Writing—review and editing, J.Z. and A.J.W.

Funding: This research was supported by the National Natural Science Foundation of China (No. 31500225), NSFC-Xinjiang key project (No. U1703233) and National Natural Science Foundation of China (No. 31870318).

Conflicts of Interest: The authors declare no conflict of interest.

Abbreviations

AP2/ERF	APETALA2/Ethylene responsive factor
DREB	dehydration-responsive element-binding protein
ERF	ethylene-responsive factor
RAV	related to ABI3/VP1
RT-qPCR	reverse transcription quantitative real-time polymerase chain reaction
TF	transcription factor
DT	desiccation tolerance

References

1. Zhang, Y.M.; Chen, J.; Wang, L.; Wang, X.Q.; Gu, Z.H. The spatial distribution patterns of biological soil crusts in the Gurbantunggut Desert, Northern Xinjiang, China. *J. Arid Environ.* **2007**, *68*, 599–610. [CrossRef]
2. Li, X.R.; Zhou, H.Y.; Wang, X.P.; Zhu, Y.G.; O'Conner, P.J. The effects of sand stabilization and revegetation on cryptogam species diversity and soil fertility in the Tengger Desert, Northern China. *Plant Soil* **2003**, *251*, 237–245. [CrossRef]
3. Hui, R.; Li, X.R.; Chen, C.Y.; Zhao, X.; Jia, R.L.; Liu, L.C.; Wei, Y.P. Responses of photosynthetic properties and chloroplast ultrastructure of *Bryum argenteum* from a desert biological soil crust to elevated ultraviolet-B radiation. *Physiol. Plant.* **2013**, *147*, 489–501. [CrossRef] [PubMed]
4. Li, J.H.; Li, X.R.; Chen, C.Y. Degradation and reorganization of thylakoid protein complexes of *Bryum argenteum* in response to dehydration and rehydration. *Bryologist* **2014**, *117*, 110–118. [CrossRef]
5. Wood, A.J. Invited essay: New frontiers in bryology and lichenology—The nature and distribution of vegetative desiccation-tolerance in hornworts, liverworts and mosses. *Bryologist* **2007**, *110*, 163–177. [CrossRef]
6. Li, J.H.; Li, X.R.; Zhang, P. Micro-morphology, ultrastructure and chemical composition changes of *Bryum argenteum* from a desert biological soil crust following one-year desiccation. *Bryologist* **2014**, *117*, 232–240. [CrossRef]

7.   Stark, L.R.; McLetchie, D.N.; Eppley, S.M. Sex ratios and the shy male hypothesis in the moss *Bryum argenteum* (Bryaceae). *Bryologist* **2010**, *113*, 788–797. [CrossRef]

8.   Gao, B.; Zhang, D.Y.; Li, X.S.; Yang, H.L.; Zhang, Y.M.; Wood, A.J. De novo transcriptome characterization and gene expression profiling of the desiccation tolerant moss *Bryum argenteum* following rehydration. *BMC Genom.* **2015**, *16*. [CrossRef] [PubMed]

9.   Licausi, F.; Ohme-Takagi, M.; Perata, P. APETALA2/Ethylene Responsive Factor (AP2/ERF) transcription factors: Mediators of stress responses and developmental programs. *New Phytol.* **2013**, *199*, 639–649. [CrossRef] [PubMed]

10.   Xu, Z.S.; Chen, M.; Li, L.C.; Ma, Y.Z. Functions and Application of the AP2/ERF Transcription Factor Family in Crop Improvement. *J. Integr. Plant Biol.* **2011**, *53*, 570–585. [CrossRef] [PubMed]

11.   Mizoi, J.; Shinozaki, K.; Yamaguchi-Shinozaki, K. AP2/ERF family transcription factors in plant abiotic stress responses. *Biochim. Biophys. Acta Gene Regul. Mech.* **2012**, *1819*, 86–96. [CrossRef] [PubMed]

12.   Bhatta, M.; Morgounov, A.; Belamkar, V.; Baenziger, P. Genome-wide association study reveals novel genomic regions for grain yield and yield-related traits in drought-stressed *Synthetic hexaploid* Wheat. *Int. J. Mol. Sci.* **2018**, *19*, 3011. [CrossRef] [PubMed]

13.   Jin, J.P.; Tian, F.; Yang, D.C.; Meng, Y.Q.; Kong, L.; Luo, J.C.; Gao, G. PlantTFDB 4.0: Toward a central hub for transcription factors and regulatory interactions in plants. *Nucleic Acids Res.* **2017**, *45*, D1040–D1045. [CrossRef] [PubMed]

14.   Jin, J.; Zhang, H.; Kong, L.; Gao, G.; Luo, J. PlantTFDB 3.0: A portal for the functional and evolutionary study of plant transcription factors. *Nucleic Acids Res.* **2014**, *42*, D1182–D1187. [CrossRef] [PubMed]

15.   Hiss, M.; Laule, O.; Meskauskiene, R.M.; Arif, M.A.; Decker, E.L.; Erxleben, A.; Frank, W.; Hanke, S.T.; Lang, D.; Martin, A.; et al. Large-scale gene expression profiling data for the model moss *Physcomitrella patens* aid understanding of developmental progression, culture and stress conditions. *Plant J.* **2014**, *79*, 530–539. [CrossRef] [PubMed]

16.   Liu, N.; Zhong, N.Q.; Wang, G.L.; Li, L.J.; Liu, X.L.; He, Y.K.; Xia, G.X. Cloning and functional characterization of PpDBF1 gene encoding a DRE-binding transcription factor from *Physcomitrella patens*. *Planta* **2007**, *226*, 827–838. [CrossRef] [PubMed]

17.   Gao, B.; Zhang, D.; Li, X.; Yang, H.; Wood, A.J. De novo assembly and characterization of the transcriptome in the desiccation-tolerant moss *Syntrichia caninervis*. *BMC Res. Notes* **2014**, *7*, 490. [CrossRef] [PubMed]

18.   Li, H.; Zhang, D.; Li, X.; Guan, K.; Yang, H. Novel DREB A-5 subgroup transcription factors from desert moss (*Syntrichia caninervis*) confers multiple abiotic stress tolerance to yeast. *J. Plant Physiol.* **2016**, *194*, 45–53. [CrossRef] [PubMed]

19.   Li, X.; Zhang, D.; Gao, B.; Liang, Y.; Yang, H.; Wang, Y.; Wood, A.J. Transcriptome-Wide Identification, Classification, and Characterization of AP2/ERF Family Genes in the Desert Moss *Syntrichia caninervis*. *Front. Plant Sci.* **2017**, *8*, 262. [CrossRef] [PubMed]

20.   Nakano, T.; Suzuki, K.; Fujimura, T.; Shinshi, H. Genome-wide analysis of the ERF gene family in Arabidopsis and rice. *Plant Physiol.* **2006**, *140*, 411–432. [CrossRef] [PubMed]

21.   Song, X.M.; Li, Y.; Hou, X.L. Genome-wide analysis of the AP2/ERF transcription factor superfamily in Chinese cabbage (*Brassica rapa* ssp pekinensis). *BMC Genom.* **2013**, *14*. [CrossRef] [PubMed]

22.   Sakuma, Y.; Liu, Q.; Dubouzet, J.G.; Abe, H.; Shinozaki, K.; Yamaguchi-Shinozaki, K. DNA-binding specificity of the ERF/AP2 domain of Arabidopsis DREBs, transcription factors involved in dehydration- and cold-inducible gene expression. *Biochim. Biophys. Res. Commun.* **2002**, *290*, 998–1009. [CrossRef] [PubMed]

23.   Duan, C.; Argout, X.; Gebelin, V.; Summo, M.; Dufayard, J.F.; Leclercq, J.; Kuswanhadi; Piyatrakul, P.; Pirrello, J.; Rio, M.; et al. Identification of the Hevea brasiliensis AP2/ERF superfamily by RNA sequencing. *BMC Genom.* **2013**, *14*, 30. [CrossRef] [PubMed]

24.   Wu, Z.J.; Li, X.H.; Liu, Z.W.; Li, H.; Wang, Y.X.; Zhuang, J. Transcriptome-based discovery of AP2/ERF transcription factors related to temperature stress in tea plant (*Camellia sinensis*). *Funct. Integr. Genom.* **2015**, *15*, 741–752. [CrossRef] [PubMed]

25.   Gao, B.; Li, X.; Zhang, D.; Liang, Y.; Yang, H.; Chen, M.; Zhang, Y.; Zhang, J.; Wood, A.J. Desiccation tolerance in bryophytes: The dehydration and rehydration transcriptomes in the desiccation-tolerant bryophyte *Bryum argenteum*. *Sci. Rep.* **2017**, *7*, 7571. [CrossRef] [PubMed]

26. Agarwal, P.K.; Gupta, K.; Lopato, S.; Agarwal, P. Dehydration responsive element binding transcription factors and their applications for the engineering of stress tolerance. *J. Exp. Bot.* **2017**, *68*, 2135–2148. [CrossRef] [PubMed]

27. Lata, C.; Prasad, M. Role of DREBs in regulation of abiotic stress responses in plants. *J. Exp. Bot.* **2011**, *62*, 4731–4748. [CrossRef] [PubMed]

28. Licausi, F.; Giorgi, F.M.; Zenoni, S.; Osti, F.; Pezzotti, M.; Perata, P. Genomic and transcriptomic analysis of the AP2/ERF superfamily in *Vitis vinifera*. *BMC Genom.* **2010**, *11*. [CrossRef] [PubMed]

29. Zhuang, J.; Cai, B.; Peng, R.H.; Zhu, B.; Jin, X.F.; Xue, Y.; Gao, F.; Fu, X.Y.; Tian, Y.S.; Zhao, W.; et al. Genome-wide analysis of the AP2/ERF gene family in *Populus trichocarpa*. *Biochem. Biophys. Res. Commun.* **2008**, *371*, 468–474. [CrossRef] [PubMed]

30. Chen, L.H.; Han, J.P.; Deng, X.M.; Tan, S.L.; Li, L.L.; Li, L.; Zhou, J.F.; Peng, H.; Yang, G.X.; He, G.Y.; et al. Expansion and stress responses of AP2/EREBP superfamily in *Brachypodium Distachyon*. *Sci. Rep.* **2016**, *6*. [CrossRef] [PubMed]

31. Lakhwani, D.; Pandey, A.; Dhar, Y.V.; Bag, S.K.; Trivedi, P.K.; Asif, M.H. Genome-wide analysis of the AP2/ERF family in Musa species reveals divergence and neofunctionalisation during evolution. *Sci. Rep.* **2016**, *6*. [CrossRef] [PubMed]

32. Ohta, M.; Matsui, K.; Hiratsu, K.; Shinshi, H.; Ohme-Takagi, M. Repression domains of class II ERF transcriptional repressors share an essential motif for active repression. *Plant Cell* **2001**, *13*, 1959–1968. [CrossRef] [PubMed]

33. Huang, B.; Liu, J.Y. A cotton dehydration responsive element binding protein functions as a transcriptional repressor of DRE-mediated gene expression. *Biochem. Biophys. Res. Commun.* **2006**, *343*, 1023–1031. [CrossRef] [PubMed]

34. Dong, C.J.; Liu, J.Y. The Arabidopsis EAR-motif-containing protein RAP2.1 functions as an active transcriptional repressor to keep stress responses under tight control. *BMC Plant Biol.* **2010**, *10*. [CrossRef] [PubMed]

35. Hao, Y.J.; Wei, W.; Song, Q.X.; Chen, H.W.; Zhang, Y.Q.; Wang, F.; Zou, H.F.; Lei, G.; Tian, A.G.; Zhang, W.K.; et al. Soybean NAC transcription factors promote abiotic stress tolerance and lateral root formation in transgenic plants. *Plant J.* **2011**, *68*, 302–313. [CrossRef] [PubMed]

36. Punta, M.; Coggill, P.C.; Eberhardt, R.Y.; Mistry, J.; Tate, J.; Boursnell, C.; Pang, N.; Forslund, K.; Ceric, G.; Clements, J.; et al. The Pfam protein families database. *Nucleic Acids Res.* **2012**, *40*, D290–D301. [CrossRef] [PubMed]

37. Marchler-Bauer, A.; Derbyshire, M.K.; Gonzales, N.R.; Lu, S.; Chitsaz, F.; Geer, L.Y.; Geer, R.C.; He, J.; Gwadz, M.; Hurwitz, D.I.; et al. CDD: NCBI's conserved domain database. *Nucleic Acids Res.* **2015**, *43*, D222–D226. [CrossRef] [PubMed]

38. Letunic, I.; Doerks, T.; Bork, P. SMART: Recent updates, new developments and status in 2015. *Nucleic Acids Res.* **2015**, *43*, D257–D260. [CrossRef] [PubMed]

39. Bailey, T.L.; Boden, M.; Buske, F.A.; Frith, M.; Grant, C.E.; Clementi, L.; Ren, J.; Li, W.W.; Noble, W.S. MEME SUITE: Tools for motif discovery and searching. *Nucleic Acids Res.* **2009**, *37*, W202–W208. [CrossRef] [PubMed]

40. Tamura, K.; Peterson, D.; Peterson, N.; Stecher, G.; Nei, M.; Kumar, S. MEGA5: Molecular evolutionary genetics analysis using maximum likelihood, evolutionary distance, and maximum parsimony methods. *Mol. Biol. Evol.* **2011**, *28*, 2731–2739. [CrossRef] [PubMed]

41. Eisen, M.B.; Spellman, P.T.; Brown, P.O.; Botstein, D. Cluster analysis and display of genome-wide expression patterns. *Proc. Natl. Acad. Sci. USA* **1998**, *95*, 14863–14868. [CrossRef] [PubMed]

42. Brock, G.; Pihur, V.; Datta, S.; Datta, S. clValid, an R package for cluster validation. *J. Stat. Softw.* **2011**. [CrossRef]

43. Saldanha, A.J. Java Treeview-extensible visualization of microarray data. *Bioinformatics* **2004**, *20*, 3246–3248. [CrossRef] [PubMed]

44. Li, X.; Zhang, D.; Li, H.; Wang, Y.; Zhang, Y.; Wood, A.J. EsDREB2B, a novel truncated DREB2-type transcription factor in the desert legume *Eremosparton songoricum*, enhances tolerance to multiple abiotic stresses in yeast and transgenic tobacco. *BMC Plant Biol.* **2014**, *14*, 44. [CrossRef] [PubMed]

International Journal of
Molecular Sciences

Article

Floral Scent Emission from Nectaries in the Adaxial Side of the Innermost and Middle Petals in *Chimonanthus praecox*

Zhineng Li [1,2,3,†], Yingjie Jiang [1,2,3,†], Daofeng Liu [1,2,3], Jing Ma [1,2,3], Jing Li [1,2,3], Mingyang Li [1,2,3] and Shunzhao Sui [1,2,3,*]

1   College of Horticulture and Landscape Achitecture, Southwest University, Chongqing 400715, China;
    znli@swu.edul.cn (Z.L.); sherneole@163.com (Y.J.); liu19830222@163.com (D.L.);
    majing427@swu.edu.cn (J.M.); acejing@126.com (J.L.); limy@swu.edu.cn (M.L.)
2   Key Laboratory of Horticulture Science for Southern Mountains Regions, Ministry of Education,
    Chongqing 400715, China
3   Chongqing Engineering Research Center for Floriculture, Chongqing 400715, China
*   Correspondence: sszcq@swu.edu.cn; Tel.: +86-23-6825-0086
†   These authors contributed equally to this work.

Received: 10 September 2018; Accepted: 18 October 2018; Published: 22 October 2018

Abstract: Wintersweet (*Chimonanthus praecox*) is a well-known traditional fragrant plant and a winter-flowering deciduous shrub that originated in China. The five different developmental stages of wintersweet, namely, flower-bud period (FB), displayed petal stage (DP), open flower stage (OF), later blooming period (LB), and wilting period (WP) were studied using a scanning electron microscope (SEM) to determine the distribution characteristics of aroma-emitting nectaries. Results showed that the floral scent was probably emitted from nectaries distributed on the adaxial side of the innermost and middle petals, but almost none on the abaxial side. The nectaries in different developmental periods on the petals differ in numbers, sizes, and characteristics. Although the distribution of nectaries on different rounds of petals showed a diverse pattern at the same developmental periods, that of the nectaries on the same round of petals showed some of regularity. The nectary is concentrated on the adaxial side of the petals, especially in the region near the axis of the lower part of the petals. Based on transcriptional sequence and phylogenetic analysis, we report one nectary development related gene *CpCRC* (*CRABS CLAW*), and the other four YABBY family genes, *CpFIL* (*FILAMENTOUS FLOWER*), *CpYABBY2*, *CpYABBY5-1*, and *CpYABBY5-2* in *C. praecox* (accession no. MH718960-MH718964). Quantitative RT-PCR (qRT-PCR) results showed that the expression characteristics of these YABBY family genes were similar to those of 11 floral scent genes, namely, *CpSAMT*, *CpDMAPP*, *CpIPP*, *CpGPPS1*, *CpGPPS2*, *CpGPP*, *CpLIS*, *CpMYR1*, *CpFPPS*, *CpTER3*, and *CpTER5*. The expression levels of these genes were generally higher in the lower part of the petals than in the upper halves in different rounds of petals, the highest being in the innermost petals, but the lowest in the outer petals. Relative expression level of *CpFIL*, *CpCRC*, *CpYABBY5-1*, and *CpLIS* in the innermost and middle petals in OF stages is significant higher than that of in outer petals, respectively. SEM and qRT-PCR results in *C. praecox* showed that floral scent emission is related to the distribution of nectaries.

Keywords: *Chimonanthus praecox*; nectary; floral scent; gene expression

1. Introduction

Nectaries are glandular structures that secrete nectar, a carbohydrate-rich solution that is composed mainly of sugars and it generally serves as a reward for pollinators or for as protectors (e.g., ants) against herbivores, or, as a lure for animal prey in carnivorous plants [1]. Nectaries are most wide spread in angiosperms, particularly within flowers, and in ferns and Gnetales [2]. Arabidopsis *CRC* is expressed in the nectary throughout its development and plays a role in the specification and/or differentiation of the nectary [3]. *CRC* is also responsible for carpel growth and fusion, and floral meristem termination [2,4,5]. *CRC* encodes a putative transcription factor of the YABBY gene family, which also includes FIL, INO (INNER NO OUTER), YAB2 (YABBY2), YAB3, and YAB5 [6,7]. The YABBY family is characterized in *Arabidopsis* and rice [2,4,6,8–14]. INO expression occurs only in the abaxial domain of the ovule integument [4]. The "vegetative YABBYs" (FIL, YAB2, and YAB5) in angiosperm are exclusively expressed in leaf-homologous organs, both vegetative and floral, they are involved in leaf development, such as the leaf margin establishment that guides laminar growth and leaflet initiation; maintenance of leaf polarity; and, activation of leaf maturation processes and repression of shoot apical meristem genes [7]. Petunia *PhCRC1/2* expressed in developing nectaries and carpels, similar to *Arabidopsis CRC* expression [15,16]. No nectary glands develop in *crc* mutants [17]. Locations of nectaries are highly variable in broader taxonomic terms [18], although their locations within flowers are constant at the family level. Nectaries tend to be associated with the perianth in basal angiosperms [19], while they are usually associated with carpels and stamens in the eudicots. Fahn argued that nectaries position within flowers trends to shift from peripheral perianth positions in basal taxa to central positions that are associated with reproductive organs in more derived taxa [20].

Wintersweet (*C. praecox*) is a unique traditional deciduous woody flower that is popularly used in floral arrangement, bonsai growing, and landscaping in many countries because of its unique flowering time and distinctive fragrance in deep winter. Wintersweet is a potential spice material because of its volatile aromatic substances and can be used in perfumery, cosmetics, aromatic tea, aromatherapy, and the food industry [21–23].

More than 30 floral scent volatiles have been detected in *C. praecox* flowers, consisting almost exclusively of volatile benzenoids and terpenoids (monoterpenes and sesquiterpenes) [24,25]. Terpenoids play a leading role among these volatiles. These compounds in wintersweet have minimal molecular-genetic characterization, except for the homologous genes of *CpFPPS* and *CpSAMT* [22,26].

Here, we reported the distribution characteristics of nectaries in *C. praecox* and the expression profiles of nectary development related gene *CpCRC*, the other four *YABBY* family genes (*CpFIL*, *CpYABBY2*, *CpYABBY5-1*, and *CpYABBY5-2*) and floral scent related genes to help understand floral scent origination and the molecular regulation of nectary development in wintersweet and other plants of scented flowers.

2. Results

2.1. Distribution Characteristics of Nectary on Petals of Different Stages

SEM analysis was performed to determine the distribution characteristics of nectary on petals in different stages of petals in different developmental stages of floral meristem (FB, DP, OF, LB, and WP), and receptacle, pistil, and stamen (Figure 1A,F,K,R,W). The nectaries were mainly located in the adaxial side of the innermost and middle petals (red asterisk shown, Figure 1B–E,G–J,L–O,S–V,X–Z2). The adaxial/abaxial side of the outer petals and the abaxial side of innermost/middle petals of different stages had no nectary distribution. Figure 1P,Q show the adaxial and abaxial sides of the outer and middle petals in the OF stages (in green box), respectively, which there is no nectary detected at all. The numbers of nectaries changed in the FB to WP during flower senescence. The numbers of nectaries were equal or slightly higher in the middle petal than that in the innermost petal under the same magnification (×400, Figure 1). Morphological difference in different stages of nectaries are shown in Figure 1. With the development of floral meristem, the length-width ratio of nectaries became small and evaginated (×1800, Figure 1). The substance of floral scent was found beside the nectaries (red arrow shown, Figure 1S,Z1,Z2). The cell size increased from FB and DP to OF with the development of floral meristem. This phenomenon is similar to that in LB and WP. Larger number of nectaries was observed in FB or DP than in OF at the same magnification (×400) because of the different cell sizes in different development stages (Figure 1B,D,G,I,L,N). The number of nectaries in OF was higher than that in LB or WP (Figure 1L,N,S,U,X,Z1), with similar cell size in three different development stages.

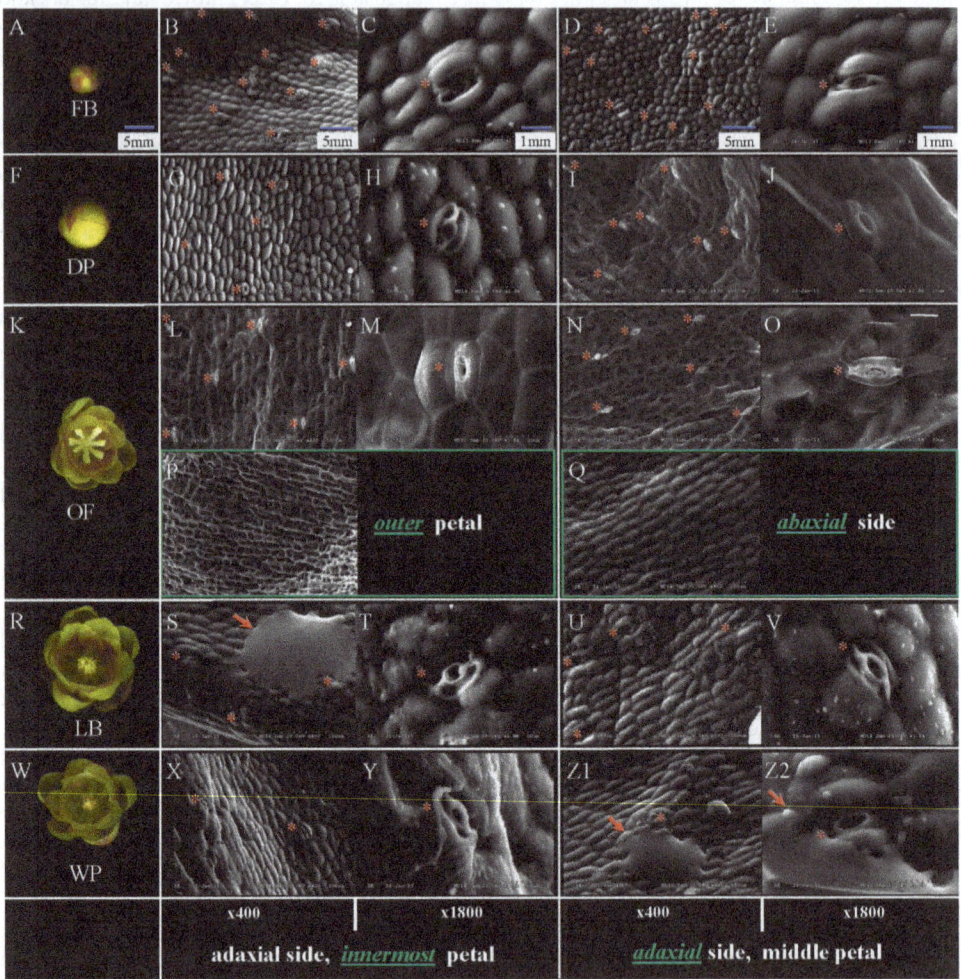

Figure 1. Morphology of the surface of petal glands on different stages in the *Chimonanthus praecox* by SEM. (**A,F,K,R,W**)floral meristem of five different developmental stages (FB, DP, OF, LB, and WP); (**B,C;G,H;L,M;S,T;X,Y**) nectaries on adaxial side of innermost petals in five different developmental stages. (**D,E;I,J;N,O;U,V;Z1,Z2**) nectaries on the adaxial side of middle petals in five different developmental stages. (**B,D,G,I,L,N,S,U,X,Z1**) surface of the glandular tissue, showing the nectary stomata (red asterisk). (**C,E,H,J,M,O,T,V,Y,Z2**) close-up of the nectary stomata. (**P,Q**) (in green box) no nectary distribution on adaxial side of outer petal and abaxial side of middle petals in OF stages. Scale bar is the same in (**A,F,K,R,W**); (**B,D,G,I,L,N,P,Q,S,U,X,Z1**); and (**C,E,H,J,M,O,T,V,Y,Z2**), respectively. Red arrow shows the substance of floral scent. The magnification was ×400 (left) and ×1800 (right), respectively.

No nectary was detected on the receptacle, perianth, stamen, and pistilin all five stages (FB, DP, OF, LB, and WP) (just show results in OF stage, Figure 2). Concentrated nectaries were found in the region near the axis of the lower part of petals, but almost none in the upper and edge of petals (Figure 3 just show the stages of FB and OF). Nectaries are not uniformly distributed in the petals (Figures 1B,D,G,I,L,N,S,U,X,Z1 and 3). These distribution characteristics of nectaries were perhaps related to the *YABBY* gene family, which controls the build of nectaries development and dorsiventral polarity.

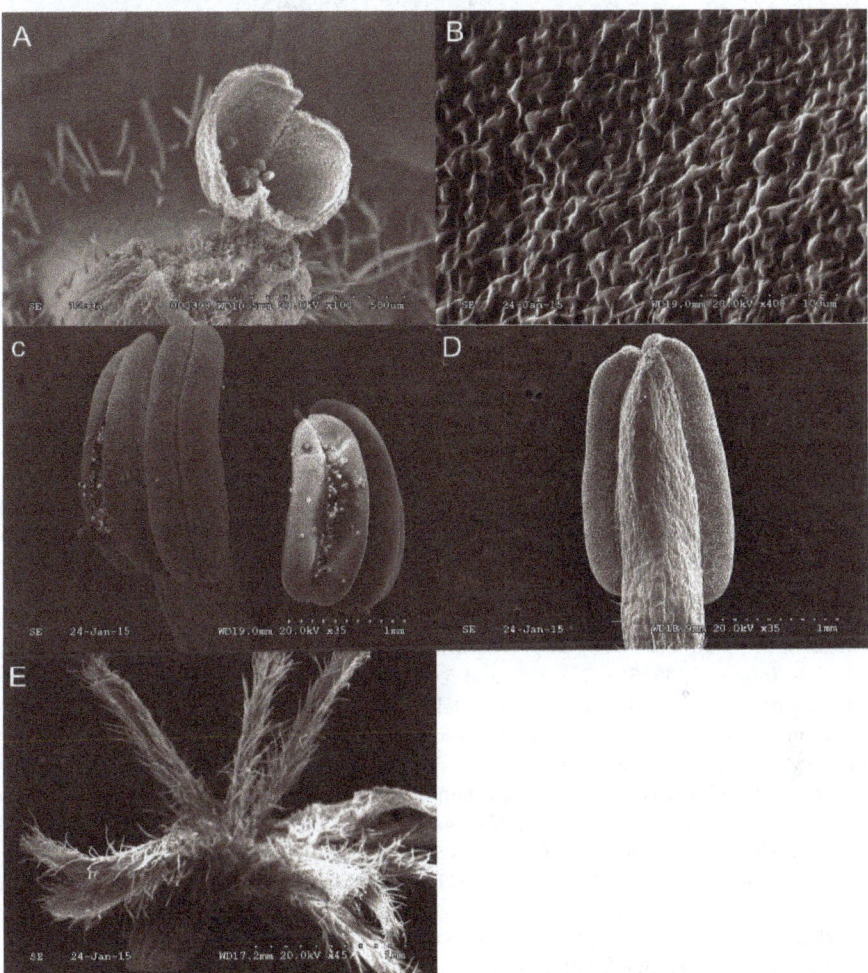

Figure 2. No nectaries distribution of the stamen and pistil in OF in the *Chimonanthus praecox* by SEM. (**A**), receptacle; (**B**), perianth; (**C**), front of stamen; (**D**), back of stamen; and, (**E**), pistil.

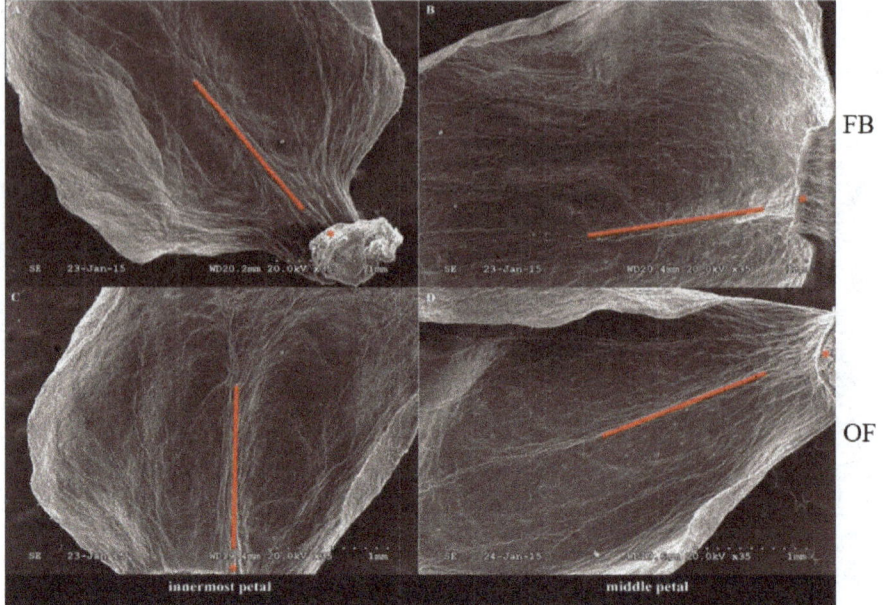

Figure 3. Distribution of the nectaries in FB and OF in the *Chimonanthus praecox* by SEM. Little white dot shown the nectaries concentrated near the axis of petals. Red line represented the axis of the petal. Red asterisks show the bottom of the petal.

2.2. Sequence Alignment and Phylogenetic Analysis

Partial or complete *CpFIL*, *CpCRC*, *CpYABBY2*, *CpYABBY5-1*, and *CpYABBY5-2* cDNAs contain open reading frame of 636, 519, 546, 555, and 552bp, respectively. The predicted CpFIL, CpCRC, CpYABBY2, CpYABBY5-1, and CpYABBY5-2 proteins of 212, 173, 182, 185, and 184 amino acid residues contains a zinc-finger domain in the N-terminus and a YABBY domain in the C-terminus (Figure 3A). The putative CpFIL protein shares 89% and 82% similarity with MgFIL from *Magnolia grandiflora* and Ny.coFIL from *Nymphaea colorata*. CpCRC protein shows 67% and 59% identity with the products of NnCRC-1/2 in *Nelumbo nucifera* and AtCRC in *Arabidopsis*. CpYABBY2 protein shares 72% and 57% similarity with NnYAB2 from *N. nucifera* and AtYAB2 from *Arabidopsis*, respectively. The putative CpYABBY5-1 protein shared high similarity (81%) with CpYABBY5-2, and they both showed 89% and 93% identity with the products of CsYAB5 in *Chloranthus serratus*, and 72% and 69% with that of AtYABBY5 in *Arabidopsis* (Figure 4A and Figure S1; Table S1).

The sequences of the five YABBYs protein homologues were aligned with the respective FIL, CRC, YABBY2, YABBY5-1, and YABBY5-2 proteins of the multiple angiosperm taxa for phylogenetic analysis. The CpYAB2 homologue clustered with non-core NnYAB2 of *Nelumbo nucifera*, DlYAB2 of *Dimocarpus longan*, Am.trYAB2 from *Amborella trichopoda* and core AtYAB2 from *A. thaliana*. The two YABBY5 sequences from *C. praecox*, CpYAB5-1 and CpYAB5-2, formed a sister group to the basal eudicot, CsYAB5 of *Chloranthus serratus*, formed a clade YABBY5 with AtYAB5 of *A. thaliana*. The CpFIL sequence aligned near to none-core basal eudicots, MgFIL of *Magnolia grandiflora*, NjFIL of *Nuphar japonica*, *Nymphaea colorata* Ny.coFIL and *A. trichopoda* Am.trFIL, and formed FIL clade. Together with *A. thaliana* AtCRC, the CpCRC, *A. trichopoda* Am.trCRC, *Oryza sativa* OsCRC, EsCRC of *Epimedium sagittatum*, EcCRC of *Eschscholzia californica*, *N. nucifera* NnCRC-1, and NnCRC-2 formed a clade of CRC (Figure 4B).

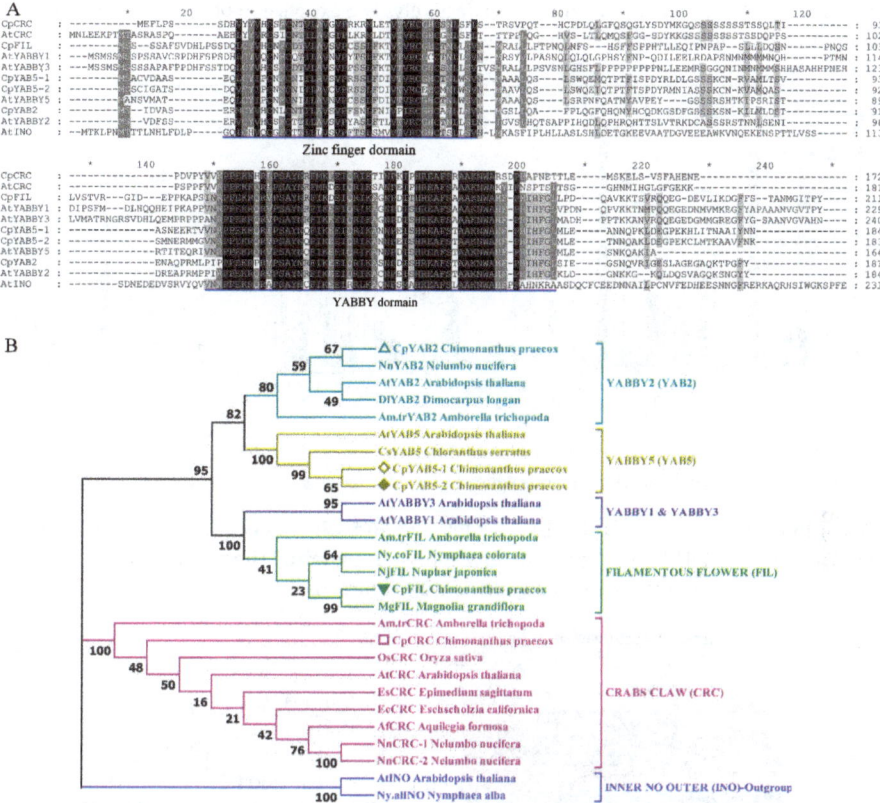

Figure 4. Sequence alignment and phylogenetic analysis of YABBY proteins. (**A**) Sequence alignment of YABBY proteins in *C. praecox* and *A. thaliana*. Conserved domains (Zinc finger domain and YABBY domain) are underlined in blue. Identical residues are highlighted in black and similar residues are highlighted in grey. Dotted line and asterisk represented the gap and the position of odd times of ten in protein sequence. (**B**) Phylogenetic analysis by Neighbor-joining (NJ) bootstrap analysis (1000 replications). AtINO (*A. thaliana*) and Ny.alINO (*Nymphaea alba*) as outgroup. The gene accession number is shown in Materials and Methods. Five different symbols in front of the protein name reprent the five YABBY protein of *Chimonanthus praecox* in this study.

2.3. Expression Analysis

In order to illustrate the correlation between nectary development and gene expression of *CpCRC* and the other four YABBY genes, also that of the nectary development and the floral scent, heatmap analysis with the RNA-Seq database in DP, OF, and WP stages [27] and qRT-PCR were conducted using cDNA derived from DP, OF, LB, and WP to determine the expression profile of five YABBY genes and 11 floral scent related genes in the flower buds of different developmental stages in *C. praecox* (Figure 1F,K,R,W). Almost no *CpDMAPP* was detected in DP, OF, LB, and WP. The relative expression of one nectary development related gene *CpCRC* and the other four YABBY family genes (*CpFIL*, *CpYABBY2*, *CpYABBY5-1*, and *CpYABBY5-2*) and seven floral scent genes (*CpSAMT*, *CpIPP*, *CpGPPS1*, *CpGPP*, *CpLIS*, *CpTER3*, and *CpTER5*) gradually increased in OF to LB and WP. The expression levels of *CpFIL*, *CpCRC*, *CpYABBY2*, *CpYABBY5-2*, and *CpTER3* were the highest, and those of *CpGPPS1* and *CpTER5* were the lowest in DP. The expression of *CpYABBY5-1*, *CpSAMT*, *CpIPP*, *CpGPP*, and *CpLIS* in DP was higher than that in OF but lower than that in WP. The highest and lowest expression levels of

CpGPPS2 and *CpMYR1* were in LB and OF, respectively. The expression of *CpFPPS* in DP was similar to that in LB, higher than that in OF, and lowest in WP (Figure 5).

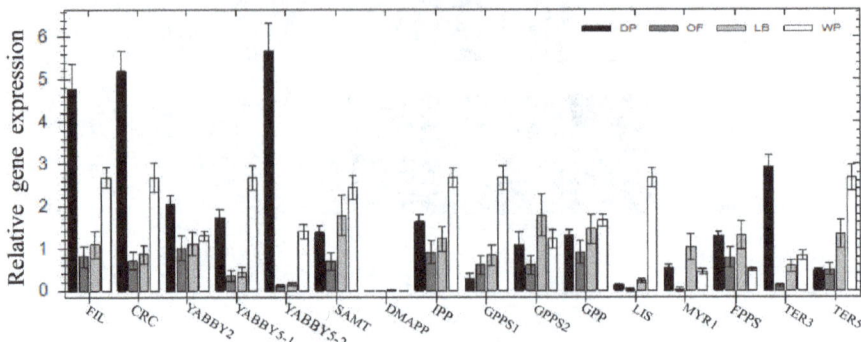

Figure 5. Quantitative real-time PCR analysis of different genes in four different developmental stages of DP, OF, LB, and WP. *Tublin* homologous gene of *Chimonanthus praecox* was used as an internal control.

According to the RNA-Seq database in DP, OF and WP stages, *CpFIL*, *CpCRC*, *CpYABBY2*, *CpYABBY5-1*, and *CpYABBY5-2* have similar expression pattern. The expression level in DP was higher than that in OF and WP and the lowest in WP (Figure 6).

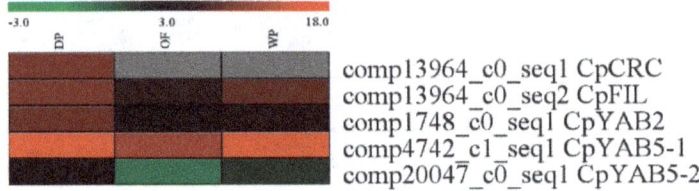

Figure 6. Heatmap analysis of 5 YABBY family gene expression in DP, OF, and WP stages.

To further clarify the correlation between the distribution characteristics of nectaries in three different round of petals and gene expression, the expression profile of five YABBY genes and five floral scent related genes in the innermost, middle, and outer petals in OF stages (Figure 1K) was detected using qRT-PCR. The expression level of 5 YABBY genes and five floral scent-related genes gradually decreased in the innermost to the middle and outer petals. The relative expression levels of *CpFIL*, *CpCRC*, *CpYABBY2*, *CpYABBY5-1*, and *CpYABBY5-2* in the innermost petals were 1.78- to 5.38-fold, and 3.24- to 12.57-fold higher than that in the middle and the outer petals, respectively. The relative expression levels of floral scent genes (*CpIPP*, *CpGPPS1*, *CpGPP*, *CpLIS*, and *CpTER5*) in the innermost petals were approximately 1.65- to 10.25-fold higher than those in the middle petals and 1.76- to 78.85-fold higher than those in the outer petals (Figure 7). They all have significant difference between the relative expression level of *CpFIL*, *CpCRC*, *CpYABBY5-1*, and *CpLIS* in middle petals and that of in outer petals (Figure 7).

Based on SEM results, for the sake of the relationship between the distribution characteristics of nectaries in the same round of petals and the gene expression, qRT-PCR was conducted using cDNA derived from the upper and lower halves in middle petals from DP, OF, and WP (Figure 1F,K,W) to further detect the expression profile of one nectary development related gene *CpCRC*, four YABBY family genes (*CpFIL*, *CpYABBY2*, *CpYABBY5-1/2*) and two floral scent genes (*CpIPP* and *CpGPPS1*). The relative expression of these five YABBY family genes and *CpIPP* had a similar expression pattern. The expression levels in the lower half of middle petals were higher than those in the upper halves in DP and OF stages, including the *CpYABBY5-1* and *CpYABBY5-2* in the WP stage and *CpGPPS1* in the

DP stage. However, the relative expression levels of *CpFIL*, *CpCRC*, *CpYABBY2*, *CpIPP*, and *CpGPPS1* in the lower half of middle petals were lower than those in the upper halves in WP stage and were similar to those of *CpGPPS1* in the OF stage (Figure 8).

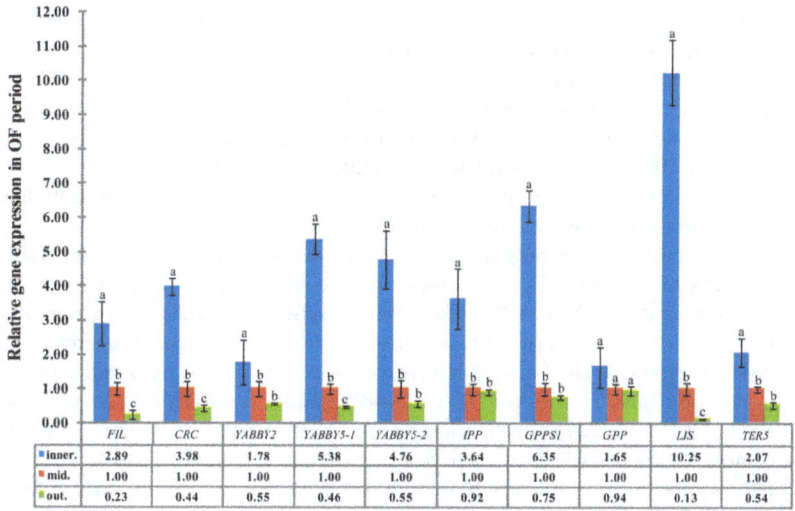

	FIL	CRC	YABBY2	YABBY5-1	YABBY5-2	IPP	GPPS1	GPP	LIS	TERS
inner.	2.89	3.98	1.78	5.38	4.76	3.64	6.35	1.65	10.25	2.07
mid.	1.00	1.00	1.00	1.00	1.00	1.00	1.00	1.00	1.00	1.00
out.	0.23	0.44	0.55	0.46	0.55	0.92	0.75	0.94	0.13	0.54

Figure 7. Quantitative real-time PCR analysis of different genes in three different rounds of petals in OF stages. Inner., mid., and out. represent innermost, middle, and outer petals, respectively. *Tublin* homologous gene of *C. praecox* was used as aninternal control. (Notes: *t*-test used for significant difference analysis; data is the means of relative expression; a, b, c show $p < 0.05$ significant level).

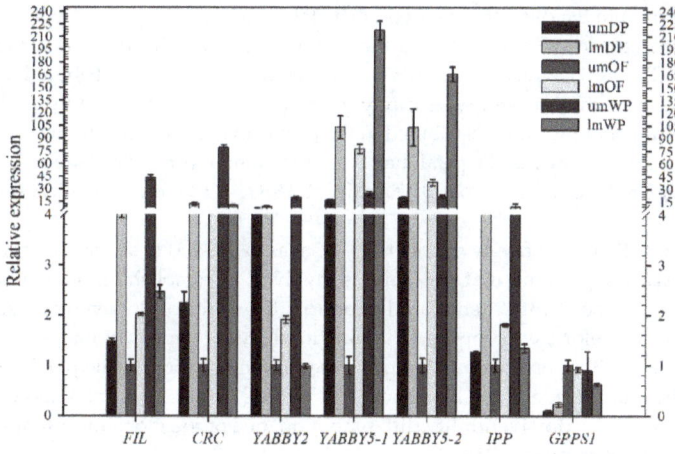

Figure 8. Quantitative real-time PCR analysis of different genes in the upper and lower halves of middle petals in DP, OF, and WP stages. *Tublin* homologous gene of *C. praecox* was used as an internal control. "um" and "lm" represent upper and lower half of middle petals, respectively.

3. Discussion

Results of SEM analysis in *C. praecox* show that nectaries were distributed on the adaxial side of the innermost and middle petals but not on the abaxial side. No nectary was detected in all five stages (FB, DP, OF, LB, and WP) on the outer petals, including in the receptacle, perianth, stamen, and pistil. The surface morphology of the innermost and middle glands of *C. praecox* is similar to that of the inner petal glands of *Alphonsea glandulosa* and *Petunia* [16,28]. The surface of the nectar glands is different from the surrounding epidermis, and nectar stomata are found across the surface of the glandular tissues [28]. The nectar stomata are raised slightly above the epidermis with an aperture for nectar secretion [29]. Although the locations of nectaries within flowers vary highly in terms of broader taxonomic terms, their locations are constant at the family level [30]. Nectaries are usually associated with carpels and stamens in eudicots, but are related to perianth in basal angiosperms [19]. *C. praecox* belongs to Calycanthaceae, Laurales, Magnoliids, and is clustered to Magnoliales, Piperales, and Canellales, which are close to Chloranthales, Austrobaileyales, Nymphaeales, and Amborellales [31]. In Magnoliidae, *C. praecox* has no nectary distribution on its receptacle, perianth, stamen, or pistil, but has some on the adaxial side of the innermost and middle petals; this finding partly supports that of a previous study, nectaries position within flowers trends to shift from peripheral perianth in basal taxa to central reproductive organs in more derived taxa [19,20].

Monoterpenes, such as myrcene, geraniol, linalool and sesquiterpene compounds, are the main aroma components of *C. praecox* [32,33]. Therefore, the concentration of universal precursor of monoterpene (GPP) and its substrate IPP can indirectly reflect the aroma production of *C. praecox*. qPCR analysis of *CpIPP* and *CpGPPS* gene in different parts of petals can indirectly determine the location of aroma substances. The expression levels of the nectary development related genes *CpCRC*, the other four YABBY family genes (*CpFIL, CpYABBY2,* and *CpYABBY5-1/2*) and five floral scent genes (*CpIPP, CpGPPS1, CpGPP, CpLIS,* and *CpTER5*) in the innermost petals of *C. praecox* were higher than those in the middle and outer petals, but they were the lowest in the outer petals. The *CpLIS* expression was increased seven-fold at the OF stage, which is responsible for α-linalool biosynthesis [27]; and α-linalool accounts for 36% of the total quantity of volatile compounds has been reported in wintersweet flowers [34]. The expression results were consistent with the characteristics of nectary distribution based on SEM analysis (Figures 1 and 7).

The expression pattern of five YABBY genes in *C. praecox* by qPCR was in accordance with the RNA-Seq in DP, OF, and WP stages (Figures 5 and 6). The expression levels of *CpIPP* and *CpGPPS* were significantly different in the different halves of the petals; those in the upper halves were significantly lower than those in the lower halves during the first two periods. This result is consistent with that of SEM analysis (Figure 1), which stated that the nectaries were mainly distributed in the lower half part of the petals near the axis and were rarely distributed on the edge and upper half part of the petals. Nectaries are not uniformly distributed in the petals, that is why the numbers of nectaries were equal or slightly higher in the middle petal than that in the innermost (×400, Figure 1). The expression characteristics of *CpFIL, CpCRC, CpYABBY2,* and *CpYABBY5-1/2* were generally similar to those of *CpIPP* and *CpGPPS*.

At least one YABBY gene family member *CpCRC* was expressed in all asymmetric above-ground organs in a polarity, suggesting that this gene is involved in establishing dorsiventral polarity in all of these organs. The YABBY gene family controls the build of dorsiventral (abaxial/adaxial) polarity [4,10,35]. Therefore, we proposed that the floral scent mainly originates from the nectaries that are distributed neither on the abaxial side of the innermost and middle petals nor on the outer petals, but on the adaxial side of the innermost and middle petals. This unbalanced distribution of the nectaries is caused by dorsiventrality differentiation, one of the most important polarities in the development of lateral organs in plants.

4. Materials and Methods

4.1. Plant Material

C. praecox plants of 21-years old were grown in the campus of Southwestern University (106°43′ E, 29°83′ N, Beibei District, Chongqing City, China) under natural photoperiod. Flower development was divided into the following five stages: FB is the stage wherein the flower bud is closed, and the petals are yellow; DP wherein the petals unroll; OF wherein the petals reach full opening, and the stamens bent toward the adaxial side of innermost petals and away from the pistils at a right angle; LB that occurs after two days of OF, where the stamens commence to move to enclose the pistils; and WP wherein the flower is pollinated, and the petals and stamens start to wilt. Floral tissue samples, such as receptacle, sepals, petals, stamens, and pistils were obtained from five different stages. Some of the petals were divided into upper and lower halves. All plant materials were harvested then fixed with FAA buffer or frozen in liquid nitrogen and stored at $-80\,^{\circ}\text{C}$ for RNA extraction.

4.2. Scanning Electron Microscope (SEM)

Fresh petals of *C. praecox* were soaked for an hour in pre-cold 2% glutaraldehyde solution and then were rinsed three to four times with 0.1 M phosphate buffer (pH 7.2) for 1 h. The buffer was discarded, and ethanol dehydration was conducted in a step-by-step gradient. Ethanol concentrations were 30%, 50%, 70%, 80%, 90%, and 100% for 25 min each. The alcohol was washed, and 1:1 mixture of isoamyl acetate to ethanol was added. Then, the solution was added with pure isoamyl acetate, soaked for 10–20 min for each step, stirred properly, and air dried before the electron microscope observation.

4.3. Sequence Alignment and Phylogenetic Analysis

The sequences of one nectary development related gene *CRC*, four other YABBY family genes (*FIL*, *YABBY2*, *YABBY5-1/2*) and 11 floral scent genes in *C. praecox* were selected from the Illumina deep sequencing [27]. Blastn of these genes were obtained and named as *CpFIL*, *CpCRC*, *CpYABBY2*, *CpYABBY5-1*, *CpYABBY5-2*, *CpSAMT*, *CpDMAPP*, *CpIPP*, *CpGPPS1*, *CpGPPS2*, *CpGPP*, *CpLIS*, *CpMYR1*, *CpFPPS*, *CpTER3*, and *CpTER5*. The sequences included in the analysis were downloaded from the NCBI GenBank (http://www.ncbi.nlm.nih.gov). The amino acid sequences of the YABBY family were aligned using ClustalX 1.83 [36]. Neighbor-joining (NJ) bootstrap analysis (1000 replications) with Poisson correction for the amino acids was performed using MEGA 4 [37]. Sequence data for analysis can be found in the GenBank/EMBL databases under the following accession numbers: *CpFIL*, *CpCRC*, *CpYABBY2*, *CpYABBY5-1* and *CpYABBY5-2* from *C. praecox*; *AtINO*, *AtCRC*, *AtYABBY1*, *AtYABBY2*, *AtYABBY3*, and *AtYABBY5* (AAF23754, NP_177078, NP_566037, AF136539, AF136540, NM_179749) from *A. thaliana*; *Am.trCRC*, *Am.trFIL*, and *Am.trYAB2* (AJ877257, AB168113, AB126654) from *Amborella trichopoda*; *AfCRC* (AY854797) from *Aquilegia formosa*; *CsYAB5* (BAF65259) from *Chloranthus serratus*; *DlYAB2* (ACN59438) from *Dimocarpus longan*; *EsCRC* (GH62810) from *Epimedium sagittatum*; *EcCRC* (CAQ17052) from *Eschscholzia californica*; *NnYAB2*, *NnCRC-1*, and *NnCRC-2* (XP_010247861, XM_010259669, XM_010259670) from *Nelumbo nucifera*; *NjFIL* (BAD83708) from *Nuphar japonica*; *Ny.alINO* (AB092980) from *Nymphaea alba*; *Ny.coFIL* (BAF65258) from *Nymphaea colorata*; *MgFIL* (BAF65261) from *Magnolia grandiflora*; and *OsCRC* (AAR84663) from *Oryza sativa*.

4.4. Gene Expression Analysis

Tissues sampled for gene expression analysis include flower buds of four different developmental stages (DP, OF, LB, and WP), three different rounds (innermost, middle, and outer) of petals in OF stages, the upper and lower halves of middle petals in DP, OF, and WP stages. Total RNA for the expression analysis was extracted using RNAprep pure kit (Tiangen, Beijing, China) according to the manufacturer's instructions. Exactly 3 μg of RQ1 RNase-Free DNase (Promega, Madison, WI, USA) pre-treated total RNA was reverse transcribed according to the instructions of the Primescript RT reagent kit (Takara, Tokyo, Japan). qRT-PCR was performed to determine the expression pattern of one nectary development related gene*CpCRC*, four YABBY family genes (*CpFIL*, *CpYABBY2*, *CpYABBY5-1*, and *CpYABBY5-2*) and 11 floral scent genes, such as *CpSAMT*, *CpDMAPP*, *CpIPP*, *CpGPPS1*, *CpGPPS2*, *CpGPP*, *CpLIS*, *CpMYR1*, *CpFPPS*, *CpTER3*, and *CpTER5*.

The primers for qRT-PCR are listed in Table 1. Reactions were performed with the Sso Fast Eva Green Supermix (Bio-Rad, Hercules, CA, USA) and analyzed using Bio-Rad CFX96 (Bio-Rad CFX Manager Software Version 1.6). Thermocycler conditions were 95 °C for 30 s, followed by 40 cycles of

95 °C for 5 s and 60 °C for 5 s. qRT-PCR products were amplified using 5 μL 2× Sso Fast Eva Green Supermix, 0.5 μL RT reaction mixture, 0.5 μL of forward and reverse primer (10 μmol/μL) each, and RNase Free dH$_2$O to a final volume of 10 μL. Relative amounts of transcripts were calculated using the comparative CT method ($2^{-\Delta\Delta Ct}$), and the values were normalized. The house-keeping gene *CpTublin* of *C. praecox* was used as internal control. Data are shown as mean values ± standard deviation (SD) from three replicates for each sample. Significant difference was carried out by *t*-test ($p < 0.05$).

The expression patterns of the five YABBY genes were estimated by FPKM values and were visualized using MultiExperiment Viewer (Broad Institute of MIT and Harvard University, Boston, MA, USA [38]).

Table 1. Primer for real-time PCR.

Gene Name	Forward Primer Sequence	Reverse Primer Sequence
Actin	AGGCTAAGATTCAAGACAAGG	TTGGTCGCAGCTGATTGCTG
CpFIL	AATCCCGACATAACCCACAGAGAG	TCCTGTTGGCGCACGCTAGTT
CpCRC	CCTCCCGTCACCTTACAAACTACAG	CTGCTACAAGGAACACTGACCGC
CpYABBY2	CCATTGTCAAGATAAAGGTAGCGATT	CTGGTGGTGGTATAGGTAGCATTCG
CpYABBY5-1	TCTCCCTCTCTATTTATCCTCGTTT	GTAAAAGGCTAAAGCAGGATCATG
CpYABBY5-2	TTTTGAACACTGGAAACTTCGTCTT	GATGCAGCTCGACATCTCACTATCT
SAMT	ACCATTTTCACATCATTGCCAGAC	CTTCCTCTTTTACCATCAAGTGCTG
DMAPP	ATCGGAGAAGAAAGTGAGCGAGAGT	GCCGTGTATCGAAGCAGCAGT
IPP	CAGACCATCTCTTTCTCCCACTTTC	GGTCGGAGAGAAGGTGGTAGAGGTA
GPPS1	GTTAGCCAACTTTCCATACCATTTC	GAGTGACAACATCATCAAAGAAGGG
GPPS2	ATGAAGATGATTAGATTTCGAGTCCAAG	ATAACCAATTTACAACCCCTGACCC
GPP	TCTACAGAAAATGGGAGAAAACGAT	TATCTGTTTCTGTCACCAAATCCAC
LIS	GGCCAAAGTTAATGAAGTGAGATCC	CGTATATGCCATCGTTGCTGCC
MYR1	TTTCACAAAAATTGCCTTCAACCTT	CAAGGTGATGGAGAACTAAAACAAAAC
FPPS	TCTTTGTCCAGTTCTTCCAGCGTT	ATCAGTGAAATCAAAGGCGGAATCT
TER3	AGAGTTGAATTGCACAGGGTGATAG	GCAGTGGATGTTGTTGATCAGCTC
TER5	CTCTCCCTCAGTCTCTTCTCCCTTT	ATCTCCATGCAACATTGGCTACAG

Supplementary Materials: Supplementary materials can be found at http://www.mdpi.com/1422-0067/19/10/3278/s1.

Author Contributions: Methodology, Z.L. and Y.J.; Software, Z.L.; Validation, Z.L. and J.L.; Formal Analysis, Y.J. and Z.L.; Resources, D.L.; Data Curation, J.M.; Writing-Original Draft Preparation, Y.J. and Z.L.; Writing-Review & Editing, Z.L.; Visualization, Z.L. and Y.J.; Supervision, Z.L.; Project Administration, S.S. and M.L.; Funding Acquisition, Z.L., S.S. and J.M.

Funding: This work was supported by grants from Fundamental Research Funds for the Central Universities (XDJK2017B031), the National Natural Science Foundation of China (No. 31500573, 31370698), Chongqing science and technology committee projects (cstc2016shmszx80112), Chongqing education committee projects (yjg20163042) and National undergraduate innovation and entrepreneurship training program (201410635065).

Conflicts of Interest: The authors declare no conflict of interest.

Abbreviations

SEM	scanning electron microscope
qRT-PCR	quantitative Reverse Transcript-Polymerase Chain Reaction
CRC	*CRABS CLAW*
FIL	*FILAMENTOUS FLOWER*
INO	*INNER NO OUTER*
YAB	*YABBY*
SAMT	*S*-adenosyl-*L*-methionine: salicylic acid carboxyl methyltransferase
DMAPP	dimethylallyl pyrophosphate
IPP	isopentenyl pyrophosphate
GPPS	geranyl diphosphate synthase
GPP	geranyl pyrophosphate
LIS	*S*-linalool synthase
MYR1	myrcenesynthase
FPPS	farnesyl pyrophosphonate synthase
TER	α-terpineol synthase

References

1. Schmid, R. Reproductive versus extra-reproductive nectaries-historical perspective and terminological recommendations. *Bot. Rev.* **1988**, *54*, 179–227. [CrossRef]
2. Lee, J.Y.; Baum, S.F.; Alvarez, J.; Patel, A.; Chitwood, D.H.; Bowman, J.L. Activation of CRABS CLAW in the Nectaries and Carpels of *Arabidopsis*. *Plant Cell* **2005**, *17*, 25–36. [CrossRef] [PubMed]
3. Baum, S.F.; Eshed, Y.; Bowman, J.L. The *Arabidopsis* nectary is an ABC-independent floral structure. *Development* **2001**, *128*, 4657–4667. [PubMed]
4. Bowman, J.L. The *YABBY* gene family and abaxial cell fate. *Curr. Opin. Plant Biol.* **2000**, *3*, 17–22. [CrossRef]
5. Gross, T.; Broholm, S.; Becker, A. *CRABS CLAW* Acts as a Bifunctional Transcription Factor in Flower Development. *Front. Plant Sci.* **2018**, *9*, 835. [CrossRef] [PubMed]
6. Goldshmidt, A.; Alvarez, J.P.; Bowman, J.L.; Eshed, Y. Signals derived from *YABBY* gene activities in organ primordia regulate growth and partitioning of *Arabidopsis* shoot apical meristems. *Plant Cell* **2008**, *20*, 1217–1230. [CrossRef] [PubMed]
7. Bartholmes, C.; Hidalgo, O.; Gleissberg, S. Evolution of the *YABBY* gene family with emphasis on the basal eudicot *Eschscholzia californica* (Papaveraceae). *Plant Biol.* **2012**, *14*, 11–23. [CrossRef] [PubMed]
8. Siegfried, K.R.; Eshed, Y.; Baum, S.F.; Otsuga, D.; Drews, G.N.; Bowman, J.L. Members of the *YABBY* gene family specify abaxial cell fate in *Arabidopsis*. *Development* **1999**, *126*, 4117–4128. [PubMed]
9. Eshed, Y.; Baum, S.F.; Bowman, J.L. Distinct mechanisms promote polarity establishment in carpels of *Arabidopsis*. *Cell* **1999**, *99*, 199–209. [CrossRef]
10. Yamaguchi, T.; Nagasawa, N.; Kawasaki, S.; Matsuoka, M.; Nagato, Y.; Hirano, H.Y. The *YABBY* gene *DROOPING LEAF* regulates carpel specification and midrib development in *Oryza sativa*. *Plant Cell* **2004**, *16*, 500–509. [CrossRef] [PubMed]
11. Liu, H.L.; Xu, Y.Y.; Xu, Z.H.; Chong, K. A rice *YABBY* gene, *OsYABBY4*, preferentially expresses in developing vascular tissue. *Dev. Genes Evol.* **2007**, *217*, 629–637. [CrossRef] [PubMed]
12. Ohmori, Y.; Abiko, M.; Horibata, A.; Hirano, H.Y. A transposon, Ping, is integrated into intron 4 of the *DROOPING LEAF* gene of rice, weakly reducing its expression and causing a mild drooping leaf phenotype. *Plant Cell Physiol.* **2008**, *49*, 1176–1184. [CrossRef] [PubMed]
13. Ohmori, Y.; Toriba, T.; Nakamura, H.; Ichikawa, H.; Hirano, H.Y. Temporal and spatial regulation of *DROOPING LEAF* gene expression that promotes midrib formation in rice. *Plant J.* **2011**, *65*, 77–86. [CrossRef] [PubMed]
14. Tanaka, W.; Toriba, T.; Ohmori, Y.; Yoshida, A.; Kawai, A.; Mayama-Tsuchida, T.; Ichikawa, H.; Mitsuda, N.; Ohme-Takagi, M.; Hirano, H.Y. The *YABBY* Gene *TONGARI-BOUSHI1* Is Involved in Lateral Organ Development and Maintenance of Meristem Organization in the Rice Spikelet. *Plant Cell* **2012**, *24*, 80–95. [CrossRef] [PubMed]
15. Lee, J.Y.; Baum, S.F.; Oh, S.H.; Jiang, C.Z.; Chen, J.C.; Bowman, J.L. Recruitment of *CRABS CLAW* to promote nectary development within the eudicot clade. *Development* **2005**, *132*, 5021–5032. [CrossRef] [PubMed]
16. Morel, P.; Heijmans, K.; Ament, K.; Chopy, M.; Trehin, C.; Chambrier, P.; Rodrigues Bento, S.; Bimbo, A.; Vandenbussche, M. The Floral C-Lineage Genes Trigger Nectary Development in *Petunia* and *Arabidopsis*. *Plant Cell* **2018**, *30*, 2020–2037. [CrossRef] [PubMed]
17. Bowman, J.L.; Smyth, D.R. *CRABS CLAW*, a gene that regulates carpel and nectary development in *Arabidopsis*, encodes a novel protein with zinc finger and helix-loop-helix domains. *Development* **1999**, *126*, 2387–2396. [PubMed]
18. Brown, W.H. The bearing of nectaries on the phylogeny of flowering plants. *Proc. Am. Phil. Soc.* **1938**, *79*, 549–595.
19. Endress, P.K. The Flowers in Extant Basal Angiosperms and Inferences on Ancestral Flowers. *Int. J. Plant Sci.* **2001**, *162*, 1111–1140. [CrossRef]
20. Fahn, A. The topography of the nectary in the flower and its phylogenetic trend. *Phytomorphology* **1953**, *3*, 424–426.
21. Kozomara, B.; Vinterhalter, B.; Radojevic, L.; Vinterhalter, D. In vitro propagation of *Chimonanthus praecox* (L.), a winter flowering ornamental shrub. *In Vitro Cell Dev. Biol. Plant* **2008**, *44*, 142–147. [CrossRef]

22. Xiang, L.; Zhao, K.; Chen, L. Molecular cloning and expression of *Chimonanthus praecox* farnesyl pyrophosphate synthase gene and its possible involvement in the biosynthesis of floral volatile sesquiterpenoids. *Plant Physiol. Biochem.* **2010**, *48*, 845–850. [CrossRef] [PubMed]
23. Zhao, K.G.; Zhou, M.Q.; Chen, L.Q.; Zhang, D.L.; Robert, G.W. Genetic diversity and discrimination of *Chimonanthus praecox* (L.) link germplasm using ISSR and RAPD markers. *Hortscience* **2007**, *42*, 1144–1148.
24. Vainstein, A.; Lewinsohn, E.; Pichersky, E.; Weiss, D. Floral Fragrance. New Inroads into an Old Commodity. *Plant Physiol.* **2001**, *127*, 1383–1389. [CrossRef] [PubMed]
25. Kram, B.W.; Carter, C.J. *Arabidopsis thaliana* as a model for functional nectary analysis. *Sex. Plant Reprod.* **2009**, *22*, 235–246. [CrossRef] [PubMed]
26. Ma, L.; Li, H.F.; Peng, C.C.; Chen, Z.Z.; Long, Z.F. Cloning of *SAMT* gene cDNA from *Chimonanthus praecox* and its expression in *Escherichia coli*. *Agric. Sci. Technol.* **2012**, *13*, 82–87.
27. Liu, D.; Sui, S.; Ma, J.; Li, Z.; Guo, Y.; Luo, D.; Yang, J.; Li, M. Transcriptomic analysis of flower development in wintersweet (*Chimonanthus praecox*). *PLoS ONE* **2014**, *9*, e86976. [CrossRef] [PubMed]
28. Xue, B.; Shao, Y.Y.; Saunders, R.M.; Tan, Y.H. *Alphonsea glandulosa* (Annonaceae), a New Species from Yunnan, China. *PLoS ONE* **2017**, *12*, e0170107. [CrossRef] [PubMed]
29. Nepi, M. Nectary structure and ultrastructure. In *Nectaries and Nectar*; Nicolson, S.W., Nepi, M., Pacini, E., Eds.; Springer: Dordrecht, The Netherlands, 2007; pp. 129–166.
30. Elias, T.S. Extrafloral nectaries: Their structure and distribution. In *The Biology of Nectaries*; Bentley, B., Elias, T.S., Eds.; Columbia University Press: New York, NY, USA, 1983; pp. 174–203.
31. Bremer, B.; Bremer, K.; Chase, M.W.; Fay, M.F.; Reveal, J.L.; Soltis, D.E.; Soltis, P.S.; Stevens, P.F.; Anderberg, A.A.; Moore, M.J.; et al. An update of the Angiosperm Phylogeny Group classification for the orders and families of flowering plants: APG III. *Bot. J. Linn. Soc.* **2009**, *161*, 105–121.
32. Deng, C.; Song, G.; Hu, Y. Rapid determination of volatile compounds emitted from *Chimonanthus praecox* flowers by HS-SPME-GC-MS. *Z. Naturforsch. C* **2004**, *59*, 636–640. [CrossRef] [PubMed]
33. Miller, E.R.; Taylor, G.W.; Eskew, M.H. The volatile oil of calycanthus floridus. *J. Am. Chem. Soc.* **1914**, *36*, 2182–2185. [CrossRef]
34. Azuma, H.T.M.; Asakawa, Y. Floral Scent Chemistry and Stamen Movement of *Chimonanthus praecox* (L.) Link (Calycanthaceae). *Acta Phytotaxon. Geobot.* **2005**, *56*, 197–201.
35. Juarez, M.T.; Twigg, R.W.; Timmermans, M.C. Specification of adaxial cell fate during maize leaf development. *Development* **2004**, *131*, 4533–4544. [CrossRef] [PubMed]
36. Thompson, J.D.; Gibson, T.J.; Plewniak, F.; Jeanmougin, F.; Higgins, D.G. The CLUSTAL_X windows interface: Flexible strategies for multiple sequence alignment aided by quality analysis tools. *Nucleic Acids Res.* **1997**, *25*, 4876–4882. [CrossRef] [PubMed]
37. Tamura, K.; Dudley, J.; Nei, M.; Kumar, S. MEGA4: Molecular Evolutionary Genetics Analysis (MEGA) software version 4.0. *Mol. Biol. Evol.* **2007**, *24*, 1596–1599. [CrossRef] [PubMed]
38. Yeung, K.Y.; Fraley, C.; Murua, A.; Raftery, A.E.; Ruzzo, W.L. Model-based clustering and data transformations for gene expression data. *Bioinformatics* **2001**, *17*, 977–987. [CrossRef] [PubMed]

International Journal of
Molecular Sciences

Article

Molecular Characterization and Overexpression of *SmJMT* Increases the Production of Phenolic Acids in *Salvia miltiorrhiza*

Bin Wang [1,2,†], Junfeng Niu [1,†], Bin Li [1], Yaya Huang [1], Limin Han [1], Yuanchu Liu [1], Wen Zhou [1], Suying Hu [1], Lin Li [1], Donghao Wang [1], Shiqiang Wang [1], Xiaoyan Cao [1,*] and Zhezhi Wang [1,*]

1   National Engineering Laboratory for Resource Development of Endangered Crude Drugs in Northwest China, Key Laboratory of the Ministry of Education for Medicinal Resources and Natural Pharmaceutical Chemistry, College of Life Sciences, Shaanxi Normal University, Xi'an 710119, China; happywangbin2003@163.com (B.W.); niujunfeng6829@126.com (J.N.); libin1989@snnu.edu.cn (B.L.); h_yoyo@snnu.edu.cn (Y.H.); hdd_1981_@163.com (L.H.); shidayuanchu@snnu.edu.cn (Y.L.); wenzhou0229@snnu.edu.cn (W.Z.); husuying0315@163.com (S.H.); shidalilin@snnu.edu.cn (L.L.); wangdonghao@snnu.edu.cn (D.W.); wsq@snnu.edu.cn (S.W.)
2   College of Chemistry, Biology and Materials Science, East China University of Technology, NanChang 330013, China
*   Correspondence: caoxiaoyan@snnu.edu.cn (X.C.); zzwang@snnu.edu.cn (Z.W.); Tel.: +86-135-7285-8369 (Z.W.)
†   These authors contributed equally to this work.

Received: 13 October 2018; Accepted: 25 November 2018; Published: 28 November 2018

Abstract: Jasmonic acid (JA) carboxyl methyltransferase (JMT), a key enzyme in jasmonate-regulated plant responses, may be involved in plant defense and development by methylating JA to MeJA, thus influencing the concentrations of MeJA in plant. In this study, we isolated the *JMT* gene from *Salvia miltiorrhiza*, an important medicinal plant widely used to treat cardiovascular disease. We present a genetic manipulation strategy to enhance the production of phenolic acids by overexpresion *SmJMT* in *S. miltiorrhiza*. Global transcriptomic analysis using RNA sequencing showed that the expression levels of genes involved in the biosynthesis pathway of phenolic acids and MeJA were upregulated in the overexpression lines. In addition, the levels of endogenous MeJA, and the accumulation of rosmarinic acid (RA) and salvianolic acid (Sal B), as well as the concentrations of total phenolics and total flavonoids in transgenic lines, were significantly elevated compared with the untransformed control. Our results demonstrate that overexpression of *SmJMT* promotes the production of phenolic acids through simultaneously activating genes encoding key enzymes involved in the biosynthesis pathway of phenolic acids and enhancing the endogenous MeJA levels in *S. miltiorrhiza*.

Keywords: *SmJMT*; transgenic; *Salvia miltiorrhiza*; overexpression; transcriptome; phenolic acids

1. Introduction

Salvia miltiorrhiza Bunge (Lamiaceae) is a well-known traditional Chinese herb with significant medicinal and economic value. Its dry roots or rhizomes (called "danshen" in Chinese) are used to treat various cerebrovascular and cardiovascular diseases in Asian countries, and are widely accepted as a health supplement in western countries [1,2]. Due to its remarkable and reliable therapeutic actions, *S. miltiorrhiza* is being developed as a potential model for research into traditional Chinese medicine [3]. Furthermore, by virtue of high-throughput technologies, the genomic sequence of *S. miltiorrhiza* was published [4].

S. *miltiorrhiza* contains two major medicinal components that are largely responsible for the observed pharmacological activities; one is a group of lipid-soluble (non-polar, lipophilic) diterpenoids, known as tanshinones, and the other is a water-soluble (polar, hydrophilic) group of phenolic acids, such as rosmarinic acid (RA) and salvianolic acid B (Sal B) [5]. *S. miltiorrhiza* is traditionally processed through extraction with water. Sal B becomes the predominant active ingredient among the phenolic acids, which is designated as a marker component of *S. miltiorrhiza* in the official Chinese Pharmacopoeia. In additional, it was reported that Sal B can provide protection against cardiovascular, neural, and hepatic diseases, as well as certain cancers [6]. Although beneficial to human health, the concentration of Sal B is low in commercial cultivars of *S. miltiorrhiza*, which currently limits its widespread use and medicinal efficiency. Furthermore, Sal B is difficult to purify from complex mixtures, resulting in inefficient chemical synthesis [7].

In *S. miltiorrhiza*, the biosynthetic pathway leading to Sal B and RA is thought to entail both the phenylpropanoid and tyrosine-derived pathways [7]. Enzymes involved in the phenylpropanoid pathway include phenylalanine ammonia-lyase (PAL), cinnamate 4-hydroxylase (C4H), and 4-coumarate/coenzyme A ligase (4CL). Tyrosine aminotransferase (TAT) and hydroxyphenylpyruvate reductase (HPPR) are active in the tyrosine-derived pathway. An additional enzyme, rosmarinic acid synthase (RAS), couples products from the two pathways. This product is then hydroxylated by cytochrome P450 monooxygenase C3'H (CYP98A) to form RA [8]. Many methods were identified for increasing the production rates of such health-promoting phenolic acids, such as genetic engineering [7], hormone induction [9], and biological fermentation [10]. Jasmonate treatment is the most commonly used method of hormone elicitation.

Jasmonates which include jasmonic acid (JA), methyl jasmonate (MeJA), and its cyclopentane derivatives, are a class of plant hormone that regulate many aspects of plant development such as root growth, production of viable pollen, fruit ripening, and senescence [11,12]. They are also involved in plant responses to biotic and abiotic stresses including insect attack, wounding, water deficiency, ultraviolet (UV) light, pathogen infection, and ozone [13–15]. Jasmonates are synthesized in plants via the octadecanoid pathway [14,16] from α-linolenic acid through a series of enzymes, beginning with an oxygenation catalyzed by lipoxygenase (LOX) [17]. The product is then converted to 12-oxo-phytodienoic acid (12-OPDA) by allene oxide synthase (AOS) and alleneoxide cyclase (AOC) [18,19]. Afterward, JA is synthesized from 12-oxo-phytodienoic acid (12-OPDA) through reduction by 12-oxo-phytodienoic acid reductase (OPR) and three steps of β-oxidation, and formation of MeJA is catalyzed by jasmonic acid carboxyl methyltransferase (JMT).

Jasmonic acid carboxyl methyltransferase (JMT) is an *S*-adenosyl-L-methionine-dependent methyltransferase of the SABATH gene family, which could specifically methylate carboxyl groups of small molecules such as jasmonic acid, salicylic acid, and benzoic acid, and named based on the first three identified genes belonging to this family, *SAMT*, *BAMT*, and that coding for theobromine synthase [20]. The *JMT* gene was first identified in *Arabidopsis* [21], and then was successively found in *Capsicum annum* [22], *Populus trichocarpa* [23], strawberry [17], and rice [24]. *JMT* acts as a cellular regulator of the level of physiologically active JA [17,24], which functions in response to several different external environmental stimuli [21] and mediates diverse developmental processes in plants [24,25]. Transgenic *Arabidopsis* lines overexpressing *JMT* had an elevated level of endogenous MeJA, and the transgenic plants showed enhanced levels of resistance against the virulent fungus [21]. Meanwhile, overexpressing *Arabidopsis JMT* in potato increased tuber yield and size [26].

In plants, MeJA, a signal molecule that acts as a second messenger, is proposed to play a role in the elicitation process [27,28], which could lead to the accumulation of secondary metabolites [29]. Furthermore, MeJA could induce plant tissues to provide a responsive system to identify and profile the transcripts and regulation factors involved in secondary metabolite accumulation [28]. When *S. miltiorrhiza* was treated with MeJA, an extensive transcriptional reprogramming of metabolism was triggered, and the biosynthesis of active ingredients was dramatically increased [30]. In addition, it was also shown that the biosynthesis of phenolic acids is stimulated by MeJA treatment [5,9,31,32], and the expression levels of most of the genes involved in the biosynthesis of bioactive compounds are induced by MeJA at different levels [8,9,28,32,33].

In addition, when a plant is subjected to biotic and abiotic stresses, it produces secondary metabolites that function as direct defenses [25], and these metabolites may have important medicinal value. Overexpression of tomato *prosystemin* (*LePS*) in *S. miltiorrhiza* enhanced resistance the pest, while the production of secondary metabolites and the level of endogenous MeJA was increased [34]. Moreover, as described above, overexpressing *JMT* in plants elevated the level of endogenous MeJA, and transgenic plants exhibited constitutive expression of jasmonate-responsive genes [21], while most genes involved in the biosynthesis of Sal B and RA could be induced by jasmonates in *S. miltiorrhiza* [8,9,28,32,33]. Thus, we inferred that, if the level of endogenous MeJA were elevated, a series of defense responses mediated by MeJA may be elicited, a group of jasmonate-responsive genes would be activated, and the accumulation of secondary metabolites could be enhanced.

Thus, we developed a novel strategy to enhance the production of phenolic acids in *S. miltiorrhiza* simultaneously using genetic manipulation and MeJA induction. On the one hand, we overexpressed *SmJMT* in *S. miltiorrhiza* aiming to activate the expression of genes responsible for biosynthesis of phenolic acids and MeJA. On the other hand, by enhancing the level of endogenous MeJA, we intended to elicit a series of biomechanisms to elevate the accumulation of secondary metabolites. After obtaining the overexpressing *SmJMT* plants, transcriptome analysis was carried out on the above transgenic plants and the untransformed control plants. The differentially expressed genes (DEGs) involved in phenolic acids biosynthesis and the α-linolenic acid metabolism pathway were identified. Finally, the contents of RA, Sal B, total phenols, total flavonoids, and endogenous MeJA in transgenic and control lines were analyzed using different biological techniques.

2. Results

2.1. Isolation and Sequence Analysis of SmJMT

Using PCR amplification, the full-length complementary DNA (cDNA) of *SmJMT* was obtained and submitted to GenBank with the accession number MH136806. The cDNA fragment contained a 1167-bp open reading frame (ORF), encoding a predicted 389-amino-acid polypeptide with an isoelectric point of 5.98 and a molecular mass of 43.5 kDa. The amino-acid sequence contains all the characteristic elements of the *S*-adenosyl-L-methionine-dependent methyltransferases [35]; two conserved binding sites of motifs I and III of *S*-adenosyl-L-methionine (SAM) [36,37], the signature of SABATH gene family members [20,21], were found using multiple-sequence alignment (Figure 1).

To determine the evolutionary relationship of the *SmJMT* with the members of JMTs from other species, an unrooted phylogenetic tree was constructed using the amino-acid sequences of *S. miltiorrhiza* JMT and 27 JMTs from other species (Figure 2). Phylogenetic analysis showed that *SmJMT* was most closely related to *SiJMT* (*S. indicum* jasmonic acid carboxyl methyltransferase) and *EgJMT* (*Erythranthe guttata* jasmonic acid carboxyl methyltransferase), both of which belong to the Lamiales order (Figure 2). Furthermore, the species belonging to same family were classed into the same clade (Figure 2), suggesting that the cluster relationship of JMT proteins from different species is consistent with the traditional taxonomy.

Figure 1. Multiple-sequence alignment of the jasmonic acid (JA) carboxyl methyltransferase (JMT) conserved amino-acid sequences from *Salvia miltiorrhiza* and selected known JMTs from other species, including the binding sites (motifs I and III are indicated) of *S*-adenosyl-L-methionine. The *S. miltiorrhiza* sequences are marked with a black arrow.

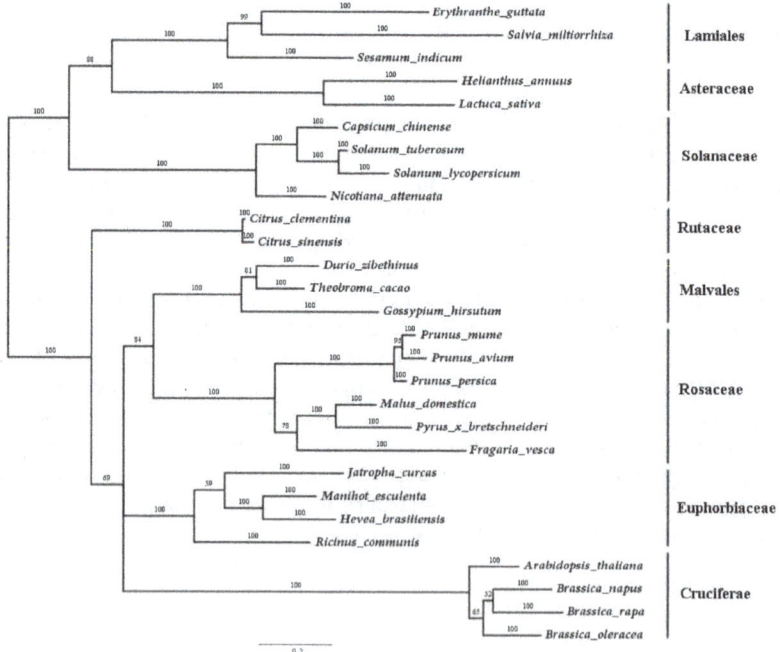

Figure 2. Phylogenetic tree based on JMT from *S. miltiorrhiza* and other species. The tree was constructed using Bayesian inference implemented in MrBayes, based on the amino-acid sequences of *SmJMT* and other species of JMTs under the model of JTT + I + G. The species taxonomy is indicated on the tree.

2.2. Generation of Transgenic S. miltiorrhiza Plants

Transgenic *S. miltiorrhiza* plants overexpressing *SmJMT* were obtained in our laboratory by *Agrobacterium*-mediated transformation. After selective culturing on a glufosinate/ammonium medium, resistant plants were confirmed through PCR amplification to contain an expected 929-bp fragment of the CaMV 35S promoter (Figure S1A). Real-time PCR analyses demonstrated that *SmJMT* was obviously overexpressed at the transcriptional level in lines OEJ-2, OEJ-5, OEJ-7, OEJ-8, OEJ-9, and OEJ-10 (OEJ stands for overexpressed *SmJMT*) (Figure S1B). Since expression was significantly higher in OEJ-7 and OEJ-10 than in the other lines and the non-transformed control, we chose them for further analysis. Due to OEJ-10 being the most highly expressed line, we chose it to do the transcriptome sequencing. However, there was no phenotypic change between transgenic lines and control lines.

2.3. Overexpression of SmJMT Enhances Production of Salvianolic and Rosmarinic Acids inTransgenic S. miltiorrhiza

To further characterize how the production of phenolic acids was modified in our *SmJMT* overexpression transgenic plants, we extracted the phenolic acids and separated them via LC/MS (Figure S2). The results show that the concentrations of both RA and Sal B, which are the two major hydrophilic active pharmaceutical ingredients in *S. miltiorrhiza*, were increased significantly compared with levels in control samples. We detected the concentrations of RA and Sal B in all the transgenic plants with overexpressed *SmJMT*. The highest concentrations were found in transgenic OEJ-10. Compared with the control, OEJ-10 showed a 3.61-fold increase in RA and a 2.00-fold increase in Sal B (Figure 3A). In OEJ-7, concentrations of RA and Sal B were approximately 1.80- and 1.15-fold higher than those of the control (Figure 3A).

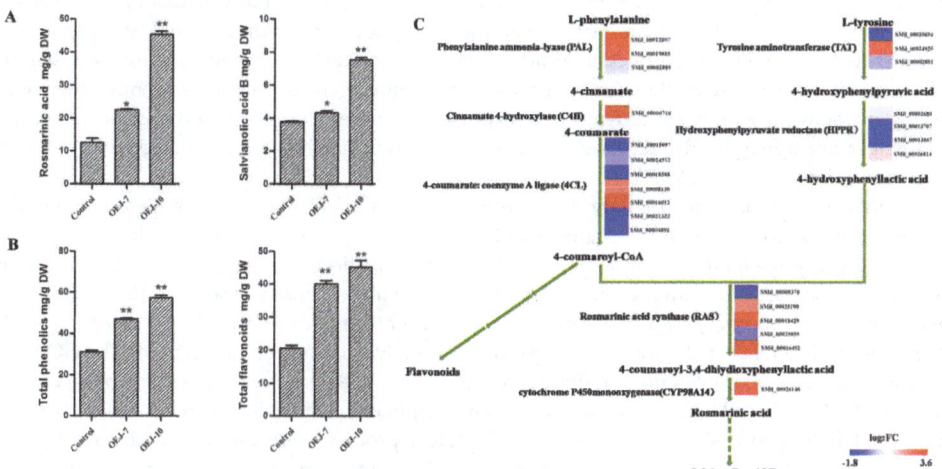

Figure 3. (**A**) Concentrations of rosmarinic acid (RA) and salvianolic acid (Sal B) in root extracts from control and transgenic OEJ-7 and OEJ-10 plants. All data are means of three replicates, with error bars indicating SD. ** Values are significantly different from the control at $p < 0.01$. (**B**) Concentrations of total phenolics and total flavonoids in root extracts from control and transgenic OEJ-7 and OEJ-10 plants. All data are means of three replicates, with error bars indicating SD. * and ** Values are significantly different from the control at $p < 0.05$ and $p < 0.01$, respectively. (**C**) Differentially expressed genes (DEGs) involved in the pathway for salvianolic acid biosynthesis between OEJ-10 and the control. For each gene, relative expression (OEJ-10 versus control) is represented as $\log_2$FC. The color scale is shown at the bottom. Higher expression levels are shown in red.

2.4. Transgenic Plants Show Higher Levels of Total Phenolics and Total Flavonoids

The results above showed that overexpression of *SmJMT* modified the accumulation of two non-flavonoid phenolic acids, RA and Sal B, while the phenolics and flavonoids share an upstream core phenylpropanoid metabolism with Sal B [38]. To determine whether the upregulation of RA and Sal B could cause activation of the phenylpropanoid pathway and provide substrates for the biosynthesis of other types of end product, global assays for phenolics and flavonoids of transgenic plants and control plants were performed. The results showed that total phenolics and total flavonoids accumulated at higher levels in the roots of the overexpression line than in the control sample. The highest concentrations of total phenolics and total flavonoids were also found in transgenic OEJ-10 (Figure 3B). Differences were significant and corresponded to a 1.85-fold increase in the total phenolics content of OEJ-10 roots and a 2.20-fold increase in the total flavonoid content. In OEJ-7, concentrations of total phenolics and total flavonoids were approximately 1.56- and 1.94-fold higher than those of the control (Figure 3B).

2.5. Transcriptomic Analysis of S. miltiorrhiza SmJMT Overexpression and Control Plants

In order to detect the differentially expressed genes between *SmJMT* overexpression transgenic and control plants, and the genes regulated by *SmJMT*, RNA sequencing (RNA-seq) experiments were performed, and the global expression profiles of OEJ-10 and the control were compared. Through Illumina deep sequencing, approximately 26.03 and 26.88 million high-quality clean reads were obtained from OEJ-10 and the control, respectively. The average length of each read was 296 bp. The Q30 values (percentage of sequences with a sequencing error rate <0.1%) for OEJ-10 and the control were 93.45% and 93.38%, respectively. Principal component analysis (PCA) showed that the two groups of samples were distributed in different regions of the spaces, and were distinguished clearly. It indicated that there are differences between the two groups of samples (Figure S3). The control samples were relatively concentrated, indicating that the biological homogeneity of the control samples was better (Figure S3). According to the criteria of differential gene expression screening, there were 2052 genes showing significant differences in expression between the OEJ-10 and control plants, with 998 genes being upregulated and 1054 genes downregulated in OEJ-10 when compared with expression in control (Figure 4 and Table S1).

The Gene Ontology (GO) analysis showed that a total of 14,375 unigenes were annotated in this manner, including 986 DEGs (Figure 5 and Table S2). The GO terms of three categories, namely biological process, cellular component, and molecular function, were assigned to categorize the function of the predicted unique sequences of *S. miltiorrhiza*. In many cases, multiple terms were assigned to the same transcript [39]. This categorization resulted in 1628 DEGs assigned to cellular component, 2573 DEGs to biological process, and 1163 DEGs to molecular function. The GO terms of "binding" (GO: 0005488) and "catalytic activity" (GO: 0003824) of molecular function; "cell part" (GO: 0044464) and "cell" (GO: 0005623) of cellular component; and "cellular process" (GO: 0009987) and "metabolic process" (GO: 0008152) of biological process were predominantly represented (Figure 5 and Table S2). Furthermore, the enriched GO terms "L-phenylalanine metabolic process" (GO: 0009694), "jasmonic acid metabolic process" (GO: 0006558), and "response to extracellular stimulus" (GO: 0009991) correlate well with the biosynthetic pathways for phenolic acids, α-linolenic acid metabolism, and plant defense.

In addition, the Kyoto Encyclopedia of Genes and Genomes (KEGG) analysis showed that a total of 379 DEGs could be to assigned to KEGG pathways, with enrichment in pathways including phenylalanine, tyrosine, and tryptophan biosynthesis (ko00400) and phenylpropanoid biosynthesis (ko00940), which are connected with the biosynthetic pathways for phenolic acids, phenylalanine, and tyrosine (Figure 6 and Table S3). Both GO terms and KEGG pathways of transcriptome analysis were correlated with the phenylalanine metabolic process, indicating that *SmJMT* could be correlated with the biosynthetic pathways for phenolic acids in *S. miltiorrhiza*.

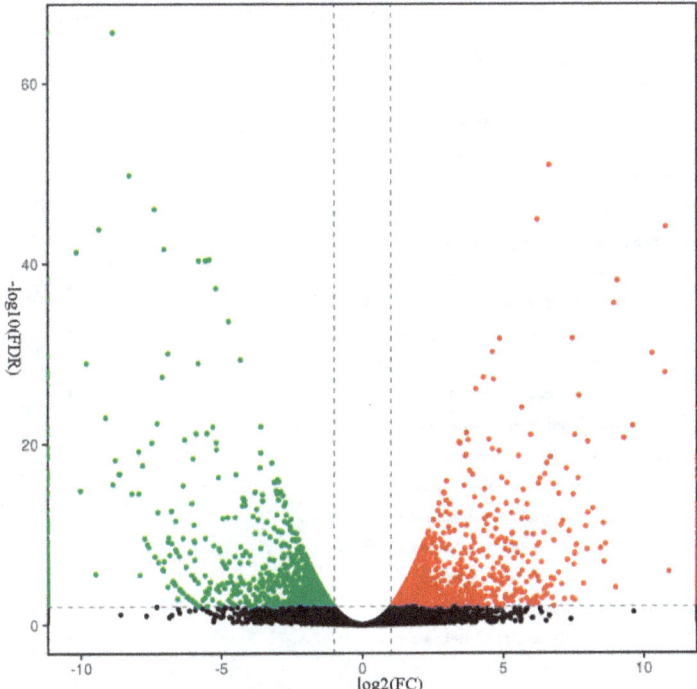

Figure 4. Volcano plot of DEGs between OEJ-10 and the control. Red points represent the DEGs that were upregulated. Black points represent the DEGs without statistically significant differences. Green points represent the DEGs that were downregulated.

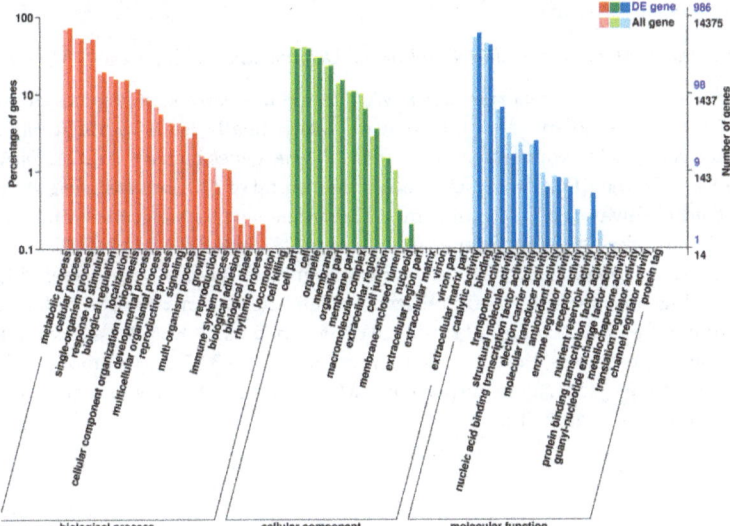

Figure 5. The second Gene Ontology (GO) classification annotation of DEGs between OEJ-10 and the control. The *X*-axis represents GO classification; red represents a biological process, green represents a cellular component, and blue represents a molecular function. The left *Y*-axis represents the percentage of DEGs with respect to all genes. The right *Y*-axis represents the number of genes.

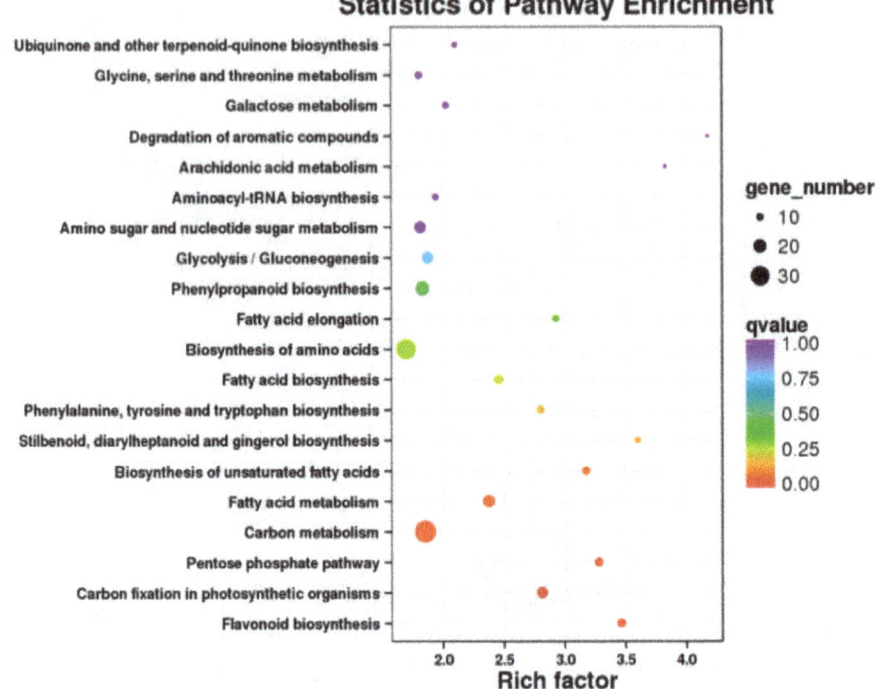

Figure 6. Kyoto Encyclopedia of Genes and Genomes (KEGG) pathway enrichment analysis of DEGs between OEJ-10 and the control. The *X*-axis represents enrichment factor; the *Y*-axis represents the pathway. The circles represent the KEGG pathways. The color of the circle represents the *q*-value. Lower *q*-values are shown in red.

2.6. DEGs Involved in α-Linolenic Acid Metabolism and Determination of Endogenous MeJA Levels

To determine whether overexpressing *SmJMT* affected the expression of genes closely associated with the pathway of α-linolenic acid metabolism, which finally leads to MeJA biosynthesis, we investigated the changes in expression of those genes. These genes include *SmLOX*, *SmAOS*, *SmAOC*, *SmOPR*, and *SmJMT*. The KEGG analysis revealed that a total of 14 genes assigned to the α-linolenic acid metabolism pathway (ko00592) and three DEGs relevant to MeJA biosynthesis were found (Figure 7A). Among the DEGs encoding putative *SmAOS*, *SmOPR*, and *SmJMT*, all of them have one unigene transcriptionally activated and being upregulated (Figure 7A and Table S4). For further detecting the endogenous MeJA levels of transgenic plants, the concentrations of endogenous MeJA in fresh leaves from OEJ-10 and control plants were determined using ELISA. According to the optical density (OD) values of samples, the concentrations of MeJA were 3.57 ± 0.08 pmol·g^{-1} for the control and 5.36 ± 0.30 pmol·g^{-1} for OEJ-10, respectively. MeJA concentrations were significantly higher for the transgenic OEJ lines (Figure 7B).

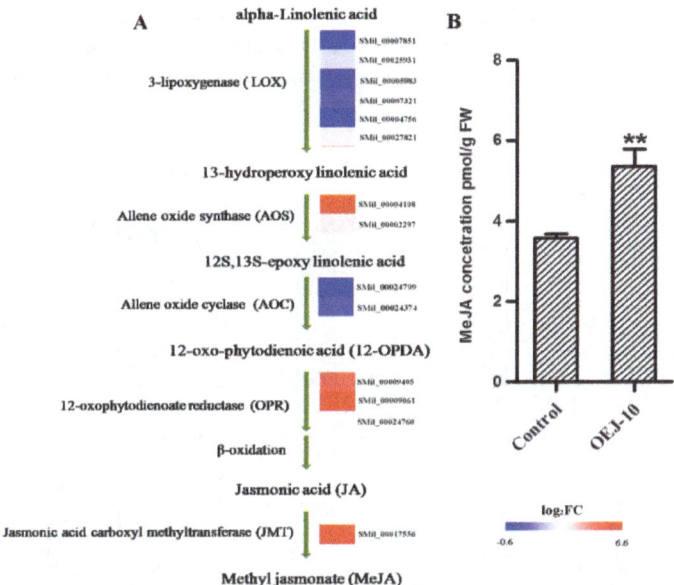

Figure 7. (**A**) DEGs involved in the pathway for MeJA biosynthesis between OEJ-10 and the control. For each gene, relative expression (OEJ-10 versus control) is represented as $\log_2$FC. The color scale is shown at the bottom. Higher expression levels are shown in red. (**B**) Concentrations of MeJA in leaf extracts from transgenic line OEJ-10 and the control. All data are means of three replicates, with error bars indicating SD. ** Values are significantly different from the control at $p < 0.01$.

2.7. DEGs Involved in the Pathway for Salvianolic Acid Biosynthesis

To evaluate whether upregulated expression of *SmJMT* could modify the activation of the key enzymes in the pathway for salvianolic acid biosynthesis, all of the putative enzyme genes in this pathway were examined through transcriptome analysis. A total of 25 unique sequences that encode seven enzymes involved in the biosynthetic pathway of salvianolic acid were present in the RNA-seq dataset, including 12 DEGs (Table S5). These enzymes were SmPAL, SmC4H, Sm4CL, SmTAT, SmHPPR, SmRAS, and SmCYP98A14 (Figure 3C). According to the transcriptome data, among the 12 DEGs in the salvianolic acid biosynthesis pathway, nine were upregulated in the OEJ-10 plants, while three were downregulated (Figure 3C and Table S5). Following comprehensive analysis of the values of counts and the annotation of the RNA-seq dataset, the expression of *SmPAL₁* (*SMil_00019885* Accession No: KF462460), *SmC4H* (*SMil_00000716* Accession No: DQ355979), *Sm4CL₃* (*SMil_00016012* Accession No: KF220556), *SmTAT* (*SMil_00024925* Accession No: DQ334606), *SmRAS* (*SMil_00025190* Accession No: FJ906696), and *SmCYP98A14* (*SMil_00026146* Accession No: HQ316179) demonstrated significant upregulation in OEJ-10, while the expression of *SmHPPR* (*SMil_00002680* Accession No: DQ09974) was not significantly upregulated.

2.8. Confirmation of RNA-Seq Data by qRT-PCR Analysis

To validate the RNA-seq data for differential gene expression between the control and transgenic OEJ lines, the expression levels of genes encoding nine key enzymes involved in the salvianolic acid biosynthesis and α-linolenic acid metabolism were analyzed by qRT-PCR (Figure 8). According to the statistical analysis of qRT-PCR, *SmPAL₁*, *SmC4H*, *Sm4CL₃*, *SmTAT*, *SmHPPR*, *SmRAS*, *SmCYP98A*, *SmAOS*, and *SmJMT* were upregulated in transgenic lines compared with the control (Figure 8). On the other hand, based on the screening conditions of DEGs, the expression of *SMil_00019885*, *SMil_00000716*, *SMil_00016012*, *SMil_00024925*, *SMil_00025190*, *SMil_00026146*, *SMil_00002680*,

SMil_00004108, and *SMil_00017556* was upregulated in transgenic lines compared with the control (Tables S4 and S5). Thus, through statistical analysis, the relative expression levels of these genes were shown to be consistent with those of the RNA-seq data, which indicated the accuracy of the results of the latter (Figure 8).

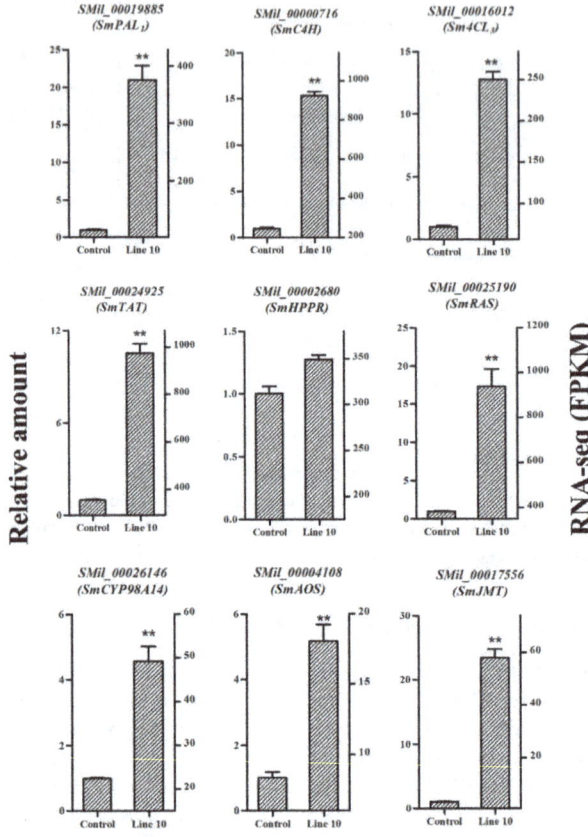

Figure 8. Validation by qRT-PCR of nine genes involved in the salvianolic acid biosynthesis and MeJA biosynthesis pathways in the control and transgenic line OEJ-10. All data are means of three replicates, with error bars indicating SD. ** Values are significantly different from the control at $p < 0.01$.

3. Discussion

Among medicinal plants, *S. miltiorrhiza* is an ideal and representative model for studying transcriptional regulation, and phenolic acid biosynthesis became a new research focus [7]. Fortunately, the *S. miltiorrhiza* genomic sequence was published [4], and this provides a good opportunity for studying the function of many valuable genes in *S. miltiorrhiza*. In this article, we report on the isolation, bioinformatics analysis, molecular characterization, and preliminary function analysis of the *JMT* gene from *S. miltiorrhiza* encoding jasmonic acid carboxyl methyltransferse. Phylogenetic analysis of the SABATH gene family of *S. miltiorrhiza* and *Arabidopsis* suggests that *SmJMT* and *AtJMT* are apparent orthologs and may have the same function [20]. Overexpressing JMT in plants can elevate the level of endogenous MeJA and improve resistance against external environmental stimuli [21,24,26]. Thus, we sought to detect the effect of JMT on the biosynthesis of MeJA in *S. miltiorrhiza* and study the function of JMT in regulating the expression of genes involved in the phenolic acid biosynthesis pathway, as well as the impact on accumulation of secondary metabolites in *S. miltiorrhiza*.

After obtaining the full-length cDNA of *SmJMT*, we constructed a *SmJMT* overexpression vector and transferred it into *S. miltiorrhiza*. Then, after obtaining *SmJMT* overexpression in transgenic *S. miltiorrhiza* lines, we compared the MeJA levels between control and transgenic lines. The results showed that overexpression of *SmJMT* significantly changed the level of endogenous MeJA in *S. miltiorrhiza* (Figure 7B). In addition, transcriptome analysis showed that the three DEGs (*AOS*, *OPR*, and *JMT*) in α-linolenic acid metabolism were upregulated in transgenic lines, while no genes were downregulated, suggesting that the α-linolenic acid metabolism pathway was obviously activated in transgenic line OEJ-10 (Figure 7A). Furthermore, AOS is the major control point of MeJA biosynthesis and the first specific enzyme [17,40,41]. The transcription level of *SmAOS* was significantly upregulated in OEJ-10 according to the transcriptome data. This strong induction of *SmAOS* may contribute to the biosynthesis of MeJA, which was consistent with the results of the ELISA experiment (Figure 7).

Jasmonic acid carboxyl methyltrans-ferase (JMT) is a key enzyme for jasmonate-regulated plant responses and defense response [17,21,24,42]. Based on previous studies, the expression of primary enzymes involved in the phenylpropanoid and tyrosine-derived pathways are upregulated by jasmonates, which could enhance the biosynthesis of phenolic acids [8,9,28,32,33]. Furthermore, the elicitation of defense responses could lead to the accumulation of secondary metabolites [25,34]. Therefore, we designed a strategy to elevate the content of phenolic acids through overexpression of *JMT* in *S. miltiorrhiza*. According to the transcriptome data, genes involved in the phenylpropanoid pathway (*SmPAL*, *SmC4H*, and *Sm4CL*) and in the tyrosine-derived pathway (*SmTAT* and *SmHPPR*), as well as *SmRAS* and *SmCYP98A*, were upregulated in transgenic lines compared with the control (Figure 3C). Meanwhile, the contents of RA, Sal B, total phenolics, and total flavonoids in transgenic lines were significantly elevated over those of the control (Figure 3A,B). The results indicated that overexpression of *SmJMT* significantly increased the contents of phenolic acids by activating the phenylpropanoid and tyrosine-derived pathways. *SmJMT* may play an important role in the regular expression of functional genes contributing to the accumulation of phenolic acids. Our results demonstrate that production of salvianolic acid could be improved by overexpression of *SmJMT* in *S. miltiorrhiza* using genetic engineering techniques. Therefore, our research on *SmJMT* overexpression in *S. miltiorrhiza* provides a good foundation for further study of the function of JMT in plants.

MeJA treatment is the most commonly used method of eliciting herbivore resistance in many different plant species, and a series of JA-mediated defense responses are quickly elicited when plants are exposed to volatile MeJA [43–45]. In brief, when treated with exogenously applied MeJA, plants could elicit most defense-resistant responses by JA [25]. However, following overexpression of the plastidic flax *AOS* cDNA in transgenic potato plants, even though the plants exhibited six- to 12-fold increased levels of JA, this increase did not activate jasmonate-responsive genes [46]. The reason for this may be that the free acid JA could not move across the cellular membrane without a carrier [21,47], while MeJA, which diffuses through the membranes, may act as an intracellular regulator and a diffusible intercellular signal transducer mediating intra- and interplant communications [21]. MeJA also plays its own role in developmental processes and defense responses. Therefore, the effects of treatment with exogenously applied MeJA may be different from those of enhanced endogenous MeJA levels in plants. Improvement of the defense ability of pharmaceutical crops through spraying of jasmonates is not feasible, while strengthening the stress resistance of medicinal herbs by elevating endogenous MeJA levels via genetic engineering would represent a new strategy. Furthermore, the plant defense is related to plant secondary metabolism. The improvement of plant defense could promote the accumulation of secondary metabolites. The defense mechanisms, resistance effects, and plant responses need to be further studied.

4. Materials and Methods

4.1. Isolation of the SmJMT Gene

Plant materials of leaves from *S. miltiorrhiza* were collected following a method described previously [20]. Total RNA was extracted from the *S. miltiorrhiza* leaf tissue with a Plant RNA Kit (OMEGA, Houston, TX, USA). RNA quantity was determined using a NanoDrop 2000C Spectrophotometer (Thermo Scientific, Wilmington, DE, USA). First-strand cDNA was synthesized using a Prime-Script RT Master Mix (TaKaRa, Beijing, China) according to the manufacturer's protocol. The full-length cDNA coding sequence for *SmJMT* was amplified from leaf cDNA with gene-specific primers *SmJMT*-F and *SmJMT*-R (Table S6), both of which were designed according to the phylogenetic analysis of the *SmSABATH* gene family [20] and the unigene sequence (*SMil_00017556*) available in the *S. miltiorrhiza* genomic database [4]. Amplification was achieved using PrimeSTAR® HS DNA Polymerase (TaKaRa, Beijing, China), and the PCR reaction was performed on a FlexCycler thermocycler (Analytikjena, Jena, Germany) in a 50-μL final volume comprising 2.5 U/μL PrimeSTAR® HS DNA Polymerase, 100 ng of first-strand cDNA, 500 nM each primer, 10 μL of 5× PrimeSTAR Buffer, and 2.5 mM deoxynucleoside triphosphate (dNTP) mixture, under the following conditions: cDNA was denatured at 94 °C for 3 min followed by 35 cycles of amplification (94 °C for 30 s, 51 °C for 30 s and 72 °C for 72 s), and then extension at 72 °C for 10 min. The PCR fragments were purified using a DNA Gel Extraction Kit (Tiangen Beijing, China), inserted into pMD19T-vector (Takara, Beijing, China), and then sequenced by Beijing Genomics Institute (Shenzhen, Guangdong, China).

4.2. Multiple-Sequence Alignment and Phylogenetic Analysis

Multiple-sequence alignment of the JMT conserved amino-acid sequences from *S. miltiorrhiza* and selected known JMTs from other species (Table S7) was performed with the DNAMAN program (Lynnon Corporation, San Ramon, CA, USA). Conserved blocks were obtained with the online program Gblocks 0.91b (http://www.phylogeny.fr/one_task.cgi?task_type=gblocks) [48]. Phylogenetic trees were constructed using Bayesian inference implemented in MrBayes [49,50] with the amino-acid sequences of the *SmJMT* and homolog from other species under the model of JTT + I + G. The model was chosen using the program ProtTest [51]. The phylogenetic tree was represented with the help of Treeview1.61 software [52].

4.3. Vector Construction and Transformation

In order to construct *SmJMT* overexpression vectors, the pMD19T-*JMT* plasmid was used as a template to amplify *SmJMT* with the primers pDONR207-*SmJMT*-F/pDONR207-*SmJMT*-R (Table S6), which contained *att*B1/*att*B2 sites. PCR amplification followed the description above. The PCR products were purified and cloned into entry vector pDONR207, using the BP recombination reaction, and then transferred into the destination vector pEarleyGate202 with the LR reaction according to the protocol from the Gateway Technology manufacturer (Invitrogen, Carlsbad, CA, United States) (Figure S4). The pDONR207-*SmJMT* and pEarleyGate202-*SmJMT* plasmid were sequenced by Beijing Genomics Institute (Shenzhen, Guangdong, China). Finally, the pEarleyGate202-*SmJMT* plasmid was transferred into *Agrobacterium tumefaciens* EHA105 using the freeze–thaw method [53].

An *Agrobacterium*-mediated gene transfer method was performed to generate transgenic plants with leaf explants from sterile *S. miltiorrhiza* cultured on Murashige and Skoog (MS) basal medium under the conditions described previously [54,55]. After transformation, MS with 1 mg·L^{-1} naphthalene acetic acid, 10 mg·L^{-1} 6-benzyl-aminopurine, 10 mg·L^{-1} glufosinate/ammonium, and 200 mg·L^{-1} cefotaxime as the selection medium was used to culture the transgenic explants. They were transferred to fresh selection medium at one-week intervals. Developing shoots were excised and placed on $\frac{1}{2}$ MS selection medium supplemented with 10 mg·L^{-1} glufosinate/ammonium, and 200 mg·L^{-1} cefotaxime for root induction [34]. After two weeks, the rooted transgenic plants were propagated through several generations to expand the culture on the MS basic medium.

4.4. PCR Detection and qRT-PCR Analysis

Genomic DNA was obtained from leaves of one-month-old transgenic and control plants using the CTAB Plant Genomic DNA Rapid Extraction Kit (Aidlab, Beijing, China) according to the manufacturer's protocol. A pair of gene-specific primers (pEarleyGate202-35S-F/R) (Table S6) was used to amplify a 929-bp fragment from genomic DNA of transgenic and control plants. PCR reactions were also performed on a FlexCycler thermocycler (Analytikjena, Jena, Germany) with a 20-μL final volume comprising 10 μL of 2× Taq PCR master Mix (Tiangen, Beijing, China), 500 nM each primer, and 100 ng of template DNA. All PCR reactions were carried out as follows: preheating at 94 °C, then 35 cycles of amplification at 94 °C for 30 s, 58 °C for 30 s, and 72 °C for 1 min, followed by a final elongation of 10 min at 72 °C. The positive control was the pEarleyGate202-*SmJMT* vector, while genomic DNA from wild-type plants served as the negative control. Amplified products were electrophoresed on a 1.0% agarose gel.

Total RNA extraction, RNA quantity determination, and first-strand cDNA synthesis from transgenic and control plants followed the method described above. Quantitative PCR was carried out on a Light Cycler 96 Instrument (Roche, Basel, Switzerland) in a 20-μL final volume comprising 10 μL of SYBR Premix Ex Taq II (Takara, Beijing, China), 20 ng of first-strand cDNA, and 500 nM each primer. The reactions were performed in triplicate under the following conditions: initial thermal cycling at 95 °C for 30 s, followed by 45 cycles of 95 °C for 10 s and 60 °C for 30 s. *Smβ-actin* (DQ243702) was selected as a reference gene [20,56]. Relative expression was calculated by the $2^{-\Delta\Delta Ct}$ method [57], and the relative expression levels were analyzed as means ± standard deviation (SD) of the biological triplicates. The lengths of the amplicons were between 100 and 250 bp. Quantitative primers are listed in Table S6.

4.5. Extraction of MeJA and Determination of Its Concentration

Fresh leaves from transgenic plants lines and control plants were used to investigate the MeJA concentration levels. Tissues (100 mg) were ground in liquid nitrogen, and 9 mL of phosphate-buffered saline (PBS; pH 7.4) was added. The extracts were then centrifuged at 8000× *g* at 4 °C for 30 min, and the upper layer was collected. Based on the method of detecting the endogenous jasmonic acid (JA) in *S. miltiorrhiza* [38], the endogenous MeJA of both transgenic and control plants was measured with a Plant MeJA ELISA Kit (mlbio, Shanghai, China) according to the manufacturer's protocol. A standard curve of optical density (OD) versus MeJA concentration at 0, 125, 250, 500, 1000, and 2000 pmol·L^{-1} was produced by testing a set of calibration standards. By comparing OD values with the standard curve, the content of MeJA in each sample was determined. The intensity of the final reaction color was measured spectrophotometrically at 450 nm to calculate the final MeJA concentration [38].

4.6. Extraction of Total Phenolics and Total Flavonoids and Determination of Their Concentrations

The roots from two-month-old transgenic and control *S. miltiorrhiza* plantlets were air-dried at 20 ± 2 °C. The dried root sample (20 mg) was ground into a powder, and mixed with 1 mL of methanol and acetone (7:3, *v/v*) in an ultrasonic bath for 1 h. The mixture was centrifuged at 6000× *g* for 3 min and the supernatant was collected.

The total phenolics content was measured using a modified colorimetric Folin–Ciocalteu method [58] with gallic acid as the standard. The extract solution (100 μL) was mixed in a centrifuge tube with 500 μL of the Folin–Ciocalteu reagent in darkness for 8 min, then incubated with 400 μL of sodium carbonate solution (7.5% *w/v*) at 40 °C for 30 min. Absorbance was measured at 765 nm against a reagent blank without the extract, and measurements were carried out in triplicate. The calibration equation for gallic acid was $y = 0.0073x + 0.1023$ ($R^2 = 0.9995$).

Total flavonoid content was determined following the procedure of Dewanto et al. [59] with epicatechin as the standard. The test sample (100 μL) was placed in a centrifuge tube, and 800 μL of 60% ethanol was added followed by 20 μL of 5% NaNO$_2$. After 6 min, 20 μL of 10% AlCl$_3$ was added. After another 6 min, 60 μL of 4% NaOH was added, and then the solution was mixed and the absorbance was measured at 510 nm against a reagent blank without the extract; measurements were carried out in triplicate. The calibration equation for epicatechin was $y = 0.0022x + 0.0737$ ($R^2 = 0.9900$).

4.7. LC/MS Analysis of Phenolic Compounds

Roots harvested from transgenic *S. miltiorrhiza* transplanted for two months and control plantlets were air-dried at $20 \pm 2°C$. Dried roots were then ground to a fine powder in a mechanical grinder with a 2-mm-diameter mesh. Samples (30 mg) were extracted with 500 μL of 75% methanol in an ultrasonic bath for 20 min, and then centrifuged at $12000\times g$ for 6 min. The residual pellet was re-extracted twice, and all supernatants were combined. Finally, the extracted samples were filtered through a 0.2-μm Millipore filter and analyzed by LC/MS.

For LC/MS, extracts were applied to an Agilent 1260 HPLC system coupled to an Agilent 6460 QQQ LC–MS system (Agilent Technologies, Palo Alto, CA, USA), an HPLC system equipped with a pump (Agilent G1312B), an auto-sampler (Agilent G7127A), and a column temperature controller (Agilent G1316A). Chromatography separation was performed with a Welch UItimate XB-C$_{18}$ column (150 × 2.1 mm, 3 μm particle size) at a flow rate of 0.4 mL·min^{-1} (temperature 30 °C) and 5 μL of sample was injected. The mobile phase comprised Solvent A (acetonitrile) and Solvent B (0.1% formic acid in deionized water), and followed a solvent gradient profile: 0–6 min, A 20–60% and B 80–50%; 6–7 min, A 60–20% and B 50–80%; 7–10 min, A 20% and B 80%. The retention times were 3.8 min for RA and 4.1 min for Sal B.

Mass spectra were acquired by a QQQ-MS instrument with an Agilent Jet Stream (AJS) electrospray ionization (ESI) source. Multiple reaction monitoring (MRM) mode was used for the quantification. For phenolic acids, the ionization mode was negative and the selected transitions of m/z were 359.1→161.1 for RA and 717.2→519.2 for Sal B. The fragmentor voltage was 130 V and the collision energy was 18 eV. The drying gas flow was 10 L·min^{-1}, and the nebulizer pressure was set to 45 psi at a capillary temperature and voltage of 350 °C and 3500 V, respectively. The sheath gas flow was 11 L·min^{-1} at a temperature of 350 °C.

4.8. Transcriptome Analysis and Identification of Differentially Expressed Genes (DEGs)

Two-month-old transgenic and control *S. miltiorrhiza* plants were collected from three biological replicates. Total RNA extraction and RNA quantity determination followed the method described above. The cDNA library preparation and construction were performed as described previously [60]. The libraries were sequenced by Biomarker Technologies Co., Ltd. (Beijing, China) with an Illumina HiSeq4000 platform. The raw transcriptome data were submitted to the National Center for Biotechnology Information (NCBI) with the accession number of SRP155681.

After sequencing of the transcriptome, clean paired-end reads were mapped to the reference genome of *S. miltiorrhiza* [4] using TopHat v2.0.12 [61]. The *S. miltiorrhiza* genome and gene model annotation files were downloaded from genome websites directly (http://www.ndctcm.org/shujukujieshao/2015-04-23/27.html) [4]. Principal component analysis (PCA) was employed to investigate the correlation of biological duplication [62]. Analysis of differential gene expression was performed by Cufflinks [63], and fragments per kilobase of transcript per million mapped reads (FPKM) was used to normalize gene expression levels [64]. The DESeq R package was used to analyze the differential expression of the transgenic and control plants [65]. The false discovery rate (FDR) was controlled by p-values, which were corrected by the Benjamini–Hochberg procedure. Genes with $|\log_2$fold-change$| \geq 2$ and an adjusted p-value < 0.01, as found by DESeq were considered differentially expressed. Gene Ontology (GO) enrichment analysis of DEGs was performed using the topGO R package [66]. GO terms with corrected p-values < 0.01 were considered significantly enriched in the DEGs. To identify significantly over-represented metabolic pathways or signal transduction pathways, all DEGs were mapped to terms in the KEGG (Kyoto Encyclopedia of Genes and Genomes) database [67], and the pathway enrichment analysis was conducted using KOBAS [68].

4.9. Verification of RNA-Seq Data by qRT-PCR

To validate the RNA-seq data, seven unigenes involved in salvianolic acid biosynthesis (*SMil_00019885, SMil_00000716, SMil_00016012, SMil_00024925, SMil_00002680, SMil_00025190,* and *SMil_00026146*) and two unigenes involved in α-linolenic acid metabolism (*SMil_00004108* and *SMil_00017556*) were selected for expression analysis through qRT-PCR experiments. Transcriptome data showed that mostly those genes had significant changes in expression between control and transgenic OEJ lines. The qRT-PCR experiment implementation and statistical analysis followed the description above. Quantitative primers are listed in Table S6.

Supplementary Materials: Supplementary materials can be found at http://www.mdpi.com/1422-0067/19/12/3788/s1. Table S1: Overview of up- and down-regulated genes; Table S2: Results of topGO enrichment analysis of DEGs; Table S3: Overview of the significant enrichment of the KEGG pathways; Table S4: DEGs involved in the α-Linolenic acid metabolism; Table S5: DEGs involved in the salvianolic acid biosynthesis; Table S6: Primer pairs used in the paper; Table S7: List of the JMT from different species; Figure S1: Expression of *SmJMT* in control and overexpression transgenic lines; Figure S2: Principal Component Analysis (PCA) about the correlation of biological duplication of the transcriptome sequencing samples; Figure S3: Mass chromatograms of standards (RA and Sal B) and samples (Control, OEJ-7 and OEJ-10); Figure S4: The *SmJMT*-overexpression vectors construction process with Gateway technology.

Author Contributions: The experiments were conceived and organized by Z.W., B.W., J.N., and X.C. B.W., Y.H., and L.H. performed the experiments. Y.L., W.Z., L.L., S.H., S.W., and D.W. contributed to the data analysis. The paper was written by B.W. and B.L. All authors discussed and approved the final manuscript.

Acknowledgments: This study was supported by the National Natural Science Foundation of China (31670299 and 31800259), the Funds of Jiangxi provincial education department (GJJ170434), the major Project of Shaanxi Province, China (2017ZDXM-SF-005), the Shaanxi province traditional Chinese medicine standardized planting base construction project (2013K14-03-01), the Natural Science Basic Research Plan in Shaanxi Province of China (2018JQ3033), and the Fundamental Research Funds for the Central Universities (GK201806006, GK201302044 and 1301031470).

Conflicts of Interest: The authors declare no conflict of interest.

References

1. Zhou, L.; Zuo, Z.; Chow, M.S. Danshen: An overview of its chemistry, pharmacology, pharmacokinetics, and clinical use. *J. Clin. Pharmacol.* **2005**, *45*, 1345–1359. [CrossRef] [PubMed]
2. Geng, Z.H.; Huang, L.; Song, M.B.; Song, Y.M. Cardiovascular effects in vitro of a polysaccharide from Salvia miltiorrhiza. *Carbohydr. Polym.* **2015**, *121*, 241–247. [CrossRef] [PubMed]
3. Wang, Q.H.; Chen, A.H. Salviae Miltiorrhiza: A Model Organism for Chinese Traditional Medicine Genomic Studies. *Acta Chin. Med. Pharmacol.* **2009**, *37*, 1–4.
4. Xu, H.; Song, J.; Luo, H.; Zhang, Y.; Li, Q.; Zhu, Y.; Xu, J.; Li, Y.; Song, C.; Wang, B. Analysis of the genome sequence of the medicinal plant Salvia miltiorrhiza. *Mol. Plant* **2016**, *9*, 949–952. [CrossRef] [PubMed]
5. Ma, Y.; Yuan, L.; Wu, B.; Li, X.; Chen, S.; Lu, S. Genome-wide identification and characterization of novel genes involved in terpenoid biosynthesis in Salvia miltiorrhiza. *J. Exp. Bot.* **2012**, *63*, 2809–2823. [CrossRef] [PubMed]
6. Ho, J.H.; Hong, C.Y. Salvianolic acids: Small compounds with multiple mechanisms for cardiovascular protection. *J. Biomed. Sci.* **2011**, *18*, 30. [CrossRef] [PubMed]
7. Zhang, Y.; Yan, Y.-P.; Wu, Y.-C.; Hua, W.-P.; Chen, C.; Ge, Q.; Wang, Z.-Z. Pathway engineering for phenolic acid accumulations in Salvia miltiorrhiza by combinational genetic manipulation. *Metab. Eng.* **2014**, *21*, 71–80. [CrossRef] [PubMed]
8. Wang, B.; Sun, W.; Li, Q.; Li, Y.; Luo, H.; Song, J.; Sun, C.; Qian, J.; Zhu, Y.; Hayward, A.; et al. Genome-wide identification of phenolic acid biosynthetic genes in Salvia miltiorrhiza. *Planta* **2015**, *241*, 711–725. [CrossRef] [PubMed]
9. Xiao, Y.; Gao, S.; Di, P.; Chen, J.; Chen, W.; Zhang, L. Methyl jasmonate dramatically enhances the accumulation of phenolic acids in Salvia miltiorrhiza hairy root cultures. *Physiol. Plant.* **2009**, *137*, 1–9. [CrossRef] [PubMed]

10. Zhou, Y.J.; Gao, W.; Rong, Q.; Jin, G.; Chu, H.; Liu, W.; Yang, W.; Zhu, Z.; Li, G.; Zhu, G.; et al. Modular pathway engineering of diterpenoid synthases and the mevalonic acid pathway for miltiradiene production. *J. Am. Chem. Soc.* **2012**, *134*, 3234–3241. [CrossRef] [PubMed]

11. Avanci, N.C.; Luche, D.D.; Goldman, G.H.; Goldman, M.H. Jasmonates are phytohormones with multiple functions, including plant defense and reproduction. *Genet. Mol. Res.* **2010**, *9*, 484–505. [CrossRef] [PubMed]

12. Cheong, J.J.; Choi, Y.D. Methyl jasmonate as a vital substance in plants. *Trends Genet.* **2003**, *19*, 409–413. [CrossRef]

13. Creelman, R.A.; Mullet, J.E. Jasmonic acid distribution and action in plants: Regulation during development and response to biotic and abiotic stress. *Proc. Natl. Acad. Sci. USA* **1995**, *92*, 4114–4119. [CrossRef] [PubMed]

14. Creelman, R.A.; Mullet, J.E. Biosynthesis and Action of Jasmonates in Plants. *Annu. Rev. Plant Physiol. Plant Mol. Biol.* **1997**, *48*, 355–381. [CrossRef] [PubMed]

15. Creelman, R.A.; Mullet, J.E. Oligosaccharins, brassinolides, and jasmonates: Nontraditional regulators of plant growth, development, and gene expression. *Plant Cell* **1997**, *9*, 1211–1223. [CrossRef] [PubMed]

16. Beale, M.H.; Ward, J.L. Jasmonates: Key players in the plant defence. *Nat. Prod. Rep.* **1998**, *15*, 533–548. [CrossRef] [PubMed]

17. Preuss, A.; Augustin, C.; Figueroa, C.R.; Hoffmann, T.; Valpuesta, V.; Sevilla, J.F.; Schwab, W. Expression of a functional jasmonic acid carboxyl methyltransferase is negatively correlated with strawberry fruit development. *J. Plant Physiol.* **2014**, *171*, 1315–1324. [CrossRef] [PubMed]

18. Sasaki, Y.; Asamizu, E.; Shibata, D.; Nakamura, Y.; Kaneko, T.; Awai, K.; Amagai, M.; Kuwata, C.; Tsugane, T.; Masuda, T. Monitoring of Methyl Jasmonate-responsive Genes in Arabidopsis by cDNA Macroarray: Self-activation of Jasmonic Acid Biosynthesis and Crosstalk with Other Phytohormone Signaling Pathways. *DNA Res.* **2001**, *8*, 153–161. [CrossRef] [PubMed]

19. Delker, C.; Stenzel, I.; Hause, B.; Miersch, O.; Feussner, I.; Wasternack, C. Jasmonate biosynthesis in Arabidopsis thaliana—Enzymes, products, regulation. *Plant Biol.* **2006**, *8*, 297–306. [CrossRef] [PubMed]

20. Wang, B.; Wang, S.; Wang, Z. Genome-Wide Comprehensive Analysis the Molecular Phylogenetic Evaluation and Tissue-Specific Expression of SABATH Gene Family in Salvia miltiorrhiza. *Genes* **2017**, *8*, 365. [CrossRef] [PubMed]

21. Seo, H.S.; Song, J.T.; Cheong, J.J.; Lee, Y.H.; Lee, Y.W.; Hwang, I.; Lee, J.S.; Choi, Y.D. Jasmonic acid carboxyl methyltransferase: A key enzyme for jasmonate-regulated plant responses. *Proc. Natl. Acad. Sci. USA* **2001**, *98*, 4788–4793. [CrossRef] [PubMed]

22. Min, S.S.; Dong, G.K.; Sun, H.L. Isolation and characterization of a jasmonic acid carboxyl methyltransferase gene from hot pepper (*Capsicum annuum* L.). *J. Plant Biol.* **2005**, *48*, 292–297.

23. Zhao, N.; Yao, J.; Chaiprasongsuk, M.; Li, G.; Guan, J.; Tschaplinski, T.J.; Guo, H.; Chen, F. Molecular and biochemical characterization of the jasmonic acid methyltransferase gene from black cottonwood (*Populus trichocarpa*). *Phytochemistry* **2013**, *94*, 74–81. [CrossRef] [PubMed]

24. Qi, J.; Li, J.; Han, X.; Li, R.; Wu, J.; Yu, H.; Hu, L.; Xiao, Y.; Lu, J.; Lou, Y. Jasmonic acid carboxyl methyltransferase regulates development and herbivory-induced defense response in rice. *J. Integr. Plant Biol.* **2016**, *58*, 564–576. [CrossRef] [PubMed]

25. Wu, J.; Wang, L.; Baldwin, I.T. Methyl jasmonate-elicited herbivore resistance: Does MeJA function as a signal without being hydrolyzed to JA? *Planta* **2008**, *227*, 1161–1168. [CrossRef] [PubMed]

26. Sohn, H.B.; Lee, H.Y.; Seo, J.S.; Jung, C.; Jeon, J.H.; Kim, J.-H.; Lee, Y.W.; Lee, J.S.; Cheong, J.-J.; Choi, Y.D. Overexpression of jasmonic acid carboxyl methyltransferase increases tuber yield and size in transgenic potato. *Plant Biotechnol. Rep.* **2010**, *5*, 27–34. [CrossRef]

27. Gundlach, H.; Müller, M.J.; Kutchan, T.M.; Zenk, M.H. Jasmonic acid is a signal transducer in elicitor-induced plant cell cultures. *Proc. Natl. Acad. Sci. USA* **1992**, *89*, 2389–2393. [CrossRef] [PubMed]

28. Luo, H.; Zhu, Y.; Song, J.; Xu, L.; Sun, C.; Zhang, X.; Xu, Y.; He, L.; Sun, W.; Xu, H.; et al. Transcriptional data mining of Salvia miltiorrhiza in response to methyl jasmonate to examine the mechanism of bioactive compound biosynthesis and regulation. *Physiol. Plant.* **2014**, *152*, 241–255. [CrossRef] [PubMed]

29. Wasternack, C. Jasmonates: An Update on Biosynthesis, Signal Transduction and Action in Plant Stress Response, Growth and Development. *Ann. Bot.* **2007**, *100*, 681–697. [CrossRef] [PubMed]

30. Ge, Q.; Zhang, Y.; Hua, W.P.; Wu, Y.C.; Jin, X.X.; Song, S.H.; Wang, Z.Z. Combination of transcriptomic and metabolomic analyses reveals a JAZ repressor in the jasmonate signaling pathway of Salvia miltiorrhiza. *Sci. Rep.* **2015**, *5*, 14048. [CrossRef] [PubMed]

31. Gao, W.; Hillwig, M.L.; Huang, L.; Cui, G.; Wang, X.; Kong, J.; Yang, B.; Peters, R. A functional genomics approach to tanshinone biosynthesis provides stereochemical insights. *Org. Lett.* **2009**, *11*, 5170–5173. [CrossRef] [PubMed]

32. Xiao, Y.; Zhang, L.; Gao, S.; Saechao, S.; Di, P.; Chen, J.; Chen, W. The c4h, tat, hppr and hppd Genes Prompted Engineering of Rosmarinic Acid Biosynthetic Pathway in Salvia miltiorrhiza Hairy Root Cultures. *PLoS ONE* **2011**, *6*, e29713. [CrossRef] [PubMed]

33. Peng, D.; Lei, Z.; Chen, J.; Tan, H.; Ying, X.; Xin, D.; Xun, Z.; Chen, W. 13C Tracer Reveals Phenolic Acids Biosynthesis in Hairy Root Cultures of Salvia miltiorrhiza. *ACS Chem. Biol.* **2013**, *8*, 1537–1548.

34. Chen, C.; Zhang, Y.; Qiakefu, K.; Zhang, X.; Han, L.; Hua, W.; Yan, Y.; Wang, Z. Overexpression of tomato Prosystemin (LePS) enhances pest resistance and the production of tanshinones in Salvia miltiorrhiza Bunge. *J. Agric. Food Chem.* **2016**, *64*, 7760–7769. [CrossRef] [PubMed]

35. Attieh, J.; Djiana, R.; Koonjul, P.; Étienne, C.; Sparace, S.A.; Saini, H.S. Cloning and functional expression of two plant thiol methyltransferases: A new class of enzymes involved in the biosynthesis of sulfur volatiles. *Plant Mol. Biol.* **2002**, *50*, 511–521. [CrossRef] [PubMed]

36. Joshi, C.P.; Chiang, V.L. Conserved sequence motifs in plant S-adenosyl-L-methionine-dependent methyltransferases. *Plant Mol. Biol.* **1998**, *37*, 663–674. [CrossRef] [PubMed]

37. Kagan, R.M.; Clarke, S. Widespread occurrence of three sequence motifs in diverse S-adenosylmethionine-dependent methyltransferases suggests a common structure for these enzymes. *Arch. Biochem. Biophys.* **1994**, *310*, 417–427. [CrossRef] [PubMed]

38. Yang, N.; Zhou, W.; Su, J.; Wang, X.; Li, L.; Wang, L.; Cao, X.; Wang, Z. Overexpression of SmMYC2 Increases the Production of Phenolic Acids in Salvia miltiorrhiza. *Front. Plant Sci.* **2017**, *8*, 1804. [CrossRef] [PubMed]

39. Wenping, H.; Yuan, Z.; Jie, S.; Lijun, Z.; Zhezhi, W. De novo transcriptome sequencing in Salvia miltiorrhiza to identify genes involved in the biosynthesis of active ingredients. *Genomics* **2011**, *98*, 272–279. [CrossRef] [PubMed]

40. Kombrink, E. Chemical and genetic exploration of jasmonate biosynthesis and signaling paths. *Planta* **2012**, *236*, 1351–1366. [CrossRef] [PubMed]

41. Wasternack, C.; Hause, B. Jasmonates: Biosynthesis, perception, signal transduction and action in plant stress response, growth and development. An update to the 2007 review in Annals of Botany. *Ann. Bot.* **2013**, *111*, 1021–1058. [CrossRef] [PubMed]

42. Jung, C.K.; Lyou, S.H.; Koo, Y.J.; Song, J.T.; Choi, Y.D.; Cheong, J.J. Constitutive Expression of Defense Genes in Transgenic Arabidopsis Overproducing Methyl Jasmonate. *J. Appl. Biol. Chem.* **2003**, *46*, 52–57.

43. Baldwin, I.T. Jasmonate-induced responses are costly but benefit plants under attack in native populations. *Proc. Natl. Acad. Sci. USA* **1998**, *95*, 8113–8118. [CrossRef] [PubMed]

44. McConn, M.; Creelman, R.A.; Bell, E.; Mullet, J.E.; Browse, J. Jasmonate is essential for insect defense in Arabidopsis. *Proc. Natl. Acad. Sci. USA* **1997**, *94*, 5473–5477. [CrossRef] [PubMed]

45. Li, C.; Williams, M.M.; Loh, Y.T.; Lee, G.I.; Howe, G.A. Resistance of cultivated tomato to cell content-feeding herbivores is regulated by the octadecanoid-signaling pathway. *Plant Physiol.* **2002**, *130*, 494–503. [CrossRef] [PubMed]

46. Harms, K.; Atzorn, R.; Brash, A.; Kuhn, H.; Wasternack, C.; Willmitzer, L.; Penacortes, H. Expression of a Flax Allene Oxide Synthase cDNA Leads to Increased Endogenous Jasmonic Acid (JA) Levels in Transgenic Potato Plants but Not to a Corresponding Activation of JA-Responding Genes. *Plant Cell* **1995**, *7*, 1645–1654. [CrossRef] [PubMed]

47. Heilmann, B.; Hartung, W.; Gimmler, H. The Distribution of Abscisic Acid between Chloroplasts and Cytoplasm of Leaf Cells and the Permeability of the Chloroplast Envelope for Abscisic Acid. *Zeitschrift Für Pflanzenphysiologie* **1980**, *97*, 67–78. [CrossRef]

48. Castresana, J. Selection of conserved blocks from multiple alignments for their use in phylogenetic analysis. *Mol. Biol. Evol.* **2000**, *17*, 540–552. [CrossRef] [PubMed]

49. Huelsenbeck, J.P.; Ronquist, F. MRBAYES: Bayesian inference of phylogenetic trees. *Bioinformatics* **2001**, *17*, 754–755. [CrossRef] [PubMed]

50. Hall, B.G. Comparison of the accuracies of several phylogenetic methods using protein and DNA sequences. *Mol. Biol. Evol.* **2005**, *22*, 792–802. [CrossRef] [PubMed]

51. Abascal, F.; Zardoya, R.; Posada, D. ProtTest: Selection of best-fit models of protein evolution. *Bioinformatics* **2005**, *21*, 2104–2105. [CrossRef] [PubMed]

52. Zhai, Y.; Tchieu, J.; Saier, M.H., Jr. A web-based Tree View (TV) program for the visualization of phylogenetic trees. *J. Mol. Microbiol. Biotechnol.* **2002**, *4*, 69–70. [PubMed]

53. Depicker, A. Transfection and transformation of Agrobacterium tumefaciens. *Mol. Gen. Genet. MGG* **1978**, *163*, 181–187.

54. Yan, Y.P.; Wang, Z.Z. Genetic transformation of the medicinal plant Salvia miltiorrhiza by Agrobacterium tumefaciens -mediated method. *Plant Cell Tissue Organ Cult.* **2007**, *88*, 175–184. [CrossRef]

55. Hua, W.; Song, J.; Li, C.; Wang, Z. Molecular cloning and characterization of the promoter of SmGGPPs and its expression pattern in Salvia miltiorrhiza. *Mol. Biol. Rep.* **2012**, *39*, 5775–5783. [CrossRef] [PubMed]

56. Yang, Y.; Hou, S.; Cui, G.; Chen, S.; Wei, J.; Huang, L. Characterization of reference genes for quantitative real-time PCR analysis in various tissues of Salvia miltiorrhiza. *Mol. Biol. Rep.* **2010**, *37*, 507–513. [CrossRef] [PubMed]

57. Schmittgen, T.D.; Livak, K.J. Analyzing real-time PCR data by the comparative CT method. *Nat. Protoc.* **2008**, *3*, 1101–1108. [CrossRef] [PubMed]

58. Yikling, C.; Jookheng, G.; Yauyan, L. Assessment of in vitro antioxidant capacity and polyphenolic composition of selected medicinal herbs from Leguminosae family in Peninsular Malaysia. *Food Chem.* **2009**, *116*, 13–18.

59. Dewanto, V.; Wu, X.; Adom, K.K.; Liu, R.H. Thermal processing enhances the nutritional value of tomatoes by increasing total antioxidant activity. *Food Sci. Technol.* **2005**, *50*, 3010–3014. [CrossRef]

60. Cao, X.; Guo, X.; Yang, X.; Wang, H.; Hua, W.; He, Y.; Kang, J.; Wang, Z. Transcriptional Responses and Gentiopicroside Biosynthesis in Methyl Jasmonate-Treated Gentiana macrophylla Seedlings. *PLoS ONE* **2016**, *11*, e0166493. [CrossRef] [PubMed]

61. Kim, D.; Pertea, G.; Trapnell, C.; Pimentel, H.; Kelley, R.; Salzberg, S.L. TopHat2: Accurate alignment of transcriptomes in the presence of insertions, deletions and gene fusions. *Genome Biol.* **2013**, *14*, R36. [CrossRef] [PubMed]

62. Schulze, S.K.; Kanwar, R.; Gölzenleuchter, M.; Therneau, T.M.; Beutler, A.S. SERE: Single-parameter quality control and sample comparison for RNA-Seq. *BMC Genom.* **2012**, *13*, 524. [CrossRef] [PubMed]

63. Trapnell, C.; Williams, B.A.; Pertea, G.; Mortazavi, A.; Kwan, G.; Van, M.B.; Salzberg, S.L.; Wold, B.J.; Pachter, L. Transcript assembly and quantification by RNA-Seq reveals unannotated transcripts and isoform switching during cell differentiation. *Nat. Biotechnol.* **2010**, *28*, 511–515. [CrossRef] [PubMed]

64. Florea, L.; Song, L.; Salzberg, S.L. Thousands of exon skipping events differentiate among splicing patterns in sixteen human tissues. *F1000Research* **2013**, *2*, 188. [CrossRef] [PubMed]

65. Anders, S.; Huber, W. Differential expression analysis for sequence count data. *Genome Biol.* **2010**, *11*, R106. [CrossRef] [PubMed]

66. Alexa, A.; Rahnenführer, J.; Lengauer, T. Improved scoring of functional groups from gene expression data by decorrelating GO graph structure. *Bioinformatics* **2006**, *22*, 1600–1607. [CrossRef] [PubMed]

67. Kanehisa, M.; Goto, S.; Kawashima, S.; Okuno, Y.; Hattori, M. The KEGG resource for deciphering the genome. *Nucleic Acids Res.* **2004**, *32*, D277–D280. [CrossRef] [PubMed]

68. Mao, X.; Cai, T.; Olyarchuk, J.G.; Wei, L. Automated genome annotation and pathway identification using the KEGG Orthology (KO) as a controlled vocabulary. *Bioinformatics* **2005**, *21*, 3787–3793. [CrossRef] [PubMed]

MDPI

St. Alban-Anlage 66

4052 Basel

Switzerland

Tel. +41 61 683 77 34

Fax +41 61 302 89 18

www.mdpi.com

International Journal of Molecular Sciences Editorial Office

E-mail: ijms@mdpi.com

www.mdpi.com/journal/ijms

Printed in July 2019
by Rotomail Italia S.p.A., Vignate (MI) - Italy